Addison-Wesley

Science Insights
Exploring Living Things

Teacher's Edition

About the Authors

Michael DiSpezio, M.A.
Michael DiSpezio is a science consultant who conducts workshops for educators throughout the United States. He taught and chaired the science departments of two independent schools in Cape Cod, Massachusetts. Mr. DiSpezio is a former marine biologist and a frequent contributor to NSTA journals. He currently resides in North Falmouth, Massachusetts.

Marilyn Linner-Luebe, M.S.
Marilyn Linner-Luebe has been a science writer and editor since 1985. She taught physical science for fourteen years and chemistry for one year at Fulton High School in Fulton, Illinois. Ms. Linner-Luebe has a master's degree in journalism, specializing in science communication, from Boston University. She currently resides in Clinton, Iowa.

Marylin Lisowski, Ph.D.
Marylin Lisowski is a professor of science education at Eastern Illinois University in Charleston, Illinois. She has taught biology, earth science, and elementary science. In addition to teaching, Dr. Lisowski leads international expeditions and field programs for high school students and teachers. Dr. Lisowski, who has been recognized as an Ohio Science Teacher of the Year, is also one of Florida's Honor Science Teachers.

Gerald Skoog, Ed.D.
Gerald Skoog is a professor of science education and the supervisor of science teachers at Texas Tech University in Lubbock, Texas. Dr. Skoog has taught biology and chemistry and has been a director of several science curriculum and training projects. He served as president of the National Science Teachers Association in 1985 and 1986.

Bobbie Sparks, M.A.
Bobbie Sparks is K–12 Science Consultant for the Harris County Schools in Houston, Texas. Ms. Sparks has taught science for sixteen years, including biology and middle school life and earth science. She has also supervised K–12 science teachers. Ms. Sparks is active in local, state, and national science organizations.

▲ *Addison-Wesley Publishing Company*

Menlo Park, California • Reading, Massachusetts • New York
Don Mills, Ontario • Wokingham, England • Amsterdam • Bonn
Paris • Milan • Madrid • Sydney • Singapore • Tokyo
Seoul • Taipei • Mexico City • San Juan

Teacher's Edition Contents

Consultants, Reviewers, and Participants	T-3
Program Rationale and Goals	T-4
Program Overview	T-5
Themes in Science	T-16
Themes in *Science Insights*	T-17
Thematic Matrices	T-18
Skills in Science	T-28
Skills Matrix	T-30
Integrated Learning	T-31
Multicultural Perspectives	T-32
Cooperative Learning in the Science Classroom	T-33
Exploring Science Visually	T-34
Concept Mapping	T-35
Portfolio Assessment	T-35
Individual Needs	T-36
Master Materials List	T-37
Science Suppliers	T-40

Project Team Acknowledgments

Editorial
Dennis Colella, Shelley Ryan, Eric Engles, Iris Martinez Kane

Design
Debbie Costello, Emily Hamilton, Kevin Berry

Market Research
Shirley Black

Photo Edit
Karen Koppel, Margee Robinson, Chris Pullo

Production
Jenny Blackburn, Steve Rogers, Don Shelonko, Dennis Horan, Nina Pohl, Ben Schroeter, Therese DeRogatis, Eric Houts, Cathleen Veraldi, Ellen Williams, Michael Burns, Bruce Saltzman, Bill Hollowell

Manufacturing
Lisa Bandini, Sarah Teutschel, Carolyn Kemp, Patsy Owens, Sheila Scott

Marketing Services
Greg Gardner

Permissions
Marty Granahan

Video/Software
Jack Hankin, Henry Shires

Front cover photographs: Telegraph/FPG International (earth); Jeff Foott/Bruce Coleman Inc. (honeycomb); Rod Planck/Tom Stack & Associates (bee); K. Murakami/Tom Stack & Associates (integrated chip wafer); Gareth Hopson for Addison-Wesley (geode).

Back cover photographs: Jeff Foott/Bruce Coleman Inc. (honeycomb); Rod Planck/Tom Stack & Associates (bee).

Copyright © 1994 by Addison-Wesley Publishing Company, Inc. All rights reserved. No part of this publication may be reproduced, stored in a retrieval system, or transmitted, in any form or by any means, electronic, mechanical, photocopying, recording, or otherwise, without the prior written permission of the publisher. Printed in the United States of America.

ISBN 0-201-25729-7
3 4 5 6 7 8 9 10—VH—97 96 95 94

Consultants, Reviewers, and Participants

Multicultural Consultants

William Bray
Stanford University
Stanford, CA

Gloriane Hirata
San Jose Unified School District
San Jose, CA

Joseph A. Jefferson
Ronald McNair School
E. Palo Alto, CA

Martha Luna
McKinley Middle School
Redwood City, CA

Peggy P. Moore
Garnet Robertson Intermediate School
Daly City, CA

Steven Oshita
Crocker Middle School
Burlingame, CA

Modesto Tamez
Exploratorium
San Francisco, CA

Content Reviewers

Antonia Adamiak
West Windsor-Plainsboro High School
Princeton Junction, NJ

Catherine K. Carlson
Parkhill Junior High School
Richardson, TX

Charles Chinn
Mortons Gap School
Mortons Gap, KY

Barbara J. Cornelius
Russell Middle School
Winder, GA

Elaine Gima
Narbonne High School
Harbor City, CA

Karel Lilly
Foshay Junior High School
Los Angeles, CA

Richard White
Austin Community Academy
Chicago, IL

Chapter Opener Participants

Steven Potter
Schuyler-Colfax Middle School
Wayne, NJ

Russ Cagle
Arundel Elementary School
San Carlos, CA

Mark Fisch
Haven Middle School
Evanston, IL

Zoe Baxter
Summit Middle School
Edmond, OK

Janice Borland
Austin Academy
Garland, TX

Dorothy G. Burns
Mason Middle School/Investigative
Learning Center
St. Louis, MO

Rich Duncan
Whitford School
Beaverton, OR

Doug Leonard
Byrnedale Junior High
Toledo, OH

Sheila Lollar
Ferry Pass Middle School
Pensacola, FL

Lisa Baumgart
Intermediate School West
Toms River, NJ

Millard Devore
Honea Path Middle School
Honea Path, SC

Janel Jones
Carl Albert Junior High
Oklahoma City, OK

Ann Harder
Antioch Junior High
Antioch, CA

Scott Hannemann
Whittier Middle School
Sioux Falls, SD

Arva Hunter
Washington Middle School
Newport News, VA

Milam Schverak
Clifton Middle School
Houston, TX

Theresa Adler
Sarah Scott Junior High
Terre Haute, IN

Elaine Gima
Fleming Junior High School
Lomita, CA

Alzena Johnson
Hilsman Middle School
Athens, GA

Gloria Nichols
Leeds Middle School
Philadelphia, PA

Virginia Thompson
Northeast Junior High School
Kansas City, MO

DuWayne Walz
Memorial Junior High
Minot, ND

Opal Lambert
Pace Middle School
Milton, FL

Kathy Berger
Kiser Middle School
Dayton, OH

Susan Middlebrooks
Southwest Junior High
Little Rock, AR

Jim Thomas
Supai Middle School
Scottsdale, AZ

Sharon Gray
Menlo Oaks School
Menlo Park, CA

Sue Trafton and Rich Marvin
Barnstable Summer School
Hyannis, MA

Science Insights *Program Rationale and Goals*

Science Insights *has been developed to meet the science learning needs of the early adolescent. To support the teacher in anticipating the special needs of this student,* Science Insights *contains a balance of features and teacher assistance that are geared to these needs.*

This middle school student...

- Needs frequent **motivation** to focus attention in an increasingly distracting environment.

- Is beginning to see **integration** of ideas and processes but is still not able to generalize easily.

- Needs **activities** and hands-on experiences to channel energy and enrich learning.

- Requires special support to learn how to work with others **cooperatively**.

- Is **self-oriented** and wants to see how learning affects personal issues and the future.

- Needs special help in learning how to make **decisions** in a world that is beginning to provide more and more options.

- Students will be motivated by the dynamic and accurate **visual learning program**, the engaging writing, and the special features that speak to their interests.

- **Integration** of the sciences and of other subjects is used throughout the program.

- An active, hands-on approach to science within the student text and within other program materials is employed to involve all students and meet the needs of different learning styles.

- Teachers are provided with strategies to encourage students to work **cooperatively**.

- Emphasis on **Science, Technology, and Society** in every section shows students that science affects them and their future.

- Questioning format and activity strategies in every section encourages **decision-making** development.

Explore Visually

Themes

Cooperative Learning

Multicultural Perspectives

Portfolio

Connections

Program Overview

Science Insights—Exploring Living Things
Student Edition

Science Insights—Exploring Living Things
Teacher's Edition

Supplementary Materials
Teacher's Resource Package

A resource of more than 700 blackline masters that can be used to review, reteach, reinforce, integrate, enrich, and assess.

- *Section Activities*
  A comprehensive series of student worksheets that includes vocabulary and section reviews, and reteach and enrich worksheets that can be used to meet the needs of different student ability levels.

- *Skills Worksheets*
  A series of student worksheets that stress science process skills, language arts and writing skills, and decision-making skills while reinforcing science content and concepts.

- *Integrated Resource Book*
  Student worksheets that encourage students to explore the thematic, interdisciplinary, and intra-science connections of science concepts.

- *Spanish Supplement*
  Spanish translations of the chapter summaries and the glossary.

- *AIDS and Other Sexually Transmitted Diseases*
  Essays and student worksheets that provide additional content and review of STDs.

- *Drugs and Alcohol Abuse*
  Essays and student worksheets that provide additional content and review of drug and alcohol abuses.

- *Assessment Program—Chapter and Unit Tests*
  Each chapter has two test versions. Chapter Test A has standard multiple-choice, true/false, modified cloze, and fill-in-the-blanks type questions. Each Chapter Test B has a skill-based and extended answer format.

Science Insights—Exploring Living Things
Laboratory Manual
Student Edition
Annotated Teacher's Edition

More than 50 hands-on exploratory and decision-making laboratory investigations and surveys.

Overhead Transparency Package

Contains 96 color transparencies, 24 blackline transparencies, and the Overhead Transparency Teacher's Guide. The Teacher's Guide includes teaching strategies for using the color transparencies, overhead blackline master transparencies, and correlated student worksheets.

Science Skills and Techniques Manual
Student Edition
Annotated Teacher's Edition

More than 35 worksheets and activities that stress basic science skills and laboratory techniques.

Computer Software

SelecTest is a flexible test-generating computer program with a bank of more than 700 questions.

Test File

A printed version of the test questions used in the SelecTest.

Spaceship Earth Video

A 25-minute video that stresses connections between people, their environment, and technology. Relevant to students from its music to its message.

Ancillary Options

Optional materials that can be used with *Science Insights—Exploring Living Things* are:

- *Multiculturalism in Mathematics, Science, and Technology*
- *One-Minute Readings: Issues in Science, Technology, and Society*
- CEPUP Modules

See the Difference, Feel the

From the beginning of each unit, Science Insights motivates and challenges students.

Motivate
Dynamic color photography stimulates student interest right from the start.

Creative Writing ▶
Lots of opportunities for you to integrate creative writing throughout each unit.

UNIT 8

UNIT OVERVIEW

Unit 8 provides an in-depth discussion of the earth's ecosystems and the way in which they are affected by people. Chapter 25 describes the earth's biosphere and the food chains and energy sources within it. The chapter discusses the chemical and physical balance of water, nitrogen, and oxygen-carbon dioxide cycles. Students explore succession and learn how organisms adapt to changes in ecosystems. Chapter 26 describes factors that affect populations. The chapter describes symbiotic relationships among organisms and behaviors that enable animals to adapt to environments. Chapter 27 discusses how precipitation, latitude, and longitude help determine climate and biomes. The chapter also describes the life-supporting elements of land biomes and the diversity of life in marine biomes. C... fies natural resources a... humans use them. The ... the sources and applic... forms of energy. Also ... chapter are pollutants ... the earth. The chapte... the need for collectiv... serve the ecosystem...

Introducing the Unit

Directed Inquiry

Have the students examine the photograph. Ask:

▶ What part of the ocean is shown here? (A part that is close to the shore)

▶ Which organisms can you name? (Angelfish, kelp, starfish, sea anemone)

▶ What advantage might these organisms gain by growing close to shore? (They get more sunlight than organisms in deep water; they are safe from many large predators, and so on. Accept all reasonable answers.)

▶ What do you think would happen to these organisms if one species were to die off? (The balance would shift. Some species might thrive for a time, but all species would eventually suffer.)

Writing About the Photograph

Have students imagine that they are scuba diving around the area in the picture. Have them describe what they see—the movements of the organisms, the colors, the effect of looking through water.

Unit 8
Ecology

Chapters

25 Organisms and Their Environment
26 Interactions Among Organisms
27 Climate and Biomes
28 Humans and the Environment

534

Excitement, Enjoy the Results

◀ **Data Analysis**
Develop students' science, reference, and research skills by using the Data Bank and its correlated skill-based unit and chapter questions.

Challenge
Challenge your students to agree or disagree with students' comments from around the country.

Follow-Up
Questions in the Chapter Review encourage students to review or to expand their thinking about each chapter photograph.

T-7

Integrate Learning in Every

Science Insights works for you by organizing each section into a three-step lesson: Motivate, Teach, and Evaluate.

Teaching Options

Opportunities to help students build a Portfolio and to develop cooperative learning skills.

Motivate

Choose the strategy that works best for students—from skills development activities to directed-inquiry questions.

◀ ## Skills Development

Skills WarmUp and Skills WorkOut activities focus students on science skills and methods

◀ ## Prior Knowledge

Find out how much science knowledge students bring to each chapter.

T-8

Lesson

Teach
Choose from a rich variety of strategies to meet individual teaching needs.

Integrated Learning
Bring themes, science and subject connections, and multicultural perspectives into each lesson.

Evaluate
Use Check and Explain to evaluate and to assess students' mastery of objectives, including skills objectives. All answers are provided and keyed right where you need them.

Science, Technology, and Society
Three-pronged content strand—Science and Technology, Science and Society, or Science and You—assures that students connect science to their world.

Build Science and Decision-

Activities
Laboratory activities promote hands-on science experiences.

Cooperative Learning
Each activity is organized by task so that you can choose to use a traditional laboratory approach or assign cooperative learning groups.

Teaching Options
Build students' understanding with a prelab discussion. Strategies and safety tips ensure safe and successful completion of each lab activity.

SkillBuilder ▶
Enrich chapter concepts with additional skills practice and hands-on activities.

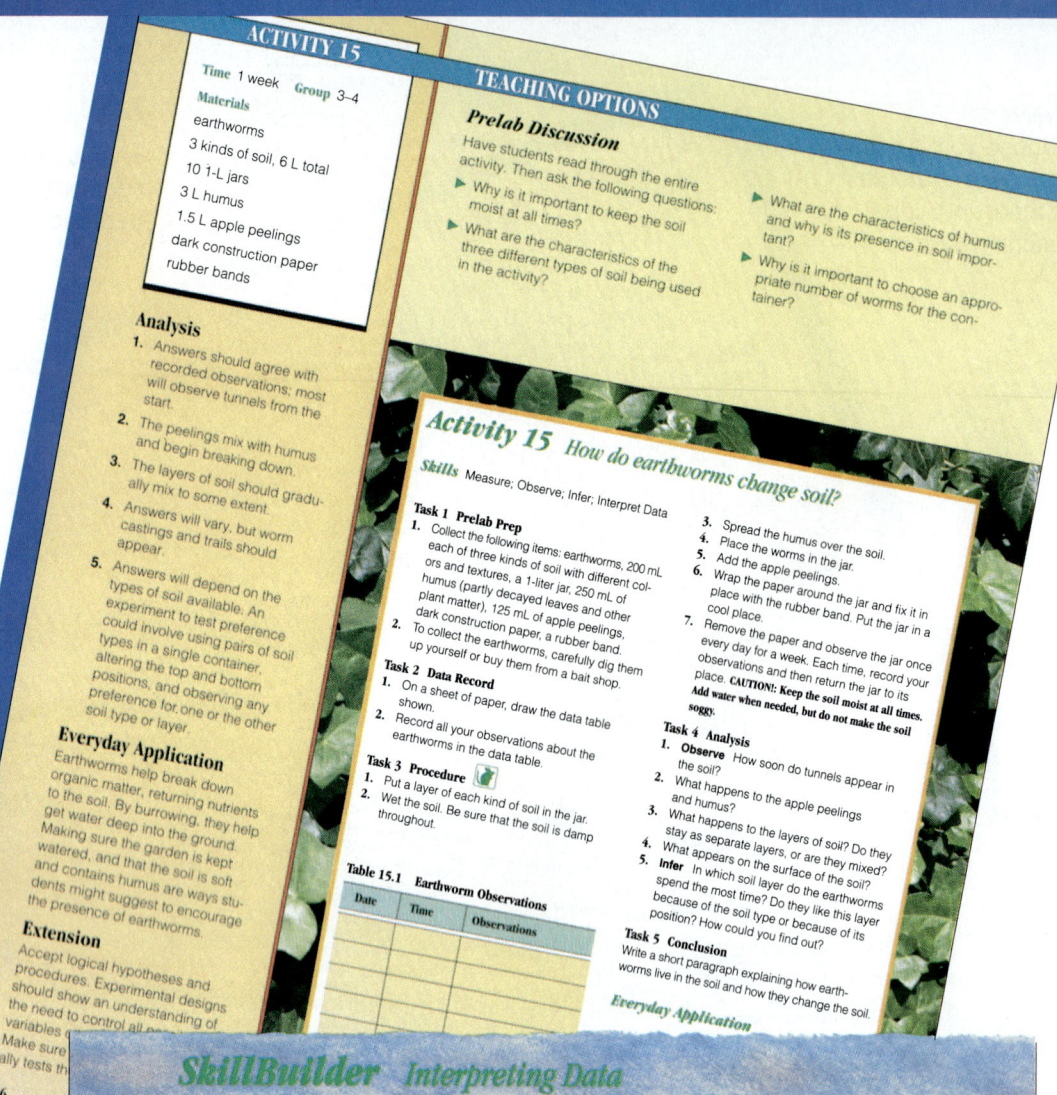

ACTIVITY 15

Time 1 week **Group** 3–4

Materials
earthworms
3 kinds of soil, 6 L total
10 1-L jars
3 L humus
1.5 L apple peelings
dark construction paper
rubber bands

TEACHING OPTIONS

Prelab Discussion
Have students read through the entire activity. Then ask the following questions:
▶ Why is it important to keep the soil moist at all times?
▶ What are the characteristics of the three different types of soil being used in the activity?
▶ What are the characteristics of humus and why is its presence in soil important?
▶ Why is it important to choose an appropriate number of worms for the container?

Analysis
1. Answers should agree with recorded observations; most will observe tunnels from the start.
2. The peelings mix with humus and begin breaking down.
3. The layers of soil should gradually mix to some extent.
4. Answers will vary, but worm castings and trails should appear.
5. Answers will depend on the types of soil available. An experiment to test preference could involve using pairs of soil types in a single container, altering the top and bottom positions, and observing any preference for one or the other soil type or layer.

Everyday Application
Earthworms help break down organic matter, returning nutrients to the soil. By burrowing, they help get water deep into the ground. Making sure the garden is kept watered, and that the soil is soft and contains humus are ways students might suggest to encourage the presence of earthworms.

Extension
Accept logical hypotheses and procedures. Experimental designs should show an understanding of the need to control all variables. Make sure the student actually tests the...

Activity 15 How do earthworms change soil?

Skills Measure; Observe; Infer; Interpret Data

Task 1 Prelab Prep
1. Collect the following items: earthworms, 200 mL each of three kinds of soil with different colors and textures, a 1-liter jar, 250 mL of humus (partly decayed leaves and other plant matter), 125 mL of apple peelings, dark construction paper, a rubber band.
2. To collect the earthworms, carefully dig them up yourself or buy them from a bait shop.

Task 2 Data Record
1. On a sheet of paper, draw the data table shown.
2. Record all your observations about the earthworms in the data table.

Task 3 Procedure
1. Put a layer of each kind of soil in the jar.
2. Wet the soil. Be sure that the soil is damp throughout.
3. Spread the humus over the soil.
4. Place the worms in the jar.
5. Add the apple peelings.
6. Wrap the paper around the jar and fix it in place with the rubber band. Put the jar in a cool place.
7. Remove the paper and observe the jar once every day for a week. Each time, record your observations and then return the jar to its place. **CAUTION!** Keep the soil moist at all times. Add water when needed, but do not make the soil soggy.

Task 4 Analysis
1. **Observe** How soon do tunnels appear in the soil?
2. What happens to the apple peelings and humus?
3. What happens to the layers of soil? Do they stay as separate layers, or are they mixed?
4. What appears on the surface of the soil?
5. **Infer** In which soil layer do the earthworms spend the most time? Do they like this layer because of the soil type or because of its position? How could you find out?

Task 5 Conclusion
Write a short paragraph explaining how earthworms live in the soil and how they change the soil.

Everyday Application

Table 15.1 Earthworm Observations

Date	Time	Observations

316

SkillBuilder *Interpreting Data*

Measuring Calories Used

Different activities use different amounts of Calories. For example, shoveling snow uses more than twice as many Calories as raking leaves. Using the data in the tables on this page, calculate the number of Calories you use in an average 24-hour period. Create your own table on a separate sheet of paper. It should include space for the estimated number of hours you spend doing each activity and space for the total Calories used for that activity.

Study your completed table, then answer the following questions.

1. What activity used the greatest number of Calories?
2. What activity used the least number of Calories?
3. Write the total number of Calories used by ten of your classmates. Then write a short statement comparing your Calorie use with their Calorie use.

Activity	Calories Burned per Hour
Sleeping	60
Reading, watching television	60
Playing a computer game	100
Playing a musical instrument	120
Doing household chores	190
Mowing grass	200
Walking	240
Bicycling	250

Using your results, write a paper discussing whether or not you get enough exercise. To support your position, give examples of the activities you do and the Calories used.

Chapter 23 Nutrition, Health, and Wellness 495

Making Skills

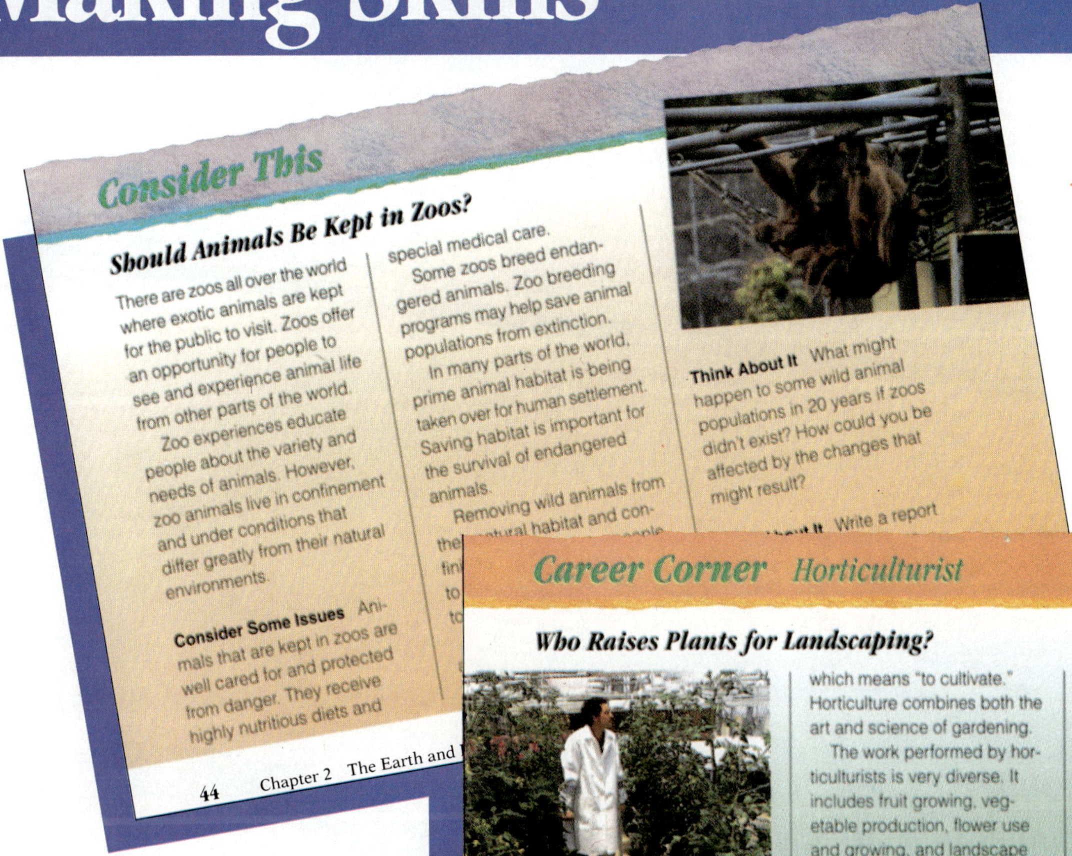

◀ **Consider This**
Decision-making skills are developed about issues related to science—encouraging students to write about them or debate them.

Career Corner ▶
Open up a wide range of career opportunities for students.

◀ **Historical Notebook**
Focus students' attention on people in science, the evolution of scientific theories, and contributions from diverse cultures.

Motivate with Features that

Unique page formats that draw in and engage students.

Explore Visually ▶

Suggested teaching questions that encourage students to explore and to study the visual elements of their texts.

Visual Learning

Carefully rendered and colorful illustrations work with the text to help sharpen students' understanding of science content and processes. Here, the visual helps students better understand the theme—Patterns of Change/Cycles.

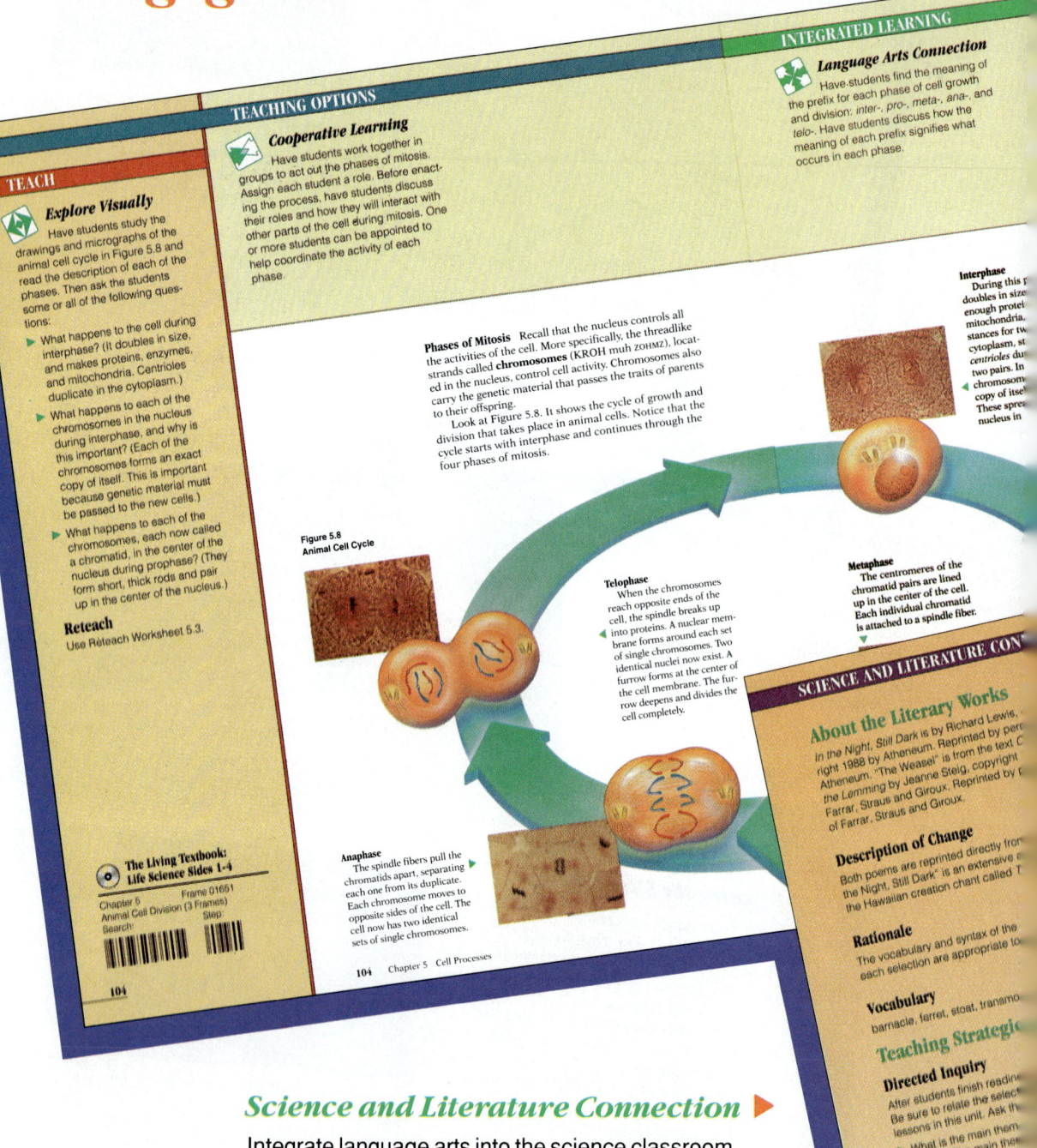

Science and Literature Connection ▶

Integrate language arts into the science classroom. Literary excerpts from notable authors and diverse cultures help students appreciate the application of science to their everyday lives.

T-12

Students Have Asked for...

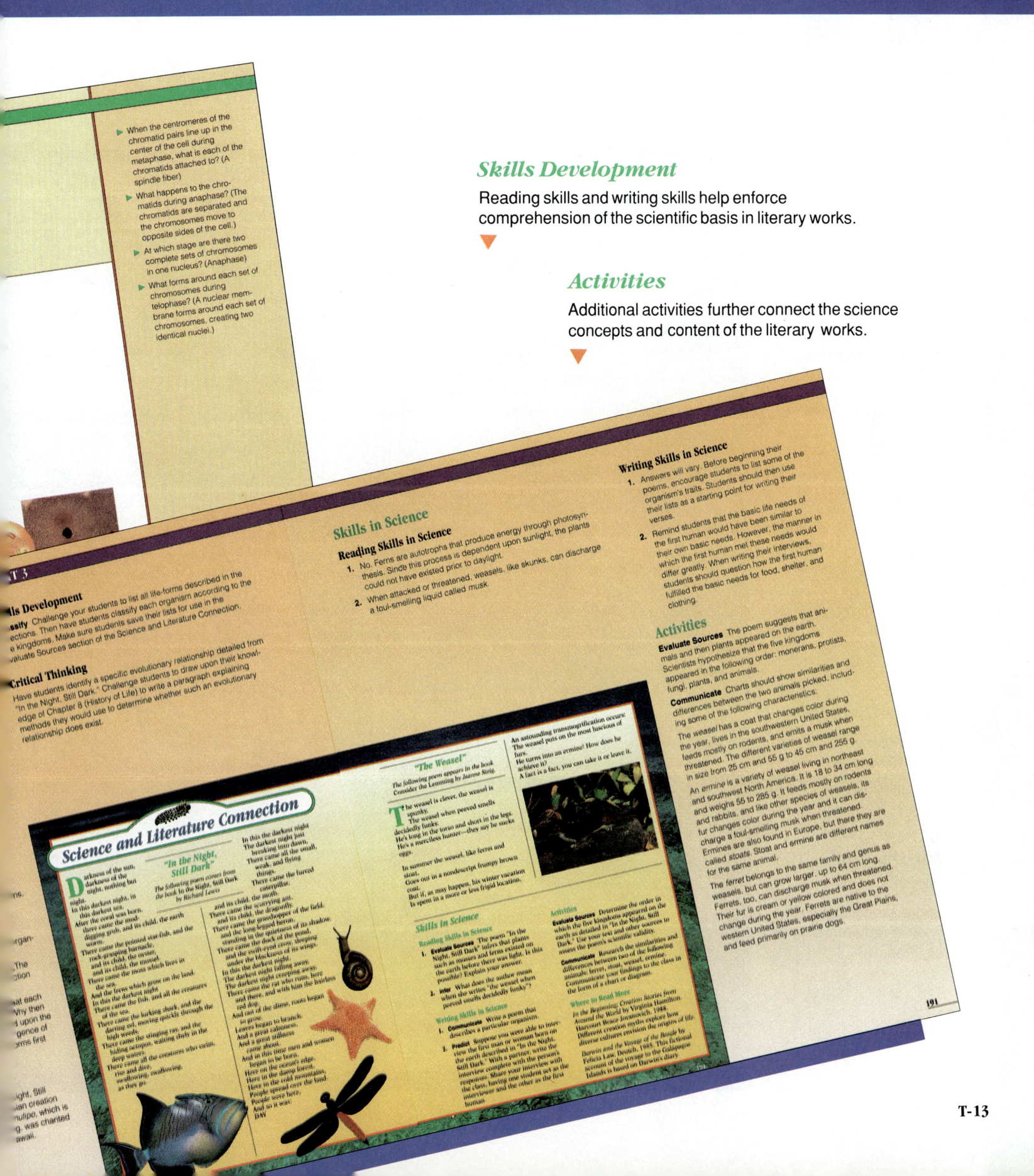

Skills Development
Reading skills and writing skills help enforce comprehension of the scientific basis in literary works.

Activities
Additional activities further connect the science concepts and content of the literary works.

T-13

Organize Lessons at a Glance

Two-page interleaf section with everything

Advance Planner

Reminders about special preparations for the chapter—keyed to the page where you'll need them.

Skills Development

An overview of the skills developed in each chapter.

Individual Needs

Suggestions for the special requirements of students.

CHAPTER 16

Overview

Students continue learning about invertebrates in this chapter. The evolution and characteristics of mollusks are presented in the first section. Arthropods are distinguished from other invertebrates in the second section. The third section provides extensive information on the insect class of arthropods, including models of body structure, development, and metamorphosis. The final section focuses on the five classes of echinoderms.

Advance Planner

- Obtain a copy of the myth about Arachne. TE page 327.
- Prepare a chart or overhead showing how ticks spread diseases and how the diseases can be avoided. TE page 327.
- Collect live or preserved grasshoppers or other arthropods (or order them from a science supplier) for TE page 332.
- Set up 15 wide-mouth jars, 8 ripe bananas, nylon stockings, small spoons, rubber bands, microscope slides, coverslips, and hand lenses for SE Activity 16, page 336.

Skills Development Chart

Sections	Classify	Collect Data	Communicate	Decision Making	Hypothesize	Infer	Model	Observe	Research
16.1 Skills WarmUp	●								
16.2 Skills WarmUp / Skills WorkOut / SkillBuilder		●	●				●		
16.3 Skills WarmUp / Skills WorkOut / Consider This / Activity			●	●	●	●		●	
16.4 Skills WarmUp / Skills WorkOut							●		●

Individual Needs

- **Limited English Proficiency Students** Provide students with a picture dictionary. Have them use the picture dictionary or the text to find definitions of the boldface terms in the chapter. Students can find and cut out pictures from magazines or newspapers that show examples of the boldface terms. Students can paste these pictures onto notebook paper in order to create their own picture dictionaries for this chapter.

- **At-Risk Students** Have students write the section headings in their science journals, leaving space between each heading. Then ask students to refer to the section objectives at the beginning of each section. Have them use their own words to make questions from each of the objectives. Then, as they work through the chapter, have students write one or two sentences that answer each objective question.

- **Gifted Students** Have students research the evolutionary development of one of the groups of invertebrates described in this chapter. Have them create a visual representation of their information by making models of fossils of several organisms related to today's invertebrate. Encourage students to arrange their fossils in chronological order.

Resource Bank

- **Bulletin Board** Have students create an information center about a very important group of arthropods, the spiders. Begin the bulletin board by having volunteers construct a spiderweb using string and pushpins. Have students label the parts of the web, indicating which parts are sticky and which are not. Have students add pictures of many different spiders and labels describing how spiders benefit human beings.

320A

you need to organize and develop your lesson plans.

CHAPTER 16 PLANNING GUIDE

Section	Core	Standard	Enriched	Section	Core	Standard	Enriched
16.1 Mollusks pp. 321–324				**16.3 Insects** pp. 331–336			
Section Features Skills WarmUp, p. 321	●	●	●	Section Features Skills WarmUp, p. 331 Skills WorkOut, p. 332 Consider This, p. 334 Activity, p. 336	● ● ●	● ● ● ●	● ● ● ●
Blackline Masters Review Worksheet 16.1 Integrating Worksheet 16.1	● ●	● ●		Blackline Masters Review Worksheet 16.3 Skills Worksheet 16.3 Enrich Worksheet 16.3	● ●	● ●	● ● ●
Laboratory Program Investigation 31		●	●	Ancillary Options One-Minute Readings, p. 28	●	●	●
Color Transparencies Transparency 45	●	●		Color Transparencies Transparencies 47, 48	●	●	
16.2 Arthropods pp. 325–330				**16.4 Echinoderms** pp. 337–340			
Section Features Skills WarmUp, p. 325 Skills WorkOut, p. 326 SkillBuilder, p. 329	● ● ●	● ● ●	● ● ●	Section Features Skills WarmUp, p. 337 Skills WorkOut, p. 339	●	● ●	● ●
Blackline Masters Review Worksheet 16.2 Skills Worksheet 16.2 Integrating Worksheet 16.2	● ●	● ● ●	● ● ●	Blackline Masters Review Worksheet 16.4 Skills Worksheet 16.4 Vocabulary Worksheet 16	● ● ●	● ● ●	
Overhead Blackline Transparencies Overhead Blackline Master 16.2 and Student Worksheet	●	●	●				
Color Transparencies Transparency 46	●	●					

◄ Chapter Planning Guide

Teaching options at a glance. Program resource materials keyed directly to section content and concepts.

Student ability recommendations help you make decisions about what materials to use to meet students' abilities and needs.

Bibliography

The following resources can be used for teaching the chapter. See page T-40 for supplier codes.

Audio-Visual Sources
(video unless noted)
Aliens from Inner Space. 25 min. 1984. FL.
The Insect Challenge. 24 min. 1985. PF.
Invertebrates. 2 filmstrips with cassettes, 14–17 min each. 1987. NGSES.
Life Cycle of the Honeybee. 12 min. 1976. NGSES.
Now You See Me, Now You Don't. 25 min film. 1977. AMP.
Riches from the Sea. 23 min. 1984. NGSES.
The World of Insects. 20 min. 1979. NGSES.

Software Resources
Animals. Apple. J & S.
Diversity of Life. Apple. J & S.
Grasshopper Dissection Guide. Apple, IBM. CABISCO.
The Insect World. Apple. DCHES.
Starfish Dissection Guide. Apple, IBM. CABISCO.

Library Resources
Allaby, Michael. The Oxford Dictionary of Natural History. New York: Oxford University Press, 1986.
Comstock, Anna Botsford. Handbook of Nature Study. Ithaca: Cornell University Press, 1986.
Durrell, Gerald. The Amateur Naturalist. New York: Alfred A. Knopf, 1989.

Themes in Science

Themes are like trellises—they provide support and structure. By using a thematic approach to teaching, you can help students relate science concepts to their lives, both in and out of the classroom. The goal of using themes to teach science is to help students understand the important ideas in science and the connections between those ideas.

Defining Themes

Using a thematic approach to teaching is not new, but different teachers may have different definitions of the word *theme*. Themes are sometimes defined as topics, such as when someone describes the theme of a lesson to be *bats*. This topical approach can be very useful to students and teachers. In this text, however, themes are defined as links between the major concepts or ideas in science. Since there are many ways to link ideas, there is no one official list of themes in science. See the chart on page T-17 for a list and descriptions of the themes used in *Science Insights*.

Using a Thematic Approach

The volume of scientific information has grown so large that it is impossible to teach students every fact that has been discovered. Too often, though, science curricula present science as a long string of unconnected facts and activities. A thematic approach to science allows students to derive a sense of how the big ideas in science relate to one another. You may wish to use themes to help students integrate different branches of science, connect science concepts with other disciplines, or relate science to their lives and society.

While it is not necessary for students to memorize themes, it is useful to pose questions that encourage students to analyze concepts in the light of specific themes. The following is a list of some questions that can be used in classroom discussions for each of the themes used in *Science Insights*:

- **Diversity and Unity** How are all these organisms alike? How are they different?
- **Energy** What is the source of energy for this organism? How has energy flowed through this system?
- **Evolution** How is this organism similar to its predecessor? What adaptations has it developed?
- **Patterns of Change/Cycles** What happens to this organism in one complete cycle? What patterns do you see in the way this organism has changed? What pattern do you see in this process?
- **Scale and Structure** How is the shape of an organism or structure related to its function? What are the levels of organization in this system?
- **Stability and Equilibrium** What is and isn't changing in the process or system we are studying? What keeps the system from changing?
- **Systems and Interactions** How do the parts of this system work together? How would changing one part of this system affect the other parts?

Themes and *Science Insights*

Themes are often used to enhance the presentation of concepts in *Science Insights*. Material in the student text is presented thematically, with various themes weaving in and out of the prose. For example, the chapters describing invertebrates, vertebrates, and plants all incorporate the theme Evolution. Students are shown how today's organisms have changed over time. Chapter 18 includes a spread that pictures and describes diversity among placental mammals. Teacher's Edition margin notes provide strategies for using themes that are appropriate for a particular concept. These include references to themes that unify an entire chapter. For example, Chapter 21, which describes control and sensing in the human body, begins with a reference to the theme Systems and Interactions. Other times themes relate specific concepts, as in Chapter 3 when the theme Patterns of Change/Cycles helps link ideas about chemical and physical changes.

Themes in *Science Insights*

Addison-Wesley's *Science Insights* series focuses on seven themes. They do not represent all possibilities.

Diversity and Unity	Throughout the sciences, diverse kinds of structures–living and nonliving–are described. Yet, despite great diversity–such as diversity of life, of chemical structures, and of geological formations–there is unity. For example, insects are widely diverse in appearance, but all have the same basic body plan.
Energy	The theme of energy is the central concept to all of science. Life processes, physical and chemical changes, interactions, and the forces that cause natural cycles and change the earth's features all involve energy.
Evolution	In the most general sense, evolution can be defined as change through time.
Patterns of Change/Cycles	Patterns of change are an essential feature of the natural world. It is useful to understand that changes occur as trends or as cycles, and that not all changes are predictable. Trends are relatively steady patterns, such as population growth and climate change. Cycles, such as the water cycle, are repeating patterns.
Scale and Structure	Scientists study the natural world at the microscopic as well as macroscopic levels. While much analysis is done by observing the smallest parts of objects, often it is necessary to observe structures as parts of systems. For example, students study organisms, such as algae, at the cellular level and also as parts of ecosystems.
Stability and Equilibrium	Stability and equilibrium refer to the ways that things do not change. All forces are balanced in a system at equilibrium. Homeostasis in living organisms and the conservation of mass in a chemical reaction are examples of stability and equilibrium.
Systems and Interactions	In science, natural systems range from body systems to the chemical systems in chemical reactions to ecosystems to the solar system. Within every system, there are many kinds of interactions.

Thematic Matrices in Science Insights

Theme	Chapter 1	Chapter 2	Chapter 3
Diversity and Unity	Although there are many areas of specialty within the body of scientific knowledge, all scientific studies are aimed at increasing human understanding of the natural world.	Although there are many diverse kinds of organisms on the earth, all living things share six common characteristics.	The many different organic compounds all contain carbon.
Energy		• All organisms use energy. • The sun is the primary source of energy.	All physical and chemical changes involve energy.
Evolution	Scientific ideas about the natural world have changed over time.	Living things have evolved adaptations that help them survive and reproduce successfully.	
Patterns of Change/ Cycles	New discoveries often effect changes in scientific theories.	All living things reproduce.	• During physical changes, physical properties of matter change, but the substances do not. • During chemical changes, new substances are produced.
Scale and Structure	Scientists gain information about the natural world by making observations at microscopic and macroscopic levels.	The cell is the basic unit of structure in living things.	• The smallest unit of an element is the atom. • The smallest unit of a compound is the molecule.
Stability and Equilibrium		Cells maintain a stable environment through a process called homeostasis.	Total mass and energy are always conserved in a physical or chemical change.
Systems and Interactions	Sharing information is essential to scientific discovery.	The earth is part of the solar system.	

Chapter 4	Chapter 5	Chapter 6
Even though there is great diversity among organisms, all organisms are made up of cells.	Plant cells differ from animal cells in the way they get energy. However, both plant and animal cells release energy through respiration or fermentation.	There are variations among the offspring of two parents. Offspring also resemble their parents in many ways.
Organelles carry on energy-storing and energy-releasing processes.	Cells store energy in a process called photosynthesis. They release energy during respiration and fermentation.	All life processes, including meiosis and DNA replication, require energy.
	The structures within cells have changed over time.	The laws of heredity help explain the mechanisms through which organisms evolve.
According to the cell theory, all cells come only from other living cells.	Cells grow and reproduce in cycles through the process of mitosis.	• Some changes or mutations in genetic material are unpredictable. • Mendel's studies provide us with insight into the patterns of inheritance.
The levels of organization in living things are cells, tissues, organs, systems, and organisms.	Cells have membranes that are permeable, semipermeable, or impermeable.	Genes are located on chromosomes, which are made of long strands of DNA. The chromosomes contain the genetic information that determines an organism's characteristics.
The cell membrane helps maintain homeostasis within cells.	The processes of osmosis and active transport help cells maintain a stable environment.	Genetic material is passed from parent to offspring.
In many-celled organisms, systems work together.	The processes of photosynthesis and respiration interact to supply the energy needs of cells.	

T-19

Thematic Matrices in Science Insights *continued*

Theme	Chapter 7	Chapter 8	Chapter 9
Diversity and Unity	Variations exist among individuals of every species. Despite variation, organisms from the same species share many similar characteristics.	The many different organisms on the earth probably evolved from single-celled organisms.	Taxonomic keys are organized according to physical similarities and differences among organisms.
Energy	Different organisms have developed adaptations that help them meet their energy needs.		Organisms can be classified according to how they get energy.
Evolution	Natural selection helps explain how organisms evolve.	The earth and organisms inhabiting the earth have changed over time.	Classification systems are based on evolutionary relationships among organisms.
Patterns of Change/ Cycles	Over time, groups of organisms develop adaptations that enable them to live and reproduce in a particular environment.	• The earth's continents were once one large landmass. • Fossils have provided clues about how organisms have changed over time.	As scientists make discoveries about cell structure, systems of classification change.
Scale and Structure	Organisms living in the same habitat often have similar structures that enable them to adapt to their environments.	By studying fossils, scientists can compare the structures of various organisms and learn more about their evolutionary relationships.	Organisms that are classified in each of the five kingdoms show characteristic cell structure and levels of organization.
Stability and Equilibrium	The offspring of organisms that reproduce asexually are identical to their parents.		
Systems and Interactions	Members of a population compete for food, water, space, and other resources. These interactions contribute to the process of natural selection.	• Changes on the earth's surface have affected the kinds of organisms that inhabit the earth. • Organisms have changed the makeup of the earth's atmosphere.	

Chapter 10	Chapter 11	Chapter 12
Viruses come in many shapes but all have capsids that surround genetic material.	There are many kinds of protists, from single-celled protozoa to large algae. However, all protists are eukaryotes, having complex cells that contain nuclei, mitochondria, and other organelles.	The many different species of plants are divided into two major groups: vascular plants and nonvascular plants.
• Anaerobic bacteria get energy through the process of fermentation. • Most bacteria meet their energy needs by ingesting nutrients. However, some bacteria make nutrients through photosynthesis.	• Protozoa ingest nutrients. • Algae meet their energy needs through photosynthesis. • Fungi absorb nutrients from their environment.	In photosynthesis, carbon dioxide and water are converted into glucose. Energy is stored in the chemical bonds in glucose.
Scientists hypothesize that the first chloroplasts were really blue-green bacteria that began to live in another bacterium.	Protozoa have evolved different ways of moving.	Plants evolved from many-celled green algae. Over time, plants have developed adaptations that allow them to survive on land.
Viruses and bacteria reproduce asexually. Offspring of bacteria and viruses are identical to the parents.	• Protists reproduce asexually. However, some protists exchange genetic material before undergoing fission. • Some algae and fungi reproduce sexually. Others reproduce asexually.	Plants play an important role in the oxygen–carbon dioxide cycle. Animals take in oxygen and release carbon dioxide. Plants use carbon dioxide and release oxygen.
Viruses are made up of RNA or DNA covered by a protein coat called a capsid. The capsid gives the virus its shape.	The cells of protists are complex. They include nuclei, mitochondria, and other organelles.	Plant cells contain the structures chloroplasts and cell walls.
Viruses and bacteria that live inside other organisms can upset the host organism's equilibrium.		Plants have a waxy coating that enables them to maintain the stable levels of moisture required for cell processes.
• Viruses cause diseases in humans and other host organisms. • Bacteria play an important role in the lives of living things. Some bacteria are harmful; most are helpful.	Single-celled algae called plankton make up the first level of the food chain. Plankton provide food for a large number of organisms.	• Chemical activities in plants allow them to interact with their environment. • Plants produce chemicals that affect other plants and animals.

Thematic Matrices in Science Insights *continued*

Theme	Chapter 13	Chapter 14	Chapter 15
Diversity and Unity	Nonflowering plants include bryophytes, horsetails, ferns, club mosses, and gymno-sperms. These plants vary greatly in appearance, but all reproduce without flowers.	Although there is great variety among flowering plants, they all produce seeds within flowers.	All animals share the same basic needs. However, animals vary greatly in the functions and structures that satisfy those needs.
Energy	Nonflowering plants meet their energy needs through photosynthesis.	The seeds of flowering plants have cotyledons that store energy in the form of sugars, starches, and oils. After germination, this energy fuels growth.	Animals meet their energy needs by ingesting foods.
Evolution	Nonflowering plants evolved special structures that enabled them to adapt to life on land.		Animals first evolved from protists. Over time, animals evolved a variety of different body plans.
Patterns of Change/ Cycles	The life cycle of a nonflowering plant includes a sexual stage and a nonsexual stage.	Flowering plants reproduce sexually. This allows for variety among offspring.	
Scale and Structure	Vascular systems in plants provide support and a means of transporting water and nutrients. This allows plants to grow tall.	The flower contains the structures necessary for plant reproduction. These specialized structures are the sepals, petals, stamen, and pistils.	Scientists classify animals on the basis of body structure. Animal body plans show radial symmetry, bilateral symmetry, or no symmetry.
Stability and Equilibrium		The vascular tissues in plants form root and shoot systems that function to maintain levels of water and minerals in plant cells.	
Systems and Interactions		Vascular plants have systems made up of cells, tissues, and organs that work together enabling the plant to live, grow, and reproduce.	Animals are many-celled organisms. Some animals have specialized cells that work together, while others have tissues, organs, and organ systems that interact.

Chapter 16	Chapter 17	Chapter 18
Although there is great diversity among invertebrates, they all share certain characteristics.	Some vertebrates are ectotherms and some are endotherms. However, all vertebrates share several traits at some stage of development, including gills, notochords, and a backbone.	The 9,000 species of birds include birds of prey, perching birds, water birds, and flightless birds. All birds, however, are endotherms. They have feathers, beaks, and wings.
Invertebrates, like all animals, get energy by ingesting food.	Fishes, amphibians, and reptiles, like all animals, get energy by ingesting food.	Birds have specialized digestive systems that help them meet their high-energy needs.
• Mollusks first evolved 570 million years ago, probably from segmented worms or flatworms. • Arthropods probably evolved from segmented worms.	The first invertebrates were water-dwelling, fishlike animals. Over time, these animals developed adaptations that enabled them to live on land.	Fossil evidence indicates that both birds and mammals evolved from reptiles.
Many insects undergo the process of metamorphosis. During metamorphosis, young organisms change in appearance as they develop into adults.	Many amphibians undergo the process of metamorphosis as they develop into adults.	Many birds migrate as the seasons change.
Invertebrates are classified into groups based on common body plans and structures. For example, mollusks have soft bodies, arthropods have jointed appendages, and echinoderms have radial symmetry.	One of the most important characteristics of invertebrates is the endoskeleton, or backbone. The backbone protects the spinal cord.	Feathers are structures that are unique to birds. They provide warmth and aid the bird in flying.
	Endotherms have body cells that produce enough heat to help them maintain a stable body temperature.	The ability to maintain a constant internal temperature is a major reason why mammals and birds have been so successful.
Many invertebrates have organ systems that interact to enable them to live, grow, and reproduce.		Mammals and birds have complex body systems that are adapted for an active life.

Thematic Matrices in Science Insights *continued*

Theme	Chapter 19	Chapter 20	Chapter 21
Diversity and Unity	The variety of specialized cell types and tissues that make up organs are necessary to the collective functions of human organ systems.		
Energy	The human body requires energy in order to move.	• The digestive system provides the body with nutrients that supply its energy needs. • The respiratory system supplies oxygen for respiration.	The nervous system uses both chemical and electrical energy to carry nerve impulses from one nerve to the next.
Evolution	Adaptation to a variety of complex needs has led to increased specialization in body structures.		
Patterns of Change/ Cycles	Skin cells are constantly dying and being replaced.	During the process of digestion, food is changed into nutrients that can be used by the body.	
Scale and Structure	• Flat bones, long bones, and irregular bones have different shapes. The shape of each is related to its function. • Muscles are made up of long, thin cells.	Structures such as villi in the intestines and alveoli in the lungs greatly increase the surface area of these organs.	Sensory receptors are located in sensory organs. The receptors allow humans to see, hear, smell, taste, and feel.
Stability and Equilibrium	The skin helps maintain body temperature and moisture levels.	The systems in the body work together to maintain levels of oxygen, water, and nutrients.	The levels of hormones in the body are kept stable by a mechanism called feedback control.
Systems and Interactions	The human skeletal and muscular systems interact to allow movement and provide support.	The digestive, circulatory, respiratory, and excretory systems all work together to allow the body to live, grow, and move.	The endocrine and nervous systems work together to respond to changes inside and outside the body.

Chapter 22	Chapter 23	Chapter 24
Sexual reproduction ensures diversity because the genetic makeup of offspring always varies from that of its parents.		
The human reproductive process requires enormous amounts of energy.	Energy is required for all human activities. Humans get most of their energy from carbohydrates.	
The female reproductive cycle is a combination of two cycles working together. In the ovulatory cycle, an egg matures and is released. In the menstrual cycle, the uterus is prepared for the fertilized egg.	Alcohol, tobacco, and drugs cause physical and emotional changes in the human body.	Usually cell division is controlled. However, the body disorder cancer causes cell division to occur in an unregulated pattern.
During human pregnancy, single-celled eggs develop into many-celled organisms.		Every pathogen has a different shape. Antibodies are specially shaped to attach to the surface of specific pathogens and destroy them.
	The human body needs a balance of nutrients, rest, and exercise to maintain health.	Usually the body's cells, tissues, and organs maintain a stable internal environment. Body disorders upset this equilibrium.
Diffusion of material across a mother's circulatory system to her developing fetus provides a means of gas exchange and a way for the fetus to obtain nutrients and remove waste.	The systems of the body work together to allow it to function. When a drug enters the circulatory system, it affects the functioning of other body systems as well.	The immune system defends the body by interacting with invading organisms.

Thematic Matrices in Science Insights *continued*

Theme	Chapter 25	Chapter 26	Chapter 27
Diversity and Unity	There are many different kinds of ecosystems on the earth, and all are affected by the interactions among biotic and abiotic factors.		The organisms in different areas of a biome vary, but they often share similar characteristics that enable them to adapt to the environment.
Energy	In ecosystems, energy flows through food chains and food webs.	Many of the interactions in an ecosystem are related to the ways in which organisms meet their energy needs.	
Evolution	Ecosystems evolve gradually. The older an ecosystem is, the more organisms it usually contains.	Ecologists who study birth and death rates of populations can predict if a species is in danger of becoming extinct.	Organisms have developed adaptations that help them live and reproduce in certain biomes.
Patterns of Change/ Cycles	In natural cycles, living and non-living factors are circulated throughout the biosphere.	Cyclic behaviors, such as migration or hibernation, generally happen as a result of changes in the environment.	A particular location has characteristic patterns of temperature and precipitation. These patterns are called an area's climate.
Scale and Structure			Ecologists have used patterns of climate, vegetation, and animal life to divide large bodies of land and water into biomes.
Stability and Equilibrium		Limiting factors in an environment help maintain stable population sizes.	Human activities have affected the stability of the earth's land, water, and air.
Systems and Interactions	Ecology is the study of the interactions of organisms and their environments.	Symbiosis is an interaction between two organisms in which at least one organism benefits.	Where an organism can live depends upon its ability to interact with different abiotic factors in the environment.

Chapter 28

Energy resources are required for human activities. There are two main types of resources—renewable and nonrenewable.

Fossil fuels were formed from the buried remains of plants and animals that lived hundreds of millions of years ago.

Often it is necessary to change energy from one form, such as solar energy, to another, such as electricity.

The health of the earth's ecosystems depends upon balancing human needs for resources with concern for the environment.

Different parts of an ecosystem interact with one another. Changes to one part affect other parts.

Skills in Science

The study of science involves more than the absorption of information. Inquiry, investigation, and discovery are also part of a scientist's work. Educators have observed that students need to develop process skills and critical thinking skills as they study science. By developing skills as well as concepts, students will be prepared for life both in and out of the classroom.

As students use process skills, they will also use particular critical thinking skills and a pattern may emerge linking these two types of skills.

Observe Observing is a skill that is fundamental to all learning. As students study science, they need to extend their senses using tools such as microscopes or hand lenses.

Classify Classifying is the grouping of objects or events according to an established scheme. When classifying, students should be able to perceive similarities and differences among objects. Classifying is both a critical thinking skill and a process skill.

Infer When students infer, they make evaluations and judgements based on past experiences. The skill of inferring may involve identifying cause-and-effect relationships from events observed, identifying the limits of inferences, and testing the validity of inferences. Inferring also relates to the critical thinking skills Find Causes and Reason and Conclude.

Predict When students predict, they formulate an expected result based on past experience. Students learn that it takes repeated observations of an event to predict the next occurrence of that event. Predicting is a critical thinking skill and a process skill.

Measure Measuring involves direct or indirect comparison of an object with arbitrary units. In science classrooms, the SI system of units is generally used. Measure is also related to the critical thinking skill Compare and Contrast.

Communicate Communicating involves an exchange of information. The exchange can involve speaking, listening, writing, reading, or creating a visual display. Communication is a common thread that runs through all critical thinking skills.

Define Operationally An operational definition is a statement about an object or phenomenon based on one's experience with it. Define operationally is related to the critical thinking skill Generalize.

Hypothesize When students hypothesize, they formulate a statement that can be tested by experiment. Hypothesizing relates to the ability to Generalize and to Form If...Then Arguments.

Make Models Models can be physical or mental representations that explain an idea, object, or event. Making models is related to the critical thinking skill Reason by Analogy.

Estimate The skill of estimating generally involves an indirect means of measuring. Estimating requires students to make mental comparisons between physical objects or lengths of time. Estimating is related to the critical thinking skill Compare and Contrast.

Control Variables As part of designing experiments, students need to identify factors that may affect the outcome of an event. They must then decide how to manipulate one factor while holding other factors constant. Controlling variables relates to the critical thinking skill Find Causes.

Collect Data The skill of collecting data includes gathering information in a systematic way and recording it.

Interpret Data Three skills are included in this category—Read a Graph, Read a Table, and Read a Diagram. When students read a graph, table, or diagram, they must explain the information presented in that graphic form and/or use it to answer questions.

Applying Skills in *Science Insights*

In *Science Insights–Exploring Living Things,* process skills and critical thinking skills are emphasized and integrated throughout the text. Students are given many opportunities to develop their skills. These opportunities range from informal to formal activities and investigations that use process and critical thinking skills. In the first section of Chapter 1, students learn how to apply nine key process skills as they study science. The following features in *Science Insights* also stress the use and application of various skills.

Process Skills and Critical Thinking Skills

Skills WarmUp
A Skills WarmUp activity appears in the margin of every section. It focuses students' attention on specific skills. The Skills WarmUp is usually a simple pencil and paper activity or discussion that leads students into the main topic of the section. For example:

▶ *Skills WarmUp, Chapter 2, page 36* Students classify familiar objects as living, once living, or non-living. The WarmUp reinforces classifying skills and can be used to assess prior knowledge.

Skills WorkOut
Skills WorkOut activities appear in the margins throughout the text. Students mainly do hands-on activities and research, extending the process skills they have learned. For example:

▶ *Skills WorkOut, Chapter 1, page 20* After students learn about graphs, they create their own graphs that show specific data. The WorkOut reinforces the skills of making graphs and doing research.

Check and Explain
Each section ends with four Check and Explain questions. The third and fourth questions require that students apply critical thinking or process skills. The questions are tied into the objectives.

Chapter Review
Two sections of the Chapter Review require students to use both process and critical thinking skills. They include Check Your Understanding and Develop Your Skills. All skills are in boldface.

SkillBuilders
There are 20 SkillBuilders in *Science Insights*. The SkillBuilder focuses on a specific process skill, but other skills are required as well. For example:

▶ *SkillBuilder, Chapter 21, page 455* Students work with partners to investigate their sense of smell and taste. The SkillBuilder reinforces the skills of collecting and recording data. In addition, cooperative learning is emphasized.

Activities
There are 28 activities in *Science Insights,* one for each chapter. All activities emphasize the process skills that students use while doing the activity.

Decision-Making Skills

Consider This
After reading about the two sides of an issue related to science in a Consider This feature, students must decide where they stand on the issue. They then communicate their opinions orally by debating the issue with classmates. Students also express their opinions in essays or editorials.

Research Skills

Historical Notebook
In the Historical Notebook feature, students read about a historical technological development in science or prominent people in science history. Students answer questions about the feature, then research a related topic or find out more about the subject of the feature.

Data Bank
Data Bank questions are presented in both the Chapter Review and Unit Opener pages. As students research answers to data bank questions, they use a variety of process skills, including predicting, inferring, classifying, and reading a diagram or table.

Skills Matrix

Key: ▲ Apply ■ Assess

Chapters	Observe	Compare and Contrast	Classify	Infer	Predict	Measure/Calculate	Estimate	Decision Making	Define Operationally	Hypothesize	Make a Model	Make a Graph	Collect/Organize Data	Interpret Data	Generalize	Find Causes	Reason/Conclude	Research	Communicate
Chapter 1	▲	■	■	■	■	▲					■	▲	▲	▲■					▲
Chapter 2		▲	▲		■			▲			■	▲■		■					■▲
Chapter 3	▲	▲	■	▲■	■	▲					▲	▲	▲■	▲■				▲	▲■
Chapter 4	■▲	▲■	▲■	▲							▲		■	▲■				■	▲
Chapter 5	▲	■▲			■		▲				■			▲■		■			
Chapter 6			▲	▲■	▲■					■	▲		▲	▲■		■	■	▲	■▲
Chapter 7	▲	■		▲■				▲■		■				■		■			■
Chapter 8	▲		■▲	▲■	■	■				▲		▲	▲■	■		■		▲	
Chapter 9	▲	▲■	▲								■▲		▲	■				▲■	▲
Chapter 10	▲		▲	▲		▲■		▲			■	■				■		▲	▲
Chapter 11	▲	■	▲■	▲■	▲■	▲				■			■	▲■	▲	■			■
Chapter 12	▲■			■	▲				▲	▲				▲■	■	■		▲	▲■
Chapter 13	▲	■	▲	■▲		▲	▲				■	■	▲	▲	▲■			▲	
Chapter 14		■	▲	▲■	▲■	▲					▲■	■		▲■	■			■	
Chapter 15	▲	■		■▲	■	▲	■	▲			▲■	■		▲■	▲			▲	▲
Chapter 16	▲	■	▲	■▲				▲			▲	▲■		▲	■			▲	▲■
Chapter 17	▲	▲■	▲■	▲■	■	▲			▲	■				■	■			▲	▲■
Chapter 18	▲	■	▲■	▲■						▲	▲			■				▲	▲■
Chapter 19	▲	■	▲■	■▲	▲	▲								▲■	■				
Chapter 20		■	▲	▲■	■▲	▲	▲			■	■		▲	▲■	▲■		■		▲■
Chapter 21	▲			▲	■			▲		■	▲			▲		▲			■
Chapter 22	▲	■	■	■	■							▲	▲■	▲	■			▲	■▲
Chapter 23	▲		▲■	▲			▲■	▲			■			▲■	▲■			■	▲	▲■
Chapter 24	▲	■	▲■	▲■						▲	■		■			▲■		▲	▲
Chapter 25			■▲	▲▲	▲■			▲		▲		■▲		▲■		■			▲
Chapter 26	■	■		▲	▲■	▲■	▲							▲	▲■	■		▲	■▲
Chapter 27			▲■		▲					▲		■	▲		▲■		■		■
Chapter 28	▲	■	▲	■	▲■					▲		▲■		■	■	■			▲

Integrated Learning

Integrated learning not only makes teaching and learning more fun, it helps students synthesize concepts and integrate skills.

Most middle schools and junior high schools are structured in such a way that fragmentation of the curriculum is obvious. Students have several teachers—each of whom specializes in teaching a particular subject in isolation of other subject areas. Teachers have observed that this approach may not be the best way to serve the early adolescent's educational needs. Instead, presenting an integrated approach to learning can help students synthesize concepts and coordinate experiences. In an integrated approach, teams of teachers work together to integrate the subjects that they teach. Integrated learning can often answer the age-old student question, "Why do I have to learn this?"

Integrating the Curriculum

A commitment to integrate the curriculum requires planning—sometimes months in advance. For integrated learning to be successful, all teachers must be involved and informed. While it may not be practical to integrate all lessons in a given school, some lessons can be integrated or individual teachers can implement integration into individual lessons within their classrooms. Ideally, integrated learning in science should occur at three levels:

▶ Integrating other sciences
▶ Connecting science to other disciplines
▶ Integrating science, technology, and society

Integrating the Sciences

Within *Science Insights—Exploring Living Things,* several strategies have been used to help teachers integrate and make connections. For example:

Section 18.1 Birds Within this section, **physics** and **life science** are integrated. After students learn the adaptations of bird flight, they learn about the physics of flight.

Section 19.2 Muscular System In Chapter 19, students are introduced to the human skeletal and muscular systems. Tying both systems together and integrating **human biology** and **physics,** lever systems in the human body are discussed. Students learn about the three classes of levers, everyday tools that are levers, and lever systems in the human body.

Margin notes throughout the **Teacher's Edition** provide numerous strategies and suggestions for integrating the sciences. For example, in Chapter 11, page 217, a chemistry integration explores bioluminescence in dinoflagellates.

Integrated Resource Book

This unique resource provides science integration worksheets that are correlated to the student text, covering topics from chemistry to physics to environmental science.

An Interdisciplinary Approach

Various components in *Science Insights* have been developed to help teachers make connections across curricula. In the **Student Edition,** features such as Skills WorkOut, Consider This, Historical Notebook, SkillBuilder, and Science and Literature connect science content, events, and concepts to language arts skills, literature, and mathematics.

Margin notes throughout the **Teacher's Edition** include strategies for making interdisciplinary connections.

Science, Technology, and Society

Within the Student Edition of *Science Insights*, each section concludes with a Science, Technology, and Society subsection.

In the Teacher's Edition, margin notes also stress STS Connections, providing additional resources for integration and relevancy.

Multicultural Perspectives

The demographics of the United States are changing rapidly. The school-age population of the nation's largest cities is becoming more and more ethnically diverse. A multicultural perspective has become essential in education as students prepare for membership in these diverse communities.

Building Multicultural Awareness

Multicultural education is a process through which students learn to respect ethnic and cultural diversity. Through this process, students can gain a global perspective on the universal role of science in the world. Multicultural education includes the history and accomplishments of people of all heritages, particularly those who have been underrepresented in the past, such as people of African, Asian, Native-American, Pacific-Islander, and Hispanic backgrounds.

Many of the techniques that you use in your classroom already are well suited to a program that includes multicultural perspectives. In particular, cooperative learning, hands-on activities, and cross-curricular strategies are effective approaches to content that combine science concepts and multicultural understanding.

Here are some suggestions to help you incorporate multiculturalism into your science curriculum:

▶ Tell students about the contribution to science and technology of people from diverse ethnic and cultural backgrounds.

▶ Invite members of the community who are scientists to speak to students about careers in science and technology.

▶ Provide opportunities for students to work in cooperative groups that are balanced with respect to gender, race, and ethnic background.

▶ Provide examples to students of how science applies to the daily lives of all people.

▶ Allow students to develop projects based on their own cultural experiences.

Multiculturalism in *Science Insights*

You will find that multiculturalism is infused throughout the *Science Insights* program.

Student Edition The Student Edition includes many text references and visuals that stress cultural diversity. In addition, Asian, African, Native-American, and Hispanic peoples are represented in illustrations and photographs. For example:

Chapter 17, page 352 In a Science and Society feature, students learn that Inuits living in the Arctic have lowered their risk of heart disease as a result of a diet high in fish. A photograph of an Inuit man fishing through a hole in the ice accompanies the text.

Teacher's Edition You will also find specific strategies for teaching multiculturalism in the Teacher's Edition margin notes, under the head Multicultural Perspectives. For example:

Chapter 4, page 81 Information about the work of African-American Charles Drew, a physician who made important discoveries about blood, is included for discussion.

Chapter 11, page 224 While studying protists, a teaching strategy suggests having students share multicultural recipes for foods made with algae.

Ancillaries References to Addison-Wesley's *Multiculturalism in Mathematics, Science, and Technology: Readings and Activities* are provided in the Teacher's Edition. This book offers a variety of multicultural readings and activities appropriate for grades 7 to 12 in a blackline master format. The lesson topics, covering a diverse multicultural spectrum, combine the vision and experiences of 12 multicultural educators throughout the United States.

Cooperative Learning in the Science Classroom

When people cooperate on a task, they often achieve surprising results.

Cooperative learning is an approach to teaching that involves building a cooperative climate in the classroom as well as structuring specific group activities. To accomplish a cooperative task, students work in small learning groups. Often students find that in sharing information with team members, they come to a better understanding of the science concepts they are studying. Process and critical thinking skills are extended as students become aware of the methods teammates use to solve a problem.

Cooperative Learning Groups

Cooperative learning groups should range in number from two to six. The ideal group size for a cooperative learning activity, however, is four. Once a cooperative learning group has been established, the cooperative group should remain together until the assigned activity has been completed. If a cooperative learning group is having difficulty working together socially or keeping on task, do not dissolve the group. It is important to keep the group intact so that students within the group will learn the social interaction skills necessary to solve problems or to complete tasks effectively through cooperation and collaboration.

Suggested Roles in Cooperative Groups

Here are some suggested roles.

Principal Investigator The Principal Investigator is responsible for managing the tasks within the activity and insuring that all members understand the goals and content of the activity. The Principal Investigator should read instructions, check results, and ask questions of the teacher. Also, the Principal Investigator should facilitate group discussions.

Materials Manager The Materials Manager is responsible for assembling and distributing the materials and equipment needed. As an activity progresses, the Materials Manager is responsible for assembling and operating equipment, as well as checking the results of the activity. The Materials Manager is also responsible for insuring that all equipment is cleaned and returned.

Data Collector The Data Collector is responsible for gathering, recording, and organizing the data. The Data Collector also is responsible for coordinating the certification of the data among all group members and reporting the results of an activity either in writing or orally to the class or to the teacher. If information is being gathered on a master table on the chalkboard, the Data Collector is responsible for recording the data on the chalkboard.

Timekeeper The Timekeeper is responsible for keeping track of time, for safety, and for monitoring noise level. The Timekeeper must also observe and record the group's social interactions and encourage group members to discuss the activity as well as check the results.

Collaborative/Social Skills

Social skills are basic to the cooperative learning process. You should assign a specific social skill for each cooperative learning task. For example, if the activity requires students to hold a debate, the social skill for the activity can be *listening carefully*. Other cooperative group skills include the following:

- Taking turns
- Sharing resources
- Encouraging participation
- Treating others with respect
- Providing constructive feedback
- Resolving conflict
- Explaining and helping without simply giving answers

Self-Evaluation

Encourage students to become actively involved in the evaluation process by providing time for them to reflect on the activity.

Exploring Science Visually

In the science classroom, visual learning strategies have become just as important as the written word. No longer are visuals just pretty pictures. They convey and expand information in a way that helps students learn.

Visual learning is the pedagogical strategy for the 1990s. The strategy employs a distinct correlation between the prose of a text and the illustrations—both photography and art. The integration of prose and visuals enables the graphics to be conveyors of content and concepts. By using visual learning in textbooks and in classroom experiences, the learning process can be enhanced. Visual learning motivates students and makes the content more relevant to them. It also helps students zero in on concepts. Besides its motivational and relevancy aspects, visual learning also helps to promote a number of skills that are essential in the educational environment and in daily life.

Visual Learning and Skills Development

The following is a list of skills that can be developed and strengthened by using visual learning strategies.

- Recognizing color cues that signal important information
- Understanding symbols and their uses
- Recognizing color cues that indicate connections between ideas
- Analyzing and interpreting observations
- Reading graphs, charts, and maps
- Comparing and contrasting
- Identifying and labeling
- Comprehending difficult concepts

Visuals in the Diverse Classroom

Think about the opportunities that visual learning offers the diverse student population in today's classrooms. In a single classroom, you may be teaching students with limited proficiency in English, as well as students who are at-risk or gifted. How do you meet the diverse needs of all these students? Visual learning is part of the answer. Because visuals are a universal language, visual learning provides opportunities for all learners. Integrated prose, diagrams, photographs, art, and maps provide students with broader educational materials, empowering students to approach the content in a way that best suits their individual needs.

Visual Learning in *Science Insights*

Color Cues In Chapter 3, The Chemistry of Living Things, the symbols for every element are color cued; for example, the symbol for oxygen is always represented in red.

Understanding Symbols and Their Uses Many symbols are used throughout the *Science Insights* text, including safety symbols and symbols in everyday life.

Analyzing and Interpreting Observations

Explore Visually is a teaching strategy used throughout the Teacher's Edition that can be used to encourage students to explore the text graphics.

Integration of Prose and Visuals Here are several examples of the various ways in which visual learning is employed in *Science Insights*:

pages 104–105 Phases of Mitosis
The phases of mitosis are presented as a two-page spread that integrates art, photography, and prose. As presented, the material emphasizes the theme Patterns of Change/Cycles.

pages 370–371 Adaptations for Flight and Feathers
In a lively visual learning spread, all aspects of bird adaptation to flight are presented in discrete content blocks directly correlated to the graphics.

Concept Mapping

Concept mapping offers a visual representation of relationships, linking concepts in a way that is highly effective in helping students synthesize new information.

Each student connects concepts differently. Therefore, constructing a concept map with a partner or a team gives students valuable experience in comprehending and communicating the meanings of scientific concepts and terms. Whether used to interpret textbook passages or as a problem-solving tool, concept mapping when paired with cooperative learning leads to lively classroom discussion.

Using Concept Maps in Your Classroom

Here are some strategies for introducing students to concept mapping and making concept maps part of the learning environment:

- Create a large concept map of the year's lessons for the bulletin board.
- List familiar and unfamiliar words from a new lesson and ask students if they can connect the words based on what they already know.
- Have students show their concept maps on the chalkboard and explain how the concepts link.
- Before testing, have students review and revise concept maps they have already made.
- Use concept mapping as a way to let students assist in planning the year's course of study.

Concept Mapping in *Science Insights*

Addison-Wesley's *Science Insights* provides concept mapping opportunities in the Make Connections section of each Chapter Review. In the first chapter of each unit, students copy a concept map on a separate sheet of paper, then complete the maps by writing the correct term in the empty spaces. With each successive chapter in a unit, the concept maps become more challenging. In the last chapter of each unit, students create their own concept maps.

Portfolio Assessment

Portfolio assessment can be an exciting, experimental, and evolving tool for both the student and the teacher.

Portfolio assessment derives its name and approach from the collection of work and achievements usually assembled by artists, designers, and architects. A portfolio is meant to illustrate the true scope of the talents of the professional. In the educational forum, portfolio assessment enables the student and teacher to work together to assemble and to complete the student's best work during a given period of time, usually for a marking period.

Using portfolio assessment, the emphasis is placed on overall achievement rather than numerical test scores. Portfolio assessment stresses what students can do rather than what they cannot do. Also, portfolio assessment allows students to remove materials from their portfolios and revise them.

Contents of the Portfolio

Portfolios can contain any items that students have negotiated with teachers to include. For each item, students should also provide an explanation of the choice. Some suggested portfolio items are:

- Photographs of any work too large to fit in a portfolio
- Rough drafts and completed work
- Diagrams, tables, graphs, or charts
- Audiotapes/Videotapes
- Projects, laboratory investigations, and activities
- Computer printouts

Portfolio and *Science Insights*

Throughout the Teacher's Edition, suggestions are provided for materials, projects, and so on that students may wish to place in a portfolio. These suggestions can be presented to students. Encourage students to plan ahead and suggest ways to revise their work if necessary.

Individual Needs

Everyday classroom teachers face the challenge of a diverse student population with individual students bringing to the educational forum unique sets of abilities, needs, and learning styles.

A Positive Learning Environment

In an effort to provide all students with a positive educational environment with optimum opportunities for success, the following strategies may help teachers structure learning for diverse student populations, especially for those students with different ability levels and those students with limited English proficiency (LEP):

- Provide tools, such as picture dictionaries and other visual learning resources.
- Encourage a variety of responses during discussions, including speaking, drawing, demonstrating, and writing.
- Use hands-on activities and demonstrations to reinforce concepts.
- Speak clearly and slowly using body language and gestures.
- Relate and incorporate various cultural content into discussions. Stress relevant examples.
- Check for understanding and comprehension frequently.
- Use cooperative learning groups for activities and investigations and actively encourage individuals to participate within their groups.
- Approach concepts in several different ways and provide relevant, common examples.

Ability Levels

In *Science Insights,* the various components of each chapter and the supplements have been designated for use by core, standard, and enriched students. The leveling of the components and supplements is presented as suggestions. Based upon individual students, classroom management, and teaching styles, the components and supplements may be used in different levels, as deemed necessary by the teacher.

Limited English Proficiency Students

Science presents a challenge to LEP students at three levels—overcoming a language barrier, achieving mastery in science, and addressing social concerns as students interact with their peers and teachers. The following teaching strategies are designed to help teachers create a positive and rewarding educational environment for LEP students:

- Create audiocassettes for each chapter that model language patterns.
- Use newspapers, magazines, and library research to provide connections between school and the larger world.
- Use illustrations with labeling exercises to build vocabulary.
- Have students keep a "dictionary" of key terms and words that includes the term in English, the term in their native language, and the definition. Attaching drawings or photographs next to the terms or words is also helpful.
- Encourage LEP students to use an English–primary language dictionary.
- Use remedial or average vocabulary and skills worksheets.
- Provide opportunities for LEP students to communicate nonverbally.

Master Materials List

Most of the activities in *Science Insights—Exploring Living Things* have been designed for use with commonly available materials. The quantities shown are based on a class of 30 students with the recommended student groupings found in the Teacher's Edition. When ordering, adjust quantities for your class size and activity groupings. Quantities shown for nonconsumable equipment used in more than one activity reflect the largest quantity needed for any of the activities.

All activities are optional. Your equipment needs will vary, depending on the activities you elect to do. To determine your material needs, each item is referenced by page number to the activity or activities requiring its use. Readily available materials such as tap water and notebook paper are not listed.

Addresses of science material suppliers follow the materials list.

Item	Quantity	Chapter Activities	SkillBuilders
antiseptic products	4 or more types	10	
apple	4		21.2, 27.3
apple peelings	1.5 L	15	
aprons, lab	30	4	
bags, self-lock plastic	32	2	
balances	6	19	
balloons, medium-sized	18	11	
bananas, ripe	8	16	
beakers, 250-mL	18	11	
beakers, 500-mL	8	20, 22	
beakers, 1-L	5	10	
beans, dark	40		7.1
beans, same variety	150–300	7	
beans, white	40		7.1
bleach, powdered	1 box	3	
bolts, 6 types	15 of each	9	
bones, beef	6	19	
bones, chicken	6	19	
bread	8 slices	2	
broth, chicken	5 packets	10	
carrot	1		21.2
clay, modeling	for 15–20 students	8, 22	
clocks, stopwatches, or watches	15	11, 24	
coverslips	100–125	1, 4, 16	
cups, measuring	15	5	
cups, small paper	170	8, 13, 19, 21, 28	
cylinders, graduated	20–25	10, 13, 22, 28	
droppers	30	1, 4, 28	

Item	Quantity	Chapter Activities	SkillBuilders
earthworms	40–50	15	
Elodea plant	1–2	1, 4, 12	
feathers, contour	30	18	
feathers, down	30	18	
fern	1		13.3
flowers (gladiolus, tulip, or lily)	30	14	
foil, aluminum	to cover 30 jars	12	
grapes	60		27.3
household objects, assorted small	15	8	
humus	3 L	15	
ice water	1.5 L	11	
indicator, cabbage juice	4 L	12	
iodine solution	100 mL	4	
jars, 1-L	10	15	
jars, small (with lids)	30	3, 10, 12	27.3
jars, wide-mouthed	15	16	
jugs, plastic drinking (clear)	6	20	
labels, adhesive	108	2, 5, 12	27.3
leaves, from 3 deciduous trees or shrubs	30 of each	28	
lenses, hand	15–20	2, 13, 16, 19	
magazines, old	15–30	25	16.2, 19.3
markers	15–30	25	
markers, wax	6	11	
microscope slides	75–100	1, 4, 16	
microscopes	30	18	
moss samples	15	13	
newspaper	1	1	19.3
nylon stocking material	to cover 15 jars	16	
onion	1		21.2
pans, cake (large)	6	20	
paper, construction (dark)	50 sheets	14, 15	
paper, graph	100 sheets	3, 27	
paper, large	60–90 sheets	25	22.1
paper, pH	30–60 strips	28	
paper, unlined	100	1, 18	
paper clips	100		1.2
paper plates	30	8	21.2
paste and/or tape	for 15–30 students	25	
pencils, colored	60	27	8.1, 22.1
pencils, grease	6–10	10, 20	

Item	Quantity	Chapter Activities	SkillBuilders
pennies	100	21, 22	
petroleum jelly	1 jar	8	
plants, nonflowering samples (not moss)	30	13	
plants, small	90	5	
plaster of Paris	for 2 L plaster	8	
plastic knives	8	2	
plastic spoons	25–30	3, 8, 11, 16	21.2, 27.3
potato	4		21.2, 27.3
pots, flower	30	5	
raisins	60		27.3
rubber bands	100	15, 16	1.2
rulers, metric	30	2, 3, 7, 17, 19, 25–27	
salt	4 oz		27.3
scalpels	30	14	21.2
scissors	30 pairs	1, 6, 25	
screws, 6 types	15 of each	9	
seashells	15	8	
seeds, assorted	1 bag		14.1
seeds, bean	100	28	
seeds, radish	100	28	
soap, liquid	1 small bottle	20	
soil, potting	2 L	28	
soil, 3 types	6 L total	15	
straws	10	12	
sugar	100 mL	11	
tape, adding-machine (optional)	45 m		8.1
tape, transparent	2 rolls		8.1
test tubes, large	20	11	
thermometers, Celsius	10	2, 3	
toothpicks, flat	30	4	
tubing, plastic	3 m	20	
turnip	1		21.2
tweezers	30	1, 4	
vinegar, distilled	3.25 L	19, 28	
water, distilled	250 mL	4	
yeast, dried	6 packets	11	

Science Suppliers

Equipment Suppliers

Carolina Biological Supplies
5100 W. Henrietta Rd.
Charlotte, NC 20391

Central Scientific Co.
11222 Melrose Ave.
Franklin Park, IL 60131

Cuisenaire Co. of America
10 Bank St.
White Plains, NY 10606

Fisher Scientific Co.
4901 W. LeMoyne Ave.
Chicago, IL 60651

Nasco
901 Janesville Ave.
Fort Atkinson, WI 53538

Sargent-Welch Scientific Co.
7300 N. Linder Ave.
Skokie, IL 60077

Audiovisual Distributors

AIMS Media
6901 Woodley Ave.
Van Nuys, CA 91405-4878

Audio Visual Narrative Arts
P.O. Box 9
Pleasantville, NY 10570

Berguall Productions, Inc.
893 Stewart Ave.
Garden City, NY 11530

Charles W. Clark Co., Inc.
168 Express Dr., S.
Brentwood, NY 11717

Churchill Films (CF)
662 N. Robertson Blvd.
Los Angeles, CA 90069

Coronet Films & Video (C/MTI)
108 Wilmot Rd.
Deerfield, IL 60015

Educational Dimensions Group
P.O. Box 126
Stamford, CT 06904

Educational Materials and Equipment Co. (EME)
P.O. Box 17, Pelham, NY 10803

Encyclopaedia Britannica Educational Corp. (EB)
425 N. Michigan Ave.
Chicago, IL 60611

Films Incorporated (FI)
5547 N. Ravenswood Ave.
Chicago, IL 60640-1199

Guidance Associates (GA)
P.O. Box 3000, Mt. Kisco, NY 10549

Human Relations Media (HRM)
175 Tomkins Ave.
Pleasantville, NY 10570

Lucerne Media (LF)
27 Ground Pine Rd.
Morris Plains, NJ 07950

Media Center
2175 Shattuck Ave.
Berkeley, CA 94704

The Media Guild
11722 Sorrento Valley Rd., Suite E
San Diego, CA 92121

National Film Board of Canada
1251 Avenue of the Americas, 16th Floor
New York, NY 10029

National Geographic Society (NGSES)
Educational Services
17th and M Sts., NW
Washington, DC 20036

Optical Data
30 Technology Dr.
Warren, NJ 07059

TVO Video
1443 W. Franklin St., Suite 206
Chapel Hill, NC 27516

WINGS for Learning
P.O. Box 66002
Scotts Valley, CA 95066

Software Distributors

CompuWare Corp.
1008 Abington Rd.
Cherry Hill, NJ 08034

CONDUIT (CO)
The University of Iowa, Oakdale
Iowa City, IA 52242

Focus Media, Inc.
839 Stewart Ave., P.O. Box 865
Garden City, NY 11530

Scholastic Software (SS)
730 Broadway
New York, NY 10003

Videodisc Distributors

Videodiscovery, Inc.
1700 Westlake Ave. N., Suite 600
Seattle, WA 98109-3012

Chapter Bibliography Supplier Codes

A-W = Addison-Wesley Pub. Co., Reading, MA.
AIT = Agcy. for Instr. Television, Bloomington, IN.
AMP = Arthur Mokin Productions, Santa Rosa, CA.
BB = Brain Bank, Gaithersburg, MD.
BF = Barr Films, Pasadena, CA.
BFA = BFA Educational Media, New York, NY.
BLS = Bio Learning Systems, Jericho, NY.
BM = Biology Media, Charlotte, NC.
CBSS = CBS Software, Greenwich, CT.
CCM = Classroom Consortia Media, S.I., NY.
CES = Cross Educational Software, Rustin, LA.
CP = Centre Productions, Inc., Boulder, CO.
CRM = CRM/McGraw-Hill Films, Del Mar, CA.
CVB = Connecticut Valley Biol., Southampton, MA.

DCHES = DCH Ed. Software, Lexington, MA.
EA = Educational Activities, Freeport, NY.
FH = Films for the Humanities, Princeton, NJ.
FL = Filmakers Library, New York, NY.
FM = Focus Media, Garden City, NY.
GH = Grandview Hospital, Westlake, OH.
HA = Hawkhill Associates, Inc., Madison, WI.
IBM = IBM, Atlanta, GA.
JF = Journal Films, Evanston, IL.
J & S = J & S Software, Port Washington, NY.
JWW = J. Weston Walch, Portland, ME.
KU = Knowledge Unlimited, Madison, WI.
LHS = Lawrence Hall of Science, Berkeley, CA.
ME = Marshfilm Ent., Inc., Shawnee Mission, KS.

MECC = MECC, St. Paul, MN.
MF = Macmillan Films, Mount Vernon, NY.
MMI Corp. = MMI Corp., Baltimore, MD.
MPL = Micro Power and Light Co., Dallas, TX.
MSP = Marty Stouffer Productions, Aspen, CO.
NOVA = NOVA, WGBH-TV, Boston, MA.
PES = Preferred Educational Software, Byron, IL.
PF = Pyramid Films & Video, Santa Monica, CA.
PFV = Phoenix Films & Video, New York, NY.
SC = Sunburst Communications, Pleasantville, NY.
SVE = Society for Visual Education, Chicago, IL.
T-L = Time-Life Video, New York, NY.
TIES = TIES, St. Paul, MN.
WN = Ward's Natural Science, Rochester, NY.

Addison-Wesley

Science Insights
Exploring Living Things

Authors

Michael DiSpezio, M.A.
Science Consultant
North Falmouth, Massachusetts

Marilyn Linner-Luebe, M.S.
Former Science Teacher
Fulton High School,
Fulton, Illinois

Marylin Lisowski, Ph.D.
Professor of Education
Eastern Illinois University
Charleston, Illinois

Bobbie Sparks, M.A.
K–12 Science Consultant
Harris County Department
   of Education
Houston, Texas

Gerald Skoog, Ed.D.
Professor and Chairperson
Curriculum and Instruction
Texas Tech University
Lubbock, Texas

▲ Addison-Wesley Publishing Company

Menlo Park, California • Reading, Massachusetts • New York
Don Mills, Ontario • Wokingham, England • Amsterdam • Bonn
Paris • Milan • Madrid • Sydney • Singapore • Tokyo
Seoul • Taipei • Mexico City • San Juan

Content Reviewers

Antonia Adamiak
Biology Teacher
West Windsor-Plainsboro High School
Princeton Junction, New Jersey

Catherine K. Carlson
Science Department Chairperson
Parkhill Junior High School
Richardson, Texas

Charles Chinn
Science Teacher
Mortons Gap School
Mortons Gap, Kentucky

Barbara J. Cornelius
Sixth Grade Science Teacher
Russell Middle School
Winder, Georgia

Elaine Gima
Biology Teacher
Narbonne High School
Harbor City, California

Karel Lilly
Science Department Chairperson
Foshay Junior High School
Los Angeles, California

Richard White
Science Department Chairperson
Austin Community Academy
Chicago, Illinois

Multicultural Reviewers

William Bray
Stanford University
Stanford, California

Gloriane Hirata
San Jose Unified School District
San Jose, California

Joseph A. Jefferson
Ronald McNair School
East Palo Alto, California

Martha Luna
McKinley Middle School
Redwood City, California

Peggy P. Moore
Garnet Robertson Intermediate
Daly City, California

Steven Oshita
Crocker Middle School
Burlingame, California

Modesto Tamez
Exploratorium
San Francisco, California

Front cover photographs: Telegraph/FPG International (Earth); Jeff Foott/Bruce Coleman Inc. (honeycomb); Rod Planck/Tom Stack & Associates (bee); Telegraph/ FPG International (integrated chip wafer); Gareth Hopson for Addison-Wesley (geode).

Back cover photographs: Jeff Foott/Bruce Coleman Inc. (honeycomb); Rod Planck/Tom Stack & Associates (bee).

Copyright © 1994 by Addison-Wesley Publishing Company, Inc. All rights reserved. No part of this publication may be reproduced, stored in a retrieval system, or transmitted, in any form or by any means, electronic, mechanical, photocopying, recording, or otherwise, without the prior written permission of the publisher. Printed in the United States of America.

ISBN 0-201-25728-9

3 4 5 6 7 8 9 10—DO—97 96 95 94

Contents

Chapter 1 Studying Science — 2
1.1	**Science Skills and Methods**	3
	Science and Technology *Safety in a Symbol*	11
1.2	**Measuring with Scientific Units**	12
	Science and You *Settling Cereals*	17
1.3	**Graphing**	18
	Science and Technology *Just a Touch of a Key*	20
1.4	**Microscopes and Lasers**	21
	Science and Technology *Optical Instruments*	25
	Activity 1	
	How do you prepare a wet mount slide?	26
	Chapter 1 Review	27

Chapter 2 The Earth and Living Things — 30
2.1	**The Living Planet**	31
	Science and Technology *Infrared Sensors*	34
2.2	**Characteristics of Living Things**	36
	Science and You *Goose Bumps*	40
2.3	**Needs of Living Things**	41
	Science and Technology *Salty Waters*	45
	Activity 2	
	How does temperature affect mold growth?	46
	Chapter 2 Review	47

Chapter 3 The Chemistry of Living Things — 50
3.1	**Matter**	51
	Science and You	
	Here an Element, There an Element	56
3.2	**Changes in Matter**	58
	Science and Technology *Practical Polymers*	62
	Activity 3	
	What happens during a chemical reaction?	63
3.3	**Molecules of Life**	64
	Science and You *Proper Proteins*	66
	Chapter 3 Review	67
	Science and Literature	
	"The Gorilla Signs Love"	70

Unit 1
Exploring the Sciences

page 1

Unit 2
Cells and Heredity

page 72

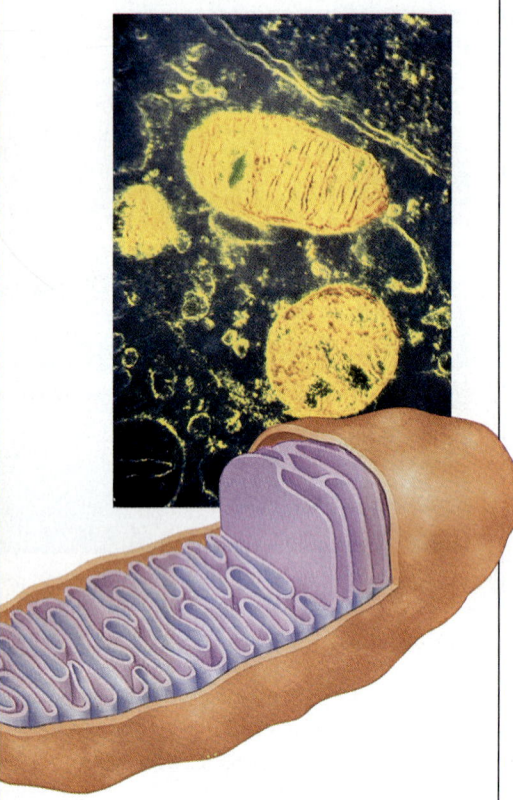

Chapter 4 Cells and Living Things — 74
- 4.1 **The Cell Theory** — 75
  - Science and Technology *Even a Closer Look* — 77
- 4.2 **Parts of a Cell** — 78
  - Science and Technology *A Living Bandage* — 84
  - **Activity 4**
    - *How do animal cells and plant cells differ?* — 85
- 4.3 **Organization of Living Things** — 86
  - Science and Technology *Pacemakers* — 88
  - Chapter 4 Review — 89

Chapter 5 Cell Processes — 92
- 5.1 **Movement of Substances** — 93
  - Science and Technology *Freezing Body Organs* — 96
- 5.2 **Energy Processes** — 97
  - Science and You *Chemistry of Sore Muscles* — 101
  - **Activity 5**
    - *How does light affect plant growth?* — 102
- 5.3 **Cell Growth and Division** — 103
  - Science and Society *Cell Division out of Control* — 106
  - Chapter 5 Review — 107

Chapter 6 Heredity — 110
- 6.1 **Basic Principles of Heredity** — 111
  - Science and You *Your Multigene Traits* — 116
- 6.2 **Chromosomes and Inheritance** — 117
  - Science and Technology *Karyotypes* — 120
  - **Activity 6**
    - *How do you make a model of gene inheritance?* — 121
- 6.3 **DNA and the Genetic Code** — 122
  - Science and You *Sunburn and DNA Damage* — 126
- 6.4 **Applied Genetics** — 127
  - Science and Technology
    - *The Human Genome Project* — 130
  - Chapter 6 Review — 131
  - **Science and Literature**
    - "The Plant People" — 134

Unit 3
Evolution and Classification

page 136

Chapter 7 Diversity of Living Things — 138
- 7.1 **Diversity and Adaptation** — 139
  - Science and You *Helping Out Your Adaptations* — 143
- 7.2 **Variation and Reproduction** — 144
  - Science and You *Variation in Humans* — 146
  - **Activity 7**
    - *How do beans vary?* — 147
- 7.3 **Natural Selection** — 148
  - Science and Technology *Germplasm Banks* — 152
  - Chapter 7 Review — 153

Chapter 8   History of Life — 156

- **8.1   Geologic Time** — 157
  - Science and Technology
    - *Clues to Continental Movement* — 161
- **8.2   The Fossil Record** — 162
  - Science and Technology   *Finding Oil* — 165
  - Activity 8
    - What is the difference between a mold and a cast? — 166
- **8.3   Human Evolution** — 167
  - Science and Society   *The Piltdown Hoax* — 170
  - Chapter 8   Review — 171

Chapter 9   Classifying Living Things — 174

- **9.1   Classification** — 175
  - Science and You
    - *Finding One Book Among Millions* — 179
- **9.2   The Five Kingdoms** — 180
  - Science and Technology   *Clues to the Past* — 182
- **9.3   Identifying Organisms** — 183
  - Science and You   *Birding* — 185
  - Activity 9
    - How do you use a taxonomic key? — 186
  - Chapter 9   Review — 187
  - Science and Literature
    - "In the Night, Still Dark" — 190

Chapter 10   Viruses and Monerans — 194

- **10.1   Viruses** — 195
  - Science and Technology
    - *Curing the Common Cold* — 199
- **10.2   Monerans** — 200
  - Science and Technology   *Foul Food* — 205
  - Activity 10
    - How effective are common antiseptics? — 206
  - Chapter 10   Review — 207

Chapter 11   Protists and Fungi — 210

- **11.1   Diversity of Protists** — 211
  - Science and Society   *Algae as Food* — 213
- **11.2   Protozoa** — 214
  - Science and Society   *Malaria* — 216
- **11.3   Algae** — 217
  - Science and You   *Polishing Up on Diatoms* — 221
- **11.4   Fungi** — 222
  - Science and Society   *Penicillin* — 227
  - Activity 11
    - How does temperature affect yeast growth? — 228
  - Chapter 11 Review — 229
  - Science and Literature
    - "The Great Mushroom Mistake" — 232

Unit 4
Simple Organisms

page 192

Unit 5
Plant Life

page 234

Chapter 12 A World of Plants — 236
- **12.1 Plant Origins** — 237
  - Science and Technology *Hydroponics* — 242
- **12.2 Chemistry in Plants** — 243
  - Science and Society *Too Few Food Plants?* — 246
  - **Activity 12**
    - Do plants breathe? — 248
  - Chapter 12 Review — 249

Chapter 13 Nonflowering Plants — 252
- **13.1 Characteristics of Nonflowering Plants** — 253
  - Science and You *Using Nonflowering Plants* — 255
- **13.2 Bryophytes** — 256
  - Science and Society *Peat Resources* — 258
- **13.3 Nonflowering Vascular Plants** — 259
  - Science and Technology *Making Paper* — 264
  - **Activity 13**
    - Which plant absorbs the most water? — 266
  - Chapter 13 Review — 267

Chapter 14 Flowering Plants — 270
- **14.1 Characteristics of Flowering Plants** — 271
  - Science and Society *Plant Medicines* — 274
- **14.2 Vascular Plant Systems** — 275
  - Science and You *Plant Edibles* — 280
- **14.3 Reproduction of Flowering Plants** — 281
  - Science and You *Pollen Allergy* — 285
- **14.4 Plant Growth** — 286
  - Science and Technology *Future Superplants* — 289
  - **Activity 14**
    - What are the parts of a flower? — 290
  - Chapter 14 Review — 291
  - **Science and Literature**
    - "Songs in the Garden of the House God" — 294

Unit 6
Animal Life

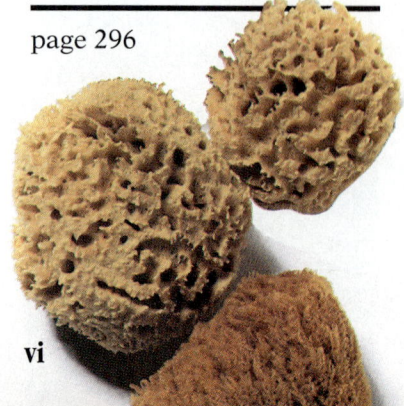

page 296

Chapter 15 Invertebrates I — 298
- **15.1 The Animal Kingdom** — 299
  - Science and Technology *Deep–Sea Exploration* — 301
- **15.2 Sponges** — 302
  - Science and You *Household Sponges* — 304
- **15.3 Cnidarians** — 305
  - Science and You *Stinging Jellyfish* — 308
- **15.4 Worms** — 310
  - Science and Society *Worms as Parasites* — 315
  - **Activity 15**
    - How do earthworms change soil? — 316
  - Chapter 15 Review — 317

Chapter 16 Invertebrates II — 320

16.1 Mollusks — 321
Science and Society *People and Mollusks* — 324
16.2 Arthropods — 325
Science and Society *Friendly Spiders* — 330
16.3 Insects — 331
Science and Technology *A Robot Honeybee* — 335
Activity 16
  What are the stages in fruit fly reproduction? — 336
16.4 Echinoderms — 337
Science and Society
  Sea Urchins, Otters, and Kelp — 340
Chapter 16 Review — 341

Chapter 17 Fishes, Amphibians, and Reptiles — 344

17.1 Common Traits of Vertebrates — 345
Science and You *Fever Biology* — 347
17.2 Fishes — 348
Science and Society *A Fishy Diet* — 352
17.3 Amphibians — 353
Science and Society *Vanishing Amphibians* — 356
17.4 Reptiles — 357
Science and You *Know Your Fangs!* — 361
Activity 17
  What changes occur during metamorphosis? — 362
Chapter 17 Review — 363

Chapter 18 Birds and Mammals — 366

18.1 Birds — 367
Science and Society *Back from the Brink?* — 374
Activity 18
  How do contour and down feathers compare? — 375
18.2 Mammals — 376
Science and Society
  Saving Endangered Mammals — 379
18.3 Diversity of Mammals — 380
Science and You *Humans as Mammals* — 384
Chapter 18 Review — 385
Science and Literature
  "The Cloud Spinner" — 388

Unit 7
Human Life

page 390

Chapter 19 Support, Movement, and Covering — 392
- 19.1 **Skeletal System** — 393
  - Science and Technology
    - *Fiber Optics! Television! Surgery!* — 399
  - **Activity 19**
    - *What gives a bone its hardness and flexibility?* — 400
- 19.2 **Muscular System** — 401
  - Science and You *Those Aching Muscles* — 406
- 19.3 **The Skin** — 407
  - Science and You *Zits and Zats!* — 410
  - Chapter 19 Review — 411

Chapter 20 Supply and Transport — 414
- 20.1 **Digestive System** — 415
  - Science and Society *Eating Disorders* — 420
- 20.2 **Circulatory System** — 421
  - Science and You *Have a Healthy Heart* — 428
- 20.3 **Respiratory System** — 429
  - Science and Society *Indoor Air Pollution* — 433
  - **Activity 20**
    - *How does lung capacity vary?* — 434
- 20.4 **Excretory System** — 435
  - Science and Technology
    - *Dialysis—The Amazing Machine* — 438
  - Chapter 20 Review — 439

Chapter 21 Control and Sensing — 442
- 21.1 **Nervous System** — 443
  - Science and Technology *PETs and CATs* — 449
- 21.2 **The Senses** — 450
  - Science and You *Too Close or Too Far?* — 455
  - **Activity 21**
    - *Are two eyes better than one?* — 457
- 21.3 **Endocrine System** — 458
  - Science and Technology
    - *Synthetic Growth Hormone* — 462
  - Chapter 21 Review — 463

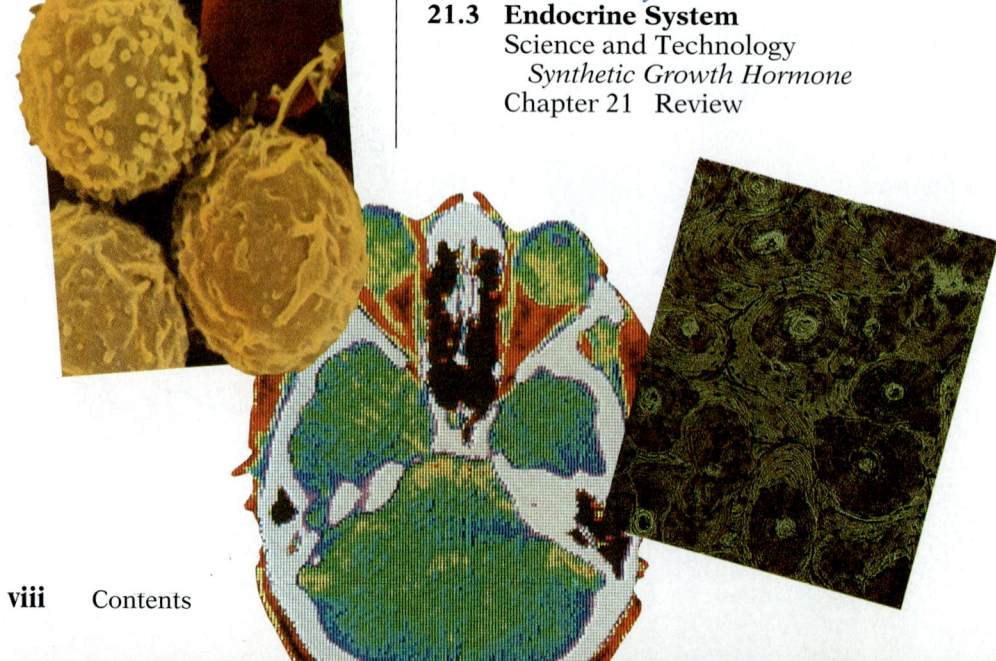

Chapter 22 Reproduction and Life Stages — 466
- **22.1 Human Reproductive Systems** — 467
  - Science and Society *STDs* — 472
- **22.2 Fertilization, Pregnancy, and Birth** — 473
  - Science and Technology *Health Before Birth* — 477
- **22.3 Human Life Stages** — 478
  - Science and Society *Our Aging Population* — 481
  - *Activity 22*
    - *Is age related to learning development?* — 482
  - Chapter 22 Review — 483

Chapter 23 Nutrition, Health, and Wellness — 486
- **23.1 Nutrients** — 487
  - Science and Society *Staple Foods* — 492
  - *Activity 23*
    - *What information is found on food labels?* — 493
- **23.2 Exercise and Rest** — 494
  - Science and Technology *Toning Up* — 497
- **23.3 Drugs and Substance Abuse** — 498
  - Science and You *Critical Choices* — 502
- **23.4 Alcohol and Tobacco** — 503
  - Science and Society *Fetal Alcohol Syndrome* — 506
  - Chapter 23 Review — 507

Chapter 24 Disease and the Immune System — 510
- **24.1 Infectious Disease** — 511
  - Science and You *Lyme Disease* — 515
- **24.2 The Body's Natural Defenses** — 516
  - Science and Technology *The Rh Factor* — 519
  - *Activity 24*
    - *Do people blink their eyes at the same rate?* — 520
- **24.3 Body Disorders** — 521
  - Science and Society *Electronic Pollution* — 524
- **24.4 Medicines That Fight Disease** — 525
  - Science and Society *Sources of New Medicines* — 528
  - Chapter 24 Review — 529
- **Science and Literature**
  - "Fantastic Voyage" — 532

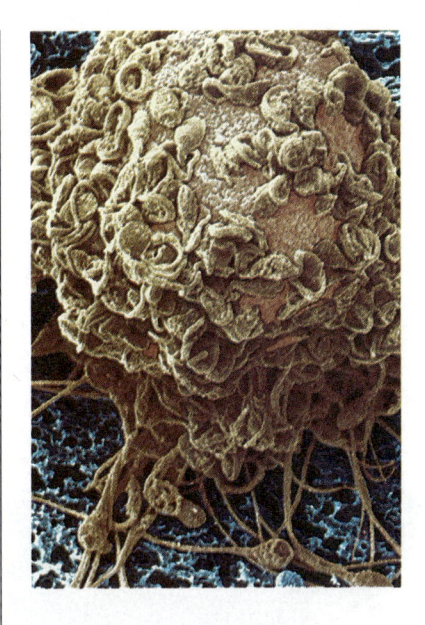

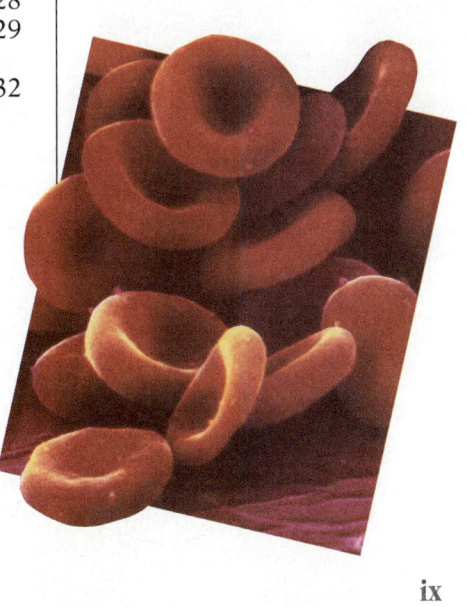

Unit 8
Ecology

page 534

Chapter 25   Organisms and Their Environment — 536
25.1   Ecosystems and Communities — 537
Science and Society   *Balance in Ecosystems* — 541
25.2   Food and Energy — 542
Science and You   *Choose Your Place!* — 546
Activity 25
   How do food chains form a food web? — 547
25.3   Cycles in an Ecosystem — 548
Science and Society   *Global Warming* — 552
25.4   Changes in Ecosystems — 553
Science and Society   *City Succession* — 555
Chapter 25   Review — 557

Chapter 26   Interactions Among Organisms — 560
26.1   Changes in Populations — 561
Science and Society   *Human Population Growth* — 564
Activity 26
   How do you measure population density? — 566
26.2   Relationships Among Populations — 567
Science and You   *Human Symbionts* — 570
26.3   Animal Behavior — 571
Science and You   *Say Good Night!* — 574
Chapter 26   Review — 575

Chapter 27   Climate and Biomes — 578
27.1   Climate — 579
Science and Society   *Know Your Microclimates* — 582
Activity 27
   How do different climates compare? — 583
27.2   Land Biomes — 584
Science and Society   *Desertification* — 591
27.3   Water Biomes — 592
Science and Society   *Living by the Yellow River* — 596
Chapter 27   Review — 597

Chapter 28   Humans and the Environment — 600
28.1   Natural Resources — 601
Science and Technology   *Alternative Fuels* — 607
28.2   Energy Resources — 608
Science and You   *Turn Off the Lights!* — 611
28.3   Pollution — 612
Science and Society   *Recycling* — 617
Activity 28
   How does acid rain affect plants? — 618
Chapter 28   Review — 619
Science and Literature
   "All Things Are Linked" — 622

Data Bank — 624–631
Glossary — 632–641
Index — 642–649
Acknowledgments — 650–653

Activities and Features

Career Corner

What Careers Use Scientific Skills and Knowledge?	10
Health Technician	81
Bread Baker	100
Scientific Photographer	184
Virologist	197
Marine Life Technician	220
Botanist	257
Horticulturist	284
Pet Store Owner	360
Exercise Specialist	403
Speech Pathologist	447
Animal Trainer	572
Agricultural Researcher	606

Historical Notebook

Disproving Spontaneous Generation	39
Solving the Mystery of the Elements	56
Looking at Lenses	76
Better Grain for More People	128
Natural Selection in Action	151
The Leakey Team	168
Plant Classification	241
Exposing the Silent Truth	368
From Free Diving to Scuba	432
The Apgar Score	476
Traditional Medicine	527

Consider This
Decision Making Skills

Should Animals Be Kept in Zoos?	44
Should Food Be Irradiated?	204
Should Protection of Coral Reefs Come First?	309
Should Pesticide Use Be Reduced?	334
Should Sharks Be Protected?	350
Should Products Be Tested on Animals?	378
Should Tobacco Advertising Be Allowed?	505
Who Should Be Tested for HIV?	514
Should Climax Forests Be Logged?	555
Should the World's Population Be Limited?	564
How Should Rain Forests Be Used?	589
How Much Should Be Done to Prevent Oil Spills?	616

SkillBuilder
Problem Solving/Process Skills

Average Mass	16
Chemical Equation Balancing Act	61
Diffusion Rate and Temperature	94
Punnett Squares	115
The Effect of Adaptations	142
Geologic Time Scale	160
Classification Systems	178
A Key to Fungi	226
Factors Affecting Photosynthesis	245
Fern Spore Cases	261
Sowing Information About Seeds	272
Design an Invertebrate	314
Arthropod Search	329
Acne Medicine Advertisements	409
Antibody Formation	426
Tasting Without Smelling?	455
Charting the Menstrual Cycle	471
Measuring Calories Used	495
Sampling a Population	540
Salt Water Versus Fresh Water	595

Science Insights

Discovering is part of the fun...

Preview
Find out the **main topics** in each unit by reading the chapter titles.

Discover
Explore a **resource section** and **develop your skills** by answering the Data Bank questions.

Explore
A **Data Bank** resource section filled with fun and interesting graphs, charts, and facts.

Unit 8 Ecology

Chapters
25 Organisms and Their Environment
26 Interactions Among Organisms
27 Climate and Biomes
28 Humans and the Environment

Data Bank

Use the information on pages 624 to 631 to answer the following questions about topics explored in this unit.

Estimating
By approximately how many British Thermal Units did the United States increase its oil usage between 1940 and 1980?

Comparing
Which continent is more densely populated, Asia or Europe?

Predicting
What type of fuel do you think will be used the most in the United States in the year 2050?

Interpreting Data
What two types of forests would you most likely find at similar altitudes?

Tallest Tree Species

California redwood	113 m
Douglas fir	93 m
Noble fir	85 m
Giant sequoia	84 m
Ponderosa pine	72 m
Cedar	67 m
Sitka spruce	66 m

Oxygen Requirements for Activities

Activity	Oxygen used per hour (L)
Baseball	70
Basketball	90
Bicycling	55
Dancing	100
Football	110

Crustaceans 3%
Arachnids 7%
Centipedes and millipedes 1%
Others 1%
Insects 88%

xii   To the Student

Preview
Use the **Chapter Outline** as a study guide. Read the **Objectives** to help you understand your learning goals.

Explore
Science is more than words. **Skills exploration** is an easy and fun way to develop skills that lead to success.

▼

Challenge
Do the **Check and Explain** questions to check your understanding.

◀

Chapter 16 Invertebrates II

Chapter Sections
- 16.1 Mollusks
- 16.2 Arthropods
- 16.3 Insects
- 16.4 Echinoderms

What do you see?

"I see a parade of colors flashing. It seems like a kind of wing—more like a butterfly's or dragonfly's wing. The colors are green, yellow, red, blue, and blue-green with black spots all over. It seems like some sort of wing because you can see the scales a little bit. You can almost see the shape of the wing, too."

Erik Gomez
Clifton Middle School
Houston, Texas

To find out more about the photograph, look on page 342. As you read this chapter, you will learn about many different kinds of invertebrates.

16.1 Mollusks

Objectives
- **Describe** the characteristics of mollusks.
- **Compare** mollusks to other invertebrates.
- **Describe** the features of each class of mollusks.
- **Make inferences** about geologic history based on mollusk fossils.

Skills WarmUp

Classify Study ten seashells. Look at the shells closely with a hand lens. Classify the shells into two groups based on one property. What property did you use? What are some other properties that you could use to divide the shells into two groups?

How fast do you move when you're getting ready for school in the morning? Do you move at a "snail's pace"? When your teacher asks you a question in class, do you just "clam up"? These expressions describe human behavior, but they also tell you something about snails and clams. Snails and clams are invertebrates called **mollusks** (MAHL uhsks). All mollusks have soft bodies, and many are covered by hard shells.

Characteristics of Mollusks

Mollusks live mostly in the ocean, but some can be found in freshwater habitats. A number of mollusks have also adapted to life on land. Snails and slugs are common in damp places in your backyard.

There are more than 100,000 species of mollusks. They range in size from tiny snails only a few millimeters across to the 20-m-long giant squid.

No matter what their shape or size, all mollusks

Figure 16.1 ▲
This oyster (top) and squid (bottom) may look very different, but they are both mollusks.

Science and Technology
Fiber Optics! Television! Surgery!

Imagine a surgeon performing knee surgery through a tiny incision, or cut, in your knee. With advances in medicine and fiber optics technology, many joint operations are performed just this way. It is called arthroscopic (AR throh SKAHP ihk) surgery.

Arthroscopic surgery is usually performed on knee, shoulder, elbow, and hip joints. Many athletes have arthroscopic surgery to help prolong their sports careers. The technique uses a straight, tubelike instrument called an arthroscope. It contains lenses and bundles of optical fibers. The lenses magnify, and the optical fibers transmit light.

During arthroscopic surgery, the arthroscope is put into a small incision. The surgeon can look through the arthroscope and see what the problem is. An image is also transmitted to a television monitor. Through a second incision, the surgeon can correct the problem by using special small instruments. The image on the monitor helps surgeons see what they are doing.

Because the incisions are so small, there is little tissue damage and little discomfort. Therefore, people heal quickly, and often can be released from the hospital on the same day as their surgery.

Figure 19.6 ▲
What are some advantages to arthroscopic surgery?

Check and Explain
1. Look at the skeleton on page 394. What are the scientific names for the breastbone, hipbone, wrist bones, and upper arm bone? Classify each bone.
2. Describe how each freely-movable joint moves.
3. **Infer** How is the shaft of

Exploring is understanding...

Explore

Photographs, diagrams, and colorful illustrations help you **discover** and **explore** the world of science.

▶

◀ Organs

A heart is one of many organs in a many-celled animal such as a dog or a human. The heart is mostly made up of cardiac muscle tissue. It also contains nerve and connective tissues. Muscle, nerve, and connective tissues work together to pump blood through the heart and body.

Connect
▲

It's easier to learn **new ideas** and **concepts** when they're closely connected to the photographs and illustrations.

xiv   To the Student

Discover
Learn skills and how science is everywhere by doing the **Activities** and **SkillBuilders**.

Activity 28 How does acid rain affect plants?

Skills Observe; Hypothesize; Control Variables

Task 1 Prelab Prep
1. Collect the following items: water, distilled vinegar, graduated cylinder, pH paper, 4 small pots or paper cups, potting soil, 2 leaves each from 3 deciduous trees or shrubs, eyedropper, radish and bean seeds.
2. Make an acidic water solution by mixing 50 mL of vinegar with 150 mL of water in a graduated cylinder. The solution should have a pH of 4.0. Test it.
3. Fill four pots or cups three-quarters full of potting soil.
4. Make the following labels: *Radish—Normal pH*, *Radish—Acidic pH*, *Bean—Normal pH*, and *Bean—Acidic pH*. Attach one label to each pot.
5. For each pair of deciduous leaves, label one *Normal pH* and the other *Acidic pH*.

Task 2 Data Record
On a separate sheet of paper, copy Table 28.1. Use it to record your data and include dates.

Table 28.1 Effects of Acid on Vegetables

	Observations
Radish—Normal pH	
Radish—Acidic pH	
Bean—Normal pH	
Bean—Acidic pH	
Leaf A—Normal pH	
Leaf A—Acidic pH	
Leaf B—Normal pH	
Leaf B—Acidic pH	
Leaf C—Normal pH	

Task 3 Procedure
1. Place a few drops of normal water on the surfaces of the leaves labeled *Normal pH*. Place water on the lea...
2. Set aside these... Continue with th... Record your ob...
3. Plant 5 radish s... 5 bean seeds i...
4. Wet the soil of ... or the acidic w...
5. Set aside thes...tions and rec...
6. Observe and... surfaces of t...

Task 4 Analysi...
1. Identify the...
2. What effect... have on th...
3. What effec... have on th...
4. Were ther... and grow...

Task 5 Con...
Write a short su... affected seed growth and leaf appearan...

Everyday Application
What is the pH of the soil in your yard or on the school grounds? Mix some soil with water and test it with the pH paper. Find out which plants like slightly acidic soil and which do not. Which plants will grow best in your soil?

Extension
Collect water samples from several sites in the area of your school. Test the pH of...

SkillBuilder Making a Model

Geologic Time Scale

The geologic time scale spans about 4,600 million years and is divided into eras. Fossils that show life forms are associated with each era. A model that shows the relative length of time in each era can help you to understand how the length of each era is different.

Collect a metric ruler, a meter stick, 4 different colored pencils, transparent tape, 1.5 m of adding-machine tape, or narrow strips of paper taped together. Copy the table below and complete the third column, using the scale of 1 cm for each 100 million years.

1. Measure out the correct amount of tape for each era. Label each era and color it a different color.
2. Use the information on pages 158–159 and the Data Bank on page 630 to draw or write in examples of life during each era.

3. Which era has the most new life forms?
4. Which era is the longest? Why is this era so much longer than the others?
5. How many times longer is the Precambrian era than the Paleozoic?
6. Does it make sense to divide the Precambrian era into two eras? If so, where would you draw the dividing line?

Geologic Time Scale

Era	Length of Time (millions of years)	Tape Length (cm)
Precambrian	3,960	
Paleozoic	395	
Mesozoic	180	
Cenozoic	66	

Connect
Learn how past discoveries have led to today's science **theories**, **laws**, and **technologies**.

Historical Notebook

Solving the Mystery of the Elements

Ancient peoples of many cultures recognized that some substances could not be broken down into simpler substances. In India, a book written around 1000 B.C. listed 5 "elements": earth, water, air, space, and light. By 300 B.C., people in China believed there were 5 elements—water, fire, wood, metal, and earth—constantly changing into one another. The ancient Greeks knew of 12 of the actual elements known today. However, they were not recognized as elements.

In the 1700s, experiments by a French chemist named Anton Lavoisier gave the concept of the *chemical* element a working definition. Lavoisier listed 33 elements. However, these were not true elements.

After chemical elements were defined, scientists began to recognize that materials suc... gold, tin, and iron were elements. In the p... years, many new elements have been...

Explore
Apply **decision-making skills** to real-world problems.

Consider This

Should Protection of Coral Reefs Come First?

The kind of climate reef corals need to grow is also the kind of climate humans like to live in. On Key Largo, an island off the coast of Florida, developers want to build a new community called Port Bougainville. Environmentalists, however, say that building the community would endanger nearby coral reefs.

...into the sea.

Coral reefs are home to an amazing array of living things, and they protect shores from erosion by waves. For these reasons and others, some people argue that reefs should be protected by limiting what humans can do in the areas around them.

Others...

Consider Some Issues Scien... tists believe t... is damaging... wide. People... polluting the oc... sewage, by an... the reefs, and b... and digging on a...

Connect
Explore real-world opportunities in a wide variety of **careers**.

Career Corner Virologist

Who Finds Out Which Viruses Cause Disease?

The word was in from San Diego. Schools were closed, and many people were forced to stay home from work. As everyone at the Centers for Disease Control in Atlanta, Georgia, had been expecting, a new virus was spreading. It had reached San Diego and was causing a flu epidemic. The virus had been identified in Asia and probably arrived on a naval ship.

Virologists have a major role in trying to keep epidemics like this one under control. In this...those most likely to become very ill from the disease.

Virologists often grow viruses in human tissue cultures. They observe what the virus does to find out if it causes disease, and if so, how. Many virologists use an electron microscope to help them see the viruses they study.

Virologists sometimes provide information to the general public about how a viral disease is transmitted. They are...agricultural animals and plants. They help prevent epidemics among sheep and cattle as well as in many crops.

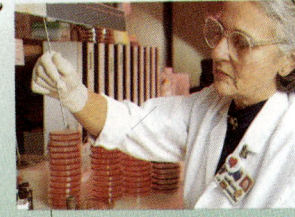

If you would like to become...

xv

Writing is understanding...

Review
Read the **Chapter Summary** and **Chapter Vocabulary** ▼

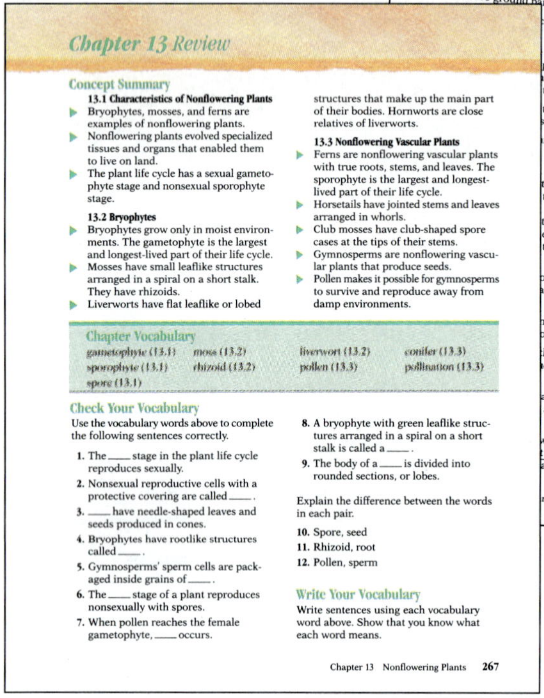

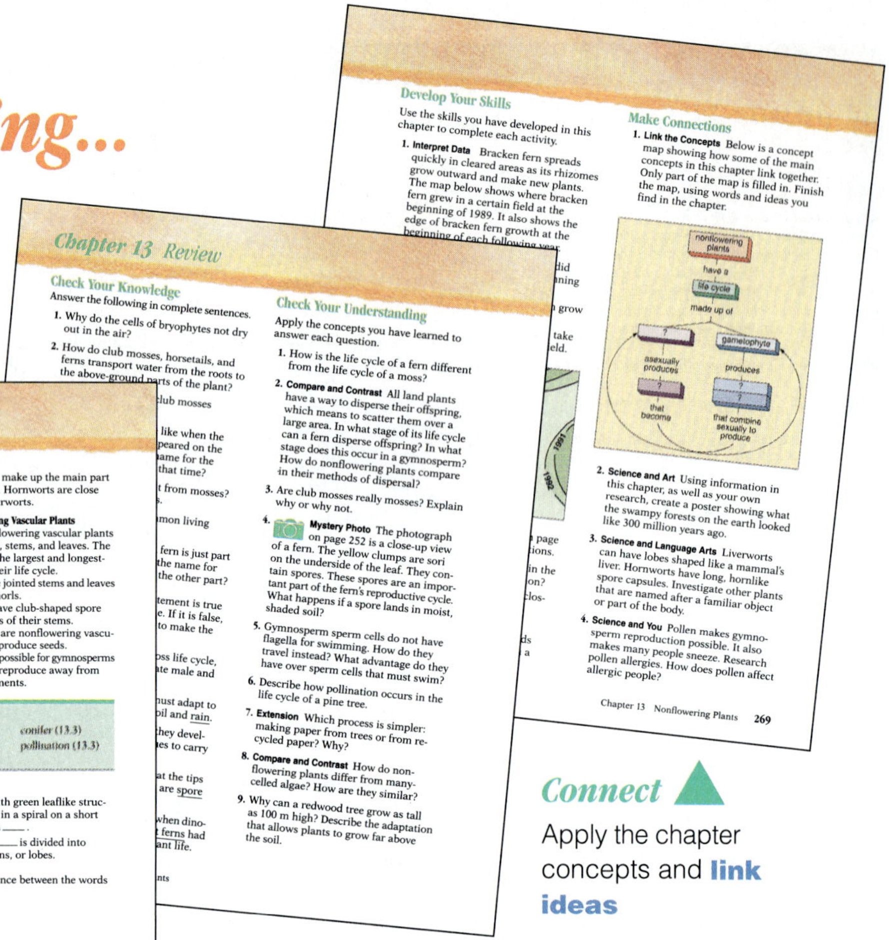

Connect ▲
Apply the chapter concepts and **link ideas**

Challenge ▲
Try the questions to **check your understanding**. Sharpen your **critical-thinking skills**

Connect
Read the literature selections to find out how **science connects** to other subjects. ▼

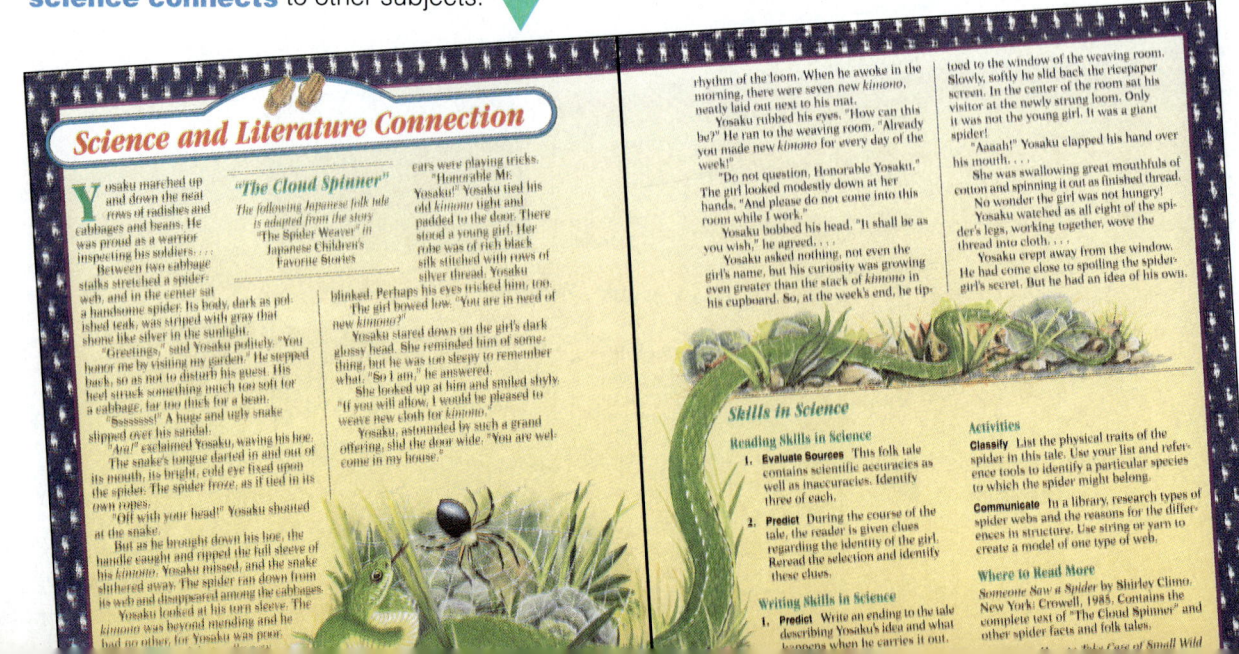

Concept Mapping...

Reading a science textbook is not like reading a magazine or a story. You usually don't have to work hard to understand a story. You probably won't remember it for a long time, either. But when you read science, you are reading to learn something new. You will need to think about what you read. You will also need to remember as much as you can. You may just *read* a story, but you will need to *study* your textbook.

Build a Concept Map

One way to help you study and remember what you have learned is to organize the information in the chapter visually. You can do this by making concept maps. In a concept map, the main ideas are identified by a word or phrase enclosed in a box. When these boxes are linked together, you can better understand the meanings of the ideas by seeing how the concepts are connected to one another. To build a concept map, follow the steps below.

Identify

1. Identify the concepts to be mapped. They may come from a short section of your book, directions from an activity, or a vocabulary list. List the concepts on a separate sheet of paper or on small cards.

Decide

2. Decide which concept is the main idea. Look for ways to classify the remaining concepts. You may want to list or rank the concepts from the most general to the most specific. For example, look at the concept map shown above. Notice that "body regions" is general, and then "head," "thorax," and "abdomen" are more specific concepts.

Organize

3. Place the most general concept at the top of your map. Link that concept to the other concepts. Draw a circle or square around each concept.

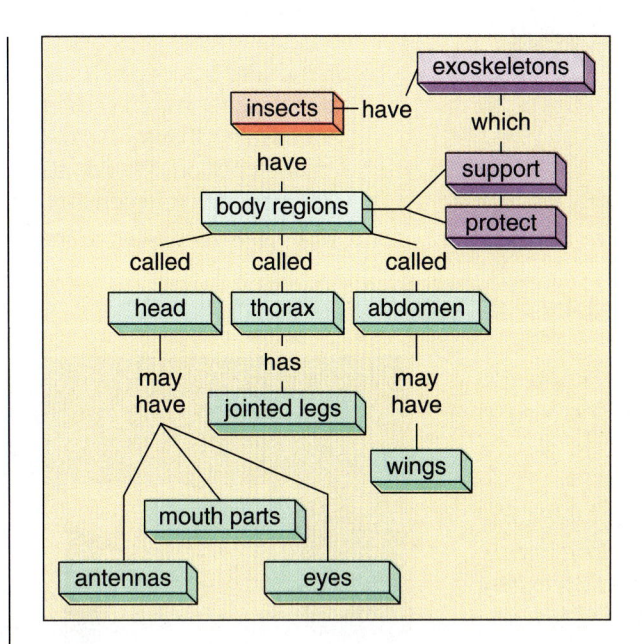

Choose

4. Pick linking words for your map that identify relationships between the concepts. Linking words should not be the concepts themselves. Label all lines with linking words that explain how each pair of concepts relate to each other.

Create

5. Start making your map by branching one or two general concepts from your main concept. Add other, more specific, concepts to the general ones as you progress. Try to branch out. Add two or more concepts to each concept already on the map.

Connect

6. Make cross-links between two concepts that are already on the map. Label all cross-links with words that explain how the concepts are related. Use arrows to show the direction of the relationship.

As you build a concept map, you are doing two things. First, you are automatically reviewing what you already know. Second, you are learning more. Once you have a completed map, you can use it to study and test yourself. You will often find that several different maps can be made from the same group of concepts.

To the Student   xvii

UNIT 1

UNIT OVERVIEW

Unit 1 provides an introduction to the study of science, a survey of living things, and a discussion of the basic characteristics of matter. Chapter 1 introduces science skills and methods. It also describes the use of microscopes and lasers. Chapter 2 explores the earth's unique position in the solar system. The life-supporting elements of the biosphere and the basic characteristics of simple and complex organisms are also explored. Finally, this chapter discusses the ways in which the needs of living things are met. Chapter 3 defines matter and its physical and chemical properties, presenting chemical symbols and formulas. The chapter also explains physical and chemical changes. In addition, the structure and function of some important organic compounds are described.

Introducing the Unit

Directed Inquiry

Have the students examine the photograph. Ask:

▶ How do you think scientists use petri dishes such as the ones shown here? (Scientists use them to grow organisms. Accept all reasonable answers.)

▶ What kinds of tools do you think scientists might need when they work in a laboratory? (They might need an apron, gloves, utensils for handling organisms, and so on. Accept all reasonable answers.)

▶ Why might it be helpful to observe organisms in an isolated laboratory environment? (It could limit the variables in an experiment.)

Writing About the Photograph

Ask students to think about the substances in this photo and imagine what they might be. Have them write a story about the substance in one of the petri dishes and its importance to the scientists who are studying it.

Unit 1
Exploring the Sciences

Chapters

1 Studying Science
2 The Earth and Living Things
3 The Chemistry of Living Things

Data Bank

Use the information on pages 624 to 631 to answer the following questions about topics explored in this unit.

Reading a Diagram

Which is larger, an atom or a molecule?

Comparing

Which planet is closest to the sun? Which planet is farthest from the sun?

Making a Graph

Make a bar graph to compare how many kilometers each planet is from the sun.

Calculating

How much larger is Jupiter than Earth?

Scientists work in laboratories and in the field, such as on ocean liners and in rain forests. They use many skills, techniques, and tools in their work. What tool is the scientist using in the photograph?

Data Bank Answers

Have students search the Data Bank on pages 624 to 631 for the answers to the questions on this page.

Reading a Diagram A molecule is larger than an atom. The answer is found in the diagram, Relative Size of Organisms, on page 625.

Comparing Mercury is the closest planet to the sun and Pluto is the farthest planet from the sun. The answer is found on page 624 in the table, Planetary Statistics.

Extension Ask students to find out which planet is the largest in the solar system and which planet is the smallest. (Jupiter is the largest planet. Pluto is the smallest.)

Making a Graph Information for students' graphs is found in the table, Planetary Statistics, on page 624.

Extension Ask students to find out which gases are most common on the earth. (Nitrogen, carbon dioxide, oxygen)

Calculating Jupiter is 130,040 kilometers larger than the earth in circumference. The answer is found on page 624 in the table, Planetary Statistics.

Extension Ask students to find out how many times Pluto would fit inside the earth. (Pluto would fit inside the earth three times.)

Answer to In-Text Question

The scientist is using petri dishes to study microorganisms. This is a preliminary step to using the microscope.

CHAPTER 1

Overview

This chapter provides an introduction to the study of science. The first section describes science skills, safety skills, and methods for designing experiments. The second section discusses base units of the SI system with everyday applications. This chapter also includes a discussion of various kinds and specific uses of graphs. The final section describes the role of microscopes and lasers in science.

Advance Planner

▶ Provide 1-liter, 2-liter, quart, and half-gallon containers for student estimation exercise. TE page 14.

▶ Use a balance to measure the masses of objects before engaging students in TE activity, page 15.

▶ Prepare Celsius thermometers, goggles, aprons, ice water, and boiling water for TE activity, page 16.

▶ Gather magazines for SE activity and TE activity, page 22.

▶ Prepare microscopes, unruled paper, scissors, droppers, water, slides, coverslips, *Elodea* leaves, newspaper, and tweezers for SE Activity 1, page 26.

Skills Development Chart

Sections	Collect Data	Communicate	Graph	Infer	Interpret Data	Measure	Model	Observe	Sequence
1.1 Skills WarmUp Skills WorkOut Skills WorkOut		●			●		●		
1.2 Skills WarmUp Skills WorkOut SkillBuilder						● ●			●
1.3 Skills WarmUp Skills WorkOut	●			●					
1.4 Skills WarmUp Skills WorkOut Activity	●		●					● ● ●	

Individual Needs

▶ **Limited English Proficiency Students** Before students read each section, have them divide a page from their science journals into three columns. Have them list boldface vocabulary terms in the first column. Have students use the glossary or dictionary to find the definition for each term. Have them write these definitions in the second column. As students read each section, have them draw a picture or diagram in the third column that represents their words. Encourage students to use the information and figures from the chapter as a resource for their drawings. For example, students could draw the tulips on page 7 for *controlled experiment*.

▶ **At-Risk Students** Show students that they regularly use scientific methods. Refer them to the description of science process skills beginning on page 3. Have students describe a situation in which they might use science skills or methods to solve a problem. It might be a situation where they find themselves in danger, have to find their way from one place to another, or need to fix a broken appliance. Have students write a description of the problem and list the science skills that would help them solve it.

▶ **Gifted Students** Have students study a picture of the Golden Gate Bridge in the San Francisco Bay area. Ask them to research the construction of the bridge. Students should consider the placement of the bridge, the geology of the area, the climate, and the materials that were used in its construction. Have students note the science skills and methods that were involved in the process of constructing the bridge. Then have them use their information to construct a scale model of the bridge.

Resource Bank

▶ **Bulletin Board** Begin this display by placing letters spelling the title *Science Is...* on the bulletin board. Arrange photographs of some famous scientists, such as Marie Curie, Albert Einstein, Yuan T. Lee, Maurice Wilkins, Rosalind Franklin, Francis Crick, James Watson, and Charles Drew. Have students add words that describe what science is. Encourage students to add drawings or other pictures to the bulletin board that illustrate the nature of science and the tools people use to study it.

▶ **Field Trip** Before beginning this chapter, arrange to take students to the school science laboratory or to another laboratory so they can observe and experiment with scientific instruments. While you're there, go over the safety rules outlined on page 6.

CHAPTER 1 PLANNING GUIDE

Section	Core	Standard	Enriched	Section	Core	Standard	Enriched
1.1 Science Skills and Methods pp. 3–11				**1.3 Graphing** pp. 18–20			
Section Features Skills WarmUp, p. 3	●	●	●	**Section Features** Skills WarmUp, p. 18	●	●	
Skills WorkOut, p. 5	●	●		Skills WorkOut, p. 20		●	●
Skills WorkOut, p. 10	●	●		**Blackline Masters** Review Worksheet 1.3	●	●	
Career Corner, p. 10	●	●	●	Skills Worksheet 1.3	●	●	●
Blackline Masters Review Worksheet 1.1	●	●		Integrating Worksheet 1.3	●	●	●
Reteach Worksheet 1.1	●	●		**Ancillary Options** Multicultural Perspectives, p. 31	●	●	●
Skills Worksheet 1.1a	●	●					
Skills Worksheet 1.1b		●	●	**Overhead Blackline Transparencies** Overhead Blackline Master 1.3 and Student Worksheet	●	●	●
Ancillary Options One-Minute Readings, p. 132	●	●	●	**Laboratory Program** Investigation 2	●	●	●
Color Transparencies Transparencies 1, 2	●	●		**1.4 Microscopes and Lasers** pp. 21–26			
1.2 Measuring with Scientific Units pp. 12–17				**Section Features** Skills WarmUp, p. 21	●	●	●
Section Features Skills WarmUp, p. 12	●	●	●	Skills WorkOut, p. 22	●	●	
Skills WorkOut, p. 13		●	●	Activity, p. 26	●	●	●
SkillBuilder, p. 16	●	●		**Blackline Masters** Review Worksheet 1.4	●	●	
Blackline Masters Review Worksheet 1.2	●	●		Integrating Worksheet 1.4	●	●	●
Skills Worksheet 1.2	●	●	●	Vocabulary Worksheet 1	●	●	
Integrating Worksheet 1.2		●	●	**Overhead Blackline Transparencies** Overhead Blackline Master 1.4 and Student Worksheet	●	●	
Ancillary Options One-Minute Readings, p. 122	●	●	●				
Laboratory Program Investigation 1	●	●	●	**Color Transparencies** Transparency 3	●	●	

Bibliography

The following resources can be used for teaching the chapter. See page T-40 for supplier codes.

Audio-Visual Sources
(video unless noted)
Laboratory Safety. 17 min. 1987. A-W.
The Microscope. 17 min. A-W.

Software Resources
Photosynthesis and Light Energy. IBM. CCM.
The Simplest Living Things. Apple. DCHES.

Library Resources
Asimov, I. *Asimov's Biographical Encyclopedia of Science and Technology.* 2d ed. New York: Doubleday and Co., Inc., 1982.
Comstock, Anna Botsford. *Handbook of Nature Study.* Ithaca: Cornell University Press, 1986.
Swertka, Eve and Albert. *Make It Graphic! Drawing Graphs for Science and Social Studies Projects.* New York: Messner, 1985.

CHAPTER 1

Introducing the Chapter

Have students read the description of the photograph. Ask students if they agree or disagree with the description.

Directed Inquiry

Have students study the photograph. Ask:

▶ What is the shape of the particles in the photograph? (The particles are spheres that seem to have a rough surface.)

▶ Do you think the particles are part of a living thing? Why? (The particles are pollen grains, which are parts of plants.)

▶ The photograph shows grains of pollen. These grains are very small. How do you think it's possible to get such a good view of something this small? (By using a microscope)

▶ What is the subject of this chapter? (Studying science)

▶ Why do you think a picture of magnified pollen opens a chapter on studying science? (To show that scientists use tools like microscopes to study the world and living things)

Chapter Vocabulary

constant
controlled experiment
cubic meter
data
density
kilogram
lens
liter
mass
meter
microscope
variable
volume

INTEGRATED LEARNING

Math Connection

Tell students that the pollen grains of ragweed shown in the photograph are magnified 325 times. Have students measure the diameter of one of the grains and calculate its actual size. (Actual size is equal to the measurement divided by 325. The actual size of the grain in the foreground is about 0.02 cm.)

Chapter 1 Studying Science

Chapter Sections

1.1 Science Skills and Methods

1.2 Measuring with Scientific Units

1.3 Graphing

1.4 Microscopes and Lasers

What do you see?

"I think the picture is pollen. These objects are smaller than a grain of salt. This was taken from under a microscope. I am highly allergic to this, so it would make me sneeze and I'll be miserable."

*Mike Scian
Schulyer-Colfax
Middle School
Wayne, New Jersey*

To find out about the photograph, look on page 28. As you read this chapter, you will learn the methods that scientists use to gain knowledge about the world you live in.

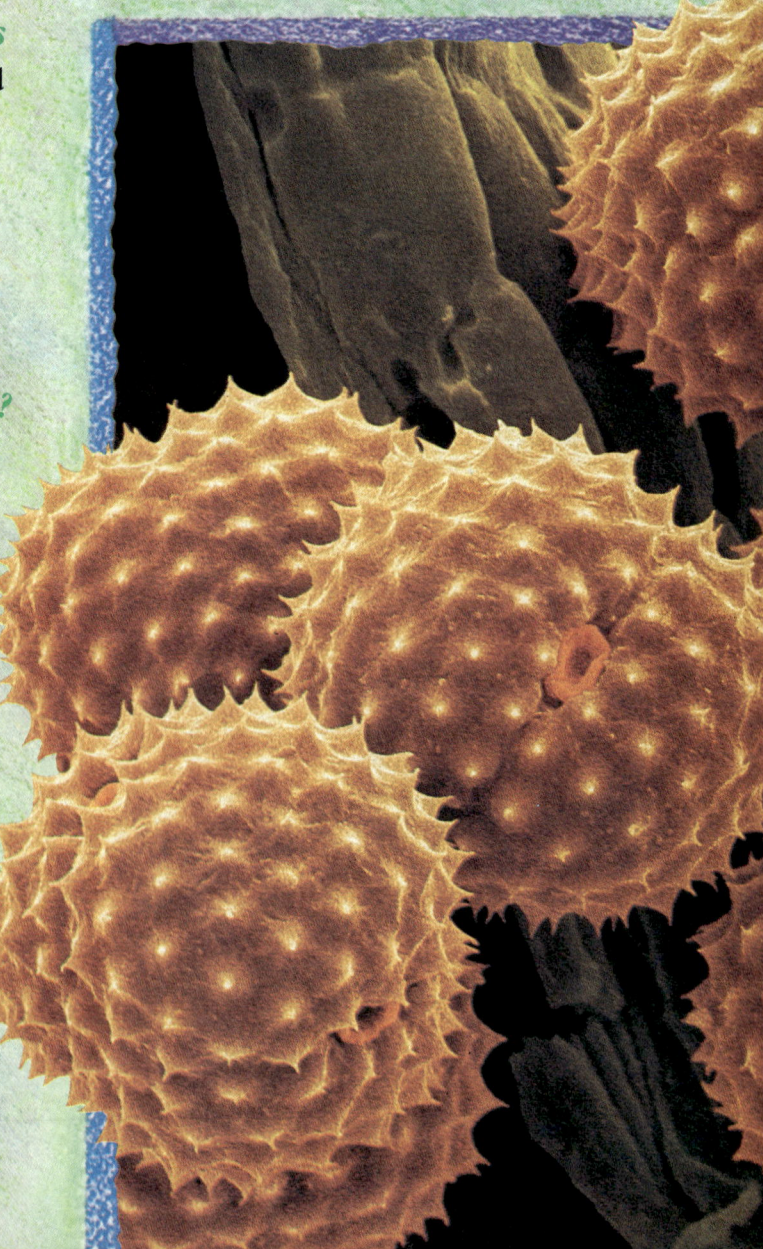

TEACHING OPTIONS

Cooperative Learning
Have students work in groups of four to find definitions for the words *knowledge, skills, systemized,* and *experiment*. Have students use these meanings to write a definition of *science*. Then have students discuss their definitions and develop one working definition for *science*. The class should then discuss all the definitions and develop one definition for *science*.

Portfolio
Have students make a list of instruments that extend their ability to observe. Have them begin with microscope, thermometer, and ruler. Students should explain how each instrument extends a person's ability to observe.

1.1 Science Skills and Methods

Objectives
- **Identify** and **use** science skills.
- **Describe** a controlled experiment.
- **Describe** a scientific method.
- **Distinguish** between theory and law.
- **Make a model** of a safety symbol.

Skills WarmUp
Communicate What is your favorite food? Picture the food in your mind and then write what you see, feel, taste, and smell. Exchange your description with a partner and see if you can guess each other's food.

What is science? If you looked up the word *science* in a dictionary, you would read many different definitions. In these definitions, you probably would read several key words again and again. Most likely, these key words would be *knowledge, skills, systemized,* and *experiment*. You may not know what systemized means. *Systemized* means "planned in an orderly way." Science is a way of gathering and organizing information about the natural world and universe. The knowledge was acquired by observing, experimenting, or studying information in an orderly way.

Science Skills

When scientists study, observe, and experiment, they gather information, or **data**. While they work, scientists use many skills. Sometimes these skills are called science process skills. As you read about these skills, you will discover a kind of secret. These skills are not some mysterious or hard way to do something. You may not realize it, but you use many of these skills every day.

Observe The most direct way of gaining knowledge about something in nature is to observe it. When you observe, you use one or more of your senses to get information about your surroundings. Your senses are sight, touch, taste, smell, and hearing. Your ability to observe can be extended through the use of tools, such as microscopes, thermometers, and rulers. Look at Figure 1.1. Write as many observations as you can about it. Which sense did you use most often? ①

Figure 1.1 ▲
You can make observations about the color, shape, and design of these traffic signs.

SECTION 1.1

Section Objectives
For a list of section objectives, see the Student Edition page.

Skills Objectives
Students should be able to:

Communicate observations about their food.

Interpret Data by examining how a phone book is organized.

Make a Model of a safety symbol.

Vocabulary
data, controlled experiment, constant, variable

MOTIVATE

Skills WarmUp
To help students understand science skills and methods, have them do the Skills WarmUp.
Answer Students should be able to describe their favorite foods in terms of smell, taste, feel, and sight.

Prior Knowledge
To find out how much students know about science skills and methods, ask the following questions:

- What do you think it means to study science?
- What kinds of tools do you use to study science?
- Describe an experiment that you've done.

Answer to In-Text Question
① The sense of sight is used most when writing observations about Figure 1.1.

Chapter 1 Studying Science 3

TEACH

Class Activity
Assign students to cooperative groups. Have them make estimates of the length of the room, the width of the room, and the height of the table or desk in front of them. Then have students take measurements and compare these with their estimates.

Skills Development
Predict Have students predict what will happen to them on the way home from school today. Have students write out their predictions and then give one or more reasons why they feel that their predictions will occur. Have them report the next day on whether or not their predictions were correct.

Answers to In-Text Questions
① The words YIELD and STOP would be written on the signs.
② There are 64 blocks in the cube and 16 flowers in the photograph.

INTEGRATED LEARNING

Integrating the Sciences

Earth Science In ancient Greece, the astronomer, mathematician, and poet named Eratosthenes estimated the circumference of the earth. Since it was physically impossible to measure such an enormous distance, Eratosthenes measured the angle at which the sun's rays hit the earth at noon on the summer solstice in two different cities. At Syene in Egypt the rays fell vertically (0°), and in Alexandria the rays fell 7° from being vertical. He compared the two angles and then used the distance between the cities to calculate the earth's circumference. Modern astronomers have found that Eratosthenes' estimate is close to being accurate. Ask students to name some other quantities that are impossible to measure exactly.

Figure 1.2 ▲
What can you infer would be written on these signs? ①

Infer When you infer, or make an inference, you suggest a possible explanation for an observation. You often can make more than one inference to explain the same observation. Look at Figure 1.2. You probably can infer several things about the drawings. For example, one inference could be that they represent traffic signs. Most likely, you also could infer that one of the drawings represents a stop sign because of its color and eight-sided shape. The shape and color help you infer its meaning without reading the word STOP on it.

Estimate When you estimate, you make careful guesses. Estimating skills are used to gather information when exact measurements are not needed or when they would be impossible or too time-consuming to get. You learn to make estimates about many things. You estimate speed, distance, size, time, and so on.

Measure When you need exact and careful information about an observation, you measure. Measurements describe the amount of an observation. Measurements include both a number and a unit. The number of things in a group, the size of an object, the speed of a car or an airplane, your height and weight—all of these are forms of measurement.

Predict When you predict, you state what you think might happen in the future. Predictions are based on past experiences and observations. Therefore, you can give the how-and-why something might occur. Knowing why a prediction is right or wrong is important.

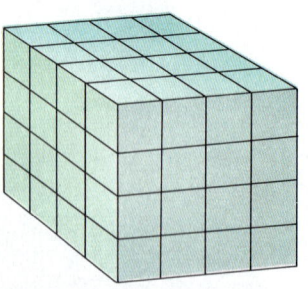

Figure 1.3 ▲
Estimate the number of blocks ▶ in the cube and the number of flowers in the photograph. Count the number of blocks and flowers for a more exact measurement. Compare your numbers. ②

4 Chapter 1 Studying Science

TEACHING OPTIONS

Cooperative Learning

Assign students to cooperative groups. Have them use the skills of observing, classifying, recording, and organizing by categorizing objects in the classroom. Have groups choose a type of object that they wish to study, such as books, notebooks, backpacks, and so on. Have them observe all the objects that are part of the group they choose. Then have groups classify the objects according to several characteristics. For example, books could be classified according to subject, size, or color. Have one spokesperson from each group report to the class on the ways that they classified their objects and the processes they used.

Skills WorkOut

To help students interpret data, have them do the Skills WorkOut.
Answers The data is organized alphabetically according to the last name. Classifying by number would only make sense if you had a person's number but could not think of his or her name.

Class Activity

Have students work in groups of three. Have one student name a problem or question and have the other two students state a hypothesis to answer it. Then have the students switch roles so that each person names a problem. For example, one problem or question might be, Why does a light shine when the switch is turned on? Have students refer to the definition of a hypothesis before they make a hypothesis to answer the question. When each group is finished, have them state how the hypothesis may be tested.

Answer to In-Text Question

③ **The shells could be classified by size, shape, color, pattern, or texture.**

◀ **Figure 1.4**
What are some different ways in which you could classify these shells? ③

Classify When you classify, you group things based on how they are alike. You may be able to group things in many different ways. Some ways to group things are by size, color, shape, texture, or any other characteristic. In science, rocks are classified by how they are formed. Living things, or organisms, are classified into groups that share common ancestors. Organisms that have common ancestors have similar characteristics.

Hypothesize When you state a hypothesis (hy PAHTH uh sihs), you suggest an answer to a problem. Your answer is based on information that you know. Think of a hypothesis as an explanation or an idea that states why something may always occur. Once you state a hypothesis, you can test your hypothesis by observing, studying, or experimenting. Your observations, research, or the results of experiments should support your hypothesis. If not, you need to look at your hypothesis again and state a new one.

Record and Organize Scientists must keep careful records of their observations. During activities and investigations, you, too, will record observations, measurements, predictions, and so on. Often you will want to organize the data you collect in some way. There are a number of ways you can record and organize data. Some ways are tables or charts, graphs, and diagrams.

Analyze Once data have been recorded and organized, you need to analyze the data. When you analyze data, you study the data to look for trends or patterns. You are looking to see if your data support your hypothesis, prediction, or inference.

Skills WorkOut

Interpret Data Examine a page from a phone book. How are these data organized? Can you uncover any pattern between name, address, and phone number? Why aren't the entries made according to numerical order of the phone numbers?

Chapter 1 Studying Science   5

TEACH • Continued

Directed Inquiry

Have students study the general safety guidelines and Table 1.1 on this page. Lead students in a discussion using the following questions:

▶ Why is it important to follow the general safety guidelines? (Eating or drinking in the laboratory is distracting to others and may lead to an accident. Knowing what to do before you do it saves time and avoids accidents. Cleaning up your work area is courteous, it prevents accidents, and it makes materials last longer.)

▶ What should you do in case of any accident? (Inform the teacher.)

▶ Why is there a symbol associated with each safety category and what is the purpose of these symbols? (The symbol provides a clear, immediate alert to the type of potential danger associated with a situation.)

TEACHING OPTIONS

Portfolio

Safety is just as important outside the laboratory as it is in the laboratory. For each of the ten categories in Table 1.1, have students write one safety rule to follow at home (in the kitchen, bathroom, or outside).

Safety Skills

Just as safety is important in everyday life, safety is important in your science classroom. When working in a laboratory, you should always follow these general safety guidelines.

▶ Do not chew gum, eat, or drink in the laboratory.

▶ Read through the activity before you start. Reread each procedure before you follow it.

▶ Clean up your laboratory work area after you complete each activity.

Study the safety symbols and guidelines in Table 1.1. By following these guidelines, you can help make the science laboratory a safe place.

Table 1.1 Laboratory Safety

Plant Safety ▶ Use caution when collecting or handling plants. ▶ Do not eat or taste any unfamiliar plant or plant parts. ▶ If you are allergic to pollen, do not work with plants or plant parts without a gauze face mask.	**Animal Safety** ▶ Handle live animals with care. If you are bitten or scratched by an animal, inform your teacher. ▶ Do not bring wild animals into the classroom. ▶ Do not cause pain, discomfort, or injury to an animal. Be sure any animals kept for observations are given the proper food, water, and living space. ▶ Wear gloves when handling live animals. Always wash your hands with soap and water after handling live animals.
Eye Safety ▶ Wear your laboratory safety goggles when you are working with chemicals, open flame, or any substances that may be harmful to your eyes. ▶ Know how to use the emergency eyewash system. If chemicals get into your eyes, flush them out with plenty of water. Inform your teacher.	
	Fire Safety ▶ Tie back long hair when working near an open flame. Confine loose clothing. ▶ Do not reach across an open flame. ▶ Know the location and proper use of fire blankets and fire extinguishers.
Heating Safety ▶ Turn off heat sources when they are not in use. ▶ Point test tubes away from others when heating substances in them. ▶ Use the proper procedures when lighting a Bunsen burner. ▶ To avoid burns, do not handle heated glassware or materials directly. Use tongs, test-tube holders, or heat-resistant gloves or mitts.	
	Glassware Safety ▶ Check glassware for chips or cracks. Broken, cracked, or chipped glassware should be disposed of properly. ▶ Do not force glass tubing into rubber stoppers. Follow your teacher's instructions. ▶ Clean all glassware and air dry them.
Clothing Protection ▶ Wear your laboratory apron. It will help protect your clothing from stains or damage.	**Electrical Safety** ▶ Use care when using electrical equipment. ▶ Check all electrical equipment for worn cords or loose plugs before using them. ▶ Keep your work area dry. ▶ Do not overload electric circuits.
Poison ▶ Do not mix any chemicals unless directed to do so in a procedure or by your teacher. ▶ Inform your teacher immediately if you spill chemicals or get any chemicals on your skin or in your eyes. ▶ Never taste any chemicals or substances unless directed to do so by your teacher. ▶ Keep your hands away from your face when working with chemicals.	**Sharp Objects** ▶ Be careful when using knives, scalpels, or scissors. ▶ Always cut in the direction away from your body. ▶ Inform your teacher immediately if you or your partner is cut.

Cooperative Learning

Scientists use many different approaches to study phenomena. However, they design experiments similarly. After students read this page and study Figure 1.5, have them design their own controlled experiments in cooperative groups. Tell them to be sure to do each step listed below the second paragraph. Groups can then present their experiments to the class.

Critical Thinking

Reason and Conclude Use Figure 1.5 to emphasize what a controlled experiment is. Have students note that all the flowers were the same and were treated in exactly the same way with the exception of the variable, which was plant food in this case. Ask students to reason and conclude why more than one plant was included in each group. (That way, one could be more certain that the results would be the same for all members of that group. In fact, if there were more flowers in each group, there would be a higher degree of certainty.)

Answer to In-Text Question

① Yes, the observations do support the hypothesis. The conclusion is that plant food does extend the time that cut flowers stay fresh because the flowers in the experimental group did not wilt and lose petals.

Experiments

What's the best way to find out about something? If you said, ask questions, you're on the right track. In your everyday life, you ask many questions to find out information. Science enables you to ask questions and then find the answers to your questions. One of the best ways to find answers in science is to experiment.

What do scientists do when they experiment? What would you do? If you said that you would carefully plan a series of activities, steps, or procedures, you would be off to a good start. Experiments must be planned and designed to:

▶ Observe how something behaves.

▶ Investigate an observation.

▶ Test an idea, prediction, hypothesis, or even an inference.

▶ Find an answer to a question.

When you plan and design an experiment, you should be sure that the experiment can be repeated. It must be able to be repeated exactly as you did it. Often, experiments need to be repeated by other people or even by you to make sure the data and conclusions are accurate. Repeating an experiment also ensures that all people doing the experiment will have the same hypothesis.

Most experiments are **controlled experiments**. In a controlled experiment, there are two test groups—the experimental group and the control group. The control group is a standard by which any change can be measured. In the experimental group, all factors are kept the same as in the control group except one.

In a controlled experiment, the factors that are kept the same are **constants**. The factor that is changed by the person doing the experiment is the **variable**. Study Figure 1.5, which shows a controlled experiment.

Figure 1.5 A Controlled Experiment ▼

Hypothesis: Adding plant food extends the time that cut flowers stay fresh.

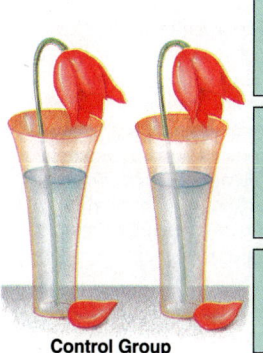

Control Group
- The same kind of flower was placed in each vase.
- Notice that the number of petals on each flower is the same.
- Each vase contains an equal amount of distilled water.

Experimental Group
- Both vases received the same amount of sunlight. The water temperatures were the same in both vases.
- An equal amount of plant food was added to the two vases in the experimental group.

After three days, the tulips in the control group wilt and begin to lose petals. The tulips in the experimental group stay fresh for one week. Do the observations support the hypothesis? Write a conclusion. ①

TEACH ▪ Continued

Directed Inquiry
Have students study the steps in Figure 1.6. Lead students in a discussion using some or all of the following questions:

▶ What is the first step in designing and planning an experiment? (State the problem.)

▶ Do you think any one step is more important than any of the others in designing and planning experiments? Explain. (No one step is more important than the others. Each step must be carried out to make the experiment truly reliable.)

▶ Why is it important that you write down what you do in each step of an experiment? (So that another person could conduct the exact same experiment and reach the same conclusion)

▶ If the data do not support the hypothesis, has the experiment failed? Explain. (Not necessarily; You've learned what does not work and that you need to solve the problem in another way.)

Skills Development
Infer How long do you think it took for the vaccines for measles and other diseases to be developed—days, months, or years? Why? (It took years. Researchers had to do many experiments before the vaccines could be developed.)

Decision Making
If you have classroom sets of *One-Minute Readings*, have students read Issue 80, "Science and Non-Science" on page 132. Discuss the questions.

Reteach
Use Reteach Worksheet 1.1.

INTEGRATED LEARNING

Language Arts Connection
Good experimental design is important to doing successful research. Poorly constructed experiments give erroneous or incomplete results. One important part of designing an experiment is stating the problem. Work with the class to brainstorm a list of problems that could be tested in an experiment. Have students practice writing problem statements. Remind them to state their problems as questions.

Methods of Science

Is there one method of approaching a problem, an experiment, or an issue? The answer is yes and no. You may have heard the phrase *scientific method* used to describe the way scientists find out about the natural world. Modern scientific method is the systemized testing of ideas, hypotheses, predictions, and inferences about the natural world. Scientific method also involves a lot of communication. Scientists constantly exchange ideas and information.

Scientific method is not a set of steps to follow like those in a cookbook recipe. Instead, there are many different ways to study, investigate, and think about scientific problems.

Read Figures 1.6 and 1.7. They show a model for designing and planning an experiment and a model for decision making. They will help you in life and as you do the activities in this book.

Figure 1.6 Designing and Planning an Experiment ▼

State the Problem
What do you want to find out? State the problem as a question.

Hypothesize or Predict
What do you think may happen? What information from past experiences or observations are you using to state your hypothesis or make your prediction?

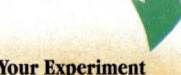

Plan Your Experiment
Do you need a control? If so, what is the variable? What are the constants? Write a step-by-step procedure that another person could easily follow. Your experiment must be able to be repeated exactly the way you did it.

Gather Data
How will you gather your data? Will you observe, estimate, or measure? Any other ways?

Record and Organize Data
How will you record and organize your data? Will you use tables and graphs? Will you include diagrams and drawings?

Analyze Data
Do you see any trends or patterns in the data? Do the data support your hypothesis or prediction? Do you need more information?

Conclude
State your conclusion based on your data. Your data should either support your conclusion or lead you to another hypothesis. Have any new questions or problems come up?

8 Chapter 1 Studying Science

Library Connection

To help students develop skills in locating information, take them to the library. Students should be introduced to the reference section, the card catalog or computer catalog, the microfilm or other periodical collections, and the *Reader's Guide to Periodical Literature*.

Directed Inquiry

Have students study the steps in Figure 1.7. Lead students in a discussion using some or all of the following questions:

▶ What is always the first step of decision making? (Think about it.)

▶ At what point should you make a decision? (When you believe you have enough information)

▶ When is a conclusion made according to this model? (A conclusion is made only after the completion of the first five steps.)

Discuss

As scientists gather more data about a topic, scientific knowledge and theories are modified. Lead a discussion about theories that have been refuted. For example, in the second century A.D., the Greek scientist Ptolemy theorized that the universe revolved around the earth. Before Jan Baptista van Helmont (1580–1644) experimented with growing a willow tree in a barrel, people thought that trees grew as they used up the soil. Encourage students to name other theories that people have had to change.

Figure 1.7 Decision Making

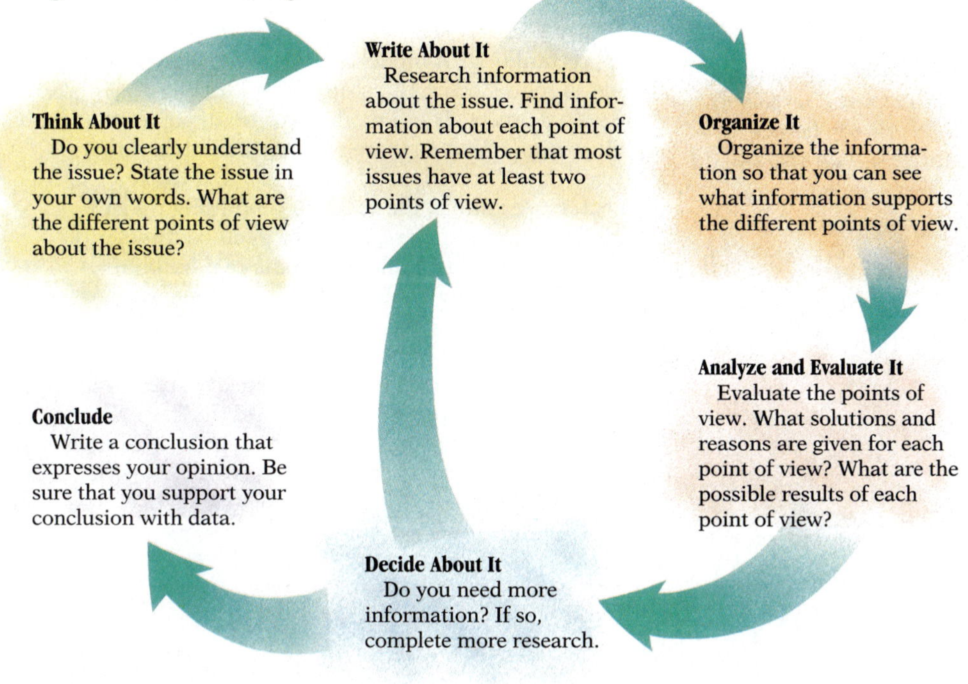

Think About It
Do you clearly understand the issue? State the issue in your own words. What are the different points of view about the issue?

Write About It
Research information about the issue. Find information about each point of view. Remember that most issues have at least two points of view.

Organize It
Organize the information so that you can see what information supports the different points of view.

Analyze and Evaluate It
Evaluate the points of view. What solutions and reasons are given for each point of view? What are the possible results of each point of view?

Decide About It
Do you need more information? If so, complete more research.

Conclude
Write a conclusion that expresses your opinion. Be sure that you support your conclusion with data.

From Facts to Laws

In science, no matter what is stated, the statement is always backed by observations, studies, tests, and experiments. Even science facts! In science, a fact is an agreement made by qualified scientists about many observations of the same thing.

If new observations are made, science facts can even be changed. Take the shape of the earth as an example. For much of recorded history, the earth was considered to be flat. It was a "fact of the times." Today, of course, everyone knows that the earth is more or less a sphere. Actually, the earth is slightly flattened at the poles and bulges a little at the equator.

When scientists want to interpret facts, they state a theory. In science, a theory is not used in the same way that a detective states a theory about a crime. A scientific theory is an idea that has been tested again and again. It is based on many separate, but related, hypotheses that have been repeatedly tested. A theory explains why and how something happens. A theory, in turn, may be stated as a scientific law. If the hypotheses of a theory are tested over and over and all the observations support the theory, a scientific law may be stated.

Just because a theory or law is stated does not mean it can't be changed. Scientific theories and laws evolve. As new hypotheses are tested, theories are often redefined.

TEACH • Continued

Skills WorkOut
To help students make a model, have them do the Skills WorkOut. **Answer** Answers may vary, but emphasize that anyone who uses the methods of science is a scientist.

Skills Development
Make Models Have students draw a scale drawing of a familiar object or place, such as their shoes or the classroom. Help them choose a scale to use in their drawings. You may wish to show them how to measure the object and use proportions to find the measurements for their models.

Career Corner
Have students work in groups of three to add more careers to the table in the Career Corner. If necessary, have students use references to find out if a particular career fits under a branch of science. Sudents can list certain careers under two or more sciences if they seem to fit there.

INTEGRATED LEARNING

Art Connection
Provide art materials such as large sheets of paper, clay, craft sticks and glue. Have students work in pairs to make a scale model of an object of their choice. The model can be either two- or three-dimensional. Students should measure the object and choose a scale before beginning to draw. They may be interested in making a model of the school building, a piece of sculpture, or a person. You might want to have them make a model or map of a country they are studying in social studies.

Skills WorkOut
Make a Model Draw a picture of a scientist. Compare your drawing with that of a classmate. In what ways are they alike? Different?

Models in Science
Have you ever seen a model train set? Or a model car, ship, or building? A model is a way to represent something. A model train, car, ship, or building looks just like the actual object. These models are small copies that are built to scale. That is, there is a mathematical ratio between the dimensions of the model and the real object. For example, the scale for a building may be 1 centimeter to 2 meters, meaning each centimeter on the model will equal 2 meters on the real building.

There are many types of models. Some models are drawn on paper, such as diagrams and building blueprints. Maps are also models. Maps do not look like the object they represent. For example, a city street map does not look like real city streets. Yet the map is very useful. Some models are mental models. In a mental model, you imagine what something would look like if you could observe it.

Career Corner

What Careers Use Scientific Skills and Knowledge?

Careers that use scientific skills and knowledge are many and varied. They may range from artists to physicists.

Table 1.2 lists several branches of science and three careers that use scientific skills and knowledge related to each branch. Notice that knowledge about different sciences is needed for some careers.

Each branch of science includes a large body of knowledge. Many of these branches have even smaller, specialized branches. For example, anatomy is a branch of biology.

Table 1.2 Careers in Science

Branch	Study	Careers
Anatomy	Structure and parts of living things	PE teacher, physician, physical therapist
Chemistry	Makeup of substances and the changes that substances undergo	Chef, food scientist, chemical machinist
Biology	All living things	Soil scientist, public health educator, medical illustrator
Meteorology	Weather and the earth's atmosphere	Air pollution technician, air traffic controller, weather reporter
Mechanics	Movement of objects and the action of forces on objects	Automobile designer, architect, underwater photographer
Physics	Properties, changes, and the interaction of matter and energy	Artist, electrician, materials scientist

Chapter 1 Studying Science

Multicultural Perspectives

Safety symbols are helpful because one does not have to know a certain language to understand them. For example, if you were in a foreign country and saw a gasoline truck with the flammable liquid symbol on it, you could probably infer what it meant. Have each student bring to class an example of an international safety symbol from a magazine or a package label.

Integrating the Sciences
Life, Earth, and Physical Science
Science is generally divided into three branches: life, earth, and physical. Remind students that science is a particular way of looking at nature's events and that all branches of science, though studied separately, are really related to one another. All scientists gather and interpret information in the same basic way.

Research
Have students do research to make lists of materials that might be hauled in tanker trucks and that may be dangerous to people or harm the environment. Ask students, Which of these materials are the most dangerous to people? Why?

EVALUATE

WrapUp
Reinforce Divide the class into groups of four. Have each group design an experiment (including all of the seven steps on page 8) from one of the three branches of science. Have each group describe its experiment. Then as a class, classify each experiment according to one of the three branches of science.
Use Review Worksheet 1.1.

Check and Explain
1. Answers may vary. Students may say that they observed what the weather was like on their way to school, used a ruler to measure something, or classified their clothes to make sure what they wore matched.
2. The two groups in a controlled experiment are the control group and the experimental group.
3. To make this decision, one should follow the six decision-making steps outlined on page 9 in Figure 1.7.
4. Answers may vary. Students should be able to design a symbol that warns of a potential danger in the home or reminds family members of home safety skills.

Scientists, engineers, architects, and certain businessmen and businesswomen are just a few people who use models in their work. They use models as plans from which real objects can be built. Models are also used to help people understand things that may not or could not be observed directly.

Science and Technology
Safety in a Symbol

A tanker truck overturns on a highway. A clear liquid spills across the road. Within minutes, the highway patrol has closed the road, and an emergency response crew has arrived. The crew looks at the side of the tanker and sees a sign with a symbol and an identification number. A quick call is made to the Chemical Transportation Emergency Center (CHEMTREC).

At CHEMTREC, the number is used to identify the substance and tell the emergency response crew the proper emergency procedures to use. CHEMTREC provides information about any health hazards, risks of fire or explosion, and cleanup procedures. With this information, spills can be handled quickly and efficiently, with few risks to people or the environment.

You have probably seen these symbols on tanker trucks. Each hazardous material is identified by a special shape, color, and identification number. For example, tanker trucks that transport gasoline will have the safety symbol for a flammable liquid. Gasoline is represented by its identification number. International safety symbols such as those used by CHEMTREC can greatly reduce chances of injury to people and harm to the environment.

Figure 1.8 ▲
CHEMTREC hazard symbols are recognized internationally.

Check and Explain

1. Name three science process skills, and explain how you used them today.
2. What are the two groups in a controlled experiment?
3. **Apply** What procedure would you use to decide if a rain forest should be cut down for farmland?
4. **Make a Model** Design a safety symbol to use in your home.

Chapter 1 Studying Science

SECTION 1.2

Section Objectives
For a list of section objectives, see the Student Edition page.

Skills Objectives
Students should be able to:

Measure parts of their bodies using a unit other than the inch.

Sequence lengths from shortest to longest.

Measure the mass of some common objects.

Measure the length, mass, and height of common objects.

Vocabulary
meter, volume, cubic meter, liter, mass, kilogram, density

MOTIVATE

Skills WarmUp
To help students understand measuring with scientific units, have them do the Skills WarmUp.
Answer Answers will vary. Explain that the two measurements should be fairly equal. Remind students to carefully align the paper clips when measuring.

Misconceptions
Students may not understand that an object's weight can vary, depending on the force of gravity. Explain to them that at the equator and on the moon objects weigh less because the pull of gravity is weaker, although the amount of matter (mass) does not change.

Integrated Learning
Use Integrating Worksheet 1.2.

INTEGRATED LEARNING

Language Arts Connection
Have students use a mnemonic device to help them remember the order and value of the SI prefixes. Mnemonics are strategies that help store information in memory. A rhyme, phrase, or mental picture can be used. An example of a mnemonic device that helps with remembering the planets in order from the sun is My Very Excellent Mother Just Served Us Nine Pizzas.

Social Studies Connection
Discuss with students the origin of the English system of measurement. (King's foot—nose-to-fingertip yard) Ask students why they think SI is used all over the world and the English system is still so prevalent in the United States. Ask students which system is more logical and why.

Skills WarmUp
Measure Place your forearm on a sheet of paper. Mark the paper at the end of your elbow and at your wrist. Then see how many paper clips can be lined up end to end between the two marks. Repeat this activity, but measure your foot instead of your forearm. How do these two measurements compare?

Table 1.3 SI Units

Measure	Unit/Symbol
Length	Meter/m
Mass	Gram/g
Area	Square meter/m^2
Volume	Liter/L, cubic centimeter/cm^3
Temperature	Degrees Celsius/°C
Time	Second/s

1.2 Measuring with Scientific Units

Objectives

▶ **Identify** SI base units and prefixes.

▶ **Compare** density, mass, and volume.

▶ **Measure** common objects using SI units.

Did you know that you get shorter during the day? While you are up and moving around, gravity pulls you toward the earth. Your body becomes more compact than it was earlier in the day. So by the end of the day, you are slightly shorter. Scientists discovered this by taking and comparing height measurements.

When something is measured, you describe it in terms of numbers and units. The system of measurement used throughout most of the world today is the *Système internationale d'unitès*, or SI. This system was developed in France in 1791 and was revised and modernized in 1960. You may have heard it called the metric system. The name comes from its base unit of length, the meter.

Base units in SI, or metric units, are shown in Table 1.3. You are most likely to use the units for length, mass, time, and temperature. SI is easy to use. Unlike other systems of measurement, SI units are based on multiples of ten. SI uses prefixes, such as *kilo-* or *centi-*, to indicate how many times a unit should be multiplied or divided by ten. These prefixes are shown in Table 1.4.

Table 1.4 Prefixes used in SI

Measurement	Unit	Symbol
kilo	1,000	k
hecto-	100	h
deca-	10	da
deci-	1/10	d
centi-	1/100	c
milli-	1/1000	m
micro-	1/1,000,000	μ

Chapter 1 Studying Science

STS Connection

There are standards for the various SI measurement units. In time measurement, seconds are measured by precise vibrations of a clock regulated by cesium atoms. Have students find out about the SI standard meter, gram and kilogram, and liter. Ask students why standards are necessary.

Multiples of ten make conversion between units very simple. You simply move the decimal point. For instance, 130 centimeters equals 1.3 meters. Think of it in terms of U.S. money: $1.00 = 100 pennies or 10 dimes. These are base 10 measurements. A centimeter is 1/100 of a meter, and a kilogram is 1,000 grams.

Length

The basic SI unit of length is the **meter** (m). A meter is about the distance from the floor to the knob of a door. Chances are that the SI measuring tool that you'll be using in class is a 1-meter-long ruler called a meter stick.

A meter stick is divided up into smaller units. A meter stick, a tape measure, or any other measuring device with a scale marked in equal divisions makes it possible for you to make exact measurements. A scale helps you find out how many units there are as you measure, without having to count them one by one.

The SI units of length are the kilometer (km), the meter, the centimeter (cm), and the millimeter (mm). Use whichever unit is most appropriate for the length you are measuring. The meter stick is marked in units called centimeters. A centimeter is 1/100 of a meter. Your little finger is about 1 centimeter across. As you can see in Figure 1.9, each centimeter is divided into ten smaller units called millimeters. A dime is about 1 millimeter thick.

Skills WorkOut

Sequence Arrange the following lengths in the correct order from shortest to longest.

a. 60 mm
b. 10 km
c. 15 dm
d. 26 dm
e. 8 km
f. 72 cm
g. 5 000 mm
h. 3 km

Figure 1.9 ▼
You will measure many distances that are less than 1 meter. The numbers on this metric ruler represent centimeters.

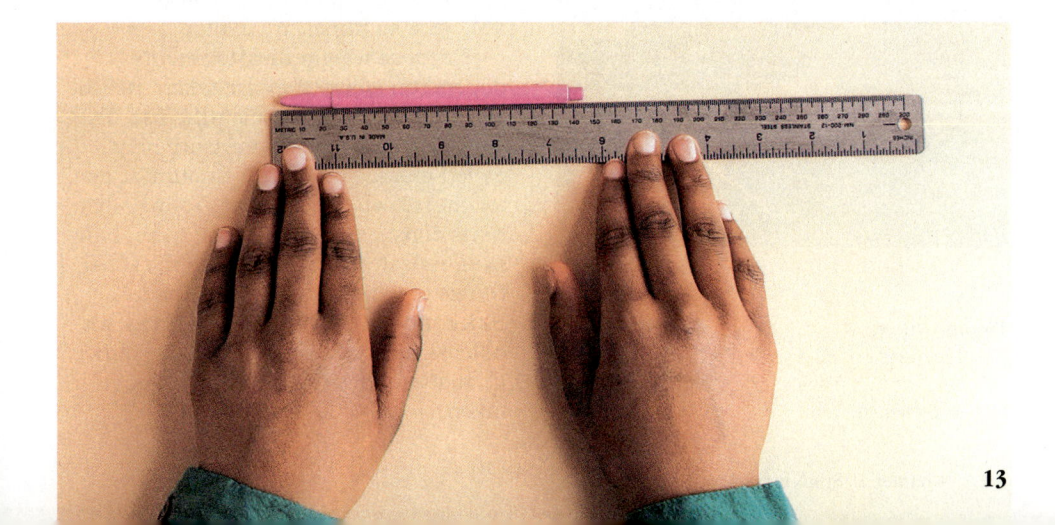

TEACH

Skills WorkOut

To help students arrange lengths in the correct numerical order, have them do the Skills WorkOut.
Answers The order from shortest to longest is the following:

a. 60 mm
f. 72 cm
c. 15 dm
d. 26 dm
g. 5000 mm
h. 3 km
e. 8 km
b. 10 km

Discuss

If students have trouble doing the Skills WorkOut, tell them how some of the lengths can be easily changed from one unit to another. For example, tell them that 80 cm is the same as 800 mm, 8 dm, or 0.8 m. You simply move the decimal point and add or discard a zero going from one unit to the next. Have students convert all the lengths in the Skills WorkOut to meters. Then have them arrange the lengths in correct numerical order.

Answers

a. 0.06 m	a.	0.06 m
b. 10 000 m	f.	0.72 m
c. 1.5 m	c.	1.5 m
d. 2.6 m	d.	2.6 m
e. 8000 m	g.	5.0 m
f. 0.72 m	h.	3000 m
g. 5.0 m	e.	8000 m
h. 3000 m	b.	10 000 m

Decision Making

If you have classroom sets of *One-Minute Readings*, have students read Issue 73, "Metric System" on pages 122 and 123. Discuss the questions.

TEACH • Continued

Skills Development

Estimate Have students think about 1-liter and 2-liter containers of soft drinks. Ask, How does a 1-liter container compare to a quart? How does a 2-liter container compare to a half gallon? Have students estimate which quantity is larger. Then have them test their estimations by filling a quart container with water and then pouring it into a liter container. Have them do a similar exercise with half-gallon and 2-liter containers.

Class Activity

Have students practice reading the water level in a graduated cylinder. Set up stations around the room. The stations will include cylinders containing various amounts of water. Divide the class into small groups. Have students take readings of the water levels at each station. When all have taken the readings at one station, have students move to the next station. When all the stations have been visited, tell students the correct readings for each. Students should revisit stations for which they have incorrect readings.

INTEGRATED LEARNING

Integrating the Sciences

Physical Science Point out that weight is a measure of gravitational attraction. Close objects exert more pull on one another and will seem to weigh more; objects farther apart will do the reverse. A person would weigh more at the poles of the earth because the poles are slightly flattened compared to the equator. As a result, at the poles an object is closer to the earth's center of gravity. Ask the following questions:

▶ What would happen to your weight if you climbed a mountain?

▶ How would your weight be affected if you went into a deep mine?

▶ Why does weightlessness occur in outer space?

▶ What is the relationship between mass, weight, and volume?

Volume

Take a deep breath. As your lungs fill with air, you can feel your chest expand. The change in lung size can be measured in terms of **volume**. Volume is the amount of space that something occupies. As your lungs expand, they occupy more space.

Because the SI base unit of length is the meter, the unit of volume is the **cubic meter**, or m^3. A cubic meter is the space occupied by a cube: $1\,m \times 1\,m \times 1\,m$. This unit of volume is used to measure large quantities, such as the volume of concrete in a building. In your science activities, you are more likely to use the cubic centimeter (cm^3).

You are probably most familiar with the metric unit of volume, the **liter** (L). Soft drinks often come in 1-liter or 2-liter containers. A smaller unit of volume is the milliliter (mL). There are about 20 drops of water in each milliliter.

In your science class, you will probably use a graduated cylinder to measure liquid volumes. As you can see in Figure 1.10, the cylinder is marked in milliliters. Notice that liquids in a graduated cylinder have a curved surface. The liquid rises slightly up the sides of the container and dips slightly in the center. That dip in the surface of the liquid is called the *meniscus*. Find the meniscus in Figure 1.10. To measure the volume of liquid accurately, you read the level of the liquid at the *lowest point* of the meniscus. The best way to do this is to look at the surface of the liquid at eye level.

Figure 1.10 ▲
You can measure liquid volume with a graduated cylinder. To read, place your eye at the surface level of the liquid.

Mass and Weight

Did you know that things are slightly lighter when weighed at the equator than at the poles? That's because an object's weight depends on the force of gravity. At the equator, gravity is measurably less, so things are lighter.

Because weight measurements depend on where they are taken, weight can be unreliable for scientific use. That is why scientists use a measurement called **mass**. Mass is the amount of matter that an object contains. Your body's mass consists of your bones, teeth, hair, skin, muscles, organs, and all the other things of which your body is made. Since it does not depend on gravity, an object's mass always remains the same, no matter where the measurement is taken.

14   Chapter 1 Studying Science

Integrating the Sciences

Physical Science One might think that objects made of materials with densities greater than water would sink when placed in water. How, then, does a steel ship float? Explain to students that the key factors are volume, the shape of the boat hull, and the amount of water displacement. A ship must displace a volume of water that weighs more than it does.

Math Connection

Ask students the following question, If you knew the volume and the density of an object, how could you figure out its mass? (The mass is equal to the volume multiplied by the density.)

Class Activity

Have students practice measuring the mass of an object. Provide balances, along with several objects to be measured. Have students weigh the objects and record their measurements. Then have students compare their measurements with those of another student. If pairs of students find that their measurements for a particular item differ, they should weigh the object again.

Skills Development

Estimate Using a meter stick, explain to students what a cubic meter is. Reinforce the meaning of the symbol m^3. Then ask students to estimate the volume of the classroom. Divide the class into groups and have them measure the length, width, and height of the room. Have students round off to the nearest meter. (Students measuring the height of the room may have to estimate if they can't measure to the ceiling.) Collect the measurements, and, as a class, calculate the volume of the room. Compare estimates to the calculation.

Answers to In-Text Questions

① **One thousand milligrams are equal to one gram.**

② **Lead is more dense than balsa wood.**

③ **An object with the same density as water will be suspended in the water.**

④ **Lead will sink in water.**

⑤ **The balsa wood will float.**

⑥ **The gram is the unit used to measure five paper clips.**

The basic unit of mass is the **kilogram** (kg). In the laboratory, however, scientists often use a much smaller unit called the gram (g). The cap of a pen has a mass of about 1 gram. For even smaller measurements, scientists use a milligram (mg). How many milligrams are equal to one gram? ①

You will use a balance to determine an object's mass. A balance lets you compare an unknown mass with a known mass. Look at the double-pan balance in Figure 1.11. When the arm of the balance is even, the two masses are equal.

Density

Take a piece of aluminum foil and crumple it into a ball. Then take some clay and make a ball the same size. Hold the two balls in your hands. Compare them. Which feels heavier? Though they are the same size, the clay ball will feel heavier than the aluminum one. The clay ball is denser.

Every substance has its own characteristic **density**. Density is the measure of how much matter is packed into a given volume. In mathematical terms:

$$\text{density} = \text{mass/volume}$$

Scientists often measure density in grams of matter per cubic centimeter (g/cm³). Lead has a density of about 11 g/cm³. This means that for every cubic centimeter of lead, there are about 11 grams of matter. Balsa wood has a density of about 0.1 g/cm³. This means that every cubic centimeter of balsa wood contains about 0.1 grams of matter. Which is the more dense, lead or balsa wood? ②

In addition to helping identify a substance, an object's density determines if the material will float or sink. The density of everything is compared to the density of pure water. Materials with densities greater than that of pure water (1 gm/cm³) always sink. Materials with densities less than that of water always float. How will an object with the same density as water behave? Will lead float ③ or sink in water? If a cubic centimeter ④ of balsa wood is placed in water, will it float or sink? ⑤

◀ **Figure 1.11**
The kilogram is the basic unit for mass in the metric system. What unit would you use to measure the mass of five paper clips? ⑥

15

TEACH • Continued

Class Activity

Have students practice using the Celsius thermometer. Set up several measuring stations for students. At each station there should be a thermometer and a beaker of water. Provide water of various temperatures from ice water to very hot water. Have students record the temperature at each station. Then have them compare measurements with a classmate. Remind students that the water will tend to become warmer if it is colder than room temperature and will tend to become colder if it is warmer than room temperature.

SkillBuilder

Answer Answers will vary depending on the size of the rubber band, the size of the paper clip, and the thickness of the notebook paper. Students may note that a larger sample provides a more precise estimate. Accuracy can be checked using a more sensitive scale.

INTEGRATED LEARNING

Math Connection

Sometimes it is useful to be able to convert Celsius to Fahrenheit or vice versa. Put the conversion equations on the board and have students practice some examples in both directions.

Use: $(F° - 32)/1.8 = C°$ or $C° = 5/9 \times (F° - 32)$

$(C° \times 1.8) + 32 = F°$ or $F° = (9/5 \times C°) + 32$

Figure 1.12 ▲
On the Celsius scale, water freezes at 0°C and boils at 100°C.

Temperature

You've probably had your temperature taken when you were sick. Chances are that your body temperature was above 98.6 degrees Fahrenheit (F), which is what it is when you are healthy. Scientists, however, do not use Fahrenheit temperature readings. Instead, they measure temperature using the Celsius scale. In fact, most of the countries of the world use Celsius as their temperature scale. On this scale, your body temperature is about 37 degrees Celsius, or 37°C.

In your classroom, you probably will use a Celsius thermometer to measure temperature. A thermometer is a thin glass tube filled with a colored liquid. As the liquid heats up, it expands and moves up the tube. As the liquid cools, it moves down the tube. The sides of this tube are marked in degrees Celsius. Two important reference marks on the Celsius scale are the freezing point (0°C) and the boiling point (100°C) of water.

SkillBuilder *Measuring*

Average Mass

The mass of a small object such as a rubber band is hard to measure directly. To get its mass, you need to determine the rubber band's average mass. First, you can count out 100 rubber bands. Then measure the total mass of the rubber bands. Divide the total mass by 100. This method gives an average mass for the rubber bands and assumes that individual masses do not vary much.

Use the method above to estimate and measure the masses of some common objects.

1. Estimate the mass of each of the following:
   a. a rubber band
   b. a small paper clip
   c. a piece of notebook paper

2. Record your estimates in a data table similar to the one shown.

3. Count out 10 of each object in step 1. Measure the mass. Divide this by 10. Record this measurement in your table.

4. Count out 100 of each object in step 1. Measure the mass. Divide this by 100. Record this measurement in your table.

How accurate were your estimates? Did the average masses vary as your sampling size increased?

Object	Estimate	Average Measurement	
		10	100

16 Chapter 1 Studying Science

Integrating the Sciences

Earth Science This is a good place to remind students that time origins for the day, month, and year are based on celestial motions. Earth's revolution around the sun determines the year; the moon's revolution around the earth determines the month; and the rotation of the earth on its axis determines the day.

Time

The basic SI unit of time is the second. In your classroom, you will use a watch with a second hand to measure time. As you know, other units such as the minute, hour, day, and year are not based on the number ten. There are 60 seconds in a minute, 60 minutes in an hour, 24 hours in a day, and 365 1/4 days in a year. This system of measuring time is used everywhere in the world. In science, however, the prefix *milli-* is used with the unit second for some short time measurements.

Science and You *Settling Cereals*

Have you ever opened a sealed cereal box to discover that the box wasn't full? Did this make you wonder what happened to the contents? Perhaps there was some mistake during packaging. You may find the answer to your questions by examining the package. Somewhere on the box you're likely to come across this statement: "Contents measured by weight, not by volume."

At the packaging plant, the boxes were filled with unbroken cereal. The amount of cereal added to each box was determined by the cereal's weight, not by the amount of space the cereal took up. Therefore, although the box might contain the right amount of cereal, it may not appear full.

During shipping, cereal flakes often crumble and settle to the bottom of the package. This settling compacts the cereal, making the box appear less than full. So when you open the sealed box, it appears as if you didn't get what you paid for . . . but you did. The cereal's weight hasn't changed at all!

Figure 1.13 ▲
Cereals and crackers are sold by weight, not by volume.

Check and Explain

1. Why do you use SI units in science?
2. Why is mass a more reliable measure than weight?
3. **Predict** Aluminum has a density of 2.7g/cm³. Will a block of aluminum sink or float?
4. **Measure** Find the following objects in your classroom: an object 5 cm long, an object with a mass of about 10 g, an object at least 2 m tall.

Discuss

Students may be unfamiliar with the very short time measurements that are made in milliseconds. Have students think of events that might be measured in milliseconds. You might start them off with the blink of an eye or a frog catching an insect with its tongue. Students may mention many others.

Discuss

Ask students what other products should be packed by weight and not volume. (Potting soil, coffee, frozen juice)

EVALUATE

WrapUp
Portfolio Have students write a description of how they would find the density of an object. Have them name the units that show density. Students can keep their descriptions in their portfolios.
Use Review Worksheet 1.2.

Check and Explain

1. The SI system is common throughout the world because of its ease of conversion between units; it's based on multiples of ten.
2. Mass is a more reliable unit because its measurement is not affected by gravity.
3. A block of aluminum will sink because its density is greater than water's.
4. Have students find and measure the objects.

Chapter 1 Studying Science **17**

SECTION 1.3

Section Objectives
For a list of section objectives, see the Student Edition page.

Skills Objectives
Students should be able to:

Collect Data about classmates and display the data using graphs.

Make a Graph of information presented.

Classify the type of graph to use in presenting data.

MOTIVATE

Skills WarmUp
To help students understand graphing, have them do the Skills WarmUp.
Answer Students should display the data using a table, bar graph, and circle graph.

Prior Knowledge
To find out how much students know about graphing, ask the following questions:

- Have you ever made a graph? When?
- Can you name some types of graphs?
- Where have you seen a circle graph before?

Answers to In-Text Questions
① Paper and paper products make up most trash.
② Plastic and food waste make up equal percentages of trash, as do metal and glass.

Integrated Learning
Use Integrating Worksheet 1.3.

INTEGRATED LEARNING

Themes in Science

Patterns of Change/Cycles Data collected during experiments and observation need to be presented in a useful way. Graphing allows comparisons that can lead to developing new theories. Data are often easier to analyze when they are presented in graph form. For example, to a scientist, population growth curves show that human populations increase exponentially, not arithmetically.

Skills WarmUp

Collect Data Count the number of boys and girls in your class. Display this information in a table, a bar graph, and a circle graph. What percentage of your class is girls? Which organizer did you use to answer the question?

1.3 Graphing

Objectives

▶ **Identify** types of graphs.
▶ **Explain** the use of variables on line graphs.
▶ **Classify** the type of graph needed to show specific kinds of data.

What is a graph? A graph is basically a picture of data. It shows numerical data as a diagram. Think of a graph as a picture of a table of data. You've probably seen many different graphs in newspapers and magazines, and on television news reports. Many businesses also use graphs. Graphs are useful tools for presenting information.

Kinds of Graphs

There are three kinds of graphs: circle, bar, and line graphs. No matter which kind of graph you look at, all graphs let you compare numerical data. Each kind of graph, however, shows the data in a different way.

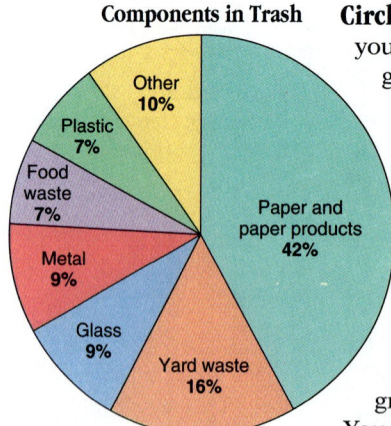

Circle Graphs Think about a pie cut into pieces, and you have a mental model of a circle graph. A circle graph is a divided circle. It shows how a part or share of something relates to the whole value. You can see what fraction, percentage, share, or proportion each part represents.

Look at the circle graph in Figure 1.14, which shows how much of our trash consists of paper and paper products. You can even compare two groups very easily. Compare the percentage of metal trash and yard waste.

Bar Graphs Just like its name describes, a bar graph has bars. The bars help you compare things. You can compare measurements—such as weight, height, and length—of the sizes of things. Bar graphs also let you compare amounts, quantities, and changes.

Look at the bar graph in Figure 1.15. It shows the shortest and longest lengths of four North American

Components in Trash
- Other 10%
- Plastic 7%
- Food waste 7%
- Metal 9%
- Glass 9%
- Yard waste 16%
- Paper and paper products 42%

Figure 1.14 ▲
① What type of waste makes up most trash? ② Which wastes make up equal percentages of trash?

18 Chapter 1 Studying Science

Geography Connection

The *x*-axis and *y*-axis are used in places other than graphs. They are also used in models, such as globes and maps. Have students locate the *x*- and *y*-axes (latitude and longitude) on a globe or world map. Show how the planet can be divided into Northern and Southern, as well as Eastern and Western hemispheres. Ask how many coordinates are needed to locate a place. The exercise could be repeated with a road map.

Math Connection

Ask students to tell the class the month and year they were born. Then have each student create a graph showing how many students were born in each month. The *x*-axis should be divided into months and cover a span of approximately two years. The *y*-axis should show the number of students.

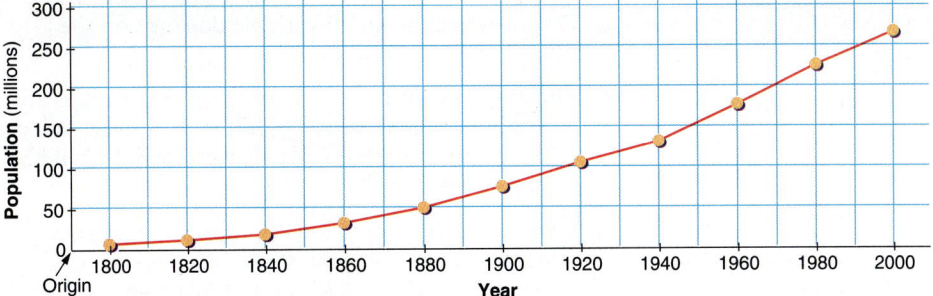

TEACH

Research

Have students draw the graph in Figure 1.15 on a sheet of graph paper. Then ask students to do research to add to the graph. For example, you could have them add diamondback rattlesnakes, timber rattlesnakes, boa constrictors, and king snakes to the graph. Students will have to add to the Y-axis for the boa constrictor, because it is much more than 140 cm in length. Remind students that the key must be changed also.

Ancillary Options

If you are using the blackline masters from *Multiculturalism in Mathematics, Science, and Technology*, have students read about Celestino Beltran on pages 31 and 32. Complete pages 32 to 34.

Answers to In-Text Questions

③ The Eastern cottonmouth has the longest maximum length, and the Eastern coral snake has the shortest.

④ The origin of the graph is at the intersection of the X-axis and the Y-axis.

---

snakes. You can see very easily which snake is the longest and which is the shortest.

Notice that the bar graph is drawn on graph paper. On the left side of the graph, there is a scale showing length in centimeters. These data are plotted on the vertical line, called the *y*-axis. Each bar is labeled on the horizontal line, called the *x*-axis. On this graph, the labels are letters keyed to the names of the snakes.

When you build a bar graph, you need to choose the scale by looking at the highest and lowest numbers in your data. On this graph, the scale ranges from 0 to 140 cm. Notice that the scale is higher than the longest snake. Also, the numbered divisions are in equal amounts of 20 cm. All bar graphs have a scale and equal divisions.

Line Graphs Line graphs let you plot several different forms of data. They also show changes. Line graphs can help you answer if-then questions. They help you see patterns or trends in data. Unlike bar graphs and circle graphs, you can also plot several different sets of data on one line graph and easily make comparisons. Figure 1.16 shows a line graph that has been drawn based on certain statistics. In this case, the population growth of the United States is measured. Notice that the population size (responding variable) is being measured every 20 years (manipulated variable).

The manipulated variables are the *x*-values. The years plotted are on the *x*-axis. The responding variable (the one that changes when the year is changed) is displayed on the *y*-axis.

Look at Figure 1.16. For the year 1880, a dot is placed at 50 million. The dot is placed directly across

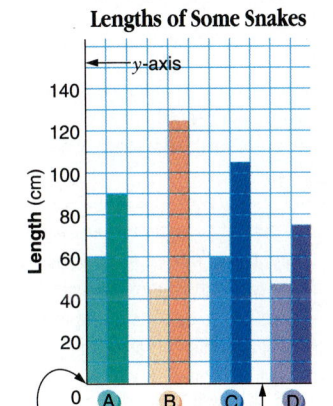

Minimum and Maximum Lengths of Some Snakes

Key
A) Southern copperhead
B) Eastern cottonmouth
C) Northern water snake
D) Eastern coral snake

Figure 1.15 ▲
Which snake has the longest maximum length? The shortest? ③

Figure 1.16
On a graph, what point is the origin? ▼ ④

United States Population Growth

Chapter 1 Studying Science

TEACH ▪ Continued

Skills WorkOut
To help students make a graph, have them do the Skills WorkOut.
Answer The graph that would best represent the information is a bar graph.

Discuss
Have students think about this question: If a computer can produce all the graphs in this book, why is it necessary for me to make a graph? (In order to better understand what they mean and how they function)

EVALUATE

WrapUp
Reinforce Ask students what their favorite television shows are. List each show and the number of students who consider it their favorite on the chalkboard. Then ask students to choose one type of graph and display the information on it.
Use Review Worksheet 1.3.

Check and Explain
1. The three types of graphs are bar, circle, and line.
2. The responding variable (*y*-axis) depends on the manipulated variable (*x*-axis).
3. The scale on the *y*-axis would show degrees Celsius from −10 to 30 in equal divisions of 5.
4. A line graph

Answer to In-Text Question
① About 100 million

② To predict the U.S. population in 2020, students will have to extend the X-axis and the Y-axis. Answers may vary, but the population should be approximately 315 million.

20

INTEGRATED LEARNING

Geography Connection
Topographic maps are actually three-dimensional graphs. Geologists use the lines and numbers on them to locate land features. Show students some examples of topographical maps.

STS Connection
Explain to students that today most graphs they see are created on a computer. Software is available that allows the user to input data and then instantly see how that data would look in the form of a circle graph, bar graph, or line graph. The axes on line graphs can be adjusted to increase or decrease the slope of the curves.

Skills WorkOut
Make a Graph Research to find out the correct numbers for one of the following: the heights of the ten highest mountains in the world; the speeds of the ten fastest animals; the area sizes of the ten largest countries. Then choose which kind of graph—bar, circle, or line—that would best represent the information. Draw your graph and present it to the class.

from the *y*-value and directly above the *x*-value. What was the population of the United States in 1920? ①

When all the values are presented as points on the graph, a line is drawn connecting all the points. Notice that the line shows a change and a trend. The rapid increase in the United States population is shown. You can even see that between certain years, the population grew much more rapidly than between other years. Predict what the population of the United States will be in 2020. ②

Science and Technology
Just a Touch of a Key

Until a few years ago, all graphs in print were drawn by hand. A graphic artist went through the same steps as you just did to produce a graph. Every time a mistake was made, it had to be erased. If the page contained too many errors, it was thrown away. As you might imagine, this could be a long and tiresome task.

Today most graphic artists have a new and powerful graph maker, the computer. All the computer needs is a graph-making program and some data, and Presto! It generates a perfect, customized graph. In fact, all the graphs in this book were made on a computer.

By taking advantage of a computer's power, complex graphing is just a keystroke away. The final graph can even be produced in several colors. In addition, the computer can calculate the exact shades needed to create the illusion of three dimensions.

Check and Explain

1. Name the three kinds of graphs.
2. On a line graph, which variable depends on the other?
3. **Analyze** The average temperatures of five cities are given in degrees Celsius: −5, 5, 15, 25, and 18. What scale would you use on a bar graph? What would be the equal divisions?
4. **Classify** Choose the kind of graph you would use if you wanted to show how many blue herons lived in a wetlands in 1970, 1975, 1980, 1985, 1990, and 1995.

20   Chapter 1   Studying Science

Integrating the Sciences

Physical Science Sunlight "powers" the earth. Point out to the class that visible light is only a part of the electromagnetic spectrum. Other wavelengths are X-rays, and ultraviolet and infrared radiation. We can't see these wavelengths but other organisms can detect some of them. Humans have developed devices to detect all wavelengths. Light must be absorbed to be seen. Ask students how they think sunglasses work.

SECTION 1.4

Section Objectives
For a list of section objectives, see the Student Edition page.

Skills Objectives
Students should be able to:

Observe and draw objects around them.

Observe the ink pattern on a magazine or newspaper.

Infer four ways that microscopes help scientists in their work.

Communicate how a compound microscope helps in the study of both living and nonliving things.

Vocabulary
microscope, lens

1.4 Microscopes and Lasers

Objectives

▶ **Distinguish** between convex and concave lenses.
▶ **Describe** the parts of a compound microscope.
▶ **Compare** compound microscopes and electron microscopes.
▶ **Infer** why scientists use microscopes in their work.

Skills WarmUp

Observe Use a hand lens to look at things around you. Look at your hand, your desk, notebook paper, a piece of hair, and other things. Make drawings of three of the things you observe. How does a hand lens change the objects you view?

There are many living things all around you that you can't see. How can you find out if these things exist? As scientists have done for many years, you can view many small living things and parts of living things by using a **microscope**. A microscope is a tool that makes things appear larger than they really are. In a microscope, light and lenses combine to bring objects into view.

Light and Lenses

The sun is the main source of light energy. Other sources of light are electric light bulbs and flames. Without light, you would not be able to see anything. Most things that you see are visible because they reflect light from other sources.

When light strikes an object, several things can happen. Some materials allow light to pass through them in straight lines. In other materials, light passes through in straight lines, but it is diffused in different directions. You can't see through these objects. In fact, most objects don't allow light to pass through them.

When light passes through glass, it bends. It bends again when light leaves the glass. The bending, or refraction, of light is caused by the difference in the speed of light in the glass and in the air. You can see how light refracts in Figure 1.17.

If glass is formed into certain shapes, it can be used to bend light. The glass can make things appear smaller, larger, closer, or farther away. A piece of glass that can bend light in some way to make an object look smaller, larger, closer, or farther away is a **lens**. There are two main types of lenses: convex lenses and concave lenses.

Figure 1.17 ▲
When light rays pass from air into water or glass, they bend. Notice how the straws appear bent in this glass of water.

MOTIVATE

Skills WarmUp
To help students understand microscopes and lasers, have them do the Skills WarmUp.
Answer A hand lens makes it possible to see small details of an object that can't be seen by the eye alone, and it causes an object to appear larger than it is.

Misconceptions
Students may not understand that light is bent (refracted) when it passes through matter. Use the example in Figure 1.17 to show how light rays bend when they pass through a glass of water, making a straw appear to bend.

The Living Textbook:
Life Science Side 1

Chapter 14   Frame 14497
How Microscopes Work (Movie)
Search: Play:

Chapter 1 Studying Science   **21**

TEACH

Skills WorkOut

To help students observe something closely by using a paper clip and water droplet as a magnifying tool, have them do the Skills WorkOut.

Answer The ink dots can be seen by using the magnifying tool.

Critical Thinking

Reason and Conclude Have students use a hand lens to look at a color photograph in a magazine. Ask students what color ink dots they see and why they were used. (Explain that different colors can be made by adjusting the percentages of ink dots that are black, magenta, cyan and yellow.)

Answer to In-Text Question

① When light rays strike a convex lens, they converge into a point. When light rays strike a concave lens, they are spread out.

Integrated Learning

Use Integrating Worksheet 1.4.

**The Living Textbook:
Life Science Sides 1–4**

Chapter 5 Frame 01534
Microscopes (21 Frames)
Search: Step:

INTEGRATED LEARNING

STS Connection

Have students examine a pair of eyeglasses to determine if the lenses are concave, convex, or some combination of the two. Discuss the terms *nearsighted* and *farsighted*. Ask students to name which condition is corrected by the two types of lenses. (Convex lenses correct farsightedness; concave lenses correct nearsightedness.)

Integrating the Sciences

Physical Science Students may be interested in knowing that millions of water drops in the earth's atmosphere act as prisms. The drops disperse light into its component colors. Ask students to infer what they may see in the sky when this happens. (Various colors of the spectrum emerge from many droplets, forming a rainbow. They may see a partial or complete rainbow, depending on the horizon line they are viewing.)

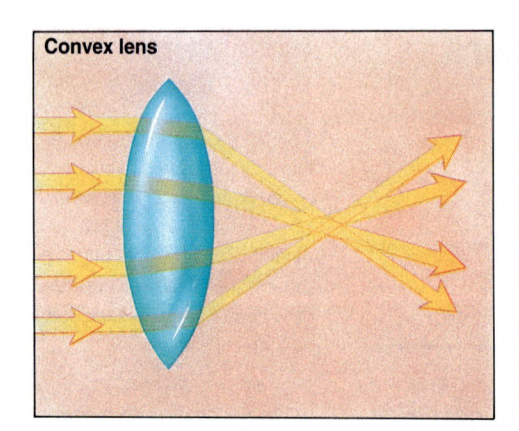

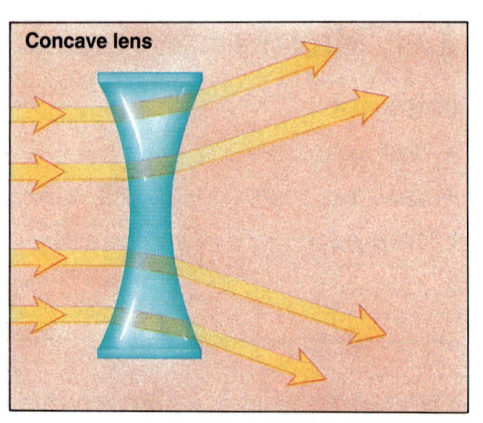

Figure 1.18 ▲
What happens when light rays strike a convex lens? What happens when light rays strike a concave lens? ①

Skills WorkOut

Observe Did you know you can build a magnifying tool with a paper clip and a drop of water? Straighten the clip, and bend back one of the ends so that a small loop forms. Drop the loop into a glass of water. Now use this tool to examine the ink pattern found in the figures shown on a magazine or newspaper page. Can you observe the ink dots used in the printing process?

Convex Lenses Convex lenses are used as magnifying lenses. They are thin on the edges and thick in the center. Look at Figure 1.18. To form images, convex lenses bring light rays together at the center of the lens. This is the thickest part of the lens. The point at which the light rays converge is the focal point.

You may be interested to know that your eyes have convex lenses in them. They are different from the hard lenses used in microscopes. Your eyes are soft, flexible lenses. With their flexibility, the muscles in your eyes can easily change the shape of the lens in each eye. The change in shape focuses light onto the back of your eye. This allows you to clearly see images of objects both near and far away.

Concave Lenses Concave lenses are thin in the center and thick on the edges. Look at Figure 1.18. Notice that concave lenses are different from convex lenses because concave lenses spread light rays out. Concave lenses are often used with convex lenses to form very sharp images.

Compound Microscopes

A microscope uses light and lenses to produce an enlarged image of an object. A simple microscope, such as a hand lens, has only one lens. A microscope with two or more lenses is called a compound microscope.

Look at the compound microscope in Figure 1.19. It has three lenses. One lens is located at the top of the viewing tube. This lens is the ocular lens. It is located in

22   Chapter 1   Studying Science

Themes in Science

Patterns of Change/Cycles Ideas about the world continue to change, and the microscope is an invention that has always had an impact on how people view the world. Have students consider the role of the microscope in the progress of medical science. Ask students why many diseases could not be successfully fought without the use of a microscope. (Many agents of disease are microbes—monerans, protists, and viruses—which are invisible without a microscope. Many clues to disease, moreover, are microscopic, such as blood cells and abnormal cells.)

---

the eyepiece. The other two lenses are located in the revolving nosepiece. These are the objective lenses. By turning the nosepiece, you can view objects under either high power or low power.

By multiplying the power of the ocular lens by the power of the objective lens, you can calculate the total magnification power of the microscope. For example, if the eyepiece is 10x and the objective is 43x, the total magnification is 430x.

Figure 1.19 Compound Microscope

Eyepiece Contains the ocular lens. You look through the eyepiece.

Body tube Holds the system of lenses.

Revolving nosepiece Contains the objective lenses and can be rotated.

High-power Objective lens

Low-power Objective lens

Stage Platform that holds slides.

Diaphragm Controls the amount of light reflected through the hole in the stage.

Light source Collects and directs light up through the hole in the stage and through the object being viewed.

Base Supports the microscope.

Coarse focus Controls movement of body tube, used only to bring the image into view.

Fine focus Controls movement of body tube, used to sharply focus the image.

Arm Supports the body tube.

Stage clips Holds microscope slides in place.

Directed Inquiry

Have students study Figure 1.19. Ask the following questions:

▶ If a low-power objective lens magnifies 10x, and the eyepiece magnifies 10x, what is the total magnification power? (100x)

▶ Can the eyepiece, the high-power objective lens, and the low-power objective lens all be used at the same time? Explain. (No, only one of the objective lenses can be used because of how the lenses are placed on the revolving nosepiece.)

▶ Which focus would you use to get the first view of the image? (The coarse focus)

▶ Why is the arm curved like an old-fashioned telephone receiver? (This shape makes the microscope easier to carry.)

▶ What is the diaphragm used for? (It controls the amount of light coming up through the hole in the stage.)

▶ What objective lens should be used first? Why? (The low-power lens should be used first. Then the high-power lens can be used to further magnify the object on the stage.)

The Living Textbook: Life Science Sides 1-4

Chapter 5 Frame 01555
Objects Under the Microscope (14 Frames)
Search: Step:

The Living Textbook: Life Science Sides 7-8

Chapter 6 Frame 01031
Taste Buds: Various Magnifications (4 Frames)
Search: Step:

TEACH • Continued

Research

Have students use references to find out more about the two kinds of electron microscopes. Have them find out why the TEM requires that objects be in thin slices. Also have them find out which kind of electron microscope will magnify one million times or more.

Discuss

Explain to students that most modern grocery stores employ scanners and UPC symbols, or bar codes. The scanner is actually a beam of light that "reads" the information on the bar code. Ask the following questions:

▶ What are some of the advantages to this system? (Items do not have to have price tags, which saves paper and work for grocers.)

▶ How would a grocery store change a price using this system? (The scanner is adjusted to read the bar code differently. The bar code itself is not changed.)

**The Living Textbook:
Life Science Sides 7-8**

Chapter 3   Frame 00388
TEM Cell Structures (10 Frames)
Search: Step:

**The Living Textbook:
Life Science Sides 7-8**

Chapter 10   Frame 01251
SEM Images of Drosophilia (12 Frames)
Search: Step:

24

INTEGRATED LEARNING

STS Connection

Because lasers are such "accurate" beams of light, they have many uses other than cutting or melting things. They can be very accurately traced, and their reflection speeds can be very accurately timed. In surveying land, measuring distances from the earth to nearby bodies such as the moon, and in tracking the speed of moving vehicles, lasers have proven quite useful.

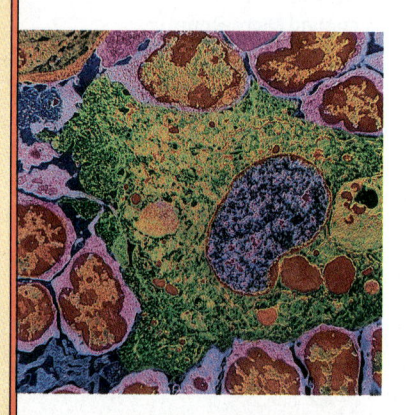

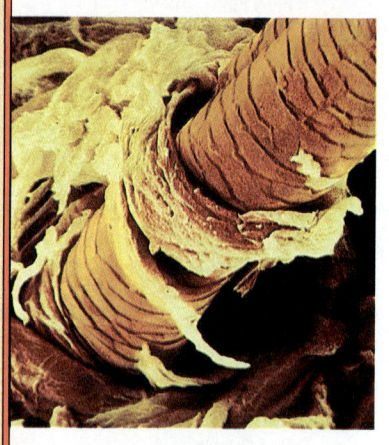

Electron Microscopes

You use a compound microscope to view objects that are magnified under 1,000x. Scientists, however, can view objects that are magnified one million times or more by using electron microscopes. Electron microscopes use tiny particles called electrons to produce enlarged images of objects.

The first type of electron microscope was developed in the 1930s. It was called a transmission electron microscope, or TEM, for short. Within a TEM, a beam of electrons is aimed at a specially prepared object. Magnets focus and direct the beam. Electrons passing through the object strike photographic film. When the film is developed, it contains a magnified image of the object.

The scanning electron microscope, or SEM, is another type of high-magnification microscope. Within an SEM, electrons are bounced off the surface of a prepared object. By recording the reflection of this beam, scientists obtain a three-dimensional view of the magnified object.

Lasers

Lasers are tools that produce a very narrow, intense beam of light. Lasers are often used by surgeons as a type of scalpel. Surgeons can use them for surgery that requires work on small, delicate body parts, such as the eye. Lasers are also used to remove cancer cells from the body.

Lasers are not only used for surgery. Welders use them to cut cloth and harden metals. You may even go to a grocery store that uses laser cash registers. These registers add up items that are marked with special codes.

Figure 1.20 ▲
A false-color TEM of cells in human tissue (top). A human hair seen through an SEM (bottom).

Figure 1.21 ▶
Lasers are used in many areas, including medicine, communication, and manufacturing.

24

STS Connection

Optical instruments have provided advances in areas of science other than biology. Explore with the class how microscopes provide information on minerals in geology, how cameras have expanded our knowledge in oceanography, and how telescopes have changed our understanding of astronomy and space science. Consider vision correction and contact lenses too. Have students make a table of optical instruments, showing the ways they are used in the life, earth, and physical sciences.

Science and Technology
Optical Instruments

A microscope is an optical instrument. An optical instrument uses mirrors or lenses to produce images. Cameras, telescopes, and binoculars are three other optical instruments that you may have used.

A camera is similar to your eyes. Both your eyes and a camera allow light to enter through a lens. However, a camera lens may be made up of several separate lenses. The image produced by a camera lens falls on photographic film that contains chemical substances that are sensitive to light.

Have you ever used a telescope to view the night sky? With a telescope, you can view objects that are very far away. Telescopes have mirrors and lenses that produce enlarged images of distant objects.

A very powerful telescope is called a refracting telescope. In a refracting telescope, a series of lenses is used to focus and magnify light from distant objects. The larger the lenses used in a refracting telescope, the greater the light-gathering power. One of the most powerful refracting telescopes is at the Yerkes Observatory in Wisconsin. This telescope has a light-gathering power that is about 40,000 times greater than the light-gathering power of the human eye!

A reflecting telescope is even more powerful than a refracting telescope. A reflecting telescope uses a mirror instead of a lens as the objective. Because they can be built larger than refracting telescopes, reflecting telescopes have greater light-gathering power.

Figure 1.22 ▲
The Lick Observatory in California has both refracting and reflecting telescopes. In the foreground, you can see one of their reflecting telescopes.

Check and Explain

1. How is a convex lens different from a concave lens?
2. Where are the lenses located in a compound microscope?
3. **Compare** How is an electron microscope different from a compound microscope?
4. **Infer** State at least four ways that you think microscopes can help scientists in their work.

Chapter 1 Studying Science 25

Enrich
If possible, show students drawings of reflective and refractive telescopes.

EVALUATE

WrapUp
Portfolio Have students use microscopes to observe prepared slides. Have them draw pictures of an object at two different magnifications. Students may keep the pictures in their portfolios.
Use Review Worksheet 1.4.

Check and Explain

1. Convex lens: used as magnifying lens, thin on the edges and thick at the center; concave lens: spreads light out, thin in the center and thick on the edges.
2. One lens is located at the top of the viewing tube, and the other lenses are located closer to the specimen.
3. Compound microscopes use lenses to bend and focus light to magnify objects under 1,000x. Electron microscopes focus a beam of electrons through an object, recording an image on photographic film at magnifications greater than 1,000x.
4. Answers will vary. Crystal structure study, medical diagnosis, laser surgery, fingerprint examination.

The Living Textbook:
Life Science Sides 1-4

Chapter 6   Frame 02255
SEM Images (31 Frames)
Search: Step:

ACTIVITY 1

Time 40 minutes **Group** 1–2

Materials
- scissors
- droppers
- microscope slides
- coverslips
- *Elodea* leaves
- newspaper
- tweezers

Analysis

1. The image of the "e" is larger, it is flipped horizontally and vertically, and is grainier.
2. The image gets larger and grainier, and possibly moves out of the field of view when switching from low to high power.
3. Answers may vary. The "e" has a definite left, right, top, and bottom, so it will appear reversed under the microscope. Students can infer that this means the image of the leaf is inverted also.

Conclusion

Conclusion should follow from students' actual work. The observer notices clues about the physical parts of the letter "e," such as the paper fiber and the variation in the ink in the print. Students might also note that a light microscope, unlike an electron microscope, allows specimens to be viewed alive.

Extension

Student observations will vary depending on the type of hair viewed. Most students will comment on the three-dimensionality and surface details of the specimen.

TEACHING OPTIONS

Prelab Discussion

Have students read the entire activity. Cover these points before students begin:

▶ Review the parts of the microscope, pointing out any differences between Figure 1.19 on page 23 and the microscopes available to the students.

▶ Review proper use and care of the microscope, including carrying it properly, illuminating the specimen, and focusing on the specimen.

Activity 1   How do you prepare a wet mount slide?

Skills Observe; Collect Data; Infer

Task 1 Prelab Prep
1. Collect the following items: unruled paper, scissors, dropper, water, slides, coverslips, Elodea leaf, newspaper, tweezers.
2. Carefully carry a microscope to your work area.

Task 2 Data Record
1. On one sheet of paper write, *Wet Mount of Letter "e"* at the top. On another sheet write, *Wet Mount of Elodea Leaf*.
2. Divide each sheet in half lengthwise. Label the upper portion *Low Power* and the lower portion *High Power*.

Task 3 Procedure
1. Use the scissors to cut out the lowercase letter "e" from the newspaper.
2. Use the dropper to place a drop of water on the center of a slide. Use the tweezers to place the letter "e," face-up, in the water drop. Add another drop of water to the slide.
3. As shown below, touch one edge of the coverslip to the water. Allow the water to spread out along this edge, then slowly lower the coverslip over the drop. Keep the "e" somewhat in the center. Be careful not to trap any air bubbles.
4. Move the low-power objective into place. Use the coarse adjustment to raise the body tube about 5 cm above the stage. Secure the "e" slide on the stage using stage clips.
5. Lower the objective so that it is about 2 mm above the coverslip.
6. Look through the eyepiece and adjust the mirror so the "e" is illuminated. **CAUTION! Do not aim the mirror directly at the sun. Direct sunlight can injure your eyes.**
7. Use the coarse focus to raise the body tube. Once you see an image, switch to fine focus. When you have the "e" in focus, draw what you see. Label the drawing *low power*.
8. Carefully click the high-power objective into place. When in high power, use only the fine focus knob. Once you have the "e" in focus, draw what you see. Label the drawing *high power*.
9. Repeat steps 2–8, using a small section of Elodea leaf instead of the letter "e."

Task 4 Analysis
1. How does viewing the letter "e" through the microscope change the appearance of the "e"?
2. What happens to the images when you switch from low power to high power?
3. How does viewing the "e" differ from viewing the leaf?

Task 5 Conclusion
Explain how a compound microscope helps you study both nonliving things and living things.

Extension
Use the wet mount technique to make slides of a human hair. Then view the slides under a microscope. Describe what you see.

CHAPTER REVIEW 1

Chapter 1 Review

Concept Summary

1.1 Science Skills and Methods
- Scientists use many skills; they observe, infer, estimate, measure, predict, classify, hypothesize, record, organize, and analyze.
- Scientists perform experiments to gather data about nature.
- Scientific facts, theories, and laws may change.
- Scientists use models to represent the parts of nature they study.

1.2 Measuring with Scientific Units
- Scientists measure length, volume, mass, density, temperature, and time with standard SI units.
- Mass is the amount of matter an object contains; it is not the same as weight.
- Density is a measure of mass per unit volume.

1.3 Graphing
- Circle graphs show how the parts of something make up a whole.
- Bar graphs compare one aspect of several different things.
- Line graphs show data defined by two variables.

1.4 Microscopes and Lasers
- Because they bend light, lenses are used in microscopes to make things appear larger.
- Electron microscopes can magnify objects more than one million times.
- Lasers are used by surgeons as a kind of scalpel.

Chapter Vocabulary

data (1.1)	meter (1.2)	mass (1.2)	microscope (1.4)
controlled experiment (1.1)	volume (1.2)	kilogram (1.2)	lens (1.4)
constant (1.1)	cubic meter (1.2)	density (1.2)	
variable (1.1)	liter (1.2)		

Check Your Vocabulary

Use the vocabulary words above to complete the following sentences correctly.

1. An object's ____ is measured as g/cm³.
2. The basic SI unit for length is the ____.
3. An object too small to be seen can be studied with a ____.
4. In an experiment, a ____ is a factor that is kept the same.
5. The SI unit often used to measure the volume of large containers is the ____.
6. An SI unit for volume based on the meter is the ____.
7. An object's ____ is a measurement of how much matter it contains.
8. In a ____, there are two test groups.
9. The information gathered by scientists is called ____.
10. A piece of glass shaped to bend light is a ____.
11. An SI unit for mass is the ____.
12. In an experiment, a ____ is a factor that is changed.
13. Liters are used to measure ____.

Write Your Vocabulary

Write sentences using each vocabulary word above. Show that you know what each word means.

Chapter 1 Studying Science 27

Check Your Vocabulary

1. density
2. meter
3. microscope
4. constant
5. liter
6. cubic meter
7. mass
8. controlled experiment
9. data
10. lens
11. kilogram
12. variable
13. volume

Write Your Vocabulary

Students' sentences should show that they know both the meanings of the words and how to apply the words.

Use Vocabulary Worksheet for Chapter 1.

CHAPTER 1 REVIEW

Check Your Knowledge

1. A number and a unit must be included in all measurements.
2. Answers may vary, but may include safety goggles, aprons, and heat-resistant or other types of gloves.
3. A controlled experiment is an experiment in which all variables are held constant except the one that is believed to have an effect on the object or process being studied.
4. A circle graph looks like a pie cut into slices.
5. Models are used to help people understand things that are not directly observable.
6. The liter (L) and the cubic centimeter (cm^3) are two SI units of volume.
7. Mass is an unchanging property of an object; weight varies depending on gravity.
8. A microscope is used to see objects that are too small to be seen with unaided eyesight.
9. False; Convex
10. True
11. False; lowest point
12. False; sometimes
13. False; 10
14. True

Check Your Understanding

1. a. bar   c. line
   b. circle   d. circle
2. a. liters
   b. centimeters
   c. degrees Celsius
   d. grams per cubic centimeter
3. Flush eyes with large amounts of water, and inform the teacher.
4. A powerful light microscope can provide a magnification of 370x. A scientist can compare pollen from different plants with the details seen in this kind of image.
5. A length of 1.5 is longer than 150 mm. $1\ m^3$ is a greater volume than 1 L.
6. 210x, 460x, and 740x

Develop Your Skills

1. a. 80%
   b. 3 times
2. a. 5.5 m (550 cm)
   b. 3 m (118 mm)
3. Answers may vary. Possible answers are given.
   a. The dog was in the pond.
   b. It is a holiday. The letter carrier has not come yet.
   c. Your friend was out in the sun. Your friend was at the beach. Your friend does not use sunscreen.
   d. Your teacher is sick. Your teacher's son or daughter is sick.

Make Connections

1.

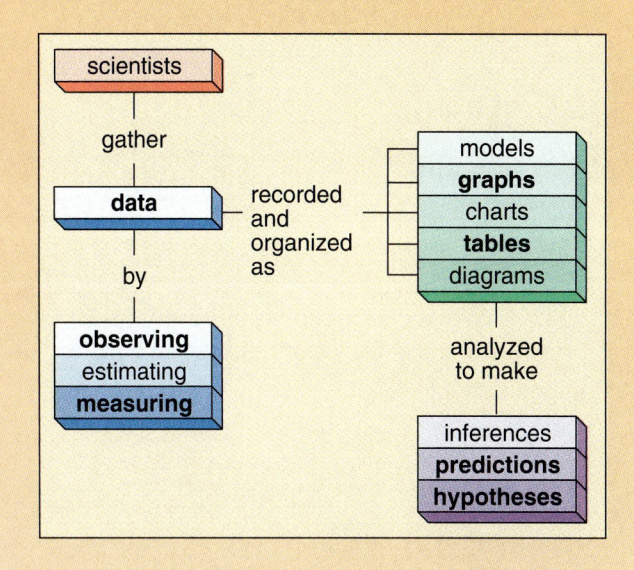

```
scientists
   │
 gather
   │
  data ──── recorded ──── models
   │         and          graphs
   by      organized      charts
   │         as           tables
observing                 diagrams
estimating                   │
measuring                 analyzed
                          to make
                             │
                          inferences
                          predictions
                          hypotheses
```

2. Answers will vary. See pages 3 to 5 in the text for a description of each of the science process skills.
3. Answers will vary, but all means for increasing magnification of light microscopes involve different combinations of lenses and/or projecting the image on a screen.
4. Student reports will vary. Mention to students any important scientists they have overlooked. Students may be interested in finding information on one of the following: Anaximander, Pythagoras, Aristotle, Archimedes, Ptolemy, Fibonacci, Albertus Magnus, Leonardo da Vinci, Copernicus, Galileo, Francis Bacon, Isaac Newton, Henry Cavendish, James Watt, Count Volta, Count Rumford.

Develop Your Skills

Use the skills you have developed in this chapter to complete each activity.

1. **Interpret Data** The circle graph below shows the makeup of a raccoon's diet over a two-week period. The amounts of each kind of food were measured in grams.
   a. What portion of the raccoon's food is made up of animals?
   b. The amount of frogs and salamanders in the raccoon's diet is how many times greater than the amount of fish?

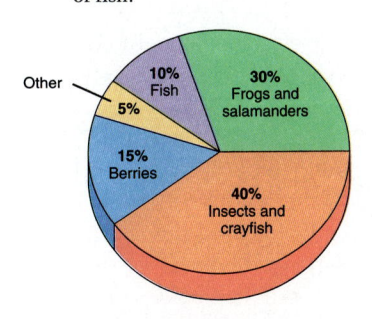

   Pie chart:
   - Other 5%
   - Fish 10%
   - Frogs and salamanders 30%
   - Berries 15%
   - Insects and crayfish 40%

2. **Data Bank** Use the information on page 631 to answer the following questions.
   a. How long is the King cobra? Convert its length to centimeters.
   b. How long is the Komodo dragon? Convert its length to millimeters.
3. **Infer** Make an inference based on each of the following observations.
   a. A wet dog walks away from a pond.
   b. Your mailbox is empty.
   c. Your friend has a sunburn.
   d. You have a substitute teacher today.

Make Connections

1. **Link the Concepts** Below is a concept map showing how some of the main concepts in this chapter link together. Only parts of the map are filled in. Copy the map. Using words and ideas from the chapter, complete the map.

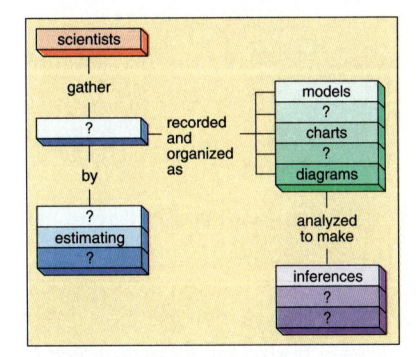

   ```
   scientists
      │
    gather
      │
      ?  ──── recorded ──── models
      │        and           ?
     by      organized      charts
      │        as           ?
      ?                    diagrams
   estimating                 │
      ?                   analyzed
                          to make
                             │
                          inferences
                             ?
                             ?
   ```

2. **Science and Living** Choose one of the science process skills and describe how you could use it to benefit your daily life.
3. **Science and Technology** Electron microscopes can magnify objects many more times than can light microscopes, but they cannot be used to study living organisms or living cells. Research ways in which high-power light microscopes can provide clear images of living cells at very high magnifications.
4. **Science and History** Many of the modern methods of science are based on the ideas or work of scientists who lived many centuries ago. Find out the names of some important scientists who lived before the 1800s. Choose one and write a report about his or her contributions to science.

Chapter 1 Studying Science

CHAPTER 2

Overview

This chapter focuses on living things and how they interact with the earth. The first section discusses the earth's unique position in the solar system and the life-supporting elements of the biosphere. The second section distinguishes between living and nonliving things in the earth's biosphere. Basic characteristics of cells and complex organisms are described. The chapter concludes with a discussion of how living things meet their basic needs and adapt to various environments.

Advance Planner

▶ Gather prepared slides of different cell types for TE activity, page 37.

▶ Make copies of an outline map of the world for TE activity, page 38.

▶ Collect magazines that contain pictures of living organisms for TE activity, page 39.

▶ Label 3 by 5 index cards with headings listed in TE activity, page 42.

Skills Development Chart

Sections	Analyze	Classify	Communicate	Compare and Contrast	Decision Making	Graph	Infer	Measure	Observe
2.1 Skills WarmUp Skills WorkOut			●	●					
2.2 Skills WarmUp Historical Notebook	●	●	●						
2.3 Skills WarmUp Skills WorkOut Consider This Activity	● ●		● ● ●		 ●	●		●	●

Individual Needs

▶ **Limited English Proficiency Students** Have students write the vocabulary terms from the chapter in their science journals, leaving several blank lines between terms. Then, as they study the chapter, have students write the figure numbers of any pictures or diagrams that describe each term. For example, students could write *Figure 2.4* next to the word *cell*. Have students use these pictures and figures to reinforce the meanings of the chapter's terms.

▶ **At-Risk Students** Invite an animal warden from your local police department to speak with at-risk students. Have students prepare for the visit by formulating questions beforehand. Students might ask why it's necessary for an animal warden to know the characteristics and needs of living things. For example, they might ask why it would be useful to know how an animal responds to a stimulus, or what adaptations make it possible for some animals to survive in urban locations.

▶ **Gifted Students** Have students use their knowledge about animals to solve a mystery. First, have students design an ideal zoo. Have them decide which animals they would like to include. Have students use the information from this chapter to plan an ideal habitat for each animal. Suggest that they draw a detailed ground plan for their zoo. Then tell students that one type of animal in their zoo is mysteriously disappearing and is in danger of becoming extinct. Have students identify the needs and adaptations of the animal to find clues to its disappearance. In this case, the culprit may be a predator or a flaw in one of the zoo's physical systems. Students can use their ground plans to show how they solved the mystery.

Resource Bank

▶ **Bulletin Board** Make headings showing the six common needs of living things and hang them on the bulletin board. Have students draw or cut out pictures that show living things meeting their needs. For example, pictures could show a plant in water, a person eating, or a rabbit digging a burrow. Have students make labels for their pictures, explaining how each organism is meeting its needs. Have students mount the pictures on the bulletin board.

▶ **Field Trip** Arrange to take students on a walk around the school. Have them look for examples of living organisms meeting their needs.

CHAPTER 2 PLANNING GUIDE

Section	Core	Standard	Enriched
2.1 The Living Planet pp. 31–35			
Section Features Skills WarmUp, p. 31	●	●	
Skills WorkOut, p. 32		●	●
Blackline Masters Review Worksheet 2.1	●	●	
Skills Worksheet 2.1	●	●	
Integrating Worksheet 2.1a	●	●	
Integrating Worksheet 2.1b	●	●	●
Integrating Worksheet 2.1c	●	●	●
Color Transparencies Transparencies 4a, b	●	●	
2.2 Characteristics of Living Things pp. 36–40			
Section Features Skills WarmUp, p. 36	●	●	
Historical Notebook, p. 39		●	●
Blackline Masters Review Worksheet 2.2	●	●	
Skills Worksheet 2.2	●	●	●
Enrich Worksheet 2.2	●	●	●
Ancillary Options One-Minute Readings, p. 114	●	●	●
Overhead Blackline Transparencies Overhead Blackline Master 2.2 and Student Worksheet	●	●	●
Laboratory Program Investigation 3	●	●	●
2.3 Needs of Living Things pp. 41–46			
Section Features Skills WarmUp, p. 41	●	●	●
Skills WorkOut, p. 43	●	●	●
Consider This, p. 44	●	●	
Activity, p. 46	●	●	
Blackline Masters Review Worksheet 2.3	●	●	
Integrating Worksheet 2.3		●	●
Vocabulary Worksheet 2	●	●	
Laboratory Program Investigation 4		●	●
Investigation 5	●	●	●
Color Transparencies Transparencies 5a, b	●	●	

Bibliography

The following resources can be used for teaching the chapter. See page T-40 for supplier codes.

Audio-Visual Sources
(video unless noted)
Adaptation to Environment. 16 min. 1983. LF.
Animal Populations: Nature's Checks and Balances. 22 min. 1983. EB.
The Building of the Earth. 55 min. 1984. T-L.
The Living Earth. 25 min. 1991. NGSES.
Our Dynamic Earth. 23 min. 1979. NGSES.

Software Resources
The Earth Through Time and Space. Apple. EA.
Odell Lake. Apple, Atari. MECC.

Library Resources
Cairns-Smith, A. G. Seven Clues to the Origin of Life: A Scientific Detective Story. New York: Cambridge University Press, 1985.
Goldin, Augusta. Water: Too Much, Too Little, Too Polluted? San Diego, CA: Harcourt Brace Jovanovich, 1983.
Horowitz, Norman H. To Utopia and Back: The Search for Life in the Solar System. New York: Freeman, 1986.
Thomas, Lewis. The Lives of a Cell: Notes of a Biology Watcher. New York: The Viking Press, Inc., 1974.

CHAPTER 2

INTEGRATED LEARNING

Writing Connection
Ask students to write a poem inspired by the photograph on page 30. Encourage them to imagine what the details and the colors represent in this scene. Have students exchange poems and discuss them with one another.

Introducing the Chapter
Have students read the description of the photograph on page 30. Ask if they agree or disagree with the description.

Directed Inquiry
Have students study the photograph. Ask:

▶ Describe the colors and textures you see in the photograph. (Reds, blacks, and blues are most obvious, but encourage students to notice the more subtle variations in color.)

▶ Where in the photo do you see sharply outlined areas? (Students should direct you to the sections where bodies of water are located.)

▶ What is the topic of this chapter? How do you think this photograph relates to the earth and living things? (Answers may vary. Students may think that the colors represent living organisms, or they may mention the importance of water.)

▶ Think of the photograph as a view of the earth's surface. Where might it be taken from? (The image is an infrared satellite photo taken of the west coast of the United States.)

Chapter Vocabulary

adaptation
atmosphere
biosphere
cell
competition
ectothermic
endothermic
homeostasis
metabolism
reproduction
response
stimulus

Chapter 2 — The Earth and Living Things

Chapter Sections
2.1 The Living Planet
2.2 Characteristics of Living Things
2.3 Needs of Living Things

What do you see?

"I think that the photograph is of the west coast. I can see the San Francisco Bay and the mountains. It looks like it was taken from space. The colors like the browns and greens represent the elevation of mountains, valleys, and flat land."

*Garrett Lamb
Arundel Elementary School
San Carlos, California*

To find out more about the photograph, look on page 48. As you read this chapter, you will learn more about why the earth is unique.

Themes in Science

Patterns of Change/Cycles The water, land, and air in the earth's biosphere are continually being changed through interactions with the sun. Have students draw a concept map or diagram that accounts for the many ways in which water is renewed in this water cycle.

Integrating the Sciences

Astronomy Johannes Kepler (1571–1630) theorized that planets move in elliptical, not circular, paths around the sun. An ellipse is an oval shape that is determined by two points called foci.

SECTION 2.1

Section Objectives
For a list of section objectives, see the Student Edition page.

Skills Objectives
Students should be able to:

Compare the relative percentage of land and water on the earth.

Communicate information about the earth using words and visual images.

Graph the distribution of land and water on the earth.

Vocabulary
biosphere, atmosphere

2.1 The Living Planet

Objectives

▸ **Define** the earth's biosphere.

▸ **Explain** why the earth is a unique planet.

▸ **Explain** the importance of the earth's land, water, and atmosphere to living things.

▸ **Design** a circle graph comparing the distribution of land and water on earth.

Skills WarmUp

Compare Use a globe. Turn the globe so that the South Pole faces you. Draw the areas of land and water. Now turn the globe so that the North Pole is in front of you. Draw the areas of land and water. Compare the two drawings. Is the amount of land and water on each half of the earth equal? Which half—south or north—has more water? More land?

MOTIVATE

Skills WarmUp
To help students compare the areas of land and water in the Northern and Southern hemispheres, have them do the Skills WarmUp.
Answer Students should observe that there is more surface covered by water in the Southern Hemisphere than in the Northern Hemisphere.

Prior Knowledge
To gauge how much students know about the biosphere, ask the following questions:

▸ How is the earth different from other planets?

▸ Where can you find living things on the earth?

▸ Why is the earth the only planet that supports life?

Integrated Learning
Use Integrating Worksheet 2.1a.

Make a list of all the places where you think living things are found. You probably named places on land, in water, and in the air. Living things can be found nearly everywhere on the earth, even in the most unlikely places. For example, several kinds of worms live so deep in the ocean that sunlight can't reach them. Most penguins live in icy Antarctica. Some kinds of lizards can even survive in the hottest deserts.

All living things exist at or near the earth's surface in a life-supporting zone called the **biosphere**. What sort of place can support life? Living things need sunlight, water, matter, and air. In the biosphere, the land, water, and air satisfy the needs of all living things.

The Third Planet

The earth is located 150 million km from the sun. The earth is part of a system of planets and other objects in space that orbit, or revolve around, the sun. This system is called the solar system. Look at Table 2.1. Of the nine planets in the solar system, the earth is the third planet from the sun and the fifth largest planet.

The earth is unique in the solar system because it is the only planet where life is known to exist. Water is found on the earth as a liquid, a solid, and a gas—unique among the planets. However, because of the earth's distance from the sun, most of the earth's water is in a liquid state. Water is needed to support life. The other planets are either closer to the sun and too hot, or farther from the sun and too cold.

Table 2.1 The Nine Planets

Position from Sun	Diameter (km)
Mercury	4,878
Venus	12,012
Earth	12,756
Mars	6,787
Jupiter	142,984
Saturn	120,536
Uranus	51,118
Neptune	49,660
Pluto	2,400

TEACH

Skills WorkOut
To help students communicate their ideas about the unique characteristics of the earth, have them do the Skills WorkOut.
Answer There will be many possible choices. Look for examples that emphasize the availability of liquid water on the earth and the widespread abundance of life.

Enrich
One career that was greatly changed when people discovered that the earth was round was ocean navigation. In 1552, Gerardus Mercator developed a way of making flat maps that enabled sailors to plot a steady course simply by drawing a straight line on the map.

Integrated Learning
Use Integrating Worksheet 2.1b.

Skills Development
Make Models Have students use the information in Table 2.1 on page 31 to compare the diameters of the nine planets. Tell them to choose a scale and calculate the approximate size that their model planets should be. (A reasonable scale would be 1 cm to 4000 km.) Ask them to construct a model of the solar system by making nine circular cutouts and arranging them in relation to the sun.

INTEGRATED LEARNING

Integrating the Sciences
Astronomy To reinforce the difference between a planet and a star, ask students which kind of body is a source of light energy and which one reflects it. (Stars are sources of light; planets are not sources of light, but reflect the light given off by the sun.)

Geography Connection
Have students use the map on pages 32 and 33 to identify the continents on which different countries are located. Explain that the equator is an imaginary line that is used to divide the earth into two halves, or hemispheres, called the Northern and Southern hemispheres. Have students identify the hemispheres in which each continent is located.

Skills WorkOut
Communicate Pretend you are in charge of collecting ten photographs to be placed in a space probe headed for distant galaxies. What photographs would you select that would best describe the earth? Explain your choices.

**Figure 2.1
Earth's Continents and Oceans**

North America This is the only continent to have every kind of climate—from the dry cold of the Arctic to the wet heat of the tropics. The abundant animal and plant life varies with the climate.

South America The greatest variety of animals on any continent live in the rivers, swamps, and tropical rain forests of South America. In fact, one-fourth of all the known kinds of mammals live in these rain forests.

Antarctica Only mosses and a few small insects can live on this ice-buried continent. However, many kinds of animals, such as fish, krill, penguins, seals, whales, and flying birds, live in the surrounding ocean.

Unlike the much larger outer planets, which are mostly gas, the earth is mostly solid. You know this solid part as land. Land provides some of the substances needed for life. It also provides a surface on which life could develop. Surrounding the land is air. The air contains the water vapor and gases also needed to support life.

Earth's Waters The most abundant substance on the earth is water. Most of the earth's water is salty and exists in the world ocean. Only 3 percent of all the earth's water is fresh. Rivers, ponds, streams, and most lakes and swampy wetlands are bodies of fresh water.

Look at the map in Figure 2.1. Notice that there is really one big body of water—the world ocean. Geographers divide the world ocean into five major oceans. Find the Pacific, Atlantic, Indian, Antarctic, and Arctic oceans on the map in Figure 2.1.

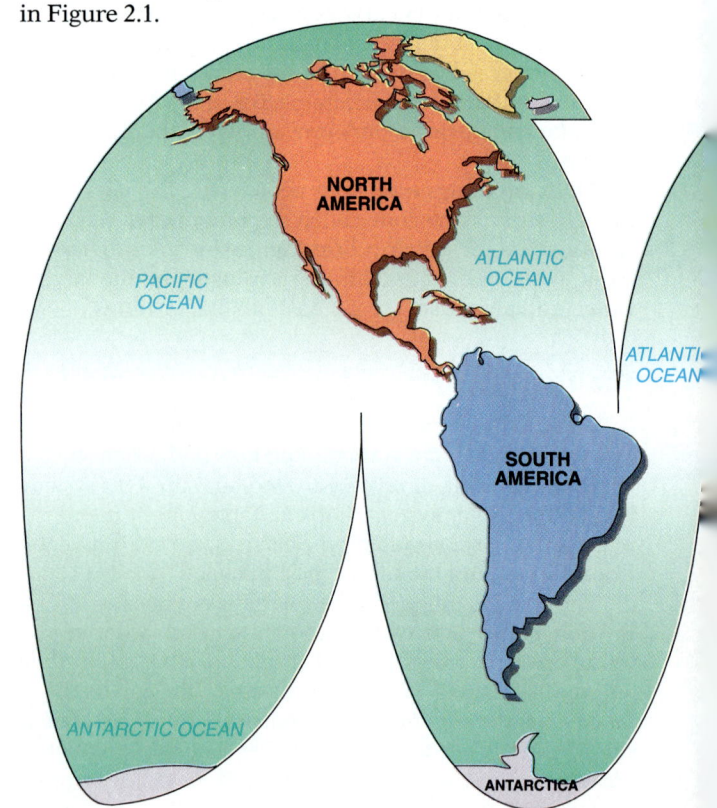

32 Chapter 2 The Earth and Living Things

Multicultural Perspectives

Explorers to the North Pole were able to survive the extreme cold conditions because they used strategies that the Inuit people taught them. Have students find out what the explorers, such as Robert Peary, learned from the Inuits. Students can work with classmates to prepare presentations based on what they find.

Critical Thinking

Reason and Conclude Point out to students that water exists in solid, liquid, and gas forms. Challenge students to name as many sources of the earth's water as they can think of. Write two headings on the chalkboard: *Location of Water* and *Percentage of Total Water*. Under the two columns mix up the answers given below. Have students try to match the locations with the correct percentages. (Frozen in ice caps and glaciers: about 2.1%; Underground water: 0.61%; Lakes, rivers and streams: 0.02%; Clouds and humidity: 0.01%; Oceans: about 97.2%)

Discuss

Ask students to examine the map on pages 32 and 33 and suggest which of the landmasses is often called an island continent (Australia). Encourage students to discuss their choices. (Australia, like an island, is completely surrounded by water.)

Of the five oceans, the Pacific is the largest and deepest. Smaller bodies of salt water in the world are seas.

The oceans and bodies of fresh water contain an amazing variety of living things. Large ocean kelp and freshwater algae provide food and shelter for some fishes. Certain animals, such as fishes, snails, and turtles, make their homes in either fresh water or ocean water. Water life ranges in size from tiny microscopic bacteria to the blue whale—the largest living animal.

Earth's Landmasses As shown in Figure 2.1, only a little of the earth's surface is actually land. The relationship is about 28 percent land to 72 percent water. The large landmasses on the earth are called continents. Geographers define continents as very large landmasses that are surrounded, or nearly surrounded, by salt water. Table 2.2 names the seven continents.

Table 2.2
The Earth's Continents

Africa
Antarctica
Asia
Australia
Europe
North America
South America

Europe Vast forests in Europe have been cut down to make way for cities and farms. Most of Europe's wildlife live in areas difficult for people to reach, or in special preserves or zoos.

Africa The world's largest desert, the Sahara, stretches across Northern Africa. Much of the continent is grassland—home for hyenas, elephants, giraffes, lions, and zebras.

Asia Asia is the largest continent. Camels, cobras, reindeer, the Giant Panda, yaks, Siberian tigers, bamboo, kudzu, beech trees, and ginkgos live in Asia.

Australia Most of Australia is dry grassland—home to many unique animals, such as kangaroos, kookaburras, wallaroos, and emu.

Chapter 2 The Earth and Living Things

TEACH • Continued

Enrich
Point out that even organisms in water need gases from the atmosphere. Ask where the oxygen comes from that fish need to breathe. (Oxygen from the atmosphere is dissolved in water.)

Research
Remote sensing satellites have given us an enormous variety of information about the surface of our planet. Satellite radar cameras, for example, have revealed glimpses of what look like irrigation canals beneath the sands of the Sahara. Have students research remote sensing images to see how a particular area on the earth changes over time, or how two different areas compare to one another.

For maps and more information about how remote sensing images are created, contact the U.S. Geological Survey, EROS Data Center, Customer Services, Sioux Falls, SD 57198.

Integrated Learning
Use Integrating Worksheet 2.1c.

INTEGRATED LEARNING

Integrating the Sciences

Environmental Science Students may have heard about concern over the ozone layer. They probably do not know that ozone (O_3) is actually one of the two forms of oxygen. It is a pale blue gas that is formed naturally by the action of lightning or ultraviolet radiation on oxygen. Ask students to look for ozone in Table 2.3 and draw a circle graph to illustrate the relatively small proportion of ozone in our atmosphere. You may also wish to have students research how the ozone layer is depleted and the progress being made in reducing the loss of this important layer.

Table 2.3 The Gases in Air

Gas	Percent
Nitrogen	78%
Oxygen	21%
Argon, carbon dioxide, neon, helium, krypton, hydrogen, xenon, and ozone	1%

Many different kinds of living things make their homes on the earth's continents. Some animals, such as snakes, live on every continent except Antarctica. The same is true of some plants. Mosses are found on every continent, including Antarctica. Some animals and plants only live naturally on one continent. For example, koalas are unique to Australia. The cacao (kuh KAY oh) bean from which chocolate is made, grows on several continents but only in areas close to the equator.

Earth's Atmosphere The earth is surrounded by an envelope of gases called the **atmosphere**. The atmosphere extends 700 km above the surface of the earth. It is constantly moving in complex, but regular, patterns. The interactions between the atmosphere, ocean, and land determine the weather and climate on the earth.

Look at Table 2.3. The atmosphere is a mixture of different kinds of gases. Notice that nitrogen is the most abundant gas in air. The next most abundant gas is oxygen. Name two other gases that make up air.

All living things need the gases in the atmosphere. For example, animals and plants use the oxygen in air with sugars to carry on certain processes. These processes provide the animals and plants with energy. Besides oxygen, plants use carbon dioxide to make fuel for growth. In the oceans, living things use the atmospheric gases dissolved in the ocean water.

Science and Technology *Infrared Sensors*

Look at Figure 2.2. It is a picture of the Monterey area in California taken from a satellite called LANDSAT I. Scientists call getting information about things at a great distance, or things one can't touch, *remote sensing*. Satellites orbiting in space are remote sensors. Instruments on board satellites provide scientists with information about the earth.

To take the photograph in Figure 2.2, the LANDSAT I satellite used heat, or infrared, sensors. All objects—even your chair, this book, and your body—give off infrared waves. As an object gets hotter, it gives off more and more infrared waves. Infrared waves are like light, but they can't be seen by the human eye. Though invisible, these waves may be detected and measured. Just as your eye can detect a yellow object as different from a blue one, infrared sensors can detect temperature changes. Especially helpful for scientists is the ability of infrared to detect many things from far away.

One specific use of infrared sensors helps scientists survey croplands and forests. Remote sensing by infrared photography provides pictures showing ground heat as shades of red and blue. Areas with thick plant life appear red; areas with thin plant life appear blue or white. By observing an area over time, scientists can detect changes in the amount of plant life in the area. Scientists can use the information to identify areas of frost damage or plant loss in tropical rain forests.

Infrared sensors can also be used to identify soil and rock types, to map natural features, to detect pollution and water flowing from rivers into oceans, and to study clouds and the ocean. The information scientists learn from remote sensors can help us use and preserve the land, air, and water on the earth.

Figure 2.2 ▲
In this LANDSAT I satellite photograph, the planted areas are red and the unplanted areas are blue.

Check and Explain

1. Define biosphere.
2. Why is the earth a unique planet?
3. **Predict** How do you think life on the earth would be affected if the earth were farther away from the sun? Explain your answer.
4. **Graph** Design a circle graph that shows the distribution of land and water on the earth. Label and title your graph.

Chapter 2   The Earth and Living Things   35

SECTION 2.2

Section Objectives
For a list of section objectives, see the Student Edition page.

Skills Objectives
Students should be able to:

Classify things as living, dead, and nonliving.

Make a Model of an organism.

Vocabulary
cell, metabolism, stimulus, response, reproduction, adaptation

MOTIVATE

Skills WarmUp
To help students begin to think about the characteristics of living things, have them complete the Skills WarmUp.
Answer Students should list the sunflower, owl, oak tree, and shark as living; the candle, shirt, radio, baseball, snow, and bicycle as nonliving; and the strand of hair as dead. It would be reasonable if students list the pear as living or dead.

Misconceptions
Students may think that growth is just an increase in size, such as the growth of an icicle. Explain to students that living things grow because their cells increase in size and divide.

Class Activity
Write the terms *icicle*, *robot*, and *car* on the chalkboard. List the characteristics of living things. Have students check whether each object has the characteristics.

Answer to In-Text Question
① Nonliving things include rocks and water. Living things include barnacles and kelp.

36

INTEGRATED LEARNING

Themes in Science
Diversity and Unity Despite the great diversity of living things, from one-celled organisms to blue whales and sequoia trees, all living things are united by six generally accepted characteristics that are common to all organisms.

Skills WarmUp

Classify Write the words *Living*, *Dead*, and *Nonliving* at the top of a piece of paper. Then place each of the following items in the correct column: pear, candle, shirt, radio, sunflower, baseball, owl, strand of hair, oak tree, snow, shark, bicycle.

Figure 2.3 ▲
Tide pools contain many living things that can be seen at low tide.

2.2 Characteristics of Living Things

Objectives
▶ **Distinguish** between living and nonliving things.
▶ **List** and **describe** the characteristics of living things.
▶ **Make a model** that shows your understanding of an organism's adaptations.

Look at Figure 2.3. Can you tell the living things from the nonliving things? Make a list of the living things and a list of the nonliving things. Then compare your lists to those of a classmate. Explain to your classmate how you decided which things were living. ①

You probably thought correctly that the rocks and water were nonliving things, but how could you tell? When you classified the living and nonliving things, you needed to identify common characteristics among the objects. Making a list of the common characteristics of living things is not as easy as it might sound. In fact, scientists still debate about the basic characteristics of life and nonlife. However, there do seem to be some generally accepted characteristics that are common to all living things. All living things:

▶ are made of one or more cells.
▶ obtain and use energy.
▶ grow and develop.
▶ respond to their environment.
▶ reproduce.
▶ are adapted to their environment.

Some nonliving things may have *some* of these six characteristics, but not *all* of them. For example, an icicle may seem to grow, but an icicle is not living. A car gets and uses energy, but it is not living—gasoline needs to be pumped into a car. A robot may move or even speak, but it is not living.

36 Chapter 2 The Earth and Living Things

Themes in Science

Scale and Structure The size and the structure of cells allow the necessary exchange of materials between the cell and its environment. The small size of cells gives them comparatively great surface area. To give students a sense of the relationship between size and surface area, compare the time it takes to peel one large potato to the time it takes to peel two small potatoes with the same total weight as the large one.

Cells

All organisms are made of one or more **cells**, and every cell comes from another cell. A cell is the basic unit of a living thing that can perform all the processes associated with life. A chemist might call a cell a chemical factory. A cell provides all the conditions needed for the chemical reactions of living things.

Cells vary in type and function. A single cell can be an entire living organism. Most of the animals and plants you are familiar with, such as cats and trees, are made of many cells. These organisms may contain thousands, millions, even billions of cells. Most cells, even those of the blue whale, are microscopic in size. The small cell size allows the necessary exchange of materials between the cell and its environment.

Nonliving things are not made of cells. Cells are found in nonliving matter only if that matter was once alive and is now dead. Wood, for example, is made of cells of the once-living tree it came from.

Figure 2.5 ▲
What is the energy source for the berries? For the mouse? ②

Energy

Think of all the things you may do in a day. Run down a hill. Eat an apple. Read this book. Or sleep. You would not be able to do these things without energy. Energy is the ability to make things move or change.

Living things get energy from their environment, or surroundings. They use that energy to grow, develop, and reproduce. All organisms need energy to build the substances that make up their cells. Plants get their energy from sunlight. Animals and humans get energy from the food they eat.

For life to be maintained, a balance must exist between an organism's energy-producing processes and its energy-using processes. The constant balancing of these two systems within an organism is called **metabolism**. Examples of metabolic processes in living things include breaking down foods, building body parts, transporting materials, and removing wastes.

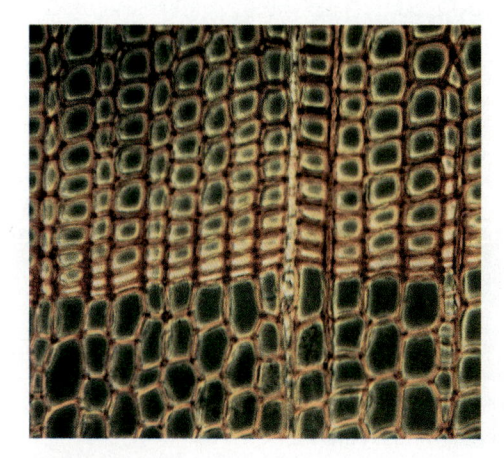

Figure 2.4 ▲
This piece of wood is no longer living. How can you tell it was once part of a living thing? ③

Chapter 2   The Earth and Living Things   37

TEACH

Critical Thinking
Reason by Analogy Explain to students that in the 1600s cells were first observed by British scientist Robert Hooke. Hooke thought the cells looked like the small rooms that monks lived in, which were called cells. Have students look at Figure 2.4, and ask them why a word that means "little rooms" is a good term to describe cells.

Class Activity
Have students use microscopes to observe prepared slides of several different kinds of cells. Students should make drawings of the cells.

Discuss
Explain to students that most living things use hydrocarbon compounds for energy. These compounds are food—made up of hydrogen, carbon, and other substances. When these hydrocarbons react with oxygen, they release energy.

Answers to In-Text Questions
② **The energy source for the berries is the sun. The mouse gets energy from the berries.**

③ **This piece of wood has cells. Cells are found only in living or once-living things.**

The Living Textbook:
Life Science Side 2

Chapter 13   Frame 14805
Living Blood Cells (Movie)
Search: Play:

The Living Textbook:
Life Science Sides 7-8

Chapter 3   Frame 00349
Plant Cells: Elodea Leaf (1 Frame)
Search: Step:

TEACH • Continued

Class Activity
Have students observe a simple stimulus response in themselves by observing how the pupils of their eyes respond to light. Have students look at the eyes of their partners and note the size of the pupils. (Tell students that the pupil is the black area at the center of their eyes.) Turn off the classroom lights. After a minute, have students look at their partners' pupils. Turn on the lights and have them observe any changes. (Students should notice that the pupils respond to the increase in light by getting smaller.)

Class Activity
On the chalkboard, write a list of stimuli and responses. Do not label them. Have students match the stimulus to the correct response. After completing the activity, discuss how certain responses help living things survive. Have students give examples.

Enrich
Explain to students that all organisms have a life span, which is the period of time that an organism lives. Within its life span, an organism goes through a cycle of change (life cycle) called development. The life span of organisms varies from several days for a mayfly to thousands of years.

Decision Making
If you have classroom sets of *One-Minute Readings*, have students read Issue 67, "Feedback: Positive and Negative," on page 114. Discuss the questions.

The Living Textbook: Life Science Side 3
Chapter 17   Frame 25668
Geotropism and Phototropism (Movie)
Search: Play:

INTEGRATED LEARNING

Geography Connection
Explain to students that migrating monarch butterflies follow a seasonal route that spans North and South America. Have students trace the migration route of monarch butterflies on an outline map of North and South America. Have students label all countries, bodies of water, and mountain ranges over which the butterflies migrate.

Stimulus and Response
How do you react when you hear a sudden loud noise? Most likely, you jump. When this happens, you are reacting to a stimulus in your environment. A **stimulus** is a change that occurs in the environment. Sound is a stimulus. Other stimuli include light, pressure, odors, and temperature changes. Reaction to a stimulus is called a **response**.

Growth and Development
All living things grow. Sometimes growth simply means getting larger—not changing. But that is not always the case. A coconut, when it sprouts, produces roots, a trunk, and leaves that continue to grow for years. As it grows, the palm tree takes in substances from the air and soil to make living tissue. After the palm tree stops getting larger, it continues to add new material to replace existing parts that wear out.

A balloon, on the other hand, may seem to "grow" if you blow it up. But a balloon grows bigger only if someone adds more air inside it. A balloon may change only if someone stretches it or twists it. A balloon won't grow bigger by just laying on a table.

As some organisms grow, most go through a cycle of change called development. Think of a caterpillar developing into a butterfly, or a small seed blossoming into a flower. Organisms grow and develop at different rates. Humans grow to adulthood in about 18 years. Some tadpoles, however, grow and become frogs in a few weeks. Some insects actually reach their adult stage in one day.

Reproduction
All organisms can reproduce. The process by which organisms produce more organisms like themselves is called **reproduction**. Although all living things can reproduce, reproduction is not essential for the survival of an individual organism. However, because all individual organisms eventually die, reproduction is necessary if a group of similar organisms is to survive.

Figure 2.6 ▶
The decreasing number of daylight hours in the fall stimulate monarch butterflies to fly to their wintering grounds where they "sleep" until the seasons change.

◀ **Figure 2.7**
During its lifetime, a butterfly grows and develops. As it develops, its appearance changes.

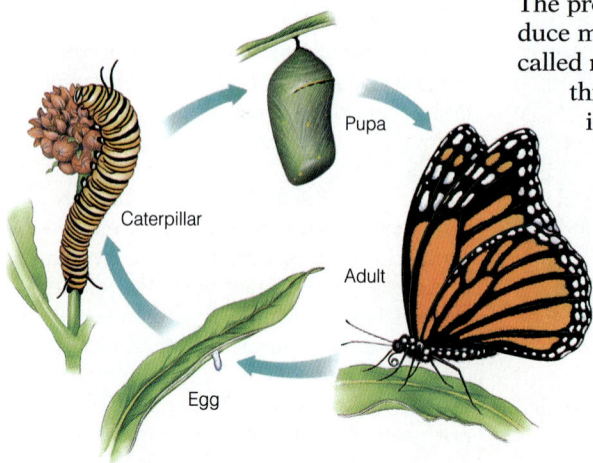

Chapter 2 The Earth and Living Things

Multicultural Perspectives

Humans have biological adaptations just like all other organisms. However, humans can also adapt to their environments behaviorally, by making their environments more fit to live in. Have students list some ways that humans have changed their environments for comfort. (For example, humans build houses, make and wear clothing, and eat foods that are best suited for the environment in which they live.) Have students research the housing, clothing, and foods that different cultural or ethnic groups use in different parts of the world. Challenge them to explain how these foods, structures, and clothes help people survive in their environments.

Adaptation

Touch each of your fingertips with your thumb. Then pick up a pen or pencil and look at your hand. You can do these actions because of the joint in your thumb. This joint is an example of an **adaptation**. An adaptation is a characteristic that enables an organism to live and reproduce successfully in its environment.

All organisms must be adapted to their enviroments to survive. For example, rattlesnakes can live in the desert. Triggerfish can survive in coral reefs. Beavers can live in forest ponds. Examples of adaptations in a beaver are shown in Figure 2.8. Look at the beaver's legs and tail. A beaver's powerful hind legs help it drag heavy loads. The flat tail serves as a rudder for swimming. These adaptations help the beaver build a home of log and mud. A beaver also has sharp teeth. The shape of its teeth helps in cutting down trees. Name some other adaptations that help a beaver survive.

Figure 2.8
Some adaptations help the beaver build its home. Others help the beaver get food, protect itself, and keep warm.

Historical Notebook

Disproving Spontaneous Generation

People have not always known that living things come from other living things. Hundreds of years ago, people thought that living things came from nonliving matter. They based this idea on their observations. For example, people sometimes saw maggots suddenly appearing on rotting meat, or mice suddenly appearing in a pile of rags. This idea, called spontaneous generation, was put to the test in the 1600s by Francisco Redi, an Italian physician.

Redi designed a controlled experiment to test the idea that maggots came from rotting meat. He put some meat into three sets of jars. One set of jars was left open. Another set of jars was sealed. The third set of jars was covered with mesh screen.

Redi thought that if spontaneous generation was possible, maggots would appear on the meat in the sealed jars. Redi's experiment and the results are shown below.

1. Observe the results of Redi's experiment shown below. Write down your observations.
2. What did Redi's experiments prove?

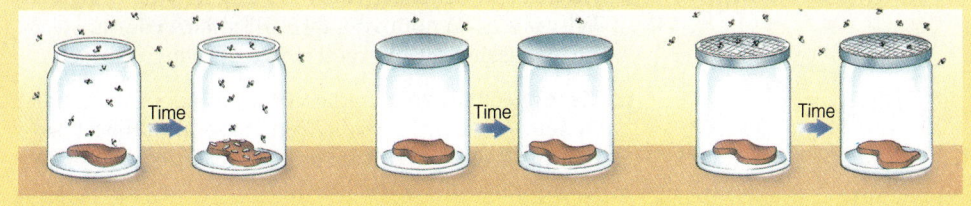

Skills Development

Hypothesize Have students choose an animal or plant that interests them. Have them find a clear photograph of the organism and list as many observable characteristics of the organism as they can. For each characteristic, students should write a hypothesis that explains how the characteristic helps the organism live and reproduce successfully in its environment.

Reteach

To reinforce the theme of unity within diversity, write the names of ten common organisms on the chalkboard. Ask students to explain how each organism has basic characteristics shared by all living things. Students should give specific examples.

Historical Notebook

Challenge students to describe another experiment that disproves spontaneous generation.

Answers
1. The jars that were left open attracted flies, and maggots appeared on the meat. The jars that were sealed attracted fewer flies, and no maggots appeared. The jars that were covered with wire mesh attracted flies, but maggots did not appear on the meat.
2. Because the maggots did not appear on the meat that was in the covered jars, Redi hypothesized that the flies were the source of the maggots.

Enrich
Use Enrich Worksheet 2.2.

The Living Textbook:
Life Science Side 8

Chapter 20   Frame 02358
Adaptation (Movie)
Search: Play:

Chapter 2   The Earth and Living Things

TEACH • Continued

Apply
Explain that many insulating materials work because they contain small air pockets. Air pockets prevent the transfer of heat by convection currents. Have students name some insulators that they use.

EVALUATE

WrapUp
Review After students have finished studying this section, ask:

▶ How are all living things alike? (Answers should include the six characteristics from page 36.)

▶ What is the difference between something that is dead and something that is nonliving? (Something that is dead was once living.)

▶ Name some characteristics of an animal that could be considered adaptations. (Answers may vary. The long neck of the giraffe is an adaptation, as are goose bumps on animals with fur.)

Use Review Worksheet 2.2.

Check and Explain
1. A cell is the basic unit of a living thing that can perform all of the processes associated with life.
2. Living things are made of one or more cells, obtain and use energy, grow and develop, respond to their environment, reproduce, and are adapted to their environment.
3. Dogs have all the characteristics of living things. Although cars use energy, they do not have the other characteristics.
4. Students' drawings should show colors that provide camouflage.

40

TEACHING OPTIONS

Portfolio
Return students' drawings from question 3 of Check and Explain. On their drawings have students write six reasons their animal can be classified as a living thing.

Figure 2.9 ▲
During cold weather the stimulus of the cold causes a bird's feathers to puff out. This helps keep in body heat, just like human goose bumps do.

Science and You *Goose Bumps*
Have you ever stepped out of a shower or bath and discovered tiny bumps on your skin? Have you noticed those same bumps when you are outside in cold weather? If so, then you have observed an adaptation called *goose bumps*.

A goose bump is caused by the contraction of a muscle located around a hair. Flex your arm. When you do this, you are contracting your upper arm muscle. The contracting muscle causes a bulge in your arm. When you are exposed to chilly air, tiny muscles in your cold skin begin to contract. These muscle contractions cause tiny bulges. Your skin gets a bumpy look, and your hairs stand on end.

Goose bumps are a helpful adaptation for animals with fur. When an animal's hairs stand on end, air is added in the fur. The hairs help hold this layer of air in place. The hair and air keep the cold air away from the animal's body. The thicker the fur, the less cold air will touch the animal's skin. This helps keep the animal's body temperature at a normal level. You don't have fur, but you do have thick hair on your head. Goose bumps on your head help keep your head warm. Look at Figure 2.9. How do goose bumps help birds?

Besides cold temperature, fear is also a stimulus for goose bumps. A scared animal with its hair on end looks larger and stronger to its enemy. This response is not very useful to you because most of your body hairs are short and sparse.

Check and Explain
1. What is a cell?
2. List and describe six characteristics of all living things.
3. **Reason and Conclude** Why is a dog a living thing? Why is a car a nonliving thing? Explain your answers by discussing the characteristics of living things.
4. **Make a Model** Draw and color an imaginary animal in a jungle environment. What colors would best protect the animal in this kind of environment? Explain your choice of colors.

Chapter 2 The Earth and Living Things

INTEGRATED LEARNING

Themes in Science
Systems and Interactions

Interactions between an organism and its environment are essential to the organism's survival.

SECTION 2.3

Section Objectives
For a list of section objectives, see the Student Edition page.

Skills Objectives
Students should be able to:

Graph class data.

Analyze the ways that people use water.

Predict the effects of changing the size of a population on an environment.

Infer an everyday application from experimental results.

Vocabulary
homeostasis, endothermic, ectothermic, competition

2.3 Needs of Living Things

Objectives

▶ **Describe** the needs of living things.

▶ **Identify** the sun as a primary source of energy.

▶ **Predict** what would happen to an organism if its needs were not met in its environment.

Skills WarmUp

Make a Graph Make a list of the five most important things you need to stay alive. Rank them in order of their importance. Then find out what your classmates wrote. Make a bar graph showing the class results.

MOTIVATE

Skills WarmUp
To help students understand the needs of living things, have them do the Skills WarmUp.
Answer Answers may vary, but should include sources of food, water, living space, and so on. Ask students to study the graph and identify the item that most people listed as most important. Ask students to explain why they did or didn't agree with this choice.

Prior Knowledge
To gauge how much students know about the needs of living things, hold a brainstorming session. Ask the following questions:

▶ How is the earth like a spaceship?

▶ How do the characteristics of an organism determine its needs?

▶ How are the needs of organisms met?

If you were to travel in space, what things would you need to survive? You probably answered air, food, and water. These are some of the things you and other living things need to live. The earth may be described as a huge "spaceship." It is the only planet in the solar system with the right environment to support the needs of millions of different kinds of organisms.

Six Common Needs

All organisms must interact with their environment to survive. There are many different environments in the biosphere. Yet each different environment provides the energy, food, water, air, proper temperature, and living space that the organisms living there need.

Energy All living things need energy. Where does this energy come from? Directly or indirectly, the sun is the main energy source for most organisms. For example, all plants use the sun's energy to make sugars and starches. Plants use these nutrients to fuel their growth. Animals, however, get their energy by eating plants or other animals.

Food All organisms need some type of food or nutrients. Food keeps the life processes going within cells. Growth and transport are processes fueled by food. Different kinds of organisms eat different kinds of food. Some organisms, such as deer, eat only plants. Some animals eat only other animals. Other animals eat both plants and animals. Some organisms do not eat their food. Plants absorb their nutrients from the soil. Some worms deep in the ocean absorb nutrients from the ocean water.

Figure 2.10 ▲
This frog gets its energy by eating other animals, such as insects and even other frogs.

Chapter 2 The Earth and Living Things 41

TEACH

Discuss
Students may want to know more about what's in blood. Tell them they can find out by looking up the terms *red blood cells*, *white blood cells*, *plasma*, and *platelets* in this book's glossary.

Critical Thinking
Reason and Conclude Why do animals require protection from freezing temperatures? (Blood is mostly water; water can freeze; frozen blood can't transport nutrients and wastes.)

Misconceptions
Students may think that living things are the only parts of the biosphere that change. Have students observe what happens to several drops of water in two drinking cups if they leave them out for a day. Seal plastic wrap over the top of one cup with a rubber band.

Skills Development
Calculate Have students calculate how much of their body mass is water. To do this have them use this formula: body mass (weight) × 70% ÷ 100 = mass of water.

INTEGRATED LEARNING

Themes in Science
Systems and Interactions Use of natural resources by humans has a powerful effect on the environment. Ask students to look at the drawing on pages 42 and 43 and to suggest ways in which human interaction with this environment might cause change.

Figure 2.11 A Pond

The sun is the main source of energy for all living things.

Water Water is a very important need of living things. Humans need water so much that you could live longer without food than without water. In fact, you can live about one or two months without food, but only about one week without water. Your body mass is about 70 percent water. This is true of most organisms.

Besides making up most of the body mass, water has two other important functions. Water is needed to dissolve and transport substances in an organism. Nutrients and wastes can be dissolved in water. Once dissolved, these substances can be transported throughout the organism's body. In animals, blood transports wastes and nutrients. Blood is mostly water. Instead of blood, trees have watery sap.

Temperature Outside temperature and internal temperature are important to living things. Few organisms can live in environments where temperatures are either extremely hot or extremely cold. Most organisms need to live within certain temperature ranges.

Freshwater plants and animals need very little water. They have special adaptations to keep too much water from entering their systems.

Oxygen and other atmospheric gases enter the surface water.

The soil on the pond's bottom contains nutrients that are absorbed by plant roots for growth.

Algae, reeds, pondweed, lily pads, and duckweed compete for a place in the sun at the pond's surface.

Writing Connection
Have students work in small groups to write a short script for a dialogue between an endothermic and an ectothermic animal in a very hot or cold climate. The dialogues can be serious or humorous, but should address the differences between the ways the two animals meet their need for maintaining a proper body temperature.

Temperature is also important to organisms because they need to maintain a proper body temperature to carry out metabolism. The ability of an organism to keep conditions inside its body the same, even though conditions in the environment may change, is called **homeostasis**.

Animals that can maintain a stable body temperature regardless of their surroundings are called **endothermic**. Birds, dogs, and humans are endothermic. Some animals, such as fishes, frogs, and turtles, can't regulate their body temperatures. As the outside temperature changes, so do their internal temperatures. These animals are **ectothermic**. They depend on heat energy from the sun for warmth.

Gas Exchange All living things must exchange gases with their environment. Land animals and plants get oxygen from the air. Organisms that live in water get oxygen that is dissolved in the water.

Skills WorkOut
Analyze You use water every day in many different ways. It is important to your general well-being. Identify the different ways that you use water. Why is each way important? Rank the ways you use water from the most important to the least important. Then write a brief essay explaining why you ranked the first use and the last use in the positions you did.

Duckweed fronds are food for snails and insect larvae.

The pond provides living space for all the pond's organisms. A male stickleback makes a nest on the pond floor and tries to attract a female to mate.

Turtles must lie in the sun to raise their body temperatures so they can become active enough to hunt for food.

Skills WorkOut
To help students understand the importance of water, have them do the Skills WorkOut.
Answer Students will probably include drinking, cleaning, swimming, and cooking on their lists of ways that they use water. Drinking water would be the most important use because human survival depends upon water. Recreational uses for water will most likely be lowest on the list.

Portfolio
Have students place their Skills WorkOut answers in their portfolios.

Explore Visually
Have students study the illustration of the pond environment. Then have them use examples from the illustration to discuss the following questions:

▶ What is the main source of energy for the pond? (Sun)

▶ Where does the stickleback fish get oxygen? (Dissolved in water)

▶ What other examples can you see of an organism meeting its needs? (Answers may vary. Some possible responses are the animals eating plants or animals that live in the pond, the turtle lying in the sun to maintain its body temperature, the lily pads getting sunlight for photosynthesis.)

▶ **Predict** How might changes in temperature, air quality, or water quality in a pond environment affect the ability of the organisms to meet their needs? (Answers may vary. For example, very cold or hot temperatures may kill plants that provide food for insects, fish, and birds; acid rain could cause the number of certain plants to increase, depriving other plants of air and sunlight.)

TEACH • Continued

Class Activity
To demonstrate the impact of competition, have students work in teams to plan for the following situation. The students and teacher from the next classroom must quickly evacuate their own room and move into yours. They are not able to bring any books, supplies, or chairs. Furthermore, they must leave their lunches behind. Ask each team to choose one classroom event, such as a science class, lunch, or taking attendance, and perform a skit demonstrating how the class would need to adjust to the increased population in the classroom.

Consider This
Research Before students write their reports, have them research the costs of maintaining zoos. Suggest that students find statistics about the number of zoos in the United States, the number of personnel working in zoos, the costs of food for zoo animals, and other costs associated with maintaining zoos.
Think About It Accept all thoughtful responses.
Write About It Students' reports should show that they have considered the conflict between preservation of animal populations and removing animals from their natural habitats.

INTEGRATED LEARNING

Integrating the Sciences

Ecology Explain to students that the earth has boundaries and is made up of many life-support systems. Explain that the wastes produced by people and industries stay on the earth forever. Also, many resources—air and water, for example—are limited; that is, there is only a given amount. Have students discuss what it means for the earth to have these limits. Ask students:

▶ How important is clean, unpolluted air and water?
▶ What steps should be taken to protect them?
▶ How important is open, undeveloped space?

Carbon dioxide is also important to living things. When you breathe out, you release carbon dioxide. Plants use carbon dioxide to make sugars. During this process, they also release oxygen into the air. Plants also release carbon dioxide as a waste product.

Living Space All organisms need living space. This space may provide food or shelter. The space may have sunlight or be dark. All living things compete for the earth's limited space. The struggle among living things to get the proper amount of water, food, and energy in a living space is called **competition**.

There are many ways in which living things compete for their space, or territory. Some birds, for example, use sounds to keep other animals away from their territory. Plants, too, compete for living space. Without sunlight or water, small plants may die in the shadow of larger plants.

Consider This

Should Animals Be Kept in Zoos?

There are zoos all over the world where exotic animals are kept for the public to visit. Zoos offer an opportunity for people to see and experience animal life from other parts of the world.

Zoo experiences educate people about the variety and needs of animals. However, zoo animals live in confinement and under conditions that differ greatly from their natural environments.

Consider Some Issues Animals that are kept in zoos are well cared for and protected from danger. They receive highly nutritious diets and special medical care.

Some zoos breed endangered animals. Zoo breeding programs may help save animal populations from extinction.

In many parts of the world, prime animal habitat is being taken over for human settlement. Saving habitat is important for the survival of endangered animals.

Removing wild animals from their natural habitat and confining them in zoos for people to view may be cruel and unfair to the animals.

International trade in wild animals may involve abusing these animals.

Think About It What might happen to some wild animal populations in 20 years if zoos didn't exist? How could you be affected by the changes that might result?

Write About It Write a report stating your position for or against keeping animals in zoos. Include your reasons for choosing that position.

Integrating the Sciences

Chemistry The process of desalination takes advantage of the physical properties of salt and water. Salt is a solid that dissolves in liquid water. Water is a liquid that can easily be changed to a gas by heating. The process of changing a liquid to a gas is called evaporation. When the water evaporates, it leaves behind solid substances dissolved in it. Therefore, salt is left behind. If the gaseous water is collected and cooled, it changes from a gas to a liquid. This process is called condensation.

Use Integrating Worksheet 2.3.

Demonstrate

Make a simple solar still to demonstrate desalination. Half-fill a large, clear cup with salt water. Cover the cup with plastic wrap and secure the plastic wrap with a rubber band. Place the cup in a sunny window and wait until water droplets appear on the plastic wrap. Have volunteers touch the water droplets with their fingertips and taste them.

Science and Technology *Salty Waters*

Beware: Deadly Water! This warning describes most of the earth's water because most of it is in oceans. Ocean water contains salts, minerals, and elements. Many kinds of living things cannot drink or live in ocean water.

Humans need some salts, especially sodium chloride, which you know as table salt. In fact, a certain amount of salt is needed in your diet for your body to function properly. Ocean water, however, contains about seven times the amount of salts that your body can withstand. People could not survive if they drank only salt water.

For centuries, people have looked for ways to remove salt from ocean water. Today salt is removed through a process called *desalination* (dee sal ih NAY shun). Most modern desalting plants use a method called flash distillation. During this process, ocean water boils or "flashes" to form steam, leaving the salt behind. When the salt-free vapor cools, it changes back into water. Once the salt has been removed and the water is purified, it is drinkable.

In some parts of the world, desalting plants are a major source of fresh water. In the desert countries of the Middle East, such as Saudi Arabia and Kuwait, desalting plants produce millions of liters of fresh water each day. Although there are some desalting plants in the United States, few places use desalting because it is a costly process.

Figure 2.12 ▲
The Yuma desalination plant desalts about 275 million liters of water each day. Desalination is one way to lower the amount of salts in some river water. It also makes ocean water drinkable.

EVALUATE

WrapUp

Reinforce Label index cards with the following headings: *Organisms Need Energy; Organisms Need Food; Organisms Need Water; Organisms Need the Proper Temperature; Organisms Need to Exchange Gases; Organisms Need Living Space.* Copy these headings onto sheets for groups of six students. Each student should choose one heading and cut out four applicable pictures from magazines and newspapers to tape down onto the sheet. Have each group bind these pages together, write captions, and make a cover.

Use Review Worksheet 2.3.

Check and Explain

1. Students' answers should include any four of the following: energy, food, water, proper temperature, gas exchange, living space.
2. The sun
3. Answers may vary, but may include: bird's nest—living space; scuba diving tank—gas exchange; pond water—water/temperature control/living space/gas exchange; rock in the shade—living space; sunflower seed—food.
4. Any change in one population would affect the availability of all resources.

Check and Explain

1. What are four needs of living things?
2. What is the main source of energy for most living things?
3. **Apply** What needs may be met with the following things: a bird's nest, a scuba diving tank, pond water, a rock in shade, a sunflower seed?
4. **Predict** Several kinds of organisms share the pond environment shown on pages 42–43. Choose one of the organisms. Describe how it would be affected if its numbers increased in this environment.

Chapter 2 The Earth and Living Things

ACTIVITY 2

Time 5 days   **Group** 4–5

Materials

32 self-lock bags
8 slices of bread
8 Celsius thermometers
8 plastic knives
hand lenses
metric rulers
labels

Analysis

1. The variable is location or temperature.
2. Mold will likely have grown in all locations but the refrigerator and the freezer.
3. Mold should have appeared first on the bread stored at the warmest temperature.
4. The largest amount of mold should have grown in the warmest location. The sample in the freezer should have no mold.
5. More mold should have grown in the warm location, less in the freezer.

Conclusion

Students' conclusions should indicate that warmer temperatures promote mold growth.

Everyday Application

Students might point out that the bread should be stored in a cool place to keep mold from growing on it.

TEACHING OPTIONS

Prelab Discussion

Have students read the activity before beginning the discussion. Ask the following questions before beginning the activity:

▶ Why is it important to begin with fresh bread?

▶ Why is it important not to remove the bread from the bag when you inspect it?

▶ Why must the bread be sealed tightly in the plastic bags?

Emphasize the importance of keeping an accurate daily log. Have students make sure that they use the same thermometer to measure the temperature each day at each location.

Activity 2   How does temperature affect mold growth?

Skills   Measure; Observe; Infer; Analyze

Task 1   Prelab Prep

1. Collect the following items: 4 self-lock bags with labels, pen, Celsius thermometer, a slice of bread, plastic knife, hand lens, metric ruler.
2. Number the labels from 1 to 4 and date each one.
3. On the first label write *Freezer*. Label the remaining bags *Refrigerator*, *Closed Box*, and *Control*.

Task 2   Data Record

1. On a separate sheet of paper, draw a data table for each location like the one shown below. You should have four separate data tables.
2. In the data tables, record all your observations about the color(s), size, and amount of any mold you see.

Task 3   Procedure

1. Measure the temperature at each location: inside a freezer, inside a refrigerator, inside a closed box in a warm location, and wherever you regularly store bread. *Note: Be sure the regular storage place is dark.* The control is your regular storage place for bread.
2. Cut the slice of bread into four equal sections. Put one section of bread in each bag.
3. Place each bag in its location.
4. Each day for five days, examine the bread for mold growth. Use the hand lens and ruler.

CAUTION! Do not remove the bread from the bag. After recording your observations, return the bag to its place.

Task 4   Analysis

1. Identify the variable in this activity.
2. At which location(s) did mold grow on the bread?
3. What was the temperature of the first location to have mold growing on the bread?
4. What were the temperature and location of the bread with the most mold? The least mold?
5. At what temperature and location did mold grow the fastest? The slowest?

Task 5   Conclusion

Write a short paragraph explaining how temperature affects the rate of mold growth on bread.

Everyday Application

Suppose you bought a month's supply of bread for lunches. Explain why the temperature of the place where you stored the bread might be important.

Extension

Develop a hypothesis about the effect of moisture on the growth of mold on bread. Write your procedures and test your hypothesis. Be sure to include a control in your experiment.

Table 2.4   Mold Growth

Date	Time	Temperature (°C)	Location (Number)	Condition of Bread

Chapter 2 Review

Concept Summary

2.1 The Living Planet
- The biosphere is the zone of water, land, and air that supports all life on earth.
- The world ocean is divided into the Pacific, Atlantic, Indian, Antarctic, and Arctic oceans.
- The landmasses on earth are divided into continents, which are the homes to many living things.
- The atmosphere that surrounds the earth contains a combination of nitrogen, oxygen, and various other gases.

2.2 Characteristics of Living Things
- Living things carry on certain life processes that nonliving things cannot.
- The cell is the basic unit of a living thing that can perform life processes.
- The processes of living things include movement, metabolism, growth and development, response, reproduction, and adaptation.

2.3 Needs of Living Things
- Living things need energy, food, water, proper temperature, air, and living space to survive in their environments.
- The sun is the primary source of energy for most living things.
- Endothermic animals maintain stable internal body temperatures. Ectothermic animals cannot regulate their body temperatures.

Chapter Vocabulary

biosphere (2.1) metabolism (2.2) reproduction (2.2) endothermic (2.3)
atmosphere (2.1) stimulus (2.2) adaptation (2.2) ectothermic (2.3)
cell (2.2) response (2.2) homeostasis (2.3) competition (2.3)

Check Your Vocabulary

Use the vocabulary words above to complete the following sentences correctly.

1. A ___ is a change in the environment that produces a response.
2. The ___ is an envelope of gases that surrounds the earth.
3. Animals that maintain stable body temperatures are ___.
4. A ___ is the basic unit of a living thing that can perform all the life processes.
5. All of the chemical changes that happen in an organism are called ___.
6. The thin zone on the earth that supports all life is called the ___.
7. An organism's ability to keep its body conditions the same although environmental conditions change is ___.
8. An ___ is a characteristic that enables an organism to survive and reproduce in its environment.
9. The struggle between living things for available living space is called ___.
10. Reacting to a stimulus in the environment is called a ___.
11. Animals that cannot regulate their body temperatures are called ___.
12. The process by which living things produce more organisms like themselves is ___.

Write Your Vocabulary

Write two or more paragraphs using five of the vocabulary words. Show that you know what each word means.

CHAPTER REVIEW 2

Check Your Vocabulary
1. stimulus
2. atmosphere
3. endothermic
4. cell
5. metabolism
6. biosphere
7. homeostasis
8. adaptation
9. competition
10. response
11. ectothermic
12. reproduction

Write Your Vocabulary

Students' paragraphs should show that students know both the meanings of the words and how to apply the words.

Use Vocabulary Worksheet for Chapter 2.

CHAPTER 2 REVIEW

Check Your Knowledge

1. All living things need energy, food, water, air, proper temperature, and living space.
2. An ectothermic animal needs energy from the sun for warmth.
3. The biosphere includes the water, the land, and the atmosphere.
4. A living thing is made up of one or more cells, obtains and uses energy, grows and develops, responds to its environment, reproduces, and is adapted to its environment. A nonliving thing would not have any of these characteristics.
5. The beaver's flat tail helps it swim. Its sharp teeth help it cut down trees. Powerful hind legs help it drag heavy loads.
6. The sun is the source of energy for living things.
7. Living things compete for food, water, shelter, and sunlight.
8. Breaking down foods, building body parts, transporting materials, and removing wastes are examples of metabolic processes.
9. Water is needed to dissolve and transport substances in an organism.
10. sunlight
11. cells
12. energy
13. metabolism
14. develop
15. nitrogen

Check Your Understanding

1. Answers may vary. For example, seeing a flash of bright light (stimulus) and blinking the eyes (response)
2. The earth is the only planet in the solar system where life is known to exist. It is the only planet where water exists as a liquid, a solid, and a gas.
3. Students' drawings should indicate several means by which the organism's needs are met by interaction with its environment.
4. Answers may vary. For example, a tree that grows tall is able to gain more sunlight than its shorter neighbors.
5. Forests and farmland look red because they contain large numbers of living things, which give off infrared energy. Towns look blue because they contain many artificial objects, such as buildings, that do not give off as much infrared energy.
6. A pumpkin seed is a living thing because it is made up of one or more cells, uses energy, is adapted to its environment, responds to its environment, and can produce a plant that can produce another seed.

7.

Need	Polar Bear	Snake
energy	from food	from food
food	hunts	hunts
water	rivers, lakes	puddles, food
air	atmosphere	atmosphere
temperature	fur, body fat	moves into or out of sun
space	finds empty space, or competes	finds empty space

Check students' paragraphs for logical responses.

Chapter 2 Review

Check Your Knowledge
Answer each of the following in complete sentences.

1. List the six common needs of all living things.
2. What is an ectothermic animal?
3. What parts of the earth's surface make up the biosphere?
4. How could you tell a nonliving thing from a living thing?
5. Give an example of an adaptation in a beaver that helps it to survive.
6. Why is the sun important to all living things?
7. Name three things that living things compete for in their living space.
8. Give an example of a metabolic process.
9. What is one function water has for living organisms?

Choose the answer that best completes each sentence.

10. Plants get energy from (sunlight, trees, people, leaves).
11. All living organisms are made of (leaves, sunlight, blood, cells).
12. To move or change, organisms need (water, matter, competition, energy).
13. Breaking down food, building body parts, and transporting materials are all processes in (metabolism, reproduction, adaptation, responding).
14. As organisms grow, they also may change or (breathe, develop, eat, compete).
15. The most common gas in the atmosphere is (oxygen, nitrogen, helium, water).

Check Your Understanding
Apply the concepts you have learned to answer each question.

1. Write an example of a stimulus. Give a response to the stimulus you chose.
2. List the unique features of the earth.
3. Draw an animal or plant. Explain how this living thing meets its needs for survival.
4. Choose a living organism and discuss how it successfully competes for its living space.
5. **Mystery Photo** The infrared photograph of the San Francisco Bay Area and the Central Valley in California on page 30 was taken by a satellite. "Warm" areas are red; "cool" areas are blue. Why do forests and farmland look red? Why do towns look blue? Predict what your community would look like from an infrared satellite photograph. Draw a satellite picture of your community.
6. **Application** Is a pumpkin seed living or nonliving? Explain your answer.
7. **Critical Thinking** Make a table listing the six common needs of living things. In the table, describe how the needs are met for a snake and a polar bear. Using the information in your table, write a paragraph comparing and contrasting how the needs are met for these two animals.
8. **Application** You read in a magazine that plants respond to the stimulus of music by growing larger. Design an experiment to test the magazine's claim. Be sure to include a control. What is the hypothesis of your experiment? What is the variable?

8. Students' experiments may vary. A possible hypothesis: If a tomato plant is exposed to music, it will grow taller than a plant not exposed to music. The variable being measured is the height of the plants.

Develop Your Skills

1. a. Hare population decreases, then increases. Fox population increases, then decreases.
   b. Hare (prey) population varies inversely with the fox (predator) population.
2. a. Earth, Mars
   b. Venus
3. Check student circle graphs. Nitrogen should occupy 78 percent, oxygen 21 percent, and other gases 1 percent.
4. Encourage students to choose examples that were not discussed in the chapter.

Make Connections

1.

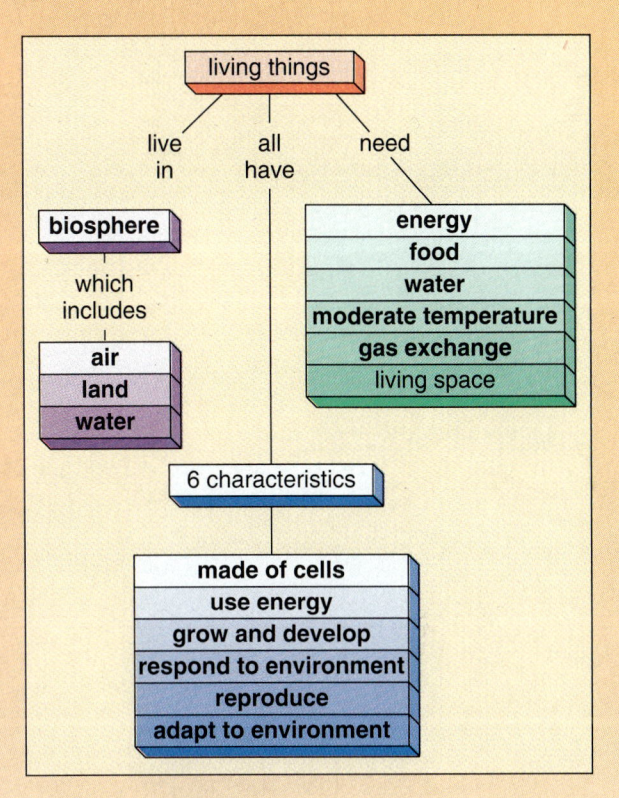

```
                    living things
           live in  | all have | need
         biosphere           energy
         which                food
         includes             water
            air          moderate temperature
            land           gas exchange
            water           living space
                     
                    6 characteristics
                    
                    made of cells
                    use energy
                    grow and develop
                    respond to environment
                    reproduce
                    adapt to environment
```

2. During the meeting, look for contributions from all students. Discussion should show knowledge of the needs of all groups, as well as an understanding of the way living things and environments interact.
3. Students' drawings should show how adaptations help living things meet specific challenges in their new environments.

Develop Your Skills

Use the skills you have developed to complete each activity.

1. **Interpret Data** The graph below shows the changing population of the arctic hare and the arctic fox over time.
   a. Describe what happens to the hare and fox population over time.
   b. What might be the cause for the hare population change?

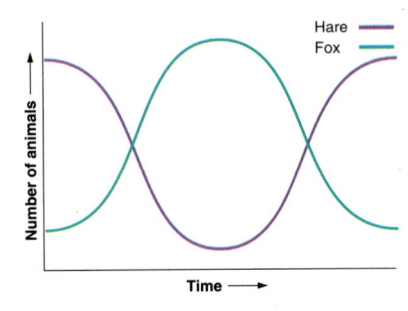

   [Graph: Number of animals vs. Time, showing Hare and Fox populations oscillating]

2. **Data Bank** Use the information on page 624 to answer the following questions.
   a. Which planets have oxygen?
   b. Which planet is the most like the earth in size?
3. **Make a Graph** Use the information contained in Table 2.3 on page 34 to make a circle graph showing the different kinds of gases in air. Remember to title your graph.
4. **Communicate** Choose a living or a nonliving thing. Have students in the class ask three questions from the list of characteristics for living things to determine whether your choice is alive or not.

Make Connections

1. **Link the Concepts** Below is a concept map showing how some of the main concepts in this chapter link together. Only parts of the map are filled in. Copy the map. Using words and ideas from the chapter, complete the map.

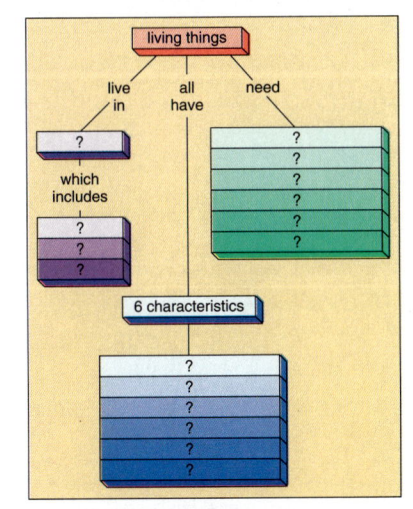

   [Concept map with "living things" at top and ? placeholders]

2. **Science and Society** Suppose a factory were built on the edge of a wilderness park in your home town. Plan a mock town meeting to discuss the impact of this factory on the plants and animals in the park. Divide the class into three groups for the mock meeting: factory owners, forest rangers, and community citizens.
3. **Science and Art** Draw or paint a picture of an environment such as a beach or a forest. On a separate piece of paper, draw a picture of an animal that could live in your environment.

Chapter 2 The Earth and Living Things

CHAPTER 3

Overview

This chapter discusses the basic characteristics of matter. The first section defines matter and its physical and chemical properties. Chemical symbols and formulas are presented to explain the composition of molecules. The second section explains physical and chemical changes, and the third section describes the structure and function of some important organic compounds.

Advance Planner

- Provide balloons for SE activity, page 54.
- Provide clay and toothpicks for TE activity, page 55.
- Prepare magnets, sugar, iron filings, stirring rods, and bowls for SE page 59.
- Collect magazines for TE activity, page 62.
- Prepare spoons, powdered bleach, water, small jars, and Celsius thermometers for SE Activity 3, page 63.
- Collect magazines containing pictures of food for TE activity, page 65.

Individual Needs

- **Limited English Proficiency Students** In their science journals, have students write definitions of the chapter terms in English and in their primary languages. Encourage students to use words that they learn early in the chapter to help them define terms appearing later in the chapter. For example, students can use the meaning of the term *atom* to define the word *molecule*.

- **At-Risk Students** Before you begin this chapter, write the titles of the three chapter sections on the chalkboard. Have students copy each title onto a page in their science journals. As students read each section, have them write the topic sentence of each paragraph in their journals.

- **Gifted Students** Have students design a set of experiments that would show how the physical and chemical properties of water are changed by pollution. Students can consider sources of pollution such as nitric acid and sulfuric acid in acid rain, or they could study water from an aquarium that contains lots of algae. If practical, encourage students to set up demonstrations of experiments for the class.

Resource Bank

- **Bulletin Board** Divide the bulletin board into three columns. At the top of the first column place the title *Atoms and Elements,* at the top of the second *Molecules and Compounds,* and at the top of the third *Molecules of Life.* As you work through the chapter, have students add pictures and examples of chemical symbols and reactions to the appropriate columns.

Skills Development Chart

Sections	Communicate	Graph	Hypothesize	Infer	Interpret Data	Measure	Observe	Research
3.1 Skills WarmUp Skills WorkOut Historical Notebook	●			●				●
3.2 Skills WarmUp Skills WorkOut SkillBuilder Activity		●		● ●		●	● ●	
3.3 Skills WarmUp Skills WorkOut			●	●				

CHAPTER 3 PLANNING GUIDE

Section	Core	Standard	Enriched	Section	Core	Standard	Enriched
3.1 Matter pp. 51–57				**Ancillary Options** CEPUP, Chemical Survey, Activities 1–3	●	●	●
Section Features							
Skills WarmUp, p. 51		●	●	**Overhead Blackline Transparencies** Overhead Blackline Master 3.2 and Student Worksheet	●	●	
Skills WorkOut, p. 54		●	●				
Historical Notebook, p. 56	●	●		**Laboratory Program** Investigation 7	●	●	●
Blackline Masters							
Review Worksheet 3.1	●	●		**Color Transparencies** Transparency 8	●	●	
Reteach Worksheet 3.1	●	●					
Skills Worksheet 3.1a	●	●	●	**3.3 Molecules of Life** pp. 64–66			
Skills Worksheet 3.1b	●	●	●				
Laboratory Program Investigation 6		●	●	**Section Features**			
				Skills WarmUp, p. 64	●	●	●
Color Transparencies Transparencies 6, 7	●	●		Skills WorkOut, p. 66	●	●	
3.2 Changes in Matter pp. 58–63				**Blackline Masters**			
				Review Worksheet 3.3	●	●	●
Section Features				Integrating Worksheet 3.3a	●	●	●
Skills WarmUp, p. 58	●	●		Integrating Worksheet 3.3b	●	●	●
Skills WorkOut, p. 59	●	●		Vocabulary Worksheet 3	●	●	●
SkillBuilder, p. 61		●	●				
Activity, p. 63	●	●	●				
Blackline Masters							
Review Worksheet 3.2	●	●					
Skills Worksheet 3.2a			●				
Skills Worksheet 3.2b		●	●				

Bibliography

The following resources can be used for teaching the chapter. See page T-40 for supplier codes.

Audio-Visual Sources
(video unless noted)
An Introduction to Chemistry. 3 filmstrips with cassettes, 16–18 min each. 1984. NGSES.
Particles in Motion: States of Matter. 15 min filmstrip with cassette. 1982. NGSES.
The Water's Edge. 28 min. FL.

Library Resources
Rosenfield, Israel, Edward Ziff, and Borin Van Loon. *DNA for Beginners.* London: Writers and Readers, 1983.
Sheeler, Philip, and Donald E. Bianchi. *Cell Biology: Structure, Biochemistry, and Function.* New York: Wiley, 1983.

CHAPTER 3

INTEGRATED LEARNING

Writing Connection
Have students write a short advertisement that features an item made from material like the object in the photograph. Ask them to describe in detail how it looks and what it feels like. Encourage them to make comparisons in their ads, such as to rainwater beading up on an umbrella.

Introducing the Chapter
Have students read the description of the photograph on page 50. Ask if they agree or disagree with the student's description.

Directed Inquiry
Have students study the photograph. Ask:

▶ What organism or part of an organism do you think the picture shows? (Most students will say it is a close-up photo of feathers.)

▶ What other details do you see in the picture? (The water droplets and the structure, color, and arrangement of the feathers)

▶ Where else have you seen water behave the way it does in the picture? (On a waxed surface, such as a car hood or a tabletop)

▶ What is the topic of this chapter? (The chemistry of living things)

▶ How do you think the picture relates to the topic of the chapter? (The way the water beads up on the feathers has something to do with the chemistry of living things.)

Chapter Vocabulary

amino acid	inorganic
atom	compound
chemical	matter
equation	molecule
chemical	nucleic acid
formula	organic
chemical	compound
symbol	product
compound	reactant
element	simple sugar

50

Chapter 3
The Chemistry of Living Things

Chapter Sections
3.1 Matter
3.2 Changes in Matter
3.3 Molecules of Life

What do you see?

"I think the picture shows duck feathers. Duck feathers have a special oil on them. This makes the water bead so that the duck will not absorb water and become heavy."

Katie Eskra
Haven Middle School
Evanston, Illinois

To find out about the photograph, look on page 68. As you read this chapter, you will learn about matter and its importance to living things.

50

Themes in Science
Diversity and Unity The physical universe is unified because it is all made up of matter. Even so, the physical properties of all the matter in the universe show great diversity. As students study this section, encourage them to look for examples of how the substances in the universe are alike and how they are different.

Math Connection
Have students use their own words to describe how they would measure the physical properties of mass and volume. They can review page 15 of Chapter 1 for an explanation of density.

SECTION 3.1

Section Objectives
For a list of section objectives, see the Student Edition page.

Skills Objectives
Students should be able to:

Communicate the physical characteristics of an object.

Infer the relationship between positively and negatively charged particles.

Make a Graph of the proportion of elements in the human body.

Vocabulary
matter, element, atom, compound, molecule, chemical symbol, chemical formula

3.1 Matter

Objectives
▶ **Distinguish** between atoms and molecules.
▶ **Distinguish** between elements and compounds.
▶ **Interpret** the parts of a chemical formula.
▶ **Graph** the proportions of elements in the human body.

Skills WarmUp
Communicate Choose one object in your classroom. Then write everything that you can think of to describe the object. How many different ways can you describe the object? See if a classmate can guess what your object is, based only on your descriptions.

What do you have in common with a glass of water, a star, and a balloon filled with air? Like you, these things are made up of **matter**. Matter is anything that takes up space and has *mass*. Mass is the amount of matter in an object. The mass of an object is not the same as the space it takes up. A balloon filled with water has a greater mass than the same balloon filled with air.

Look around you. Everything is made of matter, including the book you are reading and the chair you are sitting in. Even your own body is made of matter. In fact, the entire universe is made of matter.

Physical Properties of Matter
How would you describe a pencil? It may be the color yellow. When you touch the pencil, it probably has a smooth texture. A pencil is also long, thin, and hard. Color, texture, shape, and hardness are some of the *physical properties*, or characteristics, of matter. Physical properties can be observed or measured without changing the makeup of a substance. Mass, volume, and density are three other physical properties of matter.

Physical properties help you identify kinds of matter. For example, each different kind of matter has a different *density*. You can see this if you fill a jar with the same amount of water and oil. Oil is less dense than water, so the oil will float on top of the water. All substances that are less dense than water float on top of water. All substances that are denser than water sink in water. Before you can decide what a substance is, you need to know several of its properties.

Figure 3.1 ▲
You can see many kinds of matter in the photograph. Describe the physical properties of two substances that you see. ①

MOTIVATE

Skills WarmUp
To give students practice in describing the physical properties of matter, have them do the Skills WarmUp.
Answer Any object can be described by several different characteristics. Some examples are color, shape, hardness, and volume.

Prior Knowledge
To gauge how much students know about the characteristics of matter, ask all or some of the following questions:

▶ Where can you find matter? Do you think there is any place where you wouldn't find matter?

▶ How are liquids different from solids or gases?

▶ What are some names for the particles of matter?

Answer to In-Text Question
① Answers may vary. Students could describe a helmet, for example, as white, smooth, rounded, and hard.

Chapter 3 The Chemistry of Living Things

TEACH

Misconceptions
Students may think that particles of matter move only when they see the matter itself moving. Stress to students that movement occurs even though they cannot see it.

Demonstration
Have students observe the rate at which a drop of food coloring spreads through a glass of water. Ask how long they think it would take particles in the same drop to spread through a pool of water the size of the classroom. Then open a bottle of perfume and have students indicate when they can smell the fragrance. Ask students to compare the rates of particle movement in liquids and in gases.

Directed Inquiry
Have students study Figure 3.2 and read about solids, liquids, and gases. Ask the following questions:
▶ Which drawing shows the fastest-moving particles? (Gas)
▶ Which drawing shows particles arranged in a pattern? (Solid)
▶ How would you use the drawings to explain why gases tend to be less heavy than solids? (Students could compare the number of particles in the drawings.)

Discuss
Tell students that scientists also classify plasma as a state of matter. Matter exists as plasma in lightning bolts or stars. Gases change into plasma at extremely high temperatures.

Ancillary Options
If you are using CEPUP modules in your classroom for additional hands-on activities, experiments, and exercises, begin the Chemical Survey on page 1 in *Chemical Survey and Solutions and Pollution*.

Answer to In-Text Question
① **As particles gain energy they move more quickly.**

TEACHING OPTIONS

Cooperative Learning
Ask students to work in groups to name as many different substances as they can in one minute. One student can record the items for each group. Combine the lists, then work as a class to classify the substances as solids, liquids, or gases. Save the list for later work.

Table 3.1
The Particle Theory of Matter

All matter is made of very tiny particles.
There are spaces between particles.
All the particles in a substance are the same; different substances are made of different particles.
Forceful attractions draw particles together; the attractions may be strong or weak.
The particles move at all times; as particles gain energy, they move faster.

Figure 3.2
Matter can exist as a solid, liquid, or gas. ▼

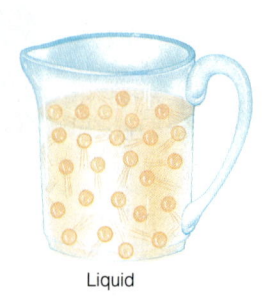

Phases of Matter

All matter is made of tiny particles. You cannot see these particles. Scientists, however, have carried out many experiments in which they have observed the behavior of particles in matter. To explain their observations, scientists have developed the *particle theory of matter*. Read Table 3.1, which lists these observations. What causes particles to move faster? ①

The particle theory of matter helps explain the differences among the *phases* of matter. Phase is a property that scientists can use to classify matter. Solids, liquids, and gases are the three phases of matter.

Solids Your pencil, book, and chair are all solids. Solids have a definite shape and a definite volume. Because the particles in a solid are very close together, the solid keeps its shape. However, the particles in a solid vibrate back and forth. They are held in fixed positions by strong forces.

Liquids A liquid has a definite volume, but it does not have a definite shape. The particles of a liquid move around freely. If you pour a liter of water into different containers, the water always takes the shape of the containers. The volume of the water, however, stays the same. A liter of water will not fit into a half-liter bottle.

Gases A gas has no definite shape or volume. The particles of a gas move constantly and rapidly in all directions. They are much farther apart than the particles in solids or liquids. A gas fills all the available space in a container. You can see this by pumping air into a tire or a balloon.

Solid Liquid Gas

Chapter 3 The Chemistry of Living Things

INTEGRATED LEARNING

Themes in Science
Scale and Structure Point out that each element has a different number of electrons, but all atoms of a particular element have the same number of electrons. Explain that the atom's electrons determines its chemical properties.

Language Arts Connection
Have students imagine themselves as journalists interviewing Marie and Pierre Curie, the French scientists who isolated the radioactive element uranium in 1911. Have students read about the Curies and list five questions they would ask in an interview.

Research
Arrange the class into small groups and have each group make a list of elements that they want to know more about. Each group should record what they already know about each element. Have students use reference materials to research where the elements are likely to be found.

Discuss
Use Figure 3.3 to emphasize that each element is made up of only one kind of atom and that each atom is made up of three kinds of particles. Have students read the information in Figure 3.3. Ask students:

▶ What features of atoms change from element to element? (The number of protons is different for each element, as are the mass and the number of electrons.)

Answer to In-Text Question

② Neutrons, electrons, and protons make up an atom.

Figure 3.3 Different Models of Atoms

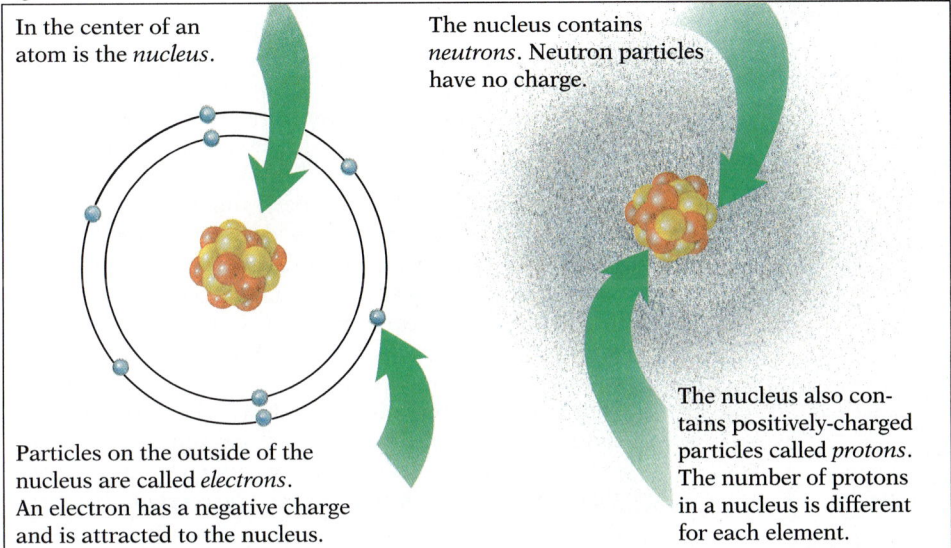

In the center of an atom is the *nucleus*.

The nucleus contains *neutrons*. Neutron particles have no charge.

Particles on the outside of the nucleus are called *electrons*. An electron has a negative charge and is attracted to the nucleus.

The nucleus also contains positively-charged particles called *protons*. The number of protons in a nucleus is different for each element.

Elements and Atoms

All forms of matter are made of one or more basic substances called **elements**. There are 91 known elements in nature. Elements cannot be changed into simpler substances by any chemical process or by heating.

Elements exist in nature as solids, liquids, or gases. Most elements are solids at room temperature. Iron, silver, gold, lead, and carbon are examples of solids. Some elements exist as gases. Oxygen, hydrogen, nitrogen, and helium are examples. These elements are present in the earth's atmosphere. However, you do not see or smell them because they are odorless, colorless gases. Only two elements exist as liquids at room temperature. They are mercury and bromine. Because they are very poisonous, mercury and bromine are normally found only in chemical laboratories.

The building blocks of elements are **atoms**. Each element is made of only one kind of atom. Each kind of atom has specific properties that make it different from every other kind of atom. An oxygen atom is different from a tin atom, helium atom, carbon atom, and so on.

Each atom is made up of different particles called *subatomic particles*. Study the models of an oxygen atom in Figure 3.3. What are the three main parts of an atom? ②

TEACH ▪ Continued

Skills WorkOut
To help students observe how positive and negative charges interact, have them do the Skills WorkOut.
Answers The pieces of paper become attached to the balloon, showing the negative and positive charges of atoms. Rubbing a student's hair with the balloon causes negative charges, or electrons, to move from the student's hair to the balloon. In this analogy, the balloon represents the atom's nucleus and the paper represents electrons.

Class Activity
Show students the periodic table of the elements. Go through the class list of substances recorded at the beginning of this section and classify items as elements and compounds.

Reteach
For hands-on work with molecular models, use Reteach Worksheet 3.1.

INTEGRATED LEARNING

Writing Connection
Have students write a story in which at least two of the following are main characters: an atom, a molecule, or a compound. The story should focus on their interactions. For example, would the characters get to know each other, or would they keep to themselves?

Integrating the Sciences
Chemistry For practice with the chemical characteristics of atoms, assign Skills Worksheet 3.1.

Skills WorkOut
Infer You can observe that atoms have positive and negative charges. You need a small balloon and about 20 small squares of notebook paper. Blow up the balloon and tie it. Then rub the balloon against your hair about 6 times. Hold the balloon close to the squares without touching them. What happens? How does this show the negative and positive charges of atoms?

Compounds and Molecules

A substance that is made of more than one kind of element is called a **compound**. In a compound, elements are chemically combined. For example, oxygen and hydrogen are elements. At room temperature, they are both gases. When oxygen and hydrogen combine chemically, they form water. Water is a compound.

Most compounds are made of **molecules**. They are the smallest part of a compound that have all the properties of the compound. A molecule is made of two or more atoms chemically bonded together.

The elements that make up a compound are always present in the same proportions. Each molecule of water consists of two atoms of hydrogen and one atom of oxygen. Each molecule of carbon dioxide consists of one atom of carbon and two atoms of oxygen.

A compound cannot be separated into the elements that form it except through a chemical reaction. For example, water can be separated into oxygen and hydrogen by passing an electric current through it. Electric energy breaks the chemical bonds that hold the molecule together.

Figure 3.4 Molecule of Water

The symbol for hydrogen is H.
The symbol for oxygen is O.

H_2O

The subscript means two atoms of hydrogen are in each molecule of water.

No subscript after oxygen means only one atom of oxygen is in each molecule of water.

Figure 3.5 Molecule of Carbon Dioxide

The symbol for carbon is C.
The symbol for oxygen is O.

CO_2

No subscript after carbon means only one atom of carbon is in each molecule.

The subscript means two atoms of oxygen are in each molecule of carbon dioxide.

Chapter 3 The Chemistry of Living Things

Integrating the Sciences

Chemistry Nitrogen is essential to all living things and is found throughout the biosphere. Have students construct models of the two-atom nitrogen molecule using clay and toothpicks and label the parts of the molecule.

Explore Visually

Have students study Figure 3.6. Point out that the formula contains both letters and numerals. Ask the following questions:

- What element does the C represent? (Carbon)
- What is the small number to the right of the C called? (A subscript)
- What does the subscript after the C mean? (12 atoms of carbon)
- What element does the O represent? (Oxygen)
- How many elements are present in the sucrose molecule? (Three—carbon, hydrogen, and oxygen)
- What is the total number of atoms in a molecule of sucrose? (45, or 12 + 22 + 11)

If students have problems arriving at the answers for the last two questions, stress that each letter is the symbol for an element and that each subscript indicates the number of atoms of that element present in the sucrose molecule.

Answers to In-Text Questions

① The chemical symbol for iron is Fe.

② The formula for carbon dioxide is CO_2.

Chemical Symbols and Formulas

Chemists all over the world use a kind of shorthand to name elements and compounds. Every element is represented by a **chemical symbol**. A chemical symbol is represented by either a capital letter or a capital letter and a small letter. Table 3.2 shows some elements and their symbols. What is the chemical symbol for iron? ①

Combinations of symbols are used to represent compounds. These symbols are called **chemical formulas**. Look at Figure 3.5. What is the chemical formula for carbon dioxide? ②

Sometimes a formula represents a molecule of an element, not a compound. For example, the symbol for nitrogen is N. Nitrogen, however, occurs naturally as a molecule containing two atoms of nitrogen bonded together. Nitrogen is a *diatomic molecule*. Therefore, the formula for a molecule of nitrogen is N_2.

Look at the formulas in Figures 3.4–3.6. Each symbol in a chemical formula represents an element. The numbers written below the line in a chemical formula are *subscripts*. A subscript indicates the number of atoms in the molecule. If there is only one atom of an element in a molecule, no subscript follows the symbol for the element. If there is more than one atom of an element in the molecule, the symbol is followed by a subscript.

Table 3.2 Common Elements

Name	Symbol
Aluminum	Al
Calcium	Ca
Carbon	C
Chlorine	Cl
Copper	Cu
Gold	Au
Helium	He
Hydrogen	H
Iron	Fe
Nitrogen	N
Oxygen	O
Sodium	Na
Zinc	Zn

Figure 3.6 Molecule of Sucrose (table sugar)

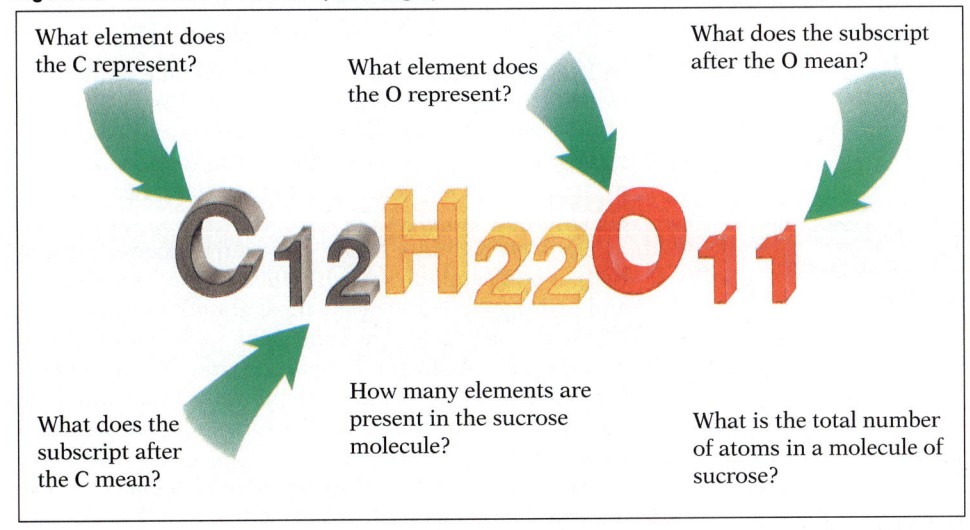

$C_{12}H_{22}O_{11}$

- What element does the C represent?
- What element does the O represent?
- What does the subscript after the O mean?
- What does the subscript after the C mean?
- How many elements are present in the sucrose molecule?
- What is the total number of atoms in a molecule of sucrose?

Chapter 3 The Chemistry of Living Things

TEACH ▪ Continued

Historical Notebook

Enrich Have students research alchemy. Ask them to write a brief description of alchemy and its relation to chemistry.

Answers
1. Answers may vary. At the time, the term *element* had no scientific meaning. The definition of chemical elements allowed scientists to make an important distinction among substances.
2. Joseph Priestley and Carl Wilhelm Scheele were involved in the discovery of oxygen, Anton Lavoisier named hydrogen, and Marie and Pierre Curie discovered radium.

Discuss

Ask students to consider how their body might compare to other animals in regard to the main elements that it has. Also ask students to consider how the main elements in squash might compare to other plants.

INTEGRATED LEARNING

Integrating the Sciences

Health While iron is found in the human body only in trace amounts (about 0.0004 percent), it is vital to healthy blood. Iron-deficiency anemia is a condition that results when people do not get sufficient iron through their diet. Have students find iron-rich foods among those that they eat. (Red meat, egg yolks, carrots, fruit, whole wheat, and grain vegetables)

Multicultural Perspectives

Have students research myths and ancient history to find out how earlier civilizations in various parts of the world described and/or named the primary materials of the universe.

Historical Notebook

Solving the Mystery of the Elements

Ancient peoples of many cultures recognized that some substances could not be broken down into simpler substances. In India, a book written around 1000 B.C. listed 5 "elements": earth, water, air, space, and light. By 300 B.C., people in China believed there were 5 elements—water, fire, wood, metal, and earth—constantly changing into one another. The ancient Greeks knew of 12 of the actual elements known today. However, they were not recognized as elements.

In the 1700s, experiments by a French chemist named Anton Lavoisier gave the concept of the *chemical* element a working definition. Lavoisier listed 33 elements. However, 7 of these were not true elements.

After chemical elements were defined, scientists began to recognize that materials such as gold, tin, and iron were elements. In the past 200 years, many new elements have been discovered. Today chemists can even change some of the elements into new ones.

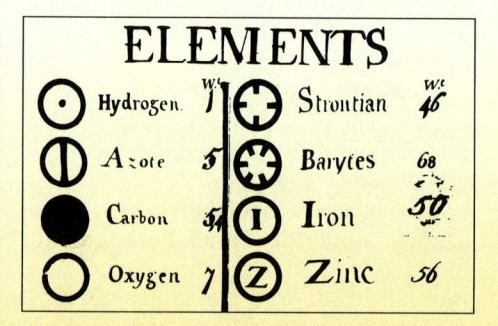

1. Why do you think it was important for scientists to define chemical elements?
2. **Research** Find out who discovered radium, hydrogen, and oxygen. Write a report about the discovery of one of these elements.

Science and You
Here an Element, There an Element

Did you know that your body contains some of the same elements as the earth's crust and a squash? Table 3.3 on page 57 shows the main elements in the human body, the earth's crust, and a squash. Notice that oxygen is the most abundant element in all three. What other elements are common to all three?

Look at the tables listing the main elements in the human body and the squash. Why do you think oxygen and hydrogen are two of the most abundant elements? If you guessed that it has something to do with water, you are right. All living things need water to carry out life processes. Notice that carbon is also abundant. It is the basis of important molecules that both you and plants need to carry out life processes.

Most living things contain the same elements in similar proportions. A table listing the elements in the body of a

Chapter 3 The Chemistry of Living Things

TEACHING OPTIONS

Portfolio

In addition to the circle graph that students made for the Check and Explain, have students make graphs for the proportions of elements found in the squash and the earth's crust. Students can identify elements common to both charts, then add this work to their portfolios. Later in this chapter, when students discuss the organic compounds in foods, they can review and revise their graphs.

Figure 3.7 ▲
What three main elements does your body have in common with summer squash? ①

Figure 3.8 ▲
Rocks are combinations of minerals, and most minerals are made of more than one element.

Table 3.3

Human Body	%
Oxygen	65
Carbon	18
Hydrogen	10
Nitrogen	3
Calcium	2
Phosphorus	1.1
Potassium	0.35
Sulfur	0.25
Chlorine	0.15
Sodium	0.15
Iodine	0.1
Trace Elements	under 0.1

Earth's Crust	%
Oxygen	46.6
Silicon	27.7
Aluminum	8.1
Iron	5.0
Calcium	3.6
Sodium	2.8
Potassium	2.6
Magnesium	2.1
Other	1.5

Squash	%
Oxygen	85
Hydrogen	10.7
Carbon	3.3
Potassium	0.34
Nitrogen	0.16
Phosphorus	0.05
Calcium	0.02
Magnesium	0.01
Iron	0.008

cat or a giant squid would be very similar to a table listing the elements in the human body.

Look at the table listing the elements in the earth's crust. Some of the elements, such as aluminum, iron, and magnesium, are metals. Your body also contains metals, but they are in very small amounts. Magnesium and iron are examples. They are part of a group called *trace elements*.

Notice that oxygen makes up nearly 50 percent of the total mass of the earth's crust. The value for oxygen does not include oxygen in the atmosphere. Oxygen combines with other elements to form many solid rock materials.

Check and Explain

1. How does an atom differ from a molecule?
2. The chemical formula for calcium carbonate, or lime, is $CaCO_3$. What elements make up $CaCO_3$? How many atoms are in one molecule of $CaCO_3$?
3. **Infer** Rust is a substance that forms when iron combines with oxygen. Is rust an element or a compound?
4. **Make a Graph** Make a circle graph showing the proportions of elements found in the human body.

EVALUATE

WrapUp

Reinforce To reinforce the concepts from this section, work with your class to develop a concept map that shows the relationships among the levels of organization of matter—compounds, elements, molecules, atoms, and atom particles.

Use Review Worksheet 3.1.

Check and Explain

1. An atom is the smallest building block of an element, while a molecule is the most basic part of a compound. A molecule is made up of two or more atoms.
2. Each molecule of $CaCO_3$ is made up of five atoms: one calcium atom, one carbon atom, and three oxygen atoms.
3. Rust is a compound because it is a combination of elements—iron and oxygen.
4. Students' graphs should indicate that the elements oxygen, carbon, hydrogen, nitrogen, and calcium make up 98 percent of the human body, in the proportions given in Table 3.3. The remaining 2 percent includes phosphorus, potassium, sulfur, chlorine, sodium, iodine, and trace elements.

Answer to In-Text Question

① Oxygen, hydrogen, and carbon are the three main elements common to both the human body and summer squash.

SECTION 3.2

Section Objectives
For a list of section objectives, see the Student Edition page.

Skills Objectives
Students should be able to:

Infer the changes in bread when it is frozen and toasted.

Observe ways to separate substances in a mixture.

Interpret Data presented in balanced chemical equations.

Communicate the meaning of an equation for a chemical reaction.

Graph changes in temperature measured in an experiment.

Vocabulary
chemical equation, reactant, product

MOTIVATE

Skills WarmUp
To give students practice in distinguishing between physical changes and chemical changes, have them do the Skills WarmUp.
Answer Frozen bread would become physically harder as the water in it changes to ice. As bread is toasted, some of the surface area is burned. Burning is a chemical change.

Prior Knowledge
To gauge how much students know about changes in matter, hold a brainstorming session. Begin the discussion by displaying a clear container filled with water. Ask all or some of the following questions:

▶ What are some of the physical properties of water? (Water is a tasteless, colorless, odorless compound that is liquid at normal temperatures.)

INTEGRATED LEARNING

Themes in Science
Patterns of Change/Cycles
There are patterns that can be observed in the ways that matter changes. Physical changes involve changes in a substance's physical properties only. Chemical changes, such as burning or rusting, involve changes in both physical properties and chemical properties.

Integrating the Sciences
Earth Science The earth and its atmosphere always contain the same amount of water. Some of it is liquid, some is solid (glaciers and ice caps), and some is gaseous, as in water vapor. Have students research the water cycle by looking it up in the index of this book.

Skills WarmUp

Infer How does a piece of bread change when it is placed in a freezer for a few days? What happens when a piece of bread is toasted? Write a paragraph describing the changes in the frozen bread and the toasted bread. How do the changes differ?

Figure 3.9 ▲
Faces have been carved into Mount Rushmore. Though the shape of the mountain changed, the granite did not.

3.2 Changes in Matter

Objectives

▶ **Explain** how water changes phases.
▶ **Distinguish** between physical and chemical changes.
▶ **Describe** the main parts of a chemical equation.
▶ **Interpret** what happens during a chemical reaction.

Matter changes in many ways. You know, for example, that water can change from a liquid to a solid. Ice is frozen water. Some of its properties are very different from those of liquid water. But is it still water? The answer is yes. You probably have seen wood burn. Is it still wood after burning? The answer is no. Eventually a burning piece of wood changes into ashes, soot, water vapor, and other gases.

Physical Changes

Matter can exist as solids, liquids, or gases. On the earth, water is a substance that exists naturally in all three phases. Energy causes water to change from one phase to another. For example, when heat energy is added to ice, the ice changes to liquid water. As more heat is added, water will eventually change to water vapor. Water vapor is a gas.

When water changes from one phase to another, it undergoes a *physical change*. A physical change is one in which some of the physical properties of a substance change, but the identity of the substance remains the same. If you cut a piece of wood in half, you get two pieces of wood. The identity of the wood does not change. It is still wood. If you bend a wire hanger, it changes the hanger's shape, but the metal does not change into something else.

Mixtures Sometimes a physical change occurs when substances combine. A mixture is two or more substances that do not combine chemically. A mixture can exist as a solid, liquid, or gas. For example, the atmosphere is made of a mixture of different gases. The gases may mix

58 Chapter 3 The Chemistry of Living Things

Integrating the Sciences

Nutrition Taro root is a major staple food for many populations in Polynesia and other Pacific islands. The fleshy root can't be eaten raw because its abrasive texture irritates the tissues of the mouth. However, cooking causes a chemical change that breaks down the root into an edible pulp. The pulp is then mashed into a paste, forming the basis of a dish called poi. Poi contains many carbohydrates. Have students discuss the chemical changes they observe in foods through cooking, such as in eggs and cake batter.

▶ What shape does water take? How would pouring it into another container affect it? (It will take the shape of the container.)

▶ How could you change some of the physical properties of this water? (Freeze or heat it)

TEACH

Skills WorkOut

To help students understand the definition of a mixture, have them do the Skills WorkOut.
Answer The magnet can pull the filings out. Also, a hand lens and tweezers can be used to separate the particles. You can tell you have a mixture if the materials that make up a substance can be separated.

Class Activity

Give each student a sheet of newspaper and have them note two physical properties of the paper. (Students might observe size, appearance, feel, and smell.) Next, ask them to produce a physical change in the newspaper. (Students might tear the paper or crumple it.) Finally, ask students how they might produce a chemical change in the paper. (One suggestion would be to burn the paper. A chemical property of the paper is that paper will burn easily, and a chemical change is that burning produces substances that are different from the paper.)

Answer to In-Text Question

① The cars are rusting. That is, iron oxide is being produced.

---

in the atmosphere, but they keep their separate chemical properties. Soil is made of a mixture of different solids such as clay and rock particles.

Solutions If you have ever tasted sea water, you know that it is very salty. Salt water is an example of a solution. A solution is a type of mixture in which one substance is evenly mixed with another substance. When salt is added to water, the taste of the water changes and the salt becomes invisible. However, neither the makeup of the water nor the salt has changed. The water is still water. The salt is still salt. In sea water, particles of water are evenly mixed with different kinds of salt.

Chemical Changes

When wood burns, it undergoes a *chemical change*. The wood is changed into many different substances. In a chemical change, the chemical identity of a substance is changed. Chemical changes occur in both living and nonliving things. Some examples of chemical changes are plants making sugars and starches for growth; the changing color of leaves; or cooking food.

During a chemical change, a chemical reaction takes place. When a chemical reaction occurs, new substances are produced. One common chemical reaction takes place when iron atoms combine with oxygen atoms to produce iron oxide, or rust. Rust is a new substance that has different properties than iron. For example, rust is a different color than iron.

Skills WorkOut

Observe Mixtures can sometimes be separated out. To see how this happens, you need a magnet, some sugar, a bowl, and some iron filings. Find two ways to separate the substances in the mixture. How can you tell that you have made a mixture?

Figure 3.10 ▲
Burning trees produce heat, light, ash, and gases. It is a chemical change because you can't get the trees back by mixing the products together. What chemical changes are happening to these cars? ①

TEACH • Continued

Discuss
Have students look at Figure 3.11. Ask the following questions:
- Name the reactants in this chemical reaction. (Carbon, oxygen)
- Name the product in this chemical reaction. (Carbon dioxide)
- How many molecules of carbon dioxide are indicated on the right of the chemical equation? (1)

Reteach
Ask students to explain what the reactants and products would be in the following chemical reactions:

$2Na + Cl_2 \longrightarrow 2NaCl$
$H_2 + Cl_2 \longrightarrow 2HCl$
$CO_2 \longrightarrow C + O_2$
$2HCl \longrightarrow H_2 + Cl_2$

Critical Thinking
Infer Ask students to suggest which parts of an atom are most affected during chemical reactions or changes. (Since electrons are on the outside of the atom, it would be far more likely for these particles to be involved than it would be for any of the particles in the nucleus.)

Enrich
Have students research fuels such as propane (C_3H_2) and methane (CH_4). What products are formed when these are burned? (CO_2 and H_2O)

INTEGRATED LEARNING

Themes in Science
Stability and Equilibrium

Another way to say that matter cannot be created or destroyed in a chemical reaction is: *Total mass is always conserved*. Physicists refer to this principle as the law of conservation of mass.

Chemical Reactions and Equations

Chemists represent chemical reactions by writing **chemical equations**. In a chemical equation, symbols represent elements and formulas represent compounds. Look at Figure 3.11. It shows the chemical equation for the formation of carbon dioxide from carbon and oxygen.

Figure 3.11
Formation of Carbon Dioxide

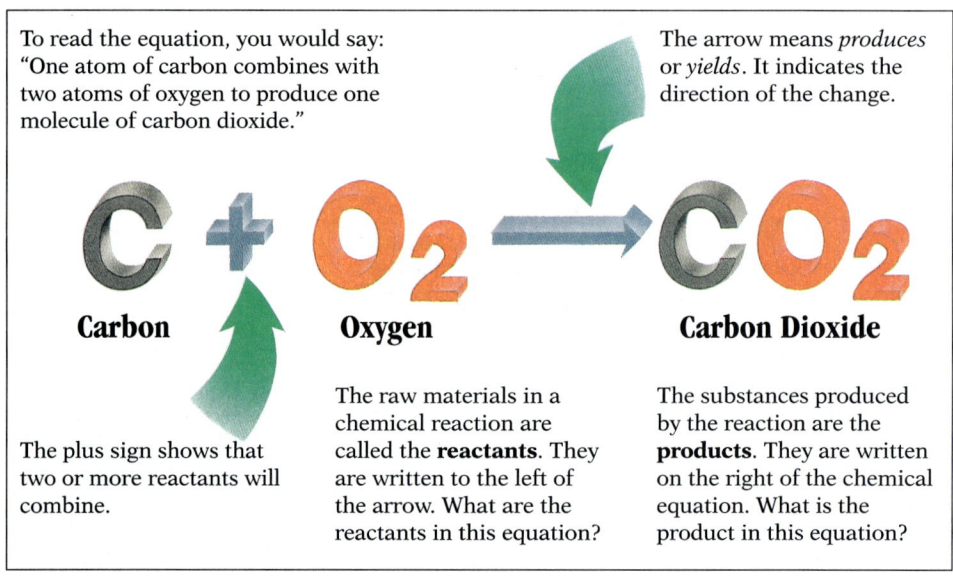

To read the equation, you would say: "One atom of carbon combines with two atoms of oxygen to produce one molecule of carbon dioxide."

The arrow means *produces* or *yields*. It indicates the direction of the change.

$$C + O_2 \longrightarrow CO_2$$

Carbon   Oxygen   Carbon Dioxide

The plus sign shows that two or more reactants will combine.

The raw materials in a chemical reaction are called the **reactants**. They are written to the left of the arrow. What are the reactants in this equation?

The substances produced by the reaction are the **products**. They are written on the right of the chemical equation. What is the product in this equation?

Balanced Chemical Equations

Look at the following chemical equation for the formation of water from hydrogen and oxygen:

$$H_2 + O_2 \rightarrow H_2O$$

In this equation, you can identify the atoms and molecules in the reaction. Notice, however, that two oxygen atoms are on the left side of the arrow but only one is on the right. One oxygen atom has disappeared.

Scientists have shown that atoms do not disappear during chemical reactions. Matter cannot be created or destroyed. To correct the equation, you must *balance* it. You balance a chemical equation by placing numbers called *coefficients* in front of chemical formulas. A coefficient is a number that shows how many molecules or atoms of a substance are involved in the reaction. The number of atoms of each element must be the same on both sides of a chemical equation.

To balance the equation, first place a coefficient before the product. You need to show that two atoms of oxygen are on each side of the equation. So write a 2 before the product:

$$H_2 + O_2 \rightarrow 2H_2O$$

Now the oxygen atoms are balanced, but the hydrogen atoms are not. There are four hydrogen atoms on the right and only two on the left. Write a 2 before the H on the left side of the equation:

$$2H_2 + O_2 \rightarrow 2H_2O$$

The left side of the equation now has four atoms of hydrogen and two atoms of oxygen. The same number of atoms are on the right side. To read this equation, you would say: "Two hydrogen molecules react with one oxygen molecule to produce two water molecules."

SkillBuilder *Interpreting Data*

Chemical Equation Balancing Act

A balanced chemical equation has the same number of atoms of each element on both sides of the equation. Try balancing the following equations.

1. Write each equation shown. Leave space below each equation in case you need to balance it.

2. Count the atoms that represent the reactants in the equation. Then count the atoms that represent the products. If the number of atoms in the reactants and product(s) are not equal, you need to balance the equation.

3. For the unbalanced equations, add a coefficient in front of a formula or symbol so the number of atoms of that element is equal on each side of the equation. Continue this process until you have balanced the atoms of each element.

4. Check your balancing act by counting the atoms of each element on both sides of the equation.

Equations

$P + O_2 \rightarrow P_4O_{10}$

$N_2 + O_2 \rightarrow NO$

$BaCl_2 + H_2SO_4 \rightarrow BaSO_4 + HCl$

$Li + Cl_2 \rightarrow LiCl$

$2K + Cl_2 \rightarrow 2KCl$

$Mg + N_2 \rightarrow Mg_3N_2$

1. What does a coefficient represent?
2. Why must chemical equations be balanced?

Using references, write two balanced equations. Then change the equations so that they are unbalanced. Exchange the unbalanced equations with a partner. See if you can balance them.

Chapter 3   The Chemistry of Living Things   61

TEACH • Continued

Discuss
Have students name items in the classroom that are made from plastic.

EVALUATE

WrapUp
Portfolio Ask students to cut out pictures that show examples of physical properties and physical changes, label them, and paste them onto a sheet of paper. Then have them also illustrate chemical properties and chemical changes. Students may wish to add these pages to their portfolios.

Use Review Worksheet 3.2.

Check and Explain
1. The addition of heat to solid water (ice) increases the kinetic energy of particles in the water, enabling them to move around freely.
2. The reactants are the raw materials. The products are the substances produced by the reaction.
3. The burning lump of sugar is a chemical change because a new substance is produced in the reaction.
4. One hydrogen molecule (two atoms of hydrogen) reacts with one molecule of oxygen (two atoms of oxygen) to produce a molecule of hydrogen peroxide.

Answer to In-Text Question

① Hard plastics would be useful in making computer bodies and electronics. Clear plastics can be used for screens. Some disk drives are made from flexible plastics.

INTEGRATED LEARNING

Integrating the Sciences
Chemistry The polymer rayon, which is used in clothing, was developed during World War II. Explain that rayon is made from a naturally occurring polymer, cellulose ($C_6H_{10}O_5$). Ask students where cellulose comes from. (Wood, cotton, and so on—plant cell walls)

Science and Technology
Practical Polymers

How many different plastic products can you name? Plastics are one of many types of *synthetic polymers*. They are formed from chemical reactions. In these types of reactions, thousands of small molecules are joined into long chains. Polymers are giant molecules. Each polymer is a compound with its own chemical properties and structure. Polymers form a variety of materials—from the material that makes up hard football helmets, to the material that makes up squeezable catsup bottles, to the material for plastic wrap.

Polymers are important substances in many ways. In the plastics industry alone, there are more than 45 different polymers in use today. These plastics are used in telephones, pipes, toothbrushes, toys, compact discs, and other items. In addition to plastics, synthetic polymers are used to make many of the fabrics for clothes that you wear. Nylon, rayon, and dacron are examples. Polymers are also used to build artificial hearts, knees, and other body parts. Even computers would not exist without polymers. The keyboard and data storage disks are both plastic or plastic coated. The microchip is embedded in plastic.

At one time, cars were made mostly of steel. Today most cars have plastic door trims, hoods, and fenders. In the future, cars will contain even more plastic parts. One day you may ride in a car made almost entirely of plastic, including the engine!

Figure 3.12 ▲
Plastics can be rubbery, stringy, clear, opaque, hard, or flexible. Which of these characteristics are useful in computers? ①

Check and Explain
1. Explain how water changes from a solid to a liquid.
2. What are the reactants and products in a chemical equation?
3. **Reason by Analogy** A lump of sugar burns. Is this an example of a physical change or a chemical change? Explain why.
4. **Communicate** In your own words, describe what the following chemical reaction for hydrogen peroxide means: $H_2 + O_2 \rightarrow H_2O_2$.

Chapter 3 The Chemistry of Living Things

TEACHING OPTIONS

Prelab Discussion

Have students read the activity before beginning the discussion. Ask the following question before beginning the activity:

▶ What evidence may be used to tell if a chemical reaction is occurring? (Heat released and/or absorbed, gas [bubbles] produced, color change, formation of a precipitate) Remind students that the new products formed have different properties than the reactants as the atoms have formed new combinations.

Safety Note: Remind students not to use the thermometer as a stirring rod.

ACTIVITY 3

Time 45 minutes **Group** 3–4

Materials
8 small jars
8 Celsius thermometers
1 box of powdered bleach
8 spoons
metric rulers
graph paper

Analysis

1. Students should observe bubbles forming after adding bleach; also that the jar becomes warmer.
2. The rising temperature is evidence that heat is being produced.
3. Answers should agree with each student's own data.
4. Temperature stops rising when all the bleach has reacted and the reaction stops.

Conclusion

Accept all logical answers; however, students should recognize that the release of bubbles (a gas) and heat indicate that a chemical reaction is occurring. The activity did not require energy—it produced heat (energy).

Everyday Application

Students may identify the type of fiber (natural or synthetic) as a common factor.

Extension

Oxidation is also a chemical reaction. Have students observe oxidation by placing pennies in a saucer on top of a paper towel that has been wet with vinegar. Have students leave the experiment for 24 hours. (The acetate part of vinegar combines with the copper to form a green coating, which is copper acetate.)

Activity 3 What happens during a chemical reaction?

Skills Observe; Measure; Graph; Infer

Task 1 Prelab Prep
1. Collect the following items: ruler, pencil, paper, graph paper, spoon, powdered bleach, water, small jar, Celsius thermometer.
2. Fill the jar three-quarters full with water.

Task 2 Data Record
1. On a separate sheet of paper, copy Table 3.4. Number up to 10 minutes.
2. Record your observations in the data table.
3. After your table is complete, graph your data. Use your ruler and graph paper to draw the line graph shown below. Add your data points. Remember to title your graph.

Task 3 Procedure
1. Add one teaspoon of bleach to the jar of water. **CAUTION! Wear gloves, a laboratory apron, and goggles when using bleach.**
2. Stir the mixture with the spoon. Observe the mixture as you stir.
3. Put the thermometer into the liquid.

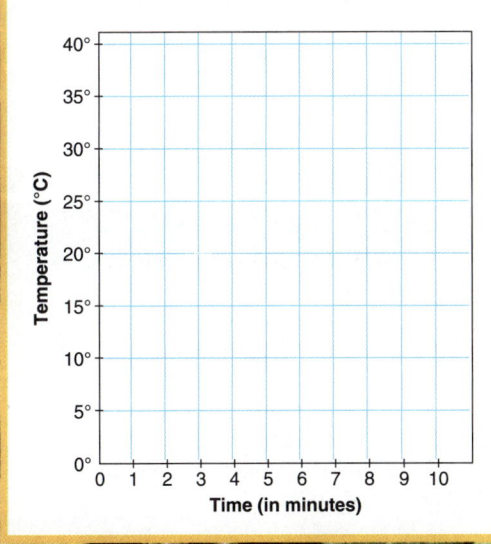

Table 3.4 Temperature Change

Time (Minutes)	Temperature (°C)	Observations
1		
2		
3		
4		
5		
6		

4. Observe the thermometer every minute for ten minutes. Record the temperature at each observation in your data table.

Task 4 Analysis
1. What happened when you added bleach to the water?
2. What does the rising temperature indicate?
3. When did the temperature stop rising?
4. Why did the temperature stop rising?

Task 5 Conclusion
Write a short paragraph explaining why this activity is an example of a chemical reaction. Did the activity require energy? Explain.

Everyday Application
Household bleach contains a chemical called sodium hypochlorite. This chemical contains oxygen that easily combines with dyes to form a colorless compound. The combination of oxygen and dyes causes colors to fade. Select ten items from your closet. Check the labels for washing instructions. Classify your clothes into two groups: those that can be washed with bleach and those that can't. What else do the clothes in each group have in common?

SECTION 3.3

Section Objectives
For a list of section objectives, see the Student Edition page.

Skills Objectives
Students should be able to:

Infer living and nonliving things that contain the elements C, H, O, and N.

Hypothesize about the chemical composition of living things.

Infer the presence of carbon compounds in specified substances.

Vocabulary
organic compound, inorganic compound, amino acid, simple sugar, nucleic acid

MOTIVATE

Skills WarmUp
To give students practice in recognizing the chemical similarity of organic molecules, have them do the Skills WarmUp.
Answer The symbols represent carbon, hydrogen, oxygen, and nitrogen. Answers may vary. Some possible responses are that air contains all these elements; water contains hydrogen and oxygen; sugars contain carbon, oxygen, and hydrogen; and proteins contain nitrogen, carbon, hydrogen, and oxygen.

Misconceptions
Students may think that all compounds containing carbon are organic. Stress that some carbon compounds, such as carbon dioxide, are not organic. Organic compounds occur naturally in the bodies, products, and remains of living things.

INTEGRATED LEARNING

Themes in Science
Diversity and Unity When learning about the molecules of life, students observe examples of unity and diversity. Although carbon is the basis for all life on earth, carbon atoms form many kinds of chemical bonds. Thus, organic compounds have tremendous variety, making up by far the majority of the chemical compounds on the earth.

Language Arts Connection
Have interested students research the origins of the words *carbohydrate* and *protein*.

Skills WarmUp
Infer What elements do the symbols C, H, O, and N represent? Write the names of these elements on a piece of paper. Then list as many things that you can think of that contain the elements. Include both living and nonliving things.

3.3 Molecules of Life

Objectives
▶ **Define** and describe carbohydrates, lipids, and proteins.
▶ **Explain** why nucleic acids are important to the cell.
▶ **Infer** whether certain compounds are organic or inorganic.

If you could see the molecules that make up your body, you would notice something similar about them. Each molecule would be made of long, twisting chains of carbon atoms. Carbon is the basis for all life on earth. In all living things, carbon atoms form many kinds of bonds that make up many molecules.

Organic Compounds

In all living things, carbon compounds are present. Carbon compounds that occur naturally in the bodies, the products, and the remains of living things are called **organic compounds**. Living things also contain **inorganic compounds**. Most inorganic compounds do not contain carbon. Water and minerals are the main inorganic compounds in living things.

Three groups of organic compounds are especially important to living things. These three groups are carbohydrates, proteins, and lipids. Most of the life processes use one or more of these groups of compounds.

Proteins Proteins are complex molecules. They are made of long chains of **amino acids**. Amino acids are simple molecules that contain carbon, oxygen, hydrogen, and nitrogen. The specific amino acids and their arrangement determine the shape of the protein.

Your cells use proteins to build and repair body parts. Proteins also help speed up chemical reactions and protect the body from disease. No single protein serves all of these functions. The human body contains more than 100,000 different types of protein molecules, each with its own function.

Figure 3.13 ▲
An octopus' skin can change color because of organic compounds called pigments.

64 Chapter 3 The Chemistry of Living Things

Multicultural Perspectives

Have students draw or cut out pictures of foods from other countries and then label them to indicate the primary organic compounds they contain. They can use the picture on this page as an example.

Social Studies Connection

Grains (cereals) provide much of the necessary carbohydrates for people all around the world. Ask students to name some common grains. Point out that all of these grains were domesticated by different civilizations to provide reliable food sources. Ask interested students to find out in what part of the world each of the following grains originated: wheat, corn, rice.

TEACH

Portfolio

Have students make lists of foods that are high in proteins, carbohydrates, and lipids. Also have students try to find pictures of these foods in newspapers and magazines and mount them on loose-leaf paper. They should label each food, showing the organic compound found in the highest amount in each. Students can staple the papers together and add the booklets to their portfolios.

Answer to In-Text Question

① People need carbohydrates in order to store energy.

Carbohydrates Carbohydrates are organic compounds that most living things use to store energy. Each carbohydrate molecule is made up of oxygen, hydrogen, and carbon atoms.

Starches and sugars are the two classes of carbohydrates. Sugars are simple carbohydrate molecules. Starches are complex carbohydrate molecules consisting of many sugars linked together. In living things, sugars and starches are broken down into **simple sugars**. Simple sugars, such as glucose, are then transported to all the cells in the body. In the cells, simple sugars provide fuel for energy.

Carbohydrates are found in many of the foods that you eat, such as potatoes, bread, rice, pasta, and bananas.

Lipids Lipids are a group of organic compounds that contain carbon, hydrogen, and oxygen. Lipids are placed in one group because they share one physical property. They do not mix with water.

Fats, oils, and waxes are three types of lipids. Foods that contain lipids include butter, vegetable oil, and animal fats.

Your body uses lipids for energy storage, drawing on them when carbohydrates are not available. A specific amount of lipids provides more than twice as much energy as the same amount of carbohydrates. In your body, lipids are broken down into simpler molecules and transported to cells.

Nucleic Acids Carbohydrates and proteins combine in living cells to form **nucleic acids**. Nucleic acids control all the activities of a cell. They also carry the instructions that enable organisms to reproduce.

There are two types of nucleic acids in living cells, DNA and RNA. DNA directs a cell to perform all of its processes. RNA copies these directions and takes them to the parts of the cell where they are carried out.

Figure 3.14 Organic Compounds in a Tostada

The corn that makes up the tortilla is a source of carbohydrates. So are vegetables such as tomatoes and lettuce. Why do you need carbohydrates? ①

Dairy products, such as cheese, are sources of fat. Oil, which is used in making a tortilla, also contains fat. Your body stores fat for energy.

Beans and cheese are sources of proteins. Your body cannot store proteins, so your diet must include foods containing proteins.

The Living Textbook:
Life Science Sides 7-8

Chapter 3   Frame 00401
Carbohydrates, Lipids,
Proteins (3 Frames)
Search: Step:

The Living Textbook:
Life Science Sides 7-8

Chapter 9   Frame 01184
DNA Structure (2 Frames)
Search: Step:

TEACH • Continued

Skills WorkOut
To help students understand the chemical composition of living things, have them do the Skills WorkOut.
Answer The cucumber slices shrink as the water in them evaporates. Students can infer that water is an important element in living things.

Discuss
Stress to students that eight essential amino acids can be gotten only in food. Have students name foods they eat that are high in protein.

Integrated Learning
Use Integrating Worksheets 3.3a and 3.3b.

EVALUATE

WrapUp
Review Use Review Worksheet 3.3.

Check and Explain
1. Organic compounds are carbon compounds in the bodies, the products, and the remains of living things. Proteins are complex molecules made up of amino acids. Carbohydrates are made up of oxygen, hydrogen, and carbon. Nucleic acids are combinations of carbohydrates and bases.
2. Nucleic acids control cell activities.
3. The body would use lipids from stored fat.
4. No: water, sand, table salt; These substances don't contain carbon. Yes: catsup, apple, butter; These substances were once living or are produced from living things.

Skills WorkOut

Hypothesize Cut part of a cucumber into slices as thin as you can make them. Place the slices on a plate near a sunny window. Observe and record what happens for several days. What conclusion can you draw about the chemical composition of living things?

Science and You *Proper Proteins*

Your body needs 20 different amino acids to build all its proteins. Your cells, however, can only manufacture 12 of them. The other eight, called *essential amino acids*, must be obtained from proteins in the foods you eat.

Most animal proteins, such as fish, meat, poultry, and eggs, contain all eight of the essential amino acids in the proper proportions. Animal proteins are called *complete proteins*. Most plants lack one or more of the essential elements. Plant proteins are called *incomplete proteins*. You can, however, get all the essential animo acids from plants alone by combining certain types of plant foods, such as beans and grain. A meal of beans and rice gives you the same protein value as a meal of steak and eggs.

The key to obtaining essential amino acids from plants is how you combine them in your diet. For example, beans supply you with a certain amino acid that is missing in corn. Corn, on the other hand, contains an essential amino acid that is not present in beans. Thus, a meal of corn and beans will provide you with all the essential amino acids that your body requires.

A lack of essential amino acids in your diet can be very dangerous to your health. It can lead to protein deficiency. Some of the symptoms of protein deficiency include lowered energy, stunted growth, and lowered resistance to disease.

Check and Explain

1. What are organic compounds? Name and describe three types of organic compounds.
2. Why are nucleic acids important to living things?
3. **Predict** What might happen if you did not eat enough foods high in carbohydrates?
4. **Infer** Decide if the following substances contain carbon compounds. Explain your conclusions.

   a. water
   b. catsup
   c. apple
   d. table salt
   e. sand
   f. butter

Chapter 3 The Chemistry of Living Things

CHAPTER REVIEW 3

Chapter 3 Review

Concept Summary

3.1 Matter
- All things are made of matter. Matter is anything that takes up space and has shape.
- Matter exists as solid, liquid, or gas. Solids have definite volume and shape. Liquids have definite volume but not shape. Gases have neither definite volume nor definite shape.
- All matter is made of elements. The building blocks of elements are atoms.
- Two or more elements joined together form a compound. A molecule is the smallest part of a compound.

3.2 Changes in Matter
- A physical change is a change in a substance's physical properties but not its chemical identity. A chemical change forms a new substance.
- Chemical equations show how raw materials react to form products.

3.3 Molecules of Life
- Organic compounds, such as proteins, carbohydrates, and lipids, contain carbon and occur naturally in living things.
- The human body contains many proteins, each with a specific function such as building or repairing body parts or speeding up chemical reactions.
- Living things use carbohydrates such as starches and sugars to store energy.
- Lipids are used for energy storage when carbohydrates are not available.
- The two types of nucleic acids, DNA and RNA, control cell activity and contain instructions for cell reproduction.

Chapter Vocabulary

matter (3.1) molecule (3.1) reactant (3.2) amino acid (3.3)
element (3.1) chemical symbol (3.1) product (3.2) simple sugar (3.3)
atom (3.1) chemical formula (3.1) organic compound (3.3) nucleic acid (3.3)
compound (3.1) chemical equation (3.2) inorganic compound (3.3)

Check Your Vocabulary

Use the vocabulary words above to complete the following sentences correctly.

1. Long chains of ____ make proteins.
2. Anything that takes up space and has mass is ____.
3. Cell activities in living things are controlled by ____.
4. A chemical reaction can be represented by a ____.
5. In living things, carbohydrates are broken down into ____ for fuel.
6. A substance that is made up of more than one kind of element is called a ____.
7. The raw materials in a chemical reaction are called ____.

Explain the difference between the words in each pair.

1. chemical symbol, chemical formula
2. mixture, solution
3. reactant, product
4. organic compound, inorganic compound
5. atom, molecule
6. element, compound
7. amino acids, nucleic acids

Chapter 3 The Chemistry of Living Things 67

Check Your Vocabulary

1. amino acids
2. matter
3. nucleic acids
4. chemical equation
5. simple sugars
6. compound
7. reactants

1. A chemical symbol is a one- or two-letter symbol for an element. A chemical formula is a combination of chemical symbols used to represent a compound.
2. A mixture is two or more compounds that do not combine chemically. A solution is a type of mixture in which the substances are evenly mixed.
3. A reactant is a substance that is used up in a chemical reaction. A product is a substance produced by a chemical reaction.
4. Organic compounds are found in living things and generally contain carbon. Inorganic compounds generally do not contain carbon.
5. Atoms are the building blocks of elements and compounds. Molecules are the building blocks of compounds.
6. Elements are the basic substances that make up all matter. A compound is a substance made up of more than one element.
7. Amino acids are molecules that make up proteins. Nucleic acids are made up of proteins and carbohydrates.

Use Vocabulary Worksheet for Chapter 3.

CHAPTER 3 REVIEW

Check Your Knowledge

1. Liquid
2. Carbohydrate
3. One
4. Atoms are rearranged to form new compounds.
5. Proteins are used to build and repair body parts, to speed up chemical reactions, and to protect the body from disease.
6. Solid, liquid, gas
7. Ice melting is an example of a physical change.
8. Organic compounds generally contain carbon; inorganic compounds do not.
9. Matter is anything that takes up space and has mass.
10. Chemists represent elements with one- or two-letter symbols.
11. False; physical change
12. False; compounds
13. True
14. True
15. True
16. False; numbers

Check Your Understanding

1. a. Al (aluminum), O_2 (oxygen)
   b. AlO_2 (aluminum oxide)
   c. 4Al, $3O_2$, $2Al_2O_3$
   d. O_2, Al_2O_3
   e. Al_2O_3
   f. Al (aluminum), O (oxygen)
   g. Al, O
   h. O_2, Al_2O_3
2. a. Physical
   b. Physical
   c. Chemical
   d. Chemical
   e. Chemical
3. Accept all logical answers; however, students should recognize that if it is not used, the extra energy in the food would be stored as body fat. If less food energy is taken in, the body would use stored energy to supply its needs.
4. Answers may vary, but students should recognize that since lipids do not mix with water, the feathers may have an oily coating that repels water.

Chapter 3 Review

Check Your Knowledge
Answer each of the following in complete sentences.

1. What form of matter has definite volume but not definite shape?
2. Starches and sugars are examples of what type of organic compound?
3. How many types of atoms can be contained in an element?
4. What happens during a chemical change?
5. How does your body use proteins?
6. Name the three phases of matter.
7. Give an example of a physical change.
8. What is the difference between organic compounds and inorganic compounds?
9. What is matter?
10. How do chemists represent elements?

Determine whether each statement is true or false. Write *true* if it is true. If it is false, change the underlined term so that the sentence is true.

11. A <u>chemical change</u> affects a substance's characteristics but does not change its identity.
12. Molecules are made of atoms and are the building blocks of <u>elements</u>.
13. Organic compounds include <u>proteins</u>, carbohydrates, and lipids.
14. In chemical equations, chemical formulas represent <u>compounds</u>.
15. A <u>physical</u> change occurs when ice melts.
16. You balance a chemical equation by placing <u>letters</u> called coefficients in front of the chemical formulas.

Check Your Understanding
Apply the concepts you have learned to answer each question.

1. Sometimes a thin, cloudy coating will form on an aluminum surface such as an aluminum pot or pan. A chemical reaction takes place between oxygen in the air and the aluminum to produce the coating, aluminum oxide. The equation for this reaction is written below.

   $$4Al + 3O_2 \rightarrow 2Al_2O_3$$

   Identify the following in the equation:
   a. reactants   e. compounds
   b. product   f. elements
   c. coefficients g. chemical symbols
   d. molecule   h. chemical formulas

2. **Classify** Identify each of the following as a physical or chemical change.
   a. Ice melting into water
   b. Drink mix dissolving in water
   c. Gasoline burning in a car's engine
   d. Hydrogen and oxygen forming water
   e. The breakdown of lipids for energy

3. **Predict** What would happen if you regularly ate large quantities of fatty foods and used up very little energy? How would your body respond if you ate less food than you needed for the energy you use? Explain your answers.

4. **Mystery Photo** The photograph on page 50 shows beads of water on pheasant feathers. Water does not penetrate the feathers; instead, it rolls off. Feathers are waterproof. Review the pages on organic compounds and look for references to water. What might help bird feathers repel water? How?

68 Chapter 3 The Chemistry of Living Things

Develop Your Skills

1. a. Carbon dioxide, nitrogen, oxygen
   b. Elements: nitrogen, oxygen; compound: carbon dioxide
   c. Both contain nitrogen and oxygen
2. a. 6
   b. 18; 6; 12
   c. 18; 6; 12
3. a. Point A
   b. Points B and C
   c. Point D
   d. Physical change

Make Connections

1. Concept maps will vary. You may wish to have students do the map below. First draw the map on the board, then add a few key terms. Have students draw and complete the map.

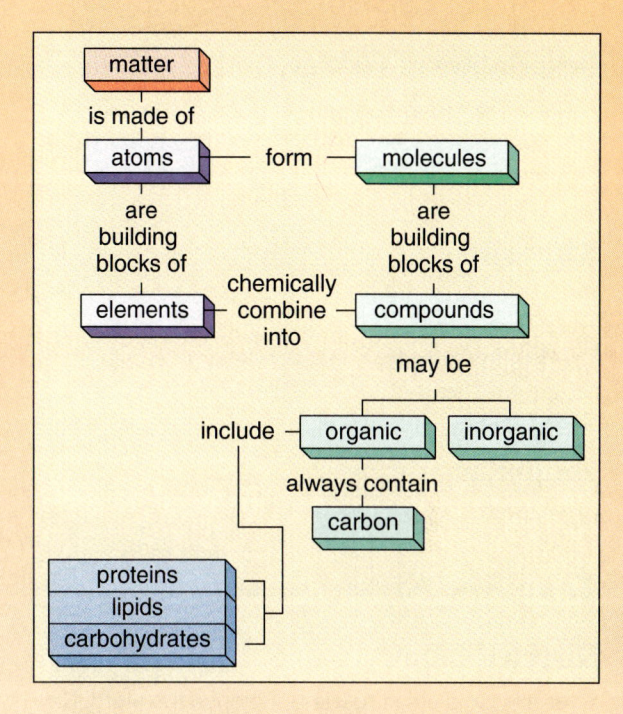

```
matter
  is made of
atoms ── form ── molecules
  are                     are
  building                building
  blocks of               blocks of
  elements ── chemically combine into ── compounds
                                              may be
               include ── organic    inorganic
                          always contain
                          carbon
proteins
lipids
carbohydrates
```

2. Answers may vary. Paragraphs should show research and an understanding of how the benefits of using synthetic polymers are balanced against the drawbacks.
3. Answers may vary; typical values are: playing tennis, 173–270; playing basketball, 191–298; walking slowly, 65–102; running, 182–284; biking, 187–291; sleeping, 23–35.
4. Answers may vary. Students should list ten events in each of two categories; for example, physical changes—peeling an orange, drying hair; chemical changes—burning fuel in a car or bus to get to school, digesting breakfast.

Develop Your Skills

1. **Data Bank** Use the information on page 624 to answer the following questions.
   a. What gases exist in the atmosphere of Mars?
   b. Which gases on Mars are elements? Compounds?
   c. How is Mars' atmosphere like Earth's?
2. The equation below shows the chemical breakdown of glucose, a simple sugar.

   $$C_6H_{12}O_6 + 6O_2 \rightarrow 6CO_2 + 6H_2O$$

   a. How many molecules of oxygen are needed to break down each molecule of glucose?
   b. In the reactants, how many atoms of oxygen are there? Atoms of carbon? Atoms of hydrogen?
   c. In the products, how many atoms of oxygen are there? Atoms of carbon? Atoms of hydrogen?
3. **Interpret Data** The graph below shows temperature changes for water.
   a. At which point(s) on the graph would water be in a solid phase?
   b. A liquid phase?
   c. A gas phase?
   d. Is the phase change of water a chemical or a physical change?

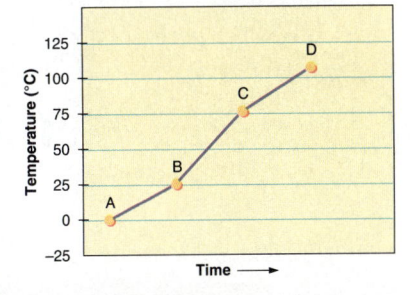

Make Connections

1. **Link the Concepts** Draw a concept map showing how the concepts below link together. Add terms to connect, or link, the concepts.

   atoms   inorganic
   elements   compounds
   molecules   carbohydrates
   compounds   lipids
   organic   proteins
     compounds   carbon
                    matter

2. **Science and Technology** Many items that people use every day are made of synthetic polymers. One drawback to synthetic polymers is that they decompose very slowly, causing waste problems. Find out more about synthetic polymers at the library. What are some positive and negative effects of polymer use? Should people limit the amount of objects made with polymers? Write a paragraph in which you state and defend your position. Include suggestions to limit polymers' negative effects.

3. **Science and P.E.** The energy your body uses to grow, play, and carry out basic functions comes from the food you eat. This energy is measured in calories. Use a book on health to find out how many calories you use during a half hour of the following activities.
   a. playing tennis   d. running
   b. playing basketball e. biking
   c. walking slowly   f. sleeping

4. **Science and You** Make a list of ten physical changes you experience daily. Make a list of ten chemical changes you experience daily. Compare your list of changes with that of a classmate.

SCIENCE AND LITERATURE CONNECTION   UNIT 1

About the Literary Work

"The Gorilla Signs Love" was adapted from *The Gorilla Signs Love* by Barbara Brenner, copyright 1984 by Lothrop, Lee, & Shepard Books. Reprinted by permission of Lothrop, Lee, & Shepard.

Description of Change

The excerpt details one experience of the author. Any changes or deletions are noted by brackets or ellipses.

Rationale

Any changes or deletions were made for the sake of space and continuity. Material that did not directly address the narrator and her relationship with Naomi was omitted.

Vocabulary

primate, Ameslan

Teaching Strategies

Directed Inquiry

After students finish reading, discuss the story. Be sure to relate the story to the science lessons in this unit. Ask the following questions:

▶ Naomi's action at the end of the selection came as a surprise to the narrator. What assumption did Naomi dispel when she shook the can? (By shaking the can, Naomi showed that she was processing her observations of the narrator, not merely imitating.)

▶ As you read the selection, how did you think it would end? (Responses may vary. Some students may have thought Naomi would attempt to drink from the empty can or that she wasn't really communicating, just imitating.)

Skills Development

Predict Emphasize that soda is not normally part of a gorilla's diet. Ask students to suppose that the narrator shared all of her meals with Naomi and that the gorilla became accustomed to "human" food. Have students predict the effect this might have on the animal. (First, we do not know whether Naomi would be able to digest and obtain the nutrients she needs for survival from "human" food. Second, the animal would eventually become dependent upon humans for food, losing her ability to seek food sources from her natural environment.)

Infer Ask students why it is important in research to avoid contact with animals in the wild. (Contact with the animals may affect how they behave, which would make the results of the research less accurate.)

Cooperative Learning

Help students understand that at the conclusion of the passage, the narrator faces a dilemma—whether to share her experience with other team members. While the experience is significant in terms of animal behavior, she did break the "no talking" rule. Have small groups of students hypothesize as to what the narrator decided to do following her visit with Naomi. Challenge students to design a skit that details what they think might have happened next.

Science and Literature Connection

The Gorilla Signs Love

The following excerpt is from the book The Gorilla Signs Love *by Barbara Brenner*

It was mostly because of my father that I was in Africa that fateful summer. I had won a prize for what now seems like a fairly dopey little high school science project on gorilla communication. But Max acted as if I had invented the wheel. He rewarded me by arranging for me to work for the summer with none other than Dr. Charlotte Wingate. . . .

Like Jane Goodall and Dian Fossey and the other primate researchers, Charlotte had come to Africa to observe apes in the wild. She had been there for two years—long enough to locate a small troop of western lowland gorillas. She was hoping to get some new information on this very rare and endangered species of gorilla, studying them with no attempt to tame them or train them or interact with them in any way. I, on the other hand, was dying to communicate with them. I would have liked to find out more about their "language" and even to try to reach them with mine. But Charlotte had laid out the ground rules during my first five minutes in camp. Trying to get friendly with the gorillas was a no-no. I was there to assist her work and to learn, not to do my own thing. . . .

[During a walk one morning] I sat down to rest and drink the can of soda I'd brought along. By this time it had stopped raining, but I was damp and sticky. My back was itching. I rubbed it against the back of a tree. Felt good. In the middle of my ecstatic back-scratching [a gorilla we called] Naomi appeared.

She came into view as she always did. Quietly, without fanfare. Suddenly she was at my side.

"Hi, baby." It was out before I remembered. No talking.

Naomi squatted next to me, seeming content just to be there. I took another swig of soda. When I glanced over at her I saw that she was watching this operation with great attention.

It occurred to me that she might like some soda. But I did not ask her. I was honoring Charlotte's no-talk rule. . . .

I sat and drank. I made up my mind that if this gorilla really wanted some soda, she was going to have to ask for it.

Impossible? Not for Naomi. All at once she turned those lively brown eyes in my direction. After she'd caught my eye, she gave me the "gimme" sign. . . . Still I waited. It was then she raised her hand to her lips in a perfect imitation of my drinking.

Skills in Science

Reading Skills in Science

1. The narrator is a high school student, most likely between the ages of 15 and 18. She is in Africa. Max is her father.
2. If the team members got friendly with the gorillas, they would influence the behavior of the animals. The purpose of the research is to study the animals and their behavior in a natural environment. As shown by the narrator's encounter with Naomi, communication can happen without words ever being spoken.

Writing Skills in Science

1. Students might describe the excitement the narrator felt when she communicated with Naomi. They might also describe her frustration at not being allowed to "do her own thing" with the gorillas. In addition, the narrator might be concerned with both Charlotte and her father's reaction to her breaking the team's rules.
2. Answers may vary. The narrator assumed that Naomi would find the soda more tasty than liquids natural to the surroundings, such as dirty rainwater.

Activities

Communicate You may wish to have students compare communicating with American Sign Language (gestures that express ideas rather than words) and the American Manual Alphabet (words that are spelled out letter by letter). Have them try communicating simple sentences to one another.

Collect Data Extend the activity by having students determine the role humans have played in reducing the world's gorilla population.

My whole body turned shaky with excitement. . . .

All of Charlotte's rules went out of my head.

"Yes, you want a drink. Sure you do. Okay. Here." I handed Naomi the can of soda. I would have given her anything.

"Drink," I said. As I said it I gave her the official Ameslan (American Sign Language) sign for it—a clenched fist and a thumb to the mouth. It was very close to what Naomi had done on her own.

She took the can and put it to her lips. She drank. Then she handed the can back to me.

Here was my chance to test whether she understood the sign.

"Drink," I said, making the sign again.

She ignored me completely. She reached out and popped a leaf into her mouth. It was as if the drinking sequence had never occurred. . . .

"Drink," I said again to my new friend, making the sign once more.

Still Naomi ignored me. I persisted. Finally she did something so clever that I had to laugh.

She leaned over, gently took the can of soda from my hand and shook it. She was showing me that it was empty.

"Love"   "Drink"

Skills in Science

Reading Skills in Science

1. **Find Context Clues** How old is the narrator? Where did her encounter with Naomi take place? Who is Max?
2. **Infer** Why did Charlotte insist that members of the team not get friendly with the gorillas? Did this rule ensure that there wouldn't be any communication between the humans and the gorillas?

Writing Skills in Science

1. **Detect the Writer's Mood** Imagine you are the narrator of this story. Write a letter to Max describing your experience with Naomi.
2. **Uncover Assumptions** Anthropomorphism is the attributing of human motivation or behavior to animals. Describe some ways the narrator anthropomorphizes Naomi.

Activities

Communicate Visit the school or local library to learn more about American Sign Language. Practice communicating with a partner using Ameslan.

Collect Data The author mentions that the lowland gorilla is an endangered species. Research population estimates of this species as well as steps taken to protect it.

Where to Read More

As Dead As a Dodo by Peter Mayle. Boston: Godine, 1982. Beautiful illustrations accompany the narrative describing sixteen vanished species.

Gorillas in the Mist by Dian Fossey. Boston: Houghton Mifflin, 1983. Famed primate researcher Dian Fossey describes her experience studying gorillas in their natural environment.

UNIT 2

UNIT OVERVIEW

In this unit, students are introduced to cell theory and theories of heredity. In Chapter 4, students learn about the historical development of cell theory. They also learn the makeup of cells, cell processes, and tissues, organs, and organ systems. Chapter 5 describes diffusion, osmosis, and active transport. Also explored are the energy processes of photosynthesis, respiration, and fermentation in organisms. The chapter closes with a detailed description of mitosis and a discussion of changes and growth in organisms. Chapter 6 explains the work of Gregor Mendel and outlines the methods by which scientists trace inherited traits. Students learn about the function of chromosomes and the process of meiosis. The chapter also discusses mutations and introduces the concepts of DNA replication, applied genetics, and genetic engineering.

Introducing the Unit

Directed Inquiry

Have the students examine the photograph. Ask:

▶ Why do you think all the sunflowers have similar characteristics? (Their genetic structure causes them to develop in the same general way.)

▶ What do you think causes the observable differences among the flowers? (Each flower has an individual genetic blueprint that comes from its parents.)

▶ How might variations, such as height, affect the flowers' need for sun and water? (Variations in height and arrangement of leaves and petals might cause different needs for sunlight and water.)

Writing About the Photograph

Ask students to imagine they are walking in this field of daisies and to write a short poem about the sights, smells, sounds, and sensations they have in the field.

Unit 2 Cells and Heredity

Chapters

4 Cells and Living Things

5 Cell Processes

6 Heredity

Data Bank

Use the information on pages 624 to 631 to answer the following questions about topics explored in this unit.

Interpreting Data

Which lives longer, a red blood cell or a skin cell?

Comparing

Which is larger, an animal cell or a frog egg?

Classifying

What things can only be seen with an electron microscope? What can only be seen with a light microscope? What can be seen with both a light and an electron microscope?

The photograph to the left is of sunflowers in an Illinois field. What characteristics do the sunflowers have in common? Do you see any differences among the sunflowers? If so, what are they?

Data Bank Answers

Have students search the Data Bank on pages 624 to 631 for the answers to the questions on this page.

Interpreting Data A red blood cell lives longer than a skin cell. The answer is found in the table, Cell Numbers and Life Spans, on page 626.

Comparing A frog egg is larger than an animal cell. The answer is found on page 625 in the diagram, Relative Sizes of Organisms.

Extension Ask students why frog and chicken eggs are larger than most plant and animal cells. (The eggs contain many cells.)

Classifying Atoms, small molecules, lipids, proteins, ribosomes, and viruses can be seen only with an electron microscope. According to the diagram, there is nothing that can be seen only with a light microscope. However, answers may vary, and students may say that euglena, grains of salt, and frog eggs may be seen more clearly with the aid of a light microscope. Mycoplasma, mitochondria, most bacteria, nuclei, and plant and animal cells can be seen with both an electron and light microscope. These answers are found on page 625 in the diagram, Relative Sizes of Organisms.

Extension Ask students to make a bar chart comparing the maximum sizes of organisms seen through an electron microscope.

Answer to In-Text Question

The sunflowers share the same petal and leaf shape, width of stem, and color. They differ in number and size of petals and leaves, height, and flower development.

CHAPTER 4

Overview

This chapter introduces students to the cell, the basic unit of function in living things. The first section explains the three main ideas about the cell that scientists call cell theory. The second section of this chapter describes the structure of plant and animal cells. The functions of the cell organelles are described also. The chapter concludes with discussion of the larger systems of cells—tissues, organs, and organ systems.

Advance Planner

▶ Provide clear plastic bags and water for use in studying refraction. TE page 76.

▶ Gather prepared slides for TE page 77.

▶ Prepare cellophane, food coloring, test tubes, and rubber bands for TE activity, page 79.

▶ Gather prepared slides of stained plant and animal cells for TE activity, page 80.

▶ Gather distilled water, droppers, clean slides, coverslips, tweezers, an *Elodea* plant, iodine, and flat toothpicks for SE Activity 4, page 85.

Skills Development Chart

Sections	Classify	Communicate	Compare and Contrast	Infer	Interpret Data	Model	Observe	Research
4.1 Skills WarmUp	●							
Skills WorkOut								●
Historical Notebook		●						
4.2 Skills WarmUp			●					
Activity			●		●	●	●	
4.3 Skills WarmUp				●				

Individual Needs

▶ **Limited English Proficiency Students** Cover the labels and photocopy the figures on pages 79, 80, 82, and 83. Draw arrows to the important parts of each figure. Have students make labels from strips of index cards and place the labels in the correct place on the photograph. Have students refer to the text to check the placement of the labels. Students can save the labels in an envelope so they can review the terms by arranging them again after they've studied the chapter.

▶ **At-Risk Students** As an extension of the analogy presented in the chapter, have students construct a map of a town. As students read the chapter, have them draw and note the parts of a town as the cell parts are described in the chapter. Students should include roads, factories, the mayor's office, schools, and houses. As students label the different parts of their towns, ask them also to write the comparable cell parts in parentheses. For example, when students label a road, have them write *cell membrane* in parentheses.

▶ **Gifted Students** Have students produce a news program announcing a fourth statement for the cell theory (see page 77). Students should use what they know about cells to come up with a reasonable new statement. For example, someone might have been able to manufacture a cell in a laboratory from nonliving substances. Students can videotape the program and show it to the class.

Resource Bank

▶ **Bulletin Board** Place outlines of a plant cell and an animal cell on the bulletin board. (See Figures 4.4 and 4.5.) Have students add labels to the bulletin board identifying the parts of the cell. As you work through Section 4.2, have students add drawings or photocopies that show enlargements of the cell organelles. Have them make labels for the organelles that describe how each structure in the cell functions.

CHAPTER 4 PLANNING GUIDE

Section	Core	Standard	Enriched	Section	Core	Standard	Enriched
4.1 The Cell Theory pp. 75–77				**Overhead Blackline Transparencies** Overhead Blackline Master 4.2 and Student Worksheet	●	●	
Section Features Skills WarmUp, p. 75	●	●	●				
Skills WorkOut, p. 76	●	●		**Laboratory Program** Investigation 8	●	●	●
Historical Notebook, p. 76		●	●				
Blackline Masters Review Worksheet 4.1	●	●		**Color Transparencies** Transparencies 9, 10, 11, 12	●	●	
Skills Worksheet 4.1	●	●		**4.3 Organization of Living Things** pp. 86–88			
Integrating Worksheet 4.1	●	●					
4.2 Parts of a Cell pp. 78–85				**Section Features** Skills WarmUp, p. 86	●	●	●
Section Features Skills WarmUp, p. 78	●	●	●	**Blackline Masters** Review Worksheet 4.3	●	●	
Career Corner, p. 81	●	●	●	Vocabulary Worksheet 4	●	●	
Activity, p. 85	●	●	●				
Blackline Masters Review Worksheet 4.2	●	●		**Ancillary Options** One-Minute Readings, p. 41	●	●	●
Reteach Worksheet 4.2	●	●	●				
Integrating Worksheet 4.2	●	●		**Laboratory Program** Investigation 9	●	●	●
Ancillary Options Multicultural Perspectives, p. 89		●	●				

Bibliography

The following resources can be used for teaching the chapter. See page T-40 for supplier codes.

Audio-Visual Sources
(video unless noted)
Cell Biology. 17 min film. 1981. C/MTI.
Learning About Cells. 16 min film. 1976. EB.
Microscope: Making It Big. 22 min film. 1981. FL.
Notes of a Biology Watcher: A Film with Lewis Thomas. 57 min. 1982. T-L.

Software Resources
The Simplest Living Things. Apple. DCHES.
Your Body—Series II. Apple, TRS-80, Commodore. FM.

Library Resources
de Duve, Christian A. A Guided Tour of the Living Cells, Vols. 1 & 2. New York: Scientific American Books, 1985.
Elting, Mary. The Macmillan Book of the Human Body. New York: Aladdin, 1986.
Fichter, George S. Cells. New York: Franklin Watts, 1986.
Thomas, Lewis. The Lives of a Cell: Notes of a Biology Watcher. New York: The Viking Press, Inc., 1974.

CHAPTER 4

INTEGRATED LEARNING

Math Connection
Have students count the number of cells in the photograph. It might help to have hand lenses available. Have students measure one of the cells with a metric ruler and estimate the magnification of the photograph. The Data Bank on page 625 will help them estimate the actual size of a cell.

Themes in Science
Scale and Structure All organisms, even the most complex, are made of cells. Similarly, all novels are made of words. Discuss this analogy with your class. Encourage students to consider how paragraphs, sentences, and words are like systems, organs, and cells in an organism.

Introducing the Chapter
Have students read the description of the photograph on page 74. Ask if they agree or disagree with the description. The photograph shows circular clusters of bone cells. The cells have been dyed red. There are no cell walls. The actual cells are the small, spidery spots in the round mineral matrix.

Directed Inquiry
Have students study the photograph. Ask:

▶ How many cells do you see in the photograph? (Students will say there are seven. Point out the actual cells.)

▶ How would you describe the arrangement of the cells? (Shaped like bundles of circles)

▶ How can nutrients and other materials get to and from the cells? (Answers may vary. The dark central circles are blood vessels.)

▶ The rest of each cluster is mostly minerals. How might this cell structure and arrangement help bones? (Accept all reasonable responses. Some students may relate a bone's cell structure to its support function.)

Chapter Vocabulary

cell	cytoplasm
cell membrane	nucleus
cell theory	organ
cell wall	organelle
chlorophyll	tissue
chloroplast	

Chapter 4 — Cells and Living Things

Chapter Sections
4.1 The Cell Theory
4.2 Parts of a Cell
4.3 Organization of Living Things

What do you see?

"This picture looks like a group of cells. It looks like there are cell walls. The nucleus is glistening or there is light shining through the nucleus. They may be animal cells. The cells look like they have been dyed red."

Christa Haynes
Summitt Middle School
Edmond, Oklahoma

To find out about the photograph, look on page 90. As you read this chapter, you will learn about different cells and the important life processes they perform.

TEACHING OPTIONS

Portfolio
Have students create a timeline that depicts the development of the cell theory. Have them keep their timelines in their student portfolios. Have them expand their diagrams after they read the Historical Notebook, and again after they read page 77.

4.1 The Cell Theory

Objectives
- **Describe** the historical development of the cell theory.
- **State** the major ideas of the cell theory.
- **Make analogies** between everyday objects and cells.

Skills WarmUp

Classify All things can be classified as either living, once living, or nonliving. Classify each of these things as living, once living, or nonliving: an autumn leaf, a cotton shirt, a tulip, a steel beam, a car, a dog, string beans, and a cow. How do you know what things are living?

Each brick in a brick house is a unit of structure for the house. In living things, the basic unit of structure is the **cell**. All living things or things that were once living are made up of one or more cells. Though you can see the bricks in a brick house, most cells are hard to see with the unaided eye. The cell is also the basic unit of function in all living things. Each cell carries out all life processes.

Development of the Cell Theory

Ideas about cells are taken for granted by most life scientists today. However, 350 years ago, no one even guessed that living things were made of units so small they couldn't be seen.

The first steps in developing modern ideas about cells were made in the late 1600s. In 1675, a Dutch shopkeeper named Anton van Leeuwenhoek (LAY vehn hook) looked at pond water through a simple microscope. He observed what he called "animalcules," or what are known today as single-celled organisms. Without knowing it, van Leeuwenhoek was probably the first person to see living cells.

Around the same time, Robert Hooke, an English scientist and inventor, used a crude compound microscope to look at a very thin slice of cork. Hooke observed that the cork was made up of "a great many little boxes." Hooke called these boxes *cells,* which comes from a Latin word meaning "little rooms." Hooke's name stuck. However, Hooke was not looking at living cells. He was actually looking at the walls of dead plant cells.

Almost 200 years passed before the importance of Hooke's and van Leeuwenhoek's observations was fully

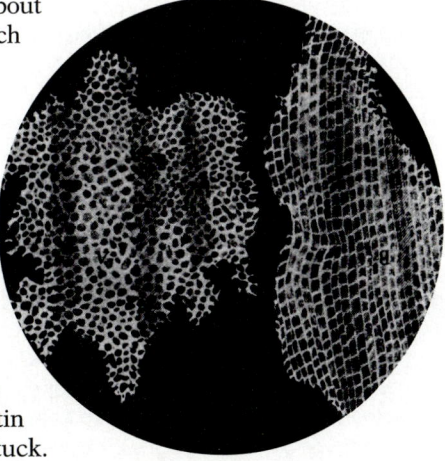

Figure 4.1 ▲
Robert Hooke made this drawing of cork cells using a compound microscope that he built.

SECTION 4.1

Section Objectives
For a list of section objectives, see the Student Edition page.

Skills Objectives
Students should be able to:

Classify things as living, once living, and nonliving.

Research the history of microscopes.

Make Analogies that refer to the characteristics of a cell.

Vocabulary
cell, cell theory

MOTIVATE

Skills WarmUp
In order to evaluate students' prior knowledge about the characteristics of living things, have them do the Skills WarmUp.
Answer Living things—dog and cow; nonliving things—steel beam and car; once-living things—autumn leaf and cotton shirt. It would be reasonable to list the string beans and tulip as living or once living. If necessary, have students review the characteristics of living things, Chapter 2.

Prior Knowledge
To gauge how much students know about the cellular basis of life, hold a brainstorming session. Ask the following questions:
- Where are cells found?
- How many cells are there in a single human being?
- What do you think might be inside cells?

Integrated Learning
Use Integrating Worksheet 4.1.

TEACH

Skills WorkOut
To help students understand more about the development of the microscope, have them do the Skills WorkOut.
Answer With the high-powered lenses of single-lens microscopes, both the object to be viewed and the viewer's eye nearly had to touch the lens. The solution involved making compound microscopes, which have a lens at each end of a tube. Later, German scientists Ernst Abbe and Carl Zeiss further refined the focusing quality of lenses, leading to the production of high-quality optical microscopes.

Historical Notebook
Activity Have students look at the photograph on page 74 through a hand lens. Then have them move a second lens under the first, keeping the second lens close to the page. Have students observe how the appearance of the image on the page changes as they vary the distance between the two lenses.

Answers
1. Hooke's microscope enabled people to see things they hadn't seen before.
2. If possible, have students observe some real cells under a microscope before they begin to write.

INTEGRATED LEARNING

Integrating the Sciences
Physical Science Microscopes and magnifying glasses rely on the principle of refraction. To illustrate refraction, have students make their own water lenses. To make a lens, fill a clear plastic bag with 2 cm of water and tightly seal the bag with a knot. Have students look at a page of newsprint through their water lenses and describe what they see.

Multicultural Perspectives
Italian physiologist Marcello Malpighi (1628–1694) made several important scientific discoveries by examining cells and tissues under a microscope. Have students research and report about Malpighi's contributions to biology, including the study of blood, embryos, and glands. Have students add an entry about Malpighi to their timelines.

Skills WorkOut
Research Find out more information about some of the advances that have led to today's scientific optical microscopes. What problems did microscope makers encounter? Look up the work of Ernst Abbe and Carl Zeiss. What were their contributions? Write a short report on your findings.

understood. During that time, new and better microscopes were developed. In the 1830s, a German botanist named Matthias Schleiden noticed that all the plants he observed under a microscope seemed to be made of tiny units. In 1838, he stated that all plants were made up of similar units, or cells.

Other scientists began to discover that cells were not only found in plants, but in other living things as well. Another German scientist, Theodor Schwann, observed animal tissue and stated that cells were the building blocks of both plants *and* animals. Schwann also pointed out many similarities between plant and animal cells. He noted that there were many different kinds of cells, each with a different function.

In 1858, German physician Rudolf Virchow made an important conclusion about the cells Schleiden and Schwann had seen in all living things. Virchow stated that cells are produced only by other living cells.

Historical Notebook

Looking at Lenses

Magnifying lenses have been used for more than 3,000 years. However, Anton van Leeuwenhoek was the first to use the type of glass lenses that are used today. Van Leeuwenhoek ground his own lenses as a hobby and made his own simple microscopes. These microscopes used only one lens, not multiple lenses as in compound microscopes. He viewed many things with his simple microscopes, including pond microorganisms and even the scum on his teeth. Some of his lenses could magnify up to 270 times.

Robert Hooke was one of the first inventors to successfully use a compound microscope by combining two or more lenses. He placed two lenses in a metal tube, one on each end. The photograph on the right shows a copy of Hooke's microscope. Although Hooke's microscope could magnify more than 100x, it was difficult to observe many details of living cells.

In the early 1800s, better glass-making methods produced lenses that revealed more details in a cell. Since that time, microscope designers have constructed some microscopes that can even show a three-dimensional view of a cell.

1. Why was Hooke's microscope an important development?
2. **Write** Imagine what it would be like to be one of the first people to discover cells through your own microscope. Write a description of your discovery. Remember that you do not yet know what a cell is.

Chapter 4 Cells and Living Things

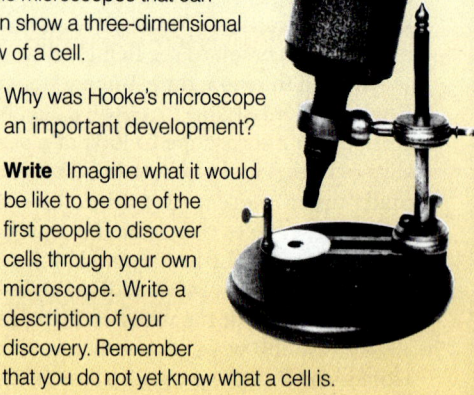

TEACHING OPTIONS

Cooperative Learning

To demonstrate the impact of new technologies on research, have the class work in groups of three and give each group a prepared slide. One member of each group should inspect the slide with unaided eyes. Another should inspect it with a hand lens. The third should inspect it through a microscope. They can then draw a composite picture of what they saw, with the first student drawing the outline and the others adding more and more detail. Afterward, allow time so that everyone has a chance to use a hand lens and a microscope.

These ideas about the cell have been put together into a set of statements called the **cell theory**. The cell theory states that

▶ All living things are made up of one or more cells.

▶ Cells are the basic units of structure and function in living things.

▶ All cells come only from other living cells.

Science and Technology *Even a Closer Look*

Since the time of van Leeuwenhoek, one of the greatest challenges for scientists has been to find a way to clearly view the parts of a cell. Another challenge has been to observe a living cell without killing it.

In a compound-light microscope, many cell parts appear larger, but not always more detailed. Details of a cell's structure are best seen in a transmission electron microscope, or TEM. A TEM can magnify an object up to a million times. However, a TEM uses beams of high-speed electrons. The beams create a vacuum. No living cells can survive in this vacuum. TEMs, therefore, can't be used to observe living cells.

In their search for a more detailed view of living cells, scientists are developing new, improved microscopes. At Cornell University, researchers have developed a microscope called the *superoptical microscope*. Lenses aren't used in a superoptical microscope. Instead, light is conducted through optical fibers in a tube. The light travels along the optical fibers to devices that produce a picture on a large screen. It's almost like watching living cells on television!

Figure 4.2 ▲
Superoptical microscopes do not damage the cells or cell parts being viewed. Scientists can use these microscopes to study living cells.

Check and Explain

1. Which scientist gave us the name "cell"?
2. What are the three parts of the cell theory?
3. **Reason and Conclude** Why is a cell considered the smallest thing that can be called living?
4. **Make Analogies** The beginning of this section describes cells as being similar to bricks in a building. What other things are like cells? Make a list. For each item, explain how it is similar to a cell.

Chapter 4 Cells and Living Things 77

Discuss

Students may ask whether microscopes have been developed that allow us to see atoms. A device used to observe atoms would not work like an optical microscope. Reflected light will not form an image of objects smaller than one-half the wavelength of visible light (such as an atom).

EVALUATE

WrapUp

Portfolio Have students expand their timelines by adding information about TEMs and superoptical microscopes. Have them write one or two sentences describing the importance of the events they included on their timelines.
Use Review Worksheet 4.1.

Check and Explain

1. Robert Hooke
2. Refer to the bulleted list on this page.
3. Each cell carries out all life processes.
4. Answers may vary, but analogies should indicate students' understanding of the cell as a structural unit, a functional unit, or both.

The Living Textbook:
Life Science Sides 7-8

Chapter 3   Frame 00388
Cell Ultrastructure (10 Frames)
Search: Step:

The Living Textbook:
Life Science Sides 7-8

Chapter 3   Frame 00407
Cellular Structure (5 Frames)
Search: Step:

SECTION 4.2

Section Objectives
For a list of section objectives, see the Student Edition page.

Skills Objectives
Students should be able to:

Compare and Contrast plants and animals.

Organize Data about the function and structure of the cell.

Observe plant and animal cells in an experiment.

Vocabulary
cell membrane, cell wall, nucleus, cytoplasm, organelle, chloroplast, chlorophyll

MOTIVATE

Skills WarmUp
To prepare students for the comparison of plants and animals at the cellular level, have them do the Skills WarmUp.
Answer Similarities include the ability to grow, reproduce, and use energy. Differences include characteristics such as appearance, mobility, and ways of getting energy. Few students will list cellular characteristics. Explain to students that many of the differences between plants and animals are at the cellular level.

Misconceptions
Students often assume that cell functions are simpler than those at a higher level of organization. Stress to students that each individual cell carries out all life processes and that each part of the cell has a special function.

INTEGRATED LEARNING

Themes in Science
Systems and Interactions The parts of a cell must work together to carry out all necessary life processes. Using the analogy of a city, have students discuss how the parts of the city interact and how the whole city system could gradually break down if only one part stopped working.

Skills WarmUp

Compare and Contrast
Make a list of five similarities between plants and animals. Make another list of five differences between plants and animals. Discuss your lists with several classmates. Did you write the same similarities and differences?

4.2 Parts of a Cell

Objectives
▶ **Describe** the structure of a cell.
▶ **Explain** the functions of a cell's organelles.
▶ **Compare** and **contrast** plant and animal cells.
▶ **Construct a table** that shows an understanding of cell functions and structure.

Imagine a small town or city. What are some of its parts? There are roads, factories, schools, and houses. There are power lines and telephone lines. Many different people work in the town, supplying needed services to the town and to surrounding areas. Each person has a special job to do. Each building has a special use. All the people, services, buildings, and other structures work together to make the town function properly.

A cell is like a small town or city. The different parts of a cell have special jobs. Each part helps the cell carry out its life processes. Each part helps to keep the cell working properly. Just as the parts of a town must work together, the parts of a cell also must work together to live.

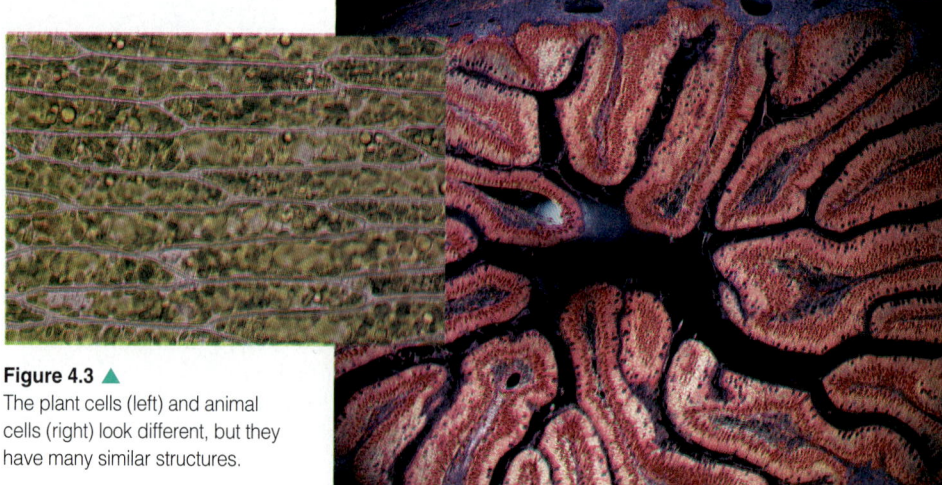

Figure 4.3 ▲
The plant cells (left) and animal cells (right) look different, but they have many similar structures.

Chapter 4 Cells and Living Things

Integrating the Sciences

Nutrition Fiber, an important part of a nutritious diet, comes from cell walls. Have students refer to page 489 for a discussion on fiber. Make a list of high-fiber foods on the chalkboard. Ask students if any of the foods come from animals. Have them relate this information to what they've learned about the differences between plant cells and animal cells.

Figure 4.4 Plant Cell

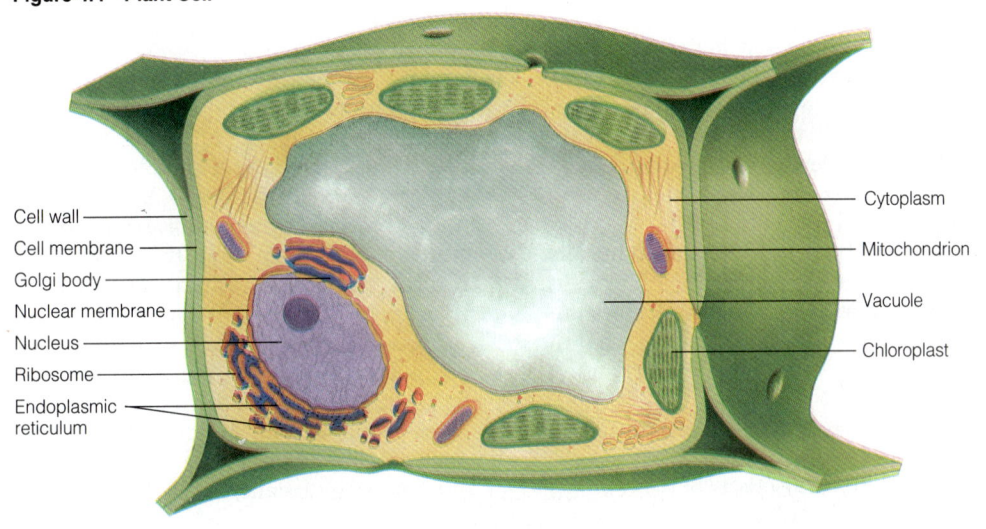

Labels: Cell wall, Cell membrane, Golgi body, Nuclear membrane, Nucleus, Ribosome, Endoplasmic reticulum, Cytoplasm, Mitochondrion, Vacuole, Chloroplast

Cell Membrane

On a map, the boundaries of each town are marked with a heavy line. Driving from town to town, you see signs that show the edge of one town and the beginning of the next.

All cells have boundaries, too. Look at the plant cell in Figure 4.4 and find the **cell membrane**. A cell membrane is a thin structure that surrounds both plant and animal cells. It is mostly made up of proteins and lipids.

The cell membrane has several functions. It protects the inside of the cell by separating the cell from its surroundings. Cell membranes also support the cell and give it shape.

The most important job of the cell membrane is controlling what enters and leaves the cell. All things that a cell needs, such as water and nutrients, enter through the cell membrane. Waste products exit through the cell membrane. The cell membrane controls this traffic. It prevents harmful substances from entering and keeps useful substances inside.

In this way, the cell maintains a stable internal environment even though conditions outside the cell may change. The ability of a cell to maintain a stable internal environment is called homeostasis (hoh mee oh STAY sihs). The cell membrane is the part of the cell that makes homeostasis possible.

Cell Wall

Plant cells have an additional structure outside the cell membrane called the **cell wall**. The cell wall is one of the characteristics that separates plant cells from animal cells.

The cell wall is stiff. It provides support and protection for the cell, allowing the plant to grow upright. The cell wall is partly made up of fibers of cellulose (SEHL yoo lohs). Cellulose is a complex compound formed from sugar molecules.

Chapter 4 Cells and Living Things

TEACH

Class Activity

Have students work in groups of three to observe the movement of materials through cell membranes. Have students place several drops of food coloring in a test tube that is about half-full of water. Next, have them secure a strip of cellophane (available as dialysis tubing) over the top of the test tube with a rubber band. Explain to students that cellophane is chemically like plant membranes. Have them place the tube upside down in a jar of water. Students should observe their jars the next day and record their observations.

Discuss

Stress to students that plants and animals differ in structure. Have students discuss how it would affect an animal, such as a cheetah, to have stiff cell walls. Ask them if the cell walls would affect the cheetah's ability to move. (Yes, it would not be able to move as easily or quickly if it had cell walls.)

**The Living Textbook:
Life Science Sides 7-8**

Chapter 3   Frame 00349
Plant Cells: Elodea Leaf (1 Frame)
Search:

TEACH • Continued

Class Activity
Have students look at prepared, stained cells of plants and animals. Have them identify the nucleus.

Directed Inquiry
Have students study Figures 4.4 and 4.5. Ask them some or all of the following questions:

▶ How do you think the shape of the plant cell shown might help the plant? (This kind of shape gives the plant some support, particularly if a number of such cells are arranged in columns.)

▶ How do you think plant cells can be distinguished from animal cells? (Plant cells contain chloroplasts and have cell walls. Animal cells do not have chloroplasts or cell walls.)

▶ Which kind of cell is more likely to become altered in shape? Why? (The animal cell; it only has a thin cell membrane as its outer boundary.)

▶ To what large structure in a single cell can you compare your brain? Why? (Answers may vary, but the nucleus is a likely answer; it is the control center for most of the cell's activities.)

The Living Textbook:
Life Science Side 7

Chapter 35   Frame 37263
Structure of DNA (Animation)
Search: Play:

INTEGRATED LEARNING

Integrating the Sciences
Physical Science The nucleus is the name given to the organelle that controls most of the cell's activities. The term nucleus is also used in physical science. Have students look at Figure 3.3 on page 53 and Figure 4.5 on this page. Ask them to compare the nucleus of a cell with the nucleus of an atom.

Language Arts Connection
Students have read how a cell is like a town or city. Have them use another metaphor to show the way a cell functions. Have them write an essay explaining how their metaphor describes the way the cell wall, cell membrane, and nucleus work together.

Nucleus

In a town, the mayor and city council make important decisions about the workings of the town. They govern the town. In a cell, the **nucleus** governs the cell. The nucleus is the control center for most of the cell's activities. It also controls cell reproduction.

Recall that molecules called nucleic acids carry instructions for cell processes and reproduction. These molecules are stored in the nucleus. They are what the nucleus uses to control the workings of the cell.

The instructions carried on nucleic acid are formed by a type of chemical code. This code tells the other parts of the cell how to perform their functions. Different parts of the DNA's code are turned "on" or "off" to control cell processes.

Most cells have a nucleus surrounded by a membrane called the nuclear membrane. It separates the nucleus from the rest of the cell. Bacteria and blue-green bacteria have a nuclear area. This area is not surrounded by a membrane. In cells with a nuclear membrane, the membrane controls the substances that enter and leave the nucleus.

You can see the nucleus of a cell with a compound microscope. It is easy to identify. The nucleus is usually round or oval shaped, and near the middle of a cell. The nucleus also appears darker than the rest of a cell. Find the nucleus in Figure 4.5.

Figure 4.5 Animal Cell

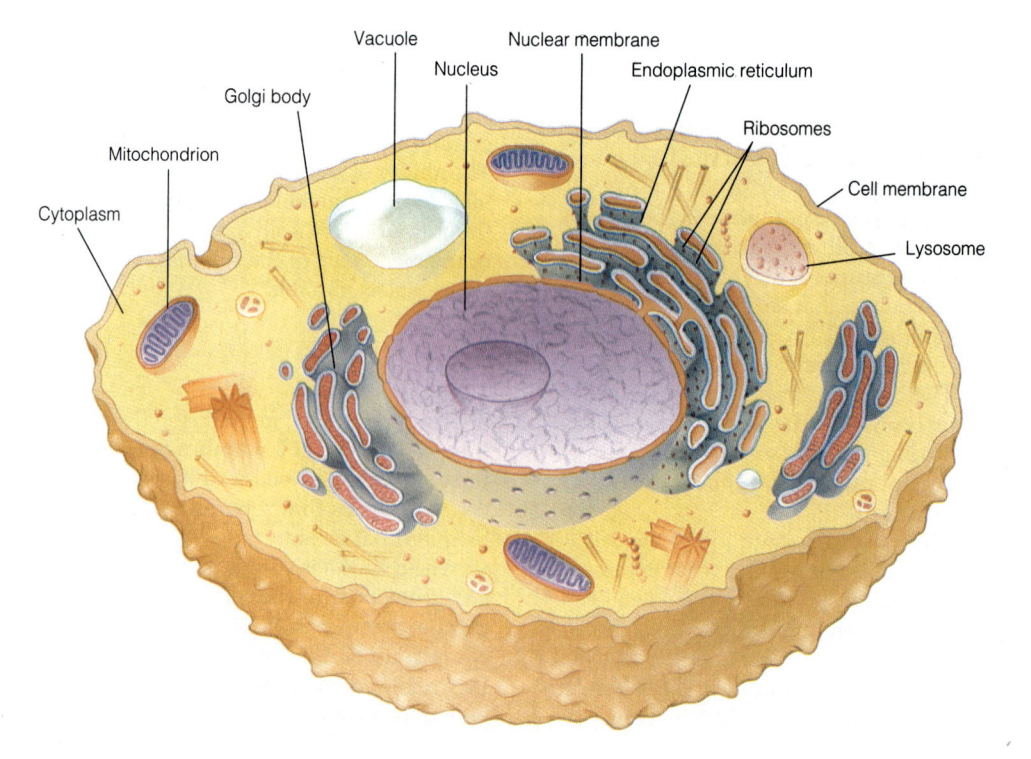

Labels: Cytoplasm, Mitochondrion, Golgi body, Vacuole, Nucleus, Nuclear membrane, Endoplasmic reticulum, Ribosomes, Cell membrane, Lysosome

80   Chapter 4   Cells and Living Things

Multicultural Perspectives

Charles Richard Drew (1904–1950), an African-American surgeon and medical researcher, discovered that blood taken from donors could be separated into red blood cells, which spoil quickly, and plasma, which doesn't. He found that patients suffering from blood loss would recover whether given plasma or whole blood. Ask students what effect they think this discovery had on the practice of blood transfusion, and how it might affect blood storage and transportation. Have students look up blood and plasma in the index of this book.

Cytoplasm

Most of the cell's processes take place in the **cytoplasm**. The cytoplasm is all the living material in a cell except the nucleus. The prefix *cyto-* means "of the cell." *Plasm* comes from the Greek word meaning "something molded." The cytoplasm is a jellylike substance that contains many different chemical compounds. Just like other living things, most of the cytoplasm is water. In fact, it is about 80 percent water. The nucleus and other special tiny parts called **organelles** float in the cytoplasm. Each organelle in a cell has a special form and function. Organelles carry out life processes such as releasing energy, making proteins, and storing food and water.

Although the nucleus controls the growth of a cell, most of the materials needed for growth are made in the cytoplasm. The cytoplasm expels waste materials through the cell membrane. In addition, the organelles in the cytoplasm carry out their functions here.

Career Corner Health Technician

Who Tests Your Blood?

Have you ever had your blood tested? If you have, you may not realize that a health technician, not a doctor, tests your blood. The health technician performs a number of different tests on the blood, such as finding the number of blood cells, typing your blood, and checking for unusual chemicals. These tests will tell a doctor whether or not your blood is healthy.

Health technicians perform almost everything that needs to be done in a lab. The actual job the technician does depends on the type of lab he or she works in. For example, *biological technicians* conduct research, such as studying and analyzing cells and microscopic organisms, to help biologists develop cures for human diseases. The technician shown in the photograph is an *optical technician*. Optical technicians make eyeglasses and contact lenses. And the technician at your doctor's office is a *medical-laboratory technician*. These technicians not only look at your blood, they also do such things as prepare solutions, keep records of tests, and clean and sterilize equipment.

If you are interested in being a health technician, you will need two years of training after high school. You may also need some college courses in math and science. Some special skills are learned on the job or through apprenticeships.

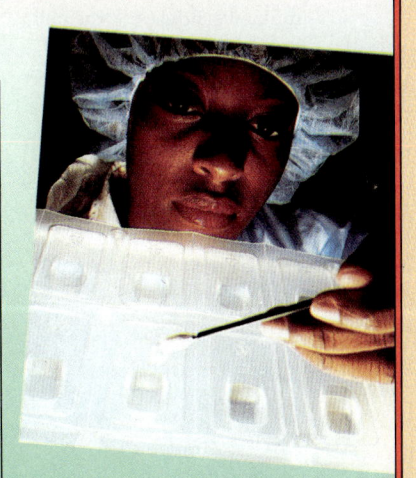

Chapter 4 Cells and Living Things

Reteach

After students have read about cytoplasm, have them fold a sheet of paper lengthwise to form two columns. On one side of the paper, have students list all the substances and structures that are contained in the cytoplasm. Opposite each structure or substance, have them list its function.

Use Reteach Worksheet 4.2.

Discuss

Stress to students that the nucleus is an organelle.

Integrated Learning

Use Integrating Worksheet 4.2.

Career Corner

Research Have students find out about other careers in health, such as dietetics, X-ray technology, physical therapy, dental hygiene, and so on. Have them give a brief oral report to the class about the kind of work done by these technicians and the training required.

Ancillary Options

If you are using the blackline masters from *Multiculturalism in Mathematics, Science, and Technology*, have students read about Ernest Just and complete pages 90 to 94.

The Living Textbook: Life Science Side 2

Chapter 10   Frame 03301
Cytoplasmic Streaming (Movie)
Search: Play: Step:

The Living Textbook: Life Science Side 2

Chapter 13   Frame 14803
Living Blood Cells (Movie)
Search: Play: Step:

TEACH • Continued

Critical Thinking

Reason and Conclude Have students explain why or why not the location of the nucleus or any organelle in the cell can be pinpointed. (Since the nucleus and organelles float in the cytoplasm, these structures are not fixed in a particular location.)

Reteach
Use Reteach Worksheet 4.2.

Explore Visually
Have students study the drawings and micrographs of the organelles in a cell on pages 82 and 83. Then ask the following questions:

▶ What are the oval-shaped, glucose-making organelles that are found in many plant cells? (Chloroplasts)

▶ Describe the structure of mitochondria. (Mitochondria have an outer membrane and a folded inner membrane.)

▶ What do mitochondria do for the cell? (They release the energy stored in nutrients.)

▶ How does the structure of endoplasmic reticulum help it perform its function in the cell? (The folded network of tubes works to transport materials throughout the cell.)

▶ Where are proteins made? (In the ribosomes)

▶ Describe the structure of the Golgi bodies. (They look like stacks of flattened sacs.)

▶ Which organelles contain powerful digestive enzymes? (Lysosomes)

▶ What materials might be found inside a vacuole? (Fluids, water, food materials, and wastes)

INTEGRATED LEARNING

Writing Connection
Making use of the analogy between cell organelles and the services in a town, have students choose one organelle and write up a bill for that organelle's services. Address this to another part of the cell.

Organelles

The organelles in a cell are like the different services in a town. In a town, each service provides a special job. Trash is collected and removed from the town. Communication is provided by the telephone company. The electric and gas company supplies energy to all the buildings in the town. In a cell, each organelle has a special job to do. Together, the organelles keep the cell working properly.

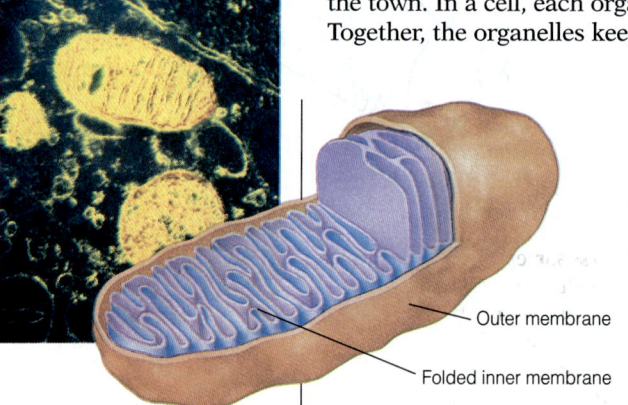

Mitochondria ▲
The structures that are the powerhouses of the cell are the mitochondria (myt uh KAHN dree uh). You can compare the mitochondria to a power plant in a town. Both structures provide energy. Mitochondria release the energy stored in food. This energy can then be used by the cell to carry out its life processes.

Ribosomes
The tiny, round dark organelles in a cell are ribosomes. Some ribosomes are attached to the endoplasmic reticulum. Others float freely in the cytoplasm. Ribosomes are the protein factories of cells. They assemble proteins, which the cell uses for growth, repair, and control. ▶

Endoplasmic Reticulum
A town has roads and highways. A cell has the endoplasmic (ehn duh PLAZ mihk) reticulum (rih TIHK yuh luhm), or ER, for short. The ER is made up of membranes. The membranes fold into a network of tubes and canals that run throughout the cytoplasm. Just as the roads and highways of a town are used to transport materials from one place to another, the ER transports materials throughout a cell. ▼

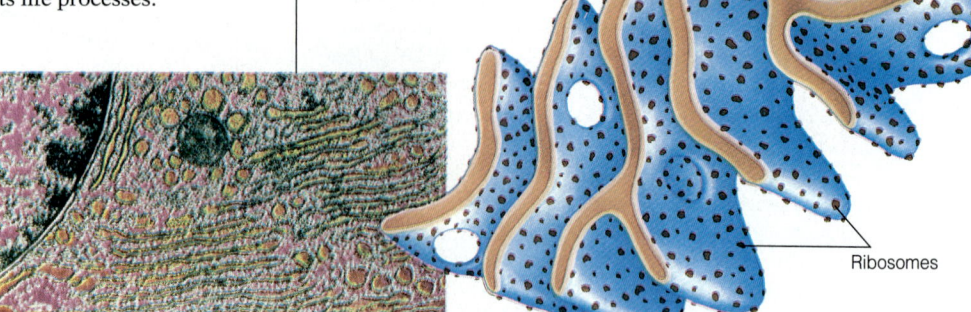

82 Chapter 4 Cells and Living Things

STS Connection

Research Explain that researchers have been able to chart the movement of substances through the Golgi bodies by using a technique called radioautography. Have students research and write a brief report describing this procedure and explaining its usefulness.

Themes in Science

Energy Chloroplasts and mitochondria are involved in cellular energy processes. Have students look up photosynthesis and respiration in the index of this book. Stress that chloroplasts and mitochondria are the sites of these processes.

▶ Point out that both the mitochondria and chloroplasts are involved somehow with energy. Ask students to distinguish between the roles played by the two organelles.

Golgi Bodies ▶

Both animal and plant cells have Golgi (GOHL jee) bodies. Golgi bodies look like stacks of flattened sacs. You can think of a Golgi body as a center of manufacturing and shipping within the cell. Materials that are transported by the endoplasmic reticulum usually stop first at Golgi bodies, where they are altered or stored before moving to other parts of a cell.

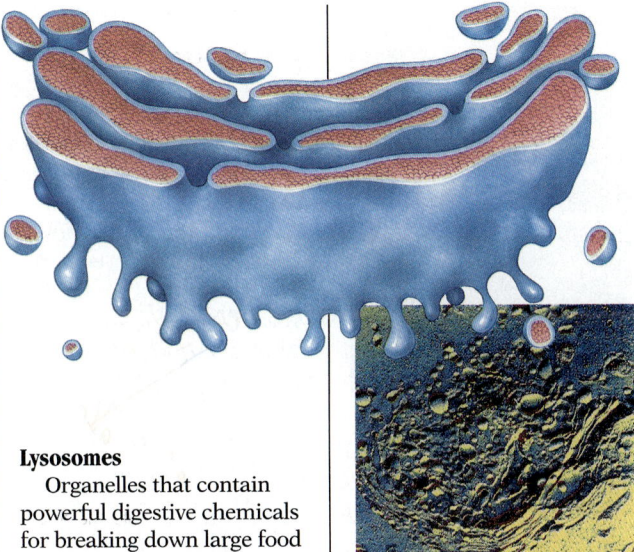

Lysosomes

Organelles that contain powerful digestive chemicals for breaking down large food molecules into smaller food molecules are called lysosomes. Lysosomes also break down waste products and old cell parts. You can think of them as the "garbage collectors" of the cell. These small, baglike structures are found in animals and a few plants.

Vacuoles

The fluid-filled sacs that float in the cytoplasm are called vacuoles (VAK yoo ohls). Vacuoles store water and food materials. They also store wastes and help the cell get rid of the wastes. Both animal and plant cells have vacuoles.

Chloroplasts

Plant cells have green, oval-shaped organelles called **chloroplasts**. Plants are the only organisms that have these structures. The chloroplast contains a compound called **chlorophyll**. Chlorophyll traps energy from the sun to make glucose. Chlorophyll also gives the chloroplast its green color.

The Living Textbook: Life Science Side 1

Chapter 19   Frame 38599
Cell Locomotion and Organelles (Movie)
Search: Play: Step:

The Living Textbook: Life Science Sides 7-8

Chapter 3   Frame 00349
Elodea: Chloroplasts (1 Frame)
Search:

Chapter 4 Cells and Living Things

TEACH • Continued

Discuss
If possible, invite a technician from a local burn center to speak with the class. Have students prepare questions about treatments for burn patients.

EVALUATE

WrapUp
Reinforce Have students draw a diagram of a town. Encourage them to include all the structures and services that help a town run smoothly. Then have them label their town diagrams with the names of the organelles that provide similar functions in a cell.
Use Review Worksheet 4.2.

Check and Explain
1. Answers should include the names and brief descriptions of five of the seven types of organelles.
2. Lettuce cells all have a cell wall and chloroplasts; mouse cells have neither. Mouse cells contain lysosomes; lettuce cells do not.
3. a. Endoplasmic reticulum—highway, because its function is to provide transportation pathways; b. Vacuole—storage warehouse, because its main function is storage; c. Cell wall—nutshell, because its main function is support and protection; d. Cell membrane—skin, because it partly protects and partly controls the traffic of substances into and out of the cell.
4. Students' tables should distinguish between name and function for each structure shown on pages 79 and 80.

INTEGRATED LEARNING

Writing Connection
Have students imagine themselves as 5 square cm of skin on a human hand. Have students write a creative one-page description of what might happen to the skin within the course of a day.

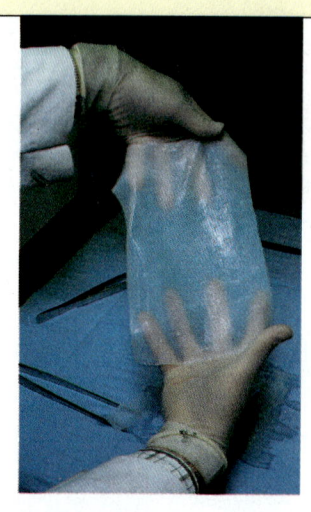

Figure 4.6 ▲
Artificial skin combined with living skin cells from a patient's own body can be used to help burn patients recover. This type of "bandage" helps prevent scars from forming.

Science and Technology *A Living Bandage*
If you have ever burned your skin, you know that it is very painful. Serious burns can also be very dangerous. To prevent infection and help speed up healing, physicians may cover serious burns with a special bandage. Unlike the bandages you use at home, this bandage is made of living skin cells!

The "skin" that forms the bandage closely resembles human skin. However, it is actually a manufactured combination of living human skin cells and proteins from cows. The skin cells are grown in a laboratory from a small portion of healthy skin. This process is known as *culturing,* or growing, cells.

To grow skin cells in the laboratory, scientists must provide the cells with the same environment as a normal skin cell. In the human body, blood provides skin cells with food, oxygen, special chemicals, and a way to remove wastes. In the laboratory, portions of healthy skin are removed from the body and placed in germ-free bags. These bags are filled with nutrients and chemicals. Like blood, this mixture meets the needs of the cells so that they can grow and reproduce.

When enough cells have grown and reproduced, they are removed from the bags. Doctors use the cultured skin cells in surgery. The cells are also used to test new drugs and surgical techniques.

Check and Explain

1. Name and describe five organelles found in the cytoplasm.
2. What organelles could you find in lettuce cells but not mouse cells? What organelles could you find in mouse cells but not lettuce cells?
3. **Reason by Analogy** Match the cell part to the object that shows the best analogy. Explain.

   a. Endoplasmic reticulum   Skin
   b. Vacuole   Nut shell
   c. Cell wall   Storage warehouse
   d. Cell membrane   Highway

4. **Organize Data** Create a table that gives the name and function of each organelle of the cell.

84   Chapter 4 Cells and Living Things

TEACHING OPTIONS

Prelab Discussion

Have students read the activity before beginning the discussion. Encourage students to look at as many cells as possible, because all of the structures may not be visible in any given cell. Mention that it takes only a very light scrape to collect the cheek cell sample. Begin a discussion of staining by asking about the reason for adding iodine to the cheek cell smear. Some students may observe a circulating motion of chloroplasts (caused by cytoplasmic streaming). Have students hypothesize why this might occur.

If you use glycerin in the Extension activity, explain that it causes moving organisms to slow down.

ACTIVITY 4

Time 45 minutes **Group** pairs
Materials
distilled water
60 microscope slides
coverslips
tweezers
droppers
flat toothpicks
iodine solution
Elodea plant

Analysis
1. nucleus, cell membrane, cytoplasm
2. cell wall, chloroplasts, vacuole
3. nucleus

Conclusion
Accept any logical answer. Students should, however, point out the most obvious differences: shape, color, chloroplasts, cell wall.

Extension
Suggest that students use a small amount of glycerin to slow down any fast-moving organisms present in the water.

Activity 4 How do animal cells and plant cells differ?

Skills Observe; Compare and Contrast; Make a Model; Interpret Data

Task 1 Prelab Prep
1. Before beginning the activity, make sure your work area is clean. Be sure to wear a lab apron.
2. Collect the following items: distilled water, dropper, 2 slides, tweezers, elodea plant, coverslips, microscope, flat toothpick, iodine solution, 3 sheets of paper.

Task 2 Data Record
1. On a separate sheet of paper, draw the data table shown. Record your observations in the data table.
2. Label a second sheet of paper *Elodea Cell*. Label a third sheet of paper *Cheek Cell*. You will draw models of the observed cells on these sheets.

Task 3 Procedure
1. Place a drop of water in the center of a clean slide. **CAUTION! Slides and coverslips have sharp edges.**
2. Use tweezers to remove a small leaf from the elodea plant. Place the leaf in the drop of water. Place a coverslip over the leaf.
3. Observe the elodea leaf through the microscope under low power. Make notes of what you see.
4. Observe the elodea under high power. Record what parts of the cells you see in the data table. Then draw a single cell and label the different parts you have identified.
5. Using the blunt end of a toothpick, gently scrape some cells from the inside of your cheek.
6. Smear the cheek cells onto a clean slide. Add a drop of water and a coverslip.
7. Observe the cheek cells under low and high power. Record your observations.
8. Remove the slide and lift up one edge of the coverslip. Add a drop of iodine solution. Replace the coverslip. **CAUTION! Iodine can stain skin and clothes.**
9. Observe the cheek cells under low power and high power. Record what parts of the cells you see in the data table. Then draw a single cell and label the different parts you have identified.

Task 4 Analysis
1. What structures did you see in both the elodea cells and the cheek cells?
2. What structures did you see only in the elodea cells? What structures did you see only in the cheek cells?
3. What structures of the cheek cells could you observe after adding iodine?

Task 5 Conclusion
Write a short paragraph explaining the similarities and differences between the elodea cells and the cheek cells. Based on your observations, how do plant cells differ from animal cells?

Extension
Collect some pond or aquarium water. Make slides using drops of water, and see how many organisms and cell structures you can identify. Draw your observations on a chart.

Table 4.1 Parts of Cells

Organelle	Elodea Cell	Cheek Cell
Cell membrane		
Cell wall		
Cytoplasm		
Nucleus		
Vacuole		
Chloroplasts		

SECTION 4.3

Section Objectives
For a list of section objectives, see the Student Edition page.

Skills Objectives
Students should be able to:

Infer from life activities of pond organisms their abilities to adapt to environmental changes.

Classify structures as tissues, organs, or systems.

Vocabulary
tissue, organ

MOTIVATE

Skills WarmUp
To help students understand the differences between simple and complex levels of organization, have them do the Skills WarmUp.
Answer Both organisms move, acquire food, and react to touch/light. The fish's activities are more varied than those of the microscopic organism. The fish is more likely to adapt to changes in its environment.

○ **The Living Textbook:**
Life Science Sides 7–8

Chapter 3 Frame 00363
Animal Cells and Tissues (25 Frames)
Search: Step:

○ **The Living Textbook:**
Life Science Side 2

Chapter 15 Frame 26460
Muscle Tissue (Movie)
Search: Play: Step:

86

INTEGRATED LEARNING

Themes in Science
Scale and Structure Explain that specialization affects every level of organization but is most obvious at the tissue level. The various structures of an organism all interact with other structures. Organisms are made up of organ systems, which in turn are made up of organs, and so on down to the cellular level.

Skills WarmUp

Infer Make a list of the life activities of a pond organism that can only be seen under a microscope, such as an ameba. Make a second list of the life activities of a fish. Compare the two lists. Which organism do you think is more likely to adapt to changes in its environment? Why?

Cells
In many-celled organisms, cells perform many different functions. The cells in your muscles work together so that you can move. Muscle cells, however, do not work alone. For example, muscle cells contract only when impulses from nerve cells set them into action. The muscle cells shown here are found in the heart. ▼

4.3 Organization of Living Things

Objectives

▶ **Identify** tissues, organs, and systems in an organism.
▶ **Explain** how organs work together in systems.
▶ **Classify** the organs, tissues, and systems in an organism.

You are a many-celled organism because your body is made up of more than one cell. In fact, the human body is made up of billions of cells. However, many organisms, such as amebas, have only one cell. They are single-celled organisms. To see most single-celled organisms, you need a microscope.

A many-celled organism carries out more complex activities than a single-celled organism. For example, you can walk, swim, run, jump, and throw. An ameba, however, can only swim. In a many-celled organism, cells are organized to work together. There are five main levels of organization: cells, tissues, organs, organ systems, and the organism.

Cells, Tissues, and Organs

Most many-celled organisms have cells that are *specialized;* that is, there are different types of cells that do different kinds of work. Each group of specialized cells is organized into **tissues**. Each kind of tissue performs a certain function.

◀ **Tissues**
Muscle cells make up three different kinds of muscle tissue in the human body: smooth, skeletal, and cardiac. The cardiac muscle tissue shown here is found only in the heart.

86

Themes in Science

Systems and Interactions Just as the organelles within cells interact to release energy, assemble proteins, produce new cells, and fulfill other needs, so the cells combine in organs and systems that interact to fill the needs of an entire organism. As you study the organization of living things, emphasize the interdependence between structures at every level.

Physical Education Connection

During exercise, the respiratory and circulatory systems work together. Have students take five minutes to do some form of aerobic exercise, such as jumping jacks, arm circles, or jogging in place. Have students notice how these exercises affect their hearts and their breathing. Can they observe signs of different systems working together?

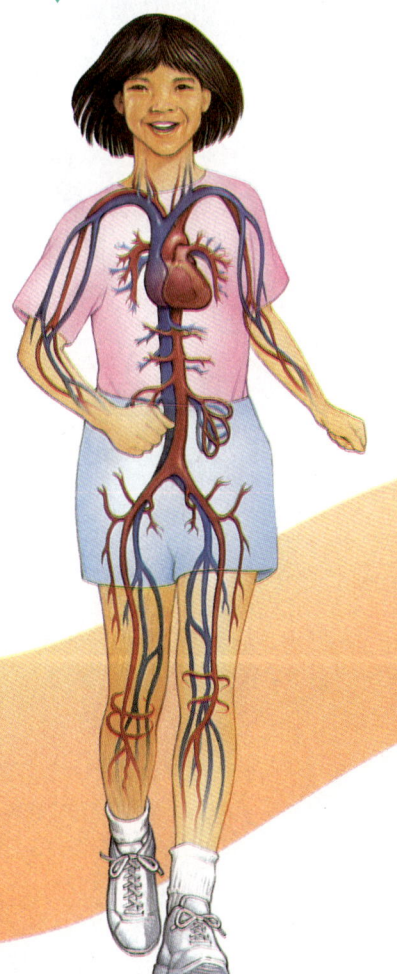

Prior Knowledge

To gauge how much students know about how the systems in their bodies interact, ask the following questions:

▶ Which one of your body systems is the most important?

▶ Why do you think scientists use the word *system* to describe your skeleton or your digestive system?

▶ Can you think of an activity you do that requires different parts of your body to work together?

In many organisms, tissues are organized into groups called **organs**. An organ is a group of tissues that work together to perform special functions. Both plants and animals have tissues and organs.

Organ Systems

A group of organs that work together make up an organ system. Some organisms, such as sponges, have no organ systems. Some organisms, such as jellyfish, have a few organ systems. And very complex organisms, such as dogs and humans, have ten organ systems.

Organisms

The highest level of cell organization is the organism. All organisms carry out life processes. In some organisms, such as humans, the different organ systems work together to keep the organism alive. One organ system, called the respiratory system, enables you to breathe. Your muscular and skeletal systems enable you to support, protect, and move your body. Your digestive system enables you to release the energy and nutrients your body needs to work properly.

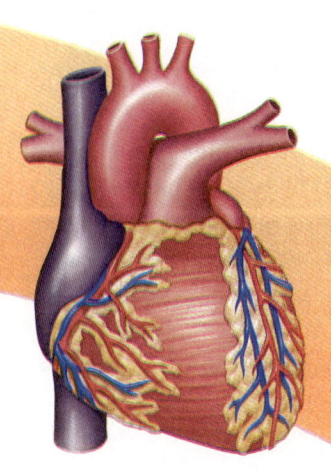

Organ Systems
The heart, blood, and a network of blood vessels make up the human circulatory system. You also have nine other organ systems:

- skeletal
- muscular
- skin
- endocrine
- reproductive
- digestive
- respiratory
- excretory
- nervous

◀ **Organs**
A heart is one of many organs in a many-celled animal such as a dog or a human. The heart is mostly made up of cardiac muscle tissue. It also contains nerve and connective tissues. Muscle, nerve, and connective tissues work together to pump blood through the heart and body.

TEACH

Skills Development
Make Models Have students imagine that the levels of organization in the body are like a pyramid in which atoms are at the lowest level and the organism itself is at the highest level. Draw a pyramid on the chalkboard and place the words *Atoms* at the bottom of the pyramid and *Organism* at the top. Ask students to supply the names of the levels of organization that lead up to *Organism* from *Atoms*. (Molecules, Organelles, Cells, Tissues, Organs, Organ systems)

Explore Visually

Have students study the words and illustrations on pages 86 and 87. Then ask the following questions:

▶ Which picture shows cells most clearly? (Far left on page 86)

▶ What do you think the term *cardiac* means? ("Relating to the heart")

▶ What system is shown in the picture of the runner? (Circulatory)

▶ What organs make up the circulatory system? (Heart and blood vessels)

▶ What are the names of different types of blood vessels? (Veins, arteries, capillaries)

Chapter 4 Cells and Living Things

TEACH ▪ Continued

Research
Have students call the public relations department of a local hospital or the American Heart Association to find out why people who use pacemakers can't be around microwaves. (The timing device in some pacemakers is sensitive to microwave signals.)

Decision Making
If you have classroom sets of *One-Minute Readings*, have students read Issue 25, "Organ Transplants" on page 41. Discuss the questions.

EVALUATE

WrapUp
Review On the chalkboard, make three columns. Label the first column *cells, tissues,* and *organs*. Label the second column *systems*, and the third column *organisms*. Have students name some living things and their parts and write their ideas in the proper column.
Use Review Worksheet 4.3.

Check and Explain
1. Answers may vary. A list might include muscle cell, cardiac tissue, heart, circulatory system, and human.
2. Each identifies a different level of organization: The heart is an organ made of cardiac muscle tissue, which is in turn made up of muscle cells.
3. No, because a tissue is defined as a group of cells.
4. a. System; b. Organ; c. Tissue

Answer to In-Text Question
① Both a pacemaker and heart tissue use an electric signal to contract the muscles of the heart.

88

INTEGRATED LEARNING

Mathematics Connection
Have students find their pulses, either in their wrists or at the base of their necks. Have them count how many times their hearts beat within 15 seconds. If their heart rates were to remain constant, how many beats would they count in a minute? (Pulse × 4 = M) In an hour? (M × 60 = H) In a day? (H × 24 = D) In a week? (D × 7 = W) In a year? (W × 52 = Y)

Figure 4.7 ▲
What kind of energy does a pacemaker and your heart tissue use to contract your heart muscles? ①

Science and Technology *Pacemakers*
Each day your heart beats about 100,000 times! With each beat, the heart contracts, pumping blood throughout your body. This continuous supply of circulating blood keeps you alive.

To work efficiently, the muscle tissues of the heart must work together. The coordination of these tissues is controlled by another tissue that produces an electric signal. When the muscle cells detect the electric signal, they tighten, or contract, in unison.

For some people, the tissue that produces the electric signal does not work properly. When this happens, the muscle tissues in the heart may not work together. The heartbeat slows down or becomes uneven. This, in turn, may decrease the amount of blood circulating through the body.

If a person's heartbeat is too slow or uneven, it can be regulated by a pacemaker. A pacemaker is an electronic device that keeps blood circulating properly. Pacemakers are made of a battery, a miniature timing device, and a set of wires. They send electric signals to the muscle cells to tell them when to contract.

Pacemakers are placed inside a person's body. The wires are attached to the heart. At regular intervals, electrical energy is sent from the battery through the wire to the heart. This electric "jolt" stimulates the heart muscles, causing them to contract at the same time.

Check and Explain
1. List at least one example of each of the following: cell, tissue, organ, organ system, and organism.
2. Describe the difference between a cardiac muscle cell, cardiac muscle tissue, and the heart.
3. **Reason and Conclude** Can a single-celled organism contain tissue? Explain.
4. **Classify** Classify the following as a tissue, organ, or system:
   a. Brain, spinal cord, and nerves
   b. Heart
   c. Group of muscle cells

88 Chapter 4 Cells and Living Things

Chapter 4 Review

Concept Summary

4.1 The Cell Theory
- The cell theory states that all living things are made of cells. Cells form the structure and function of living things, and cells only come from other living cells.
- The cell theory was developed through the work of several scientists including van Leeuwenhoek, Hooke, Schleiden, Schwann, and Virchow.

4.2 Parts of a Cell
- Cells are made of different parts that work together to meet the needs of the cell.
- The outer cell layer is called the cell membrane. It controls what enters and leaves the cell.
- The nucleus controls the cell's activities and stores the information needed for reproduction.
- Cytoplasm is the gel-like substance inside the cell that makes up the cell material.
- Organelles are the small structures inside of a cell that carry on life processes such as releasing energy, making proteins, and storing food and water.
- Chloroplasts and cell walls are found only in plant cells.

4.3 Organization of Living Things
- Multicellular organisms have specialized cells to perform different functions.
- Cells of similar structure and function join together to form tissues. Groups of tissues make up organs, and groups of organs working together make up organ systems.
- An organism, the highest level of cell organization, can carry out all life processes.

Chapter Vocabulary

cell (4.1)	cell wall (4.2)	organelle (4.2)	tissue (4.3)
cell theory (4.1)	nucleus (4.2)	chloroplast (4.2)	organ (4.3)
cell membrane (4.2)	cytoplasm (4.2)	chlorophyll (4.2)	

Check Your Vocabulary

Use the vocabulary words above to complete the following sentences.

1. A cell's activities, including reproduction, are controlled by the cell's ____.
2. In plants, sun energy is converted to food by the compound ____.
3. Specialized cells organized to perform a certain function are called ____.
4. A plant cell differs from an animal cell because it has a ____ outside the cell membrane.
5. The living material in the cell, excluding the nucleus, is called ____.
6. A set of statements called the ____ describe the basic concepts about cells.
7. A group of tissues that works together to perform a special function is called an ____.
8. The thin outer covering, or the ____, controls what enters and exits cells.
9. The tiny, special parts of a cell that carry out life processes are called ____.
10. The basic unit of structure and function for all living things is the ____.
11. The chlorophyll in a ____ gives a plant its green color.

Write Your Vocabulary

Write sentences using each vocabulary word above. Show that you know what each word means.

CHAPTER REVIEW 4

Check Your Vocabulary
1. nucleus
2. chlorophyll
3. tissues
4. cell wall
5. cytoplasm
6. cell theory
7. organ
8. cell membrane
9. organelles
10. cell
11. chloroplast

Write Your Vocabulary
Be sure that students understand how to use each vocabulary word in a sentence before they begin.

Use Vocabulary Worksheet for Chapter 4.

CHAPTER 4 REVIEW

Check Your Knowledge

1. All living things are made up of cells. Cells are the basic units of structure and function in living things. All cells come only from other living cells.
2. The cell membrane protects the inside of the cell, supports the cell, and gives it shape.
3. Cellulose is located in the cell wall.
4. Answers may vary. See pages 82–83.
5. *Cell* is the Latin word for "little room."
6. Different types of cells perform specific functions for an organism.
7. van Leeuwenhoek
8. Cell membrane
9. Mitochondria
10. Organ
11. Lysosomes
12. Cell
13. Nucleus
14. Environment
15. Living cells

7. The skin is an organ because it is made up of different tissues that work together.
8. If scientists could make cells from nonliving elements, the part of the cell theory that states that all cells come from other living cells would have to be changed.
9. Animal cells and plant cells have many common structures.

Check Your Understanding

1. Chlorophyll would allow animal cells to produce glucose, which they could then use for energy. Animals might require very little food, if any.
2. An organelle is a cell structure like a "tiny organ."
3. A group of specialized cells makes up a tissue; tissues combine to form an organ; a group of organs that work together makes up an organ system.
4. Answers may vary. Some examples are that cell walls and membranes are like the town's boundaries, the nucleus is like the town mayor, mitochondria are like its power plant, ER are like roads, and ribosomes are like protein factories.
5. This tissue is part of the skeletal system.
6. A muscle cell would have more mitochondria because it needs energy to contract.

Chapter 4 Review

Check Your Knowledge
Answer the following in complete sentences.

1. List the statements in the cell theory.
2. What is one purpose of the cell membrane?
3. Where in a cell is cellulose located?
4. List two organelles and their functions.
5. How did cells get their names?
6. Explain the statement "cells are specialized."

Choose the answer that best completes each sentence.

7. The first person to observe living cells was probably (Schleiden, Schwann, Hooke, van Leeuwenhoek).
8. The (cell membrane, lysosome, nuclear membrane, cellulose) protects, supports, and gives shape to cells.
9. The cell powerhouses are the (Golgi bodies, lysosomes, mitochondria, vacuoles).
10. Your heart is an example of an (organ system, organ, organism, organization).
11. Digestive chemicals in (chloroplasts, ribosomes, lysosomes, chlorophyll) break down food, wastes, and old cell parts.
12. The unit of structure for living things is the (organism, tissue, plant, cell).
13. The cell's control center is the (brain, cytoplasm, nucleus, scientist).
14. To grow skin cells in a laboratory, scientists must provide the cells with the same (chloroplasts, environment, hair, light) as a normal skin cell.
15. The cell theory states cells come only from (living cells, pond water, nonliving cells, cork).

Check Your Understanding
Apply the concepts you have learned to answer each question.

1. If animal cells contained chlorophyll, how would animal life be different?
2. The suffix *-elle* means "small" or "tiny" in Latin. Using this information, write a definition for *organelle*.
3. Describe how cells, tissues, organs, and organ systems are related.
4. Describe how the parts of a cell are like a city or town. Identify what parts of a town are similar to several cell organelles.
5. **Mystery Photo** The photograph on page 74 shows the bone tissue of a human thigh bone. The white spaces are channels that contain nerves or blood vessels for living bone cells. The black dots are cells in the bone matrix. Bone cells produce hard layers between one another. Fibers grow between the cells and hold them together. In what organ system would this tissue be found?
6. **Reason and Conclude** Which kind of cell would you expect to have more mitochondria, a muscle cell or a fat storage cell?
7. **Classify** What makes the skin an organ, rather than a tissue or organ system?
8. **Apply** How would the cell theory be affected if scientists could make living cells from carbon, oxygen, and other elements found on earth?
9. **Compare and Contrast** Compare the labels on Figure 4.4 and Figure 4.5. What do they tell you about the structures of these cells?

Develop Your Skills

1. a. Plant; cell wall, large vacuole, chloroplasts
   b. 3
   c. Yes
2. In designing their experiments, students should recall the characteristics of living things presented in Chapter 2.
3. a. Mitochondrion
   b. Plant cell

Make Connections

1.

```
                    cells
              contain    make up
                              tissues
        organelles        make up
          which               organs
          include          make up
        mitochondria
        ribosomes        organ systems
        vacuoles
        Golgi bodies       make up
        lysosomes
        endoplasmic        organisms
        reticulum
        chloroplasts
```

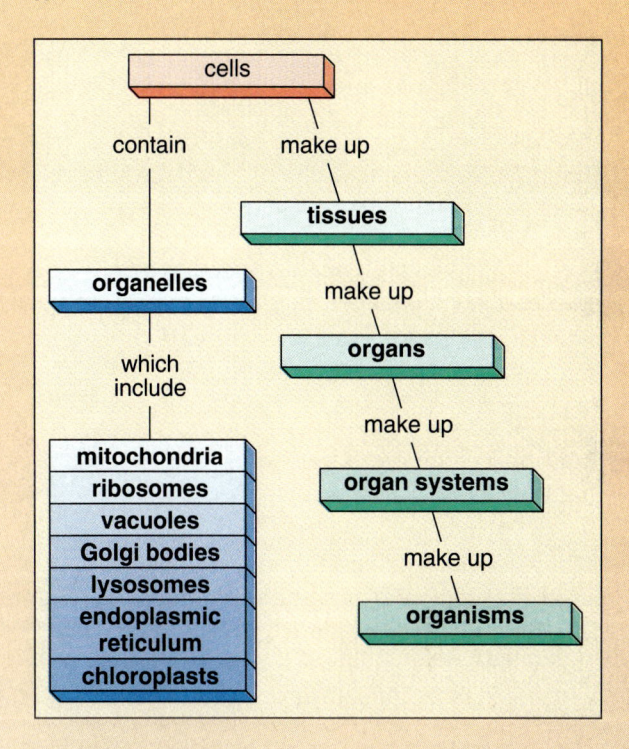

2. Ask students what questions the listener asked when the story was recited.
3. Students' research should show that they understand the challenges and problems presented by tissue culturing.
4. Any timeline should include at least the following items:

   1675 van Leeuwenhoek is first to observe living cells and Robert Hooke coins term *cell*
   1838 Schleiden observes plant cells and Schwann observes animal cells
   1858 Virchow states that all cells come from other cells

Develop Your Skills

Use the skills you have developed in this chapter to complete each activity.

1. **Observe** Look at the cell picture below to answer the following questions.
   a. Is this a plant or animal cell? How can you tell?
   b. How many mitochrondria are in the cell?
   c. Is there a vacuole?

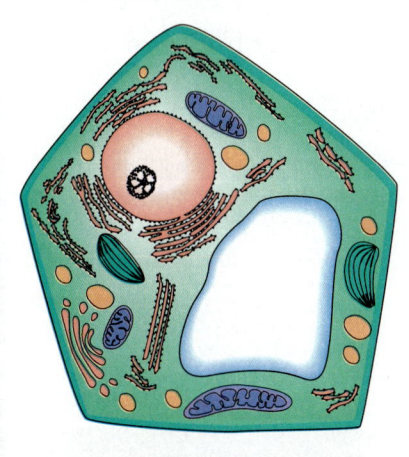

2. **Design an Experiment** Imagine you are Anton van Leeuwenhoek looking at small moving objects in your microscope. Design an experiment to determine whether these things are living or nonliving. Make sure you change only one variable.

3. **Data Bank** Use the information on page 625 to answer the following questions.
   a. Which is larger, a mitochondrion or a ribosome?
   b. Which needs a light microscope to be seen, a plant cell or a virus?

Make Connections

1. **Link the Concepts** Below is a concept map showing how some of the main concepts in this chapter link together. Only parts of the map are filled in. Copy the map. Using words and ideas from the chapter, complete the map.

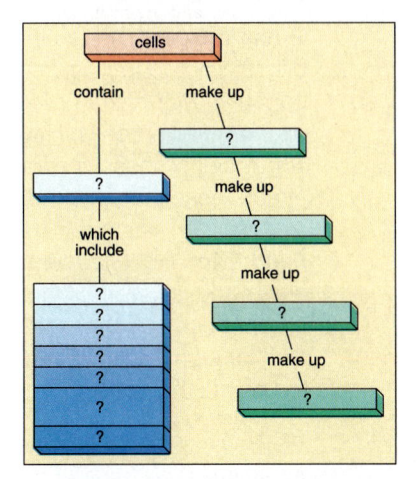

2. **Science and Language Arts** Write a children's story about the cell and its parts. Give personalities to the cell organelles. Tell the function of each organelle as part of your story. Try reading your story to a child to see if your information is clear and exciting.

3. **Science and Technology** Scientists can culture many different kinds of human cells. Choose two types of human cells. Research in your library how these cells are cultured and for what purpose.

4. **Science and Social Studies** Draw a time line that shows the events that led to the cell theory.

Chapter 4 Cells and Living Things

CHAPTER 5

Overview

The processes that take place in cells are the subject of this chapter. The first section describes diffusion, osmosis, and active transport. In the second section, photosynthesis, respiration, and fermentation are discussed. A detailed discussion of mitosis and the changes and growth in living things closes the chapter.

Advance Planner

▶ Supply carrot sticks, salt, water, and cups for TE activity, page 94.

▶ Provide eggs, jars, and vinegar for SE page 95.

▶ Gather magazines for use in identifying energy sources. TE page 101.

▶ Collect labels, planting pots, measuring cups, and small plants of the same size and species for SE Activity 5, page 102.

▶ Supply clay and string, yarn, or wire for use in making models of the phases of mitosis. TE page 106.

Skills Development Chart

Sections	Classify	Compare and Contrast	Control Variables	Estimate	Infer	Interpret Data	Observe
5.1 Skills WarmUp SkillBuilder Skills WorkOut	●					●	●
5.2 Skills WarmUp Skills WorkOut Skills WorkOut Activity	●	●	●		●		●
5.3 Skills WarmUp		●					

Individual Needs

▶ **Limited English Proficiency Students** Use a tape recorder to record the definitions and pronunciations of the chapter's boldface and italic words. Along with the definition of each term, tell students which figures in the text illustrate the term. Have them use the tape recorder to listen to the information about each term. Then have students define the terms using their own words. Have them make their own tape recordings of the terms and their definitions. They can use this tape along with the text figures to review important concepts in the chapter.

▶ **At-Risk Students** Make a list of the times each day that people are affected by photosynthesis and respiration. Some events might be breathing, eating fruit, running, looking at trees, and eating meat. Have students copy the lists into their science journals and describe which process is involved in each activity. Have students make similar lists for the other cell processes that are discussed in this chapter.

▶ **Gifted Students** Researchers at the National Aeronautics and Space Administration (NASA) have found that plants reduce the amounts of formaldehyde, carbon monoxide, and other poisonous gases in the air. Have students use what they've learned about cell processes to explain how plants might help reduce pollution. Have students prepare a display telling people about their discoveries.

Resource Bank

▶ **Bulletin Board** Place the title *Cell Processes* on the bulletin board. Photocopy Figures 5.2, 5.5, 5.6, and 5.8, covering the captions. Mount the figures on colored paper and hang them on the bulletin board. Have students make labels indicating which cell process is shown in each figure. Students can attach these labels to the bulletin board.

▶ **Field Trip** Have students visit a greenhouse at a local nursery or conservatory. Arrange to have someone from the greenhouse speak to the students. Encourage students to look for signs that all the cell processes are taking place in the plants they observe.

CHAPTER 5 PLANNING GUIDE

Section	Core	Standard	Enriched	Section	Core	Standard	Enriched
5.1 Movement of Substances pp. 93–96				**Blackline Masters** Review Worksheet 5.2 Integrating Worksheet 5.2	●	● ●	● ●
Section Features Skills WarmUp, p. 93 SkillBuilder, p. 94 Skills WorkOut, p. 95	● ●	● ● ●	● ●	**5.3 Cell Growth and Division** pp. 103–106			
Blackline Masters Review Worksheet 5.1 Skills Worksheet 5.1 Integrating Worksheet 5.1	●	● ● ●	● ●	Section Features Skills WarmUp, p. 103	●	●	●
Ancillary Options One-Minute Readings, p. 83	●	●	●	Blackline Masters Review Worksheet 5.3 Reteach Worksheet 5.3 Vocabulary Worksheet 5	● ● ●	● ● ●	● ● ●
Laboratory Program Investigation 10	●	●	●	Overhead Blackline Transparencies Overhead Blackline Master 5.3 and Student Worksheet	●	●	●
Color Transparencies Transparency 13	●	●	●	Laboratory Program Investigation 11		●	●
5.2 Energy Processes pp. 97–102				Color Transparencies Transparencies 14a, b	●	●	
Section Features Skills WarmUp, p. 97 Skills WorkOut, p. 99 Skills WorkOut, p. 100 Career Corner, p. 100 Activity, p. 102	● ● ● ● ●	● ● ● ● ●	● ●				

Bibliography

The following resources can be used for teaching the chapter. See page T-40 for supplier codes.

Audio-Visual Sources
(video unless noted)
The Cell: Basic Unit of Life. 15 min filmstrip with cassette. 1981. NGSES.
Photosynthesis: Life Energy. 22 min. 1983. NGSES.

Software Resources
Cell Respiration. Apple. J & S.
Cells. Apple. J & S.
Photosynthesis and Light Energy. IBM. CCM.
Photosynthesis and Transport. Apple. J & S.

Library Resources
Fichter, George S. Cells. New York: Franklin Watts, 1986.
Sheeler, Philip, and Donald E. Bianchi. Cell Biology: Structure, Biochemistry, and Function. New York: Wiley, 1983.
Thomas, Lewis. The Lives of a Cell: Notes of a Biology Watcher. New York: The Viking Press, Inc., 1974.

CHAPTER 5

Introducing the Chapter

Have students read the description of the photograph on page 92. Ask if they agree or disagree with the description. The green color in plants comes from chlorophyll.

Directed Inquiry

Have students study the photograph. Ask:

▶ Does this photograph show the structure at its actual size? (No. The photograph is a magnification.)

▶ What kind of object is shown in the picture? (Students may answer that a cell is shown; some may say that it is part of a cell.)

▶ Is this structure part of a plant, an animal, or both? Why do you think so? (The green color may lead students to conclude that the structure is part of a plant. Point out the cell wall in the photograph.)

▶ Read the list of Chapter Sections. What are some processes that go on inside cells? (Processes that involve the movement of substances; acquiring, storing, and releasing of energy; and cell growth and division)

Chapter Vocabulary

chromosome
diffusion
fermentation
mitosis
osmosis
photosynthesis
respiration

INTEGRATED LEARNING

Language Arts Connection

Chloroplasts belong to the group of organelles called plastids. Other plastids are chromoplasts and leucoplasts. Have students use dictionaries to find the meaning of plastid. Then have them hypothesize the meaning of chloroplast, chromoplast, and leucoplast. They can use dictionaries to check their answers.

Integrating the Sciences

Chemistry Chlorophyll is a pigment that gives chloroplasts their green color. Have students identify some of the other pigments in plant cells.

Chapter 5 — Cell Processes

Chapter Sections

5.1 Movement of Substances
5.2 Energy Processes
5.3 Cell Growth and Division

What do you see?

"I see a figure-8 shaped, lined, green structure. I think that this structure is a chloroplast. I think this because it appears to be submerged in cytoplasm inside a cell. I think it's green because of the chlorophyll inside. Chloroplasts are what give plants their green color. It is important because it aids the plant in photosynthesis, which makes the plant grow."

Andrea Winn
Austin Academy
Garland, Texas

To find out about the photograph, look on page 108. As you read this chapter, you will learn more about cells and their life processes.

Themes in Science
Systems and Interactions The cell processes described in this chapter interact and work together to supply the needs of living organisms.

Multicultural Perspectives
Inform students that another meaning of diffusion is the spread of cultural elements from one group of people to another by contact with each other. Ask students to discuss examples of cultural diffusion that they may have observed or experienced.

SECTION 5.1

Section Objectives
For a list of section objectives, see the Student Edition page.

Skills Objectives
Students should be able to:

Classify objects transported in and out of an area.

Interpret Data from a graph.

Observe cell membrane properties.

Classify everyday occurrences as examples of diffusion, osmosis, or active transport.

Vocabulary
diffusion, osmosis

5.1 Movement of Substances

Objectives

- **Describe** the role of the cell membrane in moving substances into and out of cells.
- **Explain** what it means for the cell membrane to be selectively permeable.
- **Distinguish** between diffusion and osmosis.
- **Infer** how the process of osmosis affects the cells of living things.

Skills WarmUp

Classify Make two lists. On one list, include the things you transport into your bedroom each day. On the other list, include the things you transport out of your bedroom each day.

Imagine what would happen to cities if all roads, railroads, and airports suddenly shut down. There would be no way to transport materials from one place to another. Food, clothing, medicine, or other resources could not be brought into cities. Garbage could not be carried out of cities. If all lines of transport were blocked, cities would stop functioning. Like a city, your body also needs to transport materials. Your cells must take in needed materials, such as nutrients, water, and oxygen. Your cells must also get rid of wastes.

Diffusion

Look at Figure 5.1. What happens to the hot water when the tea bag is placed in the cup? Tea from the leaves moves out of the bag and into the surrounding water, flavoring it and coloring it brown. Movement of the tea throughout the water is explained by a process called **diffusion** (dih FYOO zhuhn). Diffusion is the movement of a substance from an area of high concentration to an area of low concentration. In the water, tea moves from high concentration in the bag to areas of lower concentration in the water. Eventually, tea spreads evenly throughout the water.

Water diffuses freely through the cell membrane in the same way that tea diffuses through a cup of water. Water surrounds most cells in your body, and all your cells contain water. Many substances vital to cells are dissolved in water. Cells, therefore, are affected by the diffusion of substances that are dissolved in water.

Figure 5.1▲
Tea leaves diffuse in a cup of hot water.

MOTIVATE

Skills WarmUp
To help students understand the movement of substances, have them do the Skills WarmUp.
Answer Students should note that some of the items that are transported into their bedrooms are the same items that are transported out. Some items change; for example, clothes may be clean and folded going into their rooms, but dirty when carried out. Some of the substances that are transported in and out of cells are also changed.

TEACH

Portfolio
Have students make and label a diagram illustrating the process of diffusion. After they've read about osmosis, have them add it to their diagrams.

Chapter 5 Cell Processes 93

TEACH • Continued

Class Activity
Have students work in small groups. Provide each group with two carrot sticks, some salt, and two cupfuls of water. Have each group put a spoonful of salt in one cup and stir well. Have them observe the characteristics of the carrot sticks and then place one stick in each jar. Have students leave the carrots in the cups overnight. The next day, have them observe the carrots and note how the characteristics have changed. What does this experiment show about osmosis? (The carrot stick from the plain water is more firm because water will diffuse out of the carrot stick placed in the salt water.)

Skills Development
Hypothesize Have students suggest why a plant wilts, and why it becomes rigid again after it's watered. (Water diffuses out of cells into the soil, making the cells shrink. Adding water to the soil causes water to diffuse back into the plant.)

Discuss
Stress the difference between diffusion and osmosis. Osmosis is diffusion of water across a cell membrane.

SkillBuilder
Demonstrate Use a beaker with ice water and a beaker with hot (70°C) water. Have student volunteers mix drink mix into the two beakers so the class can observe the rates of diffusion.

Answers
1. 100°C
2. 100°C
3. The higher the water temperature, the faster the diffusion rate of a substance.
4. 100°C

94

INTEGRATED LEARNING

Multicultural Perspectives
Explain to students the custom of preserving food by drying or salting it. A high concentration of salt can cause water to diffuse out of a cell and dry out. Pemmican, or dried buffalo meat, was a staple for Native Americans. Have students research the use of dried or salted foods in other cultures.

Integrating the Sciences
Physical Science Solutions are described in terms of their concentration, which is a ratio of the dissolved substance to the overall solution. Students use the terms *concentrated* or *diluted* to describe common solutions, such as a fruit drink mix. Stress that the term *concentration* here refers to the ratio of water to dissolved substances, rather than the other way around.

Osmosis

The diffusion of water across the cell membrane is called **osmosis** (ahz MOH sihs). Osmosis and diffusion are types of passive transport. During passive transport, materials move freely through the cell membrane because they do not require energy.

Like any other substance, water tends to move from high to low concentration. Water containing dissolved substances has a lower concentration of water than pure water. What would happen, then, if an animal cell were placed in pure water? The inside of an animal cell contains many dissolved substances. Therefore, the water concentration in the cell would be less than the water concentration outside the cell. So the water from the outside would diffuse through the membrane into the cell. The cell would swell and might even burst. Your body cells are like the animal cells. To prevent your cells from bursting, your body fluids have the same concentration of dissolved substances that your cells have.

SkillBuilder Interpreting Data

Diffusion Rate and Temperature

When you put a package of drink mix in water, you usually stir it to spread the mix evenly through the water. But if you didn't stir it, eventually the drink mix would spread through the water on its own, by diffusion. The water temperature can affect the diffusion rate of a substance.

The graph shows the rate at which drink mix diffuses through water at different temperatures. Study the graph, then answer the following questions:

1. At what temperature did the drink mix take the least amount of time to diffuse?
2. At what temperature was the diffusion rate the slowest?
3. What can you conclude about the effect of water temperature on the rate of diffusion?
4. For maximum diffusion in a short amount of time, what water temperature would be best for mixing the drink mix completely?

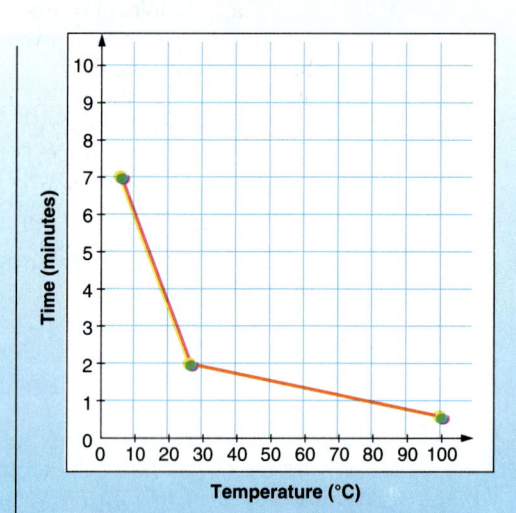

In one or two sentences, explain the methods you would use to make a packaged drink mix diffuse quickly.

94 Chapter 5 Cell Processes

Themes in Science

Scale and Structure The semipermeability of a cell membrane is important to the whole organism as well as to the individual cell. For example, oxygen moves by diffusion into a fish's blood as water passes across the cell membrane of its gills, which contain many blood vessels. The membrane is impermeable to many other substances made up of large molecules.

Integrating the Sciences

Environmental Science Explain to students that many landfills are lined with plastic. Have students discuss the usefulness of such a lining. Ask if it's like an impermeable, semipermeable, or permeable membrane. (Impermeable)

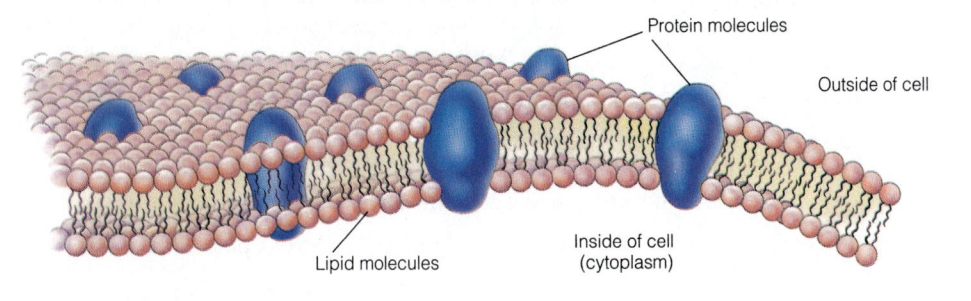

Skills WorkOut

To help students observe some of the properties of a cell membrane, have them do the Skills WorkOut.
Answers The vinegar dissolves the calcium in the shell. The egg increases in size over the three-day period. This shows that the cell membrane allows some of the vinegar to come in.

Discuss

Ask students the following questions:

▶ How are the layers of lipids in a cell membrane similar to a screen on a window? (There are small gaps, or spaces, between the lipids in the membrane and the wires in a screen.)

▶ What does a screen let in? What does it keep out? (Air can pass through the screen. Bugs can't.)

▶ What do the lipid layers in the cell membrane allow into the cell? (Water and small molecules)

▶ How can large particles get through the cell membrane? (Protein "carriers" carry large molecules through the membrane.)

Integrated Learning

Use Integrating Worksheet 5.1.

The Living Textbook: Life Science Sides 7-8

Chapter 3   Frame 00415
Animal Cell Osmosis (3 Frames)
Search: Step:

The Living Textbook: Life Science Side 2

Chapter 10   Frame 05406
Osmosis in Leaves and Onion Epidermis (Movie)
Search: Play:

Cell Membranes and Transport

Everything that moves into or out of a cell must pass through the cell membrane. The membrane is the key to how cells control the movement of life substances.

Certain substances diffuse freely into or out of cells. For these substances, the membrane is *permeable* (PUR mee uh buhl). For example, if the concentration of oxygen is lower inside a cell, oxygen will diffuse into the cell.

For many other substances, the cell membrane is a barrier. They cannot pass freely through the membrane, even if the substances are more concentrated on one side of the membrane than on the other side. For these substances, the membrane is *impermeable*.

Because a cell membrane is impermeable to some materials and permeable to others, it is described as *selectively permeable*. This characteristic of a cell membrane is explained by its structure. Look at the model of a cell membrane in Figure 5.2. Notice that the cell membrane is made up of two layers of lipid molecules. The lipid molecules slip past each other to form tiny gaps. Water and other small molecules can easily enter or leave the cell through these gaps. But many larger molecules can't fit through the tiny gaps.

How, then, do larger molecules like sugars and amino acids get into a cell? Larger molecules move through the cell membrane using special particles of protein embedded in the cell membrane. You can see these particles of protein in the model of the cell membrane in Figure 5.2. The protein molecules, or "carriers," actually carry larger molecules from one side of the membrane to the other. Carrier molecules are specialized to do certain jobs. Each carrier molecule allows the movement of only one type of molecule across a cell membrane.

Skills WorkOut

Observe You can observe some of a cell membrane's properties. Measure and record the circumference of a raw egg at its center. Place the egg, in its shell, into a jar. Cover the egg with clear vinegar. Close the jar with a lid. Over three days, observe the egg and record what you see. At the end of three days, remove the egg, measure it again, and record your observations. What happened? How does this show that a cell membrane is selectively permeable?

Figure 5.2
A Model of a Cell Membrane

Protein molecules
Outside of cell
Inside of cell (cytoplasm)
Lipid molecules

Chapter 5 Cell Processes

TEACH • Continued

Discuss
Ask students what happens to lettuce or other leafy plants when they freeze. Stress that one of the important parts of cryobiology is diffusing water from cells so that the cells are not harmed by ice crystals.

EVALUATE

WrapUp
Portfolio Give students their diagrams of diffusion and osmosis. Have them add an illustration of active transport. Suggest that students use colored pencils to highlight active and passive transport.
Use Review Worksheet 5.1.

Check and Explain
1. It allows only certain substances in and out of the cell.
2. Both allow some, but not all, materials to pass through them. Membranes have carrier molecules and strainers do not.
3. The leaves, the tea bag, and some of the water are affected by osmosis. The entire cup of water is affected by diffusion.
4. **a.** The water concentration inside the cat's body cells would be greater than the water concentration outside. So, the water inside the cells would diffuse to the outside. The cat would become dehydrated. **b.** Since the water concentration is greater outside the celery cells, water will diffuse from the outside into the cells.

Decision Making
If you have classroom sets of *One-Minute Readings*, have students read Issue 49, "Cryogenics" on pages 83 and 84. Discuss the questions.

Answer to In-Text Question
① **When the water in a cell freezes, ice crystals damage the cell.**

INTEGRATED LEARNING

Language Arts Connection
Explain to students that the root word *cry-* or *cryo-* means cold or freezing. Have them suggest the meaning of the following words: *cryobiology*, *cryosurgery*, *cryotherapy*, and *cryonics*.

Figure 5.3 ▲
Cryobiologists study the effects of cold on living things. What happens to human cells when they are frozen? ①

Active Transport
Many substances that can move freely through the membrane must be kept at a higher concentration on one side of the cell membrane than on the other. In order to move a substance to a more concentrated area, a cell uses active transport. Active transport requires energy. The large particles of protein in the cell membrane are involved in active transport.

Science and Technology
Freezing Body Organs

How can human tissue be frozen and not be destroyed? The answer to this question is important for preserving transplant organs. Cryobiology is the study of freezing living things without harming them.
Recall that freezing causes water to expand and that cell contents are mostly water. When the water in a cell freezes, ice crystals damage the cell. Some techniques that reduce the water content also upset the balance of vital substances in the cell. Cryobiologists have developed a technique for preserving thin samples of tissue. Animal tissue is cooled rapidly to a very low temperature, and water is diffused from the cells. The dry tissue is then warmed to room temperature. Cryobiologists are trying to perfect this process for making large organs usable for transplants.

Check and Explain
1. Why is the cell membrane selectively permeable?
2. In what way is a vegetable strainer like a selectively permeable membrane? In what way is it not?
3. **Reason and Conclude** A tea bag is placed in hot water. Identify the substances affected by osmosis and those affected by diffusion.
4. **Infer** What happens to the cells of the organisms in the following situations? Explain how the process of osmosis affects the cells.

   a. A cat drinks salty water.

   b. A wilted piece of celery is placed in water.

Chapter 5 Cell Processes

Themes in Science

Energy All organisms require energy in order to live. In this section, students will study three cell processes that involve energy. Photosynthesis enables organisms to store energy. Respiration and fermentation result in the release of energy.

SECTION 5.2

Section Objectives
For a list of section objectives, see the Student Edition page.

Skills Objectives
Students should be able to:

Estimate the amount of vegetables and fruits they eat in a week.

Observe oxygen on the surface of a submerged leaf.

Classify fruits and vegetables by the presence of chlorophyll and glucose.

Predict if photosynthesis would take place in specified situations.

Control Variables affecting plant growth in an experiment.

Vocabulary
photosynthesis, respiration, fermentation

5.2 Energy Processes

Objectives

▶ **Explain** how plants store energy through photosynthesis.

▶ **Describe** how plants release energy during cell respiration.

▶ **Compare** photosynthesis and respiration.

▶ **Predict** if photosynthesis will take place.

Skills WarmUp

Estimate Estimate the amount of vegetables and fruit you ate during the last week. What is the source of the energy in these foods?

Organisms get the energy they need to carry out their life processes from the chemical bonds in food. For example, when you eat a tuna fish sandwich, your body uses the energy that was stored in the cells of the tuna. The tuna obtained this energy by eating smaller fishes. The smaller fishes, in turn, may have fed on microscopic organisms or plants that lived in the ocean.

All animals depend on other organisms to obtain energy. Plants, however, do not depend on other organisms for energy. Plants get energy by a process called **photosynthesis** (foh toh SIHN theh sihs).

Photosynthesis

The word *photosynthesis* means "putting together with light." During photosynthesis, plants use the energy in sunlight and materials from air and water to join molecules of water and carbon dioxide. These materials make glucose, a simple sugar.

Photosynthesis takes place in the chloroplasts of a plant's cells. It occurs in two stages. During the first stage, energy from sunlight is absorbed by chlorophyll. Chlorophyll is a green pigment in the chloroplasts of a plant's cells. The light energy trapped by chlorophyll is used to split molecules of water (H_2O) into molecules of hydrogen (H_2) and oxygen (O_2). Some of the oxygen is a waste product that is released through tiny openings in the plant's leaves. The hydrogen remains in the chloroplasts.

During the second stage of photosynthesis, hydrogen combines with carbon dioxide (CO_2) to produce glucose ($C_6H_{12}O_6$). This completes the sugar-making process.

Figure 5.4 ▲
Chlorophyll gives this wild ginger plant its green color. Chlorophyll traps light energy from the sun for photosynthesis. Where in the plant cell is chlorophyll found? ②

MOTIVATE

Skills WarmUp
To help students understand that energy for all cell processes comes from the sun, have them do the Skills WarmUp.
Answer Answers may vary. Explain that chemical bonds in carbohydrate molecules store energy. The ultimate source of energy is the sun.

Prior Knowledge
To gauge how much students know about energy processes, ask all or some of the following questions:

▶ Why do plants need energy?
▶ Where do plants get their energy?
▶ How do you get your energy?
▶ What is respiration?

Answer to In-Text Question
② **Chlorophyll is found in the chloroplasts of a plant's cells.**

Chapter 5 Cell Processes

TEACH

Explore Visually

Have students study the chemical formula at the top of the page and Figure 5.5. Then ask some or all of the following questions:

▶ What happens to the energy that comes to the plant from the outside? (It is trapped by the chlorophyll in green plant cells and then used in the process of photosynthesis.)

▶ What happens to the oxygen produced during photosynthesis? (It is given off through the leaves.)

▶ What happens to the hydrogen produced? (Light energy combines the hydrogen with carbon dioxide to make glucose.)

▶ Where might a plant get its energy for photosynthesis other than from the sun? (The plant may be grown in the kind of artificial light that has ultraviolet rays.)

The Living Textbook:
Life Science Sides 7-8

Chapter 3   Frame 00422
Photosynthesis (2 Frames)
Search: Step:

The Living Textbook:
Life Science Side 7

Chapter 20   Frame 02361
Photosynthesis (Animation)
Search: Play:

INTEGRATED LEARNING

Language Arts Connection

Glucose, fructose, and sucrose are all sugars from which living things get energy. Glucose is derived from the Greek *glykys*, meaning *sweet*. The suffix *-ose* from glucose is used to mean *sugar* as different sugars are identified and named. Have students look at the list of ingredients on a packaged food product and identify the ingredients that are sugars.

Integrating the Sciences

Chemistry Some bacteria use sulfur, not oxygen, for photosynthesis. Write the following equations on the board:

$6CO_2 + 6H_2S + \text{Light Energy} \longrightarrow C_6H_{12}O_6 + 6S_2$

$6CO_2 + 6H_2O + \text{Light Energy} \longrightarrow C_6H_{12}O_6 + 6O_2$

Have a volunteer circle the elements that are different in the two equations.

Combined, the two stages of photosynthesis result in a chemical reaction described by the following equation:

$$\underset{\text{Carbon dioxide}}{6CO_2} + \underset{\text{Water}}{6H_2O} + \text{Light Energy} \rightarrow \underset{\text{Glucose}}{C_6H_{12}O_6} + \underset{\text{Oxygen}}{6O_2}$$

Reactants   **Products**

Sunlight provides the energy needed for the reaction to take place. The reactants, shown on the left side of the arrow, are carbon dioxide and water. These are the materials plants need to make glucose. Plants get carbon dioxide from air, and water from air and soil. The products, shown on the right side of the arrow, are glucose and oxygen.

During photosynthesis, light energy is changed, or converted, into chemical energy. This chemical energy is stored in sugar or starch in the plant's leaves and seeds. Major stages of photosynthesis are shown in Figure 5.5.

Figure 5.5 Photosynthesis

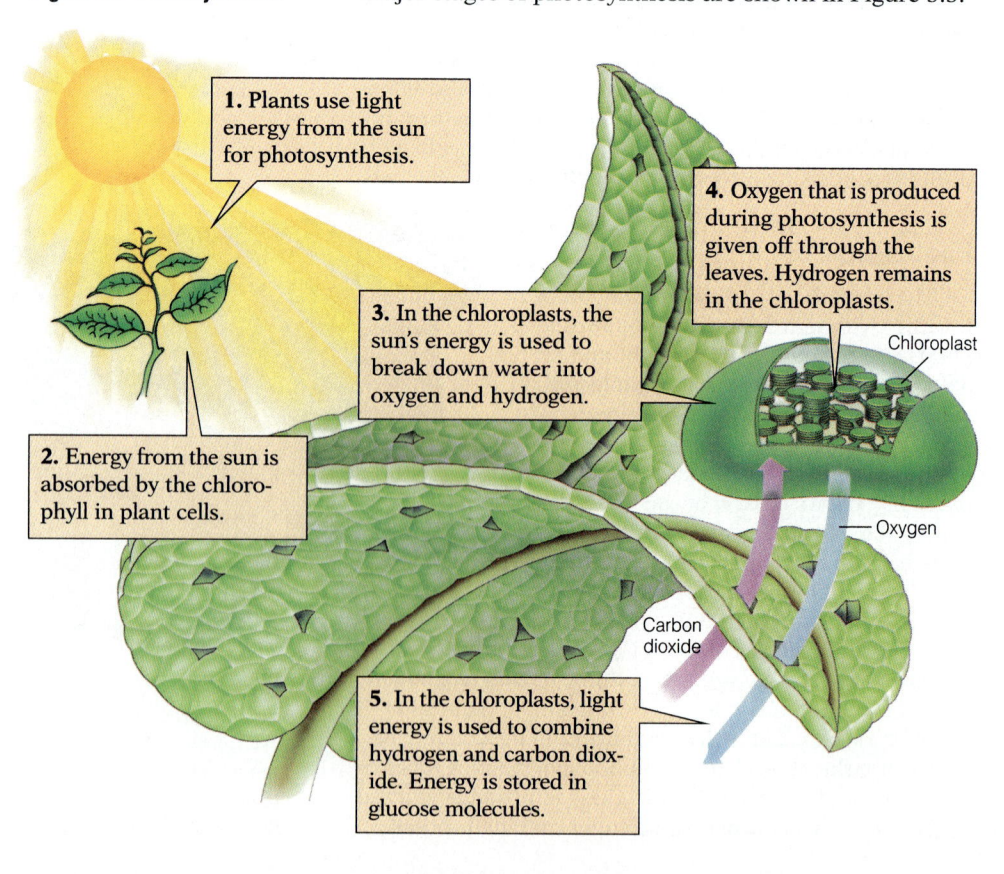

1. Plants use light energy from the sun for photosynthesis.

2. Energy from the sun is absorbed by the chlorophyll in plant cells.

3. In the chloroplasts, the sun's energy is used to break down water into oxygen and hydrogen.

4. Oxygen that is produced during photosynthesis is given off through the leaves. Hydrogen remains in the chloroplasts.

5. In the chloroplasts, light energy is used to combine hydrogen and carbon dioxide. Energy is stored in glucose molecules.

Themes in Science

Systems and Interactions

Photosynthesis and respiration are complementary reactions that allow living organisms to store and to release energy. In photosynthesis, energy is stored in carbohydrate molecules. In respiration the energy in carbohydrates is released.

Respiration

The cells of both plants and animals use the chemical energy stored in glucose. However, before any cell can use this energy, the glucose molecules must break apart. Cells combine glucose with oxygen in order to break down glucose into simpler substances. This process, called **respiration**, releases the stored energy of glucose.

During respiration, glucose breaks down into water and carbon dioxide and releases energy. The energy is carried throughout the organism by a compound called ATP. The chemical reaction for respiration is

$$\underset{\text{Glucose}}{C_6H_{12}O_6} + \underset{\text{Oxygen}}{6O_2} \rightarrow \underset{\text{Carbon dioxide}}{6CO_2} + \underset{\text{Water}}{6H_2O} + \text{Energy (ATP)}$$

What are the products of this chemical reaction? What are the reactants? How does this reaction compare with the reaction for photosynthesis? ①

Like photosynthesis, respiration takes place inside the cell. Respiration begins in the cytoplasm. Enzymes break down carbohydrates into glucose. Glucose breaks down to form acids, carbon dioxide, and two molecules of ATP. The acids break down further in the mitochondria of the cell. Here, oxygen joins with the acids to produce carbon dioxide and water. Many molecules of ATP are also released. ATP carries energy throughout the organism. The processes of photosynthesis and respiration are compared in Table 5.1.

Skills WorkOut

Observe Place a freshly picked leaf in water. Wait 30 minutes. While the leaf is still under water, observe the surface of the leaf. What forms on the leaf? Why?

Table 5.1 Photosynthesis versus Respiration

Photosynthesis	Respiration
Uses water (H_2O)	Gives off water (H_2O)
Uses carbon dioxide (CO_2)	Gives off carbon dioxide (CO_2)
Makes glucose ($C_6H_{12}O_6$)	Breaks down glucose ($C_6H_{12}O_6$)
Gives off oxygen (O_2)	Uses oxygen (O_2)
Requires light	Occurs all the time; does not need light
Takes in light energy	Releases energy (ATP) stored in glucose ($C_6H_{12}O_6$)
Occurs only in cells with chlorophyll: cells of plants and some bacteria and protists	Occurs in cells of most organisms

Skills WorkOut

To help students observe one of the products of photosynthesis, have them do the Skills WorkOut.
Answer Bubbles of oxygen form on the surface of the freshly picked leaf. The leaf is still releasing oxygen as a waste product of photosynthesis.

Discuss

Have the students review Table 5.1. Ask the following questions:

▶ Why does photosynthesis require light? (Light provides energy for the reaction.)

▶ Why can respiration occur even if there is no light? (Energy is stored in the carbohydrates' chemical bonds.)

▶ In what kinds of organisms do photosynthesis and respiration occur? (Photosynthesis occurs in plant cells. Respiration occurs in plant and animal cells.)

Enrich

The compound ATP is an adenosine triphosphate. Have students look up the structure of the ATP molecule, write it down, and circle the bonds that store energy.

Answer to In-Text Question

① The products are carbon dioxide and water. The reactants are glucose and oxygen. Respiration and photosynthesis are complementary processes.

Integrated Learning

Use Integrating Worksheet 5.2.

The Living Textbook: Life Science Sides 7-8

Chapter 3 Frame 00421
Respiration (1 Frame)
Search:

INTEGRATED LEARNING

TEACH • Continued

Skills WorkOut
To help students distinguish chlorophyll from glucose, have them do the Skills WorkOut.
Answer Answers will vary. Green foods contain chlorophyll, and sweet-tasting foods contain glucose. Some foods contain both.
Extension Have students make a Venn diagram that shows which foods are sweet, which are green, and which are both sweet and green.

Class Activity
Have students list the foods they eat that have been changed by fermentation during processing. Write these items on the chalkboard. (Examples are yogurt, sour cream, many bread products, and so on.)

Career Corner
Research Have students compare the action of yeast in baking breads and pastries to the action of baking powder or baking soda in making cakes, cookies, and quick breads (those without yeast). They can consult a baker to find out how baking powder and soda work. Students can make a diagram of each process to show the similarities and differences. Ask them to decide which attributes of the substances make them important for baking.

STS Connection
Research Have students research how the fuel ethanol is produced. Ask them to write a report that explains how fermentation is involved in the process. Students can include descriptions of the raw materials, the equipment used, and the steps in the process.

Skills WorkOut
Classify List the vegetables and fruits you ate during the last three days. Divide the list into four columns: green, nongreen, sweet, and nonsweet. Identify the foods that contain chlorophyll. Identify those that contain glucose.

Fermentation
Respiration happens only if oxygen is present. When the cells do not have enough oxygen to carry out respiration, another energy-releasing process may take place. This process is called **fermentation**.

During both respiration and fermentation, glucose breaks down to form carbon dioxide and an acid. In respiration, the acid is broken down further in the presence of oxygen to make carbon dioxide, water, and many molecules of ATP. When no oxygen is present, fermentation occurs. The acid is broken down further to produce carbon dioxide, either alcohol or lactic acid, and a small amount of ATP.

The production of either lactic acid or alcohol is determined by the type of cell. For example, your muscle cells may produce lactic acid if oxygen is not present. The cells of certain fungi and bacteria produce lactic acid used in making yogurt and cheese. Yeast cells ferment during bread baking to produce alcohol.

Career Corner   Bread Baker

Who Bakes the Bread for Your Sandwiches?

Without understanding the fermentation process, the ancient Egyptians discovered that adding yeast to flour causes bread to rise. Bakers all over the world have carried on the practice for centuries.

Most bakers use only the basic ingredients in bread: yeast, flour, water, and salt. They knead and "set" the dough, then let it rise for 3 to 16 hours at a temperature around 29°C. The yeast causes the small amount of sugar in the flour to ferment. As the sugar breaks down, alcohol and carbon dioxide are released into the dough. Flour has an elastic substance, called gluten, that traps bubbles of carbon dioxide in the bread. As the bubbles expand, the dough rises.

A final burst of fermentation, called "oven spring," occurs during baking. The heat kills the yeast and evaporates the alcohol in the dough. Without yeast fermentation, the bread you eat would be flat and hard instead of fluffy and soft.

Baking bread requires special skills. Bakers must know the right balance of ingredients. They must be able to judge when dough has been kneaded to the right consistency. A good loaf of bread is allowed to rise and bake at just the right time. Bakers receive on-the-job training and often learn skills from more experienced bakers. You can also learn to be a commercial baker through courses offered in community colleges or vocational schools.

Chapter 5 Cell Processes

Integrating the Sciences

Health Have students do research and prepare brief oral reports on the benefits of a particular exercise program. Students should indicate in their reports whether or not the exercise program they describe is aerobic.

Chemistry Write the chemical formula for lactic acid ($CH_3CHOHCOOH$) on the chalkboard. Explain that lactic acid is what causes muscles to cramp. Students will be interested to find that lactic acid is used in foods and manufacturing, as well as in life processes. Have students research lactic acid.

Class Activity

Give each student a clothespin. Have students open and close the clothespin 50 times. After students have finished, ask the following questions:

▶ How does your hand feel? (Students may say tired, warm, or sore.)

▶ What substance is making your hand feel that way? (Lactic acid)

EVALUATE

Science and You *Chemistry of Sore Muscles*

Have you ever had a muscle cramp during or after exercise? What do you think caused it? You probably never knew that the pain you felt was caused by fermentation!

In your cells, glucose is constantly combining with oxygen to provide your body with the energy it needs. When you run, swim, or do other forms of exercise, the muscle cells doing the work need an extra supply of oxygen to carry out respiration. However, even though you are breathing hard, you may not be taking in as much oxygen as your tired muscles need.

Fortunately, your muscle cells store energy in the form of a carbohydrate called glycogen. When your cells do not have enough oxygen to carry out respiration, glycogen breaks down through the fermentation process. In this process, the glycogen changes into glucose, and then into carbon dioxide and lactic acid, releasing ATP.

If you breathe deeply after exercise, more oxygen is available to your muscle cells. Respiration then continues. Normally, the lactic acid produced by fermentation changes back into glucose or glycogen in your cells. However, the lactic acid formed through fermentation can accumulate in your muscle cells. If too much lactic acid builds up, it can cause a painful cramp or sore muscles.

Figure 5.6 ▲
When you don't get enough oxygen to muscles during exercise, muscle cells begin to produce energy by lactic acid fermentation. This runner will need to breathe deeply and walk after she finishes her race to avoid cramps and sore muscles.

WrapUp

Review Have students draw a plant, a person running, a cow, a hamburger, and the sun or cut out these pictures from magazines. Tell students to identify the energy source in each picture and make a diagram showing the energy relationships among the items.

Use Review Worksheet 5.2.

Check and Explain

1. Photosynthesis requires light energy, carbon dioxide, and water.

2. Photosynthesis means "putting together with light." Photosynthesis puts chemicals together in the presence of light to make glucose.

3. Photosynthesis uses energy from the sun, water, and carbon dioxide to make glucose and oxygen. Respiration combines glucose and oxygen, making carbon dioxide, energy (ATP) and water. The products of one process are the reactants of the other process.

4. Photosynthesis would take place in *a* and *b*.

Check and Explain

1. What three things are required for photosynthesis to occur?

2. Explain how the meaning of the word *photosynthesis* is related to the process.

3. **Compare** How do the products and reactants in photosynthesis compare to those in respiration?

4. **Predict** Which of these situations meet all three requirements for photosynthesis to occur?

   a. A house plant in a well-lighted room at midnight

   b. Sprouting seedlings in an open field on a sunny day

   c. A lawn during a severe drought

   d. A cornfield on an evening with no moonlight

Chapter 5 Cell Processes

ACTIVITY 5

Time 4–5 days **Group** pairs

Materials
30 adhesive labels
90 small plants
30 pots
15 measuring cups

Analysis

1. Location (light or dark)
2. Those kept in the dark may be yellowish or have died. Those in the light will have grown larger and will be green.
3. Answers depend on the plants used.
4. Photosynthesis is the process by which plants get energy to survive. Without light, photosynthesis can't occur, so plants will not grow.
5. Conclusions should follow from actual results.
6. Possible causes include varying temperature, humidity, the health of plants at the start.

Conclusion

Students' paragraphs should include data to support their answers to Analysis question 5.

Everyday Application

Photosynthesis can't occur, and plants may get sick or die. Different plants need different amounts of sunlight, so it is important that these needs are considered when locating them. Unhealthy plants could be helped by changing their location or by varying the water supplied.

Extension

Accept all logical responses. Experimental designs should show concern for controlled variables.

TEACHING OPTIONS

Prelab Discussion

Have students read the entire activity before beginning the discussion. Ask:

▶ Why is it important to control the variables in this lab, such as the amount of water each plant gets, the location of the plants, and the species of plants?

▶ What factors other than the amount of light may affect the growth of plants?

Activity 5   How does light affect plant growth?

Skills

Observe; Infer; Control Variables

Task 1   Prelab Prep

1. Collect the following items: 2 labels, 2 pots, 6 small plants of the same species and similar in size, measuring cup.
2. Number the labels *1* and *2* and date each one.
3. Find suitable locations for the plants: a place in the sunlight and a place in darkness.

Task 2   Data Record

1. Write a different location on each label:
   1. Sunlight   2. Darkness.
2. Copy the data table below onto a separate sheet of paper.
3. Use the data table to record your observations about the appearance of the plants in each pot.

Table 5.2   Plant Data

Date	Pot Number	Observation
	1 Sun	
	2 Dark	
	1 Sun	
	2 Dark	
	1 Sun	
	2 Dark	
	1 Sun	
	2 Dark	
	1 Sun	
	2 Dark	
	1 Sun	
	2 Dark	

Task 3   Procedure

1. Label each pot. Put three plants in each one. Place the pots in their locations.
2. Water the plants according to the needs of their species. Use the measuring cup to make sure all the plants get the same amount of water.
3. Record the appearance of the plants in each pot on the data table every other day. Do not remove the plants from their locations.

Task 4   Analysis

1. Identify the variable in this activity.
2. Which plants changed the most during the time you observed them? Which plants changed the least?
3. What were the light needs of the plants you used? Check the seed package or pot label on your plants.
4. What concept from this chapter relates to the change you observed in the plants?
5. What conclusion can you draw from the changes in the appearance of the plants? Explain.
6. Can you think of any other cause(s) that might have contributed to the effect you observed?

Task 5   Conclusion

Write a short paragraph explaining how sunlight affected the plants. Explain why.

Everyday Application

What happens to plants if they are not placed near sunlight? Explain why it is important to grow plants in the right locations. Then describe some different ways that you could bring unhealthy plants back to life.

Extension

Develop a hypothesis about the effect of different amounts of sunlight on plants. Write your procedures and test your hypothesis.

INTEGRATED LEARNING

Themes in Science

Patterns of Change/Cycles Cell growth and division proceed according to specific patterns. As students study Figures 5.7 and 5.8, stress the graphic representation of these cycles.

SECTION 5.3

Section Objectives
For a list of section objectives, see the Student Edition page.

Skills Objectives
Students should be able to:

Compare their height and weight over two years as it relates to the number of cells in their bodies.

Make a Model showing how telophase differs in plant and animal cells.

Vocabulary
mitosis, chromosome

MOTIVATE

Skills WarmUp
To help students understand how cell division affects them, have them do the Skills WarmUp.
Answer Students should note higher figures for height and weight than figures of two years ago. Most of their cells increased in number; students may note that their muscle and bone cells increased the most.

Misconceptions
Students may think that they can't observe the effect of cell growth and division because it takes place deep within their bodies. Discuss some effects of cell growth and division that they can perceive, such as the healing of a scratch or the growth of hair.

5.3 Cell Growth and Division

Objectives

▶ **Explain** how cell division affects living things.
▶ **Explain** the role of mitosis in cell division.
▶ **Describe** the various phases of mitosis.
▶ **Make models** of the phases of mitosis.

Skills WarmUp

Compare Recall your height and/or weight from two years ago. Compare these figures with your present height and weight. What kinds of cells in your body increased in number? What cells do you think increased the most?

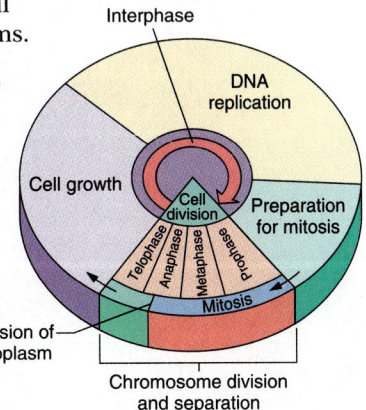

The cells that make up your body are constantly changing. For example, cells change when different substances enter and leave the cell. The size and number of cells you have also changes constantly. Individual cells get larger or they divide and make new cells. In an adult human, as many as 25 million new cells are made every second. Most new cells replace damaged or dead cells.

Cell Division

Your body makes new cells during a process called *cell division*. When cell division takes place, two identical cells are produced. In single-celled organisms, cell division results in the formation of two new organisms. In many-celled organisms, cell division increases the number of cells making up the organism. Because of cell division, a many-celled organism can grow and develop from a single cell, the fertilized egg. As the cells increase in number, the organism grows.

Figure 5.7 shows the cycle of cell growth and division. A cell may take a few hours or a few days to complete one cycle. Most of a cell's life is spent in a phase of growth and development called interphase. During interphase, the amount of DNA in the nucleus is doubled. To form a complete new cell, both the nucleus and the cytoplasm must divide. The amount of time it takes a cell to divide depends on the type of cell, the temperature, and the nutrients in the cell. The process by which a cell nucleus divides into two identical nuclei is called **mitosis** (my TOH sis). As you can see in Figure 5.7, the process of mitosis occurs in several phases.

Figure 5.7 ▲
During a cell cycle, a cell grows, prepares for division, and divides into two identical cells, each of which begins the cycle anew.

Chapter 5 Cell Processes   103

TEACH

Explore Visually

Have students study the drawings and micrographs of the animal cell cycle in Figure 5.8 and read the description of each of the phases. Then ask the students some or all of the following questions:

▶ What happens to the cell during interphase? (It doubles in size, and makes proteins, enzymes, and mitochondria. Centrioles duplicate in the cytoplasm.)

▶ What happens to each of the chromosomes in the nucleus during interphase, and why is this important? (Each of the chromosomes forms an exact copy of itself. This is important because genetic material must be passed to the new cells.)

▶ What happens to each of the chromosomes, each now called a chromatid, in the center of the nucleus during prophase? (They form short, thick rods and pair up in the center of the nucleus.)

Reteach
Use Reteach Worksheet 5.3.

**The Living Textbook:
Life Science Sides 1-4**

Chapter 5 Frame 01651
Animal Cell Division (3 Frames)
Search: Step:

104

TEACHING OPTIONS

Cooperative Learning

Have students work together in groups to act out the phases of mitosis. Assign each student a role. Before enacting the process, have students discuss their roles and how they will interact with other parts of the cell during mitosis. One or more students can be appointed to help coordinate the activity of each phase.

Phases of Mitosis Recall that the nucleus controls all the activities of the cell. More specifically, the threadlike strands called **chromosomes** (KROH muh ZOHMZ), located in the nucleus, control cell activity. Chromosomes also carry the genetic material that passes the traits of parents to their offspring.

Look at Figure 5.8. It shows the cycle of growth and division that takes place in animal cells. Notice that the cycle starts with interphase and continues through the four phases of mitosis.

**Figure 5.8
Animal Cell Cycle**

Telophase
When the chromosomes reach opposite ends of the cell, the spindle breaks up into proteins. A nuclear membrane forms around each set of single chromosomes. Two identical nuclei now exist. A furrow forms at the center of the cell membrane. The furrow deepens and divides the cell completely.

Anaphase
The spindle fibers pull the chromatids apart, separating each one from its duplicate. Each chromosome moves to opposite sides of the cell. The cell now has two identical sets of single chromosomes.

104   Chapter 5   Cell Processes

INTEGRATED LEARNING

Language Arts Connection

Have students find the meaning of the prefix for each phase of cell growth and division: *inter-*, *pro-*, *meta-*, *ana-*, and *telo-*. Have students discuss how the meaning of each prefix signifies what occurs in each phase.

Interphase
During this phase, the cell makes enough proteins, enzymes, mitochondria, and other substances for two cells. In the cytoplasm, structures called *centrioles* duplicate to form two pairs. In the nucleus, each chromosome forms an exact copy of itself, or replicates. These spread through the nucleus in threadlike strands.

Metaphase
The centromeres of the chromatid pairs are lined up in the center of the cell. Each individual chromatid is attached to a spindle fiber.

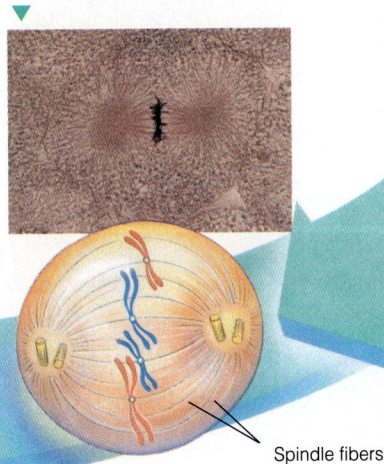

Prophase ▲
The chromosomes, now called chromatids, form short, thick rods and pair up in the center of the nucleus. A centromere connects the two halves of each double chromatid. In the cytoplasm, a fibrous structure, called the spindle, forms between the centriole pairs. The centrioles move to opposite sides of the cell, and the nuclear membrane breaks apart.

▶ When the centromeres of the chromatid pairs line up in the center of the cell during metaphase, what is each of the chromatids attached to? (A spindle fiber)

▶ What happens to the chromatids during anaphase? (The chromatids are separated and the chromosomes move to opposite sides of the cell.)

▶ At which stage are there two complete sets of chromosomes in one nucleus? (Anaphase)

▶ What forms around each set of chromosomes during telophase? (A nuclear membrane forms around each set of chromosomes, creating two identical nuclei.)

The Living Textbook: Life Science Side 2

Chapter 10   Frame 08398
Cell Division (Movie)
Search: Play:

TEACH • Continued

Class Activity
Have students use hand lenses to observe Figure 5.9. Challenge them to identify the phase of mitosis that is pictured in different cells.

EVALUATE

WrapUp
Reinforce Have students apply what they have learned in this section by asking them to make a phase-by-phase, three-dimensional model of the phases of mitosis. Have students use clay, string, yarn, wire, or other materials to construct their models. The final phase should include models of two identical cells.

Use Review Worksheet 5.3.

Check and Explain
1. A single-celled organism divides to form two new organisms. A many-celled organism grows and develops because the cell division occurring within the cell increases the number of cells making up the organism.
2. The four phases of mitosis are prophase, metaphase, anaphase, and telophase.
3. The nuclear membrane must break apart and the spindle must form to prepare for cell division.
4. Check student answers for accuracy.

The Living Textbook: Life Science Sides 1-4
Chapter 5   Frame 01657
Plant Cell Division (6 Frames)
Search: Step:

INTEGRATED LEARNING

STS Connection
Research Have students choose one type of cancer to research. Have them find out about some of the technology used to diagnose and treat that type of cancer. Ask students to prepare written reports. Display their reports so students can share information.

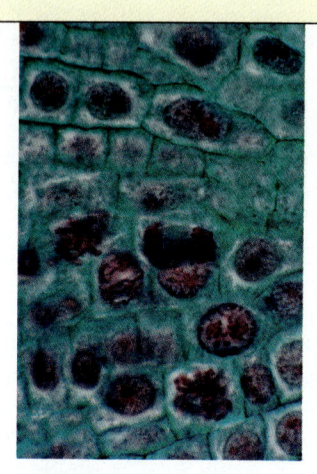

Figure 5.9 ▲
The cells at the tip of this corn root are in various stages of their cell cycles. The cell in anaphase has a cell plate beginning to form across the middle of the cell.

Mitosis in Plants Cell division differs slightly in plants and animals. In plant cells, spindle fibers don't attach to centrioles during mitosis. The rigid plant cell wall does not form a furrow in the middle as the animal cell membrane does. Instead, a structure made of liquid-filled pouches called the cell plate forms across the middle of the dividing cell. As this plate grows, its edges fuse with the old wall to divide the cell into two. Each half of the cell then forms its own cell wall. When the cell walls are complete, the newly formed cells take on nutrients and increase in size.

Science and Society
Cell Division out of Control

Have you ever wondered what cancer is? Think about how the nucleus controls the cell and tells it when to stop dividing. Scientific research shows that cancer develops when cells divide at a faster-than-normal rate. These fast-dividing cells are abnormal. They may have unusual chromosomes. Parts of the cell may be missing or out of place. The cell surface is altered. Cancer cells can invade tissue in different parts of the body.

For years, doctors treated all cancers by removing the cancerous tissue. Through research, radiation and chemotherapy were found to be useful tools for treating cancer. Scientists have identified many *agents* that may contribute to cancer in humans. Among these are excess exposure to certain chemicals, ultraviolet radiation, and cigarette smoke. However, the best cure for cancer would be learning what triggers cancerous growth in the first place.

Check and Explain
1. How can cell division affect a single-celled organism? A many-celled organism?
2. What are the four phases of mitosis?
3. **Find Causes** Why is it important for the nuclear membrane to break apart?
4. **Make a Model** Draw the telophase of a plant cell and the telophase of an animal cell. Show clearly how this phase differs in plants and animals.

Chapter 5 Cell Processes

Chapter 5 Review

Concept Summary

5.1 Movement of Substances
- Cell membranes regulate the passage of materials in and out of the cell.
- Some substances, like oxygen and water, diffuse freely through the cell membrane by osmosis.
- Large molecules, like sugars and amino acids, must pass through the membrane with carrier proteins.
- A cell membrane is selectively permeable because it is impermeable to some materials and permeable to others.

5.2 Energy Processes
- During photosynthesis, light energy is used to change carbon dioxide and water into glucose and oxygen.
- During respiration, oxygen is combined with glucose to release carbon dioxide, water, and ATP.
- Fermentation is an energy-releasing process that produces carbon dioxide, alcohol or lactic acid, and ATP.

5.3 Cell Growth and Division
- All cells have a cycle of growth and division.
- The period of growth and development in a cell is called interphase. During interphase, the cell doubles in size.
- The cell's nucleus and cytoplasm divide during mitosis. Mitosis has four phases: prophase, metaphase, anaphase, and telophase.

Chapter Vocabulary
diffusion (5.1) photosynthesis (5.2) fermentation (5.2) chromosome (5.3)
osmosis (5.1) respiration (5.2) mitosis (5.3)

Check Your Vocabulary
Use the vocabulary words above to complete the following sentences correctly.

1. The process of breaking down glucose with oxygen is called ____ .
2. The diffusion of water across the cell membrane is called ____ .
3. The process by which plants convert light energy to chemical energy is ____ .
4. Movement of a substance from a high concentration to a low concentration is ____ .
5. A cell divides into two identical nuclei during ____ .
6. Respiration releases more ATP energy from glucose than ____ does.
7. Cell activity is controlled by ____ in a cell's nucleus.

Identify the word or term in each group that does not belong. Explain why it does not belong.

8. Photosynthesis, respiration, fermentation
9. Osmosis, diffusion, mitosis
10. Chromosome, mitosis, fermentation
11. Mitochondria, centromere, chromatid
12. Carbon dioxide, water, spindle fiber
13. Mitochondria, mitosis, respiration
14. Fermentation, chlorophyll, photosynthesis
15. Interphase, anaphase, telophase

Write Your Vocabulary
Write sentences using each vocabulary word above. Show that you know what each word means.

CHAPTER REVIEW 5

Check Your Vocabulary
1. respiration
2. osmosis
3. photosynthesis
4. diffusion
5. mitosis
6. fermentation
7. chromosomes

For some of the following answers, students' responses may vary. Some logical possibilities follow.

8. Photosynthesis; others are ways energy is released from glucose.
9. Mitosis; others are ways substances move into and out of a cell.
10. Fermentation; others are involved in cell division.
11. Mitochondria; others are involved in cell division.
12. Spindle fiber; others are raw materials for photosynthesis.
13. Mitosis; respiration occurs partly in mitochondria.
14. Fermentation; chlorophyll is needed for photosynthesis.
15. Interphase; others are steps during which chromosomes are dividing and separating.

Write Your Vocabulary
Be sure that students understand how to use each vocabulary word in a sentence before they begin.

Use Vocabulary Worksheet for Chapter 5.

CHAPTER 5 REVIEW

Check Your Knowledge

1. Phases of mitosis are prophase, metaphase, anaphase, and telophase.
2. Oxygen and glucose are the reactants of respiration.
3. Diffusion refers to the movement of any substance from a region of higher concentration to a region of lower concentration. Osmosis refers specifically to the diffusion of water through a membrane.
4. The membrane has small gaps in it that are too small for some molecules to pass through.
5. The products of fermentation are alcohol or lactic acid, carbon dioxide, and a small amount of ATP.
6. The cell plate functions to separate the two newly formed cells and provide a foundation for the new cell wall between the cells.
7. Photosynthesis:
   $6CO_2 + 6H_2O + \text{Light Energy} \longrightarrow C_6H_{12}O_6 + 6O_2$
   Respiration:
   $C_6H_{12}O_6 + 6O_2 \longrightarrow 6CO_2 + 6H_2O + \text{Energy (ATP)}$
8. Accept all logical answers; examples: water, oxygen, carbon dioxide
9. Telophase
10. True
11. False; chlorophyll
12. False; present
13. False; the air
14. False; cell membrane
15. False; telophase
16. True
17. True
18. False; chromatids

Check Your Understanding

1. A given substance will pass freely through a permeable membrane, but will not pass freely through an impermeable membrane.
2. a. Cloudy weather means less sun energy—photosynthesis would slow down.
   b. Without water, photosynthesis cannot occur.
   c. At night, there is even less light energy than on a cloudy day, so photosynthesis would likely stop.
3. In plant cells, the spindle fibers do not attach to centrioles, the cell plate forms across the middle of the dividing cell, and a cell wall develops between the two newly formed cells.
4. a. Plants use photosynthesis to obtain energy for their cells to use for cell processes.
   b. The chloroplasts are the structures inside a plant cell where photosynthesis actually occurs.
5. Make sure that students include the formation of the cell plate and cell wall, in addition to the other steps described in Figure 5.8.
6. Accept all logical answers. An example would be a drop of spilled liquid spreading through cloth fibers.
7. Large molecules get into a cell with the help of protein carrier molecules embedded in the cell membrane.
8. Students' tables should show all or some of the following entries under photosynthesis (P), respiration (R), and fermentation (F): raw materials—(P) water, carbon dioxide, energy; (R) oxygen, glucose; (F) glucose; products—(P) glucose, oxygen; (R) carbon dioxide, water, ATP; (F) carbon dioxide, ATP, alcohol or lactic acid; energy—(P) sunlight; (R) ATP; (F) ATP; location—(P) chloroplasts; (R) cytoplasm, mitochondria; (F) cytoplasm, mitochondria.
9. Mitochondria are the structures in the cell where fuel (glucose) is used to produce energy (ATP) for the cell.

Chapter 5 Review

Check Your Knowledge
Answer the following in complete sentences.

1. What are the phases of mitosis?
2. What are the reactants of respiration?
3. How does diffusion differ from osmosis?
4. What makes the structure of a cell membrane selectively permeable?
5. List the products of fermentation.
6. What is the function of a cell plate?
7. Give the overall equations for photosynthesis and respiration.
8. Name two substances that diffuse freely across a cell membrane.
9. During what phase of mitosis does the cell divide into two new cells?

Determine whether each statement is true or false. Write *true* if it is true. If it is false, change the underlined term to make the statement true.

10. For substances that diffuse freely in and out of cells, the membrane is permeable.
11. The chloroplasts capture light energy from the sun.
12. Respiration can only happen if oxygen is absent.
13. Plants get carbon dioxide from soil.
14. Everything that passes in or out of a cell must move through the centrioles.
15. Cell division occurs during interphase.
16. In a cell membrane, protein molecules carry sugars into the cell.
17. Respiration begins in the cytoplasm of a cell and ends in the mitochondria.
18. During anaphase, spindle fibers pull the cytoplasm apart.

Check Your Understanding
Apply the concepts you have learned to answer each question.

1. What is the difference between a permeable membrane and an impermeable membrane?
2. How could each of the following affect photosynthesis in a plant:
   a. Cloudy weather
   b. Lack of water
   c. Nighttime
3. Describe how mitosis differs in plants and animals.
4. **Mystery Photo** The photo on page 92 shows a chloroplast in the leaf of a pea plant. The large white object in the cell is a white starch body, a product of photosynthesis.
   a. Why do plants use photosynthesis?
   b. What is the role of the chloroplasts during photosynthesis?
5. **Make a Model** Draw a picture of a plant cell cycle. Be sure to label each stage as well as the cell parts. Clearly show what is happening in the cell at each stage.
6. **Apply** Give an example of diffusion in your everyday life.
7. Describe how large molecules like sugars and amino acids get into a cell.
8. **Make a Table** Make a table comparing photosynthesis, respiration, and fermentation in the following areas: raw materials, products, energy, location.
9. Explain why the mitochondria are sometimes called the "power houses" of a cell.

Develop Your Skills

1. a. Yes; no; substance 2 is too large to pass through the gaps in the membrane, while substance 1 is not
   b. Substance 1 moves from left to right because there is a lower concentration of substance 1 on the right side.
   c. Substance 2 is too large to flow through the membrane.
   d. It is an example of osmosis. Osmosis is diffusion through a cell membrane.
2. a. Brain cells; skin cells or white blood cells
   b. Red blood cells and brain cells
3. Students should recognize that some water will be absorbed by the beans, lowering the water level.

Make Connections

1.

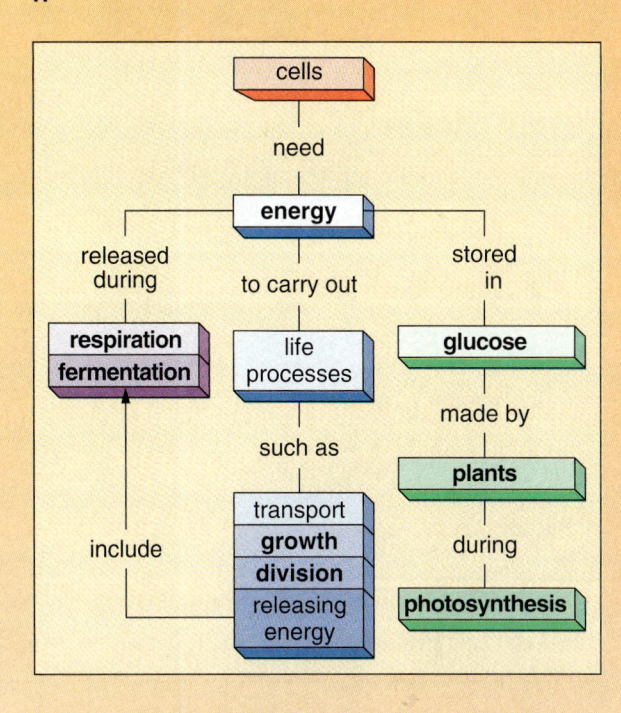

```
                        cells
                          |
                         need
                          |
                        energy
        released          |         stored
         during       to carry out    in
           |              |            |
      respiration        life        glucose
      fermentation     processes        |
                          |          made by
                       such as          |
                          |          plants
                       transport        |
        include         growth        during
           |           division         |
                      releasing    photosynthesis
                       energy
```

2. Students' answers should show that a better understanding of fermentation as a metabolic process has led to advances in its use in food making.
3. Students' writing might include points of interest that would help a feature story, such as disagreements between the researchers both then and now.
4. Bubbling in the dough causes it to rise, which can be observed. The alcohol has evaporated. Students' answers should include details that show they have actually observed the process.

Develop Your Skills

Use the skills you have developed to complete each activity.

1. **Interpret Data** The drawing below shows two substances in water. A membrane divides the water in half.
   a. Is the membrane permeable to substance 1? Substance 2? How do you know?
   b. In which direction will substance 1 flow? Why?
   c. In which direction will substance 2 flow? Why?
   d. Is this an example of diffusion or osmosis? Explain.

   ○ Substance 1   ◻ Substance 2

2. **Data Bank** Use the information on page 626 to answer the following questions.
   a. Which body cell has the longest cell life? Shortest cell life?
   b. Which cells live at least one week?

3. **Predict** Suppose you put 30 beans in a jar and then covered the beans with water. Then you marked the water level on the jar with a pen. What would happen to the water level in the jar? Explain.

Make Connections

1. **Link the Concepts** Below is a concept map showing how some of the main concepts in this chapter link together. Only parts of the map are filled in. Copy the map. Using words and ideas from the chapter, complete the map.

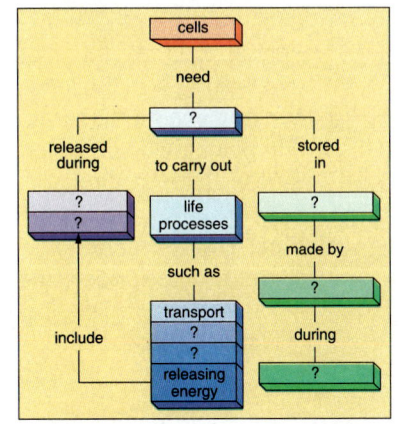

2. **Science and Technology** Throughout history people have used the process of fermentation to make certain kinds of food. Research the use of fermentation in a particular food process, such as cheese or yogurt making.

3. **Science and Language Arts** Research the story of how the steps in mitosis were discovered. Write a short story of the discovery for a magazine.

4. **Science and You** Make bread or cake following a recipe that uses yeast. Watch for signs of fermentation. How do you know that fermentation is occurring? Why do you not taste the alcohol in bread or cake made with yeast?

CHAPTER 6

Overview

This chapter explores fundamental theories and applications of heredity. The first section describes the work of Gregor Mendel and outlines methods by which scientists trace inherited traits. Chromosomes and the process of meiosis are discussed in the second section. The third section includes a description of the structure of chromosomes, introduces the concept of DNA replication, and discusses genetic mutations. The last section describes applied genetics and genetic engineering.

Advance Planner

▶ Supply colored buttons, plastic paper clips, or candies for SE page 125.

Skills Development Chart

Sections	Classify	Collect Data	Communicate	Infer	Interpret Data	Model	Predict	Research
6.1 Skills WarmUp Skills WorkOut SkillBuilder	●			●			●	
6.2 Skills WarmUp Activity		●		●	● ●	●		
6.3 Skills WarmUp Skills WorkOut				●		●		
6.4 Skills WarmUp Historical Notebook			● ●					●

Individual Needs

▶ **Limited English Proficiency Students** Stress to students that it is important to learn the vocabulary of this chapter in the order in which it is presented. Have students write all of the vocabulary words in their notebooks. Have students use their own words to define each term. Then ask them to compare the meanings of related terms. For example, have students write a sentence describing the difference between the following pairs of terms: *dominance* and *incomplete dominance, genotype* and *phenotype*.

▶ **At-Risk Students** Have students write each of the section titles on a separate page in their science journals. Have them choose four important ideas in each section and copy these onto the journal pages. Have them explain why each of these ideas is important to them.

▶ **Gifted Students** Have students develop a plan for breeding a citrus fruit that is frost resistant. Encourage them to research the latest technology for breeding hybrid fruit. Students can create a poster containing an advertisement for their new fruit and explaining the process they use to produce it.

Resource Bank

▶ **Bulletin Board** Illustrate four of the traits of pea plants shown on page 112 and indicate which traits are dominant and which recessive. (Dominant traits are yellow seed, round seed, smooth seed, and tall plant.) Make letters representing each trait and place them under the illustrations. Make a Punnet square in the middle of the board and have students use it along with the letters to predict the genotypes for the first and second generations of offspring in a cross between one plant that is pure dominant for a trait and one that is pure recessive for that trait.

▶ **Field Trip** Arrange to have students visit the produce section of a local supermarket to observe examples of hybrid fruits and vegetables. If possible, have the produce manager tell students about some of the traits that growers try to get when they grow fruits and vegetables.

CHAPTER 6 PLANNING GUIDE

Section	Core	Standard	Enriched	Section	Core	Standard	Enriched
6.1 Basic Principles of Heredity pp. 111–116				**6.3 DNA and the Genetic Code** pp. 122–126			
Section Features Skills WarmUp, p. 111 Skills WorkOut, p. 113 SkillBuilder, p. 115	● ●	● ● ●	● ● ●	Section Features Skills WarmUp, p. 122 Skills WorkOut, p. 125	● ●	● ●	● ●
Blackline Masters Review Worksheet 6.1 Skills Worksheet 6.1 Integrating Worksheet 6.1	● ● ●	● ● ●	● ● ●	Blackline Masters Review Worksheet 6.3 Integrating Worksheet 6.3	●	● ●	● ●
Overhead Blackline Transparencies Overhead Blackline Master 6.1 and Student Worksheet	●	●		Ancillary Options Multicultural Perspectives, p. 85		●	●
Laboratory Program Investigation 12	●	●	●	Overhead Blackline Transparencies Overhead Blackline Master 6.3 and Student Worksheet	●	●	●
Color Transparencies Transparencies 15, 16	●	●		Laboratory Program Investigation 13	●	●	●
6.2 Chromosomes and Inheritance pp. 117–121				Color Transparencies Transparency 19	●	●	
Section Features Skills WarmUp, p. 117 Activity, p. 121	● ●	● ●	● ●	**6.4 Applied Genetics** pp. 127–130			
Blackline Masters Review Worksheet 6.2 Reteach Worksheet 6.2 Skills Worksheet 6.2 Integrating Worksheet 6.2	● ● ●	● ● ●	● ● ● 	Section Features Skills WarmUp, p. 127 Historical Notebook, p. 128	● ●	● ●	● ●
Color Transparencies Transparencies 17, 18	●	●		Blackline Masters Review Worksheet 6.4 Vocabulary Worksheet 6	● ●	● ●	● ●

Bibliography

The following resources can be used for teaching the chapter. See page T-40 for supplier codes.

Audio-Visual Sources
(video unless noted)
DNA. 16 min. 1987. A-W.
DNA: Laboratory of Life. 21 min film or video. 1985. NGSES.
The Gene. 45 min video. 1992. HA.
Genetic Engineering. 2 filmstrips with cassettes, 17–18 min each. 1988. NGSES.
An Introduction to Heredity. 2 filmstrips with cassettes, 17–18 min each. 1983. NGSES.

Software Resources
Heredity. Apple. J & S.
Heredity Dog. Tutorial with graphics. Apple, Commodore. HRM.
Human Genetics. Apple. FM.
Meiosis. Apple, IBM. EME.

Library Resources
Asimov, Isaac. *How Did We Find Out About DNA?* New York: Walker, 1985.
Rosenfield, Israel, Edward Ziff, and Borin Van Loon. *DNA for Beginners.* London: Writers and Readers, 1983.
Shine, I., and S. Wrobel. *Thomas Hunt Morgan: Pioneer of Genetics.* Lexington, KY: University of Kentucky Press, 1976.
Stableford, Brian. *Future Man: Brave New World or Genetic Nightmare?* New York: Crown Publishers, 1984.
Watson, J. D. *The Double Helix.* New York: Atheneum, 1968.

CHAPTER 6

Introducing the Chapter

Have students read the description of the photograph on page 110. Ask if they agree or disagree with the description. Explain that not all of the colors result from breeding; some of the coloration has to do with conditions in which the plant grew—such as the mineral content of the soil.

Directed Inquiry

Have students study the photograph. Ask:

▶ How does the ear of corn in the picture differ from others you have seen? (Different colored kernels on the same ear)

▶ What does the picture have to do with the topic of this chapter? (Heredity has to do with the way traits are passed from parents to offspring.)

▶ What are some examples of inherited traits that you have seen in the plants, animals, and people around you? (Possible answers include size; types of leaves and flowers; or hair, fur, and eye color.)

Chapter Vocabulary

dominant
gamete
gene
genetics
genotype
heredity
hybrid
incomplete dominance
meiosis
mutation
phenotype
recessive
replication
X chromosome
Y chromosome

INTEGRATED LEARNING

Writing Connection

After students have studied page 110, have them write a news article featuring an imaginary interview with the person who produced this ear of corn. Questions might include the following:

▶ What do you call this kind of corn?

▶ Is there a special need for this kind of corn?

▶ What other varieties do you grow?

Encourage students to be creative and to add details about the grower, the crop, and the places where this corn is grown and used.

Chapter 6 Heredity

Chapter Sections

6.1 Basic Principles of Heredity

6.2 Chromosomes and Inheritance

6.3 DNA and the Genetic Code

6.4 Applied Genetics

What do you see?

"The photo is a close-up of some corn. The corn was bred to have different colors. Some people think that it looks pretty. They put it in a display, usually in the fall."

*Jennette Piry
Mason Middle School
St. Louis, Missouri*

To find out about the photograph, look on page 132. As you read the chapter, you will learn more about how characteristics are passed from generation to generation.

Themes in Science
Patterns of Change/Cycles
Genetic traits are passed from parent to offspring. Have students list some traits that they think they have inherited. Have them make a second list of traits that they think were not inherited. Have students compare the two lists. Ask: What do the inherited traits have in common? What do the noninherited traits have in common?

Language Arts Connection
Have students bring in photographs of themselves and make a list of adjectives to describe characteristics that appear in the photograph, and characteristics that do not appear. Students can attach the photographs to the lists. Display the photographs around the classroom.

6.1 Basic Principles of Heredity

Objectives
- **Describe** the work of Gregor Mendel.
- **Distinguish** between dominant and recessive genes.
- **Explain** the difference between phenotype and genotype.
- **Predict** the results of a simple genetic cross using a Punnett square.

Skills WarmUp

Classify Write the characteristics that best describe you. Mix your description with the others in the class. Select one at random. Match the description you select with the person it describes. What characteristic was most helpful?

How would you describe yourself so you could be found in a crowd? What characteristics help to identify you? Notice the many differences among the people in the crowd shown in Figure 6.1. The people have different skin color and hair color. Their noses and ears are shaped differently. These are just a few characteristics, called traits, that make each person unique. All living things are identified by their traits.

Where did your traits come from? You inherited each trait from your parents. The passing of traits from parents to offspring is called **heredity**. The genetic material of your parents combined to form a genetic code that determines all your traits.

Mendel's Experiments

In the 1860s, an Austrian monk named Gregor Mendel made the first detailed investigation of how traits are inherited. Mendel studied the heredity of pea plants he grew in the monastery garden. The conclusions he reached in his research form the foundation of **genetics**, the study of inheritance.

Pea plants were ideal for Mendel's study because they were easy to grow. More importantly, peas have several traits that occur in just two easy-to-recognize forms. For example, the plants are either tall or short, and the seeds are either green or yellow. The structure of pea flowers was also important because Mendel could control the pollination of each plant. He could either let a pea plant pollinate itself, or he could cross-pollinate one pea plant

Figure 6.1 ▲
Each of these people has a unique set of traits determined by his or her heredity.

SECTION 6.1

Section Objectives
For a list of section objectives, see the Student Edition page.

Skills Objectives
Students should be able to:

Classify others by characteristics.

Infer how a white-furred guinea pig could be born to black-furred parents.

Predict the results of genetic crosses using Punnett squares.

Vocabulary
heredity, genetics, gene, dominant, recessive, phenotype, genotype, hybrid, gamete, incomplete dominance

MOTIVATE

Skills WarmUp
To help students understand how heredity and characteristics are related, have them do the Skills WarmUp.
Answer Answers will vary. Most students will mention characteristics such as height, weight, hair color, or talent.

Prior Knowledge
To gauge how much students know about heredity, ask the following questions:
- What is heredity?
- Can you think of some traits that are passed from parents to offspring?

The Living Textbook: Life Science Side 7
Chapter 39 Frame 44304
Color Variation (Movie)
Search: Play:

Chapter 6 Heredity 111

TEACH

Discuss

Lead students through Mendel['s] process of achieving pure pla[nts]. Ask:

▶ How long would Mendel ha[ve] had to grow pea plants to m[ake] sure they were pure? (He w[ould] have had to grow them for [two] years.)

▶ In what way did he have to pollinate these plants? Why? (He had to self-pollinate them so he could be sure that they were pure.)

Skills Development

Predict Have students predict what might happen in this situation: A plant having pure round seeds was crossed with a plant having pure wrinkled seeds. In the first generation, all the plants produced round seeds. What kinds of seeds do you think will be produced in the second generation? Why? (Some seeds will be round and some seeds will be wrinkled. The trait for wrinkled seeds was passed on, even though it was hidden in the first generation.)

Answer to In-Text Question

① The short trait was hidden.

The Living Textbook: Life Science Side 7
Chapter 38 Frame 43731
Pea Plant Crossing (Animation)
Search: Play:

112

INTEGRATED LEARNING

Math Connection

Mendel studied seven different traits of peas, with 14 possible expressions, or forms. Seven traits were dominant and seven were recessive. Tell students to assume that each pea plant carried all seven traits. Have them calculate how many different combinations of traits Mendel could produce by mixing traits to produce a single pea plant. (2^7, or $2 \times 2 \times 2 \times 2 \times 2 \times 2 \times 2 = 128$)

Seed shape	Seed color	Seed coat color	Pod shape	Pod color	Flower position	Stem length
Round	Yellow	Colored	Smooth	Green	Side	Tall
Wrinkled	Green	White	Pinched	Yellow	End	Short

▲ Figure 6.2 Traits of Peas Studied by Mendel

Figure 6.3 ▲
What trait was hidden in the first generation of offspring? ①

(Parent generation: Tall × Short; First generation: All tall; Second generation: Tall, Tall, Tall, Short)

with another. By crossing two different plants, Mendel was able to see how the traits of two parents showed up in the offspring.

Look at Figure 6.2. Mendel studied the inheritance of seven different traits. To start, he made sure that he had pea varieties in which each of these traits stayed the same, generation after generation. For example, he grew tall plants that produced only tall plants. Mendel called such plants pure-breeding, or pure. When the pure plants were self-pollinated, they produced only offspring like themselves.

Mendel began his experiments by studying the stem length of peas. He took pure tall plants and crossed them with each other. Like self-pollinated pure tall plants, they produced only tall plants. He crossed pure short plants. Just as he expected, the pure short plants produced only plants with a stem length like their own.

Next, Mendel crossed a pure short plant and a pure tall plant. This time, the results were a surprise. All the seeds that came from crossing pure tall and pure short plants grew into tall plants. However, when these tall offspring crossed with each other, the next generation had both tall and short plants! The trait for shortness disappeared in the first generation, but it reappeared in the second. This experiment is shown in Figure 6.3.

112 Chapter 6 Heredity

STS Connection

In 1956, scientists in Brazil were working to cross an aggressive bee that made lots of honey with other bees that were more docile. They had hoped to create a docile bee that would make lots of honey. Instead, a new kind of bee appeared, which was even more aggressive than its genetic parents. To the alarm of U.S. citizens, populations of this new breed, nicknamed the "killer bee," have settled in places as far north as Dallas, Texas. Ask students to imagine the best and worst outcomes in trying to achieve desired combinations of traits in farm animals or pets.

Patterns of Inheritance

Mendel tried similar crosses with other traits and found the same pattern. In each case, crossing plants with different pure traits produced offspring with only one of the traits. The other trait was somehow hidden. When these offspring were crossed with each other, the hidden trait showed up in some of their offspring.

Mendel came up with an important hypothesis that fit these observations. He decided that a trait is determined by a pair of "factors." When two pea plants are crossed, he concluded, each parent must contribute *just one* factor to each offspring. In this way, a pair of factors passes to the offspring. Mendel also decided that a factor could be either strong or weak. When a strong factor and a weak factor paired, the strong one determined the trait in the offspring. Only when two weak factors paired could a weak factor express its trait.

Dominant and Recessive Genes Today Mendel's factors are called **genes**. A gene is a unit of genetic material that determines a trait. As Mendel discovered, many genes exist in two forms. The strong form Mendel described is called a **dominant** gene. The weak form is called a **recessive** gene. In peas the gene for tallness is dominant and the gene for shortness is recessive.

Genotypes and Phenotypes One of Mendel's important findings was that the traits a plant expressed differed from the factors, or genes, the plant inherited. Either two tall genes *or* a tall gene and a short gene could produce a tall plant. The trait that an organism actually shows is called the **phenotype** (FEE noh TYP). The gene combination that determines the phenotype is called the **genotype** (JEHN oh TYP). When a genotype is a combination of a dominant and a recessive gene, it is called a **hybrid**.

Genotypes are written with letter symbols. A capital letter is used for the dominant gene. The same letter in lowercase is used for the recessive gene. For example, T is a symbol for tallness, and t for shortness. The symbols TT, tt, and Tt represent the three different genotypes for stem length found in pea plants. Which two genotypes produce tall plants? Which is the hybrid genotype? The genotypes of the peas in Mendel's cross are shown in Figure 6.4.

Skills WorkOut

Infer A guinea pig with white fur is born to parents with black fur. However, in guinea pigs, black fur is the dominant trait over white fur. Explain how this could have happened.

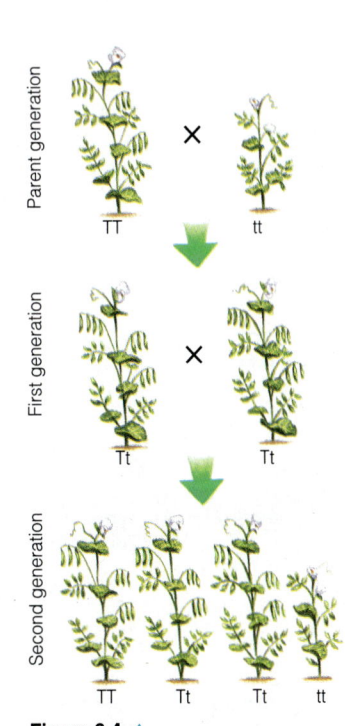

Figure 6.4 ▲
Compare the genotypes of these pea plants to their phenotypes.

Skills WorkOut

To help students infer how offspring can be different from their parents, have them do the Skills WorkOut.

Answer White fur is a recessive trait that could show up in some of the offspring only if the parents had the hidden genes for this trait.

Reteach

Explain to students that a phenotype for a trait is merely what's visible (what has been expressed) and does not necessarily mean that the trait is pure. It's possible that a recessive form of the trait is hidden and could be expressed in a later generation. Explain that the actual gene combination for the trait is the genotype. Ask:

▶ If two tall plants produce some offspring that are short, what are the genotypes of the parents? (Tt)

▶ If all the offspring of the parents are tall for generations, what are the genotypes of the parents? (TT)

Answer to In-Text Question

② The two genotypes that produce tall plants are TT and Tt. The hybrid genotype is Tt.

**The Living Textbook:
Life Science Sides 7-8**

Chapter 10   Frame 01203
Fruit Fly Genotype and Phenotype
(2 Frames)
Search: Step:

TEACH ▪ Continued

Skills Development

Communicate Ask students to explain the following in their own words: How does the combination of genes result in the different forms of offspring in Figure 6.5? (Answers may vary. For example, each parent could have one dominant gene and one recessive gene. The genes split to form gametes, resulting in four possible combinations of gene pairs that would be distributed among offspring.)

Discuss

Read through the principles of genetics with students. Stress that the word *segregate* means "separate." Help them understand that genes segregate independently. For example, the gene for tall plants can go to a gamete containing a gene for wrinkled seeds or to one containing a gene for round seeds.

Critical Thinking

Reason and Conclude Ask students what the evolutionary advantage of independent segregation might be. (If genes combine in a variety of ways, there will be a variety of traits in the offspring. This helps offspring meet challenges of the environment in different ways.)

114

INTEGRATED LEARNING

Math Connection

Remind students that only one gene from a pair is passed along to offspring from one parent. The genes are passed at random. Although the odds for each result are 50 percent, actual results might vary. Ask students to toss a penny and dime four times and record the number of heads and tails. Then have them increase the sample size by tossing the coins 40 more times. How did the results vary?

Inherited Genes

What happens when genes combine as shown in Figure 6.4? Remember that a parent passes on only one of its paired genes to each offspring. The pair of genes must separate. Gene separation occurs during the formation of sex cells, which are called **gametes** (GAM eets). In peas, the gametes are either pollen grains or unfertilized seeds, called ovules.

Look at Figure 6.5. The squares on the left stand for two of the offspring produced by Mendel's cross of a pure tall plant and a pure short plant. Notice that in both, one of the genes is T and the other is t. When each of these plants produces gametes, half the gametes will get the T gene. The other half will get the t gene. The diagram shows the two kinds of gametes produced by each parent.

When pollen from one parent fertilizes ovules from the other, the gametes join in pairs. Four different combinations of the T and t genes are possible. One combination, for example, is a T from the first parent and a T from the second parent. Another combination is a T from the first parent and a t from the second parent. What are the other combinations shown in Figure 6.5? Check to see that these are the same genotypes as the plants in the second generation in Figure 6.4.

The gametes join with each other randomly. In other words, each combination has an equal chance of occurring. An ovule containing a T, for example, has a 50-50 chance of being fertilized by a pollen grain containing a T or one containing a t.

Principles of Genetics

Mendel's work became the basis for the principles of genetics. Since Mendel's time, the work of many other scientists has also helped establish these principles. The first two principles summarize what you just learned.

▶ The traits of an organism pass from the parents to the offspring.

▶ Gene pairs determine traits.

▶ Organisms inherit genes in pairs. One gene comes from the gamete of each parent.

▶ Two members of a gene-pair separate, or segregate, at random. Half the gametes carry one of the genes and half carry the other gene.

▶ Genes are dominant or recessive. When paired, a dominant gene hides the effect of a recessive gene.

▶ The gene for a particular trait segregates independently of the genes for other traits. That is, genes do not segregate together.

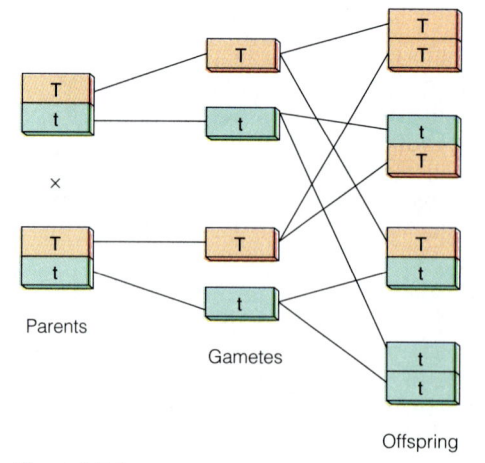

Figure 6.5 ▲
Each offspring gets a different combination of genes from the two parents. You can see in this cross that only one out of four offspring will have the tt genotype.

114   Chapter 6   Heredity

Themes in Science

Diversity and Unity The principles of heredity help explain the diversity among species. Diversity helps a species survive because it presents a variety of solutions to changing environmental conditions and factors.

Punnett Squares

An easy way to predict the genetic results of a cross between two organisms is to draw a diagram called a Punnett square. Notice that the Punnett square in Figure 6.6 has four boxes. Each box contains one of the possible combinations of genes for one trait inherited from the parents.

In this Punnett square, both parents are hybrid plants (Tt). When they cross, the result is that one out of four offspring is genotype TT. Two out of four are Tt. What is the other possible genotype? ①

Punnett squares can help you predict the numbers of different phenotypes produced by a certain cross. For the cross in Figure 6.6, three out of four boxes, or 75 percent, contain genotypes that result in tall plants. Therefore, about 75 percent of the offspring from this cross will grow to be tall. How can you tell that 25 percent will be short? ②

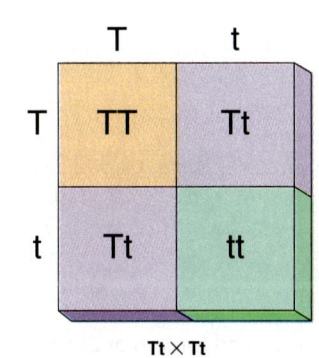

Figure 6.6
The letters on the left and top margins of a Punnett square stand for the genotypes of the two parents.

SkillBuilder *Predicting*

Punnett Squares

In dogs, dark hair (D) is dominant over light hair (d). What would happen if a dog that breeds pure for dark hair mates with a dog that breeds pure for light hair? What are the possible phenotypes of the offspring? You can use a Punnett square to answer these questions.

You begin by setting up the Punnett square as shown here. The letters on the left represent the possible genes in the sex cells of the dark-haired parent, and the letters on the top show the possible genes in the sex cells of the other parent. Copy this Punnett square on a separate sheet of paper and fill in the genotypes of the offspring.

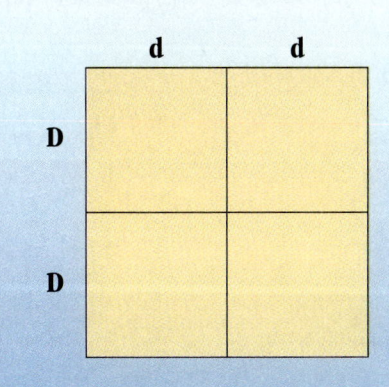

1. What genotype will all the offspring have? What phenotype is this?
2. Make another Punnett square to show a cross of two of these offspring. What are the possible genotypes that result from this cross?
3. For each puppy produced by this second cross, what are the chances that it will have light hair?
4. If a litter of eight puppies is produced, how many are likely to have dark hair?
5. Explain how Punnett squares can be used to predict the results of genetic crosses.

Chapter 6 Heredity 115

Skills Development

Predict Draw a Punnett square on the chalkboard. Have students predict the number of offspring that are tall if you cross a parent plant with genotype Tt with one that is TT. (All the offspring will be tall.)

Integrated Learning

Use Integrating Worksheet 6.1.

SkillBuilder
Answers
1. All the offspring will have the genotype Dd. The phenotype is for dark hair.
2. The possible genotypes from a cross of the offspring are DD, Dd, and dd.
3. The chances are one in four, or 25 percent, that a puppy will have light hair.
4. Two of the eight puppies are likely to have light hair.

Answers to In-Text Questions

① The other possible genotype is tt.

② You can tell that 25 percent will be short because one out of four boxes is tt.

The Living Textbook:
Life Science Sides 7-8

Chapter 10   Frame 01196
Punnett Squares (2 Frames)
Search: Step:

TEACH • Continued

Discuss
Ask students how pink flowers display incomplete dominance. (Pink flowers are a blend of traits. They show the red and white colors of both parents.)

EVALUATE

WrapUp
Portfolio Have students write a brief description of Mendel's work and draw a picture showing one of his experiments. Students should label their drawings, indicating genotype and phenotype of each plant. Students may wish to keep the papers in their portfolios.

Use Review Worksheet 6.1.

Check and Explain
1. A dominant gene codes for a trait's dominant form, which is expressed even when both dominant and recessive genes make up the genotype. The recessive gene codes for a form that is hidden unless both genes are recessive.
2. Round for both RR and Rr; wrinkled for rr.
3. By crossing pure green-pod plants with pure yellow-pod plants and noting which pod color appears in the offspring
4. An even chance for Ss or ss genotypes (with no SS genotype) and for long- or short-haired phenotypes

The Living Textbook: Life Science Sides 7-8

Chapter 10   Frame 01198
Incomplete Dominance (2 Frames)
Search: Step:

INTEGRATED LEARNING

Themes in Science
Patterns of Change/Cycles

Sometimes, many genes combine to affect one trait. Traits that students are learning about in this lesson are different in important ways from traits that they are already familiar with, such as eye color and hair color. Mendel was researching simple dominant-recessive traits that appeared in only two distinct forms. Eye color, on the other hand, is determined by no less than 23 pairs of genes in humans.

Incomplete Dominance

Many genes in plants and animals behave differently than the ones Mendel studied in peas. In one common pattern, two genes in a pair are different, but neither is dominant over the other. When the different genes pair together, they produce a trait somewhere between the traits of the two parents. For example, a cross between a red-flowered snapdragon and a white-flowered snapdragon produces offspring with pink flowers.

Genes that interact in this way show what is called **incomplete dominance**.

Figure 6.7 ▲
This snapdragon is the offspring of a red parent and a white parent.

Science and You *Your Multigene Traits*

Many inherited human traits are not as simple as stem length and seed color in peas. More than one gene determines these traits, so they are called multigene traits. None of the genes is dominant or recessive, but each gene contributes to the trait. Many different combinatons of these genes are possible. As a result, there is a wide range in these traits. For example, humans are not either tall or short. They are any height within a certain range. Another example is skin color. Three or four possible genes control skin color. A person's skin color varies from very light to very dark.

Check and Explain

1. What is the difference between a dominant gene and a recessive gene?
2. If the gene for round seeds in peas (R) is dominant, and the gene for wrinkled seeds (r) is recessive, what would be the phenotype of the following genotypes? RR, Rr, rr.
3. **Find Causes** Are green pods or yellow pods dominant in pea plants? Explain how you could find out.
4. **Predict** In dogs, short hair is dominant and long hair is recessive. Complete a Punnett square to show the results of a cross between a shorthaired dog with a genotype of Ss and a longhaired dog with a genotype of ss. What genotypes are possible? What phenotypes are possible?

Chapter 6 Heredity

Themes in Science

Scale and Structure Genes are segments of chromosomes, which are in turn segments of coiled DNA. Refer students to Table 6.1 and ask:

▶ What is the normal number of chromosomes for a human? (46)

▶ Where are chromosomes located? (In the cell nucleus)

6.2 Chromosomes and Inheritance

Objectives

▶ **Describe** the process and results of meiosis.

▶ **Explain** how sex is determined in humans.

▶ **Explain** why some traits are sex-linked.

▶ **Infer** the genotype of colorblind parents.

Skills WarmUp

Interpret Data Flip a coin ten times. Write the number of times heads and tails appear. Add your numbers to those of the rest of the class. Compare the number of times heads and tails appear. What can you determine about the chance of heads or tails appearing with any toss?

Suppose you wanted to see a gene. Where would you look? If you looked inside a cell's nucleus at the right time, you would see dense structures like the ones in Figure 6.8. These structures are called chromosomes.

Each of the chromosomes you see in Figure 6.8 is made of many genes. One gene is a very small segment of a chromosome. There may be as many as 50,000 genes on one human chromosome.

Chromosome Facts

Mendel discovered the principles of inheritance without knowing about chromosomes. His principles, however, related directly to the way chromosomes behave. Around the time of Mendel's death in 1884, the development of high-power microscopes enabled scientists to see and study chromosomes. They soon realized that chromosomes were the physical basis of inheritance.

Chromosomes normally occur in pairs because the offspring inherits one chromosome from each parent. Both chromosomes in a pair have the same appearance and length, and both contain the same types of genes. Consequently, genes also occur in pairs. Whatever its characteristic number of chromosomes, an organism has the same number of chromosomes in all of its body cells.

Every species of living things has a certain number of chromosome pairs in its cells. An organism's size or complexity doesn't determine this number. Table 6.1 shows the number of chromosome pairs for various organisms.

Table 6.1 Normal Chromosome Number

Species	Pairs of Chromosomes
Cat	19
Dog	39
Goldfish	47
Horse	32
Human	23
Onion	8
Potato	44

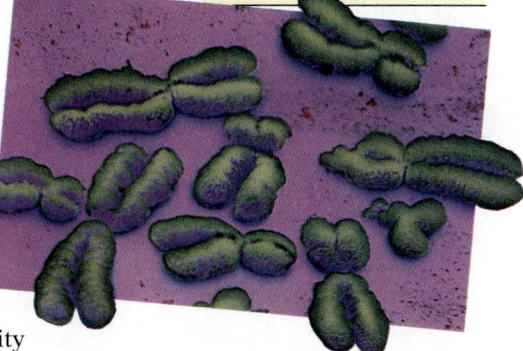

Figure 6.8 ▲ Human Chromosomes

Chapter 6 Heredity 117

SECTION 6.2

Section Objectives
For a list of section objectives, see the Student Edition page.

Skills Objectives
Students should be able to:

Interpret Data about chance events.

Infer the genotypes of a colorblind female's parents.

Make a Model of gene inheritance.

Infer why inheritance is a matter of chance.

Vocabulary
meiosis, X chromosome, Y chromosome

MOTIVATE

Skills WarmUp
To help students understand probability, have them do the Skills WarmUp.
Answer There is a 50 percent chance of either heads or tails appearing with any one coin toss.

Misconceptions
Students may think that the cells of large or complex organisms have a greater number of chromosomes than smaller, less complex ones. Explain that an organism's size and complexity are not related to the number of chromosome pairs its cells contain. Refer them to Table 6.1.

The Living Textbook: Life Science Sides 7-8

Chapter 10 Frame 01161
Chromosome Photos (6 Frames)
Search: Step:

TEACH

Directed Inquiry

Have students study the diagram and the paragraphs on this page. Lead a discussion using some or all of the following questions:

▶ How many chromosomes are in the parent cell of the organism? (6, or 3 pairs)

▶ How many chromosomes are in each gamete? (3)

▶ How many chromosomes are in the fertilized egg of this organism? (6, or 3 pairs)

▶ How many chromosomes came from each parent? (3)

▶ What is meiosis? (The process of cell division that produces reproductive cells, which have one set of unpaired chromosomes each)

▶ Will every gamete contain identical chromosomes? Explain. (No. They contain different combinations of chromosomes.)

▶ If there are 46 chromosomes in the parent cell, how many will be in each gamete? Why? (There will be 23; each gamete contributes half of the chromosomes to the fertilized egg.)

Reteach

Use Reteach Worksheet 6.2.

The Living Textbook: Life Science Sides 7-8

Chapter 7 Frame 01150
Meiosis Diagrams (5 Frames)
Search: Step:

118

INTEGRATED LEARNING

Art Connection

Have students create a model showing the stages of meiosis in an organism with three pairs of chromosomes. Have students use shape rather than color to distinguish the chromosomes. Ask students:

▶ Which chromosomes will you need to make duplicates of? (Parent chromosomes)

▶ How many different combinations of chromosomes are possible in the reproductive cells? (8)

Meiosis

Recall that each sex cell, or gamete, contributes one gene to the offspring. For each chromosome from a pair to end up in a different sex cell, the chromosome pairs must separate.

The formation of gametes involves a special kind of cell division called **meiosis** (my OH sihs). The process of meiosis produces gametes with one set of unpaired chromosomes. Follow the process of meiosis in Figure 6.9. It shows what happens in an organism with three pairs of chromosomes in its body cells.

1. The chromosomes in the parent cell condense and arrange themselves. The two chromosomes in each pair are shaded differently to make them easier to follow. Each contains different forms of the same genes.

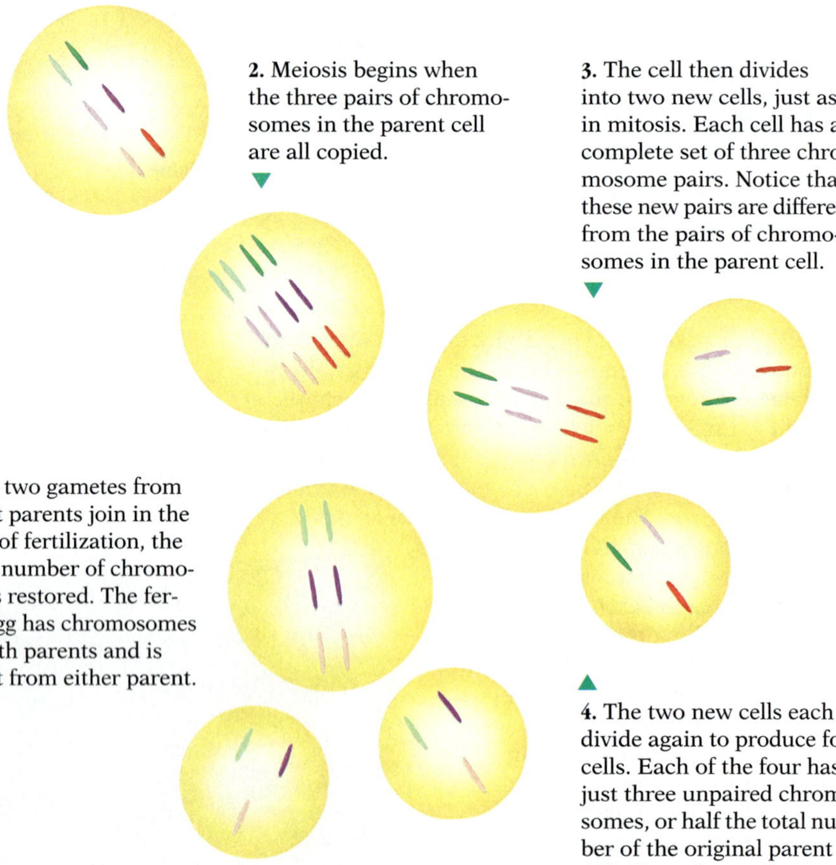

2. Meiosis begins when the three pairs of chromosomes in the parent cell are all copied.

3. The cell then divides into two new cells, just as in mitosis. Each cell has a complete set of three chromosome pairs. Notice that these new pairs are different from the pairs of chromosomes in the parent cell.

4. The two new cells each divide again to produce four cells. Each of the four has just three unpaired chromosomes, or half the total number of the original parent cell. These four cells are all gametes. Notice that they contain two different combinations of chromosomes.

5. When two gametes from different parents join in the process of fertilization, the original number of chromosomes is restored. The fertilized egg has chromosomes from both parents and is different from either parent.

Figure 6.9
The Process of Meiosis

118   Chapter 6   Heredity

Themes in Science
Patterns of Change/Cycles

Offspring inherit both characteristics and potential characteristics from parents. With sex-linked traits, characteristics may appear to skip generations because of sex as well as genotype. To help students remember how sex-linked traits are passed along, you may wish to introduce the expression *X-linked genes*. Researchers often use this name because sex-linked genes are carried on the X or Y chromosome in the gamete.

Discuss

Have students answer the question in the caption for Figure 6.10 by locating chromosome pair 23. (This is a karyotype for a male.) Now ask:

▶ What sex chromosomes are in male sex cells? (Sperm have one of either an X chromosome or a Y chromosome.)

▶ What sex chromosomes are in female sex cells? (Each egg cell has only one X chromosome.)

Critical Thinking

Reason and Conclude Have students study Figure 6.11. Remind them that colorblindness is a trait caused by genes on the sex chromosome. The male children of normal-vision parents can have colorblindness even if only the X chromosome from the mother is defective. Ask students, Can females have colorblindness? Explain. (Yes, if the father is colorblind and the mother is either colorblind or carries the genes for colorblindness.)

Sex Determination

The 23 pairs of chromosomes in a human cell are shown in Figure 6.10. Notice that the chromosomes in each of the first 22 pairs are alike. The chromosomes in pair 23, however, are not alike. The longer one is called an **X chromosome**, and the shorter one is a **Y chromosome**. These are the sex chromosomes. The combination of sex chromosomes you have in your body cells determines your sex. Females have two X chromosomes. Males have one X and one Y chromosome.

Since a female has only X chromosomes, her egg cells, or ova, all contain an X chromosome. Consequently both males and females inherit an X chromosome from their mother. Male sex cells, or sperm, carry either an X or a Y chromosome. If a sperm containing a Y chromosome fertilizes the ovum, the baby will be a boy. If the sperm has an X chromosome, the baby will be a girl. The father's sperm, therefore, determines the sex of a child.

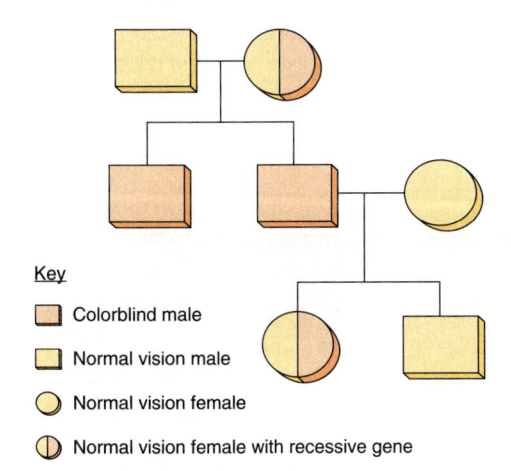

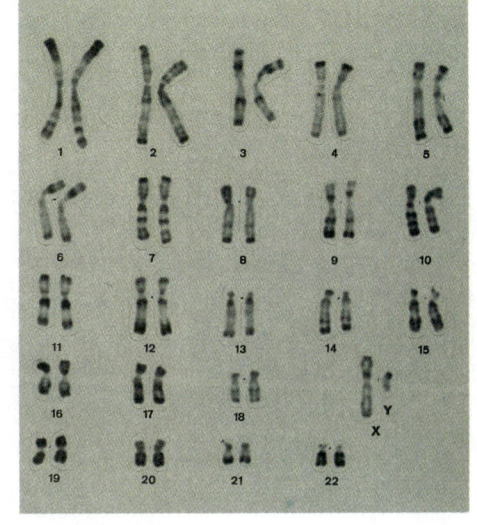

Figure 6.10 ▲
Are these the chromosomes of a male or a female?

Sex-Linked Traits

Some genes not related to sex traits are also carried on the X chromosome. In males, traits for the genes carried on the X chromosome that do not match genes on the Y chromosome are always expressed in the offspring. These traits are called sex-linked traits because the genes for them are on the sex chromosome.

A common sex-linked trait is colorblindness. The genes needed to distinguish red from green are on the X chromosome. A female with one defective and one normal X chromosome has normal color vision. However, if a male is missing these genes on his X chromosome, he is colorblind. There are no genes for normal color vision on the Y chromosome to cover for the defective X chromosome.

One way of tracing a trait through generations of offspring is to make a pedigree chart. Figure 6.11 is a pedigree showing how colorblindness was inherited in one family. A horizontal line joining a circle and a square means the couple had children.

Key
- Colorblind male
- Normal vision male
- Normal vision female
- Normal vision female with recessive gene

Figure 6.11 ▲
A pedigree shows how traits are inherited.

**The Living Textbook:
Life Science Sides 7-8**

Chapter 10   Frame 01288
Tests for Color-blindness (7 Frames)
Search: Step:

Chapter 6 Heredity

TEACH • Continued

Research

Students may ask why an excess of genetic material can lead to potentially harmful abnormalities. Use an analogy to suggest that complicated mechanisms such as automobiles work best if they have only a certain number of working parts—no more, no less. As a research project, have students find out about other conditions, besides Down's syndrome, that are caused by extra or missing chromosomes.

EVALUATE

WrapUp

Reteach Have students copy the pedigree in Figure 6.11 and add another generation to the chart.
Use Review Worksheet 6.2.

Check and Explain

1. 32 pairs

2. In males, chromosome pair 23 has one X and one Y chromosome; a female's 23rd pair has two X chromosomes. The 23rd chromosome in a sperm cell may be either an X or a Y chromosome, while an egg cell's 23rd chromosome is an X chromosome.

3. Red-green colorblindness is called a sex-linked trait because the genes for it are carried on the sex chromosome.

4. A colorblind female's parents must both supply the X chromosome that has the genes for colorblindness. The father must be X^{cb}/Y. The mother must be X^{cb}/X^{cb}.

Answer to In-Text Question

① There is an extra number 21 chromosome.

INTEGRATED LEARNING

STS Connection

Research has led to many different prenatal tests for genetic disorders. Fetoscopy and ultrasound provide images of a developing fetus. Amniocentesis and chorionic villi sampling analyze cells in the fluid surrounding the developing baby. Ask students to research one kind of test and report on the disorders it helps to diagnose.

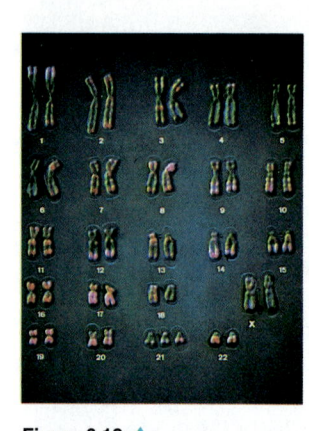

Figure 6.12 ▲
Find the extra chromosome in the karyotype (left). People with Down's syndrome can often live normal lives (right). ①

Science and Technology *Karyotypes*

The photograph of the 23 pairs of human chromosomes in Figure 6.12 is a karyotype. Karyotypes aid in counting and identifying chromosomes. Sometimes there are extra or missing chromosomes. Chromosomes are visible only in dividing cells. To make a karyotype, dividing cells are stained with a dye and then photographed. The individual chromosome pairs are cut out and organized by length on a sheet of paper. Extra or missing chromosomes are then noted.

The karyotype in Figure 6.12 shows an extra number 21 chromosome. This causes a condition called Down's syndrome. This extra chromosome prevents normal development of the eyes, mouth, hands, and brain.

Check and Explain

1. Horses have 32 pairs of chromosomes in their body cells. How many chromosomes are in their sex cells?

2. In humans, how do the chromosomes of a male differ from those of a female?

3. **Reason and Conclude** Why is red-green colorblindness called a sex-linked trait?

4. **Infer** Colorblind females are rare, but they do exist. A colorblind female's parents must have what genotypes?

Chapter 6 Heredity

TEACHING OPTIONS

Prelab Discussion
Have students read the entire activity, and then discuss the following points:

▶ Remind students of the process of meiosis and the formation of actual gametes, and emphasize that this activity is a model of fertilization.

ACTIVITY 6

Time 40 minutes **Group** pairs
Materials
paper
scissors

Analysis
1. Answers will vary by group depending on the results of the trials.
2. There was a 50 percent chance of the child being a female each time, regardless of the gender of a previous child.

Conclusion
Accept all logical conclusions. Just as the paper "chromosomes" could land with either side facing up, a child can receive either gene of the pair that each parent has for a given trait. As long as each trait is located on a separate chromosome, one trait can be passed on independently of another trait.

Extension
Students should find that only certain offspring will be colorblind, and that colorblindness will only appear in males.

Activity 6 *How do you make a model of gene inheritance?*

Skills Model; Collect Data; Interpret Data; Infer

Task 1 Prelab Prep
1. Collect 4 strips of paper, each 10 cm long and 2 cm wide.
2. Fold the strips lengthwise. Each strip of paper represents a chromosome carrying a gene pair for one characteristic. One side of a "chromosome" carries one gene, and the other side carries the other gene. Number the strips 1 through 4. Write your initials on each strip.
3. For each of the four characteristics in Table 6.2, select a genotype. You should end up with four different genotypes, such as LL, cc, Dd, and XX.
4. On one side of your first chromosome, write one letter of your chosen earlobe genotype. On the other side, write the other letter. For example, if your genotype is Ll, one side should have a L and the other side a l.
5. Repeat step 4 three times to label the genes of the other three chromosomes.
6. Work with a partner identified by your teacher. Each pair should consist of a male and a female. Your partner should have made up his or her own set of chromosomes.

Task 2 Data Record
1. On a separate sheet of paper, make a chart with a row for each characteristic. Make a column for you and your partner and three columns to represent offspring.
2. Use the chart to record the genotypes and phenotypes of you and your partner, and the results of your chromosome pairing.

Task 3 Procedure
1. At the same time as your partner, take your chromosome number 1, hold it above the floor or table, and drop it. Note the genes that face up on each chromosome. Together they are the gene pair that determine whether a child has free or attached earlobes. Record the genotype and phenotype for this characteristic on your chart.
2. Repeat step 1 with the other three chromosomes.
3. Repeat the pairing process twice to determine the genotypes and phenotypes for a second and third child.

Task 4 Analysis
1. How were the three children alike and different?
2. What were the chances of the child being a female each time? If the first child was a female, what were the chances that the second child would be a female?

Task 5 Conclusion
Use the results of this activity to explain why inheritance is a matter of chance. Is Mendel's conclusion that one trait could be passed on to offspring independently of another trait supported? Explain.

Extension
Add the trait for colorblindness to the X chromosome of the male and to one of the X chromosomes of the female. Repeat the experiment.

Table 6.2 Characteristics

Chromosome	Phenotype	Genotype
1	Free earlobes Attached earlobes	LL or Ll ll
2	Curly hair Straight hair	CC or Cc cc
3	Dimples No dimples	DD or Dd dd
4	Female Male	XX XY

SECTION 6.3

Section Objectives
For a list of section objectives, see the Student Edition page.

Skills Objectives
Students should be able to:

Infer whether traits are dominant or recessive.

Make a Model of a DNA segment and a genetic mutation.

Communicate information about mutations.

Vocabulary
replication, mutation

MOTIVATE

Skills WarmUp
To help students understand dominant and recessive traits, have them do the Skills WarmUp.
Answer Tongue folding and attached earlobes are recessive traits.

Prior Knowledge
To gauge how much students know about DNA and the genetic code, ask the following questions:

▶ What substance are genes made of?

▶ What is a mutation? What might cause a mutation to occur?

▶ What causes a sunburn? What sort of changes can too much sun produce?

The Living Textbook: Life Science Side 7
Chapter 35   Frame 37263
Structure of DNA (Animation)
Search: Play:

122

INTEGRATED LEARNING

Themes in Science
Scale and Structure Genes are located on chromosomes in every cell of the body. Point out to students that Mendel's idea of a gene was an abstract mathematical concept that explained the results of his cross-breeding peas. It wasn't until much later that workers found that the Mendelian gene is actually a segment of DNA located on a chromosome within the cell nucleus.

Skills WarmUp

Infer Can you fold your tongue? Are your earlobes attached? Infer by the number of people in your class who answer yes to each of these questions which traits are dominant and which are recessive.

6.3 DNA and the Genetic Code

Objectives

▶ **Describe** the kind of information carried by a gene.

▶ **Describe** the structure of DNA.

▶ **Explain** how DNA replication takes place.

▶ **Communicate** how mutations can be both harmful and helpful.

Look closely at the words you are reading right now. Why do they make sense to you? Could you read this page if the words were out of order? Words are a code that you know. The code is your language. Each word transmits information to you. If the words or letters were in a different order, they would make no sense or have a different meaning.

In a similar way, the genes in the cells of your body contain a code called the genetic code. The genetic code is the basis of heredity. Like a language, the genetic code transmits information. A gene is very much like a long sentence in the "language" of heredity.

The Genetic Code

How do you think a gene carries information? How does it tell an organism to have dark hair or to be tall? The language of heredity is a chemical language. A gene is actually a set of instructions for assembling a protein. The genetic code tells a cell how to put together a particular protein piece by piece. But what do proteins have to do with an organism's traits?

Proteins are very important to cells. They perform a variety of tasks depending on their structure. Proteins form most of the cell's substance. They are a part of every cell process. They carry other substances or messages to organs or tissues. Some make hair or skin a certain color. Others stimulate the body to grow longer bones. Still others make it possible for your eyes to distinguish red and green. Proteins have a role in creating all the traits of an individual organism. Thus, by carrying a code for assembling proteins, genes determine traits.

122 Chapter 6 Heredity

Themes in Science

Evolution Genetic mutations in the sex cells are one way in which a species evolves. If a mutation helps an organism survive, that organism passes the new trait to its offspring.

STS Connection

DNA matching has given law enforcement officials a new and increasingly effective way of identifying criminals. U.S. courts can convict criminals on the basis of matching DNA from body cells left at the crime scene with cell samples taken from the suspects. The samples are bar-codelike patterns made from repeated sequences of DNA bases.

TEACH

Directed Inquiry
Have students study the diagrams and paragraphs on page 123. Lead a discussion using some or all of the following questions:

▶ Which nitrogen bases fit together to make base pairs? (Adenine and thymine fit together, and guanine and cytosine fit together.)

▶ When a DNA molecule unzips to form two strands, what is added to each strand from the cell's cytoplasm, and what is made? (Bases in the cytoplasm attach to the bases on the strands so that two complete molecules of DNA are created.)

▶ What is the copying of DNA called? (Replication)

▶ What is a gene in the DNA code? (A particular sequence of base pairs)

▶ Suggest how a group of bases making up a gene are like a word. (The group is "read" by the cell as a genetic instruction code for a particular amino acid.)

Integrated Learning
Use Integrating Worksheet 6.3.

DNA

Genes are made of a substance called deoxyribonucleic (dee AHKS uh RY boh noo KLAY ihk) acid, which is commonly referred to as DNA. Genes carry coded information due to the special properties of DNA. The way DNA is put together, or its structure, forms a code that is "read" by other parts of a cell. Also, DNA makes identical copies of itself. These properties make it possible for DNA to carry all the information an organism needs to live and reproduce.

A DNA molecule "unzips" to form two strands, as shown below. Notice that as the molecule unzips, the base pairs separate. Each single strand of DNA then picks up bases present in the cell's cytoplasm. In this way, two complete molecules of DNA are created. Notice that each new DNA molecule has the same order of base pairs as the original. This copying process is called **replication**.

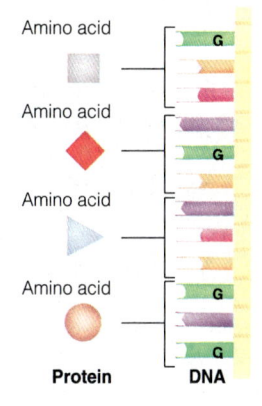

▲ A DNA molecule is shaped like a long ladder twisted into a spiral. Study the model of the DNA molecule above. The uprights of the ladder are alternating sugar and phosphate groups. The rungs of the ladder are nitrogen bases joined in pairs. There are four different nitrogen bases: adenine (A), guanine (G), cytosine (C), and thymine (T). The four bases fit together to make base pairs. A always makes a base pair with T. G always makes a base pair with C. The genetic code is set by the order of nitrogen bases in the DNA.

▲ A gene is a particular sequence of base pairs. When a cell "reads" the genetic code of a gene, it uses an unzipped segment of DNA. Each group of three bases on one of the strands is like a word in the genetic code. The sequence A-C-T means something different from T-C-A. Each of these "words" is one genetic instruction, coding for a particular amino acid.

**The Living Textbook:
Life Science Sides 7-8**

Chapter 9   Frame 01184
DNA Models (2 Frames)
Search: Step:

TEACH ▪ Continued

Discuss
Have students study Figure 6.13. Ask them to locate the base pairs that have been replaced, added, or removed. Have them identify each pair by the names of the proteins. (T-A is thymine-adenine; C-G is cytosine-guanine)

Class Activity
Have students work in pairs. Place a sequence of letters indicating DNA bases on the chalkboard. The sequence should be at least 15 bases long. Have each student write the same sequence on paper. Then have them each change, add, or replace one base pair. Have students exchange papers and try to locate and circle the "mutation."

Enrich
Explain to students that researchers often have a problem linking cancer with particular mutagens because the cancer may not occur until years after the exposure. In fact, cancer victims often are exposed to a variety of mutagens, making it difficult to target the specific cause of the cancer. An example is the asbestos worker who also smokes, then develops lung cancer later in life.

Answer to In-Text Question
① Changes in the DNA sequence may cause a different protein to be made. These changes are called mutations.

**The Living Textbook:
Life Science Sides 7–8**

Chapter 10   Frame 01246
Fruit Fly Mutations (17 Frames)
Search: Step:

TEACHING OPTIONS

Cooperative Learning
Assign students to groups of four. Have each group write down as many ten-letter sequences as possible using only the four letters A, B, C, and D. After fifteen minutes, ask the class to come back together and share the survey results. Was any one group able to come up with all possible combinations? (No—there are more than a million possible combinations.) Discuss what this implies about the amount of information that can be contained on a DNA molecule and the number of mutations possible.

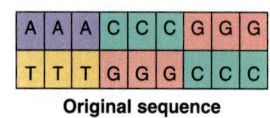

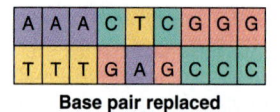

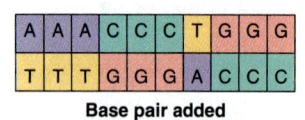

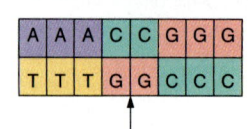

Figure 6.13 ▲
How does each of these errors change the DNA sequence? ①

Mutations
When you copy from the blackboard, sometimes you may make mistakes. In a similar way, mistakes occur when DNA and chromosomes replicate. Look at Figure 6.13 to see some common mistakes in replication. Changes in the DNA sequence may cause a different protein to be made.

Another kind of mistake results in larger changes in the hereditary material. Whole pieces of chromosomes may be left out or added when cells divide. A change in DNA or the chromosomes is called a **mutation**. Mutations occur in both body cells and sex cells. Only mutations in sex cells pass on to offspring.

Most mutations are not harmful. A mutation in a body cell might cause that cell to die but have no effect on the body. Sometimes mutations in the sex cells are useful for the species. For example, a mutation could change the size of a food plant or the number of seeds it produces. Some mutations are harmful. Gene mutations in humans may cause as many as 4,000 different diseases and disorders. Some mutations in body cells are known to cause cancer. Some mutations in sex cells can cause birth defects.

Chemical Mutagens
Substances and conditions that increase the chance of mutations are called mutagens. Mutagens can affect DNA and change the genetic code. Mutagens are everywhere in the environment. They are found in some chemicals used for killing insects and in some chemical wastes from factories. Small amounts of these chemicals may get into the air, food, and drinking water. The chemicals in cigaratte smoke are common mutagens.

The sun, some kinds of rocks, and nuclear materials all give off radiation that can increase the rate of mutations. Many devices that run on electricity, such as televisions and computer screens, also give off radiation. In 1986, the explosion of a nuclear power plant at Chernobyl in the Ukraine exposed many people in Eastern Europe to radiation. Nuclear bomb explosions in the 1940s and 1950s increased the level of radiation in the soil and atmosphere in some areas. Prolonged exposure and high doses of mutagens cause genetic changes that may result in cancer.

INTEGRATED LEARNING

STS Connection

Encourage students to create two more columns labeled *Diagnosis* and *Treatment* for Table 6.3, noting the new information for each of the listed disorders. These data are vital to parents who seek genetic counseling. For example, PKU can be treated successfully by having the individual follow a special diet. However, this treatment works only if the disorder is diagnosed early in life.

Skills WorkOut

To help students model DNA, have them do the Skills WorkOut. Students may also wish to model a stage in replication.

Enrich

Tell students that sickle-cell anemia may cause death. Also tell them that people in Africa are resistant to malaria if they carry the trait for sickle-cell anemia. Large numbers of people in the African population will not get malaria as a result of having the sickle-cell trait. Malaria is a disease that is caused by parasites in the red blood cells.

Research

Have students pick one of the disorders in Table 6.3 and use references to find out more about the specific genetics of that disorder. Have students write a report on the disorder and include it in their portfolios.

Ancillary Options

If you are using the blackline masters from *Multiculturalism in Mathematics, Science, and Technology*, have students read about Harvey Itano on pages 85 and 86. Complete pages 86 to 88.

Inherited Diseases

Many diseases are caused by the effects of inherited genes. In most cases, there is only a small difference between the DNA sequence in the defective gene and a normal one. This difference is enough to cause serious and often fatal diseases. These disease-causing genes are the result of a mutation that occurred long ago. They pass from one generation to the next.

Recessive genes cause most inherited diseases. A person with one of the genes is a carrier. A person with two genes has the disease. Cystic fibrosis is the most common genetic disease of this type in the United States. About one child in every 2,000 is born with cystic fibrosis. The disease causes thick mucus that clogs the airways of the lungs. Bacteria grow on this mucus and destroy the lungs. In 1989, scientists found the gene that causes cystic fibrosis on chromosome 7.

A recessive gene also causes sickle-cell anemia. In people with sickle-cell anemia, most of the red blood cells stiffen and become an odd shape. These diseased cells carry less oxygen than normal cells. The disease is most common among African Americans. About 1 in 500 have sickle-cell anemia and eventually die from it.

Dominant genes cause a few inherited diseases. The most common disease of this type is Huntington's disease. This disease usually strikes people between the ages of 35 and 45. Other kinds of disorders are listed in Table 6.3.

Skills WorkOut

Make a Model Use colored buttons, plastic paper clips, or candies to make a segment of DNA. Use the following code: red = A, green = T, brown = G, yellow = C. Make the following strand of DNA: T-C-C-G-T-A-T-T-T-G-G-T-T-G-G. Make this a double strand using the base-pairing rules: T pairs with A; C pairs with G. Next, model a mutation by remaking the strand with two errors.

Table 6.3 Inherited Disorders in Humans

Disorder	Description	Cause
Albinism	Inability to produce pigment in some or all organs, resulting in abnormally light skin, hair, and eye color	Recessive gene
Achondroplasia (form of dwarfism)	Condition in which person is far below normal size and lacks ability to grow normally	Dominant gene
Huntington's Disease	Progressive loss of muscle control and mental abilities, occurring from about age 35 to 45	Dominant gene
Phenylketonuria (PKU)	Mental retardation (unless treated during infancy) resulting from buildup of phenylalanine, an amino acid in proteins	Recessive gene
Tay-Sachs Disease	Brain cells do not function normally, resulting in progressive loss of motor skills and mental abilities, blindness, and seizures; babies born with Tay-Sachs disease live only for a few years	Recessive gene

The Living Textbook: Life Science Sides 7-8

Chapter 10   Frame 01286
Sickle-cell Microviews (2 Frames)
Search: Step:

TEACH · Continued

Critical Thinking
Evaluate Sources Have students bring in sunscreen products from home. As a class, compare the ingredients, SPF numbers, and claims made on the labels of the lotions.

EVALUATE

WrapUp
Reteach Have students look back at Figure 6.9 on page 118. Have them describe what is happening to the DNA in the cell nucleus in each of the numbered steps.
Use Review Worksheet 6.3.

Check and Explain
1. A gene carries information based on a chemical pattern in the DNA from which it is made, a code that is "read" by other parts of a cell.
2. A-T-T-C-C-G
3. Damage to one of only two chromosomes would lead to more mutations than would alterations to one of 46.
4. A mutation is a change in DNA or chromosomes. Though mutations sometimes result in diseases and disorders, they also lead to unique physical qualities for individual organisms and genetic diversity for populations.

Answer to In-Text Question
① People's exposure to the sun varies. Too much sun over a long period of time can cause premature aging and even skin cancer in some people.

TEACHING OPTIONS

Cooperative Learning
Many people who want a suntan use tanning machines. Proponents say that these machines are safe because the level of UV radiation is low and one can guard against overexposure. Opponents say that exposure to UV radiation is always harmful, and that such machines may malfunction. Split the class into two teams. Then have the teams research the issue and hold a debate on the use of tanning machines.

Figure 6.14 ▲
How much time do you spend out in the sun? Why is too much sun harmful? ①

Science and You *Sunburn and DNA Damage*
How many times have you gotten sunburned when you were only in the sun for "a little while"? Ultraviolet radiation given off by the sun burned your skin. Skin damage from ultraviolet, or UV, radiation builds up over a long period of time. UV radiation can damage the DNA in skin cells. Such damage can cause premature aging. With time, UV radiation can cause different forms of skin cancer in some people.

Dark skin has some protection against sunburn and UV radiation. For this reason, skin cancers are rare among people who have large amounts of melanin, such as Africans. However, too much exposure to direct sunlight can damage the skin of all humans.

The health risks posed by exposure to UV radiation increase as the ozone layer thins. Ozone is a gas in the upper atmosphere that filters out most of the UV radiation from the sun. Some kinds of commercial chemicals released into the air destroy the ozone layer.

Sunscreen and sunblock lotions give protection from UV radiation. The sun protection factor, or SPF, is a number on the label that tells you how much UV radiation is blocked out. An SPF of 10 means that a person using this sunscreen receives the same amount of UV radiation in 10 hours that she would receive in 1 hour without the sunscreen. Sunscreens with an SPF of 15 or higher offer the best protection from UV radiation.

UV radiation comes from the sun in two slightly different energy levels, called UVA and UVB. The best sunscreens block out both kinds.

Check and Explain
1. In what form does a gene carry information?
2. If a DNA molecule is "unzipped" and one strand has the code T-A-A-G-G-C, what is the code of the other strand?
3. **Predict** How would heredity in humans be different if all the genes were on two chromosomes?
4. **Communicate** What is a mutation? Explain how some mutations can be harmful, while others are helpful. Give examples.

INTEGRATED LEARNING

Art Connection

Luther Burbank (1849–1926), an American horticulturist, introduced more than 200 varieties of fruit. He also developed the Burbank potato, the most commonly grown potato in America. Show students pictures of some of the fruits and vegetables bred by Burbank, such as the pomato, a tomato/potato; the plumcot, a plum/apricot; and the white raspberry. Have students think about a hybrid fruit, vegetable, or flower that they might like to breed. Have them draw pictures of their hybrid fruits.

SECTION 6.4

Section Objectives
For a list of section objectives, see the Student Edition page.

Skills Objectives
Students should be able to:

Communicate how an ideal apple might be grown.

Hypothesize how genetic engineering and/or selective breeding might be used to solve problems.

6.4 Applied Genetics

Objectives

▶ **Explain** how selective breeding has been used to modify the traits of plants and animals.

▶ **Identify** four useful applications of selective breeding.

▶ **Describe** how DNA is changed by genetic engineering.

▶ **Predict** how genetic engineering could be useful.

Skills WarmUp

Communicate If you could invent a perfect apple to eat, what would you do? Describe your perfect apple and how you would go about growing such an apple.

How many different breeds of dogs can you name? What characteristics make one breed different from another? In Figure 6.15, you can see two examples of the different sizes and shapes of dogs.

Even though they are different, all dogs belong to the same species. Dogs can breed with each other because they all have 78 chromosomes. Many different breeds of dogs were developed deliberately over the last 12,000 years. Dog breeders chose individual dogs with desired traits and mated them.

Knowledge of how traits pass from parents to offspring is used to develop desirable traits in other organisms as well. People breed cows that produce more milk, horses that run faster, trees that produce more fruit, and flowers of a certain color.

Selective Breeding

The process of choosing and mating organisms with desired traits is called selective breeding. Most of the plant and animal products you eat are probably the result of selective breeding. Chickens, for example, have been selectively bred to lay more eggs. As a result, the best laying hens today can lay an average of 250 eggs per year. Seventy years ago, the average per year was only 120.

How would you bring out this kind of desirable trait in an organism? Say you had a flock of 100 chickens. You might use the 10 best-laying hens to parent the next generation. You would then select the best layers from their offspring to produce another generation. The ongoing process of breeding the chickens that lay the most eggs has increased egg production more than twofold.

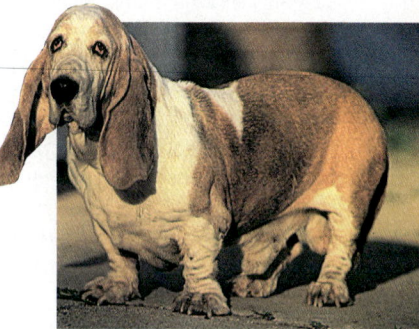

Figure 6.15 ▲
Notice the very different bodies, legs, and ears of these two dogs.

MOTIVATE

Skills WarmUp
To help students understand selective breeding, have them do the Skills WarmUp.
Answer Answers may vary. Students might suggest that the perfect apple could be grown by selective breeding of an apple tree that produces large fruit with one that produces sweet, juicy fruit.

Prior Knowledge
To gauge how much students know about applied genetics, ask the following questions:

▶ A farmer has a flock of chickens in which three prize hens lay more eggs than the other hens. What could he do to produce more prize hens?

▶ Are there any characteristics that plants have that might be helpful to animals?

The Living Textbook:
Life Science Side 7

Chapter 40   Frame 44961
Swine Breeding (Movie)
Search: Play:

Chapter 6 Heredity

TEACH

Historical Notebook

Answers

1. The process of genetic engineering is much quicker than selective breeding in developing any organism, because traits from multiple "parents" can be transferred immediately to the next generation.

2. Students may need a reference librarian to help with this question. Triticale and other "miracle" hybrids are grown in a number of developing countries.

Research

Students may have relatives or acquaintances who require insulin treatment for diabetes. (Refer to Chapter 21 for a discussion of insulin's function in regulating blood chemistry.) Explain that human insulin for diabetics would be in short supply if not for genetic engineering. Have students find out why a new source of insulin was needed in the first place. (Nonhuman insulin had to be harvested from pigs and cows, and some people had allergic reactions to it.)

The Living Textbook: Life Science Sides 7-8

Chapter 11 Frame 01314
Breeding/Genetic Engineering
(10 Frames)
Search: Step:

The Living Textbook: Life Science Side 7

Chapter 42 Frame 47299
Genetic Engineering (Movie)
Search: Play:

128

INTEGRATED LEARNING

Multicultural Perspectives

While developing such "miracle" crops as triticale, contemporary nutritionists are rediscovering ancient grains, such as amaranth and quinoa (KEEN wah) of South America. Have students make a chart comparing the growing conditions required for corn, wheat, and rice—the most common grains. They should also show where each grain comes from. Then have students find out and note where unusual grains are being planted and what the best growing conditions are for each.

Historical Notebook

Better Grain for More People

How many kinds of bread can you buy in a market? Wheat, rye, potato, and corn are just a few kinds you might choose. The grain in each kind of bread originated centuries ago in different parts of the world. For example, people ate corn in Mexico and South America 10,000 years ago. Rye originated in the Middle East and was used as grain in Ancient Rome. Wheat was grown in Asia and Europe as early as 4,000 B.C.

People soon began using methods to crossbreed, or hybridize, many different grains. These methods became especially common in the 1800s. They led to the development of plants that resist disease and produce enough grain to feed large numbers of people.

Today cross-breeding techniques are especially useful to people in developing countries, such as India and Brazil. In certain areas of these countries, populations are large, and soil is poor in nutrients. In the 1970s, scientists developed a grain called triticale. It is a crossbreed of wheat and rye. Triticale has the protein quality of rye, and it can grow in poor soil.

1. What is the advantage of using cross-breeding methods to develop new grains?

2. **Research** Find out what countries grow and use triticale. How much triticale is grown each year?

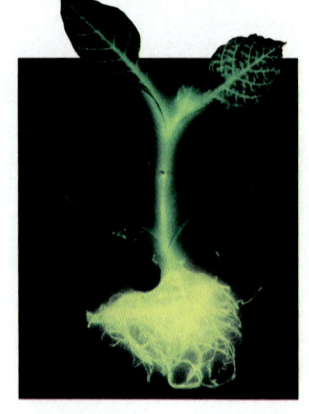

Figure 6.16 ▲
Genetic engineers transferred a firefly gene to this plant. Notice how the plant now "glows" in the dark.

Genetic Engineering

Using selective breeding, it may take a long time to develop a special trait in an organism. Selective breeding slowly changes particular genes. However, since 1975, scientists have developed an entirely new way of changing the traits of organisms. They actually take genes from one kind of organism and place them in another. This process is called genetic engineering.

Through genetic engineering, scientists can produce organisms with traits that are not possible through selective breeding. Genetic engineers have made fish grow fatter with genes taken from rats. They have developed tobacco plants that produce chemicals normally made by organs in the human body. Genetic engineers have also produced plants that are resistant to frost. They've even produced crops, such as potatoes, that make their own insect repellent. The possibility of custom-designed life forms is almost endless.

128 Chapter 6 Heredity

STS Connection

Researchers are working to engineer microbes that break down all varieties of toxic waste. Ask students how an oil-eating microbe might be useful to the petroleum industry. (The microbes could help eliminate the damage caused by oil spills.)

Recombinant DNA Normally the DNA from two different kinds of organisms could never combine. When genetic engineers move genes from organism to organism, they create new DNA. This new DNA is called recombinant DNA.

For recombinant DNA to change an organism's traits, it must be in every cell of the organism's body. For this reason, the DNA in single-celled organisms, such as bacteria, is the easiest to change. The bacterium passes on to its offspring any change made in its DNA.

Look at Figure 6.17. It shows how a human gene is added to the genetic material of a bacterium. The small ring inside the bacterium is a piece of DNA that many bacteria have in addition to their larger chromosome. It is called a plasmid. Using special enzymes, the plasmid is split open. A human gene is then "spliced" into the plasmid and the ends rejoined.

Bacteria as Chemical Factories The gene added to the bacterium in Figure 6.17 carries instructions for producing a human protein. As the bacterium and its offspring carry on their life processes, they produce this human protein.

One human gene that genetic engineers have spliced into bacteria is the gene for the hormone insulin. Insulin is normally made by the pancreas. The pancreas of a person with diabetes cannot make insulin. Diabetics can live normal lives, however, by taking the insulin made by genetically altered bacteria. Engineered bacteria produce many other useful substances, such as human growth hormone, a protein to treat anemia, and a substance that may help fight cancer.

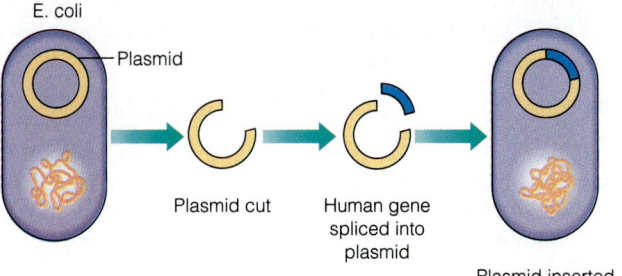

◀ **Figure 6.17**
The genetically-engineered bacterium will faithfully replicate the human gene in its plasmid. All of its offspring will carry out the instructions in the new gene.

E. coli — Plasmid
Plasmid cut
Human gene spliced into plasmid
Plasmid inserted into E. coli

Directed Inquiry

Have students study Figure 6.17 and the paragraphs on this page. Lead a discussion using some or all of the following questions:

▶ What is recombinant DNA? (DNA that has been created when genetic engineers move genes from one organism to another)

▶ Where is the DNA in the diagram? (In the plasmid and in the larger chromosome)

▶ Why are bacteria used to change an organism's traits? (The DNA must be in every cell of the organism's body. Since bacteria are single-celled, they are the easiest to change.)

▶ How is a human gene added to the genetic material of a bacterium? (The plasmid in the bacterium is split open and a human gene is spliced into the plasmid.)

▶ How can bacteria be used to make insulin and other proteins? (The gene for a protein is spliced into bacteria, and the bacteria and its offspring produce the protein according to the gene's instructions.)

▶ What advantages are there for using bacteria to produce proteins for humans? (Bacteria can produce proteins quickly and in quantity because they reproduce so rapidly.)

TEACH • Continued

Discuss
Ask students to explain why scientists began the human genome project. (One answer is that if genes that trigger diseases can be located within the human genetic code, it might be possible to remove or alter these harmful genes. Researchers may also find genes that protect people from certain diseases.)

EVALUATE

WrapUp
Reinforce Have students explain how they might use both selective breeding and genetic engineering to produce tomatoes resistant to frost and bruising.
Use Review Worksheet 6.4.

Check and Explain
1. Dog breeders mate dogs showing desired traits.
2. Genetic engineers use enzymes to open a section of DNA. A gene from another organism is then spliced onto the open DNA section.
3. Selective breeding helps farmers successfully obtain crops with desired characteristics.
4. Answers may vary. A human hormone might be supplied by splicing a gene for production of the hormone onto a bacterium that would then produce it. Mating plants that produce the most grain; mating surviving plants from a food crop destroyed by disease; mating sheep with the highest wool quality

---

INTEGRATED LEARNING

Math Connection
Students may have difficulty understanding the ambitious scope of the human genome project. Ask students how many total DNA base pairs the project workers are trying to identify, if each genome carries an average of 50,000 genes. Is it in the millions or billions? (Billions; 50 thousand times 50 thousand, or 2.5 billion)

Science and Technology
The Human Genome Project

Together all the genes of a species make up its *genome*. The human genome has 50,000 to 100,000 genes. In 1988, scientists began a project that would "read" the entire library of genetic information contained in the human genome. The project includes scientists from all over the world and will cost $3 billion or more. Scientists expect to complete the project around the year 2005.

The data from this project will be used to make a map of each chromosome. To give you an idea of how complicated this is, imagine creating a map that shows every building on every street in every city and town in the United States! To start, scientists must cut the chromosomes into smaller and smaller pieces to study them. Using computers, people in hundreds of laboratories then store and track all the accumulated information. Scientists will also use special sequencing machines to determine the sequence of base pairs in DNA.

The chromosome maps will show the location of genes that cause diseases such as cancer and heart disease. Genes that protect some humans from certain diseases may also be found. Once these human genes are mapped, it may be possible to detect and remove disease-causing genes. The human genome project, therefore, offers the possibility of curing many inherited diseases.

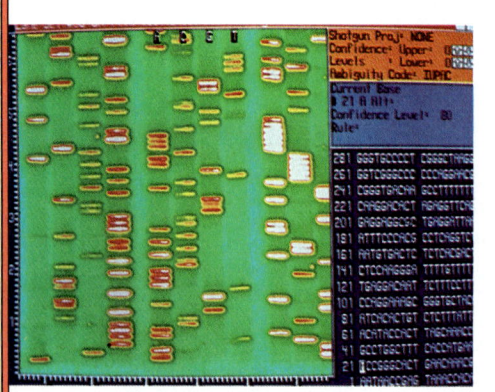

Figure 6.18 ▲
A computer helps tell scientists the order of base pairs on a short piece of human DNA.

Check and Explain
1. Explain how breeders get the most desirable traits in dogs.
2. How do genetic engineers change the genetic material on a chromosome?
3. **Infer** How does selective breeding help farmers? Explain.
4. **Hypothesize** Describe how genetic engineering and/or selective breeding might be used to solve each of the following problems: lack of a human hormone; low grain production; food plants destroyed by a specific disease; poor wool quality in a herd of sheep.

130   Chapter 6   Heredity

CHAPTER REVIEW 6

Chapter 6 Review

Concept Summary

6.1 Basic Principles of Heredity
▶ Every living thing is identified by its traits.
▶ Individual traits are determined by genes, which are either dominant or recessive. A recessive gene is masked by a dominant gene.
▶ Offspring receive one gene from each parent.

6.2 Chromosomes and Inheritance
▶ Genes are located on the chromosomes.
▶ A sex cell containing one set of single chromosomes is formed during meiosis.
▶ The X and Y chromosomes determine sex in humans.
▶ Sex-linked traits can occur in males because the Y chromosome cannot make up for the effects of genes on the X chromosome.

6.3 DNA and the Genetic Code
▶ All the information an organism needs to live and reproduce is carried by DNA.
▶ The sequence of the base pairs in DNA makes up instructions for assembling proteins.
▶ DNA can make copies of itself through the process of replication.

6.4 Applied Genetics
▶ Selective breeding can bring out desired traits in a species.
▶ Genetic engineering is the insertion of a gene from one species into the DNA of another.

Chapter Vocabulary

heredity (6.1)	recessive (6.1)	gamete (6.1)	Y chromosome (6.2)
genetics (6.1)	phenotype (6.1)	incomplete dominance (6.1)	replication (6.3)
gene (6.1)	genotype (6.1)	meiosis (6.2)	mutation (6.3)
dominant (6.1)	hybrid (6.1)	X chromosome (6.2)	

Check Your Vocabulary

Use the vocabulary words above to complete the following sentences correctly.

1. The first work in ____ was done by Gregor Mendel with pea plants.
2. A ____ is a unit of DNA inherited by an organism.
3. A ____ gene is masked if it is paired with a dominant gene.
4. A ____ is the gene combination for a single trait.
5. Of the two sex chromosomes, the ____ has fewer genes.
6. A duplicate DNA molecule forms during ____.
7. A female has two ____.
8. The trait that shows in an organism is its ____.
9. A ____ gene is always expressed.
10. During ____, cells are formed with one set of unpaired chromosomes.
11. When each parent breeds pure for a different trait, the offspring is called a ____.
12. A ____ causes a change in the DNA of a cell.
13. A trait somewhere between two others is a result of ____.
14. A sex cell is called a ____.
15. ____ is the passing of traits from parent to offspring.

Check Your Vocabulary

1. genetics
2. gene
3. recessive
4. genotype
5. Y chromosome
6. replication
7. X chromosomes
8. phenotype
9. dominant
10. meiosis
11. hybrid
12. mutation
13. incomplete dominance
14. gamete
15. Heredity

Use Vocabulary Worksheet for Chapter 6.

CHAPTER 6 REVIEW

Check Your Knowledge

1. The genotype of the offspring will be the same as that of the parent.
2. The genetic code determines an organism's traits.
3. Meiosis produces two identical cells, each with half the number of chromosomes of the parent cell.
4. The sex chromosomes: X and Y
5. Bacterial cells are commonly used in genetic engineering because they have a small amount of genetic material, a simple genome, and they reproduce rapidly.
6. Incomplete dominance occurs when two different genes in a pair interact to produce a trait somewhere between the traits of the two parents, with neither being dominant. An example is pink snapdragons produced from a cross between a white parent and a red parent.
7. A mutation is harmful when it causes a change in an organism's cells that leads to death or disease, as in cancer or birth defects. Mutations can be helpful when they give an organism and its offspring an advantage in survival, such as increasing size or the number of seeds a plant produces.
8. Punnett square
9. thymine
10. even
11. protein
12. one
13. cigarette smoke
14. genotype

Check Your Understanding

1. Mutations in sex cells can be passed on to offspring; mutations in body cells will not be passed along.
2. Answers will vary. The XXY genotype produces Klinefelter's syndrome, which is an abnormal male phenotype characterized by long legs, small testes, and breast development.
3. See page 118 in the text for a description of meiosis.
4. A person can carry a recessive genetic disease without having it if he or she also has a dominant gene to mask the recessive one.

5.

	C	c
C	CC	Cc
c	Cc	cc

6. Accept all logical, supported answers.
7. It is possible for the boy to be colorblind if his mother carries a defective gene.
8. Answers may vary, but may include fear of unintended consequences resulting from the creation of a new type of organism.
9. Answers may vary. Dog breeders want to know the pedigrees of their animals to insure that the animals and their offspring are or will be purebred.

Chapter 6 Review

Check Your Knowledge

Answer the following in complete sentences.

1. When pure-breeding plants are self-pollinated, what will be the genotype of their offspring?
2. What does the genetic code do?
3. What is produced during meiosis?
4. What set of human chromosomes is not always identical?
5. What kind of cells are most commonly used in genetic engineering? Why?
6. What is incomplete dominance? Give an example.
7. When is a mutation harmful? When is a mutation helpful? Give examples of both.

Choose the answer that best completes each sentence.

8. The results of a genetic cross can be predicted by using a (puzzle, line graph, set of equations, Punnett square).
9. Of the four nitrogen bases, adenine always pairs with (guanine, cytosine, thymine, protein).
10. The number of chromosomes in a many-celled organism is always (odd, large, greater than 20, even).
11. A gene is a set of instructions for assembling a (protein, chromosome, trait, genotype).
12. Meiosis produces gametes with (one, two, three, four) set(s) of unpaired chromosomes.
13. A combination of a dominant and recessive trait is a (phenotype, genotype, hybrid, factor).

Check Your Understanding

Apply the concepts you have learned to answer each question.

1. What is the difference between mutations in sex cells and mutations in body cells?
2. **Infer** Just as some people have an extra twenty-first chromosome, others have an extra sex chromosome. What do you think would be the gender of someone with two X chromosomes and a Y?
3. Explain what happens to the chromosomes in a cell during meiosis.
4. How is it possible for a person to carry a genetic disease without actually having it?
5. **Predict** Use a Punnett square to show the results of a cross between two parents, each with one dominant gene for curly hair.
6. **Mystery Photo** The photograph on page 110 shows an ear of ornamental corn. This kind of corn has been selectively bred to have kernels with different colors. If you looked at two ears of this kind of corn, do you think the color patterns would be exactly the same? Why or why not?
7. A boy's father is not colorblind. Is it possible for the boy to be colorblind?
8. **Application** Many scientists are concerned that genetic engineering may be dangerous. What do you think they worry about happening?
9. **Extension** Why do dog breeders want to know the pedigrees of all their animals as far back as possible?

Develop Your Skills

1. a. Either AA or Aa
   b. No
2. a. yellow
   b. 6022 yellow, 2001 green

Make Connections

1. Concept maps will vary. You may wish to have students do the map below. First draw the map on the board, then add a few key terms. Have students draw and complete the map.

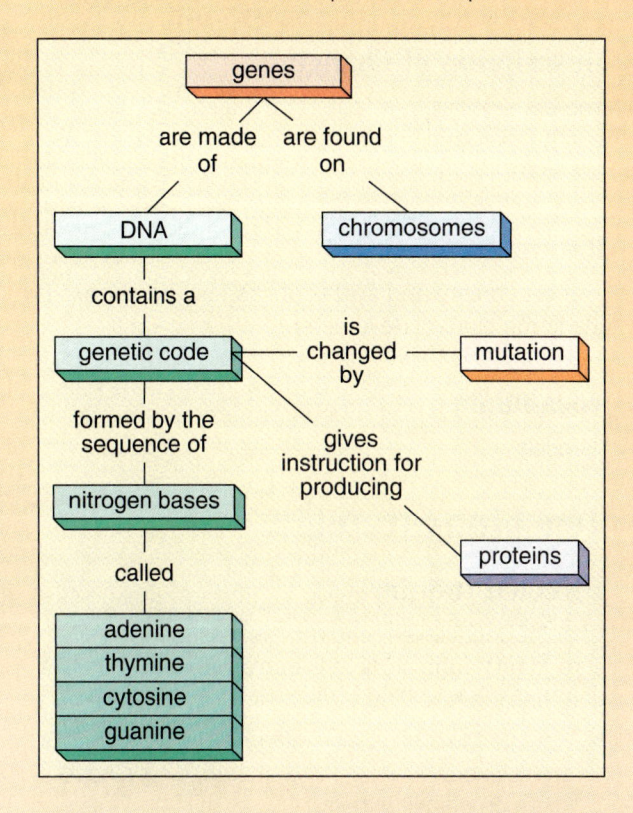

```
                        genes
                 are made    are found
                    of          on
                    |            |
                   DNA       chromosomes
                    |
                contains a              is
                    |               changed ─── mutation
              genetic code             by
                    |
              formed by the         gives
               sequence of      instruction for
                    |             producing
              nitrogen bases                ─── proteins
                    |
                  called
                    |
                 adenine
                 thymine
                 cytosine
                 guanine
```

2. In both cases, the number of possible outcomes is 2, and there is an equal chance of either outcome occurring in every trial.

3. Answers will vary. Students might infer that infrequently-occurring eye colors are due to recessive genes, while frequently-occurring ones result from dominant genes. In reality, dark eye color is caused by a dominant gene, while light eyes are recessive.

4. Answers will vary. The problems a colorblind person has are related to the inability to distinguish certain colors, especially red and green. Solutions generally involve written messages or symbols to convey the same meaning as the color signal (the word *stop* printed on a red stoplight, for example).

5. Answers will vary. Many types of bacteria have been genetically engineered to produce chemicals for human use. Examples include insulin-producing bacteria and frost-fighting bacteria.

6. There are 64 possible triplet codes (4 x 4 x 4 = 64).

Develop Your Skills

Use the skills you have developed in this chapter to complete each activity.

1. **Interpret Data** The pedigree below shows the inheritance of a trait called achondroplasia in a certain family. Achondroplasia is a form of dwarfism. It is rare, but the gene that causes it is dominant. Blue symbols in the pedigree indicate individuals with achondroplasia. Squares stand for males and circles for females.

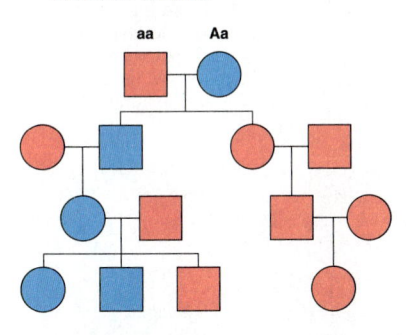

   a. What is the genotype of the achondroplasic son in the first generation?
   b. If the average–height daughter in the third generation marries an average–height man, can they have any achondroplasic children?

2. **Data Bank** Use the information on page 624 to answer the following questions.
   a. When Mendel crossed a pure yellow-seeded pea with a pure green-seeded pea, what seed color did all the offspring have?
   b. When the offspring of this cross were crossed with each other, how many yellow-seeded plants resulted? How many green-seeded plants?

Make Connections

1. **Link the Concepts** Draw a concept map showing how the concepts below link together. Add terms to connect, or link, the concepts.

   genes   genetic code
   cytosine   mutation
   chromosomes   proteins
   adenine   DNA
   nitrogen bases thymine
   guanine

2. **Science and Math** How is the chance of a human baby being a boy or girl similar to the chance of a tossed coin landing heads or tails? Use a coin toss to explain why there is a 50-50 chance of the baby being a girl.

3. **Science and You** Do a survey of the eye colors of the people in your class. Graph the results. What is the most common eye color? What can you infer about the genes that determine eye color?

4. **Science and Society** Write a paragraph discussing the problems a colorblind person might have when buying clothes, crossing the street, using makeup, or painting a picture. What could be done to solve these problems?

5. **Science and Technology** Find out what genetically engineered organisms have been produced recently. Describe what their added genes are meant to do.

6. **Science and Math** Recall that there are four nitrogen bases and that a group of three bases, called a triplet, forms a code for a certain amino acid. Determine the number of possible triplet codes.

SCIENCE AND LITERATURE CONNECTION UNIT 2

About the Literary Work

"The Plant People" was adapted from *The Plant People* by Dale Carlson, copyright 1977 by Franklin Watts, Inc. Reprinted by permission of Franklin Watts, Inc.

Description of Change

The excerpts were adapted for the sake of clarity, but the dialogue was kept as true to the original as possible.

Rationale

The vocabulary and syntax of this text are appropriate to the student level.

Vocabulary

osmosis

Teaching Strategies

Directed Inquiry

After students finish reading, discuss the story. Be sure to relate the story to the science lessons in this unit. Ask the following questions:

- How do Mike and Larry differ? (Mike's body cells are still normal while some of Larry's cells have taken on plantlike traits.)
- Why do you suppose Larry did not experience pain when he placed his skin into the spikes of the cactus plant? (It is likely that the nerve cells in his skin were destroyed.)
- What are some possible explanations for the fact that Larry needed to obtain nutrients from a cactus while other members of the community could "eat" from plants in town? (Possible responses might include that each person's body cells resembled the cells of a particular kind of plant. The person could then "eat" by touching only this kind of plant.)

Critical Thinking

Reason and Conclude Explain to the students that in the process of osmosis, water molecules move across a membrane from a place where there are many molecules to a place where there are few. Then ask the following questions:

- What does this indicate about the number of water molecules found in Larry and the cactus plant? (Since Larry had a look of pleasure on his face, water must have moved into his body. This indicates that a greater concentration of water molecules were held by the plant than by Larry.)
- Predict what might have occurred if the plant had less water molecules than Larry's body. (If this were the case, water molecules would have moved out of Larry and into the plant.)

Science and Literature Connection

The Plant People

The following excerpt is from the book The Plant People *by Dale Carlson*

Sam bent over a microscope in the school lab. He was examining a piece of Nancy Ward's skin. She hadn't minded when he cut her. She didn't mind anything, now.

"It's impossible," said Sam.

They'd all seen some terrible things. But this was a nightmare!

Mike put his eye to the microscope. Then he straightened up.

"I think I know." Mike's voice trembled.

"What?" asked Sam.

"There's been a change." Mike breathed deeply. When he spoke again, his voice was steady. "A change in the cells. Some are still human. Others look like plant cells."

"Look like?" said Sam. "They are the cells of a plant!"

Mike broke down. "How can skin change like that?!" he cried.

"I don't know," said Sam in defeat. "I just don't know enough about science, Mike. I can tell you what's happening. But I can't tell why or how. . . ."

Mike left the school. He had to phone Carson City. Every day he telephoned. "We'll send someone down" the police kept promising. But they seemed in no hurry.

People wandered the streets. More than usual, Mike thought. They seemed to be looking for something. They stopped at the sight of growing things. . . .

[Mike] felt a hand grip his arm. It was Larry Borden, [a classmate]. Mike saw that Larry's skin was really bad. He didn't, or couldn't talk. But he seemed to want something. And he made Mike understand. He looked hard into Mike's eyes. Then, Larry pointed toward the desert. Larry wanted Mike to take him there. . . .

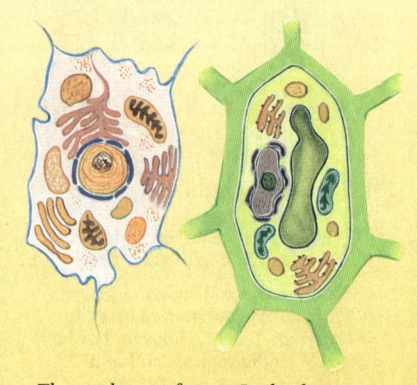

They rode out of town. In the desert, Larry began to breathe harder. He was excited. . . . Like a sleepwalker, he moved slowly. He stopped at a cactus. Before Mike could get to him, Larry plunged his hand into the thorns.

[In an instant, Mike] was by Larry's side. He wanted to help Larry deal with the pain from the thorns. But there was no pain. On Larry's face there was nothing but pleasure.

Skills in Science

Reading Skills in Science

1. The story is best defined as science fiction, a specific type of fiction. This story is highly imaginative, involving projected scientific phenomena.
2. Mike observed skin cells that resembled plant cells and watched Larry take in nutrients by osmosis.
3. Larry needed Mike to take him to the desert so that he could obtain the nutrients he needed to survive. ("Then, Larry pointed toward the desert. Larry wanted Mike to take him there.")

Writing Skills in Science

1. Students should detail the differences between plant and animal cells. Specifically, the plant-like cell would have a cell wall and chloroplasts.
2. Responses might include a need for sunlight to produce food, a lack of mobility, decline in thinking processes, and inability to communicate.

Activities

Evaluate Sources The author says osmosis had taken place when "Larry had taken food into his body through his skin." On page 94 of the text, osmosis is defined as the movement of water through a membrane from higher to lower concentration. The author is incorrect in saying that Larry had taken in *food* by osmosis.

Experiment The activity might include placing a cucumber slice in salt water, which will cause the slice to shrivel. This is due to the fact that water moves out of the slice and into the salt water.

Predict Skits may show that without certain nutrients, Larry would be unable to carry out his bodily functions.

It seemed impossible. But how could anything be impossible? Mike knew exactly what he had seen. It was called osmosis. Larry had taken food into his body through his skin!

Now Mike knew why nobody had been eating.... They couldn't eat food through their mouths.... But why had Larry needed to go out to the desert? Why not eat in his own backyard? Was there some human feeling left in him? Mike would never know. Larry Borden no longer talked.

Back in town, a terrible sight waited for Mike. The people were reaching for leaves. They were grasping for whole plants. And not just with their hands. With every part of their bodies! They ate, touching their skins to the food.

Skills in Science

Reading Skills in Science

1. **Classify** How would you best classify this text: fiction, science-fiction, or nonfiction? Give evidence for your response.
2. **Find Context Clues** What observations did Mike make that led him to conclude that Larry and other members of the town had begun to take on plantlike traits?
3. **Find the Main Idea** What was it that Larry needed Mike to do? Quote the sentence containing this information.

Writing Skills in Science

1. **Compare and Contrast** Describe the two kinds of cells in Nancy Ward's skin that Mike observed through the microscope.
2. **If...Then Arguments** How would your life be different if the cells in your body underwent the same type of change as Larry's body cells?

Activities

Evaluate Sources Compare the author's description of osmosis with the discussion of it in your textbook. Does the author of this story describe osmosis accurately?

Experiment Design an activity using thin slices of cucumber and salt water to illustrate the process of osmosis.

Predict Suppose Mike had not taken Larry out to the desert. Design a skit to show what might have happened to Larry.

Where to Read More

Cactus, the All-American Plant by Anita Holmes. Four Winds, 1982. This book examines the Sonoran cacti, provides tips for raising the plants, and includes recipes for cooking with cactus!

Cactus by Cynthia Overbeck. Lerner, 1982. A detailed study of the parts of a cactus plant and how the organism is adapted to its unique environment.

UNIT 3

UNIT OVERVIEW

This unit focuses on the diversity of living things, the history of life on the earth, and the classification of organisms. The diversity among organisms is the focus of Chapter 7. This chapter describes the survival advantages of various adaptations. Sexual and asexual reproduction processes are explained. The chapter also introduces Darwin's theory of natural selection. Chapter 8 describes how scientists measure and study geologic time. Significant events in the earth's history are explored. A description of fossils is included to illustrate how scientists study evolutionary relationships. The chapter concludes with a discussion of human evolution. Chapter 9 explains the various ways in which scientists classify living things. The chapter describes the five kingdoms and shows students how to identify organisms by using taxonomic keys and field guides.

Introducing the Unit

Directed Inquiry

Have the students examine the photograph. Ask:

▶ What part of the island is shown here? (The seashore)

▶ Why do you think these crabs are thriving? (The shore is still in its natural state; few enemies are visible.)

▶ What conditions might make these crabs unique? (The lack of natural enemies, the isolation on an island)

▶ Do you think these crabs are likely to have changed much over time? (Living in an unchanging environment, they stayed the same for many years.)

Writing About the Photograph

Ask each student to write an essay from the point of view of a crab after it encounters an animal it has never seen before.

Unit 3
Evolution and Classification

Chapters

7 Diversity of Living Things
8 History of Life
9 Classifying Living Things

Data Bank

Use the information on pages 624 to 631 to answer the following questions about topics explored in this unit.

Classifying

What is the species name of a dog? What is the species name of a polar bear?

Interpreting Data

What epochs make up the Cenozoic era? Which of these epochs is most recent?

Observing

In what ways are all the animals on the chart related on page 628? Describe some physical characteristics that these animals have in common.

Reading a Table

During which period in the earth's history did the first mammals appear? When did dinosaurs die out?

The photograph to the left is of crabs on rocks in the Galápagos Islands. What other kinds of organisms might you find on these islands?

Data Bank Answers

Have students search the Data Bank on pages 624 to 631 for the answers to the questions on this page.

Classifying The species name of a dog is *Canis familiaris* and the species name of a polar bear is *Ursus maritimus*. The answer is found in the diagram, Relationships Among Carnivora, on page 628.

Extension Ask students to describe the way in which skunks and otters are related. (They belong to the same family, known as Mustilidae.)

Interpreting Data The Paleocene, Eocene, Oligocene, Miocene, Pliocene, Pleistocene, and Holocene make up the Cenozoic era. The Holocene period is the most recent. The answers are found on page 630 in the table, Evolution of Mammals.

Observing The animals represented on page 628 are all carnivores that are also mammals. Students may note other similarities, such as the fact that they all have four legs.

Reading a Table The first mammals appeared in the Triassic period. The dinosaurs died out in the Cretaceous period. The answers are found in the table, Evolution of Mammals, on page 630.

Answer to In-Text Question

The life forms found by Darwin on these islands had not been seen anywhere else. Today, organisms such as tortoises, iguanas, and finches, to name a very few, inhabit the islands.

CHAPTER 7

Overview

The diversity among organisms is the focus of this chapter. The first section describes various adaptations that allow organisms to survive. The second section discusses the combination of genetic material in sexual reproduction that accounts for variation and genetic diversity within a single species. Darwin's theory of natural selection is discussed in the third section.

Advance Planner

- ▶ Purchase small white and dark beans (such as lima beans) for SE page 142.
- ▶ Gather old magazines or catalogs containing pictures of human-made adaptations for TE activity, page 143.
- ▶ You may want to arrange a field trip to a zoo to compare members of a species for SE page 144 (and as an adjunct to the chapter itself).
- ▶ Supply old nature or wildlife magazines with multiple photos of the same animal species for SE page 146.
- ▶ Purchase striped or textured beans (such as pinto beans) for SE Activity 7, page 147.
- ▶ If available, arrange to take students to a vacant lot for SE page 151.

Skills Development Chart

Sections	Classify	Communicate	Estimate	Infer	Measure	Observe
7.1 Skills WarmUp SkillBuilder		●		●		
7.2 Skills WarmUp Skills WorkOut Activity	● ●				●	● ●
7.3 Skills WarmUp Skills WorkOut Historical Notebook	●	●	●			

Individual Needs

- ▶ **Limited English Proficiency Students** Provide students with a picture dictionary. Have students use the picture dictionary or the text to find definitions of the boldface terms in the chapter. Have students find and cut out pictures from magazines or newspapers that show examples of the boldface terms. Also, have them find pictures showing an organism in its habitat, variation among members of the same species, and natural selection. Have students paste these pictures onto notebook paper so they can create their own picture dictionaries for this chapter.

- ▶ **At-Risk Students** Have students visit a zoo or arboretum. Have them prepare for the trip by writing questions in their science journals based on the chapter concepts. Students' questions may include: What adaptations do I see? Which of the organisms have similar natural habitats? What are some examples of diversity? What species are represented? Have at least one worker from the site speak to your students and answer any questions they have not answered on their own. Have students record answers to their questions in their science journals and share information with classmates when they return from the trip.

- ▶ **Gifted Students** Have students create a display tracing the evolution of an unusual organism. Encourage students to make three-dimensional models of each species and its environment. Have them choose an animal from a list of those found in the Galápagos Islands or Australia. Students should attempt to follow the histories of at least two strains of one population. They should refer to the chapter and note Darwin's finches as an example.

Resource Bank

- ▶ **Bulletin Board** Make a bulletin board that shows the diversity of adaptations and habitats. Hang a title on the bulletin board and hang one or two pictures showing an organism's adaptations. Attach a label explaining how the adaptation helps an organism survive in its habitat. (See page 140 for ideas.) Have students find other photographs to add to the display. Students should describe the adaptations shown in their pictures.

- ▶ **Field Trip** Take students to an aviary so they can observe some of the traits in birds that Darwin studied. Have them count the number of species of birds that they see. Arrange to have a guide talk to students about the unusual features of different birds. Have students consider which of these characteristics are adaptations.

CHAPTER 7 PLANNING GUIDE

Section	Core	Standard	Enriched	Section	Core	Standard	Enriched
7.1 Diversity and Adaptation pp. 139–143				**7.3 Natural Selection** pp. 148–152			
Section Features Skills WarmUp, p. 139; SkillBuilder, p. 142	● ●	● ●	●	**Section Features** Skills WarmUp, p. 148; Skills WorkOut, p. 151; Historical Notebook, p. 151	● ● ●	● ●	●
Blackline Masters Review Worksheet 7.1; Integrating Worksheet 7.1	●	● ●	●	**Blackline Masters** Integrating Worksheet 7.3; Enrich Worksheet 7.3; Review Worksheet 7.3; Vocabulary Worksheet 7	● ●	● ● ● ●	● ●
Color Transparencies Transparency 20	●	●					
7.2 Variation and Reproduction pp. 144–147				**Overhead Blackline Transparencies** Overhead Blackline Master 7.3 and Student Worksheet	●	●	●
Section Features Skills WarmUp, p. 144; Skills WorkOut, p. 146; Activity, p. 147	● ● ●	● ● ●	● ● ●	**Laboratory Program** Investigation 14	●	●	●
Blackline Masters Review Worksheet 7.2; Skills Worksheet 7.2	● ●	● ●					

Bibliography

The following resources can be used for teaching the chapter. See page T-40 for supplier codes.

Audio-Visual Sources
(video unless noted)
Animal Populations: Adaptation to Environment. 16 min. 1983. LF.
Did Darwin Get It Wrong? 57 min film. 1982. T-L.
Dwellings. 15 min. CABISCO.
Learning and Instinct in Animals. Filmstrip with cassette. 1983. NGSES.
Stephen Jay Gould: His View of Life. 57 min film or video. 1984. T-L.
The Wilds of Madagascar. 60 min. 1990. NGSES.

Software Resources
The Balance in Nature. Apple, Commodore. FM.
Diversity of Life. Apple. J & S.
Evolve! A Trip Through Time and Taxonomy. Apple. JWW.

Library Resources
Cork, Barbara, and Lynn Bresler. The Young Scientist Book of Evolution. Tulsa, OK: EDC Publishing, 1985.
Darwin, Charles. On the Origin of Species: A Facsimile of the First Edition. Cambridge, MA: Harvard University Press, 1975.
Gallant, Roy A. From Living Cells to Dinosaurs. New York: Watts, 1986.
Gould, S. J. The Panda's Thumb. New York: Norton, 1980.

CHAPTER 7

INTEGRATED LEARNING

Language Arts Connection

Ask students to make a list of ways in which the organism in the photograph is similar to humans. Then have them list ways in which it is different from humans. Discuss with students how the organism's particular adaptations help it to carry out its life functions. (Students' lists will vary.) They may mention that both have eyes, legs, and a mouth and that both move around to get food. When listing differences, they may say that the organism has many different body parts, is smaller, and lives in a different environment than humans.)

Introducing the Chapter

Have students read the description of the photograph on page 138. Ask if they agree or disagree with the description.

Directed Inquiry

Have students study the photograph. Ask:

▶ What kind of organism is in the center of the photograph? (An insect)

▶ Where is the organism? (On a tree)

▶ What do you notice about the color of the organism and the color of the tree it is on? (Their coloring is similar.)

▶ How do the similarities between the tree and the insect benefit the insect? (When the insect is positioned on the tree bark, it is almost invisible to predators.)

▶ How does this photograph relate to the chapter concepts of diversity and adaptation among living things? (Some students may realize that the insect's camouflaged coloration is an example of how organisms adapt to their environment.)

Chapter Vocabulary

adaptation
asexual
  reproduction
diversity
evolution
habitat
natural selection
population
sexual
  reproduction
species
variation
zygote

Chapter 7

Diversity of Living Things

Chapter Sections

7.1 Diversity and Adaptation

7.2 Variation and Reproduction

7.3 Natural Selection

What do you see?

"I see a grayish-green bug. It's on a tree, maybe a redwood. The bug looks like a piece of mold. It might use this technique for camouflage. It has four long, skinny legs and two short, big legs. The tail looks like an old samurai helmet, and the head looks like a rectangle with a circle on each short side. It has two short antennas, also."

Joshua Wragg
Whitford School
Beaverton, Oregon

To find out about the photograph, look on page 154. As you read the chapter, you will learn more about the variety of living things.

Themes in Science

Evolution Adaptations help organisms perform life functions such as food gathering, movement, protection, and reproduction. One reason there is such a diversity of adaptations among organisms is that there are many different types of habitats on the earth. Have students name different habitats on the earth. Then have them think of examples of animals that live in these habitats and discuss how their adaptations help them survive.

7.1 Diversity and Adaptation

Objectives

▶ **Define** and **explain** the term *adaptation*.

▶ **Explain** why different organisms living in the same habitat may have similar adaptations.

▶ **Infer** how organisms use their senses to adapt to their environment.

Skills WarmUp

Infer Hold your thumb straight and tape it to your index finger. Try to pick up a coin, tie a shoelace, button a shirt, and write your name. How is your thumb useful for performing these tasks? Explain how the thumb helps humans survive.

What makes one kind of organism different from another? Look at the birds in Figure 7.1. How are the ostrich and hummingbird different? You can see that each bird has a beak, wings, and feet, but each body part is a different size and shape. These structural differences in their body parts not only make them look different from one another, but they are also used differently.

In every living thing, the body structure is important to survival. The hummingbird's wings, for example, allow it to hover at a flower while it sucks nectar with its long, narrow beak. The ostrich's long, powerful legs allow it to run at fast speeds. Every kind of organism is unique because of its special characteristics and how it uses them to live.

Adaptations

A characteristic, structure, or behavior that helps an organism live is an **adaptation**. Adaptations help organisms get food, protect themselves, move, reproduce, and carry on life processes. The sharp teeth and claws of a tiger are an adaptation for catching and eating prey. A chameleon's long, sticky tongue is an adaptation for catching insects. The joints of your legs are an adaptation for walking.

An organism's adaptations help it survive in a particular **habitat**. A habitat is the area or place where an organism lives. With its adaptations, an organism is adapted to its habitat. The emperor penguin, for example, has adaptations that help it live successfully in an extremely cold climate, where other birds, such as robins and sparrows, could not live.

Figure 7.1 ▲
How are the hummingbird's adaptations different from the ostrich's? ①

Chapter 7 Diversity of Living Things 139

SECTION 7.1

Section Objectives
For a list of section objectives, see the Student Edition page.

Skills Objectives
Students should be able to:

Infer how the thumb helps humans survive.

Make a Graph showing the effect of natural selection.

Infer how senses are useful adaptations for kittens.

Vocabulary
adaptation, habitat, diversity, species

MOTIVATE

Skills WarmUp
To help students understand the idea of adaptation, have them do the Skills WarmUp.
Answer Students should realize that the thumb is useful in grasping and holding and that it enables humans to perform many specialized tasks.

Prior Knowledge
Gauge how much students know about diversity and adaptation among living things by asking the following questions:

▶ What is an adaptation?

▶ How do the adaptations of organisms relate to their habitats?

▶ What is natural selection?

Answer to In-Text Question

① The hummingbird's wings allow the bird to hover near a flower and suck nectar with its long beak. The ostrich, on the other hand, is unable to fly but can run quickly from predators with its long, powerful legs.

139

TEACH

Explore Visually
Have students study the material on this page. Ask the following questions:

▶ What is the term that refers to the variety of life in a habitat? (Diversity)

▶ What adaptation do coconuts have for floating to new islands? (Their thick, airy husks allow them to float on water to different places.)

▶ What adaptations help anteaters hunt termites? (Powerful front legs for digging and a half-meter-long tongue help them get into hard termite mounds.)

▶ What adaptation allows turkey vultures to glide and soar easily? (Long, broad wings)

▶ What adaptation allows dandelion seeds to be carried on the wind? (They are lightweight and have an umbrellalike structure to help them disperse.)

▶ What adaptation do bats have for flying in the dark? (They send out sound waves that hit an object and echo back to the bat's large ears.)

Skills Development
Infer Ask students why it is useful for bats to be able to fly in the dark. (They are nocturnal animals, which means they sleep during the day and hunt at night.)

Integrated Learning
Use Integrating Worksheet 7.1.

INTEGRATED LEARNING

Art Connection
Have students choose an animal from each of three different types of habitats, such as a desert, rain forest, and ocean. Have them draw or paint each animal in its habitat and explain how its adaptations help it survive there. For example, a seal's thick layer of blubber keeps it warm in the cold ocean waters where it lives. Have students share their artwork with the class.

Diversity of Adaptations

As you may know, many different habitats exist on the earth. Each habitat has a variety of organisms that live there in different ways. This variety of life is called **diversity**.

Below you can see some examples of the many adaptations that living things have. As you look at each organism, imagine the kind of habitat in which it lives. Think about how the organism uses all its adaptations to survive in that habitat.

▲ Coconuts, which are the seeds of one kind of palm tree, can float from island to island because of their thick, airy husks.

The long, broad wings of a turkey vulture help it glide and soar without using large amounts of energy. ▼

▲ Jackrabbits living in the desert have long ears that provide a large body surface for giving off extra heat.

▲ The powerful front legs and half-meter-long tongue of the anteater help it get into the hard mounds of termites.

▲ Dandelion seeds are carried on the wind by an umbrellalike structure. Dandelions can spread over large areas.

Bats can move freely in darkness by sending out sound waves. When the sound waves hit an object, they echo back to the bat's large ears. ▼

Chapter 7 Diversity of Living Things

Integrating the Sciences

Earth Science Explain to students that climate refers to the usual atmospheric conditions found in an area over a long period of time. Weather, on the other hand, refers to changing atmospheric conditions over short periods of time, such as a day or week. Thus, a desert has a dry climate but may have rainy weather for a day.

Themes in Science

Diversity and Unity Organisms that are not related may share similar adaptations because they live in similar habitats. For example, the snowshoe hare and the white wolf are not closely related but both have thick, white coats that blend in with their snowy, arctic habitat. Have students think of examples of organisms that are not closely related but share similar adaptations due to their similar habitats.

Critical Thinking

Reason and Conclude Tell students that plants that live in very cold climates with short warm seasons have special adaptations. Ask them:

▶ What size plants would you expect to be in very cold areas? Why? (Small and low to the ground to conserve energy)

▶ How fast would they reproduce? Why? (Quickly, because there is such a short growing season)

Discuss Explain that related organisms typically have similar adaptations. Ask students to think of other examples of related organisms with similar adaptations. Have them explain how the adaptations help the organisms. (Answers will vary but may include all land plants have roots or rootlike structures, which enable them to get water from soil; and all cats have powerful hind legs for running and jumping.)

Answers to In-Text Questions

① Some animals with short limbs and small ears are ermines, voles, and lemmings.

② Each of the structures has two "lower-arm" bones, and one "upper-arm" bone. Each structure is used for moving the body or for moving and manipulating objects.

The Living Textbook: Life Science Sides 7-8

Chapter 14   Frame 01647
Homologous Limbs/Vestigial Structures
(2 Frames)
Search: Step:

Patterns of Adaptations

Many adaptations are found in more than one kind of organism. Similar adaptations occur in organisms for two main reasons.

Organisms in the Same Habitat In order to survive the same kinds of conditions, organisms need similar adaptations. For example, animals living in cold areas tend to have rounded bodies with short limbs and small ears. Can you ① name some animals that fit this description? These body shapes expose the least amount of body surface to the cold environment and help conserve body heat.

Look at Figure 7.2. Notice the similarities in the two plants. Plants that live in very hot, dry areas have similar adaptations. Most have thick, small, hairy leaves or no leaves at all. Since plants lose water through their leaves, these adaptations help desert plants conserve water.

Figure 7.2 ▲
These two plants live on opposite sides of the world. However, they have similar adaptations because they both live in dry habitats.

Related Organisms Organisms may have similar adaptations because they are related to each other. Large groups of organisms have the same general adaptations needed for survival. For example, all plants have cell walls and almost all plants have chlorophyll to capture energy from sunlight.

Groups of more closely related organisms also have adaptations in common. All birds have wings and use beaks to get food, no matter where they live. All fishes have gills and fins. Frogs and toads both have strong back legs for hopping. Sea turtles have shells similar to the shells of pond turtles.

Organisms that are related to each other may have similar structures, even though the structures are used differently. Look at the different front limbs in Figure 7.3. How are these structures similar? How is each used? ②

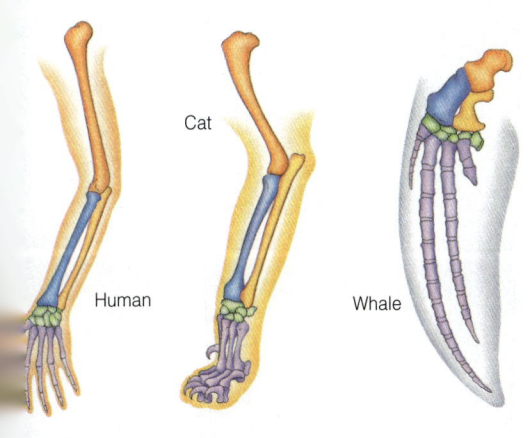

Figure 7.3 ▲
Humans, cats, whales, and all other mammals have similar front limbs. They are made up of the same kinds of bones put together in the same way. Only the shape and size of each bone vary.

Chapter 7 Diversity of Living Things **141**

TEACH • Continued

Discuss
Ask students:

▶ What is a species? (A group of living things that share a large number of characteristics and adaptations, which can mate and have offspring that are also fertile)

▶ Why aren't pandas included in the bear species? (They have a sixth finger that bears lack.)

Skills Development
Classify Have students think of characteristics of humans that distinguish them from other species. (Answers may vary. Humans have specialized hands, the ability to manipulate the environment, and fully upright posture.) Point out that humans are *not* the only species that uses tools. Chimpanzees have been observed using sticks to draw ants out of their nests.

SkillBuilder
Answers
1. Answers will vary, but the light "mice" would be picked up less often since they are better camouflaged than the dark "mice."
2. The light "mice"
3. Students' graphs should show an increase in the light "mouse" population and a decrease in the dark "mouse" population.

142

INTEGRATED LEARNING

Themes in Science
Stability and Equilibrium Two animals belong to the same species if they are able to mate and produce offspring that can, in turn, produce offspring. Tell students that people sometimes mate donkeys with horses to produce mules, which are strong and docile work animals. Ask students whether horses and donkeys belong to the same species. (They do not.)

Figure 7.4 ▲
All pandas belong to the same species.

Species

Living things that share a large number of characteristics and adaptations are identified as one kind of organism, called a **species**. A species is made up of organisms that look alike. Members of a species can mate and reproduce young that can reproduce.

All the members of a species have some adaptations that distinguish them from other species. For example, the panda has a sixth finger for stripping leaves off bamboo stems. This special finger developed from a wrist bone. It is an adaptation that is not shared by any other animal. Consequently, pandas are not the same species as bears, even though they look a lot alike.

Any two individuals in a species are never exactly the same, however. For example, you and your friends all look different from each other, but you are all human beings. Sometimes it is very difficult to tell whether two organisms belong to the same or different species

SkillBuilder *Estimating*

The Effect of Adaptations

Use small white beans and dark beans, such as lima beans, to represent mice. You'll need about 40 white beans and 40 dark beans altogether. Pretend you are an owl that preys on mice. A 10-second pickup period represents one week.

Scatter 20 dark beans and 20 white beans over a square white surface that measures 9 sq m. Pick up beans for 10 seconds. Pick up only one bean at a time and put it down before picking up another. Record the number picked up. For each bean you missed, add 2 beans of each color to the square. Repeat this procedure two more times. Be sure to record the numbers.

1. In three "weeks," how many light "mice" could the "owl" pick up from the surface? How many dark mice could the owl pick up?

2. Which color of mouse is most protected from the owl? How is color an adaptation?

3. Make a graph like the one shown below to show the trend of the decrease or increase of the mouse population over time. Use one color for white and one color for dark to show your estimations.

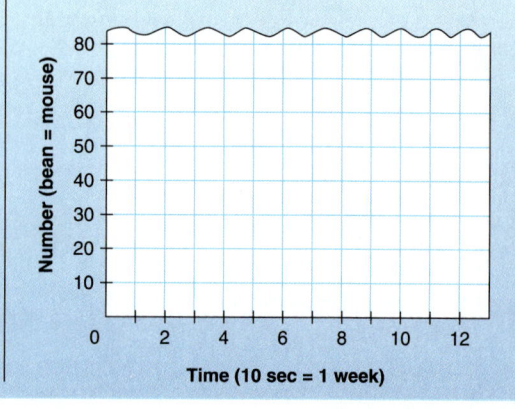

142   Chapter 7   Diversity of Living Things

STS Connection

Explain that dolphins and whales use underwater echolocation to detect obstacles and food. Humans have learned to use echolocation, in the form of sonar, to get information underwater. Sound waves are reflected back to a sonar transmitter by whatever object is in the path of the sound waves. Echolocation is used to detect icebergs, to locate enemy submarines, and to find schools of fish for fishing. Ask students how echolocation helps enhance ordinary human senses. (They may say it enhances our vision or hearing.)

---

because of individual variation. For example, German shepherds and French poodles look different, but they belong to the same species.

Look at the elephants in Figure 7.5. Are they members of the same species? They have similar adaptations, but they are not exactly the same. What differences do ① you notice? One is an African elephant and the other is an Indian elephant. You may be surprised to know that they belong to different species.

Science and You
Helping Out Your Adaptations

You share adaptations with people all over the world. All people are endothermic and have a protective coating of skin. All people see things in three dimensions, and most people see things in color.

Sometimes your adaptations are limited. You probably don't notice how you overcome these limitations. For example, your body temperature remains constant, but you still get cold. How do you adapt? You help maintain your internal temperature by wearing warm clothing and by heating your surroundings.

How do you adapt to the limitations of your eyes? For example, if you can't see something from a great distance, you may use a telescope or binoculars to overcome this limitation. Many people wear corrective lenses to see things more clearly, both up close and far away. What other limitations do you have? How do you adapt to these limitations?

Figure 7.5 ▲
The African elephant (top) and the Indian elephant (bottom) are different species.

Check and Explain

1. What is an adaptation? Give a few examples from both humans and other organisms.
2. What adaptations do humans have that make it possible for them to ride a bicycle?
3. **Compare and Contrast** Explain how the bones in Figure 7.3 on page 141 are adapted for the survival of each organism.
4. **Infer** Explain how sight, touch, and smell are useful adaptations for a family of new kittens.

EVALUATE

WrapUp
Reteach Write the section vocabulary terms on the chalkboard—*adaptation, habitat, diversity,* and *species.* Ask students to define each term and provide examples of each.
Use Review Worksheet 7.1.

Check and Explain

1. A characteristic, structure, or behavior that helps an organism survive in its habitat Examples: Tiger—sharp claws and teeth for catching and eating prey; sea turtle—shell for protection; humans—hands with thumbs for specialized tasks
2. Leg joints that allow rotation, a sense of balance while in an upright position, hands that can grasp and hold, and strong leg muscles
3. Human—for lifting, grasping, carrying; cat—for carrying body weight, walking, jumping; whale—for moving in water
4. They help kittens perform life functions such as moving, finding food, and distinguishing among family, predators, and prey.

Answer to In-Text Question

① The ears and the tusks of the African elephant are larger than the ears and tusks of the Indian elephant.

The Living Textbook:
Life Science Side 8

Chapter 20   Frame 02358
Adaptation (Movie)
Search: Play:

Chapter 7 Diversity of Living Things   143

SECTION 7.2

Section Objectives
For a list of section objectives, see the Student Edition page.

Skills Objectives
Students should be able to:

Observe and describe variation in a species.

Classify members of a species.

Estimate arm lengths, compare to actual lengths, and graph the results.

Classify bean variations.

Vocabulary
variation, sexual reproduction, zygote, evolution, asexual reproduction

MOTIVATE

Skills WarmUp
To help students understand variation in a species, have them do the Skills WarmUp.
Answer Students should observe that members of a species vary in color, size, shape, markings, and so on.

The Living Textbook: Life Science Side 7
Chapter 39   Frame 44304
Color Variation (Movie)
Search: Play:

INTEGRATED LEARNING

Themes in Science
Diversity and Unity Offspring produced by asexual reproduction are exact replicas of a single parent, receiving a copy of that parent's genetic material. In sexual reproduction, the offspring receive a complete set of genes from two different parents. This method of reproduction allows for variation among members of a species, because offspring are born with a unique combination of genes. Thus, while an offspring may share some traits of both parents, it will never share all the traits of either of them. Ask students why sexual reproduction is advantageous to organisms in a changing environment. (It allows for variations in a species, which may give that species a better chance of surviving.)

Skills WarmUp
Observe Study a group of the same species of birds, fishes, or other animal. Observe live animals or use photos from a book. Do all members of the species look alike? If not, describe how they are different from one another. Why don't all members of the same species look alike?

7.2 Variation and Reproduction

Objectives

▶ **Describe** how the members of a species may vary.

▶ **Explain** how sexual reproduction helps produce variation in a species.

▶ **Compare** and **contrast** the amount of trait variation possible in sexual and asexual reproduction.

▶ **Estimate** the amount of trait variation that can occur in a group.

Look at the kittens in Figure 7.6. How are they alike? How are they different? Even though they have similar ears and body shapes, their color patterns and sizes are not the same. Still, each kitten resembles its parents. As you know, this resemblance is the result of inheriting genes from both parents. Because each kitten looks a little different from its parents and from its brothers and sisters, this family of kittens has a variety of traits.

Variation

The differences among the kittens is an example of **variation** within a species. Variation is easy to see among humans. We are all different shapes, sizes, and colors. In what ways do you vary from your friends and the members of your family?

If you look closely, you can see that differences exist among the individuals of almost every species. Dogs, horses, apple trees, goldfish, and other species can have a great deal of variation. The variation in these species is increased intentionally by the people who breed them.

Species that reproduce in the wild also have variations. For example, individual ladybugs have different numbers of spots. Two pine trees of the same species will have different shapes. Even bacteria show differences in their ability to resist the drugs that were designed to kill them. Within a species, individual variation shows up in physical traits and in the genetic material.

Figure 7.6 ▲
Kittens in a litter often have very different color patterns.

144   Chapter 7   Diversity of Living Things

Themes in Science

Diversity and Unity A developing egg may or may not be protected by its parents, depending on the species. For example, oysters fertilize their eggs, then leave them to survive on their own. Oysters fertilize thousands of eggs to help increase the odds that some will survive. Many species of birds build nests to protect their eggs. In marsupials such as kangaroos and opossums, the young develop for a while inside the mother's womb and are then carried in the mother's pouch. Ask students how unborn babies are protected by the mother. (An unborn baby develops for a long time inside the mother's womb.)

Sexual Reproduction

Remember that when two members of the same species reproduce, each contributes genetic material to the offspring. This form of reproduction is called **sexual reproduction**. Sexual reproduction results in a new combination of genetic material that is unique in the species.

Since each new combination of genes is unique, each individual in a species varies slightly from the others. Sexual reproduction is responsible for the variation within a species.

Sexual reproduction produces new genetic combinations for two reasons. During the formation of gametes, or sex cells, the genetic material of the individual is shuffled. Recall that the gametes are the result of meiosis. During meiosis, the chromosome pairs intertwine and trade genes. In humans, the genes on the 23 pairs of chromosomes can be shuffled into at least eight million different combinations. Consequently, each gamete has a unique combination of genes.

Remember also that each parent contributes one set of genes. The two sets of genes combine when two gametes come together to form a fertilized egg, or **zygote**. The zygote contains a completely random and unique assortment of genes.

Most species reproduce sexually. However, not all living things need a partner to reproduce. A single-celled organism, for example, divides by fission to form two new individuals. The new cells get their genes from just one parent. Thus the two cells are exact copies of each other. When reproduction takes place through the copying of just one individual's genetic material, it is called **asexual reproduction**.

A large number of species that reproduce asexually can also reproduce sexually. Some have developed other ways of exchanging genetic material. The paramecium, for example, can join with another of its kind and trade pieces of DNA. This process creates variation among parameciums.

Figure 7.7 ▲
These chaffinch chicks were produced sexually. Each has a unique set of genes.

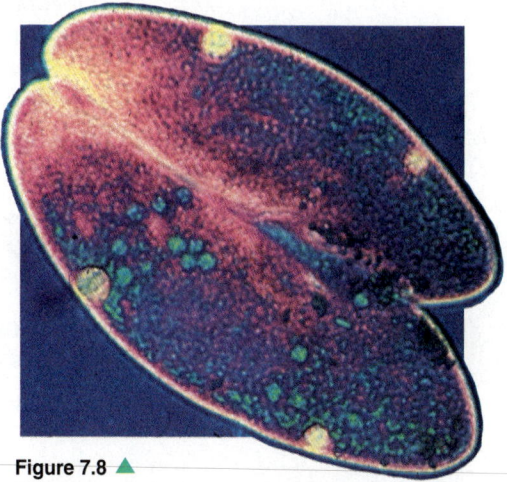

Figure 7.8 ▲
These parameciums are conjugating, or exchanging genetic material.

Chapter 7 Diversity of Living Things **145**

TEACH

Directed Inquiry
Have students study the material on this page. Ask the following questions:

▶ What are the main differences between asexual and sexual reproduction? (Asexual reproduction involves only one parent, which produces identical offspring. Sexual reproduction involves two parents who contribute an equal amount of genetic material. An offspring of sexual reproduction is different from its parents.)

▶ How does sexual reproduction produce new genetic combinations? (First, an individual's genetic material is rearranged. Second, genetic material from two individuals is combined.)

Reteach
Remind students of what they learned about genes in Chapter 6. When genes from two parents combine in an offspring, there are two genes for each trait. Genes are dominant or recessive. Some examples of dominant genes are brown eyes, color vision, and hairiness. Corresponding recessive genes are blue eyes, color-blindness, and baldness. Suggest that an offspring inherits a gene from her mother for brown eyes and a gene from her father for blue eyes. Ask students, What eye color will she have? Why? (Brown, because it is dominant)

145

TEACH • Continued

Skills WorkOut
To help students classify variations in the same animal species, have them do the Skills WorkOut.
Answer Make sure that students' classifications are logical.

EVALUATE

WrapUp
Reteach Have students list and explain the factors that contribute to diversity and variation among living things.
Use Review Worksheet 7.2.

Check and Explain
1. An organism possesses some unique genetic variations but shares most of its genetic material with the other members of its species.
2. Sexual reproduction results in a new combination of genetic material.
3. An individual's genes are reshuffled and combined with another individual's genes to produce a unique offspring.
4. Student arm lengths are likely to vary considerably yet should all be within the same general range.

INTEGRATED LEARNING

Integrating the Sciences
Chemistry Remind students of what they learned about mutations in Chapter 6. A mutation occurs when a strand of DNA is miscopied or changed in some way and passed on within the genetic code of an offspring. Genetic mutations may arise spontaneously or from the effects of chemicals, radiation, or heat. Mutations in an individual are passed on to its offspring.

Skills WorkOut
Classify Look through old magazines to find as many photographs of the *same* animal species as you can. Cut out the photos and decide on an order in which to arrange them on a posterboard. For example, you may choose to arrange the animals by size. Glue the pictures onto the posterboard. Describe the characteristics you used to arrange the individuals. In what other ways could you arrange the photographs? How much variation occurs in the species? Describe the average individual in the species.

Genetic Diversity
Variation within a species occurs when there are different forms of the same genes. You learned that sexual reproduction creates new combinations of genes, but how are different genes created in the first place? Genes change when a mutation occurs.

Mutations happen all the time in a species. A small number of the genes changed by mutation survive and are passed on. They become part of the species' gene pool. As a result of mutation and sexual reproduction, the genes of the individuals in a species show variety.

Science and You *Variation in Humans*
Look at the variety of features in human beings. Not only are eyes different colors, but often the color and texture of hair is not the same. Body parts, such as legs, arms, ears, and noses, are different sizes and shapes. Fingerprints are different. Each human is unique.

In other ways, however, all humans are very much alike. Many of the features that make individuals look different are really only "skin deep." Differences in skin color, for example, are very noticeable. However, these differences are simply the result of a greater or lesser amount of a protein called melanin.

Despite variation, every human being shares a huge number of the same genes and proteins with humans living all over the world.

Check and Explain
1. How is a single individual alike and different from the other members of its species?
2. Explain how reproduction affects variation in a species.
3. **Define Operationally** Why are so many gene combinations possible in sexual reproduction?
4. **Estimate** Measure and record your arm length from the tip of your middle finger to your elbow. Estimate the arm length of ten classmates. Find out the actual lengths and graph the results. How much do arm lengths vary?

TEACHING OPTIONS

Prelab Discussion
Have students read the entire activity. Discuss the following points before beginning the activity:

▶ Review the technique of random sampling and discuss how it can be used to draw conclusions about a population.

▶ Remind students that they are likely to see variation among the beans they observe.

▶ Discuss how variations within a species can lead to the formation of other new species.

ACTIVITY 7

Time 40 minutes **Group** 1–2
Materials
beans of the same variety
metric rulers

Analysis
1.–3. Answers will vary, but should agree with each student's recorded observations.

4. It is possible that the same unique characteristic would appear in another sample of ten beans, but the likelihood varies.

Conclusion
Conclusions will vary, but should follow logically from recorded observations. Students should note why they determined a particular trait to be a more varied trait. For example, area of markings may be more varied than color of markings because all beans in a sample may have marks that vary in area, while all may have markings of the same shade.

Everyday Application
If the largest is twice as long as the smallest, it will have approximately four times the volume. When each doubles in volume, the larger will have absorbed eight times as much water as the small one.

Extension
Students should list as many characteristics as possible.

Activity 7 How do beans vary?

Skills Measure; Observe; Classify

Task 1 Prelab Prep
1. Collect the following items: 2 sheets of paper, pencil, metric ruler, 10 beans of the same variety. Use pinto beans or any kind of striped or textured bean.
2. On a sheet of paper, draw 10 squares. Number each square from 1 to 10. Place one bean in each square. Use this grid for identifying the beans.

Task 2 Data Record
1. On the other sheet of paper, prepare a data table like the one shown.
2. Record each observation and measurement in the data table.

Table 7.1 Bean Observations

Bean	Color	Length	Markings	Texture
1				
2				
3				
4				
5				
6				
7				
8				
9				
10				

Task 3 Procedure
1. Note and record the color of each bean. Look carefully for small differences among the beans. Include any variation of color in your data.
2. Measure and record the length of each bean.
3. Examine each bean for stripes, spots, or any other kind of marking. Note all your observations of each bean in the data table.

Task 4 Analysis
1. What is the range of length for this variety of bean?
2. How many color variations did you observe? What color is the most common?
3. What kind of marking and/or texture do the beans have? Which marking is the most common? Which marking is the least common?
4. Did you find any bean with a characteristic shared by no other bean? What is this unique characteristic? If you examined 10 other beans, would you expect to see another bean with this characteristic?

Task 5 Conclusion
Write a short paragraph describing the variations that appear in this variety of bean. Include the most varied trait you observed in the beans. Identify as many variations as you can.

Everyday Application
Suppose this kind of bean must be soaked in water before cooking, and each bean doubles in volume when soaked. The beans are sorted and sold by size. What would be the difference in the volume of water for one liter of the largest-sized and one liter of the smallest-sized beans? Explain your conclusions.

Extension
Use reference books to find out how other organisms of the same species are different. You can choose a species of plant, a species of mammal, or another kind of animal. Write down all the characteristics that make the members of the species both alike and different.

SECTION 7.3

Section Objectives
For a list of section objectives, see the Student Edition page.

Skills Objectives
Students should be able to:

Infer the diet and habitat of birds based on the shape of their beaks.

Classify leaves.

Communicate how changes in traits of a species are important to its survival.

Vocabulary
natural selection, population

MOTIVATE

Skills WarmUp
To help students understand adaptation, have them do the Skills WarmUp.
Answer Answers may vary. The first and fourth birds probably eat insects and live in trees. The second probably eats cacti and lives in a dry habitat. The third bird probably eats seeds and lives mainly on the ground.

Misconceptions
Some students may believe that evolution is something that occurred in the past and has since ended. Explain that evolution is an ongoing process that is shaping life today.

INTEGRATED LEARNING

History Connection
Students may be interested to know that Charles Darwin (1809–1882) was not a good student who spent most of his time collecting insects, fishing, hunting, and reading nature books. Darwin's experiences traveling on the *Beagle* for five years proved to be the most educational of his life. Ask students to discuss experiences, other than school, that have been educational in their lives.

Geography Connection
Allow students to trace Darwin's five-year voyage on the *Beagle*. Have them locate South America, including the Galápagos Islands, on a map.

Skills WarmUp

Infer Look at the bird beaks on page 149. Make a list of the kinds of foods you think birds with these beaks eat. What kind of a habitat is most suited to each bird?

Figure 7.9 ▲
This is a portrait of Darwin about 19 years after his voyage on the *Beagle*.

7.3 Natural Selection

Objectives

▶ **Describe** some of the observations that puzzled Darwin on his voyage.

▶ **State** Darwin's theory of evolution by natural selection.

▶ **Explain** how a new species may arise.

▶ **Communicate** how changes in traits are important to the survival of a species.

Imagine that you are asked to go on a voyage around the world. Of course, the journey could be dangerous and might take several years. You're excited about going, but your parents try to discourage you. This was the situation of young Charles Darwin during the autumn of 1831.

At age 22, Darwin accepted the offer. In December, he set off on the *Beagle*, a medium-sized sailing ship. The main goal of the voyage was to study and chart stretches of the South American coastline. Darwin was to be the ship's naturalist, a person who studies the world of nature.

The voyage lasted five years. During that time, Darwin saw an incredible number of different living things. The variety of patterns in nature puzzled him. He spent the years after his return working out a theory to account for his observations. This theory eventually caused a revolution in the study of biology.

Voyage of the *Beagle*

Darwin had studied the plants and animals of England before he set sail on the *Beagle*. He was also familiar with the ideas of the scientists of his time. Most scientists believed that the earth and life on it had always been the same.

During the first part of the voyage, Darwin explored the tropical forests, the plains, and the very cold places in South America. He noticed that nearly every living thing there was very different from living things in England. Even in South American climates similar to those in

Integrating the Sciences

Earth Science Until the late 1800s, many people thought that the earth was quite young and that its landforms were created suddenly by violent upheavals such as earthquakes and floods. In the 1800s, British geologist Charles Lyell (1797–1875) supported a different and prevailing theory called uniformitarianism. He believed that the earth is very old and that geological formations were created by the same gradual processes that are still occurring, such as wind and water erosion. Darwin was familiar with this theory and applied the same reasoning to organisms. Ask students how Darwin's theory of natural selection is similar to the theory of uniformitarianism. (Both theories are based on the idea of extremely gradual change.)

England, organisms were more similar to organisms in other parts of South America than they were to English organisms. Darwin wondered why the life forms in these two places were so different when the climates were similar.

Darwin also found many fossils. They were the remains of organisms that no longer existed. Why did they die out? Many of the fossils were similar to living organisms. But curiously, they were always much bigger. One fossil find was very perplexing. High up on a mountain, Darwin found fossils of organisms that had once lived in the ocean. Why were they now so far from the sea? ①

Galápagos Islands Darwin's ideas were greatly influenced by the life forms on the Galápagos Islands. These 16 islands form a chain off the west coast of South America near the equator. Almost all the organisms Darwin found on these islands were not found anywhere else. Giant tortoises roamed the islands. Iguanas sunned themselves on the rocky shores and ate food from the ocean. Even more interesting to Darwin was that each island had many plants and animals that were different from those on the other islands.

Darwin collected 13 species of finches on the Galápagos Islands. Each had a different kind of beak, as you can see in Figure 7.10. Some had large, thick beaks for cracking seeds. Others had narrower beaks for eating insects. A few species had beaks suitable for eating cacti. The finches puzzled Darwin. In most places, finches had one kind of beak. The other kinds of beaks belonged to very different kinds of birds.

Theory of Evolution After his return to England, Darwin worked over 20 years to answer the many questions raised by his observations. Darwin had seen evidence everywhere that the surface of the earth had changed and was still changing. He concluded that if organisms in this changing world did not change along with it, they could not survive. In 1859, he published his ideas in a book, *The Origin of Species*.

Darwin claimed that organisms are the products of historical change and that new species gradually developed from previous ones. He concluded that all life shares a single and complex history. These ideas make up Darwin's theory of **evolution**. In biology, evolution is the process by which species change over time.

Figure 7.10 ▲
Notice how the beaks of the Galápagos Islands finches vary. Why is a large, thick beak most suitable for cracking seeds? ②

TEACH

Skills Development
Predict Again, refer students to the information about the Galápagos Islands finches on this page. Have students suppose that the finches had not developed any new adaptations after the islands separated. Ask students what would have happened to the finches in time. (The finches would have died out on some islands. Since different types of habitats with unique kinds of food evolved on each island, the same kind of beak would not have been successful on some islands.)

Critical Thinking
Find Causes Ask students the question, How did geological changes on the earth affect the evolution of living things? (Changes on the earth require that animals adapt to new environments. Organisms that do not adapt do not survive.)

Enrich
Use Enrich Worksheet 7.3.

Answers to In-Text Questions

① Fossils of ocean organisms on a mountain are evidence that the land was once under the ocean.

② A large, thick beak is suitable for cracking seeds because it is strong.

TEACH • Continued

Skills Development

Communicate Have students describe in their own words the four conditions that form the basis of natural selection. Ask them which of the four conditions is most characteristic of animals that lay a large number of eggs. (Overproduction)

Critical Thinking

Predict Use the following example to illustrate the idea of natural selection: In a giraffe population, some giraffes have longer necks than others. Having a longer neck allows certain giraffes to eat leaves that other giraffes can't reach during a drought. These giraffes with longer necks are more likely to survive the drought and have offspring with similar necks. If having a longer neck continues to be advantageous, these giraffes will eventually become the dominant variation in the population. Have students predict what would happen to the giraffe population if conditions changed, favoring giraffes with short necks. (Giraffes with short necks would become dominant in the population.)

Answer to In-Text Question

① Brook trout that are best adapted to their habitat are most likely to survive.

The Living Textbook: Life Science Sides 7–8

Chapter 13   Frame 01340
Natural Selection (3 Frames)
Search: Step:

150

INTEGRATED LEARNING

Themes in Science

Patterns of Change/Cycles At any given time, there are variations in a population or species. Some members of the population have adaptations that make them better equipped to survive than other members. These organisms are more likely to reproduce and pass on their successful adaptations to their offspring. In this way, populations of species change over long periods of time.

Natural Selection

In addition to claiming that life forms had evolved, Darwin hypothesized on *how* they evolved. Darwin called his hypothesis **natural selection**. Natural selection is the process by which those organisms best adapted to their environment survive and pass their traits on to the next generation. Natural selection works on a **population** of organisms, not individuals. A population is a group of individuals of the same species living in a particular area.

Darwin based natural selection on four conditions he observed in nature.

Overproduction Living things produce more young than can survive. Populations tend to stay the same size over a period of time, despite the large number of young produced.

Limited Resources Food, water, space, and other resources are limited. Members of a population must compete for these limited resources. Darwin called this the *struggle for existence*.

Variation Not all individuals in a population are exactly the same. Their traits vary slightly. These variations can be passed from parents to offspring.

Advantage of Some Variations Some variations can determine which individuals will survive the struggle for existence and reproduce.

When these conditions hold true, organisms best adapted to a particular habitat live and produce more offspring like themselves. Organisms that are not well adapted have a greater chance of dying before they can reproduce. Over generations, the number of well-adapted individuals increases. As a result, the traits and characteristics of the population as a whole change.

Darwin did not know about the genetic basis of inheritance. Today we know that natural selection changes the genetic makeup of a population. Certain genes that result in better adaptations become more common than others. Because of natural selection, a population always tends to be adapted to its habitat. When a habitat changes, individuals adapted to the new conditions survive and reproduce.

Darwin's theory of natural selection competed with other explanations of how organisms evolved. Jean Lamarck, a French biologist in the early 1800s, argued that a species changed because offspring inherited traits that their parents developed during their lives. Thus, if a giraffe's neck became longer from stretching for leaves high in trees, the giraffe's offspring would inherit the longer neck.

Many people accepted Lamarck's hypothesis. However, most scientists gradually realized that natural selection better explained evolution.

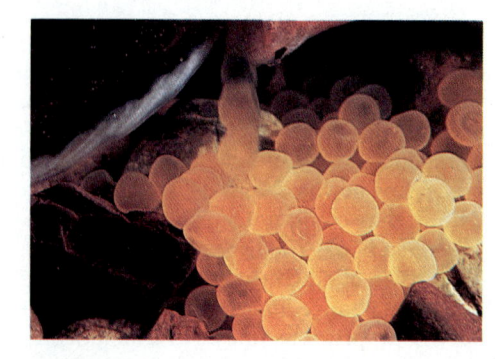

Figure 7.11 ▲
The brook trout overproduces offspring. Which ones are most likely to survive? ①

150   Chapter 7 Diversity of Living Things

TEACHING OPTIONS

Cooperative Learning
Divide the class into groups of four to explore the four conditions on which Darwin based his theory of natural selection. With one student per topic, each group should create a poster illustrating first how each of these conditions functions individually, and then how they function together to affect populations.

New Species

Just as natural selection can explain changes in a species, it can also explain how one species evolves from another. This evolutionary process is called *speciation*. Speciation begins when a portion of the population of a species becomes isolated. A population is referred to as separate when its members are not able to breed with other members of the population. Speciation can occur in several ways. Populations may be divided by a newly formed canyon or mountain range. Fish populations might be separated by the division of a lake. A pair of birds might fly to an island and start a separate population there.

Once a population is isolated, it can change in a different way from other groups. Mutations might occur in the isolated population. A gene might become more common through the process of natural selection. Over a period of time, members of the isolated population

Skills WorkOut

Classify Explore a vacant lot. Take a leaf sample from each different kind of plant growing there. Classify each leaf by color, size, and shape. Think about the climate of the lot. What characteristics do the plants have in common that are suited to the climate?

Historical Notebook

Natural Selection in Action

The peppered moth lives around the city of Manchester, England. The moth rests on the trunks of trees. Color varies in this species, so the moths can be either light or dark.

Before 1848, only light-colored moths were seen. Then people began to notice fewer light-colored moths, and began to see dark-colored ones. By 1898, nearly all of the moths around Manchester were dark-colored.

In the 1950s, H. B. Kettlewell looked for a reason why the color of the moth population had changed. He suspected that the increased burning of coal in the 1800s had changed the moths' habitat. Soot from Manchester's factories had blackened the nearby trees. Did this make the light-colored moths more visible to predators? Kettlewell set up an experiment to test this hypothesis. He found that dark-colored moths did have an advantage when the trees were darkened with soot.

1. How did the peppered moths' habitat change in the late 1800s?
2. **Write** How did the change in the moths' habitat affect the moths' traits? Write how this is an example of natural selection.

Chapter 7 Diversity of Living Things

Skills WorkOut
To help students classify types of plants, have them do the Skills WorkOut.
Answer Answers will vary depending on local climate.

Skills Development
Predict Suppose that a plant population becomes separated by a mountain range. One side of the mountain range is rainy and shady; the other is dry and sunny. Ask students:

▶ What will happen to the two plant populations? (The populations will become so different that a new species of plant will have evolved.)

Historical Notebook
Answers
1. It was blackened by industrial pollution.
2. Dark-colored moths survived better than light-colored moths and came to dominate the population.

Critical Thinking
Predict Explain that coal burning in England has been significantly reduced since the early 1800s, resulting in cleaner air. Ask students, How has this changed the peppered moth population? Why? (There are now more light-colored moths because the darker moths have become more visible to predators.)

Integrated Learning
Use Integrating Worksheet 7.3.

The Living Textbook: Life Science Sides 7-8

Chapter 14   Frame 01645
Speciation (1 Frame)
Search:

TEACH • Continued

Research
Have students write a brief report about germplasm banks. Have them find out what types of crop seeds are being stored and what living plants are preserved. They should include in their reports information about how the germplasm banks are used.

EVALUATE

WrapUp
Reinforce Ask students to explain in their own words the concept of evolution. They should include descriptions of natural selection and speciation.

Use Review Worksheet 7.3.

Check and Explain
1. Each island had unique plants and animals.
2. Natural selection—organisms best adapted to their environments survive and pass their traits on to the next generation
3. Because it proposed that all organisms share a single, complex history and that species are constantly, gradually evolving to adapt to changes on the earth
4. Traits that enable members of a species to adapt to a particular habitat provide those members with a greater chance of surviving and reproducing.

152

INTEGRATED LEARNING

Integrating the Sciences
Earth Science By reading the fossil record, paleontologists estimate that, on average, a species will die out after about one million years. The oldest fossil of a hominid is about 3.5 million years old. Ask students how much older the human species is than the average. (Humans have been around 3.5 times longer than most species.)

Figure 7.12 ▲
Australian animals have been isolated for millions of years. As a result, species like the swamp wallaby (top) and the wombat (bottom) are found nowhere else.

may look very different. Their genes may be different enough to make interbreeding impossible with the organisms that were once part of the same species. The isolated population has become a different species.

The finches on the Galápagos Islands developed into 13 separate species in this way. The population on each island was separate from the others. As each population evolved, the beaks of the finches adapted to eating the form of food most available on the island. The pouched animals of Australia are another example of organisms that evolved as separate species because of isolation.

Science and Technology *Germplasm Banks*
Just as you put money in a bank for safekeeping, there are banks for keeping safe the genetic information in seeds, or the plant's germplasm. Germplasm banks located throughout the United States make up the National Plant Germplasm System. A typical seed bank contains about a billion seeds. Almost 500,000 types of seed samples are stored throughout the United States.

The purpose of these banks is to collect, store, catalog, and distribute the germplasm of every species and variety of plant in the world. This system not only stores seeds and plant parts, but it also preserves living plants. Germplasm repositories store species of plants that lose their variation if they are stored only as seeds.

The banks are important because they ensure against losing the genetic material of wild plants. Also, they provide germplasm with traits for breeding crop plants that can resist certain kinds of pests and disease.

Check and Explain
1. What did Darwin notice that was unique about the animal and plant life on the Galápagos Islands?
2. What important hypothesis did Darwin develop by studying the beaks of finches?
3. **Reason and Conclude** Why did Darwin's concept of natural selection cause a revolution in biology?
4. **Communicate** Explain how changes in the traits of a species can be important to its survival.

152   Chapter 7   Diversity of Living Things

Chapter 7 Review

Concept Summary

7.1 Diversity and Adaptation
▶ An adaptation is a characteristic that helps an organism reproduce and survive in an environment.
▶ Adaptations are similar in different species if the adaptation is necessary for survival, or if the organisms are related to one another.
▶ The variety of organisms that live in an environment is called diversity.

7.2 Variation and Reproduction
▶ The different traits in the individuals of a species is variation within the species.
▶ Since both parents contribute genetic material during sexual reproduction, the traits of the offspring are a unique combination of genes.
▶ Asexual reproduction occurs when a single organism divides to produce another organism with identical genetic material.

7.3 Natural Selection
▶ Charles Darwin observed many plant and animal species in South America that did not exist anywhere else.
▶ The fossils of ocean organisms discovered in high mountains caused Darwin to infer that the earth's environment changes over time.
▶ Darwin based his theory of evolution on the hypothesis that species change when the organisms most adapted to an environment survive and pass their genes to the next generation.
▶ Speciation occurs when a population of organisms is isolated, or when survival depends on a certain characteristic and those individuals without it die.

Chapter Vocabulary

adaptation (7.1)	species (7.1)	zygote (7.2)	natural selection (7.3)
habitat (7.1)	variation (7.2)	evolution (7.3)	population (7.3)
diversity (7.1)	sexual reproduction (7.2)	asexual reproduction (7.2)	

Check Your Vocabulary

Use the vocabulary words above to complete the following sentences correctly.

1. A group of the same species living in a particular area is a ____.
2. The concept that new species develop from the previous ones is called ____.
3. Variation within a species is possible because of ____.
4. A structure that helps an organism survive in its environment is a(n) ____.
5. A ____ is made up of organisms that look alike.
6. An environment with a large variety of organisms has ____.
7. A fertilized egg, or ____, results from sexual reproduction.
8. A new organism is formed from one parent cell during ____.
9. An organism's adaptations help it survive in a particular ____.
10. Darwin's hypothesis about how life changed through time is called ____.
11. The differences among individuals within a species is called ____.

Write Your Vocabulary

Write sentences using each of the vocabulary words above. Show that you know what each word means.

Chapter 7   Diversity of Living Things   153

CHAPTER REVIEW 7

Check Your Vocabulary

1. population
2. evolution
3. sexual reproduction
4. adaptation
5. species
6. diversity
7. zygote
8. asexual reproduction
9. habitat
10. natural selection
11. variation

Write Your Vocabulary

Students' sentences should show that they know both the meanings of the words and how to apply the words.

Use Vocabulary Worksheet for Chapter 7.

CHAPTER 7 REVIEW

Check Your Knowledge

1. Organisms that are related to one another or that live in the same environment can have similar adaptations.
2. Darwin's hypothesis of natural selection is based on the concepts of overproduction, limited resources, variation, and the advantage of some variations over others.
3. Human traits are unique because there are different forms of the same genes.
4. A mutated gene either survives and is passed on, becoming a part of the species' gene pool, or else does not survive and does not become permanent.
5. Darwin studied finches with varying beak size and type.
6. Bat: echolocation; anteater: powerful front legs and a long tongue; jackrabbit: long ears; dandelion: seeds are carried by the wind; turkey vulture: long, broad wings
7. During the formation of gametes (meiosis), the genetic material is shuffled. Genetic material from two parents combines during fertilization.
8. False; asexual reproduction
9. False; zygote
10. True
11. False; ocean organisms
12. True
13. True

6. Students should list as many differences as possible for each trait, depending on the breed chosen. Since all dogs have the same number of chromosomes, they can still breed with each other, even though traits vary.
7. Answers will vary. A long, pointed beak will allow the bird to catch crabs and insects and eat seeds and nectar. Long legs and webbed feet would help the bird move around in the marsh.

Check Your Understanding

1. When a population is isolated, the adaptations that are most useful in the environment dominate and the species changes.
2. Variation refers to differences among individuals within a species; diversity refers to differences among species in an environment.
3. Time is important in evolution because many generations are required for speciation.
4. The movement of genes during meiosis and the combination of genes during fertilization make possible a large number of human traits.
5. Because its markings look like tree bark, the grizzled mantis is protected from its enemies. The sharp hooks on its forelegs are adapted to catch and hold prey.

154

Chapter 7 Review

Check Your Knowledge
Answer each question in a complete sentence.

1. Why do some organisms have similar adaptations?
2. On what four conditions did Darwin base his hypothesis of natural selection?
3. Why are the traits of humans unique for each individual?
4. What happens when a mutation occurs?
5. What diverse adaptation did Darwin discover in the finches of the Galápagos Islands?
6. Name one adaptation necessary for the survival of the following organisms: bat, anteater, jackrabbit, dandelion, and turkey vulture.
7. Give two reasons why sexual reproduction produces new genetic combinations.

Determine if each statement is true or false. Write *true* if it is true. If it is false, change the underlined word(s) to make the statement true.

8. <u>Overproduction</u> limits the variation of genetic material in an organism.
9. A <u>gamete</u> is a fertilized egg that forms when two sets of genes combine.
10. <u>Natural selection</u> occurs when the individuals that are most adapted to an environment are able to pass on their genes.
11. Darwin located some fossils of <u>birds</u> in the mountains of South America.
12. <u>Speciation</u> occurs when a population of organisms is isolated.
13. Within a species, individual variation shows up in physical <u>traits</u> and in the genetic material.

Check Your Understanding
Apply the concepts you have learned to answer each question.

1. Explain how isolation can play an important role in speciation.
2. **Contrast** Explain how *variation* and *diversity* differ.
3. Why is time important in evolution?
4. Explain why such a large number of gene combinations are possible for humans.
5. **Mystery Photo** The grizzled mantis pictured on page 138 eats only live prey. What adaptation does the grizzled mantis have that protects it from its predators? How is it adapted for killing its prey?
6. **Application** Use the pictures of three different breeds of dogs to identify the variation in the different characteristics given below. Explain how these differences affect the dogs' ability to breed with each other.

Body shape	Eye color
Height	Tail
Snout	Foot size
Coat	Ears

7. **Application** Use the information below to describe the beak and feet of a bird that lives in a salt marsh.
   - The water depth is 10 to 30 cm, depending on the tide.
   - Small crabs live year-round in the muddy bottom. Insects burrow in the stems of grass during spring.
   - Seeds ripen during the late autumn months. Flowers with a high amount of nectar bloom during summer.

Develop Your Skills

1. a. Narrowest: 3.00–3.24 cm; widest: ≥ 4.25 cm
   b. 3.75–3.99 cm
   c. Answers may vary. Small leaves are helpful in reducing water loss in dry climates or where sunlight is abundant. Larger leaves can help gather light in dimly lit environments.
2. a. Paleozoic era, Mississippian period; Cenozoic era, Paleocene epoch; Paleozoic era, Mississippian period; Paleozoic era, Silurian period; Cenozoic era, Oligocene period
   b. Insects

Make Connections

1.

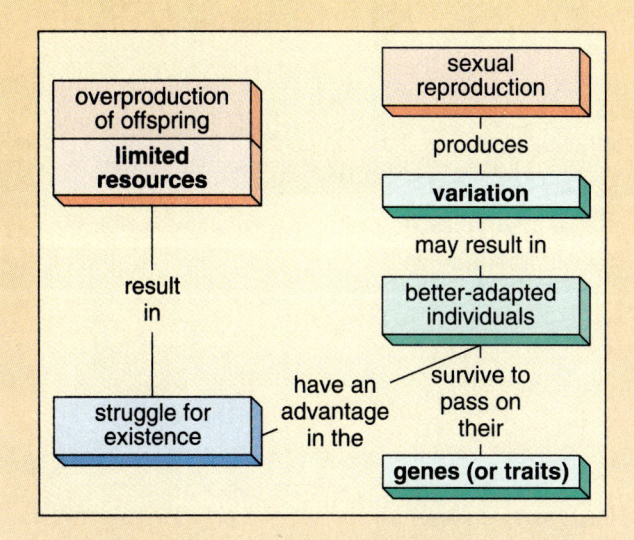

```
overproduction                    sexual
of offspring                      reproduction
limited
resources                         produces
                                  variation
   result                         may result in
    in                            better-adapted
                                  individuals
                    have an       survive to
   struggle for    advantage      pass on
   existence       in the         their
                                  genes (or traits)
```

2. Answers will vary; however, students should identify winter clothing such as coats, boots, hats, gloves/mittens. The wool, cotton, plastic, rubber, and livestock industries are a few industries that contribute to winter clothing production.

3. Seed banks get their seeds and plants from a variety of sources, including donations from plant breeders and by collecting samples from the wild. Answers may vary. It is important to collect samples from many different areas.

Develop Your Skills

Use the skills you have developed in this chapter to complete each activity.

1. **Interpret Data** The graph below shows the variation of leaf widths in one species of plant.

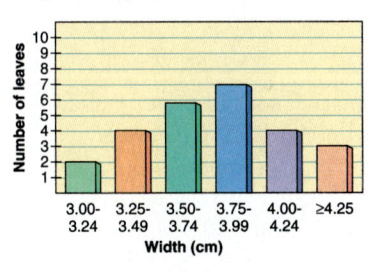

   a. What is the width of the narrowest leaf? Of the widest leaf?
   b. What leaf width occurs most often?
   c. Under what conditions might small leaves be helpful to the survival of the plant? When might large leaves be helpful to survival?

2. **Data Bank** Use the information on page 630 to answer the following questions.

   a. In which era and period or epoch did each of the following first appear on the earth: insects, primates, reptiles, sea scorpions, and mastadons?
   b. The first flowering plants appeared during the Permian period. What animals that lived during this period probably would depend on flowering plants to survive?

Make Connections

1. **Link the Concepts** Below is a concept map showing how some of the main concepts in this chapter link together. Only parts of the map are filled in. Copy the map. Using words and ideas from the chapter, complete the map.

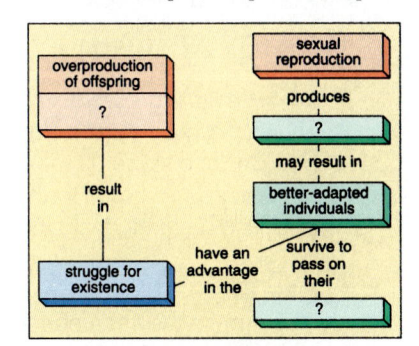

2. **Science and Social Studies** Humans are not adapted naturally to very cold weather. Write a report on the kinds of clothing made for people living in cold winter climates. Identify the industries that meet the need for winter clothing.

3. **Science and Technology** Research how seed banks get the seeds and plants that they store. Identify the locations on the National Plant Germplasm System and explain why each location is important.

CHAPTER 8

Overview

This chapter discusses how scientists measure and study geologic time. The first section illustrates significant events in the earth's history and describes the history of living things. Descriptions of fossils, molds, and casts illustrate the ways scientists study evolutionary relationships. The final section of the chapter explains the history of the development of hominids.

Advance Planner

▶ Provide almanacs or history books for SE page 157.

▶ Supply colored and uncolored clay for TE activities, page 160.

▶ Collect or purchase shells for TE activity, page 163.

▶ Prepare modeling clay, plaster of Paris (both available at a hobby or crafts store), seashells, petroleum jelly, paper cups (or plastic food containers), plastic spoons, coins, keys, leaves, washers, paper plates, and lab aprons. SE Activity 8, page 166.

Skills Development Chart

Sections	Classify	Define Operationally	Graph	Infer	Model	Observe	Research
8.1 Skills WarmUp SkillBuilder Skills WorkOut			●		●		●
8.2 Skills WarmUp Activity	●			●	● ●	●	
8.3 Skills WarmUp Historical Notebook Skills WorkOut		●			●		● ●

Individual Needs

▶ **Limited English Proficiency Students** Have students divide a page in their science journals into three columns. In the first column, have them list the chapter vocabulary. In the second column, have students write definitions of their words. Then students should refer to the concept summary and find one sentence using a vocabulary word. Have them write the concept summary sentence in the third column.

▶ **At-Risk Students** Photocopy pages 158 and 159. Write the beginning and ending dates of each geologic era in the appropriate boxes. Have students paste these pages at the top of large pieces of paper. Then have them draw vertical lines on these to make a column for each era. As they work through the chapter, have students add information or pictures about each era to these pages. For example, students could include the Grand Canyon (top) in the Precambrian Era.

▶ **Gifted Students** Have students create a puzzle of Pangaea that can be manipulated to represent the earth 200 to 250 million years ago, 65 million years ago, and today. Then have students prepare a presentation for the class that explains how the earth changed over these periods, using their puzzle to show the geographic movement. Based on information provided in the chapter and on research, students should attempt to explain the evolutionary path of one species. Students may choose as their subject either hominids or a species on the Galápagos Islands.

Resource Bank

▶ **Bulletin Board** Collect pictures of fossils and mount them on construction paper. Hang the pictures on the bulletin board. Have students make labels for the pictures describing the organism that formed each fossil and the type of fossil that is pictured.

▶ **Field Trip** Arrange to take students to a natural history museum so they can observe fossils, reconstructed skeletons of extinct animals, and other displays. Contact the museum in advance so you can have students prepare for their trip by listing exhibits that they would like to see.

CHAPTER 8 PLANNING GUIDE

Section	Core	Standard	Enriched	Section	Core	Standard	Enriched
8.1 Geologic Time pp. 157–161				**Overhead Blackline Transparencies** Overhead Blackline Master 8.2 and Student Worksheet	●	●	●
Section Features Skills WarmUp, p. 157 SkillBuilder, p. 160 Skills WorkOut, p. 161	●	● ● ●	● ●	**Laboratory Program** Investigation 15 Investigation 16	●	● ●	● ●
Blackline Masters Review Worksheet 8.1 Skills Worksheet 8.1 Integrating Worksheet 8.1a Integrating Worksheet 8.1b	● ●	● ● ● ●	● ● ●	**Color Transparencies** Transparency 22	●	●	
				8.3 Human Evolution pp. 167–170			
Color Transparencies Transparency 21a, b	●	●		Section Features Skills WarmUp, p. 167 Historical Notebook, p. 168 Skills WorkOut, p. 170	● ●	● ● ●	● ●
8.2 The Fossil Record pp. 162–166				Blackline Masters Skills Worksheet 8.3 Review Worksheet 8.3 Vocabulary Worksheet 8	● ● ●	● ● ●	
Section Features Skills WarmUp, p. 162 Activity, p. 166	● ●	● ●	●				
Blackline Masters Review Worksheet 8.2 Enrich Worksheet 8.2	●	● ●	●	Color Transparencies Transparency 23	●	●	

Bibliography

The following resources can be used for teaching the chapter. See page T-40 for supplier codes.

Audio-Visual Sources
(video unless noted)
Did Darwin Get It Wrong? 57 min. 1982. T-L.
Fossils: Clues to the Past. 23 min. 1983. NGSES.
Prehistoric Life. 6 filmstrips with cassettes. 1983. SVE.
Stephen Jay Gould: His View of Life. 57 min. 1984. T-L.

Software Resources
Dinosaur Dig. Apple, IBM, Commodore. 1984. CBSS.
The Earth Through Space and Time. Apple. EA.
Evolve! A Trip Through Time and Taxonomy. Apple. JWW.

Library Resources
Colbert, Edwin H. *The Great Dinosaur Hunters and Their Discoveries.* Revised ed. New York: Dover, 1984.
Cork, Barbara, and Lynn Bresler. *The Young Scientist Book of Evolution.* Tulsa, OK: EDC Publishing, 1985.
Gallant, Roy A. *From Living Cells to Dinosaurs.* New York: Watts, 1986.
Gould, S. J. *The Panda's Thumb.* New York: Norton, 1980.
Rydell, Wendy. *Discovering Fossils.* Mahwah, NJ: Troll, 1984.

CHAPTER 8

Introducing the Chapter

Have students read the description of the photograph on page 156. Ask if they agree or disagree with the description.

Directed Inquiry

Have students study the photograph. Ask:

▶ What organism do you see in the photograph? (Some students will recognize the insect as a cricket.)

▶ Describe the substance around the insect. (Students may mention that it is clear and yellow and contains air bubbles.)

▶ What is the substance and how do you think the insect got inside it? (Some students may realize that it is tree amber that was once soft and sticky. The insect probably became stuck and eventually encased in it.)

▶ Why didn't the insect decay over time? (The amber preserved it.)

▶ What is the topic of this chapter? (History of life)

▶ How does the photograph relate to the chapter? (It's a fossil of an organism that lived millions of years ago.)

Chapter Vocabulary

cast	geologic time
era	hominid
extinct	mold
fossil	primate
fossil record	sedimentary rock

INTEGRATED LEARNING

Writing Connection

After discussing the photograph, have students write a story or historical account of how the cricket became encased in the tree amber. Have them describe when the insect lived and what its habitat was like. They might also imagine where the fossil was found and by whom.

Chapter 8 — History of Life

Chapter Sections

8.1 Geologic Time
8.2 The Fossil Record
8.3 Human Evolution

What do you see?

"I see a cricket in some yellow material. The yellow matter is amber, which comes from trees. This cricket was probably on the tree this came from and he got stuck in it. By being in the wrong place at the wrong time, he got stuck. I think the cricket has been there for a long time. The cricket has been there since the dinosaurs ruled the earth."

*Isaac Villa
Byrnedale Junior High School
Toledo, Ohio*

To find out more about the photograph, look on page 172. As you read this chapter, you will learn about living things from the past and how life on the earth has changed through time.

Integrating the Sciences

Earth Science Approximately 200 million years ago, all of the earth's land was joined together in one mass called *Pangaea*, meaning "all earth." This large landmass then split into two smaller masses, Gondwanaland and Laurasia. These landmasses continued to break apart to form the continents that exist today. Show students on a globe that the east coast of North and South America and the west coast of Europe and Africa could almost fit together like a giant jigsaw puzzle.

Use Integrating Worksheet 8.1a.

SECTION 8.1

Section Objectives
For a list of section objectives, see the Student Edition page.

Skills Objectives
Students should be able to:

Make a Graph of important events.

Make a Model of the geologic time scale.

Research geologic eras, periods, and epochs.

Predict how life would be different if Pangaea had not split into continents.

Vocabulary
era, geologic time, sedimentary rock

8.1 Geologic Time

Objectives

▶ **Name** the major eras of geologic time.

▶ **Describe** the organisms living during each era.

▶ **Describe** methods to determine the age of rocks.

▶ **Predict** how life would be different if the earth had remained one large landmass.

Skills WarmUp

Make a Graph Make a timeline that shows at least ten important events in your life. Include events such as your date of birth, when you started school, when you moved, memorable vacations, and so on. Also include some important world events that have happened during your lifetime.

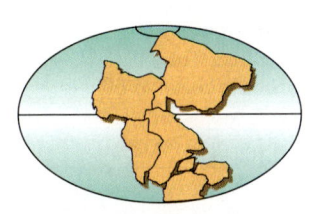

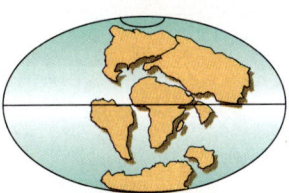

You are suddenly transported back in time 200 million years. Everything around you is different. You don't recognize any landmarks. If you went exploring, you would see that whole mountain ranges are missing. Even the continent you stand on is a different shape and in a different place.

None of the living things you see around you are familiar. The trees look very strange. Insects as large as birds fly by you. You don't see a single animal with fur.

History of Life on Earth

The earth and its life have a very long history. The earth is about 4,600 million years old. This is such an enormous amount of time that it is hard to comprehend. If 4,600 million years were condensed into one single year, a human lifetime would be only a tiny fraction of a second!

Life did not exist on the earth until about 3,500 million years ago. The organisms alive at that time were simple cells without nuclei. The world around them was very different than it is today. For one thing, the atmosphere was made up of gases that would be poisonous to most of today's living organisms!

Since that time, the earth has changed in many ways. You can see in Figure 8.1 how the continents shifted and moved apart. Mountain ranges formed. Sheets of ice covered parts of the earth and then retreated. All these changes have affected living things and the way they have evolved. Living things, in turn, have changed the earth. Through photosynthesis, organisms have filled the atmosphere with oxygen.

200–250 million years ago

65 million years ago

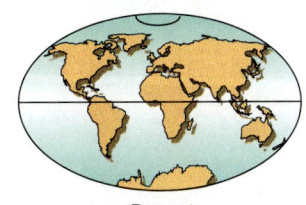

Present

Figure 8.1 ▲
At one time, the earth's continents were one large landmass (top). Notice how much the earth's land has changed since then.

MOTIVATE

Skills WarmUp
To help students understand the idea of a timeline, have them do the Skills WarmUp.
Answer Have students look in an almanac or history book to find world events to add to their timelines.

Prior Knowledge
Gauge students' understanding of geologic time by asking the following questions:

▶ How old is the earth?

▶ Do you think the area where you now live always looked this way?

Discuss
Students may be confused by figures such as 4,600 million. Explain that this is equivalent to 4.6 billion. Emphasize the advantage of keeping the unit of measurement the same (millions) when comparing various amounts.

Chapter 8 History of Life 157

TEACH

Explore Visually

Have students study the material on this page. Ask the following questions:

- How old is the earth? (4,600 million years)
- When did the first cells evolve? (About 3,500 million years ago)
- In what era did sponges, jellyfish, and worms appear? (Precambrian era)
- In what era did fishes, reptiles, and insects appear? (Paleozoic era)
- What important event ended the Paleozoic era? (Most of the animal species in the oceans died out.)

Skills Development

Estimate Have students calculate the amount of time that passed after the earth formed and before life first appeared. (About 1,100 million years, or a little over a billion years)

Answer to In-Text Question

① The Precambrian era lasted 2,860 million years.

The Living Textbook:
Life Science Sides 7-8

Chapter 14   Frame 01368
Precambrian/Paleozoic Fossils
(14 Frames)
Search: Step:

158

INTEGRATED LEARNING

Integrating the Sciences

Chemistry Life first appeared on the earth during the Precambrian era. Chemical reactions in the earth's early atmosphere are believed to have produced a "primordial soup" of chemicals in the oceans. This primordial soup probably contained amino acids from which the first single-celled organisms developed.

Earth Science The earliest evidence of life contained in the fossil record is blue-green algae that lived near the side of the ocean. These algae left imprints on clay and mud that later became rock.

Earth's Timeline

Scientists use a timeline to show the 4,600 million-year history of the earth and its life. This timeline is called a scale of **geologic time**. As you can see below, scientists divide the earth's geologic time into parts, much like a year is divided into months, weeks, and days. The largest kind of division is called an **era**. Eras are measured in millions of years. How many millions of years did the Precambrian era last? ①

Precambrian Era
After the earth formed, hundreds of millions of years passed before the first cells evolved, about 3,500 million years ago. After millions more years, fungi and algae evolved. Then, near the end of the era, the oceans began to fill with life. Many-celled sponges, jellyfish, and worms were among the simple organisms that began to appear.

Paleozoic Era
Many new life forms gradually appeared. Trilobites and other animals without backbones became common in the oceans. Then fishes evolved. Organisms began living on land about halfway through this era. Reptiles and insects appeared. At the end of the era, most of the animal species living in the oceans died out.

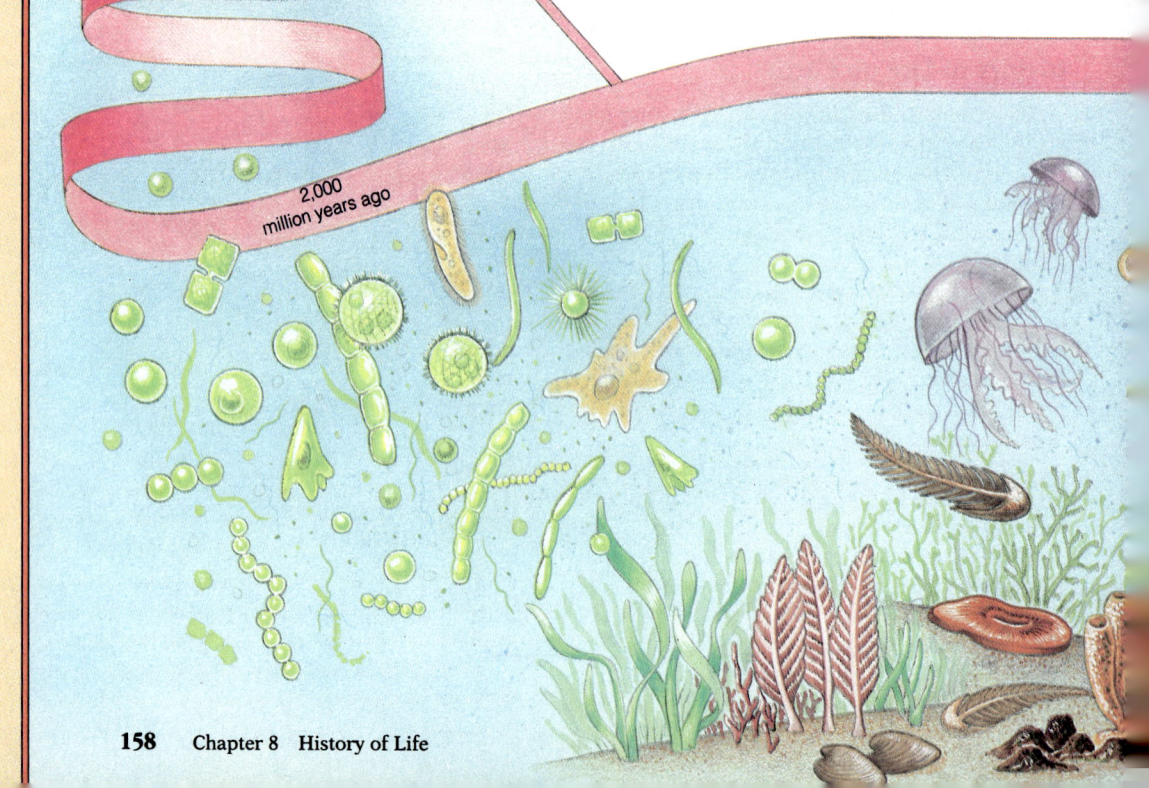

2,000 million years ago

Writing Connection
Have students pick an era before the Cenozoic and write a creative piece about what they imagine the world was like then. They may describe conditions experienced by an organism inhabiting the earth at the time or as a reporter describing the events. Students may wish to create a newspaper with articles, drawings, editorials, and so on.

There are four eras, but they are not equal lengths. Notice that the Precambrian era is much longer than the other three combined. One era ends and another begins at times in the earth's history when important changes were occurring. For example, few traces of animal and plant life from the Precambrian era exist. In fact, a variety of life forms did not appear until the Paleozoic era. Therefore, geologists separate the Precambrian era and the Paleozoic era at about 640 million years ago. What distinguishes the Paleozoic era from the Mesozoic era? ②

Mesozoic Era
Dinosaurs of all sizes were the most common form of animal for much of this era. Mammals evolved early in the era, and birds appeared about midway through. Also around the middle of the era, the earth's landmass began an important change. The one giant continent called Pangaea began to break up into separate continents.

Cenozoic Era
Mammals became very numerous and evolved into many different species. The horse and camel appeared about 37 million years ago. Large grazing animals and whales evolved. Four times in the last 2 million years of this era, huge sheets of ice spread over much of the land. During this time, the first humans appeared.

640 million years ago   245 million years ago   66 million years ago

Chapter 8 | History of Life 159

Explore Visually
Have students study the material on this page. Ask the following questions:

▶ When did the Mesozoic era begin? (245 million years ago)
▶ In which era did the dinosaurs dominate the land? (Mesozoic era)
▶ Mammals and birds appeared in what era? (Mesozoic era)
▶ In what era did mammals become abundant and diverse? (Cenozoic era)
▶ When did the first humans appear? (During the last two million years of the Cenozoic era)

Critical Thinking
Reason and Conclude
Ask students what animals did to survive when huge sheets of ice spread over the land. (They moved to areas where there were no ice sheets.)

Answer to In-Text Question
② Most of the life forms in the ocean died out at the end of the Paleozoic era. Dinosaurs, mammals, and birds appeared in the Mesozoic era.

The Living Textbook:
Life Science Sides 7-8
Chapter 14   Frame 01382
Mesozoic/Cenozoic Fossils (22 Frames)
Search: Step:

The Living Textbook:
Life Science Sides 7-8
Chapter 14   Frame 01517
Mesozoic/Cenezoic Drawings (85 Frames)
Search: Step:

TEACH • Continued

Discuss
Explain that the earth consists of many layers that have formed over time. Geologists can tell when fossils formed by the layer in which they are found.

Critical Thinking
Reason and Conclude Ask students to name some forces of erosion. (Students may mention wind, water, chemicals, and glaciers.)

SkillBuilder
Answers
3. The Paleozoic era
4. The Precambrian era was the longest because it took millions of years for life to begin and then it took millions of years more for many-celled life to evolve.
5. The Precambrian era was almost ten times longer than the Paleozoic era.
6. Some scientists divide the Precambrian era into two eras—Archeozoic era and Proterozoic era. Some students may say that the dividing line should be between life and no life or between single-celled life and many-celled life.

Answer to In-Text Question
① The rocks offer clues about how the earth has changed over time. The age of the sedimentary rock layers can be determined by how they're positioned among other layers.

160

INTEGRATED LEARNING

Integrating the Sciences
Earth Science Sedimentary rock is composed of sediments, which are small pieces of broken rocks and plant and animal remains. Some common examples of sedimentary rock are sandstone, limestone, and shale. Have students find out how each of these sedimentary rocks is formed. (Sandstone is formed by sand; limestone is composed of the skeletons of tiny marine animals; and shale is created by layers of mud and clay.)

Art Connection
Explain to students that layers of sedimentary rock are called strata. In places such as the Grand Canyon, strata are clearly observable because eroding forces, such as water, have cut down vertically through the layers, exposing them. Have students create their own canyons by layering together different colors of clay. Then, without mixing layers, they can carve away portions of the clay stack to simulate erosion.

Figure 8.2 ▲
How do the different rock layers of the Grand Canyon help scientists learn about the earth's history? ①

History Preserved in Rock
How do scientists piece together the earth's history? When landforms are worn down by wind and water, pieces of rocks called sediments are created. Sediments pile up, layer by layer. They are squeezed together to form layers of rock, as shown in Figure 8.2. Rocks formed in this way are called **sedimentary rocks**. They provide clues to how the earth has changed over time.

Layers of rocks closest to the surface formed after the layers below them. Thus, the lower layers are older than the upper layers. The *relative* age of a layer of sedimentary rock can be determined by its position among other layers.

The Grand Canyon in Figure 8.2 has many layers of sedimentary rocks exposed to view. Here the scale of geologic time is laid out on the canyon walls. The layers of rock range from 200 million years old at the top to 2,000 million years old at the bottom.

SkillBuilder *Making a Model*

Geologic Time Scale
The geologic time scale spans about 4,600 million years and is divided into eras. Fossils that show life forms are associated with each era. A model that shows the relative length of time in each era can help you to understand how the length of each era is different.

Collect a metric ruler, a meter stick, 4 different colored pencils, transparent tape, 1.5 m of adding-machine tape, or narrow strips of paper taped together. Copy the table below and complete the third column, using the scale of 1 cm for each 100 million years.

1. Measure out the correct amount of tape for each era. Label each era and color it a different color.
2. Use the information on pages 158–159 and the Data Bank on page 630 to draw or write in examples of life during each era.
3. Which era has the most new life forms?
4. Which era is the longest? Why is this era so much longer than the others?
5. How many times longer is the Precambrian era than the Paleozoic?
6. Does it make sense to divide the Precambrian era into two eras? If so, where would you draw the dividing line?

Geologic Time Scale		
Era	Length of Time (millions of years)	Tape Length (cm)
Precambrian	3,960	
Paleozoic	395	
Mesozoic	180	
Cenozoic	66	

160 Chapter 8 History of Life

Integrating the Sciences

Earth Science German meteorologist Alfred Wegener (1880–1930) advanced the theory of continental drift to explain why fossils of the same extinct organisms were found in continents on opposite sides of the ocean. Before his time, scientists hypothesized that land bridges, which no longer exist today, allowed animals to pass between continents. Wegener hypothesized that all present continents were once a part of one gigantic landmass, and that the force of the tide caused them to break off and drift apart. While his assumption that the continents have traveled is true, scientists now explain this movement by the theory of plate tectonics. Have students research and write a few paragraphs on this theory.
Use Integrating Worksheet 8.1b.

Radioactive Dating

The position of a layer of rock tells scientists its relative age. However, scientists sometimes need to know more exact ages of rocks to learn about the earth's history. To determine the actual age of a rock layer in years, they use a method called radioactive dating.

In radioactive dating, the amounts of two elements in the rock are measured. One is a radioactive element, which changes into another element at a constant rate. The other is the element the first one changes into. When the amounts of the two elements are compared, a rock's age can be determined. In the oldest rocks, most of the radioactive element has changed into the other element.

Science and Technology
Clues to Continental Movement

The earth is like a giant bar magnet. A magnetic field stretches between the North and South Poles. The magnetic field is invisible, but it can affect objects on the earth. A compass needle, for example, points north in order to line up with the earth's magnetic field.

In a similar way, particles of iron in certain rocks tend to line up with the magnetic field while the rocks are being formed. When the rocks cool, the particles are frozen in one place, "pointing" to the North Pole. Later, if the rock moves, it will no longer point to the North Pole. By determining where a rock's iron particles point now, scientists can figure out where and how the rock has moved since it was formed. This method has allowed scientists to learn how the continents have moved.

Skills WorkOut

Research The Paleozoic and Mesozoic eras are broken down further into periods. Find out the names of the periods in each era and the factors that distinguish each period. The Cenozoic era is divided into epochs. Find out the names of the epochs and the factors that distinguish each epoch. Make a chart of your findings listing the eras, periods, and epochs. Include a few highlights from each.

Check and Explain

1. List the four eras of geologic time in order.
2. Identify the era in which each of the following first appeared: humans, fishes, birds, worms, reptiles.
3. **Reason and Conclude** What were some important adaptations for animals living during the Cenozoic era?
4. **Predict** Refer to Figure 8.1 on page 157. How might life on the earth be different today if Pangaea had not split apart to form separate continents?

Chapter 8 History of Life

Skills WorkOut
To help students find out about the periods and epochs of the Paleozoic and Mesozoic eras, have them do the Skills WorkOut.
Answer Make sure that students correctly distinguish among *era*, *period*, and *epoch*.

Research
Have students use references to find out about radioactive elements in rocks and how scientists use them to date fossils.

EVALUATE

WrapUp
Portfolio Have students draw a contemporary nature scene including at least two forms of life that evolved in each of the four eras. Students should also depict layers of sedimentary rock to show the earth's changes.
Use Review Worksheet 8.1.

Check and Explain

1. Precambrian, Paleozoic, Mesozoic, Cenozoic
2. Humans, Cenozoic era; fishes and reptiles, Paleozoic era; birds, Mesozoic era; worms, Precambrian era
3. They had to adjust to environmental changes, including glaciers, and the presence of other animals, such as humans and many new mammals.
4. Such geographical variation and diversity of plant and animal life, as well as human culture, would not exist.

SECTION 8.2

Section Objectives
For a list of section objectives, see the Student Edition page.

Skills Objectives
Students should be able to:
Infer the age of fossils from their geologic layers.

Vocabulary
fossil, extinct, fossil record, mold, cast

MOTIVATE

Skills WarmUp
To help students understand the fossil record, have them do the Skills WarmUp.
Answer Students should realize that fossils are pieces that fit together like a puzzle to explain the past. Scientists study different fossils and note in which geologic layers they occur.

Prior Knowledge
To gauge how much students know about the fossil record, ask the following questions:
▶ What is a fossil?
▶ What do fossils tell us about living organisms?

The Living Textbook: Life Science Side 8
Chapter 22 Frame 05978
Dig at Site 333 (Movie)
Search: Play:

162

INTEGRATED LEARNING

Themes in Science
Evolution Information about the evolution of living things is based on the fossil record. However, the fossil record has many gaps. Most organisms decompose without leaving a trace. Furthermore, many fossils that exist have not been found. Scientists estimate that about 1 in every 10,000 species shows up in the fossil record.

Language Arts Connection
Have students look up the derivation and meaning of the word *fossil*. (*Fossil* comes from the Latin word meaning "to dig.")

Skills WarmUp
Make a Model Cut a sheet of paper into about eight differently shaped parts. Exchange pieces with a partner, and put them back together. Check each other's puzzles. How do you think looking for clues about the earth's past is like doing a puzzle? How do you think scientists go about putting pieces of the earth's "puzzle" together?

8.2 The Fossil Record

Objectives
▶ **Describe** different types of fossils and explain how they form.
▶ **Explain** how fossils provide evidence for evolution.
▶ **Generalize** about evolutionary relationships using an evolutionary tree.
▶ **Make inferences** about past life on the basis of the fossil record.

Think about how your life has changed since you were born. What records could you use to help explain the history of your life to a friend? After you have died, how can others learn about your life? People often leave behind such things as photographs, diaries, videotapes, letters, and other personal items. These records of people's lives can be used by others to understand how people lived and what they looked like.

In a similar way, scientists can use the records left by organisms that once lived to learn about past life on the earth. Bones, teeth, shells, and prints from leaves, plus animal footprints and outlines from body parts left in rocks are part of life's record.

Fossils
The remains and traces of organisms that lived in the past are called **fossils**. Fossils show that many of the organisms living in the past were different from those living today. Fossils also provide evidence that earlier forms of life have died out, or become **extinct**. Fossils provide a record of life called the **fossil record**. The fossil record extends back in time to the Precambrian era, though few fossils from this era actually exist.

Even though there are many fossils, most of the individual organisms that lived in the past have left no trace of their existence. The typical fossil was formed under fairly unusual conditions.

Figure 8.3 ▲
Scientists use fossils to reconstruct the skeletons of extinct animals.

162   Chapter 8   History of Life

Integrating the Sciences

Earth Science Plants and animals living on land have little chance of becoming fossilized because they are usually scattered by scavengers or the weather and are seldom buried before they decay. However, organisms on the continental shelf fall into the soft floor of the ocean, where they are almost buried immediately. This greatly increases the odds of their becoming fossilized.

Social Studies Connection

In A.D. 79, the Italian cities of Pompeii and Herculaneum were buried almost completely by volcanic ash from an eruption of Mount Vesuvius. Ash hardened around the bodies of humans. The bodies eventually decayed, leaving cavities. Thousands of years later, the cavities were filled with plaster to make casts, showing the bodies exactly as they were at the time of the eruption.

Fossil Formation

Most fossils are found in rocks that formed from layers of sand, mud, or silt. They show the shape or outlines of the hard parts of the organism, such as its shell or bones.

Organisms that left these kinds of fossils were covered by sediments after they died. The soft parts of the organisms decayed. Over a long period of time, the sediment covering the organism's hard parts became a layer of rock. Water seeped through the rock and dissolved the bones or shells, leaving an impression called a **mold**.

For many fossils, one last step occurred. Minerals carried by the seeping water built up in the mold to form a **cast**. You can see a cast in Figure 8.4.

Fossils formed in many other ways, too. Sometimes the soft part of an organism was preserved when it was buried where it could not be decayed by bacteria. Leaves have been found that are millions of years old and still contain chlorophyll. Organisms have also been trapped in the sap from trees or frozen in ice. They have had their entire bodies preserved. The footprints, tracks, and burrows of organisms have been preserved in rock.

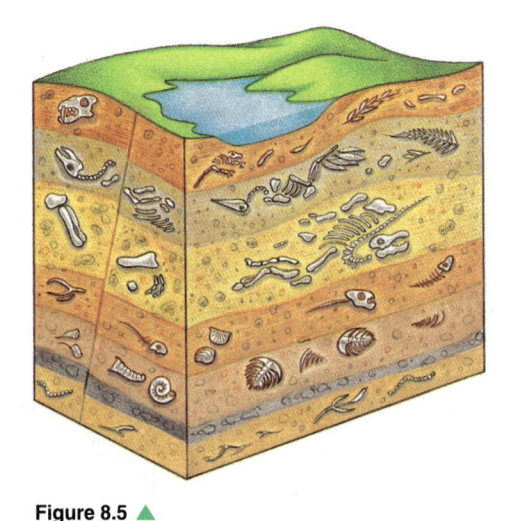

Figure 8.5 ▲
What can be learned from the fossil record? ①

Reading the Fossil Record

Fossils provide clues to how organisms and the earth have changed. They tell a story that scientists can piece together by making inferences.

The fossil record, for example, has been used to determine the order in which different forms of life have appeared. The oldest layers of rock in which reptile fossils occur are always above layers of rock containing fossils of fishes. Based on this observation, scientists conclude that fishes appeared on the earth before reptiles.

Fossils can also provide clues to how different life forms are related to each other. Imagine that fossils of fishes with leglike fins are found in layers above fossils of fishes. And above both layers, there are fossils of animals that have legs. These animals look like they lived in water part of the time. And above this layer are fossils of reptiles. What would you conclude from this evidence? It supports the idea that reptiles evolved from fishes.

Figure 8.4 ▲
A trilobite left behind this cast.

Chapter 8 History of Life

TEACH

Misconceptions
Students may not understand that fossils are not usually the organisms themselves, but rather impressions of them. Dinosaur skeletons in museums are not made of found bones; they are reproductions of bones based on information in the fossil record.

Class Activity
Have students press a shell firmly into a piece of clay to make an imprint of the shell in the clay. Then have them remove the shell to form a mold. Ask students how cast fossils form from molds. (Minerals build up in the mold, creating a three-dimensional cast.)

Skills Development
Infer Ask students to suppose that a fossil was found that showed an animal with feathers like those of a bird, but with claws on its wings and with teeth like those of a reptile. What would they conclude from this evidence? (Because a birdlike animal existed that had some traits of a reptile, it is likely that birds evolved from reptiles.)

Enrich
Use Enrich Worksheet 8.2.

Answer to In-Text Question
① The fossil record shows the order in which different life forms have appeared and how they are related to one another.

The Living Textbook:
Life Science Side 8

Chapter 23   Frame 09444
Fossil Reconstruction (Movie)
Search: Play:

TEACH ▪ Continued

Explore Visually

Have students study the evolutionary tree on this page. Ask the following questions:

▶ Which organisms did both plants and animals come from? (Complex, single-celled organisms)

▶ Which three groups of organisms did the first reptiles evolve into? (Mammals, birds, reptiles)

▶ Which groups are more closely related—the segmented worms and the mollusks or the segmented worms and the roundworms? (Segmented worms and mollusks)

▶ Which group of organisms do humans belong to? (Mammals)

▶ Which group came first, the ferns or the cone-bearing plants? (Ferns)

**The Living Textbook:
Life Science Sides 7-8**

Chapter 14 Frame 01610
Convergent/Parallel Evolution
(27 Frames)
Search: Step:

INTEGRATED LEARNING

Integrating the Sciences

Earth Science The fossil record does not show much until about 600 million years ago during the Cambrian period (the first period of the Paleozoic era), when living things began to develop skeletons and hard parts. Today scientists understand that the scarcity of fossils from before the Cambrian period is not because there was little complex life then, but because the life forms that existed were soft organisms, such as jellyfish, which did not easily become fossilized.

The Evolutionary Tree

By piecing together the clues provided by the fossil record, scientists have determined the likely ancestry of the earth's life forms. They have constructed tree-shaped diagrams that show evolutionary relationships among the different groups of organisms. A simple evolutionary tree is shown in Figure 8.6. The base of the tree represents the earliest forms of life that gave rise to all others. Each branch leads in a different evolutionary "direction."

Figure 8.6
This simplified evolutionary tree gives you an idea of how organisms are related. ▼

164   Chapter 8   History of Life

STS Connection

Ask students to name ways in which people use oil and natural gas. They may mention running cars, heating buildings, and running machinery. Point out that fossil fuels have formed over millions of years from microorganisms living in water and that these fuels now account for over half of the world's energy consumption. Have students investigate the current rate of consumption of these resources. Discuss the consequences of using up these resources at a faster rate than they are formed.

Other Evidence for Evolution

The fossil record leaves no doubt that life forms have changed over time. In addition, there are other kinds of evidence for evolution. For example, many animals look similar during their early stages of development. Look at Figure 8.7. Species with similar embryos probably have a common ancestor.

Similarities in the DNA of different organisms also show how closely they are related. And, on a large scale, the fact that all life uses the same genetic code is evidence of the common ancestry of all living things.

Science and Technology *Finding Oil*

Big pockets of natural gas and oil lie under the earth in layers of sedimentary rock. Millions of years ago, when organisms died and were buried, their bodies were slowly changed into these "fossil fuels."

Scientists use many methods to find these hidden pockets of oil and natural gas. For example, they measure the magnetic field in areas of the earth's crust. The sedimentary rocks in which oil might be found have weaker magnetic fields than other kinds of rocks. Scientists also measure the pull of gravity in different locations. Objects weigh more when they are above large masses of dense rock than when above the light, porous rocks that may contain oil deposits. If data from these tests suggest oil is present, a hole is drilled in the earth's crust to see if oil is really there.

Figure 8.7 ▲
Similar embryos mean evolutionary relationships. Here, you can see that the embryos of the fish, rabbit, and human are quite similar, even though they look very different as adults.

Check and Explain

1. How does a mold form?
2. What kinds of evidence do fossils provide about life's history?
3. **Reason and Conclude** Why do different branches of an evolutionary tree never intersect each other?
4. **Infer** Fossils of organism X are always found in rock layers below those with fossils of organism G. Both are always found in layers below fossils of organism A. Which of these organisms appeared on the earth first? Which evolved most recently?

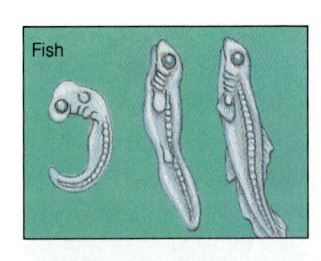

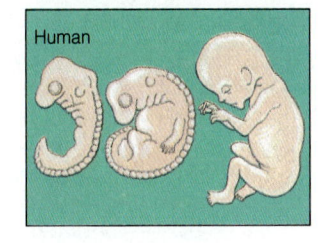

Chapter 8 History of Life

Skills Development

Infer Have students study Figure 8.7. Have them describe how the stages of the three animals are similar to one another. Ask students why similarities among species' embryos probably indicate common ancestry. (The embryonic stages might represent previous evolutionary stages in each species.)

EVALUATE

WrapUp

Reinforce Have students describe how the fossil record provides information about the evolution of life on the earth. Ask them what other forms of evidence provide information about the ancestry of living things.
Use Review Worksheet 8.2.

Check and Explain

1. Animals die and their bodies are covered by sediments, which become rock. Water seeps through the rock and dissolves the remaining bones or shells, and minerals in the water leave a mold.
2. Fossils provide clues to how the earth has changed, the order in which various life forms appeared, and how different life forms are related.
3. Organisms that share a common ancestor may represent two separate evolutionary directions that do not necessarily evolve from each other.
4. Organisms buried more deeply appeared earlier. The organisms appeared in this order: X, G, and A.

The Living Textbook:
Life Science Sides 7-8

Chapter 14   Frame 01366
Fossil Evidence of Evolution (433 Frames)
Search: Step:

ACTIVITY 8

Time 80 minutes   **Group** pairs

Materials
modeling clay
seashells
petroleum jelly
15 drinking cups
15 plastic spoons
plaster of Paris
assorted small household objects
15 paper plates
lab aprons

Analysis

1. A mold and a cast both provide a representation of an object that is the same size as the original. Molds are "negatives" of the original object, while casts take the actual shape of the original. A cast requires a mold. A mold does not require a cast.
2. The plaster of Paris represents what happens in nature: The minerals carried by seeping water form the cast.
3. An imprint will form only if the object is pressed into the clay.
4. If the clay were dried out or frozen, it would not be soft enough to make an imprint.
5. The cast gives information about the size, shape, and texture of the object. It will tell nothing about the object's weight, color, or composition.
6. Answers will vary. Students likely will have identified objects by their shapes.

Conclusion

Student conclusions will vary, but should demonstrate an understanding of the distinction between a mold and a cast.

Extension

Answers will vary depending on the region.

166

TEACHING OPTIONS

Prelab Discussion

Have students read the entire activity, and then discuss the following:

▶ Explain that this activity makes use of a technique used by naturalists, archeologists, and paleontologists in their studies. Ask students why molds and casts might be more helpful than photographs in researching animal traces, artifacts, and fossils. What do footprint impressions reveal about the animals that made them?

▶ Have students suggest some circumstances under which archeologists might make casts as opposed to molds, and vice versa.

▶ Brainstorm some familiar analogies to the casting process (ice cube tray, baking mold, concrete form, ice cream scoop). Then identify the "mold" and the "cast" in each case.

Activity 8   What is the difference between a mold and a cast?

Skills Model; Observe; Infer; Classify

Task 1 Prelab Prep
Collect the following items: modeling clay; seashell; petroleum jelly; paper cup or plastic food container; plastic spoon; plaster of Paris; objects such as coins, keys, leaves, washers; paper plates; lab apron.

Task 2 Data Record
Copy Table 8.1 on a sheet of paper. You will be identifying objects in casts that other groups will be making. Use the table to record your observations.

Task 3 Procedure
1. On a paper plate, roll a piece of modeling clay into a flat, smooth slab. Make the piece of clay about 3 by 5 cm.
2. Coat the seashell with petroleum jelly. Press the seashell into the clay. Make sure that you have a definite imprint of the shell. Remove the seashell and observe the imprint. You have made a mold.
3. Use more clay to build a low wall around the seashell imprint.
4. Use the food container and spoon to prepare a mixture of plaster of Paris according to the directions.
5. Fill the seashell imprint to the top of the wall with plaster.
6. Wait about 20 to 30 minutes for the mold to harden. Then carefully remove the clay. You have made a cast.
7. In your group, use a leaf, key, washer, or other object to make another clay imprint. Don't let anyone see the cast your group is making. Give your cast a number. After everyone is finished, go around the room and identify the object that each group's cast represents. Write the cast number in Table 8.1 and identify it.

Table 8.1 Mystery Casts

Cast Number	Object

Task 4 Analysis
1. How are the molds and casts similar to the original objects? How are they different from the original objects?
2. What does the plaster of Paris represent?
3. Explain why an imprint would not form if the seashell were just laid on the clay.
4. Explain why an imprint would not form if the clay were dried out or frozen.
5. What are some of the clues that the cast tells you about the object? What information about the object is not revealed from a cast?
6. What clues helped you identify the objects in casts that your classmates made?

Task 5 Conclusion
Write a paragraph explaining the difference between a mold and a cast. Then discuss how casts help scientists find out more about the earth's past.

Extension
Research what fossils can be found in your area. What kind are they? What era in geologic history are they from? Find out details about your area at that time, including climate, kinds of living things, and any changes that were occurring on land, such as earthquakes and volcanoes.

166

INTEGRATED LEARNING

Themes in Science

Evolution A major adaptation that distinguished hominids from other animals was bipedal locomotion, or the ability to walk upright. Large brains evolved in hominids *after* they became bipedal. Have students think of other examples of uniquely human adaptations. (Students may mention highly specialized hands, speech, and so on.)

SECTION 8.3

Section Objectives
For a list of section objectives, see the Student Edition page.

Skills Objectives
Students should be able to:

Define Operationally human adaptations for movement.

Research the work of the Leakeys.

Make a Graph showing hominid brain evolution.

Vocabulary
primate, hominid

8.3 Human Evolution

Objectives

▶ **Describe** the adaptations of primates.
▶ **Explain** the origin of hominids.
▶ **Describe** Neanderthals and Cro-Magnons.
▶ **Summarize** the changes that scientists hypothesize occurred in the process of human evolution.
▶ **Make a graph** showing the average brain sizes of extinct hominids.

Skills WarmUp

Define Operationally How does your body enable you to climb things safely, like trees? Describe some adaptations you have that help you to climb. How would your body need to be different for you to climb higher? What if you wanted to move faster? How would your body need to be different? Compare the adaptations you have for moving on the ground and moving in trees.

MOTIVATE

Skills WarmUp
To help students understand human evolution, have them do the Skills WarmUp.
Answer Students may say that humans' hands and eyesight enable them to climb. Humans could climb higher if their arms were stronger and if their feet could grip better. Humans' center of balance and limbs are built for walking, running, and jumping.

Prior Knowledge
Gauge how much students know about human evolution by asking the following questions:

▶ Under what classification of mammal do humans fall?
▶ How do humans differ from other primates?
▶ Why do physical anthropologists search for fossils of humanlike primates?

Compare your body to a dog's or a cat's. In what ways is your body different? What characteristics make you human? All your human features are adaptations that evolved over many millions of years. You share some of these adaptations with animals that have a similar ancestry. Other adaptations are unique to humans. The evolutionary history of humans can be traced back from the present through species that are now extinct.

Primate Ancestry

The human species belongs in a group of mammals called the **primates**. Primates appeared more than 65 million years ago as animals adapted to living in trees. Life in trees requires several special adaptations. The hands must be able to grasp branches. The arms must have a wide range of motion to allow swinging from branch to branch. The eyes must be able to judge distance accurately for rapid movement. This means that the eyes must face forward so that their fields of vision overlap. Even though humans do not live in trees, you will recognize these traits in your own body.

After the first primates appeared, they evolved in different directions. Some continued to be tree-dwellers and evolved into the many species of lemurs, monkeys, and other similar primates alive today. Others adapted to living at least part of the time on the ground. They evolved into the many species of modern apes, which include chimpanzees, gorillas, and orangutans.

Figure 8.8 ▲
An orangutan walks on all fours when on the ground. In the trees, it swings or walks from branch to branch.

TEACH

Misconceptions

Students may believe that there is a missing link between humans and other primates. They may be under the misconception that humans' ancestors resembled chimpanzees or other modern apes. Explain that humans and modern apes evolved from a common ancestor but did not descend from each other. Draw an ancestral tree similar to Figure 8.6 on page 164 to illustrate that apes and humans represent two divergent branches of the anthropoid tree.

Enrich

Students may question why only part of Lucy's skeleton is shown in Figure 8.9. Explain that archeologists recovered only 40 percent of her skeleton. Have students find out about Lucy and other hominids that have been discovered. Students can share information in group discussions.

Historical Notebook

Answer

1. Students should recognize that Mary Leakey's discovery was important to scientists who were researching human evolution because it added to the growing knowledge of the subject. Many scientists now believe that Africa is the cradle of human evolution. Her discovery also sparked public interest in human evolution, which led to further funding of anthropological explorations.

The Living Textbook:
Life Science Side 8

Chapter 21   Frame 03078
Lucy (Movie)
Search: Play:

168

INTEGRATED LEARNING

Writing Connection

Have students imagine they were on the site with the Leakey team in 1959 when Mary Leakey found fossils of an ancient human. Have them write a letter to a friend describing how they and the rest of the crew felt when Mary Leakey made her famous discovery.

Social Studies Connection

Donald Johanson is a famous archeologist who discovered Lucy, the oldest hominid fossil, in Ethiopia in 1974. Lucy is thought to be a young female of the species *Australopithecus* who lived 3.5 million years ago. Lucy is one of the most complete hominid skeletons and one of the greatest finds in the fossil record. Have students find out more about Johanson and his work.

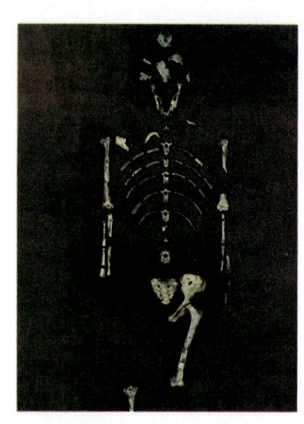

Figure 8.9 ▲
The skeleton of this female hominid was nicknamed "Lucy" by the research team that discovered her.

Human Origins

Between 4 and 8 million years ago, some apelike primates began to adapt to living entirely on the ground. A skeleton for standing and walking upright on two legs evolved. These primates were ancestors of modern humans. They were the first **hominids** (HAHM uh nihdz), or humanlike primates. In Figure 8.9, you can see one of the earliest known hominids. This 3.6-million-year-old skeleton was found in Africa in 1974.

Many other fossils of extinct hominids have been found. Scientists have classified them and determined their age. Some of these hominids are represented in Figure 8.10. Scientists are not sure how these and other species are related to each other. What they do know is that more than 2 million years ago, a species of *Australopithecus* developed a larger brain and began to use tools. This hominid was the first member of the genus *Homo*, which includes modern humans.

Historical Notebook

The Leakey Team

In 1931, anthropologist Louis Leakey began searching for hominid fossils in northern Tanzania, a country in East Africa. He found an ideal spot for his search in the canyons of the Serengeti Plain. Here the exposed rocks revealed strata dating back nearly 2 million years.

For years Leakey searched for human bone fragments. He was joined by his wife, Mary, also an anthropologist. After searching for decades, Mary made the first major discovery in 1959. She found fragments of a human skull and two large teeth! The Leakeys' first discovery turned out to be an extinct hominid whose branch had died out.

Louis Leakey died in 1972 after writing many books about the origin of humans. Mary Leakey has continued her work with her son, Richard Leakey, who is the director of the National Museum of Kenya.

1. Why was Mary Leakey's discovery important to scientists researching human evolution?
2. **Research** Find out about the research and discoveries of Richard and Mary Leakey. Write a short report about their contributions to the field of physical anthropology.

168   Chapter 8 History of Life

Language Arts Connection

Have students find out the meanings of the terms *Homo habilis*, *Homo erectus*, and *Homo sapiens*. Then ask them why they think these names were chosen for each type of human. (*Homo habilis* means "man the toolmaker," *Homo erectus* means "man the erect," and *Homo sapiens* means "man the wise.")

Neanderthals

The most widely known early hominids that are classified as *Homo sapiens* are the Neanderthals (nee AN dur THAWLZ). These early humans lived about 130,000 years ago. Even though they belong to the same species as modern humans, they looked much different. Neanderthals ranged in height from 1.5 m to 1.7 m. They had large bones and heavy brows.

Neanderthals were intelligent hunters who used tools and fire. They made their own clothes from animal hides. They lived in cooperative social groups and cared for the aged and sick. They practiced forms of art and music. Most importantly, they used language and could pass on knowledge and traditions from generation to generation. Neanderthals disappeared about 30,000 years ago.

Cro-Magnons

The first group of modern humans were the Cro-Magnons (KROH MAG nuhnz). They appeared about 100,000 years ago, possibly in Africa. Later they migrated to other areas. They arrived in Europe about 35,000 years ago. Some scientists think they caused the extinction of the Neanderthals.

Cro-Magnons looked very similar to modern humans. They had smaller teeth, longer faces, and straighter foreheads than Neanderthals. If you saw a Cro-Magnon walking down the street dressed in modern clothes, you would probably think he or she looked pretty normal.

Cro-Magnons used language and created music and art. Their colorful and expressive paintings have been found in more than 200 caves in Europe. They also created sculptures.

Figure 8.10
This timeline shows approximately when different hominids lived. Notice how the skull structures changed over time. ▼

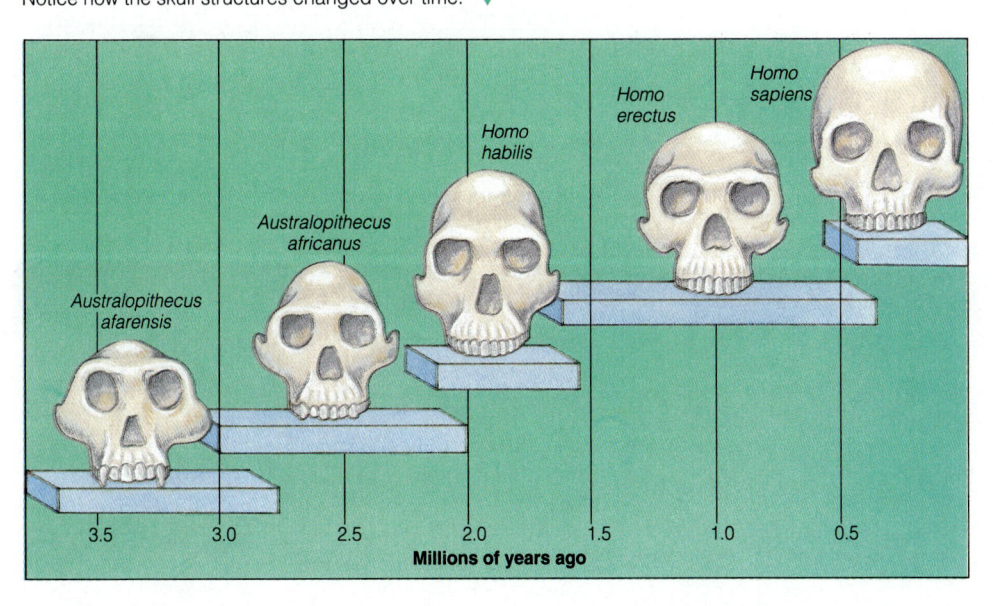

Discuss

Explain that *Homo habilis* was the earliest species of the genus *Homo*. These early hominids made tools, which they probably used to cut meat and smash open bones to obtain marrow. Because toolmaking involves using memory, planning, and abstract thinking, it marks an important development in the evolution of our species. Scientists think that *Homo habilis* used a primitive form of communication and built round huts for shelter.

Critical Thinking

Reason and Conclude Ask students how scientists know that Neanderthals were intelligent hunters who used tools and fire. (Scientists have dated many materials from Neanderthal caves, including bones of many different animals, tools, and the charcoal residue of fire pits.)

The Living Textbook:
Life Science Sides 7-8

Chapter 14   Frame 01423
A. Afarensis (39 Frames)
Search: Step:

The Living Textbook:
Life Science Sides 7-8

Chapter 14   Frame 01466
Australopithecus/Homo erectus/Neanderthal (10 Frames)
Search: Step:

TEACH ▪ Continued

Skills WorkOut
To help students find out more about the discoveries of hominid fossils, have them do the Skills WorkOut.
Answers Answers will vary because anthropologists disagree about how to classify different kinds of hominids.

EVALUATE

WrapUp
Reinforce Ask the following questions:
▶ Name some similarities between the first hominids, Neanderthals, and Cro-Magnons.
▶ Why do you think the shape of the skull and teeth of hominids evolved as they did?
▶ Has *Homo sapiens* stopped evolving?

Use Review Worksheet 8.3.

Check and Explain
1. The ability to grasp, walk on two feet, and judge distance
2. Modern humans have a larger brain and a different shaped skull and are taller.
3. Dinosaurs became extinct long before humans appeared.
4. Make sure that students' graphs contain four bars. Review the measurements in cubic centimeters.

170

TEACHING OPTIONS

Cooperative Learning
Even now, people do not know the whole story of the Piltdown forgery. Some think that Charles Dawson, the amateur paleontologist who found the remains, actually planted them. Have students work in cooperative groups to plan and present a dramatic event, such as a mock trial, about the Piltdown forgery. Have different groups research Dawson and the other scientists and choose team members to act the parts of the various characters. One group should portray the researchers who exposed Piltdown Man as a fraud. Other students can play the parts of jury members and judge.

Skills WorkOut
Research Find out more about the discoveries of hominid fossils made in the 1900s. Where was each find made? Which hominid species does each fossil belong to? What is the importance of each fossil to the understanding of human origins? Write a report describing the discoveries.

Science and Society *The Piltdown Hoax*
On December 12, 1912, newspaper headlines all over the world announced the discovery of the "missing link" between apes and humans. Pieces of a skull and jaw, tools made of bones and rocks, and remains of ancient mammals had been found in a gravel pit in Piltdown, England. The skull had room for a large brain and was the same as modern humans. However, the jaw was apelike.

The find was named Piltdown Man. The Piltdown Man was exactly what many scientists were looking for at that time. For years they had searched for a primitive human halfway between modern humans and apes. For this reason, many accepted it as real.

But in 1953, Piltdown Man was exposed as a forgery. The skull was from a modern human and the jaw was from an orangutan. Both had been stained brown so they would look old. The teeth in the jaw had been filed down to look more human. The reasons for the fraud are still unknown.

The discovery of the forgery shows how science should work. Science is self-correcting. Even though many scientists were fooled for 40 years, the truth was finally revealed. Piltdown Man was not a missing link. In fact, there was no missing link to be found, because humans are not descended from apes. Rather, modern humans and modern apes share a common ancestor.

Check and Explain
1. What characteristics do humans share with apes, such as chimpanzees?
2. In what ways are modern humans different from the first hominids?
3. **Uncover Assumptions** What is incorrect about cartoons that show humans and dinosaurs living together?
4. **Make a Graph** The average brain size of extinct hominids ranges from 400 cc in *Australopithecus afarensis* to 750 cc in *Homo habilis,* to 900 cc in *Homo erectus,* to 1500 cc in Neanderthals. Make a bar graph that shows this information.

170   Chapter 8   History of Life

Chapter 8 Review

Concept Summary

8.1 Geologic Time
- The earth is 4,600 million years old. The first living things appeared about 3,500 million years ago.
- The earth's history is divided into parts called eras. The four eras are the Precambrian, the Paleozoic, the Mesozoic, and the Cenozoic.
- Because of the way it forms, sedimentary rock preserves a record of how the earth has changed.

8.2 The Fossil Record
- Fossils are remains and traces of past life. They provide evidence that many kinds of organisms once lived on the earth but are now extinct.
- Fossils form when the shape or substance of all or part of an organism is somehow preserved after its death.
- When a fossil is formed in rock, it is often in the form of a mold and a cast.
- Fossils provide clues to how living things have evolved. They help scientists construct tree diagrams showing the ancestry of modern organisms.

8.3 Human Evolution
- Humans belong to the primates, a group of mammals that appeared over 65 million years ago.
- The basic primate characteristics are adaptations for living in trees.
- The first humanlike primates, or hominids, appeared between 4 and 8 million years ago. Hominids evolved larger brains and began using tools.
- Neanderthals lived just before modern humans evolved. The earliest modern humans were the Cro-Magnons.

Chapter Vocabulary

era (8.1)	fossil (8.2)	mold (8.2)	primate (8.3)
geologic time (8.1)	extinct (8.2)	cast (8.2)	hominid (8.3)
sedimentary rock (8.1)	fossil record (8.2)		

Check Your Vocabulary

Use the vocabulary words above to complete the following sentences correctly.

1. The remains of an organism can leave an impression, called a ____, in sedimentary rock.
2. A species is ____ when all of its members have died out.
3. Scientists use the ____ to determine the ancestry of modern species.
4. A humanlike primate is called a ____.
5. The earth's history is divided into four ____.
6. A ____ is created when mineral-containing water seeps into a fossil mold.
7. The earth's history is shown on a scale of ____.
8. Sediments collect in layers and may in time form ____.
9. Many of the characteristics of a ____ are adaptations for living in trees.
10. The footprint of an extinct animal preserved in rock is an example of a ____.

Explain the difference between the words in each pair.

11. Fossil, extinct
12. Hominid, human
13. Fossil record, evolutionary tree

Chapter 8 History of Life

CHAPTER REVIEW 8

Check Your Vocabulary

1. mold
2. extinct
3. fossil record
4. hominid
5. eras
6. cast
7. geologic time
8. sedimentary rock
9. primate
10. fossil
11. A fossil is the remains of past life; a species becomes extinct when there are no longer any individuals living.
12. All humanlike primates are hominids; humans are a specific group of hominids who are classified as *Homo sapiens*.
13. The fossil record is the collection of fossils showing the changing life on the earth; an evolutionary tree is a type of diagram showing the likely ancestors of living things.

Use Vocabulary Worksheet for Chapter 8.

CHAPTER 8 REVIEW

Check Your Knowledge

1. A primate is a member of a group of mammals that includes humans and many species of monkeys and apes. Primates are distinguished from other animals by adaptations such as forward-facing eyes, hands capable of grasping, and the ability to judge distance accurately.
2. A particular layer of sedimentary rock is older than the layers above it.
3. The bodies of these organisms were crushed or decomposed, so they could not form molds.
4. Neanderthals were similar to modern humans in that they used tools and fire, lived in social groups, practiced art and music, and used language. Differences include the smaller size of Neanderthals and several differences in appearance.
5. Many new life forms began to appear at the beginning of the Paleozoic era.
6. A mold must form first.
7. humans
8. 4,600
9. walks upright
10. Precambrian
11. embryos
12. evolutionary trees

Check Your Understanding

1. Answers may vary. Fossils of buried bones will be found, but casts will likely not have had enough time to form.
2. All of the items in the list can be considered fossils, though students might argue that the buried clamshell may not be the remains of past life.
3. To find dinosaur fossils, look in layers below where the horse fossils were found.
4. a. Answers may vary. The cricket may have died and was covered by tree sap.
   b. The cricket is a fossil, but is different from others because it was not formed in rock.
5. Answers may vary. Mammals had to compete for food and space, or were prey for dinosaurs. Once dinosaurs disappeared, the mammal populations could increase.
6. A species at the end of a branch either exists today in that form or became extinct sometime in the past, thus having no descendants. A species at the base of a branch evolved into the species beyond it on the branch.
7. Scientists believe that evolution is driven by random changes in DNA over time. The discovery of an organism without DNA would force scientists to modify or discard the current theory of evolution for a vastly different one.

Chapter 8 Review

Check Your Knowledge

Answer the following in complete sentences.

1. What is a primate? What distinguishes it from other animals?
2. What is the age of a layer of sedimentary rock compared to those above it?
3. Why did soft-bodied organisms leave few fossils?
4. In what ways were Neanderthals like modern humans? How were they different?
5. What important change occurred at the beginning of the Paleozoic era?
6. What must form first, a mold or a cast?

Choose the answer that best completes each sentence.

7. Horses evolved before (jellyfish, insects, birds, humans).
8. The earth is about (4,600, 3,500, 420, 66) million years old.
9. A primate is classified as a hominid if it (can learn, has forward-facing eyes, walks upright, has hair).
10. The oldest and longest era was the (Precambrian, Cenozoic, Mesozoic, Paleozoic).
11. When two organisms have similar (behaviors, diets, adaptations, embryos), it is evidence that they evolved from a common ancestor.
12. Scientists use (bar graphs, evolutionary trees, chemical equations, pie charts) to show how organisms have evolved and how they are related.

Check Your Understanding

Apply the concepts you have learned to answer each question.

1. **Critical Thinking** Will people living 100 years from now find fossils of the humans living now? Explain.
2. **Classify** Which of the following can be considered a fossil? Why or why not?
   a. A clamshell buried in sand
   b. An old bone
   c. An impression of a snail shell in a piece of rock
   d. A footprint in hardened mud
3. **Infer** In an area rich in fossils, you find a hillside cut away by erosion. In one layer of rock, you discover a fossilized horse jaw. Where would you look to find the fossils of dinosaurs?
4. **Mystery Photo** The photograph on page 156 shows a cricket millions of years old sealed inside a piece of hardened tree sap called amber.
   a. How do you think the cricket ended up inside the amber?
   b. Is the cricket a fossil? If so, how is it different from most fossils?
5. **Find Causes** Why do you think the mammals began to evolve rapidly about the same time that the dinosaurs became extinct?
6. **Infer** What can you infer about a species that is placed at the end of a branch on an evolutionary tree? How is it different from a species placed at the base of a branch?
7. Suppose a form of life is discovered that does not use DNA for its genetic code. How would this discovery change scientists' view of the evolution of life?

Develop Your Skills

1. a. About 4.6 billion, or 4,600 million
   b. At the top
   c. About 2.5 billion years
   d. Near the top (about 66 million years ago)
2. About 25 percent
3. a. During the Devonian period, about 408 million years ago
   b. Three
   c. Bats

Make Connections

1.

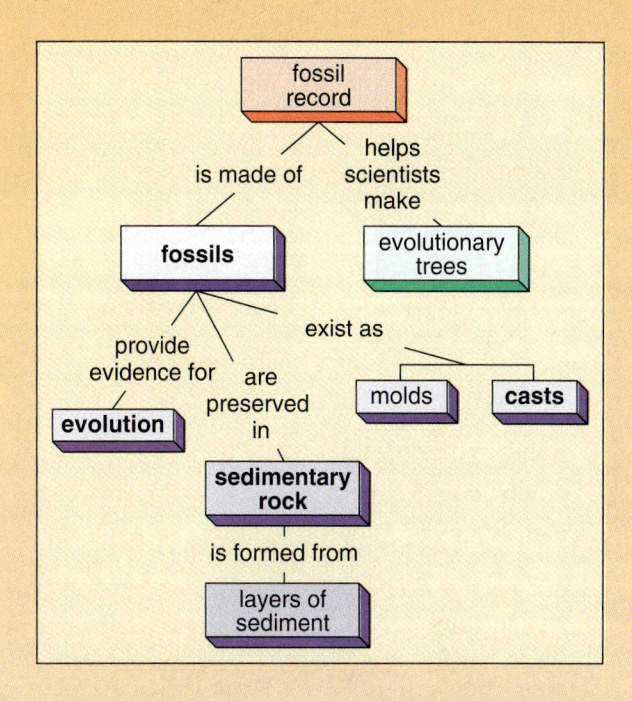

```
                    fossil
                    record
         is made of        helps
                          scientists
                            make
         fossils                evolutionary
                                   trees
    provide          exist as
    evidence for  are
                  preserved      molds    casts
    evolution      in
                 sedimentary
                    rock
                is formed from
                  layers of
                  sediment
```

2. Answers will vary. Students may find it helpful to know that they are doing research in an area of study often called anthropological linguistics.

3. Student presentations will vary. Remind students that the actual intentions of cave artists will probably never be known. Encourage students to ask such questions as: Why was this art created in places that were so hard to get to? What effect would these paintings have had on viewers at the time they were made?

Develop Your Skills

Use the skills you have developed in this chapter to complete each activity.

1. **Interpret Data** The timeline below shows some major events in the earth's history.

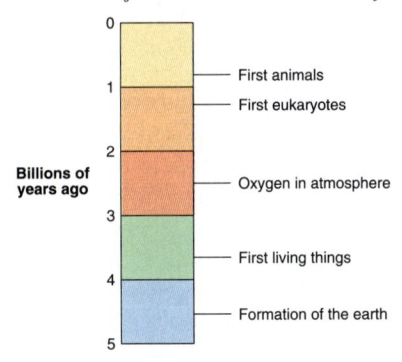

   (Timeline, Billions of years ago:
   0 —
   1 — First animals
   2 — First eukaryotes
   — Oxygen in atmosphere
   3 —
   4 — First living things
   5 — Formation of the earth)

   a. How many billions of years ago did the earth form? What is this number in millions?
   b. Where on the timeline is the present?
   c. For how long has there been oxygen in the earth's atmosphere?
   d. Where on the timeline would you mark a line to show the beginning of the Cenozoic era?

2. **Calculate** For what percentage of the earth's history was there no life?

3. **Data Bank** Use the information on page 630 to answer the following questions.
   a. When did the first amphibians appear?
   b. The Mesozoic era is divided into how many periods?
   c. Which appeared first, bats or mastodons?

Make Connections

1. **Link the Concepts** Below is a concept map showing how some of the main concepts in this chapter link together. Only part of the map is filled in. Finish the map, using words and ideas you find in the chapter.

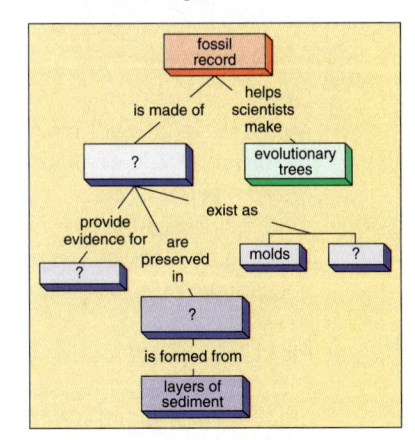

2. **Science and Social Studies** One of the most important events in the evolution of humans was the beginning of the use of language. Research the importance of language in human society. How does it make the human species different from any other? How does language shape the way humans interact with their environment?

3. **Science and Art** Find photographs of cave paintings done by Cro-Magnons and earlier humans. Study them carefully and decide what they were meant to show. Then research what scientists say they mean. Present your findings and ideas to the class.

CHAPTER 9

Overview

This chapter discusses the various ways that scientists classify living things. The first section describes the characteristics that scientists use to group organisms. The second section contains descriptions of the five kingdoms. The last section describes how taxonomic keys and field guides are used to identify organisms.

Advance Planner

▶ Provide Latin and Greek dictionaries for TE activity, page 178.

▶ Find magazines with photos or drawings of plants gathered, grown, or used by Native Americans. TE page 184.

▶ Collect screws and bolts differing in size, head type, and threading for SE Activity 9, page 186.

Skills Development Chart

Sections	Classify	Communicate	Infer	Observe	Organize Data	Research
9.1 Skills WarmUp	●					
SkillBuilder	●					
9.2 Skills WarmUp					●	
Skills WorkOut						●
9.3 Skills WarmUp			●			
Skills WorkOut		●				
Activity	●			●		

Individual Needs

▶ **Limited English Proficiency Students** Cover the captions and photocopy Figures 9.2, 9.5, and 9.7. Have students write the boldface terms for the chapter on strips of index cards and attach the terms as captions to pictures that show the meanings of the terms. They can use a term more than once. Have students meet in small groups to discuss their captions. Tell them to explain their choices.

▶ **At-Risk Students** Show students that classification is a part of their everyday lives. Have students work with a partner. To get them started, have students write down seven characteristics of their favorite food. Students should attempt to list characteristics that separate that food from other foods. Then have students write seven characteristics of their least favorite food. One student should reveal characteristics, while the partner attempts to guess the food. Students should continue supplying characteristics until their partner answers correctly. Then have students refer to Table 9.1. Have them use the same strategy to identify an organism classified in the table. One student should give a clue; the partner should attempt to guess from the clue.

▶ **Gifted Students** Have students work in a group and create two classification triangles like the one shown on page 176. Students should create triangles for two organisms from different kingdoms. Encourage students to use photographs from magazines or to draw pictures for their triangles.

Resource Bank

▶ **Bulletin Board** Divide the bulletin board into five columns. Label each column with the name of one of the five kingdoms. Have student volunteers write general descriptions of the organisms in each kingdom and add these to the bulletin board. Have them collect pictures of organisms and place them in the appropriate columns. Challenge students to find pictures of fungi, monerans, and protists, as well as animals and plants.

▶ **Field Trip** Take students to an art museum. Provide them with maps and guides to the museum and have them use the guides to figure out how the exhibits in the museum have been displayed. They can use this information to find an exhibit that they want to see.

CHAPTER 9 PLANNING GUIDE

Section	Core	Standard	Enriched	Section	Core	Standard	Enriched
9.1 Classification pp. 175–179				**Color Transparencies** Transparencies 25a, b	●	●	
Section Features Skills WarmUp, p. 175 SkillBuilder, p. 178	●	● ●	●	**9.3 Identifying Organisms** pp. 183–186			
Blackline Masters Review Worksheet 9.1 Skills Worksheet 9.1 Integrating Worksheet 9.1	● ●	● ● ●	●	**Section Features** Skills WarmUp, p. 183 Skills WorkOut, p. 184 Career Corner, p. 184 Activity, p. 186	● ● ● ●	● ● ● ●	● ● ● ●
Color Transparencies Transparency 24	●	●		**Blackline Masters** Review Worksheet 9.3 Skills Worksheet 9.3 Vocabulary Worksheet 9	● ● ●	● ● ●	
9.2 The Five Kingdoms pp. 180–182				**Laboratory Program** Investigation 17	●	●	●
Section Features Skills WarmUp, p. 180 Skills WorkOut, p. 182	● ●	● ●					
Blackline Masters Review Worksheet 9.2 Reteach Worksheet 9.2 Integrating Worksheet 9.2	● ● ●	● ● ●	● ●				

Bibliography

The following resources can be used for teaching the chapter. See page T-40 for supplier codes.

Audio-Visual Sources

(video unless noted)
Classifying Living Things. 18 min. 1987. A-W.
Classifying Microorganisms. 15 min. 1984. C/MTI.
Fungi. 18 min. 1983. BF.
Plant or Animal? 15 min. 1986. NGSES.

Software Resources

Plants. Apple. J & S.
Project Classify: MAMMALS. Apple. 1989. NGSES.
Vertebrates. Apple. 1990. J & S.
What Is This? Series. Apple. PES.

Library Resources

Allaby, Michael. *The Oxford Dictionary of Natural History.* New York: Oxford University Press, 1986.
Comstock, Anna Botsford. *Handbook of Nature Study.* Ithaca: Cornell University Press, 1986.
Margulis, Lynn, and Karlene V. Schwartz. *Five Kingdoms: An Illustrated Guide to the Phyla of Life on Earth.* San Francisco: W.H. Freeman and Co., 1982.
Teasdale, Jim. *Microbes.* Morristown, NJ: Silver Burdett, 1985.

CHAPTER 9

INTEGRATED LEARNING

Writing Connection

Classification is used to put objects and systems into orderly patterns. Lead an introductory discussion with the class to explore the characteristics used to classify things (size, color, texture, composition). Have students write an outline in which they choose a concrete or abstract category (vehicles, clothing, foods, music, buildings, and so on) and give the criteria for classifying items within that category.

Introducing the Chapter

Have students read the description of the photograph on page 174. Ask if they agree or disagree with the description.

Directed Inquiry

Have students study the photograph. Ask:

▶ How many organisms do you see in the photograph? (There is one starfish and a colony of corals. The white structure is not a plant, but is formed by the coral animals.)

▶ Would you describe the organisms as plants or animals? (Animals)

▶ How are these organisms alike? How are they different? (Both live in water and both are circular in shape. Corals live in colonies and are smaller than starfish. The starfish has an external skeleton and a spiny exterior.)

▶ What kind of group would include both organisms? What kind of group would include only one of them? (Answers may vary. For example, starfish might be grouped with animals having exoskeletons, while corals may be grouped with simpler animals.)

Chapter Vocabulary

animal	order
class	phylum
family	plant
fungus	protist
genus	species
kingdom	taxonomy
moneran	

174

Chapter 9 — Classifying Living Things

Chapter Sections

9.1 Classification
9.2 The Five Kingdoms
9.3 Identifying Organisms

What do you see?

"In this picture, I see a starfish and coral. They both are living animals. The picture shows one starfish and one coral. They both live in the water, but the starfish can move and coral doesn't. They both eat other animals in the water."

Payam Tehranian
Ferry Pass Middle School
Pensacola, Florida

To find out about the photograph, look on page 188. As you read this chapter, you will learn about the variety of living things and how they are classified.

174

Themes in Science
Diversity and Unity Share with students that despite the enormous variety of organisms populating the earth, all living things share six basic characteristics. Often, it is useful to group organisms into categories based on characteristics that are shared by members of a particular group.

Multicultural Perspectives
The ways that people classify living things have changed greatly throughout history. Have students gather information on Anaximander, Hippocrates, Aristotle, Theophrastus, and Pierre Belon, and their methods of classifying the earth's living things. Discuss how appropriate these classification ideas might be today.

SECTION 9.1

Section Objectives
For a list of section objectives, see the Student Edition page.

Skills Objectives
Students should be able to:
Classify items in a grocery store.
Make a Model that displays a system for classifying sports equipment.

Vocabulary
taxonomy, kingdom, phylum, class, order, family, genus, species

9.1 Classification

Objectives
- **Explain** why organisms are organized into groups.
- **Describe** how scientists name organisms.
- **Identify** the major classification groups.
- **Make a model** showing a classification system.

Skills WarmUp
Classify Make a list of ten different items that you might find in a grocery store. Then rewrite the list as several separate lists, so that all items on each list have at least one characteristic in common.

Did you know that a dog is more closely related to a wolf than either animal is to a fox? Since similarities in appearance can be misleading, people need a better way to classify, or organize into groups, the many different types of organisms on earth. In this way, patterns in nature can be more clearly understood. Long ago, people began developing classification systems for this purpose. These systems help make sense of the relationships among organisms by clearly defining what makes them similar and different.

History of Classification
As a small child, you learned to classify things by their use, shape, size, or color. As you grew older, you learned to classify things by smaller and more meaningful categories. Throughout time, people have classified living things in some way, usually based on how they affected the people's lives.

For instance, the earliest systems of classification probably grouped plants and animals by whether they were useful, harmful, or neither. People of the ancient Near East recognized five animal groups: domestic animals, wild animals, creeping animals, flying animals, and sea animals. More than 2,300 years ago, the Greek philosopher Aristotle defined two groups of animals: those that had red blood and those that did not. He classified plants as herbs, shrubs, or trees.

In the 1700s, the Swedish scientist Carolus Linnaeus classified about 12,000 organisms into groups based on similar characteristics. His basic system is still used today. The science of classifying living things using this system is called **taxonomy**.

Figure 9.1 ▲
Which two animals are more closely related: the gray wolf (top), Akita Inu dog (middle), or red fox (bottom)? ①

MOTIVATE

Skills WarmUp
To help students understand classification, have them do the Skills WarmUp.
Answer Students' answers should show that they can group at least a few items by their common characteristics. Lists should include items based on food groups—meat, grains, and so on.

Prior Knowledge
To gauge how much students know about classifying living things, ask the following questions:
- Why is it important to organize items into groups?
- In what ways are all living things similar?
- Do you think an insect has any of the same characteristics as a horse?

Answer to In-Text Question
① The dog is more closely related to the wolf than either is to the fox.

TEACH

Directed Inquiry

Have students study pages 176 and 177. Ask some or all of the following questions:

▶ Which category includes more organisms, the kingdom or the species? (The kingdom includes a larger number of organisms.)

▶ To what kingdom does an insect belong? How do you know? (Students can determine that the insect is an animal because of its position on the chart in Figure 9.2.)

▶ Use the description of a kingdom to identify some characteristics of insects. (They consume organisms, use energy, move about, and reproduce sexually.)

▶ To what phylum does the crow belong? (Chordata)

▶ To what class does the horse belong? (Mammalia)

▶ How does the female horse feed her young? (Like other mammals, the horse produces milk.)

▶ To what order does the cat belong? (Carnivora)

▶ To what family does the American black bear belong? (Ursidae)

▶ To what genus does the grizzly bear belong? (*Ursus*)

▶ What is one characteristic shared by bears in the genus *Ursus*? (The bears in this genus are larger than bears in other genera.)

▶ How many different kinds of organisms are in a species? (One kind of organism is in a species.)

176

INTEGRATED LEARNING

Integrating the Sciences

Physical Science Scientists have given names to various things that make up the universe, from atoms to supergroups of galaxies. In a discussion, try to get students to develop a hierarchical list of matter in the cosmos from the smallest to the largest. Impress upon them how classification applies in areas other than life science. (Neutrons, protons, and electrons; atoms and molecules; elements and compounds; stars and galaxies)

Levels of Classification

There are seven major levels of classification. Each successive level contains fewer organisms with more characteristics in common than the level above it.

Kingdom Most scientists today group all living things into five **kingdoms**. Classification by kingdom depends largely on an organism's cell structure, how it gets energy, and its movement and reproductive characteristics. A polar bear, for example, belongs to the kingdom Animalia (animals). Members of this kingdom have complex cells, consume other organisms for food, use energy, usually move about, and reproduce sexually.

Phylum The largest group in the animal kingdom is a **phylum** [FY luhm] (plural: phyla). (Plants are grouped into divisions instead of phyla.) For example, polar bears are classified in the phylum Chordata. Animals in this phylum have a flexible, skeletal rod called a *notochord*. Frogs, fishes, and birds are also classified in this phylum.

Figure 9.2
Classification of the Polar Bear

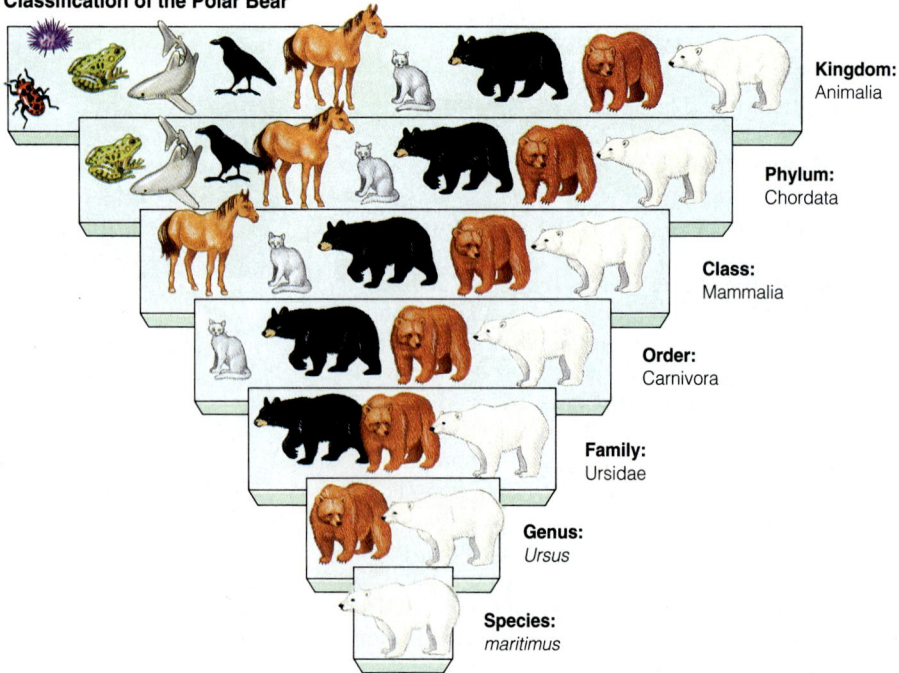

Kingdom: Animalia
Phylum: Chordata
Class: Mammalia
Order: Carnivora
Family: Ursidae
Genus: *Ursus*
Species: *maritimus*

176 Chapter 9 Classifying Living Things

Language Arts Connection

Have students create a mnemonic device that will help them remember the levels of classification. One example is "Kings play cards on fat green stools."

Table 9.1 Classification of Five Different Organisms

	Human	Yogurt-making Bacterium	Bread Mold	Ameba	Sunflower
Kingdom	Animalia	Monera	Fungi	Protista	Plantae
Phylum	Chordata	Eubacteriaces	Zygomycota	Sarcodina	Tracheophyta
Class	Mammalia	Schizomycetes	Phycomycetes	Lobosa	Anthophyta
Order	Primates	Eubacteriales	Mucorales	Amoebina	Asterales
Family	Hominidae	Lactobacilla	Mucoracease	Amoebidae	Compositae
Genus	Homo	Lactobacillus	*Rhizopus*	Amoeba	Helianthus
Species	sapiens	bulgaris	stonifer	proteus	annus

Class A phylum is divided into **classes**. Polar bears belong to the class Mammalia (mammals), endothermic animals whose females produce milk. Cats, horses, and humans are mammals. Fish and birds form separate classes in the phylum Chordata.

Order A class is divided into **orders**. Polar bears are members of the order Carnivora, animals that mainly eat meat and have well-developed tearing teeth. Cats and dogs also belong to Carnivora, while horses and humans belong to other orders in the same class.

Family An order is divided into **families**. The polar bear, like all bears, belongs to the family Ursidae. Other members of Carnivora, such as dogs and cats, form different families.

Genus Each family consists of at least one **genus** (plural: genera). The polar bear belongs to the genus *Ursus,* which includes large bears, such as the grizzly. Smaller bears, such as the American black bear, are grouped in different genera.

Species A genus contains one or more **species**. A species is made up of related organisms that are able to mate and reproduce offspring of the same type. In their natural environments, one species cannot successfully mate with another.

Figure 9.3 ▲
According to Figure 9.2, to which order does the polar bear belong? ①

Chapter 9 Classifying Living Things

Enrich

Have students use the following information to construct a chart similar to the one on page 177: lion (Animalia, Chordata, Mammalia, Carnivora, Felidae, *Felis leo*); bullfrog (Animalia, Chordata, Amphibia, Salientia, Ranidae, *Rana catesbeiana*); dandelion (Plantae, Tracheophyta, Antho-phyta, Asterales, Compositae, *Taraxacum officinale*); and paramecium (Protista, Ciliophora, Ciliata, Holotricha, Parameciidae, *Paramecium caudatum*). After students have finished their work, have them determine which of these organisms have the most and least in common with various organisms on the chart on this page, and why.

Class Activity

Have students look in dictionaries to find the meanings of the words in Table 9.1. Then have them quiz one another on the classification levels of the five organisms.

Answer to In-Text Question

① **Carnivora**

**The Living Textbook:
Life Science Sides 1-4**

Chapter 4   Frame 00747
Animals (779 Frames)
Search: Step:

**The Living Textbook:
Life Science Sides 1-4**

Chapter 7   Frame 02302
Classification of Organisms (155 Frames)
Search: Step:

TEACH • Continued

Class Activity
Remind students that scientific names are written using two words. The first word is the genus name and the second is the species name. The genus name is capitalized and italicized and the species name is lowercase and italicized. Have students write their own names in scientific form. (Names should appear as described. Students' last names should appear first.)

Apply
Have students imagine that a species of living organisms has been discovered on another planet. Discuss the characteristics of such an organism and have students describe how they would go about classifying it.

SkillBuilder
Discuss After students have completed their reports and store layouts, have a class discussion guided by the following questions:

▶ How do graphic organizers such as the one pictured help you to classify?

▶ Is the average number of classification levels you have for grocery store items more or less than the number of levels for classifying organisms?

Answers Student reports should describe the classification systems of the four students and the strengths and weaknesses of each system of organization.

INTEGRATED LEARNING

Language Arts Connection
Have each student look up the scientific names of five favorite animals (extinct or living). Then have students use dictionaries to find out the basic meanings of the parts of these names. For example, *Tyrannosaurus rex* means tyrant lizard king. A foreign language teacher could help here. Have students describe why the name fits the animal.

Use Integrated Worksheet 9.1.

Scientific Names

The name *pine* is used for hundreds of tree species. Yet silver pine and sugar pine are both common names for the same species. A single organism may have 50 different common names in 30 languages.

When Linnaeus invented his classification system, he realized the importance of using a unique name for each type of organism. He used Greek and Latin words to name organisms by genus and species. Those languages were understood by most scientists of his day. Today, biologists still use his system to name organisms. The system is called binomial nomenclature, a Latin phrase that means "two-name naming." For example, scientists of all countries know the polar bear by its scientific name, *Ursus maritimus,* which means "seagoing bear." *Ursus* is the genus, *maritimus* the species. Similarly, all pine trees belong to the genus *Pinus,* but *Pinus monticola* identifies only one species.

SkillBuilder Classifying

Classification Systems

Imagine that you are the manager of a grocery store or supermarket. You have to develop a classification system for displaying items in your store. How will you go about it?

Follow the steps below to organize your store.

1. Make a list of eight to ten different categories of items based on common characteristics.

2. Break each category down into smaller subcategories, using two or more levels based on common characteristics. Use the diagram below as an example.

3. Draw a layout of your store, showing the locations of the different items you identified in steps 1 and 2.

4. Compare your store design with the designs of three other students. Discuss why you organized your systems the way you did. Write a short report comparing the strengths and weaknesses of the different classification systems.

```
                    Dairy Products
        ┌──────────┬──────┴──────┬──────────┐
       Milk       Spread       Cheese      Yogurt
      ┌─┴─┐      ┌──┴───┐      ┌─┴──┐     ┌─┴────┐
    Whole Skim  Butter Margarine Yellow White  Plain Flavored
```

Chapter 9 Classifying Living Things

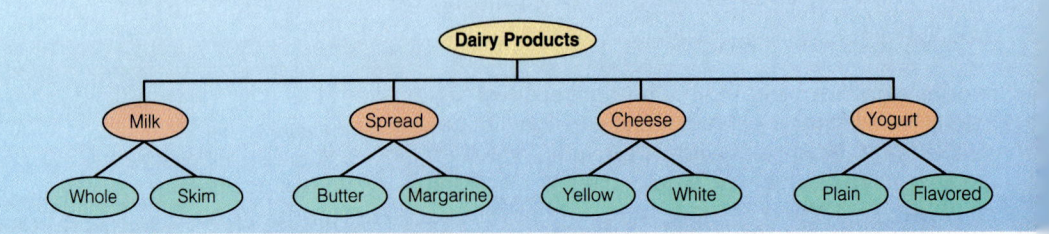

Language Arts Connection

Take students to the school library. Have them find out how the books are classified. Have each student use the classification system to help look up and locate one fiction title.

Science and You
Finding One Book Among Millions

When you look for a book in the library, you can see how a classification system works. Books can be classified one of two ways. You are probably familiar with the Dewey Decimal Classification System used in many school and public libraries. However, most large research libraries use a system developed by the Library of Congress in Washington, D.C.

In this system, two capital letters identify the category of a book. Numbers and other letters identify the individual book. Consider, for example, a book identified as QH309.A88 1984. The Q stands for science, the H for life on earth, and the 309 for environmental science. The code A88 identifies the book as *The Living Planet* by David Attenborough. The date 1984 distinguishes this edition from later editions that might include changes.

The Library of Congress classification system is not exactly like the system used to classify living things, but the two systems do have some similarities. For instance, they both provide many levels of classification and exactness in classifying items. Also, they both provide a way to include new items, such as newly discovered species or new categories of living things, or newly published books.

Figure 9.4 ▲
To help people find books, libraries group and arrange books using a classification system.

Check and Explain

1. What is the name of the science of classifying living organisms?
2. The chimpanzee belongs to the family Pongidae of the order Primata. Its genus is *Pan* and its species is *troglodytes*. What is its scientific name?
3. **Classify** List three characteristics of birds that identify them as a group and make them distinct from other types of animals.
4. **Make a Model** Develop a classification system for sports equipment. Make a chart to illustrate your classification system. Use at least three categories of equipment and at least three levels of classification within each category.

Discuss

Stress that the Library of Congress classification system and the system used to classify living things are similar in that they both allow for exactness in classifying items and provide a way for including new items. The library system uses numbers and letters, while the system used to classify living things uses Latin and Greek names.

EVALUATE

WrapUp

Reteach Have students name some objects, ideas, or events that they classify. Write their responses on the chalkboard. Ask students the following questions:

▶ Do most people classify these items in similar ways?

▶ What are some advantages of grouping items?

▶ What are some advantages of classifying living things?

Use Review Worksheet 9.1.

Check and Explain

1. Taxonomy
2. *Pan troglodytes*
3. Three distinctive characteristics of birds are feathers, wings, and beaks.
4. Answers may vary. The chart should show at least three categories and three levels of classification within each category, such as spring, summer, or winter games. Summer sports might be further divided into games played on courts, on fields, or in water.

Chapter 9 Classifying Living Things

SECTION 9.2

Section Objectives
For a list of section objectives, see the Student Edition page.

Skills Objectives
Students should be able to:

Make a Chart that illustrates relationships between people.

Research to find out about the work of zoologist Jane Goodall.

Classify an organism.

Vocabulary
fungus, moneran, plant, protist, animal

MOTIVATE

Skills WarmUp
To give students practice in distinguishing categories and relationships, have them do the Skills WarmUp.
Answer While all parts of a family are related, each person has a distinct classification: brother, sister, mother, father, uncle, and so on.

Prior Knowledge
To gauge how much students know about the five kingdoms, ask the following questions:

▶ What are some differences and similarities between cells in plants and cells in animals?

▶ Do some living things that existed millions of years ago still exist today?

▶ Are those living things the same today as they were millions of years ago?

The Living Textbook: Life Science Sides 1-4

Chapter 2 Frame 00583
Fungi (21 Frames)
Search: Step:

180

INTEGRATED LEARNING

Integrating the Sciences
Microbiology It is often difficult to observe protists, fungi, and monerans, but people encounter them daily. Have students meet in small groups to discuss where, in the course of a day, they think they encounter monerans, protists, and fungi.

Skills WarmUp
Make a Chart Make a chart that shows the relationships between people in your family.

Figure 9.5
The Five Kingdoms

9.2 The Five Kingdoms

Objectives

▶ **Name** and **describe** the five kingdoms of living things.

▶ **Explain** the relationship between classification and evolution.

▶ **Classify** a variety of living things.

For many years, biologists recognized only two kingdoms: animal and plant. Animals included all organisms that moved about, ate food, and stopped growing at a certain size. Plants included all organisms that did not move, used photosynthesis, and grew as long as they lived.

These categories were not exact. Fungi and bacteria were classified as plants even though they do not photosynthesize. The one-celled Euglena seemed to belong to both kingdoms—or neither. It has chlorophyll for photosynthesis, like a plant, but it moves about like an animal. Today's five-kingdom taxonomy is more precise. It is based on what scientists have learned about the structure of different types of cells.

Kingdom: Fungi
- Cells with cell walls, nuclei, and organelles
- Mostly many-celled
- Absorb nutrients from other organisms
- Do not move about
- Most reproduce both sexually and with spores

Kingdom: Monerans
- Simple cells with no nuclei or organelles
- Single-celled; may live in colonies
- May eat or make food
- Reproduce by cell division

Themes in Science

Evolution The ways in which organisms have evolved provide a basis for classifying them. Often different organisms in the same level of classification share common ancestors.

Classification and Evolution

Classification shows the evolutionary relationships among living things. The more groups two organisms share, the more closely they are related. This is possible because all organisms share a common history.

Humans and octopuses, for example, belong to the same kingdom. They had a common ancestor about 600 million years ago. Humans and polar bears belong to the same kingdom, phylum, and class. They had a common ancestor about 100 million years ago. Humans and gorillas had a common ancestor about 9 million years ago. They share the same kingdom, phylum, class, and order. Therefore you are more closely related to a gorilla than to a polar bear, and more closely related to a polar bear than to an octopus.

Kingdom: Animals
- Cells with nuclei and organelles
- Many-celled
- Can move about
- Eat food
- Most reproduce sexually

Kingdom: Plants
- Cells with cell walls, nuclei, and organelles
- Many-celled
- Do not move about
- Use photosynthesis
- Most reproduce sexually

Kingdom: Protists
- Cells with nuclei and organelles; some with cell walls
- Mostly single-celled
- May make or eat food
- Reproduce by cell division; some sexually

TEACH

Misconceptions
Students may think that all living organisms are either plants or animals. Stress that scientists now usually group organisms into five kingdoms. Some scientists suggest that there may be as many as 24 different kingdoms.

Explore Visually
Ask students to study Figure 9.5. Then use the following questions to lead them in a discussion about the drawing:

▶ Find and name some examples of organisms from the animal kingdom. (Birds, sheep, insects)

▶ Find and name some examples of organisms from the plant kingdom. (Trees, grasses, flowers)

▶ What are some characteristics of plants? What are some characteristics of animals? (See boxed text for possible responses.)

▶ What do you think the circular photographs represent? (Enlargements of fungi, monerans, and protists)

▶ Why do you think the images are magnified? (Because you can't see many of the organisms in these kingdoms without a microscope)

▶ How has the development of the microscope affected the way scientists classify organisms? (The microscope makes it possible to observe individual cells within organisms. Today's five-kingdom system is based on the cell structure of organisms.)

Reteach
Use Reteach Worksheet 9.2.

TEACH ▪ Continued

Skills WorkOut
To help students understand classification, have them do the Skills WorkOut.
Answer Students should mention the similarities between humans and chimpanzees: Both species exhibit complex social behavior, use tools, and are omnivorous.

Research
Have students find more information about DNA research. Challenge them to find examples of organisms that were reclassified after biologists examined strands of DNA.

EVALUATE

WrapUp
Portfolio Explain that a cladogram is a treelike diagram, similar to the one on page 628. Have students use the information on page 181 to make a diagram such as a cladogram that shows the evolutionary relationship between organisms. Students can keep their diagrams in their portfolios.
Use Review Worksheet 9.2.

Check and Explain
1. Possible answers: fungi—mushrooms; monerans—bacteria; protists—algae; plants—trees; animals—cats
2. Unlike organisms in all other kingdoms, monerans have no nuclei.
3. Many organisms that share a classification group have similar characteristics because they have evolved from common ancestors.
4. Follow the format of the chart on page 177. You may wish to have each student classify one or two organisms.

INTEGRATED LEARNING

Integrating the Sciences
Anatomy Similarities in anatomy among species lead scientists to think that species that appear to be very different may have descended from common ancestors.
Use Integrating Worksheet 9.2.

Skills WorkOut
Research Find out about the work of zoologist Jane Goodall. How has her work helped scientists learn about the close evolutionary relationship of humans and chimpanzees? Write a one-page report on your findings.

Science and Technology *Clues to the Past*
Scientists have studied the evolutionary paths of similar species for hundreds of years. Much of the research has been based on a method called *comparative anatomy*. In the science of comparative anatomy, relationships among different species are charted based on similarities in their physical structures or anatomies. Bone patterns and glands are two of the physical structures that are charted. The development of embryos is also charted. For example, birds, bats, turtles, and horses all have the same bone structures in their forelimbs. Similarities in anatomy lead comparative anatomists to hypothesize that these animals descended from a common ancestor millions of years ago.

A more recent technique for studying similarities among species has come about through DNA research. Using a technique called DNA sequencing, researchers examine long DNA strands. DNA strands are contained in the nucleus of a cell. The more similar the DNA strands are in two organisms, the more closely related the organisms probably are. Through DNA sequencing, biologists have discovered that humans and chimpanzees are more closely related than previously thought. In fact, humans and chimpanzees are each other's closest genetic cousins.

Today, the study of DNA molecules is changing classification schemes. As new DNA techniques are developed, scientists may make other startling discoveries about how species are related to one another.

Check and Explain
1. List the five kingdoms. Give an example of an organism from each kingdom.
2. What characteristic distinguishes monerans from all other kingdoms?
3. **Reason and Conclude** Why do organisms in the same classification group have characteristics that are similar?
4. **Classify** Make a chart in which you classify each of the following organisms: mushroom, sunflower, starfish, stegosaurus, dog, chimpanzee, human.

Integrating the Sciences

Chemistry Biologists study chemical similarities among organisms to find evolutionary patterns and relationships. This information is used in classifying living things. Astronomers are now suggesting that the elements themselves may have evolved in systems referred to as star furnaces. Show students a periodic table of the elements. Point out that the elements are classified according to their structure and chemical properties.

9.3 Identifying Organisms

Objectives

- **Explain** how to use a taxonomic key.
- **Compare** and **contrast** taxonomic keys and field guides.
- **Make a model** of a taxonomic key.

While walking through the woods one day, you find a clump of attractive mushrooms. You want to take them home to eat, but you do not know whether they are safe. How can you tell a delicious edible mushroom from a poisonous one?

Skills WarmUp

Infer How many clues do you need to identify the animal described below? Here are the possible choices: moth, bald eagle, sparrow, tuna, bat.

1. It has wings.
2. It has a backbone.
3. It has feathers.
4. It is longer than 20 cm.

Taxonomic Keys

A taxonomic key is a guide to identifying organisms on the basis of physical characteristics. Taxonomic keys show evolutionary relationships by arranging similar organisms in a series of numbered statements.

Figure 9.6 shows part of a key for the family of trees that includes pines, spruces, and other evergreens. Follow the steps below as you use the taxonomic key.

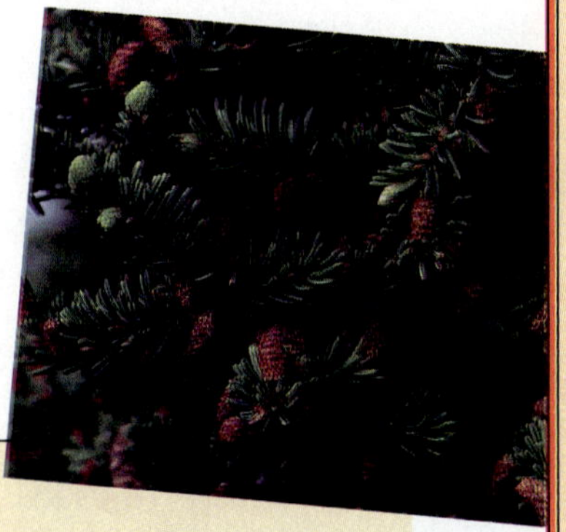

Figure 9.6 Taxonomic Key for Part of the Pine Family

1. Read parts a and b of statement 1. Choose the appropriate description. The numbers at the right, 2 and 4, tell you which statement to read next. For example, since the leaves are not in clusters, you go to statement 4.

2. Because the leaves are arranged alternately on the stem, go to statement 5.

3. Because the leaves are four-sided, go to statement 6.

4. The leaves are 6–12 mm long. You can identify the tree as a black spruce (*P. mariana*).

1a.	Leaves in clusters	2
1b.	Leaves not in clusters	4
2a.	Leaves in clusters of 5	White pine (*Pinus strobus*)
2b.	Leaves in clusters of 2 or 3	3
3a.	Leaves 8–15 cm long	Norway pine (*Pinus resinosa*)
3b.	Leaves 2–4 cm long	Jack pine (*Pinus banksiana*)
4a.	Leaves alternate or scattered	5
4b.	Leaves opposite or whorled	Juniper *(Juniperus utahensis)*
5a.	Leaves four-sided	6
5b.	Leaves flattened	7
6a.	Leaves 6–12 mm long	Black spruce (*Picea mariana*)
6b.	Leaves 15–25 mm long	White spruce (*Picea glauca)*
7a.	Leaves 12–28 mm long	Tideland spruce *(Picea stichensis)*
7b.	Leaves 18–28 mm long	Weeping spruce *(Picea Breweriana)*

SECTION 9.3

Section Objectives
For a list of section objectives, see the Student Edition page.

Skills Objectives
Students should be able to:

Infer the identity of an animal from clues.

Communicate clues that can be used to identify an organism.

Make a Model of the taxonomic key.

MOTIVATE

Skills WarmUp

To help students understand the process of identifying organisms, have them do the Skills WarmUp.

Answer You need all of the clues. The animal is an eagle.

TEACH

Directed Inquiry

Provide hand lenses. Have students use Figure 9.7 to identify the organism in the photograph on page 183. Ask the following questions:

- What are the leaves? (The needles) Are the leaves in clusters? (No, go to statement 4 in the right-hand column.)
- Are the leaves alternate or opposite? (Alternate)
- Are the leaves flat? (No, they are four-sided.)
- How long do you think these leaves are? (Students may refer to metric rulers to estimate the length, 6–12 mm.)
- What kind of tree is this? (Black spruce)

TEACH • Continued

Skills WorkOut

To help students understand how the characteristics of an organism can be used to identify it, have them do the Skills WorkOut.

Answer Answers may vary. Students' examples may be similar to the exercise described in the Skills WarmUp.

Class Activity

Have students visit the library and find different field guides. Allow time for them to study several different field guides. Challenge them to compare a page in a field guide to the page from a taxonomic key in Figure 9.7.

Career Corner

Apply After students read about scientific photography, have them look through this book to find examples of photographs that were taken using special techniques, such as time-lapse or slow-motion photography, or special equipment, such as watertight cameras or telephoto and microscopic lenses. Have students interested in photography bring their photographs to school and show them to the class.

184

INTEGRATED LEARNING

Multicultural Perspectives

Indigenous people of North America were accomplished botanists. Have students research early agriculture and make a display showing photographs or drawings of the plants gathered, grown, or used by Native Americans. Have students label each plant with a brief description of how it was used. Encourage students to indicate which plants are used by people today. Have them add photographs, drawings, or other pictures of local plants.

Skills WorkOut

Communicate Make a list of five organisms. Create a series of four clues for identifying only one organism among the five. Have each successive clue eliminate one organism as a possible answer. Then see if a classmate can use the clues to identify the correct organism from the list.

Field Guides

A field guide is another tool for identifying organisms. It is designed to be used in the field, such as on a nature hike or a trip to an aquarium or natural history museum. A field guide does not give as much scientific information as a taxonomic key. Instead, it has pictures and descriptions that help you identify what you see. Unlike taxonomic keys, field guides are not based on classification. Although guides may group organisms by order and family, they sometimes make groups on the basis of color, size, behavior, or location.

Field guides are especially popular for identifying birds, wildflowers, and other animals and plants you might encounter in nature. They usually give information on habitat, range, migration patterns, seasonal changes in appearance, and other helpful facts. Field guides are also useful for identifying nonliving natural objects, such as fossil rocks and minerals.

Career Corner — Scientific Photographer

Who Takes the Photographs for This Book?

You will probably never be near an erupting volcano or a hibernating bear. You may never see cells dividing or observe heart-transplant surgery. However, you probably are familiar with all these events through the work of scientific photographers.

Scientific photographers take pictures of everything from animals to microscopic surgery. Their work may take them to tropical rain forests, the ocean floor, or inside a laboratory. Some specialize in photographs for magazines, scientific journals, field guides, textbooks, or other books. Others make films or videotapes for research, education, or entertainment.

Scientific photographers often use special techniques and equipment. To photograph animals from a distance, they use a telephoto lens. To record cell processes, they use a special camera mounted on a microscope. To record events that happen very quickly, such as a frog capturing a fly, they use slow-motion photography. For taking pictures underwater, they use a watertight camera built to stand pressure.

Most scientific photographers specialize in a particular field, such as medicine or wildlife. They may take college courses in their field. They develop their photographic skills in college, other special courses, or on-the-job training.

184 Chapter 9 Classifying Living Things

Language Arts Connection

People who study birds are called ornithologists. Write the following words on the chalkboard: ornithologist, zoologist, paleontologist, biologist, psychologist, ophthalmologist, sociologist. Point out to students that the suffix *-ologist* can mean "one who studies." Have them hypothesize the meaning of the words on the list. Encourage them to add other similar words to the list.

Science and You *Birding*

Do you enjoy watching the birds at a backyard feeder or in a park? Do you like the outdoors and learning about nature? Then you might enjoy birding, the sport that is also a science. Birding, or birdwatching, involves observing and identifying birds. Most birders keep a log of the date and location of each species they see. They identify birds by such characteristics as size, color, shape, song or other call, distinctive body parts, and feeding and nesting behavior.

To begin birding, you need only a field guide, a pair of binoculars, and a notebook. Take a walk through your neighborhood or a local park and try to identify the different birds you see. Many areas have bird walks that are guided by experts who can help you spot and recognize a variety of birds.

Figure 9.7 ▲
Birdwatchers may use a field guide (above) to help identify the birds they see and hear.

Check and Explain

1. Explain the difference between a taxonomic key and a field guide.
2. Use the key on page 183 to identify an evergreen tree with two-leaf clusters that are 3 cm long.
3. **Classify** Lizards and snakes belong to the same order in the class Reptilia (reptiles). How would a taxonomic key distinguish between them?
4. **Make a Model** Make a list of ten animals. Create a taxonomic key that identifies all of them.

Chapter 9 Classifying Living Things 185

Class Activity
Take students outdoors so they can observe birds that live in your area.

EVALUATE

WrapUp
Reteach Have students work in small groups to classify organisms they observe on the school grounds. All group members should record detailed observations in a collective nature journal. Have them use a taxonomic key or field guide to identify the organisms.

Use Review Worksheet 9.3.

Check and Explain

1. A taxonomic key is organized in numbered statements and gives information about an organism's physical characteristics and evolution. A field guide groups organisms by location, color, size, and behavior, and includes pictures. Some field guides give information about nonliving objects.
2. Jack pine
3. A taxonomic key might distinguish between a snake and a lizard on the basis of physical characteristics such as body parts.
4. Answers will vary. Taxonomic keys should include numbered statements that follow the model in Figure 9.7.

185

ACTIVITY 9

Time 45 minutes **Group** pairs
Materials
12 different types of screws and bolts (15 of each)

Analysis
Accept all logical answers. Students should be able to use another group's key to identify one of its items.

Conclusion
Answers may vary. Explanations should contain a description of the process that each group used to develop its key.

Everyday Application
Answers may vary. A classification system used by a hardware store might be based on how the screws will be used. It might include: type of materials to be fastened, weight that must be held, location of project (indoor or outdoor), desired appearance.

Extension
Answers will vary. The key should follow the form in Figure 9.7.

TEACHING OPTIONS

Prelab Discussion
Have students read the entire activity before beginning the discussion. Ask:
▶ What kinds of characteristics are valid for the key? (Students should see that any observable characteristic is valid.)
▶ What are the advantages of listing characteristics from general to specific?

Activity 9 How do you use a taxonomic key?

Skills Observe; Classify

Task 1 Prelab Prep
1. Collect the following items: 12 or more screws and bolts, all differing in size, head type, threading, and other characteristics (see Table 9.2).
2. Divide the objects into two groups. Make sure that all items in a group share at least one defining characteristic.
3. Divide each group into two smaller subgroups. Again, make sure that all items in each subgroup share at least one characteristic.
4. Continue to separate the items into smaller and smaller groups until you end up with single items ("species").

Task 2 Data Record
1. Assign each item a name. Use real names or creative names.
2. List the characteristics you used to classify the items in each group.

Task 3 Procedure
1. Use your data to create a taxonomic key for identifying bolts and screws. You can use the one for identifying trees, on page 183, as a model.
2. Have another group use your key to identify one of your items.

Task 4 Analysis
Was the other group able to identify an item using your key? If not, discuss with your group whether the problem may have been due to inexact classification. Rethink your classification system and prepare another key.

Task 5 Conclusion
Write a short paragraph explaining the way you created your key and what the activity taught you about creating and using taxonomic keys.

**Table 9.2
Characteristics of Screws and Bolts**

Screw	
Bolt	
Wood screw	
Machine screw	
Automotive screw	
Flat-head screw	
Round-head screw	
Pan-head screw	
Slotted head	
Phillips head	
Round bolt	
Square-head bolt	
Hex-head bolt	
Eye bolt	

Everyday Application
Suppose you needed a supply of screws for a home project. Explain how a classification system would help you find them in the hardware store.

Extension
Develop a taxonomic key for something you use every day. The taxonomic key could be for identifying items of clothing, types of TV shows, or movies you have seen.

Chapter 9 Review

Concept Summary

9.1 Classification
- Taxonomy is the science of classifying living things based on similar characteristics.
- The modern system of taxonomy has seven major levels of classification: kingdom, phylum, class, order, family, genus, and species.
- The basis of modern classification was developed by Linnaeus. In this system, each organism is given a two-part name—its genus and species.

9.2 The Five Kingdoms
- The five-kingdom system of classification consists of monerans, protists, fungi, plants, and animals.
- Scientific classification outlines evolutionary relationships among living things.

9.3 Identifying Organisms
- Living things can be identified by taxonomic keys and field guides.
- Taxonomic keys identify organisms by physical characteristics. The keys show evolutionary relationships.
- Field guides identify organisms by pictures and descriptions of an organism's location, color, size, or behavior.

Chapter Vocabulary

taxonomy (9.1)	order (9.1)	species (9.1)	fungus (9.2)
kingdom (9.1)	family (9.1)	moneran (9.2)	plant (9.2)
phylum (9.1)	genus (9.1)	protist (9.2)	animal (9.2)
class (9.1)			

Check Your Vocabulary

Use the vocabulary words above to complete the following sentences correctly.

1. A shark is an ___, because it does not have cell walls, moves about, eats food, and reproduces sexually.
2. The science of classifying living things is called ___.
3. A ___ makes food by photosynthesis and has a cell wall.
4. An organism that absorbs food from other organisms, does not move, and reproduces sexually or by spores is a ___.
5. Most scientists today group all living things into five ___.
6. Plants are grouped into divisions instead of ___.
7. A class is divided into ___.
8. A ___ is made up of organisms that are able to mate and reproduce offspring of the same type.
9. Polar bears belong to the ___ of mammals.
10. A bacterium is a member of the ___ kingdom.
11. Each family consists of at least one ___.
12. An ameba belongs to the ___ kingdom.
13. All bears belong to the same ___, Ursidae.

Write Your Vocabulary

Write sentences using each vocabulary word above. Show that you know what each word means.

CHAPTER REVIEW 9

Check Your Vocabulary

1. animal
2. taxonomy
3. plant
4. fungus
5. kingdoms
6. phyla
7. orders
8. species
9. class
10. moneran
11. genus
12. protist
13. family

Write Your Vocabulary

Be sure that students understand how to use each vocabulary word in a sentence before they begin.

Use Vocabulary Worksheet for Chapter 9.

CHAPTER 9 REVIEW

Check Your Knowledge

1. two-
2. fungi
3. taxonomy
4. fungi
5. physical characteristics
6. order
7. False; polar bears
8. False; kingdom
9. True
10. False; physical appearance
11. False; genus
12. False; answers may vary, but should only include animals
13. True

Check Your Understanding

1. Monerans have simple cells with no membrane-bound nuclei or organelles; protists have complex cells with membrane-bound nuclei and organelles.
2. Answers may vary. A field guide would be used when making quick identifications in the field. A taxonomic key would be used when more detailed scientific or evolutionary information is needed.
3. Kingdom, phylum, class, order, family, genus, species
4. Answers may vary. People would have difficulty communicating about the organism.
5. Answers may vary. Examples include: both are animals, both live underwater, both feed on microscopic organisms, both have tentacles.
6. *Picea rubra* and *Picea mariana*
7. a. Juniper (*Juniperus utahensis*)
   b. Norway pine (*Pinus resinosa*)
   c. Jack pine (*Pinus banksiana*)
   d. White pine (*Pinus strobus*)
   e. Tideland spruce (*Picea stichensis*)

Chapter 9 Review

Check Your Knowledge
Choose the answer that best completes each sentence.

1. Linnaeus developed a (one-, two-, three-, four-) name system for naming organisms.
2. Bread mold belongs to the (animal, plant, fungi, moneran) kingdom.
3. The science of naming organisms and putting them in groups is called (biology, ecology, taxonomy, filing).
4. A slime mold belongs to the (fungi, moneran, protist, plant) kingdom.
5. A taxonomic key is a way of identifying organisms based on (habitat, behavior, seasons, physical characteristics).
6. Dogs belong to the (genus, order, family, species) Carnivora.

Determine whether each statement is true or false. Write *true* if it is true. If it is false, change the underlined term(s) to make the statement true.

7. Humans and <u>octopuses</u> had a common ancestor about 100 million years ago.
8. The largest group in scientific classification is the <u>species</u>.
9. In the 1700s, Linnaeus classified about 12,000 organisms into groups based on <u>similar</u> characteristics.
10. A field guide usually identifies organisms using <u>classification</u>.
11. The scientific name for an organism includes its <u>class</u> and species.
12. Humans, <u>ameba</u>, and polar bears are examples of animals.
13. Taxonomy is a science that <u>names</u> organisms and puts them into groups.

Check Your Understanding
Apply the concepts you have learned to answer each question.

1. How are protists different from monerans?
2. When might you use a field guide? When would a taxonomic key be more useful?
3. **Sequence** Order the following groups from most general to least general: class, genus, family, kingdom, phylum, species, order.
4. **Apply** Suppose an organism had three common names in different parts of the United States. How might these names lead to confusion?
5. **Mystery Photo** The photograph on page 174 shows a starfish crawling on a soft coral tree. The coral is a colony of small animals with stinging tentacles. What characteristics do these animals share?
6. Which of the following three organisms are most closely related: *Picea rubra, Morus rubra, Picea mariana*?
7. Use the key on page 183 to identify the following trees. Give the common and scientific names.
   a. Tree with opposite leaves growing on a stem
   b. Tree with 2-leaf clusters that are about 13 cm long
   c. Tree with 2-leaf clusters that are 3 cm long
   d. Tree with 5-leaf clusters
   e. Tree with flattened leaves (about 12 mm long) growing alternately on a stem

Chapter 9 Classifying Living Things

Develop Your Skills

1. a. Many organisms do not fit easily into these limited categories.
   b. Microscope
2. a. At the order level
   b. They are the same genus, but different species.
3. Accept all logical answers; however, students' designs should include all the parts of a good experiment. To prove that the grasses are different species, students would have to prove that the two grasses could not produce fertile offspring.
4. Answers may vary.

Make Connections

1. Concept maps will vary. You may wish to have students do the map below. First draw the map on the board, then add a few key terms. Have students draw and complete the map.

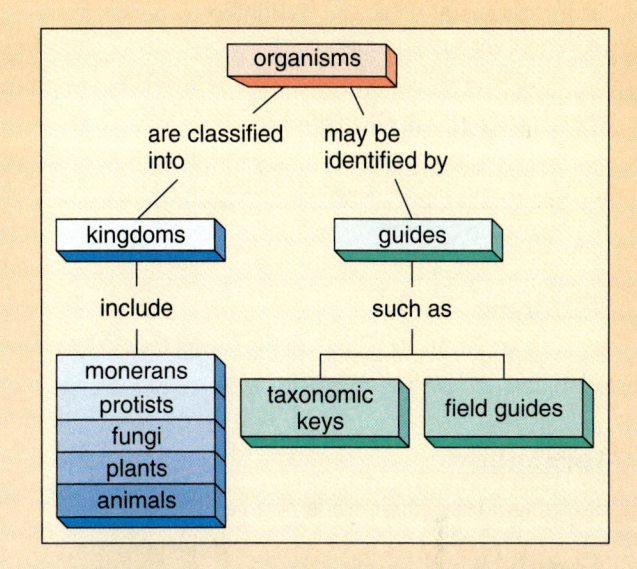

2. Answers may vary. Students' guides should follow the form of a field guide.
3. Check students' answers for accuracy. Scientific names include genus and species.
4. Answers will vary.
5. Students' drawings should indicate that they understand the process involved in DNA sequencing. Steps should show the use of new computer technology in sequencing.
6. Answers will vary. Timelines might include Aristotle's classification of living things, the invention of the microscope, the development of the Linnean system, and the later work of zoologists, anatomists, and Darwin.

Develop Your Skills

Use skills you have developed in this chapter to complete each activity.

1. **Interpret Data** The chart below shows a system of classification.
   a. What problems can you see with this system?
   b. What important invention made scientists begin to understand that this system was not accurate?

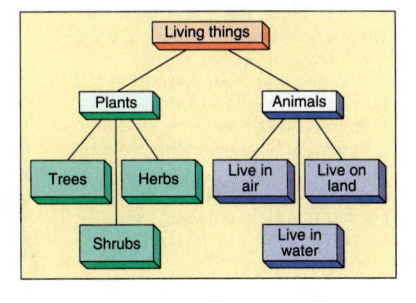

2. **Data Bank** Use the chart on page 628 to answer the following questions:
   a. At what level is the skunk related to the polar bear?
   b. According to the tree, how does the dog differ from the wolf?

3. **Design an Experiment** There are many different species of grasses. Using grass seeds or grasses, design an experiment to prove that two types of grass are different species. Describe the question, materials, procedure, control and variable, observations, and conclusion.

4. **Research** Choose five common organisms and use reference books to classify them from kingdom to species. Write their scientific names.

Make Connections

1. **Link the Concepts** Draw a concept map showing how some of the main concepts in this chapter link together. Use words and ideas from the chapter to draw your map.

2. **Science and Language Arts** Using what you have learned about field guides, develop a field guide to something nonliving. You may want to use automobiles or the utensils used in the school cafeteria.

3. **Science and You** Make a list of common plants and animals found in your area. Use a reference book to find the scientific name of each. Write the scientific names next to the common names. Include pictures or drawings of your organisms.

4. **Science and Language Arts** Many libraries use the Dewey Decimal System to group books by similarities. Visit your school or local library. Choose a science topic (such as whooping cranes, evolution, genetics). List five books related to your topic. Include the book's title, author, publisher, year of publication, and Dewey decimal number. Write a brief description about the information in each book.

5. **Science and Technology** Use your library to find out more about DNA sequencing. Draw a series of steps to explain how scientists examine the DNA strands. Describe the different uses of DNA sequencing.

6. **Science and Social Studies** Draw a timeline that shows the events that led to the modern classification system. You will need to research this topic in your library. Include advances in technology on your time line.

SCIENCE AND LITERATURE CONNECTION UNIT 3

About the Literary Works
In the Night, Still Dark is by Richard Lewis, copyright 1988 by Atheneum. Reprinted by permission of Atheneum. "The Weasel" is from the text *Consider the Lemming* by Jeanne Steig, copyright 1988 by Farrar, Straus and Giroux. Reprinted by permission of Farrar, Straus and Giroux.

Description of Change
Both poems are reprinted directly from the texts. "In the Night, Still Dark" is an extensive abridgment of the Hawaiian creation chant called *The Kumulipo*.

Rationale
The vocabulary and syntax of the original form of each selection are appropriate to the student level.

Vocabulary
barnacle, ferret, stoat, transmogrification, ermine

Teaching Strategies

Directed Inquiry
After students finish reading, discuss the selections. Be sure to relate the selections to the science lessons in this unit. Ask the following questions:
- What is the main theme of "In the Night, Still Dark"? (The main theme is the evolution of organisms.)
- Why does the weasel's fur change color? (The change in color is an adaptation for protection from predators.)
- One key belief in Polynesian culture is that each person is part of all other living things. Why then do you suppose this verse was chanted upon the birth of a child? (To celebrate the emergence of a new life by recollecting how all life-forms first came to be)

Explain
Explain to your students that "In the Night, Still Dark" is an abridged form of a Hawaiian creation chant called *The Kumulipo*. *The Kumulipo*, which is actually over two thousand lines long, was chanted for the birth of each royal child in Hawaii.

Skills Development

Classify Challenge your students to list all life-forms described in the selections. Then have students classify each organism according to the five kingdoms. Make sure students save their lists for use in the Evaluate Sources section of the Science and Literature Connection.

Critical Thinking
Have students identify a specific evolutionary relationship detailed from "In the Night, Still Dark." Challenge students to draw upon their knowledge of Chapter 8 (History of Life) to write a paragraph explaining methods they would use to determine whether such an evolutionary relationship does exist.

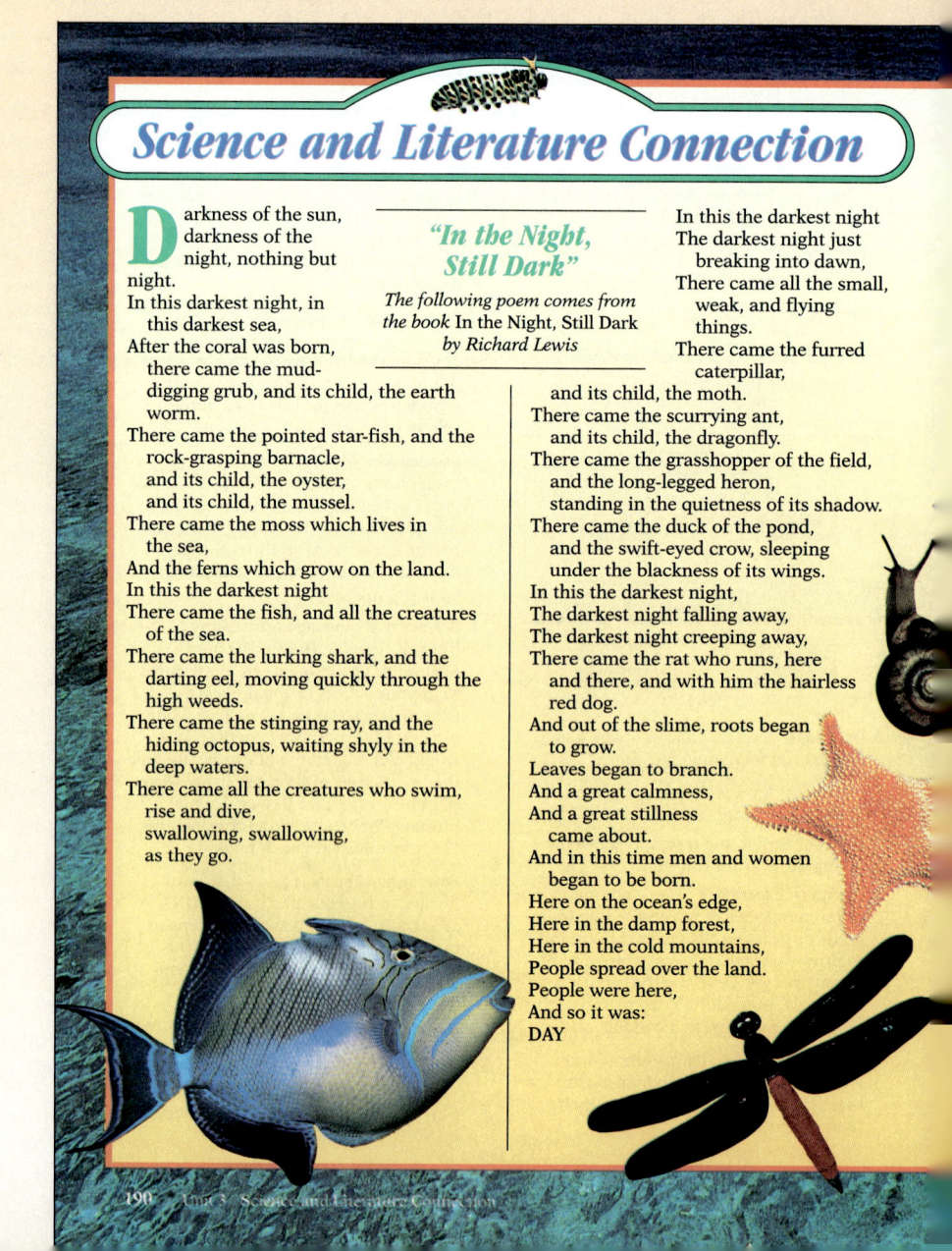

Science and Literature Connection

"In the Night, Still Dark"

The following poem comes from the book In the Night, Still Dark *by Richard Lewis*

Darkness of the sun,
 darkness of the
 night, nothing but
 night.
In this darkest night, in
 this darkest sea,
After the coral was born,
 there came the mud-
 digging grub, and its child, the earth
 worm.
There came the pointed star-fish, and the
 rock-grasping barnacle,
 and its child, the oyster,
 and its child, the mussel.
There came the moss which lives in the
 sea,
And the ferns which grow on the land.
In this the darkest night
There came the fish, and all the creatures
 of the sea.
There came the lurking shark, and the
 darting eel, moving quickly through the
 high weeds.
There came the stinging ray, and the
 hiding octopus, waiting shyly in the
 deep waters.
There came all the creatures who swim,
 rise and dive,
 swallowing, swallowing,
 as they go.
In this the darkest night
The darkest night just
 breaking into dawn,
There came all the small,
 weak, and flying
 things.
There came the furred
 caterpillar,
 and its child, the moth.
There came the scurrying ant,
 and its child, the dragonfly.
There came the grasshopper of the field,
 and the long-legged heron,
 standing in the quietness of its shadow.
There came the duck of the pond,
 and the swift-eyed crow, sleeping
 under the blackness of its wings.
In this the darkest night,
The darkest night falling away,
The darkest night creeping away,
There came the rat who runs, here
 and there, and with him the hairless
 red dog.
And out of the slime, roots began
 to grow.
Leaves began to branch.
And a great calmness,
And a great stillness
 came about.
And in this time men and women
 began to be born.
Here on the ocean's edge,
Here in the damp forest,
Here in the cold mountains,
People spread over the land.
People were here,
And so it was:
DAY

Skills in Science

Reading Skills in Science

1. No. Ferns are autotrophs that produce energy through photosynthesis. Since this process is dependent upon sunlight, the plants could not have existed prior to daylight.

2. When attacked or threatened, weasels, like skunks, can discharge a foul-smelling liquid called *musk*.

Writing Skills in Science

1. Answers will vary. Before beginning their poems, encourage students to list some of the organism's traits. Students should then use their lists as a starting point for writing their verses.

2. Remind students that the basic life needs of early humans would have been similar to their own basic needs. However, the manner in which the early human met these needs would differ greatly. When writing their interviews, students should question how early humans fulfilled the basic needs for food, shelter, and clothing.

Activities

Evaluate Sources The poem suggests that animals and then plants appeared on the earth. Scientists hypothesize that the five kingdoms appeared in the following order: monerans, protists, fungi, plants, and animals.

Communicate Charts should show similarities and differences between the two animals picked, including some of the following characteristics:

The *weasel* has a coat that changes color during the year, lives in the southwestern United States, feeds mostly on rodents, and emits a musk when threatened. The different varieties of weasel range in size from 25 cm and 55 g to 45 cm and 255 g.

An *ermine* is a variety of weasel living in northeast and southwest North America. It is 18 to 34 cm long and weighs 55 to 285 g. It feeds mostly on rodents and rabbits, and like other species of weasels, its fur changes color during the year and it can discharge a foul-smelling musk when threatened. Ermines are also found in Europe, but there they are called *stoats*. Stoat and ermine are different names for the same animal.

The *ferret* belongs to the same family and genus as weasels, but can grow larger, up to 64 cm long. Ferrets, too, can discharge musk when threatened. Their fur is cream or yellow colored and does not change during the year. Ferrets are native to the western United States, especially the Great Plains, and feed primarily on prairie dogs.

"The Weasel"

The following poem appears in the book Consider the Lemming *by Jeanne Steig.*

The weasel is clever, the weasel is spunky,
The weasel when peeved smells decidedly funky.
He's long in the torso and short in the legs.
He's a merciless hunter—they say he sucks eggs.

In summer the weasel, like ferret and stoat,
Goes out in a nondescript frumpy brown coat.
But if, as may happen, his winter vacation
Is spent in a more or less frigid location,

An astounding transmogrification occurs:
The weasel puts on the most luscious of furs.
He turns into an ermine! How does he achieve it?
A fact is a fact, you can take it or leave it.

Skills in Science

Reading Skills in Science

1. **Evaluate Sources** The poem "In the Night, Still Dark" infers that plants such as mosses and ferns existed on the earth before there was light. Is this possible? Explain your answer.

2. **Infer** What does the author mean when she writes "the weasel when peeved smells decidedly funky"?

Writing Skills in Science

1. **Communicate** Write a poem that describes a particular organism in its environment.

2. **Predict** Suppose you were able to interview a human that was among those described in "In the Night, Still Dark." With a partner, write the interview complete with the person's responses. Share your interview with the class, having one student act as the interviewer and the other as the first human.

Activities

Evaluate Sources Determine the order in which the five kingdoms appeared on the earth as detailed in "In the Night, Still Dark." Use your text and other sources to assess the poem's scientific validity.

Communicate Research the similarities and differences between two of the following animals: ferret, stoat, weasel, ermine. Communicate your findings to the class in the form of a chart or diagram.

Where to Read More

In the Beginning; Creation Stories from Around the World by Virginia Hamilton. Harcourt Brace Jovanovich, 1988. Different creation myths explore how diverse cultures envision the origins of life.

Darwin and the Voyage of the Beagle by Felicia Law. Deutch, 1985. This fictional account of the voyage to the Galápagos Islands is based on Darwin's diary.

UNIT 4

UNIT OVERVIEW

In this unit, students learn about the characteristics of simple organisms. Chapter 10 presents the characteristics of viruses and monerans. Students explore the structure of viruses, the process by which they replicate, and the ways in which they affect humans. Also, the way in which monerans live, grow, and reproduce is presented. The discussion of monerans includes the three shapes of bacteria and their effects on humans. Chapter 11 discusses the kingdoms Protista and Fungi. Cell structures, reproduction systems, and habitats of algae and protozoa are described. The chapter presents the diseases caused by protozoa and discusses the distinct characteristics and usefulness of algae. The chapter also describes the characteristics of major groups of fungi.

Introducing the Unit

Directed Inquiry
Have the students examine the photograph. Ask:

▶ What resources do you think kelp need for survival? (Water, sunlight, minerals from the rocks and water)

▶ Why do you think the kelp do not grow higher on the rocks? (They need to be close to the water.)

▶ How do you think these organisms make use of available water? (They absorb it directly into their cells.)

▶ How might the organisms high on the rocks get the water they need to survive? (They have developed roots and stems to draw water up from the earth.)

Writing About the Photograph
Ask students to look at the photograph and write a short essay telling how this shoreline might change over millions of years.

Unit 4
Simple Organisms

Chapters
10 Viruses and Monerans
11 Protists and Fungi

Data Bank

Use the information on pages 624 to 631 to answer the following questions about topics explored in this unit.

Interpreting Data

What is the approximate size of most bacteria? Most viruses?

Comparing

Which is larger, a protein or a virus?

Reading a Diagram

A virus can be seen through what kind of microscope?

Classifying

Name all the living things listed on the diagram. Then name all the nonliving things listed. On which list did you place a virus?

The photograph to the left is of a kelp bed in New Zealand. What kind of water do you think the kelp is living in, salt water or fresh water?

193

Data Bank Answers

Have students search the Data Bank on pages 624 to 631 for the answers to the questions on this page.

Interpreting Data Bacteria range from 1–10 micrometers, while viruses are under 100 nanometers. Answers are found on the chart, Relative Sizes of Organisms, on page 625.

Extension Ask students to name a simple organism early scientists were able to observe without microscopes. (Euglena could be observed by early scientists because it is visible with the unaided eye.)

Comparing A virus is larger than a protein. The answer is found on page 625 on the chart, Relative Sizes of Organisms.

Reading a Diagram A virus can be seen through an electron microscope. The answer is found on page 625 on the chart, Relative Sizes of Organisms.

Classifying Students may list viruses on either list. The answers are found in the diagram, Relative Sizes of Organisms, on page 625.

Answer to In-Text Question
Since kelp is seaweed, it thrives in salt water.

193

CHAPTER 10

Overview

The characteristics of viruses and monerans are presented in this chapter. In the first section, students learn about the structure of viruses and the process by which they replicate. Also, some ways in which viruses affect humans are described. The characteristics of monerans and how they live, grow, and reproduce are covered in the second section. Finally, three shapes of bacteria are described, and the effects of bacteria on humans are discussed.

Advance Planner

▶ Supply 3 by 5 index cards for TE activity, page 202.

▶ Provide potato slices, sugar, salt, and jars with lids for TE activity, page 204.

▶ Prepare 5 grease pencils, 5 packets of instant chicken broth, 5 1-liter beakers, 5 graduated cylinders, 25 small jars, and 4 different antiseptic products (such as mouthwash, rubbing alcohol, hydrogen peroxide, and cleaning products) for SE Activity 10, page 206.

Skills Development Chart

Sections	Calculate	Classify	Communicate	Control Variables	Decision Making	Infer	Measure	Observe	Research
10.1 Skills WarmUp Skills WorkOut		●				●			
10.2 Skills WarmUp Skills WorkOut Consider This Skills WorkOut Activity	●		●	●	●	● ● ●		●	●

Individual Needs

▶ **Limited English Proficiency Students** Have students write all of the unfamiliar words from the chapter in their science journals, leaving several blank lines between each word. Students should write one sentence from the chapter text that includes each word. Then have students find one picture that represents each word. They should write a second sentence describing the pictures and using the new words.

▶ **At-Risk Students** Mask the labels on Figures 10.1, 10.4, and 10.5 and photocopy them for students. Have them refer to the chapter vocabulary list on page 207 and label the figures with the correct terms.

▶ **Gifted Students** Have students create filmstrips showing how viruses replicate inside other living cells. If possible, encourage students to create a computer simulation or make an animated videotape, complete with a sound track.

Resource Bank

▶ **Bulletin Board** Make a title for the bulletin board, such as *Viruses and Monerans: How Do They Affect People?* Create two sections for the bulletin board so students can display descriptions or pictures of some helpful and harmful effects of viruses and monerans.

▶ **Field Trip** Arrange to take students to a food distribution center so they can see the precautions workers take to preserve food. Have students plan for the trip by preparing questions to ask a representative from the plant.

CHAPTER 10 PLANNING GUIDE

Section	Core	Standard	Enriched	Section	Core	Standard	Enriched
10.1 Viruses pp. 195–199				Consider This, p. 204		•	•
				Skills WorkOut, p. 205	•	•	•
				Activity, p. 206	•	•	•
Section Features				**Blackline Masters**			
Skills WarmUp, p. 195	•	•	•	Review Worksheet 10.2	•	•	•
Career Corner, p. 197	•	•	•	Skills Worksheet 10.2	•	•	•
Skills WorkOut, p. 199	•	•	•	Integrating Worksheet 10.2a	•	•	•
Blackline Masters				Integrating Worksheet 10.2b		•	•
Review Worksheet 10.1	•	•		Enrich Worksheet 10.2	•	•	•
Skills Worksheet 10.1	•	•		Vocabulary Worksheet 10	•	•	•
Integrating Worksheet 10.1	•	•	•	**Ancillary Options** *Multicultural Perspectives,* p. 135	•	•	•
Overhead Blackline Transparencies Overhead Blackline Master 10.1 and Student Worksheet	•	•		**Laboratory Program** Investigation 18	•	•	•
				Investigation 19	•	•	•
				Investigation 20	•	•	•
Color Transparencies Transparency 26	•	•		**Color Transparencies** Transparencies 27, 28	•	•	
10.2 Monerans pp. 200–206							
Section Features							
Skills WarmUp, p. 200	•	•					
Skills WorkOut, p. 202	•	•					

Bibliography

The following resources can be used for teaching the chapter. See page T-40 for supplier codes.

Audio-Visual Sources
(video unless noted)
Bacteria. 23 min. 1985. NGSES.
Classifying Living Things. 18 min. 1987. A-W.
Classifying Microorganisms. 15 min. 1984. C/MTI.
Introduction to the Bacteria. Filmstrip with cassette. CVB.
Neither Plant nor Animal. 2 filmstrips with cassettes, 17–18 min each. 1987. NGSES.

Software Resources
Microbe. Apple. SYN.
The Simplest Living Things. Apple. DCHES.
Understanding AIDS. Apple, IBM. HRM.

Library Resources
Magasanik, B. "Research on Bacteria in the Mainstream of Biology." Science 240 (1988): 1435–1439.
Margulis, Lynn, and Karlene V. Schwartz. Five Kingdoms: An Illustrated Guide to the Phyla of Life on Earth. San Francisco: W.H. Freeman and Co., 1982.
Teasdale, Jim. Microbes. Morristown, NJ: Silver Burdett, 1985.

CHAPTER 10

INTEGRATED LEARNING

Writing Connection
Have students write a short story about being as small as the organisms on the head of a pin. Have them write about how the world would look to them. Have students describe the dangers and excitements they would have in one morning.

Themes in Science
Systems and Interactions All the living and nonliving things in an environment are interconnected. Though they are so simple that they do not have cell structure, viruses affect the most complex living things.

Introducing the Chapter

Have students read the description of the photograph on page 194. Ask if they agree or disagree with the description. You may wish to point out that the picture shows bacteria on the head of a pin.

Directed Inquiry

Have students study the photograph. Ask:

▶ Describe what you see in the picture. Do you think this photograph shows the actual size of the objects? Why or why not? (No. This is a magnification.)

▶ What kind of instrument besides a camera was used in making this picture? (Students will probably answer a microscope and might specify an electron microscope.)

▶ Are the organisms plants, animals, or something else? (They are bacteria.)

▶ Suggest some good and bad effects that such organisms might have in our world. (They cause disease. They also aid in digestion, food making, and decomposition.)

Chapter Vocabulary

aerobe	endospore
anaerobe	flagella
bacilli	host
binary fission	moneran
cocci	spirilla
decomposer	virus

194

Chapter 10 Viruses and Monerans

Chapter Sections
10.1 Viruses
10.2 Monerans

What do you see?

"I see in this picture what appear to be many microscopic organisms. The organisms look like small bacteria or viruses of some sort. These organisms are small enough to fit one thousand of them on the point of a needle."

Jay Rhine
Intermediate School West
Toms River, New Jersey

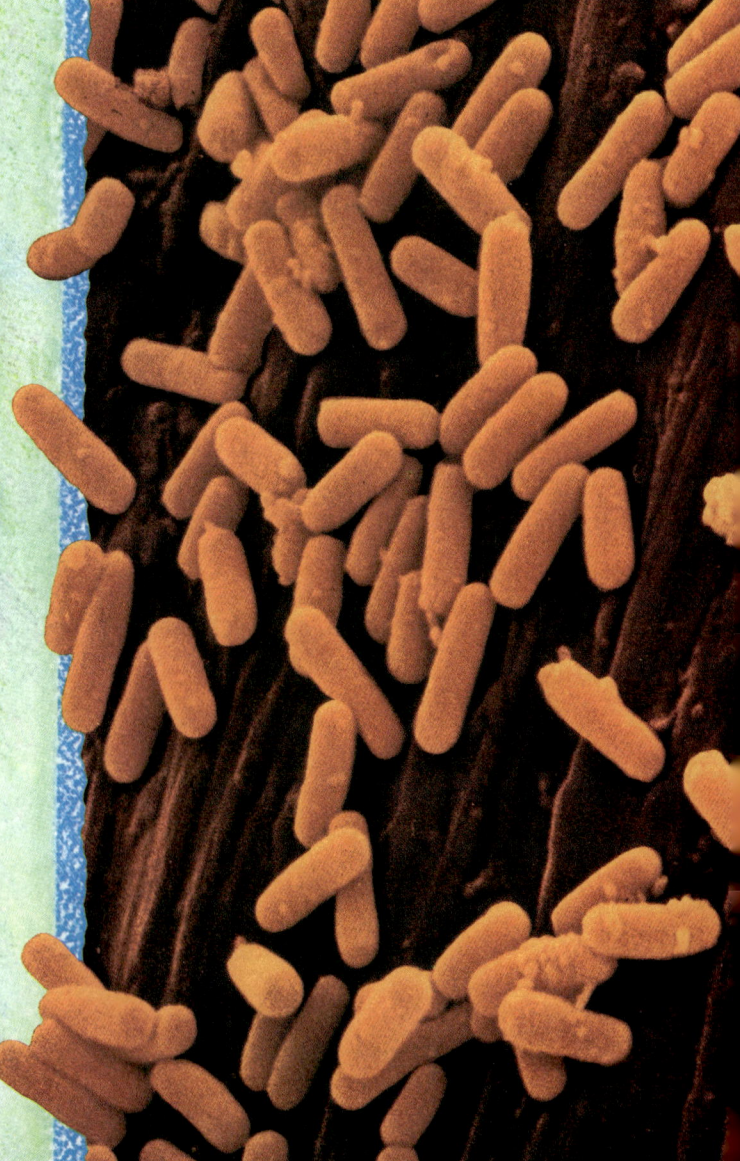

To find out about the photograph, look on page 208. As you read this chapter, you will learn about viruses, bacteria, and other monerans.

194

STS Connection

The research of French physicist Louis-Victor de Broglie (1892–1987) led to the development of the electron microscope in the 1930s. The electron microscope allowed scientists to discover viruses and treat diseases in a new way. Because of advancements in technology, scientists have invented more powerful microscopes that allow them to view things even smaller than viruses. Ask the following questions:

- What was the world like before microscopes?
- How have microscopes changed society?
- What new achievements might become possible because of microscopes?

SECTION 10.1

Section Objectives
For a list of section objectives, see the Student Edition page.

Skills Objectives
Students should be able to:

Classify common diseases into groups.

Infer how a cold virus is transmitted.

Vocabulary
virus, host

10.1 Viruses

Objectives

- **Describe** the characteristics of viruses.
- **Explain** why viruses do not fit easily into any classification system.
- **Explain** how viruses reproduce.
- **Calculate** a mathematical problem that shows an understanding of the relative size of a virus.

Skills WarmUp

Classify Make a list of all the diseases you know about. Then classify them into groups. Write on your list why you classified the diseases the way you did.

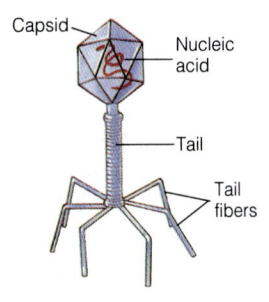

You're sneezing and coughing. Your nose is runny. You may even have a fever. You feel just plain awful. What's the problem? You have the common cold. What's the cause? Your body cells have been invaded by particles so small that they can't even be seen under a light microscope. The particles are **viruses**. A virus is a piece of hereditary material with a coat of protein.

Characteristics of Viruses

Viruses are not cells. They do not have any cell structures. There is no cytoplasm, cell membrane, or nucleus in a virus. Viruses can't move, feed, or grow outside the cell of another organism. They can't carry out any life processes. They appear to be nonliving when outside a living thing. However, they can replicate, or reproduce themselves, inside a living cell. The ability to reproduce, and the genetic code used in the process, are the only similarities that viruses share with cells.

Look at Figure 10.1. It shows three different kinds of viruses. Notice how their shapes vary. Viruses have many other kinds of shapes, too. Each virus is only a strand of RNA or DNA with a protein coat called a capsid. The capsid makes up most of the virus and gives the virus its shape.

Viruses are so small that they were not seen until the 1930s. At this time, the invention of the electron microscope enabled scientists to see viruses. To give you an idea of their size, about 100,000 average-sized viruses could be lined up end-to-end on the period at the end of this sentence.

Bacteria virus (Capsid, Nucleic acid, Tail, Tail fibers)

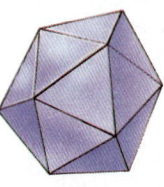

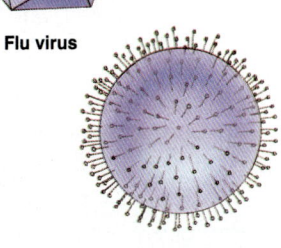

Flu virus

Polio virus

Figure 10.1 ▲
Although viruses come in many shapes, all viruses have capsids with RNA or DNA.

MOTIVATE

Skills WarmUp
To help students understand the effects of viruses on other living things, have them do the Skills WarmUp.
Answer Students' lists will probably include similar diseases, but their groupings will vary. Have students place the lists in their portfolios.

Prior Knowledge
To gauge how much students already know about viruses, ask the following questions:

- What kinds of viruses are there?
- About how big is a virus?
- What living things are affected by viruses?
- Where do you think viruses come from?

Integrated Learning
Use Integrating Worksheet 10.1.

Chapter 10 Viruses and Monerans 195

TEACH

Explore Visually

Have students study the drawings of viral replication on this page. Then ask some or all of the following questions:

▶ Why do you think the capsid can't get into the cell? (It is too large.)

▶ How would you describe the part of the bacteriophage that is left outside the cell? (It is just an empty capsid.)

▶ What is the function of the tail when the bacteriophage injects its nucleic acid into the host cell? (The tail penetrates the outer wall of the host cell so that the nucleic acid can be injected.)

▶ What happens inside the host cell after the virus invades? (The virus simply takes over the cell. It causes the cell to make new viruses.)

▶ What do you think happens to the host cell after the cell bursts open? Why? (The host cell cannot survive after it bursts; students may infer this from their knowledge of the cell membrane's importance to cell functions.)

The Living Textbook: Life Science Side 1

Chapter 17   Frame 30483
Virus Structure
Search: Play:

TEACHING OPTIONS

Cooperative Learning

Have students hold a debate to decide whether viruses should be classified as living or nonliving. Assign students to opposing positions. Have them write notes to make a case for their assigned positions. Then have students in each group pool their notes to create a three-part speech outlining the group position. Appoint three spokespersons for each group, and hold formal debates for another class. Ask the visiting class to vote on the winning position.

Classifying Viruses

Because viruses do not fit into the ideas of the cell theory, scientists have a hard time classifying them. They do not fit into any of the five kingdoms. Scientists can't even agree on whether viruses are living or nonliving.

Viruses are classified according to the **host** that the virus invades. A host is the living thing another organism lives in. Some viruses use bacteria as hosts. Other viruses infect plants, and still others live inside animals.

Viral Replication

You know that organisms reproduce by passing on their genetic material. Viruses do the same thing, except they do not have cell parts to carry out reproduction. Instead, viruses use the reproductive machinery of the cells they invade. The virus uses the cell like a factory, causing the cell to make new viruses. Figure 10.2 explains the reproduction of bacterial viruses, or bacteriophages (bak TEER ee uh FAYJ uhs).

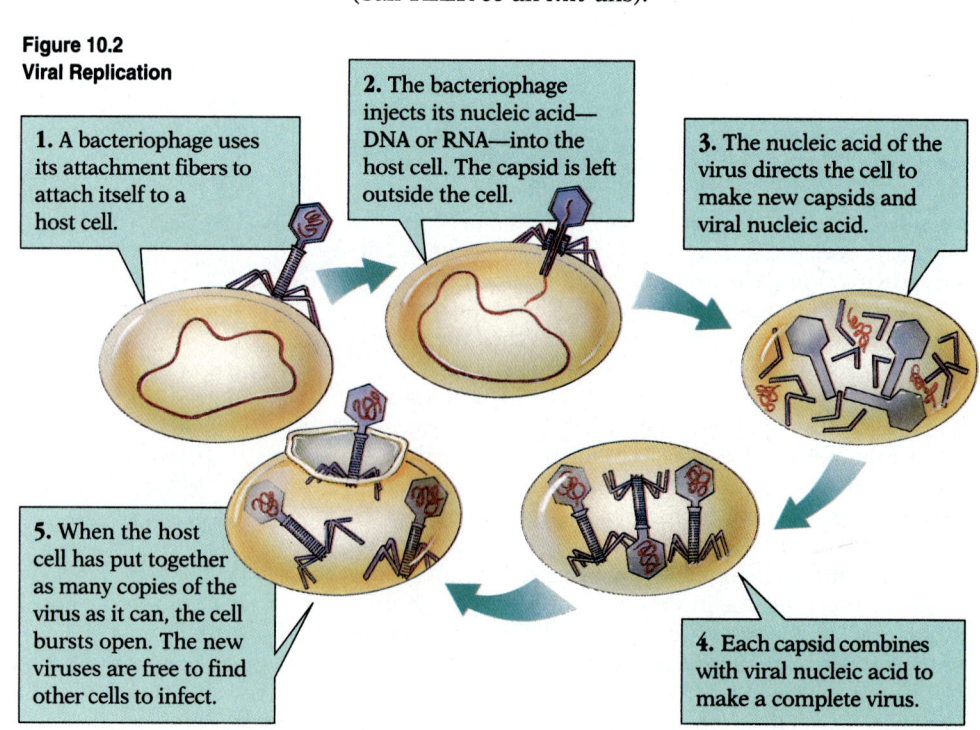

**Figure 10.2
Viral Replication**

1. A bacteriophage uses its attachment fibers to attach itself to a host cell.

2. The bacteriophage injects its nucleic acid—DNA or RNA—into the host cell. The capsid is left outside the cell.

3. The nucleic acid of the virus directs the cell to make new capsids and viral nucleic acid.

4. Each capsid combines with viral nucleic acid to make a complete virus.

5. When the host cell has put together as many copies of the virus as it can, the cell bursts open. The new viruses are free to find other cells to infect.

Chapter 10 Viruses and Monerans

INTEGRATED LEARNING

STS Connection

Research To combat viruses, the body manufactures protein substances called interferon. Scientists now are experimenting with artificial interferon to combat viruses and other diseases. Direct students to research and report on antiviral drugs such as interferon, AZT, ara-A, and acyclovir.

Themes in Science

Scale and Structure Point out to students that viruses are only 20 to 300 nanometers in width. (A nanometer is one-billionth of a meter.) Explain that viruses are often geometrically shaped. Usually viruses enter cells by attaching themselves with attachment fibers to the cell wall or membrane. Sometimes, the whole virus enters the cell. Other times, the capsid remains outside the cell membrane, and the virus DNA is injected into the cell.

Discuss

Explain that there are many viral diseases not listed in Table 10.1. Ask students if they have had chicken pox, measles, or mumps. Students may have had these viral diseases or have been vaccinated for them. Ask the question, Do you think there may be any viral diseases in people that have not yet been discovered? Explain. (Answers may vary, but it is likely that there are other diseases that will be discovered.)

Viruses and Disease

When a virus invades an organism's cells, it usually means disease or illness for the host organism. Some viral infections, such as those that cause cold sores in humans, are relatively harmless. Others, such as AIDS, are deadly.

Most viruses can only infect a specific kind of organism. The plant virus that causes disease in tobacco plants, for example, can't harm a human, or even an orange tree. The AIDS virus, called HIV (human immunodeficiency virus), only infects humans.

Humans, unfortunately, are hosts to many different kinds of viruses. Table 10.1 lists some human diseases that are caused by viruses. Most human viruses can invade only certain kinds of cells or tissues. For example, the virus that causes hepatitis infects only liver cells. The tissues affected by other human viruses are shown in the table.

Table 10.1 Viral Diseases in Humans

Disease	Tissue Infected
Smallpox	Skin
Hepatitis	Liver
Polio	Nerves
Rabies	Nerves
Common colds	Respiratory tract
Influenza	Respiratory tract
AIDS	Immune system

Career Corner

Reason and Conclude Lead a discussion about how modern transportation and people's ability to quickly move from place to place may be related to the spread of viral disease. Ask students to name other factors that may contribute to a disease spreading through populations in a large city.

Research Then have students find out more about the Centers for Disease Control. Tell them to look up what factors virologists study when they are trying to control the spread of a disease. Possible additional topics of research include the outbreak of Legionnaires' disease in Pennsylvania in 1976 and the work of epidemiologists.

Career Corner Virologist

Who Finds Out Which Viruses Cause Disease?

The word was in from San Diego. Schools were closed, and many people were forced to stay home from work. As everyone at the Centers for Disease Control in Atlanta, Georgia, had been expecting, a new virus was spreading. It had reached San Diego and was causing a flu epidemic. The virus had been identified in Asia and probably arrived on a naval ship.

Virologists have a major role in trying to keep epidemics like this one under control. In this case, virologists helped prepare a vaccine, which was sent to health centers around the country. Flu shots were given to those most likely to become very ill from the disease.

Virologists often grow viruses in human tissue cultures. They observe what the virus does to find out if it causes disease, and if so, how. Many virologists use an electron microscope to help them see the viruses they study.

Virologists sometimes provide information to the general public about how a viral disease is transmitted. They are also at the center of the search for a cure for AIDS.

Virologists also study the viruses that cause diseases in agricultural animals and plants. They help prevent epidemics among sheep and cattle as well as in many crops.

If you would like to become a virologist, you will need a college degree in biology with courses in microbiology.

Chapter 10 Viruses and Monerans

INTEGRATED LEARNING

TEACH • Continued

Directed Inquiry

Have students read the paragraphs and look at the micrographs on this page. Then ask students some or all of the following questions:

▶ In what ways are colds and herpes similar? (Both are caused by many different types of viruses.)

▶ What form of good hygiene is one way to help prevent colds? (Washing your hands before handling any food that is to be eaten)

▶ How is hepatitis commonly transmitted? (Through untreated drinking water)

▶ Which of the diseases discussed on this page can cause death? (AIDS and hepatitis)

Research

Have students use references to find out about rickettsiae. Have them find answers to these questions:

▶ How are rickettsiae like bacteria? Like viruses? How are rickettsiae different from either bacteria or viruses?

▶ How are humans affected by the diseases caused by rickettsiae?

STS Connection

Remind students that viruses can be seen only by using electron microscopes. Explain that photographs called micrographs can be made of microscopic organisms. Point out that the micrographs in Figure 10.3 were colorized and do not depict the actual colors of the viruses.

Common Cold More than 100 different viruses cause the common cold. Your body can build up cell defenses, called antibodies, against a particular cold virus. But the antibodies you build up after being exposed to one cold virus are useless against a different cold virus.

Cold viruses travel on dust and water droplets in the air. When a person with a cold sneezes or coughs, cold viruses are sprayed into the air. If you breathe in the viruses or pick them up on your hands from your surroundings, you can catch a cold.

Hepatitis The hepatitis virus causes a disease of the liver. Hepatitis makes people feel very tired. Hepatitis is spread to people by contaminated food or drinking water. When food is prepared by a person with hepatitis, the hepatitis virus can be transmitted, or passed, to others who eat the food.

Herpes Several kinds of herpes viruses cause human diseases. A fever blister or cold sore on your lips is caused by herpes virus 1. The blisters heal within a week, but the new viruses formed during the infection remain inactive within the body for a long time. Later, a fever, emotional stress, or overexposure to the sun can trigger another outbreak. Herpes virus 2 causes sores in the area of the sex organs. This herpes virus, too, can remain inactive and then become active again. Herpes virus 3 causes chicken pox and shingles.

AIDS HIV causes AIDS—acquired immune deficiency syndrome. As HIV attacks the cells of the body's immune system, the body loses its ability to defend itself against all diseases. The virus enters the body in blood or other body fluids. Scientists have not been able to prepare a vaccine against HIV because it changes constantly.

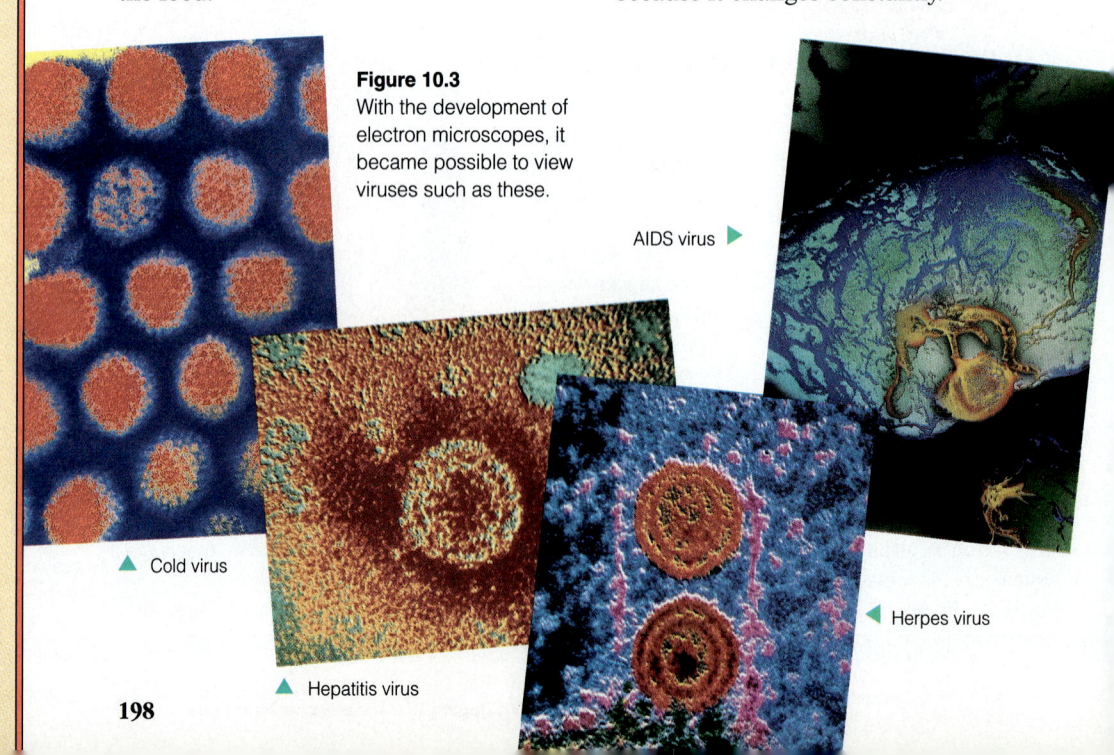

Figure 10.3
With the development of electron microscopes, it became possible to view viruses such as these.

▶ AIDS virus

▲ Cold virus

▲ Hepatitis virus

◀ Herpes virus

TEACHING OPTIONS

Portfolio

Scientists have developed vaccines against many viruses. Some vaccines are made from live but altered viruses, and others are made from killed viruses. Today, polio can be prevented by both types of vaccines. Have students research the type of vaccine Jonas Salk developed, the animals used in Salk's studies, and the date the vaccine was first used in the United States. Encourage each student to interview an older adult about his or her memory of polio or the polio vaccine. Have students either use a tape recorder or take notes during the interviews. They can then summarize the interviews in one or two pages and add them to their portfolios.

Science and Technology
Curing the Common Cold

Medical researchers have been searching for the cure for the common cold for a long time. They are experimenting with a drug called WIN 51,711. Because the researchers know about the structure of a virus and the way viruses reproduce themselves, they think WIN 51,711 may be able to be used to cure the common cold. WIN 51,711 keeps cold viruses from shedding their protein coats. If the capsid of a virus can't be left outside the host cell, the medical researchers hypothesize that the virus will not be able to inject its nucleic acid into the cell.

The researchers are not exactly sure how or why WIN 51,711 works. They know that the drug attaches itself to certain points on the virus's capsid. They hypothesize that the drug keeps the capsid from falling away from the virus. But their studies and experiments are not complete.

WIN 51,711 has not been tested on humans yet. It has only been tested on animals and in test-tube cell cultures. Will WIN 51,711 work in humans? Researchers will not know until it is tested on humans. Often, a drug works on animals or in test-tube cultures but does not work inside the human body. WIN 51,711 could also have unforeseen side effects. But who knows? Someday WIN 51,711 may be available by prescription from your doctor, and the common cold will not affect people anymore.

Skills WorkOut

Infer Think about the time of year that you most often catch colds. Is it during the winter months or summer months? Most people catch colds more often during the winter than they do in the summer. State at least two inferences that could explain this.

Check and Explain

1. List the characteristics of a virus.
2. In what ways does a virus differ from an organism? In what ways are they alike?
3. **Find Causes** Why does the reproduction of a virus in the cells of an organism cause disease or illness for the organism?
4. **Calculate** If a very large virus were enlarged 10,000 times, it would be about the size of a grain of salt. How tall would you be if you were enlarged 10,000 times?

Skills WorkOut

To help students apply what they know about how cold viruses are spread, have them do the Skills WorkOut.

Answer Students should infer that people are more likely to catch colds in winter because they are often in closed environments. Therefore, they have greater exposure to viruses.

EVALUATE

WrapUp

Portfolio Have students revise the lists they made for the Skills WarmUp at the beginning of this section. For each group, have students write the characteristics of the viruses that make that group unique.

Use Review Worksheet 10.1.

Check and Explain

1. Viruses consist of a strand of genetic material within a distinctively shaped protein coat. They have no other cell structures and are too small to be seen with light microscopes.
2. Different: lack most cell structures; depend on other organisms to grow, feed, and move. Alike: They contain protein and genetic material, and can replicate.
3. Virus reproduction usually causes disease in the host organism because host cells are used to carry out viral reproduction instead of their normal functions. Each host cell gives rise to many more viruses.
4. Probably between 12 500–15 500 m; students' calculations will vary.

Chapter 10 Viruses and Monerans

SECTION 10.2

Section Objectives
For a list of section objectives, see the Student Edition page.

Skills Objectives
Students should be able to:

Infer the relationship between bacteria and disease.

Calculate how quickly bacteria reproduce.

Research bacterial action on foods.

Infer how bacterial growth is detected in chicken broth.

Vocabulary
moneran, flagella, cocci, bacilli, spirilla, aerobe, anaerobe, binary fission, endospore, decomposer

---

MOTIVATE

Skills WarmUp
To help students understand the relationship between bacteria and disease, have them do the Skills WarmUp.
Answer Answers may vary. These precautions help discourage reproduction of microorganisms.

Prior Knowledge
To gauge how much students know about monerans, ask the following questions:

▶ Where are bacteria found?

▶ Can bacteria be helpful? How?

▶ What living things are affected by bacteria?

The Living Textbook:
Life Science Side 1-4
Chapter 5   Frame 01840
Kingdom Monera Lab Slides (3 Frames)
Search: Step:

200

INTEGRATED LEARNING

Themes in Science
Scale and Structure Monerans are single-celled organisms. Remind students that the cell is the smallest unit in living things. The cell can perform all of the activities necessary to maintain life. Monerans have cell membranes and cell walls. The DNA and ribosomes in a moneran float freely in the cytoplasm. Monerans have no internal structures.

Art Connection
Like other cells, monerans have a nuclear area that contains DNA. Yet they are unlike ordinary cells in many ways. Have students compare and contrast monerans and other cells in an illustrated chart.

Skills WarmUp

Infer Why are you told to wash your hands before you eat? Why are fresh foods kept in a refrigerator? Why don't you eat food that has fallen on the ground? Explain what microscopic organisms might have to do with these rules that people often follow.

Figure 10.4 ▲
This moneran has a cell membrane and a cell wall but no nucleus or other internal structures.

Labels: Capsule, Ribosome, Cell wall, DNA, Flagella, Cytoplasm, Cell membrane

10.2 Monerans

Objectives

▶ **Describe** the characteristics of monerans.

▶ **Classify** bacteria according to shape.

▶ **Distinguish** between harmful and helpful bacteria.

▶ **Make a graph** showing how a moneran population can increase over time.

Take a survey of the living things in your classroom. How many did you count? Compare your count to the counts of three classmates. Are your numbers in the millions? If not, they should be! All around you, on your skin, in your body, and in the food you eat, are millions of organisms so small you can't see them without a microscope. These microscopic organisms are classified as **monerans**. Monerans are single-celled microscopic organisms. Unlike other cells, they do not have a nucleus surrounded by a membrane. Instead they have a nuclear area that contains a circular molecule of DNA.

Monerans live nearly everywhere—in the air, in soil, in water, in certain foods, inside other organisms. More monerans are living at this moment than any other group of organisms. With so many of them in so many different places, monerans affect your life in a variety of ways.

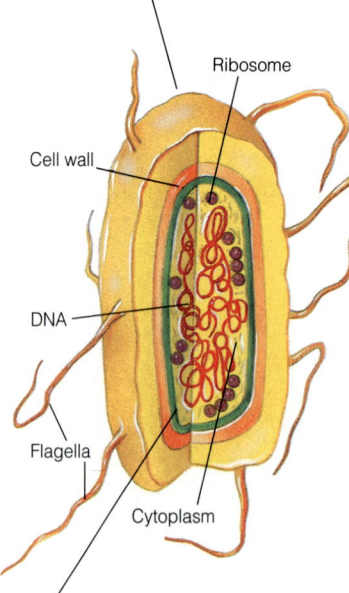

Characteristics of Monerans

Look at the moneran in Figure 10.4. Notice that the cell is very simple, with few structures. Find the nuclear area with its molecule of DNA. Look at the outside borders of the moneran. It, like all monerans, has both a cell membrane and a cell wall. This moneran also has many whiplike structures called **flagella** (fluh JEHL uh). Monerans that have flagella can move in liquids.

Although monerans are single cells, some monerans live together in groups of two or four. Some form long chains of several hundred or more. Still others form grapelike bunches.

200   Chapter 10 Viruses and Monerans

TEACHING OPTIONS

Portfolio
Have students design costumes for an educational skit on monerans. Have them design costumes for flagella, the cell wall, the capsule, ribosomes, DNA, cytoplasm, and the cell membrane. They should draw pictures and write descriptions of their designs. Encourage students to consider whether the actors will be 1) bacteria or blue-green bacteria, and 2) cocci, bacilli, or spirilla.

Cooperative Learning
Have students work in cooperative groups. Ask each group to think of a category of things containing objects that resemble the three moneran shapes. For example, in the category of sports equipment, baseballs resemble cocci, pole vaults resemble bacilli, and spirilla resemble boomerangs. Have students bring in pictures of their objects to show to the class.

Kinds of Monerans

Scientists classify monerans into two main groups: bacteria and blue-green bacteria. Blue-green bacteria are different from bacteria because they have the green pigment chlorophyll just like plant cells. Blue-green bacteria also have a blue pigment. The two pigments give blue-green bacteria their color.

Because blue-green bacteria have chlorophyll, they can get energy through photosynthesis. Unlike plant cells, however, the chlorophyll in blue-green bacteria is not in chloroplasts. In fact, scientists hypothesize that the first chloroplasts were really blue-green bacteria that began to live inside another bacterium.

Bacteria are more common than blue-green bacteria. Most bacteria get energy from food in their surroundings. Some bacteria, however, can get energy through photosynthesis. But these bacteria either don't have chlorophyll or use a simpler kind of chlorophyll than do blue-green bacteria.

One way scientists classify bacteria is by their shape. There are three common shapes. Round or egg-shaped bacteria are called **cocci** (KAHK sy). Rod-shaped bacteria are called **bacilli** (buh SIHL EYE). Spiral-shaped bacteria are called **spirilla** (spy RIHL uh). Read about these three kinds of bacteria below.

Bacteria are also classified by whether or not they need oxygen to live, grow, and reproduce. Bacteria that require oxygen are called **aerobes** (AIR ohbs). Bacteria classified as **anaerobes** (AN ur OHBS) do not need oxygen. Anaerobic bacteria get energy from food through the process of fermentation.

Cocci
Some cocci form pairs called diplococci. *Di-* means two. Others form long chains. These are called streptococci. Others form clusters called staphylococci. ▼

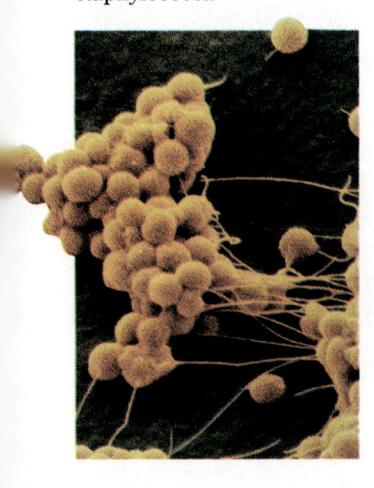

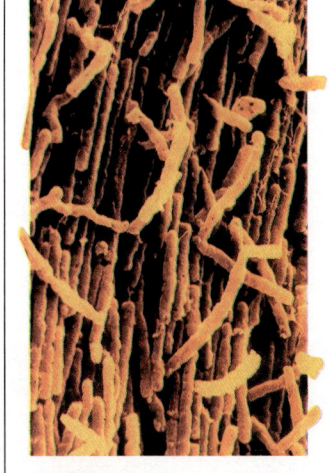

Bacilli ▲
The rod-shaped bacilli can grow in pairs or chains. They do not form clusters. Pairs of bacilli are called diplobacilli.

Spirilla
Spirilla only live as single cells. They are shaped like spirals or corkscrews. ▼

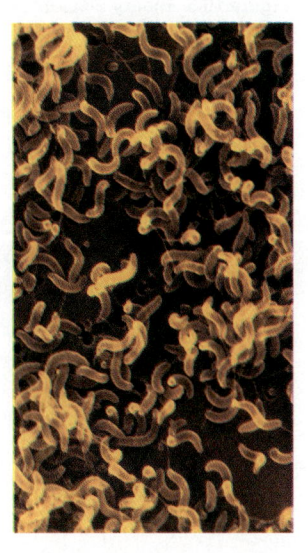

Chapter 10 Viruses and Monerans 201

TEACH

Directed Inquiry
Have students study the micrographs and text on this page. Ask some or all of the following questions:

▶ What two shapes of bacteria can be found in pairs or in chains? (Cocci and bacilli)

▶ Which two shapes of bacteria do not form clusters? (Bacilli and spirilla)

▶ Which kind of bacteria are shown in the photograph of cocci—streptococcus or staphylococcus? How do you know? (Staphylococcus; it is grouped in clusters.)

Critical Thinking
Compare and Contrast What is the difference between aerobes and anaerobes? (Aerobes require oxygen and get energy from cellular respiration. Anaerobes do not need oxygen but get their energy through fermentation.)

Ancillary Options
If you are using the blackline masters from *Multiculturalism in Mathematics, Science, and Technology*, have students read about Hideyo Noguchi on page 135 and complete pages 136 to 138.

Integrated Learning
Use Integrating Worksheet 10.2a.

Enrich
Use Enrich Worksheet 10.2.

The Living Textbook: Life Science Side 2

Chapter 13 Frame 18741
Types of Bacteria (Movie)
Search: Play:

TEACH • Continued

Skills WorkOut
To help students calculate how quickly bacteria reproduce, have them do the Skills WorkOut.
Answer There would be 6,400 bacteria after two hours.

Skills Development
Sequence Have students write the steps of virus replication on 3 by 5 index cards; each card will have one step. Collect the cards from each student. Shuffle the cards and give each student a different set of cards. Have the students correctly sequence the cards.

Directed Inquiry
Have students study Figure 10.5 on this page. Ask the following questions:

▶ What happens between the first step in binary fission and the fourth step? (The DNA is replicated, and the DNA molecules pulled apart before the cell divides.)

▶ Look at Figure 5.8 on pages 104 and 105. How does binary fission remind you of the process of mitosis in more complex organisms? (In both processes, the cell grows, duplicates its DNA, and divides to form two identical cells.)

Enrich
Use Enrich Worksheet 10.2.

The Living Textbook: Life Science Side 1
Chapter 17   Frame 31819
Bacterial Cells and Membranes (Movie)
Search: Play:

INTEGRATED LEARNING

Themes in Science
Scale and Structure The structure of a bacterium's endospore protects it so that it can live during harsh conditions. Have students make a chart comparing and contrasting the endospore or a bacterium with the capsid of a virus. Have students discuss the role the structures play in replication and reproduction.

Integrating the Sciences
Physical Science Ask students to look up the definitions of *fission*. Explain that in physical science *fission* refers to the splitting of the atomic nucleus, but that in life science the term refers to cell division in the cells of organisms such as bacteria.

Skills WorkOut
Calculate A colony of 100 bacteria begin to reproduce every 20 minutes. How many bacteria will there be after 2 hours?

Bacterial Reproduction

Bacteria reproduce asexually by a process called **binary fission**. During binary fission, one bacterial cell divides into two identical bacterial cells.

Conditions have to be right for a bacterial cell to divide. For most bacteria, this means being in a warm, moist place with a good food source. Under ideal conditions, bacterial cells can reproduce rapidly. Some bacteria can reproduce every 20 minutes or less.

Before binary fission begins, a bacterial cell usually has to grow to twice its normal size. Look at Figure 10.5. As the bacterial cell grows, the bacterium makes enough cell material, including new DNA, for each new cell. Then the cell membrane and cell wall begin to pinch inward until a cross-wall forms. Two new cells are formed, each with a set of identical hereditary material. As this process continues, large numbers of bacteria can be produced. A group of bacteria that started with one bacterium is called a colony.

Some bacteria can live through very unfavorable conditions. If the temperature is too hot or the environment is too dry, they form a thick wall around themselves. A bacterium with a thick protective wall is called an **endospore**. Some endospores can withstand boiling and freezing. When environmental conditions are favorable again, the endospore breaks open, and the bacterium becomes active again.

Figure 10.5 Binary Fission ▼

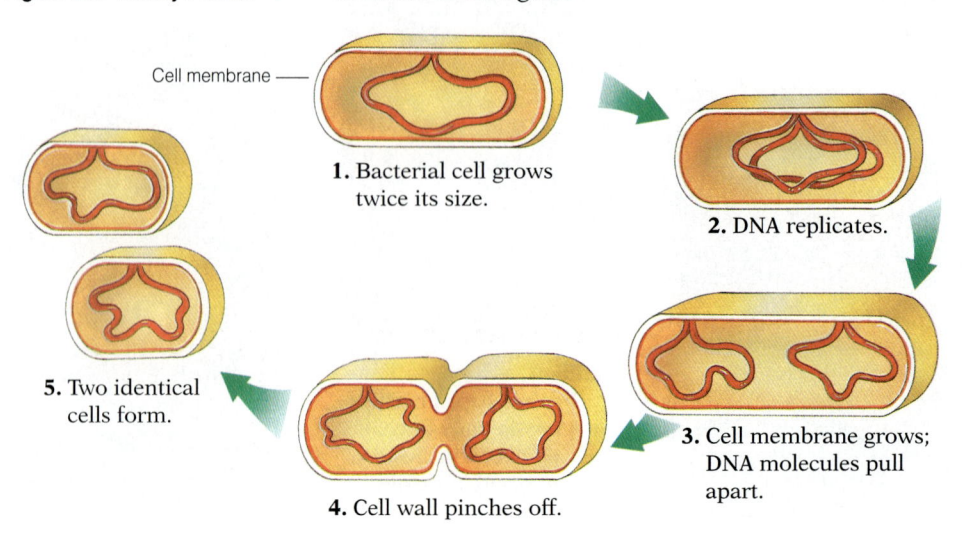

1. Bacterial cell grows twice its size.
2. DNA replicates.
3. Cell membrane grows; DNA molecules pull apart.
4. Cell wall pinches off.
5. Two identical cells form.

Chapter 10 Viruses and Monerans

Integrating the Sciences

Environmental Science Taking a cue from the decomposers in soil, technologists have attempted to create microbes that can decompose petroleum and other pollutants. Encourage students to research and report on oil-eating microbes. Students may want to keep their reports for their portfolios.

Earth Science Use Integrating Worksheet 10.2b.

Critical Thinking

Reason and Conclude Tell students that vegetarians are people who get their protein from plants like beans, peas, peanuts, and other legumes. Nitrogen-fixing bacteria change nitrogen gas into a form of nitrogen that legumes can use. Ask students how it would affect vegetarians if nitrogen-fixing bacteria were not present in the soil. (If legumes could not use nitrogen to make proteins, then vegetarians would have to get their protein from other foods.)

Discuss

Bacteria are decomposers that make a great deal of humus. Discuss the importance of humus with students. Ask students these questions:

▶ Why is humus important to the environment? (It holds the nutrients plants need to grow.)

▶ What would happen if there were no decomposers? (Minerals would not be recycled into the ecosystem, and the soil would not be fertile.)

▶ Can you describe a place that does not have much humus? (Deserts and polar regions do not have much humus and therefore little vegetation.)

Enrich

Escherichia coli (or *E. coli*) are a type of bacteria. If found in water, *E. coli* are evidence that the water was probably polluted by sewage. Ask students to do research on *E. coli*. Have them find out where these bacteria live. (The bacteria live in the intestines of animals and humans, where they aid digestion.)

The Living Textbook: Life Science Side 2

Chapter 13   Frame 19145
Antibiotics (Movie)
Search: Play:

Helpful Bacteria

Even though harmful bacteria often get all the attention, helpful bacteria far outnumber harmful bacteria. Bacteria play an important role in the lives of all living things, including you.

Bacteria and Plants Plants need nitrogen to grow. Even though the atmosphere is about 70 percent nitrogen, plants can't use it directly from the air. Some bacteria, called nitrogen-fixing bacteria, can change nitrogen gas to a form that plants can use. These bacteria grow on the roots of beans, peas, peanuts, and other plants. Others live in the soil. Animals, in turn, get usable nitrogen when they eat plants.

Bacteria and Animals Some bacteria live in the intestines and stomachs of animals. They help the animals digest food. Plant-eating animals, such as cows, sheep, and camels, feed on grasses. Bacteria help these animals digest the plant material, releasing nutrients in plants that the animals need to live.

Figure 10.6 ▲
Bacteria in the stomachs of cows digest plant matter and release nutrients.

Figure 10.7 ▲
Factory-produced cheeses are made in large molds, then cut up and packaged.

Bacteria and Food Bacteria are used widely by humans in food making, especially in producing dairy products. Milk is changed to cheese, yogurt, and sour cream through the action of bacteria. Bacteria are also involved in making soy sauce, tea, and cocoa. Many of the flavors in your food are produced by bacteria. Cheddar, Swiss, Parmesan, and other cheeses would all taste about the same without bacteria. The sour taste of vinegar and sauerkraut is also caused by bacteria.

Bacteria and Soil Soil is made up of very small rock particles and decaying plant and animal material called humus (HYOO muhs). Humus holds the large amounts of nutrients that plants need to grow. Bacteria are among the most important makers of humus. They are **decomposers**, organisms that feed on dead plants and animals for energy. As decomposers feed, they break down dead organisms into simpler substances. These substances are released into the environment. Decomposers recycle minerals into the ecosystem and help to keep soil fertile.

TEACH • Continued

Consider This
Research Sterilizing foods by irradiation can be done with one-fiftieth of the energy required for thermal sterilization. However, irradiation reduces some of the nutrients in food. Have students research food irradiation. Possible sources of information include scientific articles about food irradiation, pamphlets from the Food and Drug Administration (FDA), and food producers. Have students share information in brief oral reports.

Skills Development
Find Causes Ask students: If the human body is an ideal place for bacterial growth, how can the bacteria that cause acne and tooth decay be controlled? (Keeping skin and teeth clean will help control these two types of bacteria.)

Class Activity
Have students conduct an experiment in which they store milk in two small, tightly sealed containers. Have them keep one container in a refrigerator for three days and the other at room temperature. Ask these questions:

▶ What differences do you observe between the two containers of milk? (Milk that was kept at room temperature looks lumpy and smells sour.)

▶ What caused the milk to spoil? (Bacteria in the milk reproduced.)

Answer to In-Text Question
① **The plant is spotted and has small brown patches.**

INTEGRATED LEARNING

Integrating the Sciences
Agriculture As shown in Figure 10.8, bacteria can harm plants as well as animals. Encourage students to research and report on common bacterial pests that affect crops. Also ask students to discuss some of the measures farmers take to avoid bacterial blight.

Consider This

Should Food Be Irradiated?

The word *irradiation* sounds dangerous to many people. To irradiate means to expose to radiation. When people are exposed to radiation, it can cause cancer and birth defects. Irradiating food, however, is a very effective way of preserving it.

Consider Some Issues
Spoiled food is an expensive and dangerous problem. Those who support food irradiation say that it is a perfectly safe way to prevent food spoilage and food poisoning.

When food is irradiated, the food itself doesn't become radioactive. The radiation kills molds, bacteria, worms, and other organisms that can cause spoilage or disease.

Japan, Mexico, and Canada are currently using the food irradiation process. The United States also approved food irradiation in the 1980s. Many people in the United States, however, are worried that food irradiation may be harmful. Irradiation can create new substances in food that give it a bad taste or smell. The substances seem to be harmless, but their effects are not fully known. Another problem is that food irradiation creates dangerous radioactive waste products.

Think About It Most people feel safe eating canned, frozen, and microwaved food. Is irradiated food something people should also accept? How can scientists make sure that food irradiation is safe?

Write About It Write a paper stating your position for or against food irradiation. Include all your reasons.

Harmful Bacteria

Bacteria cause many kinds of diseases in plants and animals. In humans, tuberculosis and pneumonia are the result of bacterial infection. Bacteria also cause common problems, such as acne and tooth decay.

Once inside a living thing, bacteria can reproduce rapidly. The inside of the human body is the ideal temperature for bacterial growth. As the bacteria multiply, they produce waste products that are poisons to the living thing. The poisons cause the symptoms of the disease. In humans, the body tries to fight back by raising its body temperature. That's when you have a fever.

You get food poisoning when you eat food in which bacteria have grown. The poisons produced by salmonella food poisoning cause headache, chills, nausea, diarrhea, and stomach cramps. The symptoms are usually gone after two or three days. Salmonella bacteria

Figure 10.8 ▲
What effects from the bacterial blight do you see on the soybean plant? ①

204 Chapter 10 Viruses and Monerans

Multicultural Perspectives

In 1857, the French microbiologist Louis Pasteur discovered that fermentation was caused by microorganisms like bacteria. Pasteur demonstrated how milk sours by injecting fresh milk with living organisms. His experiment showed that milk that was not injected did not sour as the injected milk had. His studies led him to conclude that food spoils because of microbes. Pasteur invented a process that is named after him—pasteurization. This process uses intense heat to kill microbes. Ask students how they know if the milk they drink is treated. (The word *pasteurized* is on the container.)

---

most commonly infect eggs, ground meat, chicken, and sausage. Another kind of food poisoning, called botulism, can be fatal. Fortunately, botulism is not common. The bacteria that cause botulism are harmful only when they are trapped in improperly canned or jarred food.

Science and Technology *Foul Food*

Food poisoning and food spoilage are caused by the growth of bacteria in foods that are not packaged or stored properly. Using scientific knowledge about the needs of bacteria, food manufacturers can package or treat foods so that they last longer. The methods they use slow or stop bacterial growth.

Table 10.2 Methods of Preventing Food Spoilage

Method	Description
Canning	Foods are cooked at high temperatures. Then they are stored in airtight containers.
Curing	Foods are dried or cooked. Salt, vinegar, sugar, or chemical preservatives are added. These substances slow or stop bacterial growth.
Cold Storage	Many foods need to be stored at temperatures between –1°C and 10°C.
Drying	Removing most of the water from foods slows or stops bacterial growth.
Freezing	Foods are stored at temperatures below –18°C to help slow or stop bacterial growth.
Irradiation	Fruits and meats are bombarded with ultraviolet or gamma rays.

Skills WorkOut

Research Choose one of your favorite foods. Find out the different ways it can be safely preserved.

Check and Explain

1. What are the characteristics of monerans?
2. How are monerans classified?
3. **Compare and Contrast** Explain the main ways that bacteria are helpful and the ways they are harmful.
4. **Graph** One bacterium divides to form two; two divide to make four. Make a graph that shows how this pattern continues as the bacteria divide. How many bacteria will there be after five generations?

Chapter 10 Viruses and Monerans

ACTIVITY 10

Time 3 days **Group** 5–6

Materials

5 grease pencils

5 packets of instant chicken broth

5 1-L beakers

5 graduated cylinders

25 small jars

4 or more different antiseptic products

Analysis

1. Jar 5 was the control.
2. Accept all logical answers. Possibilities include: broth becomes cloudy, unpleasant odor produced.
3. Answers may vary, but should agree with recorded observations.
4. Answers may vary, but should agree with recorded observations. Bacteria might be observed in any of the jars, especially jar 5.
5. Answers will vary depending on products used.

Conclusion

Accept all logical conclusions, provided they follow from observations. Students should recognize that the less bacterial growth, the more effective the antiseptic.

Everyday Application

Answers will vary depending on products used.

Extension

Experimental designs should be complete and show good controls for all variables except the quantity of added antiseptic.

206

TEACHING OPTIONS

Prelab Discussion

Have students read the entire activity before beginning the discussion. Ask the following questions:

▶ What does it mean to "sterilize" an object or a substance?

▶ What is an antiseptic?

▶ Where does the bacteria that grows in unsterilized places come from?

▶ Where are some good places to store items that have been sterilized?

Activity 10 *How effective are common antiseptics?*

Skills Measure; Observe; Infer; Control Variables

Task 1 Prelab Prep

1. Collect the following items: grease pencil, 1 package instant chicken broth, a 1-liter beaker or 4-cup glass measuring cup, graduated cylinder, 5 small jars, 4 different everyday products claiming to have antiseptic properties (such as mouthwash, rubbing alcohol, hydrogen peroxide, cleaning products). **CAUTION! Some of these products may be dangerous, especially when mixed together; read the labels and observe all warnings and cautions**.
2. Using the grease pencil, label the jars from 1 to 5.

Task 2 Data Record

1. On a separate sheet of paper, draw the data table shown.
2. Record all your observations in the data table.

Table 10.3 Product Tests

Jar	Product Name	Condition on Day 2	Condition on Day 3
1			
2			
3			
4			
5			

Task 3 Procedure

1. Following the directions on the package, make up 3 cups of chicken broth (about 750 mL) in either the glass measuring cup or the beaker. Use very warm water from the faucet. Make sure the broth dissolves completely. Stir if necessary.
2. Using the graduated cylinder, measure 100 mL of chicken broth into each jar.
3. To each of jars 1 through 4, add 2 dropperfuls of a different "antiseptic" product. Make sure you write the name of each product in the data table before you add it to the jar.
4. Do not add anything to jar 5.
5. Place all the jars together in an open place where they will be out of direct sunlight.
6. On the second and third days, record your observations in the data table.

Task 4 Analysis

1. Identify the control variable in this activity.
2. **Infer** How can you tell if bacteria are growing in a jar of chicken broth?
3. In which jars were bacteria growing on the second day? On the third day?
4. **Infer** Which of the products prevented the growth of bacteria?
5. Read the labels from your antiseptic products. What do the labels claim these antiseptic products do?

Task 5 Conclusion

Based on your analysis of your observations, explain which of the products is the most effective antiseptic and why.

Everyday Application

What are each of the products you tested usually used for? Why would it be important to keep bacteria from growing out of control in each case?

Extension

Choose one of the antiseptics you used in this activity. Develop an experiment to test how much of the antiseptic you need to add to 100 mL of chicken broth in order to keep the bacteria from growing.

Chapter 10 Review

Concept Summary

10.1 Viruses
- Viruses are microscopic, noncellular, and unable to carry out basic life processes except inside the cells of another organism.
- Viruses are difficult to classify because they are not made of cells and appear to be nonliving when outside of a living thing.
- Viruses replicate by using the reproductive machinery of the cells they invade.
- Viral infection usually causes disease in the host. In humans, viruses can cause colds, hepatitis, herpes, and AIDS.

10.2 Monerans
- Monerans are single-celled microscopic organisms. They have both a cell membrane and a cell wall but do not have a nucleus.
- Monerans are classified as blue-green bacteria or bacteria. Bacteria are classified according to shape as cocci, bacilli, or spirilla.
- Bacteria can reproduce rapidly under good conditions by binary fission.
- Most bacteria play a positive role, such as aiding animals in digestion, decomposing dead organisms, and helping plants get nitrogen.
- Some bacteria are harmful and cause diseases in plants and animals. In humans, bacteria can cause tuberculosis, pneumonia, and food poisoning.

Chapter Vocabulary

virus (10.1)	flagella (10.2)	spirilla (10.2)	binary fission (10.2)
host (10.1)	cocci (10.2)	aerobe (10.2)	endospore (10.2)
moneran (10.2)	bacilli (10.2)	anaerobe (10.2)	decomposer (10.2)

Check Your Vocabulary

Use the vocabulary words above to complete the following sentences.

1. Bacteria reproduce asexually by means of a process called ____.
2. The living thing in which another organism lives is called a ____.
3. Organisms that feed on dead plants and animals are ____.
4. Round or egg-shaped bacteria are called ____.
5. A piece of hereditary material with a coat of protein is a ____.
6. Some monerans have a whiplike structures called ____ that allows them to move in liquids.
7. Rod-shaped bacteria are classified as ____.
8. A bacterium that requires oxygen is called an ____.
9. A bacterium with a thick protective wall is an ____.
10. Bacteria and blue-green bacteria are two kinds of ____.
11. Spiral-shaped bacteria are called ____.
12. Bacteria that do not require oxygen are classified as ____.

Write Your Vocabulary

Write complete sentences using the vocabulary words. Show that you know what each word means.

CHAPTER REVIEW 10

Check Your Vocabulary

1. binary fission
2. host
3. decomposers
4. cocci
5. virus
6. flagella
7. bacilli
8. aerobes
9. endospore
10. monerans
11. spirilla
12. anaerobic

Write Your Vocabulary

Be sure that students understand the meaning of each vocabulary word as well as how to use it in a sentence.

Use Vocabulary Worksheet for Chapter 10.

CHAPTER 10 REVIEW

Check Your Knowledge

1. by shape; by whether they require oxygen; by whether they have chlorophyll
2. Viruses are hard to classify because they share some, but not all, characteristics of living things.
3. Viruses reproduce by inserting their genetic material into a host cell and using the host cell to make new virus particles.
4. Monerans have both a cell membrane and cell wall, and circular DNA. They are microscopic; some have flagella.
5. Nitrogen-fixing bacteria convert atmospheric nitrogen to a form usable by plants. Bacteria in the digestive systems of plant-eating animals help break down plant material. Different types of bacteria are used in the production of food for humans. Bacteria break down dead organisms, recycling raw materials in the biosphere.
6. See Table 10.1 on page 197 for a list of viral diseases.
7. Blue-green bacteria have chlorophyll; bacteria do not.
8. flagella
9. virus
10. host
11. shape
12. by binary fission
13. for protection
14. need oxygen

Check Your Understanding

1. Answers may vary. Possible answers include: Both are microscopic, have genetic material (DNA or RNA), are able to reproduce. Viruses have no cytoplasm, no cell membrane or cell wall; they can't move, feed, or grow as monerans can.
2. Accept all logical answers, provided they are supported.
3. Answers may vary; however, students should recognize the beneficial contributions of bacteria and viruses.
4. Foods are preserved to slow or prevent the growth of bacteria and the decomposition of the food. Table 10.2 on page 205 describes methods of preventing food spoilage.
5. Accept all logical answers; however, students should realize that with fewer decomposers, smaller amounts of nutrients would be available to growing organisms.
6. Accept all logical answers; for example, students may suggest that the higher temperature is too hot for some bacteria to survive.
7. bacilli

Chapter 10 Review

Check Your Knowledge

Answer the following in complete sentences.

1. What are three different ways to classify bacteria?
2. Why are viruses difficult to classify?
3. How do viruses reproduce?
4. What are some characteristics of monerans?
5. What are some ways bacteria are helpful?
6. What are some human diseases caused by viral infection?
7. How do the two main groups of monerans differ?

Choose the answer that best completes the sentence.

8. Some monerans are able to move in liquid by using (capsids, flagella, spirilla, endospores).
9. AIDS is caused by a (bacterium, moneran, virus, flagellum.)
10. Viruses cannot move, feed, or grow outside the cells of a (decomposer, cow, human, host).
11. Bacteria are classified according to (size, need for oxygen, harmfulness, shape) as cocci, bacilli, or spirilla.
12. Bacteria reproduce (sexually, with the machinery of a host cell, by binary fission, using endospores).
13. A bacterium may become an endospore (for protection, to reproduce, to infect, to decompose).
14. Bacteria are classified as aerobic or anaerobic depending on whether or not they (have capsids, are round, are harmful, need oxygen).

Check Your Understanding

Apply the concepts you have learned to answer each question.

1. **Compare and Contrast** What do viruses and monerans have in common? How are they different?
2. **Reason and Conclude** Scientists disagree whether viruses are living, nonliving, or something in between. Based on what you know about viruses, how would you classify them? Give reasons to support your conclusion.
3. **Reason and Conclude** Since bacteria and viruses can cause so much disease, do you think it would be a good idea to destroy all of them? Why or why not?
4. Why are foods preserved? What methods can be used?
5. **Infer** What would happen if the population of bacteria acting as decomposers were significantly reduced?
6. When you have a bacterial infection, you often develop a fever. Why do you think that raising your body temperature is helpful in fighting the bacteria?
7. **Mystery Photo** The photograph on page 194 shows bacteria (yellow) on the head of a household pin. Based on their shape, what type of bacteria are they?

Develop Your Skills

1. The model should clearly show the following steps: bacteriophage attaches to host; bacteriophage injects nucleic acid into host; host cell is directed to make new capsids and viral nucleic acid; complete viruses are assembled; host cell bursts, releasing newly created viruses.
2. a. Approximately 1,000 times larger
   b. A grain of salt is about 1,000 times larger than a bacterium.
3. On Day 3, the person was infected with *D. pneumoniae*. On Day 5, the person began taking an antibiotic, which caused a reduction in the populations after Day 5.

Make Connections

1.

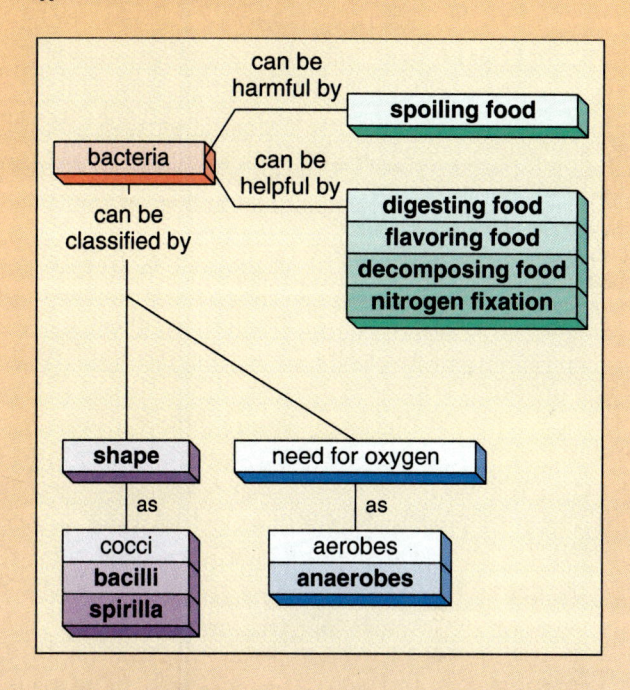

[Concept map: bacteria — can be harmful by → spoiling food; can be helpful by → digesting food, flavoring food, decomposing food, nitrogen fixation; can be classified by → shape (as cocci, bacilli, spirilla) and need for oxygen (as aerobes, anaerobes)]

2. Advertisements will vary, but should be scientifically accurate.
3. Reports will vary, but should completely address the points in the question. Student group reports should include opportunities for questions and discussion.

Develop Your Skills

1. **Make a Model** Make a model showing a bacteriophage infecting a bacterium. Label the parts of the virus and the bacterium. Describe in five steps how the bacterium is infected.
2. **Data Bank** Look on page 625 to answer the following questions.
   a. How much bigger is a bacterium than a virus?
   b. How much bigger than a bacterium is a grain of salt?
3. **Interpret Data** *Escherichia coli* is a bacteria that lives in the human large intestine. It digests some foods your body can't and even produces vitamins for you. *Dipplococcus pneumoniae* is a bacteria that causes pneumonia. Use the graph below to answer the following questions. What appears to have happened on Day 3? What happened on Day 5? What happened to the population of both types of bacteria after Day 5? Why?

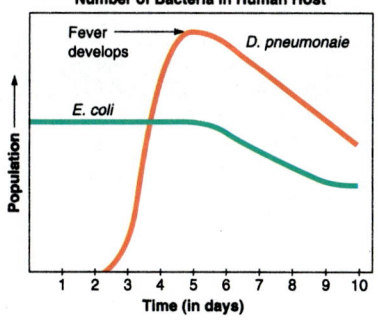

Number of Bacteria in Human Host

[Graph showing Population vs Time (in days) 1–10. E. coli line steady then declining after day 5. D. pneumoniae line rising sharply at day 3 (Fever develops), peaking around day 5, then declining.]

Make Connections

1. **Link the Concepts** Below is a concept map showing how some of the main concepts in this chapter link together. Only part of the map is filled in. Copy the map and complete it using words and ideas from the chapter.

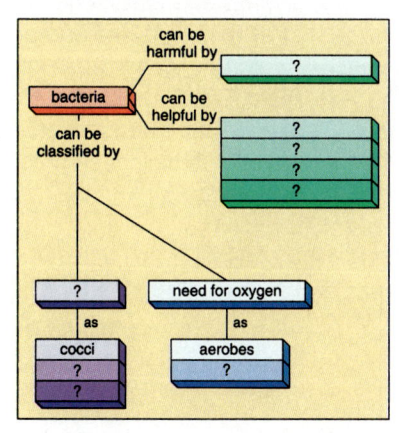

2. **Science and Art** Most types of bacteria are not harmful and do not cause disease. Design an advertisement to inform people about helpful bacteria. Draw pictures or find photographs from old magazines. Think up slogans. Include reasons why the bacteria are helpful.

3. **Science and Society** Read about one of the diseases listed in Table 10.1 on page 197. Find out as much of the following as possible: how the disease affects the body, where and when the disease was most widespread, how it affected society, how society responded to it, and if and how it was brought under control. Get into groups of three and report your findings to one another.

CHAPTER 11

Overview

The protista and fungi kingdoms are discussed in this chapter. The first section covers the cell structures, reproduction systems, and habitats of algae and protozoa. In the second section, protozoa are classified and described in more detail. The diseases that are caused by protozoa are also discussed. Algae are classified and described in the third section. The chapter closes with a discussion of the characteristics of the major groups of fungi.

Advance Planner

▶ Collect a sample of pond, river, or lake water. If none is available locally, contact a biological supply company. TE activity, page 212.

▶ Supply prepared slides of different types of protozoa, euglenas, diatoms, and red and green algae. TE activities, pages 215, 218, and 219.

▶ Contact a local college or aquarium and invite a marine life technician to visit your class, if possible. TE page 220.

▶ Compile addresses of pharmaceutical companies (contact a local library to obtain information). TE activity, page 227.

Skills Development Chart

Sections	Classify	Communicate	Find Causes	Infer	Interpret Data	Observe	Predict
11.1 Skills WarmUp		●					
11.2 Skills WarmUp	●						
11.3 Skills WarmUp		●					
11.4 Skills WarmUp Skills Workout SkillBuilder Activity	●		●	●	●	●	●

Individual Needs

▶ **Limited English Proficiency Students** Make a tape recording of each vocabulary term and its meaning. If there is a figure that illustrates the meaning of a term, direct students to look at the figure as they listen to the tape. Have students use their own words to define each term and describe the related figure. Have them add their definitions to yours on the tape. Students can use these tapes to review the section concepts.

▶ **At-Risk Students** Assign students to cooperative groups. Choose several important figures from the chapter and photocopy them. Have students write one sentence for each chapter term on strips of paper. Have students paste the copies of the figures onto posterboard and arrange the paper labels underneath the appropriate pictures.

▶ **Gifted Students** Have students study and research the life-supporting elements in the environments of protists and fungi. Ask students to choose an ecosystem, such as a pond, ocean, or bog, and explain how protists and fungi survive there and how they contribute to that ecosystem. If possible, have students build their own system in a terrarium or aquarium.

Resource Bank

▶ **Bulletin Board** Make a bulletin board that exhibits the diversity of protists. Display several diagrams or pictures of protists and fungi. Encourage students to draw or find pictures in magazines and add them to your display.

▶ **Field Trip** If possible, plan a trip to a lake or another body of water so students can observe algae. You may be able to have them collect samples of pond water to study in the classroom.

CHAPTER 11 PLANNING GUIDE

Section	Core	Standard	Enriched	Section	Core	Standard	Enriched
11.1 Diversity of Protists pp. 211–213				**Blackline Masters** Review Worksheet 11.3 Integrating Worksheet 11.3	●	● ●	●
Section Features Skills WarmUp, p. 211	●	●	●	**Overhead Blackline Transparencies** Overhead Blackline Master 11.3 and Student Worksheet	●	●	●
Blackline Masters Review Worksheet 11.1 Skills Worksheet 11.1	● ●	● ●		**11.4 Fungi** pp. 222–228			
11.2 Protozoa pp. 214–216				Section Features Skills WarmUp, p. 222 Skills WorkOut, p. 223 SkillBuilder, p. 226 Activity, p. 228	● ● ●	● ● ● ●	●
Section Features Skills WarmUp, p. 214	●	●		Blackline Masters Review Worksheet 11.4 Skills Worksheet 11.4 Integrating Worksheet 11.4 Reteach Worksheet 11.4 Vocabulary Worksheet 11	● ● ● ● ●	● ● ● ● ●	● ●
Blackline Masters Review Worksheet 11.2 Integrating Worksheet 11.2	● ●	● ●	●				
Laboratory Program Investigation 21	●	●	●	Laboratory Program Investigation 22	●	●	
Color Transparencies Transparency 29	●	●		Color Transparencies Transparencies 30, 31	●	●	
11.3 Algae pp. 217–221							
Section Features Skills WarmUp, p. 217 Career Corner, p. 220	● ●	● ●	● ●				

Bibliography

The following resources can be used for teaching the chapter. See page T-40 for supplier codes.

Audio-Visual Sources
(video unless noted)
Classifying Living Things. 18 min. 1987. A-W.
Classifying Microorganisms. 15 min. 1984. C/MTI.
Fungi. 18 min film. 1983. BF.
Neither Plant nor Animal. 2 filmstrips with cassettes, 17–18 min each. 1987. NGSES.
Protist Behavior. 11 min film. 1975. WN.

Software Resources
Microbe. Apple. SYN.
Protozoa. Macintosh, IBM, Apple. VE.
The Simplest Living Things. Apple. DCHES.

Library Resources
Gallo, R. C., and L. Montagnier. "AIDS." *Scientific American* 259 (October 1988): 47–48.
Jacobs, Francine. *Breakthrough: The True Story of Penicillin.* New York: Dodd, Mead, & Co., 1985.
Lee, J. J., S. H. Hutner, and E. C. Bovee, eds. *An Illustrated Guide to the Protozoa.* Lawrence, KS: Society of Protozoologists, 1985.
Margulis, Lynn, and Karlene V. Schwartz. *Five Kingdoms: An Illustrated Guide to the Phyla of Life on Earth.* San Francisco: W.H. Freeman and Co., 1982.
Teasdale, Jim. *Microbes.* Morristown, NJ: Silver Burdett, 1985.

CHAPTER 11

Introducing the Chapter

Have students read the description of the photograph on page 210. You may wish to point out that the grooves on the mushroom's underside are gills that produce spores. Fungi get food through underground structures called hyphae. Ask students if they agree or disagree with the description.

Directed Inquiry

Have students study the photograph. Ask:

▶ Where have you seen organisms that looked like those in the photograph? What were they? (Answers may vary; mushrooms.)

▶ Where are the grooved structures on mushrooms you have seen? (Underneath the caps)

▶ What do you notice about the arrangement of the mushrooms? (The little mushrooms are crowded under the large one; tinier mushrooms are crowded under the small ones.)

▶ What is the topic of this chapter? (Protists and fungi)

▶ Why is a photograph of a mushroom included in this chapter? (Mushrooms are a type of fungi.)

Chapter Vocabulary

alga	hyphae
budding	protist
cilia	protozoan
eukaryote	pseudopodia
fungus	spore

210

INTEGRATED LEARNING

Writing Connection
Have students write a letter to a friend describing the photograph. Have them pretend that their friend has never seen mushrooms. Encourage students to add details about where the fungi grow and about their shape, color, and texture.

Themes in Science
Diversity and Unity Although the members of a kingdom have strong similarities, there are variations among species. Help students create a chart of kingdom Protista showing its species and their different characteristics. Have students include information about each species' movement, shape and size, cell complexity, reproduction, and feeding.

Chapter 11 Protists and Fungi

Chapter Sections

11.1 Diversity of Protists
11.2 Protozoa
11.3 Algae
11.4 Fungi

What do you see?

"In this picture I see a mushroom. It looks like one big mushroom with little ones all around it. I think the grooves are so these mushrooms can get water, oxygen, and air."

Kim Rice
Honea Path Middle School
Honea Path, South Carolina

To find out more about the photograph, look on page 230. As you read this chapter, you will learn about fungi and protists.

210

TEACHING OPTIONS

Cooperative Learning
Have students work in small groups to compare monerans and protists. Have groups make large, labeled diagrams of a moneran cell and a protist cell. Then have pairs of students use their diagrams as they give brief oral reports that compare and contrast monerans and protists.

11.1 Diversity of Protists

Objectives
- **Describe** how protists differ from monerans.
- **Compare** and **contrast** protozoa and algae.
- **Explain** the role of plankton in food chains.
- **Make inferences** about how protozoa obtain their nutrition.

Skills WarmUp

Communicate Imagine that you can swim wherever you want in a pond. But you're microscopic in size. What do you see? Write about your journey.

You watch a group of submarine-shaped creatures whizz by through the murky water. The long hairs covering their bodies beat rapidly. Suddenly an even larger beast appears, its scary mouth gaping open. Is it chasing the others? Then you notice a huge jellylike blob oozing toward you.

Are you on an alien planet? No, you're still on the earth, at the bottom of a pond. But you've been shrunk down to the size of the tiniest speck. The creatures you've been observing are single-celled, animal-like organisms called **protozoa** (PROH tuh ZOH uh). Protozoa (singular: protozoan) belong to kingdom Protista.

Kingdom Protista

Members of kingdom Protista are called **protists** (PROH tihsts). Protozoa like the ones living in the pond are one major group of protists. The other major group of protists is made up of plantlike organisms, both single-celled and many-celled, called **algae** (AL jee). Seaweeds are common kinds of algae (singular: alga).

Why do scientists classify these two different groups of organisms in the same kingdom? All protists have one thing in common. They are **eukaryotes** (yoo KAIR ee OHTS), meaning their cells are complex, like the cells of plants and animals. They have nuclei, mitochondria, and other organelles. Having complex cells is what separates the protists from the monera.

However, protists aren't as complex as plants or animals. The protozoa are only single-celled. The algae are either single-celled or do not have all of the many specialized parts that plants have. Kingdom Protista, therefore, includes all eukaryotes that are not plants, animals, or fungi.

Figure 11.1 ▲
Algae are plantlike protists often found growing in masses on the surface of ponds. How do algae differ from plants? ①

SECTION 11.1

Section Objectives
For a list of section objectives, see the Student Edition page.

Skills Objectives
Students should be able to:

Communicate ideas about microscopic life in a pond.

Infer what protozoa eat.

Vocabulary
protozoan, protist, alga, eukaryote

MOTIVATE

Skills WarmUp
To help students understand the microscopic scale of protists, have them do the Skills WarmUp.
Answer Answers may vary; students might describe microscopic life forms such as protozoa and algae. Save these descriptions so students can add to them in the WrapUp for Section 11.4.

Prior Knowledge
Students will need to refer to their knowledge about monerans so they can compare protists and monerans. Ask the following questions:

- What are some organisms that are classified as monerans?
- Where do monerans live?
- What are some characteristics of monerans?

Answer to In-Text Question
① Algae differ from plants in that they are either single-celled or do not have all of the many specialized parts that plants have.

Chapter 11 Protists and Fungi 211

TEACH

Class Activity
Have students look at pond water through a microscope. Have them try to identify and draw the organisms they see, including protozoa and algae. Ask students to write down observations they make about the organisms' sizes, movements, and so on.

Skills Development
Classify Have students look up asexual and sexual reproduction in the glossary. Stress that in asexual reproduction, two identical daughter cells are produced from one parent. Sexual reproduction involves the sharing of genetic material. Have students describe how the offspring of organisms that reproduce asexually differ from offspring produced by sexual reproduction. (Offspring from asexual reproduction are identical to the parent. Offspring from sexual reproduction combine characteristics from both parents.)

The Living Textbook: Life Science Sides 1–4
Chapter 1 Frame 00528
Protist Slides (51 Frames)
Search: Step:

INTEGRATED LEARNING

Art Connection
Huge numbers of protozoa live in the sea and are eaten by other organisms. Have students do research to find out which living things feed on plankton. Then have them create a mural showing the variety of organisms that is found near the ocean's edge.

Protozoa

Animal-like protists, the protozoa, live wherever there is water. Besides ponds, you can find them in soil, in the ocean, and inside other organisms. They have no chloroplasts, so they must take in food from their surroundings.

Most protozoa are microscopic. Even so, they are much larger than bacteria. If an average protozoan were the size of a soccer field, then a bacterium would be the size of a soccer ball.

Even though it is made up of only one cell, a protozoan can do almost everything a many-celled animal can do. It moves and senses changes in its environment. It eats and reproduces.

A protozoan has a cell membrane, but no cell wall. The absence of a cell wall is another way in which protozoa differ from monerans and most algae.

Most protozoa reproduce asexually. In a process called fission, the protozoan divides into two similar daughter cells. Some protozoa exchange genetic material before undergoing fission. This is a simple form of sexual reproduction.

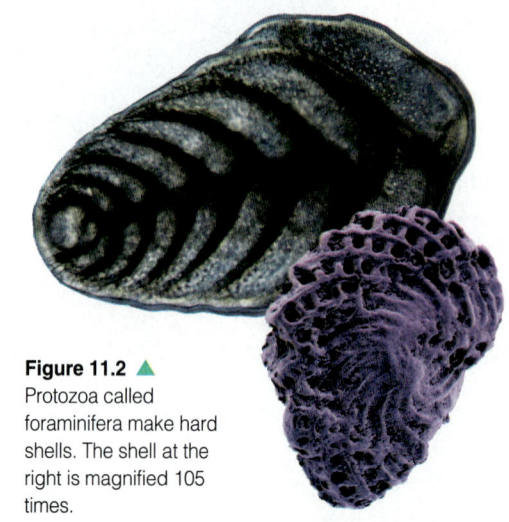

Figure 11.2 ▲
Protozoa called foraminifera make hard shells. The shell at the right is magnified 105 times.

Figure 11.3 ▲
Many-celled algae grow on rocks at the ocean's edge.

Algae

Protists with chloroplasts are classified as algae. Algae undergo photosynthesis, and most have rigid cell walls. It is because they share these characteristics with plants that they are considered plantlike.

Unlike protozoa, some algae are many-celled organisms. These algae are known as seaweeds and can grow larger than many land plants. But their simple organization and lack of transport vessels are characteristics of the protists.

Algae reproduce in a variety of ways. Some groups of algae reproduce asexually by fission. Other groups have developed forms of sexual reproduction. In some many-celled algae, special sex cells are used to exchange genetic material.

Single-celled algae living in the ocean are called plankton. They form the first level of the food chain. Chloroplasts within these organisms harness the sun's radiant energy and convert it into glucose. Animals that feed upon these protists absorb their stored energy.

Chapter 11 Protists and Fungi

Multicultural Perspectives

Aquaculturists in China have been raising seaweed and marine animals on sea farms for more than 3,000 years. Have students research the purposes and techniques of aquaculture in China. Have them share their information in brief oral reports.

Geography Connection

The alga Irish moss is a valuable commercial product used as a thickening agent in foods such as ice cream and pudding. It is raised on Prince Edward Island, Canada, and on other North Atlantic coasts. Have students locate Prince Edward Island on a map and determine different routes that manufacturers might use to transport Irish moss to your area.

Enrich

Have students find and cut out pictures of different foods that contain seaweed. Have them place the pictures on posterboard and display them in the classroom.

EVALUATE

WrapUp

Portfolio Have students imagine they are science magazine photographers who are assigned to take three photographs of protozoa and three photographs of algae. Have students draw pictures to represent their photographs. They can add the pictures to their portfolios.

Use Review Worksheet 11.1.

Check and Explain

1. Protists have complex cells with nuclei, organelles, and cell walls; monerans have simple cells with no nuclei, organelles, or cell walls.

2. Plankton are single-celled algae that live in the ocean. They form the first level of the food chain of the ocean.

3. Algae and protozoa are eukaryotes with complex cells. All protozoa are single-celled; some algae are many-celled. A protozoan has no cell wall; most algae have cell walls.

4. Answers may vary. Protozoa eat food from their surroundings, such as other protists, or bacteria. Some protists are photosynthetic.

Science and Society *Algae as Food*

When was the last time you ate algae? It was probably very recently, because many foods contain extracts of algae. Carrageen, for example, is a substance obtained from an alga called Irish moss. Carrageen is added to such foods as ice cream and puddings to make them thicker. It is also found in many toothpastes.

Algin is another substance made from algae that you may have eaten. Extracted from kelps, algin is added to ice cream to prevent its ingredients from separating. Like carrageen, algin also adds body to foods.

Many people in the world eat algae from the sea, both fresh and dried. In Japan, raw seafood and rice are wrapped in sheets of dried algae to make the delicacy called sushi. Sushi is also popular in the United States. Japanese people also use algae in soups, teas, biscuits, and as a flavoring.

Figure 11.4 ▲
Edible algae are dried on a beach in Japan (left). Sushi is a popular dish in many Japanese restaurants (right).

Check and Explain

1. How are protists different from monerans?

2. What are plankton? What role do they have in the food chains of the ocean?

3. **Compare and Contrast** Why are algae and protozoa grouped together in the same kingdom? What characteristics separate protozoa and algae?

4. **Infer** What do protozoa eat? Name a possible food and explain why protozoa are likely to eat it.

Chapter 11 Protists and Fungi   213

SECTION 11.2

Section Objectives
For a list of section objectives, see the Student Edition page.

Skills Objectives
Students should be able to:
Classify animals by how they move.
Organize information about the four major groups of protozoa.

Vocabulary
pseudopodia, cilia, spore

MOTIVATE

Skills WarmUp
To help students understand animal locomotion, have them do the Skills WarmUp.
Answer Answers will vary. Students' lists will probably note different limbs and body plans. (Wings, legs, fins, scales, and so on)

Misconceptions
Many students will use the word *germs* to mean "something small that makes you sick," but when asked to identify germs, they will say "bacteria." Tell students that, like bacteria, some protists are disease-causing organisms.

Answer to In-Text Question
① The ameba uses its psuedopodia for locomotion.

The Living Textbook:
Life Science Side 1
Chapter 19   Frame 38600
Protozoans (Movies)
Search: Play:

214

INTEGRATED LEARNING

Writing Connection
Amebas and other sarcodines behave like science fiction monsters such as The Blob. Like a blob, they change shape, ooze over their prey, and engulf and store food. Have students write a horror story about a giant ameba. Encourage them to describe in detail the way the ameba moves, engulfs, and stores its food.

Integrating the Sciences
Earth Science Many saltwater sarcodines have skeletons made of calcium carbonate. In fact, most limestone is formed from the shells of living organisms. The white cliffs of Dover in England are made up of these organisms. Show students a picture of these white cliffs and have them research organic sedimentary rock and the formation of chalk.

Skills WarmUp

Classify Write down the names of ten different animals. Then put them into groups based on how they move.

Figure 11.5 ▲
The bloblike ameba oozes around in search of food. What structures does it use to move? ①

11.2 Protozoa

Objectives
▶ **Name** the different structures protozoa use to move.
▶ **Identify** and **describe** protozoa that cause disease in humans.
▶ **Make inferences** about the habitats of protozoa based on the ways they move.
▶ **Organize data** about the four major groups of protozoa.

One characteristic that many protozoa share with animals is the ability to move. Organisms that get their energy through photosynthesis may stay in one place, but those that must take in food often have to search for their meals.

Animals have developed a variety of structures that help them move, such as wings, legs, and fins. In a similar way, protozoa have evolved different ways of moving, each using a different kind of structure. Protozoa that move in similar ways tend to have many other similarities. Just as you find it easy to classify vehicles as cars, boats, or planes, scientists have grouped the protozoa according to how they move.

Sarcodines

One group of protozoa, the sarcodines (SAR kuh deens), move using footlike projections of cell matter called **pseudopodia** (soo duh POH dee uh). A typical sarcodine extends a pseudopod, and cytoplasm begins to flow into it. The entire cell oozes into the pseudopod, and at the same time a new pseudopod forms. The cell is always changing shape.

Most sarcodines are free-living, meaning they don't live inside other organisms. A few, however, are parasites. One sarcodine causes an illness in humans called amebic dysentery (uh MEE bihk DIHS uhn TAIR ee).

Look at Figure 11.5. It shows a common sarcodine, the ameba (uh MEE buh). Sarcodines such as the ameba use their pseudopods for capturing food. The ameba engulfs the food and stores it within a food vacuole.

214   Chapter 11   Protists and Fungi

TEACHING OPTIONS

Cooperative Learning
Have students work in groups of three. Take each group aside and assign each member the role of a sarcodine, a ciliate, or a flagellate. Then have the groups perform their roles in front of the class, giving students in the audience the opportunity to identify who is acting out which type of protozoa.

Ciliates

The ciliates are another group of mostly free-living protozoa. Ciliates move using tiny, hairlike structures called **cilia** (SIHL ee uh). The cilia work like little oars, moving back and forth in the water to move the organism along.

With cilia covering much of their bodies, many ciliates are able to swim very fast, for a protozoan. Ciliates can move forward and backward, and they can turn quickly.

A common ciliate is the paramecium, shown in Figure 11.6. Its tapered body is enclosed by a hard covering. On one side of the body is a channel lined with cilia called the oral groove. Food particles are swept down this oral groove into the paramecium's mouth.

Food passing through the paramecium's mouth enters a food vacuole. As this vacuole circulates through the cytoplasm, its contents are digested and distributed around the cell.

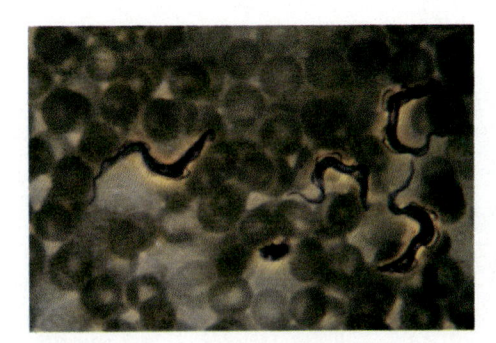

Figure 11.7 ▲
Victims of African sleeping sickness have this parasitic flagellate in their blood.

Flagellates

A third group of protozoa, the flagellates (FLA juh LAYTS), use a whiplike organelle called a flagellum to move. Lashing of the flagella creates a current of water that propels the flagellate. Look at the flagellate in Figure 11.7. Can you see the flagella? Some flagellates have one flagellum, others have many.

You may remember that monerans, too, have structures called flagella. Moneran flagella, however, are very different from the flagella of flagellates. They are made of different parts and materials and are moved in a different way. Why do you think these two different structures have been given the same name? ③

Although most flagellates are free-living, some members of this group live inside other organisms. A termite's gut, for example, contains a population of flagellates. These flagellates help the termite by digesting wood.

Some of the flagellates that live inside other organisms are parasites that can injure or kill their host. For example, African sleeping sickness is a deadly human disease caused by a flagellate. It is transmitted by the bite of the tsetse fly.

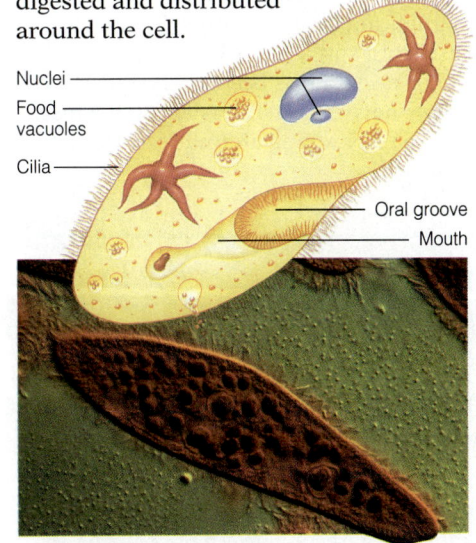

Figure 11.6 ▲
How does a paramecium move? ②

Nuclei
Food vacuoles
Cilia
Oral groove
Mouth

TEACH

Class Activity
Have students examine prepared slides of different types of protozoa. Students should identify the cilia, flagella, and pseudopodia on the organisms. Have students classify the protozoa based on their method of locomotion.

Critical Thinking
Reason by Analogy Ask students to describe how a paramecium would move forward by comparing it to a rowboat with its cilia functioning like oars. (Students should realize that the cilia would move, like oars on a rowboat, from the front of the organism toward the back.) Then ask students to explain how a paramecium would move to the right. (The cilia on the left side of its body would move from front to back, and the right cilia would move from back to front.)

Answers to In-Text Questions
② A paramecium uses its cilia for locomotion.

③ The two different structures are both called flagella because they perform the same function—they both enable the organism to move.

Chapter 11 Protists and Fungi

TEACH • Continued

Discuss

Have students research flagellates that live inside other organisms. Then discuss some of the following questions:

▶ How do flagellates get inside other organisms?

▶ Describe how some flagellates help their host organisms.

▶ Describe some flagellates that harm their hosts.

▶ How do flagellates get inside the human body?

EVALUATE

WrapUp

Reteach Write the names of the four types of protozoa on the chalkboard. Ask students what all of these organisms have in common. Then discuss how the four types differ from one another. Create a graphic organizer that displays the information.

Use Review Worksheet 11.2.

Check and Explain

1. A pseudopod is a footlike projection of cell matter.
2. Amebic dysentery; African sleeping sickness; malaria
3. Sarcodines are most common in soil.
4. Students' charts should include information about habitats, ways of movements, and effects on other animals.

216

INTEGRATED LEARNING

Multicultural Perspectives

Charles Laveran, a French physician, discovered the parasite that causes malaria in 1880. Before then, people thought that malaria was caused by poisonous swamp air. In fact, malaria is named from the Italian words *mala aria*, which mean "bad air." Have students create a class display that contains pictures of and short reports about the disease, its causes, and its treatments. Encourage them to include information about Charles Laveran and the other scientists who helped identify the cause of the disease and found a cure for it.

Use Integrating Worksheet 11.2.

Figure 11.8 ▲
An *Anopholes* mosquito infects a person with the sporozoan that causes malaria.

Sporozoa

Unlike the other groups of protozoa, sporozoa (SPOH roh ZOH uh) have no special organelles used for movement. They are all parasites, living inside the bodies of cattle, birds, humans, and other animals. All sporozoa produce special reproductive cells called **spores**, from which new sporozoa can grow.

One well-known sporozoan causes malaria. A mosquito bite transfers the organism into a person's body. There, it reproduces and infects red blood cells. Mosquitos that feed on infected blood pick up the sporozoan. Spores are produced in the insect's gut. They migrate to the mosquito's salivary glands, where they can be transferred into someone else.

Science and Society *Malaria*

In places such as Central Africa, Brazil, India, and the Far East, malaria is a very common disease. In fact, over 150 million people are stricken with malaria each year. Malaria doesn't usually cause death, but a person infected with it often has symptoms of the disease for a long time. Headaches, high fever, and violent shivering are the main symptoms.

In the 1960s, drugs and the use of insecticides helped prevent the disease from spreading. Today, however, the disease is spreading again. This is because the mosquitos that infect organisms have become resistant to the insecticides. Scientists are researching alternative vaccines to this serious disease.

Check and Explain

1. What is a pseudopod?
2. Describe two human diseases caused by protozoa.
3. **Infer** Based on what you know about how different protozoa move, which group would you say is most common in soil?
4. **Organize Data** Make four columns on a piece of paper and write the name of a group of protozoa at the top of each. Below each name, write down all that you have learned about that group.

216 Chapter 11 Protists and Fungi

Integrating the Sciences

Chemistry Light produced by living organisms, such as the light from glowing dinoflagellates, is called bioluminescence. In some organisms, energized electrons travel through an electron transport system that releases energy as light. Encourage students to research bioluminescence.

11.3 Algae

Objectives

- **Identify** and **describe** six groups of algae.
- **Explain** the causes and dangers of red tide.
- **Describe** human uses of some algae.
- **Classify** algae based on their characteristics.

Skills WarmUp

Classify In order to grow, algae need sunlight and fresh water or salt water at different temperatures. List the places in your neighborhood that you think can support algae growth.

The algae, whether single-celled or many-celled, all use chlorophyll to capture the sun's energy to make glucose. Algal cells also have other light-absorbing substances, or pigments. These extra pigments help transfer the sun's energy to chlorophyll.

Because the extra pigments can hide the green color of chlorophyll, many kinds of algae are not green. Instead they appear brown, red, or yellow. The pigments contained in algae form the basis for their classification. Of the six major groups of algae formed in this way, three have mostly single-celled members, and three contain mostly many-celled forms.

Dinoflagellates

One major group of algae is made up of the single-celled dinoflagellates. Most dinoflagellates live in the ocean. Their cells are encased in rigid plates of cellulose. Two flagella extend from a groove in the plates. As these flagella lash about, they make the dinoflagellate spin. Dinoflagellates often contain large amounts of an extra red pigment.

When the population of certain types of dinoflagellates grows rapidly, they color the water red. This condition, known as red tide, is dangerous because these protists also contain a poison. Animals that eat the dinoflagellates keep the poison in their bodies. When people eat these poison-containing animals, such as clams, they can get sick and even die.

Some dinoflagellates can change stored chemical energy into light. If you've walked on a beach at night, you may have seen glowing dinoflagellates. They make the waves sparkle as they crash on the sand.

Figure 11.9 ▲
Dinoflagellates have different shapes, but all have two flagella.

Chapter 11 Protists and Fungi 217

SECTION 11.3

Section Objectives
For a list of section objectives, see the Student Edition page.

Skills Objectives
Students should be able to:

Classify places where algae can grow.

Classify algae into two groups based on common characteristics.

MOTIVATE

Skills WarmUp
To help students understand the needs and characteristics of algae, have them do the Skills WarmUp.
Answer Answers will vary. Students' lists should note habitats with moisture and occasional sunlight.

Misconceptions
Most students think of algae as many-celled plants. In the beginning of this chapter, students learned that kingdom Protista includes both single-celled and many-celled algae. Here is a good place to stress that many types of algae are single-celled organisms.

The Living Textbook:
Life Science Sides 1-4

Chapter 1   Frame 00569
Dinoflagellates (6 Frames)
Search: Step:

TEACH

Critical Thinking
Reason by Analogy
Show students a Petri dish and challenge them to describe how it is like the shell of a diatom cell. Have students handle the Petri dish, noting how its two pieces fit together. (The top and bottom of the Petri dish fit together like the halves of the diatom shell. Like the Petri dish, the shell of the diatom cell is made of a glasslike material.)

Class Activity
If possible, have students use microscopes to look at prepared slides of euglenas and/or diatoms. If they look at euglenas, have students notice the flagella and look for the eyespots. If they look at diatoms, have them notice the variety of shell shapes and sizes. Then have students draw what they observe and compare their drawings with the micrograph on this page.

Answers to In-Text Questions

① **Euglenas are animal-like in that they eat monerans and other protists. Also, they are active swimmers, using flagella to move themselves.**

② **The eyespot helps a euglena get the light it needs for photosynthesis.**

The Living Textbook:
Life Science Sides 1-4
Chapter 1   Frame 00563
Diatoms (4 Frames)
Search: Step:

INTEGRATED LEARNING

Art Connection
Because of their symmetry and beautiful design, diatoms have often been called "jewels of the sea." Have students observe micrographs of diatoms. Review the definition of symmetry and show students how a diatom is radially symmetrical. Provide art supplies and have students create their own symmetrical designs.

Diatoms

Single-celled, gold-colored algae make up another group called diatoms (DY uh TAHMS). The gold color comes from large amounts of yellow, orange, and brown pigments. Diatoms are important producers in both freshwater and saltwater environments. In the ocean, they make up part of the plankton.

The diatom cell is covered by a shell made from a glasslike material. The shells come in two halves that fit together like a pillbox. The shells have many beautiful and delicate shapes as you can see in Figure 11.10. Each species has its own shape and markings.

When a diatom dies, the shell sinks to the sea bottom. Over millions of years, diatom shells have formed thick deposits on the ocean floor. When geologic processes raise these deposits above sea level, they can be mined. They are used for products that need a fine grit, such as cleansers and polishes.

Figure 11.10 ▲
The extra-fine details of diatom shells were once used to test a microscope's focusing ability.

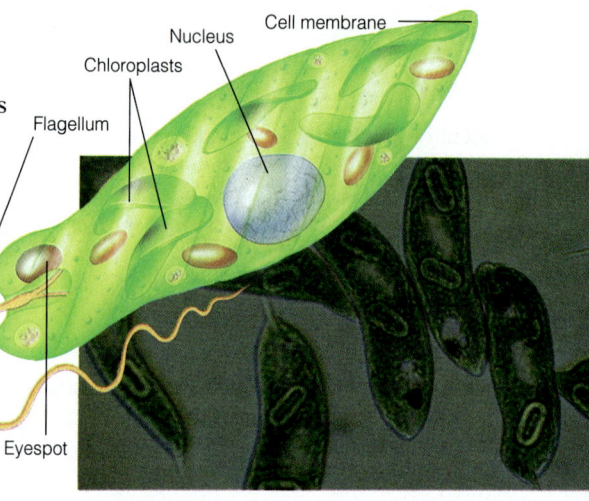

Figure 11.11 ▲
Euglenas have characteristics of both algae and protozoa. In what ways are they animal-like? ①

Euglenas

Ponds and similar environments are home to single-celled, grass-green protists called euglenas (yoo GLEE nuhs). Look at Figure 11.11. Most euglenas have chloroplasts and undergo photosynthesis. Unlike other types of algae, however, euglenas are able to live without sunlight because they can eat.

In dark conditions, euglenas dine on monerans and other protists. They are also active swimmers, using flagella to propel themselves. These animal-like characteristics show that the boundary between algae and protozoa is blurry.

Euglenas have tapered bodies encased in tough but flexible coverings. At the base of the single flagellum is an eyespot. Of course, the eyespot isn't actually an eye. Light striking the eyespot directs the action of the flagellum, causing the protist to swim toward light. Why is an eyespot a valuable adaptation for a euglena? ②

Chapter 11   Protists and Fungi

Themes in Science

Scale and Structure Different types of colonial and complex algae have a variety of specialized structures. In each case, the structure of a particular type of algae helps it meet its needs. Have students draw three types of algae, showing how their specialized structures facilitate their function and survival. Encourage students to label their drawings and include explanatory notes.

Green Algae

Both single-celled and many-celled algae are included in a large group called the green algae. Green algae live in freshwater habitats and the ocean. They are abundant in soil, and some live inside other organisms. Green algae are bright green because they contain large amounts of chlorophyll.

Green algae have a great variety of forms. The simplest green algae are single-celled. Other types of green algae form colonies, groups of cells that live together but do not show much specialization. Some of these colonial algae exist as long filaments. Another kind of colonial green algae has its cells arranged in a hollow ball.

Other types of many-celled green algae have structures adapted for certain functions. They are more complex than colonial green algae. For example, a green algae called sea lettuce has its cells arranged in thin, leaflike sheets.

Many-celled green algae are similar in many ways to plants. In fact, the cells of all green algae are more plantlike than those of any other group of algae.

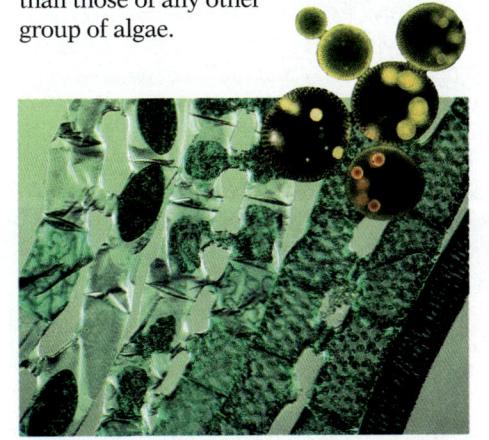

Figure 11.12 ▲
Volvox (top) and *Spirogyra* (bottom) are colonial green algae.

Brown Algae

Many of the seaweeds that carpet rocky shores belong to the brown algae. Brown algae are all many-celled, and nearly all live in the ocean. Like the many-celled green algae, their cells are grouped to form specialized structures. The cell walls of brown algae contain a jellylike material called algin. Algin's rubberiness helps protect brown algae from battering by waves.

Many kinds of brown algae look very plantlike because their specialized structures resemble those of plants. Many brown algae have a rootlike structure called a *holdfast* that anchors the algae to a stable surface. They also have stalks and leaflike blades.

Some brown algae also have air bladders, structures plants don't have. Air bladders enable brown algae to keep their bodies near the ocean's surface. Why does being at the surface benefit an alga? ③

The kelps are the largest brown algae. Their giant blades form an underwater environment much like a forest. Its canopy protects many different types of organisms.

Figure 11.13 ▲
Many kinds of brown algae are harvested for food and fertilizer.

Chapter 11 Protists and Fungi

Class Activity

If possible, have students use microscopes to look at prepared slides of green algae and red algae, such as *Chlorella, Volvox, Chlamydomonas, Spirogyra,* and *Gelidium.* Both *Chlorella* and *Chlamydomonas* are single-celled organisms, and the others are colonial forms. Ask students to describe how green algae differ from other groups of algae.

Critical Thinking

Reason and Conclude Students have read that a holdfast anchors the alga to a stable surface. Ask students to describe at least three ways in which this ability to anchor could be advantageous to an alga. (The alga is kept near the shore where there are more nutrients in the water. The alga is protected from battering waves on the shore. The alga's stability allows it to grow larger.)

Answer to In-Text Question

③ **Being at the surface keeps the alga near the light it needs for photosynthesis.**

The Living Textbook:
Life Science Sides 1-4

Chapter 1   Frame 00547
Green Algae (6 Frames)
Search: Step:

TEACH ▪ Continued

Critical Thinking

Compare and Contrast Ask students which group of algae is more likely to grow at great depths in the ocean. Encourage them to explain their answers. (Red algae are more likely to grow at greater depths. The red pigment in red algae enables them to trap light more efficiently for photosynthesis.) Ask students to explain how the red pigment is similar to and different from chlorophyll. (It is like chlorophyll in that it traps light; however, unlike chlorophyll, the red pigment is able to trap energy from very dim light.)

Career Corner

If possible, invite a marine technician from a local college or aquarium to visit your class. Have the students prepare a list of questions to ask the visitor. If there are no marine technicians in your area, have students write to a marine biology laboratory.

The Living Textbook: Life Science Sides 1-4
Chapter 1 Frame 00579
Seashore Algae (4 Frames)
Search: Step:

TEACHING OPTIONS

Cooperative Learning

Algae are an important part of the ecological balance on the earth. They undergo photosynthesis and provide food for other organisms. Yet an excess of algae can upset the balance of nature. Have students work in pairs to conduct research on the ecological pros and cons of algae. One student should research how algae are a vital part of many of the earth's ecosystems. The other student should investigate what leads to an overgrowth of algae and the consequences of such an occurrence. The student pairs should then present their findings to the class.

Use Integrating Worksheet 11.3.

Figure 11.14 ▲
Some red algae, called coralline algae, have hard calcium deposits in their cell walls.

Red Algae

Many of the brown algae's characteristics are shared by the group called red algae. All red algae are many-celled, and most live in saltwater habitats. They are large enough to be called seaweeds, and they have specialized structures such as holdfasts and stalks.

Red algae get their name from a reddish pigment that other kinds of algae do not have. Unlike other pigments, it can harness the energy of the dim light that penetrates below the ocean's surface. The amount of this pigment in a red alga depends on the depth at which it grows. Those growing near the surface do not need very much of it, so they are green. Those growing at medium depths are red. Red algae growing deep below the surface contain so much of this pigment that they appear almost black.

The bodies of red algae are divided into thin filaments. Often these filaments are branched or interwoven, creating

---

Career Corner — Marine Life Technician

Who Studies Ocean Organisms?

As you collect your last specimen, your diving buddy communicates that it is time to surface. Slowly you rise among the kelp, tracing the path made by your rising bubbles. Surfacing, you carefully place your specimens in a collection pail and head off to the laboratory.

Inside the laboratory, you catalog your collection and begin a series of tests and experiments as directed by a scientist. These observations and experiments will help scientists learn more about the life that inhabits the oceans.

Marine life technicians can be found helping scientists in many different ways. Some technicians work on research ships where they collect and catalog marine specimens. Others use scuba diving to observe organisms.

Most marine technicians, however, work in the laboratory. There, they assist scientists in studying the biology of sea-dwelling organisms. Some technicians may care for the organisms used in experiments.

To become a marine life technician, you should have a science background and enjoy working in a laboratory. A two-year college degree that includes courses in general science and laboratory techniques is recommended. Even with a college background, however, most of the laboratory skills you will need to know will be learned on the job.

Chapter 11 Protists and Fungi

lacy designs. Some red algae have hard calcium deposits in their cell walls. They are called coralline algae. Red algae are most common in warm tropical seas, where many different species live.

Red algae are often harvested by humans for use as food. Some contain substances used as food thickeners. One red alga is harvested for making agar. Agar is a jelly-like material used for growing bacteria in the laboratory.

Science and You *Polishing Up on Diatoms*

Have you ever rubbed a liquid or paste polish between your fingers? If so, you may have discovered that the polish contained very small particles of grit. When rubbed over a rough surface, these particles smooth out pits and bumps.

Many polishes, including some brands of toothpaste, contain grit made of ancient deposits of diatom shells. As you may recall, these deposits were formed on the ocean floor as the shells of dead diatoms settled to the bottom. These deposits are called diatomaceous (DY uh tuh MAY shus) earth.

Diatomaceous earth is made up of silica, the same substance that makes up the common mineral quartz. So why aren't polishes made from ground-up quartz? If you look at diatomaceous earth under a microscope, you can see the answer. The individual diatom shells are still there. The shells have many sharp edges. These edges are what make diatomaceous earth good for polishing.

Figure 11.15
The remains of diatom shells are preserved in diatomaceous earth.

Check and Explain

1. Which two groups of algae contain only many-celled species?

2. Describe one way in which humans use algae.

3. **Reason and Conclude** In some older classification systems, euglenas were placed in the animal kingdom. Explain why this classification made sense.

4. **Classify** Divide the six groups of algae you have studied into two large groups. What will you call the groups? On which characteristics do you base your classification?

Chapter 11 Protists and Fungi 221

SECTION 11.4

Section Objectives
For a list of section objectives, see the Student Edition page.

Skills Objectives
Students should be able to:

Find Causes for the way sugar affects yeast.

Predict what would happen if all the fungi and bacteria on the earth died out.

Classify protozoa using a taxonomic key.

Hypothesize the ideal conditions for fungi growth.

Vocabulary
fungus, hyphae, budding

MOTIVATE

Skills WarmUp
To help students observe a fungus that is useful to humans, have them do the Skills WarmUp.
Answer Students will see that the water with the sugar in it has become bubbly, and more bubbles are coming from the yeast. Students might note that the bubbles are related to the combination of yeast and sugar in water.

Prior Knowledge
To gauge how much students know about fungi, ask the following questions:

▶ Can you name a fungus?
▶ Where do fungi live? Why do they live there?
▶ Why do you think we categorize mushrooms as fungi?

Answer to In-Text Question

① Unlike plants and animals, fungi absorb food. Fungi can't make food through photosynthesis as plants do, nor can they ingest food as animals do.

222

INTEGRATED LEARNING

Themes in Science
Evolution Living things are classified on the basis of evolutionary relationships. As you study this section, stress to students that because fungi absorb food rather than ingest it or photosynthesize, they are classified in a different kingdom from animals, plants, monera, and protists.

Skills WarmUp

Find Causes Add a small amount of active dry yeast to two containers of warm water. Add sugar to one of the containers. Mix well and wait about ten minutes. What do you see? Explain your observations.

Figure 11.16 ▲
These mushrooms are the reproductive structures of a fungus. How do fungi differ from other kinds of organisms? ①

11.4 Fungi

Objectives

▶ **Explain** why fungi are different from members of the other four kingdoms.
▶ **Describe** the basic characteristics of each major fungus group.
▶ **Describe** the lichens and slime molds.
▶ **Hypothesize** about what fungi need to grow best.

Have you ever uncovered a slice of bread with a dark, furry coating? If so, then you observed a bread mold. A bread mold is a common example of a **fungus**, a spore-producing organism with complex cells and no chlorophyll. The plural of fungus is fungi (FUHN gee), which is also the name of the kingdom in which these organisms belong.

The fungi can be found nearly everywhere. Like monerans, they cause disease, spoil food, and damage crops. But also like monerans, fungi are important in the earth's ecology and help humans in many ways.

Characteristics of Fungi

How are fungi different from organisms in the other four kingdoms? Fungi, first of all, are eukaryotes, which separates them from the monera. Unlike plants and algae, fungi lack chlorophyll and can't photosynthesize. Unlike animals and protozoa, fungi cannot ingest food either. Instead, fungi *absorb* nutrients from their environment.

Most fungi feed upon dead organisms or other non-living organic matter. They digest this food outside their cells by releasing digestive juices. These chemicals break down the organic matter into simpler substances that fungi can use as nutrients. Organisms that obtain their nutrition in this way are called decomposers. Although most fungi are many-celled, some fungi, such as yeast, exist as single cells.

The bodies of the many-celled fungi are formed from branching, threadlike filaments called **hyphae** (HY fee). Figure 11.17 shows what hyphae look like. In most fungi, the hyphae are divided into single cells.

222 Chapter 11 Protists and Fungi

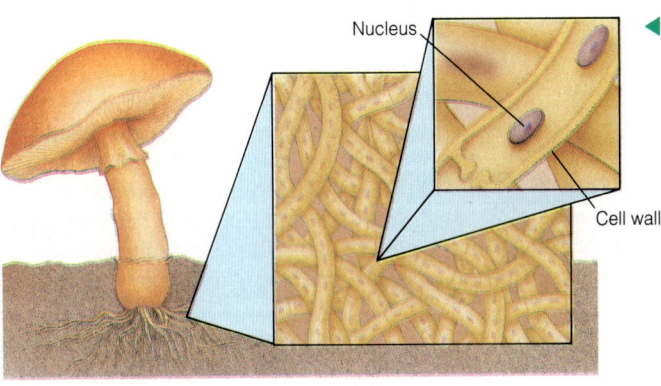

Figure 11.17
Hyphae are thin strands of tissue one cell thick. How do they help a fungus get its food? ②

Having a body made of hyphae helps a fungus obtain its nutrients. Hyphae can grow rapidly, reaching wherever there might be food to digest and absorb. And because hyphae are only one cell thick, all cells can be in contact with the food source.

Hyphae are also good building blocks. In complex fungi, the hyphae form a broad underground mat called a mycelium (my SEE lee uhm). Hyphae can also grow in closely packed bunches to form fruiting bodies. The fruiting bodies of some groups of fungi are commonly called mushrooms. Mushrooms and other fruiting bodies function to produce and release the spores that fungi use to reproduce.

Diversity of Fungi

Mycologists (my KAHL uh jihsts), the scientists who study fungi, recognize over 100,000 species of fungi. Among these many fungi, there are three different ways of producing spores through sexual reproduction. Each way of producing spores involves a different spore-producing structure. Mycologists use these different structures to classify fungi into three groups: threadlike, sac, and club.

Some fungi, however, do not reproduce sexually at all. Or at least they are not known to have sexual reproduction. So they are placed in a fourth major group, the imperfect fungi. Occasionally, a member of this fourth group is found to have a sexual stage in its life cycle and is moved to another group.

Skills WorkOut

Predict Write a paragraph describing what would happen if suddenly all the fungi and bacteria on the earth died out.

Chapter 11 Protists and Fungi 223

TEACH

Discuss
Write the terms *plant* and *fungi* on the chalkboard and have students list some of the observable characteristics of each group of organisms. Ask students the following questions:

▶ Which group reproduces by spores? (Fungi) By seeds? (Plants)

▶ How do the cells differ? (The cells of plants contain chlorophyll and can photosynthesize; fungi cells have no chlorophyll and can't photosynthesize.)

Critical Thinking
Compare and Contrast Have students look at Figure 11.17 and ask them how hyphae are similar to and different from roots. (Hyphae are like roots in that they grow and branch into the soil or whatever medium they are growing in. They are different from roots in that they give off substances that break down whatever medium the hyphae are growing in. These materials become dissolved in water and are taken into the fungus and used for its life activities.)

Skills WorkOut
To help students understand the role of fungi and bacteria in the earth's ecosystems, have them do the Skills WorkOut.
Answer Answers will vary. Fungi and bacteria digest wastes and the remains of dead organic matter. Without fungi and bacteria, these wastes would pile up. The source of food from fungi would disappear.

Answer to In-Text Question
② Hyphae can grow quickly, enabling them to reach food sources.

TEACH • Continued

Class Activity
Have students grow bread molds. Moisten a piece of bread with water, place it in a plastic sandwich bag, and store the bag in a dark, warm place. After several days, have students observe the bread for mold growth. Students should use hand lenses but should not remove the bread from the plastic sandwich bags. Have students note color and other observable differences and compare their observations with those of other groups. Ask students the following questions:

▶ Are the molds all the same color? (No)

▶ What do you think the color differences may mean? (Different kinds of molds)

Skills Development
Estimate Have students think about this problem: Suppose the number of black bread mold spore cases on a piece of bread is 1,000 per square centimeter. How many spore cases would you estimate to be on an entire piece of bread? (Answers may vary depending on the size of the bread. However, if a piece of bread is 10 by 15 centimeters, then the number of spore cases would be 100,000 to 150,000.)

The Living Textbook:
Life Science Side 3

Chapter 11   Frame 05289
Bread Mold (Movie)
Search: Play:

224

INTEGRATED LEARNING

Multicultural Perspectives
Encourage students to research and present information about popular mushrooms and sac fungi (truffles) used in the cuisine of different cultures. French, Chinese, and Japanese foods often contain fungi.

Use Integrating Worksheet 11.4.

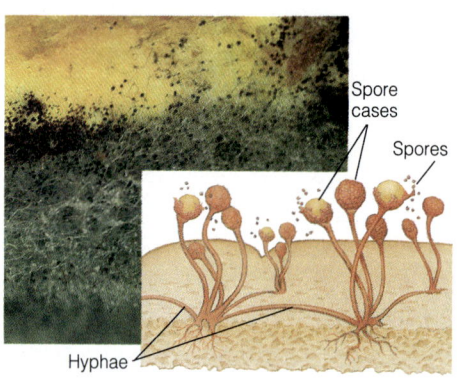

Figure 11.18 ▲
The tiny dark spots you see on bread mold are the spore cases of a threadlike fungus.

Threadlike Fungi

One group of fungi produces spores in round spore cases at the tips of hyphae. Members of this group are called threadlike fungi. Perhaps the best known threadlike fungus is the black bread mold, *Rhizopus stolonifer*.

When a *Rhizopus* spore germinates, it gives rise to a loose mat of hyphae. Some of the hyphae anchor the mold to the substance on which it grows and release digestive chemicals. These rootlike hyphae are called rhizoids.

Other types of hyphae grow upward and bear the cases in which spores are produced. You can see these spore cases in Figure 11.18. The spores released from the cases float around in the air, and a few find good places to grow.

Another kind of threadlike fungus has more control over where its spores go. This fungus lives on animal dung. It aims its spore cases toward bright light, where grass is likely to be growing. Then it shoots its spores like little cannonballs! The spores stick to the grass, and when cows eat it, they disperse the spores.

Sac Fungi

The largest group of fungi consists of the sac fungi. Sac fungi include diverse forms, such as the single-celled yeast, plant parasites, blue-green molds, mildews, and the edible mushrooms called morels and truffles. When reproducing sexually, all these fungi produce spores in saclike structures.

Yeasts have a huge impact on human society. They obtain the energy they need through a process called fermentation. During fermentation, sugar is broken down and its stored energy released. Carbon dioxide gas and alcohol are produced as wastes. Yeasts growing in bread dough produce bubbles of carbon dioxide gas, which cause the dough to rise.

In addition to reproducing sexually, yeasts use a form of asexual reproduction called **budding**. When a yeast cell buds, part of the cell wall swells and is eventually pinched off from the parent as a separate yeast cell.

Other sac fungi are harmful or annoying. Dutch elm disease is caused by a sac fungus. Humans are infected by a sac fungus when they get athlete's foot.

Figure 11.19 ▲
Morels are a kind of sac fungus. They are highly prized for cooking.

224   Chapter 11   Protists and Fungi

Themes in Science

Systems and Interactions The fungi and algae that form lichen have a special relationship. Each helps the other survive. Many of the earth's organisms have symbiotic relationships. Ask students to identify another similar relationship they have learned about in this chapter. (Flagellates live and feed inside the intestines of termites. They help termites by aiding in the digestion of the wood that termites eat. See page 215.)

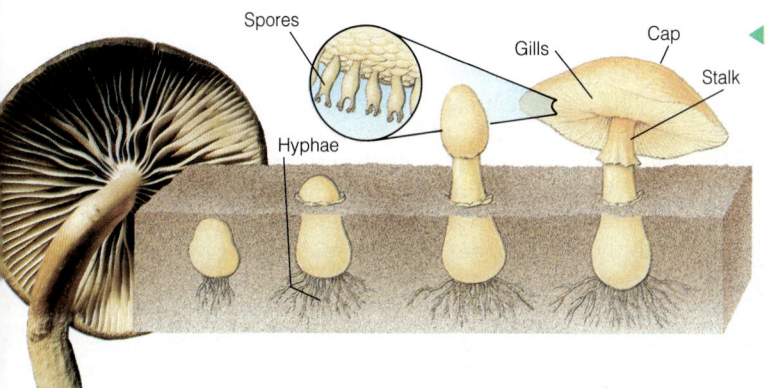

Figure 11.20
The fruiting body of a club fungus begins as a small, round mass of hyphae. The mature fruiting body can release billions of spores from the many clublike structures lining its gills.

Club Fungi

The mushrooms that spring up in lawns and top some kinds of pizza belong to the club fungi. This group also includes the rusts and smuts, both of which cause plant diseases. What these fungi have in common is a spore-producing structure shaped like a club.

In most of the common mushrooms, these club-shaped structures are found lining the gills on the underside of the mushroom's cap. When the spores are ready, they drop from the gills and are scattered by the wind.

However, a mushroom is only a small part of a larger, hidden mass. Beneath the soil's surface, a club fungus exists as a mycelium that can spread through a large area of soil.

The fruiting bodies of some types of club fungi are edible. Other types of club fungi, however, contain poisons that can cause serious sickness and death.

Imperfect Fungi

Fungi that have no known sexual reproductive cycle are classified as imperfect fungi. This group includes the fungi that produce the antibiotic penicillin and those species that give certain cheeses their distinct flavors.

Lichens

Have you noticed colorful, crustlike blotches on rocks or trees like the ones in Figure 11.21? They are organisms called lichens (LY kuhns). A lichen is made up of a fungus *and* an alga. Living together, these organisms from different kingdoms help each other survive.

The alga produces glucose for both itself and the fungus. The hyphae of the fungus anchor the lichen and keep in moisture. The hyphae are also able to absorb minerals and other necessary nutrients from the air and the surface on which the lichen grows. A lichen is able to grow in habitats where neither the fungus nor the alga could live alone.

Figure 11.21
What two kinds of organisms make up a lichen? ①

Class Activity

Have students work in small groups to make spore prints. Obtain mushrooms that are fully open and yet still fresh. Have each group cut off the stalk with scissors and then place the cap on a clean piece of paper. Then have them place the mouth of a jar or beaker over the cap and allow the cap to sit undisturbed for three or four hours. Then lift off the jar and carefully remove the cap. Have students examine the pattern made by the spores. Allow students to look at some of the mushroom spores under the microscope. Warn students not to taste the mushrooms.

Reteach

Use Reteach Worksheet 2.4.

Discuss

Point out that, because they are made up of algae and fungi, lichens can be discussed in this section or in the section on algae.

Research

Have students research the three kinds of lichens—foliose, crustose, and fruticose. Have students write a report in which they discuss the similarities and differences among the three types of lichens.

Answer to In-Text Question

① **A fungus and an alga make up a lichen.**

The Living Textbook:
Life Science Sides 1-4

Chapter 1   Frame 00592
Lichens (11 Frames)
Search: Step:

Chapter 11 Protists and Fungi

TEACH

Enrich

Have students investigate the life cycle of a slime mold and write a brief report of their findings. Ask them to define the terms *plasmodium* and *sclerotium* in their reports. Have them also explain what the slime mold takes in as nutrients in the plasmodium stage and why it is not like a fungus. (It takes in living bacteria, spores, and protists. True fungi do not get nutrients from other organisms.)

SkillBuilder

Stress to students the importance of collecting data in a systematic way.

Answers

1. The four main fungi groups are imperfect, threadlike, sac, and club.
2. The distinguishing characteristics are having a sexual stage, producing spores in upright hyphae, producing spores in saclike structures.
3. Step 1b describes the type of fungi.
4. Only threadlike fungi produce spores in cases at tips of upright hyphae. Two kinds of fungi produce spores in another kind of structure.

Students' keys should have at least three steps and should resemble the key shown.

Answer to In-Text Question

① Slime molds are found on forest floors.

The Living Textbook: Life Science Sides 1-4

Chapter 1 Frame 00544
Slime Mold (2 Frames)
Search: Step:

226

TEACHING OPTIONS

Portfolio

Challenge students to imagine that they are attending a taxonomy conference. Direct students to prepare a speech proposing a classification for slime molds and defending their opinions. After presenting their speeches, have the class vote on the best classification. Students can keep the speeches in their portfolios.

Cooperative Learning

Have the class work in groups. Assign each group a type of fungi—threadlike, sac, or club. Have students create three-dimensional models of their fungus and use the models to demonstrate the reproductive process. For example, students could make a model of a club fungus releasing its reproductive spores.

Figure 11.22 ▲
Many slime molds are very colorful. Where do you think you could find one? ①

Slime Molds

Funguslike organisms called slime molds present a problem in classification. As organisms without chlorophyll, they are like fungi. However, they are also very much like protists because they can move. In addition, the cells that germinate from some slime-mold spores have flagella. Many scientists place the slime molds with the protists, but they are studied by mycologists.

One kind of slime mold is a large mass of cytoplasm with many nuclei. It engulfs food particles and grows by extending pseudopodia through the soil. When conditions become too dry for growth, it reproduces sexually.

Another kind of slime mold exists most of the time as separate cells. For one stage of their life cycle, these cells come together in one mass and resemble a giant, many-celled ameba. This mass of protoplasm can move and even crawl over objects in its way. When it stops migrating, it produces asexual fruiting bodies.

SkillBuilder Classifying

A Key to Fungi

You know that scientists often use a taxonomic key. Recall that a taxonomic key is a series of statements that describe an organism's physical characteristics. Keys help scientists classify organisms.

Look at the simple key shown. This key can be used to place a fungus into one of the four major fungi groups. Study the key, then answer the questions.

1. What are the four main fungi groups in this key?
2. Based on information in the key, what are some characteristics that distinguish the four groups of fungi?
3. Why didn't you have to go to another step in the key after Step 1b?
4. Why did you have to go to another step in the key after Step 2b, but not Step 2a?

Key to Fungi
1a. Has sexual stage…GO TO 2
1b. Has no sexual stage…IMPERFECT FUNGI
2a. Spores produced in cases at tips of upright hyphae…THREADLIKE FUNGI
2b. Spores produced in another kind of structure…GO TO 3
3a. Sexual spores formed in a saclike structure…SAC FUNGI
3b. Sexual spores formed on club-shaped structures…CLUB FUNGI

Using this key as a guide, make your own key that can be used to place protozoa into one of the four groups discussed earlier. However, instead of placing the names of each group of protozoa on your key, leave a blank space. Then have a partner go through your key and fill in the name of each protozoa. Check your partner's answer. Is the key correct?

226 Chapter 11 Protists and Fungi

INTEGRATED LEARNING

Social Studies Connection
The Irish potato blight of the 1800s was caused by the fungus *Phytophthora infestans*. Have students research the social and economic effects of this agricultural disaster. Have students share information and present a play or documentary about how one family was affected by the blight.

Science and Society *Penicillin*

Fungi, like most organisms, have several defenses against harmful bacteria. Perhaps their best known defense is an adaptation found in a group of imperfect fungi. These fungi produce toxins, called antibiotics, that target and kill bacterial cells. The fungi that produce these poisons belong to the genus *Penicillium*. The poisons they produce are called penicillin.

Penicillin's antibacterial action was first discovered in 1928 by British scientist Alexander Fleming. Fleming was growing cultures of bacteria when one was accidentally contaminated with a *Penicillium* mold. Fleming observed that bacteria would not grow near the mold. He hypothesized that the *Penicillium* produced a substance that killed bacteria.

Ten years later, Howard Florey purified the antibacterial substance. Since this poison did not affect human cells, he proposed that it be used as a drug to fight bacterial infection. Soon, penicillin began to be used to treat bacterial diseases and infections. The drug has saved many lives and has revolutionized the practice of medicine.

Since the discovery of penicillin, many other antibiotics have been developed. Penicillin, however, remains the best treatment for many kinds of bacterial infections. Today, penicillin is obtained from special cultures of *Penicillium*. These molds are maintained in large vats of nutrient broth. The penicillin extracted from these vats is purified, then shipped to drug suppliers.

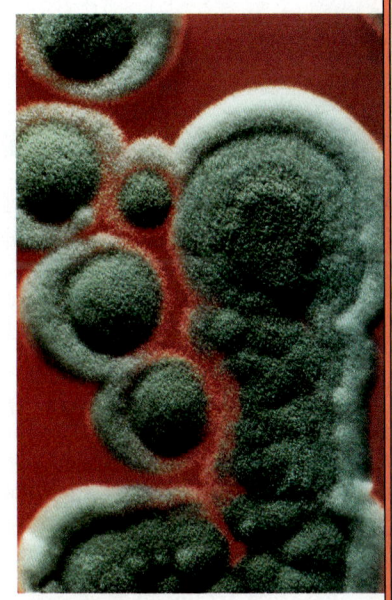

Figure 11.23 ▲
The fungus *Penicillium* is the source of one of the most important drugs ever developed.

Check and Explain

1. Why are the fungi placed in a kingdom of their own?
2. To which group of fungi do most mushrooms belong?
3. **Predict** Two slices of bread are placed on a table. One is covered. On which slice will mold first start growing? Explain.
4. **Hypothesize** What do fungi need to grow best? Form hypotheses about the ideal temperature, amount of light, level of moisture, and type of food for a bread mold. Then design experiments to test these hypotheses.

Research
Have students write to drug companies to gather information about antibiotics produced from fungi. Have them make a list of specific questions to ask about the sources of antibiotics, the methods of purifying them, and the development of new kinds of antibiotics.

EVALUATE

WrapUp
Reteach Return students' work from the Skills WarmUp, Section 11.1. Have them revise their descriptions to include new information that they've learned about protists and fungi.

Use Review Worksheet 11.4.

Check and Explain

1. Fungi are eukaryotes, which separates them from monera. They do not photosynthesize, which separates them from plants, and they do not ingest food, which separates them from animals and protozoa.
2. Sac fungi
3. Mold will probably grow sooner on the covered slice of bread because moisture is required for mold growth.
4. Answers may vary. Fungi need moisture, warmth, and little light. They get nutrients from bread. For a description of an experiment, see Activity 2 on page 46.

ACTIVITY 11

Time 90 minutes   **Group** 4–5

Materials

6 wax markers
6 packets of dried yeast
sugar
18 large test tubes
18 250-mL beakers
18 medium balloons
6 spoons
clock or watch
ice

Analysis

1. Variables: amount of yeast, amount of sugar, temperature of water bath, size of balloon; controlled variable: water temperature
2. The balloon was becoming inflated.
3. The greatest change should have occurred in the warm water.
4. Yeast cells are fermenting the sugar, releasing carbon dioxide as a product.
5. Sugar is food for the yeast cells. Without it, fermentation can't occur.

Conclusion

Students' conclusions should be based on their actual observations.

Everyday Application

Keeping it warm

Extension

Accept all logical experimental designs. Students might address the following:

▶ What result corresponds to the best temperature?
▶ What is the highest temperature at which yeast grows? The lowest?
▶ What is the lowest temperature at which yeast grows?

TEACHING OPTIONS

Prelab Discussion

Have students read the activity before beginning the discussion. Ask the following questions:

▶ Why is the balloon needed for this investigation?
▶ Why is it important for the yeast in the three tubes to come from the same packet?
▶ Why is it important to divide the yeast equally?

Activity 11   How does temperature affect yeast growth?

Skills   Observe; Infer; Interpret Data

Task 1   Prelab Prep

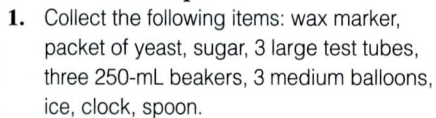

1. Collect the following items: wax marker, packet of yeast, sugar, 3 large test tubes, three 250-mL beakers, 3 medium balloons, ice, clock, spoon.
2. Use the wax marker to label the test tubes 1, 2, and 3.

Task 2   Data Record

1. On a separate sheet of paper, draw Table 11.1.
2. Record all your observations about the balloon size, the appearance of the yeast mixture, and any other observations in the table.

Task 3   Procedure

1. Divide the contents of the yeast packet into approximately three equal parts. Place one part in each of the three test tubes.
2. Use a spoon to add about 5 mL of sugar to each test tube.
3. Carefully fill each tube halfway with warm water.
4. Place a balloon over the mouth of each tube. (Make sure that all the air has been squeezed out of the balloon.)
5. Place tube 1 in a beaker filled with ice water. Place tube 2 in a beaker filled with room-temperature water. Place tube 3 in a beaker filled with warm water.
6. Observe each tube, then record your observations in the table. Repeat at the end of 5, 30, and 60 minutes.

Task 4   Analysis

1. Identify the variable in this activity.
2. How did you know when the yeast was growing?
3. In which beaker did the largest change occur?
4. What caused the observed changes? What gas did the yeast produce?
5. Why was sugar added to the test tubes? What would have happened without sugar?
6. Why did you place balloons on the test tubes?

Task 5   Conclusion

Write a short paragraph explaining what happens when yeast is placed in a warm sugar-water solution.

Everyday Application

How could you increase the rate at which baking dough rises?

Extension

Design an experiment that could be used to discover the best temperature for yeast growth.

Table 11.1   Yeast Observations

Test Tube Number	Temperature of Bath	Observations			
		At Beginning	After 5 Minutes	After 30 Minutes	After 60 Minutes

Chapter 11 Review

Concept Sumary

11.1 Diversity of Protists
- Protists have complex cells.
- The two main groups of protists are protozoa and algae.
- Protozoa are single-celled. Algae can be single-celled or many-celled.
- Single-celled algae make up plankton, which is the first level of the food chain in lakes and oceans.

11.2 Protozoa
- Sarcodines use pseudopodia to move, ciliates use cilia, and flagellates use flagella.
- Sporozoans have no special structures for movement.
- Sporozoans are parasites and can cause diseases such as malaria.

11.3 Algae
- The groups of algae include the single-celled dinoflagellates and diatoms, the single- and many-celled green algae, and the many-celled brown algae and red algae.
- Red tide is caused by dinoflagellates that grow rapidly and contain poison.
- Humans use algae for food and to make food additives and fertilizer.

11.4 Fungi
- Fungi are eukaryotes that absorb food.
- Threadlike fungi produce round spore cases at the tips of hyphae.
- Sac fungi include diverse forms such as yeast, truffles, and mildews.
- Club fungi include many mushrooms, rusts and smuts. They have a spore-producing structure shaped like a club.
- Lichens are made up of a fungus and an alga living together.

Chapter Vocabulary

protozoan (11.1)	eukaryote (11.1)	spore (11.2)	hyphae (11.4)
protist (11.1)	pseudopodia (11.2)	fungus (11.4)	budding (11.4)
alga (11.1)	cilia (11.2)		

Check Your Vocabulary

Use the vocabulary words above to complete the following sentences correctly.

1. Sarcodines use ____ to move.
2. Members of the kingdom Protista are ____.
3. A yeast cell reproduces asexually by ____.
4. Ciliates use ____ to move.
5. Organisms with complex cells are called ____.
6. A spore-producing organism with complex cells and no chlorophyll is a ____.
7. Single-celled animal-like organisms belonging to the kingdom Protista are ____.
8. Plantlike organisms of the kingdom Protista are ____.
9. Sporozoans produce reproductive cells called ____.
10. Many-celled fungi bodies are formed by filaments called ____.

Write Your Vocabulary

Write sentences using each vocabulary word. Show that you know what each word means.

CHAPTER REVIEW 11

Check Your Vocabulary
1. pseudopodia
2. protists
3. budding
4. cilia
5. eukaryotes
6. fungus
7. protozoa
8. algae
9. spores
10. hyphae

Write Your Vocabulary
Be sure students understand how to use each vocabulary word in a sentence before they begin.

Use Vocabulary Worksheet for Chapter 11.

CHAPTER 11 REVIEW

Check Your Knowledge

1. Red algae are many-celled and live below the surface of the ocean.
2. *Volvox* and *Spirogyra* are colonial green algae. Sea lettuce has many cells aligned in leaflike sheets.
3. Many scientists place slime molds with the protists because they can move, using either pseudopodia or flagella.
4. Protozoa are single-celled, animal-like protists that live wherever there is water—ponds, oceans, soil, and inside other organisms.
5. Sporozoa have no organelles for movement and reproduce by way of spores.
6. Diatoms are single-celled, gold-colored algae with glasslike shells. As a type of algae, diatoms are part of the lowest level of the food chain.
7. False; flagellum
8. True
9. False; sexual
10. True
11. True
12. True
13. False; eukaryotes
14. False; sarcodines

Check Your Understanding

1. Answers will vary.
2. Accept all logical answers. By carrying out photosynthesis, algae can contribute oxygen to the aquarium and provide shelter and a food source for other organisms in the aquarium.
3. Sporozoa, because they cannot move
4. a. Spores need to fall from the gills before they disperse.
   b. Fruiting body
   c. Club fungi
5. Answers may vary. Possible answers include: algae may be many-celled, protozoa are single-celled; algae carry out photosynthesis; algae have no transport vessels; protozoa have more complex internal organization; most algae have cell walls.
6. Decomposers feed on dead organisms. Decomposers do not have to use energy to hunt, and their food supply is assured.
7. A paramecium uses cilia to propel itself and to move food particles into its oral groove.
8. The mycelium is spread out equally in all directions, covering a circular area.

Chapter 11 Review

Check Your Knowledge

Answer the following in complete sentences.

1. Are red algae many-celled or single-celled? Where do they usually live?
2. Describe two examples of green algae. Are they many-celled or single-celled?
3. How are slime molds usually classified? What are the reasons for this classification?
4. What are protozoa? Where do they live?
5. How are sporozoans different from other protozoa?
6. Describe diatoms. Are they single-celled or many-celled? Where do they fit in the food chain?

Determine whether each statement is true or false. Write *true* if it is true. If it is false, change the underlined word to make the statement true.

7. The flagellates move using a whiplike organelle called a <u>cilia</u>.
8. The sarcodines move using footlike projections of cell matter called <u>pseudopodia</u>.
9. Fungi have three different ways of producing spores through <u>asexual</u> reproduction.
10. A lichen is made up of a <u>fungus</u> and alga.
11. Protists that have <u>chloroplasts</u> are classified as algae.
12. The two major groups of protists are the protozoa and the <u>algae</u>.
13. The kingdom Protista includes all <u>organisms</u> that are not plants, animals, or fungi.
14. The ameba is one of the best-known <u>ciliates</u>.

Check Your Understanding

Apply the concepts you have learned to answer each question.

1. Some algae are used in food. Which of the five types of algae have you eaten? Which would you never want to eat?
2. **Extension** If you had an aquarium, would you consider algae a welcome or unwanted addition? Why?
3. Which of the protozoa is the least animal-like? Why?
4. **Mystery Photo** The photograph on page 210 shows the underside of mushrooms. The grooves covering the underside are called gills. This is where mushrooms produce spores. A single mushroom can produce billions of spores.
   a. Infer why the gills are on the underside of the mushroom, pointing toward the ground.
   b. The mushroom is only part of the fungus. What part is it?
   c. These mushrooms are not morels or truffles. What type of fungi are they?
5. Describe two major differences between algae and protozoa.
6. **Infer** Fungi are decomposers. What is a decomposer? What are some of the advantages of being a decomposer?
7. How does a paramecium use cilia? Describe two uses.
8. **Application** Every year you notice a ring of mushrooms growing in the school lawn. Each year the ring gets larger. What can you infer about the location of the mushrooms' mycelium and its growth?

Chapter 11 Protists and Fungi

Develop Your Skills

1. a. There are 105,600 species.
   b. Threadlike fungi have the fewest species; sac fungi have the most.
2. Students' posters should show protists illustrated in this chapter as well as realistic impressions of protists that have only been described. Fanciful drawings are acceptable, as long as described features are referred to.
3. a. Euglena
   b. Yes
4. Animal-like characteristics: propel themselves using flagella, prey on other protists; plantlike characteristics: chloroplasts, photosynthesis; structures that fit both: nucleus, cell membrane

Make Connections

1. Concept maps will vary. You may wish to have students do the map below. First draw the map on the board, then add a few key terms. Have students draw and complete the map.

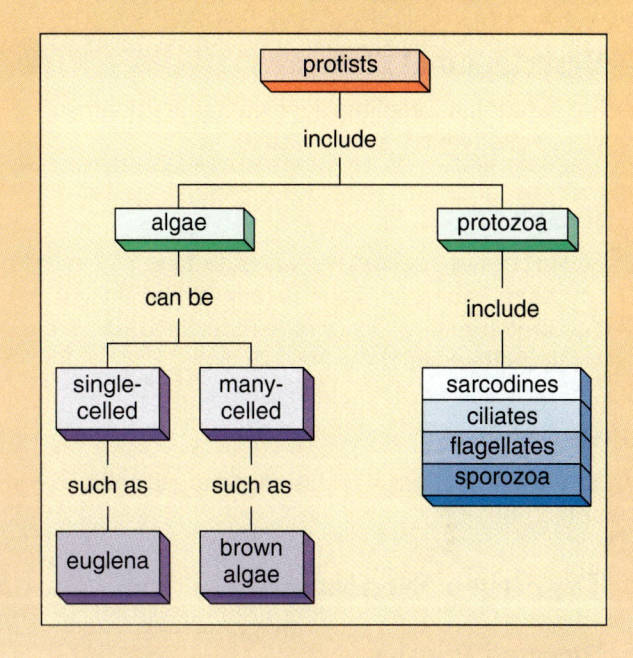

2. Answers will vary. Possible beneficial aspects include fermentation using yeast, edible fungi as food, antibiotics such as penicillin. Possible harmful aspects: some club fungi are poisonous; a type of sac fungi causes athlete's foot
3. Students' research will lead them to discover the broad range of types and uses of fungi and algae.
4. Students' research will lead them to discover that red tide occurs most often in subtropical waters. The blooms usually appear after a long period of hot, dry weather following a heavy rainstorm.

Develop Your Skills

Use the skills you have developed in this chapter to complete each activity.

1. **Interpret Data** The table below shows the number of species included in each group of fungi.
   a. How many species of fungi are there in all the groups combined?
   b. Which group of fungi has the fewest species? The most species?

Type	Number of Species
Threadlike fungi	600
Club fungi	25,000
Imperfect fungi	25,000
Lichens	25,000
Sac fungi	30,000

2. **Communicate** Using what you have learned in this chapter, create a poster that shows different kinds of protists. In addition to the examples shown in the chapter, use the descriptions of each group to infer what other protists may look like.
3. **Data Bank** Use the information on page 625 to answer the following questions.
   a. Which is larger, a typical plant cell or a euglena?
   b. Is a euglena visible to the unaided eye?
4. **Classify** Refer to Figure 11.11 on page 218. Which of the euglena's structures are animal like? Which are plantlike? Are there any structures that fit both categories?

Make Connections

1. **Link the Concepts** Draw a concept map showing how the concepts below link together. Add terms to connect, or link, the concepts.

   algae   single-celled
   protists   flagellates
   ciliates   sarcodines
   euglena   protozoa
   brown algae sporozoa
   many-celled

2. **Science and Language Arts** Write a paragraph describing how fungi can be beneficial to humans. Write another paragraph describing how they can be harmful to humans. Before writing, you may want to create a list of harmful and beneficial interactions between fungi and humans.
3. **Science and You** Fungi and algae can be used as food, or to make food additives. Research ways fungi and algae are used as food. Then describe a breakfast, lunch, and dinner, each including at least one fungus or alga.
4. **Science and Research** You have learned about the phenomenon known as red tide. Red tide is caused by a population explosion, or algal bloom, of dinoflagellates. Find out where red tide occurs most often in the world. Also research the most common environmental conditions under which red tide occurs. Include what methods, if any, are currently being used to fight red tide.

SCIENCE AND LITERATURE CONNECTION   UNIT 4

About the Literary Work

"The Great Mushroom Mistake" was adapted from the short story *The Great Mushroom Mistake* by Penelope Lively, copyright 1985 by E.P. Dutton, Inc. Reprinted by permission of E.P. Dutton, Inc.

Description of Change

Passages that were long in character development were edited for the sake of space.

Rationale

Scientifically inaccurate variables, like the inferred "unnatural" ability of Mrs. Hancock to grow things and the "unnatural" ability of Aunt Sadie to kill things, were edited from the selection.

Vocabulary

trowels, greengrocer, profusion, preened, cobbled, skirting-board

Teaching Strategies

Directed Inquiry

After students finish reading, discuss the story. Be sure to relate the story to the science lessons in this unit. Ask the following questions:

▶ What organism is the focus of this story? (Mushroom)

▶ Mushrooms are classified as what type of organism? (Fungi)

▶ What are the basic characteristics of members of this kingdom? (Fungi absorb food from their environment, are usually many-celled, and lack specialized organs and tissues.)

Skills Development

Classify   Ask students the following questions:

▶ What division of this kingdom does the mushroom belong to? (Club fungi)

▶ What are the characteristics of club fungi? (Spores are produced in the lining of the gills on the underside of the mushroom cap.)

Critical Thinking

Predict   Explain to the students that mushrooms require a cool, damp environment. Challenge students to find the part of the selection that describes this need of the organism. Then ask students, How might the Hancocks have used information about the environmental needs of a mushroom to thwart the invasion of their home? (Altering the mushrooms' environment so that it became hot and arid would impede their growth. This might be accomplished by putting a sunlamp over the container, holding a hair dryer over the container, and/or burning the existing plants.)

Science and Literature Connection

"The Great Mushroom Mistake"

The following excerpt is from the short story "The Great Mushroom Mistake" by Penelope Lively.

This year [for their mother's birthday, Sue and Alan Hancock] wanted something special—something different, something no one else's mother had. They searched the usual shops, and were not satisfied. Indeed, it was not until the very day before the birthday that they walked past the greengrocer on the corner near their school and saw the very thing.

Outside the shop were two shallow wooden boxes from which bubbled a profusion of gleaming white mushrooms: crisp fresh delicious-looking mushrooms. And alongside the boxes was a notice: GROW YOUR OWN MUSHROOMS! A NEW CROP EVERY DAY!

They looked at each other. Their mother had grown just about everything in her time, but never mushrooms. They went into the shop. They bought a small plastic bag labelled MUSHROOM SPORE, another bag of earthy stuff in which you were supposed to plant it, and an instruction leaflet.

Mrs. Hancock was thrilled. She couldn't wait to get going. The instruction leaflet said that the mushrooms like to grow indoors in a darkish place. A cellar would do nicely, it said, or the cupboard under the stairs.

The Hancocks had no cellar and the cupboard under the stairs was full of the sort of thing that takes refuge in cupboards under the stairs: old shoes and suitcases and a broken tennis racket and a chair with only three legs. Mrs Hancock decided that the only place was the cupboard in the spare room, which was used by guests only and no guest was threatening for some while. The children helped her to spread the earthy stuff out in boxes and scatter the spore. Then, apparently, all they had to do was wait.

During the night, Alan woke once and thought he heard a faint creaking sound, like a tree straining in the wind. And when, in the morning, they opened the door of the spare room cupboard there in the boxes was a fine growth of mushrooms—fat adult mushrooms and baby mushrooms pushing up under and around them. Mrs Hancock was delighted; the children preened themselves on the success of their present; everyone had fried mushrooms for breakfast.

The next day they found the cupboard door half open and mushrooms tumbling out onto the floor. "Gracious!" said Mrs Hancock. "It'll be mushroom soup for lunch today." The instruction leaflet said DO NOT SOW MORE SPORE TILL CROPPING CEASES, so they decided to wait and see what happened. . . .

That night, the creaking was more definite. Both children heard it; a sound

Skills in Science

Reading Skills in Science

1. The selection is fictional. Mushrooms do not make noise. Also, there is a limited amount of "earthy stuff" to support the growth of the organism, and the rate of reproduction presented in the story is highly unlikely.
2. Mushrooms do grow in soil, require a damp, cool environment, and reproduce by spores.

Writing Skills in Science

1. Before students begin the assignment, you may want to discuss the types of information a news article presents. To develop this concept, have students examine articles from a local newspaper and list five questions that they answer.
2. The spores that were planted grew into many mycelia that eventually sprouted caps. With an accelerated growth rate, these mushrooms may have dropped more spores from their gills into the soil.

Activities

Communicate Possible types of poisonous mushrooms include the death cap (*Amanita phalloides*), emetic russula (*Russula emetica*), fly agaric (*Amanita muscaria*), and jack-o'-lantern (*Omphalotus olearius*).

Collect Data A ring of mushrooms results from a large mycelium (the underground part of a mushroom) growing outward from a central point. Mushrooms grow from the edge of the mycelium and form an expanding circle. Ancient people called these *fairy rings* and believed that the rings were footprints left by fairies that danced in the area. The mushrooms that encircled the region were believed to be the seats upon which the tired fairies rested.

---

something between rustling and a splitting—the sound of growth. And in the morning there were mushrooms all over the spare room floor, a clump on the stairs and several clusters under the table in the hall. The Hancocks gazed at them in astonishment. "They *are* doing well," said Mrs Hancock, with a slight trace of anxiety in her voice. . . .

Over the next few days, the mushroom invasion continued. They found mushrooms in the bathroom and beside the cooker and in the toy chest. When they got up in the mornings they had to walk downstairs on a carpet of small mushrooms which squeaked faintly underfoot, like colonies of mice. It was when Mrs Hancock had to vacuum mushrooms from the sitting room carpet that they realized the situation had got quite out of control.

Skills in Science

Reading Skills in Science

1. **Classify** Classify this selection as fiction or nonfiction. Support your classification with information contained in the selection.
2. **Evaluate Sources** The selection paints a rather unique portrait of the life processes of a mushroom. Identify at least three descriptions that are scientifically accurate.

Writing Skills in Science

1. **Find the Main Idea** Write an article for a local newspaper that describes the great mushroom invasion of the Hancock home.
2. **Infer** Describe what happened in the guest room cupboard that first night while the Hancocks slept. Include information you learned in this unit about mushrooms.

Activities

Communicate Certain types of mushrooms can cause illness or even death if eaten. Visit a library to learn more about poisonous mushrooms. Make a chart to show the symptoms and effects of eating two types of poisonous mushrooms.

Collect Data Sometimes mushrooms grow in a circle in the wild. Use reference tools to explain how this occurs. What folklore has arisen about these rings of mushrooms?

Where to Read More

Uninvited Guests and Other Stories by Penelope Lively. New York: E.P. Dutton, 1985. Contains "The Great Mushroom Mistake" and other stories.

The Mushroom Hunter's Field Guide by A.H. Smith and N.S. Weber. University of Michigan Press, 1980. This field guide will help mushroom hunters classify specimens.

UNIT 5

UNIT OVERVIEW

Unit 5 introduces students to the plant kingdom and describes the characteristics, origins, and adaptations of nonflowering and flowering plants. Chapter 12 describes plant evolution and classification, including the distinct adaptations and characteristics of nonvascular and vascular plants. The chapter outlines the chemical processes of plants, including plant growth and hormones, photosynthesis, and respiration. A feature exploring the importance of plants is also included. Chapter 13 presents the characteristics, origins, and adaptations of nonflowering plants. Students are introduced to the structures of bryophytes, specifically mosses and liverworts. They also learn the characteristics and life

Introducing the Unit

Directed Inquiry
Have the students examine the photograph. Ask:
▶ How do the flowers in bloom affect the appearance of the desert? (They make it more colorful.)
▶ How do you think the flowers provide a survival advantage to the plants? (They make the plants more visible so that organisms can find and pollinate them.)

Writing About the Photograph
Ask students to write a story about falling asleep in the desert and awakening after the flowers bloom.

Unit 5 Plant Life

Chapters
12 A World of Plants
13 Nonflowering Plants
14 Flowering Plants

cycles of nonflowering vascular plants, specifically ferns and gymnosperms. The chapter also provides a look at gymnosperms in the context of geologic time. In Chapter 14, students learn about the structure and characteristics of angiosperms, or flowering plants. The chapter illustrates vascular plant systems using a transport system model to show how plant cells, tissues, and organs work to sustain plant life. Reproduction in flowering plants is also discussed. Finally, the chapter distinguishes plant growth from animal growth, explaining the action of plant hormones and describing seasonal growth patterns.

Data Bank

Use the information on page 624 to 631 to answer the following questions about topics explored in this unit.

Calculating

How much taller is the California redwood than the cedar?

Reading a Graph

How many metric tons of root crops are produced worldwide?

Making a Graph

Make a bar graph that shows the essential nutrients in plants. Your graph should show the form available for each element listed and the percent of dry weight.

Comparing

What is the tallest tree shown? What is the shortest tree shown?

The photograph to the left is of wildflowers and a yucca tree in the Mojave Desert in California. What time of year do you think this photograph was taken?

235

Data Bank Answers

Have students search the Data Bank on pages 624 to 631 for the answers to the questions on this page.

Calculating The California redwood is 46 meters taller than the cedar. The answer is found in the diagram, Tallest Tree Species, on page 627.

Extension Ask students to find out which local tree species is tallest.

Reading a Graph Six million metric tons of root crops are produced worldwide. The answer is found on page 625 in the graph, World Crop Production.

Extension Ask students to find out which crops are most commonly grown in the region of the country where they live.

Making a Graph Students' bar graphs should include the nine elements given on the chart, Essential Nutrients in Plants, on page 624.

Extension For each plant listed on the table on page 624, ask students to name which plant part is eaten—seed, stem, root, leaf, or fruit.

Extension Ask students which essential nutrient becomes especially scarce during a drought. (Hydrogen, which is available as water, becomes especially scarce.)

Comparing The tallest tree shown is the California redwood and the smallest tree shown is the Sitka spruce. The answer is found on page 627 in the diagram, Tallest Tree Species.

Answer to In-Text Question

This photograph was taken during the spring, when the desert is in full bloom.

235

CHAPTER 12

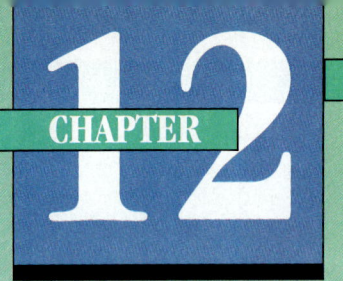

Overview

Chapter 12 is an introduction to the plant kingdom. The first section describes plant evolution and classification. Included in this description are the various adaptations and characteristics of nonvascular and vascular plants. The second section describes plant growth and the processes of photosynthesis and respiration.

Advance Planner

▶ Obtain slides of stomates or geraniums for TE page 244.

▶ Obtain a prism for TE page 245.

▶ Purchase radish seeds, clear cups, and soil for TE activity, page 246.

▶ Prepare 30 small jars with lids, sprigs of *Elodea* (or other water plant), labels, aluminum foil, straws, and cabbage juice indicator (available from science supply stores). SE Activity 12, page 248.

Skills Development Chart

Sections	Define Operationally	If ... Then Arguments	Interpret Data	Hypothesize	Observe	Predict	Research
12.1 Skills WarmUp	●						
Skills WorkOut		●					
Historical Notebook							●
12.2 Skills WarmUp				●			
SkillBuilder			●				
Activity					●	●	

Individual Needs

▶ **Limited English Proficiency Students** Photocopy pages 238, 239, and 244. Have students highlight all the places where the chapter vocabulary terms appear on these pages. They should write a definition for each term and add it to the appropriate page. They can then save these pages and use them to review the chapter vocabulary.

▶ **At-Risk Students** Assign students to cooperative groups. Photocopy the concept summary on page 249. Before students read the chapter, have them search through the text for the ten facts listed in the concept summary. When students find a fact, have them write the page number of their "find" on their group concept summary, next to the fact. Photocopy the group summary and distribute one to each student.

▶ **Gifted Students** Have students carefully read the material on page 242, which discusses hydroponics. Then ask students to create a story about a space colony and the use of hydroponics to maintain a food supply. Students' stories should include research about their chosen planet's environment and elements. Have students include information about plants' nutrients and processes on their space colony.

Resource Bank

▶ **Bulletin Board** Title your bulletin board *Riddles About Plants*. As you work through this chapter, encourage students to use the information in the text to make up riddles about plants. Have them write their riddles on squares of construction paper, asking them to add pictures that give clues to each riddle's answer. Students who know the answer to a riddle can write the answer on an index card and hang the answer next to the riddle.

CHAPTER 12 PLANNING GUIDE

Section	Core	Standard	Enriched	Section	Core	Standard	Enriched
12.1 Plant Origins pp. 237–242				**12.2 Chemistry in Plants** pp. 243–248			
Section Features Skills WarmUp, p. 237	●	●	●	**Section Features** Skills WarmUp, p. 243	●	●	●
Skills WorkOut, p. 238		●	●	SkillBuilder, p. 245		●	●
Historical Notebook, p. 241	●	●		Activity, p. 248	●	●	●
Blackline Masters Review Worksheet 12.1	●	●		**Blackline Masters** Review Worksheet 12.2	●	●	●
Skills Worksheet 12.1	●	●		Skills Worksheet 12.2	●	●	
Integrating Worksheet 12.1		●	●	Integrating Worksheet 12.2		●	●
				Reteach Worksheet 12.2	●	●	
				Vocabulary Worksheet 12	●	●	
Ancillary Options Multicultural Perspectives, p. 115	●	●	●	**Laboratory Program** Investigation 24	●	●	●
Laboratory Program Investigation 23	●	●	●	**Color Transparencies** Transparency 32	●	●	

Bibliography

The following resources can be used for teaching the chapter. See page T-40 for supplier codes.

Audio-Visual Sources
(video unless noted)
Carnivorous Plants. 12 min. 1974. NGSES.
Photosynthesis: Life Energy. 22 min. 1983. NGSES.
Plants: Parts and Processes. 2 filmstrips with cassettes. 1988. NGSES.
The Sequoia Giants of Sequoia National Park. 28 min film. HOL.
The World of Plants. 43 min. 1990. UL.

Software Resources
Basic Photosynthesis. Apple. BLS.
Photosynthesis and Light Energy. IBM. CCM.
Plants. Apple. J & S.

Library Resources
Bell, P. R., and C. L. F. Woodcock. The Diversity of Green Plants. 3d ed. London: E. Arnold (Publ.) Ltd., 1983.
Facklam, Howard and Margery. Plants: Extinction or Survival. Hillside, NJ: Enslow Publishers, Inc., 1990.
Margulis, Lynn, and Karlene V. Schwartz. Five Kingdoms: An Illustrated Guide to the Phyla of Life on Earth. San Francisco: W.H. Freeman and Co., 1982.
Pringle, Lawrence. Being a Plant. New York: Crowell, 1983.
Rahn, Joan Elma. Plants Up Close. Boston: Houghton Mifflin Co., 1981.

CHAPTER 12

INTEGRATED LEARNING

Writing Connection

Have students write a description of the plant shown in the photograph. Ask them what kind of plant they think it is and why. Have them describe its environment. Is it in a forest, a tropical island, a desert, the mountains, or a prairie? What other types of plants might live in the same habitat? Does it need sunlight or shade? How much water does it need? How can they tell? Encourage students to draw the plant in the environment they describe.

Introducing the Chapter

Have students read the description of the photograph on page 236. Ask if they agree or disagree with the description.

Directed Inquiry

Have students study the photograph. Ask:

▶ What kind of organism is shown in this photograph? (Students will probably say a cactus.)

▶ Describe the physical characteristics of this organism. (The rows of spikes and the green color)

▶ How is the plant's green color important? (It helps the plant absorb energy from sunlight.)

▶ Describe this plant's environment. (Dry, sunny desert)

▶ What kind of special characteristics do you think it needs to survive in this environment? (The ability to store water)

Chapter Vocabulary

angiosperm   nonvascular plant
Calvin cycle   stomate
gymnosperm   vascular plant

The Living Textbook: Life Science Sides 1–4

Chapter 3   Frame 00635
Plant Kingdom (112 Frames)
Search: Step:

Chapter 12 — A World of Plants

Chapter Sections

12.1 Plant Origins
12.2 Chemistry in Plants

What do you see?

" I see a cactus. It is very big and has repeating sections. The purple things are prickly thorns. The cactus uses these thorns for protection from animals and people. "

Chavon Webb
Carl Albert Junior High School
Oklahoma City, Oklahoma

To find out more about the photograph, look on page 250. As you read this chapter, you will learn more about the origins of plants and their life processes.

Themes in Science

Systems and Interactions Plants are fundamental to life on the earth not only because they provide oxygen, but also because so many species depend on them for nutrients. Most plants, however, do not rely on other organisms as a source of nutrients. Through photosynthesis, plants make glucose using carbon dioxide, water, and energy from light.

Social Studies Connection

Point out that early humans obtained food by hunting and by gathering grain, fruit, and nuts. About 10,000 years ago, people began learning how to cultivate and harvest edible plants. Primitive farming tools enabled these early farmers to produce enough food to help support large populations in the Middle East, northern China, northwest India, and the Americas.

SECTION 12.1

Section Objectives
For a list of section objectives, see the Student Edition page.

Skills Objectives
Students should be able to:

Define Operationally how people use plants.

Infer the advantages of plants being tall.

Vocabulary
nonvascular plant, vascular plant, gymnosperm, angiosperm

12.1 Plant Origins

Objectives

▶ **Describe** the origins of modern plants.
▶ **Compare** and **contrast** vascular and nonvascular plants.
▶ **Infer** the uses of plant adaptations.

Skills WarmUp

Define Operationally Make a list of plants and plant products you use every day. What do the plants on your list have in common? How many different ways do plants affect your life?

How many different plants do you know? Probably quite a few. Grasses and garden flowers, such as tulips, roses, and orchids, are plants. Pine, elm, and maple trees are plants. Mosses and ferns are also plants. Every day you probably eat parts of such common food plants as corn, rice, wheat, or potatoes.

People all over the world have long depended on plants for food, clothing, housing, medicines, and many other needs. Look at Figure 12.1. Plants from faraway places produce items that are part of your everyday life. The banana you ate for breakfast comes from a plant native to Malaysia. Bananas have been cultivated in Southeast Asia for more than 1,000 years. Vanilla, chocolate, and coffee come from plants grown in South America and Africa. Many spices, such as cinnamon and nutmeg, come from Asia and Africa. An early form of paper was made from papyrus reeds in ancient Egypt.

Characteristics of Plants

When you see a plant, you can probably tell that it is a plant. However, some organisms, such as algae, look like plants, but aren't. You need to know about the chemicals and cells inside a plant to see why it is classified as a plant. Plants have the following characteristics:

▶ All plants are many-celled.
▶ All plants are made of tissues and organs.
▶ All plants have cell walls.
▶ Almost all plant cells contain chlorophyll.

Plants, unlike most animals, continue to grow throughout their entire lives. Plants differ from many-celled algae because they have adaptations for living on land. These adaptations include a waxy coating to keep the cells moist and reproductive organs enclosed in protective tissue.

Figure 12.1 ▲
The bananas you eat are grown on plantations like this one in Costa Rica (top). Many different spices are available at this African market (bottom).

MOTIVATE

Skills WarmUp

To help students understand the usefulness and diversity of plants, have them do the Skills WarmUp.
Answer Students should give examples of plants used for food and for other purposes, such as cotton for cloth. They should also discuss how the plants enhance their lives.

Prior Knowledge

Gauge how much students know about plants. Ask the following questions:

▶ How are plants different from animals?

▶ How are plants important to you?

The Living Textbook:
Life Science Sides 1-4

Chapter 5   Frame 11873
Plant Kingdom Microviews (95 Frames)
Search: Step:

Chapter 12   A World of Plants

TEACH

Skills WorkOut
To help students understand the evolutionary importance of vascular systems in plants, have them do the Skills WorkOut.

Answers If plants never developed a vascular system, (1) the only plants on land would be bryophytes, (2) there would be fewer species of animals on land, (3) only the herbivores that eat low plants and a small number of carnivores would exist on land, and (4) the ocean would be the dominant region for life.

Skills Development
Infer Ask students whether moss usually lives in shade or direct sunlight. Why? (In shade; moss would dry out in full sunlight.)

Critical Thinking
Compare and Contrast Ask students what the difference is between nonvascular and vascular plants. (Vascular plants have tissues that transport water internally; nonvascular plants must absorb water from their environment.)

Integrated Learning
Use Integrating Worksheet 12.1.

The Living Textbook: Life Science Side 3
Chapter 12 Frame 06715
Liverwort (Movie)
Search: Play:

238

INTEGRATED LEARNING

Themes in Science

Evolution The structures that developed in early bryophytes helped support them on land. These structures distinguished them from their ancestors, which were only able to float on water. However, the bryophytes were still tied to the moist climates because they lacked structures to transport water internally, which would have been a necessary adaptation for living on land. Eventually, tracheophytes evolved, with internal tube-like structures that enabled them to move liquids. Ask students to explain how this adaptation enabled these plants to spread to more arid climates. (Plants could absorb water from roots inside the earth and the vascular system could carry water to all parts of the plant.)

Skills WorkOut

If...Then Arguments
Consider the following and write four *If...Then* statements. Nonvascular plants were the first plants to evolve. If plants never developed a vascular system, how would the world be different today? How would animal locations and feeding habits be different?

Plant Evolution and Classification

Plants evolved from many-celled green algae. Cells of green algae are very similar to plant cells. Both green algae and plant cells have the same pigments used in photosynthesis. Both have cell walls made of cellulose and store carbohydrates in the form of starch. Evidence from fossils shows that the first plants appeared about 400 million years ago.

The first plants were small and lived close to the water's edge. They had to stay very close to water because they had no system for transporting water throughout the plant. One cell would simply transfer materials to its neighbor, much like a sponge soaks up water. Plants that absorb water in this way are called **nonvascular** (NAHN VAS kyuh lur) **plants**. Nonvascular plants lack specialized tissues to transport water. Bryophytes (BRY oh FYTS), such as mosses, are living representatives of these early plants.

To grow successfully on land, plants evolved tissues with tubelike structures that can transport water. Plants with these vascular tissues are called **vascular** (VAS kyuh lur) **plants**. Over millions of years, this adaptation allowed vascular plants to colonize the land. Eventually vascular plants evolved other adaptations that let them live in a wide variety of land environments. The two most important adaptations to evolve in the vascular plants were the seed and the flower. Today, plants without vascular tissue, without seeds, and without flowers still exist along with flowering plants.

Bryophytes
Mosses are the most common bryophytes. These nonvascular plants have a variety of shapes and are generally less than 2 cm tall. More than 20,000 species are known to exist. Liverworts, shown at right, are also bryophytes. There are about 8,000 species of liverworts.

Nonvascular plants

238 Chapter 12 A World of Plants

Integrating the Sciences

Earth Science Forests of horsetails and giant club mosses, sometimes as big as 45 meters tall, covered much of the earth's landscape at one time. Over the course of 300 million years, the remains of these plants have become large coal deposits in the earth. Have students draw an ancient forest of giant club mosses and horsetails.

Themes in Science

Evolution All plants fall into one of two categories—nonflowering or flowering. Nonflowering plants include the ancient plants, such as ferns, horsetails, mosses, and gymnosperms. Flowering plants evolved later. Today, there are nearly one-quarter of a million species of flowering plants. Tell students that they will learn more about nonflowering plants in Chapter 13 and flowering plants in Chapter 14.

Explore Visually

Have students study the material on this page. Ask the following questions:

▶ Which one of the five groups of vascular plants pictured on this page is the most abundant today? (There are more than 235,000 species of angiosperms.)

▶ Which group pictured is the least abundant group today? (The horsetails)

▶ Which three groups are the most ancient vascular plants? (Ferns, club mosses, horsetails)

▶ Which group was the first one to produce seeds? (Gymnosperms)

Critical Thinking

Reason and Conclude Of the five groups of plants, which ones do not produce seeds? (Ferns, club mosses, horsetails)

Ferns ▲
Like club mosses and horsetails, ferns are an ancient group of vascular plants. More than 12,000 species of ferns are now living. Most ferns live in the tropics and other moist forest areas.

Angiosperms ▲
The first flowering plants, called **angiosperms** (AN jee oh SPURMZ), appeared about 120 million years ago. All angiosperms produce seeds enclosed in a fruit or pod. More than 235,000 species are known.

Gymnosperms ▲
The first plants to evolve seeds were the **gymnosperms** (JIHM noh SPURMZ). All gymnosperms produce naked seeds not enclosed in a fruit. Seeds and other adaptations allowed gymnosperms to live in dry places. More than 7,000 species exist today.

Club Mosses ▲
These vascular plants are one of the oldest groups. About 1,000 species exist today. Large treelike club mosses, now extinct, lived millions of years ago.

Vascular plants

Horsetails ▲
These vascular plants are considered "living fossils." They have changed very little over millions of years. Only about 15 species survive.

The Living Textbook: Life Science Side 3

Chapter 13   Frame 07798
Moss (Movie)
Search: Play:

The Living Textbook: Life Science Side 3

Chapter 14   Frame 09074
Fern (Movie)
Search: Play:

Chapter 12 A World of Plants

TEACH • Continued

Explore Visually

Have students study the material on this page. Ask the following questions:

▶ What are some adaptations of plants that live in tropical rain forests? (Some grow in the treetops where sunlight is plentiful; others grow on the forest floor and have large leaves that absorb the little light that comes through.)

▶ What adaptations allow the saguaro to survive in desert conditions? (Saguaros have a thick, green stem that stores water and captures the sun's energy. They also have a waxy coating that keeps the plant from drying out and spines for protection and shade.)

▶ What adaptations help creosotes conserve water in a desert environment? (They grow slowly, have waxy leaves that help retain water, and have a long, shallow root system that absorbs water over a large area.)

INTEGRATED LEARNING

Writing Connection

Have students create an imaginary planet in outer space on which plant evolution is occurring. Have them describe the geologic and environmental characteristics of the planet and explain what kinds of plants grow there. Students can make up new kinds of plants or discuss ones they know of on the earth. They should describe the plants' adaptations and explain how they help the plants survive in their environment. Students can write about their planet or draw a picture of it with brief descriptions.

Adaptations of Flowering Plants

Flowering plants grow in practically all land habitats—from seashores to mountain tops, from the tropics to the arctic, and from marshy areas to deserts. To survive such diverse conditions, these plants have developed special adaptations. Unlike animals, plants can't move from one place to another when conditions for growth become unfavorable. Most plants are firmly anchored to the ground and must adapt to changes in the environment.

▲ Many plants in the tropical rain forest grow in the sunny treetops, high above the ground. Their roots cling to the tree. Some have leaves that form a cup shape to store water.

Plants growing in the shade of the rain forest trees look different from plants growing in treetops. They have large leaves to absorb as much light as possible. ▼

▲ Cacti, like this saguaro (suh WAR oh), have a thick, green stem that can store water and capture energy from sunlight. A waxy coating on the stem keeps the plant from drying out. Spines protect the plant from animals and shade the stem.

Many desert plants, like this creosote bush, have adapted to dry conditions in a variety of ways. The plants use water efficiently. They grow slowly, so they need little water. The small waxy leaves of the creosote bush are adapted to slow water loss from the hot desert sun. The creosote bush also has a large, shallow root system to absorb water over a large area. ▼

240 Chapter 12 A World of Plants

Multicultural Perspectives

In the early 1920s at the age of 50, Ynez Mexia took up botanical studies at the University of California at Berkeley and began an obsession with plants that took her all over the world in search of new species. Mexia collected tens of thousands of plant specimens in remote areas of North and South America. Mexia's greatest and most daring adventure was her three-thousand-mile journey up the Amazon through Peru and Brazil. Some of her discoveries, such as the *Mexianthus mexicanus,* were named after her. Have students read more about Mexia's adventures collecting plant specimens.

Although these African euphorbias (yoo FOR bee uhz) look very similar to American cacti, they are unrelated. What environmental conditions are the euphorbias adapted for? ①

Alpine plants, like this cushion plant, are small and grow close to the ground. Plants in high mountain areas face short growing seasons with high winds and low temperatures. ▼

Historical Notebook

Plant Classification

Identifying and classifying plants has been important to people throughout history. One early classification system was developed by Greek scientist Theophrastus about 300 B.C. He classified plants into four groups: herbs, undershrubs, shrubs, and trees. Illustrated books for identifying plants were published in China in the eleventh century. These had both medical and scientific importance, because plants were widely used for medicine in China. During the Middle Ages in Europe, plants were also classified by medicinal use. Herbalists named plants after the human organ or disease they were used to treat.

The plant classification system used today is based on observations of reproductive structures of flowers and other plants. This system was developed in 1735 by Swedish botanist Carl von Linné, also known as Carolus Linnaeus. He used two words, a genus name followed by a species name, to name each plant. A drawing from his notebook is shown here. Since Latin was the most important scientific language in Linnaeus' time, scientific names are still in Latin or a Latinlike form. For example, the scientific name for red oak is *Quercus rubra*. *Quercus* is the genus of all oaks. The species name, *rubra,* is Latin for "red."

Dryas octopetala

1. Explain how plant classification changed through history.

2. Obtain a field guide to plants from the library. Look up the scientific name for three plants common in your area.

▶ Ask students what kind of environment they think African euphorbias are adapted to. (Euphorbias are like cacti and are adapted to the mainly dry climates of Africa.)

▶ What are some adaptations of plants that grow in high mountain areas? (They are small and grow close to the ground, which enables them to withstand high winds and low temperatures.)

Critical Thinking

Reason and Conclude Remind students that *Quercus rubra* is the Latin name for red oak. Ask them what *Quercus alba* means if *alba* is Latin for *white*. (White oak)

Ancillary Options

If you are using the blackline masters from *Multiculturalism in Mathematics, Science, and Technology,* have students read about Ynez Mexia on pages 115 and 116. Complete pages 116 to 118.

Historical Notebook

Answers

1. Students' answers will vary but should mention classification systems in ancient Greece, China, and during the Middle Ages in Europe, as well as the modern system established by Carl von Linné.

2. Three common plants in most parts of the United States are the dandelion (*Taraxacum officinale*), corn (*Zea mays*), and common ragweed (*Ambrosia artemisiifolia*).

Answer to In-Text Question

① **Euphorbias are adapted to desert conditions.**

Chapter 12 A World of Plants

TEACH · Continued

Critical Thinking
Reason and Conclude Ask students in what places hydroponics would be a particularly useful farming method. (In regions where there is a lack of fertile soil)

EVALUATE

WrapUp
Review Have students list the plants they have studied in this section. They should identify whether each plant is vascular or nonvascular and describe how it is adapted to its environment.
Use Review Worksheet 12.1.

Check and Explain
1. Vascular plants have vascular tissues that transport water; nonvascular plants do not have these tissues.
2. They have thick, waxy stems to store water and capture the sun's energy.
3. Plants in the rain forests have large leaves, which capture sunlight. The long, shallow roots of the creosote help it absorb water in the desert. The cushion plant grows close to the ground, which protects it from the high winds of its mountain habitat.
4. The long trunks and numerous branches of tall trees enable them to absorb more sunlight. Their height helps disperse seeds over a wide area.

Answer to In-Text Question
① Some advantages to hydroponics are that crops can be grown virtually anywhere and plant growth can be accelerated. However, not all plants grow well hydroponically.

242

TEACHING OPTIONS

Cooperative Learning
Have students work in small groups to find out where hydroponic farming methods are commonly used today. They should also investigate what crops grow well using hydroponic farming, what crops do not, and why. Some members of each group can make diagrams or models that display hydroponic farming methods. Other students can use these models as they present oral reports to the class.

Figure 12.2 ▲
These plants are being grown hydroponically. What are some advantages and disadvantages to this method? ①

Science and Technology *Hydroponics*

Hydroponics (HY droh PAHN ihks), shown in Figure 12.2, is a method of raising plants without soil. Instead, the plants are grown in water enriched with plant growth nutrients, such as nitrogen, phosphorus, and potassium. These and other plant nutrients are usually present in the soil and are essential for plant growth. Hydroponics has several advantages over traditional soil methods. More plants can be grown in a limited amount of space. Food crops mature more rapidly and produce greater yields. Some food crops that grow well hydroponically are tomatoes, lettuce, kale, spinach, and cucumbers.

The word *hydroponics* comes from two Greek words meaning "water working." The word was coined in 1936 by W. F. Gericke at the University of California. Dr. Gericke grew tomato vines more than 7 m tall in basins of water and nutrients. His experiments caused a sensation. Some people thought hydroponics could make traditional farming methods obsolete.

While hydroponic gardening didn't replace soil methods of growing plants, it has become useful in urban areas and other places with limited space. Hydroponics probably will be used to grow crops in future space stations and space colonies. Using hydroponics, every factor affecting plant growth can be regulated. As a result, the plants grow faster, taste better, and look more attractive.

Check and Explain
1. Distinguish between nonvascular and vascular plants.
2. List several ways in which cacti have become adapted to living in a dry environment.
3. **Find Causes** Environmental conditions caused different types of plants to evolve from one common ancestor. Describe three environmental conditions and the type of plant that adapted to each condition.
4. **Infer** Some vascular plants such as trees can grow to be more than 100 m tall. Discuss the advantages to being tall.

242 Chapter 12 A World of Plants

INTEGRATED LEARNING

Themes in Science

Energy Unlike animals, plants have the ability to make glucose. With the help of chlorophyll, plants convert energy from sunlight into chemical energy, which usually takes the form of ATP. Plants use the energy from ATP to convert carbon dioxide, taken from air and water, into glucose. While animals rely on energy sources, such as plants and other animals, to perform their basic life functions, plants rely on energy from sunlight.

12.2 Chemistry in Plants

Objectives

- **Describe** the two stages of photosynthesis.
- **Give examples** of chemical activities of plants.
- **Communicate** how energy is stored and used in a plant.

You can think of a plant as a chemical factory. Part of this chemical factory makes the sugar glucose. To make glucose, the plant uses energy from sunlight and raw materials from air and water. This process is called *photosynthesis*. Look at the plant in Figure 12.3. Which part takes in energy from the sun? ②

Some glucose made by the plant is stored for later use. The rest is transported throughout the plant. The glucose is used for energy to help the plant convert soil nutrients to body tissues and grow larger. The energy is also used to make other plant products, such as plant oils.

A plant's chemical factory also produces hormones. Hormones are complex chemicals that regulate growth and development. Plant hormones direct plant growth. As a result, stems grow toward light and roots toward water. Hormones also stimulate the growth of new leaves and flowers.

Skills WarmUp

Hypothesize Plants need light for photosynthesis. Hypothesize what would happen to plants grown in the dark. Consider how darkness would affect the amount of food they produced, their growth, their color, and their overall health. What experiment would be needed to test your hypothesis?

Figure 12.3
A plant is a natural chemical factory. ▼

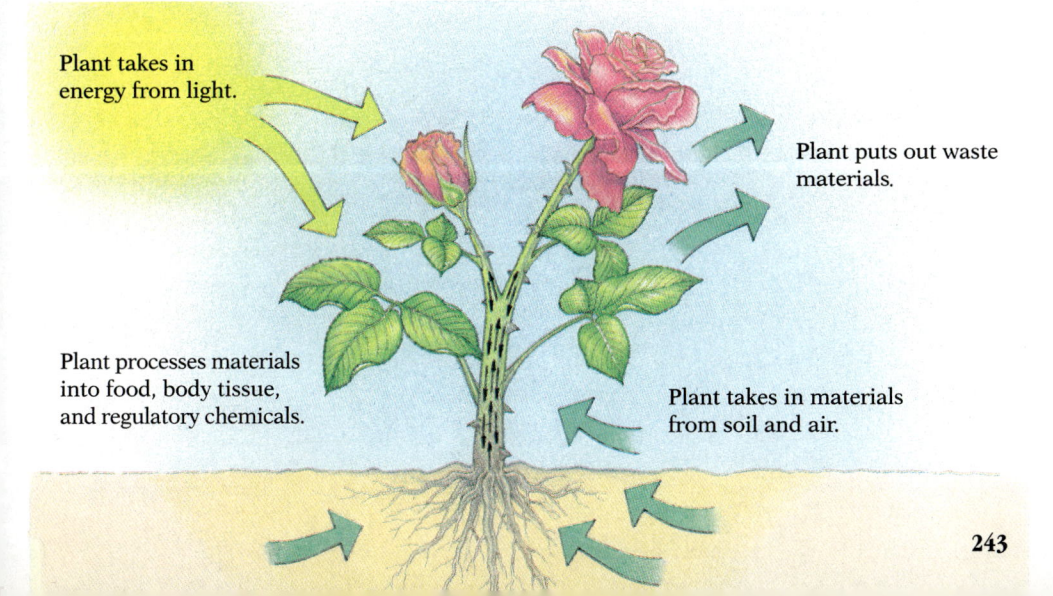

Plant takes in energy from light.

Plant puts out waste materials.

Plant processes materials into food, body tissue, and regulatory chemicals.

Plant takes in materials from soil and air.

SECTION 12.2

Section Objectives
For a list of section objectives, see the Student Edition page.

Skills Objectives
Students should be able to:

Hypothesize what would happen to plants growing without light.

Communicate how a leaf makes and uses glucose.

Vocabulary
stomate, Calvin cycle

MOTIVATE

Skills WarmUp
To help students understand plant chemistry, have them do the Skills WarmUp.
Answer Students should recognize the importance of light in photosynthesis. They could test their hypotheses by growing two plants in the same conditions, except for the amount of light they receive.

Prior Knowledge
To gauge how much students know about plant chemistry, ask the following questions:

- Where do plants get nutrients?
- How do plants and animals depend on one another?

Answer to In-Text Question
② The leaves

The Living Textbook:
Life Science Side 3

Chapter 17   Frame 26888
Stomatal Opening (Microscopic Movie)
Search: Play:

TEACH

Class Activity
While working in groups, have students observe slides of stomates with a microscope or hand lens. If slides aren't available, obtain geranium plants and have students look at the undersides of the leaves. Ask students if they can see the chloroplasts. What is their shape? Have groups record and discuss their observations.

Reteach
Review with students the material on photosynthesis and respiration in Chapter 5.

Directed Inquiry
Have students study Figure 12.4. Ask the following questions:

▶ What part of a plant absorbs light? (Chlorophyll)

▶ When is light absorbed by the chlorophyll? (During the first stage of photosynthesis)

▶ When does the plant use carbon dioxide? (In the Calvin cycle)

▶ Do plants release oxygen at night? (No)

Skills Development
Infer Ask students if they think plants release oxygen at night. (Oxygen is released during the first stage of photosynthesis, which requires light. Plants don't release oxygen if they don't receive light.)

The Living Textbook: Life Science Side 7

Chapter 20 Frame 02361
Photosynthesis (Movie)
Search: Play:

INTEGRATED LEARNING

Integrating the Sciences

Chemistry Explain that a plant's chloroplasts function like tiny solar panels to collect the sun's energy. Chloroplasts contain the green pigment *chlorophyll*. Tell students that a plant's leaves are usually rich in chlorophyll. Ask them how they can tell if a plant's leaves have a lot of chlorophyll. (Deep green indicates a high chlorophyll content.)

Use Integrating Worksheet 12.2.

Photosynthesis

The green parts of a plant make glucose for the whole plant, using light as an energy source. Light is absorbed by the green pigment chlorophyll. Chlorophyll is located inside the organelles called chloroplasts. The plant obtains the raw materials from the environment. Water is taken up from the soil. Carbon dioxide from the air enters openings on the plant surface, called **stomates** (STOH mayts). Carbon dioxide dissolves and travels into the cells that manufacture glucose. Glucose is used for plant growth or is stored. Oxygen is produced along with glucose. The oxygen that is not needed by the plant is released into the air through the stomates. The overall process of photosynthesis is shown in the following chemical reaction:

$$6CO_2 + 12H_2O \xrightarrow{\text{Light}} C_6H_{12}O_6 + 6H_2O + 6O_2$$

Carbon Dioxide and Water — Light → Glucose, Water, and Oxygen

This process actually takes place in two stages, shown in Figure 12.4.

Figure 12.4
Two Stages of Photosynthesis

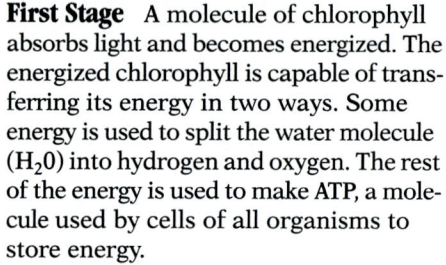

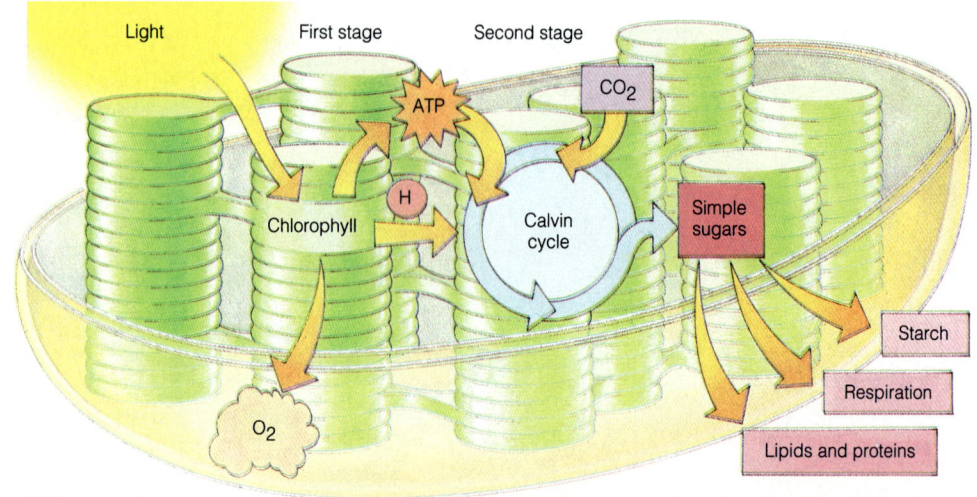

First Stage A molecule of chlorophyll absorbs light and becomes energized. The energized chlorophyll is capable of transferring its energy in two ways. Some energy is used to split the water molecule (H_2O) into hydrogen and oxygen. The rest of the energy is used to make ATP, a molecule used by cells of all organisms to store energy.

When water is split into hydrogen and oxygen, the hydrogen stays in the chloroplasts to continue more chemical reactions. The oxygen passes out of the leaf through the stomates.

Second Stage Carbon dioxide is combined with the hydrogen and ATP made in the first stage. Unlike the first stage, the chemical reactions of the second stage do not require light. These chemical reactions, called the **Calvin cycle**, produce glucose and other simple sugars. The plant uses these sugars for metabolism, growth, and other life processes. Some sugars are used for energy. Other sugars may combine chemically to form complex starches, or they may undergo chemical reactions to form oils or proteins.

Integrating the Sciences

Physical Science Sunlight is absorbed by the green pigment chlorophyll in most plants. Hold up a prism and align it with a beam of sunlight to show students the spectrum of light. Explain that sunlight is polychromatic, or contains many colors or wavelengths of light. Green light is in the center of the spectrum. When sunlight hits chlorophyll, green light is reflected and other wavelengths are absorbed.

Respiration

During photosynthesis, plant cells convert light energy into chemical energy stored in glucose. To release this energy, plant cells, like animal cells, use the reverse process known as respiration. In respiration, oxygen combines with glucose, releasing energy in the form of ATP. Respiration takes place in the cell's cytoplasm. The chemical reaction is:

$$C_6H_{12}O_6 + 6O_2 \xrightarrow{\text{Enzymes}} 6CO_2 + 6H_2O + ATP$$

Glucose and Oxygen — Enzymes → Carbon Dioxide, Water, and Energy

Plants use less oxygen during respiration than they produce during photosynthesis. The excess oxygen is released into the air, and used by animals for respiration. This oxygen is the plant's contribution to the global oxygen–carbon dioxide cycle. Notice in Figure 12.5 that animal respiration, in turn, produces waste carbon dioxide that is used by plants for photosynthesis.

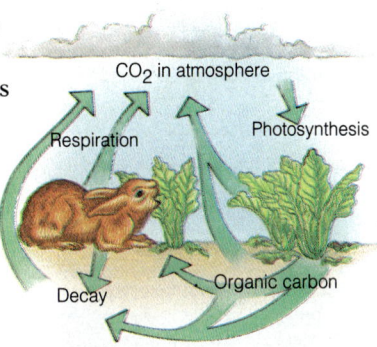

Figure 12.5 ▲
Plants play an important role in the global, oxygen–carbon dioxide cycle.

SkillBuilder Interpreting Data

Factors Affecting Photosynthesis

Light intensity and temperature affect the rate of photosynthesis in plants. The graph shows the relationship between the rate of photosynthesis, light intensity, and temperature for two plants. Use the graph to help you answer the questions.

1. What happens to the rate of photosynthesis as light intensity increases?

2. At which temperature, 20°C or 30°C, is the rate of photosynthesis higher? How can you tell by looking at the graph?

3. Based on the graph, describe the trend you see in the relationship between temperature, light, and the rate of photosynthesis.

4. Trace the graph on a sheet of paper. If the temperature is 25°C, predict the rate of photosynthesis by plotting a third line on your drawn graph.

5. Describe the conditions present when the rate of photosynthesis is almost the same for the two plants. Based on this observation, which factor affects photosynthesis more, temperature or light intensity? Explain your conclusions.

Chapter 12 A World of Plants

Misconceptions

Students may think that photosynthesis takes place only in plant cells and respiration takes place only in animal cells. Stress that both animal cells and plant cells use the process of respiration to release energy.

Skills Development

Predict Ask students what would happen to animals, including humans, if most or all plants died out. (Animals and humans could not survive without the oxygen supplied by plants.)

SkillBuilder

Answers

1. The rate of photosynthesis increases.

2. 30°C; the 30°C-line is higher than the 20°C-line on the graph.

3. The higher the temperature and light intensity, the faster the rate of photosynthesis.

4. The 25°C-line would fall midway between the 20°C-line and the 30°C-line.

5. If the rate of photosynthesis is the same for the two plants, the conditions of light intensity and temperature must be nearly the same. The graph indicates that temperature affects photosynthesis more than light intensity because the 30°C-line is higher.

**The Living Textbook:
Life Science Side 7**

Chapter 26   Frame 16823
Carbohydrate Production in Leaves
(Movie)
Search: Play:

TEACH • Continued

Class Activity

Allow students to observe how plants react to different conditions in their environment. Obtain radish seeds and have small groups of students plant them in clear cups filled with soil. Have each group design a different condition for its plant. For example, one group could put a plant in partial shade to see if it grows toward light. Another group could place its plant on its side to see in which direction its roots grow. Students should water and observe their plants regularly, recording any interesting developments. Encourage students to draw conclusions about their observations. Have groups share their observations and conclusions with the class.

Discuss

To emphasize the importance of cereals in the diet, have students name some foods that they eat during a usual day. Write their responses on the chalkboard. Then go through the list with the class and use colored chalk to underline all the foods that come from cereal grains.

Answer to In-Text Question

① **The green pigment will fade. Then the leaves will turn yellow, red, or brown and fall from the tree.**

Reteach

Use Reteach Worksheet 12.2.

INTEGRATED LEARNING

Multicultural Perspectives

Students may be interested to know about some ancestors of common plants. Many plants have been cultivated by people for hundreds and even thousands of years. Some of these plants have changed dramatically as a result of human influence. For example, the tomato's wild ancestor is a berry the size of a small grape. Wild tomatoes are much sweeter and more flavorful than their modern descendants. Wild carrots that thrive all over Europe and Asia have white roots, which are the parts that people eat. A rare breed of orange-rooted carrot grew in Afghanistan and was eventually cultivated in other parts of the world. This is the carrot many people eat today.

Figure 12.6a ▲
The sugar maple has bright green leaves during the long days of summer.

Figure 12.6b ▲
As the days get shorter, the green pigment in the leaves starts to break down. Red and yellow pigment begins to show through. What will happen to the leaves by the end of ① autumn?

Chemical Interactions of Plants

In addition to manufacturing glucose within their cells, plants have many other chemical activities. Some of these chemical activities allow plants to react to conditions in the environment. A plant in a window grows toward the light. If a potted plant is laid on its side, the stem responds to gravity and gradually grows upward while the roots grow downward.

During autumn, as the days grow shorter, the green coloring in the leaves of many trees fades and other colors become visible. You can see how leaves change color in Figure 12.6. As longer days return with the coming of spring, chemical activities stimulate the plant to develop new growth. All these responses, and many more, are controlled by hormones produced by the plant.

Plants also produce chemicals that affect other plants and animals. Some desert plants produce chemical substances in their roots that "poison" the ground around them. The poisons prevent other seeds from sprouting. This keeps other plants from growing nearby and using scarce water. When attacked by insects, many plants begin to produce small amounts of toxic chemicals. These chemicals serve to repel the insects.

Science and Society *Too Few Food Plants?*

As you know, food plants you enjoy daily come from all over the world. In fact, about 150 different species of plants are cultivated for food by people worldwide. Of these, the 12 plants listed in Table 12.1 are especially important staple foods. These 12 plants stand between human survival and starvation.

The three most important food plants are known as cereal grains: wheat, corn, and rice. These grains are among the oldest cultivated plants, going back more than 10,000 years. Each cereal grain was cultivated and developed by cultures from different regions of the world. Wheat was cultivated in Africa, Europe, and the Middle East. Corn was cultivated in North, South, and Central America. Rice was cultivated in Asia. All cereal grains are seeds and contain a high amount of starch and small amounts of oil and protein. They are low in water, are easily transported, and can be stored for long periods without spoiling.

Chapter 12 A World of Plants

TEACHING OPTIONS

Cooperative Learning

Have groups of two or three students choose a plant in Table 12.1 to research. Groups should devise a list of questions they have about the plant and research the answers. Students could investigate when the plant was first cultivated, what cultures eat it now, how it is prepared and served in different cultures, what nutrients it provides, what environment it grows in, and so on. Have groups write a report of their findings and present them in class.

Table 12.1 Twelve Essential Plants

Plant	Origin	Food Value
Wheat	Middle East, Africa	Starch, protein, oil
Corn	Central America, South America	Starch
Rice	Asia	Starch
Potato, Yam	South America	Starch
Cassava	South America	Starch
Sugar cane	Asia	Sugar
Sugar beet	Europe	Sugar
Soybeans	Asia	Protein, starch
Beans	South, North, and Central America	Protein, starch
Coconut	World tropics	Oil, protein
Bananas	Southeast Asia	Starch, sugar, protein, oil

Food plants known as root crops, such as potatoes and yams, are also from different regions of the world. The potato was cultivated for many thousands of years by Native Americans in the Andes mountain areas of South America. Some of the many varieties of potatoes they developed are shown in Figure 12.7. Potatoes were brought to Europe in 1570. They became widespread because the potato plants were easy to grow. They grew even in areas with poor soil where other food plants failed.

Figure 12.7 ▲
Many unusual varieties of potatoes are still available in this Andean market.

Check and Explain

1. What chemical activities take place in plants, beside making glucose?
2. Where does photosynthesis take place in a plant? When does photosynthesis take place?
3. **Generalize** Sometimes it is said that plants use carbon dioxide to produce oxygen, while animals use oxygen to produce carbon dioxide. In what way is this statement misleading?
4. **Communicate** In your own words, explain how a leaf makes and uses glucose.

Discuss

Have students study Table 12.1. Ask them to think of examples of foods they eat that contain these plants. Have them imagine what their diets would be like without foods from other countries.

EVALUATE

WrapUp

Review Have students write a narrative from the point of view of a plant that is undergoing photosynthesis. They should describe each step of the process in detail, specifying what is happening in each part of their bodies. Encourage students to be creative about how they feel in each chemical stage.
Use Review Worksheet 12.2.

Check and Explain

1. Respiration and regulation of growth and development by hormones
2. It occurs within molecules of chlorophyll inside a plant's chloroplasts. It begins when sunlight energizes chlorophyll molecules.
3. Answers may vary. Plants use carbon dioxide to make glucose. Plants also use oxygen during respiration.
4. Answers may vary. The cells of a leaf use solar energy, water, and carbon dioxide to produce glucose. The plant uses glucose to grow and change nutrients in the soil into tissues in its body.

Chapter 12 A World of Plants

ACTIVITY 12

Time 3 days **Group** 3

Materials

30 small jars with lids
Elodea sprigs
cabbage-juice indicator
labels
aluminum foil
straws

Analysis

1. Carbon dioxide made the indicator change color.
2. As a control—to serve as a comparison
3. There should have been no change when the sprig was placed in the jar. Some time later, the indicator should have begun to change color as the plant's cells carried out respiration, producing carbon dioxide.
4. To prevent photosynthesis, which requires light

Conclusion

Conclusions will vary, but should follow logically from each student's own observations. Evidence of carbon dioxide production (color change in the indicator) shows that plants do carry out respiration.

Everyday Application

Exposed constantly to a grow lamp, a plant would carry out photosynthesis continuously, provided water and carbon dioxide were plentiful. Plants would be likely to show increased growth.

TEACHING OPTIONS

Prelab Discussion

Have students read the entire activity. Discuss the following points before beginning the activity:

▶ Discuss the nature and use of chemical indicators. You might add that the term *indicator* comes from a Latin word that means "to proclaim."

▶ Review the chemical relationship between photosynthesis and respiration. How might the two processes affect the observations made during this activity?

Activity 12 Do plants breathe?

Skills Observe; Predict

Task 1 Prelab Prep
1. Collect the following items: 3 labels, a pencil, 3 small jars with lids, cabbage-juice indicator, sprig of *Elodea* or other water plant, aluminum foil, a straw.
2. Fill each jar full of the cabbage-juice indicator. **CAUTION! This liquid will stain clothing and skin.**
3. Make the following three labels: *Control*, *Elodea*, *Breath*.

Task 2 Data Record
1. On a sheet of paper, draw the data table shown.
2. Record all your observations about the color of the cabbage-juice indicator in the data table.

Task 3 Procedure
1. Place the *Elodea* into one of the jars filled with cabbage-juice indicator. Put the lid on the jar. Record your observations.
2. Attach the *Elodea* label to the jar. Cover the outside of the jar with aluminum foil.
3. Put the lid on the second jar filled with cabbage-juice indicator. Record your observations in the data table.
4. Attach the *Control* label to the jar. Cover the outside with aluminum foil.
5. Use the straw to blow for 2 minutes into the cabbage-juice indicator in the third jar.
6. Put the lid on the jar and record your observations.
7. Attach the *Breath* label to the jar. Cover the outside of the third jar with aluminum foil.
8. Set the three jars where they will not be disturbed.
9. After 24 hours, remove the aluminum foil and record your observations. Replace the aluminum foil.
10. After 48 hours, remove the aluminum foil and record your observations.
11. Remove the *Elodea* sprig and pour the cabbage-juice indicator down the sink. Be sure to rinse the jars.

Task 4 Analysis
1. What chemical in your breath made the cabbage-juice indicator change color?
2. Why did you prepare a jar with only cabbage-juice indicator?
3. How did the indicator react when the *Elodea* leaf was placed in the jar? What happened 24 hours later? What happened 48 hours later? Explain your observations.
4. Explain why the jars were covered with aluminum foil.

Task 5 Conclusion
Write a paragraph summarizing what you observed in this activity. Do plants breathe? Explain.

Everyday Application

Explain how a plant would be affected if you exposed it to constant light using a "grow lamp."

Table 12.2 Cabbage-Juice Observations

Jar	Prediction	Immediate	24 Hours	48 Hours
Control				
Elodea				
Breath				

CHAPTER REVIEW 12

Chapter 12 Review

Concept Summary

12.1 Plant Origins
- Plants evolved from many-celled green algae about 400 million years ago. However, plants differ from algae because they have adaptations for living on land.
- Vascular plants have systems for transporting water through the plant, whereas nonvascular plants lack these structures.
- Gymnosperms are vascular plants that produce seeds, and angiosperms are vascular, flowering plants with seeds enclosed in a fruit.
- Flowering plants have developed many special adaptations for surviving in many different environments.

12.2 Chemistry in Plants
- Photosynthesis is the process by which a plant makes simple sugars from water and carbon dioxide, using energy from the sun. Oxygen is produced as a result.
- Photosynthesis occurs in a plant's chloroplasts.
- Photosynthesis has two stages. The first stage requires light to split water into hydrogen and oxygen, making ATP. In the second stage, the Calvin cycle produces glucose and other simple sugars.
- The plant uses the simple sugars for metabolism, growth, and all its life processes.
- Plants, like animals, breathe by taking in oxygen to break down glucose.
- Chemical activities occur in plants that enable them to react to environmental conditions.

Chapter Vocabulary
nonvascular plant (12.1)   gymnosperm (12.1)   stomate (12.2)   Calvin cycle (12.2)
vascular plant (12.1)   angiosperm (12.1)

Check Your Vocabulary
Use the vocabulary words above to complete the following sentences.
1. Plants with tubelike structures for transporting water are called ____.
2. Plants with no specialized system for conducting water are ____.
3. Carbon dioxide enters a plant through ____ located on the plant's leaves.
4. The conversion of carbon dioxide, hydrogen, and ATP into glucose occurs during the ____ of photosynthesis.
5. Flowering plants are ____.
6. Plants that produce seeds not enclosed in a fruit are ____.

Find Word Relationships
Pair each numbered word(s) with a vocabulary word. Explain in complete sentences how the words are related. You may use the vocabulary words more than once.

1. Gas exchange
2. Tubelike vessels
3. Bryophyte
4. Simple sugars
5. Rose
6. Mosses
7. Spongelike
8. Flowers
9. Chemical reactions
10. Fruit

Write Your Vocabulary
Write sentences using each vocabulary word above. Show that you know what each word means.

Chapter 12 A World of Plants   249

Check Your Vocabulary
1. vascular plants
2. nonvascular plants
3. stomates
4. Calvin cycle
5. angiosperms
6. gymnosperms

Find Word Relationships
1. Stomates; gases move into and out of leaves through stomates.
2. Vascular plants; vascular plants have tubelike vessels running through their tissues.
3. Nonvascular plants; bryophytes, like mosses, are a type of nonvascular plant.
4. Calvin cycle; during the Calvin cycle, a simple sugar (glucose) is produced.
5. Angiosperm; angiosperms are flowering plants.
6. Nonvascular plant; a moss is a bryophyte, which is a type of nonvascular plant.
7. Nonvascular plant; without a system for transporting water up from roots, nonvascular plants absorb water like a sponge.
8. Angiosperm; angiosperms are flowering plants.
9. Calvin cycle; the Calvin cycle is a set of chemical reactions that are part of the process of photosynthesis.
10. Angiosperm; all angiosperms produce seeds enclosed in a fruit or a pod.

Write Your Vocabulary
Students' sentences should show that they know both the meanings of the words and how to apply the words.
Use Vocabulary Worksheet for Chapter 12.

CHAPTER 12 REVIEW

Check Your Knowledge

1. Because they can absorb water directly into cells, rather than through roots, it is an advantage for them to live near water.
2. Stomates (also *stoma*) are openings in the surface of leaves that let in carbon dioxide.
3. A plant will grow toward light; a plant will respond to gravity by sending the stem upward and the roots downward; trees' leaves change color and fall off in response to decreasing periods of daylight and cooling temperatures.
4. Bryophytes, ferns, club mosses, angiosperms, gymnosperms, horsetails
5. Glucose, oxygen, and water are the products of photosynthesis.
6. Photosynthesis takes place inside chloroplasts.
7. All plants are many celled; all are organized into tissues and organs; all have cell walls; all contain chlorophyll.
8. green algae
9. desert
10. simple sugars
11. ATP
12. oxygen
13. two
14. light

Check Your Understanding

1. Photosynthesis: takes place in chloroplasts; uses carbon dioxide, water, and light energy; and produces glucose and oxygen. Respiration: occurs in mitochondria; uses oxygen and glucose; and produces water, carbon dioxide, and energy (ATP).
2. Photosynthesis is the process by which molecules of glucose are put together using light.
3. A moss is nonvascular; a fern is vascular. A plant must have means to transport water up from the roots to enable it to gain height.
4. Spines are actually tiny leaves. In addition to providing protection from feeding animals, the spines have a small surface area, which helps the cactus conserve water.
5. A plant should take in more carbon dioxide during the day, because when light is present, photosynthesis will occur, using up carbon dioxide.
6. a. The bubbles are filled with oxygen gas produced by photosynthesis in the plant leaves.
   b. If the *Elodea* were removed, the fish might suffocate.
7. Answers may vary, but students should indicate that ferns grow in areas with plenty of moisture, so they would not survive in a desert region.

Chapter 12 Review

Check Your Knowledge
Answer the following in complete sentences.

1. Why are most nonvascular plants small and live near water?
2. What is the function of stomates?
3. Describe three ways in which plants respond to their environment.
4. Name six major groups into which plants are classified.
5. What does a plant produce as the result of photosynthesis?
6. In what plant structure does photosynthesis occur?
7. What characteristics do all plants have in common?

Choose the answer that best completes each sentence.

8. Plants evolved from (green algae, sponges, fungi, club mosses).
9. Plants that grow slowly and have fleshy leaves probably live in the (rain forest, high mountains, desert, marsh).
10. A product of the Calvin cycle is (carbon dioxide, hormones, chlorophyll, simple sugars).
11. During respiration, energy is released in the form of (ATP, hormones, enzymes, glucose).
12. Plants are important to animal life because they release (water, oxygen, carbon dioxide, nitrogen) into the atmosphere.
13. Photosynthesis has (one, two, three, six) stages.
14. The first stage of photosynthesis requires (oxygen, starches, enzymes, light).

Check Your Understanding
Apply the concepts you have learned to answer each question.

1. In a table, compare and contrast respiration and photosynthesis.
2. The word *photo* comes from the Greek word meaning "light," and *synthesis* comes from the Greek word meaning "to make." Using this information, write a definition for photosynthesis. How does this definition relate to what occurs during photosynthesis?
3. Most mosses are small plants. A fern can grow as tall as a small tree. Which one is vascular? Nonvascular? Why is the plant's size a clue to the answers?
4. **Mystery Photo** The photograph on page 236 shows a close-up view of a barrel cactus. What is the function of the spines shown in the photograph? List other successful plant adaptations to the desert environment.
5. When would you expect a plant to take in more carbon dioxide, at night or during the day? Why?
6. **Infer** The water plant *Elodea* was put in a fishbowl with a goldfish. One day later bubbles appeared underneath the plant's leaves.
   a. What gas is in the bubbles? Explain.
   b. Predict what might happen to the goldfish if you removed the *Elodea* from the fishbowl.
7. A friend of yours used to live in the tropics of Hawaii where she had a lush fern garden. She moved to the desert area of Nevada. She wanted to grow another fern garden in her new backyard. Is this a good idea? What will probably happen to her garden?

Develop Your Skills

1. a. The rate increases up to a point as CO_2 concentration increases, then levels off.
   b. It begins to level off.
   c. Answers will vary. There is a limit to how fast the reactions involved in photosynthesis can go. At higher levels of carbon dioxide, some step in the reaction reaches its maximum.
   d. When there is no CO_2 present, photosynthesis cannot occur, so the rate is zero.
2. Accept any logical answer. Make sure, however, that observations are consistent with the types of plants in the area.
3. a. Oxygen and carbon
   b. H_2O

Develop Your Skills

Use the skills you have developed in this chapter to complete each activity.

1. **Interpret Data** The graph below shows the relationship between the rate of photosynthesis and the amount of carbon dioxide in the air.

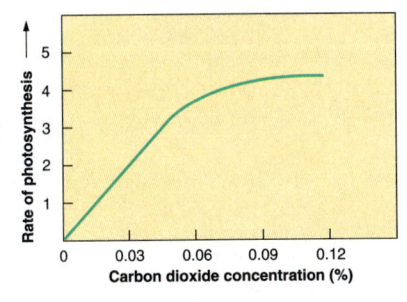

   a. Describe what happens to the rate of photosynthesis as the amount of carbon dioxide increases.
   b. What happens to the rate of photosynthesis at a carbon dioxide level of 0.09 percent?
   c. Infer why photosynthesis levels off at higher levels of carbon dioxide.
   d. What happens when there is no carbon dioxide present?

2. **Observe** Look at several plants that grow near your home or school. Record how each plant reacts to conditions in the environment.

3. **Data Bank** Use the information on page 624 to answer the following questions.
   a. What elements make up the largest percentage of dry weight in plants?
   b. Hydrogen is available to plants in what form?

Make Connections

1. **Link the Concepts** Below is a concept map showing how some of the main concepts in this chapter link together. Only parts of the map are filled in. Copy the map. Using words and ideas from the chapter, complete the map.

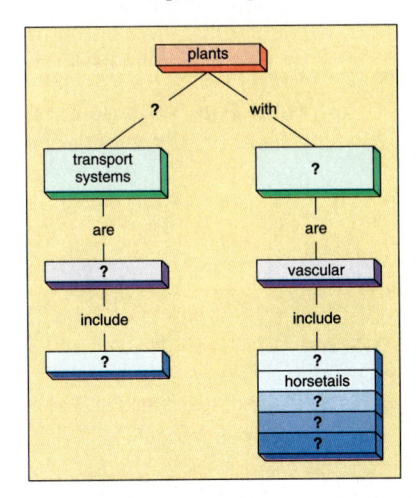

2. **Science and Math** Collect pictures of different types of plants found around the world. The entire collection is your set. Create subsets from this main set. How many different subsets did you create? What did you use as the basis for forming your subsets?

3. **Science and Technology** Research how scientists are planning to use hydroponics in space stations of the future. Find out what kinds of plants will be grown and how their needs will be met. List the problems plants may have living in space.

Make Connections

1.

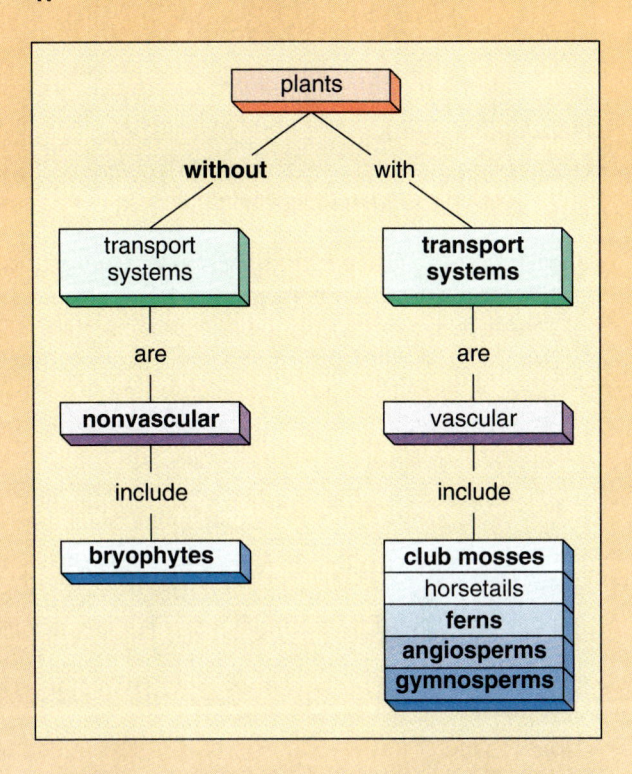

2. Answers will vary. Many students will try to organize their subsets according to the scheme given in this lesson. Accept any logical basis for classification, however, because one challenge is for students to show that classification of subsets is based on characteristics of the whole set.

3. Answers will vary. Problems include reduced gravity and difficulty in making transparent structures strong enough to withstand the conditions in space.

CHAPTER 13

Overview

Nonflowering plants are the focus of this chapter. The first section describes the characteristics, origins, and adaptations of nonflowering plants. The second section presents bryophytes, specifically mosses and liverworts, and their structures. Nonflowering, vascular plants are described in the final section, which provides details about the characteristics and life cycles of ferns and gymnosperms. The chapter closes with a look at gymnosperms in the context of geologic time.

Advance Planner

▶ Gather old nature or science magazines for SE page 253 and TE page 258.

▶ If desired, arrange for a botanist to speak to the class and answer questions. TE page 257.

▶ Obtain fern fronds for SE page 261.

▶ Contact a paper recycling center and invite a recycling expert to visit the class to discuss paper recycling methods and to answer student questions. TE page 265.

Skills Development Chart

Sections	Classify	Collect Data	Estimate	Graph	Infer	Measure	Observe	Research
13.1 Skills WarmUp	●							
13.2 Skills WarmUp								●
13.3 Skills WarmUp SkillBuilder Skills WorkOut Activity		● ● ●	●	●	●	●	●	

Individual Needs

▶ **Limited English Proficiency Students** Mask the labels and photocopy Figures 13.2, 13.4, 13.8, and 13.11. After you read about the life cycles described in each figure, have students attach labels that describe each step to the figures.

▶ **At-Risk Students** Have students work together to create a collage from pictures and actual parts of local, nonflowering plants. Collages can include drawings, photographs from catalogs or magazines, seeds, leaves, pinecones, pieces of bark, or anything made from nonflowering plants. Display the collage in the classroom.

▶ **Gifted Students** Have each student research and write about one particular fossil fuel. They should describe where the fossil grew as a plant, what nutrients and environment it needed to survive, how it died, and how it was transformed into a fuel. Students' papers should include information about the kind of habitat in which the plant grew millions of years ago. Encourage them to use diagrams and information about the chemical composition of the plant and the resulting fossil fuel. Then have students predict the availability of the fuel source. Invite them to publish their research papers and make them available for classmates to read.

Resource Bank

▶ **Bulletin Board** Have the class work together to make a bulletin board display showing the life cycle of a fern or pine. Have students create models or drawings of the various parts of the life cycle and place them in the correct sequence on the bulletin board.

▶ **Field Trip** Arrange to take students to a forest preserve or park so they can collect data for SE activity page 264.

CHAPTER 13 PLANNING GUIDE

Section	Core	Standard	Enriched	Section	Core	Standard	Enriched
13.1 Characteristics of Nonflowering Plants pp. 253–255				**Overhead Blackline Transparencies** Overhead Blackline Master 13.2 and Student Worksheet	●	●	●
Section Features Skills WarmUp, p. 253	●	●	●	**Laboratory Program** Investigation 25	●	●	●
Blackline Masters Review Worksheet 13.1 Integrating Worksheet 13.1	●	● ●	● ●	**13.3 Nonflowering Vascular Plants** pp. 259–266			
Color Transparencies Transparency 33	●	●		**Section Features** Skills WarmUp, p. 259 SkillBuilder, p. 261 Skills WorkOut, p. 264 Activity, p. 266	● ● ●	● ● ● ●	● ● ● ●
13.2 Bryophytes pp. 256–258				**Blackline Masters** Review Worksheet 13.3 Integrating Worksheet 13.3 Reteach Worksheet 13.3 Vocabulary Worksheet 13	● ● ● ●	● ● ●	●
Section Features Skills WarmUp, p. 256 Career Corner, p. 257	● ●	● ●	● ●	**Laboratory Program** Investigation 26		●	●
Blackline Masters Review Worksheet 13.2 Skills Worksheet 13.2a Skills Worksheet 13.2b	● ●	● ●	●	**Color Transparencies** Transparency 34	●	●	

Bibliography

The following resources can be used for teaching the chapter. See page T-40 for supplier codes.

Audio-Visual Sources
(video unless noted)
Ancient Forests. 25 min. 1992. NGSES.
Plants: Parts and Processes. 2 filmstrips with cassettes. 1988. NGSES.
The Sequoia Giants of Sequoia National Park. 28 min film. HOL.
The World of Plants. 43 min. 1990. UL.

Software Resources
Basic Photosynthesis. BLS.
Hothouse Planet. EME.
Photosynthesis and Light Energy. IBM. CCM.
Plants. Apple. J & S.

Library Resources
Burnie, David. How Nature Works. Pleasantville: The Reader's Digest Association, Inc., 1991.
Burnie, David. Plant. New York: Alfred A. Knopf, Inc., 1989.
Durrell, Gerald. The Amateur Naturalist. New York: Alfred A. Knopf. 1989.
Imes, Rick. The Practical Botanist. New York: Simon & Schuster Inc., 1990.
McTigue, Bernard. Nature Illustrated—Flowers, Plants, and Trees 1550–1900. New York: Harry N. Abrams, Inc., 1989.
Wilkins, Malcolm. Plantwatching. New York: Facts on File Publications, 1988.

CHAPTER 13

TEACHING OPTIONS

Cooperative Learning

Assign students to cooperative groups. Take a walk around the school and have students list and draw as many examples of plants as they can find. Have them identify the plants that they list as flowering or nonflowering. When they return from the walk, have students work with their teams to use field guides or taxonomic keys to identify each plant. Have students color their drawings and display them. Encourage them to share information with other groups to help them identify more difficult plants.

▶ Introducing the Chapter

Have students read the description of the photograph on page 252. Ask students if they agree or disagree with the description. Point out that the yellow clumps on the underside of the leaves are sori. The sori contain spores.

Directed Inquiry

Have students study the photograph. Ask:

▶ What kind of plant is shown in this photograph? (A fern)

▶ How is this plant similar to a rose? (Both have green leaves.) How is it different? (This plant doesn't have flowers.)

▶ What kind of environment would you expect this plant to grow in? (Moist)

▶ What important thing do you think you will learn about nonflowering plants in this chapter? (How they reproduce without flowers)

▶ Chapter Vocabulary

conifer
gametophyte
liverwort
moss
pollen
pollination
rhizoid
spore
sporophyte

Chapter 13 Nonflowering Plants

Chapter Sections

13.1 Characteristics of Nonflowering Plants

13.2 Bryophytes

13.3 Nonflowering Vascular Plants

What kind of plant is this?

"I think this is a plant that grows in the jungle or in a moist place. It looks like it needs a moist, humid, wet climate. The yellow clumps look like animals that grow on the plant. They could also be seeds that grow out of the plant, and when they fall they will create a new plant."

Rodrigo Figueroa
Antioch Junior High School
Antioch, California

To find out more about the photograph, look on page 268. As you read this chapter, you will learn about different nonflowering plants and how they grow and reproduce.

INTEGRATED LEARNING

Themes in Science
Energy Plants supply the energy for all life on the earth. Plants harness energy from sunlight and convert it to sugars, which they store and use for their life processes. Herbivores eat plants to sustain themselves and many, in turn, are eaten by carnivores. Ask students to explain why even animals that do not eat plants are dependent on them for survival. (The prey of these animals eat plants.)

Integrating the Sciences
Earth Science Explain to students that the fossil record indicates that primitive blue-green algae were abundant in the earth's oceans about three-and-one-half billion years ago. The first land plants evolved from a more complex form of algae and appeared much later—about 433 million years ago. These nonflowering plants differed from their algae ancestors in that they could support themselves on land.

SECTION 13.1

Section Objectives
For a list of section objectives, see the Student Edition page.

Skills Objectives
Students should be able to:
Classify nonflowering plants.
Infer why vascular plants are larger than nonvascular plants.

Vocabulary
gametophyte, sporophyte, spore

13.1 Characteristics of Nonflowering Plants

Objectives
▶ **Give examples** of common nonflowering plants.
▶ **Explain** how nonflowering plants adapted to living on land.
▶ **Compare** plant and animal life cycles.
▶ **Infer** reasons why vascular plants are larger than nonvascular plants.

Skills WarmUp
Classification How many nonflowering plants can you find? Collect leaves and other parts of plants you think are nonflowering. Also collect photographs from magazines. Arrange your collection into groups. When you finish this chapter, evaluate your method of classifying.

MOTIVATE

Skills WarmUp
To help students understand the characteristics of nonflowering plants, have them do the Skills WarmUp.
Answer Students should arrange their collections into groups that share similar characteristics. Have students save their collections to use when they've read the chapter.

Prior Knowledge
Gauge what students know about nonflowering plants by asking the following questions:
▶ What are some examples of nonflowering plants?
▶ Which came first, flowering plants or nonflowering plants?
▶ What adaptations occurred in nonflowering plants that enabled them to survive on land?

Answer to In-Text Question
① Mosses, ferns, and redwood trees are the main nonflowering plants shown.

Imagine a forest without any broad-leaf trees or other flowering plants. What kind of plants would grow in such a forest? To find out, look at Figure 13.1. A forest of nonflowering plants would include mosses, ferns, horsetails, and cone-bearing trees. When dinosaurs walked the earth, the ancestors of these plants grew in vast forests and swamps.

Origins and Adaptations

Four hundred million years ago, nonflowering land plants began to evolve from water-living green algae. Eventually, nonflowering plants completely colonized and changed the land environment. To colonize the land, nonflowering plants had to solve many problems that plants living in water don't have. A land plant must adapt to both soil and air. The air space above the soil has sunlight but no water and no support. The soil has water and minerals but no sunlight. To live on land, nonflowering plants evolved more specialized tissues and organs that satisfied some of their needs in air and some in soil.

In three separate stages of evolution, nonflowering plants solved each of the major problems of living on land. Bryophytes developed a waxy covering to keep their cells from drying out in the air. Club mosses, horsetails, and ferns evolved vascular tissue to transport water from the roots to above-ground stems. With vascular tissue they could grow upward into the air and downward into the soil. Finally the gymnosperms evolved a way to reproduce that didn't depend on water.

Figure 13.1 ▲
What kinds of nonflowering plants can you see in this photograph? ①

Chapter 13 Nonflowering Plants 253

TEACH

Discuss

Have students describe the three major adaptations that occurred in nonflowering plants as they began to grow on land. Which plants developed these adaptations? (Bryophytes developed a waxy covering to retain water; club mosses, horsetails, and ferns evolved vascular systems for transporting water; and gymnosperms evolved a way to reproduce without water.)

Directed Inquiry

After students have studied the material on this page, ask the following questions:

▶ What is a spore? (A nonsexual reproductive cell with a protective coat)

▶ What plant does a spore become? (Gametophyte)

▶ What plant does the fertilized egg develop into? (Sporophyte)

▶ What are the parts of a seed? (Embryo, stored energy supply, protective coat)

▶ As plants adapted to land, what happened to the size of the gametophytes and sporophytes? (Gametophytes became smaller and sporophytes became larger.)

The Living Textbook:
Life Science Side 3

Chapter 15   Frame 09886
Pine Life Cycle (Movie)
Search: Play:

254

INTEGRATED LEARNING

Language Arts Connection

Explain that the word *phyte* means "plant." Ask students to infer what the words *gametophyte* and *sporophyte* mean. (A gametophyte is a gamete plant, or a plant that produces gametes. A sporophyte is a spore plant, or a plant that produces spores.)

Life Cycles

Like the algae they evolved from, nonflowering plants have a life cycle made up of two different stages. Look carefully at Figure 13.2. Notice that plants alternate between a sexual stage and a nonsexual stage. In the sexual stage, a plant called the **gametophyte** (guh MEET uh FYT) produces egg and sperm cells. The sperm and egg combine to form a fertilized egg cell, which grows into an embryo. The embryo grows into a **sporophyte** (SPOH roh FYT), the plant that makes up the nonsexual stage. The sporophyte produces **spores**. A spore is a nonsexual reproductive cell with a protective coat. A spore develops into a gametophyte. The gametophyte reproduces sexually, and the cycle continues. Some plants have separate male and female gametophytes. In other plants, one gametophyte produces both eggs and sperm.

This two-stage life cycle evolved in a water environment, where the gametophyte and sporophyte could be separate organisms. The least complex plants have kept this arrangement. They have to make do with rainwater to transport the eggs and sperm. This limits where they can live.

Other plants adapted more completely to land by modifying their sexual stage. It became shorter, and the gametophytes became smaller. These two changes allowed the sporophyte to evolve ways of protecting the gametophyte and helping the eggs and sperm to come together without water.

In the most complex nonflowering plants, the gymnosperms, the sexual stage takes place almost completely within the sporophyte. This adaptation was made possible in part by the evolution of the seed, shown at right. A seed is made up of an embryo, a stored energy supply, and a protective seed coat.

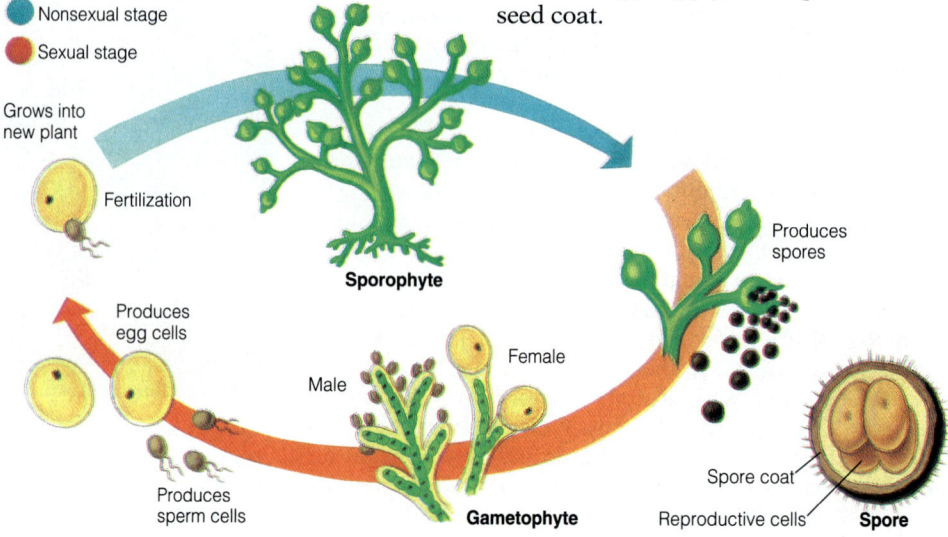

Figure 13.2 ▼
Life Cycle of a Nonflowering Plant

254   Chapter 13   Nonflowering Plants

TEACHING OPTIONS

Portfolio
Have students make a list of twenty items in their homes that are made from plant products. Then have them determine whether the items came from a nonflowering plant. Have students share their findings in class. Students may wish to keep their lists in their portfolios.

Science and You Using Nonflowering Plants

As you go through your day, you may encounter many things made from nonflowering plants. First of all, this page is paper that probably contains wood pulp from cone-bearing trees, such as pine and fir. Your pencil, your chair, your desk, and even your classroom may be constructed of wood or wood products from these same nonflowering trees. In fact, most paper products and many wood products come from cone-bearing trees. In addition, cleaning products containing pine oil may help keep your classroom clean.

If you look inside a potted plant or flower arrangement, you may find sphagnum (SFAG nuhm) moss, another nonflowering plant. Florists use sphagnum moss as a packing material because it is highly absorbent. In landscaped areas and gardens, sphagnum moss in the form of peat may be mixed in with the soil.

Are you reading this page with the help of a light bulb? Its light may come from electricity produced by the burning of coal. The coal formed from ancient nonflowering plants that grew during the Carboniferous period, about 350 million years ago. These plants were buried when they died and changed into coal by heat and pressure deep underground.

If you have ever tasted food containing pine nuts, you know that some nonflowering plants are good to eat. Pine nuts are high in protein and carbohydrates, but lower in fat than other nuts. Another nonflowering plant that people eat is the young curled frond of a fern, called a fiddlehead.

Figure 13.3 ▲
Ponderosa pines (top) are a major source of wood and paper products. Ancient plants like the seed fern (bottom) contributed to the earth's supply of coal.

Check and Explain

1. Name four common nonflowering plants. Describe how they differ from one another.
2. Describe three adaptations gymnosperms have for living on land.
3. **Compare and Contrast** How are the life cycles of nonflowering plants different from those of animals? How are they similar?
4. **Infer** Why can ferns and gymnosperms grow much larger than bryophytes?

Chapter 13 Nonflowering Plants   255

EVALUATE

WrapUp
Reinforce Write on the chalkboard the vocabulary words printed in bold in this section. Ask students to explain the role of each term in the life cycle of nonflowering plants. Have them explain how these life cycle adaptations help nonflowering plants survive on land.

Use Review Worksheet 13.1.

Check and Explain

1. Answers may vary. Mosses are nonvascular; ferns and horsetails have vascular tissue, as do pine trees, which also can survive and reproduce in dry areas.
2. Shortened sexual stage, small gametophytes, and seeds
3. Nonflowering plants alternate between a sexual and a nonsexual stage; animals progress from an immature to an adult stage. In both groups, an embryo develops after a sperm fertilizes an egg.
4. Because they have vascular tissue to transport water and nutrients

SECTION 13.2

Section Objectives
For a list of section objectives, see the Student Edition page.

Skills Objectives
Students should be able to:

Research root words of plant names.

Infer how a drought might affect mosses' reproduction.

Vocabulary
moss, rhizoid, liverwort

MOTIVATE

Skills WarmUp
To help students understand bryophytes, have them do the Skills WarmUp.
Answer The root words for *bryophyte* are *bryon* (moss) and *phyton* (plant).

Misconceptions
Students may reasonably assume that club mosses are mosses and, therefore, bryophytes. Explain that club mosses are not in the moss family, though they contain the word *moss* in their name. Club mosses evolved after bryophytes and have vascular systems that enable them to live in a wider variety of environments.

Answer to In-Text Question
① The capsules

The Living Textbook:
Life Science Side 3
Chapter 13   Frame 07798
Moss (Movie)
Search: Play:

256

INTEGRATED LEARNING

Themes in Science
Evolution Bryophytes were the first land-dwelling plants to evolve. A moss is a common type of bryophyte that still thrives on the earth today. Mosses developed two important adaptations that distinguish them from their algae ancestors. Mosses have leaflike structures that they use for reproduction and rootlike structures, called rhizoids, that enable them to cling to land. Ask students why bryophytes need to live in moist environments. (They do not have vascular systems for transporting water.)

Skills WarmUp

Research Look up the word *bryophyte* in the dictionary. Discover the root words from which it was formed.

Figure 13.4 ▲
The life cycle of a moss is shown in the diagram. What are the structures at the tips of the moss plants in the photograph? ①

13.2 Bryophytes

Objectives

▶ **Describe** where bryophytes grow and explain why they grow there.

▶ **Compare** and **contrast** the characteristics of mosses and liverworts.

▶ **Infer** how environmental conditions affect bryophyte reproduction.

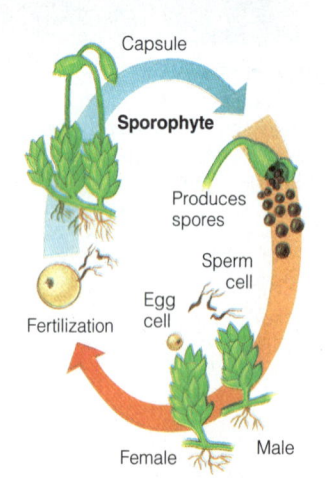

Imagine you are walking through a cool, moist forest. On the ground you see a fuzzy patch of deep green. What could it be? You bend down to examine it closely and notice that the fuzzy patch is made of many small plants like the one in the photograph in Figure 13.4. You have encountered a member of the group of plants known as *bryophytes*. Bryophytes are nonvascular plants that grow only in moist environments.

Mosses

Study the plant in Figure 13.4. The tiny green leaflike structures arranged in a spiral on a short stalk are characteristic of a bryophyte called a **moss**. A moss attaches to the ground by small **rhizoids** (RY zoydz). These are rootlike structures found in bryophytes.

Rhizoids are not true roots because they do not have vascular tissue to carry water. Water travels through a moss by osmosis, and glucose travels between cells by diffusion. These processes are slow and support only a small plant.

Look at the moss life cycle shown in Figure 13.4. In the first stage, spores develop into separate male and female gametophytes. The male gametophyte grows sperm-producing sex organs, and the female gametophyte grows egg-producing sex organs. The sperm cells use water from rain or dew to swim to the egg cell and fertilize it. The fertilized egg remains in the female sex organ and develops into a sporophyte. The sporophyte, still attached to the female gametophyte, grows a stalk with a capsule on the end. In the capsule, cell divisions produce spores that will be released to continue the cycle.

256   Chapter 13 Nonflowering Plants

Integrating the Sciences

Earth Science Many mosses serve as pioneer plants in barren places. The mosses grow in rock. Over many generations, they help to break up rock and lay down organic material, creating fertile soil for other species. If possible, show students a rock that is covered with moss. Have them note how the moss seems to be changing the rock.

Liverworts

In wet areas of a forest or bog, you may find bryophytes like the **liverwort** shown in Figure 13.5. The bodies of most liverworts have the flat, leaflike structures that you see here. Other liverworts have bodies divided into rounded sections, or lobes. Liverworts got their unusual name because the shape of the lobed liverwort looks somewhat like a mammal's liver. Both kinds of liverworts have rhizoids.

The life cycle of liverworts is similar to that of mosses. The leaflike structures belong to the gametophyte, which is the most noticeable stage. The smaller sporophyte produces capsules that contain special spring-loaded cells. These cells launch the spores and help them disperse.

Close relatives of the liverworts are the hornworts. The gametophytes of both plants look very much alike. However, their sporophytes look very different. Hornworts get their name from their sporophytes, which are hornlike.

Figure 13.5 ▲
Notice the difference between the male and female gametophytes on this liverwort.

Career Corner Botanist

Who Studies How Plants Grow and Reproduce?

Imagine cutting a trail through thick jungle vegetation in Brazil. Every so often, you find an unfamiliar plant that you carefully dig up and place in a special container. You hope one of these plants contains a substance that will kill cancer cells. You are a botanist at work.

A botanist may study any aspect of the plant kingdom. Many botanists work in laboratories, studying plant life processes and reproduction. Others may focus on developing new varieties of food crops or controlling plant diseases.

Some botanists work outside much of the time. They may explore an area to see if rare plants live there, for example. If rare plants are discovered, it may mean that the land cannot be developed.

Many research positions in botany require a doctorate degree. However, for many careers related to botany, you need only a bachelor's degree or some college-level training in botany. These careers include forester, forest ranger, landscaper, florist, environmental educator, botanical illustrator, and nursery manager.

A bachelor's degree in botany may also qualify you for an entry-level job as a technician or lab assistant. This kind of job may involve helping scientists perform research projects for private companies, universities, or government agencies.

Chapter 13 Nonflowering Plants

TEACH

Critical Thinking
Reason and Conclude
Have students explain why bryophytes need water for reproduction. (Sperm cells use water to swim to the egg cell and fertilize it.)
Predict Ask students where in a city they might find mosses, liverworts, and hornworts. Why? (These plants may be found in shady, wet areas such as the north side of large buildings, trees, or houses where they are protected from the drying effects of the sun.)

Career Corner

Discuss Tell students that sometimes botanists find plants that no one has discovered before. This is most likely to happen in a tropical rain forest because thousands of different plants live under the protection of the dense forest canopy. Discovering and naming new plants and studying their cycles is an exciting part of being a botanist.

The Living Textbook:
Life Science Side 3

Chapter 12   Frame 06715
Liverwort (Movie)
Search: Play:

TEACH • Continued

Research
Have students use references to find out what will happen to peat left in the ground. Also have students try to find out how scientists know what happens to peat over time.

EVALUATE

WrapUp
Portfolio Have students cut out pictures of liverworts and mosses and mount them on pieces of notebook paper. Have them label the rhizoids and the leaflike structures and explain how these features help the plants adapt to a land environment. Students can keep these pictures in their portfolios.
Use Review Worksheet 13.2.

Check and Explain
1. Answers may vary. Rain forests and temperate forests are environments in which unusual mosses and liverworts often grow.
2. Answers may vary. Both environments provide ample water and unique plant life.
3. Both are nonvascular, live in moist environments, have rhizoids and leaflike structures, need water to reproduce, and produce spore-releasing capsules. They differ in their shape and structure.
4. They would not be able to reproduce, because their sperm need water to fertilize egg cells.

The Living Textbook:
Life Science Sides 1-4
Chapter 3   Frame 00637
Bryophytes (5 Frames)
Search: Step:

INTEGRATED LEARNING

STS Connection
Both coal and peat are fuels that are used to provide heat and electricity. Although they are still widely used, other sources, such as petroleum and natural gas, have gained broader use. Have students create a chart naming several fuels. They should show how each is obtained and evaluate its advantages and disadvantages, such as pollution, cost, and so on.

Integrating the Sciences
Chemistry Peat mosses were used to dress wounds during World War I. Have students find out what physical properties make peat moss a useful dressing for wounds. (Peat moss absorbs liquid and has antibiotic properties.)

Science and Society *Peat Resources*

Peat has been used for thousands of years as a fuel. It forms from the compacted bodies of dead sphagnum moss plants. As the moss grows in a bog, each new generation grows on top of the others. The weight of the new plants compacts the layers beneath. The compacted moss decays slowly, forming peat. Large peat deposits occur in Europe, North America, and Asia.

Dried peat burns readily. In many places in the world, peat fires heat homes and cook meals. In Ireland and Scotland, some people still cut peat in blocks using hand tools. They dry it and store it in piles near their homes.

Large amounts of peat fuel electrical power generating plants in Europe and Asia. In the United States and Canada, mined peat is sold for landscaping and gardening. The peat is mixed into the soil to increase its ability to hold moisture and nutrients. Various techniques are used to harvest the large quantities of peat needed. For example, a steel drum with spikes is dragged across a bog, churning up the peat. After the peat dries, vacuum machines collect the top layer.

Sphagnum moss itself has other important uses. Highly absorbent, it is used for wound dressings and baby diapers by people from many different cultures. It contains sphagnol, a compound with antibacterial properties. Traditional healers in China and North America use the moss to prepare medicinal teas and salves.

Figure 13.6 ▲
In Ireland, a man stacks peat to dry it. The peat will eventually be used for fuel.

Check and Explain
1. If you were chosen to lead an expedition looking for unusual mosses and liverworts, where would you go? Describe two environments you would choose.
2. Explain why you selected each environment in the first question.
3. **Compare and Contrast** What characteristics do mosses and liverworts have in common? How are they different?
4. **Infer** How might a long drought affect the reproduction of mosses?

Themes in Science

Scale and Structure Plants that adapted to the land needed strength to support themselves against the pull of gravity. They had to be able to retain water necessary for their cell processes. And, finally, they had to be able to transport water and nutrients from the soil throughout the whole plant for photosynthesis. Have students describe the evolutionary adaptations that enabled plants to solve these three problems. (Bundles of tubelike vascular tissues gave the plants support, enabling them to grow taller than their ancestors. The vascular tissues also allowed them to transport nutrients and water from the soil. A waxy covering kept plants from drying out.)

SECTION 13.3

Section Objectives
For a list of section objectives, see the Student Edition page.

Skills Objectives
Students should be able to:

Make a Graph showing the number of species of club mosses, horsetails, and ferns.

Estimate the number of spore cases on a fern frond.

Collect Data about a species of conifer.

Make a Model of a forest landscape from the time of the dinosaurs.

Infer why some nonflowering plants absorb more water than others.

Vocabulary
pollen, conifer, pollination

13.3 Nonflowering Vascular Plants

Objectives

▶ **Describe** ferns, horsetails, club mosses, and gymnosperms.

▶ **Explain** the importance of pollen as a plant adaptation.

▶ **Compare** and **contrast** the life cycles of ferns and pine trees.

▶ **Make a model** of an ancient landscape dominated by nonflowering vascular plants.

Skills WarmUp

Make a Graph There are about 1,000 species of club mosses, 15 species of horsetails, and 12,000 species of ferns living today. Display this information on a circle graph.

MOTIVATE

Skills WarmUp
To help students understand the diversity of nonflowering vascular plants, have them do the Skills WarmUp.
Answer Students' graphs should accurately show the number of species of ferns, horsetails, and club mosses.

Prior Knowledge
Gauge what students know about nonflowering vascular plants by asking the following questions:

▶ How do vascular plants differ from nonvascular plants?

▶ What are some examples of vascular plants?

Reteach
Use Reteach Worksheet 13.3.

Imagine walking the length of a football field. Now imagine a plant that grows taller than that distance! Such a plant is a nonflowering vascular plant known as a redwood tree. Redwood trees are among the tallest of all living things, reaching heights of over 100 m.

Why can a redwood tree grow so tall? As plants evolved on land, they developed tubelike vascular tissues to carry water and nutrients. The bundles of vascular tissue gave the plants support, allowing them to grow taller. The tubelike vascular tissue could also transport water from the soil to the rest of the plant. Because of this adaptation, vascular plants could grow almost anywhere on land.

Look at Figure 13.7. By about 300 million years ago, tall vascular plants that looked like giant feather dusters, asparagus stalks, and tree-sized ferns covered the earth. Descended from those fantastic forest plants are today's horsetails, ferns, and club mosses, as well as the giant redwood.

Ferns

Nonflowering vascular plants called ferns were important members of the ancient forests. Ferns still grow today in forests around the world. Most grow in warm, tropical regions, but some species are found in regions where seasonal changes occur. A few species of tropical ferns are tree-sized like those living millions of years ago, but most of the 12,000 or so species of ferns alive today are smaller.

Unlike mosses, ferns have true roots, stems, and distinctive leaves. These organs all contain vascular tissue. The leaves of ferns are sometimes called fronds. Each is

Figure 13.7 ▲
With the evolution of vascular tissue, nonflowering plants could form forests.

Chapter 13 Nonflowering Plants 259

TEACH

Discuss
Students may wonder why they can't usually see the gametophyte of the fern. Explain that it is very small and is found only in late spring or early summer.

Skills Development
Infer Tell students that the life cycle of a fern may be said to begin with the tiny spores that are shot out of the sori of a mature fern. If a spore lands in a moist spot, it will grow into a gametophyte, which will eventually produce more sporophyte plants. Ask students to infer why it is beneficial for the spores to be shot out a distance away from the plant. (To spread the spores to new areas where plants won't have to compete for sunlight and nutrients)

Critical Thinking
Compare and Contrast Ask students how the life cycles of ferns and horsetails differ from those of mosses. (In ferns and horsetails, the sporophyte is much larger and longer-lived than the gametophyte; in mosses, the gametophyte is larger and lives longer.)

Answer to In-Text Question
① The pointed structure contains spore cases.

The Living Textbook: Life Science Side 3
Chapter 14   Frame 09074
Fern (Movie)
Search: Play:

The Living Textbook: Life Science Sides 1–4
Chapter 3   Frame 00647
Horsetails (4 Frames)
Search: Step:

INTEGRATED LEARNING

Integrating the Sciences

Earth Science The first ferns appeared during the Devonian age. Fossils of these early ferns are common in Mesozoic rocks. However, most of the 12,000 species of ferns alive today are not closely related to these early forms. Most modern ferns are descendants of two fern families that first appeared during the Jurassic period. Have students find out when these geologic periods occurred. They should create a timeline, indicating when the first plants appeared (3.5 billion years ago), when plants began colonizing the land (433 million years ago), and when the first vascular plants appeared (400 million years ago).

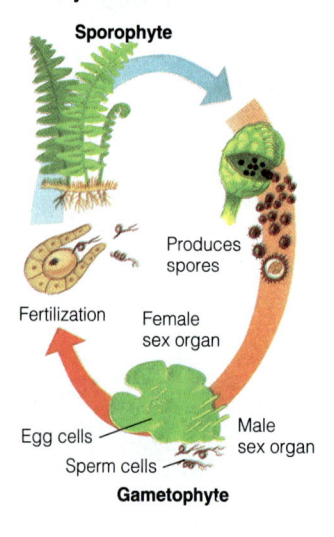

Figure 13.8 ▼
Life Cycle of a Fern

Sporophyte · Produces spores · Fertilization · Female sex organ · Egg cells · Sperm cells · Male sex organ · Gametophyte

Figure 13.9 ▲
What do you think is the pointed structure at the end of this horsetail plant? ①

---

divided into leaflets. They have a waxy coating to protect them from drying out. The stems of most ferns are underground stems called *rhizomes* (RY zohmz). Roots grow from the underside of the rhizome and leaves grow from the topside. Tree ferns have tall, upright stems.

The life cycle of a fern is shown in Figure 13.8. The plant that you know as a fern is the sporophyte part of the life cycle. On the underside of its mature fronds grow spore cases called *sori* (SOH ree) that look like small brown spots. Inside the sori, special cell divisions occur, producing spores. The sori have springlike structures that can fling the spores several meters when they are mature. Once released in this way, spores travel greater distances by wind and water.

If a spore lands in moist, shaded soil, it develops into a small, heart-shaped gametophyte with both male and female sex organs. Sperm cells produced by the male organ swim to the female sex organ containing the egg cell. Fertilization occurs, and the zygote develops into an embryo. The embryo, protected and nourished by the gametophyte, develops into the familiar leafy sporophyte.

How is the fern life cycle different from the life cycle of a moss? In a moss, the gametophyte is the largest and longest-lived generation. In a fern, it is the sporophyte that grows larger and lives longer.

Horsetails

The plants that look like giant bottle brushes in Figure 13.7 are the ancestors of the horsetails. Some grew up to 15 m tall. Modern horsetails, shown in Figure 13.9, are much smaller. They are common in many damp places, but there are only 15 species.

Horsetails have jointed stems. A circle, or whorl, of narrow leaves grows from each joint, making the plant look something like the bushy tail of a horse. Horsetails were once called scouring rushes, because people used them to scrub cooking pots. The cell walls of the stem contain hard grains of silica, making the plant abrasive.

The life cycle of a horsetail is similar to that of a fern. The sporophyte is much larger and longer-lived than the gametophyte. Spores produced by the sporophyte grow into tiny free-living gametophytes that produce both egg and sperm cells. The fertilized eggs grow into new sporophytes.

Chapter 13   Nonflowering Plants

Multicultural Perspectives

Tell students that Johann Scheuchzer (1672–1733) was a Swiss naturalist and physician. Scheuchzer studied fossils of plants and fishes from the Miocene rocks at Oeningen in Switzerland. Show students Switzerland on a map. Have them find out what kinds of plant fossils Scheuchzer studied.

Club Mosses

The plants shown in Figure 13.10 can be found growing on forest floors in temperate regions. They are members of another group of nonflowering vascular plants called club mosses. The club-shaped structures at the tips of their stems are spore cases.

Although these plants have the word *moss* in their name, they are not mosses. Unlike true mosses, club mosses have vascular tissue. Their life cycles are similar to those of the horsetails and ferns. The plants you see in Figure 13.10 are sporophytes.

Most species of club mosses grow in tropical areas on the branches and trunks of trees. Hundreds of millions of years ago, relatives of modern club mosses grew as large as trees. The treelike club mosses all became extinct, however, when the earth's climate changed. The club mosses alive today are descendants of a group of smaller club mosses that survived.

Figure 13.10 ▲
The most common club mosses in North America are these *Lycopodium*.

SkillBuilder *Estimating*

Fern Spore Cases

Ferns store spores in spore cases, or sori, located under the frond leaflets. Each sori contains thousands of tiny spores. To increase the chances of new fern growth, a fern plant has many sori.

1. Obtain a fern frond.
2. Observe the underside of the frond. Look for the sori.
3. Copy the table below on a sheet of paper.
4. Based on your observation, estimate the total number of sori on the frond. Record your estimate.
5. Count the total number of sori on ten leaflets. Record in your table. Calculate the average number of sori for each leaflet. Record.
6. To calculate the estimated number of sori on the entire frond, count the total number of leaflets on the frond. Record. Multiply the total number of fronds by the average number of sori for each leaflet.

Write a paragraph comparing your estimation to the calculated estimation for sori found on the frond. Give reasons why your observed estimate was different from the calculated estimate.

Estimation from Observation	Number of Sori on Ten Leaflets	Average Number of Sori for Each Leaflet	Total Number of Leaflets	Calculated Estimation

Enrich

Tell students that the ancestors of ferns, horsetails, and club mosses grew to be large trees millions of years ago. Only a few kinds of ferns in tropical areas grow that big today. Most modern ferns, horsetails, and club mosses are small plants.

Skills Development

Classify Ask students which of the three plants—ferns, horsetails, and club mosses—are most like mosses in appearance. Which are most like flowering plants in appearance? (Club mosses look most like mosses; ferns look most like some flowering plants.)

Discuss

Ask students the question, If there are eight fronds on a fern, 1,000 sori on each frond, and 12 spores in each sorus, how many spores are produced by the fern? (96,000 spores)

SkillBuilder

Answers Answers will vary depending on the kind of fern frond. Sample answers are as follows:

4. Observed estimate—500
5. Number of sori on ten leaflets—367
   Average number of sori on each leaflet—37
6. Total number of leaflets—27; calculated estimate—999

TEACH ▪ Continued

Directed Inquiry

Have students study the material on this page. Ask the following questions:

▶ In what important way do gymnosperms differ from ferns, horsetails, and club mosses? (They make seeds.)

▶ What is a seed? (An adaptation for protecting and nourishing a plant embryo)

▶ How are the leaves of conifers different from all the other gymnosperms? (Conifers have needles.)

▶ In what structures do gymnosperms produce seeds? (Cones)

▶ What are the four main groups of gymnosperms living today? (Conifers, cycads, gnetophytes, and the ginkgo)

▶ Which gymnosperm is most like an oak or maple tree? Why? (The ginkgo, because it loses its leaves in the autumn)

Answers to In-Text Questions

① Students may mention pines and redwoods.

② All the plants shown produce seeds.

The Living Textbook:
Life Science Side 8

Chapter 38   Frame 30622
Coniferous Forest (Movie)
Search: Play:

262

INTEGRATED LEARNING

Themes in Science

Scale and Structure Two important adaptations led to the development of trees. Tubelike vascular tissues and a material called *lignin* that made the stems of some plants tough and woody allowed plants to grow taller than their nonvascular ancestors. Because taller plants and trees get more light, they became very abundant. The tallest tree is the redwood, which can grow as tall as 112 meters.

Integrating the Sciences

Earth Science Bristlecone pines, a type of angiosperm, are among the oldest living things on the earth. These pine trees grow in North America in the Rocky Mountains. Because they live at high altitudes where there is little oxygen, bristlecones grow extremely slowly. Some specimens alive today are thought to be as much as 6,000 years old.

Gymnosperms

Nonflowering vascular plants that produce seeds are called gymnosperms. Cone-bearing trees, such as pine and redwood, are the most common gymnosperms alive today. Because of their seeds and other adaptations, these plants live and reproduce in places where ferns, horsetails, and club mosses cannot.

A seed is an adaptation for protecting and nourishing a plant embryo. Along with an embryo, a seed contains an energy supply. It is surrounded by a protective coat. The embryo inside a seed may remain dormant for a long period of time. Then, when the conditions are right, it can begin growing. Seeds can travel over a distance to establish new plants in new places.

Gymnosperms have another important adaptation to living on land. Recall that ferns, horsetails, and club mosses need standing water so that sperm cells can swim to the egg cells. Gymnosperms evolved a way to get around the need for water. Their sperm cells don't need to swim. Instead, they develop inside protective cases that can be carried by moving air. These dustlike particles are known as **pollen**. Pollen made it possible for gymnosperms to survive and reproduce away from damp environments. As a result, gymnosperms grow in deserts and on dry, windy mountaintops.

There are four main groups of gymnosperms alive today. One member of each group is shown below. Study the photographs carefully. What do all the plants have in common? ②

▲ The most common living gymnosperms, the **conifers**, have needle-shaped leaves and seeds produced in cones.
① What conifers can you name?

Cycads (SY kadz) are relics of the ancient forests. This cycad is native to the state of Florida. ▼

The ginkgo, also called the maidenhair tree, has fan-shaped leaves and fleshy cones. Unlike other gymnosperms, the leaves of a ginkgo fall from the tree in autumn. ▼

◀ This plant belongs to a group called gnetophytes (NEE toh FYTS). It has long straplike leaves and bears clusters of small cones on short stalks.

262 Chapter 13 Nonflowering Plants

Themes in Science

Scale and Structure Pine trees produce enormous numbers of pollen grains. Each grain has two tiny air-filled sacs attached to it. These air sacs act like sails to carry the pollen grains great distances through the air. Ask students to think of other examples of seeds that have aerodynamic shapes. (Students might mention maple seeds or dandelion seeds.)

Geography Connection

Have students find out which kinds of nonflowering plants grow in each climatic region of the world (tundra, forest, grassland, desert, aquatic). Have students present their findings in a map or color-coded chart of plant environments.

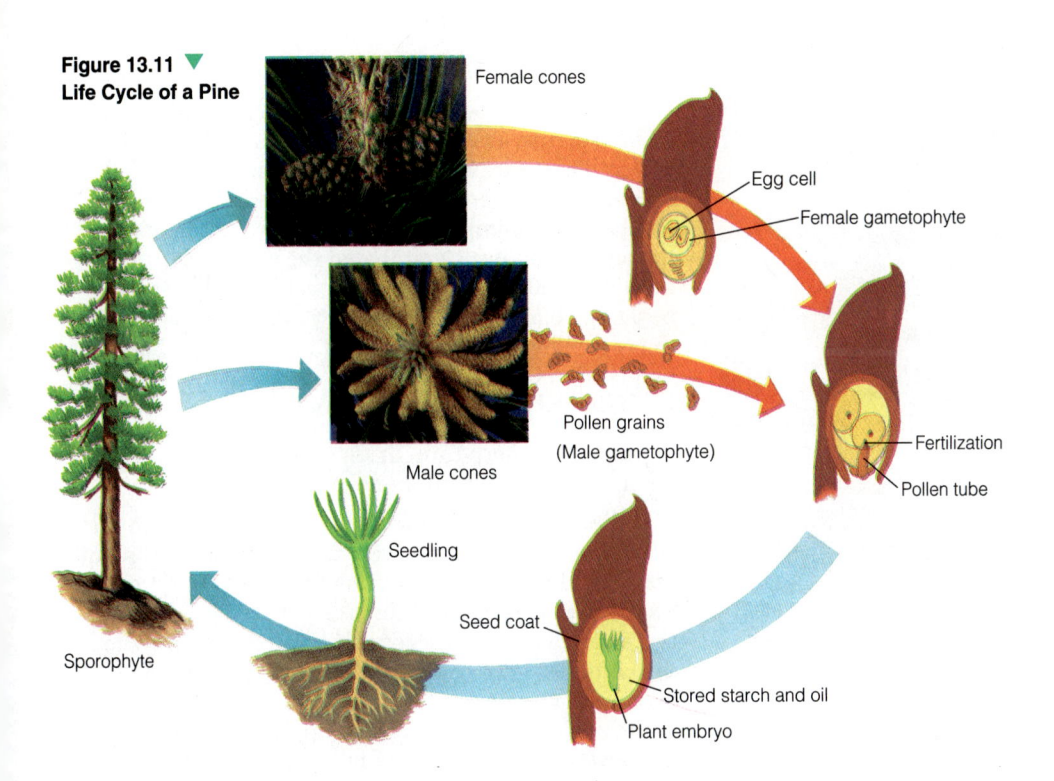

Discuss

Have students study the life cycle of a pine in Figure 13.11. Emphasize that the female and male gametophytes are very small and that the female gametophyte remains in the cone where it produces eggs. The male gametophyte is the pollen grain that is carried by the wind to the cone. Ask students:

▶ What is pollination? (Pollination occurs when the pollen grains reach a cone where there is a female gametophyte.)

▶ How does the male gametophyte reach the female gametophyte to fertilize it? (It grows a pollen tube through which sperm travel to the egg.)

Skills Development

Infer Ask students to infer how it is beneficial for a pine tree to produce so many pollen grains. (In order to increase the odds that some grains will reach the female gametophyte)

Reteach

Tell students that there are several steps in the pine life cycle. You can begin with the sporophyte, or pine tree. List the steps on the chalkboard. Have students number the steps in sequential order. Help them see that the last step is the same as the first.

Figure 13.11 ▼
Life Cycle of a Pine

Female cones — Egg cell — Female gametophyte — Pollen grains (Male gametophyte) — Fertilization — Pollen tube — Male cones — Seedling — Seed coat — Stored starch and oil — Plant embryo — Sporophyte

Pine Tree Life Cycle

The life cycles of gymnosperms differ greatly from the life cycles of ferns, horsetails, and club mosses. The pine tree provides a good example of how the gymnosperm life cycle works.

The pine tree is the sporophyte generation. It has two types of cones. The large woody cones produce spores that develop into female gametophytes. Smaller, nonwoody cones produce spores that develop into male gametophytes. Unlike the gametophytes of ferns, horsetails, and club mosses, these gametophytes are not free-living.

Look at Figure 13.11. The female gametophyte remains in the cone. It grows into a many-celled mass in which several egg cells begin to develop. The male gametophyte is packaged into a tiny pollen grain. These are released and carried by the wind.

When the wind-borne pollen reaches a cone where the female gametophyte is developing, **pollination** occurs. The male gametophyte grows a pollen tube through which sperm can reach the egg cell. Fertilization occurs as the sperm and egg fuse. Notice in Figure 13.11 that the fertilized egg develops into an embryo. The embryo is packaged into a seed, along with a food supply. When the seed sprouts and grows into a young sporophyte, the cycle is complete.

In this life cycle, the gametophyte generation is very small. For this reason, reproduction in gymnosperms *appears* to involve only one generation producing male and female sex cells.

The Living Textbook:
Life Science Side 3

Chapter 15 Frame 09886
Pine Life Cycle (Movie)
Search: Play:

The Living Textbook:
Life Science Sides 1-4

Chapter 3 Frame 00670
Conifer Cones (5 Frames)
Search: Step:

Chapter 13 Nonflowering Plants **263**

TEACH • Continued

Critical Thinking
Reason and Conclude Ask students to describe the adaptations that enable angiosperms to survive in dry conditions. (They have a vascular system to transport water and needlelike leaves with waxy coatings and sunken pores to retain water.)

Enrich
Have students use references to find out about extinct plants called *seed ferns*. These plants existed when the dinosaurs roamed the earth. Have students prepare reports about the seed ferns and present them to the class.

Skills Development
Infer Have students consider that it takes 17 trees to make a metric ton of paper. Explain that millions of trees are cut down each year in the United States to use for paper and building materials. Ask students what can be done to conserve trees. (Paper should be recycled and trees should be replanted.)

Skills WorkOut
To help students understand how environmental factors affect the needle lengths of conifers, have them do the Skills WorkOut.
Answers Conifers in cold, dry, and high-altitude areas have shorter needles.

Integrated Learning
Use Integrating Worksheet 13.3.

The Living Textbook:
Life Science Sides 7-8
Chapter 17   Frame 01815
Coniferous Forest (75 Frames)
Search: Step:

264

INTEGRATED LEARNING

Themes in Science
Stability and Equilibrium Life cycles of plants are affected by the conditions in their environments. Have students imagine that a rainy area became dry. Which nonflowering plants would dwindle? Why? (Bryophytes, because they need water to reproduce) Which would thrive? Why? (Vascular plants, because they reproduce with wind-borne pollen)

Integrating the Sciences
Chemistry The chemicals in gymnosperms make them valuable sources of tar, turpentine, and rosin. Have students find out which parts of gymnosperms are used to make turpentine, tar, or rosin. Have them share information in brief oral reports.

Figure 13.12 ▲
These fossils of cones (top) and ginkgo leaves (bottom) are the kinds of evidence scientists use to piece together the history of gymnosperms on the earth.

Gymnosperms and Geologic Time
Fossils like those in Figure 13.12 tell the history of gymnosperms. The first gymnosperms appeared about 350 million years ago. At that time, the earth's land masses were joined together in a supercontinent called Pangaea. The climate was warm and wet. Club mosses, horsetails, and ferns dominated the landscape.

About 250 million years ago, geologic processes began to change the landscape. Mountains formed, and many shallow seas drained. Conditions on land became much drier. Gymnosperms, particularly conifers, were well adapted for the dry conditions. Gymnosperms did not require standing water to reproduce. The conifers had needlelike leaves with thick waxy coatings and sunken pores to conserve water.

When dinosaurs roamed the earth, about 200 million years ago, gymnosperms were the most common plant life. The dinosaurs lived among forests of conifers and cycads the size of small palm trees. Large fernlike plants that reproduced with seeds were also common. When the climate cooled dramatically about 70 million years ago, dinosaurs became extinct, but conifers continued to thrive. Other gymnosperms were less successful under the new conditions, and many species died out. The newly evolved flowering plants took over much of the landscape. Today gymnosperms grow mostly in dry areas with poor soil.

Skills WorkOut
Collect Data Choose a species of conifer that grows in your area. Collect needles from two of these trees growing in places that differ in moisture, altitude, soil, or slope. Measure the length of ten needles from each tree. Make a table of the data. How do environmental conditions affect needle length?

Science and Technology *Making Paper*
Think of all the paper you handle in one day. Most of this paper is made of wood fibers from conifer trees. The method used to make all this paper was developed in China about 2,000 years ago. First, wood is crushed and mixed with water to form a pulp. The pulp is poured onto a flat screen, which is shaken to intermesh the wood fibers. The water is suctioned off, leaving a thin sheet of fresh paper.

Although the basic method is similar, papermaking today takes place on a huge scale. One metric ton of paper needs wood from more than 17 mature trees. The trees are cut down and trucked to a paper mill. There the bark is removed. The trees are ground with water or treated with heat and chemicals to form pulp.

264 Chapter 13 Nonflowering Plants

Integrating the Sciences

Ecology Clearcutting, or cutting down all trees in an area, is still commonly practiced by the timber industry in the United States. Encourage students to research and report on efforts made to encourage new growth in clearcut areas. Have students analyze the success or failure of various measures.

STS Connection

Invite a recycling expert to visit the class and discuss paper recycling methods. Have the expert explain why recycled paper is not more commonplace, why it is often combined with virgin paper, and why coated paper is usually not recycled. Have students prepare questions ahead of time.

Discuss

Refer students to Figure 13.14. Explain that making paper by hand is considered an art form in Japan and other countries. Most of Japan's paper, however, is made by machines like the one pictured in Figure 13.13.

EVALUATE

WrapUp

Portfolio Have students make drawings in which they compare sporophytes and gametophytes in vascular and nonvascular plants. Encourage them to illustrate the life cycles of the plants visually and graphically. Students can keep the drawings in their portfolios.

Use Review Worksheet 13.3.

Check and Explain

1. Ferns have a long sporophyte stage and waxy leaves. Horsetails have jointed stems, circles of narrow leaves, and abrasive stem walls. Club mosses grow on branches and tree trunks and are not true mosses. Gymnosperms produce seeds and pollen for reproduction and survival in drier environments.

2. A pollen grain is a dustlike, protective particle that contains and helps disperse sperm.

3. Ferns have reproductive spores and fertilize themselves. Pines have pollen that contain sperm and travel to female cones for fertilization. Both species spend most of their life cycles as sporophytes.

4. Dinosaurs lived among forests of conifers, cycads, and large fernlike plants.

Figure 13.13 ▲
Most papermaking takes place on a huge scale, as you can see in this photograph of a paper mill.

Pulp is pumped through pipes to a papermaking machine. The pulp is sprayed onto a wide, moving belt with a screen. Water drains through the screen, leaving a damp mat of wood fibers. This is the newly made paper. The paper is lifted from the end of the moving belt and dried over a series of steam-heated rollers, like the one in Figure 13.13.

If paper is recycled, the papermaking process is much simpler. Waste paper is mixed with water in a machine resembling a giant kitchen blender. The action of the blades forms a pulp that is just like the pulp made from trees.

Figure 13.14 ▲
Some paper is still made by hand. This photograph, taken in Japan, shows a man making paper in the traditional way.

Check and Explain

1. Name and describe four groups of nonflowering vascular plants.

2. What is pollen? Why is pollen a useful adaptation to gymnosperms?

3. **Compare and Contrast** How do the life cycles of pines and ferns differ? How are they alike?

4. **Make a Model** Depict a forest landscape that may have existed when dinosaurs lived.

Chapter 13 Nonflowering Plants

ACTIVITY 13

Time 40 minutes   **Group** pairs

Materials

60 small cups or petri dishes
moss samples
samples of nonflowering plants (not moss)
hand lenses
graduated cylinders
paper towels

Analysis

1. Answers depend on plant samples used. Make sure answers agree with recorded observations.
2. Students are likely to describe structural features, like a porous surface or a large surface area.
3. Students should realize, first, that all plants need water for photosynthesis and other cell processes and, second, that plants without true roots must absorb water through their surfaces.
4. Type of plant; amount of water absorbed
5. Any water left in the cup or cylinder from a previous trial would cause a higher reading in the next trial.

Conclusion

Accept all logical conclusions, provided they follow from recorded observations.

Everyday Application

The layer of moss will help keep the soil in the pot moist. Students may also point out that decaying moss might provide fertilizer.

Extension

After two to three weeks, have students evaluate their original lists of conditions. Ask them what they would change or add to their lists.

TEACHING OPTIONS

Prelab Discussion

Have students read the entire activity before beginning. Discuss the following:

▶ Ask students why a moss must absorb water over its entire surface.

▶ Ask students to suggest why they need to be very careful during the procedure for this activity. (The quantity of water being measured is relatively small; water can be lost during the procedure.)

Activity 13 Which plant absorbs the most water?

Skills Measure; Observe; Collect Data; Infer

Task 1 Prelab Prep
Collect the following items: 4 small cups or petri dishes, water, 2 samples of other nonflowering plants beside mosses, moss sample, hand lens, graduated cylinder, paper towels.

Task 2 Data Record
1. Place one sample of each plant in a cup or petri dish.
2. On a separate sheet of paper, draw the table shown to the right. Make the space for the drawing larger than it is shown. Write the names of the plant samples you chose in the table.
3. Record all your observations in the table.

Task 3 Procedure
1. Using the hand lens, observe the three nonflowering plant samples.
2. Draw each sample in your data table.
3. Measure 5 mL of water in the graduated cylinder. Add the water to one of the samples. Record this amount in your table. Repeat for each of the other two plant samples.
4. Wait 5 minutes. As you wait, wipe up any spills, and observe the samples.
5. Without disturbing the plant sample, pour off the remaining water from one of the samples into the empty cup. Measure the amount of water in the graduated cylinder. Record the amount. Dispose of the sample water. Dry the cup with the paper towel.
6. Repeat step 5 for each of the other two plant samples.

Task 4 Analysis
1. Which plant absorbed the largest amount of water?
2. What is it about the structure of this plant that makes it absorbent?

Table 13.1 Water Absorbed

Plant	Drawing	Water Added	Water Left	Water Absorbed

3. Why do you think it is important for this nonflowering plant to be able to absorb water?
4. List the variables in this activity.
5. Why was it important to dry the cup and graduated cylinder after each water sample measurement?

Task 5 Conclusion
Write a short paragraph comparing the absorption of the nonflowering plants in this activity. If the plants didn't absorb the same amount of water, explain why.

Everyday Application
Dead moss plants are often placed in the pots of indoor plants or around the base of potted trees. Explain why mosses are used for this purpose.

Extension
Try planting your own simple terrarium with various nonflowering plants. Before you make the terrarium, be sure to make a list of the necessary conditions for successful plant growth. Keep a record of any changes that occur.

Chapter 13 Review

Concept Summary

13.1 Characteristics of Nonflowering Plants
- Bryophytes, mosses, and ferns are examples of nonflowering plants.
- Nonflowering plants evolved specialized tissues and organs that enabled them to live on land.
- The plant life cycle has a sexual gametophyte stage and nonsexual sporophyte stage.

13.2 Bryophytes
- Bryophytes grow only in moist environments. The gametophyte is the largest and longest-lived part of their life cycle.
- Mosses have small leaflike structures arranged in a spiral on a short stalk. They have rhizoids.
- Liverworts have flat leaflike or lobed structures that make up the main part of their bodies. Hornworts are close relatives of liverworts.

13.3 Nonflowering Vascular Plants
- Ferns are nonflowering vascular plants with true roots, stems, and leaves. The sporophyte is the largest and longest-lived part of their life cycle.
- Horsetails have jointed stems and leaves arranged in whorls.
- Club mosses have club-shaped spore cases at the tips of their stems.
- Gymnosperms are nonflowering vascular plants that produce seeds.
- Pollen makes it possible for gymnosperms to survive and reproduce away from damp environments.

Chapter Vocabulary
gametophyte (13.1)   moss (13.2)   liverwort (13.2)   conifer (13.3)
sporophyte (13.1)   rhizoid (13.2)   pollen (13.3)   pollination (13.3)
spore (13.1)

Check Your Vocabulary
Use the vocabulary words above to complete the following sentences correctly.

1. The ___ stage in the plant life cycle reproduces sexually.
2. Nonsexual reproductive cells with a protective covering are called ___.
3. ___ have needle-shaped leaves and seeds produced in cones.
4. Bryophytes have rootlike structures called ___.
5. Gymnosperms' sperm cells are packaged inside grains of ___.
6. The ___ stage of a plant reproduces nonsexually with spores.
7. When pollen reaches the female gametophyte, ___ occurs.
8. A bryophyte with green leaflike structures arranged in a spiral on a short stalk is called a ___.
9. The body of a ___ is divided into rounded sections, or lobes.

Explain the difference between the words in each pair.

10. Spore, seed
11. Rhizoid, root
12. Pollen, sperm

Write Your Vocabulary
Write sentences using each vocabulary word above. Show that you know what each word means.

CHAPTER REVIEW 13

Check Your Vocabulary
1. gametophyte
2. spores
3. Conifers
4. rhizoids
5. pollen
6. sporophyte
7. pollination
8. moss
9. liverwort
10. A spore is a nonsexual reproductive cell with a protective coat. A seed is a plant embryo, together with stored energy and a protective coat.
11. Rhizoids are rootlike structures that function to attach bryophytes to the ground. They do not have vascular tissue to carry water, as roots do.
12. Pollen is a protective case containing gymnosperm sperm cells. Sperm cells are actual gametes.

Write Your Vocabulary
Students' sentences should show that they know both the meanings of the words and how to apply the words.

Use Vocabulary Worksheet for Chapter 13.

CHAPTER 13 REVIEW

Check Your Knowledge

1. Bryophytes have a waxy covering that functions to keep their cells from drying out.
2. These plants have vascular tissue that carries water up from the ground.
3. Club mosses grow on forest floors in temperate regions.
4. The climate was warm and wet, and the earth's masses were joined in a supercontinent called Pangaea.
5. Answers may vary, but may include: Ferns have true roots, stems, and distinctive leaves. In ferns, the sporophyte stage is the largest and longest, while in mosses, the gametophyte is the largest and longest. Fern spores develop into a gametophyte containing both male and female sex organs; moss spores develop into separate male and female gametophytes.
6. Cone-bearing trees, such as pines, are the most common gymnosperms.
7. The other stage in the fern life cycle is the gametophyte.
8. True
9. False; air
10. False; nutrients
11. True
12. True

Check Your Understanding

1. Answers may vary. In ferns, the sporophyte stage is the largest and longest, while in mosses, the gametophyte is the largest and longest stage. Fern spores are produced inside sori and develop into gametophytes containing both male and female sex organs; moss spores are produced in capsules, and develop into separate male and female gametophytes.
2. The sporophyte stage of ferns has the ability to disperse offspring. In a gymnosperm, the embryonic (seed) stage allows for dispersal of offspring.
3. Club mosses are unlike true mosses in that club mosses have vascular tissue.
4. If a spore lands in moist, shaded soil, it will develop into a gametophyte.
5. Gymnosperm sperm cells are encased in pollen grains and are carried by the wind. This allows them to be dispersed more widely than swimming sperm cells.
6. Pollination in pine trees occurs when a pollen grain reaches a cone where a female gametophyte is developing.
7. Making paper from recycled paper is much simpler because it is easier to turn paper back into pulp than it is to turn trees into pulp.
8. Answers may vary, but may include: differences—nonflowering plants have adapted to life on land, some types produce seeds, some have vascular tissue; similarities—both have life cycles that include a gametophyte and a sporophyte stage.
9. Vascular tissue, which forms a network of tubes within the tree, allows water and nutrients to be drawn up to the top of the tree.

Chapter 13 Review

Check Your Knowledge

Answer the following in complete sentences.

1. Why don't the cells of bryophytes dry out in the air?
2. How do club mosses, horsetails, and ferns transport water from the roots to the above-ground parts of the plant?
3. Where can you find club mosses growing?
4. What was the climate like when the first gymnosperms appeared on the earth? What was the name for the earth's land masses at that time?
5. How are ferns different from mosses? Name three differences.
6. What are the most common living gymnosperms?
7. The plant you know as a fern is just part of its life cycle. What is the other part of the life cycle called?

Determine whether each statement is true or false. Write *true* if it is true. If it is false, change the underlined word to make the statement true.

8. In the first stage of the moss life cycle, <u>spores</u> develop into separate male and female gametophytes.
9. A plant that lives on land must adapt to two separate substances: soil and <u>rain</u>.
10. As plants evolved on land, they developed tubelike vascular tissues to carry water and <u>soil</u>.
11. The club-shaped structures at the tips of the stems on club mosses are <u>spore cases</u>.
12. About 200 million years ago, when dinosaurs roamed the earth, <u>giant ferns</u> had become the most common plant life.

Check Your Understanding

Apply the concepts you have learned to answer each question.

1. How is the life cycle of a fern different from the life cycle of a moss?
2. **Compare and Contrast** All land plants have a way to disperse their offspring, which means to scatter them over a large area. In what stage of its life cycle can a fern disperse offspring? In what stage does this occur in a gymnosperm? How do nonflowering plants compare in their methods of dispersal?
3. Are club mosses really mosses? Explain why or why not.
4. **Mystery Photo** The photograph on page 252 is a close-up view of a fern. The yellow clumps are sori on the underside of the leaf. They contain spores. These spores are an important part of the fern's reproductive cycle. What happens if a spore lands in moist, shaded soil?
5. Gymnosperm sperm cells do not have flagella for swimming. How do they travel instead? What advantage do they have over sperm cells that must swim?
6. Describe how pollination occurs in the life cycle of a pine tree.
7. **Extension** Which process is simpler: making paper from trees or from recycled paper? Why?
8. **Compare and Contrast** How do nonflowering plants differ from many-celled algae? How are they similar?
9. Why can a redwood tree grow as tall as 100 m high? Describe the adaptation that allows plants to grow far above the soil.

Develop Your Skills

1. a. Two
   b. 1991
   c. Answers may vary; three to four years.
2. a. Answers will vary, but may include: all are gymnosperms; all are found in temperate or cooler climates.
   b. Douglas fir
3. A seed contains an embryonic sporophyte along with stored energy. A spore contains reproductive cells that develop into a gametophyte.

Make Connections

1.

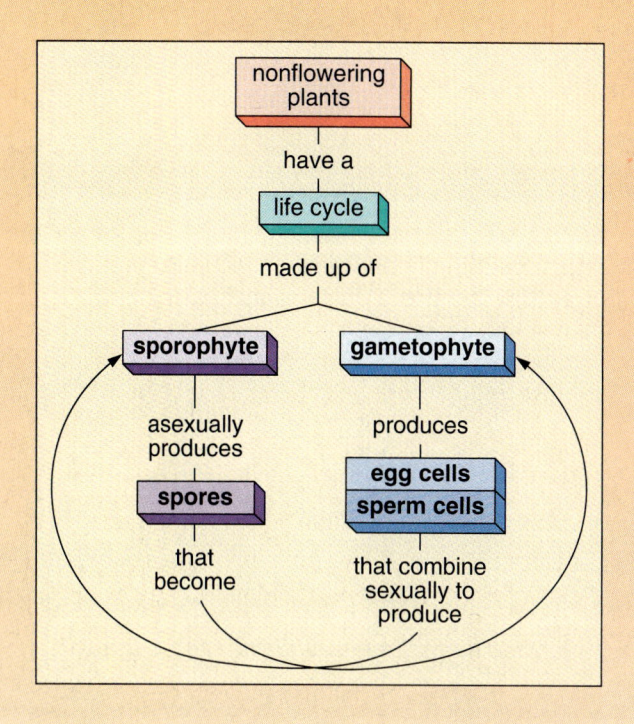

```
nonflowering plants
    have a
  life cycle
  made up of
sporophyte ─── gametophyte
asexually produces        produces
   spores           egg cells / sperm cells
that become        that combine sexually to produce
```

2. Student posters should include clues to the climate and plant and animal life of the Permian-carboniferous swampy forest.
3. Answers will vary. Some common wildflowers for investigation might include jack-in-the-pulpit, Queen Anne's lace, pearlwort, lady's slipper, bloodroot, Dutchman's-breeches, deadly nightshade.
4. Answers will vary, but should include a description of an allergic reaction. The common respiratory symptoms are sneezing, wheezing, and congestion, while more severe responses include fever, spasms, and shock.

Develop Your Skills

Use the skills you have developed in this chapter to complete each activity.

1. **Interpret Data** Bracken fern spreads quickly in cleared areas as its rhizomes grow outward and make new plants. The map below shows where bracken fern grew in a certain field at the beginning of 1989. It also shows the edge of bracken fern growth at the beginning of each following year.
   a. How many places in the field did bracken fern grow at the beginning of 1990?
   b. During which year did the fern grow the slowest?
   c. Predict how many years it will take for bracken fern to cover the field.

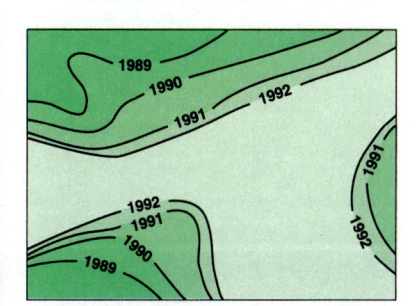

2. **Data Bank** Use the information on page 627 to answer the following questions.
   a. What do the seven tallest trees in the United States all have in common?
   b. What species of conifer grows closest in height to the California redwood?

3. **Compare and Contrast** How are seeds and spores different? How is each a part of the plant life cycle?

Make Connections

1. **Link the Concepts** Below is a concept map showing how some of the main concepts in this chapter link together. Only part of the map is filled in. Finish the map, using words and ideas you find in the chapter.

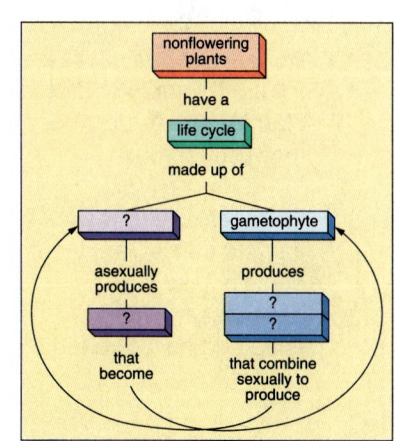

2. **Science and Art** Using information in this chapter, as well as your own research, create a poster showing what the swampy forests on the earth looked like 300 million years ago.

3. **Science and Language Arts** Liverworts can have lobes shaped like a mammal's liver. Hornworts have long, hornlike spore capsules. Investigate other plants that are named after a familiar object or part of the body.

4. **Science and You** Pollen makes gymnosperm reproduction possible. It also makes many people sneeze. Research pollen allergies. How does pollen affect allergic people?

CHAPTER 14

Overview

In this chapter, students learn about flowering plants, or angiosperms. Characteristics of flowering plants are described in the first section. The second section describes the parts of the shoot system and root system in vascular plants and emphasizes how plant cells, tissues, and organs work together. In the third section, reproduction in flowering plants is discussed. The final section distinguishes plant growth from animal growth, explaining plant hormones and seasonal growth patterns.

Advance Planner

▶ Obtain a bag of assorted seeds for SE page 272.

▶ Supply flowers for SE page 281.

▶ Provide magazines and books for TE page 282.

▶ Gather radish seeds, jars, paper towels, soil, and planters for TE activity, page 283.

▶ Purchase plants to be grafted for TE activity, page 284.

▶ Plan a field trip to a nursery, or invite a horticulturist to visit the class to talk with students. TE page 284.

▶ Prepare 30 flowers (gladioli, tulips, lilies), paper towels, black paper, and scalpels for SE Activity 14, page 290.

Skills Development Chart

Sections	Classify	Collect Data	Compare and Contrast	Hypothesize	Infer	Observe	Predict
14.1 Skills WarmUp SkillBuilder		●			●		
14.2 Skills WarmUp				●			
14.3 Skills WarmUp Skills WorkOut	●					●	
14.4 Skills WarmUp Activity	●		●			●	●

Individual Needs

▶ **Limited English Proficiency Students** Use a tape recorder to record the pronunciations and definitions of the boldface or italic words in this chapter. Along with the definition of each term, tell students which figures in the text illustrate the term. Have students use the tape recorder to listen to the information about each term. Then have students define the terms using their own words. Have students make their own tape recording of the terms and their definitions. They can use this tape along with the text figures to review important concepts in the chapter.

▶ **At-Risk Students** Invite an employee from your forest preserve or park district to speak with students. Have students prepare a list of questions about the work involved in managing a forest preserve and how park employees are affected by changing seasons.

▶ **Gifted Students** Have students design a garden. Ask students to research various nonflowering and flowering plants. Students should organize their garden based on information in the chapter—number of female and male flowers, life cycles, vegetative propagation, and so on. Students should present a visual display of their groundplan with drawings and pictures from magazines or catalogs. Students may choose to model their garden after a Japanese garden or a well-known garden such as those at Versailles or Sissinghurst.

Resource Bank

▶ **Bulletin Board** Make a bulletin board display about seeds. Include pictures or drawings of seeds, along with flowering plants. Encourage students to add pictures of seeds that they use for food.

▶ **Field Trip** As you work through the unit on plants, plan a trip to a conservatory or botanical garden. Encourage students to take field guides or taxonomic keys along. Prepare a plant search for students to complete as they observe flowering and nonflowering plants.

CHAPTER 14 PLANNING GUIDE

Section	Core	Standard	Enriched	Section	Core	Standard	Enriched
14.1 Characteristics of Flowering Plants pp. 271–274				**Blackline Masters** Review Worksheet 14.3 Skills Worksheet 14.3a Skills Worksheet 14.3b Integrating Worksheet 14.3 Enrich Worksheet 14.3	● ● ● ● ●	● ● ● ● ●	● ●
Section Features Skills WarmUp, p. 271 SkillBuilder, p. 272	● ●	● ●	●	**Ancillary Options** *Multicultural Perspectives,* p. 35		●	●
Blackline Masters Review Worksheet 14.1 Skills Worksheet 14.1	● ●	● ●		**Overhead Blackline Transparencies** Overhead Blackline Master 14.3 and Student Worksheet	●	●	●
Laboratory Program Investigation 27	●	●	●	**Laboratory Program** Investigation 29	●	●	●
Color Transparencies Transparency 35	●	●		**Color Transparencies** Transparencies 38, 39	●	●	
14.2 Vascular Plant Systems pp. 275–280				**14.4 Plant Growth** pp. 286–290			
Section Features Skills WarmUp, p. 275	●	●	●	**Section Features** Skills WarmUp, p. 286 Activity, p. 290	● ●	● ●	● ●
Blackline Masters Review Worksheet 14.2 Integrating Worksheet 14.2a Integrating Worksheet 14.2b	● ● ●	● ● ●	●	**Blackline Masters** Review Worksheet 14.4 Vocabulary Worksheet 14	● ●	● ●	
Laboratory Program Investigation 28		●	●	**Color Transparencies** Transparency 40	●	●	
Color Transparencies Transparencies 36, 37	●	●					
14.3 Reproduction of Flowering Plants pp. 281–285							
Section Features Skills WarmUp, p. 281 Skills WorkOut, p. 283 Career Corner, p. 284	● ● ●	● ● ●	● ● ●				

Bibliography

The following resources can be used for teaching the chapter. See page T-40 for supplier codes.

Audio-Visual Sources
(video unless noted)
Flowers at Work. 3d ed. 16 min. 1982. EB.
Plant Propagation: From Seed to Tissue Culture. 28 min film. 1977. MF.
Plants: Parts and Processes. 2 filmstrips with cassettes. 1988. NGSES.
Pollination. 23 min. 1983. NGSES.
The World of Plants. 43 min. 1990. UL.

Software Resources
How Plants Grow: The Inside Story. Apple, IBM. TB.
The Plant Growth Simulator. Apple. FM.

Library Resources
Burnie, David. *How Nature Works.* Pleasantville: The Reader's Digest Association, Inc., 1991.
Burnie, David. *Plant.* New York: Alfred A. Knopf, Inc., 1989.
Cochrane, Jennifer. *Food Plants.* Austin: Steck-Vaughn Co., 1991.
Imes, Rick. *The Practical Botanist.* New York: Simon & Schuster Inc., 1990.
Wilkins, Malcolm. *Plantwatching.* New York: Facts on File Publications, 1988.

CHAPTER 14

Introducing the Chapter

Have students read the description of the photograph on page 270. Ask if they agree or disagree with the description.

Directed Inquiry

Have students study the photograph. Ask:

▶ How are all of these objects alike? (They are all seeds.)

▶ Identify as many seeds as you can. (Answers may include corn, popcorn, acorns, pumpkin seeds, beans, and maple seeds.)

▶ Do all the seeds have the same texture? (No. Some are smooth, some are rough, and some are grooved.)

▶ How is the maple seed's wing-like shape a beneficial adaptation? (The wind can carry it long distances to new areas where it will not compete for light and nutrients with its parent plant.)

▶ How does the photograph relate to the chapter? (Flowering plants reproduce through seeds.)

Chapter Vocabulary

anther	sepals
cambium	stamen
dicot	stigma
epidermis	tropism
monocot	vegetative
phloem	propagation
pistil	xylem

270

INTEGRATED LEARNING

Writing Connection

Have students examine the photograph and write about how the shape or covering of each seed might help it eventually sprout and grow. For example, some shapes protect seeds, some help them sink into the earth, and others help seeds travel in air or water.

Chapter 14 Flowering Plants

Chapter Sections

14.1 Characteristics of Flowering Plants

14.2 Vascular Plant Systems

14.3 Reproduction of Flowering Plants

14.4 Plant Growth

What do you see?

"I see many seeds that are many different colors. The corn kernel came from a corn stalk and the helicopter seed came from a certain tree. Each, if planted, will grow up to be what it came from. It is kind of like a chain. First a seed must be planted and then a tree or a plant or whatever grows. Later the plant will make new seeds to be planted."

*Erica Severns
Whittier Middle School
Sioux Falls, South Dakota*

To find out more about the photograph, look on page 292. As you read this chapter, you will learn about flowering plants.

270

TEACHING OPTIONS

Portfolio

Seeds are packed with nutrition for the next generation of angiosperms. This is one reason why many seeds are popular foods. Have students prepare an informational brochure about a seed used in cooking. Encourage students to explain the nutritional value of the seed and describe its use in foods. Some seeds students could investigate are walnuts, coconuts, sunflower seeds, corn, pine nuts, cashews, and cocoa beans. Students may want to keep the brochures for their portfolios.

SECTION 14.1

Section Objectives
For a list of section objectives, see the Student Edition page.

Skills Objectives
Students should be able to:

Infer characteristics of flowering plants.

Collect Data about seed characteristics.

Organize Data about monocots and dicots.

Vocabulary
monocot, dicot

MOTIVATE

Skills WarmUp
To help students understand the characteristics of flowering plants, have them do the Skills WarmUp.
Answer Answers will vary. Students should be able to infer that flowering plants have vascular tissue and reproductive organs contained in the flowers.

Prior Knowledge
Gauge how much students know about flowering plants by asking the following questions:

▶ Name some examples of flowering plants.

▶ How do a flower's color and fragrance aid the reproduction process?

The Living Textbook:
Life Science Sides 1-4

Chapter 3   Frame 00675
Flowering Plants (72 Frames)
Search: Step:

14.1 Characteristics of Flowering Plants

Objectives

▶ **Explain** what all flowers have in common.

▶ **Discuss** adaptations of flowering plants.

▶ **Compare** gymnosperms and angiosperms.

▶ **Organize data** about monocots and dicots.

Skills WarmUp

Infer List five plants you know that have flowers. Based on the plants you know, what inferences can you make about flowering plants?

Imagine you are walking through a beautiful flower garden. What do you see? You can usually see a variety of shapes, sizes, and colors. You may wonder how there can be so many different kinds of flowers. For example, you might find a colorful, complex orchid, or a simple, white daisy. Perhaps you'll see a bee, or a butterfly landing on a flower. You might also notice hummingbirds hovering nearby. Although the flowers look very different, they all have one thing in common. Do you know what it is?

Origins and Adaptations

Fossils show that flowering plants, called angiosperms, first evolved about 130 million years ago. Scientists infer that angiosperms evolved from a gymnosperm ancestor—both have vascular tissue and both produce pollen. But unlike gymnosperms, angiosperms produce seeds within a flower.

Flowers contain the sex organs for an angiosperm. The male sex organs make pollen and the female sex organs make egg cells. Some flowers, such as tulips, contain both male and female sex organs within one flower. Other flowers, such as the squash blossoms shown in Figure 14.1, may contain either male or female sex organs.

Flowers have many adaptations for reproduction. Plants with colorful and complex flowers usually depend on insects, bats, or hummingbirds to carry pollen from one flower to another. Shape, scent, and color are flower adaptations that help plants

Figure 14.1 ▲
Squash plants have separate male and female flowers. The flower in the top photograph has male sex organs. The flower in the bottom photograph has female sex organs.

Chapter 14 Flowering Plants

TEACH

Critical Thinking

Compare and Contrast Ask students to describe the similarities and differences between angiosperms and their gymnosperm ancestors. (Both have vascular tissue and produce pollen; however, angiosperms produce flowers with seeds and gymnosperms don't.) Have students explain why pollination by animals is more effective than wind pollination. (Animals are likely to go to another flower of the same species.)

SkillBuilder

Provide students with mixtures, such as birdseed, or make your own mixture of various kinds of seeds.

Answers

Answers will vary. Sample answers follow:

1. Some seeds had wings, some had spines and barbs, some were nuts, some were feathery, some were fleshy, and some were tiny.
2. The nuts are the best seed grouping because you can eat them.
3. Wings; nuts
4. They are all seeds from flowering plants.
5. Seeds are similar in that they all give rise to new plants. They are different in the way they look and how they are dispersed.

Answer to In-Text Question

① Students may mention daisies or roses.

The Living Textbook: Life Science Sides 1-4

Chapter 5   Frame 01905
Flowering Plants (26 Frames)
Search: Step:

272

INTEGRATED LEARNING

Themes in Science

Systems and Interactions

Characteristics such as color, smell, nectar, and pollen attract pollinators to plants. Bees feed on pollen and help transport it as they fly from flower to flower. Butterflies similarly aid pollination, but they feed on nectar rather than pollen. Butterflies have a keen sense of smell. Many flowers that are pollinated by butterflies are scented. These scented flowers typically bloom in late summer when butterflies are abundant. Ask students to think of examples of scented flowers. (They may mention roses, lilacs, hyacinths, cherry blossoms, and so on.)

Figure 14.2 ▲
The world's largest flower, *Rafflesia*, grows in the forests of Malaysia. The rare flower is more than 1 m in diameter and smells like rotting meat.

attract specific pollen carriers. A specific pollen carrier is more likely to take the pollen to another plant of the same species, increasing the chances of successful reproduction. Plants that depend on wind to carry their pollen do not have this advantage.

Most angiosperms produce flowers grouped together in clusters. These flower clusters vary greatly in size, shape, and the number of flowers they contain. Other angiosperms, such as the daffodil, have only one single flower. Can you think of other angiosperms that have one flower? ①

More than 275,000 species of angiosperms are found in the world today—from the equator to the poles, and from the oceans to the mountains. They come in different forms: trees, shrubs, vines, herbs, and succulents, such as cacti. Angiosperms range in size from tiny floating plants only 10 mm long to enormous flowering trees over 100 m tall. The plant shown in Figure 14.2 has the world's largest known single flower.

SkillBuilder Collecting Data

Sowing Information About Seeds

There are probably as many seed variations as there are different types of flowers. Obtain a bag of assorted seeds. Examine the seeds very closely. Divide your seeds into groups, according to color, size, shape, or whatever characteristic you choose.

Copy the table shown on a separate sheet of paper. Record the characteristic you observed, the different variations you observed for this characteristic, and the number of groups you obtained.

After you finish, collect your seeds into one group. Divide your seeds again according to a different characteristic. Record your data in the table. Continue regrouping and sorting your seeds for several new characteristics. Study your table, then answer the following questions.

1. Describe the characteristics you observed.
2. Which seed grouping did you think was the best? Why?
3. Which characteristic had the least number of groups? The largest number of groups?
4. What seed characteristic would you use to place all seeds in the same group?
5. Explain how seeds are similar. Explain how they vary.

Observed Seed Characteristics		
Characteristics	Different Types	Number of Groups

272   Chapter 14   Flowering Plants

Integrating the Sciences

Earth Science Flowering plants, or angiosperms, are the most abundant of modern plants. There are an estimated 275,000 species of angiosperms, compared with 50,000 of all other plant species. In spite of their abundance today, angiosperms appeared relatively late in the fossil record—in the Cretaceous period after the appearance of gymnosperms.

Angiosperm Classification

Angiosperms are classified into two groups based on the number of seed cotyledons (KAHT uh LEED uhns). Cotyledons are leaflike parts of the plant embryo inside the seed.

Monocots Look at the characteristics of a **monocot** (MAHN oh KAHT) in Figure 14.3.

- *Mono-* means "one." A monocot seed has only one cotyledon.
- Monocot leaves are narrow with a parallel vein pattern.
- Inside a monocot stem, the bundles of vascular tissue are scattered.
- Monocots have fibrous roots.
- The parts of a monocot flower are found in multiples of three.

Some monocots are grasses and food plants, such as corn, and garden flowers, such as tulips. Palms are the only monocot trees.

Figure 14.3 ▲
This plant, *Lilium regale*, has the characteristics of a typical monocot.

Dicots Figure 14.4 shows the characteristics of a **dicot** (DY kaht).

- *Di-* means "two." A dicot seed has two cotyledons.
- Dicot leaves have a branching pattern of veins.
- The bundles of vascular tissue inside a dicot stem are arranged in a ring.
- Dicots have a large, thick taproot.
- The parts of a dicot flower are found in multiples of four or five.

Dicots are thought to have evolved before monocots. About five times more dicot species than monocot species exist today. Common dicots include all broadleafed trees, such as oak, maple, cottonwood, and fruit trees. Food plants such as tomatoes, potatoes, peas, squash, lettuce, and broccoli are dicots. Dicot flowers include sunflowers, roses, and violets.

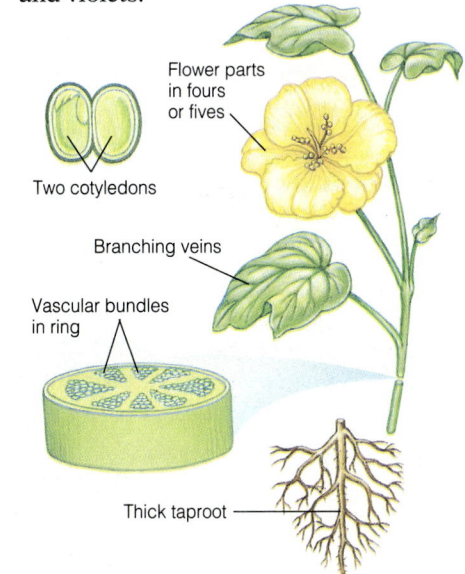

Figure 14.4 ▲
This plant, *Fremontia californica*, has the characteristics of a typical dicot.

Directed Inquiry

Have students study the material on this page. Ask the following questions:

▶ What are the two main classes of angiosperms? (Monocots and dicots)

▶ What are cotyledons? (Leaflike parts of the plant embryo inside the seed)

▶ How are monocot leaves different from dicot leaves? (Monocot leaves are narrow with parallel veins; dicot leaf veins form a branching pattern.)

▶ How do the flowers of monocots and dicots differ? (Monocots have flower parts in multiples of three; dicots have flower parts in multiples of four or five.)

Enrich

Have students prepare charts that contain information about local angiosperms. They should identify each angiosperm, state whether it is a monocot or dicot, and note whether it produces single flowers or clusters. Encourage students to use a field guide and add drawings of the angiosperms that they can include on their charts.

The Living Textbook:
Life Science Sides 7-8

Chapter 7   Frame 01146
Dicot Embryo (2 Frames)
Search: Step:

TEACH ▪ Continued

Enrich
Encourage students to research and report on other Native-American medicine, in addition to cinchona bark, that is used today.

EVALUATE

WrapUp
Reteach Have students make a concept map that includes the characteristics of angiosperms, monocots, and dicots.
Use Review Worksheet 14.1.

Check and Explain
1. Flowers have vascular tissue and produce pollen, seeds, or both.
2. Shape, scent, and color attract specific pollen carriers.
3. Both have vascular tissue and produce pollen and seeds. Gymnosperms produce seeds within cones; angiosperms produce seeds within flowers.
4. Students' tables should include most of the following: Monocots have one cotyledon, narrow leaves with parallel veins, fibrous roots, and stems with scattered vascular bundles; dicots have two cotyledons, leaves with branching veins, flower parts in multiples of four or five, a thick taproot, and stems with vascular bundles in a ring.

274

INTEGRATED LEARNING

Art Connection
Have students gather leaf and stem samples (and seed or flower samples, if possible) of an angiosperm. Have them draw the stem, vein, and vascular patterns that are characteristic of the plant and identify whether it is a monocot or a dicot. Have students label each part and explain its function. Display students' work in class.

Multicultural Perspectives
Ginseng is a root that has been cultivated in China for about 5,000 years. It has a stimulant effect and is believed to aid recovery from illness. Today, ginseng is grown commercially and sold as a powder all over the world. Students may be interested to know that they can buy a ginseng drink in many health food or grocery stores.

Figure 14.5 ▲
Tribal legends say that the medicinal properties of the cinchona tree were discovered during an epidemic. Everyone who drank water from a shallow lake with these trees growing at the water's edge got well.

Science and Society *Plant Medicines*
Did you know that aspirin and many other medicines originally came from flowering plants? For centuries, many cultures around the world used tea made from the bark of willow trees as a pain remedy. In 1827, the active ingredient of willow bark was identified. In 1899, a synthetic form of the active ingredient was made and given the name aspirin. Aspirin is still widely used to relieve pain and reduce fever.

For thousands of years, people all over the world suffered and died from a disease called malaria. Malaria is caused by a parasite and produces very high fevers. Long before the cause of malaria was understood, native peoples in South America knew an effective cure. They treated malaria and other fevers with bark from a tree they called the *quina* or *querango*. The flowers and leaves of this life-saving tree from the Andes Mountains are shown in Figure 14.5.

During the 1600s, Spanish expeditions gathered the bark, collected the flowers and seeds, and took the plant to Europe. There the bark became famous for curing many prominent people, including the Countess of Chinchón, a Spanish noblewoman. In honor of her cure, Linneaus named the plant cinchona (sihn KOH nuh).

Until the early 1800s, powdered cinchona bark was the only cure for malaria. Plantations of cinchona trees supplied the worldwide demand. About 1820, the active ingredient in the bark, called quinine (KWY nyn), was discovered. Soon after, drug manufacturers began to make synthetic quinine. Today doctors use synthetic quinine to treat malaria patients all over the world.

Check and Explain
1. What do all flowers have in common?
2. What adaptations do flowering plants have to ensure pollination?
3. **Compare and Contrast** How are angiosperms similar to gymnosperms? How are they different?
4. **Organize Data** Make a table comparing the characteristics of monocots and dicots.

274   Chapter 14 Flowering Plants

Themes in Science

Systems and Interactions The functioning of individual systems in a plant contributes to the functioning of the plant as a whole. For example, the root system brings water to the new shoots, aiding photosynthesis and other life processes.

14.2 Vascular Plant Systems

Objectives

▶ **Describe** a root system and **explain** how it works.

▶ **Explain** how each part of a shoot system works.

▶ **Compare** and **contrast** three types of roots.

▶ **Make a model** of water transport in plants.

Skills WarmUp

Hypothesize Write a hypothesis for what would happen to a plant if you removed its roots. Design an experiment to test your hypothesis.

Think about how your body systems work. For example, your circulatory system transports food, oxygen, water, minerals, and other materials throughout your body. Your circulatory system is made up of many different cells, tissues, and organs. All parts of the system work together. Your circulatory system also interacts and works with other systems of your body.

Vascular plants also have systems. The shoot system is the aboveground part of the plant. Below the ground is the root system. Each system is made of cells, tissues, and organs working together.

Water, Sugar, and Mineral Transport

Vascular tissue forms a plant's pipeline, connecting the root system and the shoot system. Follow the pipeline through the plant in Figure 14.6. One part of the vascular tissue is a set of linked cells called the **xylem** (ZY luhm). The xylem forms a pipeline for water and minerals. From the root system, water and minerals travel upward through the xylem to the shoot system. Some water is used in photosynthesis, but most of it evaporates through stomates in the leaves. The loss of water through the stomates of leaves is called **transpiration**.

A leaf is a shoot system organ. Leaves capture light and make glucose by photosynthesis. Glucose made in the leaves must travel to other parts of the plant. A second set of linked cells in the vascular tissue, called the **phloem** (FLOH ehm), transports glucose and other sugars. In the root system, the sugars provide energy for root growth. The shoot system receives sugars for growth and flower production. The plant storage organs store glucose in the form of starch for later use.

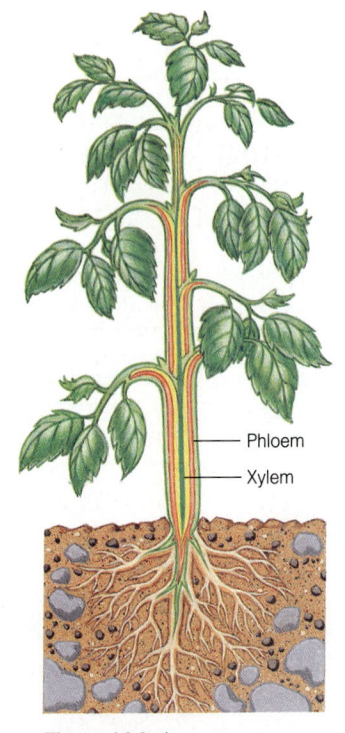

Figure 14.6 ▲
Inside every vascular plant there are xylem and phloem tissues. Water and minerals move through the xylem. Sugars, dissolved in water, move through the phloem.

SECTION
14.2

Section Objectives
For a list of section objectives, see the Student Edition page.

Skills Objectives
Students should be able to:

Hypothesize what would happen if a plant's roots were removed.

Make a Model of water transport in plants.

Vocabulary
xylem, phloem, epidermis, cambium

MOTIVATE

Skills WarmUp
To help students understand the function of root systems, have them do the Skills WarmUp.
Answer Students should realize that the plant would soon die. They could test their hypotheses by cutting off a plant's roots, replanting it, and observing the results.

Prior Knowledge
Gauge how much students know about systems in vascular plants by asking the following questions:

▶ What plant system absorbs water and nutrients from the soil?

▶ What plant systems are leaves, stems, and flowers part of?

▶ What are the main functions of a leaf?

Chapter 14 Flowering Plants

TEACH

Directed Inquiry

Have students study the material on this page. Ask the following questions:

▶ What are the two main types of vascular tissues in plants and how does each one function? (The xylem carries water and minerals up from the roots; the phloem carries sugars throughout the plant.)

▶ What are the functions of roots? (Roots support and anchor, absorb water and minerals from the soil, and store starch.)

▶ What is the function of the root cap? (It protects the end of the root and makes a slime that coats the root so it can pass easily through soil.)

▶ What do root hairs do? (They increase the root area exposed to the soil so that needed water and minerals can be absorbed.)

▶ How do water and minerals enter xylem cells inside the root? (Through osmosis)

**The Living Textbook:
Life Science Side 3**

Chapter 18   Frame 27174
Roots (Movie)
Search: Play:

276

INTEGRATED LEARNING

Themes in Science

Scale and Structure Both humans and plants have a protective outer layer, or epidermis. However, a plant epidermis performs some different functions from a human epidermis. Encourage students to compare and contrast the functions of a human and plant epidermis. Have them refer to Chapter 19 for information about human skin.

Root System

The root system of a plant is usually about the same size as its shoot system. For example, a maize plant 2 m tall often has roots that are 2 m long and extend 1 m on all sides.

Root Functions The root system has several functions.

▶ Roots support and anchor the plant.
▶ Roots absorb water and minerals from the soil.
▶ Roots store glucose in the form of starch.

Root Structure Study the root in Figure 14.7. At the tip of the growing root is a dome-shaped root cap. Tiny structures called root hairs project near the tip. Root hairs increase the area of root exposed to the soil. The greater the exposed surface area, the easier it is for the root to absorb needed water and minerals. The root hairs extend from the **epidermis** (EHP uh DUR mihs). The epidermis is an outer layer of protective cells. Beneath the epidermis is a starch storage layer. In the center are the xylem and phloem. Between the xylem and phloem is the **cambium** (KAM bee uhm). The cambium is growth tissue that makes new xylem and phloem cells.

Roots and Soil Formation In rocky environments, plant roots make soil by breaking apart solid rock. The root cap secretes slime-containing acids that dissolve rock. The force of the root growth wedges the rock apart. As fallen leaves and other plant parts decay and mix with the rock fragments, soil forms.

Figure 14.7 Root System

The root cap protects the end of the growing root. Root cap cells make a slime that coats the root, enabling it to pass more easily through the soil.

Water and minerals absorbed by root hairs enter xylem cells inside the root by osmosis.

The epidermis is an outer protective layer. It covers the entire plant like a skin.

Cambium cells divide, producing either new xylem cells or new phloem cells.

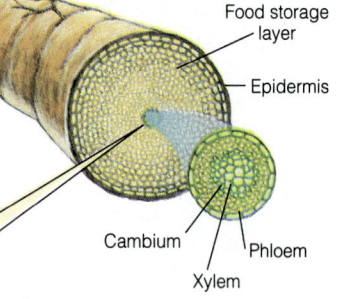

Food storage layer — Epidermis — Cambium — Phloem — Xylem

276

Integrating the Sciences

Earth Science Most trees can't grow in waterlogged soil because the soil lacks oxygen and stability. However, the tropical mangrove tree has two special adaptations that enable it to thrive on coastal mudflats. Stiltroots (or prop roots) extend in an arch from the mangrove's trunk and anchor it in the mud. Breathing roots called *pneumatophores* grow up through the mud and collect oxygen at low tide.

Math Connection

Most tree roots grow outward as well as downward. The roots of a tree 50 meters tall typically grow no more than about 2.5 meters down into the ground. However, a tree this size may have roots that spread to the sides at a distance equal to the tree's height. Provide grid paper. Have students draw such a tree and its root system to scale.

Directed Inquiry

Have students study the material on this page. Ask the following questions:

▶ How do fibrous roots help plants? (They absorb water efficiently from near the ground surface and prevent erosion by holding the soil in place.)

▶ How do taproots help plants? (They anchor plants in the soil and some store starch.)

▶ How do aboveground roots help plants? (They cling to trees and absorb water and nutrients. Some start on plant stems and grow into the soil providing support.)

Skills Development

Infer Ask students to infer why it is beneficial for many plants to have root systems with many small and hairlike roots. (This increases the amount of surface area that can absorb water.)

Portfolio

For their portfolios, have students draw and label the three types of root systems.

Root Types

The roots of a plant are adapted to the plant's environment. For example, some desert plants have roots more than 18 m long to reach water deep underground.

Although plant roots vary in size and structure, there are three general types: fibrous roots, taproots, and aboveground roots. Monocots produce a fibrous root system. Gymnosperms and dicots usually produce a taproot system. Aboveground roots are specialized roots that grow from stems or leaves. Some plants, such as corn, normally produce aboveground roots. Other plants, such as orchids, develop aboveground roots in response to their environment. Some plants develop aboveground roots in response to injuries.

Fibrous Roots

The plant shown here has fibrous roots. Fibrous roots branch out in all directions. Fibrous roots absorb water efficiently from near the ground surface. They also hold soil together, preventing erosion. ▼

Taproots ▲

This plant has a taproot. A taproot is a thick main root that grows straight into the ground. Smaller roots grow out from the sides. A taproot securely anchors a plant in the soil and may store starch. Vegetables such as carrots and beets are taproots. They contain starch stored by the plants.

Aboveground Roots

Many rain forest plants, such as the epiphyte shown below, grow high in trees. They have aboveground roots that cling to the tree and absorb water and nutrients. Other aboveground roots start on the plant stem and grow down into the soil. They support large trees. ▼

The Living Textbook:
Life Science Sides 1-4

Chapter 7   Frame 02574
Dicot Root (2 Frames)
Search: Step:

TEACH ▪ Continued

Directed Inquiry

Have students study the material on this page. Ask the following questions:

▶ What are the functions of stems? (Stems provide support, hold the leaves toward light, move water and minerals in the xylem and move sugars in the phloem, and some store water or starch and make glucose.)

▶ What kinds of plants have herbaceous stems? (Smaller flowering plants, such as wildflowers, grasses, and most vegetables)

▶ What kinds of plants have woody stems? (Dicot trees and shrubs that live many years)

▶ What adaptation do the stems of cacti have? (Green stems that store water)

▶ What adaptation does the stem of the bean plant have? (It can climb up a tree or post.)

Skills Development

Infer Ask students how the flexible stem of a herbaceous plant is a beneficial adaptation. (Its flexibility keeps it from breaking in the wind, rain, or snow.) Why are xylem cells larger in wet seasons? (They carry more water at those times.)

Answer to In-Text Question

① The smaller cells would appear as a dark band.

INTEGRATED LEARNING

Themes in Science

Scale and Structure The stem, or trunk, of a tree is protected by a tough covering of bark. Bark is made up of two layers. The inner layer is called the *bark cambium* and is made of living cells that are continuously reproducing. When these cells eventually become cut off from the tree's supply of water and sap, they die and form the outer bark.

Integrating the Sciences

Earth Science The rings of a woody stem can tell scientists about weather conditions in a given year. Have students find out how scientists get such information from plants.

Use Integrating Worksheet 14.2a.

Shoot System

The shoot system includes stems, leaves, and flowers. Some stems are flexible and slender. These are called herbaceous (hur BAY shuhs) stems. Other stems, such as those of trees, are large and woody.

Stem Functions Stems have several functions.

▶ Stems support the plant.
▶ Stems hold the leaves up toward the light.
▶ Inside a stem, water and minerals move through the xylem and sugars move through the phloem.
▶ Stems of some plants also store water or starch. Green stems make glucose through photosynthesis.

Herbaceous Stems ▲

A herbaceous stem is green, soft, and flexible. Smaller flowering plants, such as wildflowers, grasses, and most vegetables, have herbaceous stems. The top photograph shows the outside of a herbaceous dicot stem. The bottom photograph shows the inside of a dicot stem magnified 100x. Plants with herbaceous stems usually live only one or two years.

Woody Stems

A woody stem is hard, strong, and rigid. Most plants with woody stems are dicot trees and shrubs that live for many years. Look at the growth rings on this cut tree. Each year's growth has dark and light bands of xylem cells. During wet weather, large xylem cells are produced. In the dry summer, the xylem cells are small and close together. Which cells would appear as a dark band? ①

Stem Adaptations ▲

Plant stems have many unusual adaptations. Cacti, shown in the photograph, have green succulent stems that store water. The stored water is used to make glucose for the plant. Some plant stems can climb. For example, the stem of the bean plant grows in a spiral to climb up a tree or post.

Chapter 14 Flowering Plants

> **Themes in Science**
>
> **Scale and Structure** The structure of leaves varies tremendously, depending on a plant's environment. Show students a picture of the water buttercup, which starts its life underwater and later flowers above the surface. Its submerged leaves are fine and feathery and move easily in flowing water without being torn. Its upper leaves are flat and wide.
>
> **Art Connection**
>
> Have students collect several different types of leaves from flowering plants and use them to make a book of leaf prints. Have them identify the plant from which the leaf was taken and label the parts of each leaf. They should also describe how the shape of the plant's leaf helps it survive in its environment.

Leaf Functions Each leaf is a specialized organ with three main functions.

▶ Leaves capture light from the sun.
▶ Leaves make glucose through photosynthesis.
▶ Leaves exchange gases with the environment.

Leaf Structure Look carefully at Figure 14.8. Notice that a leaf is made of layers. The surface layers are the waxy cuticle and the epidermis. Inside is the moist mesophyll (MEHZ oh FIHL) layer, where photosynthesis occurs. Gases are exchanged with the environment through stomates in the plant's surface.

Leaf Adaptations The shapes and sizes of leaves vary, depending on their environment. Plants in shady forests have large leaves to capture light. Plants in dry environments have small, thick, wax-coated leaves to prevent drying.

**Figure 14.8
Leaf Structure** ▼

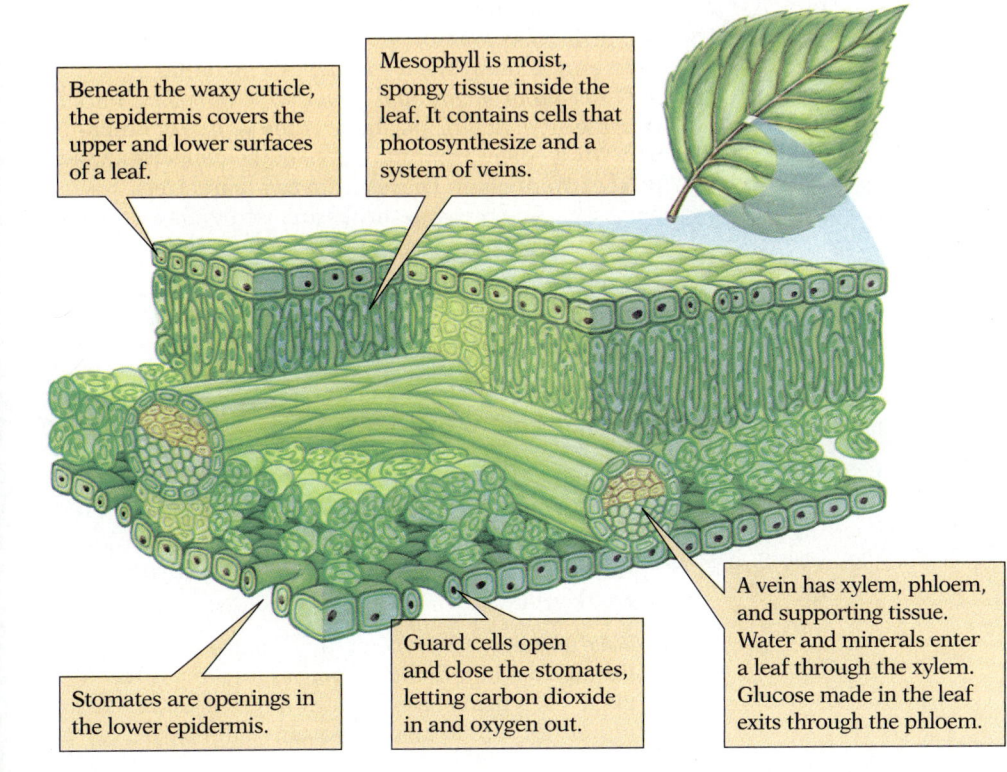

Beneath the waxy cuticle, the epidermis covers the upper and lower surfaces of a leaf.

Mesophyll is moist, spongy tissue inside the leaf. It contains cells that photosynthesize and a system of veins.

Stomates are openings in the lower epidermis.

Guard cells open and close the stomates, letting carbon dioxide in and oxygen out.

A vein has xylem, phloem, and supporting tissue. Water and minerals enter a leaf through the xylem. Glucose made in the leaf exits through the phloem.

Chapter 14 Flowering Plants

Directed Inquiry
Have students study the material on this page. Ask the following questions:

▶ What are the main functions of leaves? (Leaves capture light from the sun, make glucose, and exchange gases.)

▶ What makes up the surface layers of the leaf? (The waxy cuticle and the epidermis)

▶ What is in the mesophyll? (A system of veins and a moist, spongy tissue that is made up of cells that photosynthesize)

▶ What makes up a vein, and what does a vein do? (A vein is made up of xylem, phloem, and supporting tissue. Veins transport water and minerals into the leaf through the xylem and transport glucose out of the leaf through the phloem.)

Skills Development
Predict Ask students to predict what kind of leaves would grow on a plant that lives in a windy area, such as a mountaintop or ocean coast. (Small, tough leaves) What would happen to large, fleshy leaves in such an environment? (They would be easily torn by the wind.)

**The Living Textbook:
Life Science Sides 1-4**

Chapter 7   Frame 02570
Interior of a Leaf (1 Frame)
Search:

**The Living Textbook:
Life Science Sides 1-4**

Chapter 7   Frame 02576
Carbon Dioxide and Water Movement in Leaves (1 Frame)
Search:

TEACH ▪ Continued

Enrich
Tell students that spices from roots include turmeric, horseradish, and sassafras. Flavorings come from the leaves of citronella, bay, spearmint, parsley, and thyme.

EVALUATE

WrapUp
Review Have students draw and label the root system and shoot system of a plant and describe how its shape and structure help it survive in its environment.
Use Review Worksheet 14.2.

Check and Explain
1. Roots anchor plants, absorb water and minerals by osmosis and active transport, and store glucose starch.
2. Stems give support, hold leaves toward light, transport water and nutrients, and sometimes store water or starch or make glucose. Leaves make glucose and exchange gases. Flowers contain reproductive organs.
3. Fibrous roots branch out in all directions; taproots grow straight into the ground; above-ground roots cling to a tree or grow down into soil. All absorb water and minerals and provide support.
4. Drawings should show water moving up from root hairs, to roots, and through the shoot system.

Answer to In-Text Question
① Some answers might be spinach leaves, asparagus stems, and beet roots.

INTEGRATED LEARNING

Multicultural Perspectives
Direct students to research a salad or vegetable dish from a particular region. Ask them to list the ingredients of the salad and describe the type of plant, plant part, nutritional value, and taste of each ingredient. Encourage students to bring to class a recipe for the food or prepare the food for the class.
Use Integrating Worksheet 14.2b.

Integrating the Sciences
Health Explain that the roots of many plants are loaded with vitamins and minerals that fight disease. Root and tuber vegetables such as carrots, sweet potatoes, rutabagas, yams, and potatoes are rich in beta-carotene, vitamin C, and potassium, which help the body fight cancer and heart disease. Ask students which of these vegetables they eat.

Figure 14.9 ▲
The brightly colored edible leaves are from a plant called red chard. You probably recognize the edible turnip and red potato. What are some leaves, stems, and roots that you eat? ①

Science and You *Plant Edibles*
Think about the vegetables you ate during the past week. Chances are you tasted the roots, stems, or leaves of many different flowering plants. Maybe you even ate them raw.

In a salad, you might eat leaves from a lettuce plant, sliced radish roots, and diced celery stems. If you eat a potato, you swallow an underground part of the potato plant stem. Onion and garlic, used to flavor many foods, are the stems and the leaves from two different lily plants. You can even have roots for dessert! Carrot cake is made from the sweet, orange roots of the carrot plant. Tapioca comes from the root of the cassava plant, and spicy gingerbread is flavored with ginger root.

The plant part eaten as a vegetable is usually the plant's storage organ. Lettuce and cabbage plants save sugars in their thick leaves. Celery, rhubarb, and asparagus plants store sugars and starches in their stems. Carrot, beet, and jicama plants use their roots for storage. The stored sugars and starches give many fresh vegetables a sweet taste.

Other roots, stems, and leaves serve as spices. Bark from the stem of a tropical tree is the source of cinnamon. Mint comes from the leaves of the peppermint or the spearmint plant. Leaves from bay laurel trees, rosemary, sage, and oregano shrubs are used to flavor soups. Roots, such as ginger, produce "hot," aromatic spices.

Check and Explain
1. Describe the main functions of a plant's root system. Explain how each function is carried out.
2. What are the parts of a plant's shoot system? Explain how each part functions.
3. **Compare and Contrast** Describe three types of roots. How are the roots different from one another? How are they similar?
4. **Make a Model** Trace the path of water as it moves through a plant. Using references, first draw the plant, then use arrows to show how the water moves. Be sure to label important plant parts.

280 Chapter 14 Flowering Plants

Themes in Science
Systems and Interactions

Although some plants are able to pollinate themselves, most rely on receiving pollen from another plant of the same species. Some pollen reaches plants by the wind or by water; however, the most important pollinators are animals, particularly insects. Many plants and insects have a mutually beneficial relationship. For example, some species of yucca rely exclusively on the yucca moth for pollination. In return, the yucca moth lives in and feeds on the yucca. Have students think of other mutually beneficial relationships between animals and flowers.

SECTION 14.3

Section Objectives
For a list of section objectives, see the Student Edition page.

Skills Objectives
Students should be able to:

Observe and describe flower parts.

Classify seed-producing plants.

Infer ideal conditions for seed growth.

Vocabulary
sepals, stamen, anther, pistil, stigma, vegetative propagation

14.3 Reproduction of Flowering Plants

Objectives
- **Identify** the parts of a flower and **describe** their functions.
- **Explain** what a fruit is and how it develops.
- **Discuss** vegetative propagation.
- **Infer** how to make seeds germinate.

Skills WarmUp

Observe Carefully examine a flower. Count the number of different flower parts you see. Describe each of the different flower parts, by either drawing or writing about a sample of each part.

Have you ever watched a honeybee or a bumblebee after it has landed on a flower, such as a daisy? If you have, you probably know that the bee was searching for nectar to eat. But did you know that the daisy also benefits from the bee's visit?

If you could closely watch a bee on a daisy, you might notice a yellow dust on the bee's body. The yellow dust is an important part of the daisy's life cycle. The yellow dust is pollen from the daisy's male sex organs. The pollen is carried by wind or animals to the female part of another daisy. The pollen fertilizes the egg cells, producing a young daisy embryo that is packaged into a seed. From the seed, a new daisy plant grows.

Flower Structure

Look carefully at the flower parts in Figure 14.10. Flowers have four sets of specialized structures growing from a short stem. At the flower's base are the **sepals** (SEE puhls). Sepals are often green. They enclose and protect the flower as it develops. Just above the sepals are petals. Petals are often brightly colored.

The **stamen** (STAY muhn) is the male reproductive part of a flower. It has two parts, a long filament and a saclike **anther** (AN thur). The anther produces the pollen grains. The **pistil** is the female reproductive part of the flower. Forming the base of the pistil is a swollen ovary. From the ovary a slender stalk, called the style, arises and supports the **stigma**. The stigma is the part of the pistil that collects pollen. Its sticky or feathery quality enables it to capture pollen.

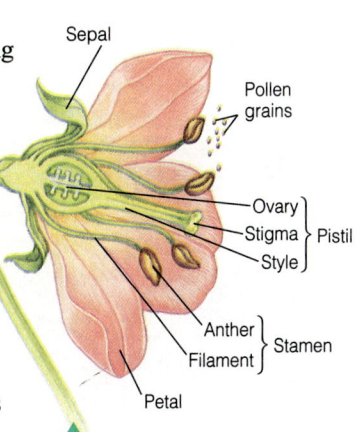

Figure 14.10 Flower Structure

MOTIVATE

Skills WarmUp
To help students understand the parts of a flower, have them do the Skills WarmUp.
Answer Students should be able to describe or draw different flower parts.

Misconceptions
Students may not realize that some plants have the ability to self-pollinate while others rely on cross-pollination. In order for plants to become pollinated, sperm must be delivered to the stigma, which contains the egg.

The Living Textbook: Life Science Side 3

Chapter 16 Frame 18191
Flowering Plant Life Cycles (Movies)
Search: Play:

The Living Textbook: Life Science Sides 7-8

Chapter 7 Frame 01058
Flower: Reproductive Organs (2 Frames)
Search: Step:

TEACH

Directed Inquiry

Have students study the material on this page. Ask the following questions:

▶ Where are the male and female gametophytes produced? (Inside the flower)

▶ What must occur before a flower can produce fruits and seeds? (Pollination)

▶ What do fertilized eggs grow into? (Plant embryos, and then eventually encased seeds)

▶ What happens to the ovary wall as seeds form? (It enlarges to become a fruit.)

▶ What is the function of fruit? (To protect and disperse seeds)

Reteach

Stress to students that not all fruit is moist and fleshy. Have students look through magazines and books to find examples of different kinds of fruit including dry pod fruit and winged fruit. Have them share the examples they find with the class.

Ancillary Options

If you are using the blackline masters from *Multiculturalism in Mathematics, Science, and Technology*, have students read about George Washington Carver on page 35. Complete pages 36 to 40.

Enrich

Use Enrich Worksheet 14.3.

The Living Textbook: Life Science Side 3

Chapter 17   Frame 25296
Fruit Ripening (Movie)
Search: Play:

INTEGRATED LEARNING

Multicultural Perspectives

The cacao bean is one of many foods native to the Americas that became popular throughout the world. In fact, cacao trees had been cultivated in Central America for centuries before Spanish explorers arrived there in the 1400s. The explorers were impressed with the drink that Native Americans made from cacao seeds. They took the beans and the recipe back to Spain. By the 17th century, small restaurants called chocolate houses were popular throughout Europe. (Explain to students that, over time, the word *cacao* became *cocoa*. The words are used interchangeably.)

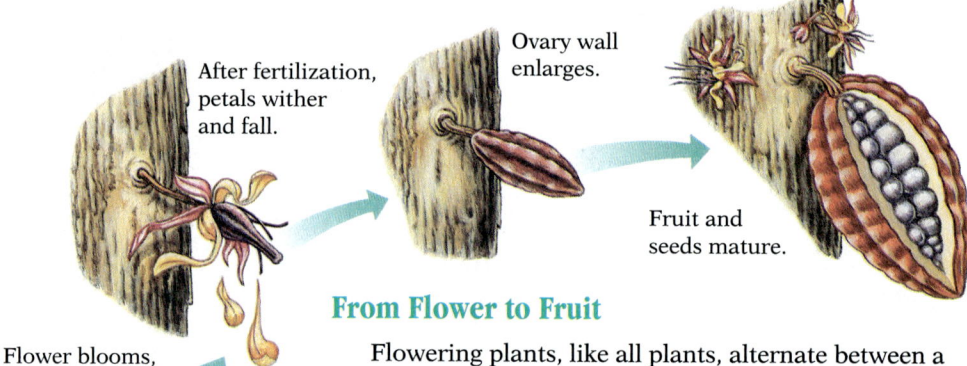

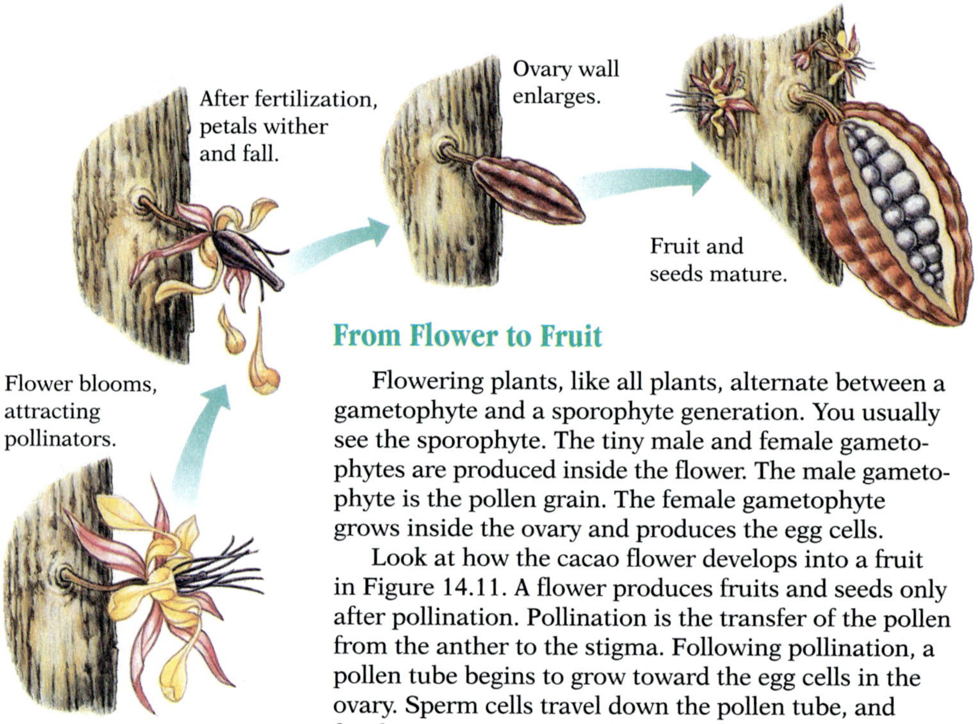

Figure 14.11 ▲
The cacao flower develops into a podlike fruit with large seeds called cocoa beans. Aztec and Maya people cultivated this plant in ancient Mexico.

Flower blooms, attracting pollinators.
After fertilization, petals wither and fall.
Ovary wall enlarges.
Fruit and seeds mature.

From Flower to Fruit

Flowering plants, like all plants, alternate between a gametophyte and a sporophyte generation. You usually see the sporophyte. The tiny male and female gametophytes are produced inside the flower. The male gametophyte is the pollen grain. The female gametophyte grows inside the ovary and produces the egg cells.

Look at how the cacao flower develops into a fruit in Figure 14.11. A flower produces fruits and seeds only after pollination. Pollination is the transfer of the pollen from the anther to the stigma. Following pollination, a pollen tube begins to grow toward the egg cells in the ovary. Sperm cells travel down the pollen tube, and fertilization occurs. Fertilized eggs grow into plant embryos. Structures develop around each plant embryo, forming seeds.

As the seeds form inside the ovary, the ovary wall enlarges to become a fruit. All fruits develop from flowers and contain seeds. Fruits also protect and help disperse seeds. Like cherries and apples, squashes and tomatoes are fruits. Not all fruits are juicy and edible. Some fruits are winged seed covers, or dry pods. The dry, podlike fruit of the cacao tree contains cocoa beans. Cocoa beans are the source of chocolate.

Figure 14.12
Kiwi, lemons, and papaya are fleshy fruits. The pea pod is a dry fruit. ▼

Writing Connection

The seeds of flowering plants do not develop unless conditions are favorable for growth. Weeks, months, and even years may pass before conditions are right for germination. Have students write a story from the point of view of a seed that is "waiting" to germinate. Have them describe the conditions that they need before they can begin to grow. Encourage them to be creative and describe what it is like in the seed's environment. Students could also add details about where the seed came from and about the germination process.

New Plants from Seeds

Just as flowers and fruits differ in shape, color, and size, so do seeds. Compare the seeds in Figure 14.12. The seeds look different, but they all share common features. All seeds have a seed coat, a plant embryo, and stored energy for the young plant. The stored energy is in the form of sugars, starches, and oils.

When a seed forms, the plant embryo usually stops growing, becoming dormant. A dormant seed can withstand freezing temperatures, or drought. Some seeds remain dormant for a few days, some for a few months, and others for many years.

If a seed lands in a suitable environment, it sprouts, or germinates. Seeds need water, oxygen, and a suitable temperature to germinate. Look at the germination process shown in Figure 14.13. Germination begins when water enters the seed coat and the seed swells. The embryo grows, using the sugars, starches, and oils stored in the cotyledons. The root tip emerges first and anchors the seed to the ground. Root hairs begin to absorb water. Then the shoot, with its tiny leaves, sprouts from the soil. The tiny leaves turn green and begin to manufacture sugars. The stored energy in the cotyledons is used, and they may shrivel and fall off.

Skills WorkOut

Classify Make a list of plants that produce seeds. Classify the seed-producing plants according to whether each plant seed is also a human food product.

Figure 14.13 Seed Germination ▼

▲ Shoots grow upward and break the ground surface.

▲ Roots grow downward. Root hairs absorb water and minerals.

▲ True leaves begin to grow. Cotyledons contain stored energy for plant growth.

▲ True leaves open. Plant begins photosynthesis.

Skills WorkOut

To help students classify plants that produce seeds, have them do the Skills WorkOut.
Answer Answers will vary.

Skills Development

Predict Ask students to suppose they saw a plant like the one pictured in four stages on this page. Ask them to predict what will happen if it receives enough water and sunshine. (The plant will continue to grow and produce more leaves. In time, it will produce flowers. If the flowers become pollinated, the eggs will be fertilized and seeds and fruits will form to begin the cycle all over again.)

Reteach

Have students fold a paper towel into thirds and moisten it with water. Have them place the towel in a shallow pan and place several bean seeds onto the moist towel. Have students cover the seeds with another moistened towel. Have them place the seeds in a warm spot for several days, making sure the towels stay moist. Have students use hand lenses to observe how the seeds germinate. Have students plant their seeds in potting soil so they can observe the development of the plants.

Integrated Learning

Use Integrating Worksheet 14.3.

Chapter 14 Flowering Plants

TEACH • Continued

Class Activity
Provide potting soil, pots, cups, and houseplants, such as geraniums or begonias, and have students experiment with propagating plants through vegetative propagation.

Enrich
Grains, nuts, legumes, berries, and pomes are among the many different kinds of fruits classified as simple, aggregate, or multiple fruits. Have pairs of students do some research to find out how different fruits are classified. Have them present their findings on an illustrated chart.

Career Corner
If possible, have a horticulturist from a local conservatory or plant nursery visit the class to talk with students. Have students prepare questions for the visitor.

TEACHING OPTIONS

Portfolio
Have students use references to find out what plants can be reproduced by vegetative propagation. Have them find out about stem cuttings, leaf cuttings, root cuttings, and grafting. Also have them find out about the vegetative reproduction of plants by means of rhizomes, tubers, bulbs, and corms. They should prepare a report about one of these topics for their portfolios.

Cooperative Learning
Have students work in teams of four and have each team bring to class a selection of four flowers. Students in each team should present each flower to the class, describing its environment and life cycle. Students should also identify the flower's parts, particularly its reproductive organs.

Reproduction Without Seeds

Skilled gardeners do not need seeds to grow many similar plants in a short time. Instead they use the method shown in Figure 14.14. They cut pieces of stems or leaves from plants and place them in a moist environment. Within a few weeks, roots develop, and the cuttings grow into mature plants. This is called **vegetative propagation** (prahp uh GAY shuhn). New plants produced this way are genetically identical to the parent plant. Vegetative propagation is possible with a root, a stem, or a leaf of the parent plant.

Some plants reproduce this way naturally. For example, strawberry plants produce runners. At the end of a runner, a new plant develops. Some grasses do this also. A small patch of grass spreads in this way until it covers hundreds of square meters. The spider plant, a common houseplant with straplike leaves, produces many small spiderlike plants with their own roots.

Figure 14.14 ▲
Under controlled conditions, vegetative propagation can be used to grow many hundreds of similar plants.

Career Corner — Horticulturist

Who Raises Plants for Landscaping?

The beautiful flowers in a floral arrangement or a botanical garden are cultivated by a horticulturist (HOR tih KUHL chur UHST).

Horticulture comes from two Latin words: *hortus*, which means "garden," and *colere*, which means "to cultivate." Horticulture combines both the art and science of gardening.

The work performed by horticulturists is very diverse. It includes fruit growing, vegetable production, flower use and growing, and landscape gardening. Horticulturists may work in orchards, commercial gardens, nurseries, greenhouses, or research facilities. Horticultural scientists experiment to determine the conditions necessary for good plant growth. They also develop ways to control plant diseases and produce new varieties of plants.

Horticulturists have a strong interest in flowers and plants. Most people in this field earn a four-year degree in biology, botany, or agricultural studies. Some two-year colleges offer a degree in architectural landscaping, which is a branch of horticulture. You can also choose to be a horticultural technician and learn about plants through on-the-job training.

If you like gardening, you can prepare to work in horticulture by taking classes in botany and by joining groups interested in gardening, such as the 4H Clubs of America.

INTEGRATED LEARNING

Writing Connection
Encourage students to imagine the wild and unpredictable trip of a grain of pollen. Then ask them to write about the adventures of a single grain of pollen. Students may write from the point of view of a naturalist, a nonscientific observer, or the grain of pollen itself.

Science and You *Pollen Allergy*

If your eyes water and you sneeze during certain times of the year, you may be allergic to pollen. Pollen allergy is sometimes called hay fever. People named it hay fever when they noticed that the pollen allergy symptoms occurred when hay grasses were common.

People suffer most from pollen allergy during the spring, summer, and fall seasons. In the early spring, flowering trees, such as oak, elm, maple, and walnut, release windborne pollen. In the summer, the small flowers of grasses shed their pollen. In the fall, ragweeds release large clouds of pollen.

Pollen allergy is a reaction of your body's immune system to pollen grains, like the ragweed pollen pictured in Figure 14.15. The tiny, dustlike pollen grains enter your nose and nasal passages as you breathe. Proteins in the pollen irritate cells in the membranes lining your nasal passages. The irritated membranes usually swell and secrete fluids.

If you have pollen allergies, you know that they are accompanied by a number of annoying symptoms. A runny nose and irritated throat are the most common symptoms of pollen allergy. Many people also get itchy, watery eyes. Sneezing and coughing, other symptoms of pollen allergy, are the responses of the body to irritations in the air passages. When particles are in the air passages, a sneeze or a cough often clears them away. But when allergy is the cause of the irritation, repeated coughing and sneezing do not help.

Figure 14.15 ▲
Ragweed pollen (bottom) has been magnified 325×. It comes from the flowers of the ragweed plant (top).

Check and Explain

1. Name the parts of a flower, and describe the function of each part.
2. What is a fruit? Describe the flower parts from which a fruit develops.
3. **If . . . Then Arguments** You notice a tiny sprout in the dirt next to your favorite plant. If this is vegetative propagation, what will happen? What will happen if it isn't?
4. **Infer** Your friend has some unusual seeds. What advice can you give on getting the seeds to grow?

Discuss
Poll the class to find out how many students are bothered by hay fever. Have students describe their reactions to pollen. Ask them during what time of year they notice these symptoms.

EVALUATE

WrapUp
Reinforce Write all the section vocabulary words on the chalkboard. Ask students to explain the role of each term in the reproduction process of flowering plants. Then ask them to name other words in the section that they think are important.

Use Review Worksheet 14.3.

Check and Explain

1. Sepals—protect developing flowers; petals— protect flowers; stamen—male reproductive structure made of filament and pollen-producing anther; pistil—female reproductive structure containing egg-producing ovary and style (slender stalk), supporting pollen-collecting stigma.
2. The ovary wall becomes enlarged and develops around seeds as they mature. The enlarged ovary wall is a fruit.
3. If yes, an identical plant will grow; if no, a completely different plant will grow.
4. Provide moist soil, a moderate temperature, and a moderate amount of sunlight.

**The Living Textbook:
Life Science Side 3**

Chapter 16   Frame 24915
Pollen (Movie)
Search: Play:

Chapter 14 Flowering Plants

SECTION 14.4

Section Objectives
For a list of section objectives, see the Student Edition page.

Skills Objectives
Students should be able to:

Predict the season of greatest angiosperm growth.

Hypothesize about how to prove that a plant is an annual, biennial, or perennial.

Classify a flower.

Vocabulary
tropism

MOTIVATE

Skills WarmUp
To help students understand seasonal plant growth, have them do the Skills WarmUp.
Answer Students should be able to predict that angiosperms will grow fastest in the summer because they get more daylight then.

Misconceptions
Many students will assume that plants grow as humans do—expanding in all directions and eventually stopping at adulthood. Emphasize that plants continue to grow larger as long as they live.

TEACHING OPTIONS

Portfolio
All plants display both primary and secondary growth. Yet some trees tend to grow tall, while others tend to grow wide and thick. Ask students to research and describe the growth patterns of several plants grown in their community. Encourage students to illustrate their findings in a comparative chart of local plant life. Students can keep their charts in their portfolios.

Skills WarmUp

Predict During which season in your area do you think an angiosperm will grow fastest? Explain your prediction.

Figure 14.16 ▲
In primary growth, the tips of roots and shoots grow longer. In secondary growth, the stem, trunk, or branches widen.

14.4 Plant Growth

Objectives

▶ **Explain** how plant growth and animal growth differ.

▶ **Describe** two plant hormones and their effects on plant growth.

▶ **Explain** why some plants bloom in summer and other plants bloom in autumn.

▶ **Hypothesize** about how to prove if a plant is an annual, a biennial, or a perennial.

Think about how you grow. As your height increases, your arms and legs get longer. As you get taller, the distance between your arms and the ground increases. If you have lived in the same neighborhood for a long time, you have probably noticed that the trees have also grown taller. However, as trees grow, you may have observed that their branches always remain the same distance off the ground. This observation tells you something about how trees grow. How is it different from the way animals and humans grow?

How Plants Grow

Plants, unlike humans, continue to grow larger as long as they live. In plants, growth occurs at special growth tissue sites. The plant's root and shoot grow longer from their tips, where growth tissues are found. Growth tissue cells are small and divide constantly, making new cells. Notice the growing tips in Figure 14.16. Growth occurs in two stages. First, more cells are made. Second, those cells increase in length. This process in plants is called primary growth.

In addition to increasing in length, a tree trunk or plant stem grows wider, or thicker. This process is called secondary growth. Examine the cross section of a woody stem in Figure 14.16 to see how secondary growth occurs. The cambium, a layer of growth tissue in the stem, continues to divide. It makes a layer of new xylem cells toward the center of the stem and a layer of new phloem cells toward the outside of the stem.

286 Chapter 14 Flowering Plants

INTEGRATED LEARNING

Language Arts Connection
Plant hormones direct plants to grow in certain ways, or tropisms. Ask students to find definitions for *geotropism* and *phototropism* and write explanations of these definitions in their own words. Have them give examples of plants or plant parts that exhibit these tropisms.

Integrating the Sciences
Chemistry Scientists have identified four other plant hormones in addition to auxins and gibberellins (cytokinins, oligosaccharins, abscisic acid, and ethylene). Have students find out what these hormones are and make a chart describing the function of each.

TEACH

Discuss
Explain that the plant hormone called *gibberellin* affects the stem length of plants and plays a role in the growth of all the plant's parts. The plant hormone *auxin* controls a plant's response to light or gravity. Ask students where the auxin of a plant is more concentrated if the plant is bending toward the light. (On the shaded side where the cells are elongated)

Skills Development
Predict Ask students to predict what would happen to a growing potted plant if it were placed on its side. (The growing stem would bend up and the root would bend down.) Have students experiment with a sturdy plant to test their predictions.

Enrich
Most plants display phototropism and geotropism. Some also display thigmotropism, or growth in response to contact with an object. Have students research and demonstrate thigmotropism.

Answers to In-Text Questions
① The plant on the right may have been given extra gibberellin to grow larger.

② The hormone auxin is making the plant bend toward the light.

The Role of Plant Hormones

Hormones play a part in regulating a plant's growth. One example is gibberellin (jihb ur EHL ihn). Hundreds of years ago, Japanese rice farmers observed that some of their rice seedlings grew abnormally tall. This was the first clue to the effects of gibberellin. Gibberellin can make dwarf plants grow tall, flowers produce seedless fruits, and plants flower prematurely. Young leaves, roots, and plant embryos make gibberellin.

Figure 14.17 shows gibberellin's effect on stem length. The hormone also plays a role in seed germination and the growth of new leaves, branches, fruits, and flowers. These effects make gibberellin useful in agriculture. For example, it is used on seedless grapes to make the fruit grow larger.

Plants in a window usually bend toward the light. You may wonder how the plant is attracted to the light. Turning in response to an environmental stimulus, such as light or gravity, is called **tropism** (TROH PIHZ uhm). When a plant turns toward light, it is called phototropism. Tropisms are controlled by a plant hormone called auxin (AWK sihn). The growing tips of plants produce auxin. The amount of auxin in the cells controls the amount of cell elongation.

The way auxin controls phototropism is shown in Figure 14.18. Experiments show that auxin is sensitive to light. As a result, auxin concentrations are always higher on the shaded side of a stem. Auxin causes the stem cells on the shaded side to grow longer, making the stem bend toward the light.

Figure 14.17 ▲
Notice the difference in the size of the plants. Why do you think the plants on the right are larger? ①

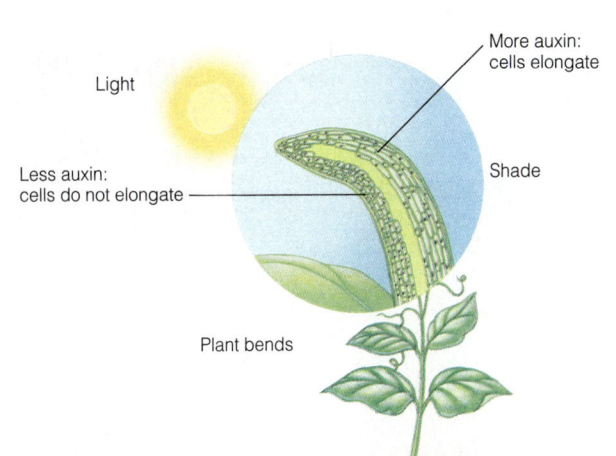

◀ **Figure 14.18**
Light affects the amount of auxin available (left). Why is the plant in the photograph bending toward the window? ②

**The Living Textbook:
Life Science Side 3**

Chapter 17   Frame 25668
Geotropism and Phototropism (Movie)
Search: Play:

Chapter 14 Flowering Plants 287

TEACH • Continued

Enrich
Tell students that short-day plants are really long-night plants. That is, it is not the length of day that is important. Instead, it is the length of night. For example, a common plant seen in December is the poinsettia. This plant will flower only when the amount of darkness is 14 hours or more. If the period of darkness is interrupted for some reason, the plant will not produce flowers.

Skills Development
Interpret Data Have students study Figures 14.21a, 14.21b, and 14.21c. Have them state the difference between annuals, biennials, and perennials. (Annuals live only one growing season. Biennials live for two growing seasons, flowering and producing seeds in the second year. Perennials live for many years and produce seeds for most of those years.)

INTEGRATED LEARNING

Geography Connection
Have students describe differences in life span and rate of growth among plants in different geographic areas. For example, magnolias grow more slowly in the Midwest than in the South. Encourage students to choose one plant and find out when and how it grows in two different states.

Figure 14.19 ▲
Long-day Plant
(16 hours light / 8 hours dark)

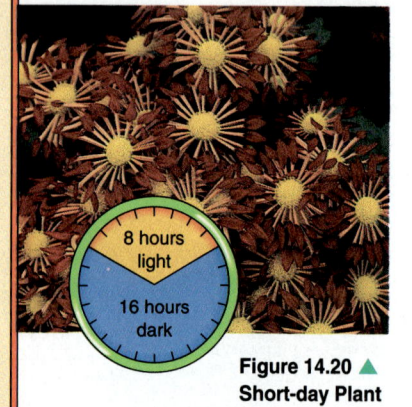

Figure 14.20 ▲
Short-day Plant
(8 hours light / 16 hours dark)

Seasonal Responses
Why do some plants always flower in summer and not at any other time of year? Plants have a light-sensitive chemical that helps them measure the length of periods of darkness and light. Examine the relationship between day length and flowering in Figures 14.19 and 14.20. Summer-blooming plants, such as columbines, are long-day plants. They need long days and short nights to trigger flowering. Autumn-blooming plants, such as chrysanthemums, are short-day plants. They flower when there are short days and long nights. Broad-leaf trees also respond to seasonal changes in day length. As days become shorter in autumn, the green chlorophyll in leaves is destroyed and other colors, such as reds and yellows, can then show through.

The life span of a flowering plant is measured in one-year growing seasons. Compare the life spans of the plants in Figure 14.21. Some plants, called annuals, complete their entire life cycle within one growing season. After setting seed, the plant dies. Plants that live for two growing seasons are called biennials.

During the first growing season, the plant germinates and develops roots, stems, and leaves. It stores energy in underground organs. During the second growing season, the plant flowers, produces seeds, then dies. Other plants, called perennials (pur EHN ee uhls), live for many growing seasons. They develop thick, woody stems and reproduce over and over again.

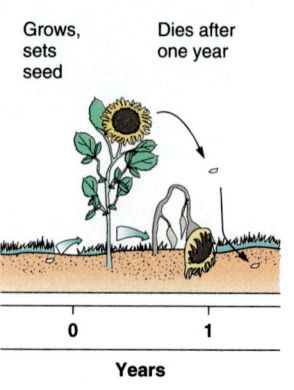

Figure 14.21a ▲
Annual Growth Pattern
(Grows, sets seed — Dies after one year)

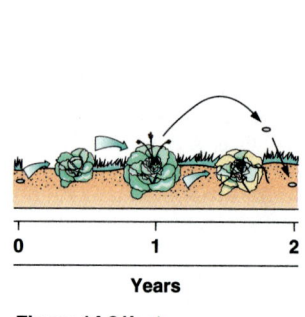

Figure 14.21b ▲
Biennial Growth Pattern
(Grows, stores food — Flowers, sets seed — Dies after two years)

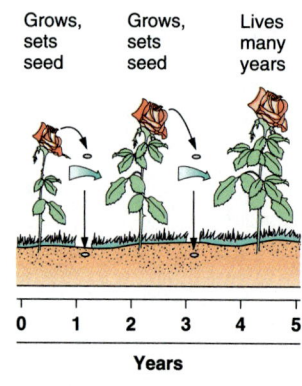

Figure 14.21c ▲
Perennial Growth Pattern
(Grows, sets seed — Grows, sets seed — Lives many years)

STS Connection
Have students find out more about some of the superplants and how they are produced. Ask them to identify the advantages of using biotechnology to produce new kinds of plants.

Science and Technology *Future Superplants*

Potatoes that resist disease; tomatoes that last for weeks without spoiling; strawberry blossoms that survive a freeze—these are just some of the potential uses for biotechnology in agriculture. Biotechnology combines genetic engineering and tissue culture. Genetic engineering takes a chosen trait from the DNA of a plant or a bacterium and splices it into host plant cells. The new cells multiply in a nutrient-rich tissue culture. These cells grow to form new plants.

For thousands of years, plant breeders all over the world introduced genes into crops through selective breeding. This process is very effective, but also very slow. Also, plant breeders can't always find the characteristics they want. Through biotechnology, scientists can insert desired traits into plants, and it takes only months to develop a desired plant strain. For example, biotechnology produced genetically altered tomatoes that do not spoil quickly. Genes from bacteria inserted into tomatoes interrupt the release of the chemical pectin. Because the release of pectin was stopped, a tomato that would spoil in one week can now last for several weeks.

In another study, tomatoes received a spliced gene from a bacterium resistant to a disease-causing virus. The plants were protected against the virus that destroys many tomato crops. Genetically altered strawberries may also be in the future. A gene from "ice minus" bacteria inserted into strawberry blossoms results in "frost-free" plants.

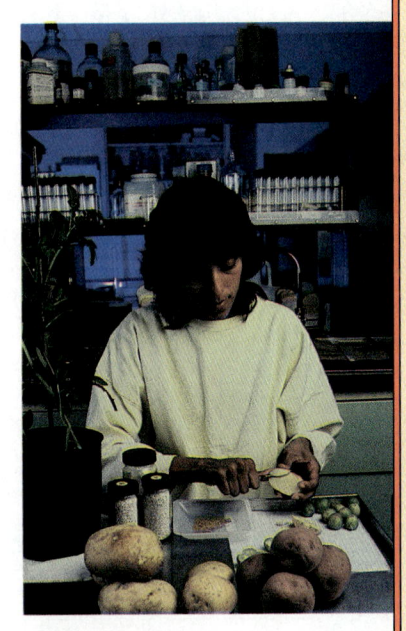

Figure 14.22 ▲
The genetic material of plants can be manipulated in the laboratory using genetic engineering techniques.

Check and Explain

1. How is plant growth different from animal growth?
2. Name two plant hormones. Describe where each one is produced and its effects on plant growth.
3. **Compare and Contrast** What do plants that bloom in the summer have in common? How do they differ from plants that bloom in the autumn?
4. **Hypothesize** Write hypotheses you could use to prove that an unfamiliar plant was an annual, a biennial, and a perennial.

EVALUATE

Discuss
Ask students to imagine that they are plant breeders. Have them describe a plant that they would like to develop.

WrapUp
Reinforce Ask students to name the factors that control plant growth and describe how they work. They should mention hormonal, seasonal, and environmental factors.
Use Review Worksheet 14.4.

Check and Explain

1. Plants grow in two stages and grow larger throughout their lives; animals grow continuously for part of their lives and then stop growing.
2. Gibberellin is made by young leaves, roots, and embryos and effects seed germination and the growth of new leaves, branches, fruits, and flowers. Auxin is made in growing tips of plants and controls cell elongation.
3. Summer-blooming plants need long days and short nights. Autumn-blooming plants need short days and long nights.
4. If annual, won't reappear after first year; if biennial, won't reappear after second year; if perennial, will appear for more than two years

Chapter 14   Flowering Plants

ACTIVITY 14

Time 40 minutes   **Group** 1–2

Materials

30 flowers (gladiolus, tulip, lily)

black paper

scalpels

hand lenses (optional)

Analysis

1. Answers may vary. Possible answers include: the sepals are green and smaller than the petals; the petals are colorful; and the sepals are located outside the petals.
2. Answers will vary depending on flowers available.
3. At the base of the pistil, students should have seen the ovary containing ovules.
4. Sepals, petals, stamens (anther, filament), pistil (ovary, stigma, style)

Conclusion

Conclusions depend on the flower observed. Refer to page 273 for characteristics of monocots and dicots.

Everyday Application

Accept any logical answer. Gardeners often use a small paintbrush to transfer pollen from the stamen to the pistil.

Extension

Answers will vary depending on the pollen available. There are a great variety of pollen shapes and sizes.

TEACHING OPTIONS

Prelab Discussion

Have students read the entire activity. Discuss the following points before beginning:

▶ Review the functions of each part of a flower.

▶ Discuss the relationship between these functions and the overall structure of a flower.

Activity 14 What are the parts of a flower?

Skills Observe; Classify

Task 1 Prelab Prep
Collect the following items: A gladiolus, tulip, or lily flower; paper towel; black paper; scalpel. A hand lens is optional. If you can't get a flower, use the photograph for your observation.

Task 2 Data Record
1. On a separate sheet of paper, draw the data table shown.
2. Record your observations about the color and number of flower parts in the table.

Table 14.1 Flower Parts

Flower Part	Number	Color
Sepals		
Petals		
Anther		
Pollen grains		
Filament		
Pistil		

Task 3 Procedure
1. Place your flower on a paper towel. Closely examine your flower.
2. Locate the outer parts, the sepals. Record the color and number of sepals in your table.
3. Inside the sepals are the petals. Examine the petals and record your observations.
4. Gently remove the sepals and the petals. Find the stalklike stamens. Remove one stamen. Observe the anther and the filament. Record your observations.
5. Tap the anther against the black paper to release pollen grains. Record the color and an estimate of the number of grains.
6. Remove the stamens. The remaining structure is the pistil. In the table, record your observations about the pistil.
7. With a scalpel, cut the pistil in half from top to bottom. **CAUTION! Be careful when using sharp instruments.** Look at the structures inside the ovary. Record your observations in the table.

Task 4 Analysis
1. What are the differences between the sepals and the petals?
2. Compare the number of stamens, petals, and sepals.
3. What did you see in the pistil?
4. List the parts of a flower.

Task 5 Conclusion
Is your flower a monocot or a dicot? Write a short paragraph explaining how you know.

Everyday Application
Flowers can be pollinated by hand. Explain the steps you would use to pollinate a flower yourself.

Extension
Collect and observe pollen grains under a microscope. What is the shape of the pollen grains?

Chapter 14 Review

Concept Summary

14.1 Characteristics of Flowering Plants
- All angiosperms have flowers containing reproductive organs.
- Shape, color, and scent are flower adaptations to attract pollen-carriers.
- Angiosperms are classified as monocots and dicots.

14.2 Vascular Plant Systems
- Vascular plants have a shoot system and root system connected by xylem and phloem tissues.
- Root systems absorb materials from soil, store glucose, and anchor plants.
- Stems and leaves are parts of the shoot system. Stems support the plant and transport materials. Leaves make glucose through photosynthesis.

14.3 Reproduction of Flowering Plants
- Flowers have petals, sepals, and reproductive organs. The male reproductive organ is the stamen. The female organ is the pistil.
- After pollination, the flower forms seeds inside a fruit. The fruit protects and disperses the seeds.
- Vegetative propagation is growing new plants from pieces of the parent plant.

14.4 Plant Growth
- Plants have primary and secondary growth.
- The plant hormones auxin and gibberellin affect plant growth.
- Angiosperms differ in their seasonal responses and in their life spans.

Chapter Vocabulary

monocot (14.1)	phloem (14.2)	stamen (14.3)	stigma (14.3)
dicot (14.1)	epidermis (14.2)	anther (14.3)	vegetative propagation (14.3)
transpiration (14.2)	cambium (14.2)	pistil (14.3)	tropism (14.4)
xylem (14.2)	sepals (14.3)		

Check Your Vocabulary

Use the vocabulary words above to complete the following sentences.

1. The outer layer of protective cells on a root is the ____.
2. A seed with one cotyledon is a ____.
3. Vascular tissue cells that transport glucose are called the ____.
4. The male part of a flower is the ____.
5. The structures growing from the stem that protect the developing flower are the ____.
6. Plant growth in response to an environmental stimulus is called ____.
7. A seed with two cotyledons is a ____.
8. The plant pipeline that transports water and minerals is the ____.
9. The growth tissue for making new xylem and phloem cells is the ____.
10. The female reproductive part of the flower is the ____.
11. Growing new plants from pieces of stems, leaves, or roots from a parent plant is called ____.
12. The male part of the flower that produces pollen grains is the ____.
13. The female part of the flower that collects pollen is the ____.

CHAPTER REVIEW 14

Check Your Vocabulary
1. epidermis
2. monocot
3. phloem
4. stamen
5. sepals
6. dicot
7. tropism
8. xylem
9. cambium
10. pistil
11. vegetative propagation
12. anther
13. stigma

Use Vocabulary Worksheet for Chapter 14.

CHAPTER 14 REVIEW

Check Your Knowledge

1. See Figure 14.3 on page 273 for a list of characteristics of monocots.
2. Leaves capture light from the sun, make glucose through photosynthesis, and exchange gases with the environment.
3. The male gametophyte is the pollen grain.
4. Seeds need water, oxygen, and the proper temperature to germinate.
5. Secondary growth refers to the widening or thickening of a tree trunk or plant stem.
6. Long-day plants require long days and short nights to trigger flowering; short-day plants flower in response to short days and long nights.
7. The lifespan of flowering plants is measured in one-year growing seasons. Plants that complete their life cycles in one season are called annuals. Perennials live for many growing seasons.
8. Fruits help protect and disperse seeds.
9. Pollination is the transfer of pollen from the anther to the stigma.
10. A cut section from a stem or leaf placed in a moist environment can produce a new plant. This is called vegetative propagation.
11. a thick taproot
12. sugars
13. xylem and phloem cells
14. herbaceous
15. mesophyll
16. filament and anther

Check Your Understanding

1. See page 282. Make sure students include all the steps.
2. Answers may vary, but may include: the cherry falls to the ground and develops where it falls; an animal eats the cherry, the seed passes through the animal's digestive system, and is deposited; and a bird carries the cherry away, eats the fruit, and drops the seed.
3. Answers may vary. By using vegetative propagation, the gardener doesn't have to take the time to collect seeds and wait for them to germinate and sprout.
4. Accept all logical answers, provided valid reasons are given.
5. Taproots store starch, which can provide carbohydrates to humans.
6. a. The seeds look different because they come from different plants; however, they all have a seed coat, a plant embryo, and stored energy.
   b. The seeds look like monocots.
7. Since the leaves are similar, one can tell that the plants are members of the same group—either monocots or dicots—and will exhibit the characteristics of the group.
8. By growing poinsettias under lights in a room without windows, one could "fool" the plants into blooming by controlling the amount of time the lights are on each day.

Chapter 14 Review

Check Your Knowledge
Answer the following in complete sentences.

1. Describe three features of a monocot.
2. List the functions of leaves.
3. What is the angiosperm's male gametophyte?
4. List the requirements for seed germination.
5. What is secondary plant growth?
6. What is a long-day plant? A short-day plant?
7. How is the life span of flowering plants measured?
8. What is the purpose of fruits for plants?
9. Define pollination.
10. How can angiosperms reproduce without seeds?

Choose the answer that best completes each sentence.

11. Dicots have (parallel veined leaves, flower parts in multiples of ten, a thick taproot, four cotyledons).
12. The phloem transports (minerals, water, starch, sugars) throughout the plant.
13. The cambium makes new (root hairs, xylem and phloem cells, epidermis cells, shoot and root cells).
14. Stems that are flexible and slender are called (filament, woody, herbaceous, taproot) stems.
15. In leaves, photosynthesis occurs in the (mesophyll, cuticle, stomates, xylem).
16. The stamen is composed of the (pistil and stigma, anther and stigma, filament and anther, filament and pistil).

Check Your Understanding
Apply the concepts you have learned to answer each question.

1. Describe how fertilization occurs in flowering plants. Start from the formation of the pollen grain.
2. Explain three possible ways that a cherry could help disperse the seed contained inside the fruit.
3. Why would a gardener choose to grow plants using vegetative propagation?
4. **Infer** Some flowers have stamens that extend far beyond their petals. Would this be an advantage or a disadvantage for such a plant? Why or why not?
5. What nutritional value do taproots have to humans?
6. **Mystery Photo** The picture on page 270 shows seeds from a variety of angiosperms, including pumpkins, corn, sunflowers, walnuts, beans, and maples.
   a. Why do the seeds look so different? What do they all have in common?
   b. From what you see in the photograph, classify the seeds as monocots or dicots.
7. **Critical Thinking** You observe two plants that have similar leaves. What conclusions can you make about these plants? Explain how you came to each conclusion.
8. **Application** Poinsettias are often seen during the winter holiday season. Poinsettias are short-day plants. Explain how you could manipulate a poinsettia plant into blooming during the summer.

Develop Your Skills

1. a. Dicot; thick taproot, leaves with branching veins
    b. The plant has a taproot system. The main root is thick and grows straight down into the ground.
    c. Seed germination
2. Accept any experimental design that is complete and thorough. Make sure students manipulate only one variable at a time.
3. a. Cereal grains
    b. Answers will vary.

Make Connections

1. Concept maps will vary. You may wish to have students do the map below. First draw the map on the board, then add a few key terms. Have students draw and complete the map.

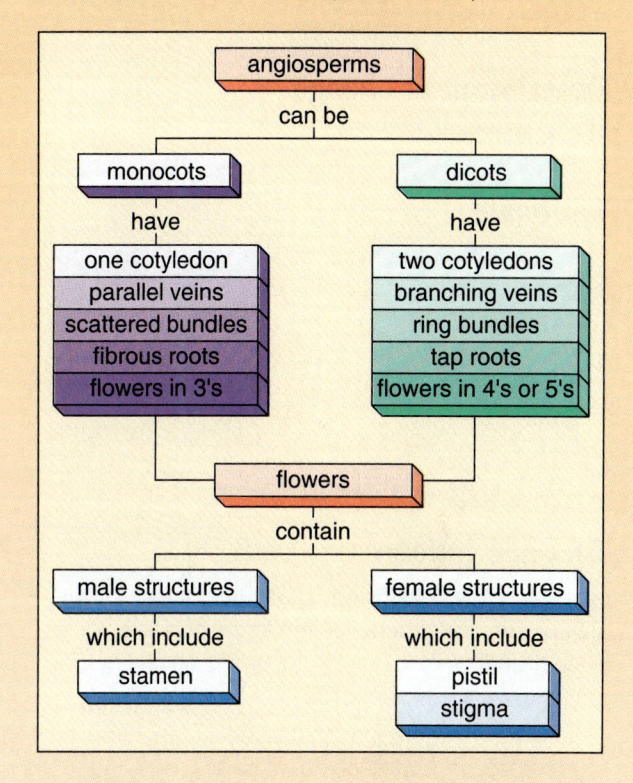

```
                    angiosperms
                       can be
         ┌───────────────┴───────────────┐
      monocots                         dicots
        have                            have
   one cotyledon                  two cotyledons
   parallel veins                 branching veins
  scattered bundles                 ring bundles
    fibrous roots                     tap roots
   flowers in 3's               flowers in 4's or 5's
         └───────────────┬───────────────┘
                      flowers
                      contain
         ┌───────────────┴───────────────┐
   male structures                 female structures
    which include                   which include
       stamen                          pistil
                                       stigma
```

2. Answers may vary. The answer for monocots should be a multiple of three and for dicots a multiple of four or five.
3. Research results will vary.
4. Answers will vary, but should include angiosperm characteristics described in the text.
5. Student reports will vary. Possible plants to research might include carrots, apples, or okra.

Develop Your Skills

Use the skills you have developed in this chapter to complete each activity.

1. **Interpret Data** The drawing below shows a plant as it changes with time.
    a. Is this plant an example of a monocot or a dicot? Describe at least two characteristics shown in the drawing that support your decision.
    b. What type of root system does this plant have? How can you tell?
    c. Which plant process is shown in the drawing?

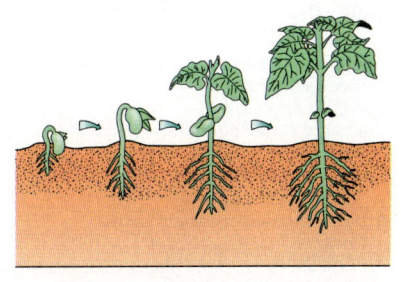

2. **Design an Experiment** New jade plants can be grown by vegetative propagation. Design an experiment to test what pieces of a jade plant will develop new plants. Remember to change only one variable at a time.

3. **Data Bank** Use the information on page 625 to answer the following questions.
    a. Which crops are most abundant in the world?
    b. **Research** Find out examples of oil-bearing and sugar-bearing crops. What countries grow these crops?

Make Connections

1. **Link the Concepts** Draw a concept map showing how some of the main concepts in this chapter link together. Use words and ideas from the chapter to draw your map.

2. **Science and Math** Calculate and record the number of possible petals for a monocot and a dicot. Estimate the maximum petal number for each group.

3. **Science and Social Studies** Before manufactured goods were available, people depended on common local plants for a variety of uses. A single species of plant could be a source of food, fiber, medicine, and building material. In the library, do research on local native plants, their uses, and the people who used them in the past or use them today.

4. **Science and You** Count the number of angiosperm varieties that you encounter on your way to school. Describe what features you used to determine whether a plant is an angiosperm.

5. **Science and Social Studies** Many of the common plants that you have probably seen in gardens came here from other countries. For example, tulips originated in western and central Asia. How did these plants travel so far? During the 19th and early 20th centuries, thousands of plant collectors travelled far distances in search of unkown plants. Many of these people were faced with obstacles during their travels. Research to find out more about the discoveries of these early plant collectors. Write a report of your findings. Include examples of the types of plants that were discovered.

SCIENCE AND LITERATURE CONNECTION   UNIT 5

About the Literary Work

"Songs in the Garden of the House God" is a poem from *In the Trail of the Wind*, edited by John Bierhorst, copyright 1971 by Farrar, Straus and Giroux. Reprinted by permission of Farrar, Straus and Giroux.

Description of Change

The selection was not altered.

Rationale

The vocabulary and syntax of the original form are appropriate to the student level.

Vocabulary

tassel, cull

Teaching Strategies

Directed Inquiry

After students finish reading, discuss the selection. Be sure to relate the selection to the science lessons in this unit. Ask the following question:

▶ In Chapter 14 you discovered the factors that cause seed germination. Based on what you have learned, what is meant by the following lines from the selection?

> Truly in the East
> The white bean
> And the great corn-plant
> Are tied with the white lightning.

(In order for a seed to germinate, it must absorb moisture. White lightning is symbolic of the water cycle that provides the soil and the white bean with this necessary substance.)

Explain

Students may be unfamiliar with the terms *tassel* and *silk*. Explain that a tassel is a cluster of small male flowers, each containing an anther. Silk is a long, threadlike substance that extends from an ovary to the tip of a husk.

Skills Development

Infer The initial lines of the selection state that the blue corn-seed will grow and flourish in one night, while the white corn-seed will grow and flourish in one day. What could students infer from this? (The two different colored seeds produce two different kinds of corn-plants—each with its own growing season.)

Critical Thinking

Reason and Conclude The selection indicates that both corn and squash were planted in the same environment. What does this indicate about the needs of these two types of plants? (The plants' environmental needs, such as soil type and climate, must be similar.)

Science and Literature Connection

"Songs in the Garden of the House God"

The following Navajo poem is from In the Trail of the Wind, *a book of Native-American poems and ritual orations edited by John Bierhorst.*

The sacred blue corn-seed I am planting,
In one night it will grow and flourish,
In one night the corn increases,
In the garden of the House God.

The sacred white corn-seed I am planting,
In one day it will grow and ripen,
In one day the corn increases,
In its beauty it increases.

With this it grows, with this it grows,
The dark cloud, with this it grows.
The dew thereof, with this it grows,
The blue corn, with this it grows.

With this it grows, with this it grows,
The dark mist, with this it grows.
The dew thereof, with this it grows,
The white corn, with this it grows.

This it eats, this it eats,
The dark cloud,
Its dew
The blue corn eats,
This it eats.

This it eats, this it eats,
The dark mist,
Its dew
The white corn eats,
This it eats.

The great corn-plant is with the bean,
Its rootlets now are with the bean,
Its leaf-tips now are with the bean,
Its dewdrops now are with the bean,
Its tassel now is with the bean,
Its pollen now is with the bean,
And now its silk is with the bean,
And now its grain is with the bean.

Truly in the East
The white bean
And the great corn-plant
Are tied with the white lightning.
Listen! It approaches!
The voice of the bluebird is heard.

Truly in the East
The white bean
And the great squash
Are tied with the rainbow.
Listen! It approaches!
The voice of the bluebird is heard.

From the top of the great corn-plant the water gurgles, I hear it;
Around the roots the water foams, I hear it;
Around the roots of the plants it foams, I hear it;
From their tops the water foams, I hear it.

The corn grows up. The waters of the dark clouds drop, drop.
The rain descends. The waters from the corn leaves drop, drop.
The rain descends. The waters from the plants drop, drop.
The corn grows up. The waters of the dark mists drop, drop.

Since the ancient days, I have planted,
Since the time of the emergence, I have planted,
The great corn-plant, I have planted,

Skills in Science

Reading Skills in Science

1. The "time of emergence" refers to the time at which humans first appeared on the earth.
2. Answers may vary. The connection between plants and the water cycle is evident in the lines "The dark cloud, with this it grows," "The dark mist, with this it grows," "The dew thereof, with this it grows," "Its dewdrops now are with the bean," "Are tied with the white lightning," "Are tied with the rainbow," "Around the roots the water foams, I hear it," and "The waters of the dark clouds drop, drop."

Writing Skills in Science

1. Student poems should show that without rain, the corn-plant seeds do not germinate.
2. Paragraphs may identify products such as cornmeal, cornbread, cornflakes, tortillas, popcorn, and other snack foods as "being with the bean."

Activities

Collect Data Collages should show such nonfood products as cosmetics, paper goods, laundry starch, textiles, and explosives.

Communicate Flowcharts should show that the anthers contained in the tassel release pollen. Pollen grains land on the silk, sending a pollen tube down through it. Sperm travel through the tube and fertilize an egg inside the ovary.

Its roots, I have planted,
The tips of its leaves, I have planted,
Its dew, I have planted,
Its tassel, I have planted,
Its pollen, I have planted,
Its silk, I have planted,
Its seed, I have planted.

Since the ancient days, I have planted,
Since the time of the emergence,
   I have planted,
The great squash-vine, I have planted,
Its seed, I have planted
Its silk, I have planted,
Its pollen, I have planted,
Its tassel, I have planted,
   Its dew, I have planted
     The tips of its leaves, I have planted,
        Its roots, I have planted.

Shall I cull this fruit
Of the great corn-plant?
Shall you break it? Shall I break it?

Shall I break it? Shall you break it?
Shall I? Shall you?

Shall I cull this fruit
Of the great squash-vine?
Shall you pick it up? Shall I pick it up?
Shall I pick it up? Shall you pick it up?
Shall I? Shall you?

Skills in Science

Reading Skills in Science

1. **Infer** What does the phrase "since the time of the emergence" refer to?
2. **Find Context Clues** Which parts of the poem describe the connection between plants and the water cycle?

Writing Skills in Science

1. **Compare and Contrast** Imagine that a drought occurs during one growing season. Write a poem that describes how this lack of rain affects the corn-plant.
2. **Predict** The poem identifies various parts of a corn-plant and states that they are "with the bean." Think about products of a corn-plant. Write a paragraph that describes other items that are "with the bean."

Activities

Collect Data Use reference texts to identify nonfood products made from corn. Share your findings with the class in the form of a collage that shows these products.

Communicate Make a flow chart to show how a corn-plant reproduces sexually.

Where to Read More

In the Trail of the Wind edited by John Bierhorst. New York: Farrar, Straus and Giroux, 1971. Contains Native-American poems and ritual orations including "Songs in the Garden of the House God."

The Lore and Legend of Flowers by Robert L. Crowell. New York: Crowell, 1982. This beautifully illustrated text describes the origins of ten flowering plants and the impact of each on history.

295

UNIT 6

UNIT OVERVIEW

Unit 6 introduces students to the animal kingdom and describes the characteristics, evolution, and adaptations of invertebrates and vertebrates. Chapter 15 focuses on animals and describes the differences between invertebrates and vertebrates. The chapter then explores the characteristics of sponges and cnidarians. The three major worm phyla are also discussed.

Chapter 16 presents the evolution of mollusks. Arthropods are distinguished from other invertebrates, and the insect class of arthropods is explored in depth. The chapter also discusses the five classes of echinoderms. Chapter 17 introduces students to vertebrates. The evolution, body features, movement, and organ systems of three classes of fish are explored. Amphibians and reptiles are described as

Introducing the Unit

Directed Inquiry

Have the students examine the photograph. Ask:

▶ Why do you think the penguins have black fur on their backs? (Black absorbs heat from sunlight.)

▶ Penguins have long flippers instead of arms. Do you think they spend most of their time on land or in water? (In water)

▶ How are penguins similar to other birds, such as ducks? (Beaks, eyes at the sides, coloring similar to that of ducks)

▶ How do the penguins' beaks help them survive in their environment? (The long beaks help them obtain fishes from the sea.)

Writing About the Photograph

Ask students to list words describing the photograph and use the descriptive words in a poem honoring penguins.

Unit 6
Animal Life

Chapters

15 Invertebrates I
16 Invertebrates II
17 Fishes, Amphibians, and Reptiles
18 Birds and Mammals

well. Chapter 18 focuses on birds and mammals. Students discover the unique characteristics of birds, exploring flight, migration, and reproduction. Also presented are the evolutionary history of mammals and reproduction among three different groups of mammals: monotremes, marsupials, and placental mammals.

Data Bank

Use the information on pages 624 to 631 to answer the following questions about topics explored in this unit.

Calculating

How much longer is a king cobra than a rattlesnake?

Reading a Table

During which period in the earth's history did reptiles begin to take over the land?

Classifying

Is the polar bear included in the same genus as the grizzly bear? Name the genus that each bear on the tree belongs to.

Making a Graph

Make a graph that shows the average life spans of the mammals listed in the table. Which mammal lives the longest? The shortest?

The photograph to the left is of king penguins on the South Georgia Island near Antarctica. Why do you think the penguins live in groups?

Data Bank Answers

Have students search the Data Bank on pages 624 to 631 for the answers to the questions on this page.

Calculating The king cobra is 3.5 meters longer than the rattlesnake. The answer is found in the diagram, Large Reptile Size Comparison, on page 631.

Extension Ask students to make a bar graph comparing the sizes of large reptiles.

Reading a Table Reptiles became dominant on land during the Permian period. The answer is found in the table, Evolution of Mammals, on page 630.

Extension Ask students how many millions of years passed between the first appearance of reptiles and the first appearance of mammals. (115 million years)

Classifying Yes. The grizzly bear and the polar bear both belong to the *Ursus* genus. The answer is found in the diagram, Relationships Among Carnivora, on page 628.

Making a Graph Female human beings are the longest living mammals. The mouse has the shortest life span. The answers are found on page 625 in the graph, Average Life Spans.

Extension Ask students to find out how the average life span of human beings has changed over the past century. (It has increased.)

Answer to In-Text Question

Living in groups provides great survival value, especially for animals living in harsh climates.

CHAPTER 15

Overview

This chapter opens the unit on animals. The first section describes the main characteristics of animals and how animals are classified. The rest of the chapter focuses on invertebrate groups. The second section explores the life cycle and characteristics of sponges. The two body forms of cnidarians are described in the third section, as well as the development of coral reefs. The final section focuses on worms, including a discussion of the three major worm phyla. Also, students learn about both the beneficial and harmful effects of worms.

Advance Planner

▶ Supply muddy or sandy water, paper towels or coffee filters, and natural sponges (available in art supply or biological supply stores) for TE activity, page 303.

▶ Purchase hydras and planaria (from a biological supply store if unavailable locally) for TE activities, pages 307 and 312.

▶ Obtain earthworms (can be purchased from a bait shop) for TE activity, page 313 and SE Activity 15, page 316.

▶ Collect 3 kinds of soil (6 L total), 10 1-L jars, 3 L of humus, 1.5 L of apple peelings, dark construction paper, rubber bands, and earthworms for SE Activity 15, page 316.

Skills Development Chart

Sections	Communicate	Decision Making	Infer	Interpret Data	Generalize	Measure	Model	Observe	Research
15.1 Skills WarmUp							●		
15.2 Skills WarmUp Skills WorkOut					●				●
15.3 Skills WarmUp Consider This	●	●			●				
15.4 Skills WarmUp SkillBuilder Activity	●		●	●		●	●	●	

Individual Needs

▶ **Limited English Proficiency Students** Use a tape recorder to record the pronunciations and definitions of the boldface or italic words in this chapter. Along with the definition of each term, tell students which figures in the text illustrate the term. Have students use the tape recorder to listen to the information about each term. Then have students define the terms using their own words. Have them make their own tape recordings of the terms and their definitions. They can use the audiotape along with the text figures to review important concepts in the chapter.

▶ **At-Risk Students** Point out to students that the text asks for their opinions and ideas. As they read each section, ask students to answer all the questions that appear in the text by first creating a list of all the questions. They should leave space for the answer after each question. Show them that some answers follow immediately after the text questions. Encourage students to ask their own questions as they read. Have them include these questions, along with answers, in their science journals.

▶ **Gifted Students** Have students select one type of organism that is described in this chapter. Have them design and construct a detailed, three-dimensional model of the organism. They should use the model to demonstrate the characteristics of the animal.

Resource Bank

▶ **Bulletin Board** Work with students to create a bulletin board display about symmetry. Have them make drawings or cut pictures from magazines that show symmetry in living things. Have them label each organism as showing radial or bilateral symmetry.

▶ **Field Trip** Arrange to take students to an aquarium where they can observe invertebrates. Have them take along a list of the characteristics of several types of invertebrates. Have them use this list to classify some of the animals they observe.

CHAPTER 15 PLANNING GUIDE

Section	Core	Standard	Enriched	Section	Core	Standard	Enriched
15.1 The Animal Kingdom pp. 299–301				**Blackline Masters** Review Worksheet 15.3 Integrating Worksheet 15.3	• •	• •	
Section Features Skills WarmUp, p. 299	•	•		**Overhead Blackline Transparencies** Overhead Blackline Master 15.3 and Student Worksheet	•	•	•
Blackline Masters Review Worksheet 15.1 Skills Worksheet 15.1	• •	• •		**Laboratory Program** Investigation 30	•	•	•
Ancillary Options Multicultural Perspectives, p. 101	•	•	•	**Color Transparencies** Transparencies 42, 43	•	•	•
15.2 Sponges pp. 302–304				**15.4 Worms** pp. 310–316			
Section Features Skills WarmUp, p. 302 Skills WorkOut, p. 303	• •	• •	•	**Section Features** Skills WarmUp, p. 310 SkillBuilder, p. 314 Activity, p. 316	• •	• • •	• • •
Blackline Masters Review Worksheet 15.2	•	•		**Blackline Masters** Review Worksheet 15.4 Skills Worksheet 15.4 Integrating Worksheet 15.4 Reteach Worksheet 15.4 Vocabulary Worksheet 15	• • • •	• • • • •	 • •
Color Transparencies Transparency 41	•	•					
15.3 Cnidarians pp. 305–309							
Section Features Skills WarmUp, p. 305 Consider This, p. 309	•	• •	• •	**Color Transparencies** Transparency 44	•	•	

Bibliography

The following resources can be used for teaching the chapter. See page T-40 for supplier codes.

Audio-Visual Sources
(video unless noted)
Animal Populations: Nature's Checks and Balances. 22 min. 1983. EB.
Classifying Living Things. 18 min. 1987. A-W.
Coral Reef. 23 min. 1979. NGSES.
Invertebrates: Conditioning or Learning? 15 min. 1975. NGSES.
The Living Earth. 25 min. 1991. NGSES.
Riches from the Sea. 23 min. 1984. NGSES.

Software Resources
Animals. Apple. J & S.
Diversity of Life. Apple. J & S.
Earthworm Dissection Program. Apple, IBM. CES.
The Plant Growth Simulator. Apple. FM.

Library Resources
Burnie, David. How Nature Works. Pleasantville: The Reader's Digest Association, Inc., 1991.
Comstock, Anna Botsford. Handbook of Nature Study. Ithaca: Cornell University Press, 1986.
Margulis, Lynn, and Karlene V. Schwartz. Five Kingdoms: An Illustrated Guide to the Phyla of Life on Earth. San Francisco: W.H. Freeman and Co., 1982.

CHAPTER 15

Introducing the Chapter

Have students read the description of the photograph on page 298. Ask if they agree or disagree with the description.

Directed Inquiry

Have students study the photograph. Ask:

▶ Do you think that the organisms shown in the photograph are monerans, protists, fungi, plants, or animals? (Animals)

▶ What characteristics of the organisms helped you make your decision? (Their complexity and their lack of green coloring)

▶ Describe the body form of these organisms. (Circular or flower-like) How is it different from the form of your body? (Humans have a more complex body structure.)

▶ How do you think the body shape of these organisms helps them get food and defend themselves? (They can detect food or danger from any side.)

Chapter Vocabulary

cnidarian
coelom
flatworm
invertebrate
larva
roundworm
segmented worm
vertebrate

INTEGRATED LEARNING

Writing Connection

Have students write a poem about the photograph. Encourage them to richly describe the appearance of the coral and the other organisms living in and around it.

Chapter 15 Invertebrates I

Chapter Sections

15.1 The Animal Kingdom
15.2 Sponges
15.3 Cnidarians
15.4 Worms

What do you see?

"In this photo there are tiny organisms called coral. Corals are invertebrates and belong to the phylum Coelenterata. Coelenterates have hollow body cavities. Have you ever wondered why corals have walls around themselves? They build the walls by using lime, a mineral found in the water. Coral can grow quite large; in some cases they form reefs."

Sherri Coates
Washington Middle School
Newport News, Virginia

To find out more about the photograph, look on page 318. As you read this chapter, you will learn about sponges, corals, jellyfish, earthworms, and other similar invertebrates.

Themes in Science
Evolution All animals are descended from common ancestors. It is possible to infer how animals evolved by studying present-day animal groups. The groups of animals from invertebrates to vertebrates show how vertebrates evolved from earlier animals.

Integrating the Sciences
Earth Science Invertebrates evolved during the Precambrian era and diversified by the Cambrian period (about 375 million years ago). Have students suggest why there are not many fossils of these early invertebrates. (These invertebrates had soft bodies.)

SECTION 15.1

Section Objectives
For a list of section objectives, see the Student Edition page.

Skills Objectives
Students should be able to:

Make a Model to demonstrate radial and bilateral symmetry.

Make a Graph to show the proportions of vertebrates and invertebrates.

Vocabulary
vertebrate, invertebrate

15.1 The Animal Kingdom

Objectives

▶ **Describe** the main characteristics of animals.

▶ **Explain** the major differences in the body plans of animals.

▶ **Distinguish** between vertebrates and invertebrates.

▶ **Make inferences** about the relationship between an animal's body plan and the way it moves.

▶ **Make a graph** comparing the number of vertebrate and invertebrate species.

Skills WarmUp

Model Draw a circle. How many ways can you fold the paper through the center of the circle to make two similar halves? Now draw a simple human figure facing forward. How many ways can you fold the paper through the center of this shape to form two similar halves? What does this tell you about the difference between the circle shape and the human shape?

Look at Figure 15.1. Does it show any animals? Which organisms do you think are animals? Why? Even though the organisms shown in the photograph do not look like the animals you are familiar with, many of them are animals. The animal kingdom includes an amazing variety of living things, from microscopic worms to giant clams to elephants to you.

Characteristics of Animals

All animals share characteristics that separate them from plants, fungi, monerans, and protists. Animals are many-celled, which makes them different from monerans and protozoa. They lack chlorophyll in their cells, so they can't make glucose through photosynthesis like plants and algae.

Animals obtain their food by *ingesting* it. They take in food from their environment and then digest it. Because they eat their food, animals differ from fungi, which digest their food outside their bodies and then absorb the nutrients.

What other characteristic do animals share? If you answered that they have the ability to move, you are right. The animals you have seen crawl, swim, jump, fly, walk, and run. But what about the animals in Figure 15.1? Do they move? Actually, many of the animals pictured stay in one place as adults. However, they were able to move at an earlier stage in their lives. For an animal, why may movement be important? ①

Figure 15.1 ▲
To which kingdom do these tidepool organisms belong? ②

MOTIVATE

Skills WarmUp
To help students understand radial and bilateral symmetry, have them do the Skills WarmUp.
Answer The circle has many lines of symmetry, but the human shape has only one. The human shape has bilateral symmetry.

Misconceptions
Many kinds of animals do not look like what students typically think of as animals. Explain that many do not have heads the way reptiles, amphibians, birds, and mammals do.

Answers to In-Text Questions
① Being able to move helps animals take in food from their environment and protect themselves from predators.
② The tidepool organisms belong to the animal kingdom.

The Living Textbook:
Life Science Sides 1-4

Chapter 7　　　Frame 02355
Survey of Invertebrates (78 Frames)
Search:　　　　Step:

Chapter 15　Invertebrates I

TEACH

Reteach
Have students study pages 86 and 87 to review the levels of organization in living things. Stress that animals are often classified according to their level of organization. Have them use this information as you discuss Figure 15.3.

Directed Inquiry
Have students study the material on this page. Then ask them the following questions:

▶ Which level of organization and which basic body type do sponges fit into? (Cells; no symmetry)

▶ Which level of organization and which basic body type do cnidarians fit into? (Tissues; radial symmetry)

▶ Which level of organization and which basic body type do worms fit into? (Organs; bilateral symmetry without a backbone)

Ancillary Options
If you are using the blackline masters from *Multiculturalism in Mathematics, Science, and Technology*, have students read "A Look at Symmetry" on page 101. Discuss the questions.

Answer to In-Text Question

① The four basic body types are as follows: no symmetry, radial symmetry, bilateral symmetry without a backbone, and bilateral symmetry with a backbone.

The Living Textbook:
Life Science Sides 1-4

Chapter 4   Frame 00780
Radial Symmetry (9 Frames)
Search: Step:

300

INTEGRATED LEARNING

Themes in Science
Diversity and Unity All animals share the same basic needs for food, shelter, and reproduction. However, animals vary greatly in their structures and functions for satisfying these needs. Encourage students to look for examples of diversity and unity among the animals discussed in this chapter.

Themes in Science
Scale and Structure Scientists classify animals based on different types of body structure. Animals with radial symmetry have bodies organized around a center, with no front or back. Animals with bilateral symmetry have bodies with two similar halves and a head that faces the direction in which the animals usually move. Ask students to find examples of radial symmetry and bilateral symmetry in the classroom.

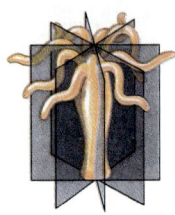

Classification of Animals

The first animals evolved from single-celled protists hundreds of millions of years ago. Since then, the animals have evolved a variety of different body plans, or kinds of overall body structures. These different body plans are the basis for animal classification.

Four levels of organization exist among animals: cells, tissues, organs, and organ systems. One major way that animal bodies differ is in their level of organization. The simplest kind of animal body has many cells but no true tissues. Animals called sponges have this kind of structure. Another group called the cnidarians (ny DAIR ee uhns) includes animals whose bodies are made up of tissues but have no organs. Several groups of animals have bodies in which tissues work together as organs. These include the flatworms and roundworms. In the earthworms, organs work together as organ systems.

Animals can also be classified by their shape. Look at Figure 15.2. Animals with *radial symmetry* have bodies organized around a center, like a wheel. They have a top and bottom, but no front, back, or head. Animals with *bilateral symmetry* have bodies with two similar halves. They have a head end that faces in the direction the animal moves. The head contains the animal's sense organs. A sponge's body can't be divided into parts. It has no symmetry.

Many of the animals with bilateral symmetry, such as dogs and birds, have an extra degree of complexity. These animals have a backbone. Animals with back-

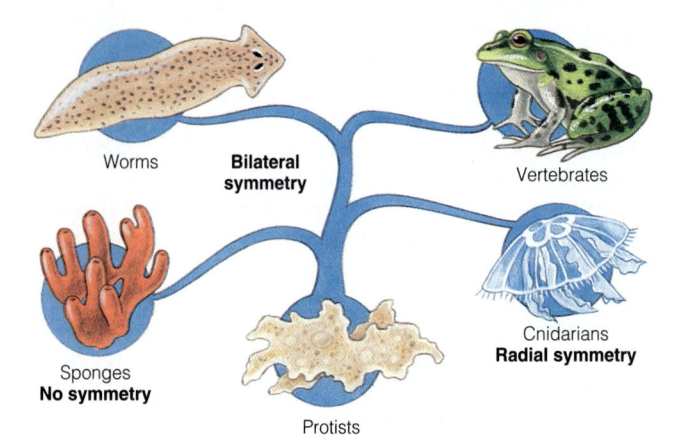

Figure 15.2 ▲
Animal bodies have either radial symmetry, bilateral symmetry, or no symmetry.

Figure 15.3 ▶
Four basic body types have evolved in the animal kingdom. What characteristics separate the four types? ①

300   Chapter 15   Invertebrates I

Geography Connection

Students may be interested to know where some hot springs, such as the one that *Alvin* discovered, are located in the ocean. Many hot springs are located in Iceland. Have students find Iceland on a world map.

---

bones are called **vertebrates** (VUR tuh brihts). You are a vertebrate. All the other animals without backbones are called **invertebrates**.

Based on these and other differences in the bodies of animals, scientists have divided the animal kingdom into 28 large groups called phyla.

Science and Technology
Deep-Sea Exploration

The ocean is rich in animal life, but most of these organisms live at or near the surface, within reach of sunlight. Algae living there are the first link in the ocean's many food chains. The deep ocean bottom, in contrast, has very little life. No sunlight reaches into these depths, so no algae exist as a food supply.

Until the 1970s, no firsthand exploration had been made of the deep sea. Then scientists built special research submarines designed to withstand the great pressures of the deep sea. The best known of these vessels is *Alvin*, owned by the Woods Hole Oceanographic Institute. Using *Alvin*, scientists found hot springs around which live many kinds of strange new invertebrates, including sea anemones, crabs, and giant tube worms.

At first, the scientists did not know how these organisms obtained food. Then they found that the water around the hot springs is filled with special bacteria. These bacteria use hydrogen sulfide from the hot spring water as an energy source. The other organisms of the community feed upon the bacteria by filtering them from the water.

Check and Explain

1. What characteristics separate the animals from the plants, algae, protozoa, monerans, and fungi?
2. What is the difference between radial symmetry and bilateral symmetry?
3. **Infer** What kind of symmetry would you expect most animals that hunt for food to have? Explain.
4. **Make a Graph** Invertebrates make up 95 percent of all known animal species. Make a circle graph that shows this and identifies the other 5 percent.

Chapter 15 Invertebrates I   **301**

Discuss
Ask students to identify the main difference between vertebrates and invertebrates. (Vertebrates have a backbone; invertebrates do not.) Explain that they will be learning about vertebrates in upcoming chapters.

EVALUATE

WrapUp
Reinforce Have students work in groups of four to look through magazines and books to find pictures of different kinds of animals. Have students in each group quiz one another on the correct classification of each animal. They should try to identify the kind of symmetry shown in each animal's body plan and the animal's level of organization.

Use Review Worksheet 15.1.

Check and Explain

1. Animals are many-celled, lack chlorophyll, and obtain their food by ingesting it.
2. If a body has bilateral symmetry, it has two similar halves. A head faces in the direction that the animal moves. If a body has radial symmetry, it is organized around a center. These bodies have a top and a bottom, but not a front, back, or head.
3. Bilateral symmetry
4. Student graphs should show that only 5 percent of all animals are vertebrates.

The Living Textbook:
Life Science Sides 1-4

Chapter 4 Frame 00807
Bilateral Symmetry (19 Frames)
Search: Step:

301

SECTION 15.2

Section Objectives
For a list of section objectives, see the Student Edition page.

Skills Objectives
Students should be able to:

Generalize about differences between many-celled animals and single-celled animals.

Research information about sponge skeletons.

Make a Model showing filter feeding.

Vocabulary
larva

MOTIVATE

Skills WarmUp
Help students understand the differences between many-celled and single-celled animals by having them do the Skills WarmUp.
Answer Answers may vary. Sponges are larger than single-celled animals. Sponges have amebalike cells that transport food and wastes to other cells. Sponges reproduce sexually and asexually.

Prior Knowledge
To gauge how much students know about sponges, ask the following questions:

▶ What makes sponges animals and not plants?
▶ How do sponges eat?
▶ How do sponges reproduce?

Answer to In-Text Question

① The collar cells move the water through the sponge. They also catch the food. The amebalike cells in the jelly layer carry food and waste to and from the other cells.

302

INTEGRATED LEARNING

Themes in Science
Scale and Structure The way animals perform their life functions can be studied at different structural levels including cells, organs, organ systems, organisms, and ecosystems. In the case of simple organisms, such as sponges, that do not have organs or organ systems, scientists study the particular specialized function of each type of cell.

Skills WarmUp
Generalize What can a many-celled animal such as a sponge do that a single-celled protist cannot? Make a list of the comparisons.

Figure 15.4 ▲
What is the function of each kind of cell in a sponge? ①

15.2 Sponges

Objectives

▶ **Describe** the main characteristics of sponges.
▶ **Identify** the functions of a sponge's three cell types.
▶ **Explain** how sponges reproduce.
▶ **Make a model** that shows how a sponge gets its food.

Examine the sponge in your kitchen or bathroom closely and you'll see it has little holes, or pores, everywhere. These pores are what make a sponge able to hold water.

Most likely your sponge was made by humans. But some sponges are natural. They are the soft skeletons of animals harvested from the ocean. These living animals are called sponges, too.

Characteristics of Sponges

Living sponges are full of pores, like your kitchen sponge. These pores are what give sponges their scientific name, Porifera (poh RIHF uhr uh).

A sponge uses its pores to get food. The sponge remains in one place, often attached to a rock. It makes water from its environment come through its pores and into the middle of its saclike body. There the sponge removes algae, tiny animals and protozoa, and bits of organic material. The water leaves the sponge through one larger hole. This way of getting food is called *filter feeding*.

Because of these characteristics, a sponge seems very plantlike. You could watch one for a whole day and not see it move. A sponge is the simplest of animals. Its cells do different jobs, but they do not form true tissues.

Look at Figure 15.4. It shows the structure of a sponge's body and its three kinds of cells. The collar cells in the inside layer use their flagella to move the water through the sponge. They also catch the food. The amebalike cells in the jelly layer carry food and waste to and from the other cells.

302 Chapter 15 Invertebrates I

Art Connection
The spicules of sponges are beautiful and delicate pieces of the sponge framework. Have students find pictures of different spicules. Some of the spicules are made of silica and others are made of calcite. Have students draw examples of each kind of spicule.

Some sponges, like the one in the diagram, are just one large sac. Others are made up of many very small sacs joined together as one organism. Most sponges have neither radial nor bilateral symmetry and no definite body shape. They are asymmetrical.

Sponges have skeletons that give support to their cells. These skeletons are made of different materials. Some are stiff and made of a glasslike substance or lime. Others are made of a flexible material.

Reproduction of Sponges

Sponges reproduce sexually and asexually. To make new sponges asexually, a sponge may bud or pinch off small sponges that will grow into new sponges. Or, if a piece of a sponge breaks off accidentally, the piece can grow to form another whole sponge.

Look at Figure 15.5. To reproduce sexually, sponges produce both eggs and sperm. Most release their sperm into the water but keep their eggs inside their bodies. When sperm cells from another sponge are drawn into a sponge's body, they are not eaten but are used to fertilize that sponge's eggs. The fertilized eggs stay inside the sponge until they have grown into tiny **larvae** (LAR vee), organisms in an early stage of development that look different from their parents.

The larvae leave the sponge through the large openings and float around. Then they drift to the bottom, attach themselves, and grow into new sponges.

Skills WorkOut
Research Find out more about sponge skeletons, including what they are made of and how they are formed. In what way are they similar to human skeletons?

Figure 15.5
In what stage of its life cycle is a sponge able to move? ②

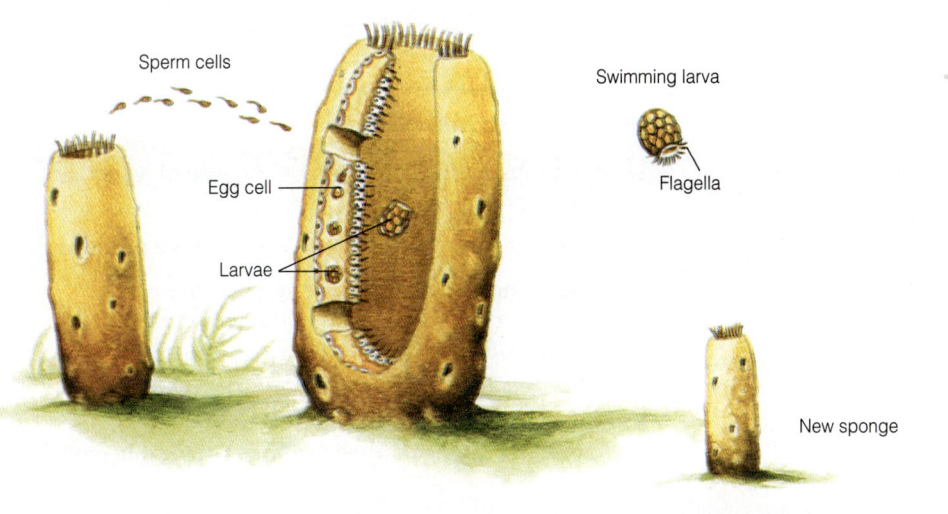

Sperm cells
Egg cell
Larvae
Swimming larva
Flagella
New sponge

Chapter 15 Invertebrates I 303

TEACH

Skills WorkOut
To help students learn more about sponge skeletons, have them do the Skills WorkOut.
Answer Some sponges are made of lime or a glasslike substance. Others are made of a flexible material. Sponge skeletons are similar to human skeletons in that they provide a framework and support for the organism.

Class Activity
Allow students to observe how efficiently sponges filter feed. Have them pour sandy or muddy water through a natural sponge and then through a paper towel or coffee filter and compare how well each filters. Have students infer why filter feeding is a helpful adaptation for sponges. (The organism is able to take in food from all directions by absorbing organic materials in moving water.)

Critical Thinking
Compare and Contrast Have students use their own words to describe how asexual reproduction in sponges is different from sexual reproduction. (In asexual reproduction, a piece from a sponge will grow into another whole sponge. In sexual reproduction, the sperm from a sponge fertilizes the egg from another. Larvae develop, leave the parent sponge, and develop into new sponges.)

Answer to In-Text Question
② Sponges are able to move at the larva stage.

EVALUATE

WrapUp

Reteach Have students use paint and natural or artificial sponges to create a painting of a sponge. Then have them label the three cell types on their sponge paintings and state how each type of cell functions. Have students use their paintings as they describe how sponges reproduce.

Use Review Worksheet 15.2.

Check and Explain

1. The cells lining the inside layer of the sponge use their flagella to move water through the sponge.

2. The sperm are released into the water and are drawn into the body of another sponge.

3. The sponge would be able to move along with the crab. The crab might have more sources of food as the sponge filters large amounts of water.

4. Answers may vary. Student models could include artificial sponges or filter paper, along with ocean or lake water.

The Living Textbook:
Life Science Sides 1-4

Chapter 4 Frame 00749
Sponges (11 Frames)
Search: Step:

304

INTEGRATED LEARNING

STS Connection

Sponges in the ocean are becoming more and more scarce. Artificial sponges, that function like natural sponges, have been developed for human purposes. Have students find out how artificial sponges are made and what materials are used to make them. Discuss the ecological advantages of using artificial sponges.

Figure 15.6 ▶
After live sponges are collected, they are often hung up on boats for a few days (right). They are then cleaned, dried, and trimmed (bottom).

Science and You *Household Sponges*

Since ancient times, people have used sponges for bathing and cleaning because sponges can hold so much water. What do you use a sponge for? Today, most people use artificial sponges made in factories.

Some people, however, still prefer natural sponges. They say that natural sponges hold more water, last longer, and are easier to clean than artificial ones. Artists and some other workers use natural sponges, too. Have you ever seen or used a natural sponge?

Natural sponges used for cleaning have skeletons made of a tough, flexible material called spongin. The living cells are removed before the sponge is sold.

Some natural sponges come from the Mediterranean Sea. Others come from the Caribbean or the Gulf of Mexico. People in boats gather them with hooks or harpoons. Divers collect sponges that live in deeper waters.

Check and Explain

1. What jobs do the cells lining the inside of a sponge's body perform?

2. What happens to the sperm that a sponge produces?

3. **Reason and Conclude** Some sponges live on the shells of crabs. How could living on a crab's shell help a sponge? How could having the sponge there help the crab?

4. **Make a Model** What familiar objects could you use to model a sponge's filter feeding? Draw your model.

304 Chapter 15 Invertebrates I

Themes in Science

Diversity and Unity Although some cnidarians swim freely and others live attached to a surface, all cnidarians share certain common features. These features include a mouth that leads to a digestive cavity, organization at the tissue level, and tentacles.

15.3 Cnidarians

Objectives

▶ **Describe** the characteristics of cnidarians.

▶ **Explain** how cnidarians reproduce.

▶ **Compare** and **contrast** the two body forms of cnidarians.

▶ **Make inferences** about the survival value of cnidarian adaptations.

Skills WarmUp

Interpret Data Choose ten small islands from a map of the South Pacific. Look up the names of these islands in an encyclopedia. How many of the islands are made of coral? What does this tell you about the importance of coral in this ocean?

You're walking along the beach with a friend. Suddenly the two of you come to a large blob of—you're not quite sure what. You start to poke it with your foot. Your friend says, "Hey, don't do that! It's a jellyfish. It might sting you."

Your friend may be right. But the blob on the beach looks nothing like the jellyfish it once was, swimming gracefully in its ocean habitat. A jellyfish is a **cnidarian**, an organism armed with stinging cells.

Characteristics of Cnidarians

Cnidarians are common in most ocean habitats, and a few live in freshwater habitats. A habitat is the area or place where an organism lives. Some cnidarians live attached to a surface and look like flowers or tiny trees. Others, like the jellyfish, swim freely in the ocean. Yet in spite of this diversity, all cnidarians have features in common.

Cnidarians have two main body forms, the *polyp* (PAHL ihp) and the *medusa* (muh DOO suh). You can see these two body forms in Figure 15.7. Medusas swim freely, and polyps usually live attached to a surface. Some cnidarians exist only as polyps. Some are medusas for most of their lives after being polyps for only a short time. In other cnidarians the two forms are equally important in the life cycle.

Both cnidarian body forms have radial symmetry. They look the same from all sides. This type of body plan helps cnidarians detect food or enemies from any direction.

Cnidarians have several features that make them different from sponges. First, cnidarians have a mouth that leads to a digestive cavity. The mouth is the cavity's only opening. Cnidarians also have tissues. The cells

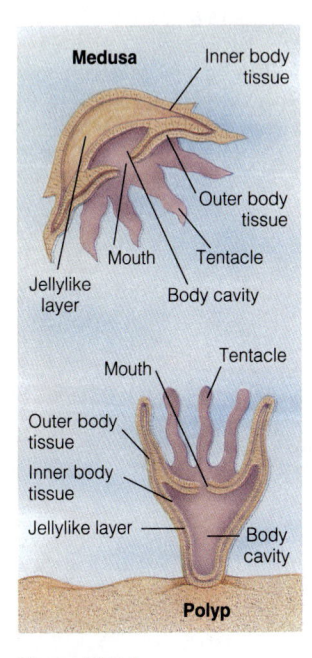

Figure 15.7 ▲
How is a cnidarian polyp similar to a medusa? ①

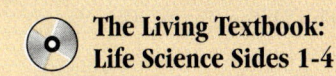

Chapter 15 Invertebrates I **305**

TEACH

Portfolio
Have students draw and label a polyp and a medusa. Have students draw lines through each figure showing whether each body type has radial symmetry, bilateral symmetry, or no symmetry. (They both have radial symmetry.) Have students place their drawings in their portfolios.

Critical Thinking
Compare and Contrast Have students compare Figures 15.4 and 15.7. Have them discuss how the physical features of cnidarians differ from the features of sponges. (Cnidarians, unlike sponges, have a mouth, tissues, tentacles, and stinging cells.)

Discuss
Have students study Figure 15.8. Ask the following questions:

▶ How many organisms are shown in the picture? (One. The picture shows a single jellyfish at various stages.)

▶ Where do the egg and sperm come from? (From a female and a male jellyfish medusa)

▶ What are some differences between the polyp and the medusa? (They differ in appearance. The medusa can move and the polyp can't.)

The Living Textbook: Life Science Side 3
Chapter 20   Frame 31821
Jellyfish/Sea Anemone (Movie)
Search: Play:

306

INTEGRATED LEARNING

Writing Connection
Have students write a poem about one day in the life of a cnidarian. How does it find food? How does it protect itself from predators? Encourage students to include descriptions of the tentacles and stinging cells, as well as how the cnidarian's body shape helps it hunt food and protect itself.

---

covering the outside of a cnidarian make up one kind of tissue. The cells lining the inside are a different kind. Both contain simple muscles and nerves. The two layers are separated by jellylike material.

Cnidarians' most striking adaptation is their tentacles, which are lined with stinging cells. Stinging cells help cnidarians defend themselves and catch food. When prey is stunned or killed by the stinging cells, the tentacles carry the food to the mouth.

Reproduction of Cnidarians

Cnidarians can reproduce both sexually and asexually. When it occurs, asexual reproduction is carried out by polyps. A small piece of the animal breaks off and grows into a new polyp. This process is called budding.

To reproduce sexually, cnidarians release eggs and sperm into the water. Each fertilized egg forms a larva. The larva grows into a new polyp or medusa.

Look at the life cycle of the jellyfish in Figure 15.8. Notice that a medusa and a polyp are part of the cycle. Many cnidarians have a life cycle that has both a medusa stage and a polyp stage. In these cnidarians, it is the medusa that reproduces sexually. In cnidarians without a medusa stage, the polyp reproduces sexually.

Figure 15.8
Jellyfish have a life cycle that includes a medusa stage and a polyp stage. Each polyp produces many young medusas.

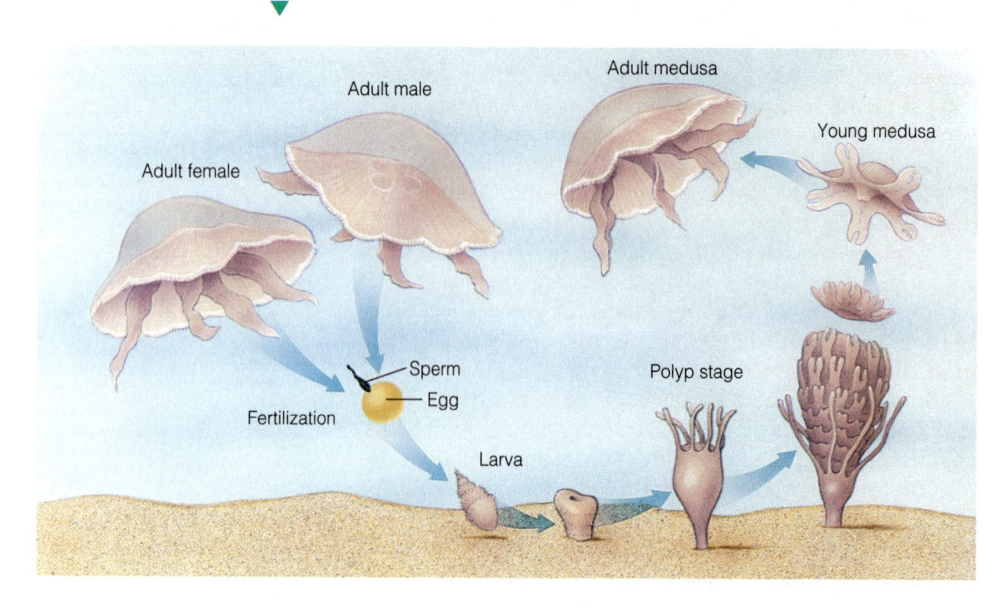

306   Chapter 15 Invertebrates I

TEACHING OPTIONS

Cooperative Learning

Have students work in groups of three or four. If possible, have each group use a hand lens or a microscope to observe a few common cnidarians such as a *Hydra*. Tell students they will need to classify the cnidarian, using this page as a guide. After all group members have observed the organism, ask each group to infer in which category their cnidarian belongs. Have each group make a large scale drawing of its cnidarian, labeling it according to the category in which it belongs and providing reasons why the category was chosen. Have students share their conclusions in a class discussion.

Explore Visually

Have students study this page. Then ask the following questions:

▶ Of the five kinds of organisms pictured, which three belong to the same main group? (Sea anemones, reef-building corals, and staghorn corals)

▶ To which group do you think the *Hydra* belongs? Why? (Students may suggest that the *Hydra* is a hydrozoan because of the similarities between the two names.)

▶ Which two groups of cnidarians have both polyp and medusa stages in their life cycles? (Hydrozoans and jellyfish)

▶ Which cnidarians pictured make hard skeletons? (Corals)

▶ How are the cnidarians divided into groups? (According to different life cycles)

▶ How do the tentacles of sea anemones protect certain neighboring fish that live safely among them? (The anemones' stinging cells do not harm the neighboring fish, but do harm other fish that prey upon the neighboring fish. These fish are thereby protected by the anemones.)

Diversity of Cnidarians

Cnidarians are divided into three main groups. Each has a different kind of life cycle. Within each group there are many shapes and sizes, and differences in habitat. Even though all the cnidarians you see on this page live in the ocean, your nearest cnidarian may be only as far away as the closest pond, river, or stream. Those are places where the *Hydra* lives. It is one of the few freshwater cnidarians.

◀ **Hydrozoans**
This sea fan belongs to a second group called the hydrozoans (HY droh ZOH uhnz). A single fan contains many polyps. Most hydrozoans have both polyp and medusa stages.

Sea Anemones
One group of cnidarians includes the sea anemones (uh NEHM uh NEES). Their scientific name means "flower animals." They are polyps all their lives. Some fish can live safely inside the tentacles of sea anemones. The anemones protect the fish. The fish, in turn, lure other fish that the anemone might eat. ▼

Reef-building Corals ▲
Tropical reefs are formed by tiny cnidarians called corals. Corals are in the same group as sea anemones. Reef corals spend their whole lives as tiny polyps living together in colonies. They make hard skeletons of calcium.

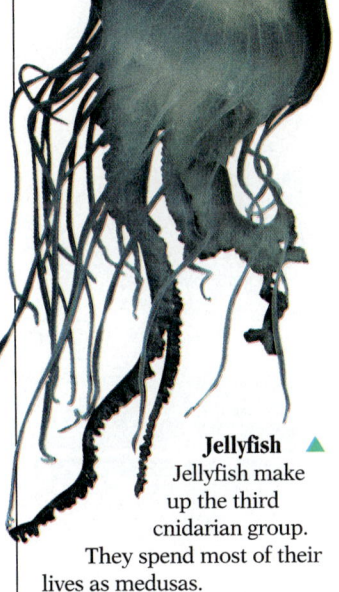

Jellyfish ▲
Jellyfish make up the third cnidarian group. They spend most of their lives as medusas.

Staghorn Coral
This staghorn coral has a shape very different from that of the reef coral. Yet it, too, is made of tiny polyps and a hard skeleton. Corals like the staghorn live on coral reefs but do not themselves help to form the reef. ▼

The Living Textbook: Life Science Side 4

Chapter 11 Frame 14046
Coral (Movie)
Search: Play:

Chapter 15 Invertebrates I

TEACH • Continued

Class Activity
Have students work in small groups to build a model of one of the three kinds of coral reefs. Provide clay, cardboard, and paints. Have them study pictures and drawings of coral reefs so that they can add details to their models. Encourage students to be creative.

Critical Thinking
Find Causes Ask students how pollutants in ocean water could keep coral reefs from forming. (Coral polyps are living organisms. Pollution may affect their ability to obtain nutrients and other substances from ocean water.)

Research
The Great Barrier Reef, which extends for 2,000 kilometers off the coast of Australia, is the largest structure ever built by living organisms. The reef is really made up of many smaller reefs that are bound together by the remains of algae and protozoa. There are at least 350 species of coral that make up the reef. Have students research the other kinds of animals found there. Students can present their findings in oral reports. Encourage them to bring pictures of the Great Barrier Reef and the animals that live there to share with the class.

Enrich
The stinging cells of most cnidarians are too weak to break through human skin. Two exceptions are the Portuguese man-of-war and the sea nettle. Have students research other harmful cnidarians and discuss how to avoid contact with these stinging organisms at the beach.

INTEGRATED LEARNING

Integrating the Sciences

Earth Science Point out to students that temperature and light are the factors that most affect the development of coral reefs. This is why coral reefs are found in the clear warm waters of the Caribbean and in the Indian and Pacific oceans. Algae cells that live in tissues of coral absorb light to produce food for the corals. These animals in turn grow and secrete calcium carbonate, which is necessary for the build-up of major coral reefs. Temperature also affects the animals' secretion of calcium carbonate.

Use Integrating Worksheet 15.3.

Figure 15.9 ▲
There are three kinds of coral reefs. A fringing reef is closer to shore than a barrier reef. An atoll is a reef atop a sunken volcanic island.

Coral Reefs

Coral reefs are found in warm ocean waters around the world. They provide homes for a greater variety of living things than any other place in the ocean. Reefs can be very large. The Great Barrier Reef off the coast of Australia is over 2 000 km long. Many tropical islands were once coral reefs that have been left dry by changing sea levels.

It is hard to believe that these large and important structures are built by tiny cnidarians. The surface of all living reefs is coated with coral polyps, each with its own hard skeleton. Below the surface are layers and layers of skeletons left by coral polyps that are now dead. The reef grows as new live corals replace ones that die on the surface.

One reason corals can produce huge reefs is that they have help from algae. Algae live inside the polyps and make part of the polyps' food. The algae also give the coral its color. Coral must therefore live in shallow water so that their algae can get enough sunlight.

In many places where there are coral reefs, scientists are worried that the coral polyps are dying off. They are not being replaced by new living corals. Scientists are not sure why this is happening, but some blame pollutants in the ocean water.

Science and You *Stinging Jellyfish*

Each cnidarian stinging cell contains a tiny harpoon at the end of a coiled, hollow thread. When something touches the cell, the long thread uncoils suddenly, and the harpoon shoots into whatever touched the cell. While jagged barbs hold the harpoon in place, a poison passes through the hollow thread. Any animal that touches a cnidarian is hit by thousands of these tiny poisoned harpoons.

The stinging cells of most cnidarians are too weak to break through human skin. Some hydrozoans and jellyfish have stinging cells that can harm people, however. The Portuguese man-of-war, a hydrozoan, and the sea nettle, a jellyfish, are examples of cnidarians harmful to humans.

If you encounter a jellyfish on the beach, it's best to avoid it. If you get tentacles from a Portuguese man-of-war on you, try to put vinegar on the tentacles as soon

308 Chapter 15 Invertebrates I

Social Studies Connection

Remind students that they learned about Charles Darwin in Chapter 7. Point out that from observing fringing reefs, barrier reefs, and atolls, Darwin formed a theory about how coral atolls developed. He suggested that the three kinds of reefs represented three stages in reef development. Corals growing along the shore of a volcanic island form a fringing reef. After many years the island sinks and the coral continues to grow upward. Eventually, this results in the formation of a barrier reef. In the final atoll stage, the island continues to sink beneath the water's surface and is covered by coral. Oceanographers have confirmed that some atolls appear to have formed by the process described by Darwin.

Consider This

Should Protection of Coral Reefs Come First?

The kind of climate reef corals need to grow is also the kind of climate humans like to live in. On Key Largo, an island off the coast of Florida, developers want to build a new community called Port Bougainville. Environmentalists, however, say that building the community would endanger nearby coral reefs.

Consider Some Issues Scientists believe that human activity is damaging coral reefs worldwide. People damage reefs by polluting the ocean water with sewage, by anchoring boats on the reefs, and by letting logging and digging on land wash dirt into the sea.

Coral reefs are home to an amazing array of living things, and they protect shores from erosion by waves. For these reasons and others, some people argue that reefs should be protected by limiting what humans can do in the areas around them.

Others point out that protecting the reefs has costs, too. Resorts such as Port Bougainville provide jobs and places for people to live. Other activities that damage reefs, such as tourism, logging, and fishing, are also important to the economy.

Think About It How can the needs of people be balanced against the need to protect the reefs?

Debate It Have a class debate in which opposing teams discuss the pros and cons of the Port Bougainville Project.

as possible. The vinegar will keep more stinging cells from firing. A paste made of baking soda and water will do the same for the sea nettle. Meat tenderizer can also stop jellyfish stings because it chemically changes the poison.

Check and Explain

1. What features do cnidarians have that sponges do not?
2. In which cnidarian body form does asexual reproduction occur?
3. **Compare and Contrast** How are the structures of polyps and medusas alike and different?
4. **Infer** How do you think each of the following help a jellyfish survive: being transparent, having stinging cells, having radial symmetry, being able to swim?

Chapter 15 Invertebrates I 309

Consider This

Research Some scientists have found evidence that many of the world's coral reefs are dying at an unnaturally rapid rate. A recent study of the Florida Keys coral reef indicated that parts of the reef are dying at the rate of 10 percent each year.

Think About It Have students investigate the current status of the Florida Keys reef. Have them find out what animals make up the reef's ecosystem.

Debate It Before students debate, you may wish to discuss some guidelines. Remind students to take turns and to listen to each other. One student may wish to act as moderator.

EVALUATE

WrapUp
Portfolio Have students draw an ocean community, including sponges, a coral reef, and all of the different types of cnidarians they have studied. Have them label the animals and briefly describe how each reproduces, eats, moves, and protects itself.

Use Review Worksheet 15.3.

Check and Explain

1. Cnidarians have a mouth that leads to a digestive cavity, tissues, and tentacles. Sponges have none of these features.
2. In the polyp
3. Polyps and medusas are both radially symmetrical. Polyps are attached to a surface while medusas can move freely.
4. Being transparent may allow some jellyfish to hide from predators; stinging cells are for self-defense and for trapping prey; radial symmetry helps it detect food or enemies from any direction; swimming helps it avoid predators and find food.

SECTION 15.4

Section Objectives
For a list of section objectives, see the Student Edition page.

Skills Objectives
Students should be able to:

Communicate observations of earthworms.

Estimate the number of earthworms in a given area of soil.

Vocabulary
flatworm, roundworm, segmented worm, coelom

MOTIVATE

Skills WarmUp
Reinforce students' prior knowledge about earthworms by having them do the Skills WarmUp.
Answer Earthworms are shiny, brown, long, and thin. They live in soil and emerge when it rains.

Prior Knowledge
To gauge how much students already know about worms, ask the following questions:

▶ How do worms differ from cnidarians?

▶ What are the three main types of worms?

▶ How are worms helpful to humans and how are they harmful?

Answers to In-Text Questions
① The body of the earthworm is shiny, brown, long and thin, and segmented.

② The earthworm belongs to the phylum known as segmented worms.

310

INTEGRATED LEARNING

Themes in Science
Evolution The evolution toward greater complexity in invertebrates is reflected in the development of specialized organs in worms. These organs, along with a defined head and a bilaterally symmetrical body, are what distinguish the structures of worms from the simpler structures of cnidarians. Such adaptations enable worms to live in a greater variety of habitats.

Skills WarmUp
Communicate Recall all the times you have seen earthworms. Based on these observations, write a paragraph describing earthworm structure and behavior.

15.4 Worms

Objectives

▶ **Explain** how worms differ from cnidarians.

▶ **Identify** and **describe** the three major worm phyla.

▶ **Explain** the importance of a segmented worm's coelom.

▶ **List** worms that live as parasites in humans.

▶ **Estimate** the size of a worm population based on a sample.

When it rains all night, what might you find on the streets and sidewalks in the morning? Worms! Have you ever wondered why they leave their homes in the soil? The rain floods their underground burrows. They escape drowning by crawling to the surface.

Earthworms are just one of the many kinds of worms in the animal kingdom. Other kinds of worms are too small to see. Some build hard shells and live in the ocean. In fact, worms live almost everywhere and have a large impact on human life.

Characteristics of Worms

Worms differ in two important ways from cnidarians. First, they have bilateral symmetry and a definite head. Nerves and sense organs are concentrated in the head. Second, instead of just two layers of tissue, they have three. And the middle layer of tissue, called the mesoderm (MEHS oh DURM), forms organs that serve particular functions.

With this added complexity, worms can live in a greater variety of habitats than cnidarians. Like cnidarian medusas, some worms swim. Like polyps, some worms attach themselves to one place. But worms also burrow in mud, crawl on surfaces, and live in other organisms. And some, like the earthworm, can live in soil.

Scientists place worms in several main groups, or phyla. A worm's body structure determines which phyla it belongs to. The three most common worm phyla are flatworms, roundworms, and segmented worms. Which group do you think earthworms belong to? ②

Figure 15.10 ▲
What do you notice about the body of this earthworm? ①

310 Chapter 15 Invertebrates I

Themes in Science

Diversity and Unity Scientists classify worms in phyla according to body type. The three main phyla are flatworms, roundworms, and segmented worms. Some worms can live in soil, some in water, some in other organisms. All worms, however, share certain characteristics such as bilateral symmetry, a head containing nerve and sense organs, and a middle layer of tissue called the mesoderm.

Flatworms

Worms with a flattened shape are called **flatworms.** Flatworms have the simplest body organization of any worm. They have organs and three layers of tissue. But like cnidarians, their digestive cavity has only one opening.

Flatworms are either free-living or parasitic. A common flatworm that lives in fresh water, the planarian, is a good example of the free-living type of flatworm. Most kinds of planaria are only a few centimeters long. They eat tiny animals in the water and are harmless to humans.

Look at Figure 15.12. It shows the structure of a planarian. The eyespots on the top of the head do not transmit images. Instead, they sense light. The nerve cells in the head form a simple brain. Food enters through the mouth, and undigested food leaves through the same opening.

A single planarian is both male and female. When two planaria exchange genetic material, each fertilizes the other. Planaria reproduce asexually by dividing their bodies in two. Each half then grows back the missing part. If you cut a planarian into several pieces, each piece will regenerate into a new worm.

There are two groups of parasitic flatworms: flukes and tapeworms. Flukes mainly live in the body tissues of animals. Some live in human tissues and cause sickness. Most tapeworms live in the intestines of different vertebrates, including humans. The tapeworm drops pieces of its body, full of fertilized eggs, into its host's waste. If another animal eats the tapeworm pieces, worms will grow in that animal's body. Tapeworms have no digestive cavity because they absorb the digested food of their host.

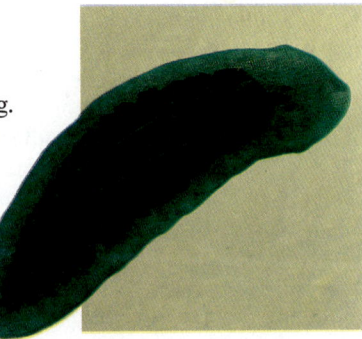

Figure 15.11 ▲
A planarian is about 2.5 cm long. What is the function of the eyespots? ③

Figure 15.12
Planaria have a simple body structure. ▼

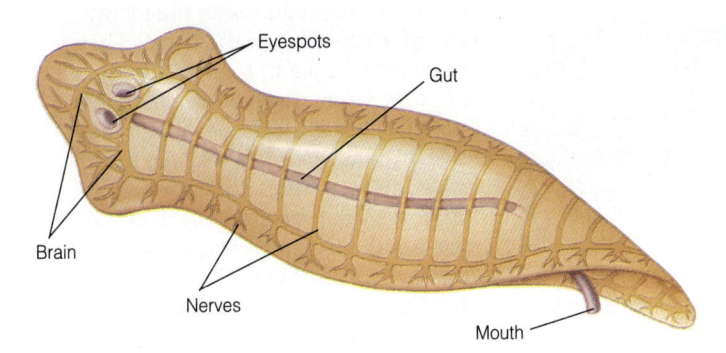

Chapter 15 Invertebrates I

TEACH

Discuss
Have students study Figure 15.12. Ask the following questions:
▶ Where is the planarian's mouth? Is it close to the head? (The mouth is at the opposite end of the body from the head.)
▶ Food enters through the planarian's mouth. How does waste leave the planaria's body? (Through the same opening)

Research
Explain that animals, including dogs and cats, commonly suffer from tapeworms and flukes. Have each student choose an animal and find out how flukes and tapeworms affect it. Ask students:
▶ What are the animal's most common parasite problems?
▶ How do they typically get inside the animal? (Tapeworm eggs must enter through the mouth and pass through the digestive system to reach the intestine.)

Integrated Learning
Use Integrating Worksheet 15.4.

Answer to In-Text Question
③ The eyespots function as light sensors.

The Living Textbook: Life Science Side 4

Chapter 11   Frame 14516
Fluke Life Cycle (Movie)
Search: Play:

The Living Textbook: Life Science Side 4

Chapter 12   Frame 23455
Marine Flatworm (Movie)
Search: Play:

TEACH • Continued

Class Activity
If possible, have students view common worms, such as planaria, with a microscope or hand lens. Have students work in groups of three or four. Group members should record and discuss their observations.

Discuss
Ask students if they think roundworms can be seen without a microscope. (They can be seen only with a microscope.) What do they eat? (Microscopic organisms and other tiny bits of organic matter in soil)

The Living Textbook: Life Science Sides 1-4

Chapter 4   Frame 00834
Roundworms (12 Frames)
Search: Step:

INTEGRATED LEARNING

STS Connection
Point out to students that some species of nematodes cause serious damage to crops. These organisms can cause severe losses in fruit, vegetable, and ornamental plant crops. There are several methods that farmers use to control harmful nematodes. One involves the use of chemicals called nematicides. Another involves the application of heat to confined areas of soil. Have students write to a representative from the Department of Natural Resources to find out if nematodes are a problem in your area and describe methods for controlling them.

Roundworms

Go outside and pick up a handful of moist garden soil. What are you holding? In addition to the soil, you may have in your hand about one million tiny worms. These worms have round bodies and belong to the **roundworm** group, also called nematodes (NEHM uh TOHDS).

Besides their round shape, roundworms have something no flatworm has: a one-way digestive tube with two openings. Look at Figure 15.13. Food enters a roundworm through the mouth located at the head. As it passes through the digestive tube, it is digested and then absorbed. Undigested food leaves the tube through the anus. In roundworms, different parts of the gut can have different functions because food moves in only one direction.

Roundworms reproduce sexually. Unlike many flatworms, the sexes are separate. In many species, females are larger than the males and can produce over 100,000 eggs in a day.

Many roundworm species attack the roots of plants. Some of their favorite plants are crops grown by humans. On the other hand, some roundworms benefit crops by killing insect pests.

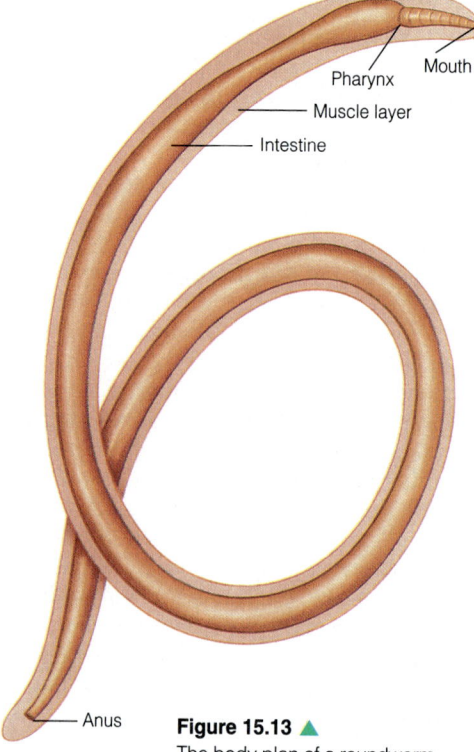

Figure 15.13 ▲
The body plan of a roundworm is often called a "tube within a tube."

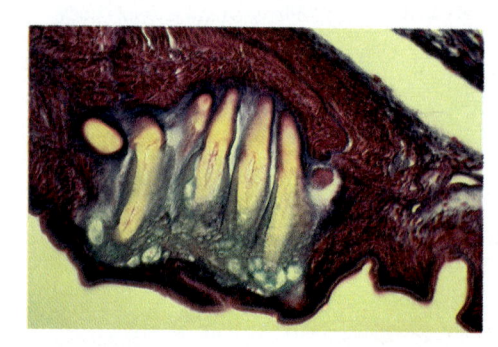

Figure 15.14 ▲
In this photo you can see a *Trichinella* inside muscle tissue.

About 50 kinds of roundworms are parasites in humans. Hookworms enter a human host by boring through the skin of the feet. Using the bloodstream, they move to the lungs. They are coughed up and then swallowed. In the intestines, they attach themselves and feed on digested food. When they have grown into adult worms, they reproduce and send eggs out with the host's waste to start the cycle over. Another roundworm parasite is *Trichinella*, which causes trichinosis. It can enter your body if you eat undercooked meat.

312   Chapter 15 Invertebrates I

Segmented Worms

Earthworms are common examples of a third worm group, the **segmented worms**. Segmented worms have bodies divided into small units or segments. A segmented body is like a beaded necklace. Just as you could thread together a large red bead, five blue beads, and eight yellow beads, a segmented worm's body is built of similar and different segments.

The segmented worms are unlike flatworms and roundworms because they have a fluid-filled space, or cavity, inside the middle tissue layer. This cavity is called a **coelom** (SEE luhm). Look at Figure 15.15 to see where the coelom is in a segmented worm's body.

A coelom provides the space needed for complex organs. The segmented worms were the first group of animals to evolve organ systems, groups of specialized organs working together.

The common earthworm is a good example of how organ systems are arranged in a segmented worm's body. Look at the earthworm in Figure 15.16. Notice the crop and gizzard. These organs help the worm digest the soil it eats. They grind up the soil and remove organic matter from it. Notice also that the earthworm has five hearts and a brain. The brain is connected to a nerve cord that runs all the way down the worm's body.

Earthworms can move in a variety of ways. Stiff bristles on the outside of the body act as anchoring points. They hold a part of the worm in place while another part pushes forward through the soil.

In these ways and others, an earthworm is well adapted to life in the soil. Its lifestyle is made possible by its segmented organization and its organ systems.

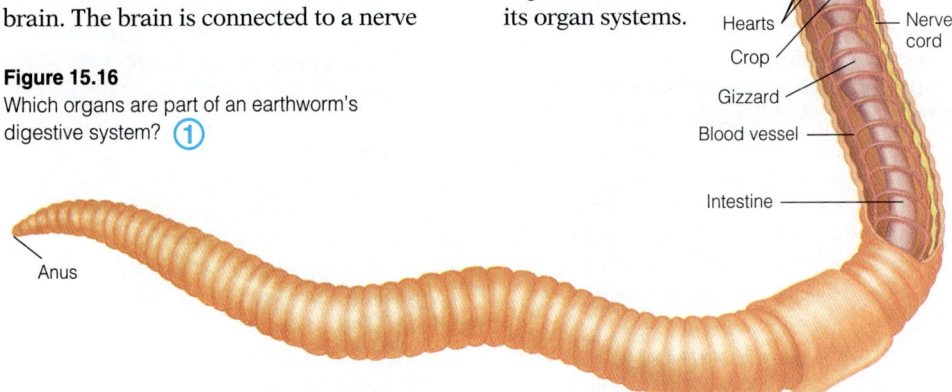

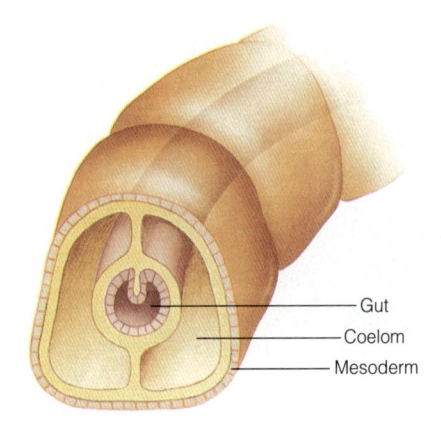

Figure 15.15 ▲
What is the function of the coelom?

(Labels: Gut, Coelom, Mesoderm)

Figure 15.16
Which organs are part of an earthworm's digestive system? ①

(Labels: Mouth, Brain, Hearts, Crop, Gizzard, Blood vessel, Intestine, Nerve cord, Anus)

Chapter 15 Invertebrates I 313

Class Activity

If possible, have students use hand lenses to examine live earthworms. Have them try to identify which end of the worm is the head. Have them note the bristles on the outside of the worm's body. Encourage students to notice the ways in which the earthworm moves.

Critical Thinking

Infer Ask students to infer why the earthworm has five hearts. (The hearts are simply enlarged blood vessels that help pump blood. Five hearts are needed to circulate the blood throughout the worm's long body.)

Compare and Contrast Have students compare the structure of flatworms, roundworms, and segmented worms. Then ask which kind of worm is the most complex of the three groups. Why? (Segmented worms are the most complex. They have a coelom and organ systems.)

Answer to In-Text Question

① The mouth, crop, gizzard, and intestine are organs that are part of the digestive system of the earthworm.

**The Living Textbook:
Life Science Side 3**

Chapter 20 Frame 35803
Proboscis/Sea Mouse Worms (Movies)
Search: Play:

**The Living Textbook:
Life Science Side 4**

Chapter 12 Frame 23676
Giant Earthworm (Movie)
Search: Play:

TEACH • Continued

Discuss
Some students may have difficulty understanding how earthworms can do so much to make soil fertile. Stress to students that fertile soil is full of living organisms. Worms help provide the organic nutrients that these organisms need.

SkillBuilder
Research Have students do some research on the sizes of invertebrates. They may be surprised to learn, for example, that tapeworms can grow to great lengths. Ask students to consider how the size of an invertebrate might affect how it adapts to its habitat. They can use this information as they create their invertebrate models.
Answers Students' model designs should address all the questions listed. In particular, the models should reflect an understanding of how the invertebrate interacts within its specific habitat.

Answer to In-Text Question
① Earthworms enrich and improve the soil.

The Living Textbook:
Life Science Sides 1-4

Chapter 4   Frame 00984
Segmented Worms (48 Frames)
Search: Step:

The Living Textbook:
Life Science Sides 1-4

Chapter 5   Frame 02139
Earthworms (18 Frames)
Search: Step:

314

INTEGRATED LEARNING

Integrating the Sciences

Earth Science Stress that if it were not for earthworms, the soil in most parts of the world would not be nearly as fertile as it is. To a large extent, a soil's fertility depends on its earthworm population. In 20 years, earthworms can increase topsoil by 10 centimeters. Have students discuss the ways in which earthworms enhance soil fertility. (Earthworms move organic matter throughout the soil, they add nitrates to the soil, and they dig tunnels, allowing air and water to reach deep into the soil.)

Earthworms and Soil

Earthworms help enrich and improve the soil. They eat organic matter in the soil that plants can't use directly. They digest this matter and take nutrients from it. What is left over are simpler substances, such as nitrates, that plants can use. Earthworms eliminate this waste from their bodies, adding it to the soil. Pellets of earthworm waste, called castings, are so full of plant nutrients that people use them as garden fertilizer.

Earthworms also improve the soil by burrowing in it. Their tunnels allow air and water to reach deep into the soil. They move organic matter from the surface down into deeper soil levels.

Many hundreds of earthworms can live under the surface of a square meter of ground. Each earthworm eats and discards its own weight in soil every day. Because it is full of living things like earthworms, soil is constantly changing.

Figure 15.17 ▲
In what ways do earthworms affect the soil they live in? ①

SkillBuilder   *Making a Model*

Design an Invertebrate

An animal is adapted to the habitat in which it lives. The characteristics of particular animals, therefore, depend on the characteristics of their habitats.

Choose a habitat. It can be a familiar one, such as a forest, a desert, or a coral reef. Or it can be a strange one, such as the deep sea, the atmosphere, or even another planet. Make notes about the features of this habitat. Is it hot or cold? Wet or dry? How much sunlight does it receive? What is there to eat? What plants grow there? Are there predators to watch out for? Are there any other dangers such as waves or high winds?

Now design an invertebrate to fit the habitat you chose. Think about the features that your animal will need to do well in this habitat. As you design your invertebrate, think about these questions:

1. Will your invertebrate have radial or bilateral symmetry, or will it have no symmetry?
2. How big will it be?
3. What adaptations will it have?
4. Where will it fit in the animal kingdom? What living animals are its closest relatives?
5. What will it eat? How will it get its food?
6. What sense organs will it have?
7. How will it move?
8. How will it reproduce?
9. How will it defend itself?

Write a short report describing your animal and explaining how it is adapted to its environment. Include pictures of your animal, showing what it looks like and what internal organs it has. Present your invertebrate design to the class.

314   Chapter 15 Invertebrates I

TEACHING OPTIONS

Cooperative Learning

Assign students to cooperative groups. Have them work to investigate the health problems caused by one of the parasitic worms. One group member can find information about the worm, its life cycle, and where it is found. Other students can investigate methods used to prevent parasitic diseases. Others can construct a diagram or three-dimensional model of the worm. Have student groups prepare brief, written reports that they can display with their models.

Science and Society *Worms as Parasites*

Several kinds of worms live as parasites in human beings. They include tapeworms, a kind of flatworm, and roundworms, such as hookworm and *Trichinella*. Some of these worms spend part of their lives in the bodies of other animals. *Trichinella* worms live in pigs, for example, and can be present in the pork eaten by people.

Parasitic worms take nutrients from a person's body. As a result, people with worms do not have enough nutrients for their own needs. Such people often feel weak and tired. They have lowered resistance to infections from microorganisms. They usually have digestive problems. *Trichinella* worms live in muscle tissue and cause muscle pains.

There are things you can do to prevent parasitic worms from entering your body. For example, if you wear shoes outside, hookworms can't bore into your feet. If you are careful to keep human and animal waste from food, you can't get tapeworms. If you eat only well-cooked meat, you can't get trichinosis.

Few people in the United States today suffer from parasitic worms. These worms still weaken people in other parts of the world, however. They are a special problem in tropical countries where, for example, most people can't afford sanitary ways to dispose of human waste. Medicine can get rid of the worms, but people quickly become reinfected.

Figure 15.18 ▲
This tapeworm was removed from the intestine of a cat.

Check and Explain

1. How do flatworms differ from cnidarians?
2. Name the three main groups of worms. For each, describe how it differs from the others.
3. **Predict** Suppose you removed all the earthworms from an acre of soil. What would happen to the soil? Why?
4. **Estimate** You dig up a small area of moist soil 10 cm wide, 10 cm long, and 10 cm deep and find five earthworms. Assuming that this number is average, how many earthworms would there be in a square plot of land 100 m on a side and 1 m deep?

Enrich

Students may be interested in knowing more about the medical uses of leeches. In the 1800s, leeches were commonly applied to the temples of persons suffering from headaches. Leeches were also used to treat mental illness, gout, skin disease, and whooping cough. Today, the anticoagulant hirudin is used medically. The drug is produced from the body tissues of the medicinal leech *Hirudo medicinalis*.

Reteach
Use Reteach Worksheet 15.4.

EVALUATE

WrapUp
Reinforce Have students make a chart of the different types of invertebrates they have studied in this chapter. (Sponges, cnidarians, and worms) Have them add the names of the species to the chart and note the main characteristics of each. Students should include details about eating, reproduction, movement, and structure.

Use Review Worksheet 15.4.

Check and Explain

1. Flatworms differ from cnidarians in that they have bilateral symmetry, a mouth, and a mesoderm.

2. The three groups are flatworms, roundworms, and segmented worms. Flatworms have the simplest body organization and have male and female organs. Roundworms have a one-way digestive tube with two openings and the sexes are separate. Segmented worms have a fluid-filled space inside their bodies called a coelem.

3. Answers may vary. Soil would become packed together and hardened. There would be fewer nitrates in the soil.

4. 50,000,000

Chapter 15 Invertebrates I

ACTIVITY 15

Time 1 week **Group** 3–4

Materials

earthworms
3 kinds of soil, 6 L total
10 1-L jars
3 L humus
1.5 L apple peelings
dark construction paper
rubber bands

Analysis

1. Answers should agree with recorded observations; most will observe tunnels from the start.
2. The peelings mix with humus and begin breaking down.
3. The layers of soil should gradually mix to some extent.
4. Answers will vary, but worm castings and trails should appear.
5. Answers will depend on the types of soil available. An experiment to test preference could involve using pairs of soil types in a single container, altering the top and bottom positions, and observing any preference for one or the other soil type or layer.

Everyday Application

Earthworms help break down organic matter, returning nutrients to the soil. By burrowing, they help get water deep into the ground. Making sure the garden is kept watered, and that the soil is soft and contains humus are ways students might suggest to encourage the presence of earthworms.

Extension

Accept logical hypotheses and procedures. Experimental designs should show an understanding of the need to control all possible variables other than plant growth. Make sure that the procedure actually tests the stated hypothesis.

TEACHING OPTIONS

Prelab Discussion

Have students read through the entire activity. Then ask the following questions:

▶ Why is it important to keep the soil moist at all times?

▶ What are the characteristics of the three different types of soil being used in the activity?

▶ What are the characteristics of humus and why is its presence in soil important?

▶ Why is it important to choose an appropriate number of worms for the container?

Activity 15 How do earthworms change soil?

Skills Measure; Observe; Infer; Interpret Data

Task 1 Prelab Prep

1. Collect the following items: earthworms, 200 mL each of three kinds of soil with different colors and textures, a 1 L jar, 250 mL of humus (partly decayed leaves and other plant matter), 125 mL of apple peelings, dark construction paper, a rubber band.
2. To collect the earthworms, carefully dig them up yourself or buy them from a bait shop.

Task 2 Data Record

1. On a sheet of paper, draw the data table shown.
2. Record all your observations about the earthworms in the data table.

Task 3 Procedure

1. Put a layer of each kind of soil in the jar.
2. Wet the soil. Be sure that the soil is damp throughout.
3. Spread the humus over the soil.
4. Place the worms in the jar.
5. Add the apple peelings.
6. Wrap the paper around the jar and fix it in place with the rubber band. Put the jar in a cool place.
7. Remove the paper and observe the jar once every day for a week. Each time, record your observations and then return the jar to its place. **CAUTION!: Keep the soil moist at all times. Add water when needed, but do not make the soil soggy.**

Task 4 Analysis

1. **Observe** How soon do tunnels appear in the soil?
2. What happens to the apple peelings and humus?
3. What happens to the layers of soil? Do they stay as separate layers, or are they mixed?
4. What appears on the surface of the soil?
5. **Infer** In which soil layer do the earthworms spend the most time? Do they like this layer because of the soil type or because of its position? How could you find out?

Task 5 Conclusion

Write a short paragraph explaining how earthworms live in the soil and how they change the soil.

Everyday Application

If you were planting a garden, why might you want earthworms in it? How could you encourage earthworms to live there?

Extension

Based on what you have learned in the activity, develop a hypothesis about how well plants grow in soil that contains earthworms as compared to soil that does not contain earthworms. Write down your procedure and test your hypothesis. Make a chart of your comparisons.

Table 15.1 Earthworm Observations

Date	Time	Observations

CHAPTER REVIEW 15

Chapter 15 Review

Concept Summary

15.1 The Animal Kingdom
▶ Animals have cells, tissues, organs, and organ systems.
▶ All animals are many-celled, ingest their food, and move during some part of their life cycle.
▶ Animals can have radial symmetry, bilateral symmetry, or no symmetry.
▶ Animals with backbones are vertebrates; animals without backbones are invertebrates.

15.2 Sponges
▶ Sponges are filter feeders that have no definite body shape and no symmetry.
▶ Sponges have cells to move water through their bodies, to catch food, and to carry food and waste to and from other cells.
▶ Sponges reproduce both sexually and asexually.

15.3 Cnidarians
▶ Cnidarians have radial symmetry in both of their two forms: polyp and medusa.
▶ Cnidarians have two layers of tissue and stinging cells.
▶ Cnidarians reproduce both sexually and asexually.
▶ Cnidarians have tentacles that help them get food. Tentacles also help cnidarians defend themselves.

15.4 Worms
▶ Worms have bilateral symmetry, a head with sense organs, and three layers of tissue.
▶ The three major worm phyla are flatworms, roundworms, and segmented worms.
▶ Segmented worms have a coelom that provides space for complex organs.

Chapter Vocabulary
vertebrate (15.1)   larva (15.2)   flatworm (15.4)   segmented worm (15.4)
invertebrate (15.1)   cnidarian (15.3)   roundworm (15.4)   coelom (15.4)

Check Your Vocabulary

Use the vocabulary words above to complete the following sentences correctly.

1. An organism armed with stinging cells is called a ____.
2. An earthworm is a ____; its body is divided into small segments.
3. Animals without backbones are classified as ____.
4. A worm with a round shape and a one-way digestive tube with two openings is a ____.
5. Organisms in an early stage of development that look different from their parents are called ____.
6. Only segmented worms have a ____, or cavity with space for organ systems.
7. Animals with backbones are classified as ____.
8. A worm with a flat shape, such as a planaria, is a ____.

Write Your Vocabulary

Write a sentence for each vocabulary word above. Show that you know what each word means.

Check Your Vocabulary
1. cnidarian
2. segmented worm
3. invertebrates
4. roundworm
5. larvae
6. coelom
7. vertebrates
8. flatworm

Write Your Vocabulary
Students' sentences should show that they know both the meanings of the words and how to apply the words.

Use Vocabulary Worksheet for Chapter 15.

Chapter 15 Invertebrates I

CHAPTER 15 REVIEW

Check Your Knowledge

1. The two main body forms are medusa and polyp.
2. Earthworms enrich and improve the soil (for plants) by breaking down organic soil matter and by boring holes deep into the soil.
3. Animals obtain food by ingesting it.
4. Answers may vary, but may include crawl, walk, swim, jump, fly.
5. Answers may vary. Hookworms enter by boring through the skin of the feet. *Trichinella* enters the body in the undercooked tissue of animals eaten as meat.
6. Coral, a type of cnidarian, is responsible for forming coral reefs.
7. Roundworms reproduce sexually; they have two separate sexes.
8. True
9. False; three
10. False; calcium
11. False; one-way
12. True
13. True
14. True

Check Your Understanding

1. Cnidarians use their tentacles to defend themselves and to catch food.
2. Segmented worms have a coelom, roundworms and flatworms do not.
3. Sponges take in food by filtering microorganisms and other organic material out of water that they draw into their bodies. This method is called filter feeding.
4. a. Sponge: invertebrate
   b. Snake: vertebrate
   c. Flatworm: invertebrate
   d. Jellyfish: invertebrate
   e. Human: vertebrate
   f. Fish: vertebrate
5. Coral polyps have radial symmetry. They are invertebrates.
6. Animal X is a cnidarian. It can be inferred that animal X also has tentacles with stinging cells.

Chapter 15 Review

Check Your Knowledge
Answer the following in complete sentences.

1. What are the two main body forms of cnidarians?
2. How do earthworms affect soil?
3. How do animals obtain food?
4. Name four ways animals move.
5. Name a roundworm that is a human parasite. How does the parasite enter a human body?
6. What type of animal is responsible for producing coral reefs?
7. Do roundworms reproduce sexually or asexually? Do they have two separate sexes?

Determine whether each statement is true or false. Write *true* if it is true. If it is false, change the underlined word to make the statement true.

8. <u>Flatworms</u> have the simplest body organization of any kind of worm.
9. Worms are classified into <u>two</u> phyla.
10. Reef corals make hard skeletons of <u>phosphate</u>.
11. Roundworms have a <u>two-way</u> digestive tube with two openings.
12. <u>Planaria</u> reproduce asexually by dividing their bodies in two. They can also reproduce sexually.
13. All worms have <u>bilateral</u> symmetry and a definite head.
14. Each cnidarian stinging <u>organ</u> has a tiny harpoon at the end of a coiled, hollow thread.

Check Your Understanding
Apply the concepts you have learned to answer each question.

1. How do cnidarians use their tentacles lined with stinging cells?
2. How is the body organization of segmented worms different from the body organization of flatworms and roundworms?
3. How does a sponge take in food? What is its method of eating called?
4. **Extension** Classify each of the following animals as a vertebrate or invertebrate.
   a. Sponge
   b. Snake
   c. Flatworm
   d. Jellyfish
   e. Human
   f. Fish
5. **Mystery Photo** The photo on page 298 shows coral polyps. Their limestone home is in the background. Each polyp is a different coral animal, though they often live in colonies like this one. Do the coral polyps have radial or bilateral symmetry? Are the coral polyps vertebrates or invertebrates?
6. **Infer** Animal X has radial symmetry. It often lives in ocean habitats. You might find it attached to a surface, or moving freely. It has tissues, a mouth, and a digestive cavity. It can reproduce sexually or asexually. What is the phyla of animal X? What is another important distinguishing feature of the phyla?

Chapter 15 Invertebrates I

Develop Your Skills

1. a. A; about 180 per cubic meter
   b. Answers will vary. It is possible that the soil was too wet for the earthworms.
2. Cnidarians are different from sponges in that cnidarians have a mouth leading to a digestive cavity, tissues, and tentacles with stinging cells.
3. a. Earthworm: bilateral
   b. Sea star: radial
   c. Sponge: neither
   d. Hydrozoan: radial
4. a. 80,000 roundworm species; 5,000 species of sponges
   b. There are many more species of roundworms because their structures have evolved to allow for wide-ranging habitats and, therefore, greater variation.

Make Connections

1.

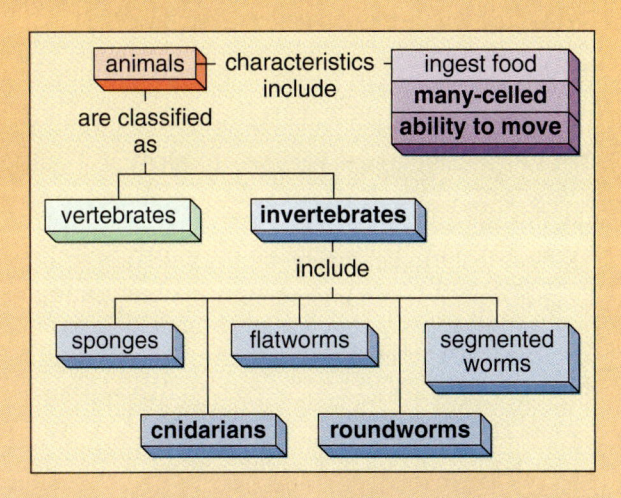

```
animals — characteristics include — ingest food
                                     many-celled
are classified as                    ability to move

vertebrates        invertebrates
                        |
                     include
                        |
        ┌───────────────┼───────────────┐
     sponges         flatwarms      segmented worms
        |               |
    cnidarians      roundworms
```

2. Student posters will vary. Coral colors are determined by the algae that live inside the polyps.
3. Answers will vary, but may include ships that lower instruments on long cables or use sonar to map the ocean and satellites that make various kinds of images of oceans.
4. Answers will vary; however, if only a few sponges are harvested from each square meter of sea floor, then there will always be sufficient numbers of individuals for sexual reproduction to occur.

Develop Your Skills

Use the skills you have developed in this chapter to complete each activity.

1. **Interpret Data** The bar graph shows the number of earthworms per cubic meter in different samples of soil.
   a. Which sample contains the most earthworms? How many are there?
   b. What can you infer about soil sample E?

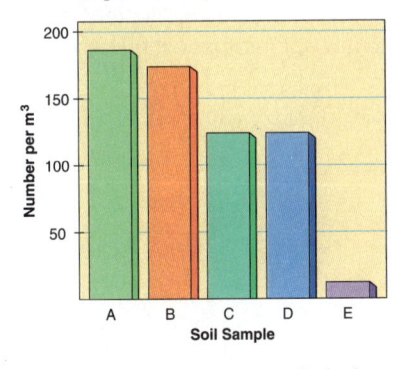

[Bar graph: Number per m³ vs Soil Sample A–E]

2. **Compare and Contrast** How is the body of a sponge similar to and different from the body of a cnidarian?
3. **Classify** Which of the following animals have bilateral symmetry? Which have radial symmetry? Which have neither?
   a. Earthworm   c. Sponge
   b. Sea star   d. Hydrozoan
4. **Data Bank** Use the information on page 629 to answer the following questions.
   a. How many species of roundworms exist? How many species of sponges?
   b. **Infer** Why do so many more species of roundworms exist than any other invertebrate shown on the graph?

Make Connections

1. **Link the Concepts** Below is a concept map showing how some of the main concepts in this chapter link together. Only part of the map is filled in. Finish the map, using words and ideas from the chapter.

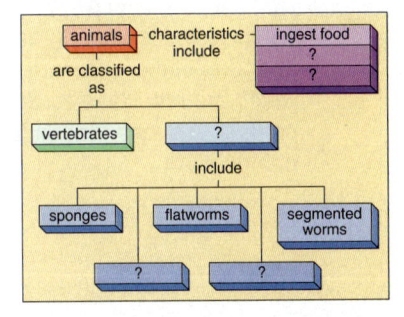

2. **Science and Art** Using what you've learned in this chapter and any additional sources from the library, design a poster of a coral reef. Include coral of different sizes and colors. What determines the different colors of coral?
3. **Science and Technology** *Alvin* is a special research submarine. What other types of vessels are used for sea exploration and research? Use the library to find out about the technology scientists use to explore the sea.
4. **Science and Society** Natural sponges are a living resource. If one type of sponge is collected too frequently, the population could be depleted. What can be done to prevent the loss of sponge diversity due to human use? Hint: you may want to review sponge reproduction.

CHAPTER 16

Overview

Students continue learning about invertebrates in this chapter. The evolution and characteristics of mollusks are presented in the first section. Arthropods are distinguished from other invertebrates in the second section. The third section provides extensive information on the insect class of arthropods, including models of body structure, development, and metamorphosis. The final section focuses on the five classes of echinoderms.

Advance Planner

▶ Obtain a copy of the myth about Arachne. TE page 327.

▶ Prepare a chart or overhead showing how ticks spread diseases and how the diseases can be avoided. TE page 327.

▶ Collect live or preserved grasshoppers or other arthropods (or order them from a science supplier) for TE page 332.

▶ Set up 15 wide-mouth jars, 8 ripe bananas, nylon stockings, small spoons, rubber bands, microscope slides, coverslips, and hand lenses for SE Activity 16, page 336.

Skills Development Chart

Sections	Classify	Collect Data	Communicate	Decision Making	Hypothesize	Infer	Model	Observe	Research
16.1 Skills WarmUp	●								
16.2 Skills WarmUp Skills WorkOut SkillBuilder		●	●				●		
16.3 Skills WarmUp Skills WorkOut Consider This Activity			● ●		● ●	 ●		● ● ●	
16.4 Skills WarmUp Skills WorkOut						 ●			●

Individual Needs

▶ **Limited English Proficiency Students** Provide students with a picture dictionary. Have them use the picture dictionary or the text to find definitions of the boldface terms in the chapter. Students can find and cut out pictures from magazines or newspapers that show examples of the boldface terms. Students can paste these pictures onto notebook paper in order to create their own picture dictionaries for this chapter.

▶ **At-Risk Students** Have students write the section headings in their science journals, leaving space between each heading. Then ask students to refer to the section objectives at the beginning of each section. Have them use their own words to make questions from each of the objectives. Then, as they work through the chapter, have students write one or two sentences that answer each objective question.

▶ **Gifted Students** Have students research the evolutionary development of one of the groups of invertebrates described in this chapter. Have them create a visual representation of their information by making models of fossils of several organisms related to today's invertebrate. Encourage students to arrange their fossils in chronological order.

Resource Bank

▶ **Bulletin Board** Have students create an information center about a very important group of arthropods, the spiders. Begin the bulletin board by having volunteers construct a spiderweb using string and pushpins. Have students label the parts of the web, indicating which parts are sticky and which are not. Have students add pictures of many different spiders and labels describing how spiders benefit human beings.

CHAPTER 16 PLANNING GUIDE

Section	Core	Standard	Enriched	Section	Core	Standard	Enriched
16.1 Mollusks pp. 321–324				**16.3 Insects** pp. 331–336			
Section Features Skills WarmUp, p. 321	●	●	●	Section Features Skills WarmUp, p. 331 Skills WorkOut, p. 332 Consider This, p. 334 Activity, p. 336	● ● ●	● ● ● ●	● ● ●
Blackline Masters Review Worksheet 16.1 Integrating Worksheet 16.1	● ●	● ●		Blackline Masters Review Worksheet 16.3 Skills Worksheet 16.3 Enrich Worksheet 16.3	● ●	● ● ●	● ● ●
Laboratory Program Investigation 31		●	●	Ancillary Options One-Minute Readings, p. 28	●	●	●
Color Transparencies Transparency 45	●	●		Color Transparencies Transparencies 47, 48	●	●	
16.2 Arthropods pp. 325–330				**16.4 Echinoderms** pp. 337–340			
Section Features Skills WarmUp, p. 325 Skills WorkOut, p. 326 SkillBuilder, p. 329	● ●	● ● ●	● ●	Section Features Skills WarmUp, p. 337 Skills WorkOut, p. 339	●	● ●	●
Blackline Masters Review Worksheet 16.2 Skills Worksheet 16.2 Integrating Worksheet 16.2	● ●	● ● ●	 ● ●	Blackline Masters Review Worksheet 16.4 Skills Worksheet 16.4 Vocabulary Worksheet 16	● ● ●	● ● ●	
Overhead Blackline Transparencies Overhead Blackline Master 16.2 and Student Worksheet	●	●	●				
Color Transparencies Transparency 46	●	●					

Bibliography

The following resources can be used for teaching the chapter. See page T-40 for supplier codes.

Audio-Visual Sources
(video unless noted)
Aliens from Inner Space. 25 min. 1984. FL.
The Insect Challenge. 24 min. 1985. PF.
Invertebrates. 2 filmstrips with cassettes, 14–17 min each. 1987. NGSES.
Life Cycle of the Honeybee. 12 min. 1976. NGSES.
Now You See Me, Now You Don't. 25 min film. 1977. AMP.
Riches from the Sea. 23 min. 1984. NGSES.
The World of Insects. 20 min. 1979. NGSES.

Software Resources
Animals. Apple. J & S.
Diversity of Life. Apple. J & S.
Grasshopper Dissection Guide. Apple, IBM. CABISCO.
The Insect World. Apple. DCHES.
Starfish Dissection Guide. Apple, IBM. CABISCO.

Library Resources
Allaby, Michael. The Oxford Dictionary of Natural History. New York: Oxford University Press, 1986.
Comstock, Anna Botsford. Handbook of Nature Study. Ithaca: Cornell University Press, 1986.
Durrell, Gerald. The Amateur Naturalist. New York: Alfred A. Knopf, 1989.

CHAPTER 16

INTEGRATED LEARNING

Writing Connection

Have students take a moment to look again at the picture of the moth's wing below. Then have them describe how they think the world appears to a moth. Have them imagine what it would be like to fly. Ask students to add details about the plants the moth visits, how it gets its food, how it protects itself from rain and predators, and so on. Encourage students to be creative.

Introducing the Chapter

Have students read the description of the photograph on page 320. Ask if they agree or disagree with the description. The photograph shows the surface of a moth's wing.

Directed Inquiry

Have students study the photograph. Ask:

▶ What do you see in the photograph? (Students will probably mention the colors and the rib-like lines.)

▶ How do you think the object in the picture feels? (Students may say that it is smooth or soft.)

▶ What kind of body part or structure do you think this is? (A wing) Why do you think so? (Students may mention that the ribs and colors make them think of a butterfly or moth's wing.)

▶ Which kind of invertebrate is shown in the picture? (An insect)

Chapter Vocabulary

abdomen	exoskeleton
antennae	gastropod
arthropod	metamorphosis
bivalve	mollusk
cephalopod	nymph
crustacean	pupa
echinoderm	thorax

Chapter 16 Invertebrates II

Chapter Sections

16.1 Mollusks
16.2 Arthropods
16.3 Insects
16.4 Echinoderms

What do you see?

❝ I see a parade of colors flashing. It seems like a kind of wing—more like a butterfly's or dragonfly's wing. The colors are green, yellow, red, blue, and blue-green with black spots all over. It seems like some sort of wing because you can see the scales a little bit. You can almost see the shape of the wing, too.❞

*Erik Gomez
Clifton Middle School
Houston, Texas*

To find out more about the photograph, look on page 342. As you read this chapter, you will learn about many different kinds of invertebrates.

Themes in Science

Diversity and Unity The phylum Mollusca is the second largest phylum in the animal kingdom. There are thousands of species of mollusks ranging from bivalves such as clams, to one-shelled gastropods such as snails, to cephalopods such as giant squids. Even so, all mollusks share the same basic body plan. This body plan includes a visceral mass, soft body parts covered by a mantle, and a muscular foot.

16.1 Mollusks

Objectives

- **Describe** the characteristics of mollusks.
- **Compare** mollusks to other invertebrates.
- **Describe** the features of each class of mollusks.
- **Make inferences** about geologic history based on mollusk fossils.

Skills WarmUp

Classify Study ten seashells. Look at the shells closely with a hand lens. Classify the shells into two groups based on one property. What property did you use? What are some other properties that you could use to divide the shells into two groups?

How fast do you move when you're getting ready for school in the morning? Do you move at a "snail's pace"? When your teacher asks you a question in class, do you just "clam up"? These expressions describe human behavior, but they also tell you something about snails and clams. Snails and clams are invertebrates called **mollusks** (MAHL uhsks). All mollusks have soft bodies, and many are covered by hard shells.

Characteristics of Mollusks

Mollusks live mostly in the ocean, but some can be found in freshwater habitats. A number of mollusks have also adapted to life on land. Snails and slugs are common in damp places in your backyard.

There are more than 100,000 species of mollusks. They range in size from tiny snails only a few millimeters across to the 20-m-long giant squid.

No matter what their shape or size, all mollusks have the same basic body plan. Like the bodies of segmented worms, mollusks' bodies contain organs organized into systems. However, their bodies are not built of segments like the segmented worms. Instead, mollusks have all their organs in one area called a visceral (VIS ur uhl) mass. The visceral mass is made up of the digestive, excretory, circulatory, respiratory, and reproductive organs.

Another characteristic of the mollusk body plan is a muscular foot. The foot is used by most mollusks for movement. In addition, a mollusk's soft body parts are covered by a skinlike tissue called a mantle. In many mollusks, the mantle produces a hard, protective shell.

Figure 16.1 ▲
This oyster (top) and squid (bottom) may look very different, but they are both mollusks.

SECTION 16.1

Section Objectives
For a list of section objectives, see the Student Edition page.

Skills Objectives
Students should be able to:

Classify sea shells.

Infer what the presence of clamshell fossils in the middle of a continent might indicate.

Vocabulary
mollusk, gastropod, bivalve, cephalopod

MOTIVATE

Skills WarmUp
To help students understand the process of classifying sea shells, have them do the Skills WarmUp.
Answer Answers may vary. Students may classify shells according to shape, color, or size.

Misconceptions
Students may think that the term *mollusk* only applies to bivalves such as mussels. Emphasize that gastropods and cephalopods are also mollusks, and explain the characteristics common to all three classes.

Portfolio
Have students look up the word *mollusk*. Ask them to draw a picture of a mollusk and write a caption explaining why the word mollusk was used to describe this group of animals.

TEACH

Critical Thinking

Compare and Contrast Have students study Figure 16.2. Have them identify the body parts that are characteristics of all mollusks. (Digestive system, mantle, shell, muscular foot) Then have students identify the body parts that are typical of gastropods. (Foot underneath body, coiled shell)

Discuss

Have students locate the snail's foot in Figure 16.2. Have them use their own words to describe how a snail moves. Ask students to explain the main difference between snails that live on land and snails that live in water. (Water snails have gills that serve as respiratory organs; land snails have evolved cavities that serve as lungs.)

Class Activity

Have students work in groups of four. Have each group observe the snails in an aquarium and make a list of all the things they observe the snails doing.

Skills Development

Infer Have students use their knowledge about what gastropods eat so they can explain why a fish tank containing snails is cleaner than a fish tank that does not contain snails. (Snails eat algae that sometimes grow on the sides of the aquarium.)

Answer to In-Text Question
① The snail lives in water.

The Living Textbook:
Life Science Sides 1-4

Chapter 4   Frame 00857
Gastropods (76 Frames)
Search: Step:

INTEGRATED LEARNING

Integrating the Sciences

Earth Science By about 225 million years ago, many mollusks had become extinct. Have students use an earth science text to discover why mollusks died out. What were the major physical events of the time that could have contributed to drastically changing the environment? (Sea levels were very low and the climate was dry.)

Evolution of Mollusks

The oldest known mollusk fossil is a clamlike organism that lived about 570 million years ago. After this time, mollusks evolved a variety of body forms. Many were common in the ancient oceans. Then about 225 million years ago, most mollusk species became extinct. The survivors are the ancestors of the kinds of mollusks alive today.

Mollusks have larvae similar to the larvae of segmented worms, or annelids. For this reason, some scientists think mollusks evolved from segmented worms. However, because of mollusks' lack of segmentation, other scientists hypothesize that mollusks evolved from ancient worms more similar to flatworms.

Diversity of Mollusks

You can tell that snails are very different from clams and that both differ from an octopus. Would you classify these three mollusks in different groups? That is what scientists do. Each of these kinds of mollusks is a member of a different group, or class.

Gastropods Mollusks that glide along on a foot underneath their bodies are called **gastropods** (GAS troh PAHDZ). Snails, limpets, slugs, and the sluglike nudibranchs are all gastropods. Most gastropods have one coiled shell. The shells of limpets, however, are flattened cones. Slugs and nudibranchs have no shell at all.

If you watch a snail moving on the ground, you might think it is scooting on its stomach. Actually, a gastropod's muscular foot spreads out under its body and is separate from the stomach. Locate the snail's foot in Figure 16.2. A rippling motion of the foot muscle moves the snail along.

Many gastropods eat plants and algae. Some eat other invertebrates. As an adaptation for getting their food, gastropods have a tonguelike organ covered with rows of teeth. It is called a radula (RAJ oo luh). A gastropod moves its radula back and forth to scrape off and scoop up food. Gastropods that live in water have respiratory organs called gills. As water flows over the gills, oxygen is extracted. Gastropods living on land have evolved a cavity inside their bodies that serves as a lung.

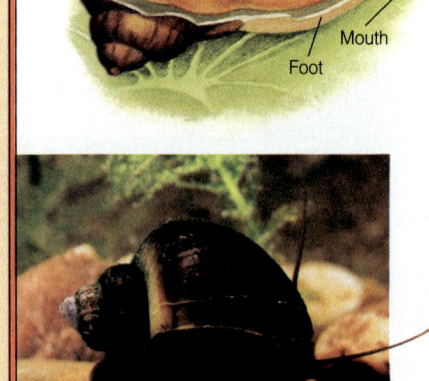

Figure 16.2 ▲
Snails are typical gastropods. Does this one live in water or on land? ①

322 Chapter 16 Invertebrates II

Integrating the Sciences

Ecology Even though most species of mussels are not harmful, one mussel, *Dreissena polymorpha*, has created great problems along the shores of the Great Lakes. This mollusk, commonly called the zebra mussel, is not native to the Great Lakes. It may have been transported in 1985 or 1986 on a transatlantic freighter. Because the zebra mussel has no natural predators in the Great Lakes, mussel populations have grown out of control. This population explosion has caused ecological and technical problems. The zebra mussels eat large amounts of phytoplankton, the food source for other animals in the Great Lakes ecosystem. Clam, walleye, and lake trout populations are particularly threatened. Zebra mussels attach themselves to any hard surface, including hulls of boats, piers, and pipes.

Bivalves Mollusks with two shells hinged together make up a second class called the **bivalves** (BY valvz). Clams, oysters, mussels, and scallops are all bivalves. Like gastropods, bivalves have a muscular foot. But the foot is shaped and used differently than a gastropod foot. It is normally hidden inside the two shells. Some bivalves move by hooking the foot in sand, and then pulling themselves along.

Bivalves have no head, and most move very little. Some remain attached to one place. Movement isn't important because they are filter feeders. Look at Figure 16.3 and find the gills of the clam. Cilia move water across the gills. Tiny food particles, such as plankton, are trapped by mucus on the gills. Cilia push the food into the animal's mouth.

The gills in bivalves have a large surface area and a rich supply of blood. Oxygen from the water passing over the gills diffuses into the blood. Carbon dioxide, a waste product of animals, diffuses from the blood into the water.

In some kinds of bivalves, water flows over the gills when the shells are open. In clams, the water enters through a muscular tube called a siphon. After it flows over the gills, it leaves through another siphon.

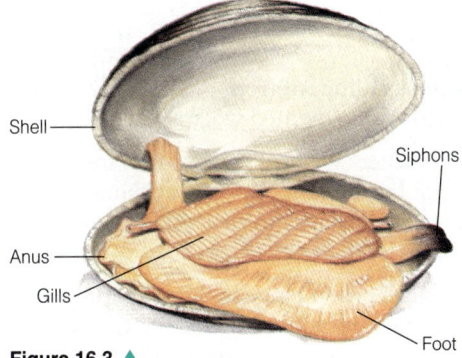

Figure 16.3 ▲
The clam is a typical bivalve.

Figure 16.4 ▲
The tentacles of a cephalopod are attached at the animal's head.

Cephalopods The octopus and squid are the major members of a third class of mollusks called **cephalopods** (SEF uh loh PAHDZ). In these mollusks, the foot is divided into tentacles. The tentacles are located at the head, away from the rest of the body. Look at the octopus in Figure 16.4. How many tentacles does it have? ②

Cephalopods are the only mollusks to have a closed circulatory system. In a closed system, the blood stays in blood vessels.

Modern cephalopods all evolved from now-extinct mollusks with external shells. In the octopus, the shell has disappeared during the process of evolutionary change. Squids have only small internal shells. Only cephalopods called nautiluses (NAWT uh luhs uhz) still have the external shells of their ancestors. Cephalopods live in the oceans and most move about freely. Unlike the filter-feeding bivalves, they are active predators. Their tentacles are covered with suction cups for grasping prey. The octopus lives mostly on the ocean floor, crawling around in search of prey. Squid swim in the open water.

Chapter 16 Invertebrates II   **323**

Portfolio

Have students find the meaning of the words *gastropod*, *cephalopod*, and *bivalve*. Have them draw a diagram of each class of mollusk and show how the organism's name describes its structure.

Critical Thinking

Compare and Contrast Have students give some examples of gastropods (snails), bivalves (clams, mussels, oysters), and cephalopods (squid, octopus, nautilus). Ask: How are cephalopods different from the other two groups of mollusks? (Cephalopods are active predators and have closed circulatory systems. Often cephalopods do not have shells.)

Enrich

Tell students that the cephalopod has both a large brain and strong sense receptors. These sense receptors include chemical sense organs and balancing organs. The cephalopod's most notable feature, however, is its eyes. The eye of a cephalopod contains a cornea, lens, and retina similar to the eye of a vertebrate. This kind of eye allows the cephalopod to see more clearly than other mollusks. Remind students that the cephalopod hunts for prey. Ask them how the cephalopod's eyes help it survive. (It can see sources of food.)

Answer to In-Text Question

② An octopus has eight tentacles.

Integrated Learning

Use Integrating Worksheet 16.1.

**The Living Textbook:
Life Science Side 4**

Chapter 11   Frame 20892
Octopus (Movie)
Search: Play:

323

◀ TEACH ▪ Continued

Class Activity
Bring in a few oysters for students to observe. Pass the closed oysters around the class and have students write their observations. Then open the oysters and have students observe the mother-of-pearl. Tell them to note the colors, luster, patterns, and so on.

◀ EVALUATE

WrapUp
Reteach Have students make a concept map that shows how the three classes of mollusks are alike and how they are different.
Use Review Worksheet 16.1.

Check and Explain

1. All mollusks have a visceral mass, which contains all the organs. All mollusks also have a muscular foot and a body covering called a mantle.

2. Mollusks differ from segmented worms in that their bodies are not segmented. Also, mollusks have a visceral mass containing their organs. A segmented worm does not have a muscular foot like a mollusk.

3. Bivalves couldn't survive on land because they filter feed on food that is suspended in water. Gastropods use their radula and teeth to eat plants and algae. Their feeding methods and diets are suited for life on land.

4. Geologists could assume that this part of the continent was once under water.

324

INTEGRATED LEARNING

STS Connection
Students may be interested in knowing that recently scientists have found new uses for mussels. Bivalves can help environmental researchers measure water pollution. As bivalves filter food in lake or ocean water, pollutants in the water become concentrated in their tissues. An examination of these tissues gives workers an idea of the level of different pollutants. Another observation of mollusks may provide new technology for treating or preventing cancer. Despite their exposure to chemicals known to cause cancer in humans, snails and other mollusks never seem to develop the disease. Researchers are trying to find out what protects these animals from cancer.

Science and Society *People and Mollusks*
Throughout the world, mollusks are known mainly as food. The oyster, a kind of bivalve, is a great delicacy to some people. Many popular dishes are made of mollusks, such as abalone, Oysters Rockefeller, clam chowder, calamari, and escargots (EHS kar GOH). Calamari is squid, and escargots are land snails. Many mollusks from all three classes are edible. In some parts of the world, people eat mollusks every day. What kinds of mollusks have you eaten?

Mollusks are also the producers of pearls. When a grain of sand gets into an oyster shell and lodges between the shell and mantle, the oyster gradually coats the particle with thousands of layers of the substance known as mother-of-pearl. In time a pearl forms. Mother-of-pearl is the same material that makes up the inside of the oyster shell. In Japan and other parts of the world, small particles are purposely placed inside oysters to make "cultured" pearls.

Bivalve and gastropod shells are used for jewelry, buttons, and other types of ornaments. Early civilizations made cooking utensils out of mollusk shells and also used them as money.

The internal shell from a type of squid called a cuttlefish is familiar to bird owners. It is the cuttlebone hung in bird cages to provide the birds with calcium. In all these ways, mollusks have been important to people throughout history.

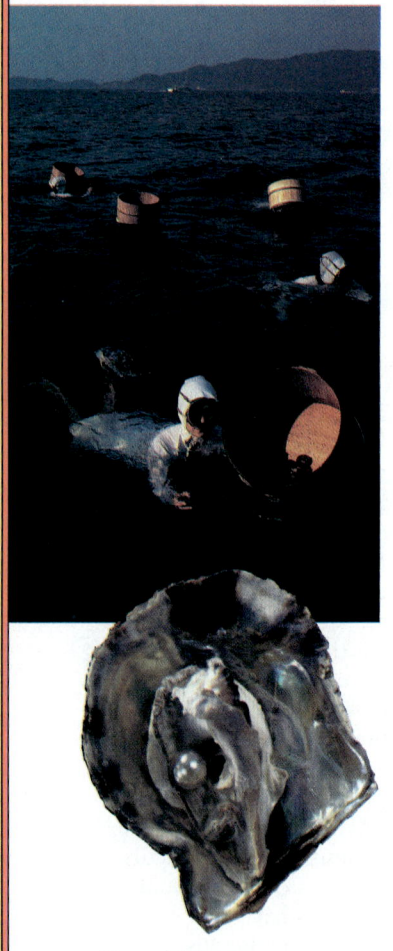

Figure 16.5 ▲
In Japan, divers collect oysters. Each oyster may contain one or more pearls.

Check and Explain

1. What are some similarities among all mollusks?

2. How are mollusks different from cnidarians? How do they differ from segmented worms?

3. **Reason and Conclude** Give an explanation for why bivalves cannot live on land. In your explanation, consider how they differ from gastropods, some of which do live on land.

4. **Infer** If geologists found clamshell fossils in sedimentary rock in the middle of a continent, what could they safely assume about the land where the fossil was found?

324 Chapter 16 Invertebrates II

Themes in Science

Scale and Structure Arthropods make up the largest phylum in the animal kingdom. They live in water, on land, and in air. They range in size from barely visible mites to spider crabs 4 m wide. The specialized body segments of arthropods and their jointed external coverings make them very strong and mobile.

16.2 Arthropods

Objectives

▶ **Describe** the characteristics of arthropods.
▶ **Identify** the major kinds of arthropods.
▶ **Explain** how arthropods are classified.
▶ **Communicate** why spiders are important organisms.

Skills WarmUp

Make a Model Based on what you know about spiders and insects, draw a general model of an arthropod and label its features.

No matter where you go on the earth, you can find invertebrates with jointed appendages called **arthropods** (AR throh PAHDZ). With more than a million different species, arthropods make up the largest phylum of animals. They include spiders, butterflies, bees, crabs, lobsters, shrimps, and millipedes.

Arthropods live in water, on land, and even in the air. Some, such as mites, are so tiny you can barely see them. Others, such as the Japanese spider crab, grow to more than 4 m across!

Characteristics of Arthropods

Arthropods are the only invertebrates with jointed appendages. Appendages are parts that extend from the body. Jointed appendages, as you have probably guessed, have joints, just as your arm has an elbow joint. Arthropods use their appendages for movement, defense, feeding, sensing, and even reproduction. A single arthropod may have a great variety of appendages, each adapted for a particular use.

Arthropods share one important characteristic with segmented worms. They both have segmented bodies. However, most arthropods have fewer repeated segments. Instead, their bodies are divided into dissimilar regions: the head, **thorax**, and **abdomen**. Look at Figure 16.6. In most arthropods, the thorax holds legs used for movement. The abdomen contains many of the animal's organs. The head has appendages that are used for sensing and feeding.

Arthropods have another important characteristic. Their bodies are covered by an outer support structure called an **exoskeleton**. An exoskeleton is much like a

Figure 16.6 ▲
The three body regions characteristic of arthropods are easy to see in this honeybee.

SECTION 16.2

Section Objectives
For a list of section objectives, see the Student Edition page.

Skills Objectives
Students should be able to:

Make a Model of an arthropod.

Communicate information about an arthropod.

Vocabulary
arthropod, thorax, abdomen, exoskeleton, crustacean, antennae

MOTIVATE

Skills WarmUp
To help students understand the characteristics of arthropods, have them do the Skills WarmUp.
Answer Drawings will vary. Students may include features shown in Figure 16.6.
Extension Have students keep their drawings in their portfolios so they can add to them at the end of this section.

Misconceptions
Because of popular depictions of spiders, lobsters, and crabs, students may believe that these animals are predators that are dangerous to humans. Explain that these animals do not generally attack humans. While there are some poisonous spiders, there are many, many more that are not poisonous.

Discuss
Have students study Figure 16.6. Have them describe the functions of the head, thorax, and abdomen.

Chapter 16 Invertebrates II **325**

TEACH

Skills Development

Infer Tell students that, because they are growing, some of the clothes they wore last year will not fit this year. Tell them there is a similar situation among arthropods. Ask students, What would happen if arthropods did not molt? Why do they need to molt? (They couldn't grow if they didn't molt. The exoskeleton does not expand.)

Research

Have students use references to find out more about trilobites. Have them try to find out about the variety of these animals, when they lived, and where fossils of these organisms might be found in the United States.

Skills WorkOut

To help students communicate information about an arthropod, have them do the Skills WorkOut.
Answer Answers may vary. Check to see that the animals chosen by students are arthropods.

Answer to In-Text Question

① The darker-looking object is the old exoskeleton.

INTEGRATED LEARNING

Integrating the Sciences

Earth Science Paleontologists make use of trilobites in dating rock layers. Trilobites evolved during the Cambrian period from 570 million years ago and became extinct in the Permian period from 280 million years ago. They evolved rather rapidly but no one species lasted for a long time. As a result, each geologic period is marked by some very specific animals. Have students refer to pages 158 and 159. Ask what other types of organisms lived during these periods. (Most animals are water animals. They are similar to present-day mollusks, medusas, sponges, and worms.)

Use Integrating Worksheet 16.2.

Figure 16.7 ▲
The compound eyes of a Mediterranean fruit fly are shown above enlarged 21 times. In the photograph on the right, a cicada has just finished molting. Can you tell which object is the old exoskeleton and which is the cicada? ①

suit of armor worn by a medieval knight. It is waterproof and helps prevent the loss of body fluids. Most of it is hard for protection, yet around the joints it is flexible.

The exoskeleton, however, does not grow with the animal. It must be shed and a new, larger exoskeleton made. The process of growing a new exoskeleton and shedding the old one is called molting. Arthropods may molt many times during their lives.

Arthropods have sense organs for sight, smell, taste, and touch. Many have eyes with multiple lenses, called compound eyes. Look at the compound eyes of the insect in Figure 16.7. Vision with compound eyes is not very sharp, but compound eyes are very sensitive to light and movement.

Skills WorkOut

Communicate Choose an arthropod to report about. Find information about the organism and a picture of it. Trace or photocopy the picture. Neatly print the information on an index card. Display your picture and information on a bulletin board.

Evolution of Arthropods

Arthropods have been on the earth a long time. Some arthropod fossils are more than 500 million years old. Arthropods called trilobites (TRY loh BYTS) were very common in the oceans for hundreds of millions of years. They became extinct about 280 million years ago.

Arthropods most likely evolved from segmented worms. In the process of evolution, groups of repeated segments were fused, or joined, to create the head, thorax, and abdomen of arthropods. Arthropods' jointed appendages probably evolved from shorter, unjointed appendages present in ancient segmented worms. With their hard exoskeleton and walking legs, arthropods were among the first animals to live successfully on land.

Chapter 16 Invertebrates II

Language Arts Connection

Arrange for each student to have a copy of the Greek myth about Arachne, or read it aloud to them. Have the class decide if there is a moral to the story and what the story says about people's attitudes toward spiders.

Integrating the Sciences

Health Ticks are a critical link in the disease cycle of Rocky Mountain spotted fever and Lyme disease. The ticks' bites generally are not painful but parasites in their blood can cause disease in humans. You may wish to refer students to page 515 and discuss with them some of the precautions they can take to avoid Lyme disease.

Class Activity

Have students use hand lenses to study the photographs in Figure 16.9. Have them find the spinnerets, cephalothorax, chelicerae, and abdomen of each spider.

Enrich

Students may wonder why spiders don't get caught in their own webs. The reason is that they instinctively know where to walk. Some of the spider silk is sticky, and some of it is not. Even if they do step on the wrong silk, they have the ability to get unstuck.

Research

Have students use references to find out about arachnids other than spiders—scorpions, ticks, and mites. They can research how scorpions, ticks, and mites differ from spiders.

Answers to In-Text Questions

② **A spider has eight legs.**

③ **The spiders differ in color, in the length and thickness of the eight legs, and in the size and shape of the cephalothorax and abdomen.**

Diversity of Arthropods

A wide variety of animals are included in the arthropod phylum, from mites to millipedes, ladybugs to lobsters. These very different animals are organized into different classes.

Arachnids Spiders, ticks, scorpions, and mites are all **arachnids** (uh RAK nihdz). Arachnids generally have two main body regions. The head and thorax are fused, forming a cephalothorax (SEF uh loh THOR aks). An abdomen is the other body region. In ticks and mites, the cephalothorax and abdomen are also fused together. Arachnids have four pairs of legs attached to the cephalothorax. Near the mouth is another pair of appendages called chelicerae (kuh LIHS ur EE).

In spiders, the chelicerae are like fangs and have poison glands. They are used to attack prey, which is mainly insects. Spiders also produce a liquid form of silk in glands and spin the silk into thread with spinnerets. When the liquid silk is exposed to air, it becomes solid and strong. Spiders make different kinds of silk for catching prey, making sacs for eggs, and anchoring webs. Not all spiders spin webs, but they all produce silk.

Spiders have thin slits in their exoskeletons that allow air into their bodies. In some spiders, the oxygen in the air diffuses directly into cells. In other spiders, the oxygen in the air diffuses into the spider's blood through book lungs. Find the book lungs in Figure 16.8. Book lungs are made up of sheets of tissue like the pages of a book. This gives them a large surface area for gas exchange. Some spiders use both for gas exchange.

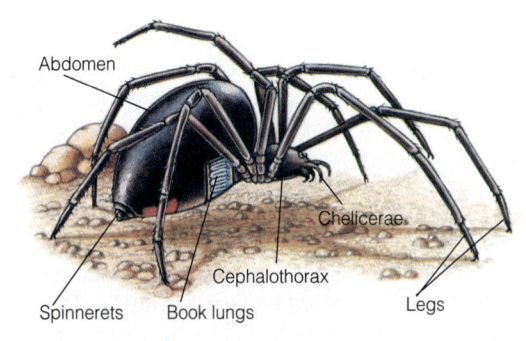

Figure 16.8 ▲
How many legs does a spider have? ②

Figure 16.9 ▲
What differences do you notice in the bodies of these spiders? ③

**The Living Textbook:
Life Science Side 4**

Chapter 17   Frame 45271
Spider (Movie)
Search: Play:

**The Living Textbook:
Life Science Side 3**

Chapter 20   Frame 37755
Bolus Spider/Diving Spider (Movies)
Search: Play:

TEACH • Continued

Discuss

Have students study the crayfish in Figure 16.10. Ask:

▶ Where are the gills of the crayfish? (They are in the cephalothorax.)

▶ What is the advantage of having a stacked compound eye? (The crayfish can see very well to catch its prey.)

▶ What are the names of the two main body parts of the crayfish? (Abdomen and cephalothorax)

Reteach

Remind students that they studied plankton in Chapter 11. Ask what organisms combine with copepods to make plankton. (Single-celled algae)

Answer to In-Text Question

① Five: eyes, antennae, mandibles, legs, and appendages for swimming. The eyes are used for seeing. The antennae are used for balance and sensing. The mandibles are used for chewing and crushing food. The legs are used for walking and jumping. The abdominal appendages are used for swimming (and for carrying eggs).

The Living Textbook: Life Science Side 4

Chapter 11   Frame 22818
Crab, Brittlestar, and Scallop (Movie)
Search: Play:

The Living Textbook: Life Science Side 4

Chapter 14   Frame 34994
Crabs (Movie)
Search: Play:

328

INTEGRATED LEARNING

Themes in Science

Scale and Structure The structure of crustaceans includes several characteristics. The multiple appendages of crustaceans are specialized. Crustaceans can have many pairs of appendages and lost appendages can be regenerated. Often, three or more pairs of appendages are modified as mouthparts, including hard mandibles. Crustaceans are the only arthropods with two pairs of antennae.

Language Arts Connection

Review the word derivations in the names *centipede* and *millipede*. Ask students why they think the names do or don't fit the animals.

Crustaceans Crayfish, barnacles, crabs, shrimps, lobsters, water fleas, pill bugs, and sow bugs are all **crustaceans** (kruhs TAY shunz). Crustaceans have many specialized appendages. At the head are jaw-like appendages called **mandibles** (MAN duh buhlz), used for chewing and crushing food. Also attached to the head are two pairs of appendages used for balance and sensing called **antennas** (an TEHN uhz). Other arthropod classes have antennas, but crustaceans are the only ones that have two pairs. Unlike spiders, which have simple eyes, most crustaceans have compound eyes.

Look at the crayfish shown in Figure 16.10. It is a typical crustacean. The crayfish uses the large claws on its thorax to grab food and to protect itself. Behind the claws are legs used for walking. Like all crustaceans, the crayfish also has appendages on its abdomen. The crayfish uses these for swimming.

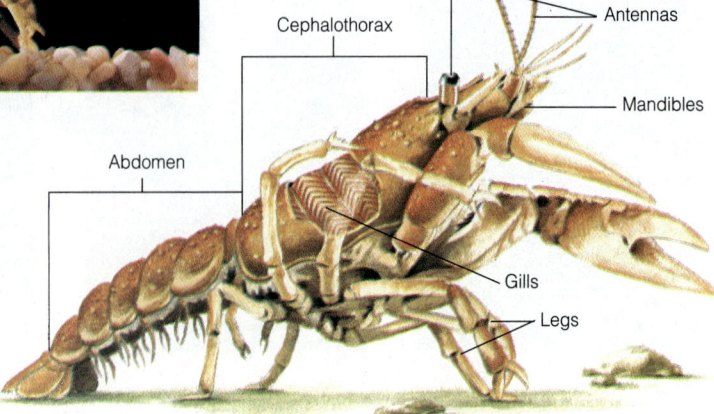

Figure 16.11 ▲
A ghost crab's claws are good weapons.

You may have seen crayfish, lobsters, crabs, or shrimps, but tiny crustaceans called copepods (KOH puh PAHDS) greatly outnumber all the other crustaceans on the earth. Copepods make up part of the plankton. They are an important part of ocean food chains.

Most crustaceans move around freely, but some crustaceans, called barnacles, remain attached to one place. Many barnacles produce hard, volcano-shaped shelters. Inside one of these shelters is a crustacean with jointed appendages. Most of these appendages are used to paddle food into the barnacle's mouth when it is under water.

Figure 16.10 ▲
How many different kinds of appendages can you see on this crayfish? What is the function of each kind? ①

328   Chapter 16 Invertebrates II

Themes in Science

Evolution The body structure of centipedes and millipedes shows the evolutionary link between arthropods and annelids (segmented worms discussed in Chapter 15). While annelids have nonspecialized segments, most arthropods have specialized segments with jointed appendages. Centipedes and millipedes have nonspecialized segments with jointed appendages. Have students use hand lenses to observe Figure 16.12, so they can see the difference between arthropods and annelids.

Centipedes and Millipedes One variation on the basic arthropod body plan is a long, wormlike body with many walking appendages. Two classes of arthropods, the centipedes (SEHN tuh PEEDZ) and millipedes (MIHL ih PEEDZ), share this type of body structure. They resemble segmented worms because of their many repeated segments. Their jointed legs, however, identify them as arthropods.

Although the word *centipede* means "100 legs" in Latin, most centipedes have about 30 legs. Each body segment has one pair of legs. Centipedes are predators. They eat insects, snails, slugs, and worms. Their adaptations for hunting include poison claws, antennas, mandibles, and the ability to move very quickly.

Millipedes, in contrast, are slow-moving and eat mainly plants and decaying organic matter. A millipede has more segments than the typical centipede. And each segment has two pairs of legs. As a millipede walks, the many legs move in a wavelike motion.

Figure 16.12 ▲
Look at these arthropods' legs. Which is the millipede? ②

SkillBuilder *Collecting Data*

Arthropod Search

Look for arthropods in your yard or in pictures in magazines. Start searching for them in both places after you copy the table below. When you encounter an organism you think is an arthropod, observe it for the characteristics in the table. Record your observations. Determine whether the organism is an arthropod. Give the organism a name, and write it down next to the number. If you are observing live organisms, release them after you complete your data collecting. Once you have observed 12 different organisms, answer the following questions.

1. How many arthropods did you observe?
2. Which characteristic of arthropods was easiest to observe? Explain why.
3. If you observed 100 different organisms, how many do you think would belong to the arthropod phylum? Explain your reasoning.
4. How might your results change if you conducted your search six months from this date?

Write a short paragraph comparing the characteristics of arthropods to other animal phyla.

	Jointed Appendages?	Segmented Body?	Exoskeleton?	Is it an Arthropod?
1.				
2.				
3.				
4.				

Chapter 16 Invertebrates II **329**

Discuss

Have students read the paragraphs on centipedes and millipedes. Lead a discussion of these organisms by asking students some or all of the following questions:

▶ If a millipede has 100 segments, how many legs does it have? (400)

▶ If a centipede has 35 segments, how many legs does it have? (70)

▶ Which animals—centipedes or millipedes—might be dangerous to handle? Why? (Centipedes could be dangerous. They have poisonous claws.)

SkillBuilder

Provide hand lenses so students can observe the organisms more closely.

Answers

1. Answers may vary depending on how many the students find.
2. The characteristic easiest to observe was jointed legs because the legs extend from the rest of the body.
3. Answers may vary but should be close to 100 percent, since the vast majority of invertebrates are arthropods.
4. Answers may vary, but the results would be about the same in six months, depending on the season.

Answer to In-Text Question

② **The photograph on the bottom shows the millipede.**

Discuss
Poll students to find out how many of them dislike spiders. Stress to students that many spiders are helpful to people.

EVALUATE

WrapUp
Review Have students repeat the Skills WarmUp from page 325. Then have them compare the two drawings.

Use Review Worksheet 16.2.

Check and Explain
1. Jointed appendages, an exoskeleton, and a body consisting of head, thorax, and abdomen
2. Lobster—crustacean; spider—arachnid; barnacle—crustacean; tick—arachnid; shrimp—crustacean; scorpion—arachnid
3. Arthropods are similar to mollusks in that they both have protective coverings and both have appendages that help them move and eat. Arthropods have jointed appendages and segmented body structures. Mollusks do not have these features. Arthropods have more in common with segmented worms because they both have segmented bodies.
4. Answers may vary. Most spiders do not harm people, so there is little to fear. Since spiders hunt and kill insects, they would reduce the number of insects in the garden.

**The Living Textbook:
Life Science Sides 1-4**

Chapter 4 Frame 01069
Centipedes (4 Frames)
Search: Step:

330

INTEGRATED LEARNING

Multicultural Perspectives
In the 1970s, Zhao Jinzao, vice president of the University of Hubei in China, helped develop a technique for protecting crops from insects. Zhao found that spiders could kill 80 to 90 percent of the harmful insects in cotton fields. He taught farmers to dig shallow pits every ten paces in their cotton fields and to throw straw into the pits. The straw protects spiders from the cold. Have students discuss the advantages of using spiders to control the insect population in crops.

Figure 16.13 ▲
Without spiders, the world might be overrun with insects.

Science and Society *Friendly Spiders*

If you saw a spider right now, what would you do? Despite popular belief, spiders are not out to get you. In fact, most spiders will bite only when frightened.

Some spider bites may cause a reaction, but most do not harm people.

Spiders are generally shy creatures who spend most of their time hunting and killing insects. In fact, spiders are insects' worst enemies. Spiders kill more insects than all the birds in the world do. Spiders also kill more insects than commercial pesticides do.

Some farmers in China know how useful spiders are for killing insects that harm their rice crops. Before these farmers flood their rice fields, they set out stacks of rice straw. Spiders crawl inside the stacks, which are then moved to save the spiders. The farmers can move the stacks to places where insects are damaging the rice crop. The spiders come out and eat the insects.

Research is now being done on the kind of poisons spiders make to capture and kill insects. These substances may be useful in the development of natural insecticides. Some pharmaceutical companies are even studying spider poisons as a source of drugs to treat heart and nervous system disorders.

Check and Explain
1. Describe at least three characteristics shared by all arthropods.
2. For each of the following arthropods, name the class in which it belongs: lobster, spider, barnacle, tick, shrimp, scorpion.
3. **Compare and Contrast** How are arthropods similar to mollusks? How are they different? Do arthropods have more in common with mollusks or with segmented worms? Explain.
4. **Communicate** Suppose you share a garden with someone who is afraid of spiders. You want to encourage spiders to live in the garden to help reduce the number of harmful insects. How would you convince your partner that this is a good idea?

330 Chapter 16 Invertebrates II

Themes in Science

Evolution Throughout their 400-million-year evolution, insects have adapted to many changes in the earth's environment. In this section, students will discover some of the variations insects have developed. These variations have given insects the ability to adapt to an incredible variety of habitats on the earth.

Integrating the Sciences

Earth Science Insects that are millions of years old are found totally preserved in amber, a form of solidified tree resin. Have students look at the photograph on page 156 or fossil field guides to find pictures of these fossils. Allow time for students to look at these pictures. Have them compare one of these fossils to a modern insect.

SECTION 16.3

Section Objectives
For a list of section objectives, see the Student Edition page.

Skills Objectives
Students should be able to:

Hypothesize how various insects respond to light.

Observe insect behavior.

Make a Model of an insect.

Vocabulary
metamorphosis, nymph, pupa

16.3 Insects

Objectives

▶ **Describe** the characteristics of insects.

▶ **Distinguish** between incomplete and complete metamorphosis.

▶ **Explain** how social insects differ from other insects.

▶ **Make a model** of the body structure of an insect.

Have you ever been on a picnic and had uninvited guests show up? Maybe there were ants crawling on the cheese, or flies landing on the potato salad. Or possibly mosquitoes bit you and bees buzzed around. Insects seem to be everywhere!

These six-legged invertebrates are arthropods. They belong to a single class, but there are more species in this class than in all the other arthropod classes put together. In fact, in their number of species, insects outnumber all other forms of life combined!

The Success of Insects

Insects appeared about 400 million years ago. By the beginning of the Cenozoic era, they had evolved an amazing variety of shapes, sizes, and behaviors. The great number of insect species is a sign that insects have been very successful as a life form. They have been able to adapt to many different habitats, from high mountains to dry deserts to rivers and wetlands to your backyard. They have been able to survive when other organisms could not. Insects owe their success to a body plan that has many advantages over others.

One important characteristic of the insect body is its flexibility. In the process of evolution, insect body parts have been modified for different ways of living. Changes in mouthparts, for example, have resulted in the adaptation of different species for eating different food. Figure 16.14 shows some of the different mouthparts insects have evolved.

Another major insect advantage is the ability to fly. Flight helps insects find food, escape predators, locate mates, and reach new places to live.

Skills WarmUp

Hypothesize Insects respond to light in different ways. Some are attracted to it. Some turn away from it. Design a short experiment that could help you check how moths, ants, houseflies, crickets, and beetles react to light.

Grasshopper

Moth

Housefly

Figure 16.14 ▲
Compare the mouthparts of these insects. What kind of food does each one eat? ①

MOTIVATE

Skills WarmUp
To help students understand insects, have them do the Skills WarmUp.
Answer Answers may vary. One approach would be to place a small light in a terrarium and observe insects' behavior with and without light.

Prior Knowledge
To gauge how much students know about insects, ask the following questions:

▶ What are some insects that you see often?

▶ What benefits do insects provide?

Answer to In-Text Question

① **Grasshoppers have mouthparts that help them cut and chew plant parts. Moths have long tongues for gathering liquid nectar from flowers. Houseflies have spongy mouthparts that help them soak up liquefied organic matter.**

The Living Textbook:
Life Science Side 4

Chapter 11   Frame 18824
Mosquito/Butterfly (Movies)
Search: Play:

Chapter 16 Invertebrates II

TEACH

Skills WorkOut
To help students observe insects, have them do the Skills WorkOut.
Answer Insects will be jumping, flying, or crawling. Some may be eating.

Class Activity
If possible, have students work in small groups to observe grasshoppers or other arthropods. The insects may be living or preserved. Have students find the three body regions. Then have them use a hand lens to look at the head to see if they can find the simple eye at the base of each antenna. Also have them look at the compound eye and the mouthparts. Then have students look for all the other parts labeled in Figure 16.15.

Enrich
Use Enrich Worksheet 16.3.

Discuss
Have students study the diagram of the grasshopper. Then ask the following questions:

▶ What are the three main regions in the grasshopper's body? (Head, thorax, abdomen)

▶ What is the function of the tympanum? (It senses sound.)

▶ What are the four front legs used for? What are the hind legs used for? (The front legs are used for walking, but the large hind legs are used for jumping.)

Answer to In-Text Question

① Answers may vary, but students should see all the parts indicated in the diagram with the exception of the simple eye.

INTEGRATED LEARNING

STS Connection
Periodically, swarms of grasshoppers devastate crops. Have students find materials that explain some of the methods of protecting crops from grasshoppers. Local Farm Bureau offices could be good information sources.

Skills WorkOut

Observe Get a piece of string about 50 cm long and a hand lens. Go outside and lay the string in a circle on a lawn, in a field, or under a tree. Sit quietly for three minutes and look for insects in the circle. What are they doing? Write your observations.

Figure 16.15 ▲
What parts of the grasshopper can you find in the photograph above? ①

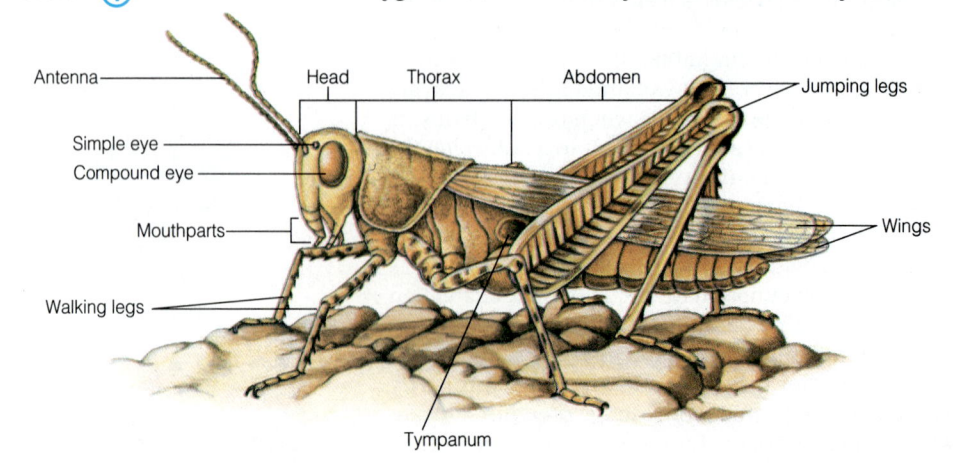

Body Structure of Insects

Insects, like all arthropods, have exoskeletons, segmented bodies, and jointed appendages. They differ from other arthropods in having three pairs of legs. In addition, they are the only arthropods with wings. All insect bodies are divided into the three basic arthropod body regions: head, thorax, and abdomen.

The grasshopper shown in Figure 16.15 is a good example of an insect. Notice where its body is divided into the three regions. Now look at the head. The grasshopper has a pair of compound eyes and three simple eyes. Its antennas are for touch and smell. The grasshopper's mouthparts are adapted for chewing.

The grasshopper's thorax is divided into three segments. It has a pair of legs attached to each segment. The first two pairs are for walking. The hind pair are powerful legs used for jumping. Notice that the grasshopper has two pairs of wings, attached to the second and third segments of the thorax. Many other insects have two pairs of wings, but some have only one pair, and others are wingless.

A grasshopper's abdomen has ten segments. Notice the tympanum (TIHM puh nuhm), which is like an eardrum; it senses sound. Other insects have tympanums on their thorax or legs.

Like other insects, a grasshopper has tiny slits in its thorax and abdomen. Air enters these slits and flows through a series of tubes called tracheae (TRAY kee EE). Oxygen diffuses from tiny tubes into the body cells.

TEACHING OPTIONS

Cooperative Learning
Have students work in small groups to create a bulletin board display about metamorphosis. Some students can collect or draw pictures of insects that go through incomplete metamorphosis; others can concentrate on complete metamorphosis. One group can draw a diagram showing the complete process. Another group can research the meaning of the word *metamorphosis* and attach information to the display. Others can research folktales or myths about it and write summaries.

Directed Inquiry
Have students study the figures and text on this page. Ask some or all of the following questions:

▶ On what kind of material does an insect lay its eggs? (On a food source)

▶ What is the name of an insect that looks like a little adult? (Nymph)

▶ What is the name that most people give to a young larva of an insect? (Caterpillar)

▶ What is the name of the process in which insects develop from an egg to an adult? (Metamorphosis)

▶ At what stage of the life cycle is the butterfly? (The adult stage)

▶ In which process—complete or incomplete metamorphosis—is there a pupa stage in which some cells are destroyed while others divide rapidly? (Complete metamorphosis)

Insect Development

All insects reproduce sexually. In most species the eggs are fertilized inside the female's body by a male's sperm. Many insects lay eggs on a food source so the young can begin to eat right after they hatch.

Most insects don't look exactly like their parents when they hatch from their eggs. As they grow, they undergo a process called **metamorphosis** (MEHT uh MOR fuh sihs). During metamorphosis, a young organism changes its appearance to become an adult.

Incomplete Metamorphosis Some insects, such as grasshoppers and lice, go through an incomplete metamorphosis. This is a series of molts in which the insect changes from an egg to a **nymph** (NIHMF) to an adult. A nymph looks like a little adult, but it has no wings and is not yet able to reproduce.

Complete Metamorphosis Look at Figure 16.17. Most insects go through complete metamorphosis, which involves a more complete change of appearance. Bees, wasps, flies, beetles, butterflies, and moths are among the insects that go through complete metamorphosis. They change from eggs to larvae to pupae to adults.

In the larva stage, the insect usually looks something like a worm. You have probably seen caterpillars, which are the larvae of butterflies or moths. During the **pupa** (PYOO puh) stage, the insect does not eat or move around. In many species the pupa is surrounded by a cocoon. During this stage, amazing changes are taking place. Most of the larva's cells are destroyed and others divide rapidly. When metamorphosis is complete, an adult insect emerges.

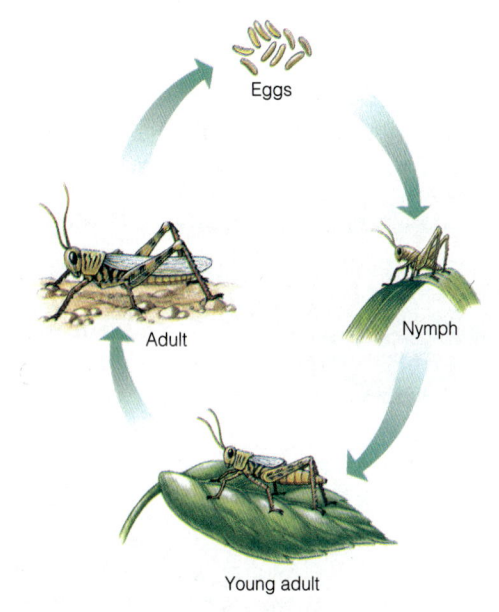

Figure 16.16 ▲ Incomplete Metamorphosis

Figure 16.17 ▲ Complete Metamorphosis

The Living Textbook: Life Science Side 3
Chapter 20 Frame 40133
Cockroach: Grooming and Mating (Movie)
Search: Play:

The Living Textbook: Life Science Side 4
Chapter 15 Frame 38063
Butterflies (Movie)
Search: Play:

Chapter 16 Invertebrates II **333**

TEACH • Continued

Enrich
Tell students that some ants in semidesert areas use other ants as living containers for food. During rainy times, some ants feed on water and nectar. As these ants store the food in their bodies, their abdomens swell. They hang upside down in the nest so that the food they hold can be used by the rest of the colony during the long, dry season.

Research
Have students use references to find out about termite colonies that form mounds. Ask them to find out how many termites may be present in a mound, where the mounds are located, and how alligators may sometimes make use of a termite mound for their developing eggs.

Consider This
Have students consider some other ideas related to pesticide use. For example, certain insects may develop a resistance to the pesticide being used in a particular area. As a result, the pesticide has to be changed to control the insects. Sometimes harmful insects can be given diseases that will cause the insects to die. Ask students how these ideas may affect their feelings about pesticide use.

Decision Making
If you have classroom sets of *One-Minute Readings,* have students read Issue 17, "Pesticide Pollution" on page 28. Discuss the questions.

INTEGRATED LEARNING

Integrating the Sciences

Chemistry One way that social insects communicate is by releasing chemicals called pheromones. Some pheromones signal members of insect colonies about the presence of food or danger. A worker ant finds food and drags its abdomen along the ground as it returns to the nest. This releases a pheromone trail that other ants can smell. The ants follow this trail to find the food source. Ask students if they ever noticed large numbers of ants gathered around a piece of food.

Social Insects

Most insects live on their own, but some species live together in colonies. They are called social insects. Social insects include termites, most ants, and some wasps and bees. A colony of social insects is in many ways like a single organism. The members of the colony have specialized functions, much like different cells. They also have systems of communication that allow them to work together smoothly.

In an ant colony, or nest, most of the ants are wingless female workers. They build the nest, gather food from outside, care for the young, and defend the nest against invaders. They communicate with each other through touch and chemicals. The worker ants do not reproduce. That job is reserved for the queen ant. She lays eggs all during her long life. The eggs are fertilized by stored sperm the queen receives after a single mating with a winged male ant.

Figure 16.18 ▲
How are these ants cooperating? Could a single ant do the job alone?

Consider This

Should Pesticide Use Be Reduced?

Crops are prime targets for hungry insects. Most farmers rely heavily on pesticides to control insect pests, which lets them produce more and better-looking crops.

Some pesticides cause health problems for the people who use them. They may remain in small amounts on the food you eat. They may kill birds, fish, and many invertebrates when they drain into rivers and lakes. Some pesticides enter the food chain.

Consider Some Issues People who want to reduce pesticide use believe the harm pesticides cause outweighs the benefits. They point out that pesticides kill beneficial insects that may help control harmful ones. They propose switching to other methods of controlling pests, including the use of organisms that attack the pests.

Supporters of pesticide use believe the alternative methods won't work as well as pesticides. They claim that reducing the use of pesticides will also reduce the amount of food that can be grown, leading to hardship for farmers.

Think About It Does the harm done by pesticides justify limiting their use? Can the problems caused by reducing the use of pesticides be overcome? What is the best policy over the long term?

Debate It Find out more about all sides of the pesticide issue. Choose a position. Debate the issue with other students in your class.

334 Chapter 16 Invertebrates II

STS Connection

Insects provide many benefits to people. Many farmers depend upon insects to pollinate crops and orchards. Bees provide honey, and moth caterpillars produce silk. Insects can also be harmful to people. Certain species carry diseases that enter food chains, affecting people. Some species destroy food crops. Research of insect reproduction cycles has helped reduce harmful insect populations.

Science and Technology *A Robot Honeybee*

All insects communicate in some way with others of their kind. Ants, for example, give off a chemical that marks a trail to a food source. They also touch each other's antennas. Fireflies communicate with flashes of light.

One of the most studied methods of insect communication is the "waggle dance" of the honeybee. The dance is used by a worker bee to communicate information about the location of a food source to other members of the hive. A bee performing the dance wiggles its abdomen, vibrates its wings, and runs around in a figure-eight pattern.

To study how information is communicated in this dance, a group of scientists in Denmark and Germany have developed a robot honeybee that can imitate the dance. Its movements are controlled by a computer. The scientists had the robot perform dances based on what they already knew about how a bee's movements communicated directions.

The scientists found that the robot could communicate enough information for some bees to find the location of a food source. But it could not do as well as a real bee. This result suggested that there is more to honeybee communication than the scientists had thought. One discovery was that the kind of sound produced by the vibration of the bee's wings seems to be important. The sound has to be just right for the correct message to be communicated.

Figure 16.19 ▲
Even though the robot honeybee looks nothing like a bee, real bees can still get information from it.

Check and Explain

1. How does oxygen reach an insect's body cells?
2. What is a social insect? Give an example, and explain why social behavior helps the group survive.
3. **Compare and Contrast** What is the difference between a larva and a nymph? How is each different from an adult insect?
4. **Make a Model** Choose an insect, other than a grasshopper, with which you are familiar. You may want to observe one for a while. Write a paragraph describing its body structure. Draw a picture and label its body parts.

Chapter 16 Invertebrates II **335**

Research
Have students use references to research how honeybees work together in a colony. Have them find out what the bees called workers, drones, and the queen do in the colony. You may wish to refer them to Figure 26.12 on page 573.

EVALUATE

WrapUp
Reteach Have students refer to the section objectives on page 331. Have them work in small groups to quiz one another on the first three objectives.
  Use Review Worksheet 16.3.

Check and Explain

1. Oxygen enters the insect's body through its thorax and abdomen, flows through the tracheae, and diffuses from tiny tubes into the body cells.
2. Social insects live in colonies. Many ants are social insects. A colony of ants can perform tasks that meet the survival needs of the individual ants. A single ant would not be able to perform all of these tasks.
3. A nymph resembles an adult insect without wings. A larva looks unlike the adult.
4. Answers may vary. A drawing and description of a butterfly should include the head, thorax, abdomen, legs, wings, mouthparts, eyes, and antennae.

**The Living Textbook:
Life Science Side 4**

Chapter 15   Frame 38467
Honeybees (Movie)
Search: Play:

335

ACTIVITY 16

Time about 2 weeks **Group** pairs

Materials

15 wide-mouth jars
8 ripe bananas
nylon stocking material
rubber bands
small spoons
microscope slides
coverslips
microscopes
hand lenses

Analysis

1. Observations will vary. The banana will spoil; tunnels may form as larvae feed.
2. Students should realize that they were looking for eggs or larvae.
3. The fruit flies might have continued to reproduce.
4. Complete; students may have observed any of the stages.
5. Answers may vary, but will likely reflect the duration of the activity, from the start to the appearance of flies in the jars.

Conclusion

Answers will vary. See page 333 for a discussion of insect metamorphosis.

Everyday Application

Answers will vary. Students may suggest that the overripe produce be packaged in such a way that it is not possible for fruit flies or other insects to lay eggs in the fruit.

Extension

Students should realize that by altering the procedure to include a jar that remains covered throughout the activity, they could show that no living thing ever appears out of the decaying banana, thereby refuting spontaneous generation.

TEACHING OPTIONS

Prelab Discussion

Have students read the entire activity. Ask the following discussion questions before beginning the activity:

▶ Where do the fruit flies in this activity come from?
▶ What is the purpose of the banana in the jar?
▶ What are the stages of insect metamorphosis?

Activity 16 What are the stages in fruit fly reproduction?

Skills Observe; Collect Data; Infer

Task 1 Prelab Prep
Collect the following items: wide-mouth jar, half of a ripe banana, nylon stocking, rubber band, small spoon, microscope slide, coverslip, microscope, hand lens.

Task 2 Data Record
1. On a sheet of paper, draw the data table shown below. Write the date you begin this activity in the first box.
2. Mark the date in each box as you record your observations.
3. Record all your observations about the changes you observe in the jar. Describe what you see with words and pictures.

Table 16.1 Fruit Fly Observations

Date	Observations

Task 3 Procedure
1. Peel the banana and put it in the jar. Leave the jar open.
2. Check the jar every day. When you see a few small flies in the jar, put the stocking over the mouth of the jar. Fasten it with a rubber band.
3. Carefully observe the flies over the next three days. Notice their features. Do not remove the stocking from the jar.
4. After three days, remove the stocking from the jar. Let all the flies go outside. Put the stocking back on the jar and fasten it.
5. For the first day or two, spoon out a tiny bit of the banana and smear it on a slide. Look at it under a microscope. Record your observations. Wash the slide.
6. Continue checking the jar every day. Use a hand lens or microscope if you want. Record your observations.
7. When you see flies in the jar again, let them go outside.
8. Clean out the jar and wash it well.

Task 4 Analysis
1. What changes did you observe in the jar?
2. What did you expect to see under the microscope in step 5? What did you see? Explain.
3. What would have happened if you had not cleaned out the jar in step 8 above?
4. Is a fruit fly's metamorphosis complete or incomplete? Explain. Which stages did you observe?
5. About how long is the life cycle of a fruit fly? Explain.

Task 5 Conclusion
Write a paragraph describing the stages of metamorphosis in a fruit fly. Explain how you know, whether you did or did not observe all the stages in this activity.

Everyday Application
Some supermarkets sell slightly overripe fruits and vegetables at bargain prices. How should this produce be packaged for sale? Why?

Extension
Many people a long time ago believed in "spontaneous generation." They thought living things could come from decaying food. How could you use what you learned in this activity to refute spontaneous generation?

INTEGRATED LEARNING

Themes in Science
Diversity and Unity Echinoderms are very different from humans yet humans and echinoderms share some characteristics. The human face shows bilateral symmetry but the human hand shows radial symmetry. Echinoderm larvae begin development much like vertebrate embryos. Stress to students that all life is related at some level.

Integrating the Sciences
Geology Many ancient echinoderms, such as crinoids, blastoids, and cystoids, became extinct at the end of the Permian period. A number of factors caused the extinction, including the formation of Pangaea and climate changes on the earth. Have students research the changes that were taking place at the end of the Permian period. Have them share information in a class discussion.

SECTION 16.4

Section Objectives
For a list of section objectives, see the Student Edition page.

Skills Objectives
Students should be able to:

Research the word echinoderm.

Infer why cutting up starfish did not control the starfish population.

Infer why some species of echinoderms are called living fossils.

Vocabulary
echinoderms

16.4 Echinoderms

Objectives
▶ **Describe** the characteristics of echinoderms.
▶ **Identify** the five classes of echinoderms.
▶ **Describe** the adaptations of starfish.
▶ **Infer** the evolutionary history of echinoderms.

Skills WarmUp
Research Look in a dictionary or encyclopedia to find out the meaning of the Greek roots that make up the word *echinoderm*. Why do you think this phylum was given this name?

What kind of stars are never seen in the sky but often found in the sea? What kind of dollars are you more likely to find on a beach than in a bank? What kind of cucumber is an animal, not a plant? If you can't guess the answers to these riddles now, you will be able to soon. They are all about invertebrates with radial symmetry called **echinoderms** (ee KY noh DURMZ).

Characteristics of Echinoderms

Echinoderms live in the ocean. Like humans, they have hard endoskeletons, which are support structures inside their bodies. Many echinoderms also have hard spines or bumps extending from the endoskeleton.

Like cnidarians, adult echinoderms have radial symmetry. They develop, however, from larvae with bilateral symmetry. In the center of an echinoderm's rounded body, or disk, is a mouth. In most echinoderms, the mouth faces downward. Some echinoderms have arms radiating out from the disk, and some don't.

Echinoderms are the only animals with tube feet, thin-walled hollow tubes or tentacles used for movement and feeding. Most echinoderms have many tube feet, each as thin as spaghetti and a few centimeters long. The tube feet are part of a system of water canals that run throughout an echinoderm's body.

Even though echinoderms have radial symmetry like the cnidarians, they are not related to cnidarians. In fact, these simple-looking organisms are more closely related to vertebrates than to the other invertebrates. Echinoderm larvae begin their development much like vertebrate embryos.

Figure 16.20 ▲
Sand dollars are pancake-shaped echinoderms that bury themselves in sand.

MOTIVATE

Skills WarmUp
To help students understand echinoderms, have them do the Skills WarmUp.
Answer Students should note that in Greek *echinos* means "spiny" and *derma* means "skin." The animals in this phylum have spiny or spiky surfaces.

Misconceptions
Students may think that starfish are a type of fish or that sea lilies and sea cucumbers are types of plants. Have students note the characteristics of these organisms so they can see why scientists classify them as echinoderms.

💿 **The Living Textbook:**
Life Science Sides 1-4

Chapter 4   Frame 01223
Echinoderms (30 Frames)
Search: Step:

Chapter 16 Invertebrates II

TEACH

Explore Visually

Have students study the pictures of echinoderms on this page and the paragraphs that describe them. Then ask some or all of the following questions:

▶ Which echinoderms are attached to the ocean floor and use their branched arms to catch food? (Sea lilies)

▶ Which echinoderms have long, flexible bodies and no arms? (Sea cucumbers)

▶ Which echinoderms have long, whiplike arms and look something like a starfish? (Brittle stars)

▶ Which echinoderms are covered with spines, but have no arms? (Sea urchins)

Answer to In-Text Question

① The brittle star can regrow its lost arms. The purpose of this adaptation is survival. If an arm breaks off, then a predator might be satisfied with the arm while the brittle star gets away.

The Living Textbook: Life Science Sides 1-4

Chapter 4 Frame 01241
Sea Urchins (10 Frames)
Search: Step:

The Living Textbook: Life Science Sides 1-4

Chapter 4 Frame 01250
Sea Cucumbers (3 Frames)
Search: Step:

INTEGRATED LEARNING

Integrating the Sciences

Oceanography Brittle stars have been found at great ocean depths, even at the bottom of some trenches. They eat decaying matter, but more interesting than this is their extremely thin body design and hairlike legs. Ask students to hypothesize why such a small surface area would be an advantage at such great depth. (The less surface area, the smaller the area for pressure to push upon.)

Diversity of Echinoderms

The 6,000 or so species of echinoderms are organized into five classes. Starfish make up the largest class. The other four classes are the brittle stars, the sea urchins, the sea cucumbers, and the sea lilies.

◀ Brittle stars look something like starfish, but they have long, whiplike arms. If you tried to pick up a brittle star, its arms would break off. What do you think is the purpose of this adaptation? ① Brittle stars eat dead or decaying matter.

▲ Sea urchins have no arms. Their bodies are covered with spines. The spines are for protection and locomotion. Sea urchins have five sharp teeth for scraping and chewing. They eat mostly many-celled algae. In Japan, people eat the reproductive organs of sea urchins. Sand dollars are also included in this class of echinoderms.

Sea lilies look like ▶ flowers, with their cup-shaped bodies attached to the ocean floor by stalks. Their branched arms are used to catch food. Sea lilies were very common hundreds of millions of years ago. The few sea lilies that exist today look very much like sea lily fossils 500 million years old.

Sea cucumbers do not ▶ look very much like other echinoderms. They have long, flexible bodies with a reduced endoskeleton and no arms. Using their tube feet, they burrow into the sand on the bottom of the sea. They move more freely than other echinoderms. Some people eat dried sea cucumbers.

338 Chapter 16 Invertebrates II

Art Connection
Provide students with supplies such as clay, paint, sand, pipe cleaners, and paper. Have students make a sculpture or model of an echinoderm.

Starfish

If you've seen an echinoderm, it was probably a starfish. They are often found clinging to rocks at the ocean's edge. Starfish have five or more arms. Look at Figure 16.21. Notice that the arms are not just appendages. They contain internal organs. The undersides of the arms are covered with tube feet.

Starfish have adapted to eating filter-feeding bivalves, such as mussels and clams. Although the hard shells of these mollusks keep most predators away, starfish have no problem with them. The starfish wraps its strong arms around the bivalve, holding on tightly with the suction cups on its tube feet. Slowly the starfish pulls open its prey's shell just a little. Then it turns its stomach inside out and pushes it into the shell. Digestive enzymes kill the mollusk, and the starfish has a meal.

Like all echinoderms, starfish have a system of water canals attached to the tube feet. Water pressure inside the canals helps keep the tube feet rigid. An opening on top of the body strains the water as it flows into the canal system.

Starfish and other echinoderms can regrow, or regenerate, lost body parts. One starfish arm including just part of the central disk can regenerate into an entire new animal.

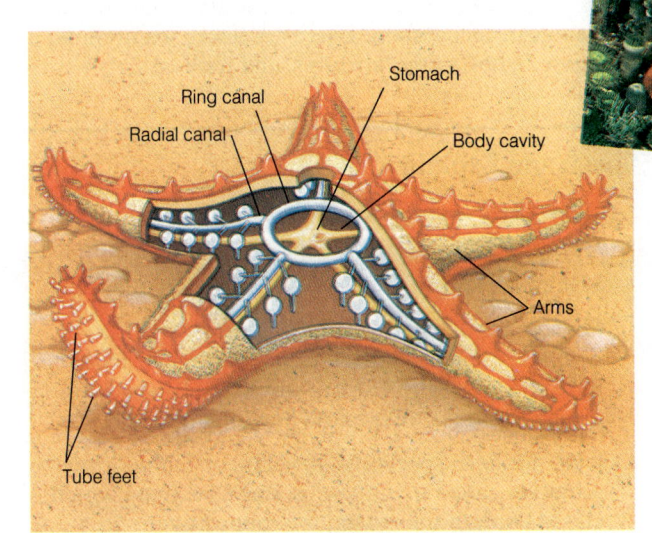

Skills WorkOut
Infer Starfish eat the mollusks harvested as shellfish. At one time, when shellfish farmers found starfish, they cut up the starfish and threw them back into the sea. But the starfish population just got larger and did even more damage. Why do you think this happened?

Figure 16.21 ▲
The red patterns on this African red starfish (above) are parts of its endoskeleton. The endoskeleton is not shown in the drawing to the left so that you can see the system of water canals inside the starfish's body.

Skills WorkOut
To help students understand starfish, have them do the Skills WorkOut.
Answer As long as the starfish has one arm and part of its central disk, it can slowly regrow all of its missing arms and other parts.

Reteach
Use a large piece of chart paper. Draw five columns on the paper and label them *Brittle Star, Sea Lily, Sea Urchin, Sea Cucumber,* and *Starfish*. Then divide the columns into four rows. Label the rows *Appearance, Food, Means of Movement,* and *Protective Adaptation*. Complete the chart in a class discussion. Display the completed chart so students can refer to it.

Discuss
Have students study Figure 16.21. Ask the following questions:
▶ What do the radial canals remind you of? (The spokes of a wheel)
▶ When water is forced into a tube foot, it expands and gives it suction. How do you suppose the suction is released? (Water is removed from the tube foot, which causes it to contract.)

Enrich
Tell students that starfish have a light-sensitive spot at the tip of each arm. They also have simple touch, taste, and smell receptors. Starfish are so flexible in the water that they often bend themselves around rocky surfaces.

TEACH • Continued

Discuss

Have students think about what happens when people interfere with the balance of nature. Ask the following questions:

▶ What would happen to the ecosystem if the sea urchins became extinct? (The otters would have to seek other food, and some would die.)

▶ What would happen if the kelp were removed? (The sea urchins would die off, and then the sea otters would disappear.)

EVALUATE

WrapUp

Review Have students make a chart listing similarities and differences among mollusks and echinoderms.

Use Review Worksheet 16.4.

Check and Explain

1. Tube feet

2. Starfish have five or more arms and tube feet. Sea lilies look like underwater flowers, and their arms catch food. Sea cucumbers have long, flexible bodies with no arms, and they burrow into sand. Brittle stars look like starfish, but their arms are longer and more fragile. Sea urchins have no arms, their bodies are covered with spines, and they have fine, sharp teeth.

3. Starfish feed on bivalves, which are blind and move quite slowly, if at all.

4. Sea lilies aren't much different from 500-million-year-old sea-lily fossils.

340

INTEGRATED LEARNING

Themes in Science

Systems and Interactions As you read about sea urchins, otters, and kelp, stress to students that living things in an ecosystem interact with one another. Changes in one part of the system have an impact on another part.

Science and Society
Sea Urchins, Otters, and Kelp

When Europeans began to settle in California in the 1700s, huge "forests" of kelp grew in a band just offshore. Kelp are many-celled algae that grow as big as trees. Air bladders keep their leaflike blades at the surface. Living in the band of kelp were many thousands of sea otters, marine mammals with thick fur.

In the 1800s the sea otters were hunted for their fur. By the early 1900s, they were nearly extinct. People noticed that while the population of otters was falling, the kelp forests were also disappearing.

The connection between the decline of both the sea otter and the kelp was the sea urchin. Sea urchins eat kelp. Normally, the population of sea urchins was kept in balance by their major predator—the sea otter. But as the sea otter began to disappear, the urchins greatly increased their numbers. They munched away on the kelp, and eventually there was little kelp left.

Then in 1911, an international treaty ended the otter fur trade. In 1938, a few sea otters were found off the California coast and immediately protected. In 1973, the Marine Mammal Protection Act was passed. These actions have helped the sea otter make a comeback.

Not surprisingly, the return of the sea otter has helped the kelp forests grow back. This is restoring the diversity of the marine ecosystem. There are fewer sea urchins, but more fishes, crabs, shrimps, and mollusks.

Figure 16.22 ▲
A sea otter floats in kelp off the California coast.

Check and Explain

1. What makes echinoderms different from other invertebrates?

2. List and describe the five classes of echinoderms.

3. **Reason and Conclude** Most predators, such as octopuses, spiders, and tigers, are fast-moving and have good eyesight. Explain why starfish, which are also predators, can be slow-moving and blind.

4. **Infer** Modern species of one class of echinoderms are often called living fossils. Can you guess which class? Explain the reason for your answer.

340 Chapter 16 Invertebrates II

Chapter 16 Review

Concept Summary

16.1 Mollusks
- Mollusks have soft, nonsegmented bodies with a mantle and a muscular foot. They have complex organ systems.
- The three classes of mollusks are the gastropods, the bivalves, and the cephalopods.

16.2 Arthropods
- Arthropods have jointed appendages and bodies divided into a head, thorax, and abdomen.
- Arthropods' bodies are enclosed in an exoskeleton.
- The major classes of arthropods are the arachnids, the crustaceans, the centipedes and millipedes, and the insects.

16.3 Insects
- Insects are six-legged arthropods. Many have wings.
- Insects go through either incomplete or complete metamorphosis to become adults.
- Termites and some ants, bees, and wasps are social insects.

16.4 Echinoderms
- Echinoderms live in the oceans. They have hard endoskeletons, radial symmetry, and tube feet.
- The five classes of echinoderms are the brittle stars, the sea urchins, the sea cucumbers, the sea lilies, and the starfish.

Chapter Vocabulary

mollusk (16.1)	arthropod (16.2)	arachnid (16.2)	nymph (16.3)
gastropod (16.1)	thorax (16.2)	crustacean (16.2)	pupa (16.3)
bivalve (16.1)	abdomen (16.2)	antenna (16.2)	echinoderm (16.4)
cephalopod (16.1)	exoskeleton (16.2)	metamorphosis (16.3)	

Check Your Vocabulary

Use the vocabulary words above to complete the following sentences correctly.

1. Arthropods have an outer support structure called an ____.
2. ____ make up a phylum of soft-bodied animals with a muscular foot.
3. An insect with incomplete metamorphosis begins life as a ____.
4. ____ are sensory organs on the heads of arthropods.
5. Crabs and lobsters belong to the arthropod class called the ____.
6. The body region between the head and abdomen of an arthropod is called the ____.
7. Insects undergo ____ to become adults.
8. A ____ has a one coiled shell.
9. The body region of an insect farthest from the head is the ____.
10. ____ make up a phylum of invertebrates with jointed legs.
11. ____ are mollusks with two hinged shells.
12. During the ____ stage, an insect undergoes most of the changes of complete metamorphosis.
13. Invertebrates with endoskeletons and radial symmetry are called ____.
14. Squid belong to a mollusk class called the ____.

CHAPTER REVIEW 16

Check Your Vocabulary

1. exoskeleton
2. Mollusks
3. nymph
4. Antennae
5. crustaceans
6. thorax
7. metamorphosis
8. gastropod
9. abdomen
10. Arthropods
11. Bivalves
12. pupal
13. echinoderms
14. cephalopods

Use Vocabulary Worksheet for Chapter 16.

Chapter 16 Invertebrates II

CHAPTER 16 REVIEW

Check Your Knowledge

1. Spiders have four pairs of legs instead of three; they have two main body regions: the cephalothorax and the abdomen.
2. Answers may vary. Possible answers include: chelicerae and claws for attacking prey or defending, mandibles for chewing and crushing food, antennae for balance and sensing, and legs for moving.
3. Answers may vary, but may include: the adaptation of the insect body to different ways of living; the ability to fly.
4. Echinoderms have tube feet, which allow them to grasp and hold prey.
5. The tentacles are similar to the muscular foot of a snail.
6. Centipedes and millipedes have jointed legs, unlike segmented worms.
7. mantle
8. echinoderms
9. pupa
10. millipedes
11. thorax
12. fast-moving

Check Your Understanding

1. Unlike arthropods, adult echinoderms have radial symmetry, hard endoskeletons, and tube feet.
2. Bivalves are filter feeders, so they may ingest the poisonous algae. Gastropods use a radula to scrape and scoop up food.
3. Answers will vary. Possible answers include: Because people found that starfish do not taste good; because they have very little "meat."
4. Accept all logical answers.
5. In order to live underwater, a spider would have to be able to extract oxygen from water and hunt for food without webs. It would be different from other spiders because of these adaptations, but would still have the same body structure as other spiders: exoskeleton, four pairs of legs, a two-part body.
6. Answers may vary. Possible answers include: The gastropod's shell does not cover its entire body, and not all gastropods have a shell.
7. All live underwater. All are primarily stationary-living, and feed on small food particles floating in the water around them.
8. Sea cucumbers are most likely to live on a muddy ocean bottom because they are burrowers. Sea urchins use thin spines for locomotion and to avoid sinking in the mud.
9. Answers may vary. A watery surrounding is vital to all echinoderms to counterbalance their internal water pressure.
10. The insect exoskeleton would not be strong enough to support a body that large.

Chapter 16 Review

Check Your Knowledge
Answer the following in complete sentences.

1. What is the difference between a spider and an insect?
2. What different functions can the appendages of arthropods serve? Name at least three functions and describe the kind of appendage used in each case.
3. Give two reasons why insects have been successful as a life form.
4. What echinoderm characteristic is shared by no other kind of animal? How is this characteristic useful?
5. An octopus's tentacles are similar to what part of a snail?
6. How do millipedes and centipedes differ from segmented worms?

Choose the answer that best completes each sentence.

7. A tissue called the (visceral mass, radula, foot, mantle) produces a shell in many mollusks.
8. Starfish are (cnidarians, echinoderms, arthropods, fish).
9. During the (pupa, egg, larva, nymph) stage, an insect is often wrapped in a cocoon.
10. Arthropods with many legs and repeated segments are (insects, spiders, millipedes, annelids).
11. An insect's wings are attached to the (head, thorax, abdomen, cephalothorax).
12. Cephalopods are the most (common, colorful, fast-moving, dangerous) of the mollusks.

Check Your Understanding
Apply the concepts you have learned to answer each question.

1. **Compare and Contrast** How do echinoderms and arthropods differ?
2. **Application** A red tide is a population explosion of dinoflagellates, one-celled algae that contain a poison. During a red tide, bivalves in the area are not safe to eat, but most gastropods are safe. Explain.
3. Why do you think starfish are not used by humans for food?
4. **Mystery Photo** The photograph on page 320 shows the surface of a moth's wing. What kinds of parts and materials do you think make up this wing? What would happen if you touched it with your finger?
5. **Extension** You discover a spider that has evolved adaptations for life in the ocean. What might some of these adaptations be? How is the spider different from other spiders? What spider characteristics does it still have?
6. Why is a gastropod's shell not considered an exoskeleton?
7. What do barnacles have in common with sponges, bivalves, and sea anemones?
8. **Application** What kind of echinoderm–sea urchins or sea cucumbers–are you most likely to find living where the ocean has a muddy bottom? Explain your answer.
9. Why have no echinoderms adapted to living on land?
10. Why is it impossible for insects to be as large as humans?

Develop Your Skills

1. a. The insect population decreases rapidly, then increases above its original level. The spider population also decreases, allowing the insects to increase in number.
    b. The initial decrease was a result of the pesticide. This also led to a decrease in the spider population. Once the pesticide had dissipated, the reduced spider population allowed the population of the insects to rebound.
    c. The reduction of the spiders (predators) allows the insect (prey) population to rise.
    d. As time passes, the spider population will increase again, leading to a reduction in the insect population, until equilibrium is reached.
2. Answers will vary. A possible answer could be that the light gives the squid an advantage in finding food or shelter.
3. a. Centipedes and Millipedes   c. About 8 times
    b. 7%

Develop Your Skills

Use the skills you have developed in this chapter to complete each activity.

1. **Interpret Data** The graph below shows the results of an experiment in a farm field.
    a. Describe what happens to the insect population and to the spider population.
    b. What seems to be the cause of these changes?
    c. Why does the insect population rise rapidly after falling for a time?
    d. What is likely to happen to the population levels as more time passes?

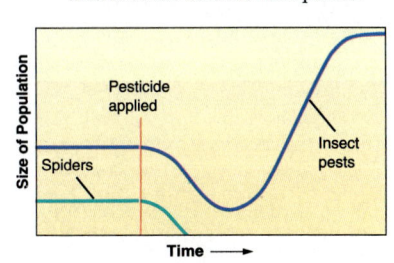

2. **Infer** Many squid that live in the ocean can produce their own light. How do you think this adaptation helps them survive? Explain.

3. **Data Bank** Use the information on page 627 to answer the following questions.
    a. Which arthropods have the fewest number of species?
    b. What percentage of the total number of arthropod species are arachnids?
    c. In number of species, insects outnumber all non-insect arthropods by how many times?

Make Connections

1. **Link the Concepts** Below is a concept map showing how some of the main concepts in this chapter link together. Only part of the map is filled in. Finish the map, using words and ideas you find in the chapter.

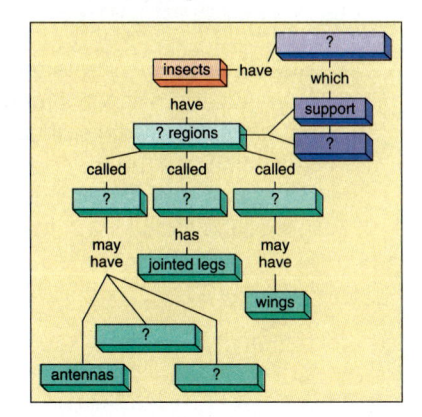

2. **Science and Society** To control insect pests, some farmers and gardeners bring in organisms that prey upon the pests. Find out what organisms are used for this purpose, and which kinds of pests each will eat. Write a report on this form of biological pest control.

3. **Science and You** Do research on the life cycle of fleas to find out why they are so hard to get rid of. Based on what you learn, explain how you would try to control fleas in your home and on your pets.

4. **Science and Art** Start a butterfly collection, using photographs instead of real butterflies. Draw your own pictures of each butterfly. Make hypotheses about the purposes of their colors and patterns.

Make Connections

1.

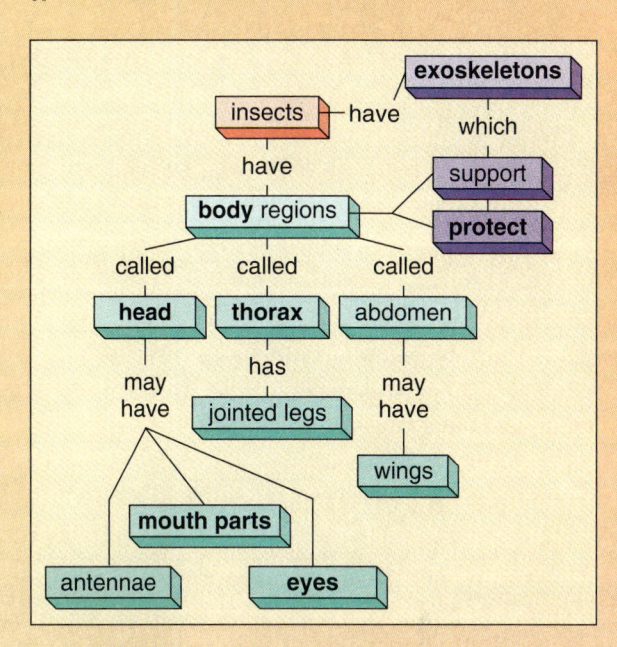

2. Student reports will vary, but should include a discussion of the advantages (such as avoiding harmful chemicals) and disadvantages (such as introduced populations reproducing out of control and becoming pests themselves) of this type of pest control.

3. Student reports will vary. A suggestion for controlling fleas might involve taking advantage of the knowledge of the life cycle of the flea.

4. Student hypotheses will vary but will likely focus on camouflage for protection or coloration for attracting mates.

Chapter 16 Invertebrates II

CHAPTER 17

Overview

This chapter introduces vertebrates. The first section explains the evolution of vertebrates, which helps to illustrate the diversity of this phylum. The second section focuses on the body features, methods of movement, and organ systems of three classes of fish. The third section describes amphibians, illustrating their unique adaptations to land and dependence on water for part of their life cycle. The chapter closes with a discussion of the evolutionary history of reptiles as well as descriptions of their characteristics and adaptations.

Advance Planner

▶ Collect old nature or wildlife magazines for student use in SE activity and TE activity, page 347.

▶ If available, set up an aquarium so students can observe fishes. If not, try to obtain photographs or a film about fishes. TE activity, page 351.

▶ If possible, set up an aquarium or terrarium containing tadpoles and/or frogs for TE activity, page 354.

Skills Development Chart

Sections	Classify	Communicate	Compare and Contrast	Decision Making	Infer	Measure	Observe	Research
17.1 Skills WarmUp Skills WorkOut	●		●					
17.2 Skills WarmUp Skills WorkOut Consider This		● ●		●	●			
17.3 Skills WarmUp					●			
17.4 Skills WarmUp Skills WorkOut Activity	●				●	●	●	●

Individual Needs

▶ **Limited English Proficiency Students** Have students write the vocabulary terms from the chapter in their science journals, leaving plenty of space between terms. Then, as they study the chapter, have them write the figure numbers of any pictures or diagrams that describe each term. Have students use these pictures and figures to reinforce the meanings of the chapter's terms.

▶ **At-Risk Students** Have students visit a pet store where they can see some of the organisms described in this chapter. Have them prepare for the visit by listing the characteristics of fishes, amphibians, and reptiles. As they observe the animals, have them write the name of an animal next to each of the characteristics on their lists. If possible, have students talk with a store employee to find out what conditions they need to provide for different animals.

▶ **Gifted Students** As an extension of the editorial activity about sharks on page 350, have students create a newspaper with various features focusing on fishes, amphibians, and reptiles. Students should use periodicals and write reports about organisms and their habitats. The newspapers could include feature articles, an editorial page, comics, and interviews. Students could review a popular movie that stars one of these animals. Have them include pictures. If possible, allow students to prepare the final layout and copy for the newspaper using a computer.

Resource Bank

▶ **Bulletin Board** Create a bulletin board display that pictures fun facts about fishes, amphibians, and reptiles. Begin by hanging photographs of each kind of organism. Then add a fact, such as "The paradoxical frog of South America grows smaller as it develops into an adult." Have students find interesting facts to add to the display.

CHAPTER 17 PLANNING GUIDE

Section	Core	Standard	Enriched	Section	Core	Standard	Enriched
17.1 Common Traits of Vertebrates pp. 345–347				**Laboratory Program** Investigation 32	●	●	●
Section Features Skills WarmUp, p. 345 Skills WorkOut, p. 347	● ●	● ●	●	**17.3 Amphibians** pp. 353–356			
Blackline Master Review Worksheet 17.1	●	●		Section Features Skills WarmUp, p. 353	●	●	
Color Transparencies Transparency 49	●	●		Blackline Masters Review Worksheet 17.3 Skills Worksheet 17.3	● ●	● ●	
17.2 Fishes pp. 348–352				Color Transparencies Transparency 51	●	●	
Section Features Skills WarmUp, p. 348 Skills WorkOut, p. 350 Consider This, p. 350	● ●	● ● ●	●	**17.4 Reptiles** pp. 357–362			
Blackline Masters Review Worksheet 17.2 Enrich Worksheet 17.2 Integrating Worksheet 17.2	● ● ●	● ● ●	●	Section Features Skills WarmUp, p. 357 Skills WorkOut, p. 358 Career Corner, p. 360 Activity, p. 362	● ● ● ●	● ● ● ●	● ● ● ●
Overhead Blackline Transparencies Overhead Blackline Master 17.2 and Student Worksheet	●	●	●	Blackline Masters Review Worksheet Skills Worksheet 17.4 Integrating Worksheet 17.4 Vocabulary Worksheet 17	● ● ● ●	● ● ● ●	●
Color Transparencies Transparency 50	●	●					

Bibliography

The following resources can be used for teaching the chapter. See page T-40 for supplier codes.

Audio-Visual Sources

(video unless noted)
About Sharks. 12 min. 1981. NGSES.
Amphibians. Revised. 14 min. 1985. C/MTI.
The Living Ocean. 25 min. 1988. NGSES.
Remarkable Reptiles. 30 min. 1986. MSP.
Riches from the Sea. 23 min. 1984. NGSES.

Software Resources

Frog Dissection Guide. Apple, IBM. CABISCO.
Operation: Frog. Apple, Commodore. SS.
Perch Dissection Guide. Apple, IBM. CABISCO.

Library Resources

Burnie, David. How Nature Works. Pleasantville: The Reader's Digest Association, Inc., 1991.
Comstock, Anna Botsford. Handbook of Nature Study. Ithaca: Cornell University Press, 1986.
Durrell, Gerald. The Amateur Naturalist. New York: Alfred A. Knopf, 1989.
Forsyth, A. "Snakes Maximize Their Success with Minimal Equipment." Smithsonian (February 1988): 159–165.

CHAPTER 17

INTEGRATED LEARNING

Writing Connection
Have students write a story or poem about the animal in the photograph. Have them consider whether they want to write from the animal's point of view or from someone else's, such as the photographer or another observer. Encourage them to be creative and to describe the animal and its surroundings in detail. What is it doing? Where does it live? What does it eat? What was it reacting to when the photograph was taken?

Introducing the Chapter

Have students read the description of the photograph on page 344. Ask if they agree or disagree with the description.

Directed Inquiry

Have students study the photograph of the chameleon's eye. Ask:

▶ What do you see in the photograph? (Students may mention scales, the large circle, and the smaller shiny circle in the middle.)

▶ What part of an animal is pictured here? (An eye)

▶ Do you think that the animal pictured is more like you or the animals that you have studied in earlier chapters? (The animal is more like a human.)

▶ To which of the three groups named in the chapter title does this animal belong? (Reptiles) Why do you think so? (It appears to have scales like those of a snake or lizard.)

Chapter Vocabulary

cartilage
ectotherm
endoskeleton
endotherm
gill

Chapter 17 — Fishes, Amphibians, and Reptiles

Chapter Sections

17.1 Common Traits of Vertebrates
17.2 Fishes
17.3 Amphibians
17.4 Reptiles

What do you see?

"What I see here is a close-up picture of a lizard's eyeball. I would describe it as colorful and scaly. I think it is a lizard because you can see the scales in the picture. It probably lives in trees in a tropical forest. The colors on the lizard would suggest a forest-type environment."

Jason Bonte
Sara Scott Junior High
Terre Haute, Indiana

To find out more about the photograph, look on page 364. As you read this chapter, you will learn about fishes, amphibians, reptiles, and some of their common characteristics.

Themes in Science

Diversity and Unity There are about 45,000 species of vertebrates. They include fishes, amphibians, birds, mammals, and reptiles. However, all vertebrates share certain features at some stage of their development. These features include gills or gill slits, a backbone, and a hollow nerve chord.

17.1 Common Traits of Vertebrates

Objectives

- **Describe** the characteristics vertebrates have in common.
- **Distinguish** between endothermic and ectothermic animals.
- **Discuss** the evolutionary development of vertebrates.
- **Classify** invertebrates and vertebrates by constructing a diagram.

Skills WarmUp

Compare and Contrast
Make a list of five important differences between an insect and a frog. Compare your list with a classmate's list. How does your list differ from your classmate's? How are the two lists the same?

Up and down the middle of your back are bones that make up your vertebrae (VUR tuh BREE). The bones in your vertebrae are joined together with cartilage to form a flexible but supportive column called the backbone.

Humans aren't the only organisms with backbones. You share a backbone with fishes, amphibians, reptiles, birds, and other mammals. Recall that these animals with backbones are called vertebrates.

Characteristics of Vertebrates

All vertebrates belong to a single phylum. The name of the phylum is Chordata. Animals in this phylum are called chordates (KOR dayts). Among the chordates are several types of animals that do not have backbones.

All chordates share several traits. One chordate trait is the presence of gills or gill slits at some stage of development. Chordates also have a flexible skeletal rod called a notochord (NOHT uh KORD). The notochord protects a chordate's hollow nerve cord.

In vertebrates, the notochord is present only in the embryo stage. As a vertebrate develops, its notochord is replaced by a backbone that surrounds and protects the spinal cord. The backbone is the central part of an internal support structure called an **endoskeleton**. A vertebrate's endoskeleton is very different from the exoskeleton of an arthropod.

Even though vertebrates share many traits, they have important differences, too. Birds and mammals are **endotherms** (EHN doh THURMS), animals with constant

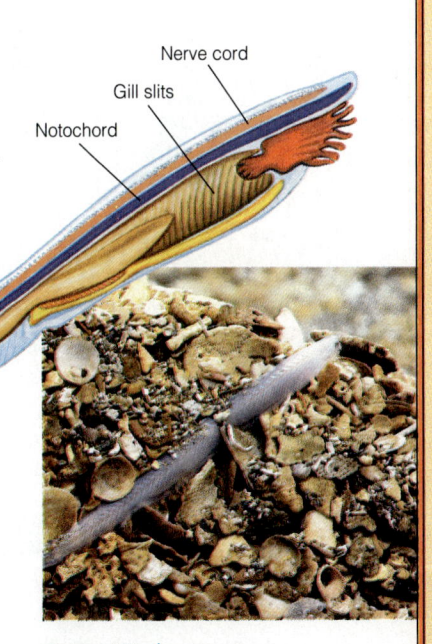

Figure 17.1 ▲
The lancelet is a chordate but not a vertebrate. What does it have in common with vertebrates? ①

Chapter 17 Fishes, Amphibians, and Reptiles 345

INTEGRATED LEARNING

TEACH

Critical Thinking

Reason and Conclude Ask students to suggest why endotherms are said to be warm-blooded and ectotherms are often referred to as cold-blooded. Then have them explain why *cold-blooded* is not a very accurate description of ectotherms. (The blood of an ectotherm is not necessarily cold. Since the temperature of its blood varies with the temperature of the environment, its blood could even be warmer than that of an endotherm.)

Explore Visually

Have students look at the evolutionary tree in Figure 17.2. Ask them the following questions:

▶ Which came first, the ancient amphibians or the ancient reptiles? (Ancient amphibians)

▶ Which came first, the ancient reptiles or the mammals? (Ancient reptiles)

▶ What are the modern groups of vertebrates? (Fishes, amphibians, reptiles, birds, and mammals)

▶ What is the ancestor of all these vertebrates? (Ancient fishes)

Reteach

Briefly review what students have learned about invertebrates. Ask them to identify what distinguishes vertebrates from invertebrates. (Vertebrates have backbones; invertebrates don't.)

The Living Textbook:
Life Science Sides 1-4

Chapter 4 Frame 01283
Fishes (27 Frames)
Search: Step:

346

Language Arts Connection

Because reptiles can't regulate their body temperatures internally, some species *hibernate* when the weather is very cold, and *estivate* or go into a state of inactivity, when it's very hot. Have students find out the definitions and derivations of these two words. Have them also name examples of animals that display this type of behavior. They can keep their notes in their portfolios.

Themes in Science

Stability and Equilibrium Fishes, amphibians, and reptiles are ectotherms. Their body temperatures depend on the external temperature. They adjust by hibernating in cold weather and remaining inactive in hot weather. Thus, they are able to maintain a stable internal temperature. Ask students whether humans are endothermic or ectothermic. (Humans are endothermic.)

---

internal body temperatures. Their body cells produce enough heat to keep their bodies warm despite the changing temperature outside their bodies. Endotherms are commonly described as warm-blooded.

Fishes, amphibians, and reptiles, in contrast, can't keep their body temperatures constant. Their body temperatures depend on the temperature of their environment. Animals with bodies that receive heat from the outside are called **ectotherms** (EHK toh THURMS). Ectotherms are sometimes described as cold-blooded. All the animals you will read about in this chapter are ectotherms.

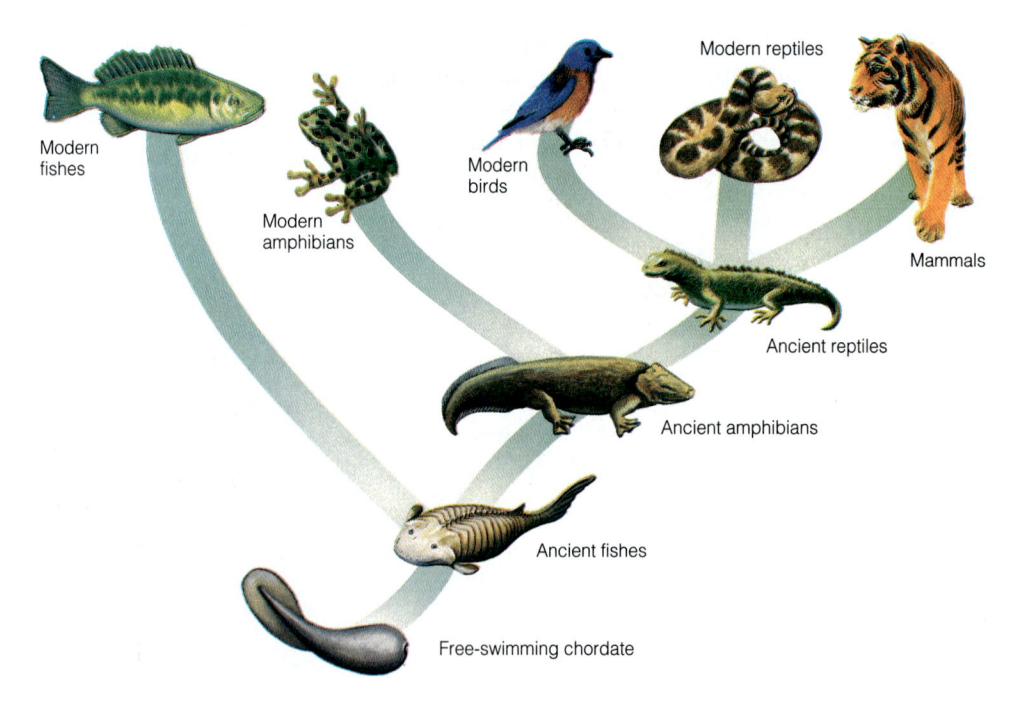

Figure 17.2
This evolutionary tree shows that fishes living hundreds of millions of years ago are the ancestors of all the kinds of vertebrates alive today. ▼

Origin of Vertebrates

The first vertebrates appeared on the earth about 500 million years ago. They were water-dwelling, fishlike animals. Over time they evolved into many different kinds of fish. For many millions of years, fish were one of the most successful life forms on the earth.

Eventually, some fishes developed adaptations that permitted them to move from the water onto the land.

346 Chapter 17 Fishes, Amphibians, and Reptiles

Themes in Science

Evolution Emphasize that the evolutionary process occurs extremely slowly. The animals in the evolutionary tree developed through many kinds of genetic mutations and variations over millions of generations. Have students review the concept of genetic mutation in Chapter 6, page 124.

Writing Connection

Ask students to imagine that they are very old amphibians who witnessed their species' transition from fish to amphibian or amphibian to reptile. Have students write a creative story of what the amphibian might say about "the good old days." Have students present their stories to the class.

Discuss

Ask students if they can remember what their temperatures were the last time they had a fever. Point out that even though 37°C is considered normal, body temperatures in healthy humans can range from 36 to 38°C (97 to 99°F).

Skills WorkOut

Have students classify seven groups of vertebrates by doing the Skills WorkOut. You may wish to write the names on the board and tell students how to pronounce them.

Answer The seven classes of vertebrates are Agnatha (jawless fishes), Chondrichthyes (cartilaginous fishes), Osteichthyes (bony fishes), Amphibia (amphibians), Reptilia (reptiles), Aves (birds), and Mammalia (mammals).

EVALUATE

WrapUp

Portfolio Have students cut out or draw pictures of vertebrates. Have them paste the pictures on notebook paper and label the characteristics of the animals that are shared by all vertebrates.

Use Review Worksheet 17.1.

Check and Explain

1. The bones that form the backbone are vertebrae.
2. Fishes, amphibians, and reptiles are ectothermic.
3. The reptile's body temperature will also increase by the same amount because it is ectothermic, unless the ectotherm does something to counteract the temperature change.
4. Diagrams should show that chordates are a separate phylum from invertebrates. Vertebrates listed should be classified according to Figure 17.2 on page 346.

One of these adaptations was strong, lobelike fins that could be used like limbs to crawl across land. Once out of the water, another adaptation—lungs—helped these animals survive on land. Lungs allowed the organisms to exchange gases with the atmosphere. These early land-dwelling vertebrates were the first amphibians.

Look at Figure 17.2. Notice that the amphibians evolved into reptiles, and the reptiles gave rise to mammals and birds. Each of these vertebrates is classified as a separate class. There are seven classes of vertebrates because the fishes are divided into three classes.

Science and You *Fever Biology*

The temperature of your home is controlled by a mechanical device called a thermostat. You may be surprised to discover that your body also contains a type of thermostat. Unlike your home's thermostat, your body's "thermostat" is made up of living cells. The cells are located in the part of your brain called the hypothalamus (HY poh THAL uh muhs).

Like that of all endotherms, your body temperature stays about the same all the time. Most often, your "thermostat" is set at about 37°C (98.6°F). Sometimes, when you are sick, a fever may cause your body temperature to rise above 37°C. You run a fever when chemicals act on the hypothalamus to "turn up" your thermostat. Doctors and scientists are not sure what function a fever has. Some have suggested that a fever may speed up the metabolism of your body. This would allow the body to produce germ-fighting particles at a faster rate.

Skills WorkOut

Classify Find out the class names for the seven groups of vertebrates. Then make a poster that lists the group's name, such as mammals, next to the class name. Use old magazines to find photos of each class of vertebrates. Tape or glue your photos next to the class of vertebrate to which it belongs. Show your poster to your class.

Check and Explain

1. What are vertebrae?
2. Which classes of vertebrates are ectothermic?
3. **Predict** If the outdoor temperature increases by 20°C, what will happen to a reptile's body temperature?
4. **Classify** Construct a diagram that shows how the following animal groups are classified: chordates, invertebrates, vertebrates, fishes, amphibians, reptiles, birds, and mammals.

Chapter 17 Fishes, Amphibians, and Reptiles

SECTION 17.2

Section Objectives
For a list of section objectives, see the Student Edition page.

Skills Objectives
Students should be able to:

Communicate the characteristics of fishes.

Infer why fishes cannot survive when they crowd tightly together.

Collect Data about the three classes of fishes.

Vocabulary
gill, cartilage

MOTIVATE

Skills WarmUp
To help students distinguish between fishes and other animals, have them do the Skills WarmUp.
Answer Lists will vary. Crayfish are invertebrates, as are silverfish, jellyfish, cuttlefish, and starfish.

Prior Knowledge
To find out how much students know about fishes, ask them the following questions:

▶ Are fishes vertebrates? Why?

▶ What are some characteristics that make fishes different from other vertebrates?

▶ What are the three classes of fishes?

INTEGRATED LEARNING

Multicultural Perspectives
German researcher Emil Heinrich Du Bois-Reymond (1818–1896) studied fish, such as eels, that are capable of generating electric currents. From experiments with these animals, he learned much about how muscles work. Have students find out what Du Bois-Reymond discovered and how he linked the disciplines of physiology and physics.

Skills WarmUp

Communicate Make a list of nonfishes that have fish names, such as *crayfish*. Then explain why each is not a fish.

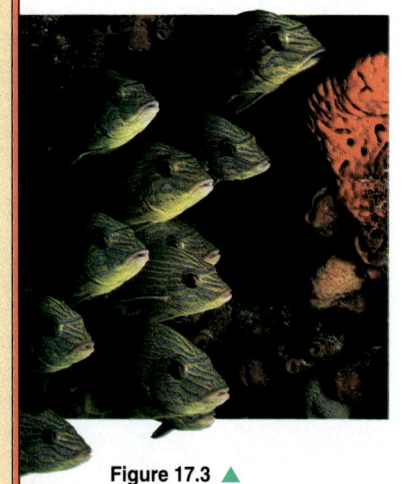

Figure 17.3 ▲
Fishes live in rivers, streams, lakes, and the ocean. These are blue striped grunts swimming off the coast of Florida.

17.2 Fishes

Objectives

▶ **Describe** the main characteristics and adaptations of fishes.

▶ **Distinguish** between the three classes of fishes.

▶ **Compare** and **contrast** the feeding methods of jawless fishes and cartilaginous fishes.

▶ **Collect data** relating to the classification of fishes.

If you have been inside a pet store, you have probably seen aquariums containing many kinds of fishes. Fishes come in many shapes, sizes, and colors. Although different kinds of fishes have different features, all fishes are adapted to life in the water. They have fins used for swimming. Most have streamlined bodies that allow them to move through the water with little resistance. What other adaptations for life in water do fishes have?

Characteristics of Fishes

Like all vertebrates, fishes have a backbone that develops from a notochord during the embryo stage. Fishes also have **gills**. Gills are organs that remove dissolved oxygen from water. The gills are located on either side of a fish's head. As a fish swims, water passes over its gills. Oxygen in the water diffuses into cells at the surface of the gills. At the same time, carbon dioxide diffuses out of blood vessels and into the passing water. Recall that all chordates have gills or gill slits at some stage of their development. Unlike most other vertebrates, fishes keep their gills throughout their lifetimes.

The 31,000 different species of fishes are grouped into three classes. Fishes in all three classes share the following characteristics:

▶ All fishes are ectotherms.

▶ Most fishes have streamlined bodies and use fins for locomotion.

▶ Most fishes have a very good sense of smell.

▶ All fishes have highly developed nervous systems.

348   Chapter 17   Fishes, Amphibians, and Reptiles

Integrating the Sciences

Physical Science Explain that an electric field surrounds the bodies of all animals. Sharks and some other fishes have the ability to detect differences in electrical fields. This ability, called electroperception, helps them locate prey. Electroperception also serves as a kind of compass for navigating in the ocean. Have students investigate electroperception in sharks and other fishes, such as catfish and elephantfish.

Jawless Fishes

The first fishes to appear on the earth were covered by thick, bony plates. Like the fishes of today, they had backbones and were considered vertebrates. However, these early fishes lacked jaws. Scientists infer that like their invertebrate ancestors, they probably were filter feeders.

A few species of fishes without jaws still exist today. They are the lampreys and the hagfish. These jawless fishes do not have the bony plates of the first fishes, but they are placed in the same class.

Lampreys and hagfish seem to have changed little since they first evolved. Unlike their more ancient relatives, however, lampreys and hagfish are not filter feeders. They are parasites. Look at the mouth of the fish shown in the upper left photograph in Figure 17.4. You can see that the mouth is a type of sucker. Lampreys and hagfish use these suckers to attach to their hosts. Once attached, these fishes rip into their host's flesh using rasp-like tongues. They then feed upon the blood and body fluids of the host.

Cartilaginous Fishes

Gently move the outer part of your ear back and forth with your fingers. Your ear holds its shape but is flexible because of a type of tissue called **cartilage** (KART uhl ihj). Cartilage is a firm but flexible connective tissue that makes up some parts of your skeleton.

Members of the second class of fishes have skeletons made up entirely of cartilage. They are called cartilaginous (KART uhl AJ uh nuhs) fishes. This group of fishes includes sharks, rays, and skates.

Unlike the jawless fishes, cartilaginous fishes have jaws. The jaws evolved from the skeletal rods that supported the gills of jawless fishes.

Most cartilaginous sharks are carnivores. Their sharp senses and sharp teeth make them well adapted for life as predators. Some of the largest sharks, however, are filter feeders. The whale shark strains water to remove the plankton it contains.

Cartilaginous fishes are more dense than ocean water, so they must keep swimming to stay above the sea floor. Swimming is also what makes a current of water pass over the gills, supplying the fish with oxygen.

Figure 17.4 ▲
The lamprey (upper left) is a jawless fish. The great white shark (lower left) and winter skate (above) are examples of cartilaginous fishes.

TEACH

Enrich
Have students find out more about jawless fishes. Why are there so few left today? (The evolution of the biting jaw in the bony fishes proved to be such a successful adaptation that the jawless fishes lost their living and feeding space to the jawed fishes.)

Discuss
After students have read about jawless fishes and cartilaginous fishes, discuss the following questions:

▶ What is cartilage? (It is a firm but flexible tissue.)

▶ How do most sharks differ from lampreys and hagfish in the way they obtain food? (Most sharks are predators; lampreys and hagfish are parasites that live on other fishes.)

▶ What and how do filter-feeder sharks eat? (They remove plankton from the water by filtering the water.)

▶ What fishes other than sharks are members of the cartilaginous fishes? (Skates and rays)

**The Living Textbook:
Life Science Side 4**

Chapter 12   Frame 27168
Ray/Eel (Movies)
Search: Play:

**The Living Textbook:
Life Science Side 4**

Chapter 16   Frame 41560
Shark (Movie)
Search: Play:

TEACH ▪ Continued

Skills WorkOut
To help students infer why schools of fishes sometimes get trapped in small areas, have them do the Skills WorkOut.

Answer Schools of fishes stay in formation by using a sense called the lateral line system, which picks up pressure waves from movement in the water. If a turn into a bay or harbor is made by the fishes in front, the whole school will follow. If the fishes become crowded together in the bay or harbor, the lack of oxygen or food could cause them to die if they can't get back to the ocean.

Misconceptions
Explain that shark attacks on humans are rare. In fact, fewer than 10 percent of shark species have been known to attack humans.

Consider This
Think About It To find out why many people believe that sharks are an important part of ocean ecosystems, students might consider the role of sharks in ocean food webs. (You may wish to refer them to Chapter 25 to read about food webs.)

Write About It Students' editorials should include an opening address to the editor, presentation of opinion supported by facts, countering of opposing views, suggestions for action, and a signature.

The Living Textbook: Life Science Sides 1–4

Chapter 4 Frame 01270
Lamprey/Shark (13 Frames)
Search: Step:

INTEGRATED LEARNING

Integrating the Sciences
Physical Science To understand *swim bladders*, students should know that hydrostatic (water) pressure increases with depth. By adjusting the amount of gas in its swim bladder, a fish is able to adjust to the changes in water pressure that occur as it swims up or down. Point out that scuba divers use devices called buoyancy compensators to adjust to changes in water pressure.
Use Integrating Worksheet 17.2.

Skills WorkOut
Infer Sometimes schools of millions of small fishes swim into small inlets or harbors where they must crowd tightly together. Unless they can find their way back to the ocean, they die. What do you think is the cause of their death? Explain.

Bony Fishes
The greatest number of fish species belong to the bony fishes. The bony fishes first appeared around the same time as the cartilaginous fishes, more than 350 million years ago. As their name implies, bony fishes have skeletons made of bone. Other adaptations, too, help them live in many kinds of water habitats.

Buoyancy Control Bony fishes have an organ called a swim bladder. This organ helps them conserve energy as they swim through the water. The swim bladder is similar to a balloon.

Fishes use their swim bladders to adjust their depth in the water. Inflating the swim bladder with just the right amount of air keeps the fish from sinking to the bottom of the ocean or floating to the surface. The swim bladder, therefore, allows the fish to maintain a *buoyancy* that is just right in water.

Consider This

Should Sharks Be Protected?

Would you swim in shark-infested waters? Probably not. Sharks are often described as terrifying creatures that kill for no apparent reason. However, this reputation is not completely deserved. Although people are killed and injured by sharks, most sharks do not attack humans. In fact, it is human attacks on sharks that have scientists concerned.

Each year about 100 million sharks are killed. As a result, the population of these predators is declining. The decline of the shark population may be causing a shift in ocean food webs. This shift could be unhealthy for many kinds of ocean life.

Consider Some Issues Unlike whales, dolphins, and seals, sharks are not protected by special laws. Some people believe that sharks should be protected because they are just as important as these other animals.

Others disagree, for a variety of reasons. Some of these people fish for sharks as part of their livelihood. Others claim that sharks do more harm than good, or simply aren't worth the fuss.

Think About It Should sharks be protected? Can you think of a compromise that might make everyone happy?

Write About It Write a newspaper editorial stating your position on the protection of sharks.

Chapter 17 Fishes, Amphibians, and Reptiles

Themes in Science

Diversity and Unity The adaptations that help animals survive in different environments result in an enormous variety of species. Discuss the characteristics that help fishes survive in their water environments. Ask students to explain how scales, fins, and gills help fishes adapt to life in water.

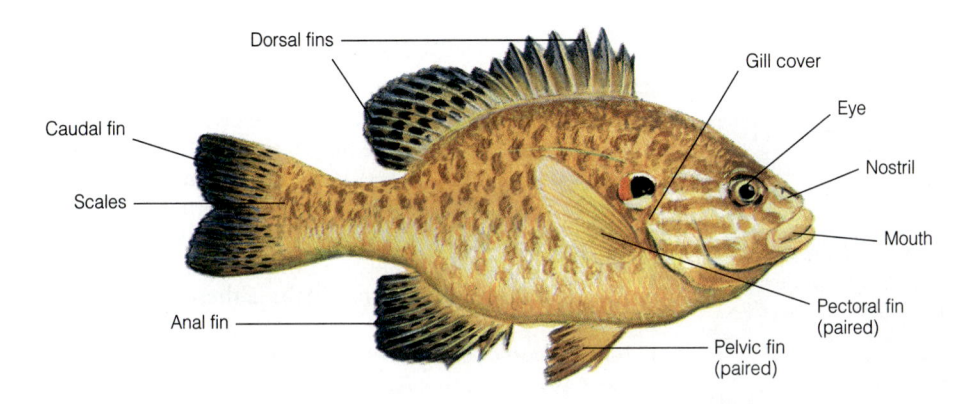

Buoyancy is the force of fluid pushing an object up. Buoyancy acts against the weight of an object. It makes an object seem like it weighs less in a fluid. If the weight of an object under water is greater than the buoyant force, the object will sink. If the weight is less than the buoyant force, the object will rise to the surface and float. If you have been swimming, you may know what it is like to float. If the weight is equal to the buoyant force, the underwater object will stay at any level. This is what happens to a fish.

Scales and Fins The outside of a bony fish is covered with a layer of smooth scales. Notice how the scales of the fish shown in Figure 17.5 overlap each other. The scales also are covered with a thin layer of mucus. The mucus and overlapping of the scales are adaptations that help the fish move through the water with little resistance. Also notice the different kinds of fins on the bony fish. The fins are adapted for swimming and guiding the fish.

Gills and Lungs Unlike the cartilaginous fishes, bony fishes do not have to swim to move water over their gills. They can pump water through the mouth and into the gill chamber. The water is pumped by movements of a special flap of tissue in the gill chamber.

Most bony fishes depend only upon gills for their gas exchange. However, some bony fishes have both gills and lungs. The lung developed from part of the swim bladder. One of these fishes, called the African lungfish, uses its lungs to help meet its oxygen needs. When it needs to take in more oxygen, the lungfish swims to the surface of the water and "gulps" fresh air.

Figure 17.5 ▲
Bony fishes have paired and unpaired fins. A fish uses its paired pectoral and pelvic fins to steer, brake, back up, and move up and down. The caudal fin propels the fish.

Figure 17.6 ▲
The African lungfish is a bony fish that lives in shallow waters and gets most of its oxygen from the air.

Class Activity
Assign students to cooperative groups. Have groups observe fishes in an aquarium and make a list of their observations. Have students identify the parts labeled in Figure 17.5.

Discuss
Ask students the following questions:

▶ How do fishes swim?
▶ How do paired fins work?
▶ How do other fins work?
▶ How do fishes "breathe"?

Skills Development
Infer Explain that fishes are able to control the balance of salt and water in their tissues. This enables them to maintain a constant internal body environment, called homeostasis. Have students look up homeostasis in the glossary. Explain that fishes living in oceans lose water to the ocean because water is more concentrated in their body tissues than in the ocean. Ask students what would happen to saltwater fishes if they could not regulate the water and salt that enters their bodies? (Their tissues would dry out, or dehydrate.)

Enrich
Use Enrich Worksheet 17.2.

**The Living Textbook:
Life Science Side 4**

Chapter 12   Frame 26902
Fish (Movie)
Search: Play:

**The Living Textbook:
Life Science Side 4**

Chapter 13   Frame 31022
Fish: Breathing and Eating (Movie)
Search: Play:

TEACH • Continued

Class Activity
Have students list their ten favorite foods. In a class discussion, combine these lists to form one list of the class's ten favorites. Ask which of the favorites are highest in fat. Have students note if foods made from fish are among the class choices.

EVALUATE

Wrap Up
Review Have students make a chart of the three types of fishes they have studied. Under each type they should list major characteristics and examples of that particular species.

Use Review Worksheet 17.2.

Check and Explain
1. Ectotherms: streamlined bodies; fins for locomotion; excellent sense of smell; highly developed nervous system
2. Swimming supplies oxygen by moving water over the gills. A swim bladder allows a bony fish to adjust its depth; a flap of tissue in the gill chamber pumps water over its gills.
3. Jawless fishes attach suckerlike mouths to a host and feed upon its body fluids. Most cartilaginous fishes use sharp teeth to tear apart the flesh of their prey; others are filter feeders.
4. Jawless: parasites with sucking mouths, rasplike tongues; Cartilaginous: skeletons made of cartilage, jaws, must swim constantly; Bony: bony skeleton, scales, swim bladder, gill pump

INTEGRATED LEARNING

Integrating the Sciences
Nutrition Have student groups research the following issues: 1) Why is high cholesterol potentially harmful? 2) What substance in fish oil helps prevent the formation of cholesterol deposits in the arteries? 3) Which other kinds of food are high in cholesterol, which are low, and which help lower cholesterol? Have group members share their findings with the class.

Figure 17.7 ▲
An Inuit man fishes through a hole in the ice. Fish is an important food in the diets of Inuits.

Science and Society *A Fishy Diet*
What are your ten favorite foods? If you eat such foods as hamburgers, steaks, ice cream, french fries, and eggs, you may be placing yourself at risk for heart disease. Heart disease is a condition associated with diets high in cholesterol. What causes a diet to be high in cholesterol? Cholesterol is naturally found in animal products. It is especially common in red meats, dairy products, and eggs. Diets that include these foods are high-cholesterol diets.

One group of people with diets especially high in cholesterol are Native Americans living in the Arctic. They call themselves Inuits (IHN oo wihts). The name given to them by Europeans—Eskimo—comes from a Native American term meaning "eater of raw meat." As this name suggests, Inuits eat much uncooked meat. Their diet includes very fatty meat taken from marine mammals, such as seals, walruses, and whales. The Inuits also eat sea birds and caribou.

With a diet high in fatty meats, Inuits should suffer from heart disease. However, a recent study performed on these people showed that Inuits actually have very healthy hearts. How can these findings be explained?

One explanation is that the Inuit diet also includes a large amount of fishes. Many fishes contain oil that seems to reverse the effects of cholesterol. A diet rich in fishes containing this oil may actually prevent the formation of deposits in the arteries, the cause of heart disease. The fishy diet of the Inuits seems to make them resistant to heart disease.

Check and Explain
1. List five characteristics of fishes that make them adapted to life in water.
2. Explain why a cartilaginous fish must swim constantly. What adaptations make it unnecessary for a bony fish to swim all the time?
3. **Compare and Contrast** How do jawless and cartilaginous fishes differ in their methods of feeding?
4. **Collect Data** In three columns, list the characteristics used to classify the three classes of fishes.

Themes in Science
Evolution The development of lobe-finned fishes into land-dwelling amphibians represents a movement toward more complex respiratory systems. Have students discuss how the ability to survive on land is a helpful adaptation for amphibians. (It allows for more diversity among species and a variety of habitats.)

Integrating the Sciences
Earth Science Paleontologists have determined that amphibians first appeared about 400 million years ago. This geological period was marked by the uplifting of landmasses. Refer students to the Data Bank on page 630, and ask when amphibians appeared. (During the Paleozoic era)

SECTION 17.3

Section Objectives
For a list of section objectives, see the Student Edition page.

Skills Objectives
Students should be able to:
Infer the function of different adaptations in frogs.
Hypothesize about how to determine whether an amphibian lives mostly on land or in water.

17.3 Amphibians

Objectives
▶ **Describe** the origin of amphibians.
▶ **Describe** the characteristics and adaptations of amphibians.
▶ **Explain** the process of metamorphosis in frogs.
▶ **Hypothesize** if an amphibian is mostly a land-dweller or water-dweller.

Skills WarmUp
Infer Frogs that live in water usually have webbed feet. Those that live in trees often have feet with pads that act as suction cups. What function do you think each of these kinds of feet serves?

MOTIVATE

Skills WarmUp
To help students understand how a frog's physical characteristics help it survive in its environment, have them do the Skills WarmUp.
Answer Webbed feet help frogs that live in water swim. Foot pads help frogs that live in trees hold onto tree trunks and branches.

Prior Knowledge
Gauge how much students know about amphibians by asking them the following questions:
▶ Name some examples of amphibians.
▶ How are amphibians different from fishes?

Answer to In-Text Question
① Frogs, like most amphibians, spend part of their lives in water, such as a pond, and part on land, such as in trees.

Have you ever touched a frog? If so, you may have observed that the frog felt slippery and cool. Frogs have a slick, slimy skin through which respiratory gases are exchanged. They are also ectotherms, which explains why they are cool to the touch.

Frogs are amphibians (am FIHB ee uhns). Other amphibians are toads and salamanders. Most amphibians spend part of their lives in water and part on land. A few, such as the mud puppy, spend their entire lives in water. Others, such as the brightly colored tree frogs that live in tropical rain forests, spend their entire lives as land-dwelling vertebrates.

Movement onto Land

Imagine seeing a fish crawl across the highway. In some places, such as parts of Florida, this sight is not uncommon. When ponds dry up or food sources become scarce, fishes called walking catfish drag themselves across the ground. They "walk" to new water holes.

Millions of years ago, this kind of ability to move over land for short periods of time led to the evolution of amphibians. Amphibians evolved from a group of lobe-finned fishes that are now extinct. The lobed fins of these fishes were strong and bony. As they evolved, the fins became stronger and more supportive. They made possible both swimming and a kind of crawling on land. Fossil records suggest that these fins evolved into the two pairs of legs that allow modern amphibians to live both on land and in water.

Figure 17.8 ▲
Study this frog's body. In what kind of habitat do you think it lives? ①

The Living Textbook:
Life Science Sides 1-4

Chapter 4   Frame 01310
Amphibians (32 Frames)
Search: Step:

Chapter 17 Fishes, Amphibians, and Reptiles

TEACH

Critical Thinking

Compare and Contrast Have students observe Figure 17.9. Ask the following question:

▶ Which parts of the lobe-finned fish are similar to the early amphibian? (The basic body shapes are similar; both have tails.)

▶ Which parts of the two animals are different? (The early amphibians didn't have bony scales. The two lower fins are replaced by legs.)

Class Activity

If possible, have students observe tadpoles and/or frogs in an aquarium or terrarium. Ask the following questions:

▶ How do tadpoles move? (By swimming)

▶ How do frogs move? (By swimming in water and hopping on land)

▶ How do tadpoles breathe? (Through gills)

▶ How do frogs breathe? (Usually through lungs)

The Living Textbook: Life Science Side 4

Chapter 12   Frame 27650
Frog Swimming (Movie)
Search: Play:

The Living Textbook: Life Science Side 4

Chapter 13   Frame 31636
Frog (Movie)
Search: Play:

INTEGRATED LEARNING

Themes in Science
Patterns of Change/Cycles

Change is a normal part of the life cycles of all living things. Some animals, including amphibians and insects, have life cycles that include distinct changes in form. This pattern of change is called metamorphosis.

Figure 17.9 ▶
The bony fins of lobe-finned fishes evolved into the four limbs of the amphibians.

Early amphibian

Lobe-finned fishes

Characteristics of Amphibians

Look at Figure 17.9. As amphibians evolved, these four-legged vertebrates continued to adapt to a land environment. The gills and simple lungs that developed in their fishlike ancestors evolved into lungs better able to exchange gases with the atmosphere. Amphibians also began to use their skin for respiration. No longer covered by bony scales, the skin could be an extra site for gas exchange. In fact, some amphibians today lack lungs and gills and use the moist surface of their skin and mouth to meet all their respiratory needs.

Although most amphibians are adapted to life on land, they remain dependent upon water. Except for some species, such as desert toads, amphibians require moist environments. Most live in or near bodies of water, or spend much of their time in moist burrows.

Amphibians also depend on water for reproduction. Their eggs lack a protective, waterproof shell. Since the eggs can easily dry out, most amphibians must return to the water to reproduce. In most species, the female lays eggs in a pond, swamp, or stream, and they are fertilized externally by the male. A few species have evolved interesting ways of adapting to living where water is scarce. Some desert toads, for example, produce a moist foam in which they lay their eggs.

The eggs of many amphibians develop into a fishlike larva stage called a tadpole. In amphibian species where the adult remains a water-dweller, the changes from the tadpole stage to the adult stage are minor. In amphibians such as frogs and toads, the tadpole changes completely as it becomes an adult.

Figure 17.10 ▲
Many amphibians, such as the leopard frog (bottom), begin life as a tadpole (top).

Chapter 17 Fishes, Amphibians, and Reptiles

TEACHING OPTIONS

Portfolio
Explain that a variety of organisms go through a great change in body form or structure during their lives (metamorphosis), and that the immature form is adapted to a different environment or lifestyle than the adult form. Have students make a list of organisms that undergo metamorphosis. They should choose one of these organisms and draw its life cycle. Students can place their pictures in their portfolios.

Frogs and Toads

Amphibians are divided into several groups. One group includes frogs and toads, the two kinds of amphibians that do not have tails as adults. Most frogs spend their lives in and around water.

When frogs reproduce, the female lays hundreds of jellylike eggs. As the eggs pass out of her body, the male fertilizes them. Within several weeks, the eggs hatch into tadpoles. Tadpoles are a water-dwelling stage in amphibian development. Like fishes, tadpoles have gills, fins, tails, and a streamlined body form.

As a tadpole develops, it undergoes great changes in appearance. These changes during development make up a process called *metamorphosis*. The stages of frog metamorphosis are shown in Figure 17.12. When metamorphosis is complete, the adult frog emerges from the water as a land-dwelling vertebrate. Most toads have a similar metamorphosis. Not all frogs and toads have a tadpole stage, however. In some species, a tiny frog or toad emerges directly from the egg.

Figure 17.11 ▲
Toads spend most of their lives on land. Here, toads bury themselves in mud to keep cool.

Figure 17.12
Stages of Frog Metamorphosis

The tadpole develops front legs. The tail begins to disappear. Lungs form. The tadpole becomes an air-breathing frog.

When the young frog becomes sexually mature, it reaches the adult stage.

The tadpole develops two small limb buds. The buds grow into hind legs.

During spawning, eggs released by the female are fertilized externally by the male.

Several days after fertilization, a tadpole emerges from the egg. It has many fishlike characteristics, including gills.

Explore Visually
After students have studied the material on this page, ask the following questions:

▶ At what stage of life do frogs release eggs or sperm into the water? (The adult stage)

▶ How are adult frogs different from frog tadpoles? (Adult frogs do not have gills, fins, tails, and a fishlike body form.)

▶ Have students briefly summarize all the stages of frog metamorphosis. (See Figure 17.12.)

Misconceptions
Students may think that they can get warts from handling toads. Explain that the skin of toads doesn't give people warts. Warts are caused by viruses.

Critical Thinking
Compare and Contrast Ask students to compare and contrast frogs and toads. (Frogs and toads both are amphibians. The adults of both animals are without tails. Both lay eggs in water. Frogs usually live in or near fresh water and have smooth, moist skin. Toads live most of their lives on land and have dry, warty skin.)

The Living Textbook: Life Science Sides 1-4

Chapter 6 — Frame 02457
Frog Development (23 Frames)
Search: — Step:

The Living Textbook: Life Science Side 2

Chapter 17 — Frame 43231
Tadpoles (Movie)
Search: — Play:

EVALUATE

WrapUp
Portfolio Have students write a story from the point of view of a tadpole undergoing metamorphosis into an adult frog. They should start with the tadpole inside its egg and discuss all the major physical changes that it experiences, including where it lives, how it moves, and what it eats. Encourage students to be creative. Students can add the stories to their portfolios.
Use Review Worksheet 17.3.

Check and Explain
1. The strong, bony, lobed fins that were used for swimming and walking by ancient fishes
2. Strong, bony legs; lungs; and use of skin for respiration are adaptations to a land environment. Methods of reproduction are adapted to water.
3. Both an adult frog and a tadpole spend their lives in and around water; however, an adult frog has lungs and four legs that allow it to exist on land, whereas tadpoles must live in water and have gills, fins, tails, and a streamlined body.
4. Answers may vary. One test would be to place the amphibian in an environment containing both land and water and observe where it spends most of its time.

The Living Textbook: Life Science Side 4
Chapter 15   Frame 40155
Newts Mating (Movie)
Search: Play:

356

INTEGRATED LEARNING

Integrating the Sciences
Ecology Explain that one reason South American rain forests are being cut down is to create pastureland for grazing cattle. There is, therefore, an economic reason to destroy the forests. However, the rain forest is the habitat for many living organisms, including amphibians. Ask students why loss of habitat is causing extinction. (A habitat provides the needs of living things, such as food, shelter, and so on.)

Figure 17.13 ▲
This salamander lives in a tropical rain forest.

Salamanders and Newts
Amphibians with tails make up another group that includes the salamanders and newts. Like frogs, salamanders and newts have a smooth, moist skin. Some live in relatively dry habitats similar to where toads live. Others prefer very moist places, and a few even live in water their entire lives.

Science and Society *Vanishing Amphibians*
How long does it take for an organism to become extinct? It may be faster than you think. As many as 100 species may become extinct each day.

The gastric brooding frog once lived in the rain forests of Australia. The female swallowed her eggs after they were fertilized and kept them inside her stomach until they developed into fully formed young. Until 1980, this frog appeared to be a hearty species. Then, within several months, it vanished.

Today many amphibians are faced with extinction. Biologists blame much of the threat on the destruction of habitats. The cutting down of rain forests, for example, is a major cause of amphibian extinction.

Amphibians are also especially vulnerable to chemical pollution. In Puerto Rico, three species of miniature frogs have vanished. Salamanders that once lived in the mountains of Colorado have also disappeared. Scientists think these species were victims of acid rain. Acid rain may affect amphibians more than other groups of organisms because amphibians' skin absorbs poisons as easily as it absorbs oxygen.

Check and Explain
1. From what structure of early fishes did the legs of amphibians most likely evolve?
2. List several characteristics of amphibians. For each, explain whether it is an adaptation to land or to water.
3. **Compare and Contrast** Write about how an adult frog is like a tadpole and how it is different.
4. **Hypothesize** Design a test to determine if an unknown amphibian lives mostly on land or mostly in water.

356   Chapter 17   Fishes, Amphibians, and Reptiles

Integrating the Sciences

Earth Science Dinosaurs inhabited the earth for about 130 million years. About 65 million years ago, they disappeared. The reason for their disappearance is still a mystery. Scientists have given a number of theories to explain this mass extinction. Some hypothesize that a giant meteorite collided with the earth, leaving a cloud of dust that blocked the sun and resulted in the death of most plant and animal life. Others hypothesize that volcanic eruptions created ash that blocked the sun. Still others hypothesize that the dinosaurs died out gradually because of slow changes in their environment. List the three theories on the chalkboard and have students discuss which theory they think is the best explanation. Encourage students to add their own theories.

SECTION 17.4

Section Objectives
For a list of section objectives, see the Student Edition page.

Skills Objectives
Students should be able to:

Classify reptiles into three groups.

Research local reptiles and amphibians and their habitats.

Infer why the largest reptiles live in warm regions.

17.4 Reptiles

Objectives

▶ **Discuss** the evolutionary history of reptiles.

▶ **Describe** the characteristics and adaptations of reptiles.

▶ **Identify** the different groups of reptiles.

▶ **Make inferences** about the relationship between the body size of reptiles and the climates they live in.

Skills WarmUp

Classify Make a list of all the reptiles you can think of, then group them into three sets. As you read this section, check on your list to see if you classified each animal correctly. Revise your list when you finish reading.

MOTIVATE

Skills WarmUp
Have students classify reptiles in the Skills WarmUp.
Answer Students should arrive at a three-way grouping of snakes/lizards, turtles/tortoises, and crocodiles/alligators.

Prior Knowledge
To gauge how much students know about reptiles, ask them the following questions:

▶ What are some of the major groups of reptiles?

▶ What are the main differences between reptiles and amphibians?

Look at the lizard in Figure 17.14. Have you seen this animal before? You've probably never found one in your backyard, but you may have seen one on your television screen. Lizards similar to this marine iguana are commonly used in science fiction movies to represent prehistoric dinosaurs.

Lizards actually are related to the extinct dinosaurs. Both are vertebrates with scaly skin adapted to life on land. But unlike dinosaurs, lizards are living today all over the earth. Together with snakes, turtles, tortoises, alligators, and crocodiles, they make up the reptiles.

History of Reptiles

Scientists can learn many things about extinct animals by studying their fossil remains. For example, scientists can infer the kind of food an animal ate by the shape of its teeth. The length and thickness of leg bones are clues about the body size of an animal. Even footprints can provide clues about an animal's movement and behavior. Using such evidence, scientists have assembled a history of reptiles. This history began when reptiles' amphibianlike ancestors developed adaptations for living their entire lives on land.

These adaptations included a body covering that reduced water loss and eggs that could be laid on land. About 300 million years ago, these adaptations appeared in a group of ancestral reptiles. These early reptiles evolved in many different directions. One path of their evolution resulted in the dinosaurs. But even before the dinosaurs evolved, the ancestors of modern turtles, lizards, and snakes had appeared.

Figure 17.14 ▲
The marine iguana is a harmless vegetarian.

💿 **The Living Textbook:**
Life Science Sides 1–4

Chapter 4   Frame 01342
Reptiles (73 Frames)
Search: Step:

Chapter 17 Fishes, Amphibians, and Reptiles 357

TEACH

Class Activity

Provide art supplies, such as paint, clay, paste, cardboard, and shoe boxes. Have students work in small groups and have each group create a diorama showing a reptile in its environment. Students should add details to their figures that show that their animal has the characteristics of a reptile. Have students write one sentence about each of the characteristics on a label. Have them attach the labels to the dioramas and display their work in the classroom.

Skills WorkOut

To help students learn about local amphibians and reptiles, have them do the Skills WorkOut.
Answer Answers may vary, but be sure students' lists show only local animals.

**The Living Textbook:
Life Science Sides 7-8**

Chapter 13   Frame 01351
Turtles and Tortoises (7 Frames)
Search: Step:

INTEGRATED LEARNING

Themes in Science

Evolution The evolution of a species is affected by its environment. Two major evolutionary adaptations allow reptiles to live on land and distinguish them from amphibians. First, tough, scaly skin protects them from drying out. Second, a waterproof amniotic egg protects and nourishes the enclosed embryo. Discuss why these adaptations are helpful on land.

Math Connection

Many dinosaurs were very large. In fact, one of the largest dinosaurs, *Brachiosaurus*, was about 22 meters long and 12 meters high. The largest land mammal now living is the elephant. Elephants average 3 meters tall and 6 meters long. Ask students how the elephant compares in size to *Brachiosaurus*. (*Brachiosaurus* was about four times the size of the elephant.)

Figure 17.15 ▲
Dinosaurs and modern reptiles share a common ancestor.

Skills WorkOut

Research Use the resources of a school or public library to make a list of local amphibians and reptiles and their habitats.

Characteristics of Reptiles

Because of the way snakes move, many people expect them to be slimy and slippery. However, if you touch a snake, you discover that the animal is not slimy at all. Unlike amphibians, whose outer skin *is* slimy, snakes and other reptiles have a dry, waterproof body covering made up of scales. This covering helps these land-dwellers conserve water. Scales also prevent the exchange of oxygen and carbon dioxide across the skin. Reptiles must therefore rely on their lungs to meet their respiratory needs.

Unlike amphibians, reptiles produce eggs that are covered by a tough, leathery shell. The shell prevents moisture from escaping, but it also keeps sperm from entering. For this reason, fertilization takes place inside the body of the female, before the eggshell develops.

Reptiles have circulatory systems more highly developed than those of amphibians. Like those of amphibians, their hearts are three-chambered. But a partial wall inside the main chamber lessens the mixing of low-oxygen blood and high-oxygen blood. The result is a heart that works nearly as well as the four-chambered hearts of birds and mammals.

Even though reptiles are ectotherms, most are very good at regulating their internal temperatures. They bask in the sun when the air is too cool and seek shade when it is too hot. Through these behavioral adaptations, many reptiles keep their body temperatures about as warm as yours throughout much of the day.

Chapter 17   Fishes, Amphibians, and Reptiles

Themes in Science

Scale and Structure Adaptations are structures and behaviors that are useful to living things in their environments. Turtles and tortoises, the most ancient group of reptiles still alive today, evolved a unique protective shell approximately 260 million years ago. The shell functions in many ways like a mobile home.

Lizards and Snakes

They may look very different, but lizards and snakes are closely related. All snakes and most lizards are carnivores. Both have a type of jaw hinge not found in other reptiles. It allows lizards and snakes to swallow their prey whole by increasing the size of their mouths.

The main difference between lizards and snakes is that snakes don't have legs. Snakes have developed ways of moving about without appendages. But some snakes still have hipbones and remnants of hind legs. These are signs that they evolved from reptiles that did have legs.

Snakes have very poor hearing and most have poor eyesight. They do, however, have keen senses of smell and taste. Snakes use their tongues to find prey and to gather other information about the environment. A snake sticks out its tongue to sample the air or the ground. The tip of the tongue picks up chemicals. The snake then places its tongue into an organ in the roof of its mouth. This sense organ detects the odors in the air or on the ground.

Turtles and Tortoises

Members of another group of reptiles all have shells. They are the turtles and tortoises (TOR tuh suhs). These reptiles have hardly changed since they appeared hundreds of millions of years ago. Their shells have proved to be important adaptations. When in danger, most turtles and tortoises can pull their heads, legs, and tails into their shells.

The shells of turtles and tortoises form from bony plates connected to their ribs and vertebrae. The shell is covered by a layer of skin. The skin is responsible for the markings and color patterns on the shell.

Turtles and tortoises differ in the structures of their shells. Most turtles have flat, streamlined shells. Tortoises have dome-shaped shells. Turtles and tortoises also live in different environments. Tortoises are land animals. Most turtles live in ponds, lakes, rivers, or the ocean. Turtles that spend most of their lives in water must return to land to lay their eggs.

◀ **Figure 17.16**
What main characteristic distinguishes snakes from lizards? ①
Why is the shell an important adaptation for the turtle? ②

Class Activity
Have students work in small groups. They can make lists of everything they know about lizards and snakes. Encourage students to share personal experiences and knowledge as well as the information they have just learned. Students should then make a list of questions about lizards and snakes. Have students do research to find answers to their questions. Then have them present their findings to the class.

Enrich
Have students investigate chameleons and their unique ability to blend in with whatever environment they are in.

Discuss
Point out that snakes shed their skin as they grow, making room for their expanding bodies. Ask students if they think turtles and tortoises shed their protective shells as they grow. (Turtles and tortoises do not shed their shells. The shells form from bony plates that are connected to their ribs and vertebrae. As a turtle or tortoise grows, the shell grows too.)

Answers to In-Text Questions
① Lizards have legs; snakes don't.
② The shell provides protection for the turtle.

The Living Textbook: Life Science Side 4

Chapter 10   Frame 04398
Sand Lizard (Movie)
Search: Play:

The Living Textbook: Life Science Side 4

Chapter 12   Frame 27846
Komodo Dragon Lizard (Movie)
Search: Play:

TEACH • Continued

Critical Thinking

Compare and Contrast Have students look closely at the photographs on this page. Ask them these questions:

▶ What are the physical differences between alligators and crocodiles? (Alligators have a wide head and a rounded snout; crocodiles have a narrow head and a more triangular snout.)

▶ What types of dinosaurs do these animals resemble? (Students might say *Tyrannosaurus*.)

Reason and Conclude Ask students how the bodies of alligators and crocodiles are adapted for life in water. (Their bodies are long and streamlined to help them swim; their eyes and nostrils are on the tops of their heads, enabling them to see and breathe while lying submerged in water.)

Career Corner

Suggest that students visit a pet store and observe the activities of the owner and staff. Students should pay attention to customer relations, management of merchandise, and pet care. Have them discuss their experiences in groups.

The Living Textbook: Life Science Side 4
Chapter 10   Frame 04142
Crocodile (Movie)
Search: Play:

INTEGRATED LEARNING

Art Connection

Have students work in pairs to create a bulletin board display about amphibians and reptiles. One student in each pair should pick an example of an amphibian and draw a picture of it, listing beside it the distinguishing characteristics of amphibians. The other student should do the same with a reptile. Have students arrange their pictures on the bulletin board.

Alligators and Crocodiles

Among the reptiles, the closest living relatives of the dinosaurs are the alligators and crocodiles. They are lizardlike in shape and can grow as long as a small boat. Their backs have large, deep scales.

Alligators and crocodiles are very similar in appearance. However, they can be distinguished by the shapes of their heads. Crocodiles have a narrow head with a triangle-shaped snout. Alligators have a broad head with a rounded snout.

Both alligators and crocodiles live in tropical areas and spend most of their time in the water. Alligators live mostly in North America and Asia. Crocodiles inhabit areas of tropical America, Africa, Asia, and Australia.

Unlike most other reptiles, alligators care for their young. The female guards her eggs, and after they hatch, both male and female protect the babies.

Figure 17.17 ▲
Both alligators (top) and crocodiles (bottom) live in the Everglades region of Florida.

Career Corner *Pet Store Owner*

Who Takes Care of All Those Pet Store Animals?

Imagine caring for 125 birds, 30 hamsters, 22 mice, 4 snakes, 25 dogs, 12 cats, and more than 1,000 tropical fish! For a pet store owner or manager, this is a typical everyday responsibility.

If you like animals, you might like working in a pet store. Pet store workers must make sure that every animal remains healthy. This means feeding the animals daily and keeping their cages and tanks clean. A pet store worker must also watch for sick or injured animals. The sooner an unhealthy animal is treated, the better its chances for recovery.

In order to tell customers how to care for their pets, pet store workers must also learn about each kind of animal. They need to be able to give advice about feeding, living quarters, treatment of diseases, and breeding. They should know what kinds of fish can live in the same tank together. A pet store worker must also be familiar with the supplies and equipment stocked by the store.

Much of this information can be learned on the job. You can also learn about the care of animals by reading books. Classes in biology at any level are very helpful.

With experience, you could take on the responsibility of managing a pet store. This includes buying and selling products, paying employees, and making sure the animals are cared for. After a while, you could be on your way to owning your own pet store!

TEACHING OPTIONS

Cooperative Learning

If a person is bitten by a snake, it's very important that she or he be able to describe the snake. Provide a reptile field guide. Have students work with partners. One student should choose a reptile and use words to describe it to the second student. The second student can use this information to draw a picture of the snake. Then have students compare the drawing to the picture in the field guide.

Science and You *Know Your Fangs!*

Most snakes are quite harmless to humans. But several snakes are poisonous. Their poison, called venom, can be deadly. You should know what snakes in your area are poisonous, or venomous, and what kind of habitats they live in.

If you are bitten by a snake, the first thing to do is find out if the snake is venomous. You can often tell whether a bite was made by a venomous snake by observing the shape of the wound. Non-venomous snakes have tiny teeth set in both the upper and lower jaws. When a non-venomous snake bites, it leaves an almost circular mark formed from the impressions of a complete set of teeth.

Most venomous snakes leave a different kind of bite mark. It is made of two small skin punctures. The punctures are from a set of two long, sharp, hollow fangs. As the snake bites, a poison produced by a special gland flows through the hollow fangs into the wounds.

If you have any reason to believe that you were bitten by a venomous snake, you must get medical attention as soon as possible. If more than one kind of venomous snake lives in the area, the doctor will want to know what kind of snake it was. Here is a situation in which your scientific skill of observation may even save your life. You should be able to describe the snake in enough detail for it to be identified. Some snakes have venom that affects the blood cells. The venom of other snakes targets the nerve cells. The doctor must know the species of snake in order to use the correct method of treatment.

Figure 17.18 ▲
This eastern diamondback rattlesnake is being "milked" to make an anti-venom serum. You can see the venom dripping out of its fangs.

Check and Explain

1. Why is it inaccurate to say that reptiles evolved from dinosaurs?
2. Identify the characteristics of reptiles that make them suited to life on land.
3. **Classify** Explain the similarities and differences among the three groups of reptiles.
4. **Infer** The largest reptiles all live in areas where it is warm year-round. Based on what you know about reptiles, suggest a reason why this is so.

Chapter 17 Fishes, Amphibians, and Reptiles **361**

Enrich

Students may be interested in knowing that snakebite remedy in the form of antivenin works by a process called passive immunization. Antibodies from a horse or rabbit injected with a weaker form of the venom attack the toxins in the victim as antigens.

Integrated Learning

Use Integrating Worksheet 17.4.

EVALUATE

WrapUp

Review Have students make a graphic organizer such as a Venn diagram that shows how the three groups of reptiles are alike and how they are different.

Use Review Worksheet 17.4.

Check and Explain

1. Ancestors of modern reptiles evolved before the dinosaurs did.
2. Waterproof scales to prevent water loss, eggs that can be laid on land, lungs, a highly-developed heart, and behavioral adaptations to regulate body temperature
3. All snakes and most lizards are carnivores and both have a jaw hinge for swallowing prey whole. Lizards have legs; snakes don't. Turtles and tortoises both have protective shells. Most turtles live in water and have flat, streamlined shells; tortoises are land animals with dome-shaped shells. Alligators and crocodiles have similar bodies, live in tropical areas, and live mainly in water. They have differently-shaped heads and only alligators care for their young.
4. Their large size would be a disadvantage in cold climates because their relatively small surface area collects proportionally little sunlight.

ACTIVITY 17

Time 45 minutes **Group** 1–2
Materials
metric rulers

Analysis

1. Check students' graphs for accuracy.
2. As body length increases, tail length decreases. At 8 to 10 weeks, the tadpole is living in water where having a tail for swimming is an advantage.
3. The lungs will have developed by around 12 to 15 weeks. By this time, the legs are long enough to allow movement onto land.
4. The hind legs grow faster than the front legs do.
5. At stages 1 and 2, the organism lives in water. By stage 3, the organism spends more time on land, and by stage 4, it is spending most of its time on land.

Conclusion
Student paragraphs will vary but should include all of the steps of metamorphosis (see Figure 17.12 on page 355) and should be based on observations made in the activity. Conclusions should mention that the metamorphosis of frogs results in limbs designed for land locomotion and lungs for air breathing.

Extension
Answers may vary; however, students should recognize that the metamorphosis of frogs parallels the evolution of land-dwelling amphibians. Limbs designed for land locomotion replaced fins for water locomotion. Also, lungs for air breathing developed in place of gills.

TEACHING OPTIONS

Prelab Discussion
Have students read the entire activity before beginning the discussion. Ask them the question, How will you measure the animal's body length for the data table? (Students should not include the tail in their measurements.)

Activity 17 What changes occur during metamorphosis?

Skills Measure; Observe; Infer

Task 1 Prelab Prep
Collect the following item: a metric ruler.

Task 2 Data Record
On a separate sheet of paper, copy Table 17.1. Make four rows, one for each drawing below.

Task 3 Procedure
1. Look at the drawings of the tadpoles and frogs shown on this page. In your data table, measure and record the animal's length and width for each stage of metamorphosis shown.
2. Identify the drawing in which the frog's hind legs first appear. Measure and record the length of a hind leg in the drawing. Repeat this step for the other drawings.
3. Identify the drawing in which the frog's front legs first appear. Then measure and record the length of a front leg. Repeat this step for each of the remaining drawings.
4. In drawings 1, 2, and 3, measure and record the length of the animal's tail section. Make your measurement from where the hind legs begin to the tip of the tail.

Task 4 Analysis
1. Make a line graph that illustrates the relationship between the length of the animal and its age. Make a similar graph for tail length.
2. Why is the tail so much longer in drawing number one?
3. In which stage might the lungs first appear? What information did you use to make this inference?
4. How does the development of the hind legs differ from the development of the front legs?
5. Tell what environment each organism lives in.

Task 5 Conclusion
Write a short paragraph explaining the changes that occur as a tadpole undergoes metamorphosis and develops into an adult frog. Include how these changes enable the frog to adapt to land.

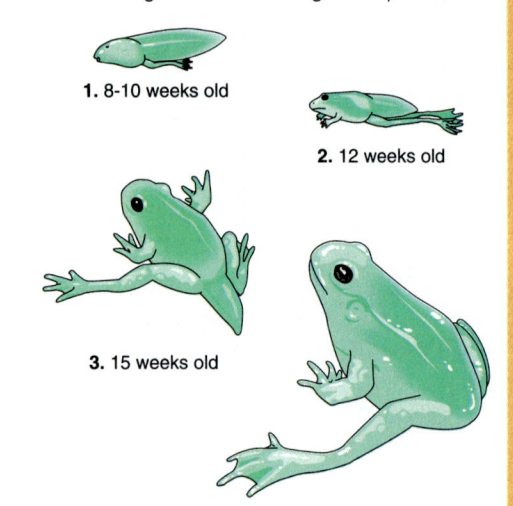

1. 8–10 weeks old
2. 12 weeks old
3. 15 weeks old
4. 1 year old

Extension
Compare and contrast the metamorphosis of frogs with the evolution of land-dwelling amphibians from fish ancestors.

Table 17.1 Frog and Tadpole Measurements

Drawing Number	Age	Body Length	Body Width	Tail Length	Hind Leg Length	Front Leg Length

CHAPTER REVIEW 17

Chapter 17 Review

Concept Summary

17.1 Common Traits of Vertebrates
- All vertebrates belong to the phylum Chordata and have endoskeletons.
- Some vertebrates are endotherms and others are ectotherms.
- The first vertebrates were fishlike animals. Amphibians evolved into reptiles, and reptiles gave rise to mammals and birds.

17.2 Fishes
- Most fishes have streamlined bodies, fins, a good sense of smell, and complex nervous systems. They are ectotherms.
- Fishes are divided into three classes: jawless, cartilaginous, and bony. The greatest number of fish species belongs to the bony fishes. All bony fishes have scales and swim bladders. Swim bladders help them conserve energy in the water.

17.3 Amphibians
- Amphibians were the first vertebrates to live on land.
- Most amphibians are dependent upon water.
- Many amphibians begin life as water-dwelling tadpoles and become land-dwelling vertebrates through metamorphosis.

17.4 Reptiles
- Reptiles are adapted to living their entire lives on land.
- Reptiles have dry, waterproof, scaly bodies. They produce eggs with a leathery coating.
- The main difference between snakes and lizards is that snakes do not have legs.
- The shells of the turtle and tortoise are important adaptations for protection.

Chapter Vocabulary

endoskeleton (17.1)   ectotherm (17.1)   gill (17.2)   cartilage (17.2)
endotherm (17.1)

Check Your Vocabulary

Use the vocabulary words above to complete the following sentences correctly.

1. Animals with constant internal body temperatures are called ____ .
2. Some parts of your skeleton are made of ____ , a firm, flexible connective tissue.
3. The central part of the ____ is the backbone.
4. Animals with bodies that receive heat from the outside are called ____ .
5. Organs that remove dissolved oxygen from water are called ____ .

Explain the difference between the words in each pair.

6. Endoskeleton, exoskeleton
7. Gill, lung
8. Ectotherm, amphibian
9. Endotherm, endoskeleton
10. Cartilage, vertebrae
11. Vertebrate, chordate

Write Your Vocabulary

Write sentences using each vocabulary word above. Show that you know what each word means.

Chapter 17   Fishes, Amphibians, and Reptiles   363

Check Your Vocabulary

1. endotherms
2. cartilage
3. endoskeleton
4. ectotherms
5. gills
6. An endoskeleton is an internal framework of bone or cartilage that supports an animal. An exoskeleton is a hard shell that provides support and makes up the exterior of an animal's body.
7. A gill is an organ adapted to take dissolved oxygen from water, while a lung is an organ that takes oxygen from air.
8. An ectotherm is an organism that cannot maintain a stable internal temperature. An amphibian is one kind of ectotherm. It spends part of its life cycle in or near water and part on land.
9. An endotherm is an organism whose *internal* metabolic processes provide enough heat to maintain a stable body temperature. The endoskeleton is an *internal* skeleton.
10. Cartilage is a firm, flexible connective tissue. Vertebrae are the inflexible bones that make up the vertebral column in vertebrates.
11. Chordates are members of the phylum chordata and have a notochord and gills or gill slits at some stage in life. Vertebrates are a subgroup of chordate and have a backbone and endoskeleton.

Write Your Vocabulary

Students' sentences should show that they know both the meaning of each vocabulary word as well as how it is used.

Use the Vocabulary Worksheet for Chapter 17.

CHAPTER 17 REVIEW

Check Your Knowledge

1. Animals with backbones are called vertebrates. Examples are fishes, amphibians, birds, reptiles, and mammals.
2. Crocodiles have narrow heads with pointed snouts, while alligators have short, wide heads with rounded snouts.
3. Both are vertebrates with scaly skin adapted to life on land. No modern lizard grows as large as some dinosaurs did.
4. Answers may vary, but should come from this list: swim bladders for buoyancy, scales, fins, gills, and bony skeletons. They first appeared more than 350 million years ago.
5. The first vertebrates appeared about 500 million years ago. They were water-dwelling fish-like animals.
6. A tadpole is an early stage in the development of the frog. Unlike a frog, it has a tail and gills and lives in water. A tadpole is like a fish because it has gills and lives underwater.
7. Salamanders and newts have tails. They have smooth, moist skin, like frogs.
8. Reptile eggs are covered by a tough, leather-like shell that prevents moisture from escaping.
9. Reptiles regulate their body temperature by moving into or out of the sunlight.
10. smell
11. amphibian
12. move water across their gills
13. seven

Check Your Understanding

1. Answers may vary; similarities include: both are reptiles, have scaly skin, are ectotherms. The main difference is that lizards have legs.
2. A snake uses its tongue to find prey and get feedback about its environment. The snake sticks out its tongue, samples the air, then touches its tongue to a special sense organ in the roof of its mouth where any odors are detected.
3. Cold-blooded; reasons may vary. Possible answer could be that the body of an ectotherm is not warm to the touch unless it is in direct sunlight.
4. Answers may vary; however, students should recognize that the chameleon is a reptile by the presence of dry, scaly skin, which helps keep the animal from drying out in a dry habitat.
5. Answers may vary; however, students should remember that alligators are found primarily in North America and Asia. An attack in an African river would likely be a crocodile.
6. Teeth can give clues about the type of food a primitive reptile ate, the size of leg bones can give clues about the body size, and footprints can tell about the animal's movement and behavior.
7. Both tortoises and turtles are reptiles with bony shells. They differ in the structure of their shells and in their habitats.
8. Sharks, rays, and skates are cartilaginous fishes.
9. Accept all logical answers; however, the smaller the organism, the smaller the ratio of its surface area to its volume, which makes for more efficient gas exchange.

Chapter 17 Review

Check Your Knowledge
Answer the following in complete sentences.

1. What are animals with backbones called? Give some examples.
2. How can you tell the difference between a crocodile and an alligator?
3. How are lizards like dinosaurs? How are they different?
4. Describe two characteristics of bony fishes. When did they first appear on the earth?
5. When did the first vertebrates appear on the earth? Where did the first vertebrates live, and what did they look like?
6. How is a tadpole different from a frog? Name two differences. How is a tadpole like a fish? Name two similarities.
7. Which amphibians have tails? Describe one characteristic of this group.
8. What kind of shell covers reptile eggs? What is one advantage of this type of egg covering?
9. Why do reptiles bask in the sun when the air is cool and seek shade when it is hot?

Choose the answer that best completes each sentence.

10. All fishes have a very good sense of (sight, hearing, balance, smell).
11. A newt is a type of (reptile, amphibian, mammal, plant).
12. Most cartilaginous fishes swim constantly to (stay warm, find food, move water across their gills, stay awake).
13. There is a total of (seven, five, four, six) classes of vertebrates.

Check Your Understanding
Apply the concepts you have learned to answer each question.

1. Describe one difference and one similarity between snakes and lizards.
2. How does a snake use its tongue? Describe the process and two uses.
3. What is a common nickname for ectotherms? Why do you think they are called this?
4. **Mystery Photo** The photograph on page 344 shows the eye and part of the head of a chameleon. Chameleons are able to change colors, primarily to camouflage themselves. When you look at the photograph, how do you know that the subject is a reptile? Is the chameleon slimy? Why or why not?
5. **Extension** If you read a story about an alligator attacking a human in an African river, would you believe it? Why or why not?
6. What are some of the fossil clues scientists have used to learn about the history of reptiles? Give two examples. What inferences are made from these clues?
7. **Compare and Contrast** What characteristics do tortoises and turtles have in common? How do they differ?
8. Which fishes are included in the cartilaginous group?
9. **Infer** Some salamanders have no lungs and use only their skin for gas exchange. How large do you think these lungless salamanders are, compared to other amphibians? Why?

Develop Your Skills

1. a. About 12°C; about 30°C
   b. The environment began to warm up at A, and began to cool down at B.
   c. Between time A and time B the lizard could have been sitting in the sun.
2. Answers may vary; however, students should realize that since many eggs will not survive to become new frogs, laying many eggs insures that some will survive to adulthood.
3. a. The reticulated python is 10 m long.
   b. The giant tortoise averages 1–3 m; the rattlesnake averages 2 m.
4. Student posters will vary. Make sure they represent all seven classes, and one each from the jawless, bony, and cartilaginous fish classes.

Make Connections

1.

```
                         vertebrate
                         classes are
   ┌──────┬──────────┬──────────┬──────────┐
mammals  amphibians  jawless    bony
                    fishes     fishes
   │         │
reptiles   birds
                              cartilaginous
                              fishes
groups are
   │                          include
┌────────┬──────────────┐
lizards &  alligators &   sharks
snakes     crocodiles     rays
                          skates
         │
       turtles &         include
       tortoises
                         lampreys
         groups are      hagfish
   ┌─────────┬──────────┐
  frogs &   salamanders &
  toads      newts
```

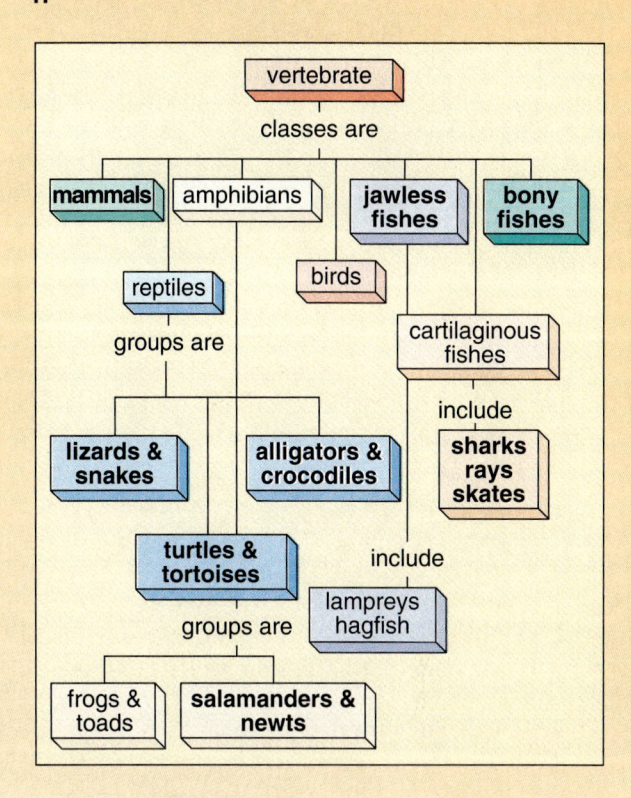

2. Reptiles walk on legs, slither, or swim. Amphibians walk, hop, or swim. Fish swim.
3. Students' reports may vary, but should include at least two examples or portrayals. Look for comments indicating that the animal was correctly or incorrectly identified as either an alligator or crocodile (if possible).

Develop Your Skills

Use the skills you have developed in this chapter to complete each activity.

1. **Interpret Data** The graph below shows the body temperature of a lizard over a 24-hour period.
   a. How cold does the lizard's body get? How warm?
   b. What happened in the lizard's environment at the time labeled A? What happened at B?
   c. What do you think the lizard was doing between time A and time B?

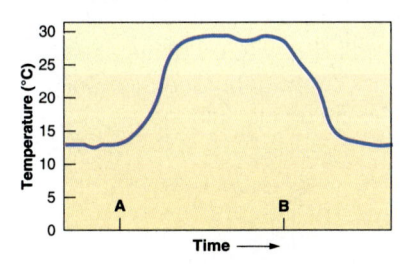

2. **Hypothesize** When frogs reproduce, the female lays hundreds of jellylike eggs. Why do you think a single frog lays hundred of eggs?

3. **Data Bank** Use the information on page 631 to answer the following questions.
   a. What is the longest reptile? How long is it?
   b. Which is longer, the rattlesnake or the giant tortoise?

4. **Communicate** Make a poster that shows examples from each of the seven classes of vertebrates. You should include one of each of the fish classes.

Make Connections

1. **Link the Concepts** Below is a concept map showing how some of the main concepts in this chapter link together. Only part of the map is filled in. Finish the map, using words and ideas you find in the chapter.

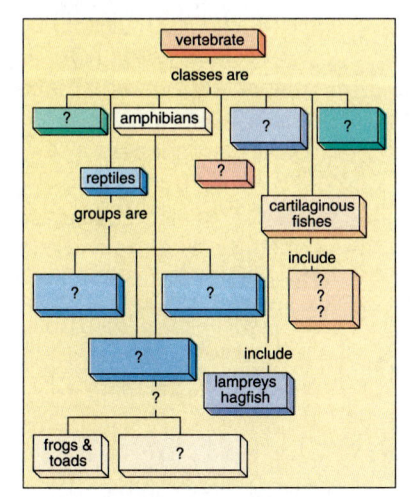

2. **Science and PE** How does a reptile move? How about an amphibian, or a fish? Practice moving as if you were a reptile, then an amphibian, then a fish. Use the information from the chapter, as well as your imagination.

3. **Science and Society** Crocodiles and alligators are often portrayed as villains in popular culture. Find examples of alligators and crocodiles in movies, comic strips, television, or books. Investigate how the alligator or crocodile is portrayed. Write a report on your findings. Include at least two examples.

Chapter 17 Fishes, Amphibians, and Reptiles

CHAPTER 18

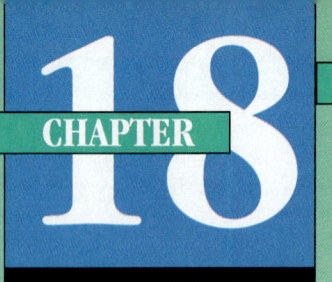

Overview

Students learn about birds and mammals in this chapter. The first section describes characteristics of birds along with a discussion of their unique ability to fly. This section also explains migration and reproduction. The second section presents a discussion of mammals, their characteristics, and their evolutionary history. The third section describes reproduction in three groups of mammals: monotremes, marsupials, and placental mammals. The diversity of placental mammals is explored through descriptions of several orders of them.

Advance Planner

▶ Order contour and down feathers from a science or biological supply company and set up microscopes for SE Activity 18, page 375.

▶ Gather books or magazines containing close-up photographs of mammals and birds for TE activities, pages 372 and 378.

▶ Use a yellow pages directory to compile addresses of medical research laboratories, cosmetics companies, and animal rights groups. TE activity, page 378.

Individual Needs

▶ **Limited English Proficiency Students** Cover the captions and photocopy pages 370, 371, 382, and 383 along with Figures 18.9 and 18.10. Have students write the boldface terms for the chapter on strips of paper and attach the terms to pictures that show the meanings of the terms. They can use a term more than once. Have students meet in small groups to discuss why they placed each term on a particular photograph.

▶ **At-Risk Students** Have students make a list of birds and mammals that they have observed in their area. Have them describe each animal, mentioning what they've noticed about its habitat, eating habits, and behavior. Then they should write a question about each animal. As they work through the chapter, have students try to answer their questions.

▶ **Gifted Students** Have students research the migration of one species of bird. Have them describe the characteristics of the bird as well as its life-style and food sources. Ask students to compare the bird's two habitats. They should draw the travel routes on an outline map. Students' reports should include information about when birds migrate, how long the migration takes, and where along the route birds stop. Display the maps and reports in the classroom or library.

Resource Bank

▶ **Bulletin Board** Create a display about reproduction in birds and mammals. Encourage students to find pictures of birds' nests, eggs hatching, marsupials, monotremes, and placental newborns. Have students label each picture with a caption saying what group of animal is shown and how that group reproduces.

▶ **Field Trip** A visit to a zoo would be very appropriate as students study birds and mammals. You might prepare an animal search for students to complete on the field trip.

Skills Development Chart

Sections	Classify	Communicate	Compare and Contrast	Decision Making	Hypothesize	Infer	Model	Observe	Research
18.1 Skills WarmUp						●			
Skills WorkOut									●
Historical Notebook									●
Skills WorkOut					●	●	●		
Activity								●	
18.2 Skills WarmUp			●						
Skills WorkOut								●	
Consider This		●		●					
18.3 Skills WarmUp	●								

CHAPTER 18 PLANNING GUIDE

Section	Core	Standard	Enriched	Section	Core	Standard	Enriched
18.1 Birds pp. 367–375				**18.2 Mammals** pp. 376–379			
Section Features Skills WarmUp, p. 367	●	●	●	**Section Features** Skills WarmUp, p. 376	●	●	
Skills WorkOut, p. 368		●	●	Skills WorkOut, p. 378	●	●	
Historical Notebook, p. 368	●	●		Consider This, p. 378		●	●
Skills WorkOut, p. 372	●	●		**Blackline Masters** Review Worksheet 18.2	●	●	
Activity, p. 375	●	●	●	Skills Worksheet 18.2	●	●	●
Blackline Masters Review Worksheet 18.1	●	●		**Ancillary Options** One-Minute Readings, p. 14	●	●	●
Enrich Worksheet 18.1		●	●				
Skills Worksheet 18.1	●	●		**Laboratory Program** Investigation 33	●	●	●
Integrating Worksheet 18.1		●	●	**18.3 Diversity of Mammals** pp. 380–384			
Overhead Blackline Transparencies Overhead Blackline Master 18.1 and Student Worksheet	●	●	●	**Section Features** Skills WarmUp, p. 380	●	●	●
Color Transparencies Transparencies 52a, b	●	●		**Blackline Masters** Review Worksheet 18.3	●	●	
				Skills Worksheet 18.3	●	●	
				Integrating Worksheet 18.3	●	●	
				Vocabulary Worksheet 18	●	●	

Bibliography

The following resources can be used for teaching the chapter. See page T-40 for supplier codes.

Audio-Visual Sources
(video unless noted)
Beyond Words: Animal Communication. 15 min. 1984. NGSES.
Birds. Revised. 14 min. 1985. C/MTI.
The Little Marsupials. 25 min. 1986. CP.
Protecting Endangered Animals. 15 min. 1984. NGSES.
Vanishing from the Earth. 3 filmstrips with cassettes. 1986. NGSES.

Software Resources
Animals. Apple. J & S.
Diversity of Life. Apple. J & S.
Vertebrates. Apple. J & S.

Library Resources
Burnie, David. How Nature Works. Pleasantville: The Reader's Digest Association, Inc., 1991.
Durrell, Gerald. The Amateur Naturalist. New York: Alfred A. Knopf, 1989.
Leen, Nina. Rare and Unusual Animals. New York: Holt, Rinehart, and Winston, 1981.
Stokes, Donald, and Lillian Stokes. A Guide to Animal Tracking and Behavior. Boston: Little, Brown, & Co., 1986.

CHAPTER 18

INTEGRATED LEARNING

Writing Connection
Have students write a poem about the photograph. They should describe the colors and patterns of the feather. Have them include details about how feathers are helpful to a bird. Ask students, Are feathers used for flying? Do they help birds stay warm? Does a bird see how colorful its own feathers are? How can feathers protect a bird? Have students read their poems to the class.

Introducing the Chapter

Have students read the description of the photograph on page 366. Ask if they agree or disagree with the description. You may wish to point out that the photograph shows the feathers of a male Asian pheasant, a relative of the peacock.

Directed Inquiry

Have students study the photograph. Ask:

▶ What part of an animal does this photograph show? Explain. (Students will probably say the feathers or outside covering of a bird because of the fringed edges, central shafts, and ridges in the vanes.)

▶ How do feathers help birds? (They keep the bird's body warm and make its body smooth and streamlined for flight.)

▶ What do the round, blue markings resemble? (Students will probably say eyes.)

▶ How might these and the other markings help the bird? (The eyelike markings could confuse predators, and the white speckles could provide camouflage. The markings also help the bird attract a mate.)

Vocabulary

cerebrum monotreme
incubate placenta
marsupial placental mammal
migrate

Chapter 18 Birds and Mammals

Chapter Sections
18.1 Birds
18.2 Mammals
18.3 Diversity of Mammals

What do you see?

"It looks like a male peacock's feathers. The purple spots are part of the design to attract females. I think it's a male peacock, because they are more colorful while the female feathers are plain. You can kind of tell they're feathers by the fringed edges."

Julie Motomura
Fleming Junior High School
Lomita, California

To find out more about the photograph, look on page 386. As you read this chapter, you will learn about birds and different types of mammals.

Themes in Science

Diversity and Unity There is great diversity among birds. They range in size from tiny hummingbirds to enormous ostriches. Some have subtle colors that blend in with their environment and some are brilliantly hued. All birds share certain unifying characteristics, however. All birds have feathers, wings, and beaks. They are all endothermic, which means they remain active even in cold temperatures.

SECTION 18.1

Section Objectives
For a list of section objectives, see the Student Edition page.

Skills Objectives
Students should be able to:

Infer the characteristics that define birds.

Research information about *Archaeopteryx*.

Make Models to compare a paper airplane and a bird.

Infer the physical characteristics of a bird based on its habitat and behavior.

Vocabulary
migrate, incubate

18.1 Birds

Objectives

▶ **Describe** the main characteristics of birds.

▶ **Explain** how birds are adapted for flight.

▶ **Interpret** how the shape of a bird's wings helps a bird fly.

▶ **Compare** and **contrast** bird reproduction and reptile reproduction.

▶ **Make inferences** about a bird's diet and lifestyle by observing its beak, wings, and feet.

Skills WarmUp

Infer You have seen birds in many places. You have probably watched them fly, look for food, and interact with each other. Based on what you have observed, describe birds as living organisms.

MOTIVATE

Skills WarmUp
To help students recognize what they already know about birds, have them do the Skills WarmUp. **Answer** Birds move, eat, reproduce, and use energy.

Misconceptions
Students may assume that birds are mammals. Explain that birds are different from mammals in several important ways: Birds have feathers instead of hair, beaks instead of teeth, and wings. Bats are the only mammals that have wings.

Answer to In-Text Question
① Birds' bodies are lightweight and streamlined for flight.

Have you ever dreamed you could fly? Imagine that you could soar over hills and valleys and flap across town to drop in on a friend. Write down what it would be like.

For most birds, flight isn't just a dream but an everyday event. As fliers, birds can easily escape enemies and cover large areas in their search for food. They can move over great distances to find good places to live. For these reasons, birds have been able to make many kinds of habitats home. As you will find out, most of what makes birds unique has to do with flying.

Characteristics of Birds

Like reptiles, fishes, and amphibians, birds are vertebrates. But birds are different from these other vertebrates in an important way. They are endotherms, organisms that have a constant internal body temperature. Reptiles and amphibians obtain heat from their external environment. When it is cold out, they are sluggish. Birds, however, can be active no matter what the temperature of their environment. The cells of their bodies stay warm and can work normally all the time.

All birds share certain other characteristics, too. You can probably guess what most of them are. Birds are the only group of animals to have feathers. They have beaks instead of teeth. Although not all birds can fly, all birds have wings. And, like reptiles, birds lay eggs with shells. How do you think each of these characteristics relates to a bird's ability to fly?

Figure 18.1 ▲
How are birds' bodies designed for flight? ①

The Living Textbook:
Life Science Sides 1-4

Chapter 4 Frame 01415
Birds (52 Frames)
Search: Step:

Chapter 18 Birds and Mammals 367

TEACH

Skills WorkOut
To give students an opportunity to learn more about *Archaeopteryx*, have them do the Skills WorkOut.
Answers Like the dinosaurs of its time, *Archaeopteryx* had a long tail and teeth and claws for catching prey. Although it had wings, scientists debate about how well it could fly. Some even question whether it was a bird at all because it lacked the strong breastbone that helps give birds the strength to fly.

Enrich
Have students research the meaning of the word *Archaeopteryx*. (It means "ancient wing.")

Historical Notebook
Before students begin their research, point out that DDT is an organic, laboratory-made (or synthetic) insecticide that was first used in 1939 against the Colorado potato beetle. It acts as a poison, disturbing the nervous systems of organisms.
Research Have students work in cooperative groups and research efforts to preserve the following bird populations in the United States: bald eagle, California condor, brown pelican, spotted owl, wood duck, osprey, and bluebird. Each group should share its findings with the class.
Answers
1. Students' answers should reflect their understanding of Rachel Carson's argument that the pesticide DDT harms organisms throughout the food web.
2. Local or state conservation offices can provide information about protection of local birds.

INTEGRATED LEARNING

Integrating the Sciences
Earth Science Explain that the evolution of birds is more difficult to trace in the fossil record than the evolution of many other animals. Remind students of what they learned in Chapter 8 about the process of fossilization. Ask them to consider the lightweight structure of bird bones and infer why they are relatively scarce in the fossil record. (Their bones are so small and light that they do not become fossilized as easily as many other kinds of land animals.)

Skills WorkOut
Research Find out more about *Archaeopteryx*. What dinosaurs lived during the same period? How was it similar to a dinosaur? How did it probably catch its prey? How well did it fly? Find out why scientists debate whether or not *Archaeopteryx* was truly a bird.

Origin of Birds
One day in 1861, a man in Europe split apart two layers of rock in a quarry. Inside he found the fossil imprint of an amazing organism. Like a reptile, it had scales, jaws with teeth, and claws on its front limbs. But also visible in the stone was the unmistakable imprint of feathers. The 150-million-year-old fossil was an ancient bird! Scientists called it *Archaeopteryx* (AR kee OP tur ihks).

The fossil of *Archaeopteryx* was convincing evidence that birds evolved from a reptile ancestor. Since then, other evidence has been collected to back up this conclusion. But scientists still don't know the exact evolutionary relationship between reptiles and birds. They know that feathers evolved from reptile scales, but they don't know why. There had to be some in-between animal that couldn't fly but that could take advantage of having feathers. Perhaps the first birds were fast runners who used their feathered wings to catch insects.

Historical Notebook

Exposing the Silent Truth
In the 1950s, an ecologist named Rachel Carson began to notice a spooky quietness around her home. She wondered why she heard so few bird songs. Puzzled, she began to do research to explain the "spring without voices."

In 1962, she published *Silent Spring*, a book that offered a reason for the decrease in the numbers of birds. She argued that when farmers used pesticides such as DDT, they destroyed more than just crop-threatening insects.

Other scientists agreed that DDT was harming organisms throughout the entire food web, including birds. One problem was that DDT killed the insects that many birds relied upon for food. But this chemical also caused the eggshells of many birds to become too thin. When the eggs were incubated, the shells broke.

Silent Spring had a tremendous impact. In 1972 Congress passed laws prohibiting use of DDT in the United States. But many dangers to birds still exist. DDT is used in countries where many birds migrate in winter. Other pesticides affect the food web, too. Birds are also threatened by the destruction of their habitats.

1. What did Carson argue in *Silent Spring*?
2. **Research** Find out what is being done in your area to protect birds.

368 Chapter 18 Birds and Mammals

TEACHING OPTIONS

Cooperative Learning

The shape of a bird's beak is specialized for eating particular types of foods. Have students explore the diversity of this adaptation. Working in groups of four, each student in a group should pick a different type of bird and make a mask of its beak. Students could choose such birds as the pelican, flamingo, coot, duck, eagle, hummingbird, parrot, crossbill, or woodpecker. Each group could then give a presentation about how its birds use their beaks. Presentations may take the form of a report, nature show, or skit.

Explore Visually

Have students study the material on this page. Ask the following questions:

▶ To which group do peregrine falcons belong? How do you know? (Birds of prey because they catch other animals for food)

▶ To which group do sparrows belong? How do you know? (Perching birds because they perch in trees)

▶ To which group do herons belong? How do you know? (Water birds because they wade in the water)

▶ To which group do ostriches belong? How do you know? (Flightless birds because they do not fly)

▶ Which of the groups of birds are probably most like *Archaeopteryx*? Why do you think so? (Students might say that birds of prey are most like *Archaeopteryx* because they have claws made for grasping and they eat other vertebrates. Students might also say flightless birds because *Archaeopteryx* probably did not fly.)

Enrich

Use Enrich Worksheet 18.2.

The Living Textbook: Life Science Side 4

Chapter 10   Frame 04589
Flamingo/Swan Feeding (Movie)
Search: Play:

The Living Textbook: Life Science Side 4

Chapter 10   Frame 05813
Penguins (Movie)
Search: Play:

Diversity of Birds

There are almost 9,000 species of birds. They are often grouped into four main types: birds of prey, perching birds, water birds, and flightless birds. Even within a group, birds have an amazing variety of shapes and sizes. Each bird is adapted to eating a certain kind of food and living in a certain kind of habitat. A bird's lifestyle is reflected in its beak, wings, and feet.

Birds of Prey ▲
Birds of prey eat mammals, fish, and other birds. They have sharp, hooked beaks good for tearing flesh, and claws made for grasping. Birds of prey include eagles, hawks, owls, falcons, and vultures. This osprey lives near the ocean and eats fish.

Water Birds
Water birds have a variety of beaks, wings, and feet. Many have webbed feet for swimming. Other water birds, like this flamingo, have long legs for wading. Some have long wings they can use to fly great distances. ▼

◀ Perching Birds
Many of the perching birds are insect eaters. They have long, pointed beaks that work like tweezers. The beaks of woodpeckers are used as drills for boring into wood to find insects. Swifts have wings that let them fly fast and turn quickly in pursuit of flying insects.

Flightless Birds ▲
Flightless birds have lost the ability to fly. They have lifestyles that make flight unnecessary. Many flightless birds, such as this rhea, have become fast runners.

Seed Eaters
Other perching birds are seed eaters. Seed-eating birds like this cardinal have thick, strong beaks for cracking open seeds. ▼

Chapter 18 Birds and Mammals

TEACH • Continued

Critical Thinking
Infer Point out that predatory birds (as well as predatory mammals) have eyes that are positioned in the front of the head, resulting in binocular vision and, therefore, greater depth perception. Birds and mammals of prey have eyes positioned on the sides of the head, affording a wider field of vision. Ask students to infer how each type of vision helps the animals adapt to their environment.

Explore Visually
Have students study pages 370 and 371. Ask the following questions:

▶ How are bird bones adapted for flight? (They are hollow for lightness, and they have thin cross-supports for strength.)

▶ How do birds avoid being hit by cars, trains, and planes most of the time? (Birds have excellent eyesight.)

▶ Why do birds need powerful muscles? (They must move their wings fast and forcefully in order to get off the ground and overcome the forces of gravity.)

The Living Textbook: Life Science Side 4
Chapter 12 Frame 28334
Hummingbird/Swan Flight (Movies)
Search: Play:

370

INTEGRATED LEARNING

Themes in Science
Energy Point out that because birds use large amounts of energy to fly, their cells are adapted to release energy from food quickly and efficiently. Ask students how the crop and the gizzard help birds obtain fast energy from food. (The crop stores food and releases it in a steady stream. The gizzard grinds it so that it's digested more quickly.)

Art Connection
Leonardo da Vinci was fascinated with birds and their ability to fly. He used his knowledge of bird anatomy to draw designs of flying machines. In his designs, he used materials such as wood to replace bones and sailcloth to replace feathers. Have students draw their own flying machines, using their knowledge of birds' bodies and the physics of flight.

Adaptations for Flight

Like airplanes, birds must be streamlined to fly. Their bones must be lightweight. In addition, they must use large amounts of energy, just as an airplane uses fuel. A bird's body releases the energy stored in food quickly and efficiently. Its cells do this work at a much faster rate than the cells of reptiles and amphibians. In the process, they create more heat. The extra heat produced by cells rapidly burning fuel is what makes birds endotherms. Adaptations for low weight and high activity are present in all parts of a bird's body.

Vision
As fliers, birds must rely on their sense of sight to avoid flying into things and to see food from far away. Most birds have large eyes and very sharp vision.

Muscles
Moving wings fast and hard enough to overcome gravity takes great muscular effort. Birds have powerful flight muscles attached to a large breastbone.

Bones
The skeletons of most vertebrates living on land are heavy. If birds had heavy bones, they couldn't fly. To save as much weight as possible, bird bones are hollow. Inside they have thin cross-supports for added strength.

370 Chapter 18 Birds and Mammals

Integrating the Sciences

Chemistry Explain that all water birds contain a special oil that repels water, allowing them to keep warm in water. Ask students how oil spills affect water birds. (The spilled oil breaks down ducks' natural oil and, therefore, their insulation. Birds in these circumstances often die of hypothermia.)

Themes in Science

Scale and Structure The structure of a bird's feathers depends on its lifestyle and environment. For example, barn owls have fringed feather edges that muffle sound, allowing them to approach prey noiselessly. Swifts have slender, curved wings that allow them great power and speed with little wind resistance. Have students discuss the two feathers in Figure 18.2.

▶ What is the function of the gizzard? (The muscular gizzard grinds up seeds and other hard food to aid digestion.)

▶ How does a bird's heart help it get the large amount of oxygen it needs to fly? (The four-chambered heart keeps blood low in oxygen separate from blood high in oxygen.)

▶ How do birds increase their intake of oxygen? (Birds have air sacs that allow them to take in more air than would be possible with lungs alone.)

Integrated Learning

Use Integrating Worksheet 18.2.

Critical Thinking

Reason and Conclude Ask students to explain how birds are able to sustain enough energy for long flights. (The crop stores food and releases it in a steady stream necessary for it to be ground up by the gizzard, moved to the intestine for digestion, and then carried to the muscle cells for energy.)

Compare and Contrast Ask students how the heart of a bird is different from the heart of a reptile. (A bird has a four-chambered heart; a reptile has a three-chambered heart.)

Answer to In-Text Question

① Contour feathers, which are both strong and light, make up most of a bird's wings and tail, the most crucial flying structures. Shorter contour feathers streamline a bird's body. The fluffiness of down feathers, which lay next to a bird's body, keep the animal warm.

Digestive System
Birds must take in a large and steady amount of food to meet their high energy needs. The crop stores food and releases it in a steady stream for digestion. A special organ, called the gizzard, grinds up the food to make it more easily digested in the intestine.

Air Supply
To provide the blood with as much oxygen as possible, birds have a special breathing system. In addition to lungs, air sacs increase the amount of oxygen a bird can take in.

Heart
Birds have four-chambered hearts that keep blood low in oxygen separate from blood high in oxygen. With this kind of heart, the flight muscles get as much oxygen as blood can carry.

Figure 18.2
How does the structure of each kind of feather match its function? ▼ ①

Feathers

Feathers are perhaps a bird's most important adaptation. Without feathers birds could not fly no matter how light their bodies. In Figure 18.2 you can see a feather called a contour feather. A contour feather is both strong and light. These feathers together make up most of a bird's wings and tail, the most important structures in flying. Shorter contour feathers cover a bird's body, making it streamlined. Contour feathers have many rows of interlocking barbs. These barbs are what make a feather strong and firm.

Under the contour feathers and next to a bird's body are down feathers, also shown in Figure 18.2. The fluffiness of down feathers traps a layer of air next to the body. This layer of air prevents heat loss. Birds could not be endotherms without the insulation provided by down feathers.

Chapter 18 Birds and Mammals

TEACH • Continued

Class Activity
Students can do this activity to observe lift in the physics of flight. Have students cut strips of paper about 3 cm wide and 12 cm long. Holding the edge of the strips up to their mouths, have them blow over the tops of their paper strips. Ask them to observe what happens. (Blowing across the top of the strip makes it lift so that it's horizontal to the floor.)

Critical Thinking
Reason and Conclude Have students read about bird migration. Then ask them to make a list of how birds prepare for migration. (One way is by eating a great deal of food and storing it in their crops.)

Portfolio
Have students collect colorful pictures of birds from newspapers and magazines. Have them mount the pictures on loose-leaf paper, label each bird, and use references to tell where each bird lives during the summer and the winter. Students can add the pages to their portfolios.

Skills WorkOut
To help students understand how birds fly, have them make models of paper airplanes by doing the Skills WorkOut.
Answer The paper airplane is like a bird in that it has wings that are streamlined for flight. The paper airplane is different from a bird in that the wings are not powered and are probably not shaped to produce lift.

Discuss
Many bird species migrate as the weather changes. Point out that when it is winter in the Northern Hemisphere, it is summer south of the equator.

372

INTEGRATED LEARNING

Integrating the Sciences
Physical Science To be able to fly, birds must have enough strength to carry their own weight. Therefore, birds' bodies must be lightweight. Their bones are small and hollow, they have lightweight beaks, and many of them have very slender legs and feet. Ask students to think of other physical characteristics of birds that make them well adapted for flight. (Feathers, streamlined body shape, excellent eyesight, and so on)

Literature Connection
In Greek mythology, Daedalus and his son Icarus flew by fastening human-sized wings made of wax and feathers to their arms. Have students find out and write a paragraph about why Icarus did not reach his destination. Then have them explain why Icarus failed based on what they know about the physics of flying.

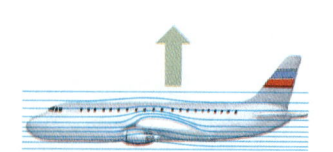

Figure 18.3 ▲
The wings of birds and airplanes both make use of the same physical principle to provide lift.

Physics of Flight
Birds use their wings to push themselves through the air. But if birds relied on flapping alone, they would tire quickly. Flying is made easier because wings provide lift, even without being flapped. The key is in the shape of the wing.

You may have noticed that an airplane's wings are not flat. The upper surface is rounded. Look at Figure 18.3 to see the shape of an airplane wing. Notice how it is similar to the shape of a bird's wing.

When a wing with this shape moves through air, the air has a longer way to go around the curved upper surface than it does across the flat bottom surface. The air above the wing must move faster to cover this longer distance in the same amount of time. This difference in air speed above and below the wing creates a difference in air pressure. The pressure under the wing is higher. So there is more force pushing up, under the wing, than there is force pushing down, on top of the wing. The result is lift. The larger the wing, the greater the lift. Birds with large wings can soar and glide for a very long time. Once they are airborne, they can cover great distances without flapping.

Migration
In places where winters are cold, many animals face a shortage of food for part of the year. Plants stop growing. Insects and other invertebrates die or bury themselves. Some animals deal with the shortage of food by becoming inactive, or hibernating. Birds, however, can go somewhere else to find food.

Taking advantage of their ability to fly, many birds **migrate**, or move to a different place during part of every year. Some birds in the northern hemisphere, for example, fly south to warmer places for the winter months.

When migrating, many birds travel along certain flying routes. Many of these routes follow coastlines, sea currents, wind currents, or land contours.

Some migrating birds travel distances that are hard to imagine. The short-tailed shearwater flies up to 32 000 km every year, circling the Pacific Ocean. The arctic tern enjoys the endless days of arctic summers, then flies all the way to Antarctica in time for summer!

Skills WorkOut
Make Models Experiment with making different types of paper airplanes. Which design will glide the longest distance? In what ways is a paper airplane like a bird? In what ways is it different?

Art Connection

Birds construct nests in a variety of ways, using a wide range of materials. Have each student create a three-dimensional representation of a nest from one of the following species: robin, tailorbird, cliff swallow.

Integrating the Sciences

Earth Science Have interested students find out the migratory routes of snow geese, whooping cranes, arctic terns, and golden plovers. Display a world map and have them trace the routes.

Skills Development

Predict Point out to students that adult birds will sometimes delay their return to a disturbed nest. Ask them to predict what would happen to the eggs under such circumstances. (The eggs might be eaten by a predator, or the embryos in the eggs might die from lack of heat.)

Critical Thinking

Reason and Conclude Have students identify the advantages of chicks being fully developed when they hatch. (They are better able to escape from enemies, find food, fight off disease, and survive in difficult climates.)

Reteach

Remind students of what they learned about reptile reproduction in Chapter 17. Have them compare reptile reproduction to bird reproduction by making lists of the major features of each.

Misconceptions

Explain to students that not all birds make nests in trees. Many birds build nests on the ground, in rocks, or even in chimneys or windowsills. Ask students where they have seen bird nests.

Reproduction of Birds

Birds reproduce much like their distant relatives, the reptiles. The male passes sperm to the female, and the eggs are fertilized inside her body. Shells form around the eggs. Then the female moves the eggs outside her body for the embryos to develop.

Unlike reptiles, however, birds have eggs that must stay warm while the embryos develop. Birds must therefore **incubate** their eggs, or use their body heat to keep the eggs warm. The need to incubate their eggs causes birds to have reproductive behavior different from that of reptiles.

Incubation can't be interrupted for long. How does one bird incubate its eggs and find food at the same time? In some species of birds, the male and female take turns sitting on the eggs. In other species, one of the parents sits on the eggs and the other finds food for both of them. Either kind of cooperation requires an attachment, called a pair-bond, between male and female. Not all birds have pair-bonds.

Figure 18.4 ▲
Many birds build nests in trees to protect the eggs from predators.

To help keep the eggs warm and protected, birds build nests. In contrast, most reptiles bury their eggs in soil. The nests of many kinds of birds are very complex structures. Look at the nest in Figure 18.4. How long do you think it took to build? In many species, nestmaking is shared by males and females. This is another reason for pair-bonding.

How do you think bird eggs and reptile eggs differ? Bird eggs have a hard shell compared to the leathery shell of reptile eggs. The hard shell keeps the embryo from being crushed during incubation.

Look at the bird egg in Figure 18.5. Along with the embryo, a bird egg contains yolk and albumen. The albumen is the "white" of the egg. Both yolk and albumen provide food for the embryo.

At the end of the incubation period, the chick breaks open the shell. The chicks of some species are fully developed when they hatch. They are ready to move and find food. In other species, the chicks are blind, helpless, and almost featherless. They must be fed and taken care of by one or both parents before they are ready to be on their own.

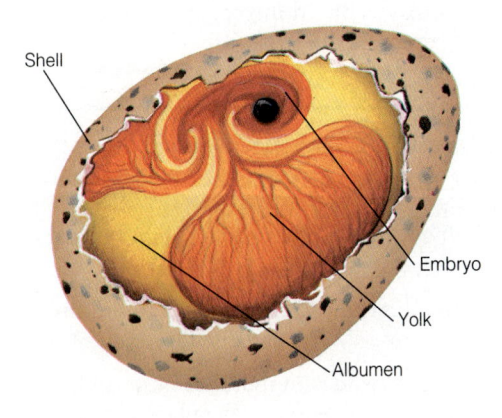

Figure 18.5 ▲
A bird egg supports a growing embryo.

The Living Textbook: Life Science Side 4

Chapter 10   Frame 05371
Ostrich and Young (Movie)
Search: Play:

The Living Textbook: Life Science Side 4

Chapter 16   Frame 44112
Grouse Courtship (Movie)
Search: Play:

TEACH • Continued

Reteach
Have students review the information they learned about Rachel Carson and DDT earlier in the chapter. Ask students how DDT enters the food chain. (DDT is still used as an insecticide in some countries. It was originally developed to help control insect damage to crops.)

EVALUATE

WrapUp
Portfolio Have students fold a piece of notebook paper to make four columns. Then have them write one bird characteristic at the top of each column. Have them draw, trace, or find a picture for each characteristic and place it in the appropriate column. Then have students write a sentence describing how that characteristic helps the bird fly.

Use Review Worksheet 18.1.

Check and Explain

1. Feathers; powerful flight muscles; and hollow, lightweight bones

2. Both have a flat bottom surface and a curved upper surface that cause the air above the wing to move faster than the air below it. This results in lift.

3. In both birds and reptiles, eggs are fertilized inside the female's body, shells form around the eggs, and the female releases the eggs. Most reptiles bury their eggs in soil; most birds build nests and use body heat to keep their eggs warm.

4. Long legs for wading; long, splayed toes for stability; a sharp, hooked beak for tearing flesh; and large wings for flying long distances with relative ease.

374

INTEGRATED LEARNING

Integrating the Sciences

Ecology Point out that while predatory birds do not live on a diet of insects, they are affected by insecticides because they eat animals that feed on insects. Such birds also eat fish that live in water that contains pesticide run-off from farms. Have students draw a picture showing how a predatory bird could ingest pesticides.

Figure 18.6 ▲
A peregrine chick is returned to its nest after being hatched in the laboratory.

Science and Society *Back from the Brink?*

The peregrine falcon was once common throughout North America. But by 1970, only two nesting pairs were counted in California. In 1978, no nesting pairs could be found in the eastern United States. The main reason for the decline was the presence of DDT in the food chain. Peregrines ate other birds with DDT in their bodies. As a result, peregrines produced eggs with shells so thin they broke when incubated.

Now peregrine falcons are on the increase. Banning DDT in the United States has helped. But peregrines still pick up DDT because they eat birds that winter in countries where DDT is still used. So for peregrines to reproduce, scientists have had to help them.

Scientists help the birds by taking peregrine eggs from the nest just after they are laid. The eggs are then incubated in a laboratory so they will hatch even if the shells are thin. Then comes the task of returning the birds to the wild. One method is to place peregrine chicks into the nests of the more common prairie falcon, who raise the foster peregrine chicks as their own.

Another successful method is bringing the chicks back to the same nest they came from. This can be done if "dummy" eggs are left in place of the real eggs. The parents continue to incubate and will accept the chicks when they are returned to the nest.

As a result of these methods, peregrines are making a comeback. But they are far from being out of danger. Many peregrine embryos die in the egg, poisoned by other chemicals present in the environment.

Check and Explain

1. Describe three adaptations that help a bird fly.

2. Why do the wings of both birds and airplanes provide lift when passing through air?

3. **Compare and Contrast** How is reproduction in birds and reptiles the same? What are the differences?

4. **Infer** A certain bird wades in marshes hunting fish. It travels long distances between the places it feeds. Describe what this bird might look like, including its wings, beak, legs, and feet.

Chapter 18 Birds and Mammals

TEACHING OPTIONS

Prelab Discussion

Have students read the entire activity before beginning the discussion. Ask the following questions:

▶ In all organisms, how is the structure of a body part related to its function(s)? What are some examples?

▶ What does it mean to say that something is insulated? In what different ways are organisms insulated against heat and cold?

▶ Why do you think it is important for contour feathers to be flexible?

ACTIVITY 18

Time 45 minutes **Group** 1–2
Materials
contour feathers
down feathers
microscopes

Analysis

1. The barbs are attached to each other by tiny parallel filaments (called barbules) with hooks that interlock each other, making a strong attachment.
2. The barbules of a down feather do not have hooks.
3. A contour feather is wide and flat, and provides a surface to push against the air as a bird flies. A down feather is fluffy and traps a layer of air against the bird to insulate it.
4. Longer contour feathers come from the wings and tail. Shorter ones cover the neck, head, and breast.

Conclusion

Students' conclusions should be based on their actual observations.

Everyday Application

Clothes made from materials that are thick and light. To copy a bird's insulation, one could wear two layers of material, creating an airy space between the two.

Extension

Coverts and inner flight feathers: help create a streamlined body shape; *outer flight feathers:* provide lift; *tail feathers:* help in controlling the direction of flight. A bird preens to "zip" together the barbs of its contour feathers, which improves both flight and insulation by keeping them airtight.

Activity 18 How do contour and down feathers compare?

Skills Observe; Infer; Hypothesize

Task 1 Prelab Prep
Collect the following items: a contour feather, a down feather, unlined paper, a microscope.

Task 2 Data Record
Record your observations of feathers on the unlined paper. Write a note about each characteristic. Make drawings to illustrate what you see.

Task 3 Procedure

1. Examine a contour feather. In one hand, hold the base of the central shaft where it was attached to the bird. Gently bend the tip of the feather with your other hand. Now, hold the shaft with one hand and wave the feather through the air. How strong is the feather? Record your observations on your paper.
2. Look at the shape of the feather. How long is it? Is it curved or straight? Is one side wider than the other side? Record your observations.
3. Examine the structure of the feather. Is the feather solid or does it come apart? Hold the tip of the feather with one hand and run the fingers of the other hand down toward the base. What happens? Now that you have "unzipped" the feather, can you "zip" it back together? What different parts does a feather seem to be made of? Make a close-up drawing of the feather's structure on your paper.
4. Now examine a down feather. Do you see a difference in the shape, size, and structure of the down feather compared with the contour feather? Record your observations.
5. Examine a contour feather under the low-power setting of the microscope. Note the individual barbs that are attached to one another. Draw what you see.
6. Examine a down feather under the low-power setting of the microscope. Are the individual barbs attached or separate from one another? Draw what you see.

Task 4 Analysis

1. How do the barbs of a contour feather seem to be attached to one another? How does this method of attachment make the feather strong?
2. **Infer** Suggest a reason why the barbs of a down feather do not fit together like the barbs that attach themselves together in the contour feather.
3. How is the structure of a contour feather related to its function? How is the structure of a down feather related to its function?
4. **Hypothesize** Based on your observations of its size and shape, hypothesize where on the bird the contour feather came from. Explain the reasons for your guess.

Task 5 Conclusion
Complete your examination of bird feathers by summarizing what you have learned in a short paragraph. Be sure to compare and contrast contour feathers and down feathers.

Everyday Application

What kind of clothes are best for keeping warm? How could you dress to copy the way a bird insulates itself with down feathers?

Extension

Find out about other feathers on a bird's body and how they contribute to a bird's flying abilities. Research the following kinds of feathers, all of which are used by pigeons when they fly: coverts, outer flight feathers, inner flight feathers, tail feathers. Make a table listing these feathers and their functions. Include contour feathers.

If you can, observe a bird for a short time. Watch the bird as it preens. What is it doing to its feathers? Why is preening necessary?

SECTION 18.2

Section Objectives
For a list of section objectives, see the Student Edition page.

Skills Objectives
Students should be able to:

Compare and Contrast a mammal and a reptile.

Observe the behavior of a mammal.

Infer what kinds of teeth different mammals have and why.

Vocabulary
cerebrum

MOTIVATE

Skills WarmUp
Help students understand the differences between reptiles and mammals by doing the Skills WarmUp.
Answer Answers may vary. Mammals have hair or fur and reptiles don't. Mammals are endotherms and reptiles are ectotherms.

Prior Knowledge
To gauge how much students know about mammals, ask the following questions:

▶ What are the closest ancestors of mammals?

▶ How are mammals similar to birds?

▶ How are mammals different from reptiles and birds?

The Living Textbook:
Life Science Sides 1–4
Chapter 4   Frame 01467
Mammals (57 Frames)
Search: Step:

INTEGRATED LEARNING

Integrating the Sciences

Earth Science Many of the characteristics of mammals, such as hair and mammary glands, usually do not become fossilized. For this reason, paleontologists look for other clues to identify mammal fossils. Scientists look for two important features of mammals—a particular kind of jaw and tiny bones in the middle ear cavity. Have students find out what the first mammals looked like and where their bones have been discovered. (They were probably small and shrewlike; England)

Skills WarmUp

Compare and Contrast
Think of one mammal you know well, like a dog or cat. Then think of one kind of reptile. Write all that you know about each animal. Then compare one with the other.

18.2 Mammals

Objectives
▶ **Describe** the characteristics of mammals.
▶ **Explain** how mammals differ from other animal groups.
▶ **Compare** and **contrast** mammals and birds.
▶ **Infer** what kinds of teeth various mammals have.

Look closely at the surface of your forearm. What do you see? The little hairs poking out of your skin are a characteristic shared by none of the animals you have studied so far. Only one group of animals has hair. They—and you—are the mammals.

Like birds, mammals are vertebrates and endotherms. They also have four-chambered hearts. What sets the mammals apart are their hair and their mammary glands. Mammary glands, active only in female mammals, produce milk to feed offspring. The milk is a balanced diet of fats, sugars, protein, minerals, and vitamins.

Origin of Mammals

Mammals evolved from ancient reptiles, just as the birds did. The early mammals and the early birds were both endotherms, adapted for a more active life than their reptile ancestors. Otherwise, the first mammals were very different from the first birds. While birds evolved feathers as a body covering, mammals developed hair. Birds evolved wings from their front limbs. The limbs of mammals, in contrast, became more efficient at moving on land or in trees.

The earliest mammals lived during the age of dinosaurs. Fossils show that they were probably no bigger than rats and similar to rats in appearance. They were very likely active at night, hiding from meat-eating dinosaurs by day. When the dinosaurs died out, however, mammals began to evolve rapidly. In taking the place of the dinosaurs, they developed a variety of shapes, sizes, and lifestyles. After a while mammals came to live in nearly every habitat on the earth.

Figure 18.7 ▲
The first mammals were small and probably ate insects.

376   Chapter 18   Birds and Mammals

Integrating the Sciences

Physical Science The dolphin's body is adapted for life in the ocean. Ask students to infer why a lack of hair might help a dolphin move more easily through the water. (Hair increases resistance. This is why some competitive swimmers shave their bodies.)

Themes in Science

Stability and Equilibrium One way mammals with fur, such as dogs and cats, maintain a stable internal body temperature is by panting when it's hot outside. Ask students how they maintain their body temperatures. (By perspiring if they are hot)

TEACH

Directed Inquiry
Have students study the material on this page. Ask the following questions:

▶ If adult dolphins have no hair, how do they keep warm? (The thick layer of fat under their skin keeps them warm.)

▶ Why do mammals need heavy coats or layers of fat? (To keep in body heat)

▶ Why do mammals sweat? (To keep cool)

▶ What kind of movement is a monkey specially adapted for? (Climbing)

Enrich
Explain that humans are omnivores, meaning they eat both plants and meat. For this reason, people have three different kinds of teeth. Ask students what these are. (Incisors, canines, and molars) Ask them to infer which are for tearing, which are for cutting, and which are for grinding. (Incisors are for cutting, canines are for tearing, and molars are for grinding.)

Characteristics of Mammals

The ability to maintain a constant internal temperature is a major reason why mammals have been so successful. In addition, mammals have several other characteristics different from those of birds, the other group of endotherms.

Hair and Skin All mammals have hair at some time in their lives. For many mammals, hair insulates the body, preventing heat loss. For others, however, hair is not important. Adult dolphins, for example, have no hair at all. They rely instead on a layer of fat under the skin to keep in body heat.

Skin is also an important insulator for land mammals that live in cold environments, such as polar bears. In addition to their heavy coats, polar bears have a thick layer of skin that keeps their bodies from losing heat. For mammals living in warm environments, the ability to sweat is an important adaptation. Sweat evaporates from their bodies, keeping them cool.

Many mammals run on land using all four limbs. This cheetah can reach speeds of almost 90 km per hour. ▼

Teeth Mammals have three different kinds of teeth: incisors, canines, and molars. Incisors are chisel-shaped teeth that are used for cutting and gnawing. Canines are long, pointed teeth that can stab prey and tear flesh. Molars are broad, flat teeth that are good for grinding food. The kinds of teeth a mammal has depends on the food it eats. Some mammals with a varied diet have all three kinds of teeth and can handle any kind of food.

Movement Mammals are active animals. They move to find food and escape enemies. All mammals, except dolphins and whales, have two pairs of limbs they use to move. The structure of these limbs, however, varies considerably. Depending on the species, they can be adapted for running, swimming, climbing, or flying. Some mammals, such as rabbits, move by hopping.

▲ Bats have limbs adapted for flight. A bat's wings, formed from flaps of skin, are very different from a bird's wings. Bats are the only mammals with wings.

◀ Dolphins are among the mammals with limbs adapted for swimming. They have adapted so completely to life in the ocean that they can't move at all on land.

The Living Textbook: Life Science Side 4

Chapter 10 Frame 08843
Bat Flight (Movie)
Search: Play:

The Living Textbook: Life Science Side 4

Chapter 10 Frame 13540
Porpoise Swimming (Movie)
Search: Play:

TEACH • Continued

Skills WorkOut
To help students understand animal behavior, have them do the Skills WorkOut.
Answer Answers may vary. Gathering food, finding a safe place to sleep, and running from predators are possible responses.

Enrich
Often scientists can observe differences among the structures of mammals' brains. For example, the olfactory lobe in a rabbit is larger than the olfactory lobe in a bird.
Use Enrich Worksheet 18.2.

Consider This

Think About It Students should consider both sides of the controversy over animal testing before they put their thoughts in writing.
Write About It Have students send for information on animal experimentation from medical research laboratories, cosmetics companies, and animal rights groups. Have students evaluate the sources. Remind them that they should be able to cite valid reasons for accepting a source as reliable. Ask students to include sources.

Decision Making
If you have classroom sets of *One-Minute Readings*, have students read Issue 8, "Animal Rights" on page 14. Discuss the questions.

The Living Textbook: Life Science Side 2
Chapter 12   Frame 12195
Mammalian Heartbeat (Movie)
Search: Play:

INTEGRATED LEARNING

Integrating the Sciences

Anatomy Stress to students that the large cerebrums of mammals often provide them with well-developed senses. Eyesight is a sense that is important to many mammals. Squirrels have particularly good eyesight. However, most mammals see in black and white. The primates, including humans and bush babies, see in color. If possible, show students photographs of mammals such as the bush baby and the squirrel.

Skills WorkOut
Observe Choose one kind of mammal, such as a friendly dog, cat, or squirrel, that you can observe off and on over a two-day period. Write all your observations of the animal's behavior. Describe what characteristics help the animal survive.

Body Systems

Like birds, mammals have organ systems adapted for an active life. The four-chambered heart transports oxygen to body cells more efficiently than the simpler hearts of reptiles, amphibians, and fishes.

A mammal's lungs are made up of millions of tiny sacs. These sacs greatly increase the surface area of lung tissue in contact with air. As a result, the lungs can transfer more oxygen into the blood. Another adaptation for breathing is a muscle in the chest called the diaphragm (DY uh FRAM). The diaphragm expands and contracts the chest cavity, causing the lungs to draw in air.

Mammals also have complex nervous systems. Their brains are large, especially the part called the **cerebrum** (SUR uh bruhm). The cerebrum is where the higher brain functions occur. With their large cerebrums, mammals are able to perform complex behaviors such as learning. Some mammals also have well-developed senses.

Consider This

Should Products Be Tested on Animals?

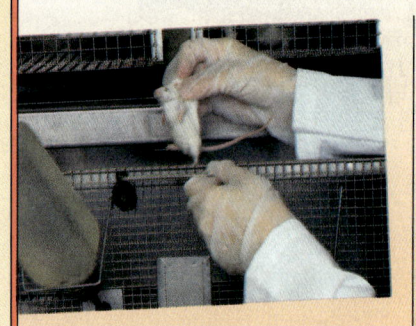

For many years, scientists and doctors have used animals to help them develop new products and inventions. They test drugs and cosmetics on animals to see if these substances have harmful side effects. They inject disease-causing microorganisms into rats to help find ways of controlling infections in humans. They try out new kinds of surgery on monkeys.

To develop many of the drugs and inventions that save human lives today, many animals were killed or made to endure pain. Each year, between 17 and 22 million animals are used for testing or experimentation.

Consider Some Issues
Opponents of animal research say that too many animals are killed or injured. They believe people do not have the right to harm animals, regardless of the benefit to humans.

Scientists, however, claim that animal research is necessary for testing medicines and discovering cures that help sick people and save lives. They say that animals in labs are treated as humanely as possible.

Think About It Is it right to kill animals for human benefit? Or, on the other hand, is it right to put limits on what can be done to save human lives?

Write About It Write a paper stating your position on animal experimentation.

Chapter 18 Birds and Mammals

STS Connection

To illustrate the complexity of wildlife conservation issues, ask students to investigate the history of whaling and write a report addressing the following questions:

▶ How long have people been hunting whales, and why is it an ecological problem?

▶ What uses do people have for whale products?

▶ Could other materials be used instead of whale products?

▶ What is being done to protect whales?

Science and Society
Saving Endangered Mammals

In the last few hundred years, humans have drastically changed the face of the earth. In the process, people have killed off many species. At least 36 species of mammals have become extinct in the last 400 years.

But today there are at least three times as many mammal species threatened with extinction. They are called endangered species. Endangered mammals include the African elephant and the blue whale. Many endangered mammals have simply been hunted so heavily there are few left. Others are in danger because their habitats have been destroyed.

Many people around the world are trying to prevent endangered species, especially mammals, from becoming extinct. This work is going on in many different areas.

Scientific research is often an important part of saving a species. Scientists try to find out, for example, what the species needs to increase in number. Another part of saving a species involves protecting the places it lives. Parks, wildlife refuges, and reserves are set aside as places where the animal can live and reproduce.

To protect species that are hunted for certain products, laws are made to ban the hunting of the animal or the buying and selling of any of its parts. It is against the law, for example, to buy or sell elephant tusks.

Some endangered species would probably not survive on their own in nature, even with protection. The only way to prevent these species from becoming extinct is to breed them in captivity.

Figure 18.8 ▲
The black-footed ferret (top), snow leopard (middle), and Florida manatee (bottom) are all endangered mammals.

Check and Explain

1. How do mammals with little or no hair stay warm?
2. Name and describe two characteristics unique to mammals.
3. **Compare and Contrast** Explain how mammals and birds are similar and different.
4. **Infer** Write down the kinds of teeth you think each of the following mammals has and why: giraffe, beaver, horse, human, elephant, mouse, lion, goat, and dog.

EVALUATE

WrapUp

Portfolio Have students draw a mammal and label its parts, indicating the specialized adaptations that make it a mammal and help it survive. They should include information about how the mammal reproduces, what kind of skin and teeth it has, how it moves, how its lungs and heart function, and so on. Students can add the drawings to their portfolios.
Use Review Worksheet 18.2.

Check and Explain

1. They rely on a layer of fat under the skin.
2. Hair and, in females, mammary glands
3. Birds and mammals are both endothermic, active, and have four-chambered hearts. Birds have feathers, beaks, and wings; mammals have hair, teeth, four limbs adapted for movement, and advanced respiratory and nervous systems.
4. Giraffes, goats, and horses have molars to grind vegetation. Beavers and mice have incisors for gnawing and molars for grinding. Elephants have molars for grinding and two large incisors (tusks) that serve as weapons. Lions and dogs have canines to hold prey, incisors for cutting flesh, and molars for grinding flesh. Humans have all three types of teeth.

Chapter 18 Birds and Mammals

SECTION 18.3

Section Objectives
For a list of section objectives, see the Student Edition page.

Skills Objectives
Students should be able to:

Classify mammals into groups.

Classify placental mammals by order and give a characteristic of each.

Vocabulary
monotreme, marsupial, placenta, placental mammal

MOTIVATE

Skills WarmUp
Help students classify different animal features and behaviors by having them do the Skills WarmUp.
Answer Answers may vary. Students may classify animals according to whether they have fur covering their bodies, how they get food, and so on.

Misconceptions
Students may think that mammals do not lay eggs. Explain that although most mammals give birth to live young, a few species do lay eggs.

The Living Textbook:
Life Science Sides 1-4
Chapter 4   Frame 01468
Monotreme (1 Frame)
Search:

380

INTEGRATED LEARNING

Integrating the Sciences

Earth Science Mammals began to diversify rapidly during the Cenozoic Era. Numerous types of monotremes and marsupials developed in the geographically isolated regions of Australia and South America, while placental mammals dominated North America. When South America joined with North America, placental mammals from the north spread south to dominate the entire continent. However, because Australia remained geographically separate, marsupials and monotremes lived with little competition from placental mammals. As a result, Australia has the largest and most diverse population of monotremes and marsupials in the world.

Skills WarmUp

Classify Write the names of all the mammals you can think of. Then classify them into different groups based on common features or behaviors. Choose any classification system you wish.

18.3 Diversity of Mammals

Objectives

▶ **Distinguish** between egg-laying, pouched, and placental mammals.

▶ **Infer** why most egg-laying and pouched mammals live in one part of the world.

▶ **Classify** the major orders of placental mammals and give examples of each.

If you've been to a zoo, then you've seen many of the different kinds of mammals that live on the earth. Elephants, giraffes, lions, zebras, monkeys, and anteaters make up a very diverse bunch. But common zoo animals don't cover all the major mammal groups. Most zoos don't have bats or shrews, few have koalas or duck-billed platypuses, and hardly any have whales. And don't forget more common mammals like dogs, mice, rabbits, goats, and cows!

Life scientists begin to make sense of this diversity by separating two kinds of mammals from all the rest. These two groups, the egg-laying mammals and the pouched mammals, reproduce very differently from most other mammals.

Egg-Laying Mammals

As you probably know, most mammals give birth to live young. Living in Australia and New Guinea, however, are mammals that lay eggs. They are the duck-billed platypus and two species of spiny anteater. These egg-laying mammals are called **monotremes** (MAHN oh TREEMZ).

Monotremes have the two unique mammal characteristics: hair and mammary glands. Their eggs are soft-shelled and hatch after ten days. When the young emerge, they are fed milk. The milk comes from glands on the mother's belly. Since the glands have no special opening, the milk seeps out and the baby sucks it off of the mother's fur. Spiny anteaters nurse their young for up to six months.

Figure 18.9 ▲
This monotreme is a short-nosed spiny anteater from Australia.

380   Chapter 18   Birds and Mammals

Multicultural Perspectives

Chinese researchers Pan Wenshi and Lu Zhi have been observing pandas in the wild since 1985. In 1989, one of the pandas gave birth, allowing the team to study the way pandas rear their young. They discovered that for more than a week after her cub was born, the mother panda did not eat. She spent nine days just nursing and caring for her cub. This surprised Wenshi and Zhi because pandas often eat for up to 12 hours a day. Encourage interested students to find out more about pandas.

Pouched Mammals

The other group of mammals with unusual reproduction is called **marsupials** (mar SOO pee uhlz). Marsupials are mammals with pouches in which the young complete their development. You have probably heard of two marsupials, the koala and the kangaroo. They both live in Australia. Another marsupial, the opossum, is common throughout North America. About 80 species of marsupials live in South America, including the rat opossum.

Marsupial eggs are fertilized inside the female's b and begin their development there. The embryos grow inside an organ called the uterus (YOOT ur uhs). They are nourished by a limited food supply that was part of the egg. When that food supply is used up, the young are born while still embryos. The tiny, blind babies then crawl into the mother's pouch. Each finds a nipple, where milk comes out of a mammary gland, and begins to suckle. The babies stay in the pouch for one to two months, until their development is complete.

Placental Mammals

Most mammals neither lay eggs nor have pouches. Their young complete their development inside the uterus. There, an embryo receives nutrients and oxygen from the mother's body through an organ called the **placenta** (pluh SEHN tuh). The placenta also takes away waste products and carbon dioxide from the embryo. Thus the placenta provides for all the embryo's needs during the entire time it is developing. Mammals with placentas are called **placental mammals**.

The length of time an embryo develops inside its mother varies, depending on the kind of placental mammal. Mice have a short, 21-day period of development. Humans are born after 9 months of growth inside the uterus. Elephants take 22 months before they are born.

After they give birth, placental mammals spend more time caring for their young than do other animals. Cnidarians, for example, release many eggs and sperm into the ocean. Placental mammals care for their young until they can be on their own. However, the degree of parental care varies. An elephant invests more time and energy into raising each offspring than a rabbit does.

Figure 18.10 ▲
What is the function of this western gray kangaroo's pouch? ①

TEACH

Skills Development

Infer Remind students that some adult placental mammals care for their young for a fairly long period of time. Ask students what some of the advantages of this practice might be. (Young mammals grow bigger and stronger and prepare for living on their own.)

Discuss

Stress to students that mammals can be classified according to how they reproduce. Ask the following questions:

▶ Why are monotremes considered mammals? (They have hair and mammary glands.)

▶ How do monotremes reproduce? (They lay eggs.)

▶ How are placental mammals different from marsupials? (The embryos of placental mammals develop in the uterus. Placental mammals don't have pouches.)

Answer to In-Text Question

① The kangaroo's offspring completes its development inside the pouch.

The Living Textbook:
Life Science Sides 1-4

Chapter 4   Frame 01469
Marsupials (6 Frames)
Search: Step:

The Living Textbook:
Life Science Side 4

Chapter 12   Frame 29044
Kangaroo (Movie)
Search: Play:

TEACH • Continued

Explore Visually

Have students study the material on pages 382 and 383. Ask the following questions:

▶ Do foxes belong to the Carnivora? How do you know? (Yes. Foxes have clawed toes, a well-developed brain, and teeth adapted for tearing flesh.)

▶ Which order of mammals has nostrils set high on the head, fins instead of forelimbs, a horizontal tail, and the absence of hind limbs? (Cetacea)

▶ What are some characteristics of Chiroptera? (They have long fingers, wings adapted to flight, and are nocturnal.)

▶ Which order of mammals are believed to be most similar to ancient mammals? Why? (Insectivora, because their bodies, habits, and diet are similar to those of the ancient mammals)

▶ What are three different kinds of primates living today? (Monkeys, apes, and humans)

▶ Which order of mammals includes the largest land mammals living today? (Proboscidea) What mammal belongs to this order? (Elephant and extinct related species)

Integrated Learning
Use Integrating Worksheet 18.3.

The Living Textbook: Life Science Side 4

Chapter 10 Frame 08547
Koala/Bat/Elephant Seals (Movies)
Search: Play:

INTEGRATED LEARNING

Art Connection

Have students create a mobile to represent the orders of placental mammals that are listed on pages 382 and 383. Students should choose one animal from each order and draw a picture of it on a small square of construction paper. Some students may want to try to make origami animals for their mobiles. On the back, they should write a description of the animal, its habitat, and its diet. Have them use sticks and yarn to construct a balanced mobile.

Diversity of Placental Mammals

Scientists classify placental mammals into nearly 20 large groups called orders. The mammals in each order share certain important characteristics and adaptations. Here are some of the orders of placental mammals.

Carnivora ▲
As you might guess from the name, these mammals are carnivores, or meat eaters. They have teeth adapted for tearing flesh, clawed toes, and a well-developed brain. This group includes not only dogs, cats, wolves, and bears, but also seals, otters, and walruses.

Insectivora
Moles and shrews belong to this group of insect-eating mammals. Most have long skulls, narrow snouts, five-clawed feet, and are smaller than 46 cm. Their bodies, habits, and diet are similar to those of the first ancient mammals. ▼

Rodentia ▲
Rodents make up the largest order of mammals. They have chisel-like front teeth adapted for gnawing. Rodents include squirrels, beavers, rats, mice, porcupines, and gophers.

Proboscidea
This order is named for the long, muscular trunks of its members, the elephants. Elephants have extra-long teeth called tusks. ▼

Cetacea ▲
Whales and dolphins belong to this group of mammals that live in the ocean. They have evolved fishlike bodies with fins and paddle-like front limbs.

382 Chapter 18 Birds and Mammals

Themes in Science

Scale and Structure Many animals have special structures and adaptations that enable them to defend themselves. Some can defend themselves by frightening enemies without actually causing bodily harm. For example, deer display their antler racks, wolves bare their teeth, cats arch their backs and raise their fur to look bigger, and squirrels make screeching noises. These harmless displays help animals avoid fighting. Ask students to think of other examples of harmless defense strategies used by mammals.

▶ Which order of mammals do you think has at least five times the number of individuals living today as any other order? (Rodentia)

▶ To what order do moose, caribou, bison, and pronghorn belong? How do you know? (Artiodactyla, because they have hooves and an even number of toes)

▶ Which order of mammals are plant eaters, good jumpers, and nest in burrows? (Lagomorpha)

▶ Which order of mammals are plant eaters, have hooves, and have an odd number of toes on each foot? (Perissodactyla)

Enrich

Have students find out about the work of animal behaviorists Jane Goodall and Dian Fossey. Explain that Goodall has lived in East Africa and observed chimpanzees in their natural habitat and that Dian Fossey studied mountain gorillas in central Africa for many years. You may want to show the National Geographic film about Goodall's work, which is available on video in most video stores.

Answer to In-Text Question

① Human beings are mammals belonging to the primates.

Primates ▲
Mammals placed in this order have thumbs adapted for grasping objects. Their eyes face forward and they eat both plants and animals. Monkeys and apes are members of this group. What other common mammal belongs to the primates? ①

Chiroptera ▲
Bats, the mammals adapted for flight, make up this order. Their wings are flaps of skin stretched between their bodies and their long fingers. Bats are active mainly at night.

◀ Artiodactyla
The familiar sheep, goats, pigs, and deer belong to this order of mammals, with hooves and an even number of toes. The giraffe is also a member. They are all herbivores, or plant eaters, and range in size from the 3.5 kg mouse deer to the nearly 4.5 tonne hippopotamus.

Perissodactyla ▶
Like sheep and goats, the mammals in this group have hooves and are plant eaters. But instead of having an even number of toes on each foot, they have an odd number. This order includes horses, zebras, and rhinoceroses. They have large, flat teeth used for grinding their food.

Lagomorpha ▲
The mammals grouped in this order have long rear limbs adapted for jumping. They include rabbits and hares.

The Living Textbook: Life Science Side 4

Chapter 10   Frame 09794
Blue and Humpback Whale/Porpoise (Movies)
Search: Play:

The Living Textbook: Life Science Side 4

Chapter 12   Frame 29550
Sloth/Elephant/Giraffe/Springbok (Movies)
Search: Play:

EVALUATE

WrapUp

Reinforce Have students make a concept map that shows similarities and differences among monotremes, marsupials, and placental mammals. Under each type of mammal, students should list examples of species.

Use Review Worksheet 18.3.

Check and Explain

1. Kangaroo and opossum

2. Placental mammals carry their young within the uterus until the embryo stage is completed. The placenta provides the embryo with nutrients and oxygen.

3. There is evidence that New Guinea was once part of the Australian continent, and would show similar patterns of evolution. Australia's separation from other land masses made it possible for unique species to evolve and thrive without competition.

4. Mouse, Rodentia order, chisel-like teeth for gnawing; tiger, Carnivora order, teeth for tearing; giraffe, Artiodactyla order, plant eater; elephant, Proboscidea order, long, muscular trunk; dolphin, Cetacea order, fishlike body with fins

Answer to In-Text Question

① Unlike dogs, humans have a hand that can grasp and a brain that can think and reason.

The Living Textbook:
Life Science Side 4

Chapter 16 Frame 44984
Female Baboon (Movie)
Search: Play:

384

INTEGRATED LEARNING

Multicultural Perspectives

Japanese conservationist Takayuki Isoyama tracks the serow, a goat-antelope that lives in the mountains of Japan. The serow, a conservation success story, was brought back from the brink of extinction. Have students find out more about Isoyama and what he has learned about the serow.

Figure 18.11 ▲
Humans and dogs are both mammals, but what makes you different? ①

Science and You *Humans as Mammals*

You are a mammal. You therefore share many characteristics with dogs, cats, rabbits, goats, even whales and moles. Your hair is one obvious sign of your membership in the mammal group. You don't call it fur, but it is present all over your body. You wear clothes and live in shelters to stay warm, but your internal body temperature is constant.

If you think about it, many mammals can do things better than humans can. For example, cheetahs can run much faster. Dolphins can swim much faster. Whales can hold their breath much longer. Dogs have a better sense of smell. Cats can see better in the dark. Monkeys are better tree climbers. Tigers have longer teeth.

What can humans claim to be the best at? As a species, humans are not specialized. Instead, humans are fairly good at many things. Not many mammals can run fairly fast *and* swim *and* climb trees *and* jump. Not many other mammals can eat and survive on so many different kinds of food. Few mammals can live in every environment from the arctic to the tropics.

Humans also rely on two important characteristics: a hand that can grasp and a brain that can think and reason. Monkeys and apes, too, have grasping hands and well-developed brains. That is why humans, apes, and monkeys are all primates. But humans alone have put brain and hand together to make many things. Humans have built cities and invented computers. Humans use written language. The ability to create makes you a very different kind of mammal indeed.

Check and Explain

1. Name two species of pouched mammals.

2. How are placental mammals different from egg-laying and pouched mammals?

3. **Infer** Find Australia and New Guinea on a map. Why do you think most egg-laying and pouched mammals live only in this part of the world?

4. **Classify** For each of the following mammals, name the order to which it belongs, and give one characteristic: mouse, tiger, giraffe, elephant, dolphin.

384 Chapter 18 Birds and Mammals

Chapter 18 Review

Concept Summary

18.1 Birds
- Birds are endothermic vertebrates. They all have beaks, feathers, wings, and lay eggs with shells.
- Birds must incubate their eggs to keep them warm. They build nests to protect the eggs.
- Birds have special adaptations for flight, including special vision, bones, muscles, and body systems.
- Bird wings are shaped to create lift in flying.
- Many birds migrate to different places during part of every year. Many birds travel along certain routes.

18.2 Mammals
- Mammals are active endotherms with hair, mammary glands, and specialized teeth.
- Mammals move in different ways because the structure of their limbs varies. However, most mammals use two pairs of limbs to move.
- Mammals have a four-chambered heart, a diaphragm, and a complex nervous system.

18.3 Diversity of Mammals
- Two types of mammals, egg-laying and pouched mammals, reproduce differently than most other mammals.
- Most mammals are placental mammals. Scientists classify placental mammals into nearly 20 different orders.
- Placental mammals have a placenta that allows the embryo to be nourished while it is developing. The mother's blood supply provides the nourishment.

Chapter Vocabulary

migrate (18.1)   cerebrum (18.2)   marsupial (18.3)   placental mammal (18.3)
incubate (18.1)   monotreme (18.3)   placenta (18.3)

Check Your Vocabulary

Use the vocabulary words above to complete the following sentences correctly.

1. Egg-laying mammals, such as spiny anteaters or the duck-billed platypus, are called ____.
2. An embryo receives nutrients and oxygen from the mother's body through an organ called the ____.
3. Many birds ____, or move to a different place during part of every year.
4. Mammals with pouches in which the young complete their development are called ____.
5. Mammals with placentas are called ____.
6. The largest part of a mammal's brain is called the ____.
7. Birds ____ their eggs, using their bodies to keep the eggs warm.

Explain the difference between the words in each pair.

8. Placenta, uterus
9. Monotreme, marsupial
10. Migrate, fly

Write Your Vocabulary

Write sentences using each vocabulary word above. Show that you know what each word means.

Chapter 18 Birds and Mammals

CHAPTER REVIEW 18

Check Your Vocabulary

1. monotremes
2. placenta
3. migrate
4. marsupials
5. placental mammals
6. cerebrum
7. incubate
8. The placenta is the organ through which a developing embryo receives nutrients and oxygen from the mother's body. The uterus is the organ in a female in which the embryo develops.
9. A monotreme is an egg-laying mammal. A marsupial is a mammal whose young complete their development inside a pouch.
10. Migration means moving to a different place during part of every year. Flying is a means of locomotion.

Write Your Vocabulary

Students' sentences should show that they know both the meaning of each vocabulary word and how it is used.

Use Vocabulary Worksheet for Chapter 18.

CHAPTER 18 REVIEW

Check Your Knowledge

1. The male and female take turns incubating the eggs and searching for food, or one parent sits on the eggs while the other finds food for both.
2. Answers will vary. Possible answers include the ability to maintain a constant internal body temperature, three different kinds of teeth, hair and skin, the ability to move.
3. Some mammals give birth to young that have developed inside a uterus attached to a placenta. Other mammals give birth to young that have completed their development in the mother's pouch. Still other mammals' young are hatched from eggs.
4. Bird eggs have hard shells and must be incubated. Reptile eggs have leathery shells and are often buried in soil. The birds' hard shell keeps the embryo from being crushed during incubation.
5. Incisors are used for cutting and gnawing. Canines can stab prey and tear flesh. Molars grind food.
6. The koala and the kangaroo are marsupials that live in Australia. The opossum is common in North America.
7. Answers will vary. See pages 382 and 383 for a discussion of mammalian orders.
8. The four main types of birds are birds of prey, water birds, perching birds, and flightless birds. Even the birds of the same general type can have a much varied appearance.
9. *Archaeopteryx*
10. endotherms
11. albumen

Check Your Understanding

1. All mammals have hair at some stage in life. For some mammals, hair helps insulate the body.
2. The earliest mammals lived during the time of the dinosaurs. They looked somewhat like rats.
3. This relationship is called a pair-bond. In addition to working together to incubate the eggs, the pair works together to find food.
4. Birds must eat often because flying requires large amounts of energy.
5. Flightless birds, such as the rhea or the ostrich, are often fast runners.
6. In addition to lungs, birds have air sacs that allow them to take in the quantities of oxygen required for flight.
7. At the end of the incubation period, the chick breaks open the egg and is hatched. Some chicks are fully developed when they hatch; some are not.
8. Answers may vary, but are likely to include differences between mammals and birds.
9. Answers may vary. Flightless birds cannot fly to escape predators or to find food. They have adapted by becoming fast runners or good swimmers.

Chapter 18 Review

Check Your Knowledge
Answer the following in complete sentences.

1. Describe two methods by which birds are able to incubate their eggs and find food at the same time.
2. List three characteristics that have helped mammals adapt to a variety of environments.
3. Describe the three ways in which mammals reproduce.
4. Describe the difference between bird eggs and reptile eggs. What is the reason for this difference?
5. Name the three different types of teeth that mammals have. What are the uses of each type of tooth?
6. List two examples of marsupials. Where do they live?
7. Name three mammal orders and give an example of an animal from each.
8. What are the four main types of birds? Do all the birds of one general type look the same?

Choose the answer that best completes each sentence.

9. The fossil of (*Archimedes, Chioptera, Archaeopteryx, Artiodactyla*) was convincing evidence that birds evolved from a reptile ancestor.
10. The extra heat produced by cells rapidly burning fuel for energy is what makes birds (endotherms, exotherms, exotic, extroverted).
11. The embryo in a bird egg gets food from the (placenta, albumen, eggshell, feathers).

Check Your Understanding
Apply the concepts you have learned to answer each question.

1. Do all mammals have hair? Describe the main function of hair.
2. When did the earliest mammals live? What did they look like?
3. **Mystery Photo** The photograph on page 366 shows the feathers of a male Asian pheasant, a relative of the peacock. The male uses the feathers for courtship displays. You may notice in the photograph that the spots of the feathers look like eyes. This may also confuse predators. Sometimes a male and female bird work together to incubate their eggs, and form an attachment. What is this relationship called? What is another function of that type of relationship?
4. Explain why birds must eat large and steady quantities of food.
5. **Application** How do flightless birds survive without flying? Give an example.
6. How does a bird get an adequate air supply?
7. What happens at the end of the incubation period? Are newborn chicks fully developed?
8. **Extension** Both bats and birds can fly. What are two differences between bats and birds?
9. **Infer** There are a few species of birds that are flightless. They include the kiwi, emu, penguin, and ostrich. Describe why being flightless is not a disadvantage to these birds. How do you think the birds adapt to being flightless?

Develop Your Skills

1. a. Bird A is larger. Smaller birds hop; larger birds walk.
   b. The hopping tracks show the marks of both feet side by side.
   c. The hopping tracks were made by a perching bird; the walking tracks by a water bird.
2. a. The first birds appeared during the Jurassic period of the Mesozoic era; however, flying insects appeared during the Mississippian period of the Paleozoic era.
   b. The first humans evolved 5 million years ago, during the Pliocene epoch of the Cenozoic era.
   c. The first birds evolved approximately 208 million years ago. Dinosaurs dominated the earth at that time.
3. Student posters will vary. Characteristics given should be related to functions.

Make Connections

1. Concept maps will vary. You may wish to have students do the map below. First draw the map on the board, then add a few key terms. Have students draw and complete the map.

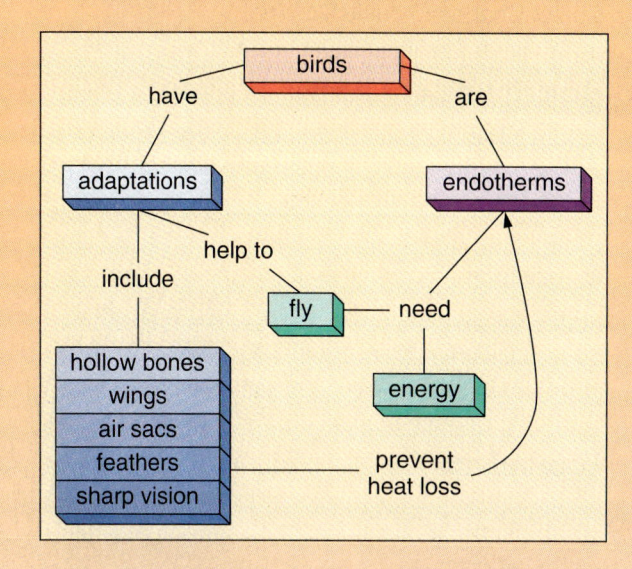

2. Students' answers on why a particular bird was chosen will vary. In many cases, the choice is made by popular vote—often by students.
3. Answers will vary. The domestication of the horse is generally accepted to have begun in Central Asia about 5,000 years ago.
4. Answers will vary. Encourage students to choose exotic organisms, and make sure they have highlighted adaptations by relating them to their animal's environment.
5. Students' stories will vary. Students may prefer not to read their writing aloud, but may be willing to let others read their work.

Develop Your Skills

Use the skills you have developed in this chapter to complete each activity.

1. **Interpret Data** The figure below shows the tracks of a hopping bird and a walking bird.

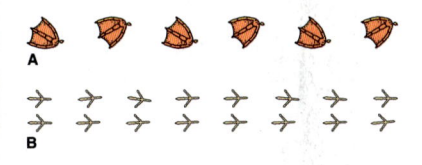

   A
   B

   a. One bird is larger than the other. Guess which is larger, and then infer the relationship between size and type of movement.
   b. How can you tell that the hopping tracks are from hopping rather than walking?
   c. Were the hopping tracks made by a bird of prey, a perching bird, or a water bird? How about the walking tracks?

2. **Data Bank** Use the information on page 630 to answer the following questions.
   a. When did the first flying animals appear? What were they?
   b. How long ago did the first humans evolve?
   c. How long ago did the first birds evolve? What were the dominant animals at that time?

3. **Communicate** Make a poster to show the characteristics of mammals. Choose one mammal, then illustrate and label the major characteristics.

Make Connections

1. **Link the Concepts** Draw a concept map showing how the concepts below link together. Add terms to connect, or link, the concepts.

   birds   fly
   endotherms   energy
   adaptations   feathers
   hollow bones   heat loss
   wings   air sacs
                  sharp vision

2. **Science and Social Studies** Most states have a state bird. Research the state bird for your state. Where does it live? Why do you think it is the bird chosen to represent your state?

3. **Science and Society** Humans have domesticated many other mammals as pets and to use their labor. Choose a pet or a work animal and find out when it was first domesticated and by whom.

4. **Science and You** Of the millions of species of birds and mammals in the world, you have probably seen only a few. Find out about an animal you have never seen. Write a report on how it is adapted to its habitat, and draw a picture of it.

5. **Science and Writing** Many stories have been written about humans and their relationships with animals. For example, in *Julie of the Wolves*, by Jean Craighead George, Julie is saved and protected by a pack of wolves. Write your own adventure story in which a person has a special relationship with an animal or group of animals. You can either be the main character in the story or narrate the story. Present the story to your class.

SCIENCE AND LITERATURE CONNECTION   UNIT 6

About the Literary Work
"The Cloud Spinner" was adapted from "The Spider Weaver" in the text *Japanese Children's Favorite Stories*, edited by Florence Sakade, copyright 1958 by Charles E. Tuttle Co., Inc. Reprinted by permission of Charles E. Tuttle Co., Inc.

Description of Change
For the sake of space, extraneous information was edited out of the tale, but the dialogue was kept as true to the original as possible.

Rationale
Edited material primarily develops the idea that Yosaku lived a rather barren life devoid of worldly goods. His "riches" were the fruits of his garden.

Vocabulary
kimono, teak, bed-mat

Teaching Strategies

Directed Inquiry
After students finish reading, discuss the story. Be sure to relate the story to the science lessons in this unit. Ask the following questions:

▶ What animal is the focus of this story? (Spider)

▶ Spiders are classified as what kind of animals? (Arachnids)

▶ How do arachnids differ from insects? (Arachnids have two main body parts; insects have three. Arachnids do not have wings or antennae, which are common to insects. Also, arachnids have four pairs of legs while insects have three pairs.)

Evaluate Sources
Explain to the students that spiders do not ingest cotton to create silk. Then ask the following questions:

▶ What does the above fact say about the accuracy of the selection? (The selection presents a fictional account of the life processes of a spider.)

▶ Reread the selection to find scientifically accurate descriptions of the spider. (Possible answers include the spider's coloring, the fact that it has eight legs, and its ability to spin a web of silk.)

Library Research
Emphasize that spiders do not spin the silk used for clothing. Tell students that silk remained a secret of Asia for many years, until silkworms were smuggled out of China. Ask students to find out more about the history of the production and trade of silk.

Science and Literature Connection

"The Cloud Spinner"
The following Japanese folktale is adapted from the story "The Spider Weaver" in Japanese Children's Favorite Stories

Yosaku marched up and down the neat rows of radishes and cabbages and beans. He was proud as a warrior inspecting his soldiers. . . .

Between two cabbage stalks stretched a spiderweb, and in the center sat a handsome spider. Its body, dark as polished teak, was striped with gray that shone like silver in the sunlight.

"Greetings," said Yosaku politely. "You honor me by visiting my garden." He stepped back, so as not to disturb his guest. His heel struck something much too soft for a cabbage, far too thick for a bean.

"Sssssss!" A huge and ugly snake slipped over his sandal.

"*Ara!*" exclaimed Yosaku, waving his hoe.

The snake's tongue darted in and out of its mouth, its bright, cold eye fixed upon the spider. The spider froze, as if tied in its own ropes.

"Off with your head!" Yosaku shouted at the snake.

But as he brought down his hoe, the handle caught and ripped the full sleeve of his *kimono*. Yosaku missed, and the snake slithered away. The spider ran down from its web and disappeared among the cabbages.

Yosaku looked at his torn sleeve. The *kimono* was beyond mending and he had no other, for Yosaku was poor. "Just so!" he said, but sadly now, and went home to his bowl of rice and bed-mat.

That night Yosaku slept soundly until the Hour of the Ox, when a tiny voice, calling from the courtyard, awakened him.

"Mr. Yosaku!"

"Eh?" mumbled Yosaku. Perhaps his ears were playing tricks.

"Honorable Mr. Yosaku!" Yosaku tied his old *kimono* tight and padded to the door. There stood a young girl. Her robe was of rich black silk stitched with rows of silver thread. Yosaku blinked. Perhaps his eyes tricked him, too.

The girl bowed low. "You are in need of new *kimono*?"

Yosaku stared down on the girl's dark glossy head. She reminded him of something, but he was too sleepy to remember what. "So I am," he answered.

She looked up at him and smiled shyly. "If you will allow, I would be pleased to weave new cloth for *kimono*."

Yosaku, astounded by such a grand offering, slid the door wide. "You are welcome in my house."

He offered tea to the girl, but she refused and went directly to the weaving room. . . . Yosaku went back to sleep on his mat again, and soon was snoring to the

Skills in Science

Reading Skills in Science

1. Inaccuracies: A spider can't change into a human; spiders do not eat cotton; spiders spin silk, not thread. Accuracies: Some snakes eat spiders; spiders do spin material (silk); some spiders live in webs.
2. Clues to the identity of the girl include the coloring of her robe, the fact that she knew Yosaku's kimono was torn, and the fact that she reminded Yosaku of something that he could not remember.

Writing Skills in Science

1. Answers may vary. Possible endings might include that the Yosaku grew rich by selling all the kimonos made by the spider-girl or that the spider-girl eventually left Yosaku to go back and live the life of a spider.
2. Some snakes feed upon spiders. Yosaku interfered with this feeding relationship when he prevented the snake from eating the spider. The snake might have caught and eaten the spider if Yosaku had not interfered.

Activities

Classify Students should use the following traits to classify the spider: shiny, black back striped with silver-grey; ability to jump; ability to spin a web.

Communicate Students' models may include an orb web, a triangular web, a dome-shaped web, a platform web, and a bowl-shaped web. Each type of web traps insects in a unique manner.

---

rhythm of the loom. When he awoke in the morning, there were seven new *kimono*, neatly laid out next to his mat.

Yosaku rubbed his eyes. "How can this be?" He ran to the weaving room. "Already you made new *kimono* for every day of the week!"

"Do not question, Honorable Yosaku." The girl looked modestly down at her hands. "And please do not come into this room while I work."

Yosaku bobbed his head. "It shall be as you wish," he agreed. . . .

Yosaku asked nothing, not even the girl's name, but his curiosity was growing even greater than the stack of *kimono* in his cupboard. So, at the week's end, he tiptoed to the window of the weaving room. Slowly, softly he slid back the ricepaper screen. In the center of the room sat his visitor at the newly strung loom. Only it was not the young girl. It was a giant spider!

"Aaaah!" Yosaku clapped his hand over his mouth. . . .

She was swallowing great mouthfuls of cotton and spinning it out as finished thread.

No wonder the girl was not hungry!

Yosaku watched as all eight of the spider's legs, working together, wove the thread into cloth. . . .

Yosaku crept away from the window. He had come close to spoiling the spider-girl's secret. But he had an idea of his own.

Skills in Science

Reading Skills in Science

1. **Evaluate Sources** This folktale contains scientific accuracies as well as inaccuracies. Identify three of each.
2. **Predict** During the course of the tale, the reader is given clues regarding the identity of the girl. Reread the selection and identify these clues.

Writing Skills in Science

1. **Predict** Write an ending to the tale describing Yosaku's idea and what happens when he carries it out.
2. **Accurate Observations** At one point Yosaku interferes in the natural relationship between two organisms. Identify Yosaku's action. What might have occurred had he not interfered?

Activities

Classify List the physical traits of the spider in this tale. Use your list and reference tools to identify a particular species to which the spider might belong.

Communicate In a library, research types of spider webs and the reasons for the differences in structure. Use string or yarn to create a model of one type of web.

Where to Read More

Someone Saw a Spider by Shirley Climo. New York: Crowell, 1985. Contains the complete text of "The Cloud Spinner" and other spider facts and folktales.

Shelf Pets: How to Take Care of Small Wild Animals by Edward Ricciuti. New York: Harper and Row, 1971. Details how to care for wild animals that are appropriate to keep as pets.

UNIT 7

UNIT OVERVIEW

In this unit, students are introduced to the functions, systems, and proper care of the human body. They also learn how the human body contracts and fights illness. In Chapter 19, the skeletal, muscular, and integumentary systems are described. Chapter 20 explores the digestive, respiratory, circulatory, and excretory systems. In Chapter 21, students learn about the nervous system, the senses, and the endocrine system. Reproduction and the human life stages are discussed in Chapter 22. Chapter 23 examines the need for proper nutrition and exercise. It also discusses the function of food Calories and the impact of stress upon body functions. The chapter concludes with a discussion about the negative effects of drug and alcohol abuse.

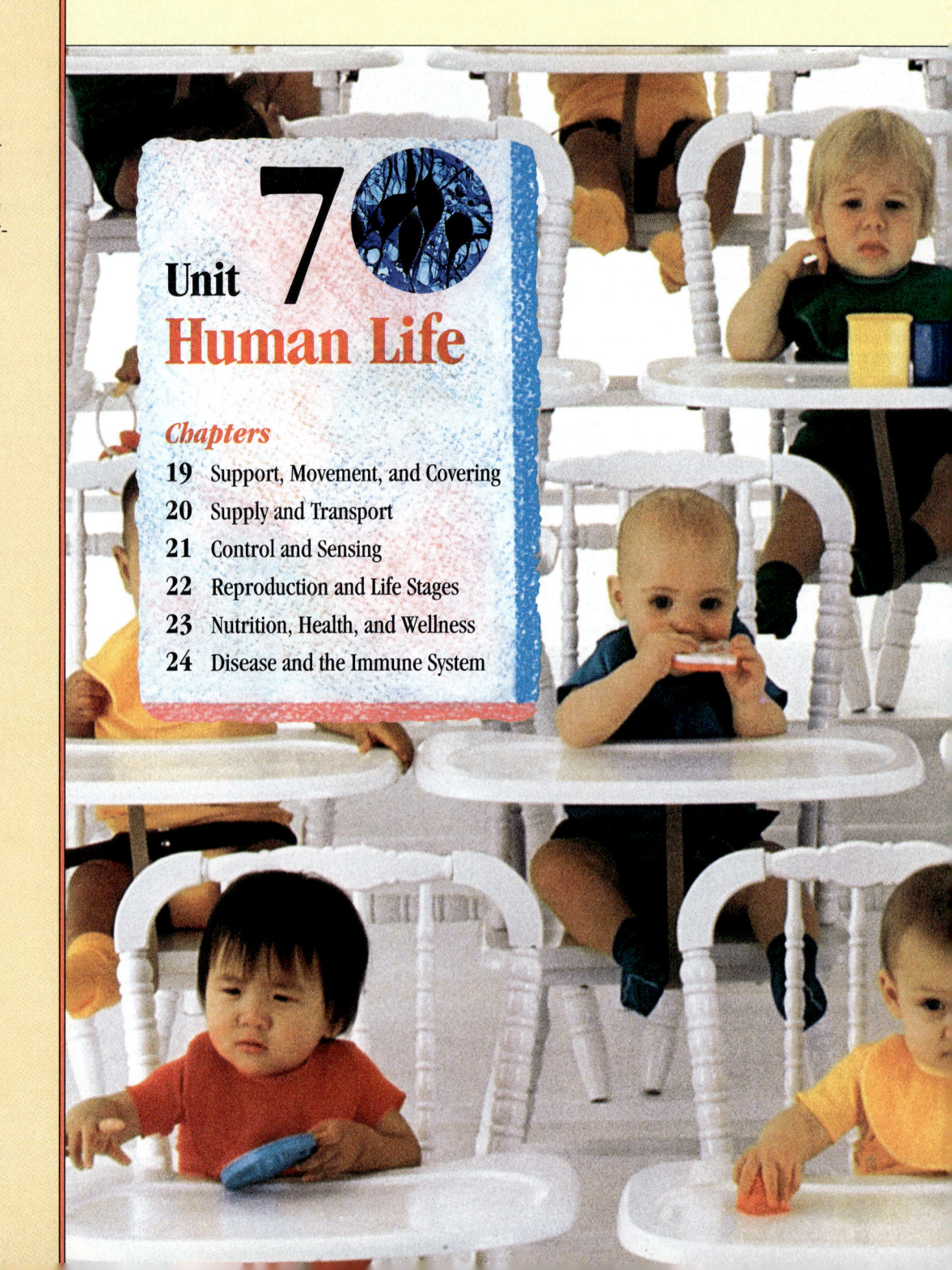

Introducing the Unit

Directed Inquiry

Have the students examine the photograph. Ask:

▶ Although they have little or no language, these infants can communicate. What methods might they use? (Laughing, crying, throwing things, posture)

▶ Why do you think infants are unable to walk? (They lack coordination; they lack muscle strength; they are top-heavy. Accept all reasonable explanations.)

▶ Name some abilities these infants will gain as they grow. (They will gain the ability to walk, talk, jump, reason, solve problems, and so on.)

Writing About the Photograph

Ask students to choose one of the infants shown in the photograph and write a monologue showing that infant's thoughts and feelings.

Unit 7 — Human Life

Chapters

19 Support, Movement, and Covering
20 Supply and Transport
21 Control and Sensing
22 Reproduction and Life Stages
23 Nutrition, Health, and Wellness
24 Disease and the Immune System

Chapter 24 presents the causes and effects of infectious diseases, including AIDS. The chapter explores medicines that fight disease and explains the functions of the skin, bodily secretions, and the immune system in protecting the body from pathogens.

Data Bank

Use the information on pages 624 to 631 to answer the following questions about topics explored in this unit.

Classifying

How many bones are in the human torso? List some of the major bones that you think are part of the torso.

Interpreting Data

What are some of the risk factors for skin cancer? What warning signals are associated with detecting skin cancer?

Reading a Table

How many red blood cells are contained in the amount of blood shown on the table? How many white blood cells are contained in the same amount of blood?

Approximately how old do you think the infants are in the photograph to the left? Do you think they can feed themselves at this stage?

Data Bank Answers

Have students search the Data Bank on pages 624 to 631 for the answers to the questions on this page.

Classifying There are 51 bones in the human torso, including the clavicle and the pelvic bones. Answers will vary. The answer is found in the table, Human Bones by Region, on page 626.

Extension Ask students to find out how the number of bones in the human hand changes from early childhood to adulthood.

Interpreting Data The answers are found in the table, Cancer Risk Factors, on page 629.

Extension Ask students to name at least one kind of cancer that is influenced by life-style choices and is therefore largely avoidable. (Answers will vary. Oral cancer is associated with tobacco use and is most likely avoidable.)

Reading a Table There are 4–6 million red blood cells per mm^3 and 4000–11 000 million white blood cells per mm^3. The answer is found on page 626 in the graph, Cell Numbers and Life Spans.

Answer to In-Text Question

These babies are close to the age of one. Judging from how they are shown grasping objects, they are probably just learning to feed themselves.

CHAPTER 19

Overview

This chapter describes main systems and structures of the human body. The first section focuses on four main types and functions of bones in the skeletal system. This section also describes cartilage. The second section discusses different muscle structures. Three lever systems of muscle contraction are modeled with everyday examples. The final section explores skin structures and functions.

Advance Planner

▶ Arrange to have a specialist in prosthetics visit your class. Refer to a yellow pages directory to obtain contacts. TE page 397.

▶ Prepare 6 chicken bones, 6 beef bones, 12 cups or beakers, 12 hand lenses, metric rulers, vinegar, balances, magazines, clay, papier-mâché, and toothpicks for SE Activity 19, page 400.

▶ Obtain prepared slides of skeletal, smooth, and cardiac muscle from a science or biological supply company for TE activity, page 403.

▶ Arrange for groups of students to interview exercise specialists working to develop fitness programs for corporations. TE page 403.

Skills Development Chart

Sections	Apply Definitions	Calculate	Classify	Evaluate Sources	Infer	Interpret Data	Observe	Predict	Relate Concepts
19.1 Skills WarmUp		●							
Skills WorkOut		●							
Activity					●	●	●	●	
19.2 Skills WarmUp	●								
Skills WorkOut			●						
Skills WorkOut									●
19.3 Skills WarmUp					●				
SkillBuilder				●					

Individual Needs

▶ **Limited English Proficiency Students** Have students write the vocabulary terms from the chapter in their science journals, leaving several blank lines between each term. Then as they study this chapter, have students write the figure numbers of any pictures or diagrams that describe or represent each term. For example, students could write *Figure 19.4* next to the word *cartilage*. Tell students they can write more than one figure number next to each word, and encourage them to add their own drawings. Have students use these pictures and figures to reinforce the meaning of the chapter vocabulary.

▶ **At-Risk Students** Before students read each section, have them write the section vocabulary terms in their science journals, leaving several blank lines between each word. Then have them write a sentence defining each term. As students read the chapter, they can write two sentences from the chapter for each term.

▶ **Gifted Students** Have students design a movable, three-dimensional model of a particular joint. Encourage students to add details to their models. They might do research to find more information about bones and joints. Suggest they find out about haversian canals, lacunae, lamellae, sinovial membranes, and hyaline cartilage. Have them show a cross section of one of the bones in their models.

Resource Bank

▶ **Bulletin Board** Have students create a display containing a life-size body map. Have them begin by tracing the body outline of one of their classmates on large paper. Then, as you work through the chapter, ask students to choose one part of the body and draw in the bones and muscles.

▶ **Field Trip** Students may enjoy the chance to visit a local gymnasium. If possible, arrange to have a trainer talk about exercise programs.

CHAPTER 19 PLANNING GUIDE

Section	Core	Standard	Enriched	Section	Core	Standard	Enriched
19.1 Skeletal System pp. 393–400				Skills WorkOut, p. 404 Skills WorkOut, p. 406	● ●	● ●	
Section Features Skills WarmUp, p. 393 Skills WorkOut, p. 396 Activity, p. 400	● ● ●	● ● ●	● ● ●	**Blackline Masters** Review Worksheet 19.2 Skills Worksheet 19.2	● ●	● ●	 ●
Blackline Masters Review Worksheet 19.1 Reteach Worksheet 19.1 Skills Worksheet 19.1 Integrating Worksheet 19.1a Integrating Worksheet 19.1b	● ● ●	● ● ● ● ●	 ● ● ●	**Laboratory Program** Investigation 35	●	●	●
				Color Transparencies Transparencies 55, 56a, b	●	●	
				19.3 Skin pp. 407–410			
Laboratory Program Investigation 34	●	●	●	**Section Features** Skills WarmUp, p. 407 SkillBuilder, p. 409	●	● ●	● ●
Color Transparencies Transparencies 53, 54	●	●		**Blackline Masters** Review Worksheet 19.3 Vocabulary Worksheet 19	● ●	● ●	
19.2 Muscular System pp. 401–406				**Color Transparencies** Transparency 57	●	●	
Section Features Skills WarmUp, p. 401 Career Corner, p. 403	● ●	● ●	 ●				

Bibliography

The following resources can be used for teaching the chapter. See page T-40 for supplier codes.

Audio-Visual Sources

(video unless noted)
The Human Body: Systems Working Together. 15 min film. 1980. C/MTI.
Man: The Incredible Machine. 28 min. 1975. NGSES.
Muscles and Joints: Moving Parts. 26 min. 1985. FH.
Muscular and Skeletal Systems. 20 min. 1988. NGSES.
Skeletal Anatomy Collection. 7 filmstrips with cassettes. BM.
Your Body: Series I. 3 filmstrips with cassettes. FM.

Software Resources

Bones, Muscles, & Skin. Apple. J & S.
The Human Body: An Overview. Apple. BB.
The Skeletal System. Apple. BB.
Your Body: Series II. Apple, TRS-80, Commodore. 1983. FM.

Library Resources

Elting, Mary. The Macmillan Book of the Human Body. New York: Aladdin, 1986.
Meredith, Susan, Ann Goldman, and Tom Lisauer. Book of the Human Body. Tulsa, OK: Educational Development Corp., 1983.
Nilsson, Lennart. Behold Man: A Photographic Journey of Discovery Inside the Body. Boston: Little, Brown, and Co., 1974.

CHAPTER 19

INTEGRATED LEARNING

Writing Connection

Have students imagine the person whose bones are shown in the photograph. Have them write a descriptive paragraph about what activity the person might be doing in such a position. Ask them if the person is sitting, standing, lying down, or bending over. How can they tell? Encourage students to be imaginative.

Introducing the Chapter

Have students read the description of the photograph on page 392. Ask if they agree or disagree with the description.

Directed Inquiry

Have students study the photograph on page 392. Ask:

- What part of an organism do you think the picture shows? (An X-ray of bones in a leg)

- What part of the chapter title relates to the picture? (Support and movement) Why do you think so? (Bones hold up other parts of the body. Bones have joints that allow parts of the body to move.)

- How many separate bones do you see? (Four—femur, patella, tibia, and fibula) What do you call the place where several bones come together? (A joint)

- What joint do you think is shown here? (The knee) Describe the way your knee moves. (Only back and forth in one direction with a great range of motion) What everyday object is a good analogy for a knee? (A door hinge)

Chapter Vocabulary

cardiac muscle	joint
cartilage	ligament
dermis	skeletal muscle
endoskeleton	smooth muscle
epidermis	tendon
extensor	vertebra
flexor	

Chapter 19 Support, Movement, and Covering

Chapter Sections

19.1 Skeletal System
19.2 Muscular System
19.3 The Skin

What do you see?

"I see a knee joint. I use it to run and to walk. Without it I couldn't do a lot of things like sports and stuff. On the knee joint is the patella, or kneecap."

Kai Bates
Hilsman Middle School
Athens, Georgia

To find out about the photograph, look on page 412. As you read this chapter, you will learn about systems that support and protect your body and allow for movement.

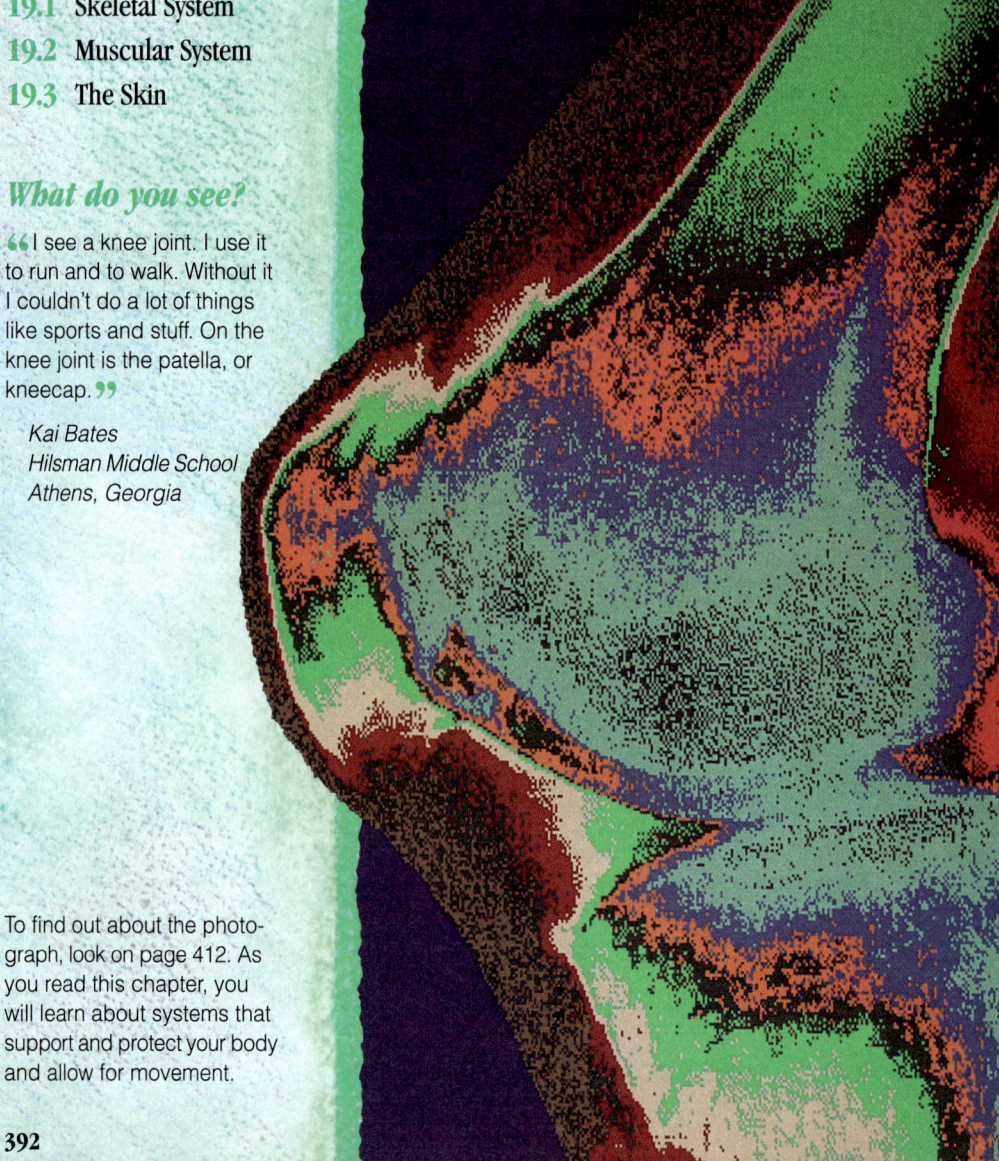

Themes in Science

Systems and Interactions The skeletal system works together with other body systems to allow an organism to function. When discussing parts of the skeleton, have students consider how the bones coordinate with other parts of the body, such as chest organs, or parts of the muscular system.

19.1 Skeletal System

Objectives

▶ **Name** and **classify** some bones in the skeleton.

▶ **Describe** the structure of a typical long bone.

▶ **Write** in your own words the functions of the bones and the skeleton.

▶ **Identify** and **compare** the different types of joints and the movements they allow.

▶ **Make analogies** between parts of the skeletal system and everyday things.

Skills WarmUp

Calculate You can find the weight of your bones by using the following formula: Your weight × 35 ÷ 100. What percentage of your body weight is the weight of your bones? What is the weight of your bones?

Can you imagine yourself without any bones? Just as the girders and beams are a building's framework, the human skeleton is the body's framework. Your bones—about 206 of them—make up this framework. The girders and beams of a building also support and give shape to a building. Your skeletal system also supports and gives shape to your body.

Unlike many organisms, such as insects, your skeleton is inside your body. An internal skeletal system is an **endoskeleton** (EHN doh SKEHL uh tuhn). All vertebrates have an endoskeleton. Insects and some other animals, such as lobsters and crabs, have exoskeletons (EHKS oh SKEHL uh tuhn). An exoskeleton is on the outside of the body.

When describing the skeleton, you can say that it is made up of two parts. One part called the axial (AK see uhl) skeleton is made of the bones of your skull, ribs, and the small bones, or **vertebrae** (VUR tuh BREE), in your backbone. Look at Figure 19.1. Notice all the vertebrae of the backbone. As an infant, your backbone had 33 separate bones. As you got older, nine bones grew together, or fused, to form larger bones. The other 24 vertebrae have remained as separate bones.

The other part of the skeleton is called the appendicular (AP uhn DIHK yuh luhr) skeleton. It includes all the bones attached to the axial skeleton. These bones make up the appendages: your arms and legs. Your collarbones, hipbones, and shoulder bones are also part of the appendicular skeleton.

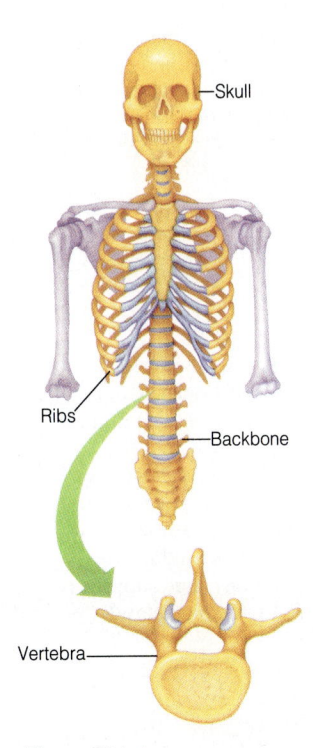

Figure 19.1 ▲
The bones of the axial skeleton are shown. The backbone is part of the axial skeleton.

SECTION 19.1

Section Objectives
For a list of section objectives, see the Student Edition page.

Skills Objectives
Students should be able to:

Calculate the weight of their bones in relation to the weight of their bodies.

Observe the number of bones and joints in the human hand.

Reason by Analogy which common objects are like parts of a skeleton.

Predict what will happen to bones placed in acid.

Vocabulary
endoskeleton, vertebra, cartilage, joint, ligament, tendon

MOTIVATE

Skills WarmUp
To find out what percentage of their body weight is bones, have students do the Skills WarmUp.
Answer 35 percent; Students' answers will vary.

Prior Knowledge
To gauge how much students know about the skeletal system, ask the following questions:

▶ What do bones do for the body? (Students are likely to say that they give support and protect organs.)

▶ How is a human's skeleton different from an insect's skeleton? (The human skeleton is internal; the insect skeleton is external.)

▶ What is inside bones? (Marrow and blood vessels)

▶ What are vertebrae? (Small bones that form the backbone)

Chapter 19 Support, Movement, and Covering 393

TEACH

Explore Visually

After students have studied the drawing and text on this page, ask them the following questions:

▶ Which large long bone is probably the strongest supporting bone in the body? Why? (The femur must be extremely strong because it acts alone in the upper part of the leg and supports the torso.)

▶ Why are long bones lightweight? (Because they are used for movement in the fingers, arms, and legs.)

▶ What is the function of flat bones? (They protect and support the body's organs.)

▶ What is the function of the cranium? (Protects the brain)

Enrich

Reiterate that bones' different shapes allow them to serve different functions. Short bones in our hands and feet allow for a wide range of specialized motion. Strong, lightweight long bones in our legs allow us to walk and run with stability and ease. Show students pictures of skeletons of other animals and have them infer how each animal's bones help serve its particular needs. For example, cats and kangaroos have large hind legs and feet that make them good jumpers.

Integrated Learning

Use Integrating Worksheets 19.1a and b.

The Living Textbook:
Life Science Sides 7-8

Chapter 5   Frame 00913
Human Skeleton (19 Frames)
Search: Step:

INTEGRATED LEARNING

Integrating the Sciences

Physical Science Discuss balance and symmetry with respect to load-bearing. Ask students why it is easier to carry loads with both hands than on just one side. Have students try this with schoolbooks.

Bones

The figure below shows some of the bones in the human skeleton. Notice that the bones have different shapes. There are four main types of bones: flat, long, short, and irregular. The shape of a bone is related to its function. Many of these bones have the same name, shape, and function as the bones of other vertebrates.

Flat Bones
The ribs, breastbone, and shoulder bones are flat bones. They protect and support body organs. ▶ For example, your ribs protect your lungs.

Long Bones
The arms and legs are long bones. These bones are strong, hollow, and light. They support weight and are used for ▶ movement. Your leg bones, for example, act like the pillars of a building. They support you when you stand.

Short Bones
There are short bones in the feet. These bones support weight and allow for ▶ many small movements.

Irregular Bones
Some bones, such as the vertebrae, are irregularly ◀ shaped. In the human ear, three tiny irregular bones conduct sound.

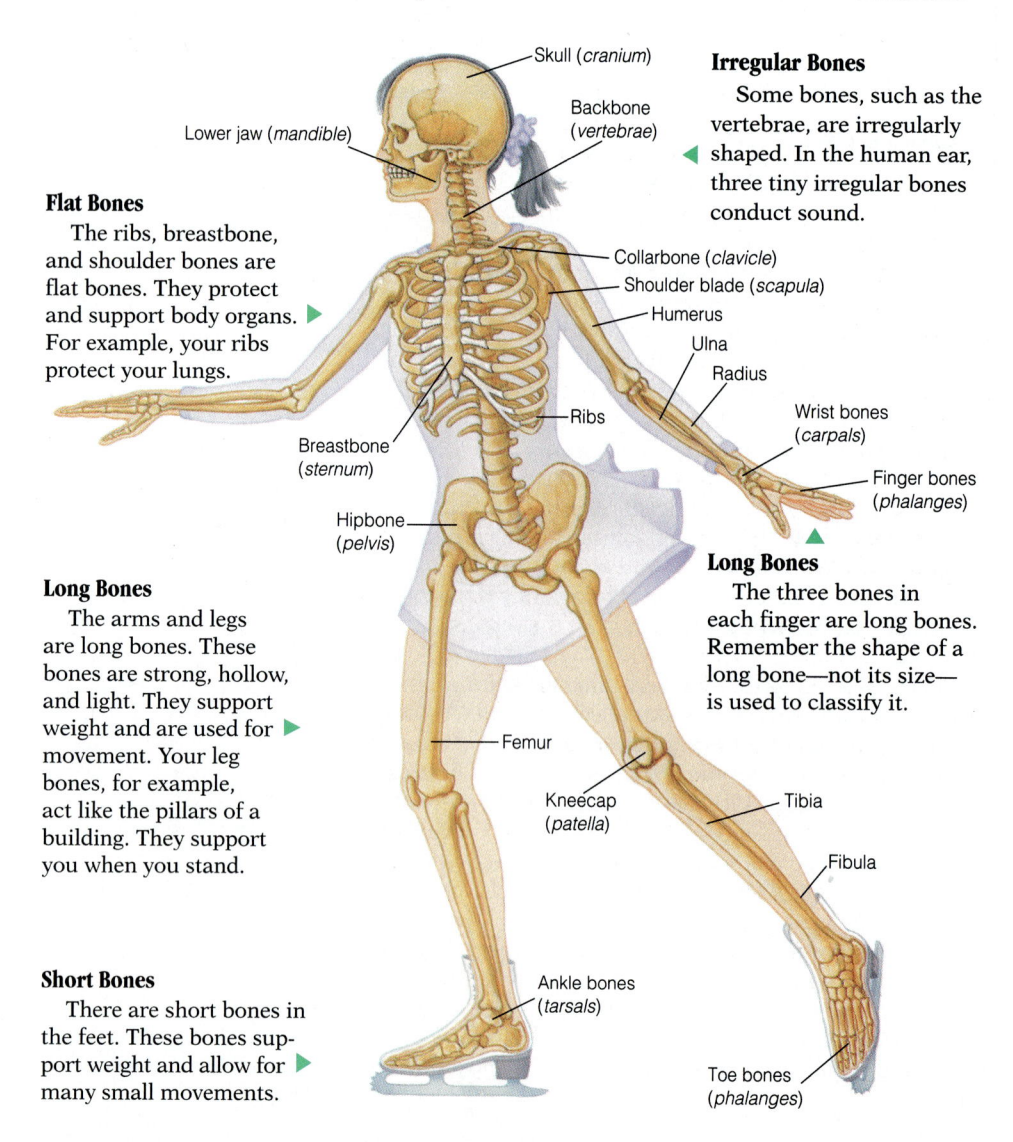

Long Bones
The three bones in each finger are long bones. Remember the shape of a long bone—not its size— is used to classify it.

Skull (*cranium*)
Backbone (*vertebrae*)
Lower jaw (*mandible*)
Collarbone (*clavicle*)
Shoulder blade (*scapula*)
Humerus
Ulna
Radius
Ribs
Wrist bones (*carpals*)
Breastbone (*sternum*)
Finger bones (*phalanges*)
Hipbone (*pelvis*)
Femur
Kneecap (*patella*)
Tibia
Fibula
Ankle bones (*tarsals*)
Toe bones (*phalanges*)

Chapter 19 Support, Movement, and Covering

Integrating the Sciences

Physical Science Tell students that human bones are able to support more weight than granite or concrete can. In fact, a block of bone the size of a matchbox is capable of supporting 10 tons. This support is necessary, because when you walk or run, you put tremendous pressure on the bones in your legs (many kilograms per centimeter). Ask students to explain how elasticity helps bones function to support body structures. Ask them to infer why rigid bones are less supportive than flexible ones. (Bones that are flexible bend slightly when under stress, and then go back to their original shape; bones that are rigid are more likely to break under pressure.)

Structure of Bones All bones are made of living bone cells surrounded by nonliving materials. These nonliving materials are protein and minerals, such as calcium and phosphorus. Protein gives bones their flexibility. The minerals calcium and phosphorus give bones their strength and hardness.

The structure of a long bone, such as the femur, is a good bone to take a close look at. Look at Figure 19.3. Notice that the bone has a long shaft and two large knoblike ends.

Think of the shaft as a cardboard tube. The shaft is made up of mostly compact bone surrounding a hollow cavity or space. Compact bone is dense and looks smooth. It helps bones withstand bangs and bumps. The cavity of the shaft contains yellow marrow. Yellow marrow is a soft tissue that contains fat.

Although the compact bone in the shaft looks solid, under a microscope you would see tiny passageways called the Haversian (huh VUR zhuhn) canals. These canals run throughout the shaft. They contain nerves and blood vessels. The blood vessels carry food and oxygen to the living bone cells and carry away wastes.

The shaft of a long bone is covered with a tough, white membrane called the periosteum (pehr ee AHS tee uhm). It is made up of connective tissue and bone-forming cells. Blood vessels in the periosteum carry oxygen and food. Nerve fibers are also present in the periosteum.

The knoblike ends of the long bone contain spongy bone. Spongy bone is softer and lighter in weight than compact bone. As you might infer from its name, spongy bone also has a lot of open spaces and holes in it. Spongy bone contains red marrow, which is where blood cells are made.

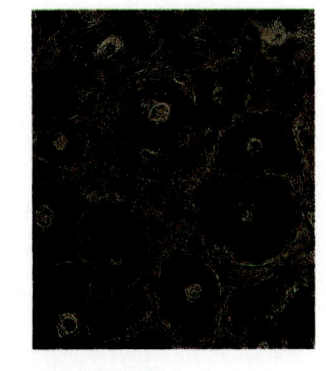

Figure 19.2 ▲
The small dark ovals are bone cells. Each cell is surrounded by larger circles of nonliving materials. What two minerals make up most of bone? ①

Figure 19.3
The femur is the longest bone in the human body.

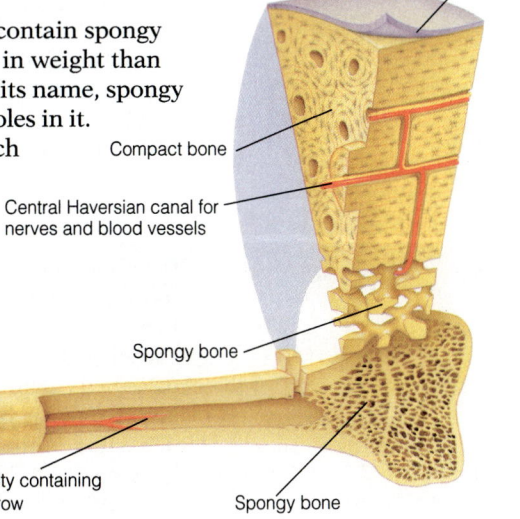

Class Activity

Ask students if they know how the cylindrical shape of bones helps support weight. Have students work in groups of four and give each group three 4 by 6 note cards. Have each group fold the first note card in half and stand it on its edges like a tent. Have students test how much weight it can support by setting books on it until it falls. Have them fold another card in half and then in half again and tape it together in the shape of a square tube. How much weight can it support? Finally, have students roll the third card into a cylinder, tape it together, and test its strength. Ask students to compare the strength of the structures. (The cylinder will hold much more weight than the other structures.)

Reteach
Use Reteach Worksheet 19.1.

Discuss
Tell students that when a bone is broken, the new bone material formed in and around the break is often stronger than the original bone. If possible, show students an X-ray of a healed bone.

Portfolio
Have students make a chart with columns for the parts of the femur. Students should write a brief explanation of the function and location of each part.

Answer to In-Text Question
① Calcium and phosphorus give bones their strength and hardness.

The Living Textbook:
Life Science Sides 7-8

Chapter 5 Frame 00951
Bone Microviews (15 Frames)
Search: Step:

TEACH ▪ Continued

Skills WorkOut
To help students determine how many bones and joints there are in each hand, have them do the Skills WorkOut.

Answer There are 14 phalanges in each hand, and there is a joint at the upper end of each. There are also 13 bones in the palm and wrist of each hand. In addition, there are many gliding joints at the wrist.

Critical Thinking
Reason and Conclude Ask students if they know any elderly people who have broken bones from a fall or an accident. Point out that elderly people break bones more easily than younger people. Ask students if they can explain why this is true. (Bones in an elderly person often do not contain the amount of hard mineral matter that they did when the person was young and, therefore, become brittle and weak.)

Discuss
Point out that the ears and the tip of the nose are made of strong, flexible cartilage. Ask students how it helps people to have these parts of the body made of cartilage instead of bone. (Since the ears and nose are extremities, they are easily bumped and injured. If they were made of rigid bone, they would tend to break easily.)

The Living Textbook: Life Science Side 7
Chapter 32 Frame 31176
Milk and the Skeleton (Movie)
Search: Play:

396

INTEGRATED LEARNING

Writing Connection
Have students imagine themselves as one of the following: an invertebrate without a skeleton, such as a clam; an invertebrate with an exoskeleton, such as an insect; or a vertebrate with a form of locomotion unlike ours, such as a fish or bird. Students should then write a description of what it is like to move. Encourage them to imagine a variety of different situations that require movement, such as eating, escaping from predators, hunting, and so on.

Skills WorkOut

Calculate and Observe
Count as many bones and joints as you can in each hand. Then look at Figure 19.4 (right). Count the number of bones and joints in the adult hand. Compare your findings.

Functions of Bones What are the functions of the bones that make up the skeletal system? With the muscle and skin systems, the skeletal system gives the body shape and support. In addition to providing shape and support, bones also have other important functions.

▶ Many bones protect your body organs. For example, the vertebrae protect the spinal cord, and the ribs protect your lungs.

▶ Many bones work with certain muscles to move the body and its parts. As a result, you can run, walk, grasp, and breathe.

▶ Most blood cells are made in the red marrow of certain bones.

▶ Bones store fat and minerals. Fat is stored in yellow marrow. Minerals, such as calcium and phosphorus, are stored in the bone itself. When the minerals are needed by the body, they are released into the blood and carried to all parts of the body.

Cartilage
Besides bone, the skeletal system has a tissue called **cartilage** (KART uhl ihj). This strong, flexible tissue gives shape to some parts of your body. For example, your ears and the tip of your nose are made of cartilage. If you bend your ears with your fingers, you can see how flexible and strong cartilage is. Cartilage also covers the ends of some bones or makes up disks between vertebrae. Cartilage keeps bones from grinding against each other. Between the vertebrae, cartilage disks act as shock absorbers. If you have ever injured your knee or suffered the pain of a "slipped disk" in your spine, you can appreciate the function of cartilage.

Bones usually develop from cartilage. Before birth, the skeleton is made mostly of cartilage. The cartilage has a covering that contains bone-forming cells. After birth, these cells begin to absorb calcium that is dissolved in the blood. The bone-forming cells change the dissolved calcium into calcium compounds that cannot dissolve in the blood. These calcium compounds are deposited in the cartilage, causing it to harden and become bone. The process of bone formation is called ossification (AHS uh fih KAY shuhn). Your bones lengthen

396   Chapter 19   Support, Movement, and Covering

STS Connection

Modern prosthetics are closely modeled after real bones, joints, and ligaments in the human body. Arrange to have a specialist in prosthetics visit your class to discuss and demonstrate how different prosthetic devices work. Have the visitor explain how technological advancements have revolutionized prosthetics in recent years.

Literature Connection

The tendon joining the calf muscle to the heel is known as the Achilles tendon. Have students find out how this tendon relates to the mythical hero.

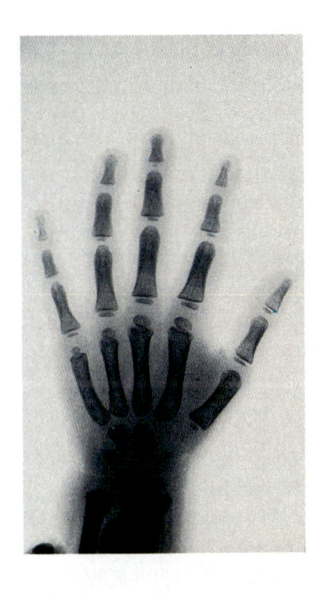

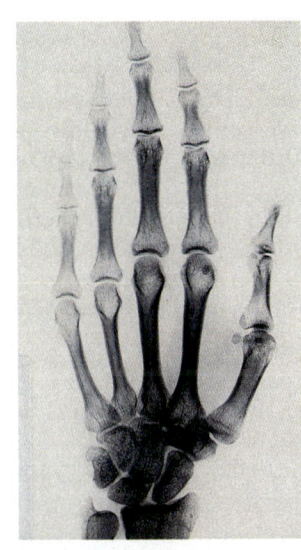

◀ **Figure 19.4**
The X-rays show the hands of a baby (left) and an adult (right). Notice that in the baby's wrist, the cartilage has not been replaced completely by bone.

until all cartilage is ossified. Ossification begins after birth and continues for about 20 years. It stops when you are between 18 and 25 years old.

Dairy products, such as milk and cheese, are rich in calcium and phosphorus. You have probably heard someone say, "Drink your milk; it's good for you." Why do you think someone would say this?

Skeletal Connections

Almost every bone in the human body forms a **joint** with at least one other bone. A joint is where two or more parts meet. The 206 bones of your body need to be held together in some way. At some joints, the bones are fused together, such as the flat bones that form the skull. In other joints, however, several bones meet but are not fused together. These joints, such as the joints at your knees and your elbows, allow you to move. The bones in these joints are connected by **ligaments**. Ligaments are connective tissues that can stretch. Bones and ligaments make up most of the skeletal system.

For the human body to move, muscles and bones need to work together. So, the muscles need to be connected to bones. Muscles are connected to bones by connective tissues called **tendons**.

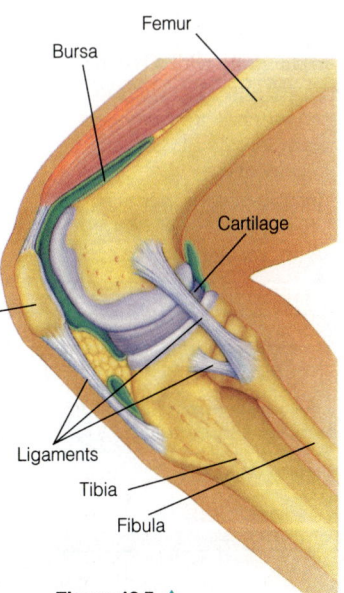

Figure 19.5 ▲
Notice that three leg bones and the kneecap are at the knee joint. The bursa is a fluid-filled sac that helps cushion bones, preventing the ends of the bones from wearing down. What three leg bones meet at the knee joint? ①

Discuss
Point out some of the advantages of drinking milk and eating other dairy products. Ask students if they know why milk is good for most babies and children. (Milk provides the calcium and phosphorus needed for growing bones.)

Enrich
Ask your class if anyone is allergic to milk. Explain that some people are allergic to a chemical in milk called lactose. Have interested students find out about milk substitutes that do not have lactose, such as soy milk.

Skills Development
Infer Have students infer why people over 50, particularly women, are often advised to supplement their diet with extra calcium. (As a person ages, calcium is lost from the bones. Since women generally have smaller bones and because they experience a loss of estrogen after menopause, their bones are more likely to become weak and brittle. Adding calcium to one's diet helps strengthen bones.)

Predict Explain that liquid in a joint acts in the same way as oil in the moving parts of a bicycle or other machines designed for movement. Explain that as people get older, the amount of liquid in their joints is depleted. Ask students what effect this might have on a person's joints. (Joints may stiffen and not work as well.)

Answer to In-Text Question

① **The femur, tibia, and fibula meet at the knee joint.**

Chapter 19 Support, Movement, and Covering

TEACH ▪ Continued

◆ Explore Visually

As students learn about and discuss the different types of joints, encourage them to experiment by moving their own joints. After students have studied this page, ask them the following questions:

▶ What joints do you use when you bend over? (The joints between the vertebrae)

▶ Besides the joint connecting the shoulder and upper arm, what other ball-and-socket joint can you name? (The joint connecting the hip with the upper leg)

▶ What kinds of joints do you think are present at your ankle? Why? (Gliding joints, because there is some movement in all directions, just as there is at the wrist)

▶ Of all the kinds of joints, which joints do you think people use most? Why do you think so? (The ball-and-socket joints and hinge joints are probably used most often. People use these joints as they run, walk, and move about.)

Enrich

Have students look for examples of joints in the classroom and in magazines. Students should make a list of the joints they find and identify which type of human joint—ball-and-socket, hinge, pivotal, or gliding—each one most resembles. Have students cut out pictures of the joints they find in magazines to include with their lists.

The Living Textbook:
Life Science Sides 7-8

Chapter 5   Frame 00989
Hinge Joint Diagrams (2 Frames)
Search: Step:

398

INTEGRATED LEARNING

Integrating the Sciences

Physical Science To reinforce students' understanding of the different types of joints, ask each student to design an imaginary machine that uses at least one of each: ball-and-socket joint, pivotal joint, hinge joint, and gliding joint. The machine may fulfill a real function or be merely decorative. Students may make a model or a drawing. Drawings should include arrows indicating the direction in which the joint moves.

Joints

There are many joints in the skeletal system. The large number of joints enables the human body to move in many different and graceful ways. Without all these joints, you would move like a robot.

At some joints, there is no movement. These are immovable joints. In adults, the joints between the flat bones of the skull are immovable joints. In a newborn baby, the joints in the skull are movable. As the child grows, the skull bones fuse.

Some joints allow a little movement. The joints between your vertebrae are slightly-movable joints. Slightly-movable joints allow some bending and twisting movements in all directions. Most immovable and slightly-movable joints are in the axial skeleton.

Most joints are freely-movable joints. These joints allow large movements, and most of them are in the appendicular skeleton.

Ball-and-Socket Joints
A ball-and-socket joint allows movement in all directions. The antenna on a portable radio is attached to the radio with a ball-and-socket joint. You can move the antenna any way you want. A ball-and-socket joint connects your shoulder and your upper arm. These joints permit you to swing your arms in a circle. ▼

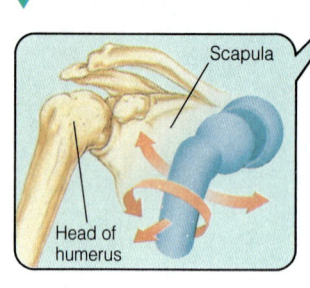

Hinge Joints
Bend your arm at the ▶ elbow. There is a hinge joint there—just like the hinge on a door. A hinge joint allows bones to move backward and forward in only one direction.

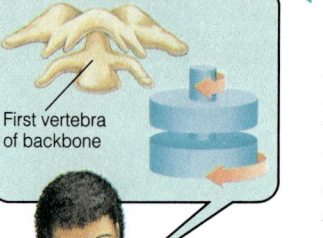

◀ **Pivotal Joints**
Nod your head. Then turn your head from side to side. The skull is joined to the first vertebra of your backbone by a pivotal joint. Pivotal joints allow two kinds of movement—side-to-side and up-and-down.

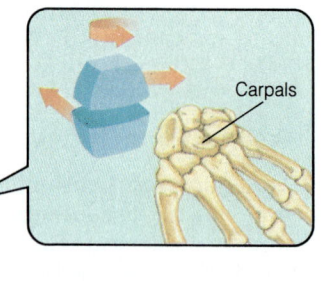

▲ **Gliding Joints**
Your wrist has many gliding joints. If you move your wrist in as many ways as you can, you will notice there is some movement in all directions. In gliding joints, the bones slide along each other. The vertebrae in your backbone are also connected by gliding joints.

398 Chapter 19 Support, Movement, and Covering

STS Connection

Biomedical implants are now available to replace parts of the body including the heart, lungs, liver, joints, tendons, ligaments, skin, and arteries. One of the biggest challenges of bioengineering is creating materials that will not be rejected by the body's immune system. Recently, biomedical implant research has begun to develop bioactive materials very similar to the natural materials they replace. These bioactive implants actually become incorporated into the living tissue of the body. Have students do research to find out more about this innovative field of medical technology.

Science and Technology
Fiber Optics! Television! Surgery!

Imagine a surgeon performing knee surgery through a tiny incision, or cut, in your knee. With advances in medicine and fiber optics technology, many joint operations are performed just this way. It is called arthroscopic (AR throh SKAHP ihk) surgery.

Arthroscopic surgery is usually performed on knee, shoulder, elbow, and hip joints. Many athletes have arthroscopic surgery to help prolong their sports careers. The technique uses a straight, tubelike instrument called an arthroscope. It contains lenses and bundles of optical fibers. The lenses magnify, and the optical fibers transmit light.

During arthroscopic surgery, the arthroscope is put into a small incision. The surgeon can look through the arthroscope and see what the problem is. An image is also transmitted to a television monitor. Through a second incision, the surgeon can correct the problem by using special small instruments. The image on the monitor helps surgeons see what they are doing.

Because the incisions are so small, there is little tissue damage and little discomfort. Therefore, people heal quickly, and often can be released from the hospital on the same day as their surgery.

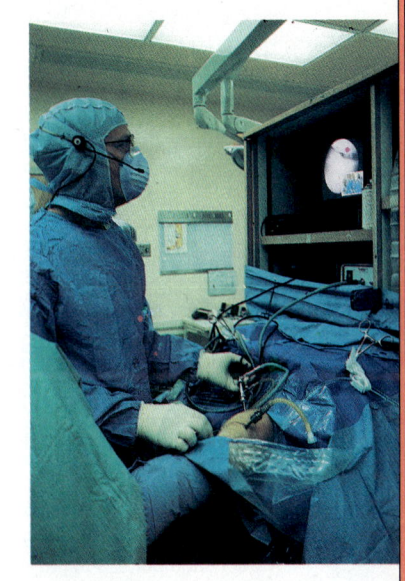

Figure 19.6 ▲
What are some advantages to arthroscopic surgery? ①

Check and Explain

1. Look at the skeleton on page 394. What are the scientific names for the breastbone, hipbone, wrist bones, and upper arm bone? Classify each bone.

2. Describe how each freely-movable joint moves.

3. **Infer** How is the shaft of a long bone related to the function of the bone?

4. **Make Analogies** Match the everyday object to the part of the skeleton that shows the best analogy. Explain your choices.

   a. Skull   Door hinge
   b. Any bone   Elastic band
   c. Ligament   A broken and glued together cup
   d. Knee joint   A storage warehouse

Chapter 19 Support, Movement, and Covering   399

Discuss
Discuss Figure 19.6. Ask students if they know anyone who's had arthroscopic surgery.

EVALUATE

WrapUp
Review Assign students to cooperative groups of four or five. Have each group make flashcards of all the different parts of the skeletal system presented in this chapter. (Refer them to the diagram of the skeleton on page 394.) Then have each group play a game in which a student draws a flashcard and must point to the corresponding skeletal part of his or her body.

Use Review Worksheet 19.1.

Check and Explain

1. Sternum, flat; pelvis, irregular; carpals, irregular; humerus, long

2. Ball-and-socket: movement in all directions; pivotal: side-to-side and up-and-down; hinge: backward and forward only; gliding: movement in all directions with sliding movement

3. Hollow shaft contains fatty marrow that provides support; periosteum covering shaft forms new bone cells and is a site for nerve fibers and blood vessels

4. a. the cup, it has nonmovable joints; b. storage warehouse, all bones store fat and minerals; c. elastic band, ligaments stretch and connect; d. door hinge, it allows back-and-forth movement.

Answer to In-Text Question
① **The injury is easily seen so that its repair is more precise. The use of small instruments decreases the trauma of surgery.**

ACTIVITY 19

Time 4 days **Group** 5–6

Materials
6 chicken bones
6 beef bones
12 cups or beakers
12 hand lenses
metric rulers
vinegar
balance

Analysis

1. a. The fact that vinegar softens the chicken bones indicates that vinegar is an acid;
   b. acetic acid
2. Each bone should have become softer or more flexible after soaking in vinegar.
3. Calcium was removed from the bones. Calcium is one of the minerals that gives bone hardness.
4. Protein remains. It gives bone flexibility, which the pieces retained after soaking.

Conclusion

1. Answers may vary; however, students should demonstrate an understanding that protein provides a flexible framework, which is filled in by calcium and phosphorus, giving the bone its hardness.
2. Students should infer from the similarities between the chicken and beef bones that all vertebrate bones are made up of protein, calcium, and phosphorus.

Everyday Application

Over a long period of time, the acid in cola drinks may erode the enamel and quickly dissolve the material of the tooth.

TEACHING OPTIONS

Prelab Discussion

Have students read the entire activity before beginning the discussion. Ask the following questions:

▶ Why are there different techniques for organizing data? (Discuss as a class Task 2, Data Record.)

▶ Why is it important for the bones to be cleaned?

▶ What major bones are shared by all vertebrates?

Safety Note: Vinegar contains acid; use safety goggles and keep away from mouth.

Activity 19 What gives a bone its hardness and flexibility?

Skills Observe; Predict; Interpret Data; Infer

Task 1 Prelab Prep

1. Collect the following items: 1 chicken bone, 1 small beef bone, 2 cups, a hand lens, vinegar, a metric ruler, and a balance.
2. In Task 3, you will need to gather a variety of art materials to make a model. You may wish to gather toothpicks, glue, drawing paper, pens, pencils, or clay.

Task 2 Data Record

1. Discuss with several other classmates the different observations that you can make about each bone, such as color, size, shape and so on.
2. Identify a way to organize the observations. You may want to use a table, a list, a chart, or some other way.
3. Make as many observations as you can about each bone. Record your observations in the table or chart.

Task 3 Illustrate by Modeling

1. Make a model of each bone you observed. You may draw pictures, use clay, papier-mâché, sticks, toothpicks, or any other material suitable for making models. Be sure to label the parts of each bone model.
2. Study the bones carefully and decide whether each bone is a long, flat, irregular, or short bone. *Hint:* The beef bone may be part of a larger bone.
3. Can you name the bones? If so, record their names.

Task 4 Procedure

1. Place the chicken bone into one cup. Place the beef bone into the other cup.
2. Pour enough vinegar into each cup to cover the bone.
3. Set the two cups in a place where they will not be disturbed for two days.
4. **Predict** Acids interact with calcium compounds. Predict what you think will happen to each bone.
5. After two days, remove the bones from the vinegar. Be sure to rinse each bone thoroughly with water.
6. Examine each bone carefully. Make as many observations as you can about each one after it was in the vinegar. Record all your observations in your table or chart.

Task 5 Analysis

1. a. Is vinegar an acid? If so, how do you know?
   b. **Name** If vinegar is an acid, use reference books to identify the acid of vinegar.
2. What happened to each bone when it was left in the vinegar for two days?
3. **Infer** What substance was removed from the bones? How do you know?
4. **Infer** What substance remains in the bones? How do you know?

Task 6 Conclusion

1. Write a short paragraph explaining how the chemical makeup of a bone is related to its strength and flexibility. Be sure to describe how the bones changed.
2. **Infer** You have worked with the bones of two vertebrates, a chicken and a cow. What can you infer about the chemical makeup of the bones of other vertebrates, including human bones?

Everyday Application

Your teeth are covered with enamel. Acids can erode enamel; however, enamel is resistant to vinegar. The acid content of a cola drink is much greater than the acid content of vinegar. What do you think might happen to your teeth if you drink a lot of cola drinks?

INTEGRATED LEARNING

Themes in Science

Systems and Interactions The muscular system works together with other body systems to allow the body to respond to both internal and external changes. As students study each kind of muscle, ask them to give an example and to explain whether an internal or an external stimulus leads to its use.

SECTION 19.2

Section Objectives
For a list of section objectives, see the Student Edition page.

Skills Objectives
Students should be able to:

Apply which definition of the word *work* is used in science.

Classify voluntary and involuntary muscle functioning.

Observe the effects of muscle fatigue in an experiment.

Classify the types of muscles found in different parts of the body.

Vocabulary
skeletal muscle, smooth muscle, cardiac muscle, extensor, flexor

19.2 The Muscular System

Objectives

▶ **State** the main function of the muscular system.
▶ **Compare** the three types of muscles.
▶ **Describe** how muscles work in pairs.
▶ **Identify** three lever systems in the body.
▶ **Classify** the type of muscle in different parts of the body.

Skills WarmUp

Apply Definitions Write a definition for *work*. Use a dictionary to find out the meaning of the word *work*. How many definitions are given? Which definition do you think is used in science? Write this definition in your own words. After you read about the muscular system, revise your definition of *work*.

Muscles are tissues that can shorten, or contract. All muscles do work this way. They pull. When a muscle pulls or contracts, it is doing work. Keep in mind that in science "work" means that a force causes something to move over some distance.

The human muscular system is the force behind the skeletal system. Without the action of muscles, bones could not move at their joints. You couldn't walk, run, or pick up things. Without muscles, you couldn't breathe. Your heart wouldn't beat. You couldn't even swallow food without muscles.

Muscle Structure

Look at Figure 19.7. It shows the scale and structure of muscle. Muscles are made up of hundreds to thousands of long, thin cells, or muscle fibers. Groups of muscle fibers are wrapped by a thin covering of connective tissue. Many bundles of muscle fibers make up a muscle. Because each muscle is an organ, there are also blood vessels and nerve fibers within each muscle.

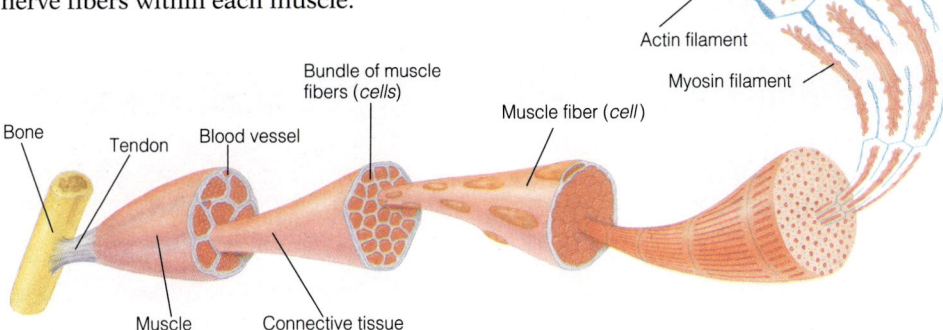

Figure 19.7
Each muscle fiber (cell) is made of smaller fibers. These fibers are made of two kinds of protein filaments. These filaments slide past one another, causing muscles to contract. What are the names of these two filaments? ①

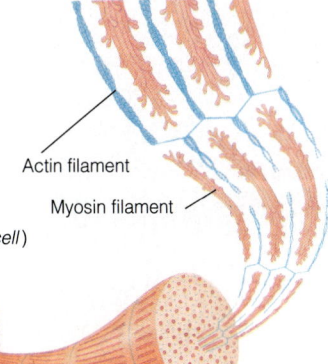

MOTIVATE

Skills WarmUp
To introduce students to the scientific meaning of *work*, have them do the Skills WarmUp.
Answer A dictionary gives several meanings. The scientific meaning relates force, effort, and distance to movement. Students' definitions will vary.

Prior Knowledge
To gauge how much students know about the muscular system, ask the following questions:

▶ What is a muscle?
▶ What are some different kinds of muscles?
▶ Where are muscles?
▶ What makes muscles move?

Answer to In-Text Question
① They are the actin filament and myosin filament.

Chapter 19 Support, Movement, and Covering 401

TEACH

Explore Visually

After students have studied this page, ask the following questions:

▶ Which muscles keep working while you're asleep—voluntary or involuntary muscles? Why? (Involuntary muscles work automatically to keep your body functioning at all times, even when you are sleeping or ill.)

▶ How does the number of muscles in your body compare to the number of bones? (There are three times as many muscles as bones—more than 600 muscles compared to 206 bones.)

▶ Name three specific muscles you would use to lift a weight above your head. (Trapezius, biceps, and triceps)

▶ What effect do the biceps muscle and triceps muscle have on the movement of the humerus? (The two muscles enable the humerus to move.)

▶ Explain how muscles help give shape to the body. (Muscles cover the bones and give the body its definition.)

The Living Textbook: Life Science Sides 7-8
Chapter 5 Frame 00966
Muscle Microviews (13 Frames)
Search: Step:

INTEGRATED LEARNING

Integrating the Sciences

Physiology A reflex is a process by which voluntary muscles are moved involuntarily, such as when a person jerks his or her hand out of a fire. Ask students to think of other examples of reflexes. Ask them to explain why reflexes are beneficial.

Types of Muscles

Trapezius! Biceps! Pectoralis! Do these words sound Greek to you? Many of the more than 600 muscles that make up the muscular system get their names from Greek or Latin words or word roots. Look at Figure 19.8. It shows some of the major muscles of the body. The white tissues that you see are tendons.

Some muscles can be moved when you want them to move. These are the voluntary muscles. Other muscles are not under your voluntary control. They are controlled automatically by your brain. These are the involuntary muscles. For example, the involuntary muscles in your stomach help you digest food, but you cannot "tell" them to make your stomach work harder to grind and digest food.

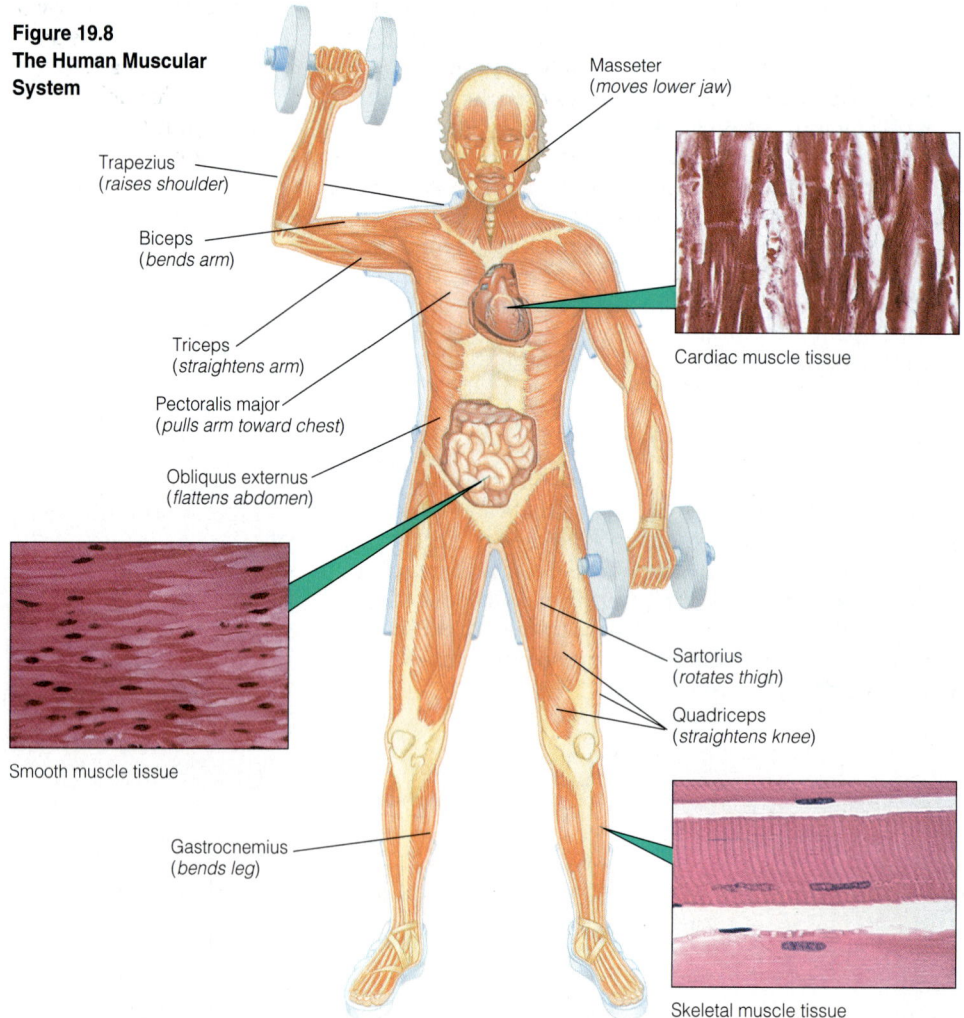

**Figure 19.8
The Human Muscular System**

- Masseter (*moves lower jaw*)
- Trapezius (*raises shoulder*)
- Biceps (*bends arm*)
- Triceps (*straightens arm*)
- Pectoralis major (*pulls arm toward chest*)
- Obliquus externus (*flattens abdomen*)
- Sartorius (*rotates thigh*)
- Quadriceps (*straightens knee*)
- Gastrocnemius (*bends leg*)

Cardiac muscle tissue

Smooth muscle tissue

Skeletal muscle tissue

Chapter 19 Support, Movement, and Covering

Math Connection

Explain that regular exercise helps lower one's at-rest heart rate, thereby reducing daily strain on the heart. Have students imagine that through daily exercise they succeed in reducing their at-rest heart rates from 72 beats per minute to 62. This means that they have reduced their heart rates by 10 beats per minute. By how many beats will they have reduced their at-rest heart rates in an entire day? (10 beats per minute × 60 minutes per hour × 24 hours per day = 14,400 beats saved per day)

Class Activity

Obtain prepared slides of skeletal muscle, smooth muscle, and cardiac muscle. Have students look closely at the three kinds of muscle under the microscope. Ask them to note the stripes in the skeletal muscle, the lack of stripes in the smooth muscle, and the branching and weaving in the cardiac muscle. Ask them which kind of muscle they think is easiest to recognize. (Most students will probably say the skeletal muscle is the easiest to identify because of its striped appearance.)

Career Corner

Have groups of students interview exercise specialists working for hospitals, health clubs, and corporations. Each group should make a presentation of its findings in class. For variety, one group could interview physical therapists or athletic trainers. Compare their training and tasks.

Skeletal Muscle You can make **skeletal muscle** move any time you want. Skeletal muscle makes movement at joints possible. It is attached to the bones by tough, elastic tendons. Look at the skeletal muscle in Figure 19.8. Its cells look striped, or striated. Skeletal muscle is sometimes called striated muscle.

Smooth Muscle Involuntary muscle is called **smooth muscle**. Look at Figure 19.8. Notice that smooth muscle doesn't have any stripes in it. Smooth muscle is in the walls of most internal organs, such as the walls of your stomach and blood vessels. You can't voluntarily control these muscles. Smooth muscles keep your internal organs working all the time, even when you are asleep.

Cardiac Muscle Did you know that your heart is a muscle? The heart is made up of **cardiac** (KAR dee AK) **muscle**. This type of muscle is only in the heart. Look at Figure 19.8. Cardiac muscle is unusual because its cells seem to branch and weave together. Also, it looks like striated muscle and acts like smooth muscle. You have no direct control over cardiac muscle in your heart. However, some people claim to be able to voluntarily slow down or speed up their heartbeat. Your heartbeat is the contraction of the heart. These contractions pump blood through the heart and to the rest of your body. Your brain will speed up or slow down your heartbeat automatically whenever your body needs a larger or smaller supply of blood.

Career Corner   Exercise Specialist

Who Sets Up Exercise Programs?

You visit a health club, and the exercise specialist asks, Do you want to tone or develop your muscles? The exercise specialist will probably measure your height and weight, and run some tests. From the information gathered, the exercise specialist will develop an exercise program just for you.

Exercise specialists perform many different tasks in their jobs. They also may work in different industries. Corporations, health clubs, and hospitals often hire exercise specialists. They may teach exercise classes or give lectures about exercise and health.

In a hospital, exercise specialists may develop special exercise programs for patients who have had heart attacks or who are overweight. These specialists will monitor a person's heart rate, blood pressure, and so on. Therefore, they must be able to use some medical instruments.

A good understanding of the parts of the body and how they work is important to exercise specialists. To become an exercise specialist, you need a four-year college degree. However, many community colleges and organizations, such as the YM/YWCA, offer certificates in exercise. You may want to write to your local college to find out more about this career.

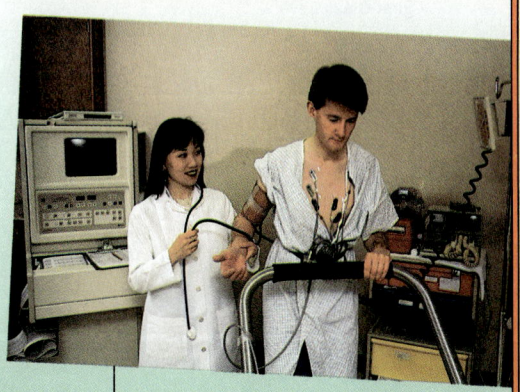

Chapter 19 Support, Movement, and Covering

TEACH • Continued

Skills WorkOut
To help students classify their eye-blinking muscles as voluntary or involuntary, have them do the Skills WorkOut.
Answer The eye-blinking muscles can be voluntary for a short time. However, they are also controlled by involuntary muscles that insure blinking to both wet and protect the eye.

Class Activity
The class can work with levers in groups of four. Give each group a rigid metric ruler to use as a lever and a pen to use as its fulcrum. Have each group tape several coins together to form a resistance force. Groups should place the coins on the 1-cm mark of the ruler and place the pen under the ruler at the 10-cm mark. Explain that in physical science the force applied to the lever is referred to as the effort force. One student should push down on the ruler at the 30-cm mark to create the effort force. Next, students should move the pen to the 20-cm mark and push down again on the 30-cm mark. Ask groups to compare the effort force when the pen is at different positions.
Discuss Ask students what happens to the effort force when the length of the effort arm (the side force is applied to) is decreased. (More force is needed when the length of the effort arm is decreased.) Compare the setup to similar lever systems (crowbar, scissors, seesaw). What kind of lever is this? (First class)

The Living Textbook: Life Science Side 7

Chapter 30   Frame 29126
Dog Movement (Movie)
Search: Play:

404

INTEGRATED LEARNING

Integrating the Sciences
Physical Science Explain to students that a lever is a type of simple machine. Most machines that we use are combinations of two or more simple machines. Have students find out the five other types of simple machines (Inclined plane, wedge, pulley, screw, and wheel and axle) and explain why levers are especially similar to muscle-and-bone systems.

Skills WorkOut
Classify
1. Blink your eyes five times.
2. Try *not* to blink for as long as you can. Time yourself and record the time.
3. Repeat step 2 four times.
4. Calculate your average time. Compare your time to the time of four classmates.

Do you think your eye-blinking muscles are controlled by voluntary or involuntary muscles? Explain.

Muscle Action
A muscle cannot contract unless it receives an electrical message from a nerve. These electrical messages are sent to nerves by the brain and spinal cord. The electrical message signals the muscle fibers to contract.

To move the body, most muscles work in pairs. When one muscle contracts, it pulls on a bone to which it is attached. At the same time, another muscle relaxes. Remember that muscles always pull; they cannot push bones apart. The contraction of one muscle can bend your leg, but another muscle is needed to straighten your leg.

To see how this works, lift one of your legs a few centimeters off the floor. Place one hand firmly on the front of your leg just above the knee. Place your other hand on the back of your leg just above the knee. Now, straighten your leg. The muscle in the front of your leg contracts. This muscle is an **extensor**. A muscle that straightens a joint is an extensor. The back muscle contracts as the knee is bent. This muscle is a **flexor**. A muscle that bends a joint is a flexor.

Figure 19.9   Lever Systems

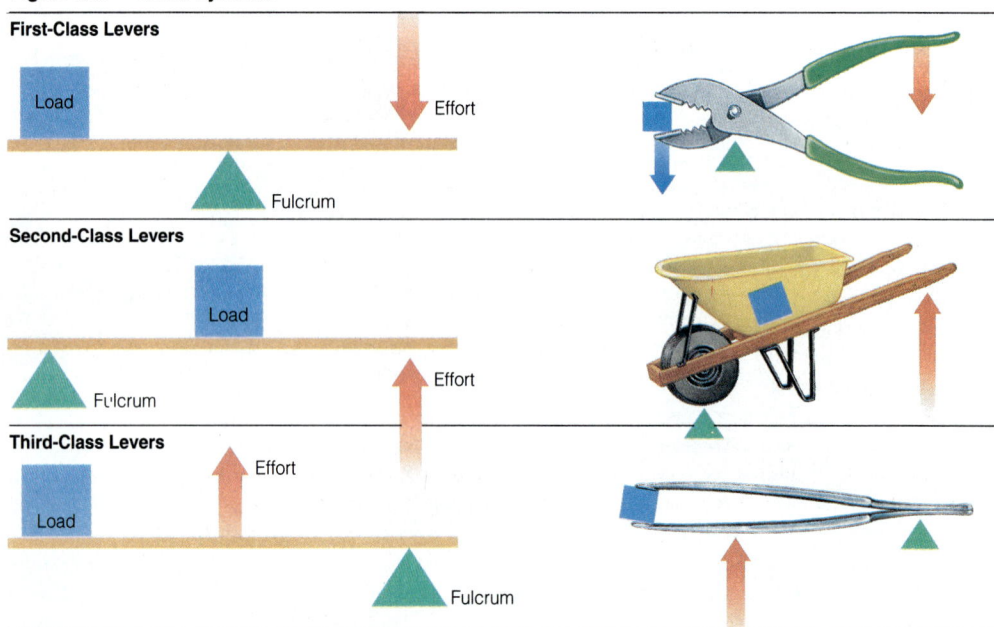

First-Class Levers

Second-Class Levers

Third-Class Levers

404   Chapter 19 Support, Movement, and Covering

Integrating the Sciences

Physical Science Explain that the mechanical advantage of a lever is the number of times the lever increases the effort force. The distance from the effort force to the fulcrum is called the effort arm. The distance from the resistance force to the fulcrum is called the resistance arm. The mechanical advantage of a lever can be calculated by dividing the length of the effort arm by the length of the resistance arm. Ask students, Would a pair of pliers with long handles or short handles have greater mechanical advantage? Why? (Long; longer effort distance in proportion to load distance)

Skills Development

Predict Ask students to predict what will happen to the triceps muscle if the biceps muscle contracts. (The triceps muscle will relax.) What will happen to the biceps muscle if the triceps muscle contracts? (The biceps muscle will relax.)

Infer Ask students which muscle is the extensor of the elbow joint and which is the flexor. Why? (The triceps is the extensor because it straightens the arm, and the biceps is the flexor because it bends the joint.)

Explore Visually

Have students study pages 404 and 405. Then ask the following questions:

▶ Which of the three types of levers requires the least amount of force to lift the load? Why? (First-class levers, because the fulcrum is between the effort and the load)

▶ Which lever in the drawings requires the least amount of muscular force to lift the load? Why? (The first-class lever system to lift the head, because the fulcrum is between the effort and the load)

▶ Which of the three uses of levers requires the greatest amount of force to lift the load? Why? (The third-class lever system, because the hand and forearm have to be lifted as well as the load)

Lever Systems in the Human Body

In the human body, most bones and muscles work together as a lever system. In a lever system, the lever is a rigid bar that moves on a fixed point called the fulcrum. A force is applied to another place, causing movement. In your body, your joints are the fulcrums. In a lever system, a force, or effort, is used to move the load. In your body, muscle contraction is the force. Force is applied where muscles are attached to the bones. The muscles pull. The load includes the bones and the tissues that are moved and anything that you are lifting or moving.

A lever reduces the amount of force needed to move something. Usually, the applied force is less than the force needed to move the object directly.

There are three kinds of levers: first-class levers, second-class levers, and third-class levers. The positions of the fulcrum, effort, and load determine how a lever works. The positions also classify a lever. Look at Figure 19.9. It shows the different kinds of levers, some everyday tools that are levers, and lever systems in the human body.

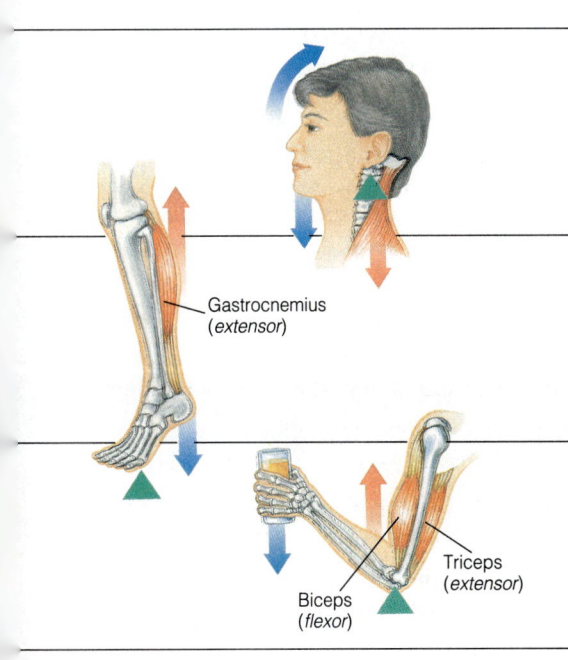

Gastrocnemius (*extensor*)

Triceps (*extensor*)

Biceps (*flexor*)

First-Class Lever: Load-Fulcrum-Effort
When you lift your head or nod your head, you use a first-class lever system. The muscles in the back of your neck are the effort. The joint at the top of your backbone and the base of your skull is the fulcrum. The bones in your face are the weight that is lifted.

Second-Class Lever: Fulcrum-Load-Effort
Standing on tiptoe activates a second-class lever in your body. The joints in the balls of your feet are the fulcrum. The muscles in your calves pull up on the heel bones of your foot. The load is the weight of your body.

Third-Class Lever: Load-Effort-Fulcrum
Whenever you use your arm to drink something, you operate a third-class lever. The biceps in your arm provide the effort. The fulcrum is the hinge joint at your elbow. The load is your hand, your forearm, the container, and the liquid.

TEACH • Continued

Skills WorkOut
To illustrate lactic acid build-up in muscles, have students do the Skills WorkOut.
Answer Students should actually experience the effects of rapid lactic acid accumulation by feeling some mild cramping. Students should relate muscle fatigue to lactic acid build-up and microscopic damage to the muscle.

EVALUATE

WrapUp
Portfolio Have students trace the skeleton on page 394 and sketch in four different muscles. Students can color these and label them according to their classification. Students should add a brief note describing the muscles' functions.
Use Review Worksheet 19.2.

Check and Explain
1. Muscles do work.
2. Skeletal: striated, voluntary skeletal movement; smooth: non-striated, involuntary organ movement; cardiac: striated, involuntary heart movement
3. Answers may vary. A flexor works with an extensor to move a joint; the flexor is the muscle that bends the joint, and the extensor straightens it.
4. Heart: cardiac; small intestine: smooth; blood vessels: smooth; toes: skeletal

---

TEACHING OPTIONS

Cooperative Learning
Assign students to cooperative groups. Each group should research the topic of exercise and demonstrate safe and effective exercise techniques to the class. For example, one group topic can be the importance of stretching before and after exercise. Another group topic can be common muscle injuries and how they can be avoided.

Skills WorkOut

Relate Concepts
1. Write your name 15 times on a sheet of paper. Record how long it takes.
2. Open and clench your writing hand as many times as you can in 45 seconds.
3. Repeat steps 1 and 2 three times.
4. Rest for 2 minutes. Then write your name 15 times.

How did your hand feel? How did your writing change? Define *muscle fatigue*.

Figure 19.10 ▲
During training, the hurdler may stretch muscles more than she would during everyday activities. This stretching may cause small tears that usually cause soreness.

Science and You   *Those Aching Muscles*

Are your muscles sore after exercise? Are they sore the next day? This soreness is caused by two different things. Some muscle fibers contract very quickly and give short bursts of energy. When these muscle fibers are working, they produce a waste product called lactic acid. Usually lactic acid is removed from the muscles by the blood. However, when lactic acid builds up and is not removed from your muscles quickly, you feel pain. Lactic acid usually builds up during heavy exercise. If you are still sore the next day, the exercise has caused microscopic tears and bruises in the muscle.

Check and Explain

1. What is the main function of the muscular system?
2. Compare the structure and function of the three types of muscle tissue: skeletal, smooth, and cardiac.
3. **Apply Definitions** Use the words *flexor* and *extensor* to explain how muscles work in pairs.
4. **Classify** Which of the three types of muscles would you find in your heart, small intestine, blood vessels, and toes?

INTEGRATED LEARNING

Themes in Science

Evolution Animals have specialized coverings that are adapted to the particular survival needs of each species. Discuss different types of skin and coverings in animals such as lizards, horses, and birds. Also discuss how each type of skin covering is adapted to the needs of each animal in its particular environment. Ask students how skin helps humans survive on the earth.

SECTION 19.3

Section Objectives
For a list of section objectives, see the Student Edition page.

Skills Objectives
Students should be able to:

Infer the functions of human skin.

Evaluate Sources of information about skin products.

Demonstrate the effect of rubbing alcohol on human skin.

Vocabulary
epidermis, dermis

19.3 The Skin

Objectives

▶ **Describe** the structure of the skin.
▶ **State** the functions of the skin.
▶ **Make analogies** between parts of the skin, their functions, and common objects.
▶ **Demonstrate** the cooling effect of evaporation.

Skills WarmUp

Infer Make a list of five objects that have an outer covering. Next to each object explain the function of the covering. Discuss your list with several classmates. Make changes or additions where necessary. Based on your discussion, what can you infer are some of the functions of your skin?

MOTIVATE

Skills WarmUp
To help students understand the skin and its functions, have them do the Skills WarmUp.
Answer Students' answers will vary. Students' lists should note functions that apply to the skin, such as protecting, supporting, watertightness, enclosing, filtering.

Misconceptions
Most students think of the skin and hair as separate body components. Emphasize that skin is an organ that includes hair, just as a skeleton includes periosteum.

Once the girders and beams of a building are in place, the outside walls are added to the framework. The inside walls are also constructed. Some of the inside walls help support the building. The outside walls protect the people who live or work in the building. Just like the walls of a building, your skin is the outer covering of your body. Your skin also helps to support and protect your body.

The Largest Organ

Skin is the body's largest organ. Not only does it cover the outside of your body, but it also covers many organs inside your body. You may not think of skin as an organ, but it is. The skin is made up of four tissues: muscle tissue, connective tissue, nerve tissue, and epithelial (ehp ih THEE lee uhl) tissue. Epithelial tissue consists of cells that cover all body surfaces.

Look at the skin cells in Figure 19.11. They make up the top layer of your skin. Notice that the cells are flattened and fit closely together like the pieces of a jigsaw puzzle. Because the cells fit so closely together, they form a protective barrier. Your skin keeps out harmful bacteria and keeps in moisture. Skin cells also contain a protein called keratin (KER uh tihn) that makes the skin waterproof.

Hair and nails are part of the skin system. Both hair and nails are made up of dead cells and keratin. If you have ever wondered why you can cut your hair and nails without pain, it is because they are made up of dead cells. Just like the keratin in animal horns, claws, and birds' beaks, keratin in your nails makes them hard.

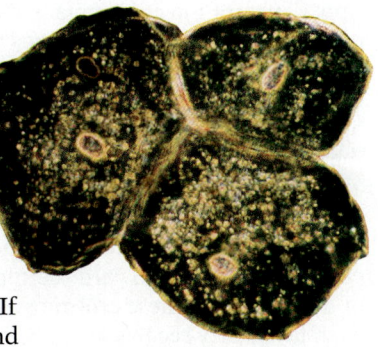

Figure 19.11 ▲
Human skin cells form a watertight and bacteria-proof covering.

Chapter 19 Support, Movement, and Covering

TEACH

Discuss

To review the model of a layer of skin in Figure 19.12, ask the following questions:

▶ What makes the dermis so much thicker than the epidermis? (The dermis contains many blood vessels, glands, and hair follicles that make up the living part of the skin.)

▶ Why are oil glands important? (They keep hair and skin from drying out.)

▶ What is the function of sweat glands? (They remove wastes and help regulate body temperature.)

Critical Thinking

Compare and Contrast Have students observe the skin on their arms, faces, the palms of their hands, and the soles of their feet. Ask them where on their bodies the epidermis seems thickest. (On the soles of their feet and the palms of their hands) Ask students how this thick epidermis is a benefit to people. (It provides extra protection.)

Class Activity

To help students learn more about the skin and its parts, have student groups make a model of a cross section of human skin. With a flat board as a base, students can use different colors of clay to make the layers of skin, oil and sweat glands, hair follicles, fat cells, sense receptors, and smooth muscles. They can use toothpicks or string for hair and wire for blood vessels. Have students label each of the parts with a letter or number and then quiz one another.

INTEGRATED LEARNING

Math Connection

The skin is the body's largest organ. In an adult, the skin makes up about 7 percent of total body weight and receives about one-third of the fresh blood pumped from the heart every minute. The skin covers a surface area of approximately 2750 square inches in the average adult. How many square meters is that? (1.77 sq. m)

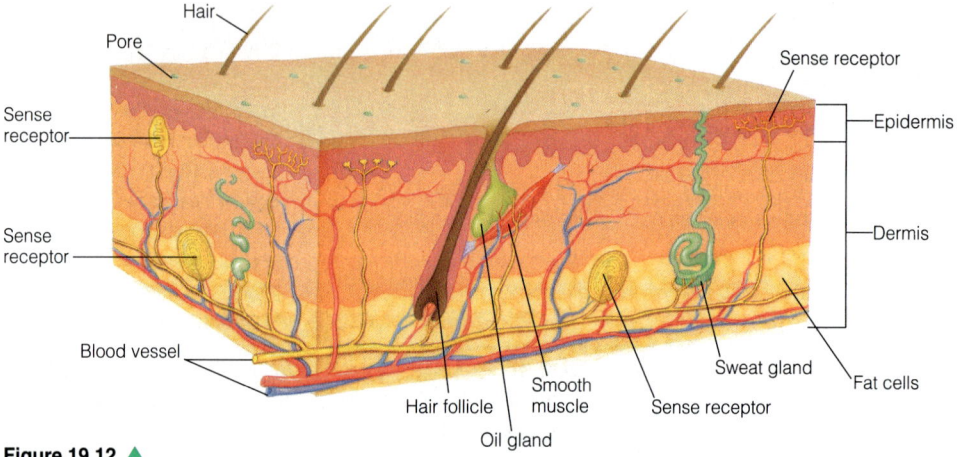

Figure 19.12 ▲
The human skin has two main layers—the epidermis and dermis—with many different structures in the dermis.

Structure of the Skin

The skin is layered. It has two main layers. The upper layer is called the **epidermis**. The lower layer is called the **dermis**. Figure 19.12 shows a model of the skin. Notice that the dermis is much thicker than the epidermis.

Epidermis Under a microscope, you would see that the epidermis has five layers. The upper layer of the epidermis is made up of dead skin cells. Rub your hands together and you will have rubbed off thousands of dead skin cells. The dead cells of the epidermis flake off or are rubbed off all the time. Therefore, they must be replaced.

The dead cells are replaced by cells in the deepest layer of the epidermis. Here a steady supply of skin cells is made. These cells move upward toward the top layer of the dermis. As these cells move through each layer, they receive less oxygen and nutrients from the blood. Eventually they die. These dead cells form the skin that you can see.

Dermis The dermis is beneath the epidermis. A thin membrane separates the dermis from the epidermis. The dermis is thicker than the epidermis: it is the living layer of skin. The dermis is made up of protein fibers and cells that form a strong network. This layer gives your skin its strength and elasticity. If you pinch the skin on the back of your hand, you will notice that it springs back just like an elastic band.

Within the dermis, there are many blood vessels, glands, and tubelike hair follicles (FAHL ih kuhlz). There are hundreds of thousands of hair follicles in the dermis. Each one grows a hair.

Oil and Sweat Glands There are oil glands and sweat glands in the dermis. The oil glands open into hair follicles. Oil passes into a hair follicle and then to the skin surface. The oil keeps both hair and skin from drying out. Your scalp and face have more oil glands than any other part of your body.

Your skin helps remove wastes from

Integrating the Sciences

Zoology Point out that sweating is not the only way that animals regulate body temperature. Have students find out how other animals, such as elephants, cats, and dogs, keep cool.

Class Activity

Have students work in groups of five or six. Obtain two Celsius thermometers per group, some cotton, and a fan. Have them wrap the bulb of each thermometer in cotton. Then have them wet the cotton on one of the thermometers. Have each group record the temperatures on both thermometers every minute for five minutes. Ask them which thermometer showed the most rapid decrease in temperature. Then ask how this experiment is similar to what happens when people sweat. (The body is similarly cooled by evaporating sweat.)

Skills Development

Infer Of all the types of sense receptors in the skin, pain receptors are the most numerous. Ask students to infer why having many pain receptors is an advantageous evolutionary adaptation. (In avoiding pain, the organism pulls away from environmental factors that might be harmful or deadly.)

SkillBuilder

Apply Ask each student to bring in an advertisement for another product and evaluate its advertising claims. Then ask the whole class to discuss how they could use this same approach to evaluate informational materials that they consult during research projects. Look for evidence that students are using their new knowledge of skin as they evaluate sources in the acne advertisements.

---

your body through the sweat glands. Sweat glands are coiled tubes that lead to a pore, or opening, on the skin's surface. Water, salt, and some body wastes are released through the pore. You know this liquid as perspiration, or sweat.

Sweating also helps control your body's temperature. During heavy exercise, on hot days, or when you have a fever, you sweat. As the sweat evaporates from your skin, your skin gets cooler. Blood flowing through the skin also cools. Heat is lost from the body. In this way, the skin acts like a built-in air-conditioning system. When you are cold, the blood vessels narrow. There is less blood flowing near the skin's surface, less heat is lost from the body, and you are kept warm.

Sense Receptors Think about all the sensations that your skin can feel. The skin is a sense organ. The dermis contains several kinds of nerve endings called sense receptors. There are special sense receptors for cold, hot, pain, pressure, and touch. Each receptor responds to a change, or stimulus, in the environment.

You do not have the same number of each kind of sense receptor. For example, there are more pain receptors than any other kind. Also, some parts of your body have more of one kind of receptor than other parts. For example, you have more touch receptors in your fingertips, the palms of your hands, on the tip of your tongue, and on your lips than any other parts of your body.

The Living Textbook:
Life Science Sides 7-8

Chapter 5 Frame 00827
Skin Microviews (12 Frames)
Search: Step:

SkillBuilder   Evaluating Sources

Acne Medication Advertisements

There are many different kinds of acne medications for sale. Most of the information that you get about these medications is from newspaper or magazine ads, and radio and television commercials. The advertisement is usually put together by someone working for the company. Before you buy a product, you should evaluate its advertisement claims.

1. Find a newspaper or magazine advertisement for an acne medication.
2. Write the name of the product.
3. Read the advertisement carefully. Then make a list of questions you would like to have answered before you decide whether or not to believe what the ad claims.
4. Analyze your questions.
   a. If a question asks for more information, write *MI* next to the question.
   b. If the question asks about the source of the information, write an *S* next to the question.
   c. If the question asks about evidence or proof of the claims in the ad, write a *P* next to the question.
5. What reasons might you have for questioning the claims in the ad?
6. What questions could you ask to help decide whether the claims in the ad are truthful?
7. What could you do to decide if the product works the way the ad claims it does?
8. What would you tell someone to think about when a person reads or hears an ad?
9. Would you buy and use the product in your ad? Explain your answer.

Chapter 19 Support, Movement, and Covering

TEACH • Continued

Apply
Emphasize that melanin is a pigment. Discuss pigments. Then ask:
▶ Look at Figure 19.12 on page 408. Why do you think skin without melanin looks pink? (It is actually translucent, but the skin itself contains small blood vessels that add a reddish cast to the skin.)

EVALUATE

WrapUp
Reinforce Have students make a list of the terms in Figure 19.12 on page 408. For each term, students can write its function in regulating the skin system.
Use Review Worksheet 19.3.

Check and Explain
1. Epidermis, dermis
2. Protecting, supporting, keeping in moisture, excreting wastes, temperature control, sense reception
3. Sweat gland, oil gland, blood vessels, melanin
4. Students' paragraphs will vary. Evaporation cools the skin because the process requires heat (more or less, depending on the substance that evaporates) and takes it from the skin.

410

INTEGRATED LEARNING

Multicultural Perspectives
Remind students that the presence of the pigment melanin affects skin color. People of all races have the same number of melanin-forming cells in their skin, but darker-skinned people have more granules of melanin itself in their skin. Melanin, which offers some protection from harmful ultraviolet rays, is formed by these cells when the skin is exposed to sunlight.

Skin and Hair Color

Skin, hair, and eye color are visible differences among people. These inherited traits are caused by a pigment, or coloring, called melanin (MEHL uh nihn). Melanin is produced by a layer of cells in the epidermis. The amount of melanin that you have in your skin determines the color of your skin. There is a wide range of skin colors. People with a lot of melanin have black or dark-brown skin. People with melanin and another substance called carotene have reddish or yellow skin. People with little melanin have light-tan or brown skin. Some people have no melanin. Their skin is white or pinkish. Their hair is also white.

Science and You Zits and Zats!

Everybody gets a pimple, but some people, especially teenagers, suffer from acne. Acne is a skin disorder caused by oil, clogged pores, and bacteria. Acne is most common in people with oily skin. The increased amount of oil produced by the oil glands clogs the skin's pores. Bacteria grow in the oil. These bacteria cause changes in the oil that irritate the surrounding skin. The result is usually a pimple. Sometimes a blackhead forms when air causes a chemical change. Table 19.1 gives some helpful hints for controlling acne. However, people who have severe acne should see a dermatologist, or skin doctor. Medical treatment can help prevent acne scars.

Table 19.1

Skin Care Hints
Wash your face regularly—2 or 3 times per day—with soap and water.
After washing, use a lotion or cream on your skin.
Avoid using oily cosmetics. Remove all makeup before going to bed.
Eat a healthful diet. Avoid sweets.
Exercise regularly.
Sleep 7 to 8 hours per day.

Check and Explain

1. What are the two main layers of the skin?
2. List the functions of the skin.
3. **Make Analogies** Name the part of the skin that shows the best analogy to the function of these common objects: an air conditioner, a lube job on a car, a thermometer, and paint.
4. **Demonstrate** Use a cotton ball and water or rubbing alcohol to show the cooling effect of evaporation on the skin. **Caution! Rubbing alcohol is poisonous.** Write a paragraph explaining how evaporation cools your skin.

410   Chapter 19 Support, Movement, and Covering

Chapter 19 Review

Concept Summary

19.1 Skeletal System
- Bones, cartilage, tendons, and ligaments all make up the human skeletal system.
- Bones are classified by their shape as flat, long, short, or irregular.
- The human skeleton has immovable, slightly-movable, and freely-movable joints. Four types of freely-movable joints are hinge, ball-and-socket, pivotal, and gliding.
- The functions of the skeletal system are support, protection, storage, and blood cell formation.

19.2 Muscular System
- The muscular system allows voluntary movement of the skeletal system and involuntary movement of organs.
- Skeletal, smooth, and cardiac muscles are made of bundles of muscle cells.
- Muscles do work only by contracting. Muscles work in pairs consisting of an extensor and a flexor.
- Skeletal muscles work together with bones as lever systems.

19.3 The Skin
- The skin is the body's largest organ. Its functions include protection, cooling, waste removal, lubrication, support, and sensory reception.
- The two main layers of the skin are the epidermis and the dermis. Sweat glands, hair follicles, oil glands, and sense receptors are all found in the dermis.
- Melanin is the pigment that gives skin, hair, and eyes their color.

Chapter Vocabulary

endoskeleton (19.1)	ligament (19.1)	smooth muscle (19.2)	flexor (19.2)
vertebra (19.1)	tendon (19.1)	cardiac muscle (19.2)	epidermis (19.3)
cartilage (19.1)	skeletal muscle (19.2)	extensor (19.2)	dermis (19.3)
joint (19.1)			

Check Your Vocabulary

Use the vocabulary words above to complete the following sentences correctly.

1. Most of the the skeleton of the newborn baby is made of ____ .
2. A chicken, a cow, and a human have an internal bone framework called an ____ .
3. Muscles are connected to bones by ____ .
4. The backbone is made of ____ .
5. Only the heart muscle is made of ____ .
6. Hinge, ball-and-socket, pivotal, and gliding are types of movable ____ .
7. The layer of skin you can see is the ____ .
8. The ____ connect bones at the joint.
9. You move your body by contracting ____ .
10. A muscle that straightens a joint is called a ____ , while a muscle that bends a joint is called a ____ .
11. Sweat glands, oil glands, and hair follicles are all found in the ____ .
12. You can't control the ____ in the walls of your internal organs.

Write Your Vocabulary

Write sentences using each of this chapter's vocabulary words. Show that you know what each word means. Check with a partner to see if your sentences are correct.

CHAPTER REVIEW 19

Check Your Vocabulary

1. cartilage
2. endoskeleton
3. tendons
4. vertebrae
5. cardiac muscle
6. joints
7. epidermis
8. ligaments
9. skeletal muscle
10. extensor; flexor
11. dermis
12. smooth muscle

Write Your Vocabulary

Be sure that students understand how to use each vocabulary word in a sentence.

Use the Vocabulary Worksheet for Chapter 19.

The Living Textbook: Life Science Sides 1-4

Chapter 5   Frame 01618
Compact Bone Microviews (3 Frames)
Search: Step:

CHAPTER 19 REVIEW

Check Your Knowledge

1. The muscle on the back of the thigh is a flexor; the one on the front of the thigh is an extensor.
2. Skeletal muscle is voluntary, striated, and moves bones by acting across joints.
3. Oil, sweat, and hair are produced by glands in the dermis.
4. Ball-and-socket: shoulder, hip; hinge: elbow, knee; pivotal: skull/vertebral column; gliding: wrist, vertebrae
5. The heart is cardiac muscle.
6. Functions of the skeleton include protection of internal organs, movement, production of blood cells, and storage of fat and minerals.
7. Bones are classified by shape—long, flat, short, and irregular.
8. The two main layers of the skin are dermis and epidermis.
9. True
10. True
11. False; dermis
12. False; skeletal
13. True
14. True
15. False; cardiac and smooth

Check Your Understanding

1. a. Human: endoskeleton
   b. Cow: endoskeleton
   c. Grasshopper: exoskeleton
   d. Lobster: exoskeleton
   e. Chicken: endoskeleton
   f. Beetle: exoskeleton
2. a. pivot, gliding
   b. ball-and-socket
3. All three act by contracting, are made up of fibers, and are controlled by the nervous system (brain).
4. The cardiac muscle is involuntary and doesn't fatigue, while the skeletal muscle is voluntary and fatigues. The skeletal muscle is voluntary and striated, while the smooth muscle is involuntary and not striated. The cardiac muscle is found in the heart and looks striated, while the smooth muscle is found in internal organs and has no striations.
5. When you get out of the shower, the water on your skin evaporates, taking away heat. You can cool yourself when you are hot by pouring water on your skin and letting it evaporate.
6. Hamstring flexes knee, quadriceps extends knee; biceps flexes elbow, triceps extends elbow; tibialis (shin) flexes foot, gastrocnemius (calf) extends foot
7. The large number of bones in the hand and foot allows for very precise movement because of the joints between these bones.
8. Drawings will vary, but students should show correct relationships among femur, tibia, fibula, and patella.

Chapter 19 Review

Check Your Knowledge
Answer the following in complete sentences.

1. Give an example of a flexor and an extensor muscle.
2. What are some characteristics of skeletal muscles?
3. What substances are produced in the dermis?
4. Give an example of a ball-and-socket, a hinge, a pivotal, and a gliding joint.
5. What type of muscle is the heart?
6. What are the functions of the skeletal system?
7. How are bones classified? What are the four classifications?
8. What are the two main layers of the skin?

Determine whether each statement is true or false. Write *true* if it is true. If it is false, change the underlined term(s) to make the statement true.

9. Skeletal muscles do work by <u>contracting</u>.
10. Many of your muscles and bones work together as a lever system in which a <u>joint</u> serves as the fulcrum.
11. Sweat glands, oil glands, and sense receptors are located in the <u>epidermis</u>.
12. Bones, cartilage, tendons, and ligaments are parts of the human <u>muscular</u> system.
13. The functions of the skin include cooling and <u>sensory reception</u>.
14. The pigment that gives skin, hair, and eyes their color is <u>melanin</u>.
15. Both <u>skeletal</u> and <u>smooth</u> muscles are involuntary muscles.

Check Your Understanding
Apply the concepts you have learned to answer each question.

1. **Classify** Which of the following organisms have exoskeletons? Which have endoskeletons?

   a. Human   d. Lobster
   b. Cow   e. Chicken
   c. Grasshopper f. Beetle

2. a. Which types of joints are named for the kind of movement they allow?
   b. Which type of joint is named for the way the bones meet at the joint?
3. **Compare and Contrast** What are three common characteristics of the three muscle types?
4. **Compare and Contrast** Identify two differences each between the following pairs of muscle types: cardiac and skeletal; skeletal and smooth; smooth and cardiac.
5. **Generalize** Apply what you know about evaporation and the body's cooling system. Identify one reason you're cold when you get out of the shower. Identify one way to cool yourself when you're hot.
6. List three flexor muscles in your body. Next to each one, describe the movement it causes. List the extensor muscle that makes a pair with each flexor listed and describe the movement each one causes.
7. **Infer** More than half of your 206 bones are in your feet and hands. Why do you think there are so many bones in human feet and hands? How is this fact important to you every day?
8. **Mystery Photo** The photograph on page 392 is a color-enhanced X-ray of a human knee joint. Make a drawing of the bones as you see them in the X-ray. Label the bones.

Develop Your Skills

1. a. There are the most pain receptors; there are the fewest hot and cold receptors.
   b. Because there are so many touch receptors, this may be skin from a fingertip, palm, or tongue.
2. a. 29; 54
   b. Student graphs should show the following: arms and legs, 9%; torso, 26%; hands, 26%; feet, 25%; head, 14%.
3. hip—ball-and-socket
   elbow—gliding
   thumb knuckle—hinge

Make Connections

1.

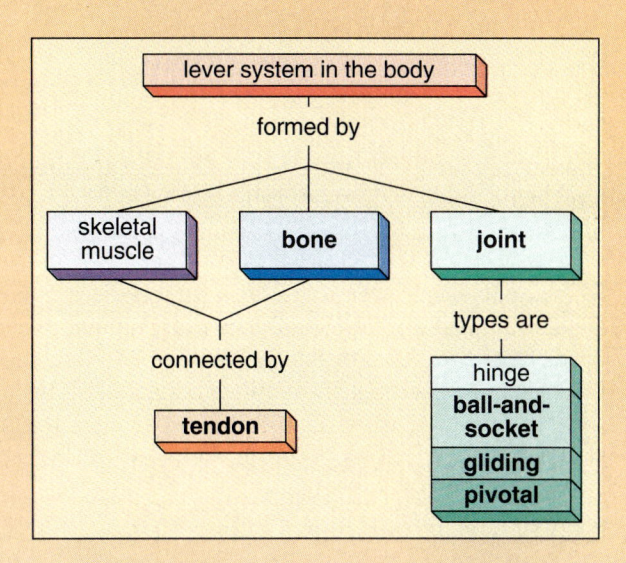

2. Drawings will vary; however, make sure students have identified major cell characteristics, structures, and shapes for each type.
3. A greenstick fracture is a partial fracture in which the bone is not broken completely through. A simple fracture is a fracture that remains covered by uninjured skin. In a compound fracture, the broken bone is exposed to the air through broken skin. A compound fracture is especially serious because of the possibilities of blood loss and infection.
4. The femur will be the longest bone for most students. Answers for the second question will vary.

Develop Your Skills
Use the skills you have developed in this chapter to complete each activity.

1. **Interpret Data** The diagram shows a map of sense receptors in a patch of skin.

   Key: ● = Cold receptor   ■ = Hot receptor
        ◆ = Pressure receptor   ▲ = Pain receptor

   a. Which type are there the most of? Which type are there the least of?
   b. What part of the body might this map be showing? Why?

2. **Data Bank** Use the information on page 626 to answer the following questions.
   a. How many bones are in the head? How many are in the hands?
   b. **Calculate** Determine the percentage of bones in each area listed. Make a circle graph of this data.

3. **Accurate Observations** Observe how you can move the following joints: hip, elbow, thumb knuckle. What type of joint do you think each is? Why?

Make Connections

1. **Link the Concepts** Below is a concept map showing how some of the main concepts in this chapter link together. Only parts of the map are filled in. Copy the map and complete it using words and ideas from the chapter.

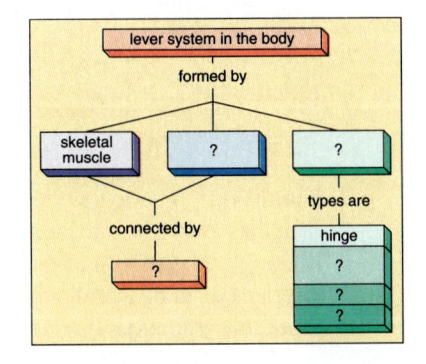

2. **Science and Art** Draw or make a model for each of these tissue cells.
   a. Skin cell
   b. Smooth muscle cell
   c. Bone cell

3. **Science and You** Have you ever fractured a bone? Three kinds of fractures are greenstick, simple, and compound. Use reference books to find out about each kind of fracture. Illustrate each type of fracture in some way, such as a drawing or a clay model. Write a description of each type of fracture.

4. **Science and Math** Measure as many long bones in your arms, legs, and fingers as you can. Organize your data in a table. Which long bone is the longest bone in your body? How many times longer is it than the shortest long bone that you measured?

CHAPTER 20

Overview

The first section in this chapter focuses on the movement of food in the digestive system. Digestive organs and processes are described and illustrated. The topic of the second section is the circulatory system. Students learn about the human heart, blood vessels, and how blood moves through the circulatory system. This section also describes the lymphatic system, blood types, and cardiovascular disease. The third section explains the movement of oxygen through the organs of the respiratory system. The final section describes the roles that the kidneys, liver, and skin play in the excretory system.

Advance Planner

▶ Provide 50-cm strips of dialysis tubing (available from a science supply company) and marbles for TE activity, page 417.

▶ Obtain an illustration of an artificial heart for TE page 421.

▶ Collect supplies for SE Activity 20, page 434.

▶ Collect filter paper, funnels, measuring cups, clear cups, water, and sand for SE page 435.

Skills Development Chart

Sections	Classify	Collect Data	Communicate	Estimate	Graph	Infer	Interpret Data	Measure	Observe	Reason by Analogy
20.1 Skills WarmUp			●							
Skills WorkOut									●	
Skills WorkOut		●								
20.2 Skills WarmUp								●		
SkillBuilder							●			
Skills WorkOut					●					
20.3 Skills WarmUp				●						
Historical Notebook						●				
Activity		●					●			
20.4 Skills WarmUp						●				
Skills WorkOut	●									
Skills WorkOut										●

Individual Needs

▶ **Limited English Proficiency Students** Photocopy for students pages 415, 418, 422, 430, 431, and 436. Have them highlight all of the places where the chapter vocabulary terms appear on these pages. They can write a definition for each term and add it to the appropriate page. Have students save these pages so they can use them to review the chapter vocabulary.

▶ **At-Risk Students** Mask the labels and make enlarged photocopies of the figures on pages 416, 423, 430, and 436. Ask students to label each part of the digestive, circulatory, respiratory, and excretory systems. As they study the chapter concepts, have them write a sentence describing the function of each organ on strips of index cards. Have them paste their sentences onto their papers.

▶ **Gifted Students** Have a group of students study how blood functions in different systems. Ask students to create four drawings or overhead transparencies showing how blood travels in each of the body systems described in the chapter. Have them describe how blood cells are changed in each system. Encourage students to make a presentation to the class.

Resource Bank

▶ **Bulletin Board** Keep the body map that students created for Chapter 19. Have students add details about the digestive, circulatory, respiratory, and excretory systems. Suggest that they use different colors of yarn or string to represent veins and arteries. Provide materials such as cotton balls or fabric so that students can create a three-dimensional effect on their display.

▶ **Field Trip** Arrange to take students to a local fire station so that they can talk with people who work as paramedical technicians and observe the equipment they use. Students can prepare a list of questions to ask the paramedics about their work and training.

CHAPTER 20 PLANNING GUIDE

Section	Core	Standard	Enriched	Section	Core	Standard	Enriched
20.1 Digestive System pp. 415–420				**Overhead Blackline Transparencies** Overhead Blackline Master 20.2 and Student Worksheet	●	●	●
Section Features Skills WarmUp, p. 415	●	●	●				
Skills WorkOut, p. 418	●	●		**Color Transparencies** Transparencies 60, 61	●	●	
Skills WorkOut, p. 420	●	●					
Blackline Masters Review Worksheet 20.1	●	●		**20.3 Respiratory System** pp. 429–434			
Enrich Worksheet 20.1			●				
Skills Worksheet 20.1	●	●	●	**Section Features** Skills WarmUp, p. 429	●	●	●
Laboratory Program Investigation 36	●	●	●	Historical Notebook, p. 432	●	●	
Investigation 37		●	●	Activity, p. 434	●	●	●
Investigation 38			●	**Blackline Masters** Review Worksheet 20.3	●	●	
Color Transparencies Transparencies 58, 59	●	●		Integrating Worksheet 20.3a	●	●	●
				Integrating Worksheet 20.3b	●	●	
20.2 Circulatory System pp. 421–428				**Color Transparencies** Transparency 62	●	●	
Section Features Skills WarmUp, p. 421	●	●		**20.4 Excretory System** pp. 435–438			
SkillBuilder, p. 426		●	●				
Skills WorkOut, p. 427	●	●		**Section Features** Skills WarmUp, p. 435	●	●	
Blackline Masters Review Worksheet 20.2	●	●		Skills WorkOut, p. 437	●	●	
Skills Worksheet 20.2a	●	●	●	Skills WorkOut, p. 438		●	●
Skills Worksheet 20.2b	●	●	●				
Integrating Worksheet 20.2		●	●	**Blackline Masters** Review Worksheet 20.4	●	●	
Ancillary Options *Multicultural Perspectives*, p. 53	●	●	●	Vocabulary Worksheet 20	●	●	

Bibliography

The following resources can be used for teaching the chapter. See page T-40 for supplier codes.

Audio-Visual Sources
(video unless noted)
Health Concerns for Today Series. 4 filmstrips. 1984. ME.
The Human Body Series: Circulatory and Respiratory Systems. 20 min. 1988. NGSES.
The Human Body Series: Nervous System. 20 min. 1988. NGSES.
The Lungs and Respiratory System. 17 min film. 1975. EB.
Man: The Incredible Machine. 28 min. 1975. NGSES.

Software Resources
Digestive System. Apple. J & S.
The Human Body: An Overview. Apple. BB.
Human Circulatory System. Apple, IBM PC. EME.

Library Resources
Elting, Mary. *The Macmillan Book of the Human Body.* New York: Aladdin, 1986.
Meredith, Susan, Ann Goldman, and Tom Lisauer. *Book of the Human Body.* Tulsa, OK: Educational Development Corp., 1983.
Nilsson, Lennart. *Behold Man: A Photographic Journey of Discovery Inside the Body.* Boston: Little, Brown, and Co., 1974.
West, J.B., and S. Lahir, eds. *High Altitude and Man.* Bethesda, MD: American Physiological Society, 1984.

CHAPTER 20

INTEGRATED LEARNING

Math Connection
Tell students that 8 percent of their total body weight is blood. Have students use their own body weights (or an average weight) and calculate how much their blood weighs. You may wish to point out that 8 percent can be written as 8 ÷ 100. (Total weight × 8 ÷ 100 = blood weight)

Themes in Science
Systems and Interactions The human body has systems that work together. These systems interact with one another to enable people to live, grow, and move.

Introducing the Chapter
Have students read the description of the photograph on page 414. Ask if they agree or disagree with the student's description. You may wish to point out that red blood cells are smaller than the period at the end of a sentence.

Directed Inquiry
Have students study the photograph. Ask:

▶ Describe the images you see in the picture. (Students will mention the flat, disklike shape, the depression in the center, and the color.) What do you think they are? (Cells)

▶ What kind of cells do you think these are? (Red blood cells) Why do you think so? (Their color and the appearance of floating suggest blood.)

▶ What is the topic of this chapter? (Supply and transport) How does the picture relate to the chapter topic? (Blood transports things that supply the body with what it needs to function.)

Chapter Vocabulary
alveoli
artery
bronchi
capillary
chemical digestion
digestion
mechanical digestion
nephron
nutrient
vein
villi

Chapter 20 — Supply and Transport

Chapter Sections
20.1 Digestive System
20.2 Circulatory System
20.3 Respiratory System
20.4 Excretory System

What do you see?

"I see large donuts flowing in some type of liquid. I think that they are red blood cells. I predict that they are the size of this period →. I think that they carry oxygen to the body."

Aneeshia Russum
Leeds Middle School
Philadelphia, Pennsylvania

To find out more about the photograph, look on page 440. As you read this chapter, you will learn about the different systems in your body that help supply and transport nutrients essential to your health.

Integrating the Sciences

Physical Science Mechanical digestion begins in the mouth. The mobility of the lower jaw enables us to chew. Explain to students that the lower jaw acts as a lever. Have them identify the part of the mouth that acts as the effort arm and the part that acts as the fulcrum. To review lever systems, refer students to pages 404 and 405. (The lower jaw is the effort arm, and the hinge where the lower and upper jaws connect is the fulcrum.) What acts as the load? (The object/mass being chewed.)

20.1 Digestive System

Objectives

▶ **Trace** the passage of food through the digestive system.

▶ **Name** and **describe** the organs used in digestion.

▶ **Communicate** how mechanical digestion helps chemical digestion occur.

Skills WarmUp

Communicate What foods did you eat recently? Choose one of the foods and describe in detail its journey from your first bite to your stomach. Read your description to a partner. Are your descriptions similar or very different?

Why do you need food? You need food because the cells of your body require a constant supply of energy and materials to build new cells and repair old ones. Food provides the materials and the energy. But your body can't use food the way it exists when you eat it. Food must be broken down into simpler parts that your body can use. This process is called **digestion.**

Digestion

The substances in food that your body needs in order to live and grow are called **nutrients.** Carbohydrates, proteins, and fats are three important kinds of nutrients in food. When your body digests food, the nutrients in it are broken down so that they dissolve in water. The dissolved nutrients are absorbed into your blood and carried to your body's cells.

Digestion occurs in a series of organs called the digestive tract. As food moves through these organs, it undergoes both chemical and physical changes.

The physical changes are a result of **mechanical digestion.** During mechanical digestion, food is broken into smaller pieces. Look at Figure 20.1. Mechanical digestion begins in your mouth as your teeth cut, grind, and mash food.

The chemical changes are a result of **chemical digestion.** Chemical changes occur when the nutrients in food are broken down into simpler molecules that dissolve in water.

Food is changed chemically by substances called acids, bases, and enzymes. Different glands throughout the digestive system produce these substances. Enzymes are molecules that speed up chemical reactions.

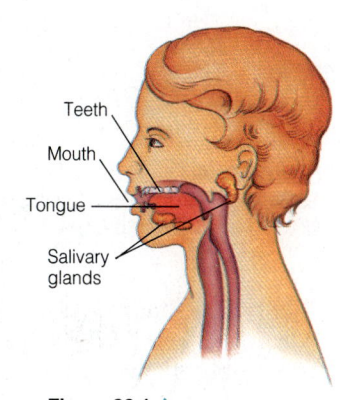

Figure 20.1 ▲
Both mechanical and chemical digestion begin in the mouth.

Chapter 20 Supply and Transport 415

SECTION 20.1

Section Objectives
For a list of section objectives, see the Student Edition page.

Skills Objectives
Students should be able to:

Communicate ideas about how the digestive system works.

Observe how a substance changes in the stomach.

Communicate ideas about how mechanical digestion aids chemical digestion.

Vocabulary
digestion, nutrients, mechanical digestion, chemical digestion, villi

MOTIVATE

Skills WarmUp
To help students understand the digestive process, have them do the Skills WarmUp.
Answer Descriptions should mention the mouth, throat, stomach, and intestines. Nutrients from food are eventually distributed throughout the body.
Extension Have students save these descriptions in their portfolios and use them during the section WrapUp.

Misconceptions

Students may think that digestion takes place only in the stomach. Have students list all the processes that they think are involved in digestion. Suggest that they keep these ideas in mind as they study this section.

The Living Textbook:
Life Science Side 2

Chapter 11   Frame 10159
Swallowing and Peristalsis (Movie)
Search: Play:

TEACH

Discuss
Have students study Figure 20.2. Ask the following questions:

▶ How is food changed in the mouth? (Teeth break food into smaller pieces, and saliva makes food softer and changes starch into sugars.)

▶ What do the muscles in the esophagus do? (The muscle action, called peristalsis, pushes food down the esophagus and into the stomach.)

▶ What chemical digestion takes place in the stomach? (Hydrochloric acid breaks down proteins and destroys microorganisms. Pepsin breaks down protein and fat.)

Enrich
Use Enrich Worksheet 20.1.

Skills Development
Infer Have students imagine they are eating a sandwich in a hurry, swallowing as fast as they can. Suddenly they feel chest pain. Ask the following questions:

▶ What is the most likely cause of the pain? (Food that is in large pieces in the esophagus)

▶ How can you prevent this problem? (Eat slowly and drink plenty of water.)

Research
Have students use reference materials to learn about the experiments of surgeon William Beaumont on his patient Alexis St. Martin. These experiments began in 1822 and lasted several years. Have students find out what Beaumont discovered about the stomach and have them present their findings to the class in an oral report. Encourage students to be creative.

INTEGRATED LEARNING

Multicultural Perspectives
Food in the stomach takes between two and six hours to be processed. In the United States, many people space mealtimes four to six hours apart, eating the largest meal in the evening. In Spain, the largest meal is often eaten at midday. Have students share their knowledge about when people in other cultures eat meals.

Process of Digestion

When you eat, food travels for about 8 m through your digestive tract. This journey can last a day or more. Each organ in the digestive tract performs different functions of mechanical and chemical digestion. As you read, follow the path of digestion in Figure 20.2.

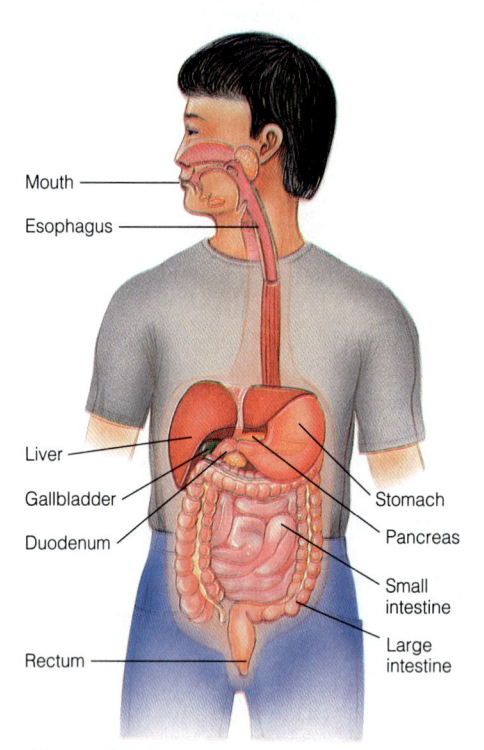

Figure 20.2 ▲
The food you eat makes a long journey through your digestive system.

Mouth Digestion begins in your mouth. When you chew a bite of food, it is broken into smaller pieces and mixed with saliva.

Saliva contains water, mucus, and an enzyme. The enzyme changes starch, a complex molecule in food, into simpler molecules of sugar. The water and mucus in saliva make the food softer and easier to swallow. Your tongue moves the food around in your mouth as it is chewed. When you swallow, the food moves into your esophagus.

Esophagus Food travels from your mouth to your stomach by a muscular tube called the esophagus (ih SAHF uh guhs). Look at Figure 20.2 and locate the esophagus. It is about 25 cm long. Food is pushed down the esophagus by the contraction and relaxation of the muscles in the esophagus. This type of muscle action is called *peristalsis* (PEHR uh STAHL sihs).

A ring of muscles at the bottom of the esophagus guards the entrance to your stomach. The muscles relax to allow food to pass into your stomach. Then they contract again to prevent food from coming back into your esophagus.

Stomach Rhythmic contractions of powerful muscles in your stomach wall mash food into a pulp and mix it with gastric juice. The gastric juice, which is made in the stomach wall, contains the enzyme pepsin, hydrochloric acid, and mucus. The thick layer of mucus protects the membranes that line the stomach from the acidic gastric juice.

The hydrochloric acid in gastric juice begins breaking down proteins and destroys microorganisms in the food. The enzymes help break down protein and fat.

The food becomes a thick, soupy mixture in your stomach. Little by little this mixture squirts out through another ring of muscles into your small intestine. After a meal, it takes about 2 to 6 hours for your stomach to empty.

Themes in Science

Energy All living things need energy to live. The source of energy for human beings is food. As you study this chapter, stress to students that the digestive system changes food into compounds that the body can use for energy.

Small Intestine The next stop in the digestive tract is a narrow, folded tube where food is further digested and absorbed. This is your small intestine, which is about 2.5 cm wide and 7 m long. Your small intestine is four or five times as long as you are tall! Find the small intestine in Figure 20.2.

The first 25 cm of the small intestine are called the *duodenum* (DOO oh DEE nuhm). As the thick, soupy mixture moves from the stomach to the duodenum, more digestive juices are added. These digestive juices are produced in the liver, the pancreas (PAN kree uhs), and the small intestine itself. The digestive juice produced by the small intestine contains enzymes and substances that neutralize the acidic food mixture.

While digestive juices are working, the thick, soupy food mixture is pushed through the small intestine by peristalsis. During this part of the journey, amino acids from proteins, simple sugars from carbohydrates, and broken-down fats are gradually absorbed into the blood. This absorption occurs through the small intestine's lining.

Liver

Your liver has many functions, including aiding in digestion. In Figure 20.3, you can see a close-up view of the liver. Find the liver in the digestive system in Figure 20.2. Where is it located? ①

The liver produces bile, a mixture of substances. Although bile doesn't contain digestive enzymes, it does contain bile salts. Bile salts aid in the digestion of fats. They are important because fats are completely undigested when they reach the small intestine.

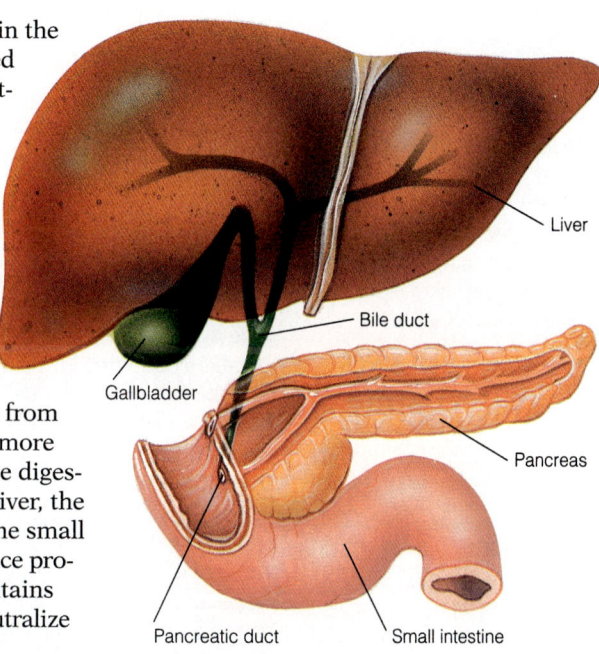

Figure 20.3 ▲
The liver, pancreas, and small intestine produce chemicals used in digestion. What is the function of the gallbladder? ②

Bile made by the liver is stored in the gallbladder. Find the gallbladder in Figure 20.3. The gallbladder moves bile into the small intestine through a duct. In the small intestine, the bile breaks down fat globules into smaller droplets of fat.

Pancreas

Figure 20.3 also shows the pancreas. The pancreas aids digestion by supplying pancreatic juices to the small intestine. Pancreatic juices are delivered to the small intestine with the bile from the gallbladder. Both enter the small intestine through the same duct. The many enzymes in pancreatic juices help to break down carbohydrates, fats, and proteins.

Chapter 20 Supply and Transport

Skills Development

Measure Provide students with yarn and a meter stick or tape measure and have them measure and cut off a piece of yarn 7 meters long. Tell them that this is how long the small intestine is. They can save these pieces of yarn to use during the section WrapUp.

Class Activity

Have students work in pairs to make a model that shows how food moves through the digestive system. Provide each pair with one 50-cm strip of dialysis tubing and one marble. Have students wet the tubing and open it by gently rubbing it between their fingers. Then have them place the marble into one end of the tubing and squeeze the tubing so the marble moves through it. After students have finished the activity, have them discuss how this model shows what happens in the digestive tract. (The tubing is like the intestines, the marble like food, the action of students' hands like the contraction of muscles along the digestive tract.)

Answers to In-Text Questions

① The liver is located above and to the right of the stomach.

② The gallbladder moves bile into the small intestine through the pancreatic duct.

Integrated Learning

Use Integrating Worksheet 20.1.

The Living Textbook:
Life Science Side 2

Chapter 11   Frame 11559
The Action of Gastric Juice (Movie)
Search: Play:

TEACH ▪ Continued

Skills WorkOut
To help students understand how bile breaks up fat, have them do the Skills WorkOut.

Answer The detergent makes the large drops of oil break into smaller droplets. Bile breaks fat globules into smaller droplets of fat.

Critical Thinking
Compare and Contrast Have students make diagrams to show the structure of the digestive systems in humans and another animal of their choice. Have them label all parts of the systems and describe their functions, noting the similarities and differences between the two digestive systems.

Discuss
Explain to students that the small intestine is made up of three parts—the duodenum, the jejunum, and the ileum. Most digestion occurs in the duodenum and most absorption occurs in the ileum. Stress to students that the many blood vessels close to the lining of the ileum carry nutrients throughout the body.

The Living Textbook: Life Science Side 7

Chapter 5   Frame 00005
Digestive System Slides (117 Frames)
Search: Step:

418

INTEGRATED LEARNING

Integrating the Sciences
Health Ask students to explain the role of fiber in the diet. Have them investigate what foods are good sources of dietary fiber and then monitor their own eating habits for a week to see if they get enough fiber in their diets.

Skills WorkOut
Observe In a medium-sized bowl of water, drip about ten drops of vegetable, peanut, or canola oil. Observe what happens for a few minutes. Then drip five or six drops of liquid dishwashing detergent into the middle of the bowl. What happens? Explain why this is similar to how bile breaks up fats in the small intestine.

Figure 20.4
Millions of villi line the small intestine. The structure of villi is shown at left. The photograph shows villi magnified 300x. ▼

Absorption

Most nutrients are absorbed from food in your small intestine. For this reason, the small intestine has a huge surface area. In fact, if the total lining of the small intestine were spread out flat, it would cover a baseball diamond!

You may think the inside of the small intestine is smooth, but it isn't. The small intestine has ridges. These ridges are lined with tiny fingerlike projections called **villi** (VIHL eye). You can see the villi in Figure 20.4. The villi and ridges are the structures that increase the surface area of the small intestine. Most nutrients would not be absorbed without this large surface area.

The lining of the small intestine is connected to a vast network of blood vessels. Simple molecules of broken-down nutrients pass through the lining of the small intestine into these blood vessels. The blood vessels carry these nutrient molecules throughout the body.

The watery leftovers that can't be digested move from the small intestine through another ring of muscles to the large intestine. The large intestine is about 1.5 m long and about 6 cm wide. The walls of the large intestine absorb the leftover water.

As the excess water is gradually absorbed, the contents of the large intestine become semisolid. This material, called feces (FEE seez), is still more than half water. Feces also contain food that can't be digested, such as apple skins and cucumber seeds. These fibrous foods are important, however, because they help to keep food moving through your digestive tract. Other parts of the feces include dead bacteria and bits of the lining of the digestive tract that rubbed off as food passed through.

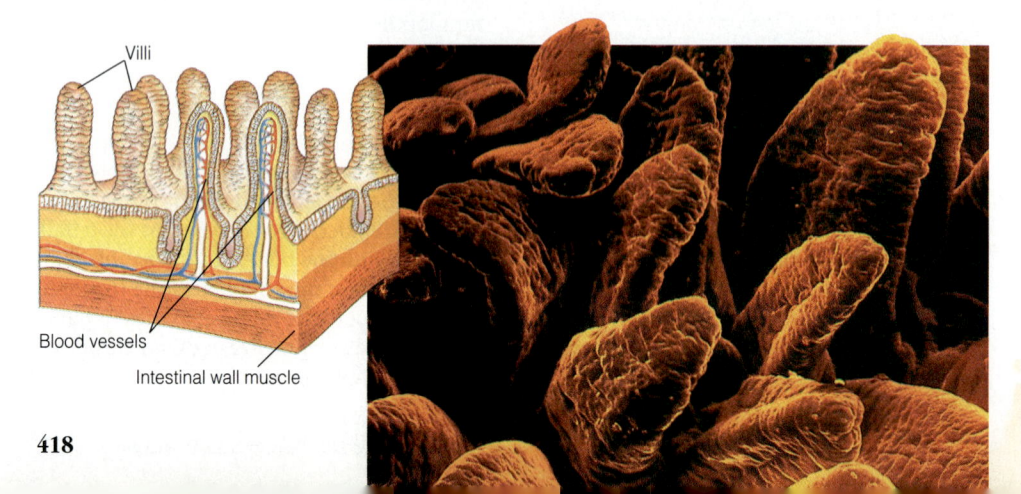

418

STS Connection

Gallstones are formed by substances such as cholesterol, bile pigments, and calcium salts. Sometimes these stones can cause considerable pain. Explain to students that one method of treating gallstones involves the use of ultrasound therapy. In this procedure, high-frequency sound waves are directed at the affected body part. The energy of the sound waves dissolves the gallstones. Have students look up the word *ultrasound* and discuss its meaning. (Ultrasound means "above sound." The high frequency of these waves is above the range that humans are able to hear.)

Directed Inquiry

Ask students the following questions:

▶ Which of the disorders in Table 20.1 have you had? (Answers will vary, but most will have experienced diarrhea, constipation, and indigestion.)

▶ What causes these disorders? (See chart.)

▶ What did you do about these disorders? (Most will have used a medication or a home remedy of some kind.)

▶ What treatment does the chart indicate is needed for appendicitis? (Surgery would be necessary.)

▶ What are some of the causes of indigestion? (See chart.)

The feces move into the rectum at the lower end of the large intestine, where they collect. At the end of the rectum is a small opening called the anus. Muscle action excretes the feces through the anus and out of the body.

Problems of the Digestive System

If you are like most people, you have experienced digestive disorders. Stomachaches, nausea, diarrhea, and other similar disorders are very common. Your digestive system has to deal with a steady stream of substances from outside your body. Every day there is a chance that these substances will contain microorganisms or chemicals that upset the normal functions of your stomach or intestines.

Table 20.1 lists some of the common disorders of the digestive system. Many of these disorders are related to a person's diet or lifestyle. Ulcers and indigestion, for example, are most common in people with high levels of stress. You probably know from experience, too, that your emotions affect your digestion. In general, the best way to prevent digestive disorders is to eat a balanced diet of the right amounts of food and to drink plenty of water.

Table 20.1 Digestive Disorders

Disorder	Description	Cause	Treatment
Appendicitis	Swollen and infected appendix	Trapped food and bacteria	Surgery
Constipation	Feces too hard	Too much water absorbed in large intestine	Medication
Diarrhea	Feces too watery	Not enough water absorbed in large intestine	Medication
Gallstones	Pain in abdomen	Hard "stones" formed in gallbladder	Medication or surgery
Heartburn	Pain below heart	Stomach contents backed up into esophagus	Change of diet; medication
Indigestion	Pain in abdomen, nausea, cramps	Food not fully digested	Change of eating habits or diet
Ulcer	Open sore in stomach or duodenum	Tissue eroded by excess stomach acid	Change of diet; reduction of stress; medication

Chapter 20 Supply and Transport

> **TEACH ▪ Continued**

Skills WorkOut

To help students understand some effects of different types and amounts of food, have them do the Skills WorkOut.

Answer Check students' descriptions to see that they reflect knowledge about how the digestive system works.

> **EVALUATE**

WrapUp

Portfolio Have students repeat the Skills WarmUp from the beginning of this chapter. Have them compare their early descriptions with the later descriptions.

Use Review Worksheet 20.1.

Check and Explain

1. The mouth performs mechanical digestion. The stomach begins chemical digestion by breaking down proteins with gastric juices. The small intestine absorbs nutrients. The liver, pancreas, and gallbladder supply digestive enzymes. The large intestine absorbs excess water and delivers undigestible materials to be excreted.

2. Digested nutrients are absorbed by blood vessels in the lining of the small intestine and are carried throughout the body.

3. People usually can survive without half of the small intestine. However, incomplete absorption could occur.

4. Mechanical digestion prepares food for chemical digestion by breaking it into small pieces.

420

INTEGRATED LEARNING

Writing Connection
Have students write a story about what happens to a particular piece of food, from an apple or a cheese sandwich, as it passes through all stages of digestion. Encourage students to be creative. For example, they could describe events from the food's point of view or tell it as a horror story.

Themes in Science
Systems and Interactions The heart and other parts of the human circulatory system work together to transport materials throughout the body.

Skills WorkOut

Collect Data Have you noticed how foods affect your digestive system? For example, how does your stomach feel after you've eaten too much food? Over a three-day period, list all the foods you eat along with approximate amounts. Then write how you feel, if anything, after you've eaten the foods. Describe how different foods and the amounts you consume affect your digestive system.

Science and Society *Eating Disorders*

Have you ever gone on a diet to lose weight? For some people, losing weight can get out of control. When this happens, they are in danger of being affected by eating disorders called anorexia nervosa (an uh REHKS ee uh nur VOH suh) and bulimia (buh LEEM ee uh). The two disorders mostly affect women in their teens and twenties. Both disorders are often related to a person's self-image.

People with anorexia nervosa are convinced they are overweight no matter how thin they are. They eat little or no food and often exercise too much. Lack of food can cause them to become nervous and act "edgy" all the time. Anorexia nervosa can severely damage a person's health. In some cases, it can lead to death.

People with bulimia actually eat large amounts of food. However, they purge their bodies of the food to avoid gaining weight. To purge, bulimics may force themselves to vomit after eating. They may take laxatives to move food through the digestive tract before it can be absorbed. Bulimia can damage the stomach and esophagus, the liver, the gallbladder, the pancreas, and even the teeth. Bulimia can also cause kidney failure or heart failure, which can lead to death.

The treatment for these eating disorders usually requires hospitalization. Patients learn about nutrition and get psychological counseling. These treatments are not really cures for the disorders, though. Instead, they are treatments for the symptoms. People who have had anorexia nervosa or bulimia may always fight the urge to lose a few more pounds or to purge after eating.

Check and Explain

1. What are the organs of the digestive system? What does each organ do in the process of digestion?

2. Explain how digested nutrients enter the blood.

3. **Reason and Conclude** What would happen if a person had half of his or her small intestine surgically removed? Would this person survive? Explain.

4. **Communicate** How does mechanical digestion help chemical digestion occur? Explain.

420   Chapter 20 Supply and Transport

STS Connection

For many years, researchers have been trying to develop a permanent artificial heart. The goal of this research is to design an instrument that can perform the pumping action of the heart for prolonged periods. Aluminum and plastic were used to produce the Jarvik 7, a device that replaces the two ventricles of the natural heart. In 1982, the first person to receive an artificial heart lived for 112 days. Have students study a diagram of the Jarvik 7 and compare its structure with the human heart shown in Figure 20.5.

20.2 Circulatory System

Objectives

- **Describe** the structure and function of the heart.
- **Trace** the path of blood through the circulatory system.
- **Identify** functions of the lymphatic system.
- **Explain** how blood types interact with each other.
- **Hypothesize** about the interaction of blood pressure and blood vessels.

Skills WarmUp

Measure Place two fingers on the pulse in your neck. Count heartbeats for 15 seconds. Multiply by four and record the number. Then run in place for one minute and count the number of heartbeats again. Did you have the same number of heartbeats both times? Explain your answer.

Think about how you move through the school building each day. You walk through the halls between your locker and the classrooms. You carry books and papers into and out of class. You are joined by many other students who move as you do through the school building.

Blood cells circulate through the body in the same way that you and your classmates circulate through the school building. However, blood can't move on its own. Blood is pumped through blood vessels in your body by the heart. Together the heart, blood, and blood vessels make up the circulatory system.

The Heart

Clench your fist. Your heart is about the same size and very similar in shape. Your heart is a strong muscle that expands and contracts in rhythmic beats as it pumps blood through the blood vessels. Your heartbeat changes in response to your body's need for blood. What happens to your heart when you exercise? What happens to it when you relax? ①

The inside of the heart is shown in Figure 20.5. Notice that each side of the heart has an atrium at the top and a ventricle at the bottom. Both atria receive blood coming into the heart, and both ventricles pump blood out of the heart. A small tissue, called a valve, opens and closes to let blood flow into the atrium and then from the atrium to the ventricle. The action of the valves keeps the blood flowing in one direction. Your heartbeat is the rhythmic sound of the opening and closing valves.

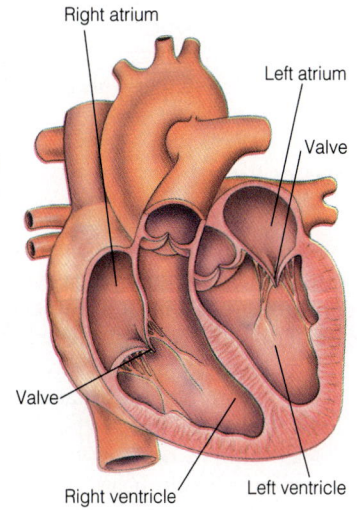

Figure 20.5 ▲
The Human Heart

Chapter 20 Supply and Transport   421

SECTION 20.2

Section Objectives
For a list of section objectives, see the Student Edition page.

Skills Objectives
Students should be able to:

Measure their heart rates before and after exercising.

Graph the heart rates of the class.

Hypothesize why it is important to have their blood pressure checked regularly.

Vocabulary
artery, capillary, vein

MOTIVATE

Skills WarmUp

To help students understand the effect of exercise on their heart rates, have them do the Skills WarmUp.

Answer Students will notice that their heart rates increase after they exercise.

Misconceptions

Students may think that blood is blue because the veins on the backs of their wrists appear blue. The blood in veins appears blue because it has been depleted of oxygen and is darker red. The blood vessel walls make the blood appear blue.

Answer to In-Text Question

① When you exercise, your heart beats fast, but it slows down when you're relaxing.

The Living Textbook:
Life Science Side 2

Chapter 12   Frame 12195
Heartbeat/Blood Flow/Heart Valves
(Movies)
Search: Play:

421

TEACH

Class Activity
Have students use the method they learned in the Skills WarmUp to monitor their heart rates at regular times each day for one week. They could do it when they wake up, right before they go to sleep, before and after gym class, or before and after meals. Have them keep a chart of their heart rates, and after a week they can write their conclusions about how their activities affect their heart rates.

Reteach
Fill a plastic container with water and seal it. Have students rhythmically squeeze the container to see how blood pressure in vessels is similarly affected by the contractions of the heart.

Skills Development
Infer Write the words *hypertension* and *hypotension* on the chalkboard. Review with students that *hyper-* means above and *hypo-* means below. Have them infer the meaning of these words. (Hypertension is high blood pressure and hypotension is low blood pressure.) Encourage interested students to research causes and effects of high and low blood pressure.

Integrated Learning
Use Integrating Worksheet 20.2.

Reteach
Write the words *artery* and *away* on the chalkboard. Underscore the letter *a* at the beginning of each word. Tell students that arteries always carry blood away from the heart.

Answer to In-Text Question
① **Veins carry blood toward the heart. Arteries carry blood away from the heart. Capillaries are very small vessels that carry blood toward and away from the heart.**

INTEGRATED LEARNING

Integrating the Sciences
Physical Science Blood pressure is an example of the force liquid exerts on the walls of a closed container. If the pressure changes in one area of the liquid, it will change in the other parts as well. This rule is stated in a theory called Pascal's principle. Have interested students plan a demonstration of Pascal's principle and describe how it applies to blood in blood vessels. They can use physical science texts for ideas about how to create a model that shows Pascal's principle. A simple model might involve filling a plastic bag or long balloon with water, sealing it tightly, and showing how all the water in the container is affected by squeezing one end.

Blood Vessels

Blood is carried through your body by more than 95 000 km of blood vessels. Look at Figure 20.6. Blood leaving the heart is forced into **arteries**, which transport it to all your organs and muscles. Arteries are tubes with thick, strong, elastic walls. They can withstand the pressure of the blood that is forced into them when the ventricles contract. Blood leaving the heart is forced through smaller and smaller arteries to all parts of your body.

The smallest blood vessels are called **capillaries**, which are finer than a human hair. Some capillaries are so small that only a single line of blood cells can move through them. Some capillaries carry blood from the heart to your muscles and organs, and other capillaries carry blood back toward the heart.

Capillaries that carry blood back toward the heart join larger blood vessels called **veins**. The heart does not push blood through veins. Valves in the veins force the blood through. Because veins are close to the surface, you can see them in your hands and feet where the skin is thin.

Heart Rate

Put two fingers on the side of your neck directly under your jaw. The rhythmic surge under your fingers is your heart pushing blood through the neck artery. Each time the ventricle forces blood into the main artery, the pressure causes every artery in your body to expand and contract. Every surge you feel is a heartbeat.

The number of times your heart beats in one minute is your heart rate, or your pulse. A normal heart rate at rest is between 65 and 75 beats per minute. Your heart rate changes throughout the day, depending on your activity. For example, it increases when you exercise. It even changes when you stand up.

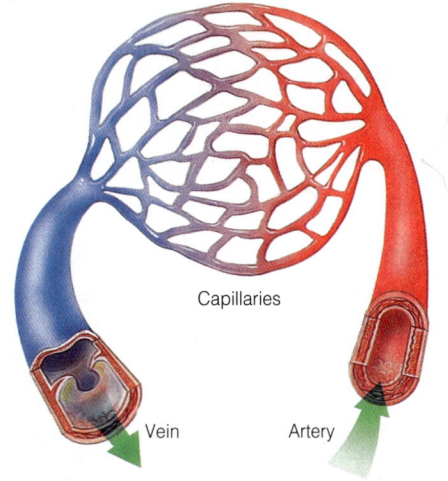

Figure 20.6 ▲
Arteries, veins, and capillaries are all blood vessels. What is the function of each blood vessel? ①

Blood Pressure

The next time you pump up a bicycle tire, hold onto the hose. Notice that with each surge of air, the hose stiffens and then relaxes. The same thing happens to your arteries when your heart pumps blood through them. The surge of blood puts pressure on the artery wall. This pressure is called your blood pressure. Blood pressure can indicate the condition of the arteries.

Blood-pressure readings are stated with two numbers. For example, normal blood pressure is 120 over 80. The first number is the amount of pressure on the artery wall when the heart contracts and forces more blood into the artery. The smaller number is the amount of pressure on the blood moving through the artery between surges.

Many people have their blood pressure checked regularly. High blood pressure can lead to strokes or heart attacks.

Chapter 20 Supply and Transport

Themes in Science

Stability and Equilibrium Blood pressure is maintained by the part of the brain called the medulla. As you study the circulatory system, stress to students that the maintenance of a stable blood pressure insures that the circulatory system will transport nutrients, oxygen, and wastes throughout the body.

Multicultural Perspectives

Explain to students that the idea that blood circulates through the body was suggested more than 2,000 years ago in China. In 200 B.C., writers of *The Yellow Emperor's Manual of Corporeal Medicine* described a dual system of circulation that, among other things, provided the body with vital essentials derived from food.

Explore Visually

Have students study Figure 20.7. Ask:

▶ When the left ventricle contracts, where does the blood go? (The blood travels through the largest artery, the aorta, and throughout the whole body.)

▶ Is the blood going into the aorta from the left ventricle oxygen-rich? (Yes, it is oxygenated before flowing into the left ventricle.)

▶ Does the right ventricle pump blood into or out of the heart? (Both ventricles pump blood away from the heart. The right ventricle pumps blood into the pulmonary artery; the left ventricle forces blood into the aorta.)

▶ Why do you think this picture shows blood in different colors? (The red blood is oxygen-rich, the blue blood is oxygen-poor.)

▶ Describe in your own words what happens during systemic circulation. (Oxygen-rich blood is pumped from the left ventricle throughout the body, delivering oxygen to the body's cells, and oxygen-poor blood is pumped back to the heart.)

▶ Describe pulmonary circulation. (Oxygen-poor blood is pumped from the right ventricle into the pulmonary artery. It is then pumped to the lungs, where it picks up oxygen.)

Circulation

The main function of the heart is to circulate blood, which transports food, oxygen, and waste products. The heart's two pumps work together in perfect rhythm to circulate your blood through your lungs and your body.

Systemic Circulation Look at Figure 20.7. Blood is forced from the heart by contraction of the left ventricle. The blood enters the aorta, which is the largest artery in the body. The aorta distributes the blood to smaller arteries and then to the capillaries.

As the cells throughout your body remove oxygen from the blood, they release carbon dioxide. Capillaries and veins carry the oxygen-poor blood back to the heart. The inferior vena cava is the final collection point for blood from the veins. Blood flows from the vena cava into the right atrium.

Circulation of blood from the heart to the body and back again is called systemic circulation. It moves blood throughout the entire body system.

Pulmonary Circulation The function of pulmonary circulation is to oxygenate the blood. This is done by passing it through the lungs.

The oxygen-poor blood entering the right atrium is pumped into the right ventricle. The right ventricle then forces this blood into the pulmonary artery connected to the lungs. In your lungs, oxygen diffuses into the blood. In addition, carbon dioxide is given up so it can be exhaled. When blood leaves the lungs, it is once again rich in oxygen. Blood enters the left atrium and repeats the cycle. The entire cycle of blood, including both pulmonary and systemic circulation, occurs in less than one minute.

Figure 20.7 ▲
The Human Circulatory System

Oxygen diffuses into, and carbon dioxide diffuses out of, capillaries in the lungs.

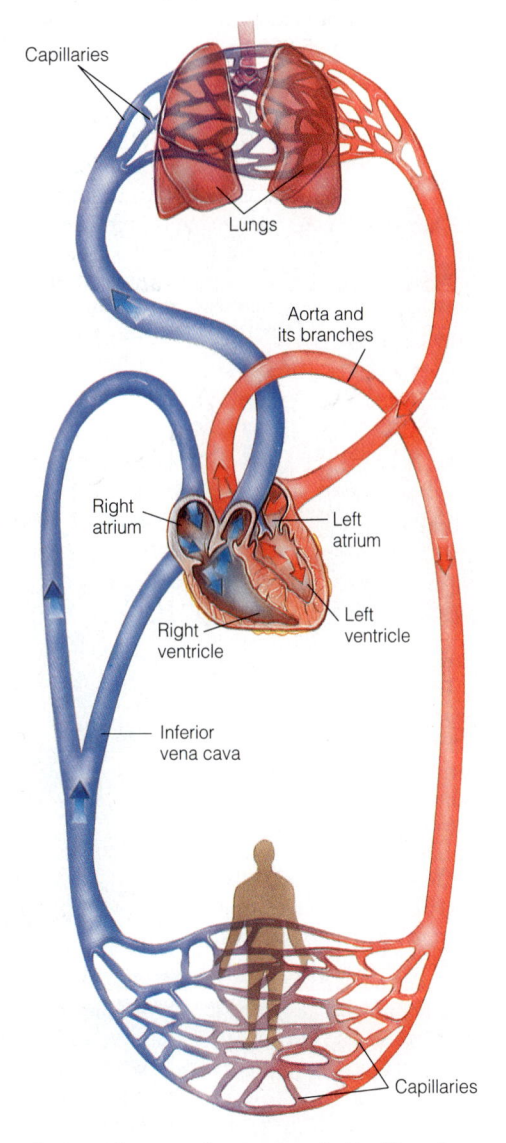

Oxygen diffuses out of, and carbon dioxide diffuses into, capillaries in the body.

The Living Textbook: Life Science Side 7

Chapter 5   Frame 00512
Circulatory System Slides (64 Frames)
Search: Step:

TEACH • Continued

Explore Visually

Have students study the micrographs, diagrams, and paragraphs on this page. Then ask students the following questions:

▶ How does the shape of the red blood cells help them function? (The smooth, round shape enables the cells to flow easily through blood vessels. The large surface area enables them to absorb substances.)

▶ How are white blood cells different from red blood cells? (White blood cells have mitochondria and nuclei. They also live longer.)

▶ In what form are salts and nutrients present in plasma? (They are dissolved in the plasma.)

▶ How and why do platelets become trapped in an injured blood vessel? (Proteins in the plasma form sticky threads that trap the platelets. The platelets then form a clot.)

Ancillary Options

If you are using the blackline masters from *Multiculturalism in Mathematics, Science, and Technology*, have students read about Charles Richard Drew on page 53 and complete pages 54 to 56.

The Living Textbook: Life Science Side 2

Chapter 13 Frame 14805
Living Blood Cells (Movie)
Search: Play:

INTEGRATED LEARNING

Integrating the Sciences

Chemistry Explain to students that hemoglobin is the substance in blood that carries oxygen. The hemoglobin molecule contains four iron atoms. These atoms each attach to an oxygen molecule, producing oxygenated hemoglobin, which is bright red.

Blood

What are you giving away when you donate half a liter of your blood? Actually you are sharing many elements that are vital to life. Blood is a complex substance made up of different kinds of cells, particles of protein, and dissolved salts and nutrients.

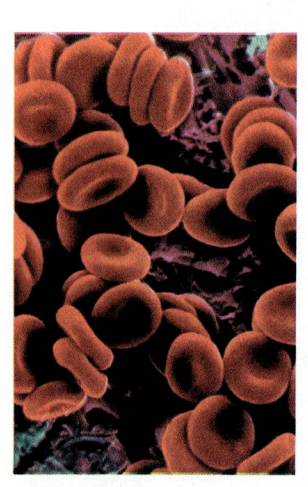

Red Blood Cells ▲
Oxygen is carried in your blood by red blood cells. These cells are specialized for this function. Unlike other cells in your body, they have no mitochondria or nuclei.
How do red blood cells carry oxygen? They contain hemoglobin, a protein that attracts and holds oxygen molecules.
Red blood cells are made in the bone marrow. Each lives only a few months.

White Blood Cells
A whole army of cells in the blood protect your body against disease. These are the white blood cells. When you get sick, the number of white blood cells increases.
White cells begin their development in the bone marrow and mature in the lymph organs and nodes. ▼

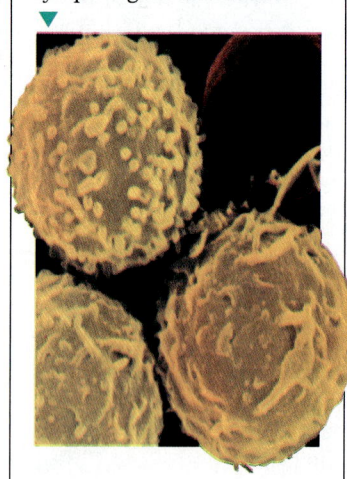

Plasma ▶
When red blood cells, white blood cells, and platelets are removed from blood, what's left is a watery yellow fluid called plasma. Plasma contains important salts that help your muscles function. It also carries dissolved nutrients.

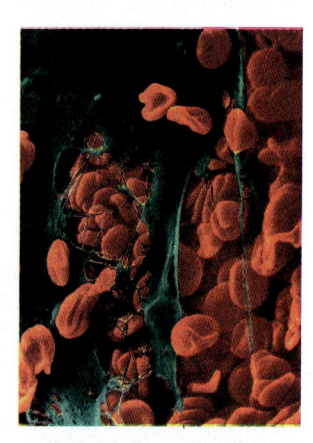

Platelets ▲
Colorless bits of cells, called platelets, cause your blood to clot. When a blood vessel is injured, proteins in the plasma form long sticky threads called fibrin. The fibrin traps platelets, which then collect, form a clot, and plug the hole.

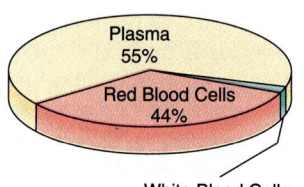

Plasma 55%
Red Blood Cells 44%
White Blood Cells; Platelets 1%

424   Chapter 20 Supply and Transport

STS Connection

Explain to students that hospitals need a constant supply of fresh blood from donors. The reason is that blood deteriorates after three weeks. Before being transported to hospitals, blood is sometimes separated into its components by a machine called a centrifuge. Blood may also arrive whole at a temperature of about 4° to 6° C.

Once the blood is separated and dehydrated, hospital workers freeze the plasma. Plasma can then be kept indefinitely as powder. The plasma supply is easier to maintain than the supply of whole blood, because a donor can give blood 40 times a year instead of the normal 3, receiving back his or her red cells from the previous visit.

Lymphatic System

Do you ever notice small, sore lumps in your neck when you are sick? These little lumps are swollen lymph nodes that are helping you get well. Lymph nodes are part of a network of lymph vessels that extend throughout your body. The main function of this network, called the lymphatic system, is to return to the bloodstream small amounts of fluid and proteins that seep out of the capillaries. Figure 20.8 shows the network of lymphatic vessels in the body.

The clear yellow fluid that flows through the lymphatic vessels is called lymph. Lymph travels through your body to a location near your shoulders. Here it drains into the circulatory system. During its journey, lymph is filtered through the lymph nodes, where bacteria and viruses can be trapped. These nodes are filled with white blood cells, called lymphocytes. Lymphocytes fight infections. When bacteria and viruses are present, the lymphocytes multiply rapidly and the lymph nodes are tender and swollen. Lymphocytes are also manufactured in the spleen, which is part of the lymphatic system. In addition to fighting infection, the lymphatic system helps keep the right level of fluid and proteins in your blood.

Blood Types

Everyone has blood made up of the same physical components. There are, however, differences among people in the chemical makeup of their blood. Your red blood cells belong to a group that identifies an important molecule, called an antigen, on its surface. Individuals with the A antigen on the surface of their red cells have type A blood. Those with the B antigen have type B blood. If the red cell surface has both A and B antigens, then the blood is type AB. If the red cell surface has neither the A nor the B antigen, the blood is type O.

Blood typing is important if you give or receive blood. The plasma of your blood can produce special proteins, called antibodies, that will bind to the antigens on red cells that are not the same type. For example, type A blood plasma makes an antibody that binds to type B blood cells. This antibody is referred to as an anti-B antibody. If type B blood is mixed with type A blood, the anti-B antibodies

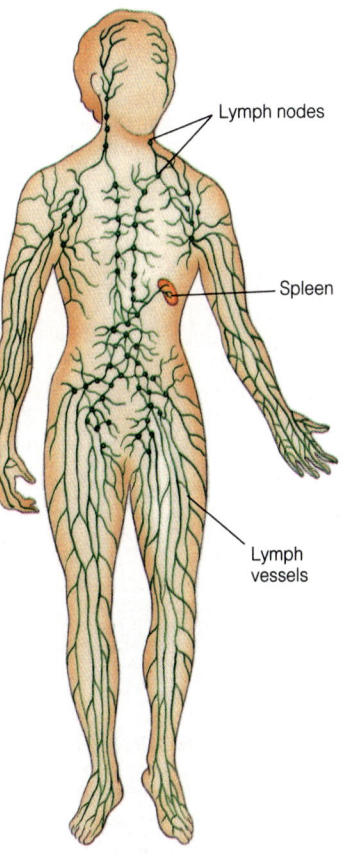

Figure 20.8 ▲
The lymphatic system includes lymph nodes, lymphatic vessels, and the spleen. How does the lymphatic system help fight infections? ①

Apply
Have students press gently against the back of their necks to see if they can feel their lymph nodes. Have students talk about times when they have noticed soreness in their lymph nodes. Ask them to use their own words to describe the reason for the soreness they felt. (Answers may vary. Lymph nodes contain white blood cells, or lymphocytes. They become swollen when white blood cells multiply to fight infection.)

Discuss
Stress to students that there is no pump in the lymphatic system as there is in the circulatory system. One of the reasons that exercise is important is that it helps keep lymph moving through the body.

Answer to In-Text Question
① **The lymphatic system traps infection-causing bacteria and viruses in the lymph nodes where the lymphocytes can attack and destroy the harmful organisms.**

TEACH • Continued

Class Activity
Have all the students in your class find out their blood types if possible. Record this information on a class chart. Have students refer to Table 20.2 and determine the people to whom they could give blood, and those from whom they could receive blood.

SkillBuilder
Before students answer questions 1 through 6, have them work with a partner to check their tables.

Answers
1. Type AB blood has neither anti-A nor anti-B antibodies.
2. Type O blood has both anti-A and anti-B antibodies.
3. All blood types
4. Type O or type A; type O or type B
5. AB blood does not have any antibodies because AB cells have both A and B antigens, so people with type AB blood can receive all types of blood.
6. Type O blood has antibodies against both A and B antigens. It would form clots if types A, B, or AB were donated to a type-O recipient.

The Living Textbook:
Life Science Side 2
Chapter 13 Frame 18273
Blood Typing and Agglutination (Movie)
Search: Play:

INTEGRATED LEARNING

Themes in Science
Diversity and Unity All people have blood that is made up of plasma, red blood cells, white blood cells, and platelets. However, the chemical makeup of blood varies. The surfaces of red blood cells have antigens that affect the chemistry of the blood.

Table 20.2 Blood Types, Donors, and Recipients

Blood Type	Cell Surface Antigen	Can Donate Blood To	Can Receive Blood From
A	A	A, AB	A, O
B	B	B, AB	B, O
AB	A, B	AB	A, B, AB, O
O	None	A, B, AB, O	O

cause the type A cells to clump and form a clot. Clots are dangerous because they prevent the flow of blood through the blood vessels. They can cause death.

The same thing happens with the other blood types. Type B blood contains anti-A antibodies that bind to type A blood. Type AB blood cells have both A and B antigens, so AB blood contains neither anti-A nor anti-B antibodies. People with type AB blood can receive all

SkillBuilder *Interpreting Data*

Antibody Formation
Before blood transfusions are done, the donor blood is cross-matched with the blood of the person receiving it. The purpose of cross-matching the two blood types is to ensure that the blood transfusion will work. If blood clotting occurs, then a new donor must be found.

Copy the table shown on a piece of paper. Use Table 20.2 and the information on blood types to complete the table. For each blood type, place an X in the antigen row if antibodies form. Study the table and answer the following questions:

1. Type AB blood has antibodies against which antigens?
2. Type O blood has antibodies against which antigens?
3. Donated blood is safe only if it is given to a person whose blood does not contain anti-

Antigen	Blood Types			
	A	B	AB	O
A				
B				

bodies against it. Which blood type(s) can be donated to a person with AB blood?
4. Which blood can be donated to a person with type A blood? Type B blood?
5. Although type AB blood is rare, donor blood does not have to be type AB. Why do you think this is so? Explain.
6. Type O blood is the most common. However, the number of compatible blood types is limited. Why do you think this is so? Explain.

Chapter 20 Supply and Transport

Multicultural Perspectives

Explain to students that a current study of heart disease rates among North American Native Americans has emphasized that many factors combine to affect coronary illness. Researchers have noticed that North American Native Americans have fewer cases of coronary artery disease. However, the modern-day Sioux seem to have more coronary illness than other people of North American Native American heritage. Several factors may account for this, such as health and life-style choices and changes in the gene pool. You may wish to have interested students read about the study (called Strong Heart Project, and funded by the National Heart, Lung, and Blood Institute).

Skills WorkOut

To help students graph the heart rate of everyone in the class, have them do the Skills WorkOut.

Answer The lowest heart rate will probably be about 60 beats per minute. The average range for teenage students is 70 to 80 beats per minute.

Skills Development

Predict Ask students to use the information they collected about heart rates to predict what a typical person's heart rate would be while running. Ask the question, Why does a person's heart rate increase during exercise? (The rate might be about 120 beats per minute. Working muscle cells need extra oxygen and nutrients to function properly.)

Answer to In-Text Question

① The build-up of fat deposits can narrow an artery and eventually close it or cause it to bulge and then burst. Blood clots can also form around fat deposits and then break apart into pieces and block blood vessels.

---

types of blood, but they can donate blood only to other people who are type AB. Type O blood cells have neither A nor B antigens, so O blood contains both anti-A and anti-B antibodies. People with type O blood can donate to anyone, but they can receive only type O blood.

Use the information in Table 20.2 to determine blood type, donors, and recipients. Remember, two blood types can't be combined if one contains antibodies that conflict with the surface antigen(s) of the other.

Cardiovascular Disease

Do you know someone who has had a heart attack or a stroke? If you do, you know that heart attacks and strokes are very serious. Disease of the heart and blood vessels is called cardiovascular disease. It is the most common cause of death in the United States.

Cardiovascular disease starts in the blood vessels. Normally the blood flows freely through the arteries. However, with time, fatty deposits of cholesterol can build up on the walls of the arteries. The opening inside the artery is narrowed by these deposits. This condition is called atherosclerosis (ATH ur oh skluh ROH suhs). It becomes serious when a large number of the major arteries are narrowed.

As the arteries become narrower, the amount of blood flow is reduced, and the heart and blood vessels are affected. The heart has to work harder to force blood through the arteries, so the blood pressure increases. Increased pressure inside the arteries causes them to lose their elasticity and bulge. If the bulge is great, the artery wall becomes thinner and can rupture.

Blood clots are likely to form at the fatty deposits. These clots can break loose and completely close a blood vessel anywhere in the body: the lung, the brain, or the heart muscle itself. When a clot lodges in the heart muscle, a person has a heart attack. When a brain artery ruptures or is clogged by a clot, a stroke occurs.

Just about everyone, even children, has some amount of fatty deposits in their blood vessels. As you age, these deposits increase in size and amount if you don't take care of yourself. Controlling your weight, not smoking, eating a diet low in animal fats, exercising regularly, and controlling your blood pressure are all important in preventing cardiovascular disease.

Skills WorkOut

Make a Graph Record the heart rate of everyone in your class. What is the highest heart rate? What is the lowest heart rate? Graph the results. What is the average range for most students?

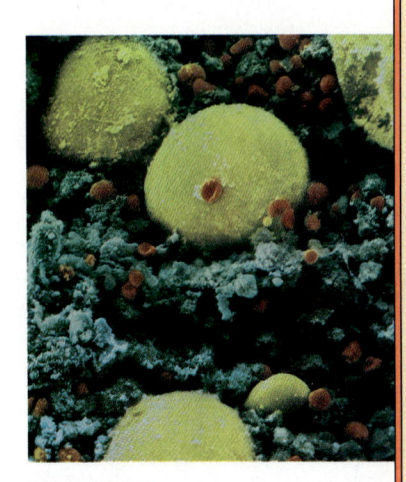

Figure 20.9 ▲
The yellow shapes are fat deposits. How does the buildup of fat deposits affect the arteries? ①

The Living Textbook:
Life Science Side 7

Chapter 24   Frame 09766
Atherosclerosis (Movie)
Search: Play:

The Living Textbook:
Life Science Side 7

Chapter 25   Frame 15572
Blocked Blood Vessels (Movie)
Search: Play:

Chapter 20 Supply and Transport

EVALUATE

WrapUp
Reinforce Have students make a bulletin board display of the circulatory system. Have them draw a figure of a person that shows all the parts of the circulatory system. Then have them make and label an enlarged diagram that shows how blood travels inside the heart. Students can trace the path of blood on the body diagram.

Use Review Worksheet 20.2.

Check and Explain

1. Blood is pumped into the arteries when the ventricles in the heart contract. Blood is pumped through the veins by valves in the veins themselves.
2. Lymph nodes contain lymphocytes, which are white blood cells that fight bacteria and viruses. When there is an infection in the body, lymphocytes multiply rapidly. Bacteria and viruses are carried to the lymph nodes where the lymphocytes destroy them.
3. Blood is forced from the heart into the aorta from the left ventricle. This blood must travel throughout the body.
4. If arteries become clogged, blood pressure increases. This increased blood pressure causes the blood vessels to lose their elasticity. Checking blood pressure helps people to monitor cardiovascular health.

Answer to In-Text Question

① Regular exercise strengthens the heart muscle and generally keeps the cardiovascular system in good shape.

INTEGRATED LEARNING

Integrating the Sciences
Health Poll the class to find out what types of and how much exercise students engage in. Explain that only some forms of exercise, such as running, swimming, biking, and brisk walking, provide a cardiovascular workout. Emphasize the importance of doing these kinds of exercise at least three times a week for 20 minutes each session.

Figure 20.10 ▲
How can regular exercise help lower the risk of heart disease? ①

Science and You Have a Healthy Heart

What is your risk of ever having a heart attack? If very few people in your family have had heart attacks, you might be at low risk. However, if your family has a long history of high blood pressure and cardiovascular disease, you could be in a higher risk group. You can't change what you inherited from your family, but you can control some factors that contribute to cardiovascular disease.

Medical studies show a link between heart disease and a high-fat diet. Limiting the amount of such foods as butter, eggs, red meat, and cheese reduces the amount of fat in your diet. You can make decisions about what you eat.

Exercise is another important factor in lowering your risk of heart disease. Swimming, biking, running, hiking, and brisk walking are all good for your cardiovascular system. They are aerobic exercises in which you must breathe deeply. However, it's important that you do them consistently for at least 20 minutes at a time. To get the best results, you can decide to do aerobic exercises three times a week.

Weight control and not smoking are also important in lowering your risk of heart disease. You can control the amount of food you eat, how much you exercise, and whether or not you smoke. No one knows exactly what your risk of cardiovascular disease is. However, good health habits beginning now will more than likely lower your risk.

Check and Explain

1. How is blood moved through the arteries? Through the veins?
2. Explain how the lymphatic system fights infection. Describe one other function of the lymphatic system.
3. **Find Causes** The left ventricle of the heart is the largest chamber. How is this important to systemic circulation?
4. **Hypothesize** Explain how blood pressure and the health of your blood vessels are related. Why is it important to periodically have your blood pressure checked?

Themes in Science

Energy Human beings require energy. In cellular respiration, chemical energy is transformed into heat and mechanical energy. This process requires oxygen and produces carbon dioxide as a waste product. The respiratory system provides the gas exchange that is necessary for cellular respiration.

20.3 Respiratory System

Objectives

- **Explain** the function of the ribs in inhaling and exhaling.
- **Describe** the flow of air through the respiratory system.
- **Predict** how the respiratory system is affected by cigarette smoking.

Whether you are playing football, cheering in the stands, or playing in a band, you need energy. The food you eat is a source of energy. In order for your body to release energy from the food you eat, you need oxygen. Your cells use oxygen to "burn" the molecules that your digestive system supplies.

Breathing and Air Pressure

The air around you is about one-fifth oxygen. You get the oxygen you need by breathing in air. Organs that help with breathing make up the respiratory system.

Put your hands on either side of your ribs. Take a deep breath and hold it a few seconds. What happens? When you inhale, your ribs move up and out as air moves into your lungs. When you exhale, your ribs move down and in as gases move out of your lungs.

Look at Figure 20.11. Your ribs move up and out when you inhale because the muscles between your ribs are contracting. Below your ribs and lungs is a large, flat muscle called the diaphragm (DY uh FRAM). When you inhale, your diaphragm contracts and moves down. The contracting diaphragm works with the contracting muscles between your ribs to increase the space in your lungs. When the space inside your lungs increases, the air pressure inside your lungs decreases. Because the air pressure outside your lungs is greater than inside your lungs, air moves into your lungs.

When you exhale, the contracted rib muscles and diaphragm relax. Your ribs move down and in, the diaphragm moves up, and the space in your lungs decreases. The air pressure inside your lungs is greater than the air pressure outside. In order to equalize the air pressure, air moves out of your lungs.

Skills WarmUp

Estimate Count the number of times you breathe in one minute right now. Then estimate your breaths for the same amount of time for other daily activities, such as after you eat, when you are studying hard, and after you do some exercise. Write down your estimates. Later, find out the exact numbers. How close were you?

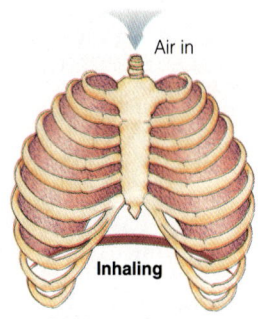

Air in

Inhaling
Diaphragm contracts

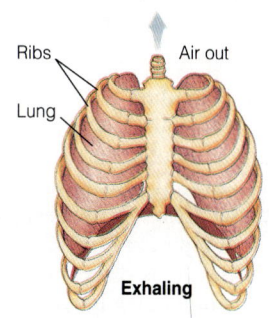

Ribs — Air out
Lung

Exhaling
Diaphragm relaxes

Figure 20.11 ▲
The diaphragm contracts when you inhale. What happens to the diaphragm when you exhale? ②

SECTION 20.3

Section Objectives
For a list of section objectives, see the Student Edition page.

Skills Objectives
Students should be able to:

Estimate the number of breaths they take during different activities.

Compare photosynthesis in plants to gas exchange in animals.

Predict the effects of smoking on the respiratory system.

Measure their lung capacities.

Vocabulary
bronchi, alveoli

MOTIVATE

Skills WarmUp
To help students understand how the number of breaths they take in a minute is affected by activity, have them do the Skills WarmUp.
Answer Students will notice that their respiration rates are affected by their activity level.

Prior Knowledge
Begin by asking students to put their hands on their ribs and take a deep breath. Ask the following questions:

- What do you feel when you take a deep breath?
- Which parts of your body help you breathe?
- What happens to the air you inhale?
- Has it ever been hard for you to breathe? Talk about how that felt.

Answer to In-Text Question
② When you exhale, the contracted rib muscles and diaphragm relax.

Chapter 20 Supply and Transport 429

TEACH

Directed Inquiry

Have students study Figure 20.12. Ask:

▶ How does air enter the respiratory system? (Through the nose and/or the mouth)

▶ How is air filtered in the nasal passages? (Hairs and mucous membranes filter particles that are in the air.)

▶ What structure connects the nose and mouth to the trachea? (The pharynx)

▶ Place your hand on your throat. Hum softly. Describe what you feel. What is it? (Students should feel vibrations from the larynx.)

▶ What is the purpose of the epiglottis? (It closes the trachea when a person swallows. This keeps food from entering the lungs.)

▶ What is the trachea lined with? (Cilia)

▶ How does air pass from the trachea into the lungs? (It passes through the bronchi.)

Skills Development

Infer Ask students how oxygen gets from the lungs to the cells throughout the body. (Gas exchange takes place between the alveoli and the capillaries surrounding the alveoli.)

**The Living Textbook:
Life Science Side 7**

Chapter 5 Frame 00624
Respiratory System Slides (44 Frames)
Search: Step:

430

INTEGRATED LEARNING

Language Arts Connection

Often the derivation of a particular word gives insight into its current use. Have students find the roots and origins of the words *pharynx, trachea, bronchi, alveoli,* and *larynx*. They could share their findings.

Path of Air

Each time you take a breath, air travels through the organs of your respiratory system. Figure 20.12 traces the path of air, which begins in the nose.

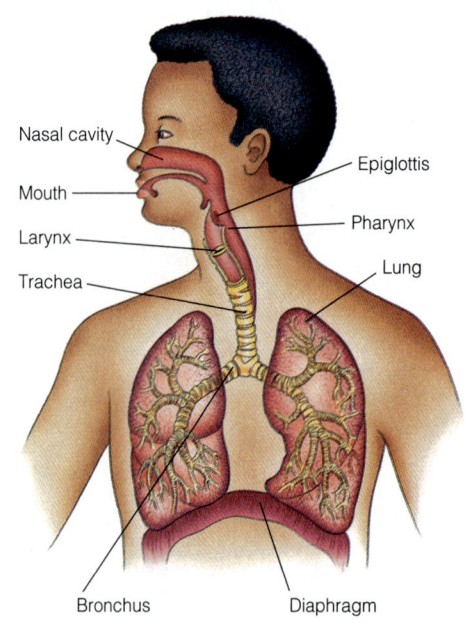

**Figure 20.12 ▲
The Human Respiratory System**

Nose The main entrance and exit to your respiratory system is your nose. The nose is connected to the winding passages that warm and filter the air as it passes through the respiratory system.

When you breathe in, hairs in your nasal passages filter dust, pollen, and other small particles from the air. These particles also stick to the mucous membranes that line your nasal passages. If they build up, sneezing forces them out.

Tiny blood vessels in the nasal passages warm the air as it passes through. The air also collects moisture in the nasal passages.

Sometimes breathing through your nose can be difficult, such as when you have a cold. So you breathe through your mouth. When you run or exercise, you need to take in air at a faster rate. Your air supply is increased by breathing through your mouth and nose.

Pharynx Another name for the throat is the pharynx (FAIR ihnks). The pharynx is like a funnel that leads from your nose and mouth to your windpipe. Locate the pharynx in Figure 20.12. The pharynx is lined with cilia. Cilia are tiny hairlike parts that help filter air on its way to your lungs.

Trachea Air travels from your pharynx to a tube called the trachea (TRAY kee uh). Another name for the trachea is the windpipe. This ringed tube is about 10 cm long and 1.5 cm wide.

The trachea is a tough, flexible passageway that air can move through all the time. When you move your head and neck, the trachea twists and stretches. Notice the epiglottis in Figure 20.12. The epiglottis closes off the trachea when you swallow. This keeps food from entering your lungs. The epiglottis opens when you breathe or talk.

At the top of the trachea is the larynx (LAR ihnks), or voice box. Air passing over the vocal cords in the larynx produces sounds. You form these sounds into speech with your mouth.

Like the pharynx, the trachea is lined with cilia. The cilia continue to clean air as it passes by. The bottom of the trachea branches into two narrow tubes called **bronchi** (BRAHN kee). Each bronchus leads into one lung. Locate the bronchi in Figure 20.12.

430 Chapter 20 Supply and Transport

Themes in Science

Scale and Structure The structure and number of alveoli greatly increases the surface area of the lungs. Even though one lung is the size of a football, the 300 million alveoli make its surface area much greater. This large surface area is important for gas exchange.

Integrating the Sciences

Earth Science Explain to students that atmospheric air, which is the air they breathe, is a mixture of gases. Nitrogen makes up 78 percent of atmospheric air, oxygen makes up 21 percent, and trace gases such as carbon dioxide, water vapor, helium, methane, and argon make up a total of 1 percent. Have students show the percentages in a circle or bar graph.

Critical Thinking

Compare and Contrast Remind students that the process that releases energy in cells is called respiration. Have them look up cellular respiration in the index and review the process. Have students use their own words to describe the difference between breathing and respiration. (Breathing is the process that allows for the exchange of carbon dioxide for oxygen in the lungs. Respiration, which takes place in the cells, transforms chemical energy into heat and mechanical energy.)

Class Activity

Have students locate where their ribs end on each side of their chests. Tell them that the diaphragm is here and that the lungs are just above it. Have them breathe deeply and feel their ribs move upward and outward. Explain that the ribs act to encase and protect the lungs. Have them feel their ribs again in front and in back as they breathe deeply. Ask them to identify the parts of the body that move during breathing. (The diaphragm, lungs, and ribs)

Integrated Learning

Use Integrating Worksheet 20.3a.

Class Activity

Obtain some limewater. Have each student exhale into limewater through a straw and observe what happens. (It becomes cloudy.) Ask them if they can explain what caused the water to become cloudy. (Carbon dioxide in the exhaled air caused the cloudiness.) Ask them where the carbon dioxide comes from. (From the body's cells)

Lungs

The oxygen you inhale with air is exchanged for waste gases from the body in the lungs. Inside your lungs, the bronchi branch into thousands of tiny bronchioles (BRAHN kee OHLS). Each bronchiole ends in a cluster of tiny air sacs called **alveoli** (al VEE uh LY).

The bronchioles and alveoli look like the branches of a tree. The biggest branches are the bronchi. Bronchi are covered by cilia and a thin film of mucus. Dust and pollen are trapped by the mucus before they can reach the alveoli.

There are about 300 million alveoli in each lung. Although each lung is only about the size of a football, the large number of alveoli greatly increases the surface area of the lungs. If the total surface of the lungs were spread out flat, it would cover a tennis court! This large surface area is important for the quick exchange of oxygen and waste gases.

Exchange of Gases

As oxygen moves into the bloodstream, carbon dioxide moves into the alveoli. The gases exchange places by diffusion. Look at Figure 20.13. Notice that the capillaries surround the alveoli. The alveoli are coated with a moist film. Oxygen dissolves in this film and diffuses into the capillaries. The red blood cells in the capillaries release the carbon dioxide and replace it with oxygen. The carbon dioxide diffuses out of the capillaries and into the moist film on the alveoli. The carbon dioxide is then exhaled from the lungs.

Every cell of your body needs oxygen for respiration to take place. After the blood receives oxygen from your lungs and releases carbon dioxide to the lungs, the blood flows back to the heart. The heart pumps this oxygen-rich blood into all your cells. Carbon dioxide diffuses from the cells into the blood.

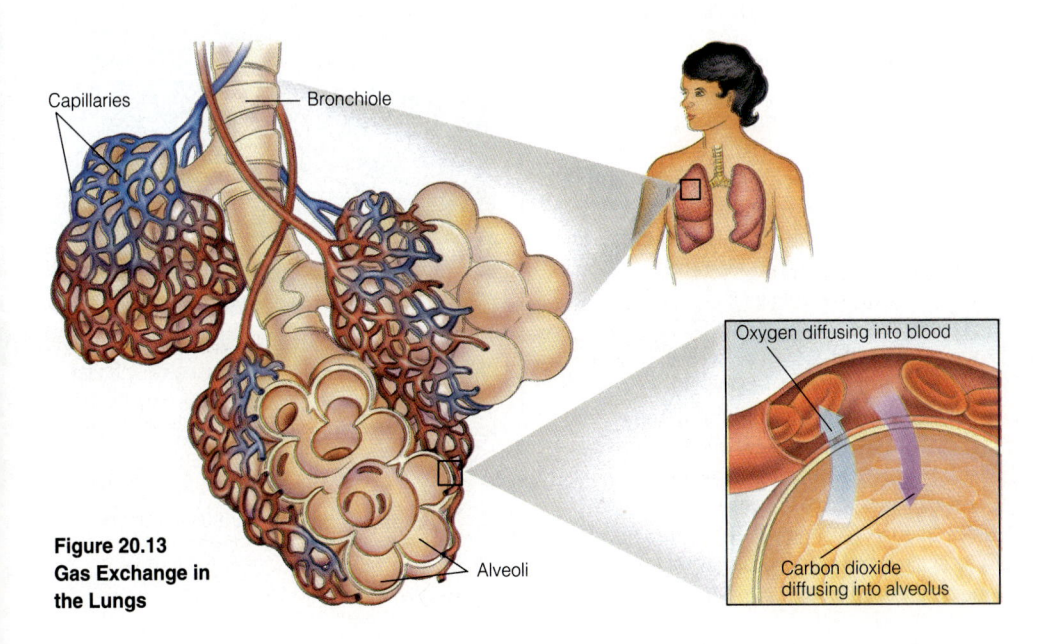

Figure 20.13 Gas Exchange in the Lungs

Chapter 20 Supply and Transport

TEACH • Continued

Discuss
Point out that respiratory disorders are quite common. Ask for a show of hands to see how many students have had a cold in the past year. (Probably everyone will have had a cold.) Find out how many have had asthma, bronchitis, or pneumonia. (Some will have or have had these disorders.) Explain that people can become seriously ill with these disorders. Ask students what conditions can make respiratory problems especially dangerous. (Such ailments are dangerous if not properly treated or if the lungs and body are in a weakened condition because of smoking, drug abuse, poor eating habits, lack of sleep, or illness.)

Historical Notebook
Inform students that snorkeling was popularized in America in the 1930s and the aqualung was invented in 1943.

Answers
1. Snorkel, diving bell, aqualung
2. A snorkel is a long hollow tube that is curved at the bottom and fitted with a mouthpiece. The diver breathes through this submerged end, being sure to keep the top of the snorkel above water.

Portfolio
Have students write an imaginative personal narrative in which they tell about a snorkeling, diving bell, or scuba diving experience. Ask them to use descriptive language and to describe how they are able to observe underwater activities through the aid of their equipment.

INTEGRATED LEARNING

Integrating the Sciences

Oceanography Although many areas of the oceans have not been explored, technological advances have made exploration of the oceans easier. Many different ocean exploration vehicles have been invented throughout the years. Have students find out when each of the following ocean exploration vehicles was invented and how each works: bathysphere, bathyscaphe, diving bell, Alvin, Jason, and scuba. During a class discussion, have students place the vehicles in chronological order of invention.

Respiratory Disorders

The function of the hairs, cilia, and mucous membranes in the respiratory system is to filter out substances that might be harmful to your lungs. Sometimes harmful substances get past these filters and cause respiratory problems. The most common respiratory problem is the common cold, caused by any one of 100 viruses.

Asthma is a respiratory problem that occurs when muscles in the bronchi and bronchioles contract. When this happens, the size of the airways into the lungs is reduced. This narrowed airway makes breathing difficult and causes wheezing. Asthma can be brought on in a number of ways, including allergies and extreme cold.

Bronchitis is an inflammation of the mucous membranes that line the bronchi. Bronchitis causes coughing and high fevers. Pneumonia is a respiratory disorder that infects the lungs. It is usually caused by bacteria. Symptoms include chest pains and coughing.

Historical Notebook

From Free Diving to Scuba

Finding methods of underwater exploration has interested people for many centuries. As early as 4500 B.C., divers in the Mediterranean Sea held their breath and dived for pearls and sponges. By 100 A.D., divers in Greece and Italy were using hollow reeds as snorkels. The diver could search the sea floor underwater for a long time by breathing through a snorkel, which has one end above water.

The first apparatus used for breathing underwater was the diving bell. In 1690 Edmund Halley built the first diving bell that held more than one person. Halley's wooden bell had glass portholes. Oxygen was supplied by two barrels.

The first safe underwater breathing device was the *aqualung*, which was invented by Jacques-Yves Cousteau, a naval officer, and Emile Gagnan, an engineer. The aqualung, which is an air tank and regulator, is part of the scuba equipment used today.

Scuba, which stands for self-contained underwater breathing apparatus, has an air tank, goggles, snorkel, fins, and a weight belt. Scuba equipment enables a diver to explore the ocean depths for long periods of time.

1. Name two kinds of apparatus used for breathing underwater.
2. Explain how a snorkel works.

Chapter 20 Supply and Transport

STS Connection
Many schools have had asbestos-removal programs. Have students find out how workers safely remove asbestos from buildings.

Integrating the Sciences
Botany Indoor pollutants come from plastics, paints, and perfumes. Explain to students that researchers have found that houseplants not only remove carbon dioxide from the air but also remove other, more harmful chemicals. You may wish to encourage students to bring houseplants to keep in the classroom.

Use Integrating Worksheet 20.3b.

Science and Society *Indoor Air Pollution*

In the late 1970s, many people found ways to make their homes and businesses more energy-efficient. However, they did not know that they were contributing to air pollution. For example, many people put better insulation and tighter seals around windows and doors to save energy. Unfortunately, these devices also lessen the flow of natural air into and out of a building.

Studies by the Environmental Protection Agency (EPA) show that indoor air often contains more pollutants than outdoor air. Examples of pollutants include aerosol sprays, cleaning products, paint products, and air fresheners. Concentrations of these pollutants can be as much as 100 times higher inside the building than in the open air.

Some indoor pollutants cause itchy eyes, nausea, and headaches. Other pollutants are linked to diseases that may take years to develop, such as cancer. Thirty of the indoor air pollutants identified by the EPA are known to cause cancer. These substances include asbestos and a radioactive gas called radon. Radon seeps into buildings from underground. There are ways to check for radon and asbestos in buildings. In many communities, asbestos is being removed from schools.

To combat indoor air pollution, you can be more careful about using household products that contain harmful chemicals. A lot of people are using natural products instead. Many office workers are getting better ventilation in their offices by using desktop air purifiers. Smoking is no longer allowed in many buildings.

Figure 20.14 ▲
Asbestos is a material that has been used for insulation and fireproofing in many older buildings. It has been linked to lung cancer. Here, workers begin removing asbestos.

Check and Explain

1. Describe how air moves through the respiratory system.
2. What happens to the diaphragm when you breathe in and out? What do you think happens to the diaphragm when you have the hiccups?
3. **Compare** Write how the process of photosynthesis in plants compares to the exchange of gases in animals.
4. **Predict** What respiratory problems might a person who smokes be more likely to develop than a nonsmoker? How do you think smoking can make some respiratory problems worse?

EVALUATE

WrapUp
Reinforce Have students work in small groups to make three-dimensional models of the respiratory system. Provide art supplies and items such as milk containers, bags, balloons, and paper-towel tubes. Have students use their models to demonstrate how the respiratory system works.

Use Review Worksheet 20.3.

Check and Explain

1. Air enters through the nose and travels through the pharynx and trachea to the bronchi and lungs.
2. When a person inhales, the diaphragm contracts, which increases the space in the lungs. The diaphragm relaxes when a person exhales, decreasing the space in the lungs. When a person has the hiccups, the diaphragm contracts quickly, causing the person to take a short gasp of air.
3. In photosynthesis, plants take in carbon dioxide and give off oxygen. Animals take in oxygen and release carbon dioxide.
4. People who smoke are more likely to develop bronchitis, emphysema, and lung cancer. Many people are allergic to smoke, so asthma can be made worse by smoking or by being around someone else who is smoking.

ACTIVITY 20

Time 60 minutes **Group** 4–5

Materials
- 6 clear plastic drinking jugs
- 6 large cake pans
- 3 m plastic tubing
- 6 grease pencils
- 6 500-mL beakers
- liquid soap

Analysis

1. Answers may vary but should agree with student data tables.
2. Answers may vary but should be consistent with class data. Average capacity for 12-year-olds is in the range of 2 liters.
3. Answers may vary but should be accurate based on group data.
4. Answers may vary but should be consistent with collected data.

Conclusion

Lung capacity is affected by body size—a larger person will have greater lung capacity; level of fitness—a person who exercises regularly will have greater capacity; and health—asthma or other respiratory diseases can decrease lung capacity, as can smoking.

Extension

Answers may vary. Comparing lung capacity measurements over a period of time could expose lung problems not detectable by other means or could show progress during the treatment of an illness. A device called a spirometer is commonly used to measure lung capacity.

TEACHING OPTIONS

Prelab Discussion

Discuss the following questions:

▶ Why it is important that the jug be completely filled with water before each trial?

▶ What happens when a person exhales into the tube? Why does this allow measurement of lung capacity?

▶ What are some other methods of determining lung capacity? How do they compare with this method?

▶ Can all the air in one's lungs be exhaled at one time?

Activity 20 How does lung capacity vary?

Skills Measure; Collect Data; Interpret Data

Task 1 Prelab Prep
1. Collect the following items: a large, clear plastic drinking jug, a large cake pan, a half-meter length of plastic tubing, a grease pencil, a 500-mL beaker, liquid soap.
2. Find a large sink or tub.

Task 2 Data Record
1. On a sheet of paper, draw the table shown. Write your name at the top. Also write the names of people whose lung capacity you plan to measure.
2. After you measure the lung capacity of each person listed, record the volume in the table.

Table 20.3   Measuring Lung Capacity

Name	Lung Capacity

Task 3 Procedure
1. Fill the jug to the top with water.
2. Put the cake pan in the sink. Completely fill the pan with water.
3. Hold your hand tightly over the opening of the jug.
4. Turn the jug upside down over the sink. Put the upside-down jug in the cake pan so the opening is underwater. Remove your hand.
5. Slip one end of the tubing into the jug. Hold on to the jug so it doesn't fall over. Pinch the other end of the tubing.
6. Take a deep breath. Release the pinch at the end of the tubing. Exhale all the air in your lungs into the tubing. Quickly pinch the tubing shut again.
7. Use a grease pencil to mark the water level on the side of the jug.
8. Put your hand over the opening of the jug and remove it from the pan.
9. Pour out the water in the jug. Refill the jug with water up to the line you marked.
10. Use the beaker to measure the amount of water in the jug. Record this volume amount as your lung capacity.
11. Repeat steps 1 to 10 to determine the lung capacity of each person. Be sure to wash the tubing with soap and water between each use.

Task 4 Analysis
1. What is your lung capacity?
2. How does your capacity compare to the lung capacities of your classmates?
3. Find the average lung capacity of your group.
4. Create a chart that shows the lung capacity of each person in your group.

Task 5 Conclusion
What factors affect a person's lung capacity? For example, how might the lung capacity of a person who exercises a lot be different from someone who doesn't exercise very often? Explain why lung capacity varies among people.

Extension
Explain how measuring lung capacity could be used in the diagnosis and treatment of some respiratory illnesses. Find out actual methods used to measure lung capacity.

INTEGRATED LEARNING

Themes in Science

Systems and Interactions The parts of the circulatory system function to circulate materials throughout the body. The organs in the digestive system function together to break down food into nutrients and waste products. The parts of the respiratory system work together to provide oxygen and remove carbon dioxide from cells. The parts of the excretory system rid an organism of waste. As you finish this chapter, stress that all of the systems in the human body interact with one another to support life.

20.4 Excretory System

Objectives

▶ **Describe** how the skin rids the body of wastes.

▶ **Describe** how the kidneys remove wastes from the body.

▶ **Define operationally** the path of urea from the liver through its excretion from the body.

You know that the lungs remove carbon dioxide from your body. Your cells produce other wastes as well. For example, when your tissues contain too much salt and water, they become wastes. If wastes are not removed, or excreted, they build up to dangerous levels in your body.

Your liver and skin are excretory organs. They play an important part in removing certain wastes from your body. However, the main organs of excretion are the kidneys.

Liver

You learned about some of the functions of the liver when you read about the digestive system. During metabolism, wastes with nitrogen are released. These wastes are toxic to the body. Your liver combines the wastes with carbon dioxide to form urea. Urea is a nitrogen compound that is poisonous in large amounts and must be removed from your body. Urea enters the bloodstream through the capillaries in the liver. The circulatory system carries urea to your kidneys, where it is separated from the blood and flushed from your body in urine. Urine is made up of urea, water, and excess salts.

Skin

You know that your skin has several layers. Within your skin layers are sweat glands and hair follicles. Wastes and excess water are removed from your blood by the sweat glands. The waste and water form sweat, which is released onto the skin surface through pores in your skin.

Did you ever notice that if sweat from your face gets into your mouth, it tastes salty? Along with excess water, salt is removed from your blood in the form of sweat.

Skills WarmUp

Infer Set up this simple activity. You'll need a piece of filter paper that is cone-shaped, a funnel, a measuring cup, a clear cup, water, and sand. Put sand and about 200 mL of water into the measuring cup. Put the filter paper into the funnel. Pour the sand-water mixture into the empty cup through the filter. What happens? How do you think this is like your kidneys?

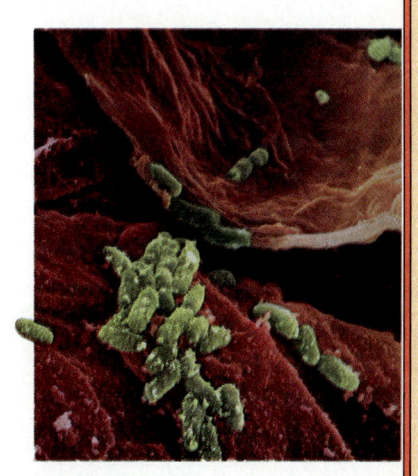

Figure 20.15 ▲
The green objects shown are bacteria in a pore of skin. Why is the skin an excretory organ? ①

SECTION 20.4

Section Objectives
For a list of section objectives, see the Student Edition page.

Skills Objectives
Students should be able to:

Infer how the kidneys work by using a model.

Classify organs according to which systems they are part of.

Collect Data about lung capacities.

Define Operationally the process by which the waste from the liver is excreted.

Vocabulary
nephron

MOTIVATE

Skills WarmUp
To help students understand how the kidneys function, have them do the Skills WarmUp.
Answer The kidneys act as filters, removing wastes from the blood.

Prior Knowledge
To gauge students' understanding of the excretory system, ask the following questions:

▶ How does your body get rid of carbon dioxide?

▶ What other wastes does your body need to remove?

▶ How does your body remove those wastes?

Answer to In-Text Question

① **The skin is an excretory organ because some wastes and excess water are released from the body through pores in the form of sweat.**

Chapter 20 Supply and Transport

TEACH

Discuss

Have students study Figure 20.16 and trace the flow of materials from the kidney to the outside of the body. Ask students the following questions:

▶ Why don't blood cells and proteins enter the capillaries around the renal artery? (They are too large.)

▶ What substances enter the capillaries around the renal artery? (Water, some salts, and nutrients)

▶ What does the bladder do? (It stores up to 1 liter of urine.)

Critical Thinking

Reason and Conclude Point out that the renal artery branches off into a network of capillaries. Blood cells and proteins are too large to enter these capillaries. Ask students how they think the capillaries in the kidneys compare in size to the capillaries in a muscle or in a lung. (The capillaries in the muscles and in the lungs are larger.)

Answers to In-Text Questions

① Water, nutrients, and some salts must reenter the bloodstream to supply the cells of the body with these materials.

② Urine leaves the body through the ureter. The function of the nephron is to filter wastes out of the blood coming to the kidney.

The Living Textbook:
Life Science Side 7

Chapter 5 Frame 00668
Excretory System Slides (28 Frames)
Search: Step:

INTEGRATED LEARNING

STS Connection

Many mechanical systems involve systems of filtration. Divide the class into work groups and have each group make a poster or chart comparing the kidney filtration system to a mechanical filtration system. Student groups could investigate sewage treatment systems, air conditioning systems and ionizers, automobile engine lubricant filter systems, or water softening systems. Charts and posters should include information about each system's structure, what is being filtered, and how it is being filtered.

Kidneys

Reach back and touch your lower back, just above your waist. You have found approximately where your two kidneys are located. They are about 10 cm long and shaped like large beans.

Each time your heart beats, about 20 percent of your blood enters the kidneys through the renal artery. As you can see in Figure 20.16, the artery branches off into a network of capillaries. Some parts of blood, such as blood cells and proteins, are too large to enter the capillaries. They remain in the bloodstream. Other substances, such as water, some salts, and nutrients, enter the capillaries. They are filtered by tiny structures called **nephrons** (NEHF rahnz). You can see a diagram of a nephron in Figure 20.16. The substances pass through the collecting tubes in the nephron. They are then absorbed back into the bloodstream and leave through a vein. Why is it necessary for water, nutrients, and some salts to reenter your bloodstream? ①

The liquid that remains in the collecting tube of the nephron is urine. Notice the ureter in Figure 20.16. Here urine flows from the core of the kidney into the urinary bladder. The bladder is a muscular sac of tissue. The bladder will store about 1 L of urine. When the bladder is full, its strong muscular walls squeeze urine

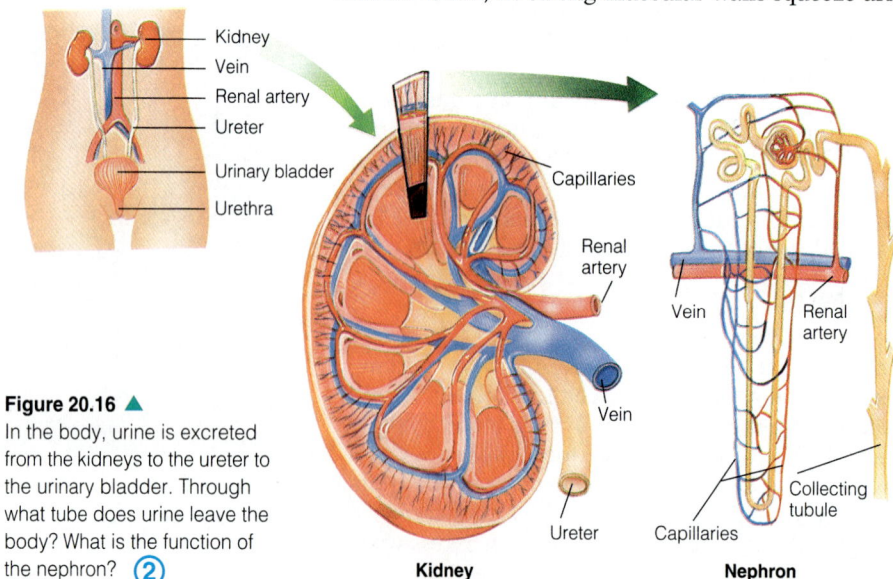

Figure 20.16 ▲
In the body, urine is excreted from the kidneys to the ureter to the urinary bladder. Through what tube does urine leave the body? What is the function of the nephron? ②

436 Chapter 20 Supply and Transport

Integrating the Sciences

Physical Science Have students work in small groups to make a classroom display about the uses of ultrasound. Some students can report on the nature of sound waves and how ultrasound waves are different from the sound waves that people can hear. Others can report on the medical uses of ultrasound waves. Encourage students to write brief reports and provide pictures and make diagrams for the display.

Discuss

Spend some time talking with students about the symptoms of excretory disorders. Remind them about the location of their kidneys. Any pain in their kidneys could be a symptom of an excretory disorder. Difficulty in urinating or a feeling of burning when urinating can also be symptoms. Health workers can test for urinary tract infections through a process called urinalysis. Stress the importance of treating these disorders.

Skills WorkOut

To help students make analogies about the structure and function of the kidneys, have them do the Skills WorkOut.

Answers The filter paper represents the nephrons in the kidneys. The water to be filtered represents the blood that enters the kidneys through the renal artery. The clear water that passes out of the filter represents the substances that pass through the collecting tubes in the nephrons, which are soon thereafter reabsorbed into the bloodstream.

---

out through the urethra. The urethra is a tube that leads from the bladder to the outside of the body. Look at Figure 20.16 and trace the flow of urine to the outside of the body.

Excretory Disorders

Sometimes microorganisms get into the organs of the excretory system. Microorganisms may enter the body through the urethra, other body organs, or the blood. When microorganisms get into the urinary tract, they cause infections. Some infections can be treated by drinking a lot of water, resting, changing your diet, or taking antibiotics. Other infections are more serious.

When the nephrons of the kidneys become infected, a disease called *nephritis* may result. Repeated infections can damage the kidneys to the point that they can't remove wastes anymore. A person with severe kidney damage may have to be connected to an artificial kidney. If the kidneys no longer function, a transplanted kidney may be necessary. Fortunately, one kidney can do the work of two.

Another kidney disorder occurs when salts in the urine form crystals, or kidney stones. They are usually the result of frequent bacterial infections. If the kidney stones are too large to pass in the urine, they can lodge in the ureter and keep urine from leaving the kidney. Kidney stones can be very painful. They are often treated with a technique called *ultrasound*. The ultrasound vibrations break up the kidney stones the same way that a certain sound pitch can shatter glass. Once broken up into small crystals, the kidney stones can easily pass through the ureter.

Skills WorkOut

Make Analogies Refer back to the activity you did in the *Skills WarmUp* on page 435. What part of the human body best represents the filter paper? What body part best represents the water to be filtered? What body part best represents the clear water that passes out of the filter?

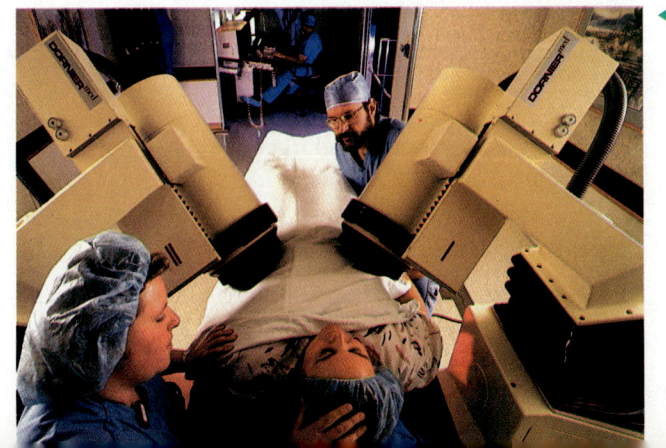

◀ **Figure 20.17**
Today, ultrasound is often used to break apart most large kidney stones so they can be passed out of the body.

TEACH • Continued

Research
Have students use reference materials to find out why some people have to have dialysis to remove wastes from their blood. Have students find out what can be done to prevent a person from needing dialysis.

Skills WorkOut
To help students understand how to classify organs, have them do the Skills WorkOut.
Answer Answers may vary. Lists could include organs such as the heart, lungs, stomach, kidneys, and others named in this chapter. Organs that are part of the same system can be grouped together. One organ that can be grouped with two groups is the mouth, because it aids in the digestion of food and it is a passageway for air.

EVALUATE

WrapUp
Portfolio Have students draw a concept map that shows how water, oxygen, and carbon dioxide are changed in the human body. They can keep these pages in their portfolios.
Use Review Worksheet 20.4.

Check and Explain
1. Urea, water, and salt; the kidney keeps a correct balance of needed salts and minerals.
2. Wastes such as ammonia are poisonous.
3. Water combines with urea to form urine and aids in carrying wastes out of the body.
4. Waste from the liver goes into the kidneys, where it is separated from the blood and flushed from the body in the form of urine.

INTEGRATED LEARNING

STS Connection
Many people believe that there is a serious health care problem in the United States. Have students use newspaper and magazine articles to write a report about the pros and cons of a national health care system. Have them include information about the present costs of major medical procedures, acute as well as chronic (such as dialysis), and have them offer their personal opinions as to which viewpoints they support and why.

Skills WorkOut
Classify Make a list of body organs you have read about in this chapter. Find ways to group the organs based on how they are alike and different. Think about their locations, functions, and so on. Which organs fit in more than one group? Why?

Science and Technology
Dialysis—The Amazing Machine

Sometimes a person's kidneys fail to function at all. However, a person whose kidneys fail is still able to lead a normal life, thanks to the kidney dialysis machine. A kidney dialysis machine takes over the functions of the kidneys and removes all waste products from the blood.

During dialysis, a person's blood flows into special tubing. The tubing is made of a material that allows only certain substances to diffuse through its walls. Inside the dialysis machine, the tubing is surrounded by a special fluid. Wastes and other unneeded substances in blood diffuse from the tubing into the fluid. Certain substances from the fluid diffuse into the blood. The blood then flows back into the person. The dialysis process takes six to eight hours. A person might need dialysis two or three times a week.

In some cases, temporary kidney failure is caused by sudden trauma or illness. The patient may have dialysis done in the hospital until the kidneys heal and no longer need the dialysis.

When kidney failure is permanent, dialysis may be done in a hospital or elsewhere. A new type of dialysis machine was developed in recent years. This machine is cheaper, less traumatic, and requires less time. Most importantly, kidney dialysis patients are able to perform the procedure themselves. They hook up to the machine at bedtime and let the machine work through the night. This amazing machine allows people to go on with their daily lives.

Check and Explain
1. What substances does the kidney remove from the body? What substances does it keep?
2. Explain why death would occur if a person's kidneys no longer functioned, and treatment on a dialysis machine was not possible.
3. **Infer** Why is water important for the excretory system to function properly?
4. **Define Operationally** Explain the process by which the waste from the liver is excreted from the body.

Chapter 20 Review

Concept Summary

20.1 Digestive System
- Digestion is the process of breaking down food into particles small enough to cross cell membranes.
- Mechanical digestion involves mashing and grinding food. Chemical digestion occurs when food is broken down into simpler molecules with the aid of enzymes, acids, and bases.
- The main parts of the human digestive tract are the mouth, esophagus, stomach, small intestine, large intestine, and rectum.

20.2 Circulatory System
- The heart serves as a muscular pump that moves blood through the vessels of the circulatory system.
- The left side of the heart pumps blood to the organs of the body. The right side of the heart pumps blood to the lungs.
- Blood has liquid and solid parts. Blood transports oxygen, carbon dioxide, nutrients, and wastes. Blood also helps the body fight disease and repair damaged blood vessels.
- The lymph system returns fluid to the capillaries and aids in nutrient absorption and body defense.

20.3 Respiratory System
- The diaphragm and rib muscles work together during breathing to change the air pressure in the lungs. Air moves into and out of the lungs through the nose or mouth, pharynx, trachea, and bronchial tree.
- Gas exchange occurs in the alveoli and body cells by diffusion.

20.4 Excretory System
- In humans, the lungs, liver, skin, and urinary system all aid in excretion.
- The kidneys remove wastes carried by the blood and restore nutrients.

Chapter Vocabulary

digestion (20.1)	chemical digestion (20.1)	capillary (20.2)	alveoli (20.3)
nutrient (20.1)	villi (20.1)	vein (20.2)	nephron (20.4)
mechanical digestion (20.1)	artery (20.2)	bronchi (20.3)	

Check Your Vocabulary

Use the vocabulary words above to complete the following sentences correctly.

1. Blood leaving the heart travels through ____ to all organs and muscles.
2. In the kidneys, a ____ is a system of tubes acting as a filter.
3. Gas exchange occurs in the tiny air sacs of the lungs called ____.
4. Carbohydrates, proteins, and fats are three important ____ in food.
5. Your teeth cutting, grinding, and mashing food are examples of ____.
6. The smallest kind of blood vessel is a ____.
7. Tiny fingerlike projections, or ____, line the walls of the small intestine.
8. Blood leaves the organs and travels through ____ back to the heart.
9. The breakdown of food into simpler parts that your body can use is called ____.
10. Air enters a lung after having passed through the ____.
11. Acids, bases, and enzymes change food during ____.

CHAPTER REVIEW 20

Check Your Vocabulary

1. arteries
2. nephron
3. alveoli
4. nutrients
5. mechanical digestion
6. capillary
7. villi
8. veins
9. chemical digestion
10. bronchi
11. digestion

Use Vocabulary Worksheet for Chapter 20.

CHAPTER 20 REVIEW

Check Your Knowledge

1. Food gets broken into smaller pieces through mechanical digestion. Food molecules are converted into simpler molecules by chemical digestion.
2. The first number, 120, is the pressure when the heart is contracting; the second, 80, is the pressure when the heart is relaxed.
3. Answers may vary. Possible answers are controlling weight, refraining from smoking, decreasing fat intake, exercising regularly, and controlling blood pressure.
4. Oxygen and carbon dioxide dissolve in this moist film and diffuse between the alveoli and the capillaries.
5. Answers may vary. Possible answers appear in Table 20.1.
6. After leaving the right side of the heart, blood flows to the lungs.
7. The lungs, kidneys, and skin all remove wastes from the body.
8. The four blood types are A, B, O, and AB.
9. pulmonary circulation
10. digestive
11. respiratory
12. urine
13. plasma
14. small intestine

Check Your Understanding

1. a. Muscle action mixes the food and moves it along the digestive tract.
   b. The villi in the small intestine provide a large surface area for absorption of nutrients.
   c. Enzymes are necessary for the chemical digestion of many nutrients.
2. a. Ureter—carries urine to the bladder
   b. Platelet—clots blood
   c. Lung—exchanges gases
   d. Stomach—digests proteins
   e. Lymph node—traps viruses and bacteria
3. Carbon dioxide is produced by cellular respiration. It leaves cells, diffuses into plasma, flows through blood vessels to the heart, moves on to the lungs where it diffuses into air sacs, and is finally exhaled.
4. Red blood cells carry oxygen. White blood cells protect the body against disease. Platelets take part in blood clotting. Plasma is the watery fluid that acts as a solvent and contains many proteins.
5. Inhaled air is richer in oxygen. Exhaled air is richer in CO_2.
6. A pulse is difficult to feel in a vein because blood flowing through veins does not have sufficient pressure to cause the vein to expand.
7. The correct order of steps is d, b, f, c, a, e.

Develop Your Skills

1. a. urine
   b. More water is excreted during heavy exercise in the form of sweat as the body attempts to cool itself.
   c. You will become dehydrated because you lose approximately 3 L of water per day in hot weather.
2. a. Sleeping; jogging and swimming
   b. Sleeping; jogging and swimming
   c. As energy expenditure increases, so does the need for oxygen.

Chapter 20 Review

Check Your Knowledge
Answer the following in complete sentences.

1. Name the two ways food may change during digestion.
2. Suppose your blood pressure was 120 over 80. What do the two numbers mean?
3. Name three things you can do to help prevent cardiovascular disease.
4. Why are alveoli covered with a moist film and surrounded by capillaries?
5. Describe three digestive disorders and list their treatments.
6. Where does blood flow after it leaves the right side of your heart?
7. Name three organs that remove wastes from the body.
8. What are the four blood types?

Choose the answer that best completes each sentence.

9. Blood is oxygenated during (digestion, pulmonary circulation, systemic circulation, excretion).
10. Gallstones, heartburn, and ulcers are all disorders of the (digestive, circulatory, respiratory, excretory) system.
11. The lungs, pharynx, and trachea are all parts of the (digestive, circulatory, respiratory, excretory) system.
12. Excess salts, water, and nitrogen wastes form (sweat, urine, nephrons, urea).
13. Nutrients are carried in the blood by (red blood cells, white blood cells, platelets, plasma).
14. Most nutrients are absorbed in the (gallbladder, small intestine, stomach, liver).

Check Your Understanding
Apply the concepts you have learned to answer each question.

1. How is each of the following important in food processing?
   a. Muscle action
   b. Villi
   c. Enzymes
2. Match the body part on the left with its function on the right.
   a. Ureter   Digests proteins
   b. Platelet   Clots blood
   c. Lung   Traps viruses and bacteria
   d. Stomach   Exchanges gases
   e. Lymph node   Carries urine to bladder
3. **Sequence** Where in the human body is carbon dioxide produced? Describe the path it follows until it leaves the body.
4. **Mystery Photo** The photograph on page 414 shows red blood cells. What is the function of these cells? What are the other parts of blood and their functions?
5. Which is richer in oxygen—inhaled air or exhaled air? Which is richer in carbon dioxide?
6. Can you feel a pulse in your veins? Why or why not?
7. **Sequence** Order the basic steps for food digestion and absorption below.
   a. Water is absorbed.
   b. Food is liquified.
   c. Nutrients pass into bloodstream.
   d. Food is broken into smaller pieces.
   e. Wastes are removed from the body.
   f. Food is digested into nutrient forms that cells can use.

3.

Organ	Function	Nutrients
mouth	mechanical digestion, starch to sugar	starch
esophagus	carries food from mouth to stomach	none
stomach	mixes food with acids, enzymes	proteins
small intestine	continues chemical digestion, absorption	amino acids, sugars, fats
large intestine	absorbs water	water
rectum	collects feces	none

Develop Your Skills

Use the skills you have developed in this chapter to complete each question.

1. **Interpret Data** Use the table below to answer the following questions.
   a. During normal weather, what causes the major loss of water to the body?
   b. How does water excretion change during heavy exercise? Why?
   c. On a normal day, the average person takes in about 2.2 L of water. What will happen to you if you do not increase this amount during hot weather?

Table 20.4 Daily Loss of Water in Humans (in mL)

	Normal Weather	Hot Weather	Extended Heavy Exercise
Lungs	350	250	650
Urine	1400	1200	500
Sweat	450	1750	5350
Feces	200	200	200
Total	2400	3400	6700

2. **Data Bank** Use the information on page 627 to answer the following questions.
   a. Which activity uses the least amount of oxygen per hour? The most oxygen?
   b. Which activity probably needs the least amount of energy? The most energy?
   c. **Infer** Explain the relationship between oxygen use and energy need.

3. **Make a Table** Design a table that lists in order the organs of the digestive system. List the functions of each organ. Identify the types of nutrients digested or absorbed at each stage.

Make Connections

1. **Link the Concepts** Below is a concept map showing how some of the main concepts in this chapter link together. Only part of the map is filled in. Copy the map. Using words and ideas from the chapter, finish the map.

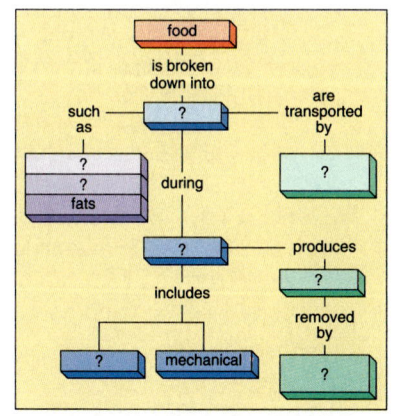

2. **Science and Language Arts** Write a series of metaphors and similes for various human body organs. For example: "The heart is like a geyser that explodes every few seconds."

3. **Science and You** Hold one arm up in the air and let the other arm hang down at your side for two minutes. Compare the color of each hand and how much your veins stand out. What do you think causes the difference?

4. **Science and Technology** Astronauts in space must be able to eat, breathe, and remove wastes from their bodies. Research how scientists meet the oxygen and food needs of astronauts as well as reduce or remove carbon dioxide and food waste buildup.

Chapter 20 Supply and Transport   441

Make Connections

1.

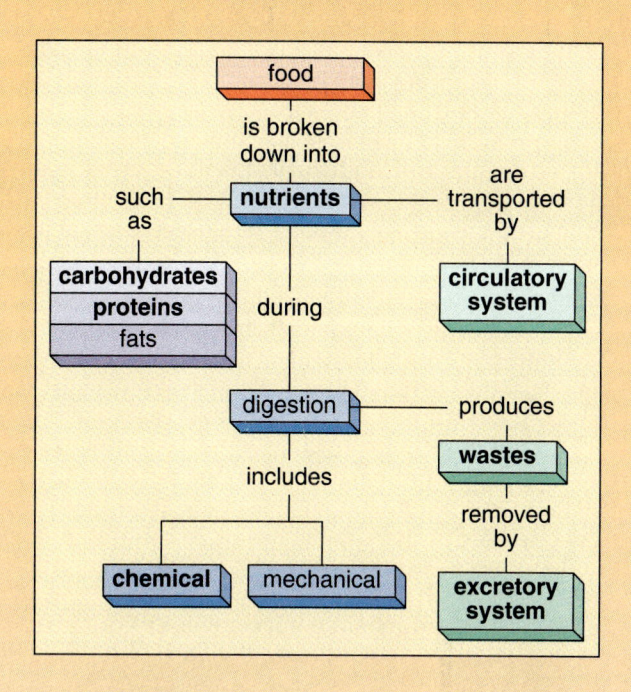

2. Answers will vary.
3. The hand held in the air may become pale in color and the veins should collapse as blood leaves. The hand hanging down may begin to look pinkish, and the veins will begin to bulge as blood collects. The difference is caused by the effect of gravity on the blood; gravity pulls the blood in the raised hand down, while preventing blood from leaving the lowered hand as rapidly.
4. Answers may vary; however, students' answers should take into account the limited atmosphere for gas exchange and the limited space for storage of food and collection of wastes.

CHAPTER 21

Overview

In this chapter, students learn about the human nervous system, senses, and endocrine system and how they work together. The first section explains the information and communication systems of the central and peripheral nervous systems. This section includes descriptions of the parts of the nervous system and information about how these parts function. The second section describes how the senses help the body interact with the environment. Models and discussions of each of the senses included. The last section explores the endocrine system, describing the glands and their functions.

Advance Planner

▶ Invite a speech pathologist to visit the class. TE page 447.

▶ Supply sugar, salt, vinegar, and bittersweet chocolate for TE activity, page 453.

▶ Collect differently textured items and blindfolds for TE activity, page 454.

▶ Supply blindfolds and small, peeled, same-sized cubes of apple, potato, carrot, onion, and turnip for SE page 455.

Skills Development Chart

Sections	Calculate	Collect Data	Hypothesize	Infer	Interpret Data	Observe	Research
21.1 Skills WarmUp Skills WorkOut		●		●			
21.2 Skills WarmUp Skills WorkOut SkillBuilder Activity			●	●	●	●	
20.4 Skills WarmUp Skills WorkOut			●				●

Individual Needs

▶ **Limited English Proficiency Students** Have students write all of the unfamiliar terms from the chapter into their science journals, leaving several blank lines between each term. Have them each find one figure in the text that represents each term. Photocopy these figures and have students paste them into their journals next to the correct term. They can write captions that include the vocabulary terms for each figure.

▶ **At-Risk Students** Invite a doctor or a nurse from a local hospital to talk with students about the electromyelogram (EMG). In this procedure, electrical impulses in the body are measured. Encourage them to prepare for the visit by looking through the chapter summary so they can ask your guest how knowledge about body systems relates to administering and interpreting EMGs.

▶ **Gifted Students** Have students invent a machine that can perform the functions of one of the systems in this chapter. For example, a machine that models part of the nervous system may include electrical wires and batteries. Students can send an electrical impulse through the wires. A small light that flashes can represent a response in the brain. Have students display their machines. Encourage them to write descriptions of how their machines function and how they are similar to the endocrine or nervous system.

Resource Bank

▶ **Bulletin Board** Begin a bulletin board display about the senses. Hang diagrams of the brain, eye, ear, nose, mouth, tongue, and fingertip on the board. Have students cut pictures from magazines that show something about how each sense functions, and add the pictures to the bulletin board.

▶ **Field Trip** Sometime during your study of the human body, you may wish to take students to a museum of science and technology so they can observe exhibits showing how the body functions and how modern technology extends the senses.

CHAPTER 21 PLANNING GUIDE

Section	Core	Standard	Enriched	Section	Core	Standard	Enriched
21.1 Nervous System pp. 443-449				**Blackline Master** Review Worksheet 21.2 Integrating Worksheet 21.2a Integrating Worksheet 21.2b	● ● ●	● ● ●	● ●
Section Features Skills WarmUp, p. 443 Skills WorkOut, p. 444 Career Corner, p. 447	● ● ●	● ● ●	●	**Laboratory Program** Investigation 40		●	●
Blackline Masters Review Worksheet 21.1 Enrichment Worksheet 21.1 Integrating Worksheet 21.1	●	● ●	● ●	**Color Transparencies** Transparencies 65, 66, 67	●	●	
				21.3 Endocrine System pp. 458-462			
Laboratory Program Investigation 34 Investigation 39		● ●	● ●	**Section Features** Skills WarmUp, p. 458 Skills WorkOut, p. 462	● ●	● ●	●
Color Transparencies Transparencies 63, 64	●	●		**Blackline Masters** Vocabulary Worksheet 21 Section Reviewsheet 21.3 Skills Worksheet 21.3	● ● ●	● ● ●	
21.2 The Senses pp. 450-457							
Section Features Skills WarmUp, p. 450 Skills WorkOut, p. 454 Skill Builder, p. 455 Activity, p. 457	● ● ●	● ● ● ●	● ● ● ●	**Laboratory Program** Investigation 41		●	●
				Color Transparencies Transparencies 68	●	●	

Bibliography

The following resources can be used for teaching the chapter. See page T-40 for supplier codes.

Audio-Visual Sources
(video unless noted)
3-2-1 Contact: The Five Senses. 45 min. 1985. GA.
The Brain: Its Wonders and Mysteries. 2 filmstrips with cassettes, 16 min each. 1982. NGSES.
Health Concerns for Today Series. 4 filmstrips. 1984. ME.
The Human Body Series: Nervous System. 20 min. 1988. NGSES.
Marvels of the Mind. 23 min. 1980. NGSES.
The Sense Organs: Our Keys to the World. Filmstrip with cassette. 1990. CABISCO.

Software Resources
The Ear. Apple. MPL.
The Human Body: An Overview. Apple. BB.
Your Body—Series II. Apple, TRS-80, Commodore. FM.

Library Resources
Elting, Mary. *The Macmillan Book of the Human Body.* New York: Aladdin, 1986.
Kuffler, S. W., J. Nicholls, and A. Martin. *From Neuron to Brain.* 2d ed. Sunderland, MA: Sinauer Associates, 1984.
Nilsson, Lennart. *Behold Man: A Photographic Journey of Discovery Inside the Body.* Boston: Little, Brown, and Co., 1974.
Restak, Richard. *The Brain.* New York: Bantam Books, 1984.

CHAPTER 21

INTEGRATED LEARNING

Themes in Science
Systems and Interactions The nervous system works with other body systems to control reflexes and reactions. Have students consider how many systems work together when they sneeze. (The muscular system, nervous system, and respiratory system)

Writing Connection
Have students write a haiku poem in which they compare the photograph to another object, such as a telephone receiver, an ameba, an interstate highway, or a computer chip.

Introducing the Chapter
Have students read the description of the photograph. Ask if they agree or disagree with the description.

Directed Inquiry
Have students study the photograph. Ask:

▶ What do you think this photograph shows? (Students may suggest cells connected by strings or branches.)

▶ How are the cells here different in shape from other cells you have seen? (They have branches that connect them with other cells.)

▶ What everyday things does this picture remind you of? (Students may suggest a road map, plant roots or branches, or a computer circuit board.)

▶ How do you think this photograph is related to the topic of the chapter? (These are cells that are involved with control and sensing.)

Chapter Vocabulary

cochlea	neuron
eardrum	pupil
endocrine gland	retina
hormone	sensory neuron
iris	synapse
motor neuron	

Chapter 21 Control and Sensing

Chapter Sections
21.1 Nervous System
21.2 The Senses
21.3 Endocrine System

What do you see?
"This picture seems to have a bunch of nuclei with a stringy substance going to and from each one. I think the surroundings of the picture give it away, because I think they are magnified nerve cells."

*Mike Eaton
Northeast Middle School
Kansas City, Missouri*

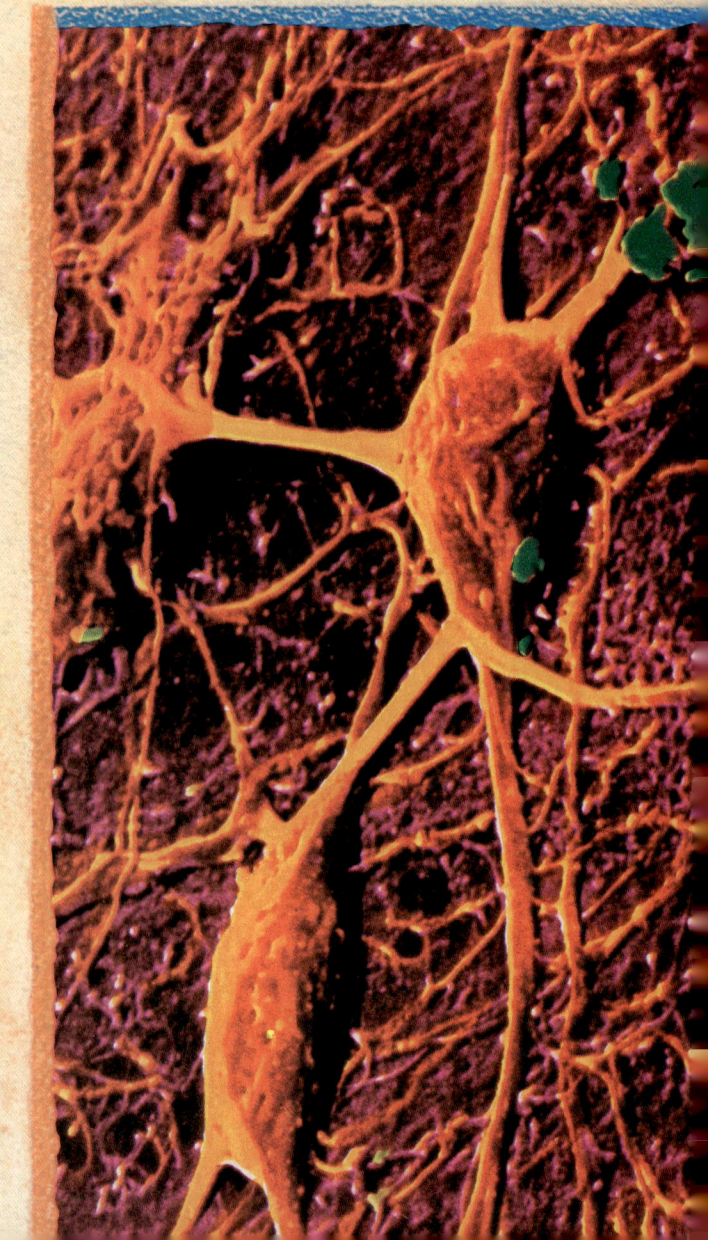

To find out more about the photograph, look on page 464. As you read this chapter, you will learn about how you control your body and how your senses work.

STS Connection

A sensory deprivation chamber is a device that is sometimes used in stress management therapy. The chamber is a tub filled with salt water that is completely enclosed in a chamber. A person floats in the tub from one to four hours or more. The water is kept at body temperature. No light, sound, or other stimuli are present in the chamber. Ask students to describe what it might be like to be in a sensory deprivation chamber. Have students infer how the chamber might help people who are stressed.

21.1 Nervous System

Objectives

- **Name** the parts of the nervous system and **describe** their functions.
- **Explain** how neurons carry impulses throughout the body.
- **Describe** the parts of the brain and their functions.
- **Make a model** of a reflex.

Skills WarmUp

Infer Have a partner hold a ruler vertically above the floor. Position your thumb and forefinger around the bottom of the ruler, but don't touch it. Your partner will then drop the ruler at any time. Catch the ruler as soon as you see it falling. Try this a few times. Did you catch it each time? Why did you react to the falling ruler?

Has this ever happened to you? The muffin you heated in the oven smells so good, you can't wait to try it. But as soon as you pick it up, you drop it. Your fingers can't hold the muffin. It's too hot! Even before you knew it was hot, your fingers received the message to let go. What might have happened had you held on to the muffin? What made your fingers let go of the muffin?

Functions of the Nervous System

The reason you'll let go of anything hot, such as the muffin fresh out of the oven, is because of your nervous system. Your nervous system is a communication and control system made up of the brain, spinal cord, and billions of nerves. The nervous system receives information from your environment and from inside your body. The system interprets this information and causes the body to respond to it.

Your nervous system lets you know if something is hot or cold, sweet or bitter, rough or smooth. It controls your movements. It protects you from harm by letting you feel pain. The nervous system even lets you solve problems and learn music. In addition, the nervous system controls reactions that involve emotion. It lets you be happy or sad, angry, or calm.

You wouldn't be wrong if you thought that the nervous system is the most important system in your body. You couldn't even raise an eyebrow without the nervous system. But remember, all human body systems work together to keep you alive.

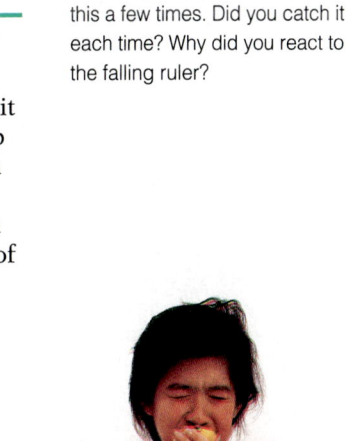

Figure 21.1 ▲
Reacting to the taste of a bitter lemon is possible because of your nervous system.

Chapter 21 Control and Sensing   443

SECTION 21.1

Section Objectives
For a list of section objectives, see the Student Edition page.

Skills Objectives
Students should be able to:

Infer how nerve branches throughout the body allow quick reactions to stimuli.

Calculate the average time it takes to react to a falling ruler.

Make a Model of a nerve impulse traveling through a neuron and across a synapse.

Vocabulary
neuron, synapse, motor neuron, sensory neuron

MOTIVATE

Skills WarmUp

To help students understand how the nervous system carries impulses throughout the body, have them do the Skills WarmUp.
Answer Answers may vary. Although some students may not catch the ruler each time, each will react to the falling ruler because the nervous system is sending messages to the parts of the body that cause them to react.

Misconceptions

Students may think that the central nervous system is not related to body movement and the function of organs. Explain that the nerves branch out to all parts of the body as well as to the organs. The nervous system not only controls all body movement but also is involved in the functions of all other body systems.

443

TEACH

Skills WorkOut
To help students calculate their reaction times, have them do the Skills WorkOut.
Answer The difference in average times should not vary greatly.

Explore Visually
Have students study the drawing and the paragraphs. Ask the following questions:

▶ When someone throws a ball, what body organ coordinates all the activities? This organ belongs to what part of the nervous system? (The brain coordinates these activities. It's a part of the central nervous system.)

▶ When messages from the brain travel to your arm, through what parts of the nervous system do they travel? (They travel down the spinal cord to the spinal nerves and then to the nerves in your arm.)

▶ From where do the spinal nerves branch out? (They branch out from between the vertebrae.)

▶ If you look carefully at the spinal cord and the spinal nerves, how are the spinal nerves arranged? (They are arranged in pairs, one going to each side of the body in a branching pattern.)

Integrated Learning
Use Integrating Worksheet 21.1.

INTEGRATED LEARNING

Themes in Science
Stability and Equilibrium The human nervous system coordinates all responses to the environment. This provides stability for the entire body. The central nervous system processes information and coordinates the body's actions. The peripheral nervous system senses changes in the environment and causes actions, but processes little information.

Multicultural Perspectives
The Chinese medical art of acupuncture works in harmony with the nervous system to relieve pain. Many scientists believe that acupuncture interrupts pain signals that the nervous system sends to the brain. Encourage students to read about the history of acupuncture and its use in the United States.

Skills WorkOut
Calculate Try the ruler activity again with a partner. This time, have your partner hold the ruler with the zero end closest to the floor. Catch the ruler five times. Record in centimeters the distance the ruler falls past zero before you catch it. Then calculate your average reaction time. How does it compare with your classmates' averages?

Parts of the Nervous System

In the study of the human body, the nervous system is divided into two parts. Look at Figure 21.2. The central nervous system, or CNS, is made up of the brain and spinal cord. The peripheral (puh RIHF uhr uhl) nervous system, or PNS, connects the central nervous system to the rest of the body. It is made of the spinal nerves and the many nerves that branch from the brain and spinal nerves.

Think about the two words that describe the nervous system. *Central* means the main part or most important part of something. *Peripheral* means lying on the outside or away from the central part. By looking at these definitions, you can tell how the two parts of the nervous system were named.

Brain
The brain is the main control center of your body. It directs and coordinates all body processes, thoughts, behaviors, and emotions. It is often compared to a computer. Yet no computer has as much memory or as many connections.

Spinal Nerves
Thirty-one pairs of spinal nerves branch out into both sides of the body from between the vertebrae. Spinal nerves carry messages to and from other nerves. Each nerve contains thousands of nerve fibers.

Spinal Cord
Extending from the base of your brain down your back is a bundle of nerve fibers called the spinal cord. If you want a first-hand idea of where your spinal cord is, try this. Feel the vertebrae in the back of your neck and down your spine. The vertebrae cover and protect the spinal cord.

Nerves
Many nerves branch to all parts of your body. Some go to body organs, such as your heart and lungs. The actions of these nerves are controlled automatically by the brain. Some nerves connect the spinal cord to muscles. These nerves cause muscles to contract.

Figure 21.2 ▶
The Human Nervous System

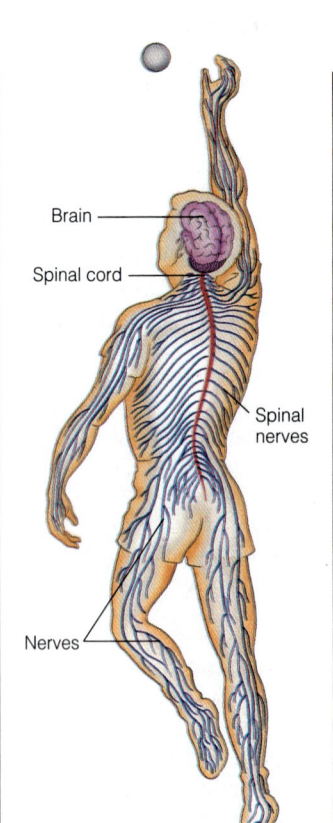

Chapter 21 Control and Sensing

Integrating the Sciences

Chemistry Point out to students that neurons often need chemical energy to carry nerve impulses from the axon of one neuron to the dendrite of another. There are more than 50 different chemicals that can serve as neurotransmitters in animals.

Neurons

Your brain, spinal cord, and nerves are made of nerve cells, or **neurons** (NOO rahns). Each neuron receives and sends electrical and chemical messages.

Look at Figure 21.3. It shows the parts of a neuron. The cell body is like a telephone switchboard that receives and relays incoming messages. Notice that the cell body has a nucleus just like all cells.

The cell body receives electrical messages from branched cell parts called dendrites. Notice that the many short dendrites are close to neighboring neurons. The cell body relays the messages to the long axon. The axon carries nerve impulses away from the cell body. The axon ends in branches, too. These branches of the axon transmit nerve impulses to the dendrites of another neuron.

Nerve Impulses

A neuron uses both electrical and chemical energy to send a nerve impulse. Look again at Figure 21.3 to follow the process. An electrical impulse travels from the dendrites to the cell body. From the cell body, the electrical impulse travels down the axon. But the end of the axon doesn't touch the next neuron. There is a small space between one neuron and the next one. This space between neurons is called a **synapse** (SIHN aps). The electrical impulse stops at the gap. That's when the axon releases a chemical into the synapse.

Chemical energy is used to carry the nerve impulse to the dendrite of the next neuron. The chemicals travel across the synapse. They join with molecules in the dendrites of the next neuron. The chemical energy causes an electrical impulse in the dendrites of this neuron.

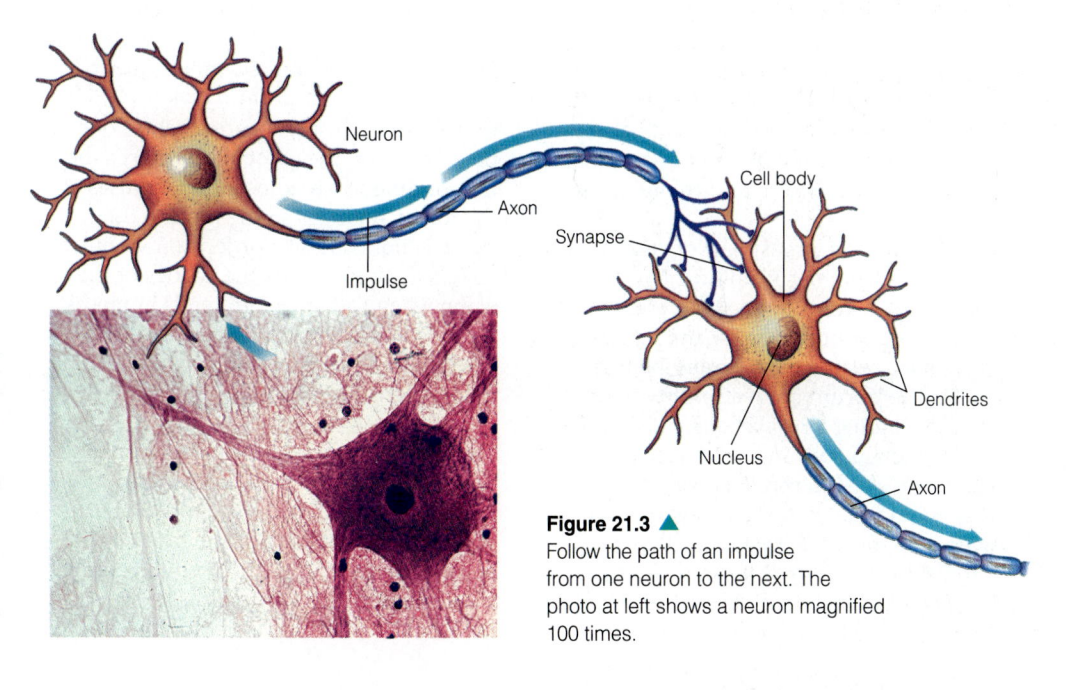

Figure 21.3 ▲
Follow the path of an impulse from one neuron to the next. The photo at left shows a neuron magnified 100 times.

Critical Thinking

Reason by Analogy Have students compare the neuron to a light switch by discussing the following:

▶ When you turn on a light switch, the electricity travels through the circuit to the light. Ask students how this relates to the neuron. (The nerve impulses travel to the cell body.)

▶ When you turn off the light switch, the light goes off. Have students relate this action to the neuron. (The nerve impulses no longer travel to the cell body.)

Discuss

Explain to students that nerve impulses always move in one direction: *from* the dendrites, *to* the cell body, then *to* the axon. Only the tip of the axon makes the chemical that stimulates the next neuron. Write the word *axon* on the chalkboard. Underline the letter *a*. Tell students that it might help them remember that axons always carry impulses *away* from the cell body of the neuron if they notice that *axon* and *away* both start with the letter *a*.

Skills Development

Hypothesize Since nerve impulses always move in one direction, ask students how messages can travel both to and from the brain. (Some neurons carry messages to the brain and some carry messages away from the brain.)

**The Living Textbook:
Life Science Side 2**

Chapter 12 Frame 20738
Neurons and Impulse Conduction (Movie)
Search: Play:

Chapter 21 Control and Sensing

TEACHING OPTIONS

TEACH • Continued

Class Activity
Have students brainstorm a list of simple movements, such as peeling an orange, swallowing, reading, and so on. Then ask them to identify the part of the brain that controls each function. (When you peel an orange, the cerebrum interprets nerve impulses from the eyes, enabling you to see the orange. The cerebrum directs your hands and arms to pick up the orange and tear apart the rind. The cerebellum makes the movements smooth. While the cerebrum and the cerebellum function, the medulla controls the automatic body processes that keep us alive.)

Skills Development
Infer Explain that the brain takes in about 50 mL of oxygen each minute. Ask students what would happen if a person's blood supply to the brain were cut off. (Within 10 seconds the person would become unconscious from a lack of oxygen. Without oxygen, the brain cells would be permanently damaged. The damaged cells would die and could not be replaced.)

The Living Textbook:
Life Science Sides 7-8

Chapter 4   Frame 00511
Nerve Cell (1 Frame)
Search:

The Living Textbook:
Life Science Sides 7-8

Chapter 4   Frame 00492
Brain Microviews (14 Frames)
Search: Step:

Portfolio
Have students make a list of the major activities they perform each day. Then ask them to chart the daily activities controlled by each part of the brain. Point out that complex activities might require more than one part of the brain. Students may want to keep the charts in their portfolios.

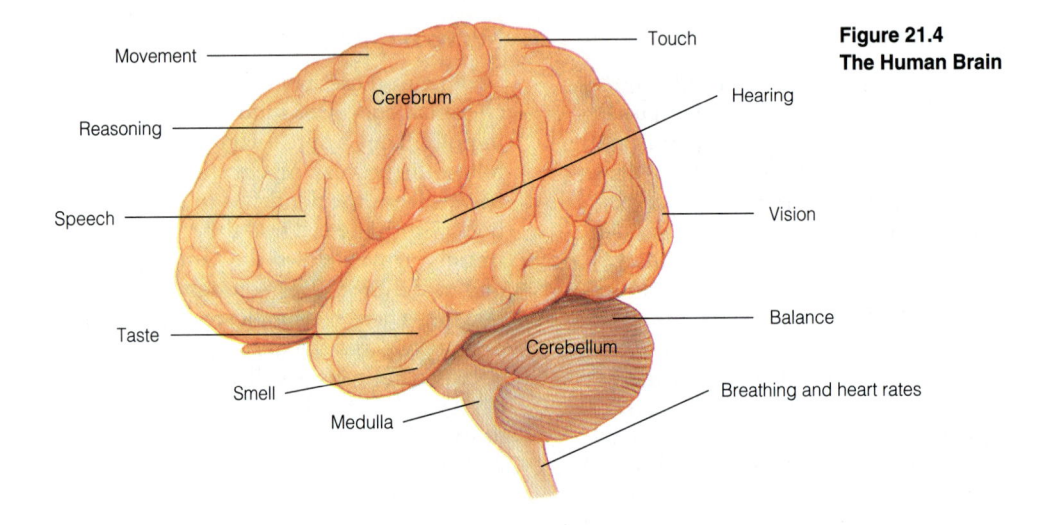

Figure 21.4
The Human Brain

The Brain

The brain is made up of more than 10 billion neurons, forming a spongy nerve tissue. The nerve tissue is surrounded by membranes that nourish and protect the brain. The main job of the brain is to receive and interpret messages from outside and inside your body. You can see the three main parts of the brain in Figure 21.4. They are the cerebrum, the cerebellum, and the medulla.

Cerebrum The largest part of your brain is the cerebrum (SUR uh bruhm). The folds in the cerebrum greatly increase the surface area of this part of the brain. More surface area means more neurons.

The cerebrum controls many functions, including all conscious body movement, such as running in a race. The cerebrum also interprets nerve impulses that come from your sense organs—the eyes, ears, nose, tongue, and skin. The cerebrum enables you to see this page.

The cerebrum is divided into two halves. The left half of the cerebrum controls your ability to speak, use mathematics, and think logically. The right half is the center for musical ability, the creation of art, and the expression of emotions. Nerve paths between the two halves allow them to communicate.

Cerebellum The second largest part of the brain is the cerebellum (SUR uh BEHL uhm). When the cerebrum directs your body to move, the cerebellum makes the motion smooth. It adjusts the impulses of the motor neurons so the motion they cause is not robotlike or jerky. A violinist moves the bow across the strings of the violin when the cerebrum sends impulses to the musician's hands and arms. The cerebellum fine-tunes that motion so the violinist creates a soothing sound.

Medulla The brain stem is a bundle of nerves connecting the cerebrum to the spinal cord. The medulla (mih DUHL uh) is the lowest part of the brain stem. Nerve impulses from the medulla control many automatic body processes, such as heartbeat, breathing, and blood pressure.

Chapter 21 Control and Sensing

INTEGRATED LEARNING

Themes in Science

Systems and Interactions

Explain to students that the brain, spinal cord, and nerves function together as a system to enable humans to receive information and respond to it. For example, if a person dances, the cerebrum enables the person to hear the music and move in a certain way. The cerebellum adjusts the impulses of the motor neurons so the motions are smooth. If the beat of the music speeds up and the dancer moves more quickly, the medulla signals the body to take in more oxygen.

The Spinal Cord

The spinal cord connects the brain to the peripheral nervous system. It acts as a highway for the passage of nerve impulses both to and from the brain. Nerve impulses pass through the spinal cord when your brain tells your body to move, such as when you reach for a glass of water. As you grasp the glass, nerve impulses pass through the spinal cord in the other direction to tell the brain that the glass feels cold.

The spinal cord connects with nerves all over your body through spinal nerves. These nerves are made of two kinds of nerve fibers. Some of the fibers lead from the spinal cord to the muscles and cause the muscles to contract. They are made up of neurons called **motor neurons**. Other nerve fibers carry sense information to the spinal cord, which then sends it on to the brain. These fibers are made up of **sensory neurons**. Impulses are carried between sensory and motor neurons by association neurons.

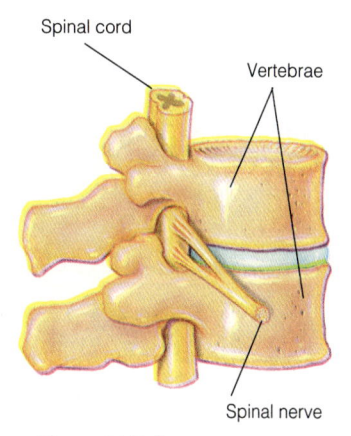

Figure 21.5 ▲
The nerves in the spinal cord are protected by vertebrae.

Career Corner Speech Pathologist

What Does a Speech Pathologist Do?

A ten-year-old boy is struggling to conquer stuttering. A hearing-impaired teenager is learning to speak. A six-year-old girl is working to overcome a lisp. An adult who has suffered a stroke is slowly regaining his speech. The person helping each of these people with problems in speaking is called a speech pathologist. The work of speech pathologists is as varied as the speech disorders they treat.

When treating patients, the pathologist uses various methods. The methods used depend on many factors, including age, type of speech disorder, and case history. A pathologist may use video and audio recording machines. This way, patients can see their lips move on the television screen and hear their speech errors on the tape recorder. The patients learn to speak correctly by watching and listening to the pathologist pronounce the sounds and words correctly. The pathologist will eventually have patients do tongue exercises and drills to help them improve their speech.

If you would like to help people with speech problems, many universities offer training in speech therapy. You need a master's degree in the field to work as a speech pathologist. Then you might work in schools, hospitals, or university speech clinics. Speech pathologists with doctorate degrees may teach in colleges or universities. They may also engage in research.

Discuss

Draw two columns on the chalkboard. At the top of one column, write the term *motor neurons,* and at the top of the other, write *sensory neurons.* Give students examples of various messages that are carried by nerves, such as *feels hot, tastes sweet, run, breathe, chew, looks red,* and so on. Have students identify which kind of neuron carries the message. Write each message in the appropriate column.

Research

Have students use references to find out where in the body the three kinds of neurons—sensory, motor, and association—are likely to be found.

Enrich

Use Enrich Worksheet 21.1.

Career Corner

If possible, invite a speech pathologist to visit the class. Have students prepare a list of questions to ask the visitor. Encourage students to ask about the way different parts of the brain interact to enable a person to speak.

The Living Textbook:
Life Science Sides 7-8

Chapter 4   Frame 00506
Brain Diagrams (4 Frames)
Search: Step:

The Living Textbook:
Life Science Sides 7-8

Chapter 4   Frame 00447
Spinal Column (5 Frames)
Search: Step:

TEACH • Continued

Explore Visually

Have students study Figure 21.6 and the paragraphs on this page. Then ask some or all of the following questions:

▶ If you picked up a second hot muffin a few minutes later, would you expect to have the same reaction? Why? (Yes. The reaction should be the same as long as the receptors in the skin are not damaged. The reflex action would simply be repeated.)

▶ Which is faster—the nerve impulse traveling to the spinal cord or the nerve impulse traveling to the brain? How do you know? (The nerve impulse to the spinal cord is faster because it has less distance to cover. By the time the pain message is registered in the brain, the body has already responded.)

▶ If the nerve impulse travels 1 m to the spinal cord at 90 m per second and the nerve impulse travels at the same speed and the same distance to the muscles in the hand, how quick would the reaction be? (1/45 second)

▶ Would you expect the reaction to be faster or slower if the heat from the muffin is detected in a place other than the hand, such as in the bottom of the foot? Why? (The reaction depends on the body part involved, since the density of nerve cells varies throughout the body.)

448

INTEGRATED LEARNING

Integrating the Sciences

Physical Science Tell students that light travels at a fixed rate, just as nerve impulses do. Direct students to find out how fast light travels. (Light travels about 299 792 km per second.)

Reflex Actions

Remember the hot muffin? It was too hot to hold, so you immediately dropped it without having to think about dropping it. Dropping the muffin was a reflex action.

A reflex action is a simple response to a stimulus. Sneezing is a reflex. It is automatic. Blinking is also a reflex. It can be automatic or it can be controlled. Reflex actions are controlled by the spinal cord.

Look at the diagram of the reflex in Figure 21.6. The incoming message is switched in the spinal cord directly to outgoing motor neurons. The nerve impulse then travels quickly to the hand or to another part of the body in danger. It causes the body part to move away from the source of the stimulus. The message from the sensory neurons is also sent to the brain. But by the time the message of pain is registered in the brain, the body has already responded.

Figure 21.6
Some quick movements happen as a result of short impulse pathways. ▼

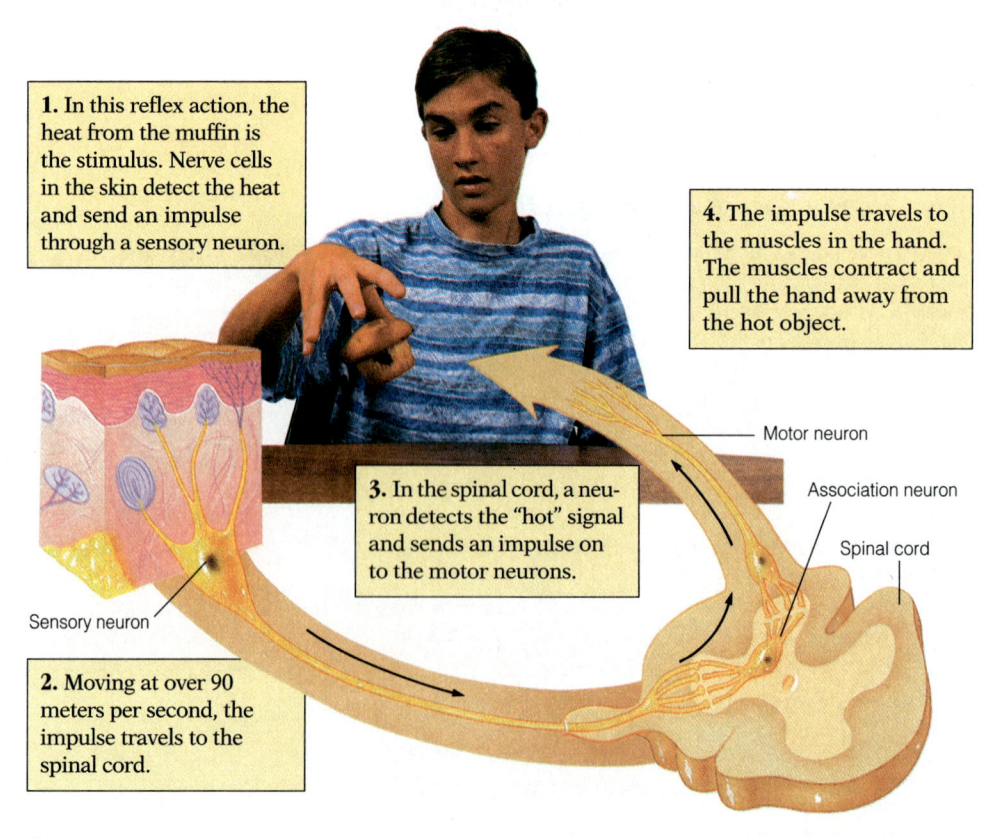

1. In this reflex action, the heat from the muffin is the stimulus. Nerve cells in the skin detect the heat and send an impulse through a sensory neuron.

2. Moving at over 90 meters per second, the impulse travels to the spinal cord.

3. In the spinal cord, a neuron detects the "hot" signal and sends an impulse on to the motor neurons.

4. The impulse travels to the muscles in the hand. The muscles contract and pull the hand away from the hot object.

Sensory neuron
Motor neuron
Association neuron
Spinal cord

448   Chapter 21   Control and Sensing

STS Connection

PET scans provide pictures of the brain. They record changes in blood supply in areas of the brain by tracing radioactive sugar in the bloodstream. The patient may be told to think of a person's name, then of that person's face. From readings of radioactivity in the brain, a computer creates color pictures of the specific areas that were stimulated by the task. Comparing the results with those of normal brain activity, doctors can diagnose problems without having to perform surgery.

Science and Technology PETs and CATs

Did you know that it is possible to take a picture of the brain? Pictures of the brain are achieved through a process called computerized tomography, or CT. At one time, X-rays were the only method for obtaining information about the inside of a human body. This method often gave insufficient results because X-rays are most sensitive to bony structures. CT imaging is different. It is like taking a slice out of a loaf of bread without cutting the bread. However, a person is much more complex than bread.

There are two main types of computerized tomography: computerized axial tomography, or CAT, and positron-emission tomography, or PET. A CAT machine is shaped like a human-sized doughnut. When a patient is placed inside the tube, the tube shoots an X-ray beam into the skull. The different tissues in the brain collect the radiation in different ways. The computer takes this information and translates it into a detailed picture of the scanned brain slice. Often many different slices of the brain are imaged. CAT is used to scan for tumors, brain damage from accidents, and cancer.

PET patients ingest a small dose of a radioactive substance. The PET scanner picks up the radiation sent out from the brain. A computer creates pictures of the brain's activity in blues and yellows. PET scans have been used for research on people affected by mental illness, Alzheimer's disease, and epilepsy.

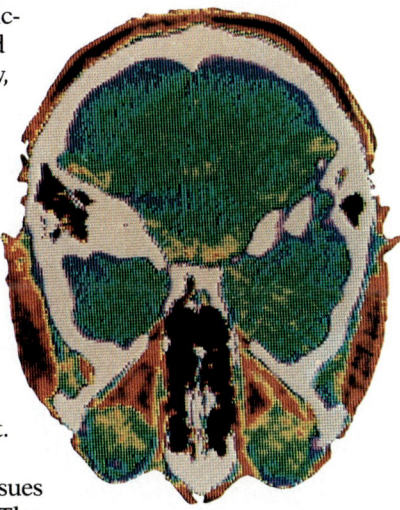

Figure 21.7 ▲
CAT scans are used to detect problems in the brain, such as tumors.

Check and Explain

1. List the parts of the nervous system and describe the function of each.
2. Which part of the brain did you use in answering the first question?
3. **Infer** Drugs called anesthetics are used during surgery. An anesthetic stops the transmission of nerve impulses at the synapses. How does this help a patient being operated on?
4. **Make a Model** Create a model that shows how a nerve impulse travels through a neuron and across a synapse.

EVALUATE

WrapUp
Review Have students make flash cards of the parts of the nervous system. On the back of each card, have students describe how that part of the nervous system functions. If helpful, have students include simple drawings on their flash cards. Students can use the cards as they review the concepts from this section.
Use Review Worksheet 21.1.

Check and Explain
1. The CNS is made up of the brain and spinal cord. The CNS directs and coordinates all body processes, thoughts, behaviors, and emotions. The PNS is made up of spinal nerves and nerves. The PNS connects the CNS to the rest of the body.
2. The cerebrum
3. The patient does not feel pain since the electrical impulse stops at the synapse.
4. See Figure 21.3 on page 445.

Chapter 21 Control and Sensing

SECTION 21.2

Section Objectives
For a list of section objectives, see the Student Edition page.

Skills Objectives
Students should be able to:

Infer how the lens in the eye changes its shape and affects vision.

Observe how the fingers detect temperature changes very quickly.

Collect Data during an experiment.

Infer possible causes of deafness.

Vocabulary
retina, pupil, iris, eardrum, cochlea

MOTIVATE

Skills WarmUp
To help students understand how the eye functions, have them do the Skills WarmUp.
Answer The words on the page become magnified and focused. The lens in the eye changes its shape. The lens then focuses the rays of light so they form a clear image on the retina.

Prior Knowledge
To gauge how much students know about the senses, ask the following questions:

▶ What kind of information do you get through your senses?

▶ How does this information help you?

Answer to In-Text Question

① The sensory receptors in your eyes, ears, and skin provide the feedback that helps you experience a roller coaster ride.

INTEGRATED LEARNING

Themes in Science
Systems and Interactions To meet the demands of the environment, the senses work together with the brain. They enable people to find food and shelter and avoid danger. Encourage students to discuss the role of the senses in human adaptations. How do the senses help them survive?

Integrating the Sciences
Physical Science Point out to students that cameras produce an inverted image of an object rather than the image the eye sees. Ask students to research and report on the similarities and differences between a camera lens and the eye.

Skills WarmUp

Infer Find a page in this book with small print, such as the index. Then roll your forefinger in tightly to form a very tiny hole. Hold the printed page close to your eyes so that it is barely blurred. Close one eye. Then look at the page through the hole with the open eye. What happens? Why?

21.2 The Senses

Objectives

▶ **Describe** the function of sensory receptors.

▶ **Identify** the different stimuli the senses can detect.

▶ **Outline** the processes of seeing, hearing, smelling, tasting, and touching.

▶ **Make inferences** about possible causes of deafness.

How would your life be different without your sense of sight? How would you communicate with people if you couldn't hear? Would you enjoy eating if you couldn't taste your food? Your senses work together with your brain to keep you informed about your environment. They provide pleasure, keep you from danger, and help you interact with others.

Sensory Receptors

You know that you use your eyes to see, your ears to hear, your nose to smell, your tongue to taste, and your skin to feel. Each of these sensory organs is sensitive to certain parts of the world around you. Each is able to detect a certain kind of stimulus. Your eyes see, for example, by picking up the stimulus provided by rays of light.

How are your sensory organs able to perform their functions? They contain special nerve cells called sensory receptors. The function of a sensory receptor is to take in a stimulus and convert it into nerve impulses. These impulses can then be transmitted to the brain. Here they become the experience of sight, sound, taste, touch, or smell.

The sensory receptors in your eyes detect light rays. The receptors pick up different colors and brightnesses and create nerve impulses to match. The sensory receptors in your tongue and nose sense different chemicals. In your ears, sensory receptors respond to sound waves. One kind of receptor in your skin detects pressure and another ① senses heat.

Figure 21.8 ▲
How do your senses help you experience a roller coaster ride? ①

450   Chapter 21 Control and Sensing

Integrating the Sciences

Physical Science The human eye is designed to capture light. Light reaches the eye in two ways: directly from a source like the sun or a light bulb, or indirectly when light bounces off something, like the moon. We see trees, buildings, and each other's faces due to reflected light.

Language Arts Connection

Humans rely on vision much more than other animals do, and it shows in our language. Wise people are said to have "insight." People say, "You see?" when they mean, "You understand?" Have students find other examples of the ways we use vision words to signal knowledge. (Other examples would be "foresight" or "revision.")

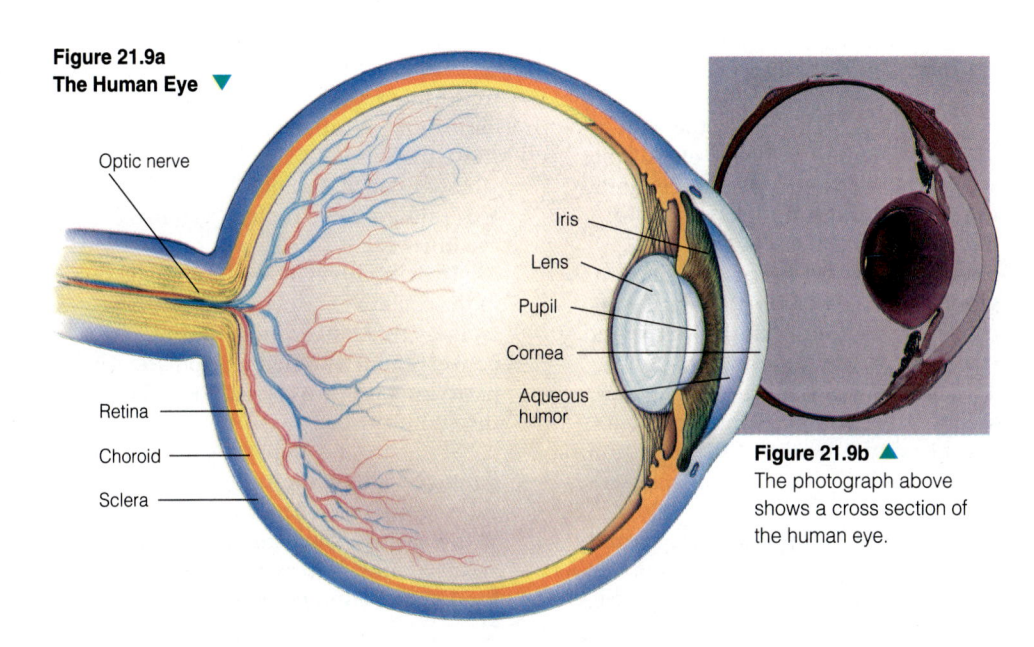

Figure 21.9a
The Human Eye ▼

- Optic nerve
- Retina
- Choroid
- Sclera
- Iris
- Lens
- Pupil
- Cornea
- Aqueous humor

Figure 21.9b ▲
The photograph above shows a cross section of the human eye.

Light and Seeing

Light doesn't just let you see, it *is* what you see. Light rays enter your eyes. Your eyes and your brain work together to form and interpret images of the objects that you see.

Structure of the Eye Look at Figure 21.9a. Your eye is made up of three layers. The tough, white, outer covering of most of the eye is called the sclera (SKLUR uh). The middle layer is the choroid (KOR oyd) layer. Lining the back and sides of the inner eye is the **retina** (REHT ih nuh).

In the front of the eye, the choroid layer has an opening called the **pupil**. Surrounding the pupil is a round, colored disk called the **iris**. The iris controls how much light enters the eye by changing the size of the pupil. Behind the pupil is a piece of clear tissue called the lens. The lens is attached to muscles that can change its shape.

The retina is a layer of nerve tissue made up of receptors called rods and cones that detect light and color.

Light and Image Light bounces off objects in the world. Some of it enters your eye. The lens then focuses, or gathers together, the rays of light. By changing its shape, the lens changes how the light rays bend when they pass through it. In this way, the lens focuses the rays of light so that they form a clear image when they strike the retina. This image is an upside-down likeness of the object the light rays are bouncing off of.

The rods and cones in the retina detect the image. They convert it, piece by piece, into nerve impulses sent to the brain along the optic nerve. The brain uses the impulses to remake the image. In the process, it turns the image right-side up. It also combines the two slightly different images from each eye to make a three-dimensional image.

TEACH

Directed Inquiry
Ask students to study the drawings and the paragraphs on this page. Lead them in a discussion using the following questions:

▶ How do you suppose messages travel from the retina to the optic nerve? (They travel through the sensory neurons of the retina to the optic nerve.)

▶ Is the pupil a structure that is part of the eye? Explain. (No. *Pupil* is the name of the opening through which light enters the eye.)

Demonstrate
Use a hand lens and sheet of paper. Darken the room. Have a volunteer stand about 3 meters from a window, holding up the paper. Place the hand lens between the paper and the window. Move the lens until a focused image is projected on the paper. Tell students that the image is similar to those appearing on the eye's retina.

Skills Development
Infer Why are the images seen by each eye slightly different? (One eye is looking at the object from one side of the head and the other eye is looking at it from the other side of the head. Students can see the difference by holding a pencil at arm's length and then looking at it with one eye at a time.)

The Living Textbook:
Life Science Sides 7-8

Chapter 6   Frame 01048
Vision Diagrams (5 Frames)
Search: Step:

TEACH • Continued

Enrich
Tell students the eardrum is so thin that it vibrates at even the smallest sound. As it vibrates, it moves a tiny bone called the hammer against a tiny bone called the anvil. The anvil then moves a third tiny bone called the stirrup (see Figure 21.10). The bones amplify the vibrations to the inner ear, which transmits the signals to the brain.

Class Activity
Have students amplify vibrations to their ears. Each student needs a metal spoon and a 60-cm piece of string. Have students tie the spoon to the middle of the string. Then have them tie each end of the string to each of their two index fingers. Have students lower the spoon so it hits the tabletop. Then have them put their index fingers in their ears and knock the spoon against the table again. Have students note how loud the sound is. This shows that the sound waves are amplified as they travel through the string.

Discuss
Refer students to Figure 21.10. Have them find the semicircular canals. Tell students that balance is controlled by three semicircular canals set at right angles to one another. These canals detect head movement in any direction. Head movement moves the fluids. The attached nerves signal the brain about the head's position. Ask students to infer why spinning in a circle makes a person feel dizzy. (The fluids continue to move after the person stops spinning.)

The Living Textbook:
Life Science Sides 7-8
Chapter 6   Frame 01055
Hearing Diagrams (2 Frames)
Search: Step:

INTEGRATED LEARNING

Integrating the Sciences
Physical Science Emphasize to students that sound travels in waves. Have them describe the sound they hear when a siren on a vehicle passes by them. (The sound is higher as it approaches and lower after it passes.) Explain that the pitch changes because sound waves are closer together (making a higher pitch) when the siren is approaching, and are farther apart (making a lower pitch) when the siren is moving away.

Themes in Science
Diversity and Unity Humans can hear only those sounds in which the pitch falls within a certain range. Bats hear sounds that are higher than the human range; whales hear sounds that are lower. The protruding forehead of a porpoise is a sound sensor, sending out high-pitched sounds and picking up echoes when the sounds bounce off objects.
   Use Integrating Worksheet 21.2a and 21.2b.

Sound and Hearing

Extend your ruler over the edge of your desk. Lightly flick the end of the ruler. What do you hear? Notice that as the ruler vibrates, it makes a sound.

Sound Waves All sound comes from vibration. When an object vibrates, it causes the molecules in the air around it to push together and stretch apart. This movement of air molecules is a sound wave. Sound waves move outward from the source of the vibration. They can travel through solids, liquids, and gases.
   The ear picks up sound waves and converts them into nerve impulses. With the help of the brain, you experience sound waves as sound.

Hearing Look at Figure 21.10. As you can see, most of your ear is hidden inside your head. The external ear is only a funnel that helps collect sound waves. Sound waves pass through the ear canal and strike a thin, round, tightly stretched membrane called the **eardrum.** When sound waves strike the eardrum, it vibrates.

A chain of three tiny bones is attached to the inner surface of the eardrum. The eardrum's vibrations are transmitted to these three bones. The bones pass the vibrations on to the **cochlea** (KAHK lee uh). It is a fluid-filled structure that makes up the inner ear.
   The vibrations move through the fluid in the cochlea. The cochlea contains nerve endings that detect the vibrations and convert them into nerve impulses. The nerve impulses are carried to the brain by the auditory nerve.

Balance The semicircular canals you can see in Figure 21.10 are not involved with hearing. They enable you to keep your balance. The canals are filled with fluid and motion receptors. When you move or rotate your head, the fluid in the canals shifts, affecting the receptors. The receptors send nerve impulses to the brain, which detects the way you moved and coordinates muscle movements that keep you balanced.

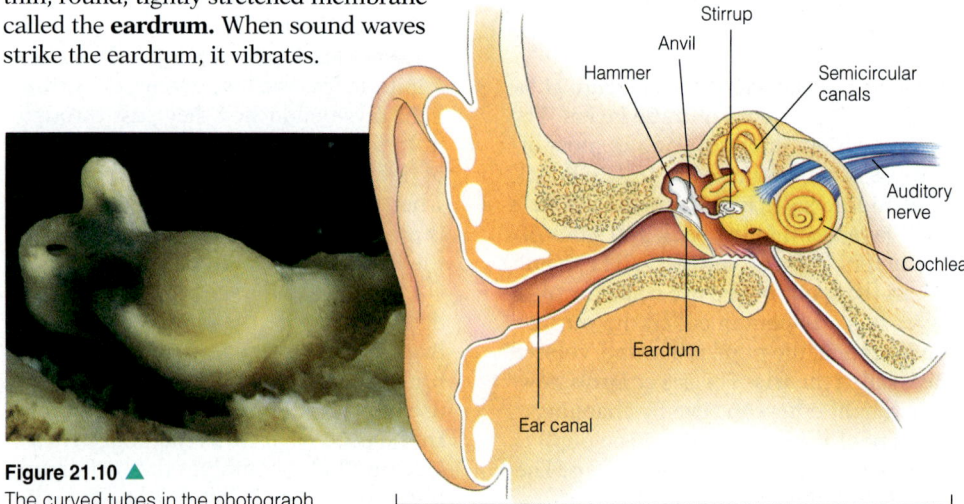

Figure 21.10 ▲
The curved tubes in the photograph are the semicircular canals.

452   Chapter 21   Control and Sensing

STS Connection

Smells can send all kinds of signals. Biologists are now using smells as a natural way to protect plants from predatory insects. For example, gypsy moths are very destructive. Biologists know that male moths locate females by a special female scent called a *pheromone*. So biologists make special traps, using the female scent; the males investigate and are caught and destroyed.

Multicultural Perspectives

In many countries, herbs and spices are used to achieve various flavors. For example, Indian curry contains turmeric, fenugreek, coriander, cumin, mace, and ginger. Have students interview neighbors, family, or local chefs to learn about the herbs and spices of their ethnic heritage. Encourage students to bring samples of spices to class.

Discuss

Have students answer the following questions about the sense of smell:

▶ Can you smell odors as well when the mucus lining the tissues inside your nose is very dry? Explain. (No. The mucus has to be moist to dissolve scent molecules.)

▶ Why is it sometimes hard to smell odors when you have a cold? (A cold often causes the nose to be stuffed up and the tissues inflamed. Both conditions keep odors from reaching the sensory receptors, which help odor stimuli to reach the brain.)

Class Activity

You might want to have your students try some or all of the different tastes on their tongues. Have them try grains of sugar, grains of salt, drops of vinegar, and shavings of bittersweet chocolate on different parts of the tongue. Have them moisten their tongues between tastings. Ask students to compare the taste buds on their tongues to the labeled drawing on page 453.

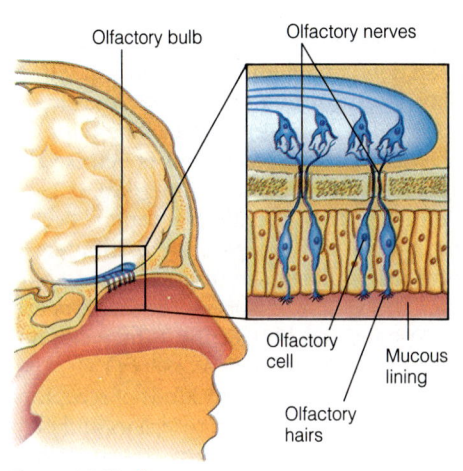

Figure 21.11 ▲
Smell receptors in the nose

Figure 21.12 ▲
Taste buds in the tongue (left); photograph of clusters of taste buds (right)

Smell

Think about what happens to you when a meal is being cooked near you. What are the odors, and how does your brain detect them? All odors are chemicals carried in the air.

The food you'll eat at a meal gives off different chemicals as it is cooked. These chemicals are molecules in the gaseous state. When you breathe, you take in some of these molecules. The molecules dissolve in the mucus that lines the tissues inside your nose. They stimulate sensory receptors located in this tissue. The receptors change odor stimuli into nerve impulses. Look at Figure 21.11. These nerve impulses are carried by the olfactory nerves to the brain.

It is not clearly understood how you can tell one odor from another. It may be that different odor receptors respond to different kinds of chemicals. At least 7, and perhaps as many as 50, basic odors can be distinguished. Most smells are different combinations of several of these basic odors.

Taste

The aroma of a food you like makes you want to taste it. Like smell, taste is the result of chemicals being detected by sensory receptors. The sensory receptors on your tongue, however, respond to molecules in the liquid state or molecules dissolved in liquid.

Look at Figure 21.12. Your tongue's sensory receptors are called taste buds. Taste buds give you four different kinds of taste sensations: salty, sweet, bitter, and sour.

Your taste buds are all very similar, but some are more sensitive to each of the four taste sensations. This may explain how areas of the tongue specialize in sensing either saltiness, sweetness, sourness, or bitterness. Find where these areas are located in Figure 21.12.

Have you noticed that when you have a cold it is harder to taste food? That is because most "tastes" are actually combinations of tastes and smells. When your nose is blocked, it can't detect the odors of food very well.

The Living Textbook:
Life Science Sides 7-8

Chapter 6   Frame 01031
Taste Bud Microviews (4 Frames)
Search: Step:

The Living Textbook:
Life Science Sides 7-8

Chapter 6   Frame 01054
Taste and Smell Diagrams (2 Frames)
Search: Step:

TEACH • Continued

Skills WorkOut
To help students observe how their fingers react to different water temperatures, have them do the Skills WorkOut.
Answer The sensory receptors in the fingers got used to the same strong sensation and stopped sending reports to the brain. When the sensory receptors were exposed to a change in sensation—the warm water—they sent new messages to the brain.

Critical Thinking
Reason and Conclude When people are exposed to very cold temperatures, their fingers, toes, ears, and nose may feel numb. When these people go into a warm place and slowly warm up, their fingers and toes begin to hurt. Ask students to suggest why this happens. (The receptors detect a change in sensation and begin to send messages to the brain. In fact, warming these parts too fast will produce severe pain.)

Enrich
You may wish to tell students that, of the parts of the body, the fingertips are not the most sensitive to pain. The neck, elbows, scalp, and backs of the hands and knees are more sensitive to pain than the fingertips.

Answer to In-Text Question
① There are large numbers of touch receptors on the tip of the tongue.

The Living Textbook:
Life Science Sides 7-8
Chapter 6 Frame 01023
Touch Receptor Microviews (5 Frames)
Search: Step:

454

TEACHING OPTIONS

Cooperative Learning
Assign students to cooperative pairs. Distribute items with various textures around the room. Blindfold Partner A. Have Partner B lead Partner A to the objects, then record Partner A's impressions as he or she examines the objects by touch. Reverse roles (and, if possible, change some of the items or their order) and repeat the exercise. Have students list the objects they were able to identify by touch.

Skills WorkOut

Observe Fill one jar or bowl with cold water, another with warm water, and a third with water just hot enough to touch. Label each jar. Place one forefinger in the hot water and the other forefinger in the cold water for one minute. Then put both fingers in the warm water jar. Describe what happens. What does this tell you about your sense of touch?

Figure 21.13 ▶
There are different kinds of receptors in the skin. Touch receptors are located near the surface of the skin. Pain receptors are located deep within the skin.

Touch
What are the different sensations you can feel with your skin? You can detect cold and heat. You can tell the difference between the touch of a feather and the touch of someone's hand. You can feel pricking pain and aching pain. Each of these sensations is detected by a certain kind of sensory receptor in the skin.

You can see in Figure 21.13 that these receptors occur at different levels in the skin. They are also distributed unevenly. You have many more pain receptors than cold receptors, for example. Touch and pressure receptors are concentrated on the hands and fingertips. Where else do you have large numbers of touch receptors? ①

Even though the skin contains many kinds of receptors, they fall into three main groups. Touch receptors and pressure receptors both detect movement or pressure. Cold and heat receptors both respond to temperature changes. Pain receptors detect many kinds of stimuli but are placed in a group of their own.

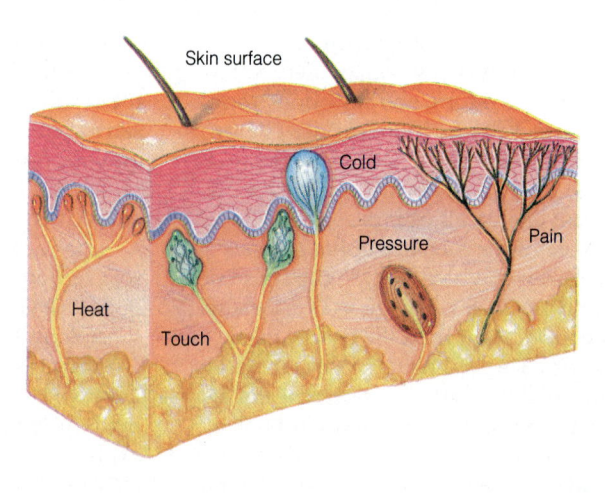

Pain is an especially important sense for survival. Imagine, for example, if you couldn't feel the burning sensation of a hot muffin. Pain is felt in different ways. For example, if you've cut yourself, you know that the result is an intense, quick pain. If you've burned yourself, the painful feeling usually develops slowly and is long lasting. Have you ever had an aching pain? What does it feel like?

454 Chapter 21 Control and Sensing

INTEGRATED LEARNING

STS Connection

Different areas of the brain interpret different elements of vision. One interprets form, another interprets motion, still another interprets depth, and a final area interprets color. Recently scientists have discovered that certain people who have become totally blind in an accident can still "see" color. They can pick out colored cards during tests. The way they do this is still a mystery.

SkillBuilder Collecting Data

Tasting Without Smelling?

Prepare small, peeled cubes of an apple, potato, carrot, turnip, and onion—all the same size. Place the pieces on a plate. Blindfold a partner, so she can't see what kind of cube you choose. Tell your partner to hold her nose while you place a cube on her tongue with a spoon.

Copy the table at right. After your partner chews the cube, ask which food she has eaten. Make a check mark on the chart under that food. Draw a star under what the cube really is. If your partner guesses correctly, the check and the star will be in the same box. After your partner tastes each food, switch roles. This time you will taste while your partner records.

Trial	Apple	Potato	Carrot	Turnip	Onion
1					
2					
3					
4					
1					
2					
3					
4					

1. Why was it difficult to determine which food you were eating?
2. Which foods were the hardest to distinguish? Why?
3. What clues did you use to determine what you were eating?
4. When have you not been able to taste food? Write why it is important to be able to smell food.

Science and You *Too Close or Too Far?*

How many people do you know who wear glasses or contact lenses? What are some reasons people wear glasses? ②

People's eyes are not always able to function normally. Glasses or contact lenses correct many people's vision problems. Two of the most common vision problems are nearsightedness and farsightedness.

When you view a distant object, your lens flattens to project the image of the object farthest from the lens. For a person with normal vision, the image falls on the retina. If the eyeball is too long, the image forms in front of the retina, as you can see in Figure 21.14. Nearsighted people can't see distant things clearly. Nearsightedness can be corrected by wearing lenses that are thinner in the middle than at the edges. These lenses bring the image back on the retina, where it should be, and into focus.

SkillBuilder

Hypothesize Have students design another experiment to test how the flavors of foods are affected by smell. Have students phrase their hypotheses in the form of questions.

Answers Students' answers should indicate that most flavors combine taste and smell. Students may name texture as an important clue used to identify foods without the aid of their sense of smell.

Enrich

Point out to students that, as people grow older, their vision often changes. People who are nearsighted sometimes have trouble seeing close objects. To solve this problem, people often wear special glasses called bifocals. These glasses have two sections to each lens. One part corrects the near vision and the other part corrects the distant vision. Sometimes contact lens wearers wear one contact lens that corrects farsightedness and another that corrects nearsightedness.

Answer to In-Text Question

② **People wear glasses or contact lenses to help them see better. The glasses commonly correct the problems of nearsightedness and farsightedness.**

Chapter 21 Control and Sensing

EVALUATE

WrapUp

Portfolio Have students draw or cut out pictures of stimuli that can be detected by the senses, and mount the pictures on notebook paper. On each page, have students describe the process involved in seeing, hearing, smelling, tasting, or feeling the stimulus.

Use Review Worksheet 21.2.

Check and Explain

1. A sensory receptor takes in a stimulus and converts it into nerve impulses.

2. The sensory receptors in the eyes detect light rays. The receptors create nerve impulses to match the different colors and levels of brightness detected.

3. Both have a lens that focuses on objects. The camera and the eye differ in that the eye, together with the brain, turns the image right-side up and produces a three-dimensional image.

4. Sound could be blocked at any one of the three tiny bones on the inner surface of the eardrum.

456

INTEGRATED LEARNING

Art Connection

What we see determines how we interpret the environment. Encourage students to draw two pictures or write two descriptions of the same scene or object under different lighting. How does the lighting affect their vision and, therefore, their perceptions?

Integrating the Sciences

Physical Science Point out to students that there are two shapes of lenses. Convex lenses are thicker in the center than at the edges. Concave lenses are thicker at the edges and thinner at the center. Convex lenses bring light rays to a point and are called converging lenses. Concave lenses make light rays spread apart and are called diverging lenses.

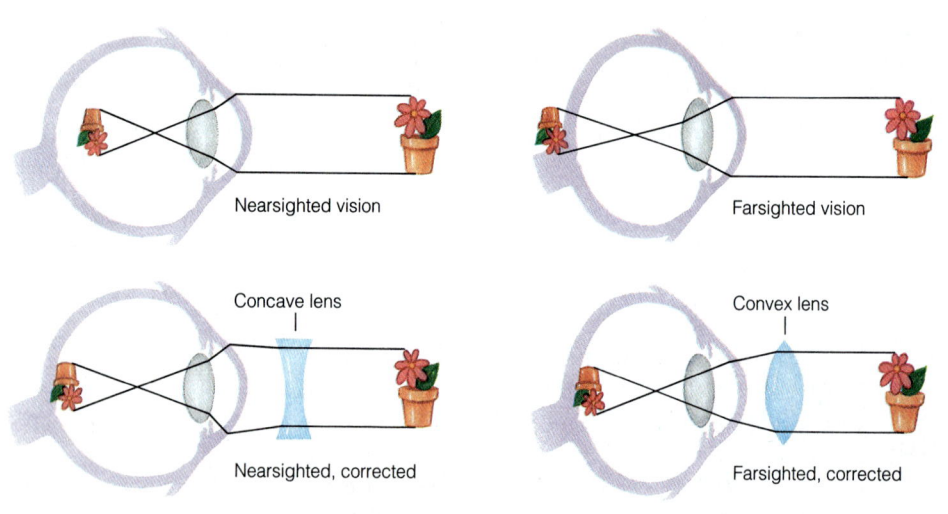

Figure 21.14 ▲
A concave lens (left) corrects nearsighted vision. A convex lens (right) corrects farsighted vision.

When you view a close object, your lens thickens to project the image of the object closer to the lens. If the eyeball is too short, the image forms behind the retina. People are farsighted if they can't see near things clearly. A farsighted person wears lenses that are thicker in the middle than at the edges. These lenses bend the light so the image is focused somewhat sooner, making it fall on the retina.

Focusing on close objects often becomes more difficult as you grow older. The reason is that the lenses in your eyes harden and lose their ability to change shape. The lenses cannot become thick enough to see close objects.

Check and Explain

1. What is the function of a sensory receptor?

2. Explain how you can see the words on this page.

3. **Compare and Contrast** How is the eye like a camera? In what ways is it different from a camera?

4. **Infer** At which places could the transmission of sound from object to brain be blocked, causing deafness?

456 Chapter 21 Control and Sensing

TEACHING OPTIONS

Prelab Discussion
Have students read the entire activity before beginning the discussion. Ask the following questions:

▶ What are some differences between the images produced by binoculars and telescopes?

▶ How do 3-D glasses work?

▶ What factors do you think will affect the ability to drop the penny in or near the cup?

ACTIVITY 21

Time 45 minutes **Group** pairs
Materials
100 pennies
15 cups

Analysis
1. Variables: distance from penny to cup, distance from student to penny/cup, distance of penny above cup
2. A single sight line allows little depth perception.
3. Answers may vary; however, students should have noticed some improvement.
4. Students may suggest that the experiment needs to control for varying eyesight between the two eyes.

Conclusion
Accept all logical answers. Students should understand that the brain judges distance by perceiving slightly varying images of an object based on the slightly different vantage points of the two eyes.

Everyday Application
Answers will vary. Possible answers include activities involving throwing, catching, or riding a bike. Clues to help judge depth are the relative size and position of familiar objects.

Extension
Answers will vary. Primates have depth perception because at some time during their evolution, binocular vision gave them a survival advantage—perhaps in detecting and escaping from predators—that has been maintained to the present.

Activity 21 Are two eyes better than one?

Skills Observe; Infer; Interpret Data

Task 1 Prelab Prep
1. Collect the following items: a cup and several pennies.
2. Put the cup on a table.

Task 2 Data Record
1. On a sheet of paper, draw the data table shown. Write the number of each trial.
2. Use the table to record your observations about where the penny falls.

Table 21.1 Guiding the Pennies

One Eye Closed	
Trial	Result
1	
2	
3	
4	
5	

Both Eyes Open	
Trial	Result
1	
2	
3	
4	
5	

Task 3 Procedure
1. Sit at least a meter away from the cup. Cover one of your eyes while a friend holds a penny at arm's length above the cup. The penny should be held slightly in front of the cup.
2. Watch only the cup and the penny. Tell your friend where to move his or her hand so the penny appears to be directly above the cup.
3. Remind your friend to move the penny exactly as you say—no more, no less. When you think the penny will fall into the cup, say, "Drop." Check how close you came. Record your results.
4. Repeat steps 1–3 four more times with the same eye closed.
5. Repeat steps 1–3 five times with both eyes open.

Task 4 Analysis
1. Identify the variable in this activity.
2. With one eye closed, why were you not able to judge the right position from which your friend should drop the penny?
3. How much did your accuracy improve with two eyes open instead of one?
4. Why weren't you told to cover the other eye for some of the drops?

Task 5 Conclusion
Write a short paragraph explaining how two eyes help you judge distances and depth.

Everyday Application
Suppose the eye doctor orders you to wear a patch over one eye for three weeks. What are some activities you should avoid? What are a few clues that might help you to judge depth?

Extension
Most mammals have eyes set at the sides of their heads. The primates, however, have eyes that face forward, and as a result they have stereoscopic vision, or the ability to perceive depth. Develop a hypothesis to explain why primates have depth perception and other mammals do not.

SECTION 21.3

Section Objectives
For a list of section objectives, see the Student Edition page.

Skills Objectives
Students should be able to:

Hypothesize that one of the endocrine glands is involved with the growth rate.

Research to find out more about diabetes.

Predict the effects of HGH on a normal adult.

Vocabulary
endocrine gland, hormone

MOTIVATE

Skills WarmUp
To help students infer about the purpose of the endocrine glands, have them do the Skills WarmUp.
Answer The pituitary gland regulates growth.

Prior Knowledge
To help students understand the endocrine system, briefly review information about the nervous system from section 21.1.

▶ How does your body get information about your environment?

▶ How does your body respond to information?

▶ How does information travel from one neuron to another?

Answer to In-Text Question

① You might feel a boost from adrenaline when you're excited about winning a big vacation or nervous about giving a speech.

458

INTEGRATED LEARNING

Themes in Science
Systems and Interactions The endocrine and nervous systems work together to respond to changes inside and outside the body. Epinephrine, or adrenaline, is often called the fight-or-flight hormone. In an emergency, it speeds up the heartbeat and respiration, raises blood pressure, slows digestion, increases the blood supply to muscles, and raises sugar levels for quick energy. Have students discuss a time when they noticed the effects of adrenaline on their nervous, skeletal, or digestive systems.

Skills WarmUp

Infer One gland in your body was very large when you were a baby. By the time you are an adult, it will almost have disappeared. Infer the purpose of the gland.

21.3 Endocrine System

Objectives

▶ **Describe** the major endocrine glands and their functions.

▶ **Explain** how hormone levels in the body are regulated.

▶ **Identify** several disorders of the endocrine system.

▶ **Predict** the effects of extra growth hormone on a normal person.

When can you run the fastest—racing in a 100-meter dash, or trying to escape an attacking dog? Most likely, the fear of the dog would boost your speed more than the desire to win a race. Why can you run faster when you are afraid?

The answer is that a substance called adrenaline (uh DREHN uh luhn) gives you a boost when you are frightened. Adrenaline causes many quick changes in your body. These changes make you faster and stronger for a short period of time. At what other times do you feel a boost from adrenaline? ①

Adrenaline is produced by a body system called the endocrine (EHN doh krihn) system. The endocrine system works closely with the nervous system. Like the nervous system, it functions to control many body processes. But the endocrine system uses chemicals, not nerve impulses, to perform its function.

Endocrine Glands

A gland is an organ that produces a chemical needed somewhere in the body. You have already learned about the glands that have ducts. They distribute their chemical directly to another organ. Another kind of gland, an **endocrine gland**, does not have a duct. An endocrine gland releases the chemicals it makes directly into the bloodstream.

Endocrine glands are the major organs of the endocrine system. The chemicals made by endocrine glands are called **hormones** (HOR mohnz). Hormones cause changes in other organs and regulate many body activities. There are 7 major endocrine glands in the human body, and more than 50 different hormones.

458 Chapter 21 Control and Sensing

TEACHING OPTIONS

Cooperative Learning

Have students work in groups of four to create a display about one of the endocrine glands. One member of each group can research the way the gland functions in the body, another can find out about disorders that occur when the gland does not function, and another can find out about methods for treating disorders. The fourth group member can collect and organize the information. Students can decide if they want to use drawings, diagrams, or three-dimensional models to represent the way the gland functions in the body.

Pituitary Gland Look at Figure 21.15. At the base of the cerebrum is a pea-sized gland called the pituitary (pih TOO uh TAIR ee) gland. Despite its small size, some people call it the "master gland" because it releases hormones that control several other glands. The pituitary has two lobes, each of which produces and stores a different set of hormones.

Among the many hormones produced by the pituitary is human growth hormone, or HGH. HGH controls how fast your muscles, bones, and organs grow. It determines the height you will reach when you are an adult.

The pituitary also stores hormones made by a part of the brain called the hypothalamus (HY poh THAL uh muhs). These hormones regulate blood pressure, water absorption in the kidneys, and the contraction of some smooth muscles in the body. The hypothalamus and the pituitary are closely connected and serve as a link between the nervous system and the endocrine system.

Thyroid Gland One gland that the pituitary controls is the thyroid (THY royd). Find this gland in Figure 21.15. The thyroid gland produces the hormone thyroxine (thy RAHKS een), which regulates your body's metabolism. Remember that metabolism is all the chemical reactions that provide energy for the body and allow it to grow. The thyroid is located at the base of the neck beneath the larynx. It lies on the trachea. It also produces calcitonin (KAL sih TOH nihn), a hormone that controls the levels of calcium and phosphorus in the body.

Pancreas You have already learned about the pancreas as a digestive gland. Locate the pancreas in Figure 21.15. A part of the pancreas is an endocrine gland as well. The pancreas releases a hormone called insulin (IHN seh lihn). This hormone helps transport sugar from the blood into the body cells, where it can be used for energy. This lowers the level of sugar in the blood.

The pancreas also produces glucagon (GLOO kuh GAHN). Its function is the opposite of insulin. Glucagon raises the blood sugar level. It does this by causing the liver to release the glucose stored there, making it available to the cells.

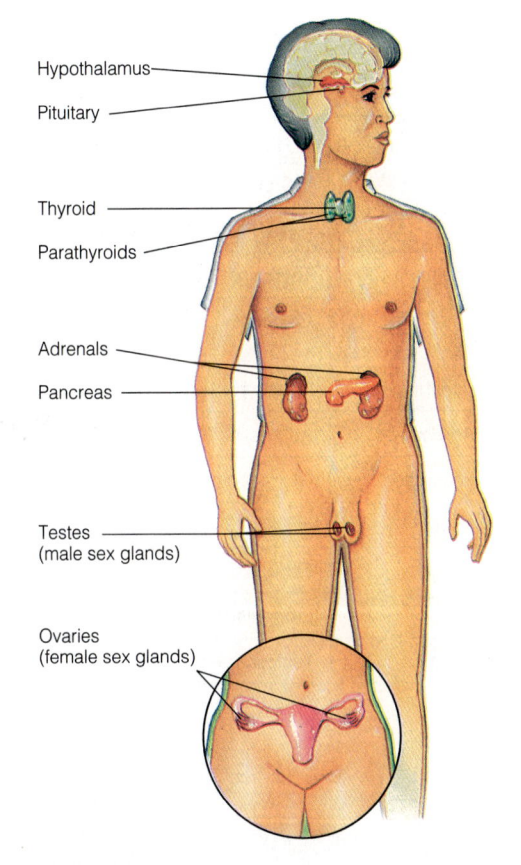

Figure 21.15 ▲
The endocrine systems of males and females are somewhat different. What endocrine glands are found only in females? ②

Labels: Hypothalamus, Pituitary, Thyroid, Parathyroids, Adrenals, Pancreas, Testes (male sex glands), Ovaries (female sex glands)

TEACH

Discuss

Have students study pages 458 and 459. Ask:

▶ Does the endocrine system work independently of all the systems in the body? Which system does the endocrine system work closely with? (The nervous system)

▶ Where is the hypothalamus located? (In the brain)

▶ What is metabolism? (All the chemical reactions that provide energy for the body and allow it to grow)

▶ Which gland regulates metabolism? (Thyroid)

▶ What endocrine glands are found only in males? (testes)

▶ What are the hormones produced by the ovaries? (Estrogen and progesterone)

Critical Thinking

Reason and Conclude

Have students review the digestive function of the pancreas (page 417). Ask students why the pancreas is considered part of the endocrine system and part of the digestive system. (It releases insulin and glucagon which regulate blood sugar levels, and it releases pancreatic juices for digestion.)

Answer to In-Text Question

② **The ovaries**

The Living Textbook:
Life Science Sides 7-8

Chapter 4   Frame 00426
Glandular Microviews (24 Frames)
Search: Step:

Chapter 21 Control and Sensing   459

TEACH • Continued

Directed Inquiry

Ask some or all of the following questions about Table 21.2:

▶ What is the function of the gonadotropic hormone released by the pituitary? (It affects the development of sex organs.)

▶ Which hormone regulates the amount of calcium in bone? (parathyroid)

▶ What is the function of adrenaline? (It stimulates organs of the body to respond to emergencies.)

▶ What two hormones are produced by the ovaries? What are their functions? (Estrogen produces female secondary sex characteristics. Progesterone promotes the growth of the uterine lining.)

▶ What hormone is produced by the testes? What is its function? (Testosterone produces male secondary sex characteristics.)

The Living Textbook:
Life Science Side 2

Chapter 12   Frame 14051
Effect of Epinephrine (Movie)
Search: Play:

INTEGRATED LEARNING

Integrating the Sciences

Botany Plants, like animals, have hormones to regulate growth. Have students create a table explaining the roles of gibberellins, auxins, and cytokinins in plant growth.

Table 21.2 Endocrine Glands

Gland	Hormone	Function
Pituitary	Growth hormone	Regulates growth of bones
	Thyroid-stimulating hormone	Stimulates thyriod to secrete hormones
	Gonadotropic hormone	Affects development of sex organs
Thyroid	Thyroxine	Increases metabolism
	Calcitonin	Controls level of calcium and phosphorus in the blood
Parathyroids	Parathyroid hormone	Regulates amount of calcium in bone
Adrenals	Adrenaline	Stimulates organs to respond to emergencies
Pancreas	Glucagon	Stimulates release of glucose from liver
	Insulin	Stimulates storage of glucose in liver
Ovaries	Estrogen	Produces female secondary sex characteristics
	Progesterone	Promotes growth of uterine lining
Testes	Testosterone	Produces male secondary sex characteristics

Feedback Control of Hormone Levels

Too much or too little of a hormone can cause serious health problems. Hormone levels are tightly controlled by a feedback system that automatically turns an endocrine gland on or off. You can compare feedback to a thermostat in a building. When the temperature inside the building drops below a set temperature, a signal turns the furnace on. What happens when the building temperature reaches the set temperature? The thermostat then signals the furnace to turn off.

The feedback mechanism in your body works like this. When a hormone is needed, a chemical signal tells the right gland to make more of it. When there is enough hormone to do the job, it causes a certain effect in the body. This effect acts as a signal for the gland to stop producing that hormone.

Look at Figure 21.16, which shows how the feedback mechanism controls blood sugar levels. When blood glucose levels fall below normal, the pancreas responds by releasing glucagon into the bloodstream. Glucagon acts on the liver to increase the rate at which the stored glucose is

460   Chapter 21 Control and Sensing

Themes in Science

Stability and Equilibrium Stress to students that hormone levels are regulated by the body to keep it in a state of equilibrium. For example, the hormones calcitonin and parathyroid are affected by a feedback control system that keeps the calcium levels in the blood balanced. Drinking milk causes the level of calcium in the blood to rise. The rise in calcium signals the thyroid to release calcitonin, which causes calcium to be deposited in bone tissue. Falling levels of calcium signal the parathyroid gland to release parathyroid hormone, which signals the bones to release calcium into the bloodstream.

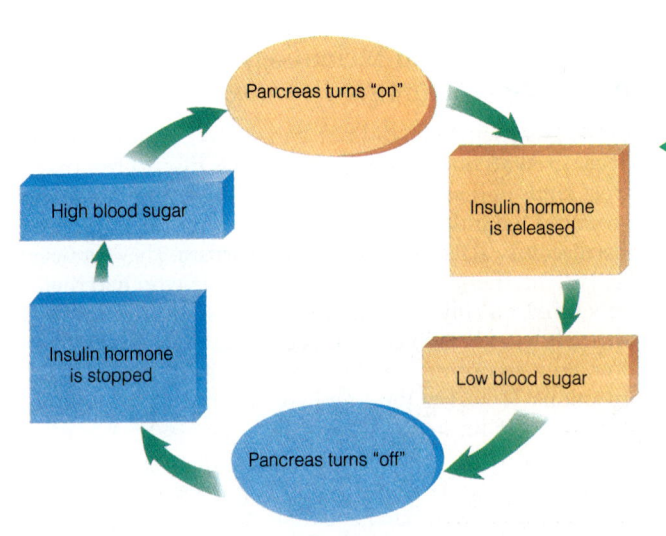

◀ **Figure 21.16**
In this diagram, you can see how feedback controls the blood sugar levels in the body.

changed to sugar. As the level of sugar rises in the blood, insulin is released by the pancreas. Insulin increases the transport of glucose from the blood into cells, causing the level of sugar in the blood to fall again. The level of sugar in the blood, therefore, is balanced by the two hormones.

Disorders of the Endocrine System

The endocrine system is in a state of delicate balance. Sometimes one of the endocrine glands fails to work properly, and the balance of the system is upset. The result is one of the endocrine disorders listed in Table 21.3.

Table 21.3 Endocrine Disorders

Disorder	Description	Cause
Diabetes	Inability of cells to properly use glucose	Too little insulin
Dwarfism	Below-normal growth	Too little human growth hormone
Gigantism	Above-normal growth	Too much human growth hormone
Goiter	Enlargement of thyroid and neck	Too little iodine in the diet
Hyperactivity	Weight loss and high level of nervousness	Too much thyroxine

Discuss

Ask students the following questions:

▶ How is the endocrine feedback system similar to involuntary muscles? (Both work without our conscious control.)

▶ How is the endocrine feedback system similar to the nervous system? (Both work without our conscious control.)

Critical Thinking

Compare and Contrast Ask students to compare the pancreas to a thermostat. (Both work to maintain balance. The thermostat maintains temperature and the pancreas maintains blood sugar levels. Encourage students to be specific with their comparisons.)

Directed Inquiry

Have students study Table 21.3. Ask the following questions:

▶ In general, what is the cause of each disorder? (Too much or too little of a hormone)

▶ What can students do to prevent endocrine disorders? (Nothing. People have little or no control of the hormones produced by the body. But today, certain drugs, hormone therapy, and medical procedures help many people with endocrine disorders lead normal lives.)

**The Living Textbook:
Life Science Side 2**

Chapter 12   Frame 14547
Effect of Acetylcholine (Movie)
Search: Play:

Skills WorkOut
Encourage students to find out more about the disease diabetes by doing the Skills WorkOut.
Answer Students' reports should include the latest advances in treatment.

EVALUATE

WrapUp
Reteach Have students fold a piece of notebook paper in half. Have them title their columns *Hormone* and *What It Does*. Ask students to list at least ten hormones in the first column and write a brief description of their functions in the second column.
Use Review Worksheet 21.3.

Check and Explain
1. Pituitary gland produces HGH, which controls growth in muscles, bones and organs; thyroid gland produces thyroxine, which regulates metabolism, and produces calcitonin, which controls calcium and phosphorus levels; pancreas produces insulin and glucagon, which maintain blood sugar levels.
2. A chemical signal tells the gland to make more of a hormone when it is needed or less of a hormone when there is enough.
3. The hormone thyroxine contains iodine.
4. An adult probably would experience no change. A teenager would grow quickly.

INTEGRATED LEARNING

STS Connection
Insulin, the hormone that diabetics need, comes from cattle and sheep. However, some diabetics are allergic to these animals, so they can't use the insulin. So, biologists spliced the human gene for insulin to a common bacteria. As a result, this bacteria produces a safe, plentiful supply of human insulin. You may wish to refer students to Chapter 6, page 129, to review this genetically engineered procedure.

Skills WorkOut
Research Find out more about the disease diabetes. How is diabetes currently being treated? What advances have been made in the search for a cure? Write a one-page report on your findings.

Science and Technology
Synthetic Growth Hormone

In the past, scientists did not know why certain people never grew to normal size. Then it was discovered that most kinds of dwarfism occurred when the pituitary gland produced too little human growth hormone (HGH). With this discovery, scientists could treat dwarfism. They extracted HGH from the pituitary glands of people who had died. This HGH was injected into the bodies of children whose pituitary glands did not produce enough HGH. The children grew to normal height if they received regular HGH injections. However, the supply of HGH collected in this way was very limited.

Then in 1979, scientists succeeded in programming bacteria to make HGH through the process of genetic engineering. The human gene that carried the instructions for making HGH was spliced into the DNA of the bacteria. Large numbers of bacteria containing the HGH gene were cloned. The HGH produced by the bacteria was collected and purified. HGH produced in this way was approved for use in 1985. It is used in treating children whose growth is far below normal.

The synthetic hormone has helped many children grow to a normal height. Taking the hormone is not without some risk, however. Regulating the dose given to each child is very difficult. Also, the hormone can stimulate the leg bones to grow so fast that they slip out of the hip socket. Most people, however, think that using HGH is worth the risk.

Check and Explain
1. List the major endocrine glands, the hormones they produce, and the function of each hormone.
2. What does it mean for hormone levels to be controlled by feedback?
3. **Infer** Knowing that lack of iodine in the diet causes a disorder affecting the thyroid gland, what can you infer about the hormone thyroxine?
4. **Predict** A normal adult takes a dose of human growth hormone. What might happen? Why? What might happen if a normal teenager took HGH?

Chapter 21 Review

Concept Summary

21.1 Nervous System
- The nervous system is the communication and control system of your body. It includes the brain, spinal cord, and nerves.
- Nerve impulses travel through neurons using electrical and chemical energy.
- The brain consists of the cerebrum, cerebellum, and medulla. The spinal cord connects the brain to the rest of the body.

21.2 The Senses
- Sensory receptors in your body allow you to convert stimulus to sense.
- The senses allow you to smell, taste, touch, see, and hear.
- Your senses work with your brain to keep you informed about your environment.

21.3 Endocrine System
- Endocrine glands release chemicals needed by the body directly into the bloodstream.
- The major endocrine glands are the pituitary, thyroid, parathyroid, adrenal, pancreas, ovaries (in females) and testes (in males).
- Hormone levels in the body are regulated by endocrine glands. Hormones cause changes in other organs and regulate many body activities.
- Diabetes, dwarfism, gigantism, goiter, and hyperactivity are disorders of the endocrine system.

Chapter Vocabulary

neuron (21.1)	sensory neuron (21.1)	iris (21.2)	endocrine gland (21.3)
synapse (21.1)	retina (21.2)	eardrum (21.2)	hormone (21.3)
motor neuron (21.1)	pupil (21.2)	cochlea (21.2)	

Check Your Vocabulary

Use the vocabulary terms above to complete the following sentences correctly.

1. The three tiny bones of the ear transfer sound vibrations to the ___.
2. Hormones are released directly into the bloodstream by ___.
3. A nerve cell, or ___, receives and sends electrical messages.
4. Nerve tissue in the ___ is made of light and color receptors.
5. Light entering the eye is controlled by the ___, a round, colored disk.
6. Thyroxine is a ___ that regulates your body's metabolism.
7. Nerve fibers that lead from the spinal cord to the muscles are called ___.
8. Nerve fibers carrying sense information to the spinal cord are called ___.
9. In the front of the eye, the choroid layer has an opening called the ___.
10. Chemicals are released into the small space, or ___, that exists between one neuron and the next.
11. Sound waves strike the ___, causing it to vibrate.

Write Your Vocabulary

Write sentences using each vocabulary word above. Show that you know what each word means.

CHAPTER REVIEW 21

Check Your Vocabulary

1. cochlea
2. endocrine glands
3. neuron
4. retina
5. iris
6. hormone
7. motor neurons
8. sensory neurons
9. pupil
10. synapse
11. eardrum

Write Your Vocabulary

Be sure that students understand how to use each vocabulary word in a sentence before they begin their stories.

Use Vocabulary Worksheet for Chapter 21.

CHAPTER 21 REVIEW

Check Your Knowledge

1. The nervous system is the communication and control system for the body. It receives information from the environment and from inside the body.
2. The two main parts of the nervous system are the central nervous system (CNS) and the peripheral nervous system (PNS).
3. Table 21.2 on page 460 lists and describes the major endocrine glands.
4. Sensory receptors detect stimuli in the internal and external environments and carry impulses to the spinal cord. For example, chemicals present in the air are detected by receptors in the nose.
5. A reflex action is an automatic response to a (usually harmful) stimulus, and is controlled in the spinal cord.
6. Table 21.3 on page 461 lists and describes disorders of the endocrine system.
7. The parts of the brain are the cerebrum, cerebellum, and medulla.
8. True
9. False; endocrine
10. False; chemicals
11. True
12. False; pituitary
13. True
14. True

Check Your Understanding

1. Pain receptors are triggered and send an impulse through a sensory neuron to the spinal cord. In the spinal cord, the impulse in the sensory neuron triggers an impulse in an association neuron. This impulse triggers an impulse in a motor neuron, which signals the appropriate muscles to contract, moving the foot away from the sharp object.
2. The perception of flavor involves the detection of odor by receptors in the nose in addition to taste buds on the tongue. If the nose is stuffed up by a cold, it is difficult to detect odors.
3. Both control body functions by sending signals to various organs and tissues. While the endocrine system sends chemical signals through the bloodstream, the nervous system sends electrochemical signals along neurons.
4. a. Neurons are specialized nervous system cells that carry messages via electrical impulses through the body.
   b. The space between neurons is a synapse. An impulse reaching the end of an axon causes the release of chemical molecules that diffuse across the gap and trigger an impulse in another neuron.
5. Answers may vary. For example, reading would not be possible without a sense of sight (or of touch, for braille). Enjoying a favorite food would not be possible without taste and smell.
6. Extreme heat can cause serious harm very quickly. A quick response to heat helps protect the body from burning.
7. Answers may vary. For example, adrenaline from the adrenal glands stimulates various body organs and tissues to respond to the demands of the race. Also, glucagon from the pancreas can stimulate the liver to release glucose into the bloodstream for use in muscles.
8. The sense of balance is affected because the fluid in the semicircular canals takes a few moments to stop moving after the ride.
9. The iris contracts and relaxes, changing the diameter of the pupil.

Chapter 21 Review

Check Your Knowledge
Answer the following in complete sentences.

1. What is the function of the nervous system?
2. Name the two main parts of the nervous system.
3. Name three endocrine glands and their role in the endocrine system.
4. Describe what sensory receptors do. Give an example.
5. What is a reflex action? Where are reflex actions controlled?
6. Describe two disorders of the endocrine system, and their causes.
7. List the parts of the brain.

Determine whether each statement is true or false. Write *true* if it is true. If it is false, change the underlined word to make the statement true.

8. The largest part of the brain is the <u>cerebrum</u>.
9. Hormones are produced by the <u>nervous</u> system.
10. Taste is the result of <u>motion</u> being detected by sensory receptors.
11. The ear picks up <u>sound waves</u> and converts them into nerve impulses.
12. Human growth hormone is produced by the <u>thyroid</u> gland.
13. Hormone levels are tightly controlled by a feedback system that automatically turns a(n) <u>endocrine</u> gland on or off.
14. The brain is called the main <u>control center</u> of the body because it directs and coordinates all body processes, thoughts, behaviors, and emotions.

Check Your Understanding
Apply the concepts you have learned to answer each question.

1. **Critical Thinking** Describe the path of the reflex that occurs when a person steps on a sharp object.
2. Why can having a cold affect your sense of taste?
3. **Critical Thinking** Compare the endocrine system and the nervous system. How are they similar? How are they different?
4. **Mystery Photo** The photo on page 442 shows human nerve cells, or neurons. The photograph is a false-color electromicrograph: a photograph taken through an electron microscope.
   a. What are neurons and what do they do?
   b. What is the space between neurons, and how do impulses cross the gap?
5. **Application** Describe three things you could not experience without your senses. Identify the sense necessary to fulfill that experience.
6. **Extension** Reflex actions that involve heat receptors are faster than any other kind of reflex action. Why do you think this is important for your body?
7. What endocrine gland(s) could help you win a swimming race? Describe how the gland(s) can help you in this situation.
8. **Infer** After riding on a roller coaster that takes you on a series of loops, and comes to a stop, you still feel like you are spinning. What sense is related to this sensation? Why?
9. When you are in bright light, your pupils become small. When you move into dim light, they become large. What part of the eye controls this change?

Develop Your Skills

1. a. Fish, Bird, Snake
   b. Dolphin; the wide hearing range helps the dolphin avoid obstacles and find prey by echolocation.
2. a. Up to 100 years
   b. Answers may vary. Students may suggest that skin cells are constantly being replaced. When brain cells die they are not replaced.
3. a. reflex
   b. voluntary
   c. voluntary
   d. reflex

Make Connections

1.

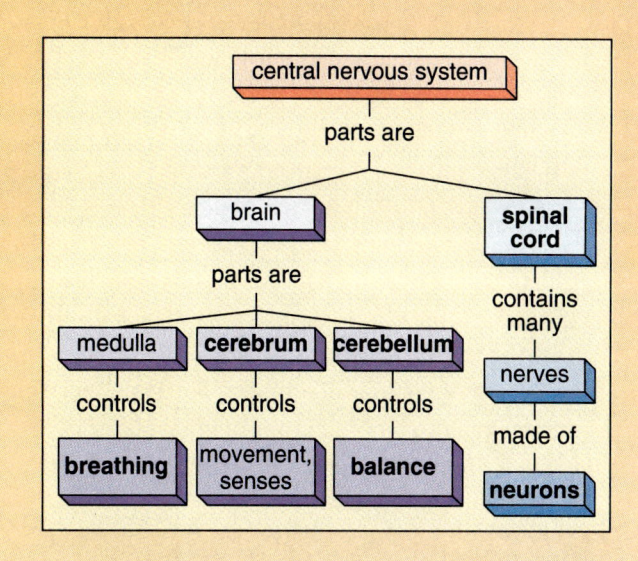

```
                central nervous system
                       parts are
                   ┌──────────┴──────────┐
                  brain              spinal cord
                parts are             contains many
         ┌────────┼────────┐             nerves
      medulla  cerebrum cerebellum      made of
      controls controls controls        neurons
      breathing movement, balance
                senses
```

2. Student-created illusions will vary. Optical illusions make use of discrepancies between the physical visual stimuli received by the eye and the perception of the stimuli by the observer.
3. Students' answers should explain how the substitute senses fulfill the needs of the sense that is lost.

Develop Your Skills
Use the skills you have developed in this chapter to complete each activity.

1. **Interpret Data** Sound is measured in hertz (Hz). Humans cannot hear sounds that measure much more than 10,000 Hz. The graph below shows the hearing ranges for various organisms. Study the graph, then answer the following questions.
   a. Which organisms have similar hearing ranges?
   b. Which organism is most sensitive to sounds? Explain why you think this organism has a wide hearing range.

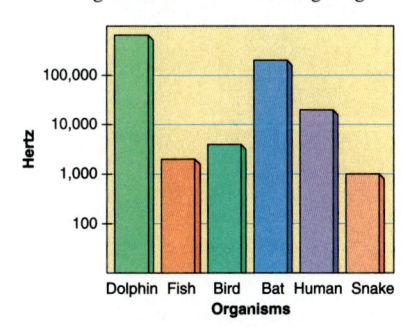

2. **Data Bank** Use the information on page 626 to answer the following questions.
   a. How long do brain cells live?
   b. **Infer** Why does a brain cell live so much longer than a skin cell?

3. **Define Operationally** Decide whether each of the following is a reflex action or a voluntary action:
   a. Coughing after eating spicy food.
   b. Climbing a tree.
   c. Blinking when a bug flies in your eye.

Make Connections

1. **Link the Concepts** Below is a concept map showing how some of the main concepts in this chapter link together. Only part of the map is filled in. Copy and finish the map, using words and ideas from the chapter.

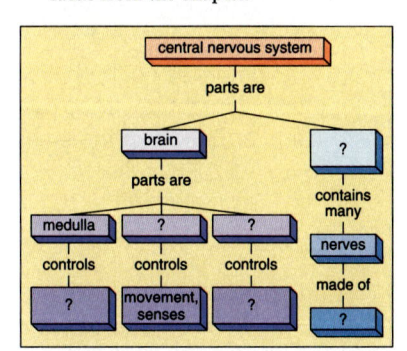

2. **Science and Art** Look through art reference books and find examples of optical illusions. For example, drawings and paintings by M.C. Escher are famous for their optical effects. After studying some optical illusions in drawings and paintings, see if you can draw some of your own. Test them out on a partner. See if your partner can "figure out" your illusion. Why do you think you see optical illusions?

3. **Science and You** Think of some things you do every day, like brushing your teeth or eating your lunch. Which of your senses do you use to do these activities? Imagine what it would be like to do the same activities if you had lost one of your senses, such as vision. What other senses might you use instead?

CHAPTER 22

Overview

This chapter explains human reproduction and life stages. The first section describes the reproductive structures of the male and female and the process of sexual maturation. Fertilization and the major stages of reproduction, including pregnancy and birth, are presented in the second section. The last section traces the primary life stages of human development.

Advance Planner

▶ Provide old magazines for SE page 473.

▶ If desired, arrange to have students visit a nursing home or have a group of elderly visit the class.

▶ Collect clay, labels, pennies, measuring cups, identical clear containers, and taller, thinner clear containers.

Skills Development Chart

Sections	Collect Data	Communicate	Graph	Model	Observe	Research
22.1 Skills WarmUp SkillBuilder Skills WorkOut	●		●	●		
22.2 Skills WarmUp Skills WorkOut Historical Notebook		●			●	● ●
22.3 Skills WarmUp Activity	●	●	●		●	

Individual Needs

▶ **Limited English Proficiency Students** Cover the labels and arrange write-on lines beside different parts of each of the figures on pages 468, 469, and 474. Make enough copies of the figures for students. Have them make labels from strips of index cards and place them in the correct places on the figures. Have them refer to the text to check the placement of the labels. Students can save the labels in an envelope so they can review the terms by arranging the labels again after they've studied the chapter.

▶ **At-Risk Students** Have students create their own concept summaries. First, give each student a photocopy of the concept summary at the end of the chapter. Then, as they read the chapter, have them use their own words to describe the concepts in the chapter summary. Have students meet in cooperative groups to compile a group concept summary. Photocopy this summary and allow students to use it to review the material in this chapter.

▶ **Gifted Students** Ask students to research basic life stages as studied by at least two major psychologists (Erikson, Piaget, and so on). Have them create a report that includes photographs of family members who are at various life stages.

Resource Bank

▶ **Bulletin Board** Have students create personal timelines that show the major events in their lives up to the present. Encourage them to add photographs or drawings. Arrange the timelines on the bulletin board. As you study the last section, it might be fun to have students predict future events and add them to the timelines.

CHAPTER 22 PLANNING GUIDE

Section	Core	Standard	Enriched	Section	Core	Standard	Enriched
22.1 Human Reproductive Systems pp. 467–472				**Blackline Masters** Review Worksheet 22.2	●	●	●
Section Features Skills WarmUp, p. 467 SkillBuilder, p. 471 Skills WorkOut, p. 472	● ●	● ● ●	● ●	**Ancillary Options** One-Minute Readings, pp. 2, 38	●	●	●
				Laboratory Program Investigation 42		●	●
Blackline Masters Review Worksheet 22.1 Reteach Worksheet 22.1	● ●	● ●	●	**22.3 Human Life Stages** pp. 478–482			
Laboratory Program Investigation 42	●	●	●	**Section Features** Skills WarmUp, p. 478 Activity, p. 482	● ●	● ●	● ●
Color Transparencies Transparencies 69, 70, 71	●	●		**Blackline Masters** Review Worksheet 22.3 Skills Worksheet 22.3 Integrating Worksheet 22.3 Enrich Worksheet 22.3 Vocabulary Worksheet 22	● ● ● ● ●	● ● ● ● ●	● ● ● ● ●
22.2 Fertilization, Pregnancy, and Birth pp. 473–477							
Section Features Skills WarmUp, p. 473 Skills WorkOut, p. 476 Historical Notebook, p. 476	● ● ●	● ● ●	●				

Bibliography

The following resources can be used for teaching the chapter. See page T-40 for supplier codes.

Audio-Visual Sources
(video unless noted)
Growing Up Female. 12 min film. 1983. JF.
Human Biology: Aging. 26 min. 1985. FH.
The Human Body Series: Reproductive Systems. 20 min. 1988. NGSES.
The Human Body Series—Stages of Life: Reproduction, Growth, and Change. Filmstrips with cassettes. 1980. NGSES.
Human Reproduction. 3d ed. 20 min film. 1981. CRM.
The Miracle of Life. 60 min. NOVA.

Software Resources
The Human Systems—Series III: The Reproductive System. Apple. FM.
Practicing Sexual Decision Making. Apple. CABISCO.
Reproduction and Growth. Apple. J & S.
Understanding AIDS. Apple, IBM PC. CABISCO.

Library Resources
Elting, Mary. *The Macmillan Book of the Human Body.* New York: Aladdin, 1986.
Mayle, Peter. *What's Happening to Me?* Secaucus, NJ: Lyle Stuart, Inc., 1975.
Meredith, Susan, Ann Goldman, and Tom Lisauer. *Book of the Human Body.* Tulsa, OK: Educational Development Corp., 1983.
Miller, Jonathan, and David Pelham. *The Facts of Life.* New York: Viking Penguin, 1984.

CHAPTER 22

INTEGRATED LEARNING

Writing Connection
Have students take a moment to observe the photograph. Then have them write a haiku poem that describes the experience of beginning life.

Themes in Science
Patterns of Change/Cycles
Human beings, like all organisms, develop according to predictable patterns. Each individual begins life as a fertilized egg. The egg develops into an embryo, and then a fetus. After birth, human beings continue to grow and develop. As adults, they are able to reproduce, continuing the cycle.

Introducing the Chapter
Have students read the description of the photograph on page 466. Ask if they agree or disagree with the description.

Directed Inquiry
Have students study the photograph. Ask:

▶ What is the object in the photograph? How do you know? (Most students will recognize the human fetus because of the toes, fingers, and head.)

▶ Where is the fetus in the picture? (Inside the uterus)

▶ Do you think that the baby in the picture is fully developed? Explain. (No, because the head is much larger in proportion to the rest of the body than a baby's head is at birth)

▶ What is the topic of this chapter? (Reproduction and life stages)

▶ Why do you think a photograph of a fetus begins this chapter? (Answers will vary. The fetus is the earliest life stage during which the human form is recognizable.)

Chapter Vocabulary

fertilization	penis
fetus	scrotum
labor	semen
menstruation	testes
ovaries	uterus
oviduct	vagina

Chapter 22 Reproduction and Life Stages

Chapter Sections

22.1 Human Reproductive Systems

22.2 Fertilization, Pregnancy, and Birth

22.3 Human Life Stages

What do you see?

"I see the fetus of a human baby. It is in the mother's uterus. It has been developing for about 3¾ months. The fetus has all its toes and fingers. Its body is catching up to its head in size. This image was created by ultrasound."

*Tim Kelley
Memorial Junior High
Minot, North Dakota*

To find out more about the photograph, look on page 484. As you read this chapter, you will learn about growth and reproduction in humans.

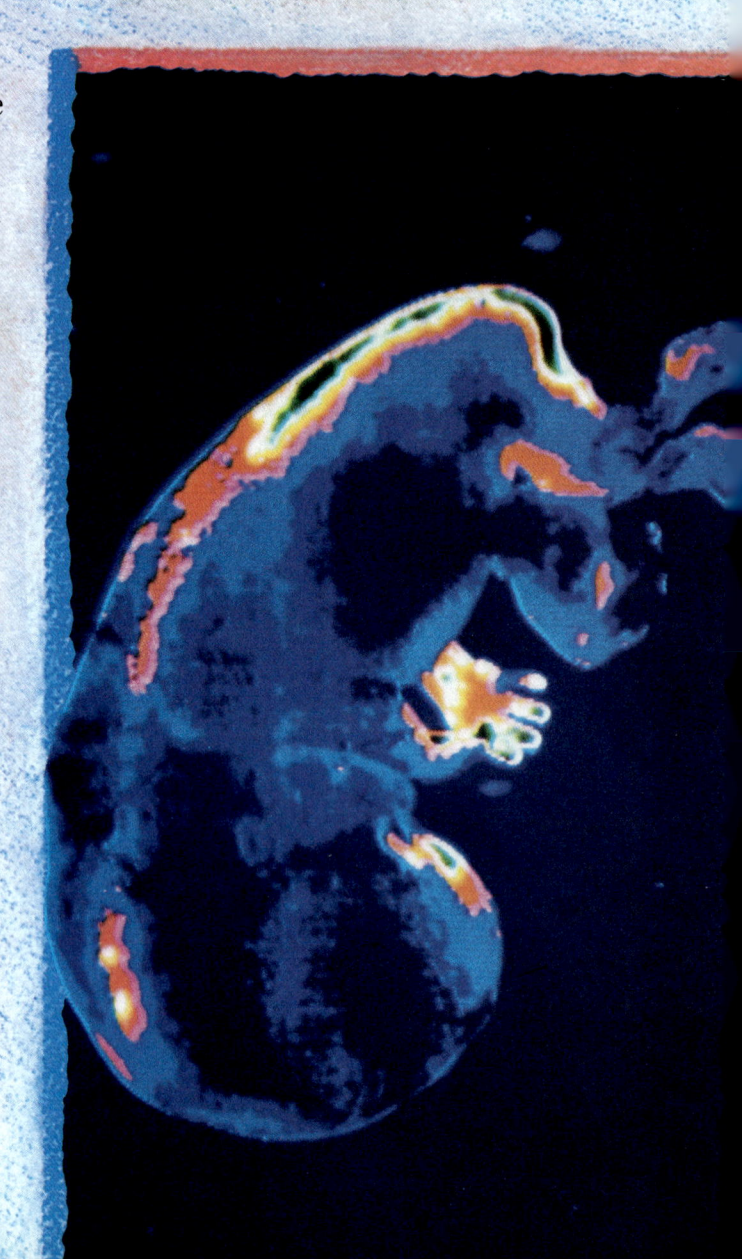

Multicultural Perspectives

Many cultures recognize puberty as an important stage in an individual's development. In parts of Latin America, a special party called a Quinceañera is given for girls when they turn 15. In India, people who have reached puberty traditionally become engaged. Many Jewish people celebrate their bar or bas mitzvah when they turn 12 or 13. Ask students what other kinds of cultural traditions exist in the United States that recognize the onset of puberty.

22.1 Human Reproductive Systems

Objectives

▶ **Name** and **describe** the male and female reproductive organs and their functions.

▶ **Explain** the process of menstruation.

▶ **Communicate** how hormones are part of the reproductive systems in males and females.

Skills WarmUp

Make a Graph Work in small groups and measure the leg length of each person in the group. Measure from the knee to the floor. Record the information. Then make a bar graph that shows the different leg lengths. Do you think that the bar graph would look very different if you measured everyone's legs next year? Predict how much each person's measurement might change.

Do you ever look at your old school pictures and wonder who that little kid was? You have been growing and changing since you were born. You'll change and grow up in different ways over the next few years, too. Sometime between ages 10 and 15, girls and boys usually experience a growth spurt. They not only get taller, but their bodies also begin to change in other ways. In girls, the mammary glands in their breasts develop and their hips widen. Boys begin to develop facial hair. Their voices deepen, and their shoulders broaden.

These changes in females and males are caused by the hormones produced in the reproductive system. They signal the beginning of sexual maturity, which is the ability to reproduce. The time when the human body becomes sexually mature and capable of reproducing is called puberty. Puberty usually occurs between ages 9 to 14 for females, and 11 to 16 for males. However, it can occur earlier or later, too.

Figure 22.1
◀ Females and males both experience many physical and emotional changes during puberty.

SECTION 22.1

Section Objectives
For a list of section objectives, see the Student Edition page.

Skills Objectives
Students should be able to:

Make a Graph of classmates' leg lengths.

Make a Model of the menstrual cycle.

Collect Data on STDs and write a report on findings.

Communicate the role of hormones in reproduction.

Vocabulary
testes, scrotum, semen, penis, ovaries, oviduct, fertilization, uterus, vagina, menstruation

MOTIVATE

Skills WarmUp
To help students understand human growth, have them do the Skills WarmUp.
Answer Next year's leg lengths would be significantly longer because of the growth spurt students are now experiencing at puberty.

Prior Knowledge

Gauge how much students know about human reproduction by asking the following questions:

▶ What happens during puberty?

▶ What occurs during fertilization?

▶ How do hormones affect human reproduction?

Chapter 22 Reproduction and Life Stages

TEACH

Directed Inquiry

Have students study Figure 22.2 and read the page. Then ask the following questions:

▶ Where are sperm produced and stored? (Produced in the testes and stored in sperm ducts)

▶ What fluids pass through the urethra? (Sperm and urine)

▶ What is the function of the scrotum? (To protect the testes)

▶ What is the difference between sperm and semen? (Sperm are male sex cells. Semen is a nutrient-rich fluid that carries and protects sperm.)

Skills Development

Infer Have students compare the number of seeds produced by seed-bearing plants and the number of sperm produced by the human male. Have them infer why it is advantageous for a species to produce a large number of male sex cells. (Higher numbers improve the chances of fertilization.)

The Living Textbook:
Life Science Sides 1-4

Chapter 7 Frame 02668
Male Reproductive System (1 Frame)
Search:

The Living Textbook:
Life Science Side 2

Chapter 17 Frame 33467
Sperm (Movie)
Search: Play:

468

INTEGRATED LEARNING

Social Studies Connection

Have students recall what they learned about Anton van Leeuwenhoek in Chapter 4, Section 4.1. Tell them that Leeuwenhoek also used his microscopes to view human sperm. He believed that each sperm contained a tiny human being that would grow to normal size after fertilization. Discuss Leeuwenhoek's misconception about sperm, and encourage students to explain why his theory was incorrect.

The Male Reproductive System

The changes in the body of a male are triggered by the hormone testosterone (tehs TAHS tur OHN). Testosterone is produced in the male sex glands, the **testes** (TEHS teez). Look at Figure 22.2 and find the testes. Notice that they are inside the **scrotum** (SKROHT uhm), a protective sac.

Because the scrotum is outside the male's body, the testes can maintain a cooler temperature than the rest of the body. This is important for the production of sperm cells. Sperm cells are male sex cells. Sperm can only survive at a temperature that is a few degrees cooler than the body. Production of sperm begins at puberty. During the rest of a male's lifetime, the testes will make hundreds of billions of sperm.

Notice in Figure 22.2 that the testes connect to two sperm ducts. Sperm are stored in these ducts after they are produced in the testes. The two sperm ducts connect to the urethra. As sperm move from the sperm ducts to the urethra, they are mixed with fluids. These nutrient-rich fluids are produced by glands. The fluids protect the sperm, and provide sperm with energy. With this energy, sperm can swim as well as survive outside the male's body. The mixture of sperm cells and fluids is called **semen** (SEE muhn).

As you can see in Figure 22.2, the urethra extends through the **penis**. The penis is the external male organ through which both semen and urine flow out of the body. However, they do not mix. Only one fluid at a time enters the urethra. The penis also releases semen into the female's body during sexual intercourse.

Figure 22.2
The male reproductive system; a close-up view of a testis is shown. ▼

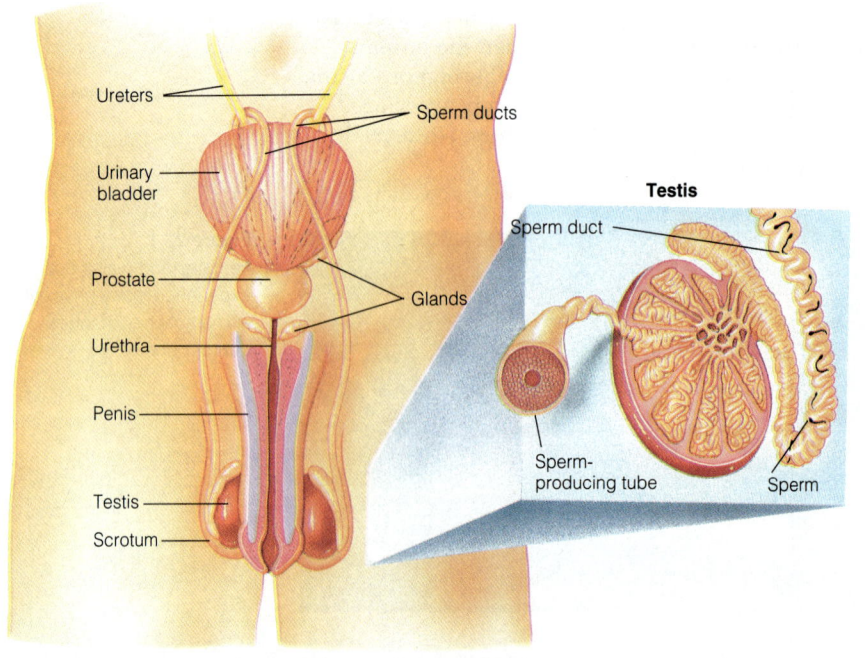

468 Chapter 22 Reproduction and Life Stages

Language Arts Connection

Have students look up the origins of the terms *ovary* and *duct*. (*Ovary* is from the Latin *ovarium*, meaning "egg"; *duct* means "to lead.") Ask them to explain the meaning of the word *oviduct*, based on the Latin definitions. (Eggs are led through oviducts.)

Themes in Science

Scale and Structure Explain to students that the human egg, or ovum, is about a tenth of a millimeter in diameter. Despite its small size, it has almost 200,000 times the volume of a single sperm. The largest animal ovum is found in some species of shark and is about 15 centimeters long.

Directed Inquiry

Have students study Figure 22.3 and read the page. Then ask the following questions:

▶ Where are egg cells produced and stored? (In the ovaries)

▶ What happens during fertilization and where does it occur? (A sperm cell unites with a mature egg in an oviduct and forms a new cell called a *zygote*.)

▶ What is the function of the uterus? (It is where the zygote develops into a baby.)

▶ What happens to a mature egg that is not fertilized? (It dies and passes out of the body.)

Critical Thinking

Compare and Contrast Ask students to compare and contrast the urethra in the female and male. (Students should recognize that the male urethra carries both urine and sperm and is located in the penis; whereas, the female urethra carries only urine and is a separate passage from the vagina.)

The Female Reproductive System

The changes in the body of a female are triggered by the hormone estrogen (EHS truh juhn). Estrogen is produced in the female sex glands, the **ovaries** (OH vuh reez). Find the ovaries in Figure 22.3.

The ovaries contain egg cells. When a female is born, her ovaries already hold all the eggs she will ever have. At birth, these eggs are tiny. During the years that a female can have children, one egg will mature each month. A mature egg is the largest cell in the human body.

Look at Figure 22.3. When an egg matures in one of the two ovaries, it is released into an **oviduct** (OH vih DUHKT). The oviducts are close to the ovaries, but they aren't actually connected to them. In an oviduct, **fertilization** occurs. Fertilization is the process in which an egg cell and sperm cell unite to form one cell, called a zygote.

The egg is swept through the oviduct by cilia into the **uterus** (YOOT uhr uhs). The uterus is a hollow organ with muscular walls. A zygote will embed itself into the lining of the uterus. There the zygote begins to divide and grow. Eventually, the egg will develop into a baby. If the egg is not fertilized, it dies and passes out of the female's body.

Notice that the lower end of the uterus opens to a canal called the cervix (SUR vihks). The cervix is connected to the **vagina** (vuh JY nuh), a passageway leading to the outside of the female's body. The vagina receives the male penis during sexual intercourse. The penis releases semen into the vagina. A female has a separate opening for releasing urine.

The vagina is also called the birth canal. When a baby is born, it passes from the uterus, through the cervix, and out of the body through the vagina.

Figure 22.3
The female reproductive system; a front and side view is shown. ▼

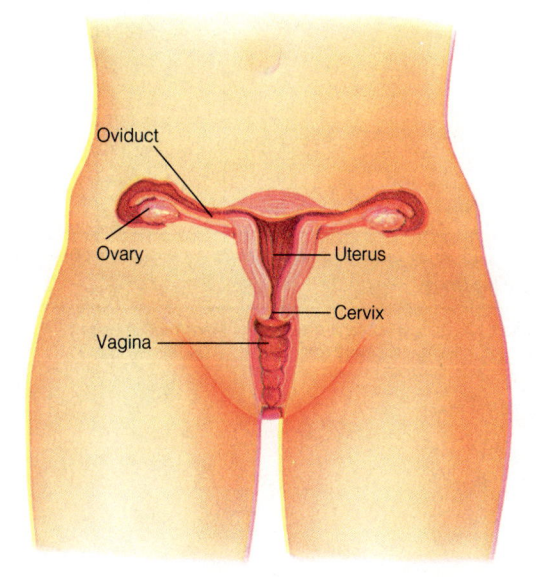

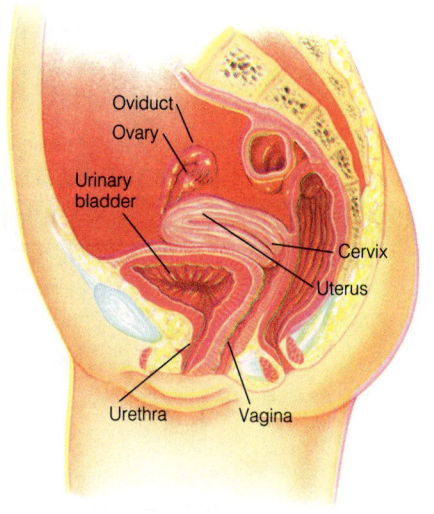

The Living Textbook:
Life Science Sides 1-4

Chapter 7 Frame 02669
Female Reproductive System (1 Frame)
Search:

Chapter 22 Reproduction and Life Stages **469**

TEACH • Continued

Directed Inquiry
Have students read pages 470 and 471. Ask the following questions:

▶ What are the two reproductive cycles in the female and which hormones control them? (The menstrual cycle is controlled by progesterone and the ovulatory cycle is controlled by estrogen.)

▶ About how long does each cycle last? (28 days)

▶ What is the purpose of the uterine lining if an egg is fertilized? (It nourishes and protects the fertilized egg as it develops into an embryo.)

▶ What hormone causes the uterine lining to thicken? (Progesterone)

Skills Development
Interpret Data Have students study Figure 22.4. Ask them to describe what happens during the first five days of the menstrual cycle if an egg is not fertilized. (The lining of the uterus breaks down and leaves the body through the vagina.)

Answer to In-Text Question
① Ovulation begins on day 14.

The Living Textbook: Life Science Side 2

Chapter 17   Frame 34543
Human Uterus and Ovary (Movie)
Search: Play:

The Living Textbook: Life Science Side 2

Chapter 17   Frame 35139
Ovulation & Egg Transport (Movie)
Search: Play:

470

INTEGRATED LEARNING

Themes in Science
Patterns of Change/Cycles

Sometimes a woman releases two eggs during her menstrual cycle. The second egg is always released within 48 hours of the first. Unfertilized eggs can survive only 12 to 24 hours in the uterus. Sperm can live in the female body for up to five days. Ask students to determine the maximum amount of time in which fertilization can occur during a normal monthly cycle. (Seven days)

The Female Reproductive Cycle

The female reproductive cycle is actually two cycles and two hormones working together. Both cycles, the menstrual (MEHN struhl) cycle and the ovulatory (AHV yoo luh TOR ee) cycle, occur each month within 14 days of each other. During the ovulatory cycle, which is controlled by the hormone estrogen, an egg matures and is released. The menstrual cycle, which is controlled by the hormone progesterone, prepares the uterus for the fertilized egg. Look at Figure 22.4 as you read about the interaction of the menstrual and ovulatory cycles.

Together, both cycles take about 28 days to complete, although it may vary from month to month. A female's first menstrual cycle usually begins between the ages of 11 and 14.

Menstrual Cycle During the menstrual cycle, progesterone causes the lining of the uterus to thicken. The blood vessels and fluids that develop there will nourish the fertilized egg. If the egg is not fertilized, the lining breaks down and **menstruation** (MEHN STRAY shun)

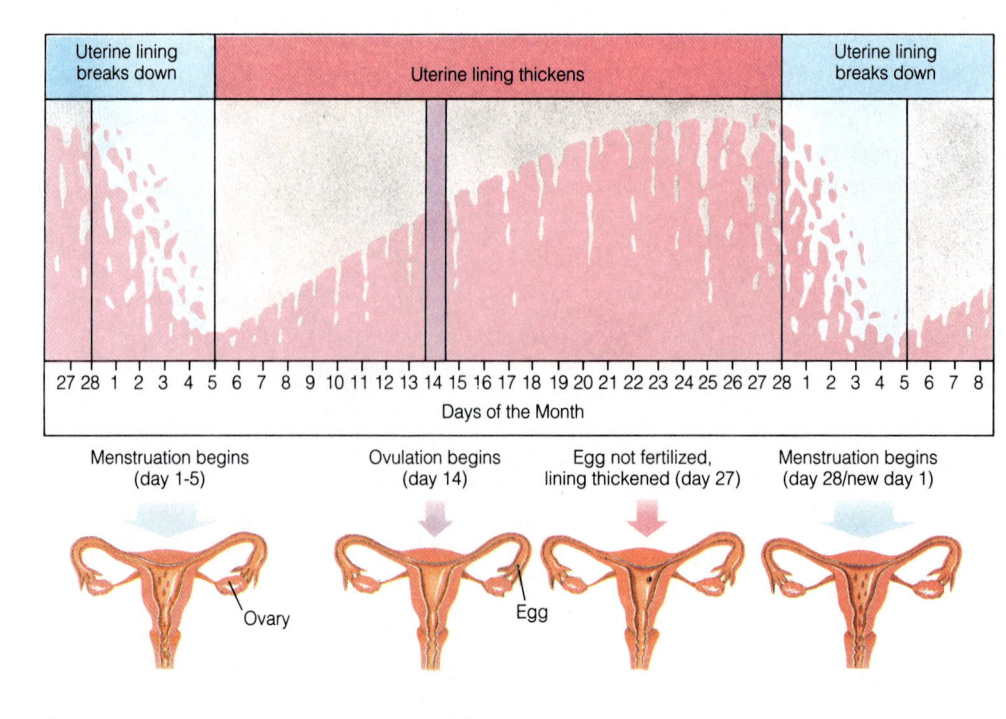

Figure 22.4 ①
In the female reproductive system, the menstrual cycle occurs each month. On what day does ovulation begin? ▼

470   Chapter 22 Reproduction and Life Stages

Integrating the Sciences

Health Many advertisements for vitamin supplements emphasize that they contain iron. Ask students to infer why many women need to supplement their diets with iron. (During menstruation, women lose about 60 to 75 milliliters of fluids, which contain iron. Iron supplements help replace this lost iron.)

begins. During the menstrual period, which lasts for three to seven days, the blood and tissue of the uterine lining and the unfertilized egg flow from the uterus. The flow leaves the body through the vagina.

These bodily changes that occur during the menstrual cycle sometimes cause abdominal cramps. In the week before menstruation, some women also experience premenstrual syndrome, or PMS. PMS can bring on a number of symptoms, including headaches, weight gain, tension, or depression. Although many women are not affected by PMS, it can be a serious problem for some.

Ovulatory Cycle About 14 days after the onset of menstruation, a drop in the estrogen level triggers the release of a mature egg from the ovary. This marks the beginning of the ovulatory cycle. If the egg is fertilized, it attaches itself to the lining of the uterus and pregnancy begins. If the egg is not fertilized, menstruation occurs about 14 days after ovulation.

SkillBuilder *Making a Model*

Charting the Menstrual Cycle

You will be modeling a female's 28-day menstrual cycle. Choose any 2-month period. Use a large sheet of paper and colored pencils to make a timeline showing all the dates of these 2 months. Be sure to label the days and the dates on your timeline.

Assume that the menstrual period lasts 5 days, that an egg is released from the ovary on the 14th day of the month, and that the egg is not fertilized. Refer to Figure 22.4 on page 470. Use the illustration to label what is happening each day of the month. Be sure to mark the following events on the timeline:

▶ Day when the uterine lining breaks down.

▶ Days when the uterine lining builds up.

▶ Day when the menstrual cycle begins.

▶ Day when ovulation begins.

After labeling the dates, make drawings of the uterine lining that show what is happening to it during the cycle. Draw the oviducts, the ovaries, and the movement of the egg. Then label all of these parts. Compare your drawing to Figure 22.4.

1. When could fertilization occur in this cycle?
2. When will ovulation begin again?
3. When will the first menstrual period begin? The second? Mark the start of the second period on your timeline.
4. Why does the uterine lining thicken? Why does it break down?

Sometimes a female's menstrual cycle is affected by factors such as stress and strenuous exercise. In a paragraph, infer how these factors might affect the cycle.

Chapter 22 Reproduction and Life Stages **471**

Discuss

Explain that the number of days between menstrual cycles is usually 28, but can vary from 21 to 35 days, depending on the individual. In addition, cycles can be interrupted or can occur irregularly if a woman is ill, under stress, or very active physically. Cycles are also sometimes irregular in girls when they first begin menstruating.

Misconceptions

Females who get their periods or become sexually mature earlier or later than most of their classmates may think that something is wrong with them. Explain that it is normal for sexual maturity to occur at different ages and that it is a gradual process that can be quite different for each person.

SkillBuilder

Answers

1. Between day 14 and day 21
2. Day 14 of the next 28-day cycle
3. The first period begins on day 28; the second begins on the next day 28.
4. The lining thickens to nourish the fertilized egg. It breaks down when the egg is not fertilized.

Students should realize that these factors can interrupt a woman's cycle or cause it to occur irregularly.

TEACH • Continued

Skills WorkOut
To help students learn about STDs, have them do the Skills WorkOut.
Answers Have students call or visit the local library for information on where they can get statistics about STDs.

EVALUATE

WrapUp
Reteach Ask students to compare and contrast the female and male reproductive systems. Have them explain how they function together to create a fertilized egg.
Use Review Worksheet 22.1.

Check and Explain

1. Testes produce sperm and testosterone; ovaries produce and store hormones and eggs.
2. The discharge of the uterine lining and an unfertilized egg; menstruation occurs when the egg is not fertilized.
3. Both eggs and sperm are sex cells. Sperm production begins at puberty and continues throughout a male's life. All of the female's few hundred eggs exist in her ovaries at birth. After puberty, one egg matures and is released monthly until menopause.
4. At puberty, the male hormone testosterone triggers the development of facial hair, broadened shoulders, deepened voice, and sperm production. The female hormone estrogen causes mammary glands to develop, hips to widen, and eggs to mature monthly.

INTEGRATED LEARNING

STS Connection
Chlamydia is the most common STD in the United States, infecting approximately three to four million people each year. Chlamydia is a bacterium that lives on host cells. In women, the disease can cause irregular menstrual cycles and sometimes sterility. In men, chlamydia causes testicular pain and can lead to arthritis. A mother can pass on the disease to her child during birth as the infant passes through the birth canal. Have students find out how chlamydia is diagnosed and treated.

Skills WorkOut
Collect Data Find out how many new cases of each of the following STDs were reported in the United States in the last three years: gonorrhea, AIDS, chlamydia, syphilis, genital herpes. Make a chart of your findings. Which disease occurred most often? Which disease occurred least often? Is any disease declining in numbers? Is any disease increasing in numbers? Write a report on your findings.

Science and Society *STDs*
A serious problem in many countries today is the spread of diseases by sexual contact. A sexually transmitted disease, or STD, is caused by either a virus or a bacterium. For example, chlamydia (kluh MID ee uh) and gonorrhea (GAHN uh REE uh) are two STDs caused by bacteria. Both of these diseases can cause painful urination or a cloudy discharge from the sex organs. However, there are often no symptoms, especially in the early stages. Complications, such as severe infections, may eventually occur if an infected person does not seek treatment. If caught in time, chlamydia and gonorrhea can be cured with antibiotics.

Syphilis (SIHF uh luhs) is also a bacterial disease that causes serious problems if it is not treated. Symptoms include sores on the mouth and sex organs, rashes, and a fever. Although these symptoms disappear, the bacteria remain in the blood. Years later the bacteria can cause blindness, paralysis, mental disorders, or heart failure.

Genital herpes (HUR peez) is a viral disease that causes painful, contagious blisters on the sex organs. Although the blisters go away after a few weeks, the virus continues to live in the body. The blisters may reappear again and again for years. Although the symptoms of genital herpes can be treated, there is no cure.

AIDS is usually transmitted through sexual contact. However, people can also get AIDS from contaminated blood and intravenous drug use. AIDS is caused by HIV, which attacks and weakens the cells of the immune system. A person with AIDS becomes less resistant to infections and diseases. Many people have died as a result of AIDS. So far, there is no cure.

Check and Explain

1. Describe the two main functions of the testes and two functions of the ovaries.
2. What is menstruation? What causes it to occur?
3. **Compare and Contrast** In what ways are eggs and sperm different? In what ways are they alike?
4. **Communicate** Explain how hormones are involved in a male's and female's reproductive developments.

Math Connection

For humans, the gestation period (the time in which a fetus develops) lasts about nine months. Different species have different gestation periods. Have students create a bar graph showing the gestation periods of the following animals: horse—11 months; cat—nine weeks; rabbit—30 days; mouse—20 to 30 days; elephant—18 to 23 months; and dog—nine weeks. (Students should use *months* for the y-axis.)

SECTION 22.2

Section Objectives
For a list of section objectives, see the Student Edition page.

Skills Objectives
Students should be able to:

Observe and describe a newborn baby.

Research to find out how fraternal and identical twins develop.

Infer how knowledge of the birth process might be useful to a parent.

Vocabulary
fetus, labor

MOTIVATE

Skills WarmUp
To help students make observations about newborn babies, have them do the Skills WarmUp.
Answer Newborn babies are dependent on adults, can't walk or talk, and sleep a lot.

Prior Knowledge
To find out how much students know about fertilization, pregnancy, and birth, ask the following questions:

▶ What is fertilization?
▶ What is a placenta?
▶ How long is pregnancy?

Decision Making
If you have classroom sets of *One-Minute Readings,* have students read Issue 1, "Definition of Life" on page 2. Discuss the questions.

22.2 Fertilization, Pregnancy, and Birth

Objectives

▶ **Describe** the process of fertilization.
▶ **Describe** the development of the fertilized egg, embryo, and fetus.
▶ **Summarize** the stages of labor.
▶ **Infer** the importance of understanding the birth process.

Skills WarmUp

Observe Look for a photograph of a newborn baby in a book or magazine, or think of a newborn you may have actually seen recently. Write down everything you observe about the baby. What makes a newborn baby so special?

Is there a newborn baby in your family or a friend's family? If so, you may know that pregnancy and birth are special times for both women and men. Pregnancy is the period of time between fertilization and birth. Pregnancy lasts for about nine months in humans. It is a time of dramatic changes in a woman's body.

Fertilization

During sexual intercourse, the male's penis enters the vagina of the female. The penis releases semen, which is filled with millions of sperm. The sperm swim rapidly from the vagina, through the cervix and uterus, and on into both oviducts.

Depending on the female's menstrual cycle, an egg may be present in one of the oviducts. If it is, the egg sends out a chemical that attracts and guides the sperm to the egg. Often many sperm will surround the egg and stick to it. Although many sperm may attach to the jellylike coating of the egg, only one sperm can penetrate through the egg's cell membrane beneath this jelly coating. The nuclear material of this single sperm then enters the cytoplasm of the egg cell and fertilizes it.

Immediately following fertilization, the egg releases chemicals that form a protective membrane around the fertilized egg, or zygote. The membrane prevents other sperm from entering.

The zygote now contains 46 chromosomes: 23 from the egg and 23 from the sperm. The zygote will begin to develop into a human being.

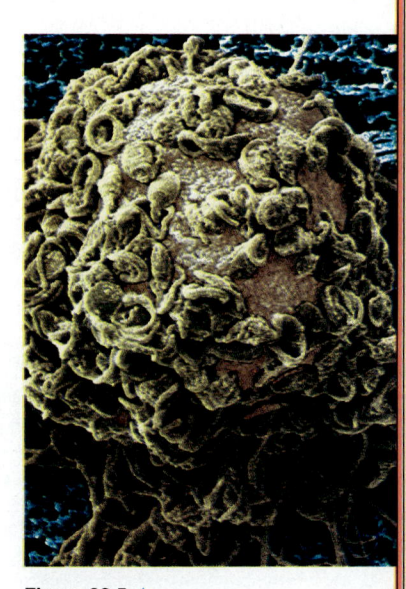

Figure 22.5 ▲
This photograph shows sperm on an egg. The sperm and egg have been magnified 3,400 times.

Chapter 22 Reproduction and Life Stages **473**

TEACH

Directed Inquiry

Have students study the material on pages 473, 474, and 475. Ask the following questions:

▶ How does the egg prevent more than one sperm from entering during fertilization? (It releases chemicals that form a protective membrane around it.)

▶ How many chromosomes are contained in an egg and a sperm? (23 in each)

▶ What is the function of the placenta? (It brings nutrients from the mother to the embryo and takes away wastes.)

▶ What is the umbilical cord? (A soft rope of blood vessels that connects the embryo to the placenta)

▶ When does most of the growth of the fetus occur? (During the last three months of pregnancy)

Reteach

Remind students of what they learned in Chapter 6 about DNA. Ask them what makes each sperm cell unique. (Each has a random pattern of genes that is encoded in DNA. Each gene carries an individual set of instructions to developing cells.)

**The Living Textbook:
Life Science Sides 1-4**

Chapter 7   Frame 02670
Fertilization (1 Frame)
Search:

**The Living Textbook:
Life Science Side 7**

Chapter 33   Frame 31802
Fetal Development (Movie)
Search: Play:

INTEGRATED LEARNING

Themes in Science

Systems and Interactions Cells divide through the process of mitosis. One cell divides into two, two into four, four into eight, and so on. For the first few divisions, all the cells divide at about the same time. However, as the cells begin to diversify, this synchronization ends and cells take on specialized functions. Some cells become part of the nervous system, others become part of the skeletal system, and so on.

Development Before Birth

The zygote, smaller than a grain of salt, will undergo dramatic changes and growth. Over the next nine months, the zygote will develop into a human baby that is made of billions of cells. As the zygote travels down the oviduct, cell division begins. One cell becomes two, two cells become four, and so on, until it forms a small ball. It continues to divide and change, reaching the uterus about six days after fertilization. The zygote attaches to the inner wall of the uterus. In Figure 22.6b, notice that the developing individual, called an embryo, does not yet look much like a human.

Within the uterus, a placenta forms. The placenta is an organ that brings nutrients to the embryo and takes away wastes. Connecting the embryo to the placenta is the umbilical cord, a soft rope of blood vessels. The umbilical cord allows the embryo to float freely within its fluid-filled sac. Although materials can pass back and forth between mother and embryo, their blood does not mix. The blood is kept apart by a thin layer of cells.

After nearly eight weeks, the embryo begins to look like a human. Fingers, toes, and eyelids have developed. All of the major internal organs have formed. From eight weeks until birth, the developing baby is called a **fetus**. You can see the fetus at ten weeks in Figure 22.6c.

From the third to sixth month of pregnancy, the fetus grows rapidly. The mother begins to feel its kicks and somersaults as it moves around. At this time, the fetus has a little hair on its head, and its fingernails have grown. It begins a regular pattern of sleeping and waking.

Most of the fetus's growth takes place during the last three months of pregnancy. A layer of fat develops under the skin, and the bones harden. Near the end of the ninth month, the fetus is about 45 cm long and weighs 3 to 3.5 kg.

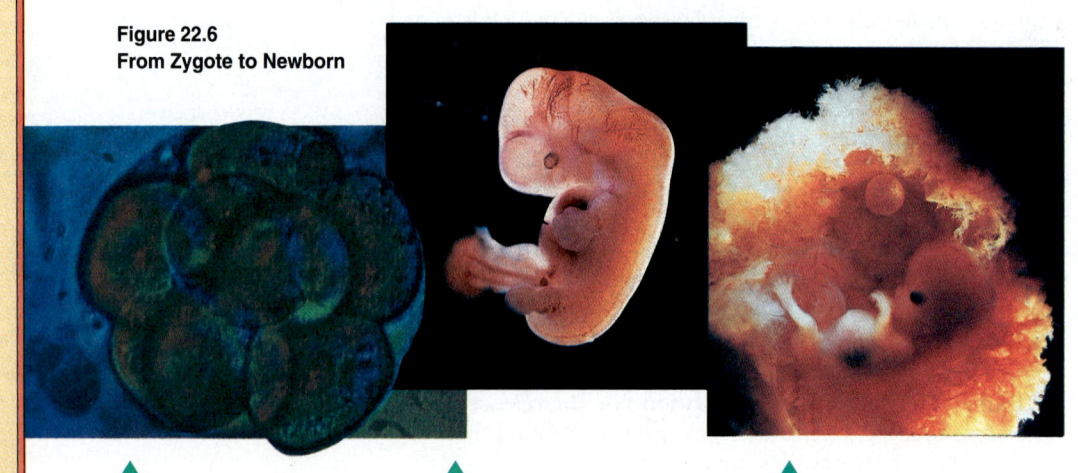

**Figure 22.6
From Zygote to Newborn**

▲ **a.** In this phase of mitosis, the zygote has divided eight times.

▲ **b.** At 5 weeks, the 1 cm long embryo has a head, buds where arms and legs will form, and a beating heart.

▲ **c.** At 10 weeks, the fetus has a human shape and is about 5 cm long. Its mass is about 25 g.

474 Chapter 22 Reproduction and Life Stages

Integrating the Sciences

Chemistry The hormone oxytocin causes the uterus to contract during childbirth. When labor is not progressing quickly enough, doctors may administer an artificial oxytocin called *pitocin*. Have students research pitocin and share their findings with the class.

STS Connection

Have students interview an obstetrician or read information on cesarean section births.

Discuss

Ask students the following questions:

▶ When is the fetus developed enough to survive outside the mother's body? (At about nine months)

▶ When does the birth process begin? What is the process called? (When the muscles of the uterus contract and relax at regular intervals; labor)

▶ What happens during the first stage of labor? (At first, the contractions are 15 to 20 minutes apart. Later, the contractions are only one or two minutes apart and are much stronger. At the end of this stage, the cervix widens to about 10 cm.)

▶ What happens during the second stage of labor? (Strong contractions push the baby toward the birth canal and out of the mother's body.)

▶ What happens during the third stage of labor? (Contractions push fluid, blood, the rest of the umbilical cord, and the placenta out of the mother's body.)

The Birth Process

After about nine months of development and growth inside the uterus, the fetus is large, strong, and developed enough so life is possible outside the mother's body. Near the end of the pregnancy, the fetus is positioned for birth, usually upside down in the mother's body. The head points toward the cervix. The birth process begins when muscles of the uterus start to contract and relax at regular intervals. The muscular contractions, called **labor**, push the baby out of the uterus and through the vagina, or birth canal.

There are three stages of labor. The first stage lasts from about 2 to 24 hours. At the beginning, contractions are about 15 to 20 minutes apart. By the end of the first stage, the contractions have gotten much stronger and are only 1 or 2 minutes apart. The cervix widens from its normal 1 cm opening to about 10 cm. Contractions push the baby's head against the cervix. Crowning, or the appearance of the baby's head in the cervix, marks the beginning of the second stage of labor.

The baby is born during the second stage of labor, which usually lasts from two minutes to one hour. The mother uses the muscles in her abdomen to push the baby toward the birth canal. Strong contractions, along with this pushing, squeeze the baby out of the mother's body.

After birth, the baby is still connected to its mother by the umbilical cord. The cord is then clamped and cut. This does not hurt the baby or the mother. Within a few days, the rest of the cord falls off and the navel, or belly button, marks the spot where the cord was attached.

In the third stage of labor, contractions push fluid, blood, the rest of the umbilical cord, and the placenta out of the mother's body. This stage lasts about 15 or 20 minutes.

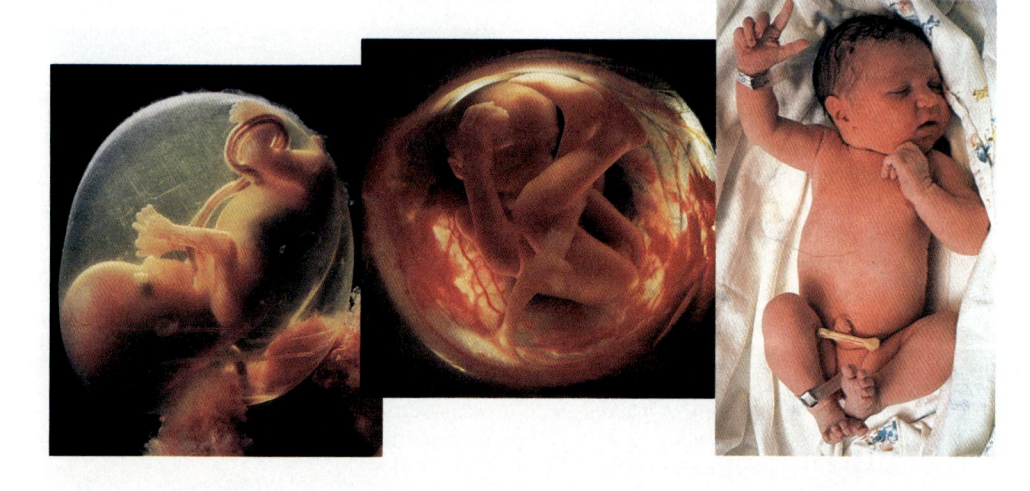

▲ **d.** At 16 weeks, the fetus begins moving around inside the uterus. Notice the umbilical cord.

▲ **e.** At 22 weeks, the fetus is nearly 35 cm long and weighs about 600 g. Its body is covered with fine hair.

▲ **f.** A baby soon after birth. Compare the newborn's development to the photographs of the embryo and fetus.

Chapter 22 Reproduction and Life Stages

TEACH ▪ Continued

Skills WorkOut
To help students learn about fraternal and identical twins, have them do the Skills WorkOut.

Answers Twins develop side by side in the uterus. Fraternal twins develop from two separate, fertilized eggs. They are not identical and may even be different sexes. Identical twins develop from the same fertilized egg that splits in two. Each identical twin has the same set of genes.

Skills Development
Infer Ask students the following questions:

▶ Why are prematurely born babies in danger of dying without intensive care? (Premature babies do not have fully developed organ systems.)

▶ Why are premature babies kept in incubators? (Incubators simulate the warm, protective environment inside the uterus.)

Historical Notebook
Have students explain what Dr. Apgar meant when she said that "Birth is the most hazardous time of life."

Answer It is important to test newborns right away to make sure the babies have no serious problems.

Discuss
Tell students that about 160,000 Americans are born each year with major birth defects. The sooner a defect is discovered, the sooner treatment can begin.

The Living Textbook:
Life Science Sides 7-8

Chapter 7   Frame 01158
Amniocentesis (1 Frame)
Search:

476

Skills WorkOut

Research Find out about fraternal and identical twins. How do they develop? How are fraternal twins different from identical twins? Write about your findings in a one-page report.

Problems of Reproduction

Human reproduction is a complex process made of many steps. Problems can occur at each step. For example, there may be problems fertilizing the egg due to a blocked oviduct or too few sperm in the semen.

Other problems can occur later in a pregnancy. Occasionally an embryo begins to develop inside an oviduct instead of in the uterus. This condition is called an ectopic, or tubal, pregnancy. It requires immediate surgery. In other cases, a pregnancy may end suddenly. This is called a miscarriage. The causes of miscarriages are sometimes unknown.

Sometimes a baby is born prematurely—before it is fully developed. Premature babies are very small. Often their organ systems are not fully developed. These babies may suffer from lung problems, brain damage, and blindness. They may not survive without intensive hospital care.

Historical Notebook

The Apgar Score

"Birth is the most hazardous time of life," wrote Dr. Virginia Apgar. When Dr. Apgar began studying childbirth, she learned that newborns were typically wrapped in blankets and examined later. Because serious problems went unnoticed in these first few moments after birth, many babies died. So Dr. Apgar devised a quick way of checking a newborn's health before it was given a thorough exam by a pediatrician. This quick check is called the Apgar Score. Dr. Apgar, who became one of the first female graduates of the Columbia University Medical School, introduced the Apgar Score in 1952.

Using the Apgar Score, a baby's health can be quickly rated at one and five minutes after birth. For example, a healthy baby breathes easily, shows good muscle tone and reflexes, and has a normal heart rate and rosy-colored skin. Each of these newborn characteristics—respiratory effort, muscle tone, reflexes, heart rate, and skin color—is given a score of 0 to 2, with a total score of 10 showing excellent health.

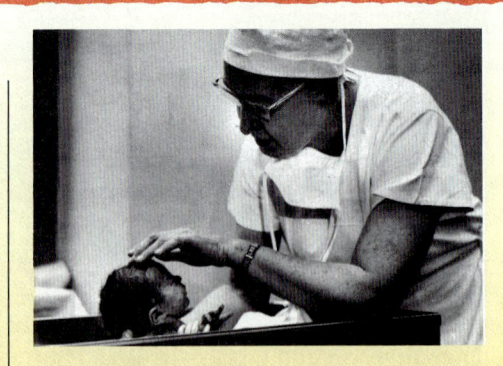

1. Why is it important to test newborns right away?

2. **Research** Find out what other kinds of medical advances have been made in recent years to test a newborn's health.

476   Chapter 22   Reproduction and Life Stages

INTEGRATED LEARNING

STS Connection

In addition to ultrasound and amniocentesis, chorionic villi sampling (CVS) has quickly become a popular way for parents and doctors to identify fetal problems. CVS is performed in the first three months of pregnancy. One reason that CVS has become a popular diagnostic tool is that it detects problems early, while they often can still be treated. Have students research and report on the risks and benefits associated with CVS.

Science and Technology
Health Before Birth

Did you know that some children have baby pictures taken before they are even born? These "pictures" are actually ultrasound scans. The scans are often used to make an image of a fetus in the mother's uterus. Scans are done by moving a special device over the mother's abdomen. The device sends out sound waves that echo off the fetus. The sound waves are changed to electrical impulses and are shown on a video monitor as a picture of the fetus.

Ultrasound scans are used to check the fetus's organs for defects and to diagnose ectopic pregnancies. Sometimes these defects can be surgically corrected while the fetus is still in the uterus, or, in other cases, shortly after the baby is born. Ultrasound scans are taken at any time during pregnancy; they do not hurt the fetus or the mother.

Another useful tool for identifying problems before birth is amniocentesis (AM nee OH sehn TEE sihs). As the fetus develops, it sheds cells into the amniotic fluid, the fluid that surrounds and cushions the fetus. A doctor uses a hollow needle to withdraw a small amount of amniotic fluid. Fetal cells in the fluid are checked to determine if the baby has certain diseases or genetic disorders. The cells and fluid also indicate the age and sex of a fetus. When a baby is expected to be premature, amniocentesis may be used to find out if the baby's lungs are fully developed. In most cases the best time for this kind of test is around the sixteenth week of pregnancy.

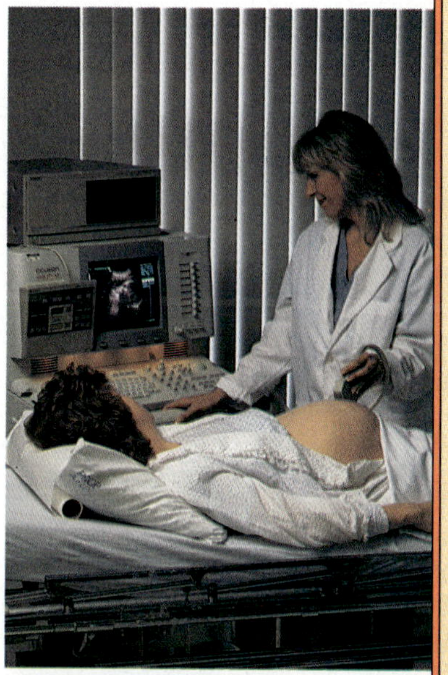

Figure 22.7 ▲
Using ultrasound, a doctor checks a pregnant woman. They are looking at an outline of her baby on the video monitor.

Check and Explain

1. In a paragraph, summarize the steps of fertilization of a human egg.
2. Describe the development of an individual from fertilized egg to birth.
3. **Evaluate** What are the possible advantages and disadvantages of testing a fetus or embryo with ultrasound or amniocentesis? Explain.
4. **Infer** How might prior knowledge of the birth process be useful to a woman in labor or a father attending the birth?

Enrich
Another test that doctors may perform on the fetus to identify problems is called *fetoscopy*. In this test, a doctor inserts a tube into a pregnant woman's uterus and examines the fetus through a lens. The doctor can also take skin and blood samples from the fetus and tissue samples from the placenta.

EVALUATE

WrapUp
Reinforce Have students make simple drawings of three of the stages of development presented on pages 474 and 475.
Use Review Worksheet 22.2.

Decision Making
If you are using classroom sets of *One-Minute Readings*, have students read Issue 23, "Fetal Medical Examination" on pages 38 and 39. Discuss the questions.

Check and Explain

1. Sperm travel through the cervix and uterus and into the oviducts. A single sperm penetrates the egg's jellylike coating. Genetic material from egg and sperm form a zygote.

2. The fertilized egg attaches itself to the uterine wall, where a placenta and an umbilical cord form. Its organ systems gradually develop, its bones harden, and it grows a layer of fat.

3. Testing allows a doctor to diagnose ectopic pregnancies, diseases, or genetic defects, which can sometimes be corrected. However, some procedures pose risks to the mother.

4. Answers will vary. They could take actions that would ease the birthing process.

Chapter 22 Reproduction and Life Stages

SECTION 22.3

Section Objectives
For a list of section objectives, see the Student Edition page.

Skills Objectives
Students should be able to:

Make a Graph showing a life timeline depicting life stages.

Predict how a growing population of elderly might affect a community.

Interpret Data collected about the relationship between age and learning.

MOTIVATE

Skills WarmUp
To help students understand human life stages, have them do the Skills WarmUp.
Answer Timelines should include major life events such as learning to walk, talk, read, and write.

Misconceptions
Because elderly people often have physical problems and may move and react slowly, young people often think that they are also mentally impaired. Explain that although diseases sometimes affect the minds of older adults, most remain fully capable of thinking, feeling, creating, communicating, and contributing to society.

INTEGRATED LEARNING

Themes in Science
Patterns of Change/Cycles All organisms change and develop in response to internal and external changes in their environment. As human beings develop from infancy through adulthood and old age, they pass through a number of stages. Encourage students to name some of the stages they have passed through. (Students may mention that they were once embryos, fetuses, babies, and children, and are now entering adolescence.)

Skills WarmUp

Make a Graph Draw a timeline showing each year of your life up to now. Mark the major events in your life, such as learning to talk and walk, going to school, learning to read and write, and anything else that is important to you.

Figure 22.8 ▲
All infants are dependent on the care provided by an adult.

22.3 Human Life Stages

Objectives

▶ **Identify** the main life stages in humans.

▶ **Describe** the characteristics of each life stage.

▶ **Predict** the impact of the aging population in the United States.

Look around you the next time you are in a crowd. All the people you see began life in the same way—as a single cell that divided and developed into a baby. As you know, those first nine months of development involved amazing changes. After birth, a new period of growth and change began. Everyone you see is still in the process of change. Change occurs throughout a person's lifetime.

Infancy

When you were born, you could cry, suck, grasp objects, hear, and follow light with your eyes. You probably spent about 16 hours a day sleeping. And you depended on other people to feed, diaper, clothe, shelter, and protect you. These first 12 months of life are called infancy.

Infancy is a time of rapid growth. The lower body, legs, and arms grow fastest. An infant's mass also increases greatly. Infants begin to react to stimuli, such as light and noise, just a few weeks after birth. They begin to respond to familiar voices. As their muscles grow stronger, they shows signs of controlling body movements. Infants can usually lift their heads after about 3 or 4 months and soon they can grasp objects. Within the first 7 or 8 months, the infant usually learns to crawl, sit up, and roll over. Baby teeth develop.

By the age of 8 months, infants can often speak simple words. By 12 months, infants can usually understand simple commands, and many can take their first steps. From 1 to 3 years of age, infants enter the toddler stage. During this stage, most toddlers learn to walk. They are very curious about their environment, touching and grabbing at objects, and putting things in their mouths. It is a very active time of life.

478 Chapter 22 Reproduction and Life Stages

Art Connection
As a period of transition, adolescence offers a unique opportunity to participate in two stages of life—childhood and adulthood. Have students draw or paint an image of the passage from childhood to adolescence. Encourage students to include both their positive and negative reactions to the stage they are approaching and to the one they are leaving behind.

Figure 22.9 ▲
Why is play an important part of childhood? ①

Childhood

After infancy, physical growth slows down, and mental and emotional growth take the lead. This period of growth, from infancy to about age 11 or 12, is called childhood. During childhood, most body systems mature, and the brain reaches its full size.

By age 2, a child has a vocabulary of a few hundred words and can speak in two-word phrases. By age 3, most children speak in sentences. Later, children learn to draw, read, and write. They begin feeding and dressing themselves.

Children also learn that they are part of a community and that their actions can affect others. They learn to cooperate and to make friends. They learn many skills through play, which is an important part of their lives.

Adolescence

You are in a stage of life called adolescence. Adolescence occurs from about age 11 to 20. It is a period of transition from childhood to adulthood.

Many cultures celebrate the transition to adulthood with special ceremonies. For example, some Native-American girls are initiated into their tribe in a ceremony. Each teenage girl shares a blanket with an older woman who promises to protect her for life.

During adolescence, the body grows to its full height, and the reproductive organs develop. As you know, the time when a person's body becomes capable of reproducing is called puberty. At this time, males and females go through many changes, both physically and emotionally. The hormones that trigger puberty also affect the emotions and may cause sudden mood swings.

Just as with other life stages, puberty does not occur at the same age for everyone. Those who mature first may feel uncomfortable with their looks. Those who mature later often worry that they are not developing as quickly as everyone else. The changes brought on by adolescence can be very confusing. Keep in mind, however, that all the changes are a natural part of growing up.

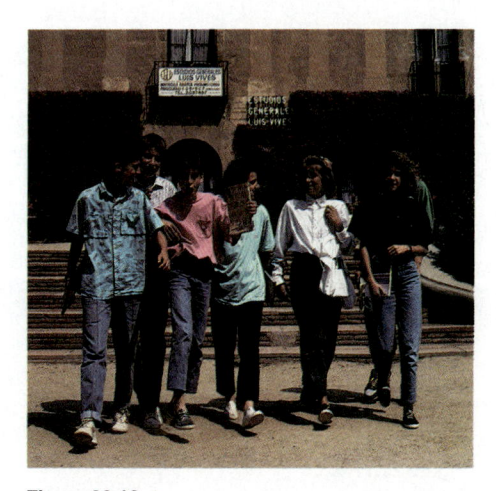

Figure 22.10 ▲
During adolescence, young people spend a great deal of time socializing.

Chapter 22 Reproduction and Life Stages

TEACH

Directed Inquiry
Have students study the material on this page by asking the following questions:

▶ What are the major human life stages before adulthood? (Infancy, childhood, and adolescence)

▶ What usually occurs during infancy? (The infant experiences rapid growth, learns to speak simple words and sentences, and learns to walk.)

▶ How long does infancy last? (About three years)

▶ What occurs during childhood? (Physical growth slows down and mental and emotional growth speed up.)

▶ What major developmental events occur in adolescence? (Full body height and sexual maturity are reached.)

Discuss
Explain that girls often enter puberty about two years before boys do. Girls are often taller and look older than boys at this age.

Answer to In-Text Question

① Play is an important part of childhood because many skills are learned through play.

Enrich
Use Enrich Worksheet 22.3.

TEACH • Continued

Directed Inquiry

Have students read the material on this page. Ask the following questions:

▶ What are some of the major events of adulthood? (Raising families and establishing careers)

▶ What happens to the body after age 30? (It begins a gradual decline that can be slowed by exercise and a healthy diet.)

▶ What happens in a woman's body at menopause? (The ovaries stop releasing eggs and women lose the ability to reproduce.)

▶ How does a person's body change in old age? (Body systems, eyesight, and hearing weaken.)

Integrated Learning

Use Integrating Worksheet 22.3.

INTEGRATED LEARNING

Multicultural Perspectives

Encourage students to discuss the extent to which the United States is or is not a "youth culture." Have students research and report on the status, roles, and treatment of elderly people in Native-American, Japanese, African, or island cultures.

Adulthood

Once the body has grown to its full height and has sexually matured, a person enters adulthood. At this time of emotional and physical maturity, usually starting between age 18 and 21, all body systems are fully developed. Bones and muscles are large and strong, coordination is good, and the anxieties of adolescence are generally over.

During adulthood, there is no additional growth, only replacement of damaged cells. Although the body starts a gradual decline after age 30, exercise and a healthy diet can help slow the decline.

The reproductive systems of men and women age differently. Men continue to produce sperm throughout most of their adult lives. But usually sometime after age 40, women experience menopause, a time when the menstrual cycle stops. After menopause, no more eggs are released from the ovaries, and a woman can no longer reproduce.

Old Age

As people grow older, their body systems may change. It may become more difficult for people to fight off disease and heal from physical injuries. Many older people have problems with their eyesight and hearing.

In the past, a normal life span was only about 40 years. Today people live longer and healthier lives thanks to better nutrition, exercise, and advances in medicine. In addition, staying interested in activities and other people is a key to staying healthy. Some older people continue their careers. Others retire and devote time to special interests, such as swimming, gardening, reading, or travelling. Many enjoy spending time with their grandchildren.

Figure 22.11 ▲
Adulthood is often a time of many new responsibilities, such as rearing families and establishing careers.

Figure 22.12 ▲
Even in later years, many people continue to work. In Poland, a man creates beautiful violins.

Chapter 22 Reproduction and Life Stages

Social Studies Connection

Life in rural America 100 years ago included large families with live-in aunts, uncles, and grandparents. The urban and mobile lifestyle of Americans today has often had the effect of separating family members from one another. Encourage students to discuss how developments of the last century have changed the shape of the family and the roles of elderly people.

Science and Society *Our Aging Population*

As medical advances help people live longer, the number of people in the United States who are over 65 is rapidly increasing. In addition, there are many more very old people, those over 85. How will our society respond to a population that includes more old people than ever before?

The very old often have physical problems. As a result, they generally have more health care needs than younger adults. In previous generations, these needs were often met by adult children. But today many adult children work full time, live far away from their parents, or have little money to spare.

Another problem facing the elderly is the financial burden of rising health care costs. Most working people contribute to social security insurance, and many have company retirement funds. However, the cost of living is rising faster than the increase in social security, and many retirement benefits are inadequate. As a result, elderly people are finding it harder to pay their medical bills and maintain healthy lifestyles.

To respond to the financial burdens and needs of the elderly, many communities now provide day-care services. These services are especially useful for adult children who work and care for their parents. In addition, some employers today assist employees making arrangements for care of their aging parents. And many older people are taking action. Large organizations of senior citizens offer special services at reduced costs.

Figure 22.13 ▲
In the United States, many elderly people live in nursing homes. As the aging population increases, more and more communities are faced with a shortage of nursing homes.

Check and Explain

1. What are the main life stages?
2. Explain why adolescence is sometimes a difficult stage.
3. **Classify** Make a table showing some of the events that occur in infancy and childhood. For example, include when a baby begins to talk and walk.
4. **Predict** How might a growing population of older people affect your community? How could the community help them? Explain.

Class Activity
Have students talk with older people to find out what their lives were like when they were younger and what they are like now. If possible, visit a nursing home or have a group of older adults come to the school to talk with students. Have students make a list of questions to ask.

EVALUATE

WrapUp
Portfolio Have students refer to the life timeline they made for the Skills WarmUp. Have them extend the timeline and make predictions of future events in their lives.

Use Review Worksheet 22.3.

Check and Explain

1. Infancy, childhood, adolescence, adulthood, and old age
2. It is a period of great physical and emotional changes.
3. Answers may vary. The table might include the following events: Infancy—lift head at three to four months, crawl and sit up at eight months, and take first steps around 12 months; childhood—speak in two-word phrases at around age two and learn to draw, read, and write between the ages of five and ten.
4. Answers may vary. A growing population of elderly may increase the need for health-care facilities. The community could help by making buildings and public transportation systems more accessible to the elderly.

Chapter 22 Reproduction and Life Stages

ACTIVITY 22

Time 60 minutes   **Group** 1–2
Materials
modeling clay
30 beakers
15 graduated cylinders
120 pennies
labels

Analysis
1. Answers will vary. Older children would usually be able to recognize that: (1) the amount of clay stays the same; (2) all three containers have the same amount of water; and (3) the two rows have the same number of pennies.
2. Comparison would show that younger children would probably answer incorrectly more often than older children would.

Conclusion
Conclusions will vary, but should show that around the age of six or seven, children are able to recognize that the amount of matter stays the same in each case.

Everyday Application
Answers will vary. Students should recognize that the material being taught should be appropriate for the developmental stage of the student.

Extension
Student reports will vary, but should present a basic outline of Piaget's stages of learning development.

TEACHING OPTIONS

Prelab Discussion
Have students read the entire activity, and then discuss the work of Jean Piaget. Tell students that he developed tasks like the ones in this activity to determine a child's learning development.

Activity 22   Is age related to learning development?

Skills Communicate; Collect Data; Observe

Task 1 Prelab Prep
1. Collect the following items: blank labels, ball of clay, 2 identical clear containers, a taller and thinner clear container, 8 pennies, measuring cup.
2. Arrange individual meetings with at least one child between age 2 and 7 and one child between age 7 and 11.
3. Before each meeting, label the identical containers *A* and *B*. Label the other container *C*. Fill A and B with exactly the same amount of water, about halfway full.

Task 2 Data Record
On separate sheets of paper, draw the data table shown. Make one for each child you will be meeting with. Circle the children's responses in your data tables.

Table 22.1   A Child's Responses

Child's Name and Age			
Which has more clay?	Ball	Sausage	Same
Which has more water?	A	B	Same
	A	C	Same
Which has more pennies?	Short row	Long row	Same

Task 3 Procedure
1. Meet with one child. Record the child's name and age.
2. Show the child a ball of clay. Then roll the ball of clay into a sausage shape.
3. Ask the child whether the ball or the sausage contains more clay, or if they are the same. Record the response.
4. Show the child containers A and B. Ask which contains more water, or if they are the same. Record the response.
5. Pour all of the water from container B into container C. Ask whether A or C contains more water, or if they are the same. Record the response.
6. Show the child a row of eight pennies. Then spread out the pennies a little to make a longer row. Ask the child which row has more pennies, or if they are the same. Record the response.
7. Repeat steps 1 through 6 with another child.

Task 4 Analysis
1. Compare the responses and ages of the children. What differences do you notice? What trends do you notice?
2. Compare your results to those of your classmates. How are they the same? How are they different?

Task 5 Conclusion
You know that the amount of matter stays the same in an object regardless of changes in its shape or position. Based on this activity, write a paragraph describing when children seem to develop an understanding of this concept. Does the activity show that age is related to learning? Explain.

Everyday Application
Why is it important to know a child's developmental stages when trying to teach difficult concepts, such as matter and energy, to a child?

Extension
Jean Piaget was a French scientist who developed a theory that identified different stages of learning development. Write a report about Piaget and his theory.

Chapter 22 Review

Concept Summary

22.1 Human Reproductive Systems
- Male sex organs include the testes and penis. The testes produce sperm.
- Female sex organs include the ovaries, uterus, and vagina. The ovaries release mature egg cells.
- In sexually mature females, ovulation and menstruation take place in a cycle that is usually about 28 days long.
- The major sex hormones are testosterone in males and estrogen and progesterone in females.

22.2 Fertilization, Pregnancy, and Birth
- Fertilization occurs when a sperm penetrates the egg and combines its genetic material with that of the egg.
- Over a nine-month period, called pregnancy, the fertilized egg grows into an embryo and then a fetus. It is nourished by the placenta, which is attached to the uterus by the umbilical cord.
- A baby is born during the process of labor.

22.3 Human Life Stages
- Infancy is a time of rapid growth during the first 12 months of life.
- During childhood, body systems mature and the brain reaches its full size.
- The life stage between childhood and adulthood is adolescence. It includes puberty, the time when a person reaches sexual maturity.
- After a long period of maturity, called adulthood, a person reaches old age.

Chapter Vocabulary

testes (22.1)	penis (22.1)	fertilization (22.1)	menstruation (22.1)
scrotum (22.1)	ovaries (22.1)	uterus (22.1)	fetus (22.2)
semen (22.1)	oviduct (22.1)	vagina (22.1)	labor (22.2)

Check Your Vocabulary

Use the vocabulary words above to complete the following sentences correctly.

1. An unfertilized egg and the uterine lining leave the body during ____.
2. Estrogen is produced in the ____.
3. The mixture of sperm cells and fluid is called ____.
4. Testosterone is produced in the ____.
5. After the third month of pregnancy, the growing embryo becomes a ____.
6. Fertilization occurs in an ____.
7. The testes are enclosed in the ____.
8. A baby is born during the second stage of ____.
9. Sperm cells are delivered into the vagina through the ____.
10. When a sperm cell penetrates the lining of the egg, a process called ____ occurs.
11. Another name for the ____ is the birth canal.
12. The development of the embryo and fetus takes place inside an organ called the ____.

Write Your Vocabulary

Write sentences using each of the vocabulary words above. Show that you know what each word means.

CHAPTER REVIEW 22

Check Your Vocabulary

1. menstruation
2. ovaries
3. semen
4. testes
5. fetus
6. oviduct
7. scrotum
8. labor
9. penis
10. fertilization
11. vagina
12. uterus

Write Your Vocabulary

Students' sentences should show that they know both the meanings of the words and how to apply the words.

Use Vocabulary Worksheet for Chapter 22.

CHAPTER 22 REVIEW

Check Your Knowledge

1. The uterus is the place where the fertilized egg develops into a baby.
2. Infancy is the shortest life stage; adulthood is the longest.
3. Puberty is the time when a person's body becomes capable of reproduction. It occurs during adolescence.
4. Sperm cannot survive at a temperature as high as body temperature.
5. During menstruation, an unfertilized egg and the broken-down lining of the uterus are shed. It is a cycle because it occurs regularly, every 28 days or so.
6. The muscular contractions of the uterus that function to push the baby out are called labor. Refer to page 475 for a description of the birth process.
7. Female: estrogen, progesterone; male: testosterone
8. after
9. ovaries
10. 3 years
11. cervix
12. tension
13. umbilical cord

Check Your Understanding

1. Both are produced by meiosis and have half the number of chromosomes that other body cells have. All egg production takes place before a female is born, while a male produces sperm from puberty on. Eggs are much larger than sperm. Many more sperm than eggs are produced.
2. Answers will vary depending on age of student. The changes are those associated with puberty.
3. Answers will vary. Possible answer: The social and intellectual characteristics of humans take time to develop.
4. Answers will vary. The fetus is about 12 weeks old. It has fingers and toes, but its head still looks somewhat like an embryo's.
5. The placenta does the work of the respiratory, digestive, and excretory systems. The circulatory and nervous systems function on their own in the fetus.
6. Answers may vary, but should include the fact that the brain is still developing.
7. The fetus's blood and the mother's blood never mix across the placenta.

Chapter 22 Review

Check Your Knowledge

Answer the following in complete sentences.

1. What is the purpose of the uterus in reproduction?
2. Which is the shortest human life stage? Which is the longest?
3. What is puberty? During which life stage does it occur?
4. Why are the testes held in the scrotum away from the rest of a male's body?
5. What occurs during menstruation? Why is it called a cycle?
6. What is labor? What occurs during the birth process?
7. What hormones are produced in the female reproductive system? What hormones are produced in the male reproductive system?

Choose the answer that best completes each sentence.

8. Menstruation occurs (after, before, just before, during) ovulation.
9. A male does not have (testes, hormones, a urethra, ovaries) in his body.
10. A child usually learns to talk in complete sentences by age (6 months, 1 year, 3 years, 5 years).
11. Between the uterus and the vagina is the (urethra, placenta, zygote, cervix).
12. A common symptom of premenstrual syndrome is (weight loss, tension, a runny nose, drowsiness).
13. In the uterus, the embryo is connected to the placenta by the (fetus, umbilical cord, zygote, oviduct).

Check Your Understanding

Apply the concepts you have learned to answer each question.

1. **Compare and Contrast** What are the differences and similarities between sperm production and egg production?
2. **Extension** Describe how your body will have changed by the time you reach adulthood.
3. **Infer** Of all the animals, humans take the longest to reach maturity. How do you think humans' long period of childhood relates to the other characteristics that make the human species unique?
4. **Mystery Photo** The photograph on page 466 shows a human fetus in the uterus of its mother. The image was created through a process called ultrasound. How old do you think the fetus is? On what did you base your answer?
5. The placenta does the work of three different body systems for the developing fetus. What are these three body systems? Which important body system is functioning on its own in the fetus?
6. **Application** Why do you not remember anything about the time you spent in your mother's uterus or your whole first year or two of life?
7. **Application** A person with type A blood cannot receive a transfusion of type B blood. The combination causes the blood to clot. However, a woman with type A blood can safely have a baby with type B blood. Explain.

Develop Your Skills

Use the skills you have developed in this chapter to complete each activity.

1. **Interpret Data** The graph below shows the mass of a developing baby at different stages of pregnancy.

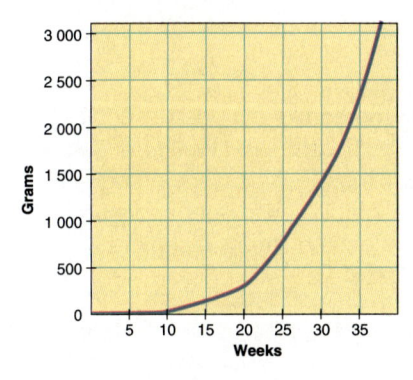

   a. How many weeks old is the fetus when it reaches 1 000 g?

   b. When does the baby increase its mass most rapidly?

2. **Make a Graph** Refer to pages 474 and 475 to make a graph showing the length of a developing baby at different stages of pregnancy. Add the following statistics to the graph: 14 weeks—18 cm; 26 weeks—38 cm; 32 weeks—41 cm. Use the same scale of weeks as in the graph above. How does the increase in length compare to the increase in mass?

3. **Data Bank** Use the information on page 625 to answer the following questions.

   a. What animal has a longer life span than humans?

   b. Which animal lives longer, a lion or an elephant?

Make Connections

1. **Link the Concepts** Below is a concept map showing how some of the main concepts in this chapter link together. Expand the map by adding the following concepts: embryo, placenta, vagina, fetus, and labor.

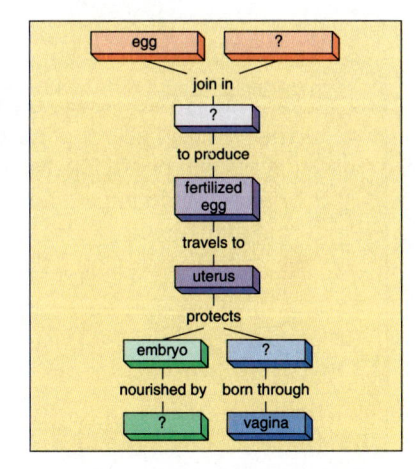

2. **Science and Society** Birth is a special event that can be difficult. Many different methods of making childbirth easier for the mother and better for the baby have been developed. Research different methods of childbirth and choose one to report on.

3. **Science and Social Studies** Find out how life expectancies differ in countries around the world. In which countries do people tend to live the longest? What are the causes of short life expectancies?

Develop Your Skills

1. a. 25 weeks
   b. During the last 12 weeks

2. The slope of the graph showing increase in length starts out steep, then becomes more gradual prior to birth, while the graph showing increase in mass starts out with a gradual slope, then becomes steeper during the last three weeks.

3. a. Tortoise
   b. Elephant

Make Connections

1.

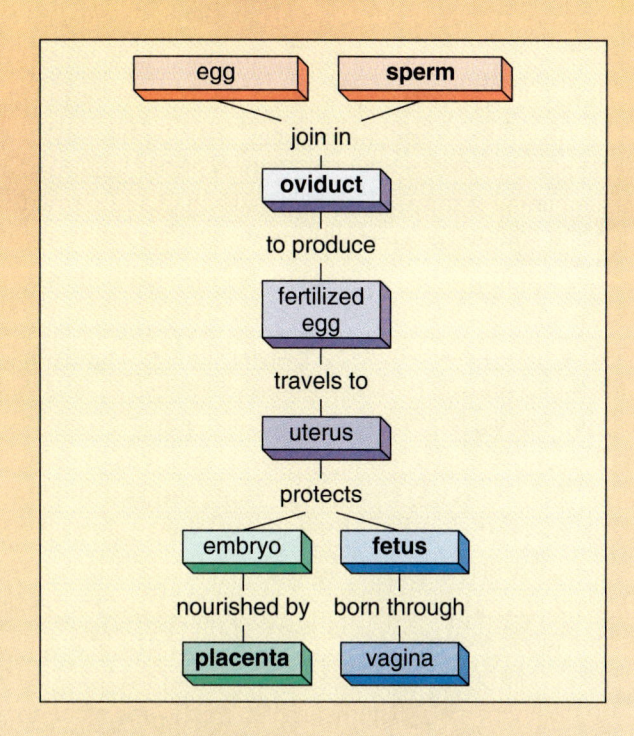

2. Answers will vary. Lamaze is an example of a method of childbirth.

3. Answers will vary. Life expectancies are shorter in countries where, on average, it is more difficult to obtain adequate nutrition and medical care.

CHAPTER 23

Overview

This chapter discusses human health and nutrition. The first section presents the essential nutrients for a balanced diet. Different food types illustrate the substances important in all life processes. The second section explains the importance of exercise for body functions and systems. Calories in food are explained. A discussion of the effects of stress is also included in this section. Students are presented with the dangers of drug abuse in the third section. The final section explores the effects of alcohol and tobacco.

Advance Planner

▶ Supply old magazines for TE activity, page 488.

▶ Purchase celery sticks and provide clear cups for SE page 490.

▶ Gather paints, old magazines, paste, and other art supplies for TE page 496.

▶ Arrange for students to visit a health club for TE page 497.

Skills Development Chart

Sections	Classify	Collect Data	Communicate	Decision Making	Estimate	Infer	Interpret Data	Observe	Research
23.1 Skills WarmUp	●								
Skills WorkOut						●			
Skills WorkOut	●								
Activity						●	●	●	
23.2 Skills WarmUp					●				
SkillBuilder							●		
Skills WorkOut						●			
23.3 Skills WarmUp			●						
Skills WorkOut									●
23.4 Skills WarmUp			●						
Skills WorkOut		●							
Consider This			●	●					

Individual Needs

▶ **Limited English Proficiency Students** Use a tape recorder to record the pronunciation of the boldface or italic words in this chapter and their definitions. Along with the definition of the terms, tell students which figures in the text illustrate certain terms. Not all terms can be matched with a figure. Have students use the tape recorder to listen to the information about each term. Then have them define the terms using their own words. Have students make their own tape recordings of the terms and their definitions. They can use this tape along with the figures to review important concepts in the chapter.

▶ **At-Risk Students** Have students copy all of the questions that appear in the text into their science journals, leaving several blank lines between each question. As students read each section, have them write the answers in their science journals. Show students that some answers appear immediately after the text question. Encourage them to ask their own questions as they read. Have them include these questions and answers in their science journals.

▶ **Gifted Students** Have students write two "day in the life" scripts—one for the "Perfectly Healthy Person" and one for the "Unhealthy Person." Students should use information in all four sections of the chapter and enact the diets and life-styles of these two characters. Encourage students to use the vocabulary provided in the chapter. Students may enjoy videotaping their skits.

Resource Bank

▶ **Bulletin Board** Make a title for the bulletin board, such as *The Dos and Don'ts of Nutrition, Health, and Wellness*. Divide the bulletin board into two columns. As you work through this chapter, encourage students to add newspaper and magazine articles that discuss nutrition and life-style choices. Have them hang the articles on the bulletin board.

CHAPTER 23 PLANNING GUIDE

Section	Core	Standard	Enriched	Section	Core	Standard	Enriched
23.1 Nutrients pp. 487-493				**23.3 Drugs and Substance Abuse** pp. 498-502			
Section Features				*Section Features*			
Skills WarmUp, p. 487	•	•		Skills WarmUp, p. 498	•	•	•
Skills WorkOut, p. 490	•	•		Skills WorkOut, p. 499	•	•	
Skills WorkOut, p. 491	•	•		*Blackline Masters*			
Activity, p. 493	•	•	•	Review Worksheet 23.3	•	•	
Blackline Masters				Skills Worksheet 23.3	•	•	
Review Worksheet 23.1	•	•	•	*Laboratory Program*			
Enrichment Worksheet 23.1		•	•	Investigation 45		•	•
Skills Worksheet 23.1	•	•		**23.4 Alcohol and Tobacco** pp. 503-506			
Overhead Blackline Transparencies				*Section Features*			
Overhead Blackline Master 23.1 and Student Worksheet	•	•		Skills WarmUp, p. 503	•	•	•
Ancillary Options				Skills WorkOut, p. 505	•	•	•
CEPUP				Consider This, p. 505		•	•
One-Minute Readings	•	•	•	*Blackline Masters*			
Laboratory Program				Vocabulary Worksheet 23	•	•	
Investigation 34		•	•	Review Worksheet 23.4	•	•	
Investigation 43		•	•	Skills Worksheet 23.4	•	•	
Investigation 44		•	•	*Ancillary Options*			
23.2 Exercise and Rest pp. 494-497				One-Minute Readings	•	•	•
Section Features				*Color Transparencies*			
Skills WarmUp, p. 494	•	•	•	Transparencies 72, 73	•	•	
Skill Builder, p. 495	•	•					
Skills WarmOut, p. 496	•	•					
Blackline Masters							
Review Worksheet 23.2	•	•					
Integrating Worksheet 23.2		•	•				

Bibliography

The following resources can be used for teaching the chapter. See page T-40 for supplier codes.

Audio-Visual Sources

(video unless noted)
Being Obese. 24 min. 1984. GH.
Choices: Alcohol, Drugs, or You. 24 min. 1983. BF.
Diets for All Reasons. 20 min. 1982. CF.
The Feminine Mistake. 24 min. 1978. PF.
Maintaining Your Health. 19 min. 1987. A-W.
Nutrition for Better Health. 15 min. 1985. EB.
Smokeless Tobacco: The Sean Marsee Story. 15 min. 1986. C/MTI.

Software Resources

Health Risk Appraisal. Apple, IBM PC, TRS-80. CABISCO.
Nutrition—A Balanced Diet. Apple, TRS-80. EME.
The Smoking Decision. Apple. SC.
What's in Your Lunch? Apple, Atari, Commodore. LHS.

Library Resources

Boston Children's Hospital, with Susan Baker and Robert R. Henry. *Parent's Guide to Nutrition and Healthy Eating from Birth Through Adolescence.* Reading, MA: Addison-Wesley, 1986.
Dixon, Bernard, ed. *Health, Medicine, and the Human Body.* New York: Macmillan, 1986.
Woods, Geraldine. *Drug Use and Abuse.* Revised ed. New York: Franklin Watts, 1986.

CHAPTER 23

INTEGRATED LEARNING

Themes in Science
Energy While studying this chapter, students will learn that, like all living things, humans need energy to perform their life activities. They get energy from the food that they eat. The primary energy sources for humans are carbohydrates and lipids (fats).

Writing Connection
Refer students to the photograph. Have them write a poem or essay about the squash and add details about how squash look and feel. Have students describe how squash smell when cooking or how they taste. Ask them to describe their favorite squash dishes.

Introducing the Chapter

Have students read the description of the photograph on page 486. Ask if they agree or disagree with the description.

Directed Inquiry

Have students study the photograph. Ask:

▶ What type of organism do the objects in the photograph come from? (Plants)

▶ What would you see if you cut one of the vegetables in half? (Seeds would be surrounded by green, orange, or yellow fleshy material.)

▶ What parts of these vegetables do you eat? (The fleshy layer and some seeds are edible.)

▶ Why is a picture of squash opening this chapter? (Vegetables are important for good nutrition.)

Chapter Vocabulary

addiction   lipid
carbohydrate   narcotic
cholesterol   protein
depressant   stimulant
hallucinogen

Chapter 23 Nutrition, Health, and Wellness

Chapter Sections

23.1 Nutrients
23.2 Exercise and Rest
23.3 Drugs and Substance Abuse
23.4 Alcohol and Tobacco

What do you see?

"In this picture, I see fruits and vegetables. Some are round, some are long. They vary in color. Some have bumps and some are smooth. I think some of them are good to eat. They are good for your body. Some have protein and vitamins C and B. They give the body energy to run, play, and participate in sports and other school activities."

Paula Lee
Pace Middle School
Milton, Florida

To find out more about the photograph, look on page 508. As you read this chapter, you will learn about human nutrition and health.

Themes in Science

Systems and Interactions In this section, students learn that food is the source of organic and inorganic compounds. Food is necessary for growth, movement, development, and energy storage in humans.

Integrating the Sciences

Chemistry Foods are made up of large molecules. During digestion, these molecules are broken into smaller units. Breaking the chemical bonds in these polymers releases the energy stored in them and provides building blocks for bone, muscle, and other tissues.

SECTION 23.1

Section Objectives
For a list of section objectives, see the Student Edition page.

Skills Objectives
Students should be able to:

Classify foods as nutritious or unhealthy.

Infer the importance of water in their diets.

Classify the foods they eat according to the food groups.

Make a Model of a healthy menu.

Interpret Data on food packages.

Vocabulary
carbohydrate, lipid, cholesterol, protein

23.1 Nutrients

Objectives

▶ **Explain** the body's need for protein, carbohydrates, and fats.
▶ **Describe** the importance of fiber.
▶ **Describe** the characteristics of a balanced diet.
▶ **State** the dangers of improper nutrition.
▶ **Make a model** of a menu that provides a balanced diet.

Skills WarmUp

Classify Write the words *Nutritious* and *Unhealthy* at the top of a piece of paper. Then place each of the following foods in the correct column: apple, corn, french fries, salad dressing, milk, soft drinks, rice, doughnut, orange juice, chocolate bar. Explain your reasoning.

What did you eat for dinner last night? For breakfast this morning? What are your favorite snacks? The foods you eat affect your health and physical fitness in many ways.

Food provides energy for you and your body's cells. It also supplies raw materials that cells need for growth and repair. The substances in food that your body needs are called nutrients. Many nutrients are necessary for good health. Foods vary in the kinds and amounts of nutrients they contain. That's why paying attention to what you eat is necessary for good health. A balanced diet is one that includes all the nutrients your body needs.

Most nutrients can be classified as organic nutrients. Organic nutrients are essential compounds found in living things. They include carbohydrates, fats and oils, proteins, vitamins, and fiber.

Carbohydrates

Your body gets most of its energy from organic nutrients called **carbohydrates**. They are made of carbon, hydrogen, and oxygen. Starch is a carbohydrate made of large molecules. Corn, potatoes, and rice are good sources of starch. So are foods made from wheat, such as bread.

When starches are broken down into smaller molecules, they form sugars. Table sugar, or sucrose, is only one of many sugars. During digestion, sucrose splits into two smaller molecules of a simple sugar called glucose. In the cells, glucose combines with oxygen during the process of respiration. During respiration, the stored energy of glucose is released.

Figure 23.1 ▲
The grain of the rice plant is high in starch. In some cases, the grain is ground into a flour, but usually it is husked, boiled, and eaten in its original state.

MOTIVATE

Skills WarmUp
To help students think about which foods are nutritious, have them do the Skills WarmUp.
Answer The nutritious foods are: apple, corn, milk, rice, and orange juice. The unhealthy foods are: french fries, salad dressing, soft drinks, doughnut, and chocolate bar.

Misconceptions

Students may think that it's better for them to eat salads than other foods, such as steak. Point out to students that if they eat a salad with croutons, bacon bits, salad dressing, and other foods that are high in fat and Calories, they may be consuming more fat and Calories than in a meal consisting entirely of meat.

Chapter 23 Nutrition, Health, and Wellness 487

TEACH

Explore Visually

Have students study Figure 23.2. Ask:

▶ Which foods do you see that are sources of protein? (Chicken, shrimp, beans, tofu, hummus, deviled eggs)

▶ Find three sources of carbohydrates. (Pita bread, pasta, whole-grain bread)

▶ Which foods in the picture are good sources of vitamins? (Fruits, cheeses, margarine, eggs, breads)

Discuss

Encourage students to think about foods in Figure 23.2 that come from other countries and cultures. Ask:

▶ Which foods have you eaten? (Answers will vary.)

▶ Where do some of these foods come from? (Pita and hummus: Middle East; tofu: Far East; artichoke: Mediterranean. Pasta may have originated in China, rather than Italy.)

▶ Which of these foods would you like to try? (Answers will vary.)

Discuss

After students have read the page, ask:

▶ How does your body use fats? (Fats store energy and insulate the body.)

▶ What is cholesterol? (Cholesterol is a fatty substance found in animal tissues. Too much cholesterol can cause heart disease.)

▶ How can you get the oils your body needs without getting too much cholesterol? (By getting oil from plants, rather than animals)

INTEGRATED LEARNING

Multicultural Perspectives

Archaeologists have discovered that around 18,000 years ago, when glaciers still covered much of the earth, African peoples were raising crops of wheat, barley, lentils, chick-peas, capers, and dates. Ask students to name prepared foods that come from these crops.

Geography Connection

In the past, food stores could provide only seasonal fruits and vegetables for consumers. Today, many of the fruits and vegetables we eat are imported from other countries. Have students list their favorite fruits and vegetables and investigate the sources of these foods during each season. Encourage students to interview the produce manager of a local food store.

Fats and Oils

What makes foods greasy? Greasy foods contain either fats or oils. At room temperature, fats are solid, while oils are liquid. In general, animals produce fats and plants produce oils. Together, fats and oils are called **lipids**. Lipids are made up of carbon, hydrogen, and oxygen.

In your body, deposits of fat serve as a source of stored energy. Gram for gram, fats supply more than twice as much energy as carbohydrates. Fat deposits stored under the surface of your skin also help insulate your body.

Eating too much fat can be harmful. High-fat food, such as french fries, hot dogs, and chocolate bars, are high in Calories and tend to make you gain weight. Moreover, fats seem to increase the amount of **cholesterol** (kuh LEHS tuh ROHL) in the blood. Cholesterol is a fatty sustance found in all animal tissues. High cholesterol levels can cause solid deposits to clog the arteries, resulting in heart disease. To cut down on the amount of fat in your diet, substitute low-fat foods like fruits and vegetables or cereal grains.

Figure 23.2
Nutrients and Food

Chicken, shrimp, and beans are sources of protein.

Proteins

Your cells need nutrients called **proteins** for growth and repair. Your body uses the proteins you eat to build its own proteins. These are used to make new cell material and substances important in all life processes. Within your body are thousands of different proteins, each doing a different job.

Proteins are large, complex molecules made up of smaller units called *amino acids*. Twenty different amino acids, combined in different ways, make up all the different proteins in your body. Just as the 26 letters of the alphabet form an endless number of words, amino acids join together to create many kinds of proteins.

Proteins in food are classified as complete or incomplete. Complete proteins contain all the amino acids your body needs to make its own proteins. Incomplete proteins are missing one or more of the essential amino acids. Cheese, eggs, meat, fish, and milk are good sources of complete proteins. Grains, nuts, and dried peas and beans contain incomplete proteins. Grains and beans eaten together, however, combine to provide you with complete protein.

Chinese noodles are a source of carbohydrates for your diet. Find three other carbohydrate foods.

Multicultural Perspectives

All people need food in order to live. However, there are many different kinds of food and many ways of preparing it. Have students choose a culture whose food they enjoy. Some possibilities are Korean, German, African, Creole, and Middle Eastern peoples. Have them work in small groups to list their favorite dishes. Have students decorate their lists with maps or other symbols of the culture they've chosen. Display the lists in the classroom. You may also wish to have an ethnic food day on which a few volunteers bring in and share favorite ethnic foods.

Vitamins

Your body also needs organic nutrients called vitamins. Vitamins help regulate the chemical reactions that convert food into energy and living tissue. Vitamins are needed only in small amounts. If you don't have enough of any one vitamin in your diet, however, disease will result.

Because your body can't produce most of the 13 vitamins it needs, they must be supplied daily. Look at Table 23.1 to see which foods contain the vitamins you need. Find some of these foods in Figure 23.2.

Fiber

Plant cell walls contain indigestible substances called *fiber*. Fresh fruits and vegetables, dried beans and peas, and whole grain cereals are all good sources of fiber. Fiber helps move food through your intestines more quickly. This may help to prevent cancer from occurring in the walls of the intestines. Because most high-fiber foods are low in Calories, substituting these foods for high-Calorie foods can help you lose weight.

Table 23.1 Vitamins

Vitamin	Sources	What It Does
A	Milk, eggs, green and yellow vegetables, liver	Promotes healthy bones, teeth, eyes, skin
B-complex	Whole-grain breads and cereals, most vegetables, eggs, milk products, meat	Helps body cells use energy and oxygen; needed for healthy skin, nerves, blood, and heart
C	Citrus fruits, tomatoes, potatoes, strawberries, cabbage	Promotes healthy bones, teeth, healing of wounds
D	Fortified milk, eggs, tuna, salmon, liver	Promotes healthy bones and teeth
E	Grains, fish, meat, lettuce, vegetable oils, margarine	Protects cell membranes
K	Green leafy vegetables, tomatoes, pork	Essential for blood clotting

Vitamins are often added to foods such as milk, cereals, and bread. What vitamins are found in leafy green vegetables and citrus fruit? ①

Whole-grain breads and vegetables add fiber to your diet. What other foods contain fiber? ②

Rub a peanut on a brown paper bag. You can see the oils these nuts contain.

Discuss

Have students refer to Table 23.1. Ask the following questions:

▶ Which vitamins are important for healthy skin? (Vitamin A, and vitamin B complex)

▶ What are good sources of these vitamins? (Milk, eggs, green and yellow vegetables, whole-grain breads, cereals, and meats)

▶ What vitamins are present in whole-grain breads, tomatoes, and green leafy vegetables? (Vitamin K, and vitamin B complex)

▶ What vitamin is important for blood clotting? (Vitamin K)

Class Activity

Have students bring in empty packages from prepared foods. Have them read the labels on each package to find the ingredients. Ask students to group their packages according to which foods provide vitamins, protein, carbohydrates, and so on. Have students make a fourth group of foods that taste good, but are not nutritious.

Decision Making

If you have classroom sets of *One-Minute Readings*, have students read Issue 9, "Vitamin C and History" on page 17. Discuss the questions.

Answers to In-Text Questions

① **Vitamin C is found in citrus fruits and vitamin K is found in green leafy vegetables.**

② **Fresh fruits and beans and peas also have fiber.**

TEACH • Continued

Skills WorkOut
To help students understand the importance of water in their diets, have them do the Skills WorkOut.
Answer The celery becomes limp when left in the open air. It becomes firm again when it is put in a glass of water for a day. This shows how necessary water is for a living thing to function properly.

Discuss
Have students study Table 23.2. Ask:

▶ What does calcium do in the human body? (Helps keep bones and teeth strong)

▶ What minerals are provided in milk? (Zinc, calcium, magnesium)

▶ What does sodium do? (Helps regulate water balance inside and outside body cells)

▶ What foods provide phosphorus? What does it do? (Grains and protein-rich foods; energy-rich molecules such as ATP need phosphorus.)

Research
Space missions require astronauts to remain in space for days. Have students investigate the type of food that astronauts eat. If students have sampled space food, ask them to describe its texture, color, and taste. Ask students if they would want to eat space food for long.

The Living Textbook: Life Science Side 7
Chapter 32 Frame 31176
Milk and the Skeleton (Movie)
Search: Play:

INTEGRATED LEARNING

Integrating the Sciences

Chemistry Students may not realize that much of the water their bodies need comes from food. On the chalkboard, write the following foods and the percentage of water in each: raw egg, 74%; peanut butter, 3%; watermelon, 92%; cornflakes, 5%; American cheese, 4%; salmon, 70%. Have students make a bar graph of these percentages.

Skills WorkOut
Infer Leave a crisp celery stick in the open air for 24 hours. What change occurs? Next put the celery in a glass of water for 24 hours. What happens? What does this show about the importance of water in your diet?

Inorganic Nutrients
Your body needs certain substances, such as water and minerals, that are not manufactured by living things. These substances are called inorganic nutrients.

Water You could live without food for up to a month, but only about a week without water. Your body can't function properly without water. It needs to take in about 2 liters of water a day.

Most of the chemical reactions in your body take place in water solutions. Water carries dissolved nutrients to the cells and carries away soluble wastes. Water makes up most of the saliva that helps you swallow food.

Minerals A supply of minerals is necessary for building many important substances in your body. Your bones, for example, are made in part from the minerals phosphorus and calcium. Like vitamins, minerals are needed only in small amounts for your body to function properly. But some minerals are required in greater amounts than others. They include calcium, phosphorus, potassium, sodium, and magnesium. Table 23.2 lists sources and functions of minerals.

Table 23.2 Minerals

Mineral	Sources	What It Does
Calcium	Milk, cheese, eggs, leafy vegetables	Promotes strong bones and teeth; necessary for blood clotting
Phosphorus	Protein-rich foods, grains	Promotes strong bones and teeth; used in energy-rich molecules like ATP
Magnesium	Milk, cheese, cereals, vegetables, nuts	Aids nerve and muscle contractions
Potassium	Most foods, especially fruits and vegetables	Helps regulate water balance inside and outside body cells
Sodium	Table salt, most foods	Helps regulate water balance inside and outside body cells
Iron	Liver, red meat, grains, eggs, leafy vegetables, dried beans	Enables blood to carry oxygen
Iodine	Iodized salt	Controls normal functioning of thyroid gland
Zinc	Meat, fish, eggs, milk, oysters	Helps heal wounds; aids growth and reproduction

Chapter 23 Nutrition, Health, and Wellness

Geography Connection

Explain that climate and geography are the two main factors that determine the kinds of crops that grow in a region. For instance, rice grows in water, so there are no rice paddies in the dry Sahara desert in Africa. Corn, on the other hand, grows well with less water than rice does, while wheat requires even less water. You may wish to use a world map to show the locations of the world's major crops.

Math Connection

Students probably know that there are more Calories in a cup of ice cream than in a cup of brown rice. Have them use the following information to calculate the number of Calories in 200 grams of carbohydrates and 200 grams of fat: 1 gram of fat yields 9 Calories; 1 gram of carbohydrate produces 4 Calories of energy. (In 200 grams of fat there are 1,800 Calories; and in 200 grams of carbohydrates there are 800 Calories.)

Skills WorkOut

To help students classify the foods they eat, have them do the Skills WorkOut.

Answers Answers may vary, but student responses should agree with the information in Table 23.3.

Portfolio

You may wish to have each student continue the classification of foods from the Skills WorkOut to cover what they have eaten for an entire week. Then have students see how well they did according to the four guidelines suggested on this page. Discuss whether students would modify their diets to fit these guidelines if they could. (Answers may vary, but some students may enjoy the high-fat, high-sugar, and high-salt foods at the fast-food restaurants and would not want to change.) Students may add these pages to their portfolios.

Enrich

Point out that Table 23.3 is based on a new food guide pyramid that was researched and released by the U.S. Department of Agriculture in 1992. You may wish to draw the pyramid on the chalkboard, showing the bread group at the bottom (6 to 11 servings per day); the fruit group (3 to 5 servings) and vegetable group (3 to 5 servings) above the bread; the milk group and meat group (2 to 3 servings) on the next level; and the fats and sweets (use sparingly) at the top.

Ancillary Options

If you are using CEPUP modules in your classroom for additional hands-on activities, experiments, and exercises, begin Chemicals in Foods: Additives.

A Healthy Diet

Your body needs varying amounts of more than 40 different nutrients. If you choose your diet by taste or convenience alone, you risk getting too little of some nutrients and too much of others. To have a healthy diet, follow the guidelines below.

▶ Eat a variety of foods. Most foods are rich in just a few nutrients. The best way to make sure your body gets all the many nutrients it needs is to eat many different kinds of foods. Table 23.3 shows one way in which foods are divided into groups according to the nutrients they contain. Eating foods from all these groups ensures a varied diet.

▶ Reduce your intake of fat and cholesterol. The diets of many Americans include too much fat and cholesterol. Eggs and red meat contain a large amount of cholesterol. Red meat, butter, ice cream, and all fried foods are high in fat. The bulk of the lipids in your diet should be from unsaturated oils, such as olive oil.

▶ Eat plenty of fruits, vegetables, grains, and dried beans. Most foods that come from plants are naturally low in fat and contain no cholesterol. In addition, most contain large amounts of fiber. For these reasons plant foods are good replacements for foods high in fat. If you greatly reduce the amount of meat, milk, and cheese in your diet, dried beans and peas and foods made from them can provide you with needed proteins.

▶ Use sugar and salt in moderation. Sugar and salt help many foods taste better. But both are unhealthy in large amounts. Many snack foods and packaged and processed foods contain much more salt or sugar than you may realize.

Skills WorkOut

Classify Write down all the foods you ate yesterday. Then place each food into one of the food groups in Table 23.3. You may need to divide some dishes into the foods they are made from. If you ate spaghetti and meatballs, for example, the spaghetti goes into the grain group, the meatballs into the meat group, and the sauce into the vegetable group.

Table 23.3 Food Groups and Their Nutrients

Fruits	Vegetables	Breads, Cereals, Rice, Pasta	Meat, Poultry, Fish, Eggs, Dry Beans, and Nuts	Milk, Cheese, and Yogurt
Vitamins A and C, minerals, carbohydrates	Vitamins A and C, minerals, carbohydrates, fiber	Carbohydrates, B vitamins, iron, fiber	Protein, iron and other minerals, B vitamins	Protein, B vitamins, vitamin D

Chapter 23 Nutrition, Health, and Wellness

EVALUATE

WrapUp

Reinforce Have students analyze and evaluate their lunch menus for one week. Direct students to determine whether each day's menu includes foods from each of the food groups, whether the foods contain the needed nutrients and vitamins, and whether the foods are low in fat. If students find a menu to be unhealthy, ask them to suggest foods that could be substituted.

Use Review Worksheet 23.1.

Check and Explain

1. Carbohydrates provide most of the body's energy. Proteins make new cell material and substances that are important for growth and repair in all the body's processes. Vitamins help regulate the chemical reactions that convert food into energy and living tissue.

2. Fiber helps move food through the intestine more quickly.

3. A junk food diet is high in fat, cholesterol, and Calories, which may result in weight gain. In addition, the lack of nutrients may make the body susceptible to illness or disease.

4. Answers may vary, but menus should contain foods from each food group as well as necessary nutrients and vitamins.

INTEGRATED LEARNING

Multicultural Perspectives

Tell students that the world production of rice totals 287.5 million tons. Have students use the following information to create a bar graph that shows, in millions of tons, how much of the world's rice is grown in each country: United States, 5; Port Timor, 5.5; South Korea, 6; Brazil, 7; Burma, 8; Vietnam, 11; Thailand, 13; Japan, 16; Bangladesh, 17; Indonesia, 23; India, 61; China, 115. Have students locate these countries on a world map.

Science and Society *Staple Foods*

Think about the food you eat. You probably eat a variety of foods. However, you and your family may eat one kind of food more often than others. A food that makes up the bulk of a diet is called a staple. What might be your family's staple food?

About half the world's people eat rice as their staple food. Southeast Asians were probably the first to use rice. Today, thousands of kinds of rice are grown in warm, wet regions around the world.

In many countries of Latin America and Africa, corn is the staple food. It was first used about 10,000 years ago by Native Americans living in what is now Mexico. Many varieties are now grown in temperate and tropical regions.

Rice and corn are versatile foods: that is, they can be prepared many ways. Greeks wrap rice in grape leaves, for example, while the Chinese fry it with pork. Some Latin Americans eat a mixture of beans and rice called *gallo pinto* for breakfast. Corn can be eaten fresh or made into cornmeal. Tortillas and tamales are popular Mexican foods made from cornmeal.

Grains, such as rice and corn, are staple foods in developing countries because they are easy to grow and cost less than other foods. Depending too much on these single foods, however, has a serious disadvantage. While grains are good sources of starch, vitamins, and fiber, they supply only small amounts of incomplete protein. People who have no other source of protein in their diet will suffer from malnutrition.

Figure 23.3 ▲
Rice (top) is grown in areas where there is enough water to flood the fields, or paddies, in which it is planted. Corn products can be found in open-air markets (bottom) of Mexico, where it is a staple crop.

Check and Explain

1. Name the three main organic nutrients and explain why your body needs them.

2. Why should you eat foods that contain fiber?

3. **Predict** How might your body be affected by a diet made up mainly of junk food?

4. **Make a Model** Make up a menu for an entire day that provides a balanced diet.

TEACHING OPTIONS

Prelab Discussion

Have students read the entire activity before beginning the discussion. Have some actual food labels available, or have students bring in labels from foods they often eat. Ask the following questions:

▶ What is the difference between a list of ingredients and a table of nutritional information?

▶ Why is it important to note the serving size when reading nutritional information on a food package? How does the serving size relate to the size of the package?

▶ Why is it a good nutritional goal to get as few Calories from fat as possible?

ACTIVITY 23

Time 40 minutes   **Group** 1–2

Materials
Paper
Pencils or pens

Analysis

1. The soup and the milk have about the same amount of fat; the milk has more than twice as many carbohydrates.
2. The milk is the better source of both carbohydrates and protein.
3. The milk has slightly less fat than the soup has.

Conclusion

Answers will vary, but students should mention information given on the label shown. Students should also name items not shown here, such as additives, recommended daily allowances, warnings for consumers with special dietary needs, storage precautions, and preparation instructions.

Everyday Application

Answers will vary. Besides nutrient tables, the only way to keep track of dietary intake is to read food labels and consult nutrition tables for fresh foods. The information from these sources can then be compared to recommended daily allowances for individuals. Point out that sodium is an important nutrient that is available in salt. Too much salt, however, can cause high blood pressure, weight gain, and water retention. Since many popular food items contain large amounts of salt and other sodium compounds to enhance flavor, it is a good idea for students to monitor salt intake, just as they should monitor fat and sugar intake.

Activity 23 What information is found on food labels?

Skills Observe; Infer; Interpret Data

Task 1 Prelab Prep
1. Collect the following items: 2 sheets of paper, pencil or pen.
2. Read the information on the labels shown on this page.

Task 2 Data Record
On a separate piece of paper, copy Table 23.4 and use it to compare the nutrients in the milk and chicken noodle soup.

Task 3 Procedure
1. Enter the data from the labels into your table. Note: Always use a capital *C* when referring to food Calories.
2. Calculate the number of Calories in one serving of milk that come from protein by multiplying the number of grams of protein listed by 4.
3. Calculate the number of Calories in the milk that come from carbohydrates by multiplying the number of grams of carbohydrate by 4.
4. Calculate the number of Calories that come from fat by multiplying the number of grams of fat by 9. Enter these figures for milk in your table.
5. Do the same set of calculations for one serving of chicken noodle soup. Enter these figures in your table.

Table 23.4 Comparing Nutrients

Nutrients in 1 Serving	Milk	Soup
Total Calories		
Grams of protein		
Calories that come from protein		
Grams of carbohydrate		
Calories that come from carbohydrates		
Grams of fat		
Calories that come from fat		

Task 4 Analysis
1. Compare the fats and carbohydrates provided by one serving of each food.
2. Which food is the best source of carbohydrate? Of protein?
3. Which food contains the least amount of fat?

Task 5 Conclusion
Write a short paragraph describing the information available on a food label.

Everyday Application

How can reading food labels help you to control the amount of fats and carbohydrates in your diet? Read the labels of all the packaged foods you ate in one day and add up how much fat and carbohydrate you consumed.

Look at the labels again and add up how much sodium, or salt, you consumed. Compare this figure to the daily recommended allowance of 1 100 to 3 300 mg of salt. Did you consume too much salt? Why is a low-salt diet important to your health?

Chicken Noodle Soup
Serving Size............250g
Calories...................131
Protein....................6.5g
Carbohydrate............9g
Fat..........................7.4g

Milk
Serving Size............250g
Calories...................200
Protein.....................13g
Carbohydrate...........20g
Fat............................7g

SECTION 23.2

Section Objectives
For a list of section objectives, see the Student Edition page.

Skills Objectives
Students should be able to:

Estimate the average number of hours per day they perform ten different activities.

Interpret Data from a table.

Infer that their bodies need rest in order to function properly.

Estimate the amount of stress caused by a given stimulus.

MOTIVATE

Skills WarmUp
To help students estimate the average number of hours they perform different activities each day, have them do the Skills WarmUp.
Answer Answers will vary.

Prior Knowledge
To gauge how much students understand about how exercise and rest affect health, ask the following questions:

▶ What do you think it means to say you're healthy?

▶ Name some kinds of things you do to exercise.

▶ How do you feel after you exercise?

▶ Describe what it feels like when you don't get enough rest.

INTEGRATED LEARNING

Integrating the Sciences
Physiology Endorphins are natural pain relievers that are found in the brain. There is strong evidence that athletes and people who follow a regular exercise program experience a sense of well-being due to the endorphins released by the brain during strenuous exercise. Discuss whether students have experienced a release of endorphins during exercise.

Themes in Science
Stability and Equilibrium A healthy life-style includes a balance of good nutrition, sleep, and physical as well as mental activity. Encourage students to discuss how their bodies feel when they don't get enough sleep or exercise.

Skills WarmUp

Estimate Make a table you can use to estimate the average number of hours per day you perform ten different activities, such as walking to school, riding a bicycle, playing a sport, playing a musical instrument, helping around the house, playing computer games, and so on.

Figure 23.4 ▲
Exercise that strengthens muscles is not limited to any age group.

23.2 Exercise and Rest

Objectives

▶ **Explain** the role of exercise in maintaining a healthy body.

▶ **State** how stress affects the body.

▶ **Interpret** the role of rest and relaxation in maintaining health.

▶ **Estimate** the stress level of events and situations.

Every day, you make personal choices about eating, exercise, and rest. You may also face choices about whether to smoke, drink alcoholic beverages, or use drugs. The decisions you make will determine your lifestyle. How does your lifestyle affect your health and well-being?

Exercise

Regular exercise is part of a healthy lifestyle. Exercise does more than strengthen muscles. It improves your flexibility and builds endurance. Actually, endurance is the most important result of exercise. Building endurance strengthens your body's most important muscle—your heart muscle.

The best way to improve your endurance is to do aerobic (air OH bihk) exercise. Aerobic exercise is vigorous activity that increases your muscles' need for oxygen. Oxygen is used in the chemical reactions that provide muscles with energy. Because of the increased demand for oxygen, your lungs work harder, and more oxygen is delivered to the blood. Your heart also works harder to pump more blood to the muscles. The object is not to make the heart beat as fast as it can but to make it beat somewhat faster than when the body is at rest. Aerobic exercise benefits your heart and lungs because the hard work improves their strength. As a result, they work more efficiently when your body is at rest.

Aerobic exercises include bicycling, dancing, jogging, rowing, skating, and swimming. Even ordinary activities like walking or mowing the lawn can become aerobic when you do them at a fast pace. Exercise also helps you maintain a healthy body weight.

Chapter 23 Nutrition, Health, and Wellness

Math Connection

Have students take their *resting heart rates*, the rate at which their hearts beat when they are sitting down. (They should time the throat pulse, just beneath the jaw, or the wrist pulse; count for 10 seconds and multiply by 6 to get the minute rate.) Then have them perform some exercise (in or outside the classroom) and time their pulses again to determine their *exercise pulse rates*. Then have students multiply their exercise rates by 65 percent and by 75 percent to find their safe *maximum heart rate ranges*. If their exercise rates fall within the safe maximum heart rate range, they are in good physical condition. If the numbers are higher than the maximum range, the students may need more exercise.

As you probably know, food is rated in Calories. A Calorie is a measure of the energy in food. A food Calorie (spelled with a capital *C*) is actually a kilocalorie (1000 Calories). The number of Calories in a serving of food tells how much heat energy is available in the food for the body to use. Foods high in Calories provide a lot of energy. Foods low in Calories provide less energy.

Like wood, food can be "burned." The heat given off by burning food can change the temperature of water. Scientists define a calorie as the amount of heat needed to raise the temperature of one gram of water 1°C. Of course, food energy in your body is not used to heat water. It is used to keep your body temperature close to 37°C. What else does food energy help you do? ①

Your body weight depends on the balance between Calories taken in and Calories used. Burning Calories as you exercise can help you lose weight. Which activities in Table 23.5 use the most Calories? ②

Table 23.5 Burning Calories

Exercise	Burned per Hour
Dancing (moderate)	210
Recreational swimming	220
Softball	230
Tennis	336
Bowling	336
Basketball	348
Football	414
Soccer	480

Based on body weight of 56.8 kg

SkillBuilder Interpreting Data

Measuring Calories Used

Different activities use different amounts of Calories. For example, shoveling snow uses more than twice as many Calories as raking leaves. Using the data in the tables on this page, calculate the number of Calories you use in an average 24-hour period. Create your own table on a separate sheet of paper. It should include space for the estimated number of hours you spend doing each activity and space for the total Calories used for that activity.

Study your completed table, then answer the following questions.

1. What activity used the greatest number of Calories?
2. What activity used the least number of Calories?
3. Write the total number of Calories used by ten of your classmates. Then write a short statement comparing your Calorie use with their Calorie use.

Activity	Calories Burned per Hour
Sleeping	60
Reading, watching television	60
Playing a computer game	100
Playing a musical instrument	120
Doing household chores	190
Mowing grass	200
Walking	240
Bicycling	250

Using your results, write a paper discussing whether or not you get enough exercise. To support your position, give examples of the activities you do and the Calories used.

Chapter 23 Nutrition, Health, and Wellness   495

TEACH

Discuss
Have students study Table 23.5. Ask the following questions:

▶ According to the chart, which kind of exercise uses the least Calories? (Dancing)

▶ How many Calories are burned in an hour of playing basketball? (348)

▶ Would a person use up more Calories running for an hour or walking for an hour? Why? (Running would use more Calories. The table shows that exercise that involves greater activity levels uses up more Calories.)

SkillBuilder
Analyze Have the students construct nutritional guidelines for three different people: one with a low, another with a medium, and a third with a high physical activity level. Remind students that they will first have to determine how many Calories the three people expend in a day before they can determine what their nutritional intake should be. Stress to students that a diet that includes no carbohydrates or fats can be just as unhealthy as one that includes too many.

Answers Answers to questions 1 to 3 should reflect data in the tables students created.

Integrated Learning
Use Integrating Worksheet 23.2.

Answers to In-Text Questions

① **Food energy is necessary for all life processes and physical activities.**

② **Soccer, football, and basketball burn the most Calories.**

TEACH • Continued

Skills WorkOut
To help students infer how the lack of rest affects the body, have them do the Skills WorkOut.
Answer Lack of rest from physical activity causes you to feel weak and to have less control over your coordination.

Discuss
Discuss the following situation: Suppose you have a friend who seems to be worried about something and who can't concentrate on schoolwork or other daily activities. If the friend continues to worry for more than a day, what should you do? (Answers may vary. Point out to students that sometimes it helps just to talk about things that are worrying them.)

Class Activity
Have students create a mural that illustrates some of the ways they release stress. Provide paints, magazines, scissors, paste, and other art supplies. Encourage students to be creative. Ask the question, How much does enjoying an activity help relieve stress?

Answer to In-Text Question
① Art classes may take your mind off the source of your stress.

INTEGRATED LEARNING

Writing Connection
Have students write a journal entry describing an event or situation that causes them to feel stress. Have them write some ideas for handling the stress. Encourage interested students to share their descriptions with you or with a classmate.

Themes in Science
Stability and Equilibrium The human body is made up of delicately balanced systems that work together to maintain order and to adapt to changes. Stress can upset this balance, sometimes leading to illness and other problems. Have students discuss whether they think stress affects them personally, what strategies they use to cope, and the effectiveness of those strategies.

Skills WorkOut
Infer Choose an exercise you can do repeatedly, such as jumping rope, dribbling a basketball, or running in place. Repeat the activity until you start to feel tired. Write down any body changes you observe. Then rest for a few minutes and observe how you feel. How does lack of rest affect your body?

Stress and Your Body

Do you sometimes feel tense? Pressured? Depressed? Everyone has these feelings at times. If you feel this way very often, however, you are probably experiencing a lot of stress. Stress is an unbalancing of your emotions or mental state. Learning to handle stress is necessary for a healthy lifestyle.

Stress is a normal part of everyday life. It is usually caused by unhappy events, but stress can also stem from happy events, such as getting a job. Stress can also build up gradually, without being caused by any single event. For example, you may feel a need to always excel in school or sports.

Your body's built-in stress response equips you to handle dangerous situations. Your heart beats faster, your blood pressure increases, and your nervous system goes on the alert. When triggered too often, however, this response can cause chronic headaches, depression, loss of sleep, and other health problems.

What can you do about stress? Reducing stress might be as simple as getting up 15 minutes earlier than you usually do in the morning, or writing down school assignments so you don't forget them. One of the best ways to relieve stress is to exercise. Sometimes stress is brought on by holding in your feelings. If you are troubled, don't bottle up your feelings. Find someone you can talk to—a family member, counselor, teacher, or friend.

Figure 23.5 ▲
The strenuous activity of competitive sports releases tension and keeps your body in good condition. How might art classes relieve stress? ①

Chapter 23 Nutrition, Health, and Wellness

Multicultural Perspectives

Since cycling is both a healthy activity and well liked in the Netherlands, the Dutch started the Ride-a-Bike Promotion Van. The van holds about 70 bicycles and can be rented by businesses and private groups for recreation. Have students investigate similar health-promoting ideas in other countries.

Integrating the Sciences

Physical Science Exercise machines change the size and direction of the effort force as well as the distance over which it is exerted. Have students visit a nearby health club and analyze how people using the exercise equipment change the effort force, distance, or direction depending on their goals in exercising.

Relaxation and sleep also help reduce stress. Any activity that differs from your normal routine can be relaxing. Plan to spend a little time each day doing something you enjoy. Growing teenagers need more sleep than adults. The amount of sleep needed varies from person to person, but your body probably does best on at least eight hours of sleep a night.

Science and Technology *Toning Up*

For many people, using an exercise machine is the most convenient way of staying fit. For one thing, exercise machines can be used at home. People who own machines can exercise at any time, in any kind of weather.

One reason people use exercise machines is to strengthen the heart and lungs through aerobic exercise. Another goal is to tone specific muscle groups.

There are many types of exercise machines that help people design their own fitness programs. For example, many stationary bicycles, called ergometers, calculate your work output as you pedal. This makes it possible to carefully monitor your exercise program. Small trampolines, or rebound exercisers, are great for your heart, lungs, and leg muscles. Some runners use treadmills for exercising indoors during bad weather. Treadmills are revolving belts that can be powered either manually or by a motor. Weight machines help a person concentrate on toning specific muscle groups. There are also exercise machines that imitate the movements used in rowing, cross-country skiing, or climbing stairs.

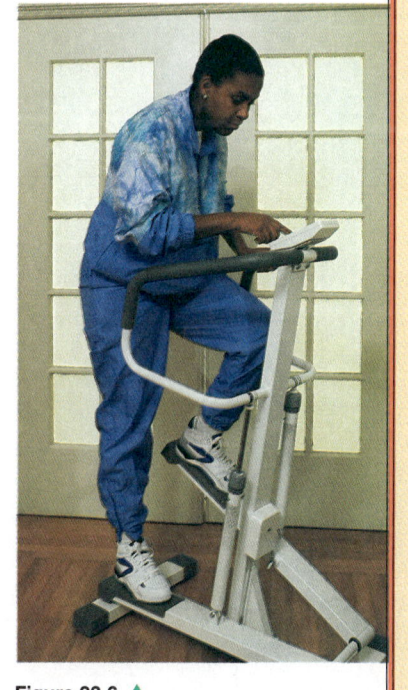

Figure 23.6 ▲
How does a stair-stepping machine help you keep fit? ②

Check and Explain

1. Why should a person exercise regularly?
2. List several ways stress affects the body.
3. **Reason and Conclude** Does regular exercise increase or decrease the number of hours of sleep you need? Why?
4. **Estimate** Make a list of situations that cause stress. Then give each item a number from 1 to 10 according to how much stress you think it causes. Compare your list with those of your classmates.

EVALUATE

WrapUp
Reinforce Have students discuss how much exercise and rest they get. Ask the following questions:

▶ What did you learn about exercise and rest from reading this chapter?

▶ How would what you've learned about exercise, rest, and stress affect your life-style choices?

Use Review Worksheet 23.2.

Check and Explain

1. Exercise strengthens muscles, improves your flexibility, and builds endurance.
2. Stress can cause tension, pressure, and depression.
3. Regular exercise increases the amount of sleep you need because your tired muscles need more rest.
4. Answers may vary. Possible factors include: death of a family member, drugs or alcohol abuse, a change in relationships with peers, divorce or remarriage of a parent, serious illness, moving. Even positive experiences can be stressful.

Answer to In-Text Question

② **It strengthens the heart and lungs through aerobic exercise, and also tones the legs.**

SECTION 23.3

Section Objectives
For a list of section objectives, see the Student Edition page.

Skills Objectives
Students should be able to:

Communicate their feelings about drugs.

Research the penalties for drug-related crimes in their state.

Vocabulary
addiction, stimulant, depressant, narcotic, hallucinogen

MOTIVATE

Skills WarmUp
To help students communicate their feelings about drugs, have them do the Skills WarmUp.
Answers Answers will vary.

Prior Knowledge
To gauge how much students know about drugs and their effect on the human body, ask the following questions:

▶ What is a drug?
▶ What are drugs used for?
▶ Can you remember any time when something you ate changed the way you felt?

Answer to In-Text Question
① Always read the label, and use the drug exactly as recommended. Pills should never be mixed, nor should they be stored in anything other than the original container.

INTEGRATED LEARNING

Themes in Science

Systems and Interactions The human body systems work together. For example, the respiratory system provides oxygen to the blood and the circulatory system delivers the oxygen throughout the body so cells can produce the energy required for movement and growth. Explain to students that when a drug enters the circulatory system, it travels to every part of the body.

Skills WarmUp

Communicate Think about the following questions, then write a short paragraph that communicates your feelings about drugs. Do you know anyone who uses drugs? What happens to someone who gets "hooked" on drugs? What happens to a drug abuser's family?

Figure 23.7
Finding the right medication in this pile of pills could be a problem. What steps can you take to guard against taking the wrong drug? ▼ ①

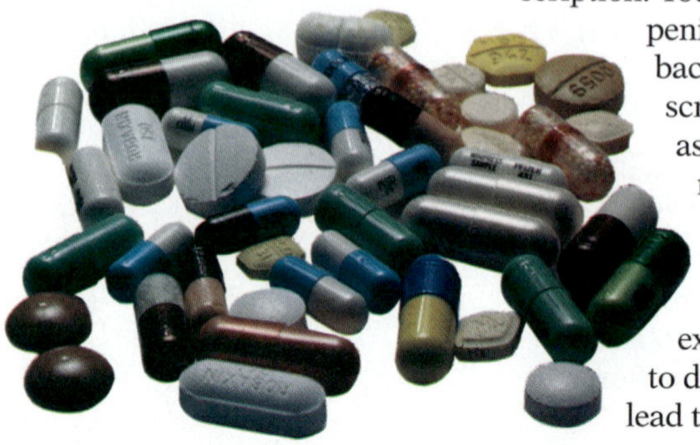

23.3 Drugs and Substance Abuse

Objectives

▶ **Describe** the major kinds of drugs.
▶ **Describe** the effects of drug abuse.
▶ **Classify** commonly abused drugs.
▶ **Analyze** the reasons why people abuse drugs.

What do you know about drugs? Are all drugs harmful? Write your own definition of a *drug*. Then make a list of all substances that you think are drugs. Did you classify alcohol, nicotine, or caffeine as drugs?

Any substance that changes the way your body or mind works is a drug. Medicines are drugs used to relieve pain and cure diseases. Generally, drugs are helpful when used for their intended purposes. But any drug can be harmful when not used properly. Any harmful or nonmedical use of a drug is called *substance abuse*.

Types of Drugs

Drugs you can buy at the supermarket are over-the-counter, or nonprescription, drugs. Aspirin, cold remedies, and insulin are just a few of the hundreds of over-the-counter drugs available. While they are fairly safe, you still need to use these drugs carefully. Always read the label and use the drug exactly as recommended.

Other drugs can be purchased only with a doctor's prescription. You may have taken a prescription drug such as penicillin or tetracycline. These drugs fight bacterial infections like strep throat. Other prescription drugs are used to treat such illnesses as heart disease and cancer. Because they are more powerful than over-the-counter drugs, prescription drugs should always be taken under a doctor's supervision. Misuse of prescription drugs can be very dangerous. For example, some athletes use drugs called steroids to develop bigger muscles. This practice may lead to serious health problems and even death.

Chapter 23 Nutrition, Health, and Wellness

Writing Connection

Have students write a story about a boy or girl their own age who is being pressured to try marijuana, crack cocaine, or another illicit drug. Tell them the story should be about how the student stands up to the peer pressure.

You may be surprised to learn that caffeine, alcohol, and nicotine (NIHK uh TEEN) are also classified as drugs. Nicotine is found in tobacco. As you know, only adults can legally buy and use products containing alcohol and nicotine.

The possession, use, or sale of certain drugs is forbidden by federal law. These drugs, such as cocaine, crack, and marijuana, have the potential to greatly harm the user.

Dangers of Drug Abuse

Improper use of drugs can damage your health, threaten your safety, interfere with your performance at school or work, and hurt your personal relationships. Abusers may spend thousands of dollars a year on drugs. Anyone convicted of drug charges faces strict penalties and a criminal record.

Long-term abuse of certain drugs causes physical dependence, or **addiction**. An addict's body gets so used to the drug that a painful condition called withdrawal results if the drug is withheld. The body also develops a drug tolerance. This means that users need to take more and more of a drug to get the same effects.

Whereas some drugs cause a physical dependence, regular use of any drug can lead to psychological dependence. This means that using the drug is a habit because the mind craves it. A drug that your mind craves can be as hard to stop using as a drug your body is addicted to.

There is a limit to how much of a drug a person can take safely at one time. When this limit is exceeded, an overdose results. An overdose may lead to death. Mixing drugs increases the risk of an overdose. For example, using alcohol and sleeping pills together greatly increases the danger of a sleeping pill overdose.

Commonly Abused Drugs

Five types of drugs are commonly abused: stimulants, depressants, marijuana, narcotics, and hallucinogens. Some of these drugs cause physical dependence and all can cause psychological dependence. While many commonly abused drugs are illegal, quite a few can be obtained with a doctor's prescription, and some can even be purchased over-the-counter.

Skills WorkOut

Research Find out the penalties for drug-related crimes in your state. You can interview a police officer, judge, or district attorney. Or you can find the information in a library.

TEACH

Skills WorkOut
Encourage students to research the penalties for drug-related crimes by doing the Skills WorkOut.
Answer Answers will vary.

Critical Thinking
Compare and Contrast Have students consider the difference between simply wanting something and being addicted to it. Have them name the physical effects of addiction. (People who are addicted to substances suffer withdrawal symptoms if they don't get the substance. Also, addicts need to take more and more of a drug to get the desired effect from it.)

Skills Development
Classify Make five columns on the chalkboard. Label the columns with the words *stimulants, depressants, marijuana, hallucinogens,* and *narcotics*. Ask students which drugs are legal and which are illegal and label them. Stress that legal drugs can be dangerous. (Be sure that you include caffeine under the list of stimulants, and point out that it is found not just in coffee but also in chocolate, tea, and many soft drinks.)

Discuss
Ask students why teenagers might begin using alcohol, nicotine, or any drug when they are aware of the possible danger. Encourage students to share their feelings as well as their ideas. Stress the importance of listening politely to one another's responses. (Answers may vary. Teenagers might like the way drugs make them feel, and they probably think that they can stop at any time, which may not be true.)

Chapter 23 Nutrition, Health, and Wellness

TEACH • Continued

Class Activity

Have students name as many products that have caffeine as they can think of. List these on the chalkboard. Then poll students to find out how much caffeine each has on a typical day. Have students describe the way caffeine makes them feel. Explain that even mild stimulants such as caffeine can be used to excess.

Discuss

Have students read the information on this page. Ask:

▶ Why do people use stimulants? (To stay awake, lose weight, feel more energetic or euphoric)

▶ What are some dangers of using stimulants? (Temporary mental illness, irritability, sleeplessness)

▶ Why do people use depressants? (To relax, get to sleep, relieve anxiety)

▶ What are some ways to feel more energetic without using stimulants? (Exercise can make a person feel more energetic. Getting more sleep can also help.)

▶ What are some ways to relieve anxiety without using depressants? (Exercise can relieve anxiety. Doing an activity you enjoy can also relieve anxiety.)

Answer to In-Text Question

① Misuse of drugs can cause injury or death.

INTEGRATED LEARNING

Integrating the Sciences

Chemistry Refer students to Chapter 21, Figure 21.3. Remind them about the role of chemical signals that cross synapses between one nerve cell and another. Explain that mood-altering drugs affect the transmission of those chemicals across synapses.

Stimulants Does sipping a cola pep you up? Cola contains a **stimulant**, a substance that stimulates the nervous system. Stimulants also raise blood pressure and make the heart beat faster. The stimulant in cola is caffeine. Caffeine is also found in coffee and tea. Tobacco contains nicotine, a mild stimulant. Amphetamines (am FEHT uh MEENZ) are a group of stimulants often referred to as "speed." They are very powerful stimulants.

Amphetamines are popular mood-enhancing drugs. Small doses create a feeling of limitless energy and a sense of pleasure called euphoria. Amphetamines also curb the appetite and reduce the desire for sleep. Doctors prescribe small doses of these drugs to treat depression and for supervised weight-loss programs.

Cocaine is also a stimulant. On the street, a form of cocaine is often sold as "crack." Long-term abusers of crack and cocaine may become tense and suspicious. They may suffer from depression and other kinds of temporary mental illness. Heavy use causes permanent brain damage, and sometimes death.

Depressants Another group of drugs relaxes the central nervous system, causing the heart rate to slow and blood pressure to fall. These drugs are **depressants**. The most commonly used depressant is alcohol. Others include barbiturates (bar BIHCH ur ihts) and tranquilizers. Some drug abusers routinely take depressants to relax and sleep. They then take stimulants to get going again. This creates a dangerous cycle.

Barbiturates are highly addictive to both the body and the mind. They change how people view things. They also slow down normal responses. Users may become confused about how many pills they have taken and die of an accidental overdose. Sudden withdrawal causes painful cramps, nausea, convulsions, and sometimes death.

Tranquilizers are widely used prescription drugs that calm people down without making them sleepy. Doctors use these drugs to treat such illnesses as depression and anxiety. Abuse of some tranquilizers can result in addiction. Mixing tranquilizers and alcohol is especially dangerous and can cause death.

◀ **Figure 23.8**
Why is it important to spread the word about the dangers of using drugs? ①

Chapter 23 Nutrition, Health, and Wellness

Integrating the Sciences

Physiology Point out that one of the most dangerous effects of drug use occurs when one type of drug is taken with another. For example, a prescription drug, such as a barbiturate, has an ingredient that adversely affects the body when it's taken with alcohol. This combination can be harmful to the central nervous system and can even cause death.

Marijuana A popular illegal drug is marijuana. It is prepared by shredding the dried leaves and flowers of the hemp plant and then smoked in cigarettes or in pipes. Smokers experience a disoriented state. Moods can vary from excitement to depression.

Marijuana contains a compound called THC, which acts on the brain and nervous system. THC reduces a person's ability to concentrate and react. There is evidence that prolonged use of marijuana damages brain cells. Marijuana's effect on the lungs is similar to that of tobacco smoke.

Figure 23.9 ▲
The leaves, flowers, and resin from the hemp plant are used to produce marijuana.

Narcotics Heroin, morphine, and codeine are powerful drugs made from the opium poppy. These drugs are called **narcotics**. The word *narcotic* means "to numb." Morphine and codeine are used legally as painkillers.

When injected into a vein, heroin produces a warm, pleasurable sensation followed by a dreamy high. Heroin is physically and psychologically addictive. Heroin addicts risk getting hepatitis or AIDS if they share needles.

Hallucinogens The drug LSD causes hallucinations, or dreamlike "trips," in which a person sees, hears, smells, or feels things that do not exist. LSD and other drugs that cause hallucinations are called **hallucinogens** (huh LOO sih nuh juhnz). Another hallucinogen is mescaline, a drug obtained from peyote cactus.

After taking LSD, a person feels the effects of the drug for 10 to 12 hours. In the unreal world of a "good trip," colors seem more brilliant and patterns unfold. Senses merge into one another: music may appear as a color, or colors may even seem to have a taste. Sometimes it seems as if flying or floating is possible. Some people under the influence of LSD have fallen to their deaths thinking they could fly.

LSD trips are unpredictable. One trip may be good, the next frightening. During a "bad trip," people may appear to be horrible monsters. These visions can repeat themselves in flashbacks. Flashbacks may occur without warning—days, even months—after an individual stops using LSD.

Figure 23.10 ▲
Narcotics are extracted from the seed pods of the opium poppy.

Skills Development

Infer Outline the effects of different drugs on perception. Ask students to infer how being under the influence of these drugs would affect driving a car, studying for a test, debating a topic, playing a sport, or other activity. (Include considerations about reaction time, ability to concentrate, and ability to retain information; the last is particularly true with marijuana.)

Discuss

Ask the following questions:

▶ How do people take marijuana? (Usually by smoking it)

▶ What are some effects of smoking marijuana? (Mood changes; often users experience a dreamy "high")

▶ How do people take heroin? (By injecting it with a needle)

▶ What are the effects of using heroin? (Heroin numbs pain and provides a warm pleasurable sensation. It is physically addictive, which means addicts need to keep increasing the dose that they take to feel the drug's effects. Sharing hypodermic needles can spread serious disease.)

▶ What are some effects of hallucinogens? (Hallucinations, delusions, flashbacks)

TEACH • Continued

Discuss
Read through the Science and You section with the class. Encourage students to respond to the questions raised in the text.

EVALUATE

WrapUp
Reinforce Work with students to make a concept map that divides the five categories of substances into three groups: those that relax, those that stimulate, and those that expand awareness.
Use Review Worksheet 23.3.

Check and Explain
1. Stimulants stimulate the nervous system, causing high blood pressure and making the heart beat faster. Depressants relax the central nervous system, causing the heart rate to slow and blood pressure to fall. Marijuana reduces a person's ability to concentrate and react, and may cause damage to the brain and lungs. Hallucinogens cause a person to hallucinate. Narcotics numb pain and are physically and psychologically addictive.
2. Drug abuse can damage your health, threaten your safety, interfere with your performance at school or work, and hurt your personal relationships.
3. Answers may vary, but the drugs may be classified using the information presented in this chapter.
4. Answers may vary. Students' letters should show that they understand some of the effects of drugs.

Answer to In-Text Question
① Singing in a chorus, playing with a pet, playing games like chess

---

INTEGRATED LEARNING

STS Connection
Explain to students that although overcoming dependency on drugs is extremely difficult, it is not impossible. Describe the following treatments:

▶ Detoxification is the supervised withdrawal from drug dependence. During detoxification, people are sometimes given medication to treat withdrawal symptoms.

▶ Therapeutic communities are specially organized places in which drug abusers can live. These communities are drug-free and provide support to people recovering from the effects of drug abuse.

▶ Outpatient programs provide support through various forms of counseling.

Figure 23.11 ▲
Helping others is a rewarding and productive way to spend time. Activities like these give teens an opportunity to meet new people and learn new things.

Science and You *Critical Choices*
One of the most important decisions you will ever make is whether or not to use drugs. It's a fact that most people who abuse drugs begin using them in their early teens or early twenties. For many teens, the pressure to use drugs comes mainly from their friends or peers.

Think about how peer pressure affects you. For example, what would you do if you went to a party where some of your classmates were smoking marijuana or drinking alcohol? Imagine that you were offered one of these substances. What is the best way to react to this kind of situation?

There are effective ways to resist the pressure to "join in." First, decide how you feel about the situation and then stick to your opinion. You can politely say "No thanks." You can walk away. You can seek out someone who feels the same way you do and supports your feelings.

There are many alternatives to using drugs. Below is a list of some healthy alternatives. What activities can you add? ①

▶ Exercise, outdoor hobbies, sports
▶ Talking to friends, family members, teachers
▶ Volunteer work
▶ Reading books, writing stories and poems, drawing, painting, playing an instrument
▶ Joining a club or an organization

Check and Explain
1. Name the major types of drugs and describe how each affects the body.
2. What may happen to someone who abuses drugs?
3. **Classify** Make a list of commonly abused drugs. Then classify each drug as illegal, legal with a prescription, or legal over-the-counter.
4. **Communicate** Write a letter to a friend, real or imaginary, whose use of drugs is having a bad effect on his or her life. Explain the dangers that person faces and suggest ways of dealing with the problem.

Chapter 23 Nutrition, Health, and Wellness

Social Studies Connection
Discuss the movement to prohibit alcohol in the United States during the 1800s, the prohibition amendment of 1920, and the repeal of the amendment in 1933.

23.4 Alcohol and Tobacco

Objectives
- **Describe** the effects of alcohol on the body.
- **Explain** alcohol abuse.
- **Name** the chemicals present in tobacco smoke.
- **Describe** the long-term effects of tobacco on the body.
- **Collect data** about the alcohol content of beverages.

Skills WarmUp
Communicate Imagine that a friend is trying to get you to smoke or drink. Think about what you would say to explain your feelings to him or her. Then role-play the situation with a classmate. Take turns expressing your feelings out loud.

Why do you think alcohol and tobacco are sometimes called *social drugs*? It is because people often use them in social gatherings. People tend to forget these substances are drugs because they see others using them so often. However, just because alcohol and tobacco are legal drugs does not mean they are safe to use.

Alcohol Abuse

Any person who drinks to the point of drunkenness is abusing alcohol. People who abuse alcohol may not be alcoholics, however. Alcoholics are addicts—they have lost control over their drinking. Alcohol addiction is called alcoholism. At first, many alcoholics are unaware of their drinking problem. They may even boast that they can handle their drinking better than their friends. Actually, a high tolerance for alcohol is one of the early symptoms of alcoholism. Many alcoholics cannot get through the day without alcohol. Often they do not remember what happened during a drinking bout. They often dismiss their abuse of alcohol as "social drinking."

Alcoholism strikes men and women of all ages and in all walks of life, including teenagers. It is often hard to identify alcoholics because they learn to hide their drinking problem. Like other kinds of addiction, however, alcoholism disrupts a person's life.

The abuse of alcohol can have very serious results for both the drinker and society. Among teenagers, alcohol-related accidents are the leading cause of death. Look at Table 23.6. In what other ways do alcohol and alcoholism affect people's lives? ②

Table 23.6
Alcohol's Costs to Society

In the United States alone . . .
▶ Drunk drivers are responsible for about half of the 46,000 automobile accidents that occur every year.
▶ Someone is killed by a drunk driver every 21 minutes.
▶ Alcohol is connected to as many as 2,800 drowning deaths each year.
▶ Alcoholism costs at least $100 billion a year in treatment, property damage, lowered worker productivity, and other losses.

SECTION 23.4

Section Objectives
For a list of section objectives, see the Student Edition page.

Skills Objectives
Students should be able to:

Communicate their feelings about peer pressure.

Collect Data about alcoholic content in beverages.

Collect Data about the substances in cigarette smoke.

MOTIVATE

Skills WarmUp
To help students imagine and communicate their feelings to a friend who is trying to get them to smoke or drink, have them do the Skills WarmUp.

Misconceptions
Students may think that drinking coffee helps a person to become sober. Point out that it takes the body about an hour to remove one drink's worth of alcohol from the bloodstream. Drinking coffee does not speed up this process.

Answer to In-Text Question
② Alcoholism costs billions in treatment, property damage, and lowered worker productivity.

TEACH

Directed Inquiry

Ask students to study the drawings and the paragraphs about the effects of alcohol and tobacco.

▶ What body systems do alcohol and smoking affect? (All body systems are affected by alcohol and smoking.)

▶ What effect does alcohol have upon the brain? The heart? The liver? (See Figure 23.12.)

▶ What effect does smoking have upon the brain? The mouth and throat? The lungs? The heart? (See Figure 23.13.)

Misconceptions

Students may think that drinking a can of beer (5 percent alcohol) or a glass of wine (12 percent alcohol) is not as much of a problem as drinking a shot of hard liquor because liquor has a higher concentration of alcohol (50 percent alcohol). In actuality, a 12-ounce serving of beer, a 4-ounce glass of wine, and a 1-ounce shot of hard liquor all contain the same amount of alcohol.

INTEGRATED LEARNING

Social Studies Connection

Have students investigate changing attitudes toward smoking in the United States. They may look at advertisements in newspapers and magazines from the 1940s to the present. They could also talk to parents, grandparents, and friends to get information. Lead students in a discussion about their discoveries.
Use Integrating Worksheet 23.4.

Effects of Alcohol

The blood absorbs alcohol directly from the stomach and intestines without digestion. Once in the bloodstream, alcohol moves quickly throughout the body. The first reactions take place in the brain.

Remember that alcohol is a depressant. It dulls the senses and weakens the brain's control centers. A drinker may become confused and lose control of speech, movements, and actions. Heavy drinking over time permanently damages brain cells.

The heart and liver are also affected by alcohol abuse. Heavy drinking increases blood pressure and eventually weakens the heart muscle. The liver breaks down alcohol and removes it from the body. Excessive drinking overworks the liver and can damage it, possibly leading to a condition in which the liver is severely scarred. Figure 23.12 summarizes the major effects of alcohol on the human body.

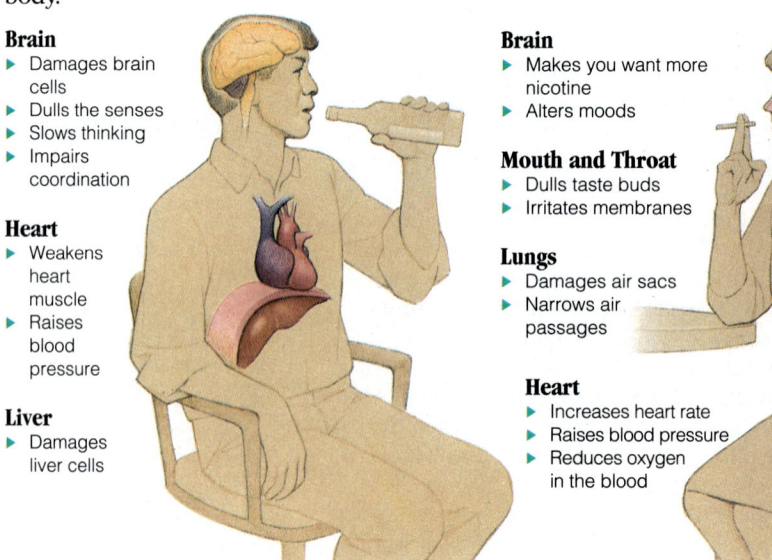

Brain
▶ Damages brain cells
▶ Dulls the senses
▶ Slows thinking
▶ Impairs coordination

Heart
▶ Weakens heart muscle
▶ Raises blood pressure

Liver
▶ Damages liver cells

Figure 23.12 ▲
How does alcohol affect the liver?

Effects of Tobacco

Nicotine is a poisonous compound found in tobacco leaves. It acts as a mild stimulant on the heart and other organs. Because nicotine is very addictive, heavy smokers find it hard to quit.

Cigarette smoke contains more than 3,000 substances. Some, such as carbon monoxide, are poisons. Others, such as tars, have chemicals that cause cancer.

In the brain, nicotine stimulates cravings for more nicotine. It can also affect a person's moods. Nicotine dulls the sense of taste and irritates membranes in the nose and throat.

In the lungs, nicotine damages the air sacs and increases mucous secretions in the bronchial tubes. When this happens, normal breathing is affected. Nicotine also increases the heart rate. Figure 23.13 summarizes the major effects of nicotine on the human body.

Brain
▶ Makes you want more nicotine
▶ Alters moods

Mouth and Throat
▶ Dulls taste buds
▶ Irritates membranes

Lungs
▶ Damages air sacs
▶ Narrows air passages

Heart
▶ Increases heart rate
▶ Raises blood pressure
▶ Reduces oxygen in the blood

Figure 23.13 ▲
How does nicotine affect the lungs?

Chapter 23 Nutrition, Health, and Wellness

Integrating the Sciences

Chemistry People are growing concerned about the effects of second-hand smoke on nonsmokers. Some studies show that there is more tar, nicotine, and carbon dioxide in second-hand smoke. Also there is 50 times as much ammonia and three times as much of a substance called "3-4 benzopyrene," which is thought to cause cancer.

STS Connection

A popular aid to people who are trying to quit smoking is the nicotine patch. It releases a small amount of nicotine through the skin in smaller and smaller doses over a month, in order to break the addiction gradually. Doctors recommend that patch users also have counseling to break smoking-related behavior patterns. Ask students why it is probably better to use both methods.

Skills WorkOut

To help students understand the substances in cigarette smoke, have them do the Skills WorkOut.

Discuss

Stress to students that once a person begins to smoke, it is often difficult to quit. Smoking is addictive. Nicotine reinforces a person's desire to smoke.

Consider This

Evaluate Sources Have students bring in three examples of cigarette ads from teen or general interest magazines. Ask the following questions:

▶ What are the hidden messages in the ads? (That smoking will keep you young, happy, popular, carefree, active, and healthy)

▶ Compare the images in the ads with the Surgeon General's warnings. Are the warnings less effective than the ads? Why or why not? (Answers will vary. The ads are usually more effective emotionally because of their bright colors and attractive images; the warnings are effective because they appeal to logic.)

Decision Making

If you have classroom sets of *One-Minute Readings*, have students read Issue 5, "Tobacco" on page 8. Discuss the questions.

Enrich

Use Enrich Worksheet 23.4.

Diseases and Tobacco

Studies show conclusively that smoking is the leading cause of preventable death in the United States. Smoking has been linked to a number of serious diseases. For example, smokers are much more likely to get heart disease than nonsmokers. The buildup of tars in the lungs causes a respiratory disease called emphysema (EHM fuh SEE muh). Smokers are ten times more likely to die from lung cancer than nonsmokers.

Recent studies show that nonsmokers who inhale tobacco smoke are also more likely to get heart disease or lung cancer. Moreover, children of smokers have more respiratory infections than do children of nonsmokers.

For people who already smoke, there is some good news. All of these risks can be dramatically reduced by quitting smoking. Two years after quitting, for example, a person's risk of heart attack returns to normal. After ten years, the risk of lung cancer is nearly normal.

Skills WorkOut

Collect Data Use the library to find out the kinds and amounts of gases, tars, and other substances in cigarette smoke. Make a table that lists 20 of these substances. Next to each, briefly explain why they are harmful.

Consider This

Should Tobacco Advertising Be Allowed?

The health hazards of smoking have been firmly established. In the United States, federal law requires that all cigarette packages carry a statement that smoking is harmful to your health. The law also forbids cigarette advertising on radio and television. However, tobacco companies can still advertise their products on billboards and in newspapers and magazines.

Consider Some Issues
Some people believe that cigarette ads should not be allowed anywhere. They claim that such ads persuade many people to start smoking. They feel that tobacco ads project a sophisticated image that attracts teenagers. They claim that some advertising is aimed at groups that may not be aware of the problems caused by using tobacco and are easily swayed by the ads.

The tobacco companies argue that tobacco products are legal in this country, so they have a right to advertise. Tobacco companies claim that their ads are designed for adult audiences. They say ads can't make people buy cigarettes and that adults are free to choose whether or not they want to smoke.

Think About It Do tobacco companies have a right to advertise a product known to be unhealthy for people to use? How influential are cigarette ads? Do these ads affect your own attitude about smoking?

Write About It Write a paper stating your position for or against cigarette advertising. Use the ads from magazines and newspapers to support your position.

Compare your position with those of your classmates. Form two teams to debate whether or not tobacco advertising should be allowed.

TEACH • Continued

Research

Have students find newspaper and magazine articles written by doctors or nurses who have treated babies with fetal alcohol syndrome. Ask what treatments are used for infants born with FAS.

EVALUATE

WrapUp

Reinforce Very often people who smoke are also drinkers. To review the effects of tobacco and alcohol, make a concept map that concentrates on the combined effect of smoking and drinking upon an individual.

Use Review Worksheet 23.4.

Check and Explain

1. Nicotine
2. Tobacco damages the lungs, mouth, throat, heart, and brain. Using tobacco may lead to lung cancer, emphysema, or heart disease. Alcohol affects the heart, blood pressure, and liver cells. Alcohol may cause heart disease, liver damage, or pancreatitis.
3. Answers may vary, but students' answers may be based on stereotypes. Alcoholism affects people regardless of race, class, or career choice.
4. Answers may vary. (Beer contains 5% alcohol, wine contains 12% alcohol, hard liquor contains 50% alcohol)

Answer to In-Text Question

① Women should refrain from drinking alcohol during pregnancy.

506

INTEGRATED LEARNING

STS Connection

You may wish to provide students with more details about fetal alcohol syndrome. During the first weeks of pregnancy, an embryo develops all of its organ systems. It is during the first three months, or trimester, of pregnancy that fetuses are most likely to be harmed by alcohol and drugs in the mother's system. Fetal alcohol syndrome occurs when the developing fetus is exposed to ethanol (alcohol) or its by-product, acetaldehyde. Babies with fetal alcohol syndrome show signs of mental retardation and delayed mental development. They can also have abnormalities of the face, head, and respiratory system. Stress to students that fetal alcohol syndrome is unpredictable. Although it's associated with heavy drinking, the syndrome has been known to develop from lighter alcohol intake.

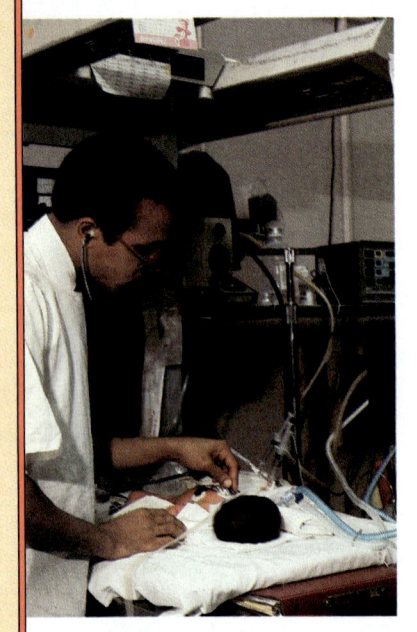

Figure 23.14 ▲
Babies of alcoholic mothers start life at risk. How can fetal alcohol syndrome be prevented? ①

Science and Society
Fetal Alcohol Syndrome

When pregnant women drink, the alcohol passes through the placenta to the baby. It affects the baby's fast-growing tissues by killing cells or slowing their growth. The brain is affected most by alcohol.

Babies born to women who drink heavily during pregnancy may have a condition known as *fetal alcohol syndrome*. This condition is actually a mixture of physical, mental, and behavioral defects. It is one of the leading known cause of mental retardation in the United States.

Babies with fetal alcohol syndrome are smaller than normal at birth. Many of these babies show some degree of permanent mental retardation. The brain damage also affects their behavior. As children, they have poor coordination and short attention spans. Many are jittery and impulsive. They often are unable to consider the consequences of their actions. As a result, even those who have normal intelligence often have difficulty succeeding at school or living on their own.

Scientists are not sure how much alcohol is harmful to an unborn baby. Studies suggest, however, that even moderate drinking may retard fetal growth. Babies of teens who drink heavily are doubly at risk since these babies are likely to be born too small or too soon.

Check and Explain

1. What is the substance in tobacco smoke that causes addiction?
2. What are the long-term effects on the body of using alcohol and tobacco?
3. **Uncover Assumptions** Who is most likely to be an alcoholic—a business executive who coaches little league, a single mother on welfare, or an unemployed carpenter? What assumptions did you make in reaching this conclusion? Can these assumptions cause you to make the wrong conclusions about people?
4. **Collect Data** Find information about the alcohol content of different kinds of alcoholic beverages. Create a circle graph or chart that displays this information.

506 Chapter 23 Nutrition, Health, and Wellness

Chapter 23 Review

Concept Summary

23.1 Nutrients
- Your body needs nutrients for energy, and for growth, replacement, and repair of cell parts.
- Carbohydrates and lipids provide energy; proteins provide materials for growth and repair.
- Vitamins, fiber, and minerals help the body to function normally.
- Water is needed for transport of nutrients and wastes. Most of the chemical reactions in your body take place in water solutions.

23.2 Exercise and Rest
- Regular exercise can help keep you healthy. Aerobic exercise improves endurance.
- Stress is a normal part of life, but sometimes stress can overwhelm your built-in stress response or trigger it too often.
- Relaxation, sleep, and exercise help reduce stress.

23.3 Drugs and Substance Abuse
- Any substance that changes the way your body or mind works is a drug.
- Alcohol, nicotine, and caffeine are drugs.
- Substance abuse is the harmful or non-medical use of drugs.
- Drug abuse can lead to physical and psychological dependence.
- Abused drugs include stimulants, depressants, marijuana, narcotics, and hallucinogens.

23.4 Alcohol and Tobacco
- Alcohol and nicotine are addictive drugs.
- Alcohol depresses the central nervous system. It affects the brain, as well as the heart and liver.
- People who become alcoholics have lost control over their drinking.
- Tobacco smoke contains nicotine, carbon monoxide, tars, and other substances that can harm the body.

Chapter Vocabulary

carbohydrate (23.1)   protein (23.1)   stimulant (23.3)   narcotic (23.3)
lipid (23.1)   addiction (23.3)   depressant (23.3)   hallucinogen (23.3)
cholesterol (23.1)

Check Your Vocabulary

Use the vocabulary words above to complete the following sentences correctly.

1. Physical and psychological dependence on a drug is called ____.
2. Fats and oils are nutrients called ____.
3. Morphine and codeine, both ____, numb feelings.
4. Drugs that cause dreamlike "trips" are called ____.
5. Starch is a ____ made of carbon, hydrogen, and oxygen.
6. Caffeine is the ____ in coffee.
7. Alcohol is a ____ because it relaxes the central nervous system.
8. A fatty substance found in all animal tissues is called ____.
9. Cells need ____ for growth and repair.

Write Your Vocabulary

Write sentences using each vocabulary word above. Show that you know what each word means.

Chapter 23 Nutrition, Health, and Wellness 507

CHAPTER 23 REVIEW

Check Your Knowledge

1. Lung cancer, emphysema, heart disease
2. Stimulants cause the central nervous system to raise blood pressure and heart rate. Cola, coffee, tea, and tobacco contain stimulants.
3. Answers may include building muscles, strengthening the heart, increasing lung capacity.
4. Answers may include fruit, vegetables, beans and peas, and whole grain breads and cereals. Fiber helps food move more quickly through the intestines, and may help lower cholesterol and prevent cancer.
5. Narcotics abuse can lead to addiction, and hepatitis or AIDS can result from sharing needles.
6. Eat a variety of foods; reduce intake of fats and cholesterol; eat plenty of fruits, vegetables, and grains; and use salt and sugar in moderation.
7. Alcohol slows down brain functions, including reflexes and reactions, which are necessary for safe driving.
8. The number of Calories tells the amount of energy present in a particular food.
9. True
10. False; 65
11. False; more
12. False; impairs
13. True
14. True
15. False; low
16. False; inorganic

Check Your Understanding

1. Accept all logical answers; however, students should recognize the importance of variety, including fruits and vegetables; moderate intake of fat, cholesterol, salt, and sugar.
2. Answers may vary. Possible answers include: getting adequate rest, planning ahead, getting regular exercise, talking to someone about problems.
3. Dependence develops when a person has taken a drug for so long that he or she needs the drug just to feel normal. Addiction is a strong physical or psychological craving for a drug. Withdrawal is the physical illness that results when an addicted person is denied the drug.
4. a. Most energy: potato
   b. Protein: fish
   c. Vitamin C: potato
5. Vitamin A: milk, eggs, green and yellow vegetables, liver; vitamin C: citrus fruits, tomatoes, potatoes, strawberries, cabbage
6. During aerobic exercise, muscles use more oxygen, which makes the heart and lungs work harder, strengthening them.
7. Answers will vary; however, students should realize that the teenager described has probably become addicted to nicotine.
8. Stimulants cause the central nervous system to raise blood pressure and heart rate, while depressants have the opposite effect on the body.
9. Answers will vary, but could include presenting scientific evidence of the health dangers of smoking; calculating the financial cost of the cigarettes; describing the harm to others from second-hand smoke; reminders that a smoker can lose the ability to taste and smell.
10. Athletes eat large amounts of carbohydrates to store up energy for the events.

Chapter 23 Review

Check Your Knowledge
Answer the following in complete sentences.

1. Name three diseases linked to smoking.
2. How do stimulants affect the human body? Name three products that contain a stimulant.
3. Name at least two benefits of aerobic exercise.
4. Name two foods that are good sources of fiber. How is fiber beneficial?
5. What are two dangers associated with narcotics abuse?
6. Name four things you can do to ensure that you have a healthy diet.
7. Why is it dangerous for a person to drive after she or he has been drinking?
8. What does the number of Calories in a serving of food tell you?

Determine whether each statement is true or false. Write *true* if it is true. If it is false, change the underlined word to make the statement true.

9. Soybeans, cheese, meats, and fish are sources of <u>incomplete</u> proteins.
10. About <u>25</u> percent of your body is water.
11. Cross-country skiing uses <u>fewer</u> Calories than baseball.
12. THC <u>improves</u> judgment, coordination, and concentration.
13. An <u>overdose</u> is when a person takes more of a drug at one time than is safe.
14. Walking at a fast pace can be an <u>aerobic</u> exercise.
15. Foods that come from plants are naturally <u>high</u> in fat.
16. Water and minerals are <u>organic</u> nutrients.

Check Your Understanding
Apply the concepts you have learned to answer each question.

1. In planning your diet, what three factors do you think are most important? Explain.
2. Describe three methods for reducing stress.
3. Explain what is meant by dependence, tolerance, and withdrawal.
4. Assume that the three foods listed in the right column are available in equal amounts. Which food would you eat to best satisfy each of the needs in the left column?

   a. Most energy   Potato
   b. Protein   Butter
   c. Vitamin C   Fish

5. **Mystery Photo** The photograph on page 486 shows a variety of winter squash. Squash are rich in potassium and vitamins A and C. What other foods are rich in vitamins A and C?
6. Explain what happens to your body during aerobic exercise.
7. **Critical Thinking** A teenager began smoking to be sociable. After a time, smoking became a habit. Explain this change in terms of the effect of smoking on the body.
8. **Critical Thinking** Compare the effects of stimulants and depressants.
9. **Application** Suggest ways you could help a friend decide not to start smoking, and an adult to quit smoking.
10. **Application** Before athletic events, many runners and other athletes eat large amounts of foods high in carbohydrates. Why do athletes eat these foods?

Develop Your Skills

1. a. Steaming
   b. Not cooking, steaming, blanching, boiling, heating (canned)
2. a. Blood alcohol concentration
   b. Coordination and balance deteriorate and judgement is impaired.
   c. Any activity that could cause physical harm if one's concentration were impaired, like chopping vegetables; driving

Make Connections

1.

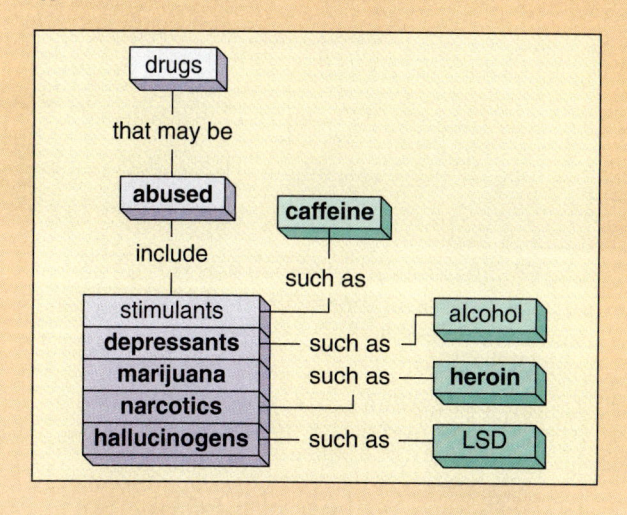

```
                drugs
                  |
             that may be
                  |
               abused          caffeine
                  |              |
               include        such as
                  |
           ┌─ stimulants ─┐
           │  depressants │── such as ── alcohol
           │  marijuana   │── such as ── heroin
           │  narcotics   │
           └─ hallucinogens ── such as ── LSD
```

2. Answers will vary, but should exhibit understanding of the major detrimental physical effects of each type of drug; may also include psychological and social effects.

3. Answers will vary, but should focus on the effects on the heart, lungs, and on coordination and balance. Steroids allow an athlete to develop greater muscle mass, but have harmful side effects on the heart, skin, and reproductive system.

4. Answers will vary. Students should compare methods, including the success rates and any drawbacks, such as cost.

Develop Your Skills

Use the skills you have developed in this chapter to complete each activity.

1. **Interpret Data** The chart below shows how the percentage of vitamin C in peas changes as peas are processed for eating.

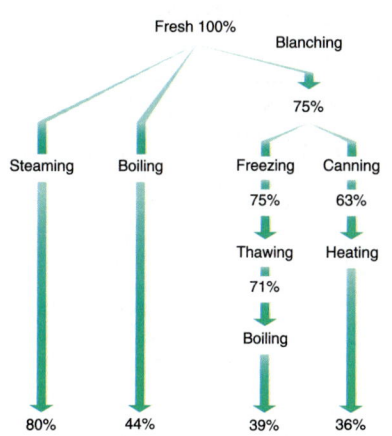

   Fresh 100%
   - Steaming → 80%
   - Boiling → 44%
   - Blanching 75%
     - Freezing 75% → Thawing 71% → Boiling → 39%
     - Canning 63% → Heating → 36%

   a. How should you prepare peas so that you get the most vitamin C per serving?
   b. Order the methods of preparing peas from 1 to 5, with 1 being most nutritious and 5 being least nutritious. Be sure to include *not* cooking the peas as a method.

2. **Data Bank** Use the information on page 628 to answer the following questions.
   a. What does BAC mean?
   b. What happens to a person in this weight range after four drinks?
   c. What are some activities that might be unsafe after two drinks? After five drinks?

Make Connections

1. **Link the Concepts** Below is a concept map showing how some of the main concepts in this chapter link together. Only part of the map is filled in. Finish the map, using words and ideas from the chapter.

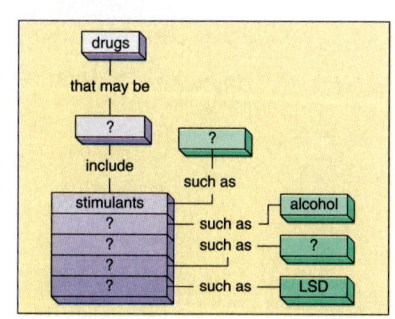

   ```
              drugs
                |
           that may be
                |
                ?          ?
                |          |
             include    such as
                |
           ┌─ stimulants ──┐── such as ── alcohol
           │      ?        │── such as ── ?
           │      ?        │
           └──   ?    ─────── such as ── LSD
   ```

2. **Science and Art** Design an advertisement for a billboard or magazine that teaches people about the dangers of smoking, or drug or alcohol abuse.

3. **Science and PE** Drugs, smoking, and alcohol have an effect on physical abilities. Find out how each of these would affect your ability to perform in a chosen sport or activity. You may also research the use of steroids by athletes to improve muscle strength. What effect do steroids have on the body?

4. **Science and Technology** Cigarette smoking is addictive and difficult to quit. Many people who want to quit simply cannot give up smoking. Research current methods to quit smoking, such as nicotine gum, patches, acupuncture, hypnotism, and other programs. Report on the effectiveness of each program.

CHAPTER 24

Overview

This chapter focuses on how the human body contracts and fights illness. The first section explains the causes of and defenses against infectious diseases. This section also explains the HIV virus and AIDS. The second section describes how the body protects itself from pathogens in the environment. Skin, bodily secretions, and the immune system are presented. The third section distinguishes body disorders from other diseases and discusses cancer. The chapter closes with medicines that protect people from disease.

Advance Planner

▶ Provide a stopwatch or clock with a second hand for SE Activity 24, page 520.

Skills Development Chart

Sections	Calculate	Classify	Communicate	Decision Making	Find Causes	Infer	Observe	Research
24.1 Skills WarmUp Consider This			●	●	●			
24.2 Skills WarmUp Skills WorkOut Activity	●		●			●	●	
24.3 Skills WarmUp		●						
24.4 Skills WarmUp Skills WorkOut Historical Notebook			●		●			● ●

Individual Needs

▶ **Limited English Proficiency Students** Have students write all of the vocabulary terms and any other unfamiliar words in their science journals, leaving several blank lines between each word. Have students use their own words to define each term. Then have them write sentences describing the difference between the following pairs of terms: *cancer* and *carcinogen*, *virus* and *bacteria*.

▶ **At-Risk Students** Invite a healthcare professional to speak with students. Have them prepare a list of questions about the challenges to professional people in the health industry. Students may ask about the knowledge and skills required for this type of work.

▶ **Gifted Students** Have students create an audio-visual program that shows the course of a bacterial or viral infection. They can create a series of drawings on paper or overhead transparencies that show how the pathogens enter the body, how they affect it, and how the body responds. Students should write a script to accompany their visuals. Encourage them to present their program to the class.

Resource Bank

▶ **Bulletin Board** Create a display about the body's defense systems. Have some volunteers study the photograph on page 510 and draw or paint pictures showing T-cells attacking a pathogen. Then, as you work through the chapter, have students add drawings or written descriptions of other kinds of natural defenses.

CHAPTER 24 PLANNING GUIDE

Section	Core	Standard	Enriched	Section	Core	Standard	Enriched
24.1 Infectious Disease pp. 511–515				**24.3 Body Disorders** pp. 521–524			
Section Features Skills WarmUp, p. 511 Consider This, p. 514	●	● ●	● ●	Section Features Skills WarmUp, p. 521	●	●	●
Blackline Masters Review Worksheet 24.1 Skills Worksheet 24.1 Integrating Worksheet 24.1	● ● ●	● ● ●	● ● ●	Blackline Masters Review Worksheet 24.3 Integrating Worksheet 24.3	●	● ●	● ●
Ancillary Options One-Minute Readings, p. 40	●	●	●	Ancillary Options *Multicultural Perspectives*, p. 49	●	●	●
Laboratory Program Investigation 46	●	●	●	**24.4 Medicines That Fight Disease** pp. 525–528			
24.2 The Body's Natural Defenses pp. 516–520				Section Features Skills WarmUp, p. 525 Skills WorkOut, p. 526 Historical Notebook, p. 527	● ● ●	● ● ●	●
Section Features Skills WarmUp, p. 516 Skills WorkOut, p. 519 Activity, p. 520	● ●	● ● ●	● ● ●	Blackline Masters Review Worksheet 24.4 Enrich Worksheet 24.4 Vocabulary Worksheet 24	● ● ●	● ● ●	● ● ●
Blackline Masters Review Worksheet 24.2	●	●	●	Ancillary Options CEPUP, Risk Comparison, Activities 1–6	●	●	●

Bibliography

The following resources can be used for teaching the chapter. See page T-40 for supplier codes.

Audio-Visual Sources
(video unless noted)
The AIDS Epidemic: Is Anyone Safe? 50 min. GA.
The Clinical Story of AIDS: An Interview with Dr. Paul Volberding. 28 min. CVB.
Health Concerns for Today Series. 4 filmstrips. 1984. ME.
The Human Body Series—Staying Healthy: The Body's Defenses. Filmstrips with cassettes. 1980. NGSES.
Our Immune System. 25 min. 1988. NGSES.
Your Immune System. 2 filmstrips, 12–15 min each. 1991. NGSES.

Software Resources
Health Risk Appraisal. Apple, IBM PC, TRS-80. CABISCO.
Pathology: Disease and Defenses. IBM PC. CCM.
Understanding AIDS. Apple, IBM PC. CABISCO.

Library Resources
Dixon, Bernard, ed. *Health, Medicine, and the Human Body.* New York: Macmillan, 1986.
Elting, Mary. *The Macmillan Book of the Human Body.* New York: Aladdin, 1986.
Lambert, Mark. *Medicine in the Future.* New York: Bookwright, 1986.
Taylor, Ron, with Isaac Asimov, ed. *Health.* New York: Facts on File, 1985.

CHAPTER 24

Introducing the Chapter

Have students read the description of the photograph on page 510. Ask if they agree or disagree with the description. Point out that the lymphocytes are attacking a cancerous growth.

Directed Inquiry

Have students study the photograph. Ask:

▶ What organisms or parts of organisms are shown in this photograph? (Many students will recognize the white objects as magnified white blood cells, which they first saw in Chapter 20. They may say that the other objects are harmful cells or foreign bodies.)

▶ Where is the activity in the photograph taking place? (Inside the body)

▶ What do you think is happening here? (The white cells are attacking the other cells.) Do you think this activity is harmful or helpful? Why? (Helpful, because the white cells are destroying cells that might be causing disease)

▶ Remind students that the foreign bodies are cancer cells. Have them suggest some analogies for the process pictured. (Students might suggest war or predators eating their prey.)

Chapter Vocabulary

active immunity
antibiotic
antibody
cancer
carcinogen
immune system
infectious disease
passive immunity
pathogen
vaccine

INTEGRATED LEARNING

Writing Connection

After discussing the photograph, have students write a brief science fiction account describing the action that is taking place inside the body. Have students read their stories to the class.

Chapter 24
Disease and the Immune System

Chapter Sections

24.1 Infectious Disease
24.2 The Body's Natural Defenses
24.3 Body Disorders
24.4 Medicines That Fight Disease

What do you see?

“I think that this is inside the human body. The white things are probably white blood cells. The white blood cells are attaching themselves to a harmful disease. This is important because without white blood cells, diseases would infect the body and it could not function normally.”

Ben Wilkins
Kiser Middle School
Dayton, Ohio

To find out more about the photograph, look on page 530. As you read this chapter, you will learn about diseases, what causes them, and how your body functions to prevent them.

510

Themes in Science

Systems and Interactions The immune system is the human body's internal system of defense. White blood cells produced in the immune system surround invading organisms and produce specific types of antibodies that attack the invaders. Antibodies either deactivate the invaders or immobilize them so that white blood cells can destroy them.

24.1 Infectious Disease

Objectives

- **Distinguish** between infectious diseases and noninfectious diseases.
- **Identify** three kinds of organisms that can cause disease in humans.
- **Describe** ways that disease-causing agents can be spread from person to person.
- **Make a graph** of the projected numbers of AIDS cases worldwide.

Skills WarmUp

Find Causes Think about the last time you were sick. Then list all the possible causes for the illness. Compare your list to a classmate's. Are they similar? How do they vary?

The feeling is all too familiar. You wake up with a tightness in your chest or a sore throat. Maybe your nose is runny and your head and muscles ache. You feel awful. You know you're getting sick.

Everyone gets sick. People catch colds and other illnesses, such as the flu, that usually go away in a few days. Unfortunately, some people also come down with illnesses that are more serious, such as cancer. When a person's body no longer functions properly, he or she may have a disorder or a disease. Learning about the causes of disease can help you to prevent them and get over them faster.

Causes of Disease

What causes a person's body not to work the way it's supposed to? You know that microorganisms, such as bacteria and viruses, can invade the body and cause disease. A disease that can be transmitted among people by harmful organisms is called an **infectious disease**. The organisms that cause infectious diseases are called **pathogens**.

Another kind of disease occurs when something goes wrong with the body itself. A body part may wear out or stop working the way it is supposed to. This kind of illness is called a noninfectious disease, or a body disorder. Some body disorders are inherited. Others are the result of old age or an unhealthy lifestyle. Whatever their cause, body disorders differ from infectious diseases because they can't be spread from person to person.

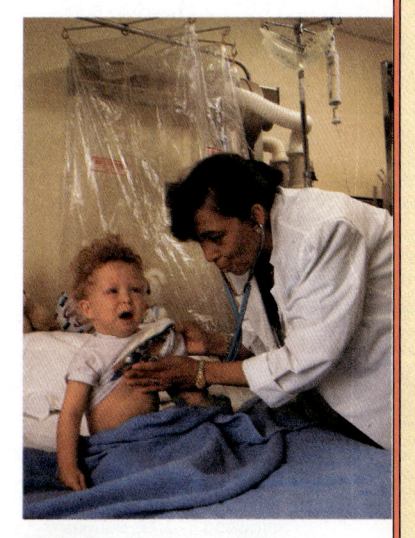

Figure 24.1 ▲
Sometimes when you're sick, a doctor may find the cause and help your body defend itself.

SECTION 24.1

Section Objectives
For a list of section objectives, see the Student Edition page.

Skills Objectives
Students should be able to:

Find Causes for a previous illness.

Communicate how bacteria harm the body.

Make a Graph of the projected number of AIDS cases in the world by 1995.

Vocabulary
infectious disease, pathogen, immune system

MOTIVATE

Skills WarmUp
To help students understand the nature of infectious disease, have them do the Skills WarmUp.
Answer Answers may vary but may include: invasion of the body by bacteria, viruses, or other pathogens; and lowered immune system functioning due to stress and/or lack of sleep.

Prior Knowledge
To find out how much students know about infectious disease, ask the following questions:

- What are the main types of infectious diseases?
- How do people get infectious diseases?
- Can you get AIDS from shaking someone's hand?

Chapter 24 Disease and the Immune System 511

TEACH

Skills Development

Infer Ask students to think about the last time they took cough syrup, antihistamines, or aspirin for a cold, and what happened when the medication wore off. Have them infer whether cold remedies treat the disease or the symptoms. (The symptoms)

Classify Have students make lists of infectious diseases that they have had, the age at which they had them, and the treatment given. If they are not sure about some of this information, have them check with parents or guardians. Students should then classify each disease as viral, bacterial, fungal, and so on.

Research

Have students work in pairs. Ask each pair to choose a type of disease from Table 24.1 to research. They should make a poster with written descriptions and drawings of the disease they researched. Encourage them to answer the following questions:

▶ How is the disease spread?
▶ What are its symptoms?
▶ How can it be avoided?
▶ Where is it common?
▶ How is it treated?

Exhibit the posters in class.

Integrated Learning

Use Integrating Worksheet 24.1.

The Living Textbook:
Life Science Side 1

Chapter 17   Frame 31819
Bacterial Cells and Membranes (Movie)
Search: Play:

INTEGRATED LEARNING

Multicultural Perspectives

Hideyo Noguchi (1876–1928), a Japanese bacteriologist, discovered how to grow the organisms that cause syphilis and other diseases. Ask students to describe some advantages of cultivating pathogens in the laboratory. Encourage students to find out more about Hideyo's research.

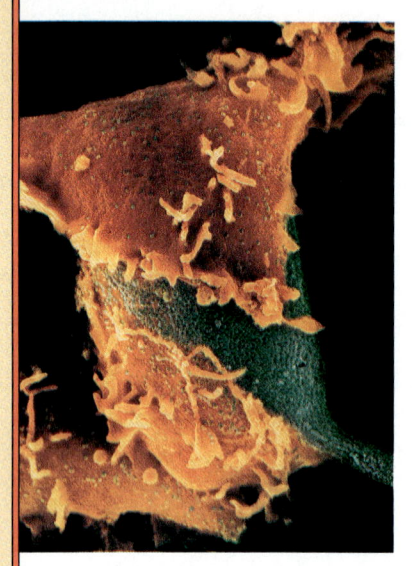

Figure 24.2 ▲
Newly-assembled viruses burst from a cell, ready to infect others.

Table 24.1
Some Infectious Diseases

Disease	Pathogenic Organism
AIDS	Virus
Common cold	Viruses
Gonorrhea	Bacteria
Herpes	Viruses
Influenza	Viruses
Malaria	Protozoan
Pneumonia	Viruses or bacteria
Polio	Virus
Rabies	Virus
Ringworm	Fungus
Strep throat	Bacteria
Trichinosis	Roundworm
Tuberculosis	Bacteria

Kinds of Infectious Disease

For most of human history, people had no idea that living things too small to be seen were responsible for their illnesses. As scientists discovered the pathogens that caused diseases, they found that each one infected the body in a different way. Today the treatment and prevention of an infectious disease depend on knowing what kind of organism is the pathogen at work.

Viral Diseases As you have learned, viruses harm the body because they use the body's cells to reproduce. Cells infected by the virus are destroyed in order to make new viruses, which infect other cells. Nearly everyone has had a viral disease called the common cold. Other kinds of viral diseases are influenza, chicken pox, hepatitis, herpes, and AIDS.

Bacterial Diseases Bacteria, as you know, are in the air, on your skin, and in the food you eat. Fortunately, most bacteria are not harmful. Those bacteria that are harmful are usually destroyed or reduced in number by the natural defenses of a healthy body.

When they cause disease, bacteria harm the body's cells in two ways. In most bacterial diseases, the pathogenic bacteria produce toxins, or poisons, that harm cells. Food poisoning, for example, is caused by toxins from bacteria growing in spoiled food. Other pathogenic bacteria harm cells directly by growing around them and preventing the normal flow of substances in and out of the cells.

Other Infectious Diseases Some infectious diseases are caused by other kinds of organisms. Certain kinds of fungi, for example, cause disease by infecting the body and producing enzymes that digest body cells. Ringworm and athlete's foot are examples of fungal diseases.

Some protozoa are pathogens. Malaria is caused by a protozoan that lives inside liver and blood cells. A kind of ameba causes a serious form of diarrhea called amebic dysentery.

A number of worms also cause disease. Parasitic tapeworms infest the digestive system and rob the host of food. Flatworms called flukes infest the liver. Trichinosis is caused by a roundworm called *Trichina,* which lives in the skeletal and intestinal muscles.

Integrating the Sciences

Chemistry Explain to students that heat can be used to prevent infectious diseases in two ways: sterilization and cooking. Then have them find examples of pathogens that can easily be destroyed by heat sterilization or cooking. (*Salmonella* and trichinosis)

STS Connection

Cholera is spread mainly by contaminated drinking water, and typhoid by both contaminated food and water. Have students find out where in the world these diseases are most common and why. (Central and South America and parts of Asia and Africa; the tropical climates and poor sanitary conditions in certain areas of these nations are two reasons why these diseases continue to spread.)

Directed Inquiry

Have students study the material on this page. Ask the following questions:

▶ What is similar about malaria, Lyme disease, and rabies? (They are all transmitted through the stings or bites of infected insects or other animals.)

▶ How do people get chlamydia, herpes, gonorrhea, and syphilis? (Through unprotected sexual intercourse with an infected person)

▶ Why shouldn't needles be reused? (Diseases such as hepatitis and AIDS can be spread through previously used needles.)

▶ Most people get food poisoning several times a year. Why do you think this is true? (Without knowing it, many people eat contaminated or spoiled food.)

Skills Development

Infer Some researchers think that the most common way people get infectious diseases is through their hands. Ask students:

▶ How do you think this happens? (Uninfected people touch things that infected people have touched and then touch their mouths or their food, allowing pathogens to enter their bodies.)

▶ What is a good way to avoid spreading infectious diseases? (Wash hands frequently, especially before eating)

Spread of Infectious Disease

In order for a pathogen to cause disease, it must enter the body and reproduce. Most of the time, your body's natural defenses keep pathogens from entering your body or from reproducing. Some pathogens, however, can easily overwhelm your body's defenses. Since dangerous pathogens can be almost anywhere, you should be aware of how they are transmitted to you.

People with an infectious disease may have the pathogen in their saliva, mucus, or excrement. If they get these substances on their hands, the pathogen can be transferred to anything or anyone they touch. A kiss or handshake, therefore, can spread some infectious diseases. In addition, sneezes and coughs may spray droplets of contaminated saliva into the air, which other people may breathe. ▼

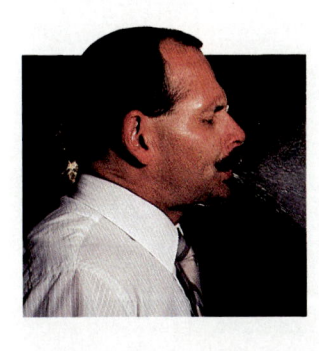

Infectious diseases that are spread by the close contact of sexual intercourse are called sexually transmitted diseases. These diseases include chlamydia, herpes, gonorrhea, AIDS, and syphilis.

▲ Food and drinking water can also carry pathogens. The pathogens in food may originally come from an infected person who handled the food. Sometimes people are infected by food that has been contaminated by bacteria such as *Salmonella*. *Salmonella* bacteria live in such farm animals as chickens. Diseases, such as cholera, are commonly spread by water contaminated with sewage.

Some diseases are spread mainly by the bites of insects and other animals. A certain mosquito carries the protozoan that causes malaria. Lyme disease is carried by ticks, and mammals infected with rabies can spread the disease by biting people. ▼

Blood from a person with a viral disease will probably contain the pathogenic virus. For this reason, needles that have been previously used can spread diseases, such as hepatitis and AIDS. ▼

The Living Textbook: Life Science Side 2

Chapter 13   Frame 18741
Types of Bacteria (Movie)
Search: Play:

TEACH • Continued

Misconceptions
Remind students that the HIV virus is the cause of the disease called AIDS. Emphasize that HIV does not cause death. Rather, it leads to death indirectly by attacking the immune system and rendering it incapable of fighting even the most minor illnesses. Thus, people with HIV end up dying of other diseases, such as pneumonia.

Reteach
Emphasize to students that the HIV virus is at epidemic proportions in many parts of the world, including the United States. Ask them the following questions:

▶ Why is AIDS spreading so quickly? (A lot of people are having unprotected sexual intercourse and injecting drugs with previously used needles.)

▶ What is the best way to avoid exposure to the virus? (Abstaining from sexual intercourse and from intravenous drug use is the best way. HIV can also usually be avoided through protected sex.)

Consider This
Discuss Before they debate the issue of mandatory HIV testing, have students make a list of the reasons for mandatory testing and the reasons against it. Ask students why this issue is so important today. (Many people are dying of AIDS.)

**The Living Textbook:
Life Science Sides 7-8**

Chapter 5 Frame 00822
AIDS-Infected Lymphocytes (1 Frame)
Search:

514

INTEGRATED LEARNING

Social Studies Connection
Explain to students that an infectious disease known as the Plague, or the Black Death, ravaged Europe and parts of Asia in the 14th century, killing up to three-quarters of the population in some regions. In just three years, between 1347 and 1351, the disease killed 75 million people. People in the midst of the epidemic did not know how the disease was transmitted or how to cure it. It was later discovered that the disease is caused by a bacterium and transmitted to humans by fleas from infected rats. Today bubonic plague is still prevalent in some parts of the world but is treatable with antibiotics.

AIDS

The majority of people are healthy most of the time because their bodies have natural defenses against disease. These defenses include a complex set of cells and tissues called the **immune system**. The immune system attacks invading pathogens and prevents them from growing inside the body.

Most pathogens cause disease by overwhelming the immune system on their way to infecting some other part of the body. But one pathogen, the human immunodeficiency virus, or HIV, attacks the immune system itself. HIV is the cause of the disease called AIDS, or acquired immune deficiency syndrome. Because it harms the immune system, AIDS leaves the body open to infection by many pathogens. People with AIDS suffer from infections that their bodies would normally fight off. When people with AIDS die, other infections, not HIV, kill them.

Figure 24.3 ▲
The AIDS quilt is a memorial to the people who have died from the disease. It is now much larger than a football field.

Consider This

Who Should Be Tested for HIV?

Over a million people in the United States alone are infected with HIV. Many do not even know that they are carrying the virus.

Many public-health officials would like to know who has HIV. However, getting the answer raises difficult issues about public health and an individual's right to privacy.

In order to determine who carries the AIDS virus, tests would be required of nearly everyone. Should people be tested against their will? Should an HIV test be required for jobs, medical insurance, or a marriage license?

Consider Some Issues
People who favor widespread testing for HIV believe that the AIDS threat is serious enough to justify testing. They say that accurate data are needed on how many people are carrying the virus now. The HIV test, in their view, will provide the information needed to slow the spread of the virus.

Those opposed to testing believe it threatens peoples' basic rights. They are concerned that people who are found to carry HIV may be forced to leave their jobs. Their health insurance may be canceled. They could be shunned by others. Opponents to testing believe that no one should be forced to take an HIV test against his or her will. They feel that the test results should be given only to the patient and his or her doctor.

Think About It Who should be tested for HIV? Who should know the test results? Is HIV testing a good way to slow the spread of the virus?

Debate It Have a class debate in which you and your classmates role-play people with opposite points of view on the issue of testing for HIV.

514 Chapter 24 Disease and the Immune System

> **Writing Connection**
> Ask students if any of them know anyone who has AIDS or has died from it. Have them write about what it would be like if they or someone they love were to become infected with the HIV virus. How would they feel? How would their lives change? How would other people treat them? Have students share their essays with the class. Use students' writings as a vehicle for discussing the realities of AIDS and the importance of AIDS awareness and compassion.

HIV has infected millions of people all over the world, and it continues to spread. So far, there is no cure for AIDS. Researchers predict that by the year 2000, two million people will have died of AIDS. Look at Table 24.2. It shows cumulative AIDS cases in adults. How many AIDS cases are expected in North America by 1995?

Every person can take steps to keep from coming in contact with HIV. The virus can be transmitted through body fluids, such as blood, semen, vaginal fluid, and breast milk. Therefore, you can't get HIV from hugging, shaking hands, or touching toilet seats or shower stalls. You can't get HIV from a cough or sneeze. HIV is transmitted in four ways—through sexual contact, intravenous drug use, contaminated blood, and breast milk. If people abstain from sex and do not inject drugs, they greatly reduce their chances of getting AIDS.

Science and You Lyme Disease

Have you walked through tall grass or brush lately? Did you check yourself for ticks afterward? In many parts of the United States, it is a good idea to do all you can to avoid being bitten by a tick. The deer tick can infect you with the bacteria that cause Lyme disease.

Lyme disease usually begins with headaches, backaches, fever, and chills. If the disease is not treated, the heart and nervous system can be permanently damaged. More and more cases of Lyme disease are being reported, and many doctors consider it an epidemic.

If you find a tick attached to you, remove it immediately. Grasp it with tweezers where it meets the skin and pull straight up. Quick removal of the tick may prevent transfer of the bacteria that cause the disease.

Table 24.2
AIDS Cases in Thousands

Region	1992 Estimate	1995 Projection
N. America	257	534
W. Europe	99	279
Australia/ Oceania	4	11
Latin America	173	417
Africa	1,367	3,277
Caribbean	43	121
E. Europe	2	9
NE Asia	3	14
SE Asia	65	240

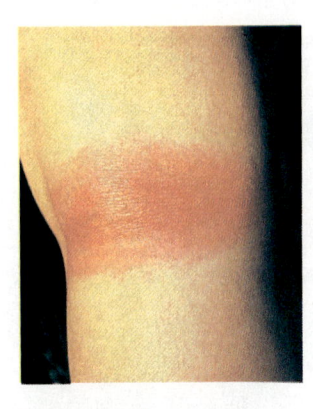

Figure 24.4 ▲
This rash on the back of the knee is one sign of Lyme disease infection.

Check and Explain

1. Name three kinds of organisms that can be pathogens.
2. Describe three different ways you could get a disease from a friend.
3. **Find Causes** How do bacteria harm the body's cells?
4. **Make a Graph** Use the data in Table 24.2 to make a bar graph showing the projected number of AIDS cases in the world in 1995.

Discuss
Explain that ticks infected with Lyme disease are most common in wooded areas. People who live in such areas should be particularly careful about checking for ticks after being outside. Dogs should also be checked, since they are susceptible to the disease, too. Explain that deer ticks are very small (the size of a pin head), so extra care must be taken to find them.

EVALUATE

WrapUp
Reinforce Have students trace Table 24.1 on page 512 and add an additional column labeled *How Organism Enters Body*. They should use this column to describe how the organism responsible for each disease enters the body. They may want to add other diseases to the chart.

Use Review Worksheet 24.1.

Check and Explain

1. Viruses, bacteria, and protozoa
2. Through a kiss, a handshake, or by being near them when they sneeze
3. By releasing harmful toxins or by growing around them and preventing the normal flow of substances in and out of the cells
4. The graph should plot the data (from Table 24.2) for 1992 and the projection for 1995 in each region.

Decision Making
If you have classroom sets of *One-Minute Readings*, have students read Issue 24, "AIDS and Society's Responsibility" on page 40. Discuss the questions.

SECTION 24.2

Section Objectives
For a list of section objectives, see the Student Edition page.

Skills Objectives
Students should be able to:

Infer four ways to keep a cut from becoming infected.

Communicate how bacteria in different parts of the body are destroyed.

Infer what would happen if people didn't blink.

Vocabulary
antibody

MOTIVATE

Skills WarmUp
To help students understand how to prevent infection, have them do the Skills WarmUp.
Answer Washing it and using peroxide to kill pathogens, protecting the cut with a bandage, and keeping it clean all help to avoid infection.

Prior Knowledge
To find out how much students know about the body's natural defenses, ask the following questions:

▶ How does your skin protect you?

▶ What are some ways in which your body prevents pathogens from entering it?

▶ What is the immune system?

The Living Textbook: Life Science Sides 7-8

Chapter 10 Frame 00827
Skin, Lymph Nodes, and Spleen Microviews (33 Frames)
Search: Step:

516

TEACHING OPTIONS

Portfolio
Reinforce the fact that having healthy skin is vital to resisting infection. Briefly review the section about skin in Chapter 19. For their portfolios, have students draw a cross section of the dermis and epidermis. Have them include a description of how skin repairs itself and how perspiration helps prevent infection.

Skills WarmUp
Infer Name four things you could do to keep a cut from becoming infected. Explain how each precaution is effective.

Figure 24.5 ▲
In addition to cooling the body, perspiration helps the skin be a better barrier against infection.

24.2 The Body's Natural Defenses

Objectives
▶ **Describe** how body secretions prevent infection.

▶ **Describe** what happens during the inflammation process.

▶ **Explain** how the body can become immune to a particular pathogen.

▶ **Hypothesize** about why organ transplants may be rejected.

Usually when you cut your finger, it's no big deal. As long as the cut isn't too deep, soap and water, a bandage, and time are all that are needed to make your finger as good as new.

You probably aren't aware of all the work your body does to prevent infection the minute your skin is cut. Many lines of defense come into play. They work together to make sure bacteria that enter the cut do not grow inside your body.

The way your body fights infection in a cut is just one example of how it protects itself in an environment full of pathogens. Your body's natural defenses constantly prevent bacteria, viruses, and other pathogens from giving you diseases.

Skin

Every organism has a barrier between itself and its environment. Your barrier is your skin. Skin is thick and tough. Like bricks in a wall, the cells that make up skin are packed tightly together. Skin prevents pathogens from entering the body, where they would find ideal growing conditions.

Your skin is not only a physical barrier, it is a chemical one, too. Glands in the skin produce perspiration, or sweat. Perspiration contains enzymes that can destroy the cell walls of many kinds of bacteria.

Your skin gets a lot of wear and tear. The outer layer is constantly being rubbed, scraped, and washed away. New skin cells are always being created in the inner layer to replace them. The rapid cell division in your skin also helps it repair itself quickly when skin is damaged.

516 Chapter 24 Disease and the Immune System

INTEGRATED LEARNING

Themes in Science

Systems and Interactions The body has many systems, in addition to the immune system, to defend itself against disease. The skin acts as a protective barrier between the body's internal environment and the outside world. Fluids such as saliva, stomach juices, and tears contain enzymes that help destroy pathogens. Inflammation fights bacteria that enter the body through skin wounds. Ask students to explain the function of mucus in defending the body against harmful invaders.

Body Secretions

Skin does not cover your body completely. Openings such as your mouth, nose, and eyes can be doorways for pathogens. To guard against infection, the tissues in these openings produce substances that kill pathogens or discourage their growth.

Saliva In your mouth, saliva contains strong enzymes that can destroy some kinds of bacteria.

Stomach Juices Many pathogens that reach your stomach are killed by the acids used to digest foods.

Mucus The tissues lining your nose, sinuses, throat, and windpipe produce mucus. Mucus is sticky enough to trap many of the pathogens you inhale. In the lungs and windpipe, cilia help push the mucus upward toward the mouth. There it is swallowed and carried to the stomach, where any pathogens it contains are killed by stomach acid. Pathogen-containing mucus in your nose may be removed when you blow your nose.

Tears Glands in your eyes constantly produce tears that wash away pathogens. In addition, tears contain enzymes that destroy some bacterial substances. Each time you blink, your eyelids bathe your eyes with tears and keep them free of pathogens.

Inflammation

Have you observed what your body does when you cut or scrape yourself? You bleed, of course, but the area around the wound also gets red and puffy.

When the skin is broken, bacteria can enter the body and reproduce. To prevent infection, the body begins to set up a line of defense. The injured cells in the wound produce a substance called histamine (HIHS tuh MEEN). The histamine widens blood vessels and increases blood flow to the area. At the same time, the injured tissues leak fluid. The result is a swelling and reddening called inflammation.

Inflammation of the injured area raises its temperature, which helps slow the growth of bacteria. The extra flow of blood also brings with it white blood cells. These cells seek out bacteria, which are engulfed and digested. You can see white blood cells in action in Figure 24.6.

Inflammation is a response that fights any kind of bacteria. It is a general line of defense. Your body cells have other general lines of defense, too. For example, when cells are attacked by viruses, they produce a substance called interferon. Interferon helps other cells resist further infection by the virus.

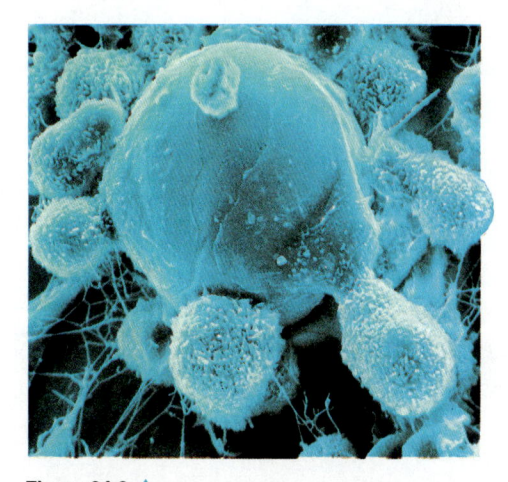

Figure 24.6 ▲
White blood cells gang up to destroy a foreign substance in the body. What do you think is the function of the bumps on the surface of the white blood cells? ①

Chapter 24 Disease and the Immune System

TEACH

Discuss
Have students identify three places on the body where pathogens can enter. (Mouth, nose, and eyes) Then have students name the fluids that help prevent pathogens from entering through these passageways. (Saliva, stomach juices, mucus, and tears) Ask them what sticky material lines all the air passages to trap pathogens. (Mucus)

Misconceptions
Many people believe that it is necessary to use antiseptic solutions or antibiotic ointments to treat minor cuts and scrapes. Explain that these remedies usually do not aid the healing process and may actually burn or irritate the skin. Simply cleaning the wound is usually adequate treatment.

Critical Thinking
Reason and Conclude Ask students to recall the last time they skinned a knee or other part of the body. Ask them what their bodies did to prevent infection. (Bled and/or became inflamed)

Skills Development
Infer Ask students if they have ever seen a pet dog or cat licking its own wound. Ask them to infer how this instinctive behavior can be beneficial. (The saliva contains bacteria-killing enzymes.)

Answer to In-Text Question
① **The bumps on the white blood cells help the cells detect and destroy bacteria.**

The Living Textbook:
Life Science Side 7

Chapter 28   Frame 20340
Wound Healing (Movie)
Search: Play:

TEACH • Continued

Reteach
Remind students that when they have a cold they become immune to the particular virus that caused the illness; however, there are many different strains of cold-producing viruses. When they catch another cold, it is caused by a different strain.

Skills Development
Infer Have students infer what would happen to a family if its members had no memory cells and one family member caught a cold. (Family members would probably pass the same cold back and forth indefinitely.)

Discuss
Have students think of antibodies as pieces of a puzzle; only a certain piece will fit in a certain place in the puzzle. Have them look at Figure 24.7, and remind them that an antibody fits onto a pathogen at a certain place on the pathogen's surface. But, unlike a puzzle, there are many antibodies of the same shape in the body that fit on many pathogens of the same shape.

The Living Textbook:
Life Science Side 2

Chapter 13   Frame 19769
Antibodies (Movie)
Search: Play:

The Living Textbook:
Life Science Side 7

Chapter 29   Frame 28105
Antibody Formation (Movie)
Search: Play:

518

INTEGRATED LEARNING

Integrating the Sciences

Chemistry People are constantly exposed to pathogens in the environment. The body has about two trillion differently shaped cells in its immune system that match an equal number of different pathogen cells. Thus, there are usually antibody cells in the immune system that match most pathogens the body encounters and prevent infection.

The Immune System

In addition to your body's general lines of defense, a complex defensive system protects your body against specific pathogens. This defensive system is called the immune system.

The immune system is made up of several types of cells that circulate through the body. They are always on the lookout for anything that does not belong in the body. These cells are able to destroy an invading pathogen and "remember" it in case it enters the body again.

Recognition of Pathogens Imagine that you are reaching into a jar of jellybeans high on a shelf. All the jellybeans feel the same, but then you feel something you know is different. It's a piece of candy wrapped in cellophane. In a similar way, the cells of your immune system can tell the difference between your body cells and anything that is different. Cells called lymphocytes "feel" the surface of whatever they contact in the body. When lymphocytes contact a pathogen, they recognize it as a foreign substance because its surface is different.

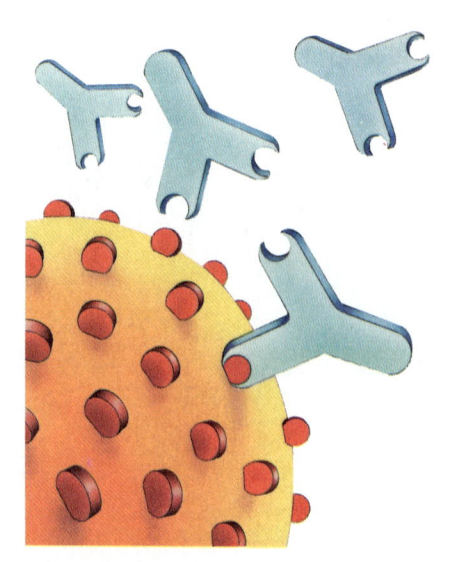

Destruction of Pathogens When the immune system recognizes a pathogen, certain lymphocytes begin producing **antibodies** for the pathogen. An antibody is a specially shaped molecule that matches a certain part of the surface of a particular pathogen. Look at Figure 24.7. Notice how the antibodies fit onto the surface of the pathogen.

Millions of antibodies are produced. They circulate through the blood and lymph, sticking to the pathogen wherever the two come together. Once a pathogen has an antibody sticking to it, it can't infect a body cell.

Pathogens are also "tagged." Cells called phagocytes seek out tagged pathogens and eat them. Tagged pathogens may be destroyed in other ways, too. For example, certain proteins may break open the pathogen's cell membrane.

Immunity Once the immune system destroys most of the pathogens that cause a disease, you get well. Some of the lymphocytes that produced the antibodies to the pathogen stay in the blood, however. They are called *memory cells* because they remember how to produce the antibodies. If the same pathogen enters your body again, the memory cells produce antibodies right away. The antibodies destroy the pathogen before it can cause disease. You are therefore *immune* to that disease. You can become immune to a variety of diseases caused by both viruses and bacteria.

Figure 24.7 ▲
Every pathogen has a different kind of surface. The "shapes" on its surface are determined by the chemicals that make it up. Antibodies are made to fit the surface shapes like a lock and key.

STS Connection

Tell students that about 85 percent of the population is Rh-positive and about 13 percent of all marriages are made up of an Rh-negative wife and an Rh-positive husband. Tell students that at 28 weeks during the pregnancy and at 24-36 hours after delivery, the mother of an Rh-positive baby is given an injection of immunoglobulin which prevents the development of antibodies to Rh-positive cells in her bloodstream. Her future babies are not at risk.

Science and Technology *The Rh Factor*

Although antibodies generally work to fight disease in the body, they can also cause problems. This is the case when a pregnant mother produces antibodies for a certain substance in the blood of her fetus. This substance is the Rh factor.

Some people, called Rh-positive, have the Rh factor on their red blood cells. Other people do not have the Rh factor and are called Rh-negative. An Rh-negative woman can become pregnant with an Rh-positive baby if the baby inherits the Rh gene from the father.

When the mother is exposed to small amounts of the fetus's blood, she develops antibodies to the Rh factor. For the first baby, this is usually no problem. But the mother's immune system "remembers" how to produce anti-Rh antibodies. If she has a second Rh-positive baby, she produces the antibodies again. The antibodies can cross the placenta in the final weeks of pregnancy and cause the baby's blood to clump. If this happens, the baby's life is in danger.

The Rh factor was a serious problem for some women during the 1950s. The blood of their newborns had to be replaced with donor blood that did not contain antibodies. Some babies died.

Fortunately, the problem of the Rh factor is easily avoided today. The mother is given an injection of anti-Rh antibodies immediately after an Rh-positive baby is delivered. The Rh-positive blood cells in the mother's blood are destroyed before her immune system can produce antibodies for them. Any Rh-positive babies she has after that will be safe.

Skills WorkOut

Communicate Tell what would destroy bacteria that are on or in the following places: skin, eye, nose, stomach, lung, bloodstream.

Check and Explain

1. Name four body substances that fight infection.
2. Describe how the body prevents infection in a cut.
3. **Infer** What is one danger a person faces if he or she has a severe burn on a large area of the body?
4. **Hypothesize** When an organ is transplanted from one person to another, the body of the person receiving the organ produces antibodies against it. Explain why the immune system reacts this way.

Skills WorkOut

To help students understand how different parts of the body destroy bacteria, have them do the Skills WorkOut.
Answer Salt for skin, tears for eye, mucus for nose and lung, acid for stomach, and white blood cells for bloodstream

EVALUATE

WrapUp
Portfolio Have students make a simple sketch of the human body and label all of the areas through which pathogens can enter. At each point of entry, ask them to also list the defense used by the body and write a brief description of how it works.
Use Review Worksheet 24.2.

Check and Explain

1. Saliva, mucus, tears, and stomach acids
2. By releasing histamine to increase blood flow and cause inflammation; extra blood flow brings white blood cells to destroy bacteria; inflammation increases temperature, which slows the growth of bacteria.
3. A severe burn greatly increases risk of infection because much of the protection provided by the skin has been destroyed.
4. The body of a person receiving a transplanted organ produces antibodies against the organ because it is perceived as a foreign, invading body.

The Living Textbook: Life Science Sides 7-8

Chapter 3 Frame 00419
Phagocytosis (1 Frame)
Search:

ACTIVITY 24

Time 20 minutes **Group** pairs
Materials
stopwatches

Analysis

1. Answers may vary, but should agree with recorded data. Humans normally blink every 2 to 10 seconds.
2. Students should observe whether the difference between the extremes in their samples is large or small.
3. Averages should agree with recorded data. Students should compare their blinking rates to the group average.
4. Answers may include: health of eyes and eyesight, general health (allergies, colds), variability of the room's lighting.

Conclusion
Intervals between blinks range from 2 to 10 seconds. Lighting and eye wear might also affect blinking rates. The blinking rate affects the lubrication and cleansing of the eyes.

Everyday Application
Increased blinking might result from eye irritants or dry conditions. Blinking and reflex tears help rid eyes of irritants. Inability to blink would lead to eye damage from infection and dehydration.

Extension
The class average should be comparable to group averages.

TEACHING OPTIONS

Prelab Discussion
Have students read through the entire activity, then discuss the following points:

▶ What are the functions of eyelashes, tears, and blinking as they relate to the immune system?

▶ Suggest that students record *all* factors that might affect the blinking rate, such as eyeglasses or contact lenses.

Activity 24 Do people blink their eyes at different rates?

Skills Observe; Calculate; Infer

Task 1 Prelab Prep
1. Collect the following items: a stopwatch or clock with a second hand, a sheet of paper and a pencil for every other student, and a sheet of paper for each group.
2. Name one person in the class to serve as timekeeper.

Task 2 Data Record
1. Copy Table 24.3 onto a sheet of paper.
2. Use the table to record the observations of each person.

Task 3 Procedure
1. Watch your partner's eyes for three minutes. Count the number of times the person blinks. Note: Try not to think about blinking when you are being watched.
2. Write down the number of times your partner blinks. Calculate the number of blinks per minute for your partner.
3. Repeat steps 1 and 2 for each person in your group.

Task 4 Analysis
1. What is the difference between the smallest and the largest number of blinks per minute for your group?
2. How many more times per minute did the fastest blinker blink compared to the slowest blinker? Is this a big difference or a small difference?
3. Calculate the average number of blinks per minute for your group. How does your blinking rate compare to the group average?
4. What variables could explain the differences between people?

Task 5 Conclusion
Based on your data analysis, write a paragraph explaining how the rate of blinking differs among individuals and among groups of people. Infer how the blinking rate is important in protecting your eyes.

Everyday Application
What could cause your blinking rate to increase? Explain how an increased rate could protect your eyes. What would happen if you blinked less than most people? Predict what would happen if you didn't blink at all.

Extension
Graph the results for the entire class by showing the number of people at each blinking rate. How does your group's blinking-rate average compare with the other group averages? Where is your blinking rate on the graph?

Table 24.3 Blinking

Name	Length of Time Observed	Number of Blinks	Blinks per Minute	Other Observations

INTEGRATED LEARNING

Themes in Science

Stability and Equilibrium Usually the body's cells, tissues, and organs work together to maintain a stable internal environment. Sometimes, however, one or more of these systems break down, causing a body disorder. Body disorders can be hereditary or can be caused by aging, malnutrition, or exposure to radiation or harmful chemicals in the environment. Ask students to help you make a list of harmful things in the environment that can cause body disorders. (Examples might include asbestos, pesticides, carbon monoxide, sewage, radon, cigarette smoke, and so on.)

SECTION 24.3

Section Objectives
For a list of section objectives, see the Student Edition page.

Skills Objectives
Students should be able to:

Classify the causes of diseases.

Infer the connection between lung cancer and smoking.

Classify body disorders and their causes.

Vocabulary
cancer, carcinogen

24.3 Body Disorders

Objectives

▶ **Identify** three different kinds of body disorders.

▶ **Explain** what cancer is and how it harms the body.

▶ **Infer** the causes of different kinds of cancer.

▶ **Classify** body disorders and their causes.

Skills WarmUp

Classify Write down the names of ten diseases. Put a check mark by the diseases that are *not caused by bacteria or viruses*. Then discuss what might cause these diseases.

Do you have older relatives with health problems? They may have arthritis, heart disease, diabetes, Alzheimer's disease, or cancer. Many people suffer from one or more of these diseases as they get older. How are these diseases different from infectious diseases such as pneumonia, influenza, and the common cold?

Body disorders, as you have learned, are not caused by pathogens. They are diseases that result from a breakdown or defect in some part of the body. Many body disorders commonly occur as the body ages. Others, however, can happen at any age. Sometimes, too, a person is born with a body disorder.

Causes of Body Disorders

There are many types of body disorders, each with a different cause. One kind of body disorder occurs when the immune system turns against the body. In diseases such as lupus and rheumatoid arthritis, the immune system attacks the body's own cells. These are called autoimmune diseases because the body treats its own cells as foreign invaders. Scientists do not fully understand what causes the immune system to malfunction in this way.

Some body disorders are caused by poor nutrition. Disease can occur if the body doesn't get enough protein, certain minerals, or vitamins. Scurvy, for example, occurs when people don't have enough vitamin C in their diets. When children have nutritional diseases for a long time, their brains and other body parts may be permanently damaged.

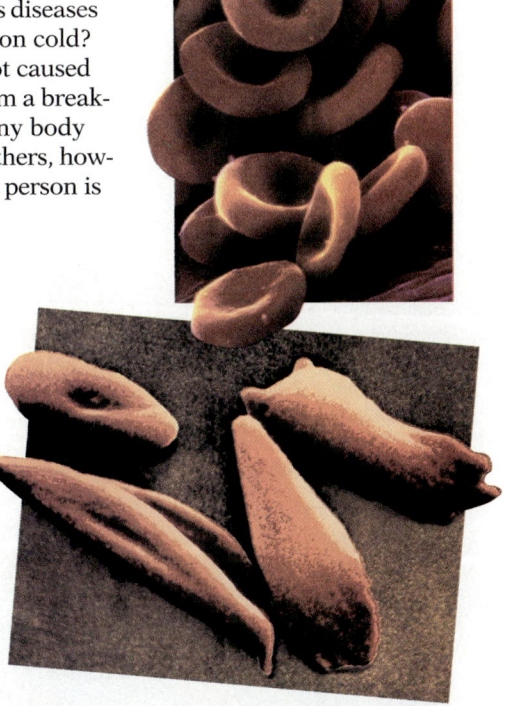

Figure 24.8 ▲
A body disorder called sickle-cell anemia causes normal red blood cells (top) to become sickle-shaped (bottom).

MOTIVATE

Skills WarmUp

To help students understand the different causes of diseases, have them do the Skills WarmUp.

Answer Many diseases, such as cancer, diabetes, heart disease, Down's syndrome, arthritis, lupus, and Alzheimer's, are not caused by bacteria or viruses. All of these body disorders are caused by a breakdown or defect in some part of the body.

Misconceptions

Students may think that cancer is an infectious disease that can be spread from one person to another. Explain that cancer can't be transmitted. It is abnormal cell growth in the body that occurs as a result of heredity, environment, or a combination of both.

Integrated Learning

Use Integrating Worksheet 24.3.

💿 **The Living Textbook: Life Science Side 2**

Chapter 13 Frame 16113
Sickling Blood Cells (Movie)
Search: Play:

Chapter 24 Disease and the Immune System

TEACH

Critical Thinking

Predict Point out that many teenagers smoke, take drugs, eat poorly, and do not get enough exercise. Have students predict how these factors may affect one's health. (Students may mention cancer and cardiovascular disease.)

Compare and Contrast Have students list the similarities and differences between a cancer cell and a pathogen. (Both reproduce in the body and cause disease. The immune system treats both as harmful invaders and tries to destroy them. However, pathogens are outside invaders usually caused by either bacteria or viruses, whereas cancer cells are body cells that reproduce abnormally.)

Discuss

Explain that about half of the 400,000 cancer deaths each year in the United States could have been prevented through lifestyle changes. Have students infer some of the steps that can be taken to prevent cancer. You may wish to list them on the board. Examples include: Don't smoke; don't chew tobacco; avoid exposure to radiation (sun, X-rays, radon); avoid handling or breathing chemicals; avoid fatty, salt-cured, or smoked foods; avoid drinking alcohol; don't eat foods that have become moldy; eat fruits, vegetables, and grains; avoid sexually transmitted diseases; and avoid obesity.

The Living Textbook: Life Science Side 7

Chapter 24   Frame 09766
Atherosclerosis/Blocked Blood Vessels (Movies)
Search: Play:

INTEGRATED LEARNING

Multicultural Perspectives

About one out of every 800 African Americans in the United States is born with sickle-cell anemia. The disease is an inherited disorder in which the red blood cells become distorted sicklelike shapes. The disease can result in circulatory complications including chronic anemia, fever, abdominal and joint pains, and jaundice. Currently no cure for sickle-cell anemia has been found. Have students find out what treatments are available for the disease and share their findings with the class.

The most common body disorders today are caused mainly by the way people live in modern industrial societies. Cardiovascular disease, which affects the heart and blood vessels, can result from unhealthy diets, lack of exercise, and too much stress. Cancer is often caused by cigarette smoking and exposure to harmful chemicals and radiation.

Some body disorders are the result of a genetic defect. In sickle-cell anemia, for example, the body can't produce normal hemoglobin, because the wrong instructions are carried in the DNA. Genetic diseases, such as sickle-cell anemia, are hereditary because the defective gene is passed on from parent to child. Other genetic diseases, such as Down's syndrome, are not passed from generation to generation.

Cancer

Most of your body cells have the ability to divide to make new cells. Usually cell division is kept under strict control. Sometimes, however, a cell loses control and starts to divide at a rapid rate. All the cells produced also lack any control over their rate of division. A mass of cells, called a tumor, develops and continues to grow. The tumor robs the surrounding healthy cells of nutrients and oxygen. Healthy cells are crowded out and destroyed. This is what happens in the body disorder called **cancer**.

Although cancer is a body disorder, you can compare the behavior of a cancer cell to that of a pathogen. A cancer cell causes disease in the body as it grows and reproduces. The main difference between a cancer cell and a pathogen is that the cancer cell is the body's own. It's not an outside invader.

Figure 24.9
A cancer tumor in the brain shows as an orange spot in this CAT scan (left). In the cancer called leukemia, white blood cells become cancerous (right).

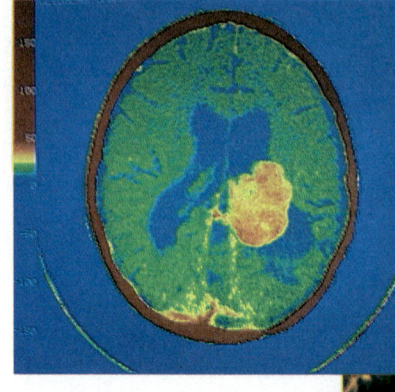

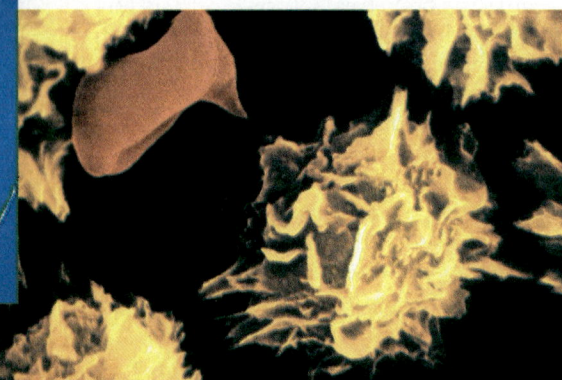

> ### Math Connection
> Have students research the five most common types of cancer found in the United States. Then have them find out the total number of cancer cases in this country and calculate what percentage of the total each of the five most common types represents. Have students make a circle graph of their findings, including a category for all other forms of cancer combined.

> ### Social Studies Connection
> Students may be interested to know about the physicist Marie Sklodowska Curie (1867–1934). She was the first person to win two Nobel prizes, which she received for her work with radioactivity and for her discovery of polonium and radium. Curie's work with radium led to X-ray and isotope methods of diagnosing and treating cancer.

Discuss
Have students consider the following situation: Two people the same age smoked the same number of years, had the same kind of job, and were exposed to the same environment. One got lung cancer, and the other did not. Ask students to offer a likely reason for this difference. (Genetic factors might be the cause of cancer developing in one person and not in the other.)

Skills Development
Infer Remind students that some cancer cells are triggered by viruses. Based on what they know about the body's use of interferon to fight viruses, ask them to infer what a successful method of controlling the growth of virus-induced cancer cells might be. (Adding additional interferon to the body to kill the virus might help the body resist further cancerous growth.)

Reteach
Remind students of what they learned about the lymph system in Chapter 20. Discuss how the lymph system works and why it is so important to prevent cancer cells from spreading into it.

Ancillary Options
If you are using the blackline masters from *Multiculturalism in Mathematics, Science, and Technology*, have students read about Jewel Plummer Cobb on pages 49 and 50. Complete pages 50 to 52.

Causes of Cancer Many cancers are caused by unhealthy substances in the environment. Different substances that cause cancer are called **carcinogens**. Carcinogens include many kinds of chemicals, pollutants in the air, cigarette smoke, exposure to radiation, and substances in some foods.

Each kind of carcinogen is linked to one or more particular kinds of cancer. For example, cigarette smoke is linked to lung and throat cancer. Certain foods are associated with colon cancer. Excessive radiation can cause a cancer of the white blood cells called leukemia. Even though most cancers are probably caused by something from outside the body, heredity sometimes plays a role, too. Scientists think that certain genetic factors may increase the chances of breast, colon, and stomach cancer developing in some people.

Development of Cancer How does a carcinogen cause a cell to become cancerous? This is one major question cancer researchers are trying to answer. They think the carcinogen somehow changes the DNA that controls cell division.

Once a cell becomes cancerous, the immune system acts to stop its growth. Recall that the immune system recognizes anything that is not one of the body's normal cells. Since a cancer cell differs from other body cells, the immune system treats it like an invading bacterium or virus and tries to destroy it.

The immune system is not always successful in destroying cancer cells. If it is not, the tumor grows. One reason cancer is so deadly is that cells from the tumor spread the cancer to other parts of the body through the lymph system.

Treatment of Cancer Many cancers are treated by removing the tumor through surgery. However, doctors may try other ways of destroying the tumor. They can use radiation or chemicals that kill cancer cells. The use of chemicals for cancer treatment is called chemotherapy.

In some cases, the cancer patient recovers completely. However, if the cancer is discovered after it has spread to other parts of the body, the patient's chance of recovery is greatly reduced. Scientists are developing new and more effective ways of treating cancer. Cancer, however, is still a major killer.

Figure 24.10 ▲
A scientist works to develop better anticancer drugs.

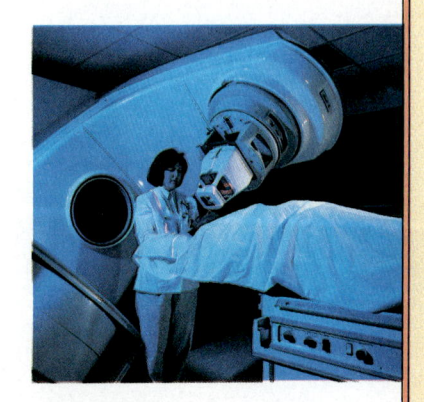

Figure 24.11 ▲
Radiation therapy is commonly used to destroy some types of tumors.

The Living Textbook: Life Science Sides 7-8
Chapter 10   Frame 01311
Lung Cancer Cells (1 Frame)
Search: Step:

TEACH ▪ Continued

Portfolio
Explain that one out of every thirteen Americans suffers from hay fever. Oddly, the symptoms of hay fever do not involve hay or fevers. Have students write brief reports about the actual causes and symptoms of hay fever.

EVALUATE

WrapUp
Reinforce Have students make a diagram grouping the body disorders they have studied according to their possible causes. One group should be labeled *autoimmune*, another *environment and behavior*, and the third *genetic*. Remind students that some groups will overlap.
Use Review Worksheet 24.3.

Check and Explain
1. Scurvy occurs because of poor nutrition. Autoimmune disorders, such as arthritis, occur when the immune system attacks the body's own cells. Cardiovascular disease is caused by lifestyle choices as well as by genetic factors.
2. Cancer is a body disorder because it is caused by the body's own cells, not an outside invader. It is often fatal once it has been spread throughout the body by the lymph system.
3. Smoking, exposure to radiation, and pollution, for example
4. Cancer: carcinogens, genetic factors, or life-style; Heart disease: life-style or genetic factors; Scurvy: lack of vitamin C; Sickle-cell anemia and Down's syndrome: genetic defects; Arthritis: malfunctioning immune system

TEACHING OPTIONS

Cooperative Learning
Assign students to cooperative groups. Have each group work together to list electronic devices they use regularly, such as computers, video games, electric blankets, and televisions. Have students create a display about the possible effects of spending a lot of time close to such devices. One member of each group can do research about some effects, another can find ways to reduce exposure to the radiation emitted from electronic devices. Others can create charts or other visual aids to use in the display.

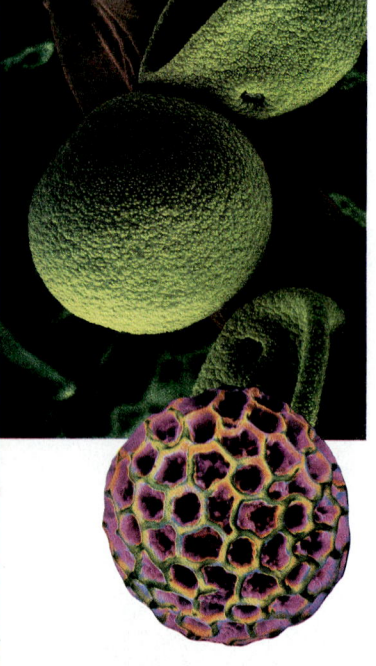

Figure 24.12 ▲
Grains of pollen, shown here magnified hundreds of times, cause an allergic reaction in some people.

Allergies
You probably know some people who have to stay away from dogs and cats or who get hay fever in the spring. These people have an allergy. Allergies are caused by the immune system responding to harmless substances as if they were harmful pathogens. These substances are called allergens. They can be almost anything, from pollen, to the chemicals in insect bites.

A substance becomes an allergen when a person's body produces antibodies against it. Whenever the person comes in contact with the allergen, the antibodies cause an allergic reaction. The body releases histamine, which causes the symptoms of the allergic response.

Science and Society   *Electronic Pollution*
Scientists have known for a long time that air pollution is harmful to the lungs and may have a role in causing cancer. Now evidence suggests that there is another kind of pollution that may be a health hazard. It is a form of weak radiation produced by electrical appliances, power lines, radio transmitters, and video display terminals.

Studies show that in areas where people live near high-voltage power lines, there are more cases of leukemia than in the general population. Studies also indicate that some people who work at video display terminals for long periods of time report other kinds of health problems, such as eyestrain and headache. Scientists are doing more research to determine how people's health may be affected by electronic pollution.

Check and Explain
1. Name three different kinds of body disorders, and explain how each one occurs.
2. Why is cancer a body disorder? When is it fatal?
3. **Infer** If a person develops lung cancer, what might be the cause?
4. **Classify** Make a chart of the following body disorders and their causes: cancer, heart disease, scurvy, sickle-cell anemia, Down's syndrome, arthritis.

Chapter 24 Disease and the Immune System

INTEGRATED LEARNING

STS Connection
In 1796, Edward Jenner, an English physician, discovered that people who had been infected with cowpox could not become infected with the far more serious smallpox. He intentionally infected healthy people with the lesser disease in order to prevent them from contracting the more serious one. Have students find out why he named this method of disease prevention "vaccination."

Multicultural Perspectives
Explain that some African tribes practiced inoculation for many generations before it was adopted in European and North American cultures. For example, in some African communities, parents exposed their children to a less serious form of smallpox in order to prevent them from contracting a deadly form of the disease later.

SECTION 24.4

Section Objectives
For a list of section objectives, see the Student Edition page.

Skills Objectives
Students should be able to:

Find Causes for infectious diseases they've had.

Research the techniques used by Pasteur and Salk to isolate pathogens.

Infer the effect of booster shots.

Vocabulary
vaccine, active immunity, passive immunity, antibiotic

24.4 Medicines That Fight Disease

Objectives
- **Describe** how a vaccine works.
- **Compare** active and passive immunity.
- **Distinguish** between antibiotics and antibodies.
- **Infer** the purpose of a booster shot.

Skills WarmUp

Find Causes Write down all the infectious diseases you have had in your life. Compare your list with that of others in your class. Discuss possible reasons why everyone has not had the same diseases.

When your grandparents were your age, many children came down with infectious diseases such as measles, whooping cough, and mumps. These diseases were so common that they were called childhood diseases. Although they were not considered to be life-threatening, they were still serious illnesses, which could have side effects.

You have heard of some of these diseases. Most people don't get them anymore. Many diseases that once threatened people's lives are now easily prevented or cured. Medicines developed during the last 100 years now protect people from many infectious diseases.

Vaccines

Remember that you become immune to some types of infectious diseases once you have had them. You can't catch the disease again. Your body produced antibodies to fight off the pathogen. If the pathogen appears again later, your immune system is ready to prevent another infection.

A medicine called a **vaccine** makes it possible to become immune to a disease without having the disease. A vaccine contains dead or weakened pathogens, which can't cause disease in the body. However, the white blood cells still treat the pathogens as harmful invaders and make antibodies against them. So, when the body comes into contact with the live pathogen, infection cannot occur.

A vaccine put into the body is called a vaccination. Vaccinations are usually given to young children. You probably received vaccinations for measles, mumps, polio, diphtheria, and whooping cough.

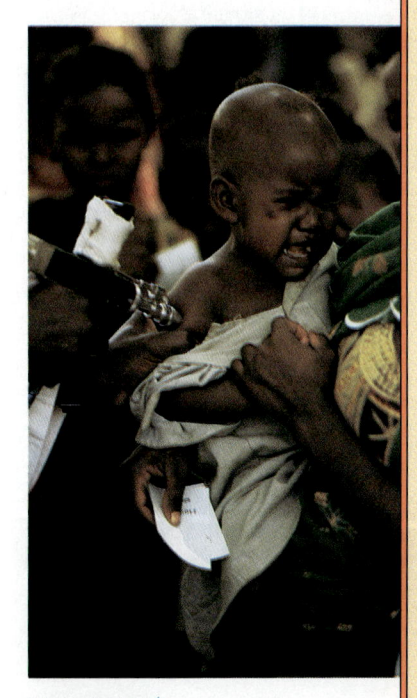

Figure 24.13 ▲
A child in the African nation of Chad receives a measles vaccination.

MOTIVATE

Skills WarmUp
To help students understand how diseases can be prevented, have them do the Skills WarmUp.
Answer Not all students were exposed to the same pathogens; not all students received the same vaccinations.

Misconceptions
Students may think that a vaccine is a medicine used to treat a disease. Explain that a vaccine is preventive rather than curative. A vaccine contains a dead or weakened version of the pathogen responsible for a disease, which is injected into the body. White blood cells, by making antibodies against the pathogen, prevent future infection.

Chapter 24 Disease and the Immune System

TEACH

Skills WorkOut
To help students understand vaccine research, have them do the Skills WorkOut.

Critical Thinking
Compare and Contrast Have students compare active immunity and passive immunity. Ask them to explain which kind of immunity to disease is most effective, and why. (Active immunity is most effective because it triggers the body to produce antibodies that will be ready to act against a particular pathogen before it can cause disease. Passive immunity is achieved through borrowed antibodies and is only temporary.)

Reason and Conclude Emphasize that active immunity can be established through direct contact with a pathogen or from a vaccine. Ask students why a vaccine is the best way to obtain active immunity. (Active immunity from a vaccine enables one to acquire immunity to a disease without actually contracting it.)

INTEGRATED LEARNING

Multicultural Perspectives
German bacteriologist Paul Ehrlich (1854–1915) observed that certain chemicals attack disease without attacking healthy tissue. Ehrlich found a cure for syphilis in a form of arsenic, creating the first synthetic drug. Have students discuss the impact of synthetic drugs on modern medicine. How do they differ from vaccines?

STS Connection
Inform students that millions of children in developing nations die each year of diseases that are preventable through vaccination. Have students find out why vaccination is not consistently used in areas of Latin America, Africa, Asia, and the Caribbean.

Skills WorkOut
Research Write a one-page report comparing the techniques used by Louis Pasteur and Jonas Salk to isolate pathogens. Find out why Jonas Salk developed a live vaccine for polio.

Active Immunity A vaccine is a medicine that *prevents* disease. It can't cure a disease once infection has begun. The vaccine does not act on the pathogen itself. It "fools" the immune system into action, making it produce antibodies. When the body produces its own antibodies for a particular pathogen, **active immunity** occurs. Both vaccines and pathogens cause active immunity.

Passive Immunity In addition to vaccination, another way to prevent certain diseases is to inject a person with antibodies that were produced by a donor. The person who receives the antibodies gains **passive immunity** against the disease. Passive immunity is only temporary. Since the lymphocytes of the person receiving the vaccine did not make the antibodies, they can't make more antibodies at a later time. Injections of antibodies are given to people who need protection from a disease for a short time.

Viruses and Vaccines Vaccines are available for both viral and bacterial diseases. In the case of viral diseases, vaccines are especially important. Viral diseases can't be easily cured with medicines once the infection starts, so it is better to prevent them. One way a medicine can treat a viral infection is to interfere with the virus's reproduction. Development of drugs that will accomplish this task is a slow process. However, scientists recently developed some antiviral drugs. Two of these drugs are medicines that help prevent the flu. Scientists do not yet know why the drugs are effective.

Figure 24.14
Many countries have a goal of vaccinating all their children against common childhood diseases. This poster is from Burma.

Multicultural Perspectives

Many herbal remedies traditionally used by Native Americans were adopted by Western medicine because of their effectiveness. Have students find out what the following natural substances have been used for: ipecac, cinchona (the source of quinine), pinkroot, alum, witch hazel, hellebore, and medicinal curare.

Antibiotics

Although viral diseases are difficult to treat once the body's cells are infected, many bacterial diseases can be cured. The drugs that make it possible to stop bacterial infections are called **antibiotics**. Unlike vaccines, antibiotics act directly on the pathogen. They kill the bacterial cells or prevent them from reproducing.

Antibiotics have saved many lives by stopping infections that the immune system could not handle. Penicillin was the first antibiotic used commercially. It is still widely used today for such illnesses as strep throat and certain types of pneumonia. Many other antibiotics are now available. Some are created in the laboratory. Others are produced in nature by other living things.

Some people are allergic to certain antibodies. If you ever have a bad reaction to an antibiotic, it is important to remember which one. Any doctor treating you in the future will need to know.

Historical Notebook

Traditional Medicine

Hospitals, X-rays, blood tests, and miracle drugs have been around for only a short time. Not too long ago, people cured illness and disease differently than we do today. They obtained medications from flowers, grasses, seeds, roots, and animals. They also thought of illness much differently. They believed it was caused by some sort of imbalance. This kind of medicine, often called traditional medicine, is still practiced today by many people around the world.

The Chinese developed very complex and effective ways of healing beginning thousands of years ago. The goal in Chinese medicine is to restore the body's balance of vital energy, called Qi (CHEE). Imbalances in Qi are corrected with methods such as acupuncture, shown at right.

Native Americans also have a long history of traditional healing practices. In addition to herbs, they use chanting and rituals to help create the conditions under which the sick can become well. Today traditional medicine is being studied by modern doctors. They have seen that traditional healing can sometimes work when modern medicine can't.

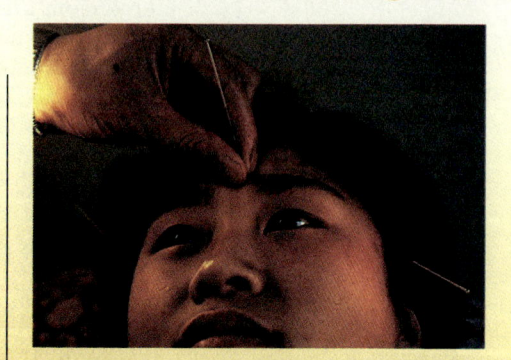

1. How does traditional medicine differ from modern medicine?

2. **Research** Do research on Chinese medicine or acupuncture, and write a report.

Reteach

After students have read about antibiotics, have them compare antibiotics and vaccines. Ask them to explain in what circumstances each type of treatment is given. (Antibiotics are given when a person already has a pathogen; vaccines are given before a pathogen enters the body to prevent future infection.) Ask students to identify the type of illness antibiotics are used for and compare it with what vaccines are used for. (Antibiotics are used to fight bacterial infection; vaccines prevent both viral and bacterial infections.)

Historical Notebook

Research Explain that in acupuncture, very fine metal needles are inserted into the skin at any of 800 designated points on the body. Acupuncture is used as a pain reliever and to treat many common ailments including arthritis, hypertension, and ulcers. Have students find out more about the history and uses of acupuncture. They can share their findings with the class.

Enrich

Use Enrich Worksheet 24.4.

The Living Textbook: Life Science Side 2

Chapter 13 Frame 19145
Antibiotics (Movie)
Search: Play:

Chapter 24 Disease and the Immune System

TEACH • Continued

Class Activity
Have students list aloud some of the illnesses that they have had. Write their responses on the chalkboard. Have students describe how they were treated for each illness. Ask them to identify which treatments come from nature.

EVALUATE

WrapUp
Portfolio Have students make a list of all the diseases they have learned about in this chapter. Then have them pretend they are doctors and prescribe the most appropriate treatment for each disease.
Use Review Worksheet 24.4.

Check and Explain

1. A vaccine contains dead or weakened pathogens incapable of causing disease, for which the body produces antibodies that protect it against possible future infection.

2. Active immunity occurs when the body produces antibodies for a pathogen, preventing later infection by it. Passive immunity, which is temporary, occurs when a person is injected with donor antibodies.

3. Antibodies are produced by the body to fight a pathogen and prevent later infection by the same type of organism; antibiotics are drugs that kill bacterial cells or stop them from reproducing.

4. A booster shot causes additional antibodies to be released, making future infection by the pathogen even less likely.

INTEGRATED LEARNING

STS Connection
The skin of the African clawed frog contains natural antibiotics called *magainins*. Dr. Michael Zasloff discovered that magainins help fight disease by preventing wounds from becoming infected. Have students find out more about how magainins are currently being used.

Figure 24.15 ▲
The skin of the African clawed frog produces a new and powerful antibiotic.

Science and Society
Sources of New Medicines

Nature has always been an important source of medicines for human beings. A bread mold contains the antibiotic penicillin. The bark of willow trees provided people with an aspirinlike substance. Many other living things are also sources of medicines.

Scientists think that far more useful drugs are still to be discovered in such organisms as tropical trees, corals, and frogs. These organisms may produce substances that can fight cancer, stop viral infections, and boost the immune system. Excited by these possibilities, some scientists are conducting studies in rain forests and ocean environments.

Organisms living in the ocean may be one of the largest untapped sources of new medicines. Researchers discovered a substance in an alga, for example, that greatly increases the ability of the immune system to fight infection. This substance could help AIDS patients live much longer. Promising substances were also discovered in certain sponges and tunicates.

Another important source of new medicines may be organisms living in tropical rain forests. Many of these organisms are still unknown to humans. Unfortunately, rain forests are being destroyed at an alarming rate.

Almost any organism might contain a useful drug. An African frog, for example, produces a chemical in its skin that kills bacteria, fungi, and protozoa. This chemical is an entirely new kind of antibiotic.

Check and Explain

1. How does a vaccine make you immune to a disease without itself causing the disease?

2. What is the difference between active and passive immunity?

3. **Compare and Contrast** What is an antibiotic? How does it differ from an antibody?

4. **Infer** Some kinds of vaccines are given more than once. The second vaccination is called a booster shot. What do you think is the effect of a booster shot on the immune system?

Chapter 24 Review

Concept Summary

24.1 Infectious Disease
- Infectious diseases are caused by organisms called pathogens.
- Pathogens may be bacteria, viruses, fungi, protozoa, or worms.
- Pathogens may be spread through water and food and by contact with infected animals or people.
- AIDS is an infectious disease caused by a virus that attacks the immune system.

24.2 The Body's Natural Defenses
- Skin and body secretions help protect the body against pathogens.
- Inflammation is a general line of defense that protects against infection when the skin is damaged.
- The body's immune system can recognize and destroy specific pathogens. It can remember a pathogen and prevent it from infecting the body a second time.

24.3 Body Disorders
- Body disorders result from a breakdown or defect in some part of the body.
- Cancer is an often deadly body disorder with many causes.
- Allergies occur when the immune system treats a harmless substance, such as pollen, as a harmful pathogen.

24.4 Medicines That Fight Disease
- A vaccine makes a person immune to a certain disease even though he or she has not had the disease.
- Antibiotics are drugs that kill bacteria or prevent them from reproducing inside the body.

Chapter Vocabulary
infectious disease (24.1) antibody (24.2) vaccine (24.4) passive immunity (24.4)
pathogen (24.1) cancer (24.3) active immunity (24.4) antibiotic (24.4)
immune system (24.1) carcinogen (24.3)

Check Your Vocabulary

Write the word or term from the list above that best matches each of the phrases below.

1. Molecules produced by lymphocytes that attach to the surface of pathogens.
2. A body disorder in which body cells divide uncontrollably, destroying healthy cells.
3. A disease caused by harmful organisms.
4. The body system that attacks invading pathogens.
5. A medicine that prevents disease by making the body produce antibodies.
6. When the body produces antibodies for a particular pathogen.
7. The name for organisms that cause infectious diseases.
8. Drugs that stop bacterial infections.
9. Substances that cause cancer.
10. A temporary immunity to a disease.

Explain the difference between the words in each pair.

11. Active immunity, passive immunity
12. Antibody, antibiotic
13. Pathogen, carcinogen
14. Infectious disease, body disorder

Write Your Vocabulary
Write a sentence using each vocabulary word above. Show that you know what each word means.

CHAPTER REVIEW 24

Check Your Vocabulary
1. antibodies
2. cancer
3. infectious disease
4. immune system
5. vaccine
6. active immunity
7. pathogens
8. antibiotics
9. carcinogens
10. passive immunity
11. Active immunity occurs when the body produces antibodies for a particular pathogen. Passive immunity is temporary and results when antibodies are introduced into a body from a donor.
12. Antibodies are molecules produced by lymphocytes that attach to pathogens. Antibiotics are drugs that stop bacterial infections.
13. A pathogen is any organism that causes disease. A carcinogen is a cancer-causing pathogen.
14. Infectious diseases are caused by pathogens; body disorders result from a breakdown or defect in some part of the body.

Write Your Vocabulary
Students' sentences should show that they know both the meanings of the words and how to apply the words.

Use Vocabulary Worksheet for Chapter 24.

CHAPTER 24 REVIEW

Check Your Knowledge

1. Answers will vary, but should be based on the discussion on page 517.
2. Bacterial infections can be treated with antibiotics.
3. Memory cells are special types of lymphocytes that remain in the blood after an infection. These cells can respond quickly if the same pathogen enters the body again.
4. Sickle-cell anemia results from a genetic defect that causes red blood cells to be malformed.
5. Carcinogens include: many chemicals, pollutants, cigarette smoke, radiation.
6. A vaccination is a weakened or destroyed form of a pathogen which, when injected into the body, causes antibodies to be produced. If the pathogen enters the body subsequently, no infection will occur.
7. An allergen is a substance that causes an allergic reaction.
8. strep throat
9. AIDS
10. histamine
11. cancer
12. poor nutrition
13. antibiotic
14. immunity

Check Your Understanding

1. There are many different strains of cold and flu viruses, so even if a person develops immunity to a particular strain, the same person can still be infected by another.
2. Accept all logical answers. Both are examples of illness; the distinction is between the causes of the illness.
3. Answers may vary. The skin is a physical barrier preventing the entry of foreign substances and pathogens.
4. Answers may vary. Chicken pox could be spread by someone, even if that person is immune, if contact occurs before the carrier's immune system has a chance to destroy the virus.
5. Answers will vary, but may include measles, mumps, polio, flu, diphtheria, whooping cough, tetanus.

6. Answers should include all or some of the following: recognition of cancerous cells by the immune system, production of antibodies, binding of antibodies to cancer cells, and destruction of cancer cells by phagocytes.
7. An antibiotic is prescribed after a person has come down with a bacterial infection. A vaccine is given before a person has been exposed to a pathogen.
8. Answers will vary. Table 24.1 on page 512 lists some common infectious diseases and their causes.

Chapter 24 Review

Check Your Knowledge
Answer the following in complete sentences.

1. List two types of body secretions that guard against infection and describe how they do their job.
2. What kind of infectious diseases can be treated with antibiotics?
3. What are memory cells? How do they help your immune system?
4. What causes sickle-cell anemia?
5. List three carcinogens.
6. What is a vaccination?
7. What is an allergen? Give two examples of allergens.

Choose the answer that best completes each sentence.

8. A disease caused by bacteria is (influenza, malaria, strep throat, polio).
9. The human immuno-deficiency virus, or HIV, causes the disease (AIDS, herpes, gonorrhea, tuberculosis).
10. During the inflammation response, (an antibody, histamine, saliva, mucus) is released and causes swelling.
11. Chemotherapy is used to treat (cancer, the common cold, diabetes, pneumonia).
12. Some body disorders, such as scurvy, are a result of (bacterial infection, allergens, stress, poor nutrition).
13. Penicillin is one example of an (antibiotic, infection, immunity, allergen).
14. If you cannot catch a disease again because your body has antibodies for the pathogen, you have developed (chemotherapy, an antibiotic, an immunity, a vaccine) to the disease.

Check Your Understanding
Apply the concepts you have learned to answer each question.

1. Explain why you continue to get colds and the flu, even though your body produces antibodies against the pathogens that cause these diseases.
2. **Compare and Contrast** How are infectious diseases and body disorders alike? How are they different?
3. Discuss how your skin prevents pathogens from entering your body.
4. **Application** You visit a friend with chicken pox, knowing you're safe from getting the disease because you have already had it. Is it possible, however, that you could spread the chicken pox virus to your brother or sister who has not had it? Explain your answer.
5. List two vaccinations that you received as a child.
6. **Mystery Photo** The photograph on page 510 shows magnified lymphocytes, or white blood cells, attacking a cancerous growth. Imagine the photograph is part of a movie. Describe the chemical and physical methods the immune system would use to destroy the cancer cell.
7. When would a doctor prescribe an antibiotic to a patient? When would a doctor give a vaccination?
8. **Extension** Make a list of the various types of infectious diseases you have had. Next to each illness, write the type of pathogen that probably caused the disease. You may want to use reference books to locate the cause of some illnesses.

Develop Your Skills

1. a. measles
   b. polio
   c. The graph clearly shows the decline in the number of cases of each disease directly after the introduction of vaccines.
2. a. Lung cancer and oral cancer
   b. Answers will vary, but should include as little exposure as possible to known carcinogens.

Make Connections

1. Concept maps will vary. You may wish to have students do the map below. First draw the map on the board, then add a few key terms. Have students draw and complete the map.

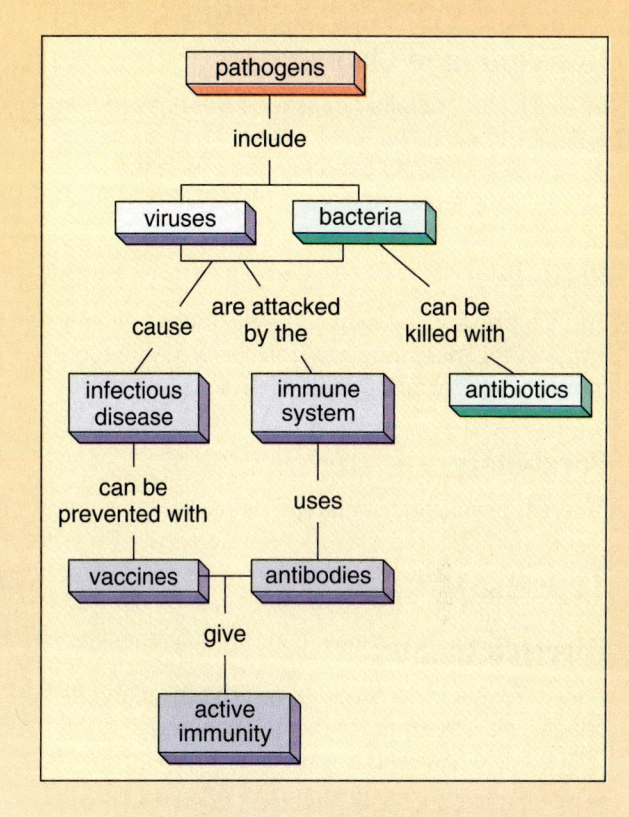

2. Student reports will vary. Check to see that students have references for statistics and dates.
3. Student maps will vary somewhat, depending on when information is obtained. Lyme disease is carried by a tick that is commonly associated with deer, so the disease will tend to be most prevalent in regions with large deer populations.
4. Graphs will vary. Students may see links between particular types of cancers and other statistics; for example, cigarette smoking and lung cancer, or meat consumption and intestinal cancer.
5. Answers will vary. Students may identify things such as cleaning chemicals, cigarette smoke, electromagnetic radiation.
6. Answers will vary. Medical researchers are working on antiviral drugs to cure the disease and vaccines to prevent the disease.

Develop Your Skills

Use the skills you have developed in this chapter to complete each activity.

1. **Interpret Data** The graph below shows the occurrence of polio and measles since 1950. The dot on each line indicates when a vaccination for that disease was introduced.

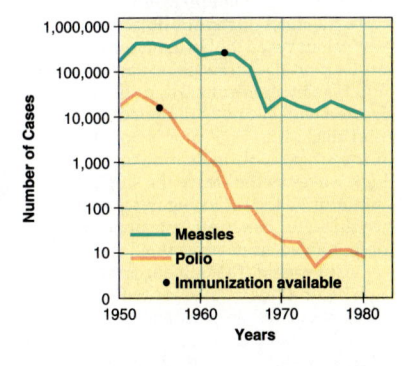

   a. Which disease was, and still is, the most common?
   b. Which disease has been most effectively controlled by its vaccine?
   c. How does this graph visually display the effectiveness of vaccines?

2. **Data Bank** Use the information on page 629 to answer the following questions.
   a. What types of cancer are linked to smoking?
   b. Describe a lifestyle that would reduce your chances of getting cancer.

Make Connections

1. **Link the Concepts** Construct a concept map to show how the following concepts from this chapter link together: pathogens, active immunity, immune system, vaccines, infectious disease, antibodies, viruses, bacteria.

2. **Science and History** Research how infectious diseases have had an effect on human history. Choose a particular time in history when there was an epidemic, or uncontrolled spread of a disease. Find out the numbers of people killed during the epidemic and the results. Write a short report.

3. **Science and You** In which areas of the United States is Lyme disease most common? Find out the answer with the help of a doctor, public-health official, or librarian. Then draw a poster-sized map showing where the most and fewest cases have been reported.

4. **Science and Society** Research the occurrence of cancer in different countries. Make a graph showing the number of cases per 100,000 people. What do you think causes the differences in the cancer rate?

5. **Science and You** What things in your home may help to cause cancer? Research the kinds of chemicals, building materials, and sources of radiation that are suspected as carcinogens. Then survey your household to find if you have any of those items. If so, what can be done to make your home safer?

6. **Science and Technology** Find out what progress scientists are making in the search for a cure for AIDS. What are the most effective drugs being used? What kind of treatments do scientists have the most hope for?

SCIENCE AND LITERATURE CONNECTION UNIT 7

About the Literary Work

Fantastic Voyage is a novel written by Isaac Asimov based on the screenplay by Harry Kleiner, copyright 1966 by Houghton Mifflin Company. Reprinted by permission of Houghton Mifflin Company.

Description of Change

Sophisticated, detailed passages were edited from a short section of the text for the sake of readability, but the dialogue has been kept as true to the original as possible.

Rationale

This vocabulary and syntax are too difficult for the entry-level Limited English Proficiency student, but the story itself is appropriate for the student level.

Vocabulary

faceted, physiology, melange, detritus

Teaching Strategies

Directed Inquiry

After students finish reading, discuss the story. Be sure to relate the story to the science lesson in this unit. Ask the following questions:

▶ What components of the blood does the crew see? (Red blood corpuscles, platelets, white blood cells)

▶ Which body systems are referred to in the selection? (Circulatory system and immune system)

Critical Thinking

Reason and Conclude Ask students the following questions:

▶ Why do white blood cells pose a threat to the *Proteus* and its crew? (The immune system will identify the vessel as a foreign body and attempt to destroy it.)

▶ From the information given in the selection, where do you think the *Proteus* is? (Because the crew sees the oxygen-rich blood and an artery wall, the *Proteus* must be near the neck close to an artery that carries blood to the head.)

Science and Literature Connection

Fantastic Voyage

The following excerpt is from the novel Fantastic Voyage *by Isaac Asimov.*

Duval looked about with exultation. "Conceive it," he said. "Inside a human body; inside an artery. Owens! Put out the interior lights, man!" . . .

The interior lights went off, but a form of ghostly light streamed in from outside, the spotty reflection of the ship's miniaturized light beams fore and aft.

Owen's had brought the Proteus into virtual motionlessness with reference to the arterial blood stream, allowing it to sweep along with the heart-driven flow. . . .

Grant turned to the window. Almost at once he was lost in amazement at the wonder of it all.

The distant wall seemed half a mile away and glowed a brilliant amber in fits and sparks, for it was mostly hidden by the vast melange of objects that floated by near the ship.

It was a vast, exotic aquarium they faced, one in which not fish but far stranger objects filled the vision. Large rubber tires, the centers depressed but not pierced through, were the most numerous objects. Each was about twice the diameter of the ship, each an orange-straw color, each sparkling and blazing intermittently, as though faceted with slivers of diamonds.

Duval said, "The color is not quite true. If it were possible to deminiaturize the light waves as they leave the ship and miniaturize the returning reflection, we would be far better off. It is important to obtain an accurate reflection." . . .

"But even if it's not an accurate reflection," said Cora in an awed tone, "surely it has a beauty all its own. They're like soft, squashed balloons that have trapped a million stars apiece."

"Actually, they're red blood corpuscles," said Michaels to Grant. "Red in mass, but straw-colored individually. Those you see are fresh from the heart, carrying their load of oxygen to the head and, particularly, the brain."

Grant continued to stare about in wonder. In addition to the corpuscles, there were smaller objects; flattened plate-like affairs were rather common, for instance. (Platelets, thought Grant, as the shapes of the objects brought up brightening memories of physiology courses in college.)

[Michaels cried out,] "See that!"

Grant looked off in the direction of the pointing finger. He saw small, rodlike objects, pushing fragments and detritus and, above all, red corpuscles, red corpus-

Skills in Science

Reading Skills in Science

1. The *Proteus* is a miniaturized vessel that was inserted into the patient's circulatory system.
2. Statements that accurately describe the circulatory system include references to the make-up of the blood, discussion of the structure and function of red and white blood cells, and descriptions of the motion of the bloodstream.

Writing Skills in Science

1. Letters will vary and may include comparisons of the various kinds of blood vessels and descriptions of the components of whole blood.
2. A miniaturized vessel might be able to scrape fatty buildup from artery walls, seek out and destroy harmful bacteria, break apart blood clots, and repair damaged body tissue.

Activities

Communicate Students' charts should show differences in the shapes of the components of blood and how they work together.

Compare and Contrast Answers will vary according to the disease selected. Responses should reflect an understanding of the effects of the disease.

---

cles, red corpuscles. Then he made out the object at which Michaels was pointing.

It was huge, milky, and pulsating. It was granular and inside its milkiness there were black twinkles, flashing bits of black so intense as to glow with a blinding non-light of its own.

Within the mass was a darker area, dim through the surrounding milkiness, and maintaining a steady, unwinking shape. The outlines of the whole could not be clearly made out, but a milky bay suddenly extended in toward the artery wall and the mass seemed to flow into it. It faded out now, obscured by the closer objects, lost in the swirl.

"What . . . was that?" asked Grant.

"A white blood cell, of course. There aren't many of those; at least, not compared to the red blood corpuscles. There are about six hundred fifty reds for every white. The whites are much bigger, though, and they can move independently. Some of them can even work their way out of the blood vessels altogether. They're frightening objects, seen on this scale of size. That's about as close as I want to be to one."

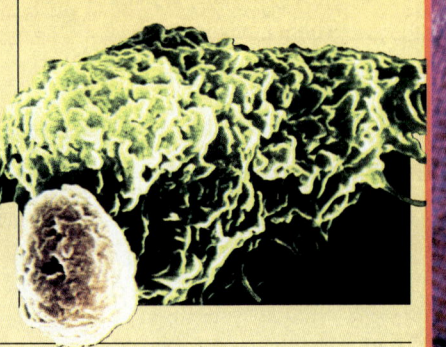

Skills in Science

Reading Skills in Science

1. **Infer** What is the *Proteus*? Where is it?
2. **Evaluate Sources** Although this story is fictional, some information it presents is scientifically accurate. Identify at least three statements that correctly describe the circulatory system.

Writing Skills in Science

1. **Predict** Imagine you are a member of the *Proteus's* crew. Write a letter to a classmate describing your journey through the human circulatory system.
2. **If . . . Then Arguments** Suppose it were possible to miniaturize a vessel to transport a crew through the passageways of the circulatory system. Write a paragraph that describes how this might affect treatment of various diseases of this body system.

Activities

Communicate Use reference tools and information from this story to make a chart that shows the components of blood, their functions, and their physical characteristics.

Compare and Contrast Research a disease that affects the blood or circulatory system. Describe how the crew's view of the blood vessel would differ if the *Proteus* were in a person who had this disease.

Where to Read More

Fantastic Voyage by Isaac Asimov. Boston: Houghton Mifflin Company, 1966. The crew of a miniaturized submarine ventures through a man's body to try to save his life.

The Body Victorious by Kjell Lindqvist and Stig Nordfeldt. New York: Delacorte, 1987. Astonishing photographs taken with an electron microscope detail the inner workings of the human body.

UNIT 8

UNIT OVERVIEW

Unit 8 provides an in-depth discussion of the earth's ecosystems and the way in which they are affected by people. Chapter 25 describes the earth's biosphere and the food chains and energy sources within it. The chapter discusses the chemical and physical balance of water, nitrogen, and oxygen-carbon dioxide cycles. Students explore succession and learn how organisms adapt to changes in ecosystems. Chapter 26 describes factors that affect populations. The chapter describes symbiotic relationships among organisms and behaviors that enable animals to adapt to environments. Chapter 27 discusses how precipitation, latitude, and longitude help determine climate and biomes. The chapter also describes the life-supporting elements of land biomes and the diversity of

Introducing the Unit

Directed Inquiry

Have the students examine the photograph. Ask:

▶ What part of the ocean is shown here? (A part that is close to the shore)

▶ Which organisms can you name? (Angelfish, kelp, starfish, sea anemone)

▶ What advantage might these organisms gain by growing close to shore? (They get more sunlight than organisms in deep water; they are safe from many large predators, and so on. Accept all reasonable answers.)

▶ What do you think would happen to these organisms if one species were to die off? (The balance would shift. Some species might thrive for a time, but all species would eventually suffer.)

Writing About the Photograph

Have students imagine that they are scuba diving around the area in the picture. Have them describe what they see—the movements of the organisms, the colors, the effect of looking through water.

Unit 8
Ecology

Chapters

25 Organisms and Their Environment
26 Interactions Among Organisms
27 Climate and Biomes
28 Humans and the Environment

life in marine biomes. Chapter 28 identifies natural resources and describes how humans use them. The chapter explains the sources and applications of various forms of energy. Also presented in the chapter are pollutants and how they affect the earth. The chapter also emphasizes the need for collective efforts to preserve the ecosystems on the earth.

Data Bank

Use the information on pages 624 to 631 to answer the following questions about topics explored in this unit.

Estimating

By approximately how many British Thermal Units did the United States increase its oil usage between 1940 and 1980?

Comparing

Which continent is more densely populated, Asia or Europe?

Predicting

What type of fuel do you think will be used the most in the United States in the year 2050?

Interpreting Data

What two types of forests would you most likely find at similar altitudes?

The photograph to the left is of a coral reef in the Caribbean. What kind of interactions do you see between the different living things in this picture?

Data Bank Answers

Have students search the Data Bank on pages 624 to 631 for the answers to the questions on this page.

Estimating Answers will vary. The United States increased its oil usage by approximately 35 quadrillion British Thermal Units between 1940 and 1980. The answer is found on page 631 in the graph, Energy Use in the United States by Type of Fuel.

Comparing Asia is more densely populated than Europe. The answer is found on page 626 in the table, Population Density of the Continents (1991).

Extension Ask students to research how the population density of North America has changed over the past century.

Predicting Answers may vary. Helpful information is found in the graph, Energy Use in the United States by Type of Fuel, on page 631.

Interpreting Data A deciduous forest and a coniferous forest would most likely be found at a similar altitude. The answer is found in the diagram, Vegetation Variations Near the Equator, on page 626.

Answer to In-Text Question
The coral reef is a good example of the interactions of organisms because a variety of organisms together form the basic reef structure. Also, other organisms, such as fishes, depend on coral reefs for food.

CHAPTER 25

Overview

An in-depth discussion of the ecosystems of the earth is provided in this chapter. The first section describes living and non-living interactions within the earth's biosphere. The second section uses familiar organisms to explore the energy sources of various food chains. The third section presents the chemical and physical balance of cycles in an ecosystem. Water, nitrogen, and the oxygen–carbon dioxide cycles are discussed and illustrated. The final section explores succession and how organisms function when changes occur in various ecosystems.

Advance Planner

▶ Have students collect cardboard boxes and other supplies for making models of food webs. TE page 544.

▶ Collect magazines for SE Activity 25, page 547.

▶ Gather sand, potting soil, pebbles, a graduated cylinder, three beakers, and water. A graduated cylinder and beakers can be purchased or ordered from a biological supply company. TE page 549.

Skills Development Chart

Sections	Classify	Collect Data	Communicate	Decision Making	Estimate	Infer	Model	Predict
25.1 Skills WarmUp SkillBuilder					●	●		
25.2 Skills WarmUp Skills WorkOut Activity	●	●	●			●	●	
25.3 Skills WarmUp						●		
25.4 Skills WarmUp Consider This			●	●				●

Individual Needs

▶ **Limited English Proficiency Students** Provide students with a picture dictionary. Have them use the picture dictionary or the text to find definitions for the boldface terms in the chapter. Have students find and cut out pictures from magazines, newspapers, and catalogs that show examples of the boldface terms and paste these pictures onto loose-leaf paper so they can create their own picture dictionaries.

▶ **At-Risk Students** Tell students that they encounter the concepts presented in this chapter in their everyday lives. For example, a snowstorm is part of the water cycle. Have students write the title of the first three section heads at the top of a page in their science journals. Ask students to list examples of other ecosystems and populations for Section 25.1, food chains for Section 25.2, and other parts of natural cycles for Section 25.3.

▶ **Gifted Students** Have students create a display comparing the ecosystems in their area 100 years ago and today. Students might collect physical specimens, such as leaves, soil, and pictures of plants and animals. They may even want to build a terrarium that contains a living part of the local ecosystem.

Resource Bank

▶ **Bulletin Board** Begin to create a bulletin board display on ecosystems by listing some abiotic and biotic factors. Have students collect pictures of these factors from magazines. As you work through the concepts in this chapter, have them use their pictures to create collages of different ecosystems. Arrange their collages on the bulletin board.

▶ **Field Trip** Arrange to take students on a walk around a park, forest preserve, beach, or other location in your area where they can observe the parts and the interactions in an ecosystem. Have them find examples of organisms that are part of a food web.

CHAPTER 25 PLANNING GUIDE

Section	Core	Standard	Enriched	Section	Core	Standard	Enriched
25.1 Ecosystems and Communities pp. 537–541				**25.3 Cycles in an Ecosystem** pp. 548–552			
Section Features Skills WarmUp, p. 537 / SkillBuilder, p. 540	● ●	● ●	●	**Section Features** Skills WarmUp, p. 548	●	●	●
Blackline Masters Review Worksheet 25.1 / Skills Worksheet 25.1	● ●	● ●	●	**Blackline Masters** Review Worksheet 25.3 / Skills Worksheet 25.3 / Integrating Worksheet 25.3	● ●	● ● ●	●
Color Transparencies Transparencies 74a, b	●	●		**Ancillary Options** *Multicultural Perspectives*, p. 119 / *One-Minute Readings*, p. 19	●	● ●	● ●
25.2 Food and Energy pp. 542–547				**Laboratory Program** Investigation 47	●	●	●
Section Features Skills WarmUp, p. 542 / Skills WorkOut, p. 546 / Activity, p. 547	● ● ●	● ● ●	●	**Color Transparencies** Transparencies 78, 79, 80	●	●	
Blackline Masters Review Worksheet 25.2 / Skills Worksheet 25.2 / Integrating Worksheet 25.2 / Reteach Worksheet 25.2	● ● ●	● ● ● ●	●	**25.4 Changes in Ecosystems** pp. 553–556			
				Section Features Skills WarmUp, p. 553 / Consider This, p. 555	●	● ●	● ●
Overhead Blackline Transparencies Overhead Blackline Master 25.2 and Student Worksheet	●	●	●	**Blackline Masters** Review Worksheet 25.4 / Skills Worksheet 25.4 / Vocabulary Worksheet 25	● ● ●	● ● ●	●
Color Transparencies Transparencies 75, 76, 77	●	●		**Color Transparencies** Transparency 81	●	●	

Bibliography

The following resources can be used for teaching the chapter. See page T-40 for supplier codes.

Audio-Visual Sources
(video unless noted)
Adaptation to Environment. 16 min. 1983. LF.
Animal Populations: Nature's Checks and Balances. 22 min. 1983. EB.
The Building of the Earth. 55 min. 1984. T-L.
Earth: Its Water Cycle. 11 min. 1974. C/MTI.
Exploring an Ecosystem. 17 min. 1987. A-W.
Pond-Life Food Web. 10 min. 1976. NGSES.
Seas of Grass. 55 min. 1984. T-L.
Water: A Precious Resource. 23 min. 1980. NGSES.

Software Resources
The Environment I: Habitats and Ecosystems. IBM PC. IBM.
The Environment II: Cycles and Interactions. IBM PC. IBM.
Hothouse Planet. Apple, IBM PC. EME.
Odell Lake (Science Volume 3). Apple, Atari. MECC.

Library Resources
Bendick, Jeanne. *Ecology: Science Experiences.* New York: Franklin Watts, Inc., 1975.
Diether, Vincent G. *The Ecology of a Summer House.* Amherst: University of Massachusetts Press, 1984.
Nilsson, Lennart. *Close to Nature: An Exploration of Nature's Microcosm.* New York: Pantheon, 1984.
Rockwell, Jane. *All About Ponds.* Mahwah, NJ: Troll Associates, 1984.
Schwartz, George and Bernice. *Food Chains and Ecosystems.* Garden City, NJ: Doubleday, 1974.

CHAPTER 25

INTEGRATED LEARNING

Themes in Science

Systems and Interactions While studying this chapter, students will discover ways in which all the living and nonliving things on the earth are interconnected. They will also discover that there are many interactions within ecosystems, and that changes to one part of an ecosystem affect the other parts.

Writing Connection

Have students imagine that another photograph was taken five minutes after the one on this page. Have them describe what changes may be observed in the environment and what the organism in the photograph might be doing. Ask them to describe any other organisms that might appear in the second picture.

Introducing the Chapter

Have students read the description of the organism on page 536. Ask if they agree or disagree with the student's statement.

Directed Inquiry

Have students study the photograph. Ask:

▶ Do you think that the organism in the photograph is close to its home? (Yes. It builds nests in tall reeds near marshy areas.)

▶ How do you think the bird meets its needs? What do you think it eats? (Students may suggest that the bird eats bugs, plants, or fish.)

▶ What are the living and nonliving things that the bird uses? (The grasses provide camouflage, fish provides food, and the marsh provides water.)

▶ Why do you think this photograph is in a chapter on organisms and their environment? (It shows the bird in its environment, which is a wetland environment.)

Chapter Vocabulary

abiotic factor
biotic factor
community
condensation
consumer
ecosystem
energy pyramid
evaporation
food chain
food web
niche
nitrogen fixation
population
precipitation
producer
succession

Chapter 25 Organisms and Their Environment

Chapter Sections

25.1 Ecosystems and Communities

25.2 Food and Energy

25.3 Cycles in an Ecosystem

25.4 Changes in Ecosystems

What helps this animal live here?

"This is some kind of water bird in some kind of marsh. It is easy for it to blend in and hide from predators. This bird has a beak that is used for spearing fish."

Kevin Brown
Southwest Junior High School
Little Rock, Arkansas

To find out more about the photograph, look on page 558. As you read this chapter, you will learn how organisms interact with each other and their environment.

Integrating the Sciences

Earth Science The biosphere is composed of land, air, and water. In geology, these three components are called the lithosphere, the atmosphere, and the hydrosphere. Have students refer to earth science texts to find descriptions of these components and create drawings or models that show the three spheres.

Language Arts Connection

Write the prefixes *hydro-*, *atmos-*, and *litho-* on the chalkboard and ask volunteers to find the meanings of each. Then have students find other words in the dictionary that use these prefixes. Have them use their own words to define two such terms in a way that relates to the prefix.

SECTION 25.1

Section Objectives
For a list of section objectives, see the Student Edition page.

Skills Objectives
Students should be able to:

Infer where an organism lives and how it survives.

Estimate the number of organisms in a population by sampling.

Infer how adaptations help determine an organism's niche.

Vocabulary
ecosystem, biotic factor, abiotic factor, community, population, niche

25.1 Ecosystems and Communities

Objectives

▶ **Describe** an ecosystem.
▶ **Identify** the abiotic and biotic parts of an ecosystem.
▶ **Distinguish** between habitats and niches.
▶ **Infer** how adaptations help determine an organism's niche.

Skills WarmUp

Infer In your mind, picture organisms that you think may live in a forest. Then write about one of the organisms. Tell what part of the forest the organism may live in and how it survives.

Think of how many nonliving things you need or use to get through a day. You need air for breathing. You need water to brush your teeth and to take a shower or bath. What other nonliving things in your environment are a necessary part of your day?

You interact with nonliving things in your environment each day. You also interact with many living things. This is true of all organisms. The study of the interactions of organisms and their environments is called ecology (ee KAWL uh jee).

Scientists who study these interactions are called ecologists. They may study organisms living on a small leaf or organisms living in a huge volcanic crater.

The Biosphere

Imagine a jigsaw puzzle with more than a thousand pieces to put together. At first it may seem like an impossible task. Gradually you get all the pieces to fit together, and you have a complete picture! All the different living things and nonliving things on the earth are like pieces of a huge puzzle. Like a puzzle, they fit together, or interact to form one working system.

All organisms live at or near the earth's surface in a life-supporting zone called the *biosphere*. The biosphere includes the earth's waters, its surrounding air, and a portion of the earth's land surface. Within the biosphere are many different environments that organisms inhabit.

Figure 25.1 ▲
What are the parts of the earth's biosphere? ①

MOTIVATE

Skills WarmUp

To help students understand some of the ways that organisms interact with their environment, have them do the Skills WarmUp.
Answer Answers may vary. One example would be an owl that makes its nest inside a hollow tree, eats small rodents that occupy the forest floor, and hides from larger birds of prey underneath branches.

Misconceptions

Students may think that communities always include people and the structures that they create. As you work through this chapter, you may wish to stress that there are many different types of populations that can be considered parts of communities.

Answer to In-Text Question

① The biosphere includes the earth's water, its air, and the earth's land surface.

TEACH

Class Activity
Have students work in small groups, each building a model of one particular ecosystem. The models can be realistic or symbolic. Some possible ecosystems would be a log, an aquarium, or a prairie. Encourage students to represent both biotic and abiotic factors in their ecosystem.

Explore Visually
Have students study the wetland environment pictured on these two pages. Then ask some or all of the following questions:

▶ What organisms of this wetland environment can you name? (A heron, a deer, a dragonfly, reeds, cattails, turtle, muskrats, spurtina grass, red-winged blackbirds, ducks, and raccoons are shown.)

▶ How could you find out how many different populations are in the wetland? (Students should realize that they need to count only the number of different organisms to find the number of populations.)

▶ Find some examples of organisms interacting with biotic and abiotic factors in this ecosystem. (Answers may vary. For example, the ducks swim in and drink the water, which is an abiotic factor, and eat plants and fish, which are biotic factors.)

**The Living Textbook:
Life Science Sides 7-8**

Chapter 17   Frame 01754
Biosphere (1 Frame)
Search:

INTEGRATED LEARNING

Themes in Science
Scale and Structure The earth contains many ecosystems that in turn can be divided into smaller categories such as communities, populations, habitats, and niches.

Social Studies Connection
Have students find the five countries with the largest populations of human beings.

Ecosystems

Look at the wetland environment below. Within this environment, many kinds of interactions take place. An area in which living things interact with one another and with their environment is called an **ecosystem**. Ecosystems may be large, such as a wetland, the ocean, or a desert. They may also be small, such as a puddle or a piece of a rotting log.

The living parts of an ecosystem are called **biotic factors**. Animals are examples of biotic factors. The nonliving parts of an ecosystem are called **abiotic factors**.

Ecosystems do not happen overnight. They evolve gradually. The older an ecosystem is, the more organisms it usually contains.

**Figure 25.2
A Wetland Ecosystem**

Reeds are common plants in wetlands. Some reeds grow up to three meters tall.

Communities

How many different organisms do you recognize in the wetland ecosystem below? When two or more groups of different organisms live together and interact with each other in the same area, they form a **community**. The plants and animals shown are just a few of the types of organisms that live in a wetland ecosystem. All the organisms in a wetland are a community.

The biotic relationships in an ecosystem hold a community together. These relationships are always changing. For example, in a wetland ecosystem that has seasonal changes, there are more plants during the summer. More insects enter the community because many depend on plants for food and shelter. With more insects, the number of frogs also increases because there is a large supply of food.

Abiotic factors also contribute to changes in communities. For example, fewer plants will grow in the winter because there is less sunlight and the climate is cold. Some organisms that depend on plants will leave the community to seek food and shelter elsewhere.

Organisms interact closely with one another in the wetland. Cattails serve as landing pads for dragonflies. When danger is near, herons point their long bills and necks upward to blend in with tall marsh plants.

538   Chapter 25 Organisms and Their Environment

Themes in Science

Stability and Equilibrium The interactions between biotic and abiotic factors in an ecosystem maintain a delicate balance that supports the organisms living there.

▶ Describe an example of one population getting food from another. (One example is insects eating plants.)

▶ Describe an example of two populations competing. (Herons and raccoons both eat fish.)

▶ How would this ecosystem look in the winter? (Fewer plants would be growing, so animals may have moved elsewhere to find food and shelter. Areas of the wetlands may be frozen, so ducks may have moved in search of water.)

▶ Describe the niche of one of the populations in the picture. How does this niche differ from the niche occupied by another organism? (Answers may vary, but should indicate that students are considering all the activities and behaviors that satisfy an organism's need for food, water, living space, and reproduction. For example, the duck meets its needs by eating insects and water organisms, lays its eggs on land, and lives by the shore. Its niche differs from the deer which is a terrestrial animal.)

Populations

All the organisms of the same species living in a certain place make up a **population**. Notice that in the wetland community, there are many different populations. All the raccoons make up one population. Among the plants, cattails make up another population. Your school and the area around it may include populations of people, ants, grass, plants, and even spiders. The size of each population may change depending on conditions, such as available food, space, and water.

In a community, the different populations interact with one another in many ways. Some populations eat food that is produced by other populations. Some get their homes from others, and some get protection from others. Populations may also compete for food and shelter.

Habitats and Niches

In a wetland ecosystem, each species lives in a certain area. The area or place where an organism lives in an ecosystem is called a *habitat*. You can think of a habitat as an organism's "address." In a wetland, the habitat of the cattails is the edge of the water. Here cattails grow well because they get plenty of water and a lot of direct sunlight.

Each organism in a wetland has a different job, or role. The specific role an organism has in its habitat is called its **niche** (nihch). A niche includes all the activities and behaviors that satisfy an organism's need for food, water, living space, and reproduction.

Each population in a habitat has a different niche. For example, herons and deer share the same wetland habitat and have similar abiotic needs. They do not, however, have the same niche. Herons live near water and eat fish. Herons make their nests high in trees. Deer have no nesting sites. They roam the wetlands in search of food, eating only plants and plant parts.

Wetlands support plants and animals that are adapted to living in a watery environment. Water is an abiotic factor in the wetlands. Other abiotic factors include soil, air, and sunlight.

Wetlands attract a diverse group of ducks and other wildfowl. Ducks have webbed feet for swimming, and long, limber necks for finding food in the water and the water's muddy bottom.

Wetland plants take up and use nutrients in soil. Many can absorb pollutants from the soil.

The Living Textbook:
Life Science Side 8

Chapter 26 Frame 11519
Analogous Niches (Movie)
Search: Play:

TEACH • Continued

Discuss
Have students choose an organism pictured in Figure 25.2. Describe how the organism is adapted to its niche in the ecosystem.

SkillBuilder

Activity Fill a clear jar with beans. Have students estimate the number of beans inside the jar. Allow them to measure the jar, count the beans in a given area, or lift the jar to test its weight. Do not let them open the jar until all of the students have made an estimate. Then let students work together to count the beans. Have students compare their estimates with the actual number.

Answers
1. Answers will vary. A reasonable estimate would be 100 ants.
2. Answers are 14, 15, 16, or 17.
3. Students may make their initial estimate by counting the number of ants in one square and multiplying by six. Students will observe that the estimate becomes more accurate as they proceed with the process.
4. Answers will vary, but must be 84, 90, 96, or 102.
5. Answers will vary, but should fall between 14 and 17.
6. The actual number is 91. Answers will vary but may include sampling can help scientists if an area is very large.

The Living Textbook: Life Science Side 8

Chapter 20 Frame 02358
Adaptation (Movie)
Search: Play:

INTEGRATED LEARNING

Social Studies Connection
Have students research the methods that were used to take the most recent census of persons living in the United States. Have volunteers present brief oral reports to the class. Ask students to compare the method they used to sample the ant population with the methods used for the U.S. census.

Niches and Adaptations

All organisms have characteristics that help them survive in ecosystems. Characteristics that help organisms get food and water, protect themselves, and build homes are adaptations.

Adaptations help determine an organism's niche. For example, a heron waits in water until fish come within reach. As fish dart through the water, the heron cranes its neck, spears a fish, and tosses it around. The heron will then eat the fish whole. The heron's long legs and neck and sharp beak enable it to catch fish. These adaptations help the heron live and reproduce successfully in its niche in the wetlands.

Populations with similar adaptations occupy similar niches. They eat foods that are alike. They construct homes and seek protection in much the same way. If more than one population has the same niche, they will compete for food and shelter.

Figure 25.3
The heron's long neck, long legs, and sharp beak for catching food are adaptations for life in its wetland niche.

SkillBuilder *Estimating*

Sampling a Population

Scientists sometimes need to know the size of a population. However, a population is often much too large to count each organism. The size of a whole population can be estimated by counting a small part, or sample, of the population.

1. Estimate the number of ants you think are in the population shown. Record your estimate. *Note: Do not count the ants.*
2. Choose any square in the grid. Count and record the number of ants in the square.
3. Now estimate the total number of ants in the grid by multiplying the number of ants you counted by the number of squares in the grid. Record your answer.
4. Repeat steps 2 and 3 two more times. Record your answers.
5. Add the three sample estimates, then divide the total by three. This is the average. Record the average.
6. Count all the ants in the grid. Record the total.

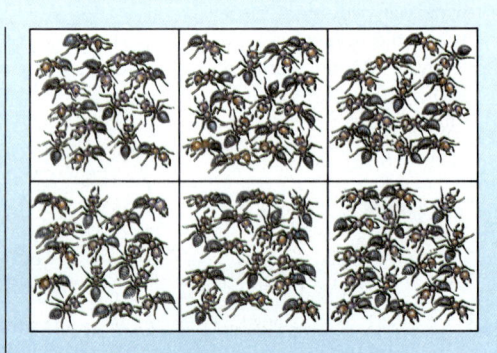

How do you think sampling can help scientists estimate an entire population's size?

Themes in Science

Systems and Interactions In a balanced ecosystem, living and nonliving parts interact successfully. Furthermore, ecosystems, such as wetlands, interact with other ecosystems, such as bays. By changing one part of an ecosystem, people often cause changes to other parts.

Science and Society *Balance in Ecosystems*

In any ecosystem, organisms need a balanced environment. A balanced ecosystem is one in which all living and nonliving things are interacting successfully. If any part of the ecosystem is disturbed, other parts will also be disturbed. People sometimes cause these disturbances. In some wetland ecosystems, people have affected the balanced environment.

At one time, people did not consider wetland areas to be important ecosystems. In many countries, wetlands were drained for farmland. Some wetlands were filled in so that houses could be built. Other wetland ecosystems were destroyed by pollution.

How does the destruction of wetlands cause imbalance in ecosystems? If a wetland area is destroyed, so are important plants and the habitats of many kinds of animals. You may think that the destruction of a wetland ecosystem stops there. However, wetlands protect other ecosystems, such as bays, by absorbing pollutants from industries and other sources. Wetlands also help prevent flooding in coastal areas.

Today, many people recognize the value of wetland ecosystems. Around the world, important steps have been taken to protect and preserve wetlands. In Tunisia, Africa, a wetland nature center has been built. In El Salvador, the habitat of white-winged tree ducks has been preserved because people protested the destruction of a wetland area.

Figure 25.4 ▲
This wetland ecosystem in Botswana, Africa, was saved from possible destruction. Why is it important to preserve wetland ecosystems? ①

Check and Explain

1. What is an ecosystem? Give an example of an ecosystem.
2. Explain the difference between an organism's habitat and its niche.
3. **Classify** Make a list of abiotic and biotic factors inside and outside your classroom. Be sure to include factors, such as insects, that may not be clearly visible.
4. **Infer** Describe a turtle's niche in a wetland ecosystem. Then describe how some of the turtle's physical features have helped it adapt to the wetland.

Chapter 25   Organisms and Their Environment   **541**

Discuss
Stress that there are many ecosystems that need to be protected. Have students describe an ecosystem that they would like to preserve.

EVALUATE

WrapUp
Portfolio Have students create a concept map or Venn diagram that shows how the various elements of an ecosystem relate to one another. They should include the following: individual organisms, populations, niches, habitats, and communities.
Use Review Worksheet 25.1.

Check and Explain

1. An ecosystem is an area in which living things interact with one another and with their environment.
2. An organism's habitat is the area where it lives within the ecosystem. Its niche includes all of the activities and behaviors that satisfy its needs.
3. Answers may vary. Some abiotic factors are air, water, sun, cement, chalkboard, clothes, and desks. Some biotic factors are people, trees, grass, insects, and birds.
4. A turtle eats insects and plants that live in and around the water. It has flipperlike feet designed for swimming. A turtle can cool off in the water or use its shell to absorb the sun's heat. Its hard, camouflaged shell helps to conceal the turtle from predators.

Answer to In-Text Question

① **When a wetland is destroyed, so are its inhabitants. Wetlands protect other ecosystems and help prevent flooding.**

541

SECTION 25.2

Section Objectives
For a list of section objectives, see the Student Edition page.

Skills Objectives
Students should be able to:

Classify foods according to whether they came from plants or animals.

Infer how food chains and food webs were named.

Make a Model of a food chain.

Vocabulary
food chain, producer, consumer, food web, energy pyramid

MOTIVATE

Skills WarmUp
To help students understand food chains, have them do the Skills WarmUp.
Answer Students should indicate that fruits, vegetables, grains, and nuts come from plants. Meats and dairy products come from animals. Animals eat plants and sometimes other animals. Animals depend on plants for food, shelter, and oxygen. Plants depend on animals for carbon dioxide and for help with pollination and seed dispersal.

Prior Knowledge
To gauge how much students know about how organisms in an ecosystem meet their energy needs, ask the following questions:
▶ What things provide your energy needs?
▶ How do animals meet their energy needs?
▶ How do plants meet their energy needs?

542

INTEGRATED LEARNING

Themes in Science
Energy All living things require energy to live. The energy in an ecosystem is passed from one organism to another through a series of interactions described as a food chain.

Writing Connection
Suggest to students that they write a fictional account of how it might be if the sun went out tomorrow. How long could life go on? Would technology preserve life on the earth?

Skills WarmUp

Classify List some of the foods that you usually eat. Classify the foods that came from plants and the foods that came from animals. If the foods came from animals, name the foods that you think the animals ate. Tell why plants and animals depend on one another.

Figure 25.5 ▲
A vulture is a bird that is an example of a scavenger.

25.2 Food and Energy

Objectives
▶ **Identify** the producers and consumers in an ecosystem.
▶ **Distinguish** between a food chain and a food web.
▶ **Interpret** an energy pyramid.
▶ **Make a model** of a food chain.

What would the earth be like without energy from the sun? It would be a very cold and dark place. No organisms could live on the earth without energy from the sun.

All organisms get energy from some type of food, and the sun provides the energy needed for food crops to grow. Food provides you with energy for all your activities, including walking, eating, and even reading.

Food Chains

Food energy is passed from one organism to another through a sequence of events called a **food chain**. Plants are the only organisms that can capture the sun's energy, which they use to fuel their growth by the process of photosynthesis. Because they can fuel their own growth, plants are called **producers**. On a prairie, grasses are the main producers. In a forest, trees are the main producers. Algae carry on photosynthesis, so they are also producers.

Many organisms can't make their own food. They eat plants, animals, or other organisms. Organisms that eat other living things are called **consumers**. A food chain may include more than one consumer. For example, in a food chain in which a rabbit eats grass and an owl eats the rabbit, both the rabbit and the owl are consumers.

Some food chains have consumers that feed only on the bodies of dead organisms. These consumers are called scavengers. After a scavenger has fed on a dead organism, millions of tiny organisms called decomposers take over. Decomposers, such as bacteria and molds, break down the tissues of dead organisms.

542 Chapter 25 Organisms and Their Environment

Language Arts Connection
Have students compare the everyday uses of the words *producer* and *consumer* with the scientific meanings of the words.

A Food Chain Example

A food chain always begins with the sun, which provides energy for the producers. Producers are the next link in the chain. Look at the food chain below. It shows the following sequence: sun → plant → grasshopper → harvest mouse → marsh hawk.

Consumer
In this food chain, a marsh hawk swoops down to feed on the mouse. Animals that feed only on other animals are called carnivores (KAHR nih vorz). Carnivores have special adaptations, such as sharp teeth, for eating flesh.

Decomposers
The last link in a food chain are the decomposers. After the marsh hawk has died, decomposers feed on the hawk and break down its tissues. In the process, decomposers release wastes into the soil. These wastes are then used by other organisms.

Consumer
In one link of the food chain, a harvest mouse eats the grasshopper. The mouse also feeds on the plants. Consumers that eat both plants and animals are called omnivores (AHM nih vorz).

Consumer
The plants are eaten by a grasshopper. Animals that eat only plants are called herbivores (HUR bih vorz). Herbivores have adaptations that enable them to eat and digest plants.

Producers
In this food chain, salt-marsh plants and algae are the producers.

TEACH

Directed Inquiry
Have students study the food chain. Ask:

▶ What are the producers shown in the illustration? (Salt–marsh plants, and algae)

▶ What are the three kinds of consumers, and what do they feed on? (Herbivores eat only plants, omnivores feed on both plants and animals, and carnivores feed only on other animals.)

▶ Name several things that decomposers feed on. (Decomposers feed on all the dead organisms as well as all the wastes that come from living organisms.)

▶ Why are decomposers called the last link in the food chain? (Decomposers produce nutrients that are used by the producers.)

Class Activity
Have students construct a food chain that includes organisms in your locale. Have students consider what would happen if one organism were removed. Have them think about what would happen to the organisms above and below it on the chain.

The Living Textbook: Life Science Sides 7-8
Chapter 16   Frame 01659
Herbivores, Carnivores, Scavengers, and Decomposers (25 Frames)
Search: Step:

The Living Textbook: Life Science Side 8
Chapter 27   Frame 11960
Food Chain (Movie)
Search: Play:

TEACH • Continued

Explore Visually
Have students study Figure 25.6. Ask some or all of the following questions:

▶ Which way does energy in this food web flow from the producers? (Toward the snail, smelt, harvest mouse, and shrimp)

▶ Describe one of the food chains that is shown in this figure. (There are at least seven food chains. Students can describe them by looking at the points where arrows start and then following the direction of the arrows.)

▶ How can you tell if an animal is a part of more than one food chain? (Any animal that has more than one arrow pointing to it and/or away from it is a part of more than one food chain.)

▶ Which population do you think would be larger—song sparrows or grasshoppers? Explain. (The grasshopper population would be larger. A greater number of animals lower on the food chain is needed to supply the animals higher on the food chain.)

Critical Thinking
Predict What would happen to the short-eared owl if the rodent population decreased? (The owl population would also decrease.)

Reteach
Use Reteach Worksheet 25.2.

Answers to In-Text Questions
① **The short-eared owl feeds on Norway rats.**
② **Besides grasshoppers, other first-level consumers are snails, harvest mice, smelt, and shrimp.**

INTEGRATED LEARNING

Art Connection
Have students work in small groups to build models of the food web that includes humans. Supply materials such as yarn, clay, paints, small cardboard boxes, crayons or markers, paper, and paste.
Use Integrating Worksheet 25.2.

Food Webs

In an ecosystem, there are many different food chains. The same food source can be part of more than one food chain. Different food chains connect to form a network of feeding relationships called a **food web**. In food chains, the flow of energy happens in one direction. In food webs, the flow of energy branches out in many directions.

In many food webs, producers and consumers can be eaten by more than one kind of organism. Look at Figure 25.6, showing an example of a salt-marsh food web in the winter. The Norway rat, vagrant shrew, song sparrow, and harvest mouse all feed on grasshoppers. The marsh hawk, in turn, feeds on the Norway rat, harvest mouse, and vagrant shrew. On what animals does the short-eared owl feed? ①

In a food web, there are different feeding levels. In Figure 25.6, the grasshoppers are first-level consumers because they eat only plants. What other animals are first-level consumers? ② The shrew is a carnivore and eats first-level consumers. The shrew is a second-level consumer. The marsh hawk and short-eared owl are third-level consumers. No animals feed on the hawk and owl. They are the top consumers in the food web.

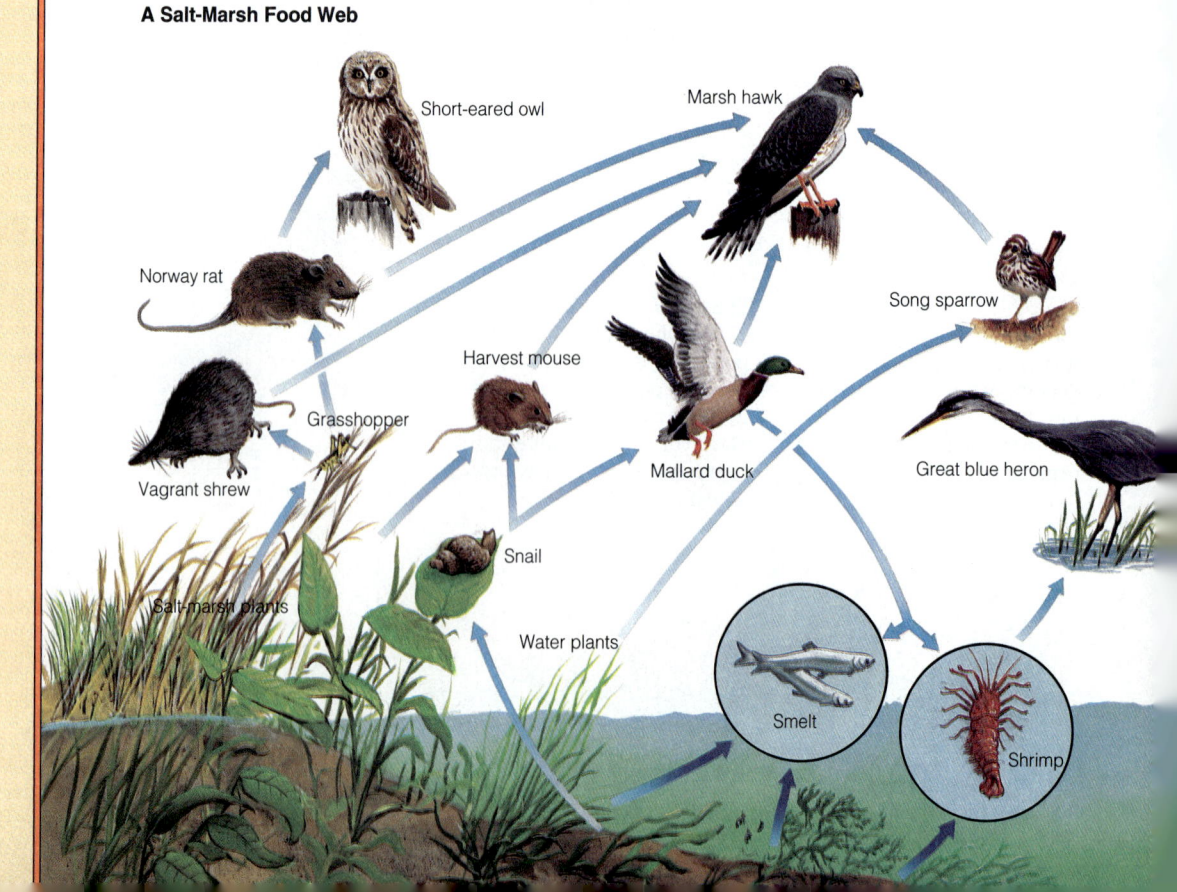

Figure 25.6
A Salt-Marsh Food Web

544

Math Connection

Only 1 to 2 percent of the solar energy that reaches the earth is actually used by plants for photosynthesis. In most food chains, only 10 percent of the energy at each stage is passed on to the next stage. Have students calculate the percentage of the energy available to producers that is available to third-level consumers. (100% × .1 × .1 × .1 = .1%)

Energy and Energy Pyramids

You have probably heard the word *energy* over and over. But what does it mean? Energy is not an easy word to define because it has many different forms, such as heat, light, sound, water, and wind.

As you know, the sun provides energy for all ecosystems on the earth. You may be surprised to learn that only 1 to 2 percent of the solar energy that reaches the earth is actually used by plants for photosynthesis.

In a food chain, the energy stored in the food is passed along from one organism to the next. An **energy pyramid** is a diagram that shows the amount of energy available at each level of the food chain. However, not all the energy is moved from one link in the chain to the next. Energy is lost at each stage because it is used by the organisms for life processes. In almost all food chains, 10 percent of the energy is transferred at every link.

Look at Figure 25.7. Each level of the pyramid represents the amount of energy stored at that level. At the base of the pyramid are the producers. This is the level where most of the energy is stored. Each level above the producers contains less available food energy than the level below it. The second level represents the first-level consumers. The third level represents the second-level consumers. On the top are the third-level consumers. At this level, the least amount of energy is available because energy has been lost as it passes through each organism. On what level would you place yourself in an energy pyramid? ③

**Figure 25.7
An Energy Pyramid**

- Third-level consumers
- Second-level consumers
- First-level consumers
- Producers

Misconceptions

Often students will assume that larger animals are higher on the food chain. Discuss the size variation in organisms at each level in the food chain. For example, herbivores, which are second-level consumers, include small animals such as slugs and large animals such as elephants.

Reteach

Have students use the information in Figure 25.7 to construct their own energy pyramids using organisms from other ecosystems.

Answer to In-Text Question

③ Most humans are second-level consumers because they feed on both producers and first-level consumers. Vegetarians would be first-level consumers. A person who eats shark meat is a third-level consumer.

**The Living Textbook:
Life Science Side 8**

Chapter 25 Frame 10183
Energy Pyramid (Movie)
Search: Play:

**The Living Textbook:
Life Science Side 8**

Chapter 28 Frame 12621
Food Web (Movie)
Search: Play:

TEACH ▪ Continued

Skills WorkOut
To help students understand food chains, food webs, and energy pyramids, have them do the Skills WorkOut.

Answers The food chain describes a linear sequence of relationships. The flow of energy in a food chain travels in only one direction. The arrows in a food web show the flow of energy traveling in more than one direction. A pyramid grows narrower at the top, which illustrates that there is less available energy at the top of the pyramid than there is at the bottom.

Apply
Have students find and record the names of foods that are good sources of protein. Have them circle foods that they could eat if they chose to be first-level consumers.

EVALUATE

WrapUp
Review Use Review Worksheet 25.2.

Check and Explain
1. Wheat is the only producer. The grasshopper, field mouse, snake, and owl are consumers.
2. In a food chain, the flow of energy happens in one direction. In a food web, the flow of energy branches out in many directions.
3. One of the reasons that energy is lost as it passes through an organism at each level of a food chain is that the organism uses energy for life processes.
4. Answers will vary.

546

INTEGRATED LEARNING

Multicultural Perspectives
Encourage students to do research on how people in different countries meet their nutritional needs. Ask students to choose a particular country and find information about the diets of the people living there. Have them consider whether the people are predominantly first-, second-, or third-level consumers.

Skills Workout
Infer Look at the food chain on page 543, the food web on page 544, and the energy pyramid on page 545. Infer how each got its name. What other names can be used to describe food chains, food webs, and energy pyramids?

Science and You *Choose Your Place!*
Most animals and plants can't change their places in a food web. A hawk, for example, will never one day become a first-level consumer and start eating only plants because a hawk is a carnivore. You, however, can change your place. You can choose to be a first-level consumer for a while, then change to a second-level consumer. You can also choose to be a third-level consumer in a food web. You can be any one of these consumers because you can survive on plants or animals, or both. You are an omnivore.

Because you can choose what kind of consumer you would like to be, you can also consider which consumer uses the sun's energy most efficiently. Suppose you could set up your own farm. If you decided to be a second-level consumer and only eat meat, you would need to keep a herd of about seven to ten cattle to support yourself. The cattle, in turn, would need about 5 acres of alfalfa for their feed.

What would happen if you decided to change from a second-level to a first-level consumer? You might, for example, choose to eat only wheat. On this diet, you could support yourself on less than an acre of wheat.

As you can see, if you choose to be a first-level consumer, you use the sun's energy more efficiently than if you are a second-level consumer. However, the disadvantage of eating only one kind of producer, such as wheat, is that wheat can't provide all the proteins you need in your diet. You would have to supplement your diet by eating other producers high in proteins, such as beans and peas.

Check and Explain
1. Identify the producers and consumers in the following food chain: Sun → Wheat → Grasshopper → Field mouse → Snake → Owl
2. How is a food chain different from a food web?
3. **Find Causes** Explain how energy is lost to a community at each level of a food web.
4. **Make a Model** Construct a simple food chain that you would expect to find in a stream community. Use any of the following organisms: frogs, algae, dragonflies, turtles, beetles, bass fish, water lilies.

546 Chapter 25 Organisms and Their Environment

TEACHING OPTIONS

Prelab Discussion
Make sure students understand the difference between a food chain and a food web. Remind students that the arrows represent the flow of energy, so they should point from the "eaten" to the "eater."

Activity 25 How do food chains form a food web?

Skills Model; Communicate

Task 1 Prelab Prep
1. Collect the following items: 2 large sheets of paper, markers, metric ruler, paste and/or tape, old magazines.
2. Cut one sheet of paper into fourteen 5-cm squares. **CAUTION! Be careful when using scissors.**
3. Study Table 25.1. On each square of paper, write the names of the organisms in the table. On the last square, write *Sun*. These squares are your cards for Task 3.

Task 2 Gathering Data
1. Use reference books to read about the organisms with which you are not familiar. Find out what each organism listed in the table eats.
2. On a piece of paper, use arrows to form as many food chains in the salt-marsh ecosystem as you can.

Task 3 Procedure
1. Refer to your food chains. Use the cards you made to construct a food web. Arrange the cards on a large sheet of paper. *Note: Be sure to include the card labeled* Sun. Leave a lot of space around each card. Attach your cards.
2. Refer to your food chains again. Construct a food web by drawing arrows from each organism to all the other organisms that eat it.
3. Label each organism *Producer, First-level Consumer, Second-level Consumer,* or *Third-level Consumer.*

Task 4 Analysis
1. Why is a food web a better model than a food chain of feeding relationships in an ecosystem?
2. Are there any omnivores in your diagram? If so, name them.
3. If you were to remove one of your organisms from the food web, how would other organisms be affected? Explain.
4. Which organisms do you think are the most abundant in the salt-marsh community? The least abundant? Explain your answer.

Task 5 Conclusion
Compare your food web model to other models in the class. Why do ecosystems have a variety of feeding relationships?

Extension
Think of a meal that you've eaten recently. On posterboard, create a food web that shows you as the top consumer for the meal.

Table 25.1 Salt-Marsh Organisms

Producers	First-level Consumers	Second-level Consumers	Third-level Consumers
Water plants	Eel	Billfish	Osprey
Marsh plants	Fluke	Cormorant	Tern
	Mud snail	Merganser	Green heron
	Bay shrimp	Redwing blackbird	
	Cricket		

ACTIVITY 25

Time 90 minutes **Group** 1–2
Materials
large sheets of paper
markers
metric rulers
paste and/or tape
old magazines

Analysis
1. A food web shows that many organisms eat, or are eaten by, more than one organism. A food chain shows only a single series of "eater–eaten" relationships.
2. See Table 25.1. Second- and third-level consumers that also eat plants are omnivores.
3. Removing one organism from the food web would have two effects: (1) If that population were prey for another population, the predator population would decrease; and (2) If the population were a predator over another population, the prey population would increase. These changes would affect all the organisms in the web.
4. The producers (marsh plants and water plants) would be the most abundant because they must exist in sufficient quantity to support the energy needs of all other organisms in the community. The amount of energy available to them at the top of the energy pyramid is relatively small, so the third-level consumers (osprey, tern, heron) would be the least abundant.

Conclusion
Answers may vary, but should be consistent with the class results.

Extension
Answers may vary.

SECTION 25.3

Section Objectives
For a list of section objectives, see the Student Edition page.

Skills Objectives
Students should be able to:

Infer how a shortage of water would affect an ecosystem.

Make a Model showing the water cycle.

Vocabulary
evaporation, condensation, precipitation, nitrogen fixation

MOTIVATE

Skills WarmUp
To help students understand the water cycle, have them do the Skills WarmUp.
Answer Water could enter the area's water supply in the form of precipitation or could be pumped into the area from a nearby body of water. Students' answers should reflect the fact that water can't be manufactured.

Prior Knowledge
To gauge how much students know about the earth, ask all or some of the following questions:

▶ Where does the water in rain and clouds come from?

▶ What kinds of cycles can you see in living and nonliving things around you?

▶ Do cycles play a part in helping you meet your needs?

Answer to In-Text Question

① The terrarium provides everything the organisms in it need for survival.

548

INTEGRATED LEARNING

Integrating the Sciences
Geology Discuss the geologic cycles that give rise to the earth's landscapes, and describe the three types of rocks involved in the rock cycle—igneous, metamorphic, and sedimentary.
Use Integrating Worksheet 25.3a.

Skills WarmUp

Infer What would it be like if the area in which you live was suddenly low on water? Name some ways in which you think more water could enter the area's water supply.

Figure 25.8 ▶
How is a terrarium like the earth's biosphere? ①

25.3 Cycles in an Ecosystem

Objectives

▶ **Describe** the cycles in nature.

▶ **Explain** the importance of natural cycles to organisms.

▶ **Construct** a model of the water cycle.

Look at the terrarium in Figure 25.8. It is a self-enclosed ecosystem. To keep the terrarium ecosystem working properly, all the biotic and abiotic factors interact in a balanced way. A terrarium is like a small model of the biosphere because the terrarium supplies all the materials the organisms in it need in order to survive.

All organisms need certain substances to live, grow, and reproduce. Many substances, such as water, pass through natural cycles in which they are used and reused. These substances circulate through both living and nonliving things.

Some cycles include mainly nonliving things. For example, forces and pressures within the earth cause rocks to change from one type to another. This process is called the rock cycle. In space, many stars that you see are in different stages of their life cycle. The sun, a medium-sized star, has a life cycle of about 10 billion years. The sun is now about 5 billion years old. At the end of its life cycle, the sun will become a white dwarf.

548　Chapter 25　Organisms and Their Environment

Themes in Science

Patterns of Change/Cycles Many of the components in an ecosystem are affected by changes that occur in a cyclical pattern. As students study the concepts in this chapter, they will learn to identify some of these cycles. They will also see visual representations of cycles.

Figure 25.9 The Water Cycle

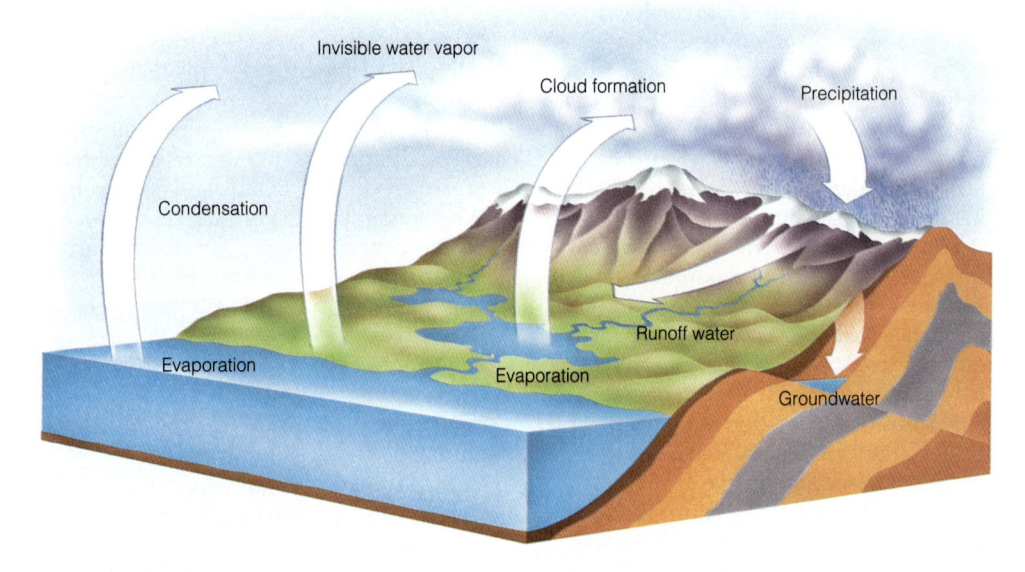

The Water Cycle

All living things need water. Water moves continually throughout the earth by means of the water cycle. Some of the water molecules that once flowed in the Amazon River could fall on you as rain. You also could be drinking some of the water that once existed as snow on the Himalaya Mountains.

Look at Figure 25.9 and follow the pathway of water between the earth and the atmosphere. Energy from the sun changes water from a liquid to a gas through the process of **evaporation**. In its gaseous state, water is called water vapor. Water evaporates from bodies of water, such as oceans, rivers, and lakes. Water also evaporates from soil, plants wet with dew, and even the bodies of animals.

Water vapor is held in the atmosphere. The amount of water vapor the air can hold depends on the air temperature.

When the atmosphere's temperature cools, water vapor changes back to a liquid by the process of **condensation**. The condensed water forms clouds, steam, or fog, and eventually falls back to the earth as **precipitation**. Rain, snow, hail, and sleet are forms of precipitation.

Most precipitation falls back into bodies of water. Some precipitation first strikes land and then flows into bodies of water. Some water droplets that fall on land may soak through the soil and become part of the groundwater. The groundwater may then flow back to the surface as a spring, or it may be pumped to the surface through wells. The water cycle continues when water that has returned to the earth's surface evaporates again.

Plants use water from the ground for life processes. They also release some water into the air. Animals also use water for life processes and release it during respiration and excretion.

TEACH

Demonstrate

Groundwater supplies often depend on the type of soil in a particular area. Students can get a sense of what this means by contrasting the porosity of different types of soil. Place sand, potting soil, and pebbles into three separate beakers. Use a graduated cylinder to pour water into the first beaker until the sample is saturated. Record the quantity of water you poured and repeat with the remaining two beakers. Explain that porosity is the ratio of the space between the particles of a substance to the mass of the substance. Ask students to compare the porosity of the three materials based on the demonstration results. Encourage students to draw some conclusions about infiltration rates in various types of the earth's surface materials and the local groundwater supply.

Misconceptions

Some students may think that the same water falls in a given area again and again. Have them look at the drawing and imagine that the water in the air, on the surface of the earth, and in the ground is constantly moving. Emphasize that water vapor moves great distances in the air all over the earth; that water in lakes, rivers, and streams is always moving; that currents throughout the world's oceans are constantly moving; and that groundwater may move great distances.

Ancillary Options

If you are using the blackline masters from *Multiculturalism in Mathematics, Science, and Technology*, have students read The Native Americans I on page 119 and complete pages 120 to 122.

TEACH • Continued

Directed Inquiry
Have students study Figure 25.11. Ask the following questions:

▶ Name all the places in the figure where some form of nitrogen is present. (Raccoon, raccoon wastes, plants, proteins, ammonia, nitrates, decomposers, air)

▶ Why is nitrogen important to the organisms in the picture? (Nitrogen is used to build proteins.)

▶ What is growing on the roots of the plant? (Nitrogen-fixing bacteria)

▶ What do nitrogen-fixing bacteria do? (Take nitrogen and change it to ammonia)

▶ What do the decomposers do? (Change the proteins in animal wastes into ammonia)

▶ What do denitrifying bacteria do? (Break down nitrogen compounds and convert them into nitrogen gas)

Enrich
Tell students about *Clostridium* and *Azotobacter*, which are two free-living genera of nitrogen-fixing bacteria in the soil. Bacteria in the genus *Rhizobium* are nitrogen-fixing bacteria that live in the nodules of legumes. Have students research agricultural practices that take advantage of these bacteria.

Apply
Have students hypothesize what the earth would be like if there were no decomposers. (Organisms that died would not be broken down. There would be bodies of animals and dead plants littering the earth's surface.)

550

INTEGRATED LEARNING

Integrating the Sciences

Chemistry Write the chemical formulas for ammonia (NH_3), ammonium (NH_4^+), and nitrogen gas (N_2) on the chalkboard. Point out that N_2 is a gas that makes up 72 percent of the earth's atmosphere. Soil bacteria convert N_2 into ammonia, which dissolves in water to produce ammonium (NH_4^+). The nitrogen in ammonium, called "fixed nitrogen," can be used by plants. Ammonium is also used to make amino acids and proteins.

Use Integrating Worksheet 25.3b.

The Nitrogen Cycle

Nearly 80 percent of the earth's atmosphere is made up of nitrogen. Nitrogen is a necessary element for organisms because it helps build proteins and other important body chemicals. Although there is a lot of nitrogen in the air, organisms can't directly use it. Nitrogen becomes available after it has been changed into nitrogen compounds.

The process of changing atmospheric nitrogen into usable compounds is called **nitrogen fixation**. Nitrogen fixation is a necessary part of the nitrogen cycle; without it, nitrogen would not be usable to living things. Nitrogen is fixed, or converted, either through the action of lightning or by bacteria. Some of the bacteria live freely in the soil. Other bacteria live in the root nodules of legumes. They carry on most of the nitrogen fixation in land ecosystems. Beans, peas, clover, and soybeans are examples of legumes. When legumes are planted, the bacteria in the nodules provide nitrogen for plant growth.

Look at Figure 25.11. Nitrogen-fixing bacteria change the soil's nitrogen gas into nitrogen compounds. The plants use the compounds to produce proteins. Plants need proteins for growth. When animals eat plants, proteins are passed through the organisms in food chains and food webs. When organisms die or release wastes, bacteria change the proteins to the nitrogen compound ammonia. Bacteria called denitrifying bacteria then break down the ammonia and other nitrogen compounds and convert them into a nitrogen gas.

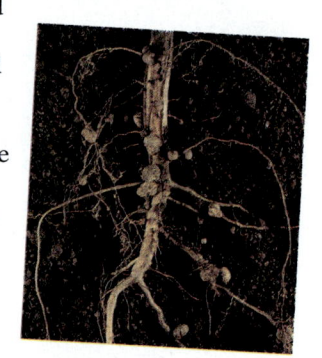

Figure 25.10 ▲
Bacteria in root nodules provide nitrogen for plant growth.

Figure 25.11 The Nitrogen Cycle

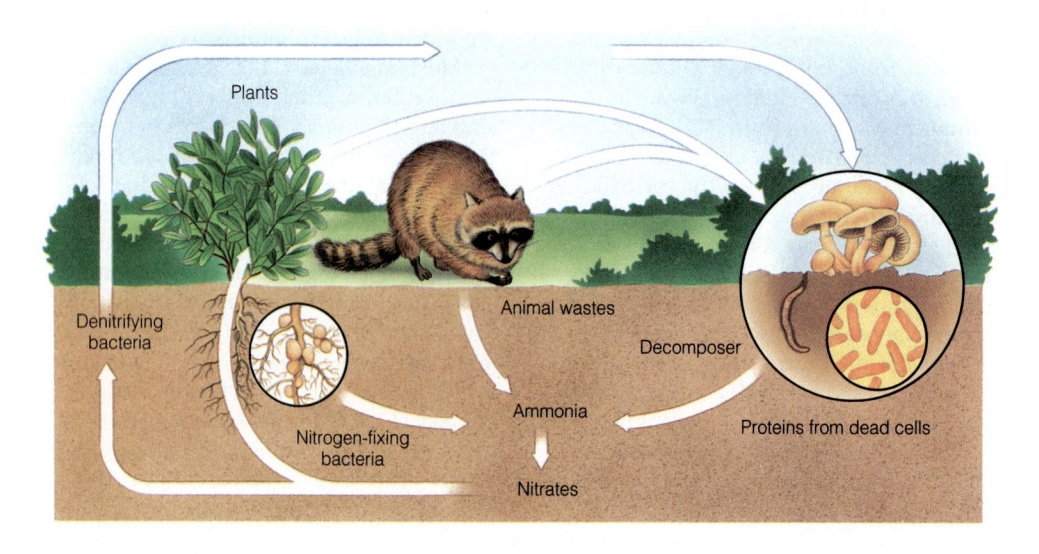

550   Chapter 25   Organisms and Their Environment

STS Connection

The oxygen–carbon dioxide cycle is affected by the use of fossil fuels for energy, because burning these fuels releases carbon dioxide into the atmosphere. Have students work in small groups. Each group can research one alternate energy source and compile a brief report that includes diagrams. Suggest that students study wind, solar, geothermal, hydrothermal, tidal, or nuclear power, as well as hydropower.

Discuss

Ask students where they would put themselves in Figure 25.12. (People would occupy a position similar to that occupied by the rabbit in the figure.)

Reteach

You may wish to review the oxygen–carbon dioxide cycle by asking students the following questions:

▶ What is the source of the carbon in the carbon dioxide you exhale? (This carbon comes from other organisms, both plant and animal, that people use for food.)

▶ Name the process that produces carbon dioxide and state where in the body that process takes place. (The process is respiration and it takes place in cells.)

▶ Where does the oxygen in carbon dioxide come from? (The oxygen comes from the air. People take oxygen into their bodies as they breathe, and it is used by the body during respiration.)

Figure 25.12 The Oxygen–Carbon Dioxide Cycle

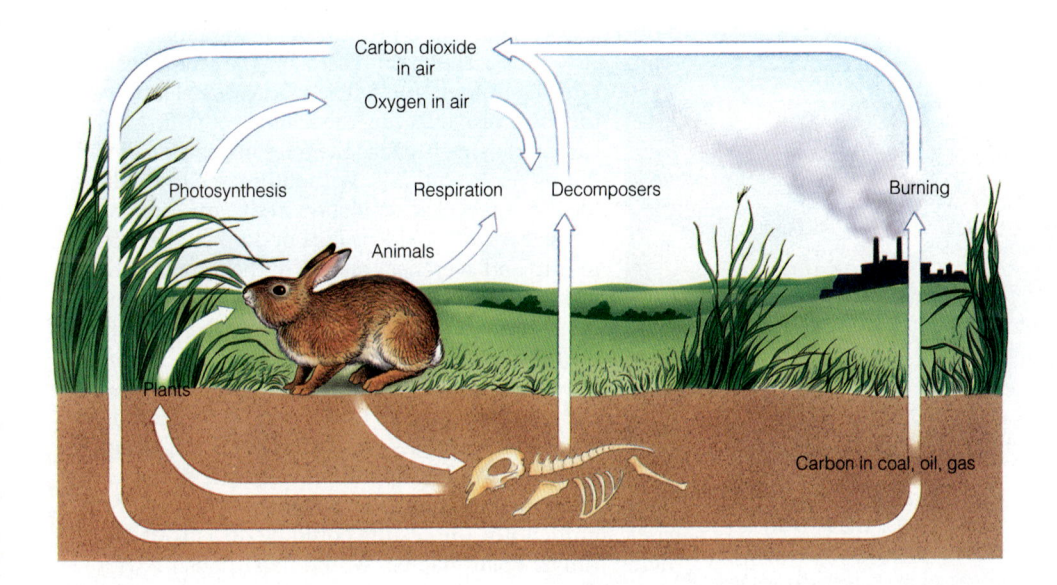

The Oxygen–Carbon Dioxide Cycle

Take a deep breath, then slowly exhale. You have just contributed to the oxygen–carbon dioxide cycle. When you inhaled, you took in air that contained oxygen. What happened when you exhaled? You released carbon dioxide back into the atmosphere.

Look at Figure 25.12. It shows the oxygen–carbon dioxide cycle. During the process of photosynthesis, plants absorb carbon dioxide from the air and release oxygen into the air. Plants are a main source of oxygen in the earth's atmosphere.

There is more than enough oxygen available in land environments. Oxygen becomes scarce at high altitudes. Here the air is thin, so there is less available oxygen. Oxygen is also scarce deep in the ground or in water-soaked soil.

Most organisms use oxygen from the air for the process of respiration. During respiration, food is broken down. Organisms get their energy from respiration and release carbon dioxide in the process.

Another way in which carbon dioxide enters the atmosphere and becomes part of the cycle is through the decay of dead organisms. Dead organisms contain carbon compounds. When decomposers break down the tissues of dead organisms, carbon dioxide is released into the air.

Sometimes the bodies of dead organisms do not decompose. Over thousands of years, the bodies are compressed underground. Millions of years later, the bodies are changed into oil, coal, and gas. People burn these fossil fuels for their energy needs. When they are burned, the fuels release carbon dioxide into the atmosphere.

The Living Textbook: Life Science Sides 7-8

Chapter 16   Frame 01654
Cycles Diagrams (4 Frames)
Search: Step:

Discuss

Ask students how recycling newspapers can reduce the amount of carbon dioxide in the atmosphere. (Carbon dioxide buildup is complicated, but trees take in carbon dioxide from the atmosphere and produce oxygen during photosynthesis.)

EVALUATE

WrapUp

Portfolio Ask students to write a narrative of a day in the life of an organism in its environment. The narrative should describe how an organism contributes to the cycles of nature and depends on them to fulfill its needs. Students may wish to include their essays in their portfolios.

Use Review Worksheet 25.3.

Check and Explain

1. A natural cycle that occurs as a substance circulates through living and nonliving things to be used and reused.
2. Organisms require food, air, and water in order to thrive. These substances are obtained by organisms through natural cycles.
3. Soil can be fertilized with animal wastes and the planting of legumes.
4. Students' models may vary but should include the three main stages of the water cycle.

Decision Making

If you have classroom sets of *One-Minute Readings,* have students read Issue 11, "Photosynthesis: Greenhouse Effect" on page 19. Discuss the questions.

552

INTEGRATED LEARNING

Geography Connection

Have students use an atlas to find the elevations of several coastal areas. Have them trace a map and use crayons or colored pencils to show where the U.S. coastlines would change if polar-glacier melting occurred. Have students choose an area, possibly a coastal area near where they live, and write a sentence or two describing how the ecosystems in that area might be affected by polar-glacier melting.

Figure 25.13 ▲
During the Gulf War in 1991, the smoke from huge oil fires in Kuwait added carbon dioxide to the atmosphere.

Science and Society *Global Warming*

Scientists have found that the earth's atmosphere holds nearly 25 percent more carbon dioxide than it did in the early 1900s. The main reason for the buildup is the burning of fossil fuels. The worldwide clearing of forests is also affecting the balance of atmospheric carbon dioxide because the trees use carbon dioxide during photosynthesis. Without trees, carbon dioxide is not recycled, so it stays in the atmosphere.

As carbon dioxide builds up in the air, it acts to trap and keep heat energy in the atmosphere. Although there is disagreement about how much the buildup of heat is actually affecting the earth, some scientists predict that the earth will eventually become much warmer than it is now. The term *global warming* refers to the worldwide warming trend that scientists are currently investigating.

What would happen if the earth's average temperature increased by 3°C? If the earth's temperature rises significantly, the results could be disastrous for living things. Polar glaciers would melt, causing widespread flooding in coastal areas. Climates around the world would change. Many agriculture regions could become too hot and dry to grow crops.

Scientists are currently studying ways to slow down the buildup of carbon dioxide in the atmosphere and reduce the risk of global warming. One way you can help is to recycle newspapers. Recycling will help preserve trees.

Check and Explain

1. What is a natural cycle?
2. Why are natural cycles important to organisms?
3. **Infer** Some farmers wanted their soil to be richer in nitrogen, but they didn't want to add chemicals. How can they increase the amount of nitrogen in the soil?
4. **Make a Model** Draw a model of the water cycle. Include any or all of the following: plants, trees, animals, clouds, soil, ponds, groundwater, mountains, rain, or snow. Also, label the processes of evaporation, condensation, and precipitation.

552   Chapter 25   Organisms and Their Environment

Themes in Science
Patterns of Change/Cycles
Ecosystems change in ways that can be predicted. In the process of succession, each community affects the environment, sometimes creating the conditions that result in its own replacement.

SECTION 25.4

Section Objectives
For a list of section objectives, see the Student Edition page.

Skills Objectives
Students should be able to:

Predict how a lack of water would affect an ecosystem.

Predict how forest ecosystems would be changed by a fire.

Vocabulary
succession

25.4 Changes in Ecosystems

Objectives

▶ **Describe** the stages of succession in nature.

▶ **Explain** how a pioneer community differs from a climax community.

▶ **Predict** how natural disasters can affect an ecosystem.

Skills WarmUp

Predict Imagine a shallow pond that is about the length of your classroom. Predict what might happen to the pond ecosystem if it eventually dried up. What kind of plants might grow there? What animals might live there?

What changes have occurred in your neighborhood within the past year? You may have noticed that a new house or building was built. You may have noticed that a road was recently paved, or that a new stop sign was placed at a busy intersection.

Ecosystems in nature also change over time. Some changes are hardly noticeable. Other changes have major effects on an ecosystem. In some cases, entire communities within the ecosystem are replaced.

Succession

If you have been to a forest, you may have wondered what the area looked like before the trees and other plants grew. You may be surprised to know that hundreds of years ago, the forest may have been a lake. The communities that once made up the lake ecosystem were gradually replaced by other communities that formed a forest ecosystem. The replacement of one community by another over a period of time is called **succession**. The complete process of succession may take hundreds, even thousands, of years.

In the process of succession, communities will grow and replace one another until a climax community forms. A climax community is a stable, almost permanent community that is not easily replaced by other communities. Once the climax community becomes established, fewer changes occur in the area. However, disturbances such as fire, volcanic eruptions, severe weather, or some human activities can change an ecosystem or destroy a climax community.

On page 554, you can follow the formation of a climax community.

Figure 25.14 ▲
Mt. St. Helens in 1981 (top) and in 1990 (bottom). How has Mt. St. Helens changed? What caused the changes? ①

MOTIVATE

Skills WarmUp
To help students understand how changes in an environment affect the populations living there, have them do the Skills WarmUp.
Answer Answers may vary. Populations of plants and animals that thrive in water would be replaced by those that thrive on land.

Prior Knowledge
To gauge how much students know about the effects of change on ecosystems, ask all or some of the following questions:

▶ What changes can you observe in your area every day?

▶ What changes can you notice at different times of the year?

▶ Are there changes you notice that are different from daily or seasonal changes?

Answer to In-Text Question
① The bottom picture shows plants and other organisms that have grown on the side of the mountain. The eruption of the volcano first killed organisms. After the volcanic ash cooled, it enriched the soil.

Chapter 25 Organisms and Their Environment

TEACH

Directed Inquiry

Have students study the pictures and descriptions on this page. Ask some or all of the following questions:

▶ What are the sizes of the first organisms in a newly formed lake? Where do these organisms come from? (The sediments that flow with rivers into the new lake include the spores of small bacteria, fungi, and protozoa.)

▶ Which organisms come first—the plants or the animals? Explain. (The plants had to come first because they are the producers upon which the consumers feed.)

▶ What causes the lake to be filled in over time? (Soil, fallen leaves, and material from dead organisms slowly fill up the lake and cause it to become a marsh.)

▶ What is the climax community in the area where you live? (Answers may vary, but it is likely that the climax community is a forest in most areas.)

Reteach

Have students fold a sheet of paper lengthwise to make four columns. Ask them to write at the top of each column the titles Pioneer Community, Lake Community, Marsh, and Climax Community. Have students list the organisms living in each community.

The Living Textbook: Life Science Sides 7-8

Chapter 16   Frame 01715
Succession (28 Frames)
Search: Step:

INTEGRATED LEARNING

Art Connection

Around Lake Michigan there are areas where several stages of succession in a sand dune environment can be observed as a person walks from the water's edge, across the dunes, and to the forest. Explain to students that closest to the lake shore is sandy area. There is usually no life on the lower beach because of pounding waves. The middle and upper beaches have sparse vegetation and few pieces of driftwood. The first sand dunes mark the line between the beach and the pioneer community, which is covered by grasses. Grasshoppers, spiders, and decomposers also live in the beach grass community. The next stage is the shrub community, which gives way to the pine woods community, which gives

---

Pioneer Community At first, the newly formed lake has no living things. In time, sediments are carried into the lake by rivers that feed it. In the sediment are airborne spores of algae, bacteria, fungi, and protozoa. These organisms form a pioneer community because they are the first to settle in the area.

Lake Community The algae and other organisms add nutrients, such as carbon, to the lake. The lake can now support small plants. Insects that eat the plants enter the lake ecosystem. A variety of plants eventually grow in and around the lake. Different kinds of animals that eat plants and insects become part of the lake community.

Marsh Ducks, dragonflies, mosquitoes, cattails, raccoons, and frogs are among the many organisms that live and interact in the lake community. As the lake ages, it slowly fills with soil, fallen leaves, and material from dead organisms. As these materials fill up the lake's bottom, it becomes shallower. The lake gradually turns into a marsh.

Climax Community As succession continues, the marsh dries up. Land plants, such as pine trees or oak trees, develop in the area. Over time, the pine or oak trees are replaced by larger trees, such as spruce or birch. The spruce–birch forest and all the populations of plants and animals living in the forest form a stable climax community.

way to the climax community, a woodland area where beech and maple trees grow. Provide art materials and have students make murals showing the stages of succession along Lake Michigan.

Integrating the Sciences

Botany As students do research to prepare for writing their papers for the Consider This feature, have them consider tree farms. Students should discover how long it takes to grow harvestable trees. A phone call to a branch of the United States Forest Service could provide some information.

Research

Explain to students that often animals that leave an area when a city is built return later. Two animals that have returned to city environments are the coyote and the peregrine falcon. In New York City, peregrine falcons have built nests inside the letters on signs outside buildings. Have students find out about one animal that has become part of the city community. Have them write a brief description of the animal and how it survives in a city environment.

Science and Society *City Succession*

Succession is a natural process; however, succession can also be caused by human actions. When people settle and develop an area, many changes take place within the communities. For example, when a city is built, the natural vegetation in the area is cleared away. Concrete structures and increasing populations of people gradually take over the area. In this process, a new environment is created.

To create a new city environment, the original topsoil is often bulldozed away. Much of the land is then covered with concrete and asphalt. People replace native plants with ornamental plants. When you picture plants in a city, you may think of tree-lined streets, green parks, and planter boxes of flowers. This introduced vegetation helps create the city environment.

Just as the vegetation changes, so does the animal population. A city environment, with its buildings, people, and roadways, is not suited for most native animals, such

Consider This

Activity Recycling paper is used as a way of limiting the number of trees that people cut down for paper. Give students an opportunity to become familiar with the recycling process.

Think About It Students should think about the long- and short-term effects of logging.

Write About It Encourage students to cite the arguments opposed to their opinions and to counter these arguments with facts.

Consider This

Should Climax Forests Be Logged?

The logging of trees is a major industry in the United States. Trees provide important products for people. However, the number of trees in the United States is not endless. In fact, the Wilderness Society estimates that 95 percent of the United States' original forests have been logged.

Consider Some Issues Many populations and communities live in forests. The roots of trees help control flooding and prevent landslides. Trees are also important in the oxygen–carbon dioxide cycle. In addition to their role in nature, forests are valued for their beauty.

Trees are considered one of the United States' most valuable natural resources. The wood from trees has provided many products, services, and jobs for people. If logging is stopped, jobs would be lost, and the prices for houses, paper, and furniture would increase.

Think About It Should trees in climax forests be cut down? What might happen to the organisms living in climax forests if logging continues? If forests are no longer logged in the United States, how will this affect people who live in areas where logging is a major industry?

Write About It Write a paper stating your position for or against the logging of trees in the United States. Write your reasons for choosing your position.

The Living Textbook: Life Science Side 8

Chapter 44   Frame 39407
Tropical Rainforest (Movie)
Search: Play:

Chapter 25   Organisms and Their Environment

TEACH ▪ Continued

Discuss
Ask students to name animals that they think would thrive in a city environment. Have them suggest what parts of the city environment help these animals meet their needs.

EVALUATE

WrapUp
Reinforce Have students work in pairs to research how succession has been affected by recent environmental disturbances. Encourage students to research events such as the 1988 fires in Yellowstone National Park, the eruption of Mt. St. Helens in 1980, and the Alaskan oil spill in 1989.
Use Review Worksheet 25.4.

Check and Explain

1. In the process of succession, communities grow and replace one another until a climax community forms.
2. A pioneer community will always be replaced by another community. A climax community is stable.
3. Answers may vary. The vegetation develops first, providing the environment that draws and sustains the animal populations.
4. Answers may vary. Students should include a general description of how the area will develop new growth typical of a pioneer community with few organisms and then follow a pattern of natural succession.

Answer to In-Text Question

① Animals who benefit from human activity thrive well in a city. These include pigeons, barn swallows, chimney swifts, and house sparrows.

556

TEACHING OPTIONS

Cooperative Learning
Have students work in small groups to present a TV documentary about logging climax forests. Have students meet in groups of four and choose a particular forest to feature. One or two students could do research, another could write a script, and another draw charts or pictures. One of the students can take the role of TV announcer, and others can be guests who are interviewed.

Figure 25.15 ▶
A city is an example of succession caused by people. Why do only certain native animals survive in a city? ①

as deer, rabbits, and snakes. They cannot survive in a city, so they migrate to more suitable places, or die off. Other animals—that benefit from human activity—move in. For example, pigeons thrive well because there are few predators. There is also a large supply of food from human scraps. Barn swallows, chimney swifts, and house sparrows often make their nests on building eaves and road overpasses. Bird nests and bird feeders placed near homes also extend the habitat range of many species.

Some mammals migrate to city environments. In the city, rats and mice find an abundance of waste food. They can nest in and under buildings. Skunks, possums, and raccoons can adapt to environments where people live.

Check and Explain

1. Briefly explain the process of succession.
2. How is a pioneer community different from a climax community?
3. **Find Causes** Why do you think climax communities are identified by their plants rather than by the animals that live in them?
4. **Predict** A forest was struck by lightning and a large fire resulted. The trees, bushes, and grasses were all burned. Predict how the area will change. What kinds of communities and populations would you expect to find in the area in the future?

556 Chapter 25 Organisms and Their Environment

Chapter 25 Review

Concept Summary

25.1 Ecosystems and Communities
▶ The biosphere is a zone near the earth's surface that supports all organisms on the earth.
▶ Organisms interact with each other and the environment in an ecosystem.
▶ Ecosystems have biotic and abiotic factors.
▶ Within each ecosystem, each organism has a habitat and a specific niche within its habitat. Adaptations help determine an organism's niche.

25.2 Food and Energy
▶ The food chain shows how food energy moves from one organism to another within an ecosystem.
▶ Organisms have a role in the food chain as producers, consumers, or decomposers. Food chains connect together to form feeding relationships, or food webs.
▶ Energy pyramids show how much energy is available at each level of the food chain. At each level in a food chain, energy is lost.

25.3 Cycles in an Ecosystem
▶ Natural cycles include the water cycle, the oxygen–carbon dioxide cycle, and the nitrogen cycle. They allow organisms to live, grow, and reproduce.
▶ The water cycle has three stages: precipitation, condensation, and evaporation.

25.4 Changes in Ecosystems
▶ Communities succeed each other, beginning with a pioneer community and ending with a climax community.

Chapter Vocabulary

ecosystem (25.1)	population (25.1)	consumer (25.2)	condensation (25.3)
biotic factor (25.1)	niche (25.1)	food web (25.2)	precipitation (25.3)
abiotic factor (25.1)	food chain (25.2)	energy pyramid (25.2)	nitrogen fixation (25.3)
community (25.1)	producer (25.2)	evaporation (25.3)	succession (25.4)

Check Your Vocabulary

Use the vocabulary words above to complete the following sentences correctly.

1. Plants are ____ because they can fuel their own growth.
2. Rain and snow are forms of ____.
3. The specific role an organism has in its habitat is called its ____.
4. The process of changing atmospheric nitrogen into usable compounds is called ____.
5. A cat eats a sparrow that ate a beetle that ate a blade of grass. The cat is a ____ in the food chain.
6. The replacement of communities over a period of time is called ____.

Explain the difference between the words in each pair.

7. pioneer community, climax community
8. condensation, evaporation
9. habitat, niche
10. scavenger, decomposer
11. community, population
12. biotic factor, abiotic factor
13. ecosystem, ecology
14. food web, food chain

CHAPTER REVIEW 25

Check Your Vocabulary

1. producers
2. precipitation
3. niche
4. nitrogen fixation
5. consumer
6. succession
7. A pioneer community is made up of organisms that settle new ground. A climax community is the final stage of succession.
8. Evaporation: the change from the liquid to the gaseous state; condensation: gas to liquid.
9. An organism's habitat is the physical area in which it lives; its niche is the role it plays in its environment.
10. A scavenger is an animal that feeds off the tissues of dead organisms; a decomposer is a microorganism that does the same.
11. A community is the entire collection of organisms, of all species, in a geographical area; a population is the group of organisms of a particular species in an area.
12. Biotic factors are living parts of an ecosystem; abiotic factors are nonliving parts.
13. An area in which living things interact with one another and with the environment is an ecosystem; ecology is the name for the study of living things and their ecosystems.
14. A food chain is a way of representing feeding relationships among organisms; a food web represents the combination of many food chains.

Use Vocabulary Worksheet for Chapter 25.

CHAPTER 25 REVIEW

Check Your Knowledge

1. A habitat is the surroundings in which an organism lives.
2. Answers may vary; examples include fire, volcanic eruptions, severe weather, human activities.
3. Answers may vary; examples include soil, sunlight, water, air.
4. Plants, which are involved in the process of photosynthesis, are a main source of oxygen in the oxygen–carbon dioxide cycle.
5. Organism, population, community, ecosystem, biosphere
6. Answers may vary; examples include soil bacteria, molds.
7. Answers may vary; examples include long, slender bill and neck that can blend in with tall marsh plants; long legs to stand in shallow water; webbed feet to keep from sinking in soft mud.
8. Producers get their energy from the sun.
9. Evaporation of liquid water into water vapor; condensation of water vapor into clouds; precipitation as rain, snow, hail, and sleet; flow of precipitation into larger bodies of water
10. False; Nitrogen
11. False; first (or base)
12. True
13. False; consumer
14. True
15. False; climax
16. False; condensation

Check Your Understanding

1. Answers may vary. Possible answers include farming, forestry, real estate development, pollution.
2. Answers may vary. Possible answers include abiotic factors—sunlight, climate, water, buildings, roads; biotic factors—humans, birds, rodents, insects, pet animals.
3. Answers may vary. Abiotic factors may include water—temperature, mineral content; pebbles, rocks, or other objects in tank; sunlight. Natural cycles: oxygen–carbon dioxide (if aquatic plants are present), water (if tank has a lid)
4. In a field, grasses give way to pine trees, which are eventually replaced by hardwood forests.
5. a. corn —> chicken —> human
   b. leaf —> beetle —> spider —> owl
   Third-level consumer: owl
   Second-level consumers: human, spider
   First-level consumers: chicken, beetle
   Producers: corn, leaf
6. The bittern is a second- or third-level consumer. Its sharp beak is adapted for spearing fish and its body shape is adapted to blend into its surroundings.
7. Answers may vary. Examples follow:
   25.1: All the living things and nonliving things in an organism's surrounding environment influence that organism and its role in its surroundings.
   25.2: All the energy available to living things comes, initially, from the sun and is transferred from one organism to another along a food chain.
   25.3: The substances necessary for life (water, nitrogen, oxygen, and carbon dioxide) are continuously cycled through the environment and the organisms living in it, insuring that an adequate supply is available.
   25.4: Succession is the process by which, due to natural and human factors, an ecosystem changes and certain species appear while others disappear.
8. Accept all logical answers.

Chapter 25 Review

Check Your Knowledge
Answer each of the following in complete sentences.

1. What is a habitat?
2. Name three ways a climax community can be destroyed.
3. List three abiotic factors.
4. What is the main source of oxygen in the oxygen–carbon dioxide cycle?
5. List the following in order from smallest to largest: community, population, organism, ecosystem, biosphere.
6. Give one example of a decomposer.
7. What adaptations have helped the heron to establish a niche in the wetlands?
8. Where do the producers in a food chain get their energy?
9. List the stages of the water cycle.

Determine whether each statement is true or false. Write *true* if it is true. If it is false, change the underlined term to make the statement true.

10. <u>Potassium</u> is a necessary element for organisms because it helps build proteins.
11. Most energy is stored at the <u>third level</u> of the energy pyramid.
12. A heron's sharp beak and long legs are <u>adaptations</u>.
13. A rabbit is an example of a <u>producer</u>.
14. Nitrogen is fixed through the action of <u>lightning</u> and bacteria.
15. A <u>pioneer</u> community is a stable, almost permanent community.
16. Clouds, steam, and fog are forms of <u>precipitation</u>.

Check Your Understanding
Apply the concepts you have learned to answer each question.

1. Describe three human activities that might disturb a balanced ecosystem.
2. Describe a city environment. Include biotic and abiotic factors.
3. Design a fish tank. What animals and plants will you include? What abiotic factors will be part of the fish tank? What natural cycles will take place in your fish tank?
4. What specific changes in a field mark the succession of a pioneer community as it eventually changes to a climax community?
5. **Extension** Draw food chains using the following:
   a. corn, human, chicken
   b. beetle, spider, leaf, owl
   Now create an energy pyramid using the items in the food chains above. Identify each organism as a producer or a consumer.
6. **Mystery Photo** The photograph on page 536 shows a bittern, a close relative of the heron. The bittern is difficult to see in the tall marsh reeds because of its coloring and tall, slim shape. Bitterns feed on fish and small animals in the marsh. What kind of consumer is a bittern? How is it adapted to life in a marsh?
7. **Main Ideas** Using the outline on page 536, write the main idea for each section in the chapter.
8. **Application** Identify four factors that could limit the number of flowers growing in a garden. Describe how each factor might limit the population growth.

Develop Your Skills

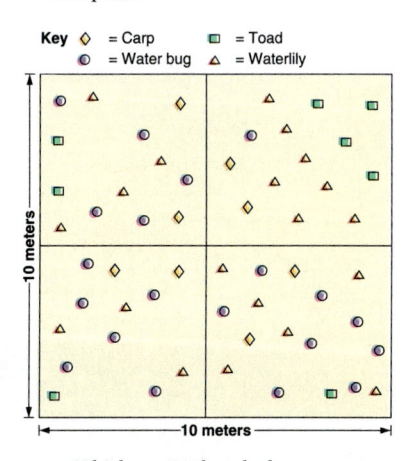

1. Answers may vary. Student diagrams should show the following: The pine tree, blueberry bush, and oak tree are producers; the caterpillar, squirrel, and finch all eat the producers and the finch may also eat the caterpillar; the hawk and the mountain lion eat the smaller animals.
2. a. waterlily and water bug
   b. Answers may vary; however, each species will be influenced by biotic factors, especially other organisms with which each competes for space, food, and mates; and by abiotic factors, which include size and depth of pond and amount of sunlight.
3. The ocean supplies 84 percent of the water evaporated into the atmosphere. Land supplies 16 percent.

Make Connections

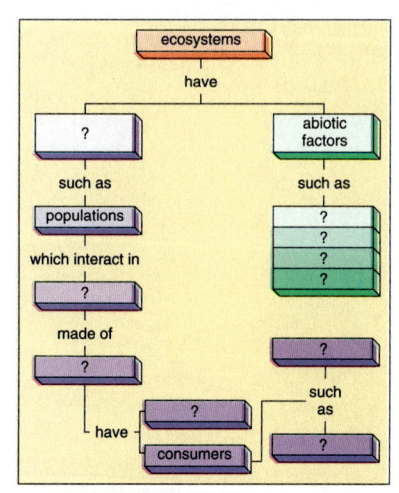

1.

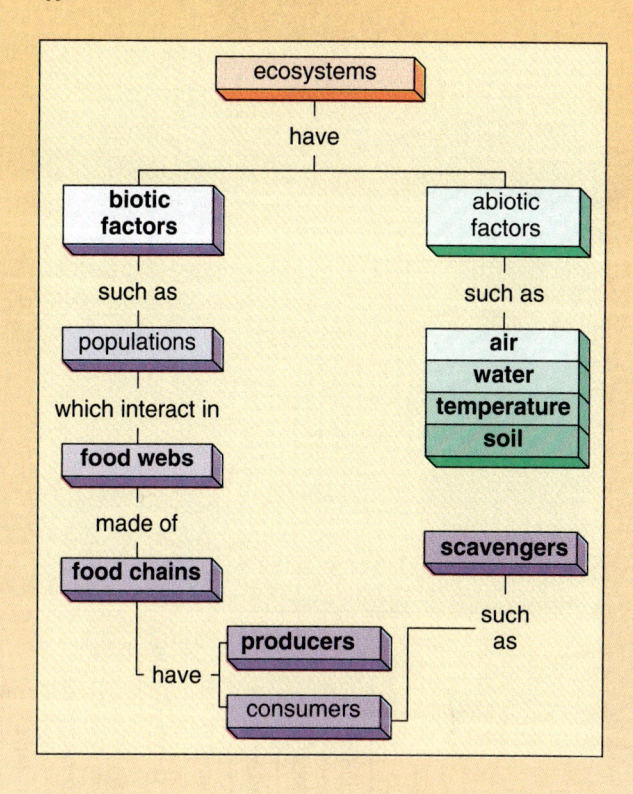

```
                    ecosystems
                        |
                       have
              _____|_____
             |                     |
       biotic factors         abiotic factors
             |                     |
          such as               such as
             |                     |
        populations               air
             |                   water
      which interact in       temperature
             |                    soil
        food webs
             |
          made of
             |                  scavengers
        food chains                |
             |                  such as
           have _____     |
                           |      |
                       producers
                       consumers
```

2. Answers will vary. Student plays should include the biotic and abiotic roles mentioned in the question.
3. Answers may vary. Alternative power sources include methanol, geothermal, and nuclear power.

CHAPTER 26

Overview

This chapter extends the discussion of organisms and how they interact and function in an ecosystem. The first section describes the factors that affect populations of organisms. The second section presents relationships that maintain balance in an ecosystem, including symbiosis and parasitism. The third section explains behaviors that enable animals to function in their environment.

Advance Planner

▶ Collect magazines for students to use in making collages.

Skills Development Chart

Sections	Calculate	Classify	Collect Data	Communicate	Decision Making	Infer	Interpret Data	Measure	Predict	Research
26.1 Skills WarmUp Consider This Skills WorkOut Activity			●	● ● ●	●		 ●	 ●	●	
26.2 Skills WarmUp Skills WorkOut		●				●				
26.3 Skills WarmUp Skills WorkOut				●						●

Individual Needs

▶ **Limited English Proficiency Students** Have students divide their science journal pages into three columns. Ask them to write all of the chapter terms in the first column, leaving several blank lines between each word. Have students write definitions in the second column. Then ask students to find pictures that represent their words in the third column. Have them meet in small groups to discuss how the pictures they chose describe the terms.

▶ **At-Risk Students** Have students use pictures from magazines and catalogs to create a collage of the chapter terms. Encourage students to create one unified picture that includes an example of each term. They should write the definitions on the borders of their collages.

▶ **Gifted Students** Have students conduct in-depth research of the complex social interactions of such animal groups as apes, chimpanzees, or wolves. They can include with their reports the interactions of their animal groups of choice in detailed paintings or drawings.

Resource Bank

▶ **Bulletin Board** Have students write haiku poems describing *mutualism*, *parasitism*, and *commensalism*. Then have students exchange poems and find or make a drawing for their classmate's poem. Display these poems and drawings on the bulletin board. You might title your bulletin board display *Symbiosis*.

CHAPTER 26 PLANNING GUIDE

Section	Core	Standard	Enriched	Section	Core	Standard	Enriched
26.1 Changes in Populations pp. 561–566				**Blackline Masters** Review Worksheet 26.2 Skills Worksheet 26.2	● ●	● ●	
Section Features Skills WarmUp, p. 561 Consider This, p. 564 Skills WorkOut, p. 565 Activity, p. 566	● ● ● ●	● ● ● ●	● ●	**Laboratory Program** Investigation 48	●	●	●
				26.3 Animal Behavior pp. 571–574			
Blackline Masters Review Worksheet 26.1 Skills Worksheet 26.1	● ●	● ●	●	**Section Features** Skills WarmUp, p. 571 Career Corner, p. 572 Skills WorkOut, p. 574	● ● ●	● ● ●	● ●
Ancillary Options One-Minute Readings, p. 27	●	●	●	**Blackline Masters** Integrating Worksheet 26.3a Integrating Worksheet 26.3b Review Worksheet 26.3 Enrich Worksheet 26.3 Vocabulary Worksheet 26	● ● ● ●	● ● ● ● ●	● ● ●
Overhead Blackline Transparencies Overhead Blackline Master 26.1 and Student Worksheet	●	●	●				
26.2 Relationships Among Populations pp. 567–570				**Color Transparencies** Transparency 82	●	●	
Section Features Skills WarmUp, p. 567 Skills WorkOut, p. 570	● ●	● ●	●				

Bibliography

The following resources can be used for teaching the chapter. See page T-40 for supplier codes.

Audio-Visual Sources
(video unless noted)
Adaptation to Environment. 16 min. 1983. LF.
Animal Camouflage. 9 min film. 1977. BFA.
Animal Populations: Nature's Checks and Balances. 22 min. 1983. EB.
Canyon Creatures. 26 min. 1984. MSP.
Konrad Lorenz: Science of Animal Behavior. 14 min. 1975. NGSES.
Predators of North America. 12 min. 1981. NGSES.

Software Resources
Balance. Apple, IBM PC. DEE.
Coexist. Apple. CO.
Ecological Modeling. Apple, IBM PC. CO.

Library Resources
Bendick, Jeanne. Ecology: Science Experiences. New York: Franklin Watts, Inc., 1975.
Graham, Ada and Frank. The Changing Desert. San Francisco: Sierra Club Books, 1981.
Rickleff, R. E. The Economy of Nature. 2d ed. New York: Chiron Press, 1983.
Van Lawick, Hugo. Among Predators and Prey. San Francisco: Sierra Club Books, 1986.

CHAPTER 26

INTEGRATED LEARNING

Writing Connection

After students have studied the photograph, have them read the chapter title and consider what the chapter is about. Then ask them to think about an interaction between two or more organisms and write a description of it. For example, the interaction could be between a human and a cat, between a herd of musk-oxen and a wolf pack, or between a cardinal and a maple tree. Students may use any example that involves an interaction between organisms. Encourage them to consider whether or not the organisms help each other.

Introducing the Chapter

Have students read the description of the photograph on page 560. Ask if they agree or disagree with the description. Explain to students that the sea anemone is *not* eating the fish.

Directed Inquiry

Have students study the photograph. Ask:

▶ What organisms are in this photograph? (Students will probably recognize the sea anemone and fish.)

▶ How does the picture relate to the topic of this chapter? (The organisms are interacting.)

▶ Do you think that this fish is safe or in danger? (This fish is safe.)

▶ How might the fish be protected by the sea anemone? (The anemone might hide the fish or sting and kill its enemies.)

▶ How might the fish be helpful to the anemone? (The fish might drop some of its food for the anemone to eat, attract prey that the anemone could eat, or drive off the anemone's enemies.)

Chapter Vocabulary

innate behavior
limiting factors
population density
predator
prey
symbiosis

Chapter 26 Interactions Among Organisms

Chapter Sections

26.1 Changes in Populations

26.2 Relationships Among Populations

26.3 Animal Behavior

What living things do you see?

❝ I see a fish and a sea anemone. I think the fish is either hiding from another fish or is being eaten by the sea anemone. I think they swim and eat together. I think their relationship is pretty simple. ❞

Kelly Reed
Supai Middle School
Scottsdale, Arizona

To find out more about the photograph, look on page 576. As you read this chapter, you will learn more about relationships among organisms.

STS Connection

The Japanese serow (mountain goat), once a threatened species, has been granted protected status by the Japanese government. The species is now thriving, but population sizes have swelled to the point that the serows have become a problem for farmers, who find their fences broken and their crops damaged by the serows. Have students write one or two sentences about how the problem might be resolved so that both serows and farmers benefit. (Students may say that controlled hunting of serows would help reduce their population. Another alternative would be to relocate some of the serows to another area.)

26.1 Changes in Populations

Objectives

▶ **Describe** the factors that affect population growth and size.

▶ **Classify** organisms as predators or prey.

▶ **Communicate** how limiting factors and population density are related.

Skills WarmUp

Predict An area is set aside as a park preserve, and no hunting is allowed in the park. A number of deer are placed in the area. There are plenty of plants to serve as food for the deer population. The herd is healthy and begins to grow faster than expected. Predict what will happen if the deer population continues to increase in the park area.

Picture yourself walking in a dense jungle. How many different kinds of organisms do you see? A jungle has many populations, or groups of similar organisms. Some populations, such as ants, fungi, and ferns, can be very large in number. Other populations, such as jaguars and boa constrictors, have fewer members. Why do population sizes vary among organisms?

Population Size

Population sizes change when new members move into an ecosystem. They decrease when members move out of an ecosystem. The number of births, or birth rate, and the number of deaths, or death rate, also affect a population's size. For example, on an ocean island, few new members enter the isolated populations. Few members leave the island. Therefore, if more births than deaths occur in a population from year to year, the population will increase. If there are more deaths than births, the population will decrease. If the birth rate and death rate are equal, the population remains steady.

Ecologists study the birth rates and death rates of populations so they can measure population growth. They can then predict the future of a population, including whether or not it is in danger of becoming extinct. For example, if the size of the caribou population shown in Figure 26.1 suddenly decreased, ecologists would try to find out the cause. They might find that a certain disease was affecting the herd. Steps could then be taken to ensure the survival of the herd.

Figure 26.1 ▲
The size of this population of caribou is influenced by many factors, including the number of births and deaths.

Chapter 26 Interactions Among Organisms 561

SECTION 26.1

Section Objectives
For a list of section objectives, see the Student Edition page.

Skills Objectives
Students should be able to:

Predict what will happen to a deer population that increases rapidly.

Collect Data to find the average household size.

Communicate how limiting factors and population density relate.

Interpret Data about population density.

Vocabulary
population density, limiting factors, predator, prey

MOTIVATE

Skills WarmUp
To help students understand changes in populations, have them do the Skills WarmUp.
Answer Some deer will either starve to death or be forced to search for food in a different location.

Prior Knowledge
To find out what students know about changes in populations, ask the following questions:

▶ What is a population?

▶ What might cause a population's size to change?

▶ How can predators help prey populations?

The Living Textbook:
Life Science Side 8

Chapter 32 Frame 16560
Population Control (Movie)
Search: Play:

561

TEACH

Skills Development

Infer Tell students that the term *population density* can refer to any kind of living population, including pine trees, robins, humans, mushrooms, bobcats, and bacteria. The population density of a species may vary greatly from place to place or from season to season. Have students name examples of factors that cause differences in seasonal and regional population density. (Weather, pollution, disease, availability of food, availability of space, and competition)

Directed Inquiry

Have students study the material on this page. Ask the following questions:

▶ What do animals in the same area compete for? (Food, water, living space, sunlight)

▶ How do some animals avoid competition? (Some can move to another area, eat different food, hunt at different times, live in a different niche of the same environment, and so on.)

Answer to In-Text Question

① The population density is 12.5 ants per sq m.

The Living Textbook: Life Science Side 8

Chapter 29   Frame 13453
Competition (Movie)
Search: Play:

The Living Textbook: Life Science Side 8

Chapter 34   Frame 20546
The Lemmings (Movie)
Search: Play:

562

INTEGRATED LEARNING

Themes in Science
Systems and Interactions

Point out that different species can avoid competition by occupying different niches within the same ecosystem. In a rain forest, some organisms live in the ground, on the ground, on the lower levels of the trees, and high up in the forest canopy. Have students paint a mural showing the zones in a rain forest, including animals that occupy each zone.

Math Connection

Have students work out the following problems:

▶ If 40 buffalo live in a 1 1/2 sq km area, what is their population density per sq km? (about 27 per sq km)

▶ How many squirrels would have to live in a 3/4 sq km area to have the same population density as the buffalo? (20)

Population Density

Populations can be the same size, but they may have different densities. The number of individuals in an area is called **population density**. Population density is calculated by dividing the number of organisms in an area by the size of the area in which they live. Suppose 60 ants live in a 4 sq m plot of grass. The population density of the ants would be 15 ants per sq m because 60 ants divided by 4 sq m equals 15. What would the population density be if 100 ants lived in an 8 sq m plot of grass? ①

Population density can affect the growth rate of a population. For example, if the population density of plant seedlings is too great, there may not be enough water to support the growth and development of all the seedlings. Many will not survive, thereby reducing the size of the population.

Limiting Factors

Certain environmental conditions called **limiting factors** keep a population from increasing in size and help balance ecosystems. Examples of limiting factors are the availability of food, water, and living space, and the spread of disease. Light, temperature, and types of soil are also limiting factors because they help determine the types of organisms that can live in an ecosystem.

Limiting factors are often related to population density. The greater the population density, the greater effect limiting factors have on a population. For example, plants may be a limiting factor to deer living in a forest. If the population of deer is too dense, there may not be enough food for each one. Some deer will not survive, and the population will decrease.

Competition

Populations can grow when their needs are met. But when many populations share an ecosystem, the resources needed by one population may be the same as those needed by other populations. When two or more species in an ecosystem must use the same limited resources, competition occurs.

Living space is a major cause of competition. All populations need a certain amount of living space to find food, to reproduce, and to interact. For example, two species of birds may compete for nesting sites in the same tree.

Because they can move about easily, animals often move to another habitat to avoid competition. Other animals, such as owls and hawks, adapt to the ecosystem by changing their feeding habits. Both owls and hawks feed on some of the same animals. To reduce competition, hawks hunt during the daytime, while owls hunt at night.

Figure 26.2 ▲
These two different species of birds are competing for the same food source.

562   Chapter 26   Interactions Among Organisms

Themes in Science

Stability and Equilibrium The populations of predators and their prey exist in a balance. Predators limit prey numbers. Conversely, the size of the prey population limits the predator population that can survive in an area. Both populations help keep the other healthy by removing the weak animals in each group. Weak or unhealthy prey are more likely to be caught by predators, while weak or unhealthy predators are more likely to die because they can't obtain food. Ask students to think of examples of predator-prey relationships.

Predators and Prey

Many interactions among organisms are related to their feeding patterns. For example, some animals obtain food by killing and eating other organisms. Animals that obtain their food in this way are called **predators**. Some common predators are wolves, lions, robins, and flesh-eating sharks. The organisms that are eaten by a predator are called **prey**.

Predator-prey relationships limit population density. In this way, they help maintain balance in an ecosystem. In fact, predators may actually be beneficial to populations of prey animals by removing members who are weak or sick. This allows only the stronger members of the prey population to survive and reproduce. Predator-prey relationships also prevent too large a population of prey organisms from developing. If there are not enough predators and too many prey, the prey population may starve or die of disease.

Look at the graph in Figure 26.3. It shows the changing populations of Canadian lynxes and snowshoe hares over an 80-year period. The lynx is the predator. The hare is the prey. As the number of hares increases, so does the number of lynxes. Then what happens? The increased number of lynxes causes a drop in the hare population. The lynxes then begin to starve to death, which allows the hare population to increase again.

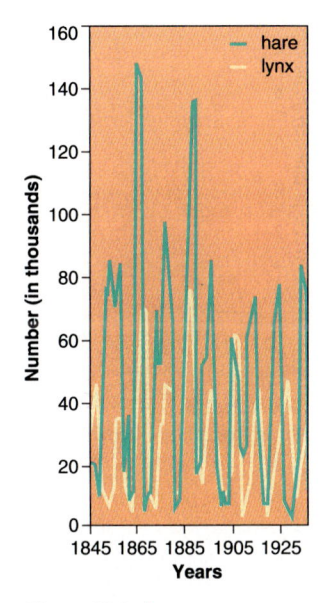

Figure 26.3 ▲
Lynx-Hare Population Cycle

◀ **Figure 26.4**
The predator-prey relationship of the Canadian lynx and snowshoe hare helps balance the sizes of both populations. What other factors might account for the changing population sizes of both animals? ②

Discuss
Explain to students that the size of the deer population in many areas of the United States has been increasing so rapidly that deer are often seen in urban areas. Many of these deer have no place to go and are killed by traffic. One reason for the increase in the deer population is the lack of a predator population in many areas.

Critical Thinking
Reason and Conclude Explain that human communities have been developed in many areas where deer and their predators once lived. Ask students what might be done to solve the problems arising from humans and deer occupying the same area. (Students should suggest ways to control the deer population and/or housing developments.)

Skills Development
Interpret Data Have students look at Figure 26.3. Ask them what determines the size of the predator population. (The size of the prey population)

Answer to In-Text Question
② Competition, pollution, disease, and the availability of food, water, and space

The Living Textbook:
Life Science Side 8

Chapter 36 Frame 25821
Predator and Prey On the Tundra (Movie)
Search: Play:

Chapter 26 Interactions Among Organisms

TEACHING OPTIONS

TEACH ▪ Continued

Skills Development

Infer The doubling rate for the human population used to be 1,500 years about 100 centuries ago. Ask students, Why did the human population grow very slowly until recently? (People didn't live long because diseases were widespread, food was not stored successfully, and clothing and shelter were often poor.)

Critical Thinking

Find Causes Ask students, What factors have caused the population explosion in recent decades? (Students may mention innovations such as modern medicine, refrigeration, improved sanitation, and so on.)

Consider This

Research Explain to students that many of the poorest countries in the world have growth rates that are much higher than those of industrialized nations such as Switzerland, Germany, Japan, Canada, and the United States. Have students find out which countries have the highest growth rates and why. Students should use this information as they evaluate whether births should be restricted.

Decision Making

If you have classroom sets of *One-Minute Readings,* have students read Issue 16, "U.S. Birthrate" on page 27. Discuss the questions.

Cooperative Learning

Emphasize the fact that human overpopulation is the major cause of many of the world's most serious problems. It leads to food shortages, energy shortages, the destruction of plant and animal habitats, species extinctions, pollution, overcrowding, and so on. Divide the class into groups of four and have each group discuss how human overpopulation is affecting the environment and quality of life for humans and other living things all over the world. Have each group choose one problem associated with overpopulation and find out more about it. Have each group make a presentation to the class about its topic.

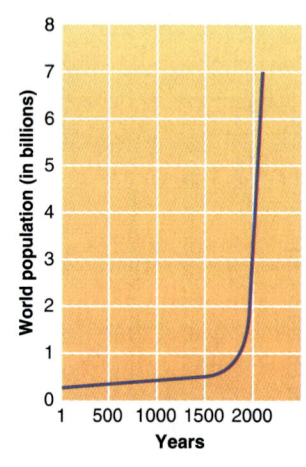

Figure 26.5 ▲
World Population Growth

Science and Society
Human Population Growth

If all the people living on the earth were lined up side by side, they would produce a line longer than the distance from the earth to the moon and back! Although the human population started to grow at a slow rate, it has grown very rapidly in the last two centuries. Today more than five billion people live on the earth.

There are different ways to measure human population growth rates. One way is to find the population's doubling rate, or the length of time it takes for the population to double in size. In 8000 B.C., the doubling rate was 1,500 years. In 1975, the doubling rate was estimated to be 35 years. If this rate is accurate, there could be 10 billion people on the earth by the year 2010.

When the growth pattern for human population is graphed, it produces a J-shaped curve. Look at Figure 26.5. Notice that the curve starts slowly. Then, when the bend

---

Consider This

Should the World's Population Be Limited?

Can the earth's environment handle as many as 10 billion people in the future? Many people fear that it can't and have suggested that governments impose limits on family sizes. They feel that reducing birth rates would help relieve the strain on the environment.

As shown in the graph at right, if each family in the United States only had two children, instead of three, for two or three generations, the population of the United States would stay at today's level. This concept is called *zero population growth.*

Consider Some Issues Large families can be a source of wealth in some cultures. Young people take care of the old, and young workers strengthen a nation's economy.

Smaller families could have a better standard of living. There would be less disease and hunger. Fewer people would put less strain on the environment.

Think About It How will the earth be affected if its population continues to grow as predicted? How can population growth be controlled? Should governments be involved in determining the sizes of families?

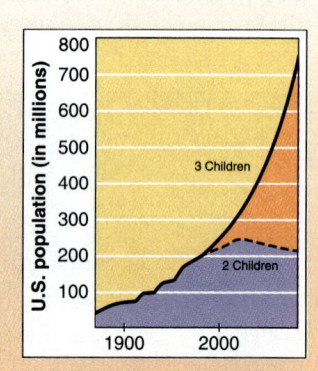

Write About It Your legislator is going to vote on imposing limits on family size. Write a letter telling the legislator how you'd like him or her to vote. Explain your opinions.

Chapter 26 Interactions Among Organisms

INTEGRATED LEARNING

✳ *Multicultural Perspectives*

Japan has a serious space shortage. People in Japan are always looking for new ways to utilize space. Recently, urban planners have begun to look into the idea of creating underground complexes containing places such as stores, theaters, offices, hotels, libraries, and museums. One plan is to build an underground network that would support as many as 500,000 people about 50 meters below the surface of the earth. Ask students what they think of the idea of living and working underground. What else could Japan do to solve its space problem?

◀ **Figure 26.6**
Although there are more than 14,000 people per sq km living in Tokyo, Japan, the population growth rate in Japan is far less than that of developing nations, such as India.

in the J becomes rounded, the line on the graph becomes almost vertical.

Why has the human population grown so rapidly? The steady increase in the human population is partly because humans think of ways to adapt to their surroundings in order to meet their needs. For example, humans can grow food and control some diseases.

Although the human population is increasing, the growth rate varies regionally. For example, the growth rate is much higher in developing regions, such as India, than in industrialized nations, such as Canada. In fact, nine out of ten babies born today are born in developing nations. Unfortunately, many of these nations do not have adequate resources to feed and provide shelter for all these people.

Skills WorkOut

Collect Data The U.S. Census Bureau collects information about the number of people who live in a household. You will be collecting similar data. Find out how many people live in the household of at least ten of your classmates. Include your own household. Graph this data. What is the average number of people in a household?

Check and Explain

1. Name factors that limit a population's size.
2. Owls often kill and eat mice. Which organism is the prey? Which organism is the predator?
3. **Infer** If the number of students in your school doubled, what effect would this have on the resources of your school? Explain.
4. **Communicate** In a population, limiting factors and population density are often related. Explain how. Give examples of limiting factors that may affect the density of a population.

Chapter 26 Interactions Among Organisms

Skills WorkOut
To help students collect and evaluate data, have them do the Skills WorkOut.
Answer Answers will vary, but the average number of people in U.S. households is 2.62.

EVALUATE

WrapUp
Reteach Have students choose a plant or animal population and describe its environment, competitors, and limiting factors. Students should identify whether the species is predator, prey, or both. They should also discuss how the species is affected by human populations.
Use Review Worksheet 26.1.

Check and Explain
1. Disease, sunlight, temperature, competition, pollution, and availability of food, water, and space
2. Owls are the predators and mice are the prey.
3. Competition for resources would increase as resources declined.
4. The greater the population density, the greater the effect of limiting factors. For example, when population density is high, there is more competition for food, water, and space. Population density is then lowered as members die or move away.

ACTIVITY 26

Time 30 minutes **Group** 1–2
Materials
metric rulers

Analysis
1. The dandelion population is the densest.
2. Answers may include: availability of sunlight, nutrients, or water; presence of other organisms that feed on the plants; the space available to each individual plant; terrain.
3. Answers will vary. Any change in the factors listed in question 2 could lead to a change in the population density.

Conclusion
Population density patterns are based on the relationship between the needs of individual species and a complex of limiting factors. Population density figures can be applied to similar populations in different places. For example, given the same soil, terrain, climate, and grazing animals, population densities for ground cover in different areas ought to be the same. Differences might point to potential problems such as pollutants or disease.

Everyday Application
Answers will vary depending on the rooms chosen. Class assignments affect population density, because administrators will put only as many students into a room as will comfortably fit.

Extension
Answers for your class will depend on statistics for your state, and on statistics for the state chosen for comparison.

TEACHING OPTIONS

Prelab Discussion
Have students read the entire activity. Discuss the following points before beginning the activity:

▶ Discuss the techniques used to study populations in the wild—counting individuals in a small plot, then extrapolating to the whole area.

▶ Discuss Figure 26.7: First discuss factors that all the individuals in the population have in common (they are all leafy flowering plants; they are varieties of ground cover). Then ask for possible explanations for the way the distribution of individuals varies over the plot.

Activity 26   How do you measure population density?

Skills   Measure; Communicate; Interpret Data; Calculate

Task 1   Prelab Prep
1. Collect the following items: metric ruler, paper, pencil.
2. Study the three patterns of population distribution in Figure 26.7.
3. Review the formula for calculating density:

$$\text{density} = \frac{\text{number of individuals}}{\text{size of area}}$$

Task 2   Data Record
On a separate sheet of paper, copy Table 26.1. Record each population's name in the table.

Task 3   Procedure
1. Calculate and record the area of Figure 26.7.
2. Select one population in Figure 26.7. Count the members present. Record this number in Table 26.1.
3. Calculate the density of that population. Record this number in the table.

Figure 26.7 ▼

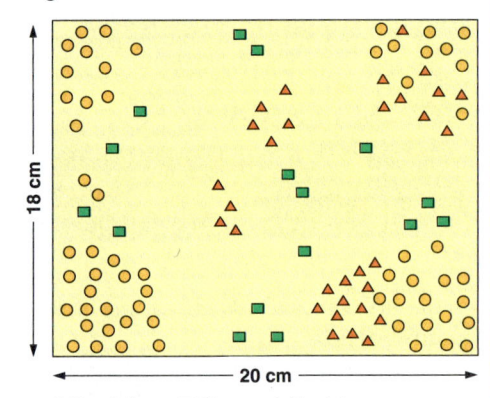

18 cm × 20 cm

○ Dandelions   ■ Clover   ▲ Plantains

Table 26.1   Population Density

Population name	Number of organisms	Density

4. Count the members and calculate the density for the other two populations. Record these numbers.

Task 4   Analysis
1. Compare the distribution patterns of the three populations. Which population has the greatest density?
2. What factors could be responsible for the differences in the population densities?
3. Name a factor or condition that could change the density of any of the populations.

Task 5   Conclusion
Write a short paragraph describing how measuring a population's density can be used to learn about the needs and characteristics of a population.

Everyday Application
Count the number of people in your classroom. Then measure the area of the classroom. What is its population density? Find the approximate population densities of other occupied rooms in your school, such as the study hall, the gym, or the lunchroom. Make a table listing the densities. How do they compare? How could knowing the population densities of these rooms affect the scheduled classes or lunch periods?

Extension
Find the population density of your state. A local library can supply you with the number of people living in your state and its area. Compare your state's density with that of another state.

INTEGRATED LEARNING

Themes in Science
Systems and Interactions
Interactions between organisms can be classified into three categories of symbiosis according to how one species is affected by the other. Have students look up the definitions of *commensalism, mutualism,* and *parasitism* to enhance their understanding of these terms.

SECTION 26.2

Section Objectives
For a list of section objectives, see the Student Edition page.

Skills Objectives
Students should be able to:

Infer how organisms interact.

Classify symbiotic relationships.

Predict what will happen to a soybean crop and the fungus that covers it.

Vocabulary
symbiosis

26.2 Relationships Among Populations

Objectives

▶ **Define** symbiosis.

▶ **Compare** and **contrast** the three major kinds of symbiotic relationships among organisms.

▶ **Explain** the helpful and harmful effects of a parasitic relationship.

▶ **Predict** if two different species can develop a symbiotic relationship.

Skills WarmUp

Infer Think about how organisms relate to each other. For example, a dog scratches its fleas; a bird builds a nest in a tree. List ten organisms. Next to each organism, write about some type of interaction that you think it has with another organism. In each case, describe whether the organisms were helped, harmed, or not at all affected by each other.

Go back to your imaginary jungle scene and picture many types of animals. Do these animals live alone? Or do they have contact with one another? In most ecosystems, different species of organisms interact with one another. These interactions help populations survive. They also help balance the ecosystems in which they live.

Symbiosis

Few organisms live alone. Populations that live in close physical contact with each other interact in a variety of ways. Many relationships between organisms involve the transfer of energy or food. Some organisms also provide services to each other such as protection, transportation, or support. A relationship between species in which at least one member benefits is called **symbiosis** (SIHM by OH sihs).

An example of a symbiotic relationship can be seen in Figure 26.8. Notice that the cattle egret is living peacefully on top of the cape buffalo. As the cape buffalo moves and grazes, it flushes out insects from the vegetation. This benefits the egret because an egret eats insects. The cape buffalo is not harmed in any way by the relationship.

There are three major types of symbiotic relationships. They are classified according to how one species is affected by the other.

Figure 26.8 ▲
The cape buffalo and the cattle egret have a symbiotic relationship. Which organism benefits from this relationship? ①

MOTIVATE

Skills WarmUp
To help students understand relationships among populations, have them do the Skills WarmUp.
Answer Human beings and the bacteria *E. coli* have a mutually beneficial relationship. The relationship between fleas and dogs helps fleas and harms dogs. Barnacles benefit from their relationship with whales without harming them.

Misconceptions
Students may think that symbiosis refers only to relationships in which organisms help each other. Stress that three types of symbiotic relationships exist.

Answer to In-Text Question
① The cattle egret

The Living Textbook:
Life Science Side 4
Chapter 14   Frame 31833
Species Interactions (Movies)
Search: Play: Step:

Chapter 26 Interactions Among Organisms

TEACH

Explore Visually

Have students study pages 568 and 569. Ask the following questions:

▶ How are animals and the nests they build in trees examples of commensalism? (The nesting animals benefit from the relationship and the trees are unaffected by it.)

▶ Explain how lichen is an example of mutualism. (Both organisms in the lichen benefit. The alga supplies the fungus with nutrients and water; the fungus protects the alga from intense sunlight and extremes of temperature.)

▶ How do orchids benefit from living in trees? (The orchids gain easy access to sunlight.) What type of symbiotic relationship is this? Why? (The relationship is an example of commensalism because the orchids benefit and the tree neither suffers nor benefits.)

The Living Textbook: Life Science Sides 7-8

Chapter 16 Frame 01695
Mutualism/Commensalism (10 Frames)
Search: Step:

568

INTEGRATED LEARNING

Integrating the Sciences

Oceanography Refer students to the photograph of the barnacle-covered whale. Barnacles begin life in a drifting larval stage. As they enter their final adult stage, they search for a place to attach themselves, such as a ship or a whale. Once attached, barnacles will remain in a chosen site for the rest of their lives. Barnacles attach by secreting a glue from their cement glands. They then grow a cone-shaped shell for protection from predators and waves. Barnacles obtain their food from the surrounding water and do not harm the animals they become attached to. Ask students why whales are good homes for barnacles. (Whales carry them to new sources of food.)

Commensalism A symbiotic relationship in which one organism benefits while the other organism is unaffected is called commensalism. Trees are often involved in relationships of commensalism. For example, some kinds of birds and squirrels build their nests in trees.

Mutualism A symbiotic relationship in which both organisms benefit is called mutualism. Organisms that share a mutualistic relationship often give each other protection and food.

An example of commensalism is the relationship between trees and flowers called orchids. The orchids live by attaching to the branches of the tree. The tree is neither benefited nor harmed. However, the orchid benefits because it gains easy access to sunlight. ▼

A lichen is an organism ▶ made up of a green alga or a blue-green alga and a fungus. It is an example of mutualism. Together these organisms live as one, often on the surface of rocks. The alga supplies the fungus with sugar, other nutrients, and water. The fungus protects the alga from intense sunlight and temperature extremes. Neither of these organisms could survive in its habitat without the other.

You can also see com- ▶ mensalism in action here, as a whale transports attached barnacles throughout the ocean. Because barnacles can't move on their own, their relationship with whales increases the barnacles' chances of finding food. The whales are not affected by the barnacles.

568 Chapter 26 Interactions Among Organisms

Art Connection
Butterflies and flowers have a mutualistic relationship. Have students draw and color a diagram illustrating how the flower and the butterfly benefit each other.

Parasitism A symbiotic relationship in which one organism benefits and the other is harmed is called parasitism. Parasites are organisms that live on or in another organism. The organism that is harmed by the parasite is called the host. Both plants and animals can act as parasites and as hosts.

A parasite uses its host as a source of food. Usually the parasite does not kill its host. However, the parasite may weaken or sicken the host. Why would the death of the host be a disadvantage to the parasite? ①

Some parasites live outside the body of the host. Fleas, ticks, lice, and mites are examples of external parasites. Each of these parasites feeds on animals, including humans. Here, you can see a louse, magnified many times, on a human hair. ▼

▲ Mistletoe is a plant that is a parasite of trees. The roots of the mistletoe tap into the tree's tissues to get food and water. The loss of nutrients to the tree can weaken the tree.

▲ Butterflies get food from the nectar of flowers. As the butterfly feeds, pollen from the flower sticks to the butterfly's body. The butterfly carries the pollen to the next plant it feeds on. In this way, the butterfly helps pollinate the plant, ensuring the survival of the plant species. What kind of symbiotic relationship do the flower and butterfly have? ②

The ants and acacia trees ▶ in Central and South America have an unusual mutualistic relationship. The ants live on the thorns of the acacia. They also eat food provided by its leaves. The acacia benefits, too. Ants protect their territory by attacking predators that try to eat the acacia. They will even destroy other plants that shade sun from the acacia.

▶ What are some parasites that live outside the body of their hosts? (Fleas, ticks, lice, and mites) Have students name some animals that are affected by these parasites. (Humans, deer, mice, dogs, and cats)

▶ Why is mistletoe considered a parasite? (The roots of the mistletoe grow into trees' tissues to get food and water.)

▶ What kind of a relationship do the ants and acacia trees in Central and South America have? Why? (It is a mutualistic relationship. Ants live on the thorns of the acacia and eat food provided by its leaves. In turn, the ants protect the trees by killing competitors that eat the acacia and by destroying vegetation that competes with the acacia.)

▶ Where do butterflies get their food, and how does this activity benefit plants? (Butterflies feed on the nectar from flowers; butterflies help pollinate flowers by carrying pollen between flowers.)

Answers to In-Text Questions

① **The death of a host would either lead to the death of the parasite or force it to look for another host.**

② **A flower and a butterfly have a mutualistic relationship.**

**The Living Textbook:
Life Science Sides 7-8**

Chapter 1   Frame 01685
Parasitism (10 Frames)
Search: Step:

Chapter 26 Interactions Among Organisms

TEACH • Continued

Skills WorkOut
To help students classify the symbiotic relationships they have with other organisms, have them do the Skills WorkOut.
Answers Examples will vary.

EVALUATE

WrapUp
Review Have students make a chart of the three types of symbiotic relationships and identify examples of organisms that are involved in each type.
Use Review Worksheet 26.2.

Check and Explain

1. A relationship between species in which at least one member benefits is symbiotic. Commensalism—one organism benefits while the other is unaffected; mutualism—both organisms benefit; parasitism—one organism benefits and the other is harmed
2. Two different species live in the same ecosystem without competing for food and space, providing benefits that help at least one of the species survive.
3. In both predator-prey relationships and parasitism, one species uses a second species as a food source at the second's expense. In a predator-prey relationship, the predator kills the prey. Parasites usually don't kill the host. Examples may vary.
4. The soybeans might be killed by the fungus, which would most likely also die; the relationship is parasitic.

570

INTEGRATED LEARNING

STS Connection
Humans have always had a symbiotic relationship with the bacteria that help them to digest food. Contemporary industrial use of bacteria is also symbiotic. Have students find out in what way people use bacteria in industry. (In producing food such as cheese, vinegar, and sauerkraut; in tanning leather; and in treating sewage) Refer them to Chapter 10 for more information.

Skills WorkOut
Classify How many symbiotic relationships do you have with other organisms? Give an example of commensalism, mutualism, and parasitism. Compare your examples to those of a classmate. Then create a chart that lists a variety of these different relationships. Label the type of relationship for each example.

Science and You — Human Symbionts
Did you know that you are never alone? Living within your body are entire ecosystems of organisms! Some of these organisms form mutualistic relationships with you. Others are parasites that use your body to meet their nutritional needs.

Millions of bacteria make their home in your intestines. The bacteria contribute to your health by making vitamin K, which is necessary for blood clotting. The bacteria also assist in changing foods that enter your intestines so that wastes can be eliminated from your body. Both you and the intestinal bacteria benefit from this association.

You know that not all types of bacteria are helpful to humans. Many parasitic bacteria cause disease. Tuberculosis, for example, is a respiratory disease caused by parasitic bacteria. A person with this disease tires easily, loses weight, and coughs persistently. Today, drugs are used to treat people infected by tuberculosis. Cholera is also a disease caused by parasitic bacteria. The bacteria that cause cholera enter the body in contaminated food or water. Once in the body, the cholera bacteria may cause a person to lose large amounts of fluids needed by the body. With adequate care, most people can recover from cholera.

Check and Explain

1. What is symbiosis? Name and explain the major types of symbiosis.
2. How do symbiotic relationships help keep an ecosystem in balance?
3. **Compare and Contrast** Explain how predator-prey relationships and parasitism are alike but different. Give an example of each.
4. **Predict** In a farm field, soybeans were the only crop that was planted on several acres. A farmer noticed that a fungus was growing all over the soybean fields. Predict what might eventually happen to the soybeans. What could possibly happen to the fungus itself? What kind of symbiotic relationship do the soybeans and fungus have?

570   Chapter 26   Interactions Among Organisms

Themes in Science

Evolution Complex patterns of behavior have evolved in animals as they compete in a variety of environments. Animal behavior is classified as either innate or learned. Innate behaviors are genetically passed from parent to offspring. Learned behavior is acquired after birth. Ask students to consider their own behavior and make a list of innate behaviors that they engage in. (Some examples of innate behaviors are shivering from cold or withdrawing one's hand from extreme heat.) Encourage them to describe how these differ from learned behaviors.

SECTION 26.3

Section Objectives
For a list of section objectives, see the Student Edition page.

Skills Objectives
Students should be able to:

Communicate about their own learned behaviors.

Research unusual animal behaviors.

Observe human behaviors and classify them as innate or learned.

Vocabulary
innate behavior

26.3 Animal Behavior

Objectives

▶ **Distinguish** between learned behavior and innate behavior.

▶ **Describe** social behavior among organisms.

▶ **Observe** behaviors and **classify** them as innate or learned.

Skills WarmUp

Communicate Think of at least 20 things you have learned since your birth, such as walking or tying your shoe. Make a list, including approximately how long it took you to learn each task. What was the hardest thing to learn? What was the easiest? Compare your list with that of a classmate.

Each year, millions of monarch butterflies travel thousands of miles to Mexico. When it gets very cold, ground squirrels curl up and become inactive. A dog may turn around several times before it lies down. All these actions show ways in which animals respond to their internal or external environments. The way an organism responds to its environment is called behavior.

Innate Behavior

Sea turtles lay their eggs on beaches. Newly hatched turtles immediately run toward the water and begin swimming. Why does this happen? The turtles were not taught this behavior. A behavior that is natural for an organism and does not need to be learned is called an **innate behavior**. Innate behaviors protect an organism and help in its survival.

Innate behaviors are inherited, or genetically passed from parent to offspring. Many innate behaviors are simple. For example, a newborn chick will peck at particles on the ground without ever seeing another chicken perform this behavior. You also have innate behaviors. For example, if something comes close to your eyes, you blink. What other innate behaviors do you have?

Some innate behaviors are complex and involve many responses. For example, some species of birds migrate great distances to warmer places in the fall. Migration is a complex animal behavior that occurs in a repeated pattern. This type of behavior is called cyclic behavior. Cyclic behaviors generally happen as a result of changes in the external environment. Another example of cyclic behavior is hibernation. Hibernation is a state of inactivity that occurs in some animals when outside temperatures are cold. Some mammals and many reptiles and amphibians hibernate.

Figure 26.9 ▲
What makes this newly hatched sea turtle move toward water? ①

MOTIVATE

Skills WarmUp
To help students understand animal behavior, have them do the Skills WarmUp.
Answer Answers may vary. Students should be able to list 20 tasks and determine their levels of difficulty.

Misconceptions
Students may think that all behavior is learned. As you read through this section, point out examples of innate behaviors.

Answer to In-Text Question
① Innate behavior

The Living Textbook:
Life Science Side 4

Chapter 16   Frame 41000
Innate Behaviors (Movies)
Search: Play: Step:

Chapter 26 Interactions Among Organisms   571

TEACH

Career Corner

Research Have students interview a veterinarian or obedience trainer. Students should prepare a list of questions to ask about what it's like to work in such a profession and what kind of experience or schooling is necessary. Encourage them to read about animal behaviorists such as Charles Darwin, Jane Goodall, Dian Fossey, Farley Mowat, and so on. Students may want to read *All Creatures Great and Small*, the first in a series of novels about the experiences of a country veterinarian.

Class Activity

Have students think of examples of their own learned behaviors. Encourage them to make a chart with different categories of behavior and the reasons behind them. For example, some students may say that eating nutritious foods and exercising regularly are learned behaviors that aren't always fun, but they manage to repeat in order to stay healthy. On the other hand, learning to ride a bike or play a sport may be easy behaviors to repeat because they are very enjoyable. Have students consider how they acquired their learned behaviors.

Integrated Learning
Use Integrating Worksheet 26.3a.

Enrich
Use Enrich Worksheet 26.3.

The Living Textbook: Life Science Side 4

Chapter 14   Frame 33725
Trapdoor Spider and Sowbug (Movie)
Search: Play:

INTEGRATED LEARNING

Multicultural Perspectives

The Australian zoologist and ethologist Konrad Lorenz (1903–1989) discovered a form of learned behavior in birds called *imprinting*. In an experiment with newly hatched goslings, Lorenz found that, for a short time during their development, young birds have a tendency to form a strong attachment to a moving object. Ordinarily, this object is the mother leaving the nest. However, Lorenz found that birds can form an attachment to other things, including humans. Have students find out more about Lorenz's discoveries about imprinting.

Career Corner *Animal Trainer*

Who Teaches Animals New Behaviors?

Have you ever taught a pet to beg for food, fetch a ball, or perform some other trick? If so, you may have a skill that you can turn into a career. Animal trainers are people who specialize in working with animal behaviors.

Many animal trainers work with animals for entertainment purposes. For example, trainers teach dolphins to perform in amusement parks and aquariums. Animal trainers also teach animals to carry out tasks that help people. These tasks may include offering protection, assisting in daily chores, and guiding or assisting people who are physically challenged.

Animals have been trained to be companions and helpers for people with disabilities. Seeing-eye dogs are often used as guides for blind people. In Thailand, trainers teach macaques, a type of monkey, how to pick and transport coconuts.

Animal trainers must learn about the innate behaviors of animals. Patience, skill, and knowledge about the animals are needed in this career. To obtain more information about this kind of animal training, you might speak with such people as obedience trainers, veterinarians, or staff who train dogs to work with the blind.

Figure 26.10 ▲
Chimpanzees learn to use sticks to get food that is hard to reach, such as termites living in mounds.

Learned Behavior

Behaviors that are acquired after birth and are a result of changes in experience are called learned behaviors. Since your birth, you have learned many things. You may have learned to ride a bicycle. You have probably learned to solve different kinds of math problems. What other things have you learned?

Animals also learn to respond to certain conditions to produce new behaviors. This helps them survive. Carp that are fed by people will swim to the surface when people are near, even if the people don't have food. Have you seen a dog bark or get excited when a leash is picked up? This, too, is a learned behavior.

Because learned behavior takes a long time to develop, it mainly happens among animals that have long life spans, such as mammals. In addition, learned behavior requires a complex nervous system. So the more advanced an organism's brain, the more it is capable of learning.

Chapter 26 Interactions Among Organisms

Art Connection

Have each student create a poster depicting a honeybee society. They should draw and label all the members, explaining their roles in the colony. Display students' posters in the classroom.

Social Behavior

Humans all over the world live in groups called societies. Many animals also form organized societies. Each member in a society has a specialized role or niche. The role may be to protect other members of the society, to get food, or to raise the young. In a society, the cooperative behavior and specific jobs of the members increase the survival chances of the group.

Social behavior is common in many insect groups, such as termites and honeybees. In a termite society, some termites build and repair the nest. Other termites produce a substance that defends the nest. Only the queen can reproduce, but she is dependent on the other members for raising the young. Each member contributes only one service, but all members benefit from each of the tasks in the society. Honeybees have similar tasks in a hive. Look at Figure 26.12. What is the role of a worker bee?

Figure 26.11 ▲
Honeybees form organized societies in hives.

Figure 26.12 A Honeybee Hive

Queen
Each honeybee colony has only one queen. The queen mates just once and spends the rest of her life laying eggs. She coordinates all activities in the colony by sending out chemical messages.

Workers
In a strong honeybee colony, there may be as many as 80,000 worker bees. Worker bees are all female. They carry out special tasks inside the colony, such as feeding the larvae.

Eggs and Larvae
The queen lays about 1,500 eggs a day. Each egg has its own cell. The queen determines whether the eggs are fertilized. If they are, they develop into workers or queens. If they are not, they develop into drones.

Drones
Drones are male bees that mate with the queen. They die soon after mating. There are several hundred drones in each honeybee colony.

Explore Visually

Ask students to study the material on this page. Ask the following questions:

▶ How many active queens are in each termite society and each honeybee society? (One)

▶ Which members of the honeybee society are the vast majority? (Workers; they carry out all the special tasks inside the colony.)

▶ How is the honeybee colony controlled? (The queen sends out chemical messages that coordinate all the activities within the colony.)

▶ What is the sole function of the drone bee? (To mate with the queen)

Critical Thinking

Reason and Conclude Can any of the members of a honeybee society get along without the other members? Explain. (No. Each member contributes only one service, but all benefit from that service.)

Compare and Contrast Have students compare the honeybee society to human society, listing similarities and differences. (Answers may vary. Students should recognize that both societies are interdependent and that human society has some similar roles as bee society but is much less structured.)

Integrated Learning

Use Integrating Worksheet 26.3b.

**The Living Textbook:
Life Science Side 4**

Chapter 15   Frame 38467
Honeybees (Movie)
Search: Play:

TEACH • Continued

Skills WorkOut
To help students learn about unusual behaviors among animals, have them do the Skills WorkOut.
Answer Answers will vary.

EVALUATE

WrapUp
Reteach Have students make a chart listing 20 human behaviors. Have them indicate if each behavior is innate or learned, and then if it benefits human society, the individual, or neither.

Use Review Worksheet 26.3.

Check and Explain
1. An organized, cooperative group in which members have specialized roles; a honeybee hive is a society.
2. Answers may vary. Blinking when an object comes close to the eye; checking to see if cars are approaching before crossing the street
3. The fact that chimpanzees can learn human sign language suggests that the learning process gives animals the flexibility to adapt to new situations by learning new behaviors, which helps them to survive.
4. Students should realize that most human behaviors are learned and only a small number are innate.

The Living Textbook: Life Science Side 8
Chapter 33   Frame 18001
Monarch Migration (Movie)
Search: Play:

INTEGRATED LEARNING

Integrating the Sciences
Chemistry Tell students that sleep allows the tissues and organs of the body to repair themselves from the stress of all the day's activities. Because the body's needs slow down during sleep, there is more energy available for tissue repair. Sleep also allows the body to get rid of waste materials that have accumulated. One of these waste materials is lactic acid. The body converts most of the lactic acid back into blood sugar so that it can be used for energy once again.

Skills WorkOut
Research Many animals have unique behaviors. Research to find out about some unusual behaviors among animals. You can start by looking in books about birds, bats, insects, and primates, such as baboons. Choose one animal to write about. Write a description of the animal's behavior and present it to the class.

Science and You *Say Good Night!*
In your lifetime, you will spend about a third of your life asleep. Is that much sleep necessary? Why do you get drowsy at certain times?

Sleep is an example of an innate cyclic behavior. It is controlled by the brain. Many scientists think that a sleep-inducing chemical builds up in your body while you are awake. When the chemical reaches a certain level, you get sleepy.

Why do you need sleep? Scientists are conducting studies to find the answer to this question. Some think that sleep may allow your brain time to process all the information it received during the day. Sleep also allows the body time to rest. For example, during sleep your heart rate and metabolic rate slow down. Your muscles also relax.

When people don't get enough sleep, they often experience difficulty in concentrating. Memory can also be affected. Lack of sleep may interfere with learning ability and coordination. In what ways does lack of sleep affect you?

Of course, people differ in the amount of sleep that they need. For example, studies have shown that you need more sleep than an adult. What does your biological clock tell you about your sleep needs? When and how often are you ready to snooze?

Check and Explain
1. What is a society? Give an example.
2. Describe two actions: one in which you use an innate behavior, and one in which you use a learned behavior.
3. **Generalize** Some animals can be trained to communicate with people. For example, sign language is exchanged with chimpanzees. What does this tell you about the learning process? Could you apply this information to other species? Explain.
4. **Observe** Select one person to watch. Carefully observe five of their actions. Name their behaviors and identify each of them as either innate or learned. Compare your findings with those of your classmates. What does this information tell you about human behaviors?

Chapter 26 Interactions Among Organisms

Chapter 26 Review

Concept Summary

26.1 Changes in Populations
- Population sizes vary among organisms. They change with number of births and when new members move into an ecosystem. They also change when members die or move out of an ecosystem.
- Limiting factors are environmental conditions that keep a population from increasing in size and help balance ecosystems.
- Predators are animals that eat other organisms. The organisms they eat are prey.

26.2 Relationships Among Populations
- Symbiosis is a relationship between species in which at least one member benefits.
- Commensalism benefits one organism; the other organism is unaffected.
- Mutualism benefits both organisms.
- Parasitism benefits one organism and harms the other.

26.3 Animal Behavior
- Innate behaviors are genetically passed from parent to offspring. Innate behaviors protect an organism and help it survive.
- Learned behaviors are acquired after birth and are a result of changes in experience. Learned behavior usually takes a long time to develop. It mainly happens among organisms with long life spans and complex nervous systems.
- Like humans, many animals also form organized societies. Members of societies have special roles and cooperate with each other. Cooperation increases the group's chances for survival.

Chapter Vocabulary

population density (26.1)   predator (26.1)   symbiosis (26.2)   innate behavior (26.3)
limiting factors (26.1)   prey (26.1)

Check Your Vocabulary

Use the vocabulary words above to complete the following sentences correctly.

1. An organism that is eaten by a predator is called ____.
2. A behavior that is natural for an organism and does not need to be learned is ____.
3. The number of individuals in an area, or ____, is calculated by dividing the number of organisms by the size of the area.
4. A relationship between species in which at least one member benefits is called ____.
5. An animal that obtains its food by killing and eating other organisms is called a ____.
6. Availability of food, water, and living space, and the spread of disease are examples of ____.

Identify the word or term in each group that does not belong. Explain why it does not belong.

7. Predator, prey, symbiosis
8. Limiting factors, competition, commensalism
9. Learned behavior, parasitism, symbiosis
10. Population density, limiting factors, cyclic behavior
11. Commensalism, prey, mutualism

CHAPTER REVIEW 26

Check Your Vocabulary

1. prey
2. innate behavior
3. population density
4. symbiosis
5. predator
6. limiting factors
7. Symbiosis; Predators eat prey for food.
8. Commensalism; Different species living in the same area often compete for limiting factors.
9. Learned behavior; Parasitism is a type of symbiotic relationship.
10. Cyclic behavior; Population density is a limiting factor.
11. Prey; Commensalism and mutualism are both examples of symbiotic relationships.
12. Competition; The graph of human population over time is a J-curve.

Use Vocabulary Worksheet for Chapter 26.

CHAPTER 26 REVIEW

Check Your Knowledge

1. Answers may vary, but may include birthrate, death rate, and migration of individuals into or out of a population.
2. Commensalism, mutualism, and parasitism are the three types of symbiotic relationships.
3. Riding a bicycle is a learned behavior.
4. Population density is calculated by dividing the number of individuals in a geographical area by the size of the area in which they live.
5. Limiting factors prevent a population from growing beyond a certain limit. Examples may include availability of food, water, or living space; the presence of disease; and the presence of a predator population.
6. Predators kill and eat prey for food. Examples: lions/gazelles, snakes/mice, frogs/flies
7. Cyclic behavior is an innate behavior that occurs in a repeated pattern at regular time intervals. Examples include migration and hibernation.
8. competition
9. J
10. limit
11. mutualism
12. limiting factors

Check Your Understanding

1. 1 ladybug/sq m
2. Answers may vary; however, students should recognize that the robin population would likely decrease due to a reduction in food or from eating poisoned worms.
3. Answers may vary. Examples of innate behaviors include blinking, sleeping, and yawning. Examples of learned behaviors include walking, brushing teeth, and reading.
4. a. Mutual
   b. The clownfish is protected and the anemone gets food.
5. The bees are exhibiting social behavior. The drones mate with the queen, the queen lays eggs and coordinates the activities of the hive, and the workers care for the larvae and do other jobs in the colony.

Chapter 26 Review

Check Your Knowledge

Answer the following in complete sentences.

1. What kinds of factors might affect the size of a population? List two examples.
2. What are the three types of symbiotic relationships?
3. Is riding a bicycle an innate or learned behavior?
4. How do you calculate population density? Give one example.
5. How do limiting factors affect a population? Give two examples of limiting factors.
6. Describe the relationship between predator and prey. Then give one example of a predator and its prey.
7. What is cyclic behavior? Give one example.

Choose the answer that best completes each sentence.

8. When two or more species in an ecosystem must use the same limited resources, (parasitism, competition, expansion, extinction) occurs.
9. When the growth pattern of the human population is graphed, it produces a (K, M, Z, J) -shaped curve.
10. Predator-prey relationships (limit, increase, exaggerate, prevent) population density.
11. A lichen is an example of (commensalism, mutualism, parasitism, predator-prey relationship).
12. Certain environmental factors called (growth factors, death factors, birth factors, limiting factors) keep a population from increasing and help balance ecosystems.

Check Your Understanding

Apply the concepts you have learned to answer each question.

1. **Extension** What is the population density of 50 ladybugs living in a 100 sq m plot?
2. **Critical Thinking** Robins are predators and worms are prey for robins. Describe what might happen to the population of robins if an area where they hunted for worms was sprayed with pesticides.
3. **Application** List two activities you did today that were innate behaviors. Then list two activities that were learned behaviors.
4. **Mystery Photo** The photo on page 560 shows a clownfish swimming among the tentacles of a sea anemone. While the tentacles of most sea anemones are very poisonous, the clownfish is immune to the sting of most varieties. So the tentacles provide protection for the clownfish, since its predators either avoid the sea anemone or are stopped by its sting. The clownfish drops small pieces of food as it feeds, and the sea anemone absorbs this food. The clownfish may also lure prey into the tentacles so that the sea anemone can feed.
   a. What type of symbiotic relationship do the clownfish and sea anemone have?
   b. How does the clownfish benefit from the relationship? How does the sea anemone benefit?
5. Describe the role of a drone, a worker, and the queen bee in a honeybee hive. What kind of behavior are the honeybees in a hive exhibiting? Explain.

Develop Your Skills

1. a. The paramecium moves in essentially a straight line until it runs into something.

   b. When it encounters an object, it changes direction. Students may or may not see a pattern. One such pattern might be the degree of change in direction after running into an object.

2. a. 22 persons per square kilometer

   b. Asia

   c. Asia is about nine times more densely populated than North America.

3. Student models will vary, but should show an understanding of the symbiotic relationship chosen and labeled.

Make Connections

1.

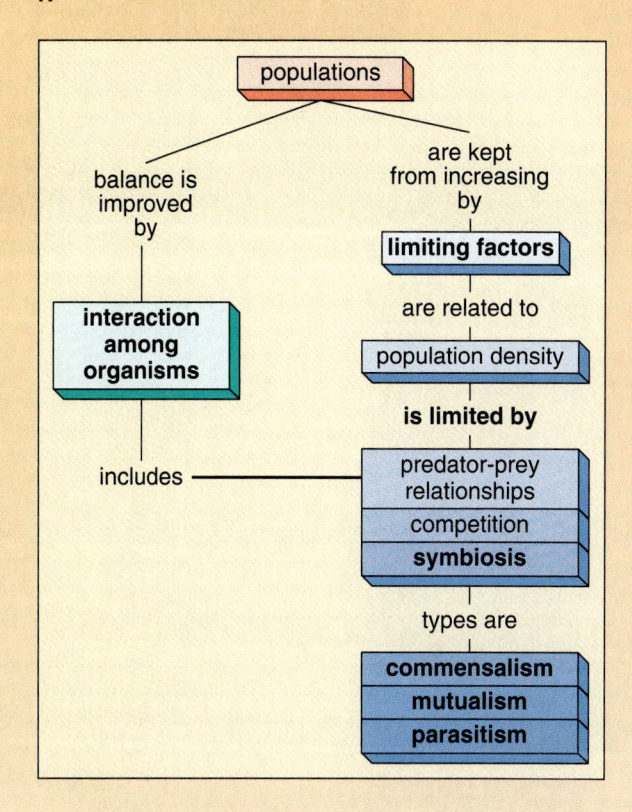

2. Answers may vary, but students should use the correct mathematical relationship: number of individuals divided by the size of the area. Answers should use square units unless the living area is best described in volume (as in an aquarium).

3. Reports will vary, but should include population data, a bar graph, comparisons, and inferences about causes of differences in population growth.

Develop Your Skills

Use the skills you have developed in this chapter to complete each activity.

1. **Interpret Data** The diagram below shows the movement of a paramecium in a petri dish. The geometric shapes are solid objects in the dish.

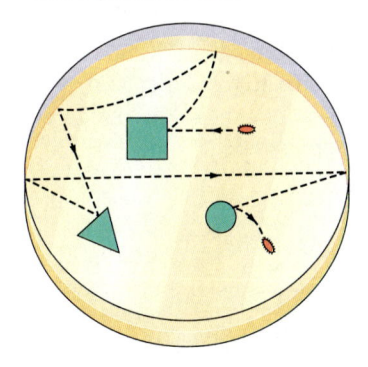

   a. How would you describe the paramecium's movement?

   b. What does the paramecium do when it encounters an object? Is there a pattern in its behavior?

2. **Data Bank** Use the information on page 626 to answer the following questions.

   a. What is the population density of Africa?

   b. Which continent has the greatest population density?

   c. How much more dense is the population in Asia than in North America?

3. **Make a Model** Create a model of a symbiotic relationship. You may create your own organisms, or use examples of real organisms. Label your model commensalism, parasitism, or mutualism.

Make Connections

1. **Link the Concepts** Below is a concept map showing how some of the main concepts in this chapter link together. Only part of the map is filled in. Copy the map. Using words and ideas from the chapter, finish the map.

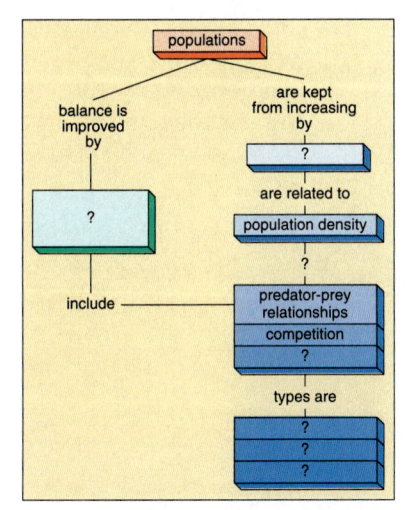

2. **Science and Math** Use the formula for calculating density to find out the population density in a place that you often visit, such as a recreation center.

3. **Science and Society** Human population growth generally varies from country to country. Choose one country in Asia, one country in Europe, and one country in Africa. Find out the growth rate of each country over the past ten years. Infer any factors that might have influenced the differences in growth rates among the countries. Write your findings in a report.

CHAPTER 27

Overview

Students learn in this chapter the factors that determine climates and help to classify biomes. Precipitation and temperature are presented in the first section to explain climatic conditions. The second section illustrates the life-supporting elements of land biomes. Plant and animal life are described within each environment. The chapter closes with a description of marine and freshwater biomes and the diversity of life within these environments.

Advance Planner

▶ Supply modeling clay, flashlights, and pencils for TE activity, page 580.

▶ Provide colored pencils, graph paper, rulers, and calculators for SE Activity 27, page 583.

▶ Prepare glass jars, salt, spoons, and small equal-sized pieces of apple, potato, grape, and raisin for SE page 595.

Skills Development Chart

Sections	Communicate	Compare and Contrast	Decision Making	Graph	Infer	Interpret Data
27.1 Skills WarmUp					●	
Skills WorkOut					●	●
Activity				●	●	●
27.2 Skills WarmUp					●	
Consider This	●		●			
27.3 Skills WarmUp		●				
Skills WorkOut					●	
SkillBuilder		●				

Individual Needs

▶ **Limited English Proficiency Students** Photocopy pages 580, 584, 592, and 593. Ask students to highlight all the chapter terms that appear on these pages and underline the sentences that explain the definitions.

▶ **At-Risk Students** Have students write each of the section titles on pages in their science journals. Have them choose four important ideas in each section and copy these onto the journal pages. Have them explain why each of these ideas is important to them.

▶ **Gifted Students** Have students create a travelogue that features a location with a climate different from that in your area. Students should find out about the area, including such elements as climate patterns, types of biomes, dominant plants, and native animals. They can draw pictures of the area or find examples in magazines and add them to the travelogue. Students can then present their travelogues to the class. They might enjoy videotaping it.

Resource Bank

▶ **Bulletin Board** Hang a picture of a world map on the bulletin board. As you work through this chapter, have students add details, such as longitude and latitude lines, and topographical features such as mountains. As you read about various biomes, have students find photographs from magazines showing characteristics of each. Have them add these photographs near the areas on the map where they would expect to find a particular biome.

▶ **Field Trip** Many zoos or conservatories have displays showing the characteristics of various biomes. You may wish to arrange a field trip to one of these exhibits.

CHAPTER 27 PLANNING GUIDE

Section	Core	Standard	Enriched	Section	Core	Standard	Enriched
27.1 Climate pp. 579–583				**Laboratory Program** Investigation 49	●	●	●
Section Features Skills WarmUp, p. 579 Skills WorkOut, p. 581 Activity, p. 583	● ● ●	● ● ●	● ● ●	*Color Transparencies* Transparency 84	●	●	
Blackline Masters Review Worksheet 27.1 Reteach Worksheet 27.1 Skills Worksheet 27.1 Integrating Worksheet 27.1a Integrating Worksheet 27.1b	● ● ● ● ●	● ● ● ● ●	● ●	**27.3 Water Biomes** pp. 592–596			
				Section Features Skills WarmUp, p. 592 Skills WorkOut, p. 594 SkillBuilder, p. 595	● ● ●	● ● ●	● ●
Color Transparencies Transparency 83	●	●		*Blackline Masters* Review Worksheet 27.3 Skills Worksheet 27.3 Integrating Worksheet 27.3a Integrating Worksheet 27.3b Enrich Worksheet 27.3 Vocabulary Worksheet 27	● ● ● ●	● ● ● ● ● ●	● ● ● ●
27.2 Land Biomes pp. 584–591							
Section Features Skills WarmUp, p. 584 Consider This, p. 589	● ●	● ●	● ●				
Blackline Masters Review Worksheet 27.2	●	●		*Color Transparencies* Transparencies 85a, b	●	●	
Ancillary Options One-Minute Readings, pp. 22, 33	●	●	●				

Bibliography

The following resources can be used for teaching the chapter. See page T-40 for supplier codes.

Audio-Visual Sources
(video unless noted)
Ancient Forests. 25 min. 1992. NGSES.
The Ecology of a Stream. Filmstrip with cassette. 1983. CABISCO.
The Living Earth. 25 min. 1991. NGSES.
The Living Ocean. 25 min. 1985. NGSES.
The Margins of the Land. 55 min. 1984. T-L.
Riches from the Sea. 23 min. 1984. NGSES.
Seas of Grass. 55 min. 1984. T-L.

Software Resources
The Environment I: Habitats and Ecosystems. IBM PC. IBM.
The Environment II: Cycles and Interactions. IBM PC. IBM.
Hothouse Planet. Apple, IBM PC. EME.

Library Resources
Carson, Rachel. *The Sea Around Us.* Racine, WI: Golden Press, 1958.
Graham, Ada and Frank. *The Changing Desert.* San Francisco: Sierra Club Books, 1981.
Rockwell, Jane. *All About Ponds.* Mahwah, NJ: Troll Associates, 1984.
Rydell, Wendy. *All About Islands.* Mahwah, NJ: Troll Associates, 1984.

CHAPTER 27

Introducing the Chapter

Have students read the description of the photograph on page 578. Ask if they agree or disagree with the description.

Directed Inquiry

Have students study the photograph. Ask:

▶ Where was the photographer who took this photograph? (Somewhere above, probably in an airplane)

▶ What do you notice first about the photograph? (Most students will say the bright colors.)

▶ How would the same scene look in a picture taken three months later? Six months later? (Three months later, the trees would have no leaves and might be covered with snow. Six months later, new green leaves would be starting to open.)

▶ What do you think the weather was like when this picture was taken? (Sunny and cool)

▶ Why do you think this picture was used to begin a chapter on climate and biomes? (Leaves change color only in a particular climate and biome.)

Chapter Vocabulary

biome	marine
climate	polar
coniferous	temperate
deciduous	tropical
desert	tundra
grassland	

INTEGRATED LEARNING

Language Arts Connection

Write the word *biome* on the chalkboard. Underline the first two letters. Ask students what they think the root *bi-* means. (Life) Have students use this information, along with the information and photograph on page 578, to infer the meaning of biome.

Chapter 27 Climate and Biomes

Chapter Sections

27.1 Climate
27.2 Land Biomes
27.3 Water Biomes

What do you see?

"This picture was taken in the fall because of the inspiring scenery and the wonderful colors of nature. Some of the animals that might live in this place are small gopher snakes, deer, gophers, squirrels, owls, mice, blue jays, and all kinds of birds. This is a good habitat for them because they can find the food they need for the winter."

Ramón Plancarte
Menlo Oaks School
Menlo Park, California

To find out about the photograph, look on page 598. As you read this chapter, you will learn about climate and biomes.

578

STS Connection

Ask students to help you create a list of local climatic conditions. Then have them compose a list of the human devices that have been made to help people adapt to such a climate. Examples might include air conditioners, furnaces, solar panels, and so on. When students have completed their lists, have them make another list for a place with a radically different climate.

Themes in Science
Systems and Interactions

Organisms interact with their environments to meet their needs. Where an organism can live depends on the abiotic factors in an area, such as rainfall, latitude, and altitude.

SECTION 27.1

Section Objectives
For a list of section objectives, see the Student Edition page.

Skills Objectives
Students should be able to:

Infer the kinds of plants that would grow well in a rain-drenched area.

Interpret Data to find the longitude and latitude of several cities.

Make a Graph showing average temperatures for a particular city.

Vocabulary
climate, biome, tropical, polar, temperate

27.1 Climate

Objectives

▶ **Describe** factors that determine the climate of a biome.

▶ **Explain** how rainfall and temperature affect vegetation.

▶ **Describe** how latitude and altitude determine the characteristics of a biome.

▶ **Interpret data** from a graph of global temperatures.

Skills WarmUp

Infer Imagine that you are traveling through a rain-drenched area. Not only is it pouring on you now, but it also rains every day here. What kind of plants would grow well in an area that has so much rain? Would there be many plants that could adapt to this environment? What would these plants look like?

Have you ever wanted to live on a tropical island or a place where it's always cold? You could probably survive in either of these places. People are very adaptable. Look at Figure 27.1. As you can see, people can live in many places, including hot deserts. Not all living things, however, can adapt to environmental conditions as well as people can. For example, orchids are adapted to an environment with lots of rain. They can't survive in a desert environment. Why? Because they can't change the desert environment to meet their needs. Where an organism can live depends on nonliving, or abiotic, factors.

Climate Zones

Weather may vary daily, but from year to year each area has average weather patterns. These patterns are known as **climate**. Climate influences what type of plant life, or vegetation, can live in an area. The vegetation supports specific animal life. So different climates support different living communities.

Ecologists have used patterns of climate, vegetation, and animal life to divide large portions of land and water into **biomes**. Biomes are large communities with similar biotic and abiotic factors. Each biome has a certain climax community of plants and animals. Land biomes include communities such as deserts and forests. There are only two types of water biomes: marine, or ocean, and freshwater.

The different climates in biomes can be classified according to two main factors. One factor is precipitation. Precipitation is moisture in the atmosphere that condenses

Figure 27.1 ▲
How have the humans living here adapted to the hot, dry conditions? ①

MOTIVATE

Skills WarmUp
To help students understand climate and biomes, have them do the Skills WarmUp.
Answer Plants that need a lot of water grow in rainy areas. Examples are ferns and plants that have cup-shaped leaves.

Prior Knowledge

To gauge how much students know about climate and biomes, have them look at a globe and ask them the following questions:

▶ Where is the equator?

▶ Which places do you think have the coldest temperatures?

▶ Where is a place that has a climate similar to the one in which you live?

Answer to In-Text Question

① Human beings can alter conditions to meet their needs. In a hot, dry climate, they can use irrigation and build shelter in such a way as to counteract the heat and dryness.

INTEGRATED LEARNING

TEACH

Class Activity
Have students work in groups of four to observe the effect of the earth's axial tilt on the seasons.

▶ Give each group a ball of modeling clay, two pencils, and a flashlight.

▶ Have students insert a pencil through the center of the ball of clay (earth's axis) and use the other pencil to draw a line around the ball (the equator) at a right angle to the pencil.

▶ Have students place the ball on a table with the pencil's eraser leaning to the right.

▶ Darken the room. Have students shine the flashlight on the ball about 15 cm from its left side. Have students observe and record where the light hits the ball.

▶ Have students shine the light the same distance from the right side of the ball and record where it strikes. (The half below the line, the Southern Hemisphere, receives the most light when the eraser points away from the light; the half above the line, the Northern Hemisphere, is brighter when the eraser points toward the light.)

Explain to students that the pencil represents the earth's axis; the line, the equator; the halves, the hemispheres; and the flashlight, the sun. The direction of the earth's axis shifts during its revolution around the sun. The hemispheres receive different amounts of light rays, creating seasonal changes.

Answer to In-Text Question
① Tropical, polar, or temperate climate

Integrating the Sciences
Earth Science Tell students that climatic changes on the earth depend on how directly sunlight falls on the hemispheres as the earth revolves around the sun. The amount of sunlight falling on the hemispheres changes as the earth orbits the sun because of the tilt of the earth's axis (23.5°). Have students draw a diagram of the earth's orbit around the sun.

Geography Connection
Pose the following situation to students: A person wants to live in a tropical climate during the winter and in a high temperate or low polar climate during the summer. Ask students where in North America that person might live during each season. (Answers will vary. A possible answer is in Ontario, Canada, during the summer and Florida during the winter.)

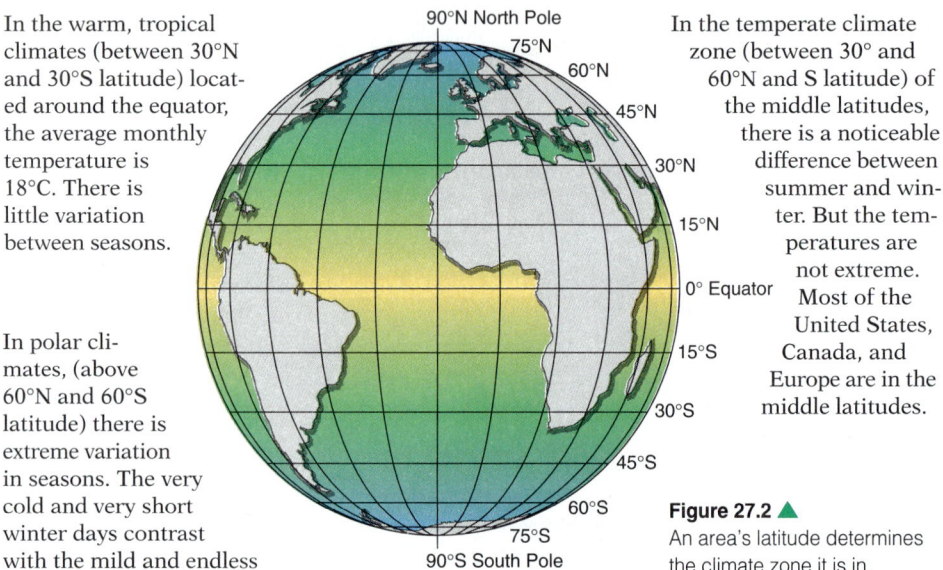

In the warm, tropical climates (between 30°N and 30°S latitude) located around the equator, the average monthly temperature is 18°C. There is little variation between seasons.

In polar climates, (above 60°N and 60°S latitude) there is extreme variation in seasons. The very cold and very short winter days contrast with the mild and endless days of summer.

In the temperate climate zone (between 30° and 60°N and S latitude) of the middle latitudes, there is a noticeable difference between summer and winter. But the temperatures are not extreme. Most of the United States, Canada, and Europe are in the middle latitudes.

Figure 27.2 ▲
An area's latitude determines the climate zone it is in.

and falls to the earth. Sleet, snow, hail, and rain are forms of precipitation. The other factor is temperature. Each climate has different temperature patterns. Some areas have periods of hot and periods of cold. Others have almost the same temperatures all year.

Different combinations of hot and cold or wet and dry make up different types of climates. In general, there are areas of the earth with particular climates. These are known as climate zones. Find the three major climate zones in Figure 27.2. What are they? What climate zone do you live in? ①

Each of the climate zones in Figure 27.2 is described by how the seasons vary. The **tropical** climate has warm temperatures year-round. The **polar** climate has a long season of very cold temperatures and a short warm season. The **temperate** climate has a cold season and a warm season of about equal length.

Precipitation

Scientists use rain gauges like the one in Figure 27.3 to measure precipitation in centimeters or inches. The amount of precipitation influences what plants will grow

Integrating the Sciences

Earth Science During winter the sun's rays are less direct than in summer because the earth is at more of an acute angle in relation to the sun. Explain that direct rays are hotter because more are absorbed by the earth's surface. Ask students to infer how Antarctica can be one of the most sunny places on the earth and also one of the coldest. (The rays reach the earth's surface at an angle.)

Earth Science A front is a dividing line between a warm air mass and a cold air mass. Fronts usually bring on a sharp change in weather. Typically, a cold air mass moves into an area and slides under warmer air, causing clouds and often rain or snow. Ask students what they think happens when a warm air mass moves into a cold area. (Warm air rises above cold air causing precipitation.)

Skills Development

Interpret Data Have students look at Figure 27.2 on page 580. Ask the following questions:

▶ What climate zone do you live in?

▶ What other climate zones have you lived in? What was the weather like?

▶ What are the two main factors that determine climate? (Precipitation and temperature)

▶ What latitudes do the tropical climates cover? (From about 20°N to 20°S)

▶ What latitudes do the polar climates cover? (From about 60°N to 90°N and 50°S to 90°S)

Skills WorkOut

To help students interpret data about latitude and longitude, have them do the Skills WorkOut.

Answers
Tokyo—36°N 140°E
Paris—49°N 2°E
Moscow—56°N 38°E
Mexico City—19°N 99°W
Lima—12°S 77°W
Nairobi—1°S 37°E
Bombay—19°N 73°E
Sydney—34°S 151°E

Integrated Learning

Use Integrating Worksheets 27.1a and b.

Answer to In-Text Question

② The climate 15° south of the equator is a tropical savannah.

---

in an area. Most plants have specific water needs. For example, cacti usually grow in dry, desert conditions. The amount of precipitation determines whether grasses, trees, or desert plants will grow in an area.

Different kinds of precipitation also influence vegetation. For example, snow protects plants by providing insulation. Some plants exposed to a cold winter die without insulation. Therefore, certain plants may only be able to survive cold winters where there is snow.

Temperature

While precipitation determines what type of vegetation is in an area, temperature determines what specific plants grow there. For example, a biome may have the right amount of rain to grow grasses. Some grasses can't survive if the temperature falls below freezing. So an area that has below-freezing temperatures will only have certain kinds of grasses.

Two factors influence temperature. One is latitude, or how close an area is to the equator. The other factor is altitude, or how high an area is above sea level.

Latitude One system used to describe a location on the earth's surface is a set of imaginary lines that are parallel to the equator. This system compares locations to the equator. The north-south location of any place on the earth's surface is known as its latitude. Latitude is measured in degrees, and the equator is 0° latitude. Figure 27.2 shows parallel latitude lines. The climate is moderate at 30° north of the equator. What do you think the climate is like 15° south of the equator? ②

Temperature is influenced by latitude. Areas at or near the equator receive the direct rays of the sun, so they have a warm climate. For example, the Amazon Basin along the equator in South America has a very hot climate. Regions such as northern Canada are far from the equator and are very cold in the winter.

Altitude An area's height above sea level is called altitude. If you climbed Mount McKinley in Alaska, or one of Colorado's Rocky Mountains, you would feel changes in temperature as you climbed to the top. Areas at high altitudes have lower average temperatures than areas at lower altitudes.

Figure 27.3 ▲
Raindrops fall into the open end of a rain gauge. The amount that collects can be measured.

Skills WorkOut

Interpret Data Find latitude and longitude lines on a map or globe. Locate where you live. What are the latitude and longitude? Then find the latitude and longitude of the following cities: Tokyo, Paris, Moscow, Mexico City, Lima, Nairobi, Bombay, and Sydney.

The Living Textbook:
Life Science Side 8

Chapter 35   Frame 23229
Global Temperature Change (Movie)
Search: Play:

Chapter 27 Climate and Biomes

TEACH • Continued

Discuss
Ask students to think of possible factors that might create different microclimates in the same area. Encourage them to predict how such environmental factors could affect weather conditions and local wildlife. (Possibilities include altitude, urban development, proximity to water, and so on.)

EVALUATE

WrapUp
Reteach Write the word *climate* on the chalkboard. Have students look through this section to find the boldface terms. List them on the chalkboard. Have students make a study web that shows how each of these terms relates to climates.

Use Review Worksheet 27.1.

Check and Explain
1. A biome is a large community with similar biotic and abiotic factors.
2. The climates around the equator are tropical, with average monthly temperatures around 18°C. The rain forests around the equator receive a lot of precipitation.
3. Answers may vary. Average temperatures should be consistent with latitude and altitude.
4. City A does not have cold winters. The average temperature in January is 7°C, which is above freezing. The warmest season is in September.

582

INTEGRATED LEARNING

Integrating the Sciences
Physical Science A drop in air pressure is another signal of precipitation. Explain that air tends to flow away from high pressure to low pressure areas to maintain equilibrium. This movement can bring on precipitation. Ask students if they know what instrument meteorologists use to measure air pressure. (A barometer)

Use Integrating Worksheets 27.1a and 27.1b.

Figure 27.4 ▲
This area in the foothills of a mountain range receives the right amount of rain for growing pear and apple trees.

Science and Society
Know Your Microclimates

You and your neighbor live in the same biome and plant the same kind of grass, but your neighbor's grass is healthier than yours. Why? It might be because of different microclimates.

Microclimates are created by local differences in environmental conditions. The differences can be related to wind currents, hills, bodies of water, or even the shadows of houses. Your grass may be in shadows, preventing it from getting enough light. To grow something in the microclimate in your house's shadow, pick a plant that doesn't need direct sunlight.

Just as you can grow different plants in your yard, farmers can use microclimates to grow different crops. For example, when moist air from the ocean moves over a coastal mountain range, the air rises and cools. The moisture in the air condenses and falls as rain. The farther up the mountain, the greater the amount of rainfall. Pear trees grow best in the wet region partway up the mountain. Apricot trees grow best in the drier region at the base of the mountain.

If a farmer tried to plant the same trees in all of the microclimates, what would happen? Obviously, the crops would not be very successful. As for improving your lawn, maybe it's simply a matter of figuring out the microclimates in your neighborhood.

Check and Explain
1. What is a biome? Give an example.
2. What type of climate would you find near the equator? Explain.
3. **Make a Model** Create a model of a climate. What is the average temperature? How much precipitation is there, and what type is it? What are its latitude and altitude?
4. **Interpret Data** Create a graph of temperatures for City A using the following average temperatures: January, 7°C; March, 11°C; May, 20°C; July, 23°C; September, 24°C; and November, 10°C. Does City A have cold winters? What season is warmest?

582 Chapter 27 Climate and Biomes

TEACHING OPTIONS

Prelab Discussion
Have students read the entire activity, then discuss the following points:

▶ Discuss how average monthly temperatures are measured and calculated.

▶ Remind students that 0°C is the temperature at which water freezes.

▶ Discuss the differences between the seasons of the Northern and Southern hemispheres.

ACTIVITY 27

Time 40 minutes   Group 1–2

Materials
graph paper
calculators
colored pencils
rulers

Activity 27 How do different climates compare?

Skills Make a Graph; Interpret Data; Infer

Task 1 Prelab Prep
Collect the following items: graph paper, calculator, 2 different-colored pencils, ruler.

Task 2 Data Record
1. You will be making 2 graphs. Label one piece of graph paper *Precipitation* and another piece *Temperature*.
2. Select separate colors that will represent the data for City A and City B. Record this legend on each of the graphs.
3. On the horizontal axis of each graph, list the months of the year.
4. Refer to Table 27.1 and write appropriate numbers on the vertical axis for each graph.

Task 3 Procedure
1. On the precipitation graph, plot the precipitation data for City A with the chosen color. Repeat the procedure for City B.
2. On the temperature graph, plot the temperature data in the same manner using the same colors as in the precipitation graph for cities A and B.
3. Calculate the annual precipitation for each city and record this number on the graph.
4. Calculate the annual average temperature for each of the cities and record this number on the graph.

Task 4 Analysis
1. Describe each city's precipitation pattern and temperature pattern. Does City A have a wet season and a dry season? How about City B?
2. Can you infer which form of precipitation occurs in what months for each of the cities?
3. Describe the climate of each city based on the precipitation and temperature data.
4. Are precipitation and temperature data sufficient to identify the type of climate or biome of an area? Why or why not?
5. What kind of vegetation do you think grows around each of the cities?

Task 5 Conclusion
Write a paragraph describing how temperature and precipitation determine climate. How does climate affect the animal and plant life?

Extension
Find out your region's current precipitation and temperature pattern. Determine if there have been any changes in these patterns in the past five years. Make a table listing average temperatures and precipitation over the past five years.

Table 27.1 Climate Calculations

Month	Precipitation (cm) A	Precipitation (cm) B	Temperature (°C) A	Temperature (°C) B
Jan.	1.30	2.3	7.0	1
Feb.	1.30	2.8	10.0	3
March	1.00	5.6	12.5	10
April	0.50	7.9	17.5	15
May	0.50	10.0	22.5	18
June	0.25	13.0	29.0	20
July	1.30	10.0	32.0	20
Aug.	1.30	9.4	30.5	22
Sept.	0.77	8.9	26.5	19
Oct.	0.77	7.0	19.0	16
Nov.	1.00	4.6	12.0	12
Dec.	0.77	3.3	7.0	5

Analysis

1. City A's precipitation is low and is fairly constant throughout the year. City B's precipitation goes up during the summer months. Both cities show higher temperatures during the summer months, with City B somewhat cooler. City B has a wet season and a dry season. Temperature graphs will be more similar than precipitation graphs.

2. Since the monthly averages never actually drop below freezing, it could be inferred that both cities get rain all year round. City B probably gets some snow during January and possibly December and February.

3. City A has a desertlike climate; City B has a temperate climate.

4. Answers may vary. A biome is defined by the organisms living in an area as well as by climate.

5. Answers will vary, but should be consistent with the students' answers to question 3.

Conclusion
Accept all logical answers. Students should recognize that animal and plant life must be adapted to the particular climatic characteristics of the area.

Extension
Students will find that graphing their data makes patterns easier to notice.

SECTION 27.2

Section Objectives
For a list of section objectives, see the Student Edition page.

Skills Objectives
Students should be able to:

Infer the kind of biome that they live in.

Make a Model that shows a plant's adaptations to a particular biome.

Vocabulary
tundra, deciduous, coniferous, grassland, desert

MOTIVATE

Skills WarmUp
To help students understand land biomes, have them do the Skills WarmUp.
Answer Answers will vary. Have students refer to Figure 27.5 to help identify their biome.

Misconceptions
Explain that although the biome map shows distinct divisions, biomes generally blend together. For example, many coniferous forests have deciduous trees and vice versa.

The Living Textbook:
Life Science Sides 7-8

Chapter 17　　Frame 01753
Biomes (13 Frames)
Search:　　　　Step:

INTEGRATED LEARNING

Themes in Science
Diversity and Unity Variations among climates have created a variety of biomes. Ranging from polar to equatorial extremes, each biome has its own unique collection of organisms that is specially adapted to that environment. Differences among biomes are largely the result of varying amounts of solar energy and moisture, which are determined by latitudinal position and altitude.

Integrating the Sciences
Earth Science The characteristics of soil influence biome formation. Soil consists of varying ratios of rocks, minerals, and decomposing organic matter called *humus*. Desert soil contains very little humus; deciduous forest soil contains large amounts of humus.

Skills WarmUp
Infer Describe the area that surrounds where you live. Write about each of the following: soil, rainfall, temperature, and kinds of plants. Based on your description, infer what kind of biome you think you live in.

Figure 27.5
Major Land Biomes of the Earth ▼

27.2 Land Biomes

Objectives

▶ **List** locations of the land biomes.

▶ **Describe** the major characteristics of the land biomes.

▶ **Make a model** of a biome.

If you went to a desert in Africa, do you think you would see the same animals and plants as those in a desert in Asia? Probably not. Organisms vary in different areas of the same type of biome. However, the available niches are very similar. The general structure of the communities is also similar.

Figure 27.5 shows the locations of major land biomes. Unlike the borders on the map, biomes don't have sharp boundaries. Each biome blends with the next. On your tour of the following biomes, you'll learn about some adaptations that plants and animals have to live and reproduce in their biomes.

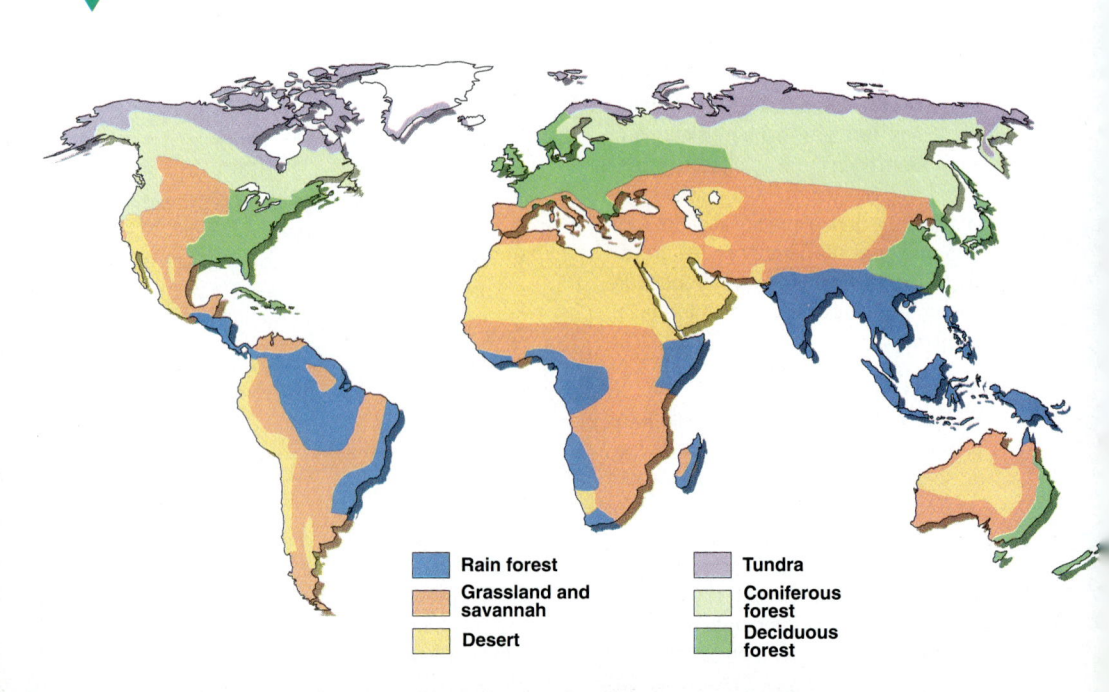

- Rain forest
- Grassland and savannah
- Desert
- Tundra
- Coniferous forest
- Deciduous forest

584　Chapter 27　Climate and Biomes

Geography Connection
Ellesmere Island, Canada, is situated 800 kilometers from the North Pole. The environment there is so harsh that the permafrost is nearly a kilometer thick in most areas and temperatures range from 21°C in the summer to −21°C in the winter. Arctic wolf, arctic fox, ermine, Peary caribou, and collared lemming live in the tundra. Have students discuss the adaptations that enable these animals to survive in the tundra.

Multicultural Perspectives
Inuit peoples have lived in the polar biome of the Arctic tundra for thousands of years. Students could do some research on Inuits and how they have adapted to such a harsh climate.

TEACH

Explore Visually
Have students study pages 584 and 585. Ask the following questions:

▶ How much of the earth's land surface is covered by tundra? Where is most of it? (One-tenth; above the Arctic Circle)

▶ What kind of animals live in the tundra? (Animals that hibernate or migrate in the winter months and a few hardy species, such as snowshoe hare and caribou, that are adapted to extreme cold)

▶ What kinds of plants live in the tundra? (Mosses, lichens, and seed-bearing plants that mature and reproduce during the short growing season)

▶ What is the climate like in the tundra biome? (Very cold and dry)

Skills Development
Classify Remind students of the three major climate zones: polar, temperate, and tropical. Ask them to identify the climate of the tundra biome. (Polar)

Tundra

The biome with low average temperatures, little rainfall, and a very short growing season is the **tundra**. Beneath the soil surface in the tundra is a permanently frozen layer of soil known as permafrost. In some areas of the arctic tundra, permafrost can be more than 500 m thick! This layer prevents trees from putting down roots. Only hardy pioneer species and the animals they support can live in the tundra.

The tundra covers about one-tenth of the earth's land surface. Most tundra is near and above the Arctic Circle, but there is also alpine tundra at the tops of mountains.

▲ Many animals in the tundra hibernate or migrate in the winter months. This snowshoe hare is a year-round resident.

▼ Plants in the alpine tundra are low and small. They often grow very slowly to conserve energy.

▲ The caribou is one of the few large animals that live in the North American tundra. Its large body size helps maintain enough heat to survive.

The Living Textbook: Life Science Side 8

Chapter 36   Frame 25821
Predator and Prey On the Tundra (Movie)
Search: Play:

The Living Textbook: Life Science Side 8

Chapter 37   Frame 29649
Tundra in Summer (Movie)
Search: Play:

Climate
Temperature: Ranges from −40°C to 10°C.
Precipitation: Less than 30 cm per year.

◀ Seed-bearing plants in the tundra must mature and reproduce during the short growing season. Non-flowering plants such as mosses and lichen are common.

TEACH • Continued

Directed Inquiry

Have students study the material on this page. Ask the following questions:

▶ What are deciduous trees? (Trees that lose their leaves during one season and grow new ones during another season)

▶ What are the dominant deciduous trees in northern areas? (Beech and maple) In southern areas? (Oak and hickory)

▶ How does the soil in deciduous forests help support a large and diverse wildlife community? (It is fertile and moist. Many layers of vegetation can grow there and support a wide variety of animals on the forest floor, in the treetops, and in the soil.)

▶ How do the fungi and bacteria of the forest floor contribute to the health of the entire biome? (Decompose and recycle materials that keep the soil fertile)

Skills Development

Infer Are most of the deciduous forests of the United States in the western or eastern half? Why? (The eastern half, because deciduous trees require a temperate climate)

**The Living Textbook:
Life Science Side 8**

Chapter 39 Frame 31307
Deciduous Forest (Movie)
Search: Play:

586

INTEGRATED LEARNING

STS Connection

Explain that much of the world's forests have been and continue to be destroyed to create land for grazing cattle or to use the wood for fire. For example, most of the countryside in England was densely forested before it was cleared for pasture land. Many people want to use the land for farming instead of grazing, since farm crops produce more food than cattle do. Other people believe that the forests should be preserved in their natural state. Ask students what they think.

Deciduous Forest

Trees that lose their leaves during one season and grow new leaves during another are called **deciduous** (dee SIHJ oo uhs) trees. The biome with deciduous trees, moderate rainfall, six months of mild temperatures, and a cold or dry season is deciduous forest. The soil in deciduous forests is rich in organic material and can hold a lot of water. This means that many plants can live on the forest floor.

Animals and plants live in different layers of the forest. Some live in the treetops, or canopy, while others live on the trees. Deciduous forests are located in parts of North America, South America, Europe, and Asia.

Squirrels, chipmunks, and numerous species of birds inhabit the canopy in the deciduous forest biome. ▼

This raccoon eats ▶ smaller animals and plants. Night vision and feet adapted to climbing or walking the forest floor make it an efficient predator.

▲ The deciduous forest floor supports a variety of fungi and bacteria that help decompose and recycle nutrients.

Climate
Temperature: Varies greatly with region; ranges from −20°C to 30°C.
Precipitation: 50 to 150 cm per year.

Dominant tree species ▶ vary in deciduous forests. In northern areas, beech and maple trees are common. In the south, oak and hickory trees are the dominant species.

586

Themes in Science

Scale and Structure The structure of most trees in coniferous forests helps them survive in the cold and snowy climate typical of this biome. The leaves on most trees are needle-shaped and waxy. Ask students to predict how the shape of these leaves helps keep the branches from breaking during a heavy snowfall. (The snow slides off the needles.)

Integrating the Sciences

Earth Science Ten thousand years ago, the land that is now covered by coniferous forests was under ice. When the ice melted, a land bridge between Russia and Alaska was formed. This land bridge was covered a few thousand years ago by the Bering Straits. This may partially explain why many similar plants and animals live on both continents. Have students find the Bering Straits on a globe.

Directed Inquiry

Have students study the material on this page. Ask the following questions:

▶ What are coniferous trees? (Trees that bear their seeds in cones and do not shed all of their leaves at one time)

▶ How are the branches of coniferous trees well-adapted to long, cold winters? (Their branches are flexible and can bend under the weight of snow.)

▶ How does the forest floor in a coniferous forest differ from the forest floor of a deciduous forest? (The soil of a coniferous forest is less fertile.)

▶ How is the climate of a coniferous forest different from that of a deciduous forest? (A coniferous forest is somewhat colder and drier.)

Skills Development

Classify How would you classify the climate of the coniferous forest biome? (It is a cool, temperate climate.)

Coniferous Forest

The biome with long cold winters of heavy snowfall and three to six months of growing season is the **coniferous** (koh NIHF ur uhs) forest. Coniferous trees, such as spruce, fir, pine, and hemlock, bear their seeds in cones. They do not shed their leaves at one time.

The soil of coniferous forests does not support many decomposers. Smaller animals live in the canopy and trees, while larger animals live on the forest floor. Large areas of Canada and northern Europe have coniferous forests. These forests also occur at high elevations in the western United States and in such states as Maine and Michigan.

Seed-eating birds, such as this evening grosbeak, can find their food in cones. ▼

▲ The porcupine lives on the ground and eats plants. Other plant-eaters include deer, moose, and elk.

Forest predators include the timber wolf, Canadian lynx, grizzly bear, and red fox. The fox is a fast runner, since some of its prey can escape to the trees. ▼

Climate
Temperature: Ranges from –30°C to 20°C.
Precipitation: 40 to 125 cm per year.

◀ Coniferous trees are well adapted to long cold seasons. Their branches are flexible and can bend under the weight of snow.

The Living Textbook:
Life Science Side 8

Chapter 38   Frame 30622
Coniferous Forest (Movie)
Search: Play:

TEACH • Continued

Directed Inquiry

Have students study the material on this page. Ask the following questions:

▶ What are the climatic conditions of a rain forest biome? (Heavy rain, high humidity, and warmth)

▶ How does the soil of a tropical rain forest differ from the soil of a deciduous forest? (The soil in a tropical rain forest is less fertile because most forest nutrients are stored in the biomass, and dead organisms are recycled very quickly.)

▶ Where in North America are temperate rain forests? (The Pacific Northwest)

Skills Development

Infer Why are epiphytes found more often in rain forests than in deciduous or coniferous forests? (There is more moisture in the air in rain forests.)

Decision Making

If you have classroom sets of *One-Minute Readings,* have students read Issue 20, "Destruction of Tropical Forests" on page 33. Discuss the questions.

**The Living Textbook:
Life Science Side 8**

Chapter 44   Frame 39407
Tropical Rain Forest (Movie)
Search: Play:

588

INTEGRATED LEARNING

Integrating the Sciences

Earth and Physical Science On the Pacific coast of the United States, heavy rainfall supports a rare temperate rain forest. Some of the world's largest trees, including redwoods and Douglas firs, flourish here because of year-round moisture from a high-precipitation weather belt. This microclimate, between the Pacific Ocean and inland mountains, is created by a weather phenomenon known as a *rain shadow.* Air approaching from the west is forced up over the mountains. As the air rises, it cools and condenses into rain. Have students locate the Pacific coast on a map. Then have them draw a diagram of the rain shadow weather pattern.

Rain Forest

The biome with heavy rainfall, high humidity, and a warm environment is the tropical rain forest. The soil in the rain forest has little organic matter. The heavy rains wash away most of the nutrients in the soil. But this biome contains the greatest diversity of life. The moisture and warmth create a climate for millions of plant and animal species.

Tropical rain forests are located near the equator in regions of Central and South America, Africa, Australia, and central Asia. There are also rain forests in cool climates in regions that have a lot of rainfall and fog. These forests are called temperate rain forests.

There are more different kinds of organisms living in and on a single rain forest tree than in an entire coniferous forest! Many trees grow to more than 45 m high.

▲ Three-toed sloths use their claws to hang from branches in tropical rain forests. They spend most of their lives in the trees and rarely come down to the ground. They climb high into the canopy of the rain forest. The canopy provides food and a place to sleep.

▲ At the tops of the rain forest trees are epiphytes like these orchids. Epiphytes are plants that gather nutrients and water from the air.

Climate
Temperature: Average is 18°C.
Precipitation: 200 to 500 cm per year.

▶ Along the western coast of North America there is a temperate rain forest. The conifers here grow much larger than trees in a coniferous forest. Lichens, mosses, and algae grow on nearly every surface.

588

Multicultural Perspectives

Grassland biomes cover about one-fourth of the earth's land surface. Grassland biomes occur at about the same latitude as deciduous forest biomes, but do not receive enough moisture to support trees. Grasslands are known under various names in different regions of the world. They are called pampas in South America, prairie in North America, steppes in Asia, and veldt in South Africa. The grassland biomes in these parts of the world support a wide variety of animal life. For example, the dominant grazing mammals of the African veldt are antelope, giraffes, and elephants. Ask students what large grazing mammals inhabited the North American prairie. (Students may mention buffalo and antelope.)

Grassland

Regions that have hot summers, cold winters, rich soil, and unevenly distributed rainfall support the growth of different species of grasses. Trees can't grow in these areas because there is not enough rain. Biomes that are characterized by grasses are known as **grasslands**.

There are different grasslands all over the world. They include the Great Plains in the United States, the pampas (PAHM puhz) in South America, and the veldt (VEHLT) in South Africa.

Each type of grassland supports different species of grasses and animals. The primary consumers may include birds, mice, rabbits, and gophers. There are also predators, such as coyotes in the United States and lions in Africa. Grazing animals also live in grasslands, such as these sheep in New Zealand shown below.

Climate
Temperature: –8°C to 28°C.
Precipitation: 25 to 75 cm per year.

Consider This

How Should Rain Forests Be Used?

More than two-thirds of the people on the earth live in countries with tropical rain forests. More than 10 000 sq km of rain forests, per year, are cleared for building materials, to supply wood for fuel, to build new towns, and for other reasons.

Consider Some Issues
People who live near rain forests depend on them for food, water, shelter, and medicines. Farm animals graze on the cleared land. People plant crops that provide food. The forests also provide work for farmers, ranchers, and loggers. As populations increase in these countries, more and more space and food are needed.

Rain forests are home to more than half of the earth's plants and animals. Habitats are lost when forests are cleared. Rain forests help cycle the earth's available fresh water. The plants in rain forests produce much of the earth's oxygen. Some countries will lose all their tropical rain forests within a few decades if the rate of destruction continues.

Think About It How do you think rain forests should be used? What steps can people take now to save rain forests?

Write About It Write a paper stating your position on rain forest use. Include your reasons for choosing your position.

Chapter 27 Climate and Biomes

Directed Inquiry
Have students study the material on this page. Ask the following questions:

▶ What are the climatic conditions of a grassland biome? (Hot summers, cold winters, and unevenly distributed rainfall)

▶ Why can't forests grow in these areas? (Not enough rain)

▶ Name some states in the United States that have grassland biomes. (Kansas and Nebraska)

Consider This

Think About It Explain to students that after rain forests are cleared, the soil quickly loses its productivity. Ask students why this happens. (Heavy rains carry away the soil's nutrients.)

Write About It In considering the issues of rain forest preservation, students should find out about the greenhouse effect. Have them explain in their papers how the destruction of the world's rain forests is contributing to this serious global problem.

The Living Textbook:
Life Science Sides 7-8

Chapter 17 Frame 01982
Grassland (44 Frames)
Search: Step:

The Living Textbook:
Life Science Side 8

Chapter 40 Frame 32402
Grasslands/Savannah (Movies)
Search: Play:

TEACH • Continued

Directed Inquiry

Have students study the material on this page. Ask the following questions:

▶ What are the climatic conditions of a desert biome? (Scarce rainfall, high air pressure, and high winds)

▶ What are the main factors that limit life in the desert? (Heat and lack of water)

▶ Describe some adaptations of desert plants. (Some plants reproduce quickly after periods of rainfall. Cacti have shallow roots that take up rainwater quickly before it evaporates. Cacti also lack leaves, which reduces water loss from evaporation.)

▶ What are some adaptations of desert animals? (Animals such as kangaroo rats and jackrabbits avoid the heat by staying underground or in the shade during the day. Animals such as lizards and snakes are light in color to reflect sunlight and thereby absorb less heat.)

The Living Textbook: Life Science Side 8
Chapter 42   Frame 34553
Desert Formation (Movie)
Search: Play:

The Living Textbook: Life Science Side 8
Chapter 43   Frame 35586
Life in the Desert (Movie)
Search: Play:

INTEGRATED LEARNING

Themes in Science

Evolution The desert climate is one of hot days and cold nights with very little precipitation. Animals of the desert biome have a variety of adaptations that enable them to survive in such harsh conditions. For example, skinks and lizards often hibernate in the cold winter months and estivate (sleep) during the hottest periods of the summer. Scorpions have sand-colored bodies and can survive for more than a year without eating. Desert gerbils and warty toads never need to drink because they obtain all the water they need from food. Encourage students to find out more about the unique adaptations of desert wildlife.

Desert

Areas that have greater rates of evaporation than rainfall are known as **deserts**. Desert biomes have scarce rainfall, high air pressure, high winds, and poor soils. Heat and water are the main factors that limit life in the desert. Each species of plant and animal has methods of conserving water and staying cool.

Hot deserts are located between latitudes of 20° north and 30° south of the equator. Cold deserts are located near the earth's poles. Currently more than one-third of the earth's land area is desert.

Thorny bushes and cacti are typical desert plants. Cacti have shallow root systems that take up water rapidly when it rains. The lack of leaves decreases water loss from evaporation. ▼

▲ Kangaroo rats, pocket mice, and jackrabbits avoid the heat by remaining in the shade or underground during the day.

Lizards, skinks, and snakes are desert predators. They eat seed-eaters, such as ants, rodents, and birds. They are often light in color so they can stay cool by reflecting sunlight. ▼

Climate
Temperature: Ranges from 5°C to 60°C.
Precipitation: Less than 25 cm per year.

To reproduce, some ▶ desert plants bloom quickly after periods of rainfall. Springtime in the Mojave Desert in California, for example, has a brilliant-colored landscape.

Social Studies Connection

In the early 1800s, the fertile farmland of the present-day Great Plains was known as the Great American Desert. It wasn't until the middle of the century that people began to see that this so-called desert was excellent farmland. News got around and homesteaders flocked to the area to plant crops. By the early 1930s, the natural vegetation of much of the land had been plowed up by farmers. A four-year drought beginning in 1933 killed crops and dried up the soil. With no strong grass to hold the soil, millions of tons of fertile topsoil were blown up into black dust clouds by the wind. For four years, dust storms battered the area that came to be known as the Dust Bowl. Ask students how this might have been prevented. (By using irrigation and alternative methods of plowing)

Science and Society Desertification

Worldwide, deserts claim more than 70 000 sq km of land per year. Although deserts expand and shrink naturally, human activity can also cause desert expansion, or desertification. Desertification destroys land that can be used for farming or grazing.

One way people cause desertification is through soil erosion. When forests are cut down or lands are overgrazed, the soil can dry out and be blown away. Scientists have calculated that between 1972 and 1981, more than 400 million tons of African soil were blown over the Atlantic Ocean each year.

Some farming practices expose soil. This is what happened in the Great Plains of the United States in the 1930s. Farmers plowed the plains to plant wheat, then high winds carried most of the soil into the air. The dust clouds were estimated to contain 355 million tons of topsoil.

Another farming practice that has led to erosion problems is the slash-and-burn technique. In this process, trees and plants are burned, leaving the soil unprotected. When rain falls on unprotected soil, huge amounts of the soil wash away.

Throughout the world, people are working to stop the spread of deserts. Some people are working to stop deforestation. Others are trying to teach people different ways of managing the land they farm.

Figure 27.6 ▲
When winds carried away its topsoil in the 1930s, this area was named the Dust Bowl.

Check and Explain

1. Describe how the tundra differs from the desert.
2. Name two biomes that you might expect to see if you climbed a very high mountain.
3. **Reason and Conclude** Tropical rain forests contain plants that grow on trees. Give reasons why these plants can survive even though they do not obtain their nutrients from roots in the ground.
4. **Make a Model** Select a biome and design a plant that could survive in that environment. Explain the plant's adaptations to the biome.

Research
Have students write a short research report about ways to prevent or reverse desertification.

EVALUATE

WrapUp
Reinforce Have students make a circle graph depicting the median temperature and precipitation for each of the six land biomes. The graph should have one line labeled *Temperature* and one line labeled *Precipitation*.
Use Review Worksheet 27.2.

Check and Explain
1. The tundra's temperature range is from –40°C to 10°C; the desert ranges from 5°C to 60°C. The tundra has a permafrost layer; the desert has a greater rate of evaporation than of rainfall and high winds and air pressure.
2. One might see deciduous forest at lower altitudes and alpine tundra at the peak.
3. Rain forest plants, such as orchids, get nutrients from water and air. The heavy rainfall, humidity, and warmth in rain forests produce air that is rich in nutrients.
4. Answers may vary. Possibilities include a tree in a coniferous forest biome that has thin needlelike leaves that conserve moisture.

Decision Making
If you have classroom sets of *One-Minute Readings*, have students read Issue 13, "Trees vs. Desert: A Project" on page 22. Discuss the questions.

SECTION 27.3

Section Objectives
For a list of section objectives, see the Student Edition page.

Skills Objectives
Students should be able to:

Compare fresh and salt water.

Infer what type of water biome an insect lives in based on its body structure.

Compare and Contrast still and fast-moving water environments.

Vocabulary
marine

MOTIVATE

Skills WarmUp
To help students understand water biomes, have them do the Skills WarmUp.
Answers The ocean contains salt water and has stronger tides and waves. Both bodies of water have similar surface temperatures and some similar animals and plants.

Prior Knowledge
Gauge how much students know about water biomes by asking the following questions:

▶ What are the main differences between marine and freshwater biomes?

▶ Why do whales live in oceans and not in lakes?

▶ Why are organisms in lakes different from those in streams?

592

INTEGRATED LEARNING

Themes in Science
Systems and Interactions
Variations in ocean surface temperatures influence climates around the world. From late in 1982 to the middle of 1983, unusually violent weather, including hurricanes, floods, and severe droughts, swept the globe. Scientists traced these events to ocean temperature changes. A warm Pacific current known as El Niño flows eastward every three to ten years as a result of weakening trade winds. As El Niño moves into normally cool ocean waters, it can create atmospheric changes that affect as much as 70 percent of the earth's weather. Have students find out more about El Niño and its influence on global weather patterns.

Skills WarmUp

Compare Imagine that you just completed training in scuba diving. Describe the kind of water you would swim in if you went scuba diving in ocean water. What would it be like if you then swam in Lake Superior? Describe the similarities and differences of the two bodies of water.

Figure 27.7
Major Zones of the Marine Biome ▼

27.3 Water Biomes

Objectives

▶ **Describe** the characteristics of marine and freshwater biomes.

▶ **Identify** adaptations of organisms living in water biomes.

▶ **Compare** and **contrast** two freshwater environments.

Why do you think the earth is often called the water planet? If you said it's because most of the earth is covered with water, you are right. More than 70 percent of the earth's surface is water.

Although most of the earth's water is in the oceans, a small percentage is fresh water. What is the main difference between ocean water and fresh water? The answer is salt. The oceans contain nearly 3.5 percent salt. Fresh water, however, contains less than 0.006 percent salt.

Littoral Zone ←→ **Sublittoral Zone**

Littoral Zone
The place where land and ocean join is called the littoral (LIHT uh ruhl) zone, or seashore. Organisms in this zone must adapt to constant changes caused by the tides. They must also survive temperature changes and the pounding action of waves.

Sublittoral Zone
From the shore, the land slopes downward and the water becomes much deeper. This area is called the sublittoral zone, or the continental shelf. Many types of organisms live here because there is an abundance of sunlight and nutrients.

592

Integrating the Sciences

Chemistry Salinity is the amount of dissolved salts present in water. As the surface of an ocean evaporates, its salinity level rises because there is more salt left in relation to the amount of remaining water. However, the salinity of ocean water remains stable because "fresh water" with low salt levels is constantly flowing into ocean water from rivers and in the form of rain. Salinity differs slightly in different areas of the ocean, but never becomes greatly imbalanced because all substances eventually become evenly spread throughout the water. Ask students why salinity levels might differ slightly in different parts of the ocean. (Some areas are hotter and drier and therefore experience more evaporation and less rain.)

Marine Biome

The ocean environment is called the **marine** biome. Look at the marine biome in Figure 27.7. Marine life varies from the microscopic plankton, which are a food source for many animals, to the giant sperm whale, which grows to more than 15 m long. How many organisms do you recognize?

Marine life is limited by a number of factors. One major factor is salinity, or salt composition. Marine organisms must adapt to the salt in water. They have the ability to remove excess salts.

Water pressure is also a limiting factor. As you go deeper into the ocean, the pressure increases greatly. Some organisms can live only in surface waters, while others can easily dive to great depths.

Light and temperature also limit marine life. These factors vary throughout the ocean. In the deepest parts of the ocean, some organisms have adapted to a very dark and cold environment.

Pelagic Zone
Most of the ocean is in the pelagic (pih LAYJ ihk) zone. Here, sunlight is only visible to a depth of about 200 m. Below 200 m, the ocean becomes very cold and dark. Water pressure increases greatly. Animals that live deep in this zone have special adaptations for survival.

TEACH

Explore Visually
Have students study the material on pages 592 and 593. Ask the following questions:

▶ What is the littoral zone, and what are some animals that live there? (The littoral zone is the place where land and ocean join. Animals such as crabs, mussels, and starfish live there.)

▶ What is the sublittoral zone, and what are some animals that live there? (The sublittoral zone is where the land slopes downward away from the shore and the water becomes deeper. Sponges, sea urchins, jellyfish, and a variety of other fishes live there.)

▶ What is the pelagic zone, and what are some animals that live there? (The deep part of the ocean; Plankton, herring, dolphins, cod, sperm whales, hammerhead sharks, stingrays, hatchetfish, and deep-sea anglerfish live there.)

Integrated Learning
Use Integrating Worksheet 27.3a.

Enrich
Use Enrich Worksheet 27.3.

The Living Textbook:
Life Science Side 7-8

Chapter 17 Frame 02243
Marine Communities (112 Frames)
Search: Step:

The Living Textbook:
Life Science Side 8

Chapter 45 Frame 42076
Deep Ocean Vents (Movie)
Search: Play:

TEACH • Continued

Skills WorkOut
To help students infer where insects come from based on their body structures, have them do the Skills WorkOut.

Answer The insects came from a stream. Their streamlined bodies and suctionlike appendages enable them to cling to rocks or other objects in fast-moving water.

Directed Inquiry
Have students study the material on this page. Ask the following questions:

▶ How are zones in marine biomes and stillwater biomes similar? (In both stillwater and marine biomes there is a shore, a shallow area off shore, and then a drop-off into deep water. Fish feed on plankton near the shore and the water surface.)

▶ Describe some adaptations of animals in fast-moving water. (Fishes such as salmon and trout are strong and have streamlined bodies to move easily through fast currents.)

▶ Why do scavengers and bottom-feeding organisms generally live in slow-moving water? (Organic matter for feeding settles to the bottom of slow-moving water, whereas it is washed away in fast currents.)

Answer to In-Text Question
① **Plankton, fishes, insects, and other organisms that live on the bottom**

INTEGRATED LEARNING

Language Arts Connection
Limnology is the branch of biological science that involves investigating lakes and ponds. Two terms describing the condition of a lake are *eutrophic* and *oligotrophic*. Students can look up the meanings of these two terms and share them with the class. (The prefix *eu-* means "well." *Oligo-* comes from Greek and means "few." The root *-trophic* comes from a word meaning "nourishment.")

Skills WorkOut
Infer A friend of yours just returned from a vacation and brought you insects for an aquarium. The insects had streamlined bodies and suction-like appendages. Infer if the insects came from a pond, a lake, or a stream. Explain your choice.

Figure 27.8 ▲
What organisms do you think live in this lake? ①

Figure 27.9
Plants grow in a shallow pond ▶ in Grand Teton National Park.

Freshwater Biomes
Organisms living in freshwater biomes have adaptations for regulating the amount of water in their bodies. A fish's gills remove most excess water. A plant releases water through openings in the undersides of its leaves. Some freshwater organisms adapt to life in running water. Some adapt to life in still water.

Still Water In the still water of lakes and ponds, light penetrates only to a certain depth. Most of the oxygen, too, is at the surface. The surface of a stillwater biome, therefore, is where most photosynthesis occurs. Deeper waters are rich in nutrients because plants and algae do not grow there and use the nutrients.

Organisms in lakes and ponds almost always have a large supply of food. Because still waters have little current, nutrients are not washed away, and plankton can easily grow.

Lakes have life zones that are similar to the marine biome. Fishes that feed on plankton live close to the surface of lakes. Plants with roots, birds, and frogs all live near the shore. In the deeper part of a lake, near the bottom, live worms, bacteria, and fungi.

Running Water Streams and rivers contain running water. Certain factors, such as light, water current, temperature, and the availability of nutrients, determine the kinds of organisms that can live in running water. For example, fishes living in fast-moving streams are strong swimmers. They generally have large fins to guide their movements. Some fishes, such as brook

STS Connection

Slow-moving rivers and stillwater lakes are easily polluted. This is because there is less dissolved oxygen in the water to break down pollutants, and there are no currents to sweep away pollutants. Have students investigate the cleanup of Lake Erie. They should find out what the major pollutants were and how the cleanup was undertaken.

SkillBuilder   *Comparing and Contrasting*

Salt Water Versus Fresh Water

Have you ever spent a long time in the water? What happens to your skin? Does the ocean affect your body differently than a lake? You will be investigating how fresh water and salt water affect organisms. Collect the following items: 8 glass jars, salt, spoon, 2 small equal-sized pieces each of apple and potato, 2 raisins, 2 grapes.

1. Fill 8 jars with tap water. Label 4 jars *Salt Water* and 4 jars *Fresh Water*. Put 2 spoonfuls of salt into each of the saltwater jars and stir.
2. Copy the data table to the right. Your version may be larger.
3. Observe the size and texture of each fruit and vegetable. Predict the effects of salt water and fresh water for each. Record your observations and predictions in the table.
4. Place one sample of each fruit and vegetable in a saltwater jar and one sample in a freshwater jar.
5. Observe each of the samples after one hour. Record these observations in your table.
6. What happened to each fruit and vegetable? Why do you think you got those results?
7. Describe the adaptations you think are needed to survive in a saltwater environment.

Data Table

	Potato	Apple	Grape	Raisin
Before				
Salt prediction				
Salt water				
Fresh prediction				
Fresh water				

---

trout, have streamlined bodies that give less resistance to fast currents. Other animals, such as snails, have sticky undersides that help them cling tightly to rocks.

In most fast-moving water, tumbling action and cool water provide much dissolved oxygen. Also, light can reach the bottom. However, little photosynthesis takes place because plants are swept away. So are many nutrients, preventing the growth of plankton. Organisms must feed on insect larvae and on plants and algae that are anchored to the bottom or attached to rocks. Fallen land plants are also a source of food.

In slow-moving streams and rivers, organisms require less oxygen and warmer water. Organic matter is more likely to settle to the bottom. This provides food for scavengers and bottom-feeding organisms, such as tube-dwelling worms, crayfish, and numerous kinds of insects. Such fishes as muskies and pike, as well as turtles, move more easily in streams with a slow current.

Figure 27.10 ▲
An Alaskan brown bear catches a salmon in a fast-moving stream. What other kinds of organisms live in fast-moving water? ②

Chapter 27   Climate and Biomes   595

SkillBuilder

Answers
The following would happen to each fruit and vegetable:
Potato—softens in salt water; no change in fresh water
Apple—softens in salt water; no change in fresh water
Grape—wrinkles in salt water; no change in fresh water
Raisin—no change in salt water; swells in fresh water
Water leaves the fruit or vegetable in salt water except for the raisin, which is already dehydrated. The raisin swells in fresh water because it absorbs water.

Reteach
Have students work in cooperative groups of four. Ask each group to make a list of organisms that live in a slow-moving stream or river and a fast-moving stream or river. Have groups compare their lists and determine if they are accurate.

Answer to In-Text Question
② Snails, trout, certain insects, and algae and other anchored plants live in fast-moving water.

TEACH • Continued

Misconceptions
Students may think that floods always cause disasters. Explain that flooding is a natural process that helps revitalize the surrounding land. Flood waters leave behind organic materials that fertilize the soil.

EVALUATE

WrapUp
Reinforce Have students compare a brook trout to a sperm whale. Ask the following questions:
▶ How are the animals alike and how are they different?
▶ How are their biomes different?
▶ How has each animal adapted to its biome?

Use Review Worksheet 27.3.

Check and Explain
1. It contains 3.5 percent salt; water pressure increases with depth; its temperature and light change with depth.
2. It would die because its tissues are adapted to a salt water environment.
3. Pollution could possibly block light needed for photosynthesis. Fewer plankton would grow, resulting in the death of plankton-eating fish.
4. Organisms in still water rely on plankton. In fast-moving water organisms must feed on insect larvae and on organisms that are anchored to the bottom.

Answer to In-Text Question
① Rivers provide water for irrigation and a means of transportation.

TEACHING OPTIONS

Cooperative Learning
Divide the class into pairs and have each choose a river to research. Students might investigate any of the following rivers: Nile, Missouri, Colorado, Ganges, Congo, Amazon, Albany, Jordan, Volga, Mississippi. Have pairs make a poster with drawings and descriptions of the river's physical characteristics, such as its location and size. Students should also include information about the human communities that live along the river and how they depend on it for survival.

Figure 27.11 ▲
Why do people tend to live along rivers? ①

Science and Society
Living by the Yellow River

Rivers are important not only to organisms such as fishes. They are important to people as well. One of the most populated river regions in the world is the Yellow River Valley in China. More than 100 million people live there.

The Yellow River's name comes from its color. The river passes through land areas where the soil erodes rapidly. This erosion fills the river with yellowish sand and silt. Most of the sand flows into the sea. However, the silt builds up in the riverbed. When the riverbed gets very high and a storm comes, the water floods out of the river. The floods are so frequent and disastrous that the river has been nicknamed China's Sorrow.

Why do people live near a river that floods? One major reason is that the silt and water from the river create very fertile cropland. Even in times of drought, the land near the river produces a lot of grain and cotton. The water is also used for water-power stations and industry.

Since 1949, China has invested much money in projects to control the river. These projects include dikes to block overflow and gates to change the water's direction. The government also built a canal called the People's Victory Canal. The canal has provided effective crop irrigation. The efforts to control the river have been successful so far. There have been no serious floods of the Yellow River for more than 40 years.

Check and Explain
1. Describe three characteristics of the marine biome.
2. If you place an organism that lives in salt water into a freshwater environment, will it survive? Explain.
3. **Reason and Conclude** In the past, garbage was regularly dumped into the oceans and the Great Lakes. If this practice happened in smaller lakes or ponds, what would be the effect on these stillwater systems?
4. **Compare and Contrast** Explain why the food sources for organisms living in ponds differ from the food sources for organisms living in fast-moving streams. How does sunlight affect these two environments?

Chapter 27 Review

Concept Summary

27.1 Climate
- Climate is an area's pattern of weather and temperature.
- Latitude and altitude influence climates.
- Different climates support different vegetation.
- A biome is a large land or water area with similar climate, vegetation, and plant and animal life.

27.2 Land Biomes
- The earth's major land biomes are tundra, deciduous forest, coniferous forest, tropical rain forest, grassland, and desert.
- Land biomes are characterized by climate and dominant plant species.

27.3 Water Biomes
- Most of the earth's water is in the marine biome.
- The freshwater biome is made up of running water and stillwater environments.
- Organisms have special adaptations for the different water biomes.

Chapter Vocabulary

climate (27.1)	polar (27.1)	deciduous (27.2)	desert (27.2)
biome (27.1)	temperate (27.1)	coniferous (27.2)	marine (27.3)
tropical (27.1)	tundra (27.2)	grassland (27.2)	

Check Your Vocabulary

Use the vocabulary words above to complete the following sentences correctly.

1. An area's average rainfall and temperature determine its ___.
2. A ___ rain forest biome has high rainfall, high humidity, and constant warm temperatures.
3. If you were in a biome with high air pressure, little precipitation, high winds, and poor soil, you'd probably be in a ___.
4. Trees that lose their leaves in the autumn and grow new ones in the spring are called ___.
5. You can find permafrost in the ___ biome.
6. The ___ biome is made up of the ocean.
7. Winters are long and cold in the ___ biome.
8. Trees can't grow in the ___ biome.
9. Large regions with distinctive weather, temperature, and plant and animal life are called ___.
10. Spruce, fir, and pine are examples of ___ trees.
11. The climate of the middle latitudes is ___.

Explain the difference between the words in each pair.

12. tropical, polar
13. latitude, altitude
14. deciduous forest, coniferous forest
15. littoral zone, sublittoral zone

Write Your Vocabulary

Write sentences using each vocabulary word above. Show that you know what each word means.

CHAPTER REVIEW 27

Check Your Vocabulary

1. climate
2. tropical
3. desert
4. deciduous
5. tundra
6. marine
7. tundra
8. tundra/marine
9. biomes
10. coniferous
11. temperate
12. Tropical refers to the area around the equator, while polar refers to the areas around the earth's poles.
13. Latitude refers to the distance north or south of the equator; altitude refers to height above the surface of the earth.
14. A deciduous forest contains trees that lose their leaves each year; a coniferous forest contains trees that bear their seeds in cones.
15. The littoral zone refers to coastal area; the sublittoral zone is the continental shelf.

Write Your Vocabulary

Students' sentences should show that they know both the meanings of the words and how to apply the words.

Use Vocabulary Worksheet for Chapter 27.

CHAPTER 27 REVIEW

Check Your Knowledge

1. Water biomes include marine, still water, and running water. People get most of their water from freshwater sources. Most of the water on the earth is located in oceans.
2. The land biomes are: rain forest, grassland, desert, tundra, coniferous forest, and deciduous forest.
3. Coniferous forests are made up of trees that bear cones and do not shed their leaves at one time. Deciduous forests are made up of trees that lose their leaves once a year.
4. Precipitation is measured using a rain gauge.
5. The higher the latitude (closer to the poles), the colder the climate.
6. Freshwater organisms lack mechanisms for removing excess salt from their bodies.
7. They would be washed away by the current.
8. The tropical zone has warm average monthly temperatures, with little variation. In the polar zone, there is a very cold winter season and a mild summer season. In the temperate zones, there is a definite cold season and warm season, but without the extremes of the tropical or polar zones.
9. Different kinds of plants need different amounts of water to survive, and cannot adapt to other climates.
10. False; can
11. True
12. True; (or *identical*)
13. True
14. False; rich
15. False; altitude

Check Your Understanding

1. Answers may vary, but could include: precipitation, temperature, presence of other organisms, availability of water.
2. Answers will vary. See pages 584 to 591 for descriptions of the land biomes.
3. The factors are salinity, water pressure, light, and temperature.
4. The leaves are changing color in response to decreasing periods of daylight and lowering temperature. Squirrels, chipmunks, and birds might live in the canopy. Raccoons might live on the forest floor, along with small plants, bacteria, and fungi.
5. Answers may vary. Possible answers include: People may not understand the effects of their actions; it may be the easiest way for them to make a living, so they do not care about the effects.
6. Students should list several reasons why the biotic factors, abiotic factors, and adaptations of organisms in their biome are compatible.
7. Answers will vary. Make sure students correctly identify characteristics of the biome they choose.

Chapter 27 Review

Check Your Knowledge
Answer the following in complete sentences.

1. What are the water biomes? Where do people get most of the water they use, and where is most of the water on the earth located?
2. What are the land biomes?
3. Compare and contrast coniferous and deciduous forests.
4. How is precipitation measured?
5. How is climate influenced by latitude?
6. Explain why most freshwater organisms cannot live in saltwater environments.
7. Explain why most rooted plants cannot survive at the bottom of fast-moving streams.
8. Name and describe the three major climate zones on the earth.
9. Why does precipitation influence the kinds of vegetation that grow in an area?

Determine whether each statement is true or false. Write *true* if it is true. If it is false, change the underlined word to make the statement true.

10. People <u>cannot</u> live in deserts.
11. Water covers more than <u>70</u> percent of the earth.
12. Where there are similar environmental conditions, there are <u>similar</u> biomes.
13. Tundra is one of the <u>largest</u> biomes.
14. Grassland biomes have hot summers, cold winters, and <u>poor</u> soils.
15. An area's height above sea level is called <u>latitude</u>.

Check Your Understanding
Apply the concepts you have learned to answer each question.

1. Describe at least three factors that influence where organisms can live.
2. **Compare and Contrast** Compare the climate and animal and plant species of two different land biomes. Give examples of where each biome is found.
3. Describe the four factors that limit life in the marine biome. Using Figure 27.7 on pages 592 and 593, give examples of how organisms have adapted to each of these factors.
4. **Mystery Photo** The photograph on page 578 was taken during autumn on the upper peninsula of Michigan. The photograph shows a deciduous forest, which supports a wide variety of plant and animal life. What is happening to make the leaves change color? What animals and plants might live in the deciduous forest canopy? What lives on the forest floor?
5. Why do you think some people continue to farm in ways that may cause desertification?
6. **Extension** Create an imaginary land biome. Describe its temperature, precipitation, latitude, and altitude. What kinds of animals and plants live there? How are they adapted to the biome? What will you call your biome? How is it different from others?
7. **Application** Choose the biome in which you would most like to live. Explain the reasons for your choice.

598 Chapter 27 Climate and Biomes

Develop Your Skills

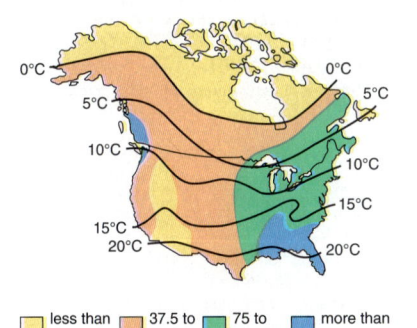

1. a. Biomes noted on this map should include rain forest, grassland and savannah, tundra, coniferous forest, deciduous forest, and desert. Students should locate biomes on this map according to the color-keyed map on page 584.
   b. Answers will vary by region.
   c. Answers will vary by region.
2. a. Coniferous forest
   b. Tropical rain forest
3. Student posters will vary, but should accurately present the plants and animals of a particular biome.

Make Connections

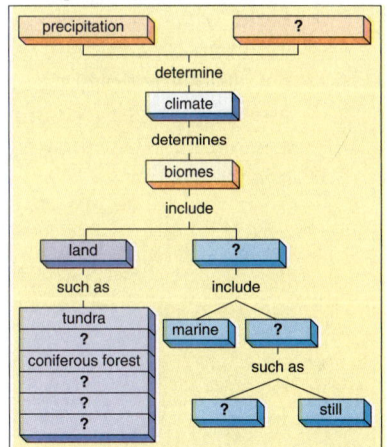

1.

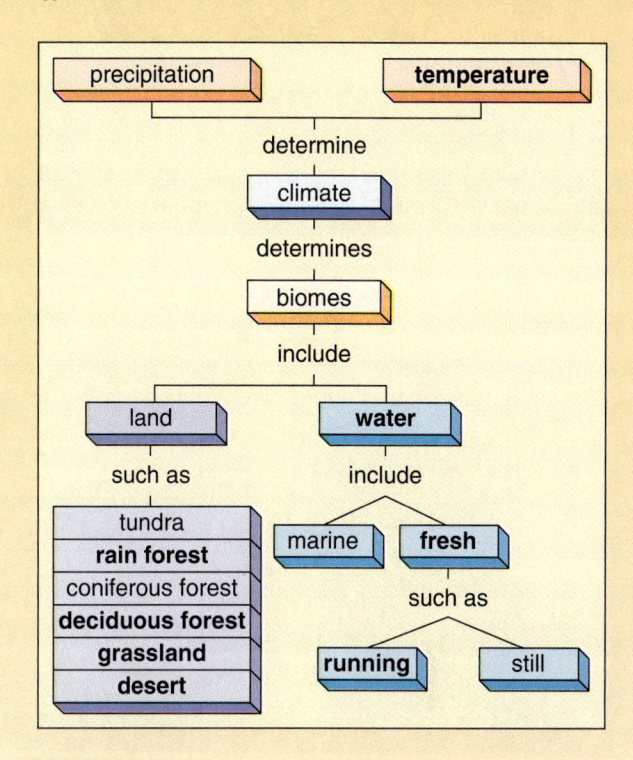

[Concept map: precipitation and temperature determine climate; climate determines biomes; biomes include land and water; land such as tundra, **rain forest**, coniferous forest, **deciduous forest**, grassland, desert; water include marine and **fresh**; fresh such as running and still.]

2. Answers will vary. Students should realize that people living in the desert have to deal with very limited water and food and large daily fluctuations in temperature. In a tropical rain forest, people have to adapt to large amounts of moisture, in the forms of both rainfall and humidity, and must learn to distinguish among a large variety of plants and animals that may be helpful or harmful to them.
3. Answers will vary, but may include wood, animal products, plant products (oils, rubber), and medicines.

CHAPTER 28

Overview

The focus of this chapter is people's uses and misuses of natural resources. The first section describes various resources and methods to conserve and manage them. The second section explains the sources and applications of various forms of energy. Included here is a model and explanation of a solar energy house. The third section describes solid, liquid, and gas pollutants and their effects. This section emphasizes the collective effort that must be made to preserve the earth.

Advance Planner

▶ Provide soil and spray bottles for TE activity, page 603.

▶ Contact the National Aeronautics and Space Administration to obtain photographs of the surface of Venus taken by the Galileo and Magellan probes. TE page 613.

▶ If possible, obtain a videocamera for TE activity, page 614.

▶ Gather distilled vinegar, pH paper, 100 radish seeds, 100 bean seeds, 15 graduated cylinders, 60 small paper cups, 30 leaves each from 3 deciduous trees or shrubs, and 15 eyedroppers for SE Activity 28, page 618.

Skills Development Chart

Sections	Classify	Communicate	Control Variables	Decision Making	Hypothesize	Measure	Model	Observe	Predict
28.1 Skills WarmUp Skills WorkOut							●		●
28.2 Skills WarmUp Skills WorkOut		●						●	
28.3 Skills WarmUp Skills WorkOut Consider This Activity		● ●	●	●	●	●		●	

Individual Needs

▶ **Limited English Proficiency Students** Provide students with 11 index cards and 11 sheets of paper. Have them write one of the chapter terms and its definition on each of the index cards. Then have students make a drawing for each term on a sheet of the paper. Ask them to paste the index cards to the correct drawings.

▶ **At-Risk Students** Photocopy the chapter concept summary for students. Have them cut out each bulleted statement and paste it into their science journals. Then ask them to write one sentence about what each statement means to them, giving an example from their home life or from the newspaper.

▶ **Gifted Students** Have students research the effects of human activities, such as deforestation, that disturb environmental conditions. Students should focus on the imbalance of predator-prey relationships for any evidence of disturbed parasitism, commensalism, and mutualism. Have them share their information in written reports.

Resource Bank

▶ **Bulletin Board** Have students plan and create a bulletin board display about things they can do to help protect the earth. Encourage them to include photographs or original artwork that visually communicate the ideas they wish to present.

▶ **Field Trip** This would be an appropriate time to arrange a trip to an area that shows the effect humans have had on the environment. It might be a good idea to plan one trip to an area that is used for farming, building, or manufacturing, and another to an environment that is being preserved, such as a conservation area.

CHAPTER 28 PLANNING GUIDE

Section	Core	Standard	Enriched
28.1 Natural Resources pp. 601–607			
Section Features			
Skills WarmUp, p. 601	●	●	●
Skills WorkOut, p. 606	●	●	●
Career Corner, p. 606	●	●	●
Blackline Masters			
Review Worksheet 28.1	●	●	
Enrich Worksheet 28.1		●	●
Skills Worksheet 28.1	●	●	●
Integrating Worksheet 28.1a	●	●	●
Integrating Worksheet 28.1b	●	●	●
Ancillary Options			
One-Minute Readings, pp. 31, 33	●	●	●
Overhead Blackline Transparencies			
Overhead Blackline Master 28.1 and Student Worksheet	●	●	●
28.2 Energy Resources pp. 608–611			
Section Features			
Skills WarmUp, p. 608	●	●	●
Skills WorkOut, p. 611	●	●	
Blackline Masters			
Review Worksheet 28.2	●	●	
Integrating Worksheet 28.2a		●	●
Integrating Worksheet 28.2b	●	●	●
Color Transparencies			
Transparency 86	●	●	

Section	Core	Standard	Enriched
28.3 Pollution pp. 612–618			
Section Features			
Skills WarmUp, p. 612	●	●	●
Skills WorkOut, p. 616	●	●	●
Consider This, p. 616	●	●	●
Activity, p. 618	●	●	●
Blackline Masters			
Review Worksheet 28.3	●	●	
Skills Worksheet 28.3	●	●	●
Vocabulary Worksheet 28	●	●	
Ancillary Options			
CEPUP, Toxic Wastes, Activities 1–7	●	●	●
One Minute Readings, pp. 57–59, 60, 64	●	●	●
Laboratory Program			
Investigation 50	●		●
Investigation 51	●	●	●
Color Transparencies			
Transparency 87	●	●	

Bibliography

The following resources can be used for teaching the chapter. See page T-40 for supplier codes.

Audio-Visual Sources
(video unless noted)
Acid Rain. 57 min. 1985. T-L.
Air Pollution: A First Film. Revised ed. 12 min. 1984. PFV.
Energy: The Problems of the Future. 23 min. 1978. NGSES.
Nuclear Energy: The Question Before Us. 26 min. 1981. NGSES.
Our Fragile Atmosphere: Current Issues. 2 filmstrips with cassettes, 16 min each. 1991. NGSES.
Pollution: World at Risk. 25 min. 1989. NGSES.
Recycling: The Endless Circle. 25 min. 1992. NGSES.

Software Resources
Air Pollution. Apple, IBM PC, TRS-80. EME.
Energy House. Apple. TIES.
Hothouse Planet. Apple, IBM PC. EME.
Water Pollution. Apple, IBM PC, TRS-80. EME.

Library Resources
Goldin, Augusta. Water: Too Much, Too Little, Too Polluted? San Diego, CA: Harcourt Brace Jovanovich, 1983.
Kiefer, Irene. Poisoned Land: The Problem of Hazardous Waste. New York: Atheneum, 1981.
Southwick, C., ed. Global Ecology. Sunderland, MA: Sinauer Associates, 1985.
Woods, Geraldine and Harold. Pollution. New York: Franklin Watts, 1985.

CHAPTER 28

Introducing the Chapter

Have students read the description of the photograph on page 600. Ask them if they agree or disagree with the student's description.

Directed Inquiry

Have students study the photograph. Ask:

▶ How do you think that people have changed the land in the picture? (It has been plowed and planted with crops.)

▶ How is this field different from one that grows wild? (A single food crop is planted and carefully cultivated here. A similar wild area of land might support many different types of plants and animals.)

▶ What kind of living organisms are supported by farmland? How are these different from the organisms supported by wilderness areas? (Farmland supports humans; wilderness supports a diversity of living organisms.)

▶ What type of farming is used on this land? (Some students may recognize it as contour farming.)

Vocabulary

acid rain	natural resource
biomass energy	nonrenewable resource
fossil fuel	nuclear energy
geothermal energy	renewable resource
hazardous waste	smog
	solar energy

INTEGRATED LEARNING

Writing Connection

Have students imagine that they are working on the land shown in the picture. Have them describe the crops, the machinery they use, and how they are affected by weather. Encourage students to be creative as they try to imagine a typical day for one of these workers.

Chapter 28 — Humans and the Environment

Chapter Sections

28.1 Natural Resources
28.2 Energy Resources
28.3 Pollution

What do you see?

"I see a cornfield. Somebody plowed it. It serves as a food source. It looks like a cornfield because it has yellowish-green and green in a continuous pattern. The yellow looks like corn."

Christina Dickson
Barnstable Summer School
Hyannis, Massachusetts

To find out more about the photograph, look on page 620. As you read this chapter, you will learn more about humans and the environment.

INTEGRATED LEARNING

Integrating the Sciences

Earth Science Explain to students that it will take millions of years for today's plants to become coal. There is a large supply of coal, about 430 billion tons, in the ground. Although the United States recovers and burns over 900 million tons of coal each year, coal supplies are so plentiful that there is enough coal to last hundreds of years. Supplies of oil are not nearly so plentiful. Even though there are still oil reserves in the United States and other countries according to some estimates, many of the world's reserves of oil could be emptied if people continue to use oil at the present rate.

SECTION 28.1

Section Objectives
For a list of section objectives, see the Student Edition page.

Skills Objectives
Students should be able to:

Predict how humans affect an environment.

Make a Model of a water source.

Organize a chart about natural resources.

Vocabulary
natural resource, renewable resource, nonrenewable resource

28.1 Natural Resources

Objectives

▶ **Distinguish** between renewable and nonrenewable resources.

▶ **Identify** ways that soil and water can be conserved.

▶ **Describe** the changes that may take place when a wild area is taken over for human use.

▶ **Organize** information relating to natural resources and their conservation.

Skills WarmUp

Predict Imagine that you and your classmates will be sent to explore two Pacific islands. The islands are very similar in size, age, and location. But one has a human population and the other does not. Predict what you would expect to see on each island. List ways in which the environment of the populated island has been changed by its human inhabitants.

Have you drunk any water or eaten food today? Have you turned on a light, ridden in a car, or used a computer? When you do any of these things, you are using one or more **natural resources**. Natural resources are materials from the environment that people use to carry on their daily lives.

Every human society that has ever existed has used natural resources. But societies vary in the kinds and amounts of resources they use. People in modern industrial societies use oil, natural gas, and coal for energy. They mine metals for making cans, cars, and many other objects. They use soil for growing food, trees for making paper, and water for keeping their bodies clean. Nearly everything on, above, and below the earth's surface is being used as a natural resource today. And these resources are being used faster than ever before.

Renewable and Nonrenewable Resources

Natural resources can be divided into two types. Resources that can be replaced by natural cycles or processes are called **renewable resources**. For example, when trees are cut for lumber, seedlings can grow up to take their place. Other renewable resources include water and soil. Even though these resources can replace themselves, the cycles that replace them can be destroyed.

Resources that can't be replaced by natural processes are called **nonrenewable resources**. The supplies of nonrenewable resources, such as oil and natural gas, get smaller as people use them.

Figure 28.1 ▲
The photograph shows copper ore being mined. Copper is a nonrenewable resource that is used for many things, including electrical wiring.

MOTIVATE

Skills WarmUp

To help students understand natural resources, have them do the Skills WarmUp.
Answer On the island inhabited by humans, land would be cleared for housing and farming. There might be a decreased plant and animal population and some pollution.

Prior Knowledge

Gauge how much students know about natural resources by asking the following questions:

▶ What are some resources that should be conserved?

▶ Is it possible for the earth's soil to "disappear"?

▶ Why are many animals in danger of extinction?

Integrated Learning

Use Integrating Worksheet 28.1a.

Chapter 28 Humans and the Environment

INTEGRATED LEARNING

Themes in Science

Patterns of Change/Cycles The earth's water moves in natural cycles. Every molecule of water is part of a continuous cycle. Water molecules evaporate into the atmosphere, forming water vapor and clouds. The water molecules then return to the earth as rain, snow, or ice. You may wish to review the water cycle shown in Figure 25.9 on page 549.
Use Integrating Worksheet 28.1b.

STS Connection

The most common industrial use for water is cooling. Water is circulated around hot material to cool it. When the heated water is returned to rivers or streams, it can be damaging to the environment even though it may not contain any toxic chemicals. Have students infer what can happen when an environment's temperature changes. (Many plants and animals can't survive such a change.)

TEACH

Discuss

Ask students the following questions:

▶ What are some ways to conserve resources? (Reduce consumption of resources, reuse, and recycle.)

▶ What are some ways you can reduce the amount of water you use? (Take shorter showers, use water saving cycles on washing machines, and turn off water while washing dishes or brushing teeth.)

Misconceptions

Students may believe that baths use less water than showers. Explain that taking a short shower uses less water than filling a bathtub.

Conservation of Resources

Have you ever shut off a light when it was not needed? If so, you were helping conserve the earth's natural resources. To *conserve* means to protect from being lost or reduced. It is the root of the word *conservation*.

Conservation of resources can be accomplished in two major ways. One way is by reducing the amount of natural resources each person uses. This means recycling and cutting down on waste. The second way is by using and managing renewable resources wisely.

Water Conservation

How much water do you think you use each day? If you are typical, you probably use at least 228 liters of water a day. That's enough water to fill several bathtubs. It includes the water used to wash your clothes and dishes, the water you bathe with, and the water you drink.

The 228 liters per person is only a small part of the water used each day in the United States. Factories and power

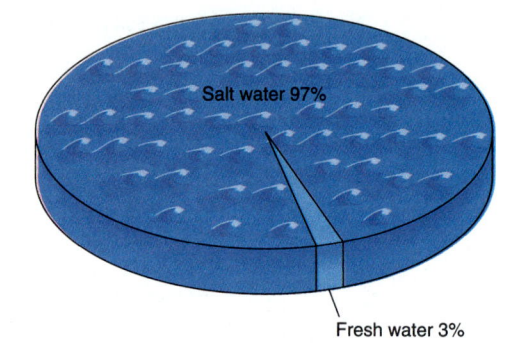

Figure 28.3 ▲
As you can see in this circle graph, only 3 percent of the earth's water is fresh. People, therefore, must continue to find ways to conserve this valuable resource.

plants use tremendous amounts of water for cooling and other purposes. Farmers use even more, especially in dry climates.

Where does all this water come from? About 75 percent of the water used in the United States comes from rivers and lakes. The other 25 percent comes from ground water. Both of these sources are limited. Look at Figure 28.3. It shows that only a small amount of the water on the earth is fresh water.

In some places, water supplies are still plentiful. In other places, however, there is not nearly enough water to meet the demand. So much water is taken from some rivers and streams that plant life and animal life are threatened. And in some places, ground water is being used faster than it can be replaced. This can cause permanent damage to the underground areas that store water, called aquifers.

A simple way you can help conserve water is by paying attention to your daily use. For example, you can take shorter showers and turn off the faucet while you brush your teeth.

Figure 28.2 ▲
Water is stored in the Roosevelt Dam in Arizona for use in farming, electrical power, and recreation.

602 Chapter 28 Humans and the Environment

Multicultural Perspectives

At one time, nearly half of China's land was covered with forests. As China's population grew, much of this land was cleared in order to grow crops. As a result, valuable topsoil was lost through erosion. Today, many farmers in China construct terraces like the one shown in Figure 28.4. The terraces prevent excessive erosion. They are supported by walls that hold water in the soil.

Soil Conservation

Soil is one of the most important natural resources as well as the most abused. It is estimated that each year the amount of topsoil blown or washed away would fill a train of freight cars long enough to encircle the earth 138 times! At the same time, much of the soil that is left becomes less fertile. New soil forms naturally but slowly. It can actually take hundreds of years for 20 cm of topsoil to form.

Causes of Soil Loss In a natural setting, soil, plants, and animals are connected as an ecosystem. The soil and its fertility are renewed by natural cycles. When crops are grown, cycles are disturbed.

Soil often erodes during periods when farmers leave it fallow, with no plants growing in it. The soil surface is exposed to wind and rain. There are no roots to help hold the soil in place.

Growing crops also reduces the soil's fertility. Crops remove large amounts of plant nutrients from the soil, which are then themselves removed. Chemical fertilizers restore these nutrients, but only for a short time. Continued use of chemical fertilizers also reduces the soil's ability to hold moisture and support organisms.

Soil Management Some farmers prevent soil erosion and loss of fertility by managing their farm fields as ecosystems. In a natural ecosystem, plants die, decompose, and return to the soil as nutrients. Farmers copy this process by plowing back into the soil the parts of crops not harvested or by leaving the remains of the last crop standing.

Another characteristic of a natural ecosystem is that a variety of plants grow in one area. Farmers copy this pattern by growing different kinds of crop plants together in one field. One of these methods is called strip cropping, shown in Figure 28.5. A related method, called crop rotation, is to plant a different crop every year. These methods also help decrease pest damage.

Contour farming of crops can also prevent erosion. Rows or crops are plowed across the slopes of hills rather than up and down the hill.

Figure 28.4 ▲
Terracing is a type of contour farming that slows down the runoff of water and helps keep the soil on hills from being washed away.

Figure 28.5 ▲
Strip cropping is the planting of alternate strips of crops in areas where soil is easily eroded by wind and water.

Discuss

Ask students the following questions:

▶ How can farming damage soil? (It depletes the soil of nutrients and makes it vulnerable to erosion when it is left fallow.)

▶ What does it mean to manage a farm as an ecosystem? (To use wise farming methods, such as plowing back into the soil parts of crops not harvested, or growing different kinds of crops together in one field)

Class Activity

Have students demonstrate the principle of terrace farming. They should form two small "hillsides" out of moistened soil—one with a flat incline and one with a series of level steps, or terraces, cut into the incline. They should use a spray bottle filled with water to simulate rain falling on the "hillsides." Have them observe the movement of the water on the "hillsides."

TEACH • Continued

Directed Inquiry

Have students study the material on this page. Ask the following questions:

▶ What are some of the consequences of using land for human development? (Valuable farmland is lost, animal and plant species are destroyed or forced to leave, and local air and water becomes polluted.)

▶ What are some reasons that forests have been cut down? (To obtain timber for fuel and making wood products, to clear land for farming and housing developments)

▶ How does the environment change when forests are cut down? (The habitats of many plants and animals are destroyed, erosion occurs, and land dries up because it loses much of its ability to store water.)

Decision Making

If you have classroom sets of *One-Minute Readings*, have students read Issue 20, "Destruction of Tropical Forests" on page 33. Discuss the questions.

604

INTEGRATED LEARNING

Geography Connection

Almost half of the world's tropical rain forests have been cleared to create farmland, grazing land, timber, and wood for fuel. If destruction continues at this rate, scientists estimate that nearly all the tropical rain forests of the world will be gone by the year 2035. Today, deforestation is occurring in Brazil, Indonesia, Colombia, and Mexico. Have students locate these countries on a world map.

Land Use

As the human population increases, more and more land is developed to meet human needs. Land is paved over to make highways and parking lots. It is graded to make room for housing developments. In the process, valuable farmland is lost and open space is reduced.

When land is taken over for human use, it is changed dramatically. Natural cycles are modified or destroyed. The number and kinds of species the land will support are reduced. The land is no longer a habitat for most of the plants and animals that once lived there.

Human development also affects the surrounding area and its life. For example, large mammals, such as deer, no longer have space in which to roam. Water becomes polluted. Paved and built-up areas may also change the local climate, raising temperatures or reducing rainfall.

Decisions about how to use land are increasingly more difficult. Voters, elected officials, and planners must choose between uses that may seem equally important. They have to weigh the needs of the present with those of the future.

Figure 28.6 ▲
When housing developments are built, valuable farmland is sometimes lost.

Figure 28.7 ▲
Trees are often replanted after forests are cut down. Here, a technician checks a planted tree seedling.

Forest Management

People use the wood from trees to build houses, to make furniture, to produce paper and cardboard, and to burn as fuel. For thousands of years, people have harvested timber from forests for these purposes. But in the last few hundred years, the demand for wood has grown as rapidly as the population. Huge areas of forest have been cut down as a result. Forests have also been cleared to make room for towns and cities and to provide land for farming.

The environment changes in many ways when forests are cut down. The habitats of many forest-dwelling plants and animals are destroyed. Erosion increases because there are fewer plants to hold the soil in place. Less water is stored in the soil, and the quality of the surface water is reduced.

The forests that are left must be managed very wisely if they are to continue to supply human needs. Forests are not just stands of trees to be harvested. They have value as wilderness, as animal habitat, and as recreation areas.

604 Chapter 28 Humans and the Environment

STS Connection

The Endangered Species Act was passed in the United States in 1973. The act was designed to help halt human-caused extinctions. An endangered species is defined as a plant or animal species that is in danger of extinction throughout all or most of its range, or habitat. A species listed as endangered is given protection under the act, and efforts are made to restore its population. Recovery steps may involve restoring a habitat, cleaning up polluted areas, providing protection from hunters, and breeding the species in captivity. Have students explain why it is important to protect endangered species. (Answers will vary. Some may say that humans have a responsibility to prevent human-caused extinctions. The destruction of one species can affect others that remain, upsetting the ecological balance.)

Wildlife Conservation

Many species of animals are endangered, or facing extinction. Some, like the mountain lion, have been hunted for sport or because they are seen as dangerous. Others have been victims of poachers who sell their skins or body parts. African elephants, for example, are hunted for their tusks, which are ivory. But the greatest single threat to most species of wildlife is the destruction of their habitats. Whenever a salt marsh is drained, a river dammed, or a forest cut down, valuable habitats for one or more species are lost forever.

In 1948, fewer than 20 whooping cranes were living. Many had been shot by hunters and much of their breeding habitat had been destroyed. Today there are over 150 whooping cranes, but the species remains close to extinction. These birds spend the winter on the coast of Texas. ▼

Blue morpho butterflies live among the trees in the tropical rain forests of Central and South America. As their habitat is logged, their population is threatened. ▼

Green sea turtles live in the warm waters of the Indian and Pacific oceans. When they return from the ocean to their nesting grounds, they can be hunted very easily. ▼

▲ Grizzly bears were once found in most of western North America. Today most grizzly bears live in western Canada and Alaska. The population has dwindled because of loss of habitat and conflict with humans.

◄ Pronghorns, or the American antelope, came close to extinction in the 1920s because of habitat loss and hunting. Now, due to herd management, their population has increased greatly.

Chapter 28 Humans and the Environment

Explore Visually

Have students study the material on this page. Ask the following questions:

▶ Why has the population of grizzly bears in North America dwindled? (Loss of habitat and conflict with humans)

▶ Why did whooping cranes almost become extinct in 1948? (Many had been shot and much of their breeding habitat had been destroyed.)

▶ Why did the population of pronghorns increase greatly after coming very close to extinction? (Concerned people are managing their herd populations to help preserve them.)

▶ Why is the population of blue morpho butterflies threatened? (Their forest habitat in Central and South America is being destroyed.)

▶ Why are the green sea turtles threatened? (Hunters kill them when the turtles leave the water to return to their nesting grounds.)

Portfolio

Have students research other animals that are currently on the endangered species list. They can then draw a picture of one and list the reasons why the animal is endangered. They should include the steps being taken to save the species. Students may keep the papers in their portfolios.

Enrich

Use Enrich Worksheet 28.2.

Decision Making

If you have classroom sets of *One-Minute Readings,* have students read Issue 19, "Destruction of Species" on page 31. Discuss the questions.

TEACH • Continued

Skills WorkOut
To help students make a model of where the water in a drinking fountain comes from, have them do the Skills WorkOut.
Answer Answers will vary depending on locale. Check students' maps for accuracy.

Career Corner
Research Have students use references, such as agricultural magazines, to find out about some of the latest findings of agricultural researchers. Also have them find out how biotechnological discoveries are communicated to farmers. Have students share information in brief oral reports.

INTEGRATED LEARNING

Math Connection
Explain that one cord of wood, or about three pine trees, can be made into 7.5 million toothpicks. The same amount of wood goes into 300 copies of a typical Sunday newspaper. Have students compute how many trees might be saved if all the households in their state recycled one Sunday newspaper each week for a year. (Example: 10,000 families multiplied by 52 newspapers divided by 100 equals 5,200 trees.)

Themes in Science
Systems and Interactions Stress to students that life on the earth is interconnected. For example, some North American birds may spend the winter in tropical areas. If the plants in those areas are gone, the birds may die off. Explain also that plants all over the world provide many uses, such as potential cures to diseases including cancer and AIDS.

Skills WorkOut
Make a Model Where does the water in the nearest drinking fountain come from? How does it get there? Find the answers to these questions, and then draw a map that traces the pathway of water from its source in your community to the water fountain.

Plant Conservation
Over one-tenth of all the known plant species on the earth are in danger of becoming extinct. Scientists estimate that many thousands of plant species will become extinct before they are even discovered and named. Many of these plants live in tropical rain forests, which are being cleared at a rapid rate. Among these endangered plant species are many with life-saving medicinal qualities. Some may be good sources of food. In addition, many endangered plants may be sources of valuable genetic material because they can contribute important traits to other species of plants.

The extinction of plant species is not just a loss to humans. Plants are primary producers. They are a vital part of the relationships that exist in nature. When plant species are lost, the balance of ecosystems can be upset. As many as 30 other species can depend on one plant for survival.

Career Corner Agricultural Researcher

Who Helps Farmers Grow More and Better Food?

Modern agriculture depends on knowledge of soil, climate, insects, chemistry, plant growth and nutrition, plant genetics, and plant diseases. An agricultural researcher may be involved in any one of these areas.

Agricultural researchers work to develop new plant varieties. They help solve soil erosion problems. They study insect life cycles to determine how to control these pests without using dangerous chemical pesticides. Some agricultural researchers experiment with planting different mixtures of crops, or with letting certain weeds grow with crops to attract beneficial insects.

Some agricultural researchers work mostly outdoors, but many spend much of their time in laboratories or offices. However, any new treatments or plant varieties developed in the lab must be tested in the field.

Since food production is a global concern, some agricultural researchers travel to other countries to assist with their agricultural needs. Advances in biotechnology may provide many new opportunities and challenges for people in agricultural research.

Many businesses and government agencies hire agricultural researchers. Some also work at universities. For most jobs, a doctorate degree in a biological science or agronomy is required.

Integrating the Sciences

Physical Science Hydrogen is produced from water by four methods: electrolysis, thermochemical water-splitting, photolysis, and algal photosynthesis. Write these four methods on the chalkboard. Have students use dictionaries or physical science texts to find the meanings of these terms.

STS Connection

Students may be interested to know that technicians at the National Aeronautics and Space Administration have used liquid hydrogen to power spacecraft. Explain that one of the difficulties with using hydrogen is storage. The hydrogen used to propel astronauts must be maintained at low temperatures, about –383°C, so that it remains in liquid form.

Discuss

Students may wonder why hydrogen power is not used more often. Explain that it takes an enormous amount of energy to obtain hydrogen from water. Also, hydrogen fuel is explosive and therefore has to be handled and stored carefully.

EVALUATE

WrapUp

Portfolio Have students write a description of the changes caused by a human population settling in a location such as a rain forest.
Use Review Worksheet 28.1.

Check and Explain

1. Lumber, water, and soil are renewable resources because they can be replaced by natural cycles. Minerals, oil, and natural gas are nonrenewable because they take millions of years to form.
2. Plant and animal habitats are destroyed, soil erosion increases, quality of surface water is reduced.
3. These methods "trick" pests by moving the crops they feed on and preventing them from reproducing and staying for years.
4. Students' charts could include water, soil, land, forests, animals, plants, natural gas, oil, and coal.

Science and Technology *Alternative Fuels*

In the United States, people depend on natural gas and oil for their energy needs. However, there isn't an infinite supply of these fuels, and they are nonrenewable resources. One day, we may run out of these fuels. Another problem is that they cause pollution.

Today scientists are searching for alternative fuels for people to use. One example is synfuels, or synthetic fuels. Synfuels are liquid and gaseous fuels that are made from oil shale, tar sands, and coal. All three, especially oil shale and tar sands, are found in great quantities in the earth. However, none can be used yet on a large scale. Major obstacles include high production costs, low energy yields, and pollution problems.

Another alternative fuel is hydrogen. Where would hydrogen be obtained? If you said water, you're right. Because there is so much water on the earth, hydrogen use could be a major energy-saving method. It is also a clean fuel. In order to use hydrogen as a fuel, however, it must be isolated from water. One method of obtaining hydrogen is to "split" water molecules into hydrogen and oxygen.

There are already some hydrogen-powered vehicles in cities around the world, but they are mostly experimental models. In order to make hydrogen fuel available to the public, the water-splitting process would have to be set up on a large scale. This would mean building huge hydrogen plants and places for storage.

Figure 28.8 ▲
Synthetic fuels are made in this plant in New Zealand (top). This hydrogen-powered automobile was built in 1992.

Check and Explain

1. Give examples of renewable and nonrenewable resources. Explain how these resources are different.
2. What ecological changes may result when a forest is cleared to make room for a housing development?
3. **Find Causes** How can crop rotation and strip cropping help reduce the damage caused by pests?
4. **Organize Data** Make five columns, and then place the name of an important natural resource at the top of each. Complete the chart by describing the importance of each resource to human society and ways that it can be conserved.

Chapter 28 Humans and the Environment

SECTION 28.2

Section Objectives
For a list of section objectives, see the Student Edition page.

Skills Objectives
Students should be able to:

Observe their use of electricity, natural gas, and gasoline.

Classify ideas for conserving energy.

Make a Model of an energy-efficient house.

Vocabulary
fossil fuel, solar energy, biomass energy, geothermal energy, nuclear energy

MOTIVATE

Skills WarmUp
To help students understand energy resources, have them do the Skills WarmUp.

Answer Students should note that our society is heavily dependent upon energy resources for thousands of daily activities. Have students save their papers in their portfolios to use during the WrapUp for this section.

INTEGRATED LEARNING

Themes in Science
Energy The sources of usable energy are limited. As you read through this section, stress to students that in order to transport, store, and use energy, people often transform it from one form, such as fossil fuel energy, into another, such as electricity, and eventually into another form, such as light.

Integrating the Sciences
Earth Science The Carboniferous period dates from 345 million years ago. Paleontologists have found fossils of organisms, such as *Lepido dendron,* a giant club moss, and *Asterophylites,* a carboniferous horsetail fern, in coal fields. If possible, show students pictures of these fossils. Stress that the coal deposits were forming at the same time as these fossils.

Skills WarmUp

Observe Make a list of everything you have done today in which electricity, natural gas, or gasoline was used. What does this tell you about the society in which you live?

28.2 Energy Resources

Objectives

▶ **Identify** types of renewable and nonrenewable energy sources.

▶ **Compare** and **contrast** alternative energy sources.

▶ **Explain** the importance of energy conservation efforts.

▶ **Make a model** of an energy-efficient house.

If you have ever experienced a power outage or blackout, you know how much people depend on energy. Energy is required to cook your food, heat and cool the place where you live, warm the water you bathe with, and transport you from home to school. Energy is even needed to make most of the products you use every day. Where does this energy come from? Will there always be enough for everyone?

Fossil Fuels

Most of the energy used in modern industrial societies is provided by coal, oil, and natural gas. These energy sources are known as **fossil fuels**. Fossil fuels were formed from the buried remains of decayed plants and animals that lived hundreds of millions of years ago. These remains were covered with many layers of mud, sand, and rock. Their complex molecules slowly changed into simpler molecules made of chains of carbon. Under certain conditions, the solid fossil fuel, coal, was formed. Under other conditions, either liquid petroleum or natural gas was created.

Because it has taken millions of years for fossil fuels to form, they are nonrenewable. Some day, the earth's supply of fossil fuels will run out. Currently humans are burning fossil fuels so rapidly that this day may not be very far in the future.

To replace fossil fuels, both today and in the future, scientists are working to develop several alternative energy sources. Many of these alternative sources have great advantages over fossil fuels. Most are nonpolluting and based on renewable resources.

Figure 28.9 ▲
Oil is extracted from wells deep in the ground (above) and at offshore sites.

Chapter 28 Humans and the Environment

Integrating the Sciences

Physical Science Explain to students that solar cells produce a small electric current. Many solar cells are required for a larger current. They are most commonly used on spaceships and satellites. Recently, solar cells have been installed in power plants. Mirrors in the plants focus sunlight onto the cells and directly generate electricity. Although solar-electric plants cost less to build than fossil-fuel plants, they require vast areas of land.

Solar Energy

Did you know that the sunlight striking the earth on a typical summer day may contain twice as much energy as the United States uses in an entire year? The sun's energy is free, nonpolluting, and will not run out for millions of years. Why, then, doesn't everyone just use the sun's energy? The problem is that there are no practical methods of collecting and using it on a large scale. The sun, therefore, isn't yet a practical alternative to fossil fuels.

Energy from the sun is called **solar energy**. One kind of solar energy is the heat produced when sunlight strikes a surface. Figure 28.10 shows a house designed to use the sun's heat energy.

The sun's heat energy can also be collected and stored in devices called solar collectors. Water is passed through dark-colored pipes that absorb the heat energy in sunlight and transfer it to the water. Today over one million homes in the United States use solar collectors as their source of hot water.

Solar energy is also produced with solar cells. They convert sunlight into electricity. If solar cells can be made cheaper and more efficient, they may become important energy sources.

Energy from the sun can also be used indirectly. Plants convert solar energy into plant tissue. When plants and plant products are burned, they release energy that people use. Energy that comes from plants is called **biomass energy**.

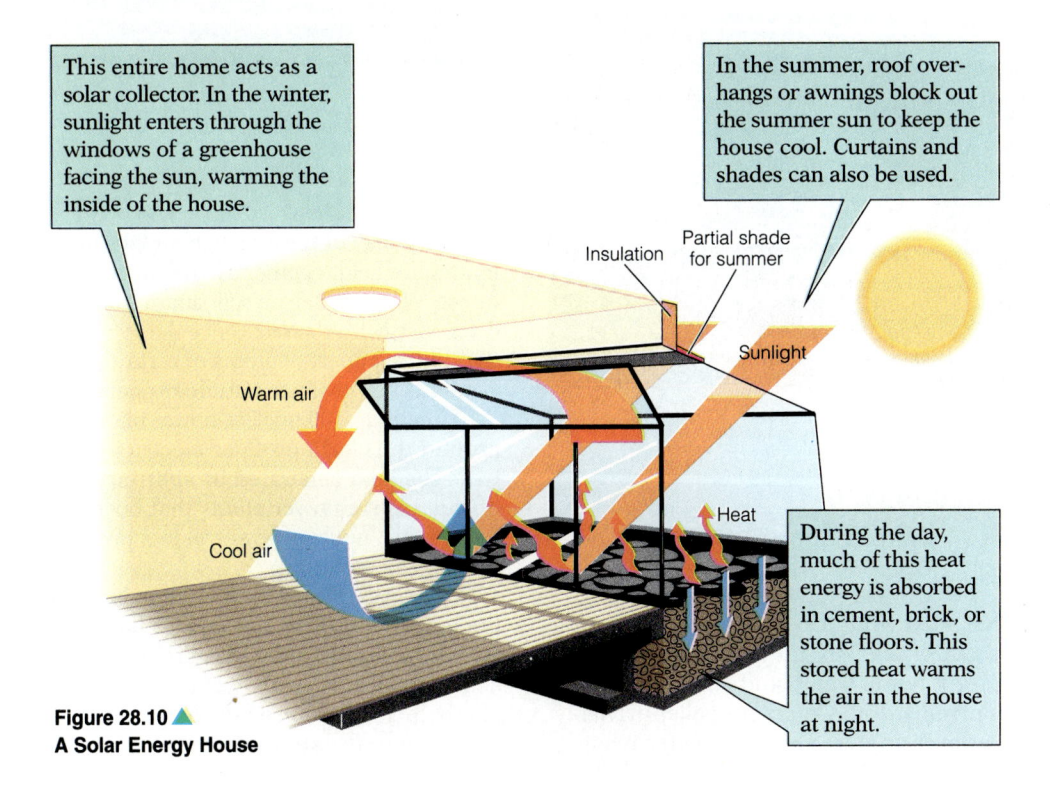

This entire home acts as a solar collector. In the winter, sunlight enters through the windows of a greenhouse facing the sun, warming the inside of the house.

In the summer, roof overhangs or awnings block out the summer sun to keep the house cool. Curtains and shades can also be used.

During the day, much of this heat energy is absorbed in cement, brick, or stone floors. This stored heat warms the air in the house at night.

Figure 28.10
A Solar Energy House

Chapter 28 Humans and the Environment   609

TEACH

Directed Inquiry
Have students study the material on this page. Ask the following questions:

▶ How can sunlight be used to heat a solar energy house in the winter? (Sunlight enters through windows of a greenhouse room facing the sun. This heat energy is collected and circulated throughout the house.)

▶ How can a solar energy house be cooled in the summer? (Roof overhangs or shades can be used to block the sun.)

▶ How can a solar energy house be heated at night? (During the day, energy is absorbed in the cement, brick, or stone floors. This stored heat warms the air in the house at night.)

▶ How are biomass energy and solar energy related? (Plants use solar energy to grow and create plant body tissue. When plants are burned, they release biomass energy, a form of solar energy.)

Enrich
Explain to students that one of the disadvantages of solar energy is that it isn't constant. Solar cells don't absorb much solar energy on cloudy days. Seasonal weather changes often cause uneven heating. Point out that a solar energy house is dependent on warm, sunny days in order to stay consistently warm at night.

Integrated Learning
Use Integrating Worksheets 28.2a and b.

609

TEACH • Continued

Critical Thinking

Reason and Conclude Explain that heat in the earth often heats up underground water sources. The pressure of the boiling water and steam sends the water shooting up through holes in the earth's surface. Ask:

▶ Which kind of energy is produced by geysers and hot springs? (Geothermal)

▶ Why isn't geothermal energy used more commonly? (Geysers and hotsprings are found only in certain areas of the world.)

Enrich

Explain that some people have proposed creating giant wind farms in the Arctic. Have students investigate the pros and cons of this idea and report their findings to the class.

Discuss

Explain to students that the energy released from one kilogram of uranium is one million times greater than the energy released by one kilogram of fossil fuel. Since nuclear energy comes from splitting the nucleus of the uranium atoms, no carbon dioxide is released into the atmosphere. However, nuclear reactors produce hazardous wastes. Safe removal and storage of the waste is difficult.

610

INTEGRATED LEARNING

STS Connection

While heat from the sun's rays can be used directly to produce energy, heat can also be harnessed through ocean thermal energy conversion. Energy is produced by floating electric power plants, which are anchored in ocean currents. Although this process uses a renewable resource, it requires that large, expensive systems be built. Have students find out more about how thermal energy can be harnessed by humans.

Geothermal Energy

Deep below the surface, the earth is very hot. In certain places, the hot interior is close enough to the surface to come in contact with ground water. The water is heated and turned to steam. The steam is released at the surface through hot springs, geysers, and wells. When people use this heat from the earth as an energy source, it is called **geothermal energy**.

In a geothermal power plant, steam from the earth drives turbines that produce electricity. Nothing is burned, so geothermal energy is pollution-free. However, geothermal energy use is limited because not many places on the earth are suitable for producing it.

Figure 28.11 ▲
A geyser in Iceland releases hot water and steam.

Water and Wind Power

One of the oldest sources of energy is the wind. Wind has been used for hundreds of years to sail ships, grind grain crops, pump water, and cut wood at sawmills. Today huge propeller-driven turbines produce electricity. Often many hundreds of wind turbines are set up in areas known as wind farms, as shown in Figure 28.12.

Water has also been a source of power for centuries. Falling water can spin turbines to produce electricity. Energy produced in this way is known as hydropower.

Figure 28.12 ▲
Windmills generate electricity in California.

Nuclear Energy

The nuclei of the atoms that make up matter are held together by very strong forces. When nuclei are split apart, tremendous energy is released. Since the 1950s, the splitting apart of atoms of certain elements has produced **nuclear energy**.

More than 400 nuclear power plants around the world produce electricity from the heat generated by splitting nuclei. Unlike power plants that burn fossil fuels, nuclear power plants produce no air pollution.

However, many people worry about nuclear energy. No safe way has been found to dispose of the dangerous radioactive waste produced. Accidents at the plants can also release radioactive substances into the atmosphere.

610   Chapter 28   Humans and the Environment

Themes in Science

Energy The unit of energy most often used by people in the United States is the kilowatt-hour, which is used to measure electricity consumption. This is the amount of energy expended by a one-kilowatt source in operation for one hour. Explain that one kilowatt-hour is used by a light bulb burning for ten hours or a TV transmitting for four hours.

Art Connection

Have students refer to the list of energy-saving suggestions on this page. Then instruct them to draw a detailed picture of a house or apartment where energy-conscious people live. Have them label the energy-saving steps these people have taken in their home.

Skills WorkOut

To help students classify the energy-saving suggestions listed, have them do the Skills WorkOut.

Discuss

After students have read the suggestions for energy efficiency, poll students to find out which steps they already take. Then find out which ones they plan to follow in the future.

EVALUATE

WrapUp

Portfolio Have students take out the lists they made for the Skills WarmUp. Have them revise and expand their lists and tell when today they conserved an energy resource.

Use Review Worksheet 28.2.

Check and Explain

1. Answers may vary. Solar energy is renewable because it is replaced by natural processes.
2. Answers may vary. Make sure students understand that alternative energy sources are those other than fossil fuels.
3. Answers may vary. Energy conservation saves money, reduces pollution, and prolongs the time the earth will remain inhabitable for humans and other life forms.
4. Answers may vary. The house should have a greenhouse and/or solar panels facing the sun.

Science and You *Turn Off the Lights!*

There are many reasons for each person to cut down on energy use. The United States uses much more than its share of the world's fossil fuels. Fossil fuels also affect the environment. Extracting and transporting them can cause environmental damage, such as oil spills. Processing fossil fuels pollutes the air, and burning them for energy pollutes it even more. Here are some suggestions to help you and your family become more energy efficient.

▶ Turn off lights and appliances when not in use.
▶ Open curtains when you need light during the day.
▶ Close the curtains when the weather is cold.
▶ Dust the light bulbs. Dusty ones waste more energy.
▶ Use fluorescent bulbs instead of incandescent bulbs.
▶ Check for drafts around windows and doors. Fix seals around the windows and doors, if necessary.
▶ Check to see if your water heater is insulated.
▶ If you have a furnace, change the air filter regularly.
▶ Cover pots of boiling water.
▶ Open the refrigerator only when you know what you want from it.
▶ Recycle cans, paper, and plastics.
▶ Encourage others to become more energy efficient.

Skills WorkOut

Classify Study the suggestions on this page for becoming more energy efficient. Classify these suggestions into three or more groups. Give each group a name and explain how the items in the group are related.

Check and Explain

1. Give an example of a renewable energy source and explain why it is renewable.
2. Which alternative energy sources would you recommend be developed for wider use? Explain.
3. **Generalize** Explain why energy conservation efforts can improve the quality of life for people and help protect the environment.
4. **Make a Model** Design an energy-efficient house. Make sure it uses the heat and light of the sun wisely. Include ways to get energy from wind, water, biomass, or solar cells.

SECTION 28.3

Section Objectives
For a list of section objectives, see the Student Edition page.

Skills Objectives
Students should be able to:

Communicate the effects of pollution.

Measure how much landfill space would be saved by recycling school paper.

Predict what the earth's ecosystems will be like in the future.

Vocabulary
smog, acid rain, hazardous waste

MOTIVATE

Skills WarmUp
To help students understand pollution, have them do the Skills WarmUp.
Answer Answers may vary. Students should realize that pollution is harmful to all living things.

Prior Knowledge
Gauge how much students know about pollution by asking the following questions:

▶ When have you noticed pollution around you?

▶ Can you see pollution? How do you know if air and water are polluted?

Answer to In-Text Question
① There are two kinds of smog. One kind results when gases from burning oil and coal combine with atmospheric moisture. The other kind is photochemical smog, which forms when car exhaust and sunlight interact in sunny, dry climates.

INTEGRATED LEARNING

Writing Connection
Remind students that the amount of pollution in the air has increased greatly during the last century. Have them imagine life in the United States 100 years from now—will there be much more or much less pollution? Taking the perspective of a person from the future, they should write a creative story about what the earth's environment will be like at the end of the next century.

Skills WarmUp
Communicate You have probably seen in books or magazines how pollution has affected the earth's air, water, or land. Maybe you've even seen pollution in your own neighborhood. Describe your reaction to the pollution.

Figure 28.13 ▲
Smog surrounds the city of Denver, Colorado. What causes smog? ①

28.3 Pollution

Objectives

▶ **Describe** some of the harmful effects and sources of air, land, and water pollution.

▶ **Explain** how pollution can be reduced.

▶ **Predict** the condition of the earth's ecosystems ten years from now.

In what ways is the environment around you polluted? Make a list of all the places where you have seen pollution. Pollution is any physical, chemical, or biological change in the environment that is harmful to living organisms. A substance is a pollutant if it makes the environment unfit or undesirable for humans and other organisms.

Air Pollution

Would you choose to eat rotten food or drink waste water? Of course not. But people often have no choice about the quality of air that they breathe. Huge amounts of solids, liquids, and gases are added to the earth's air by humans every day. Many of these materials not only make breathing difficult but cause harm to other living organisms and the environment.

Smog One of the results of air pollution is **smog**. Many cities are covered with a blanket of smog during much of the year. One type of smog results from the burning of oil and coal. It forms when tiny particles and gases from the burning combine with moisture in the air. Particles from the smog cause health problems and reduce visibility.

Areas that have sunny, dry climates and a lot of exhaust from automobiles often have another type of smog. It is called photochemical smog. This smog forms when chemicals from car exhaust react in the presence of sunlight. Ozone and other harmful substances produced in these reactions combine with nitrogen oxides and carbon monoxide from the exhaust. The result is air that is very unhealthy to breathe.

612 Chapter 28 Humans and the Environment

Integrating the Sciences

Astronomy Students may be interested to know that an extreme example of the greenhouse effect at work has been observed on the planet Venus. Boiling-hot temperatures are common on the planet's surface both at night and during the day, not because the planet is so close to the sun, but because the atmosphere is dense with carbon dioxide and sulfur dioxide that trap the sun's reflected radiation.

Chemistry Point out that ozone (O_3) is a form of oxygen produced when a single oxygen atom joins a molecule of O_2. The ozone diffused in the stratosphere is known as the ozone layer. The ozone layer screens out about 99% of the sun's ultraviolet radiation. Thus, the ozone layer protects organisms from harmful radiation. Explain that some substances, such as fluorocarbons, destroy the ozone layer.

Acid Rain The gases produced when fossil fuels are burned can create more than smog. These gases combine with the moisture in clouds to form acidic water droplets. The droplets then fall to the earth as **acid rain**. Acid rain is precipitation with a pH below 5.6.

Acid rain can damage forests and freshwater ecosystems. In many parts of the eastern United States, lakes have become so acidic from acid rain that they can no longer support most forms of life. Acid rain also corrodes buildings, metals, and painted surfaces.

Ozone Depletion At ground level, ozone is a harmful pollutant. In the upper atmosphere, however, ozone forms a layer that protects humans, plants, and animals from the sun's harmful ultraviolet rays. It also helps stabilize the earth's climate.

Recently scientists have detected a thinning of the ozone layer. You can see this in Figure 28.15. It is believed that a number of gases produced by humans are responsible for breaking down this protective layer. Scientists worry that ozone depletion will cause an increase in skin cancer.

Figure 28.14 ▲
How has acid rain affected this forest? ②

Greenhouse Effect Greenhouses trap solar energy in the form of heat to help plants grow. Look at Figure 28.16. The earth is like a giant greenhouse because certain gases in the atmosphere trap the sun's heat and keep the earth warm. However, these gases have been building up in the atmosphere for over 100 years because of the burning of fossil fuels. Scientists believe that the increasing levels of these "greenhouse gases" could cause the earth to warm, changing the climate and melting the polar ice caps.

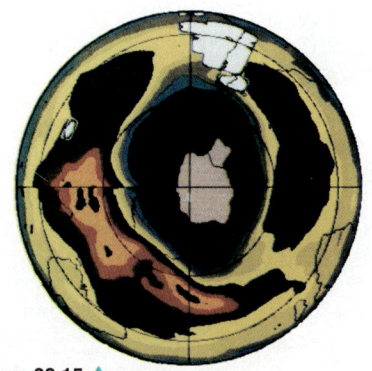

Figure 28.15 ▲
The ozone hole is shown in pink in this satellite map.

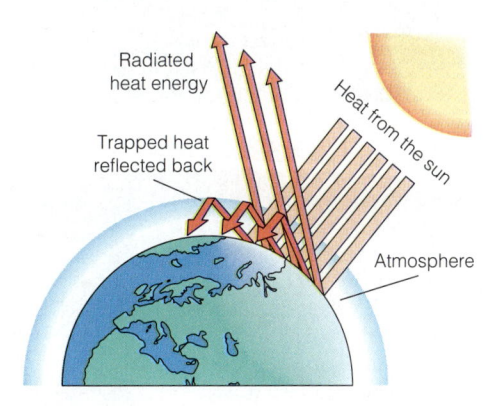

Figure 28.16 ▲
How is the earth like a giant greenhouse? ③

Chapter 28 Humans and the Environment   613

TEACH

Misconceptions
Students may think that only water is affected by acid rain. Point out that acid rain is part of a pollution cycle that affects air, water, and land.

Critical Thinking
Reason and Conclude Ask students to suggest how ozone can be a harmful pollutant and also protect life on the earth. (Direct contact with ozone is harmful; however, ozone gas in the upper atmosphere creates a protective layer that blocks cancer-producing ultraviolet rays and helps stabilize the earth's climate.)

Decision Making
If you have classroom sets of *One-Minute Readings,* have students read Issue 36, "Acid Rain" on page 60. Discuss the questions.

Ancillary Options
If you are using the CEPUP materials, have students do activities 1 to 7, Toxic Waste: A Teaching Simulation.

Answers to In-Text Questions
② The acid rain has harmed the trees in the forest.
③ Atmospheric gases trap the sun's heat, serving to warm the earth.

The Living Textbook: Life Science Side 8
Chapter 35   Frame 23229
Global Temperature Change (Movie)
Search: Play:

TEACH ▪ Continued

Discuss

After students have studied the material on this page, ask the following questions:

▶ What substances pollute the land? (Litter, garbage, solid wastes, and hazardous wastes)

▶ Where does most of the trash people throw away end up? (In landfills)

▶ What are hazardous wastes? (Radioactive, corrosive, poisonous, or disease-causing substances that are harmful to living organisms, including people)

▶ What is the problem with many old hazardous waste dumps? (Containers of hazardous wastes are buried in the ground and later leak into groundwater that animals and people drink.)

Enrich

Explain that many household cleaners contain toxic chemicals. There are many new products on the market that are just as effective and do not contain harmful chemicals. Have students investigate the cleaning products in their homes to see what kinds of ingredients they contain.

Decision Making

If you have classroom sets of *One-Minute Readings*, have students read Issues 33, 34, and 35, "Waste Disposal," "Underground Toxic Wastes," and "Radioactive Waste Disposal" on pages 57 to 59. Discuss the questions.

TEACHING OPTIONS

Cooperative Learning

Assign students to cooperative groups. Have them plan a TV documentary about how waste disposal is managed in your area. Some students may be reporters, contacting the local sanitary district to find out how household and industrial waste is disposed of. Others may prepare visual aids, such as charts, or a bag containing 3.6 kg of classroom waste. Some students may wish to find information about recycling efforts in your area. When they've finished gathering information and preparing for the documentary, have teams make their presentations to the rest of the class. If possible, students might enjoy videotaping their reports.

Land Pollution

Land is polluted by litter, garbage, solid waste, and other substances that humans dispose of. Household garbage is one of the largest sources of land pollution. Every day, each person in the United States throws away an average of 3.6 kg of garbage. Most of this enormous amount of trash ends up in landfills, places where garbage is dumped and then covered with soil. A large amount of garbage never makes it to landfills, instead becoming litter that fouls the landscape.

A more dangerous kind of land pollution can occur when **hazardous wastes** are disposed of. Hazardous wastes are substances that are very harmful to humans and other organisms. They can be radioactive, corrosive, poisonous, or disease-causing. Most hazardous wastes are produced by industry, but the typical household creates hazardous wastes, too.

Until recently, most hazardous wastes were put in barrels or other containers and buried in the ground. But many of the containers have since leaked. As a result, old hazardous waste dumps have become environmental nightmares. In hundreds of places, toxic chemicals from these dumps have seeped into the ground water and threaten to poison drinking water.

◀ **Hazardous Wastes**
Some companies dispose of their hazardous wastes illegally rather than incur the costs of proper disposal. The photograph shows workers cleaning up hazardous waste barrels left by a paint manufacturer who had gone out of business. Locating the companies who dump illegally is usually difficult, since most abandoned barrels are not labeled with company names.

Solid Wastes ▶
Solid wastes are mainly produced by agriculture, manufacturing, mining, and people's daily lives. Wastes produced by daily life include materials discarded by schools, stores, and homes. In the United States, these wastes are increasing at a rate of 2 to 4 percent a year.

Portfolio

Have students make a concept map representing all the problems associated with human use of fossil fuels. Students should include acid rain, the greenhouse effect, oil spills, water pollution, air pollution, wildlife destruction, drinking water contamination, and so on. Have students discuss their concept maps with the class and brainstorm a list of alternatives to using fossil fuels.

Water Pollution

Is there a stream, river, or lake near where you live? Most likely it is polluted. According to the Environmental Protection Agency, all but a few of the water basins in the United States are polluted.

Water pollution has many sources. Pollutants are sometimes dumped directly into rivers and lakes by factories. Fertilizers and pesticides used by farmers can enter waterways in the runoff from farm fields. Litter, oil, and sewage can flow into rivers from storm drains. Water pollution kills fish and other freshwater organisms. Polluted water can harm people if they drink it or bathe in it.

The slow seepage of pollutants from landfills, toxic waste dumps, and farm fields also pollutes the ground water. These pollutants are often hard to detect because they are tasteless and odorless. Ground water makes up the drinking water supply of many people in the United States.

The pollutants in rivers and streams eventually end up in the earth's oceans. Even though the oceans are very large, they are affected by all the waste that has been dumped into them. One serious pollutant in the oceans is oil. Each year millions of tons of oil are added to the oceans.

◀ **Freshwater Pollution**
Industrial wastes are the major source of pollution in rivers, lakes, and streams. The water cannot break down the polluting substances naturally. The poisonous substances accumulate, killing plants and animals. Here, the wastes from a paper mill have severely polluted a nearby river.

Ocean Pollution ▶
Each year millions of tons of oil are added to the world's oceans. Here, workers begin the huge task of cleaning the Alaskan shoreline in 1989. The oil tanker *Exxon Valdez* had struck a reef. More than 240,000 barrels of oil leaked into the waters, polluting the surrounding shoreline.

Directed Inquiry

Have students study the material on this page. Ask the following questions:

▶ Where does water pollution come from? (Waste water from factories; fertilizers and pesticides from farms; and litter, oil, and sewage from storm drains)

▶ How does groundwater become polluted? (By seepage of pollutants from landfills, toxic waste dumps, and farm fields)

▶ Where do pollutants in rivers and streams end up? (In the oceans)

▶ Where does oil in the oceans come from? (From oil tanker accidents, from ships that clean out fuel tanks, from land sources that pollute the rivers, and from dumping waste oil into the oceans)

Research

Have students find out more about the *Exxon Valdez* tragedy and how it affected Alaska's wildlife and human populations.

Discuss

Point out to students that plastic litter can be especially harmful to wildlife. In fact, about 100,000 mammals and 2 million birds die as a result of eating or being tangled up in plastic from fishing lines, balloons, plastic bags, ring-style beverage carriers, and milk containers. Stress the importance of disposing of plastics properly.

Decision Making

If you have classroom sets of *One-Minute Readings*, have students read Issue 38, "Oil Pollution" on page 64. Discuss the questions.

TEACH • Continued

Skills WorkOut
To help students calculate how much landfill space could be saved by recycling paper, have them do the Skills WorkOut.
Answer Answers will vary. The amount of paper will be considerable.

Class Activity
Have students work in cooperative groups of four to observe how oil reacts with water. Have each group fill a cup with water. Then have students pour a small amount of vegetable oil into the cup and stir gently. The cups should sit for 20 minutes. Then have them check the cups again and observe the oil layer on the water's surface. Have students discuss their observations with one another.

Consider This
Answer More restrictions to prevent oil spills will probably mean higher costs for people who are dependent on oil.

Skills Development
Infer Have students consider the requirements of having oil companies use tankers that have a double hull, which provides an additional barrier between the oil and the ocean. Ask them what effect they think this would have on the number of oil spills. (It would reduce the number because some accidents may not be severe enough to breach both hulls.)

INTEGRATED LEARNING

Multicultural Perspectives
Many countries have turned to bicycling as an alternative to driving in an effort to reduce air pollution and traffic congestion. Bicycling is now a major form of transportation in countries such as Germany, China, and Japan. In many of these nations' cities, half of all trips are made by bike. In the German city of Erlangen there are bike paths covering the length of about half the city's streets. So many commuters in Japan ride bicycles to train stations that there are parking towers for bicycles. Poll students to find out how many of them ride bicycles.

Skills WorkOut
Measure Calculate how much landfill space would be saved monthly if all the paper and paper products in your classroom were recycled. Begin by collecting in a box all the paper that would be discarded in your classroom for one week. Calculate the volume of this paper. Multiply this by four. How much landfill space would be saved in a month? In a year?

Protecting the Earth
There are no easy or quick solutions to environmental problems. Today, however, more and more people and governments are becoming aware of the dangers of polluting the earth. Since 1970, many laws have been passed in the United States to control the amount of pollution produced by factories and automobiles. The Environmental Protection Agency has been created. Endangered species have received legal protection.

People are helping by recycling more paper, bottles, cans, and plastic than ever before. Many people are making their homes energy efficient and cutting down on the use of gasoline by carpooling and bicycling.

But preserving the quality of life on the earth and its ecosystems may require more than small changes in people's lifestyles. Everyone can begin by trying to better understand the complex interactions of living things and the earth's nonliving environment.

Consider This

How Much Should Be Done to Prevent Oil Spills?

People in the United States depend greatly on oil for their energy needs. Oil accounts for nearly half of the total energy consumption in the United States. However, demand for oil has also harmed the environment in a number of ways. One way is by oil tankers. Accidents from these vessels cause serious spills in waters all over the world.

Consider Some Issues Since the 1970s, the U.S. government has created many new safety standards for oil tankers. Imposing regulations on oil companies is costly, however. Companies raise prices for fuel to pay for modern equipment. People are then burdened with these rising costs when buying gasoline and paying home heating bills.

Environmentalists argue that the government needs to create even more restrictions that will make oil tankers less likely to accidentally spill oil. They also feel that oil companies need to do a better job of cleaning up the oil spills by spending more money on cleanup efforts.

Think About It What action should the government take to help prevent oil spills? How will more restrictions affect people dependent on oil?

Write About It Write a proposal that describes what you think the government should do about oil spills. Include reasons for your recommendations.

STS Connection

Discuss with students the following ways that everyone can reduce waste:

▶ Buy products packaged in recyclable glass, metal, or paper rather than unrecyclable plastic and polystyrene.

▶ Buy goods packaged in recycled material whenever possible. (Look for the recycling symbol.)

▶ Donate items such as clothing and furniture to places like the Salvation Army, Goodwill, and other agencies instead of throwing them away.

▶ Maintain and repair the products you own instead of throwing them away and buying new ones.

Science and Society *Recycling*

Each day in the United States over 400,000 tons of materials are discarded and dumped. What can be done about all this waste? Recycling is one option. Recycling reduces the amount of waste that goes into landfills, *and* it saves natural resources. For example, 95 percent less energy is needed to make an aluminum can from recycled aluminum than to produce one from aluminum ore.

Fewer pollutants are created when materials are recycled. Producing a new product from raw materials often releases harmful pollutants into the environment. In addition, disposing of the product after it is used can also cause pollution. But if the product is recycled, both kinds of pollution can be eliminated.

In the United States, only about 10 percent of all the waste is recycled. The national goal for recycling is 25 percent. If you think this goal is difficult, consider that some European countries recycle 40 percent or more of their waste. To help increase recycling, many communities have started curbside recycling programs. Cans, bottles, and newspapers are picked up every week. In addition, some states have passed laws requiring deposits on beverage containers. When the containers are returned for recycling, the deposit is returned to the consumer.

Examine how much waste you create in a day. What percentage do you recycle? Does your family and school recycle over 25 percent of its wastes? Does your community have a curbside recycling program? If not, think about how you could help start one. Your actions do make a difference.

Figure 28.17 ▲
These aluminum products will be recycled and used again. How does recycling help the environment? ①

Check and Explain

1. How does acid rain affect the environment?
2. Describe three sources of water pollution.
3. **Generalize** What is an important action an individual can take to protect the environment?
4. **Predict** Describe what the earth's ecosystems will be like in ten years. What about in 50 years? Give reasons for your prediction.

Enrich
Students may have seen labels or statements like "This book printed on recycled paper." Ask students to bring examples of labels or statements to school.

EVALUATE

WrapUp
Reinforce Have students make a chart listing the ways in which they see air, land, and water being polluted in the local community. Have them propose ways to reduce this pollution.

Use Review Worksheet 28.3.

Check and Explain
1. Acid rain damages forests and freshwater ecosystems, kills fish, and corrodes buildings.
2. Students may list fertilizers, pesticides, litter, oil, sewage, toxic waste dumps, and industrial wastes.
3. Answers may vary. Emphasize to students that just one person's actions can play an important role in protecting the earth.
4. Answers may vary. Students should consider current trends in recycling and pollution control and predict how those trends will change.

Answer to In-Text Question

① Recycling reduces the amount of waste in landfills, saves natural resources, and reduces pollutants.

Chapter 28 Humans and the Environment

ACTIVITY 28

Time: 1–2 weeks   **Group:** 2–3

Materials
- distilled vinegar
- pH paper
- potting soil
- 100 radish seeds
- 100 bean seeds
- 15 graduated cylinders
- 60 small paper cups
- 30 leaves each from 3 deciduous trees or shrubs
- 15 eyedroppers

Analysis
1. The controlled variable in this experiment is the pH of the acidic solution.
2.–4. Answers will vary, but should reflect each group's recorded observations.

Conclusion
Students' summaries should follow from the group's actual observations, and include accounts of the progress of both types of potted seeds as well as changes in the appearance of the leaves.

Everyday Application
Answers will vary by region. A gardener's guide can provide information about soil requirements for various types of plants.

Extension
Answers will vary by region. Most normal samples will be slightly acidic.

618

TEACHING OPTIONS

Prelab Discussion
Have students read the entire activity before beginning. Discuss the following points:

▶ Talk about the effects of acid rain on the leaves of plants. Discuss how damaged leaves can affect the plant as a whole.

▶ Have students predict the effect of acid rain on growing plants (this experiment looks only at germinating seeds and detached leaves) and design an experiment.

▶ Discuss the fact that the procedure assumes that tap water has the same pH as normal rainwater. If possible, have students collect rainwater to use.

Activity 28   *How does acid rain affect plants?*

Skills   Observe; Hypothesize; Control Variables

Task 1   Prelab Prep
1. Collect the following items: water, distilled vinegar, graduated cylinder, pH paper, 4 small pots or paper cups, potting soil, 2 leaves each from 3 deciduous trees or shrubs, eyedropper, radish and bean seeds.
2. Make an acidic water solution by mixing 50 mL of vinegar with 150 mL of water in a graduated cylinder. The solution should have a pH of 4.0. Test it.
3. Fill four pots or cups three-quarters full of potting soil.
4. Make the following labels: *Radish—Normal pH*, *Radish—Acidic pH*, *Bean—Normal pH*, and *Bean—Acidic pH*. Attach one label to each pot.
5. For each pair of deciduous leaves, label one *Normal pH* and the other *Acidic pH*.

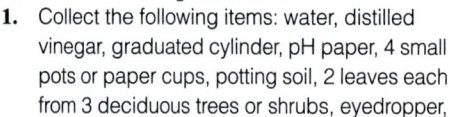

Task 2   Data Record
On a separate sheet of paper, copy Table 28.1. Use it to record your data and include dates.

Table 28.1   Effects of Acid on Vegetables

	Observations
Radish—Normal pH	
Radish—Acidic pH	
Bean—Normal pH	
Bean—Acidic pH	
Leaf A—Normal pH	
Leaf A—Acidic pH	
Leaf B—Normal pH	
Leaf B—Acidic pH	
Leaf C—Normal pH	
Leaf C—Acidic pH	

Task 3   Procedure
1. Place a few drops of normal water on the surfaces of the deciduous leaves labeled *Normal pH*. Place the same amount of acidic water on the leaves labeled *Acidic pH*.
2. Set aside these leaves for a few minutes. Continue with the second part of the activity. Record your observations later.
3. Plant 5 radish seeds in each radish pot and 5 bean seeds in each bean pot.
4. Wet the soil of each pot with the normal water or the acidic water as shown on the pot's label.
5. Set aside these pots and make daily observations and records of their growth.
6. Observe and record any changes on the surfaces of the deciduous leaves.

Task 4   Analysis
1. Identify the variable in this activity.
2. What effect did the different water samples have on the surfaces of the leaves?
3. What effect did the different water samples have on the seeds' abilities to germinate?
4. Were there differences in the germination and growth patterns of radishes and beans?

Task 5   Conclusion
Write a short summary of how the pH of water affected seed growth and leaf appearance.

Everyday Application
What is the pH of the soil in your yard or on the school grounds? Mix some soil with water and test it with the pH paper. Find out which plants like slightly acidic soil and which do not. Which plants will grow best in your soil?

Extension
Collect water samples from several sites in the area of your school. Test the pH of each sample. Collect rain or snow samples from different areas in the community and test the pH of each.

Chapter 28 Review

Concept Summary

28.1 Natural Resources
▶ Natural resources are materials people take from the environment to meet their needs. Natural resources are either renewable or nonrenewable.
▶ Wise use of water, land, and forests helps conserve these limited resources.
▶ Managing farm fields as ecosystems helps conserve their soil.
▶ Many plant and animal species are endangered because their habitats have been destroyed.

28.2 Energy Resources
▶ Fossil fuels were formed from the remains of organisms that lived long ago. They are the major source of energy in modern industrial societies.
▶ Solar energy, geothermal energy, water power, and wind power are alternative energy sources based on renewable natural resources.
▶ Many people are concerned about the use of nuclear energy because it poses several dangers.

28.3 Pollution
▶ Pollution of the air causes smog, acid rain, and destruction of the ozone layer, which may warm the earth through the greenhouse effect.
▶ Disposal of solid wastes and hazardous wastes can pollute the land. Many sources of chemicals and wastes pollute streams, rivers, lakes, the ocean, and ground water.
▶ Steps are being taken to preserve the earth's ecosystems, but much remains to be done.

Chapter Vocabulary

fossil fuel (28.2)   natural resource (28.1)   nuclear energy (28.2)   geothermal energy (28.2)
acid rain (28.3)   renewable resource (28.1)   solar energy (28.2)   hazardous waste (28.3)
smog (28.3)   biomass energy (28.2)   nonrenewable resource (28.1)

Check Your Vocabulary

Use the vocabulary words above to complete the following sentences correctly.

1. Automobile exhaust reacts in the air with sunlight to produce a kind of ____.
2. Energy from the sun is called ____.
3. Humans use materials from the environment called ____ to meet their needs.
4. Corrosive, poisonous, and radioactive wastes are examples of ____.
5. No safe way has been found to dispose of the waste produced by ____.
6. The burning of ____ can cause air pollution.
7. Supplies of a ____ cannot be replaced.
8. Energy that comes from plants is called ____.
9. Trees, water, and soil are examples of ____.
10. Freshwater ecosystems and forests are damaged by ____.
11. In some places, ____ can be produced because superheated ground water rises to the surface.

Write Your Vocabulary

Write sentences using each vocabulary word above. Show that you know what each word means.

CHAPTER REVIEW 28

Check Your Vocabulary

1. smog
2. solar energy
3. natural resources
4. hazardous wastes
5. nuclear energy
6. fossil fuel
7. nonrenewable resource
8. biomass energy
9. renewable resources
10. acid rain
11. geothermal energy

Write Your Vocabulary

Students' sentences should show that they know the meaning of each word as well as how it is used.

Use Vocabulary Worksheet for Chapter 28.

CHAPTER 28 REVIEW

Check Your Knowledge

1. Solar energy can be used for heating and for generating electricity with solar cells.
2. Answers will vary. See page 615 for a discussion of water pollution.
3. Fossil fuels are formed from the buried remains of plants and animals. They are nonrenewable.
4. Forest-dwelling plants and animals lose their habitats; erosion increases; the quality of surface water is reduced.
5. Destruction of habitat is the main cause of extinction.
6. alternative
7. the ocean
8. the greenhouse effect
9. medicine
10. heat energy inside the earth
11. more

Check Your Understanding

1. Plants and animals could be used for food; trees for cooking and warmth; rivers or streams for water.
2. a. Aluminum, fossil fuels
   b. Trees
   c. Plant and animal life
   d. Water
3. Answers will vary. Fossil fuels must be burned to release their energy, while the energy from the sun is used directly as heat, or as light in solar cells. They are similar in that both can be converted to do useful work.
4. If rows were planted along the slope instead of across, there would be great soil erosion.
5. Possible answers include recycling, encouraging adults to drive less, turning off lights, avoiding products with excessive packaging that is not recyclable.

6. Answers may vary. Fossil fuels produce various types of pollution when they are burned to release energy. Cutting down their use would reduce the amount of these pollutants.
7. The sun, wind, and water are energy sources that could be used by individuals without the large-scale development required for geothermal or nuclear power production.
8. Answers will vary; however, distance from the source does not eliminate the risk because pollutants can be carried far by wind and clouds and by water in rivers and streams.

Chapter 28 Review

Check Your Knowledge

Answer the following in complete sentences.

1. What are two ways of using energy from the sun?
2. Describe one source of water pollution. What effect might this pollutant have on the environment?
3. What are fossil fuels? Are they renewable or nonrenewable?
4. How is a forest ecosystem affected when its trees are cut down?
5. What is the main cause of the extinction of plant and animal species?

Choose the answer that best completes each sentence.

6. Many (nonrenewable, polluting, alternative, expensive) energy sources are being developed to replace fossil fuels.
7. Pollutants dumped into rivers and streams eventually find their way to (the ocean, the atmosphere, groundwater supplies, Chicago).
8. The average temperature of the earth's atmosphere may rise as a result of (the greenhouse effect, water pollution, strip cropping, nuclear energy).
9. Some endangered plants in tropical rain forests may be sources of (ozone, fossil fuel, hazardous waste, medicine).
10. A geothermal power plant uses water heated by (fossil fuels, heat energy inside the earth, volcanoes, the greenhouse effect) to produce electricity.
11. Humans are using (fewer, different, more, better) natural resources than they did 100 years ago.

Check Your Understanding

Apply the concepts you have learned to answer each question.

1. **Application** If you were lost in a wilderness area, what natural resources might you use to survive? Where would you look for them?
2. Each of the following is designed to conserve what natural resource?
   a. recycling soft drink cans
   b. reusing paper bags
   c. a protected wilderness area
   d. low-flush toilets
3. **Compare and Contrast** As energy sources, how are fossil fuels and the sun different? Are they similar in any way?
4. **Mystery Photo** The photograph on page 600 shows crops that have been planted along level lines that follow the contour of the land, called contour farming. What would happen if this land were farmed without using contour farming methods?
5. **Extension** Make a list of ten things that you can do to help protect the earth's ecosystems. Explain the purpose of each action.
6. **Predict** In what ways would the global environment benefit if fossil fuels were replaced as a source of energy?
7. **Infer** Which of the alternative energy sources you learned about in this chapter could be used by people to supply their own energy needs? Which require large-scale development by industry or government?
8. If you lived far from any cities, would you be safe from water and air pollution? Explain your answer.

Develop Your Skills

1. a. About 700 kilograms
   b. 1970–1975
   c. 850 kilograms, judging from the five-year increases in the graph; people are generating more waste all the time.
2. If each drop contains about 1 mL of water, then about 3 L would be saved each hour, 68 L every day, and 25 000 L in a year.
3. a. 1850: wood; 1920: hydropower, geothermal, and other; 1980: nuclear energy
   b. About 7 times

Make Connections

1. Concept maps will vary. You may wish to have students do the map below. First draw the map on the board, then add a few key terms. Have students draw and complete the map.

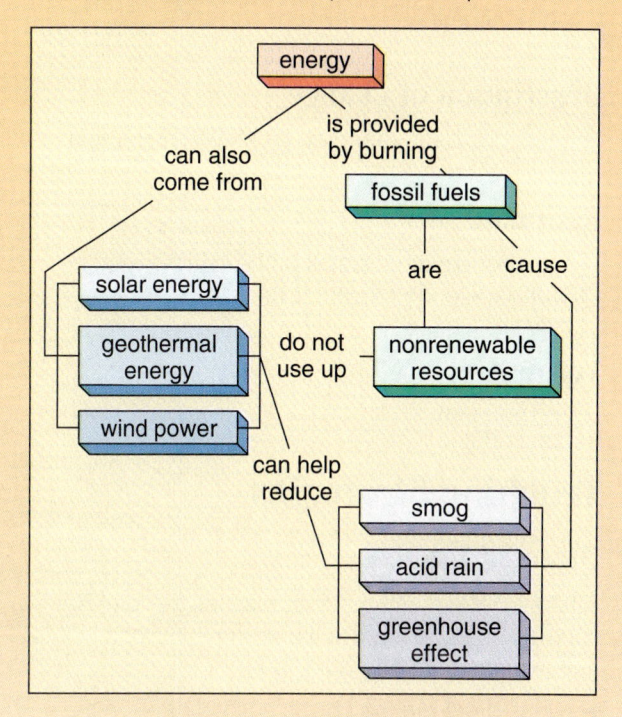

2. Answers will vary depending on the state. Encourage students to make realistic plans for protecting the species.
3. Answers will vary; however, sewage treatment generally involves filtration, sedimentation, biological or chemical purification, as well as chlorination and fluoridation.
4. Student reports will vary.
5. Student reports will vary. Most recycling centers accept newsprint, glass, and aluminum. Many also accept plastic bottles, magazines, and junk mail. Glass and aluminum is recycled back into glass and aluminum containers; plastic containers are either recycled into containers or are used for other plastic products, such as trash bags and playground equipment.

Develop Your Skills

Use the skills you have developed in this chapter to complete each activity.

1. **Interpret Data** The bar graph below shows the average amount of garbage produced per person each year in the United States.

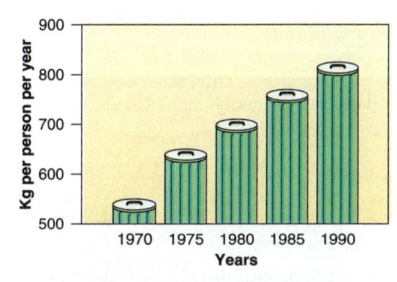

   a. About how much garbage did each person produce during 1980?
   b. During what five-year period did the amount per person increase the most?
   c. How much garbage do you predict the average person will produce in 1995? What factors should you take into account in making this prediction?

2. **Calculate** A leaky faucet drips 47 times per minute. If you fix it, how much water do you save every hour? Every day? In a whole year?

3. **Data Bank** Use the information on page 631 to answer the following questions.
   a. What type of fuel provided most of the energy used in the United States in 1850? In 1920? In 1980?
   b. Between 1900 and 1970, by approximately how many times did the United States increase its energy use?

Make Connections

1. **Link the Concepts** Construct a concept map to show how the following concepts from this chapter link together: fossil fuels, energy, smog, nonrenewable resources, acid rain, solar energy, greenhouse effect, geothermal energy, wind power.

2. **Science and Conservation** Find out what endangered plant and animal species live in your state. Choose one, and research where it lives and why it is endangered. Then outline a plan for protecting it.

3. **Science and Technology** Contact your local sewage treatment plant. Find out how sewage is treated in your community to make it safer for the environment.

4. **Science and Social Studies** Many communities are working to reduce the use of automobiles in order to save energy and cut down on air pollution. Investigate what is being done in your area to encourage bicycling, carpooling, and the use of mass transit. Then write a report on alternative transportation, including your ideas for what should be done and why.

5. **Science and You** Contact your nearest recycling center to find out how it works. What materials does it accept? Then ask where these materials are sent to be recycled. What are the glass bottles you recycle made into? What do recycled plastic containers become? Make a poster that shows the information you uncover.

SCIENCE AND LITERATURE CONNECTION UNIT 8

About the Literary Work

"All Things Are Linked" is a tale that appears in the text *The Crest and the Hide* by Harold Courlander, copyright 1982 by Coward, McCann & Geoghegan, Inc. Reprinted by permission of Coward, McCann & Geoghegan, Inc.

Description of Change

The tale has been reprinted directly from the text.

Rationale

The vocabulary and syntax of the original form are appropriate to the student level.

Vocabulary

larvae, plague

Teaching Strategies

Directed Inquiry

After students finish reading, discuss the tale. Be sure to relate the story to the science lessons in this unit. Ask the following questions:

▶ What effect did the chief's actions have on the ecosystem? (Interfered with a food chain; left vacant the niche filled by the frog population)

▶ Could this interruption in the food chain have possibly affected the food supply of the human population of this ecosystem? Explain. (Yes. Populations of consumers that fed upon frogs would eventually diminish, and it is likely that some of these higher-level populations were used as a food source by the humans.)

Social Studies Connection

Ask your students to identify the type of government by which the village was ruled. (Dictatorship) Discuss problems associated with this type of rule. (Ruler has absolute authority; needs of the ruler come before the needs of the community) Challenge students to predict how the story might have differed if the village had been a democracy.

Science and Literature Connection

"All Things Are Linked"

This story comes from the Lega people of Zaire and appears in the book The Crest and the Hide *by Harold Courlander.*

There was a certain chief in a certain village. He had many slaves. Whatever he wanted to be done, he ordered it. If it was a wise thing he wanted, his various counselors said to him: "Yes, it is good." If it was not a wise thing, they said, just the same, "Yes, it is good," because if they disagreed with him he grew angry, saying, "What! Do you say the chief doesn't know what he is doing?" But the lowest of his counselors never said yes or no. If the chief asked him about a certain thing he would think for a while and then reply: "All things are linked."

It happened one time that the chief could not sleep at night because of the croaking of frogs in the marshes. Night after night he could not sleep, and decided at last that the frogs would have to be exterminated. He told his counselors what he intended to do. One by one, as usual, they applauded him, saying, "Yes, it is good." Only the lowest of the counselors did not speak. The chief said: "You, counselor, have you no tongue in your mouth?" The man thought for a while, then he said: "O chief, all things are linked." The chief thought: "This man knows nothing else to say."

The chief sent his slaves out to exterminate the frogs in the marsh. They killed frogs until no more frogs remained. They returned, saying, "Sir, the frogs are done with." That night the chief slept well, and he slept well for many nights thereafter. He was pleased with life.

But in the marshes, the mosquitoes began to rise in swarms because there were no frogs to eat their larvae. They came into the village. They came into the chief's house and bit him. They made his life a misery. The people of the village suffered. So the chief ordered his slaves to go out and kill mosquitoes. The slaves went out, they tried, but the mosquitoes were too numerous. They continued to plague the village. The chief called his counselors. He scolded them, saying, "When I asked you about killing the frogs, you answered, 'It is good.' Why did you not say, 'If the frogs are killed the mosquitoes will multiply?' Only one of you said something for me to think about. He said, 'All things are linked,' but I did not understand his words."

622

Skills in Science

Reading Skills in Science

1. All parts of an ecosystem are interconnected. Even the slightest change in one aspect of the ecosystem will cause a subsequent change in another.
2. Other organisms in this food chain include mammals, birds, and snakes.
3. Populations of consumers that normally fed on frogs were reduced in number due to the lack of a food source.

Writing Skills in Science

1. Students' journal entries might describe preventative measures for protection from the biting mosquitoes and an increase in disease spread by the insect. Entries may also describe reduction in populations that feed on frogs.
2. Answers may vary. To quiet the noise, students may suggest a change to the chief's home.

Activities

Collect Data Answers may vary. Students should identify the method's positive and negative effects on the environment.

Communicate Answers will vary. Charts should show other organisms that feed on frogs and upon which mosquitoes feed.

The mosquito hordes made life unlivable. People left their houses and fields and went away. They went to distant places, cleared new fields, and began living again. The old village became deserted except for the chief and his family. Finally, the chief, too, took his family and went away.

Because of what happened there came to be a saying:
" 'Yes, it is good' caused a village to become deserted."

Skills in Science

Reading Skills in Science

1. **Drawing Conclusions** What did the counselor mean when he said "All things are linked"?
2. **Find Context Clues** The story identifies part of a food chain. What might the other organisms in the food chain be?
3. **Predict** What do you think happened to the population of consumers that fed on frogs?

Writing Skills in Science

1. **Infer** Suppose you lived in the village depicted in this story. Write a diary entry that describes how the extermination of frogs affected daily life in your village.
2. **Reason and Conclude** Imagine you are one of the chief's counselors. Rather than agree with his decision to exterminate the frogs, you offer him an alternative way of handling the situation. Write a new ending to the story describing your plan, the reasons you give the chief to follow your plan, and his reaction to it.

Activities

Collect Data In a library, research two different pests that are controlled by pesticides. Who are the people who want to control the pests? Why do they want to use pesticides? What are the negative effects of the pesticides? Is there another way to control the pests that you think is better?

Communicate Use reference books to identify at least two different food chains that involve both mosquitoes and frogs. Design a chart illustrating your findings.

Where to Read More

A Caribou Alphabet by Mary Beth Owens. Brunswick, Maine: Dog Ear Press, 1988. The interdependence of animals and nature is explored in this text, which was originally inspired by a project to ward off extinction of the caribou.

The Pied Piper of Hamlin by Mercer Mayer. New York: Macmillan Publishing, 1987. A colorfully illustrated contemporary retelling of this famous tale, which explores the problems of solid waste disposal, disease, and environmental relationships.

Data Bank

Planetary Statistics

Name	Distance from Sun (millions of km)	Diameter (km)	Gases in Atmosphere
Mercury	58	4 880	None
Venus	108	12 104	Carbon dioxide
Earth	150	12 756	Nitrogen, oxygen
Mars	228	6 794	Carbon dioxide, nitrogen, oxygen
Jupiter	778	142 796	Helium, hydrogen
Saturn	1 427	120 660	Helium, hydrogen
Uranus	2 870	50 800	Helium, hydrogen, methane
Neptune	4 496	48 600	Helium, hydrogen, methane
Pluto	5 900	about 3 000	Methane

Essential Nutrients in Plants

Element	Form Available	% of Dry Weight
Oxygen	O_2, H_2O	45.0
Carbon	CO_2	45.0
Hydrogen	H_2O	6.0
Nitrogen	NO_3^-, NH_4^+	1.5
Potassium	K^+	1.0
Calcium	Ca^{2+}	0.5
Magnesium	Mg^{2+}	0.2
Phosphorus	$H_2PO_4^-$, HPO_4^-	0.2
Sulfur	SO_4^-	0.1

Results of Mendel's Crosses of Peas

Cross	First Generation	Second Generation
Round X wrinkled seeds	All round	5474 round, 1850 wrinkled
Yellow X green seeds	All yellow	6022 yellow, 2001 green
Colored X white seed coats	All colored	705 colored, 224 white
Smooth X pinched pods	All smooth	802 smooth, 229 pinched
Green X yellow pods	All green	428 green, 152 yellow
Side X end flowers	All side	651 side, 207 end
Tall X short stems	All tall	787 tall, 277 short

Relative Sizes of Organisms

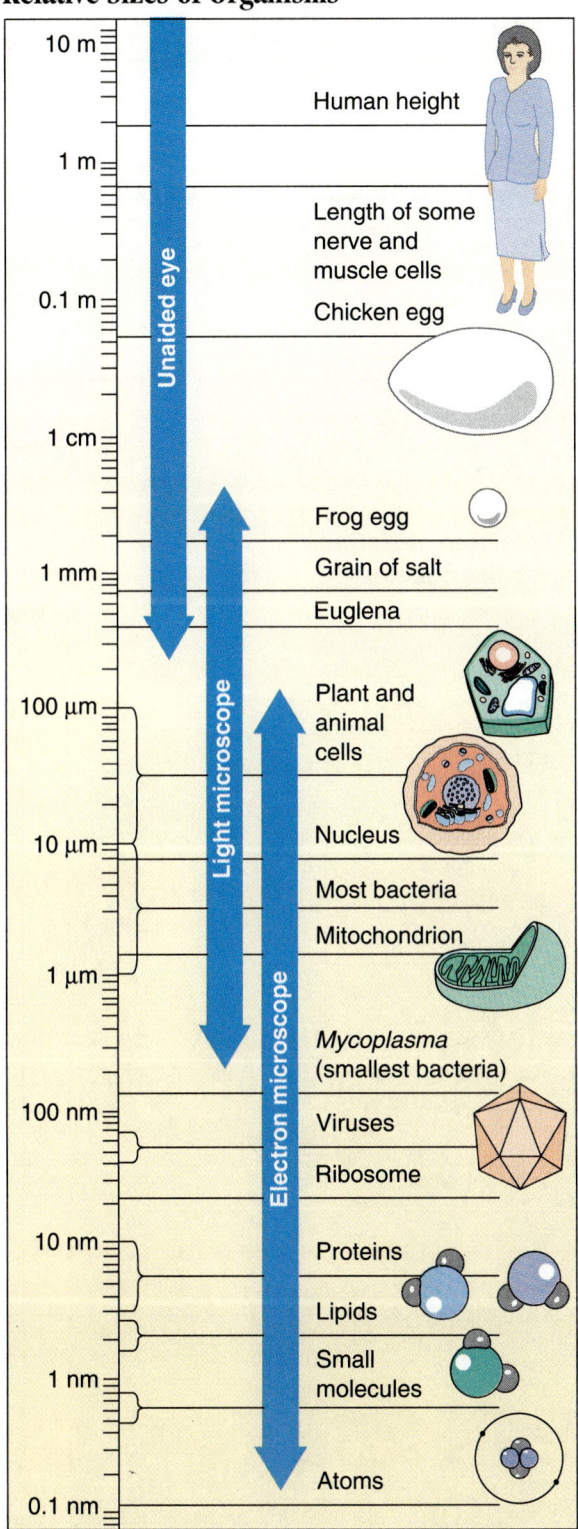

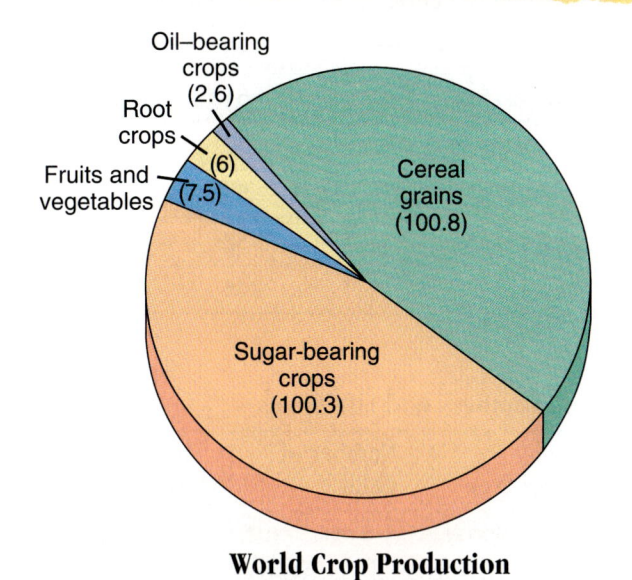

Scale		Organism/Structure
10 m	Unaided eye	Human height
1 m		Length of some nerve and muscle cells
0.1 m		Chicken egg
1 cm		
		Frog egg
1 mm		Grain of salt
		Euglena
100 µm	Light microscope	Plant and animal cells
10 µm		Nucleus
		Most bacteria
		Mitochondrion
1 µm	Electron microscope	*Mycoplasma* (smallest bacteria)
100 nm		Viruses
		Ribosome
10 nm		Proteins
		Lipids
1 nm		Small molecules
		Atoms
0.1 nm		

World Crop Production in Millions (metric tons)

- Cereal grains (100.8)
- Sugar-bearing crops (100.3)
- Fruits and vegetables (7.5)
- Root crops (6)
- Oil–bearing crops (2.6)

Average Life Spans

Species	Years
Mayfly	1 day
Mouse	2-3
Trout	5-10
Squirrel	11
Rabbit	12
Sheep	10-15
Cat, Dog	13-17
Rattlesnake	18
Owl	24
Lion	25
Horse	30
Hippopotamus	40
Pelican	45
Ostrich	50
Alligator	55
African elephant	60
Macaw	63
Dolphin	65
Raven	69
Rhinoceros	70
Man, Woman (USA)	68, 76
Tortoise	100

Data Bank

Human Bones by Region

Region of Skeleton	Number of Bones
Head	29
Torso	51
Arms and legs	20
Hands	54
Feet	52
Total	206

Cell Numbers and Life Spans

Cell Type	Number of Cells	Average Life Span
Red blood cell	4-6 million (per mm^3 of blood)	100 to 120 days
Brain cells	30 billion	Up to 100 years
Skin cells	6.5 million (per sq cm of skin)	Hours to days
White blood cells	4 000–11 000 (per mm^3 of blood)	A few hours to 9 days, depending on type

Population Density of the Continents (1991)

Asia	190 persons per sq. mi. (73 per km^2)
Europe	172 persons per sq. mi. (66 per km^2)
Africa	57 persons per sq. mi. (22 per km^2)
North America	46 persons per sq. mi. (18 per km^2)
South America	44 persons per sq. mi. (17 per km^2)
Australia	6 persons per sq. mi. (2 per km^2)

Vegetation Variations near the Equator

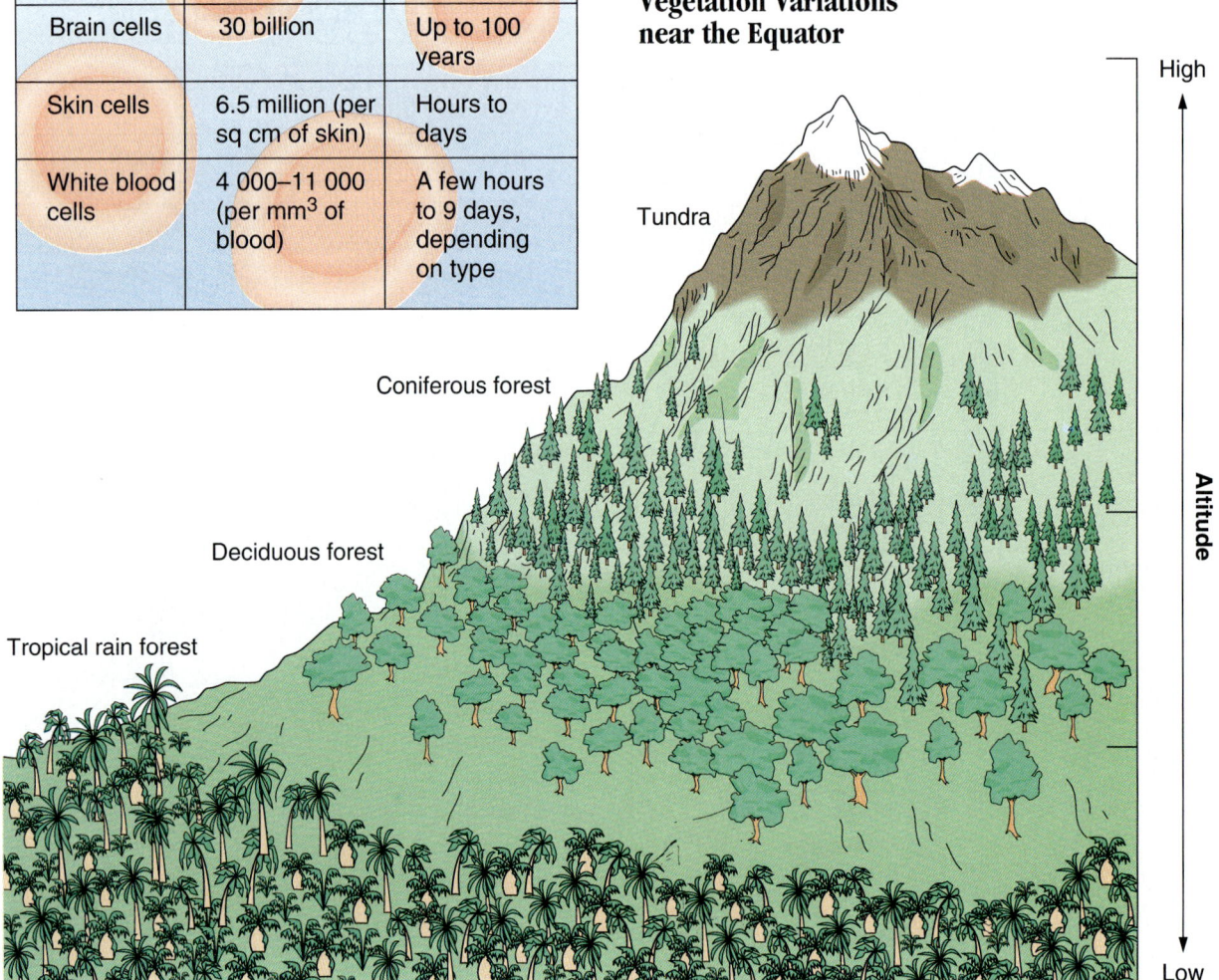

Tundra
Coniferous forest
Deciduous forest
Tropical rain forest

Altitude: High — Low

Data Bank

Tallest Tree Species

California redwood	Douglas fir	Noble fir	Giant sequoia	Ponderosa pine	Cedar	Sitka spruce
113 m	93 m	85 m	84 m	72 m	67 m	66 m

Oxygen Requirements for Activities

Activity	Oxygen used per hour (L)
Baseball	70
Basketball	90
Bicycling	55
Dancing	100
Football	110
Jogging	120
Playing piano	35
Sitting	25
Sleeping	14
Soccer	115
Swimming	120
Tennis	96
Walking	60

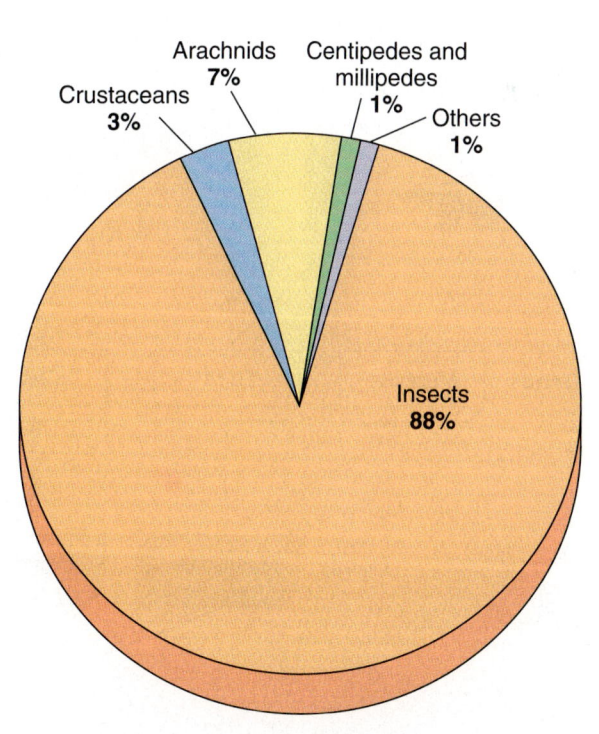

Relative Number of Species in the Major Arthropod Classes

- Insects 88%
- Arachnids 7%
- Crustaceans 3%
- Centipedes and millipedes 1%
- Others 1%

Data Bank

Effects of Blood Alcohol Concentration (BAC) on the Body*

Number of Drinks	BAC	Effects
🍸	.02–.03	Slight relaxation and mood change
🍸🍸	.05–.06	Relaxation and warmth; some loss of coordination
🍸🍸🍸	.08–.10	Increased loss of coordination; slightly impaired balance, vision, speech, and hearing
🍸🍸🍸🍸	.10–.13	Coordination and balance increasingly difficult; lack of judgment
🍸🍸🍸🍸🍸	.13–.16	Mental and physical abilities impaired; slurred speech; blurred vision; lack of motor skills
🍸🍸🍸🍸🍸🍸	.16–.19	Loss of motor control; assistance required for moving about; mental confusion
🍸🍸🍸🍸🍸🍸🍸	.19–.23	Severe intoxication; little mind and body control
🍸🍸🍸🍸🍸🍸🍸🍸	.22–.26	Unconscious; near coma state

*Figures based on body weight of 100 to 120 lbs, and drinks consumed in the span of 1 hour.

Relationships Among Carnivora

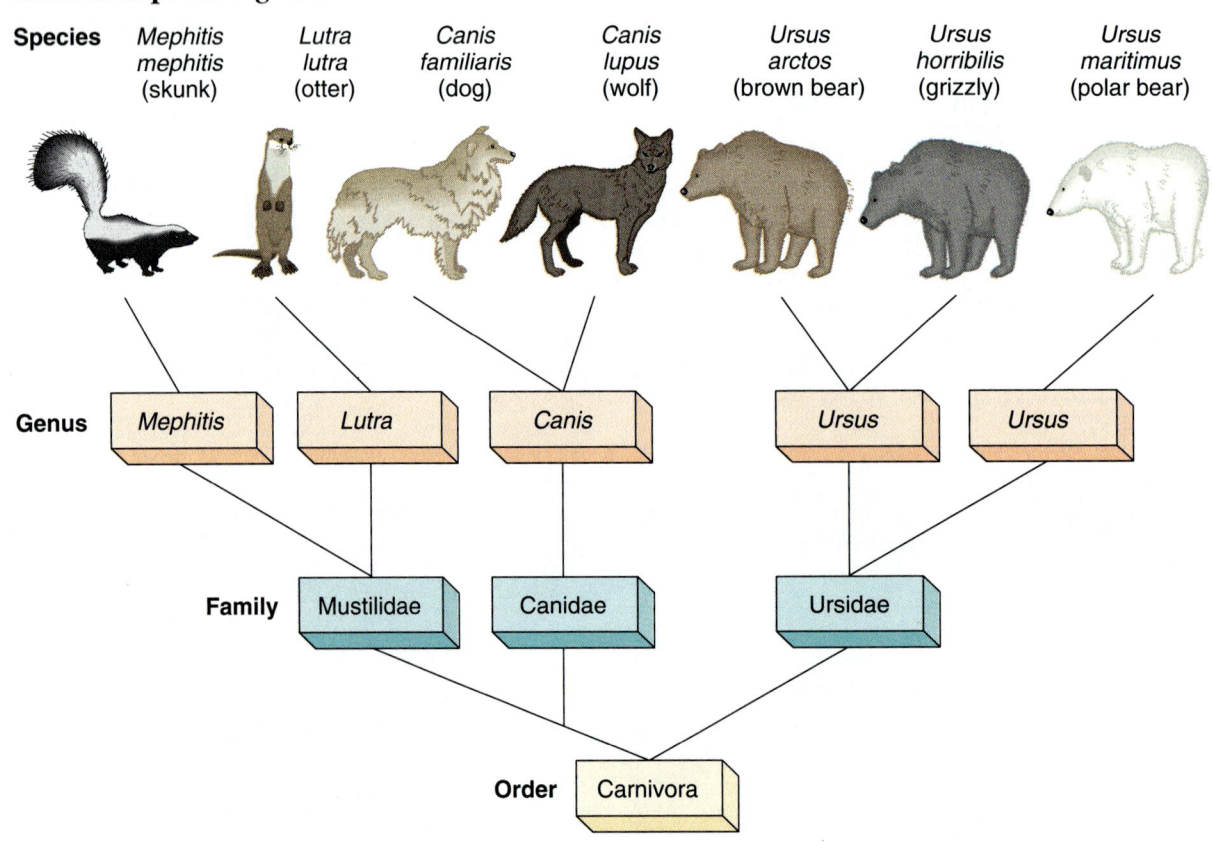

Species: *Mephitis mephitis* (skunk), *Lutra lutra* (otter), *Canis familiaris* (dog), *Canis lupus* (wolf), *Ursus arctos* (brown bear), *Ursus horribilis* (grizzly), *Ursus maritimus* (polar bear)

Genus: Mephitis, Lutra, Canis, Ursus, Ursus

Family: Mustilidae, Canidae, Ursidae

Order: Carnivora

Global Cycling of Water

Process	Annual % leaving atmosphere	Annual % into atmosphere
Evaporation from land	N/A	16
Precipitation over land	30	N/A
Surface runoff	N/A	7
Precipitation over ocean	84	N/A
Evaporation from ocean	N/A	84

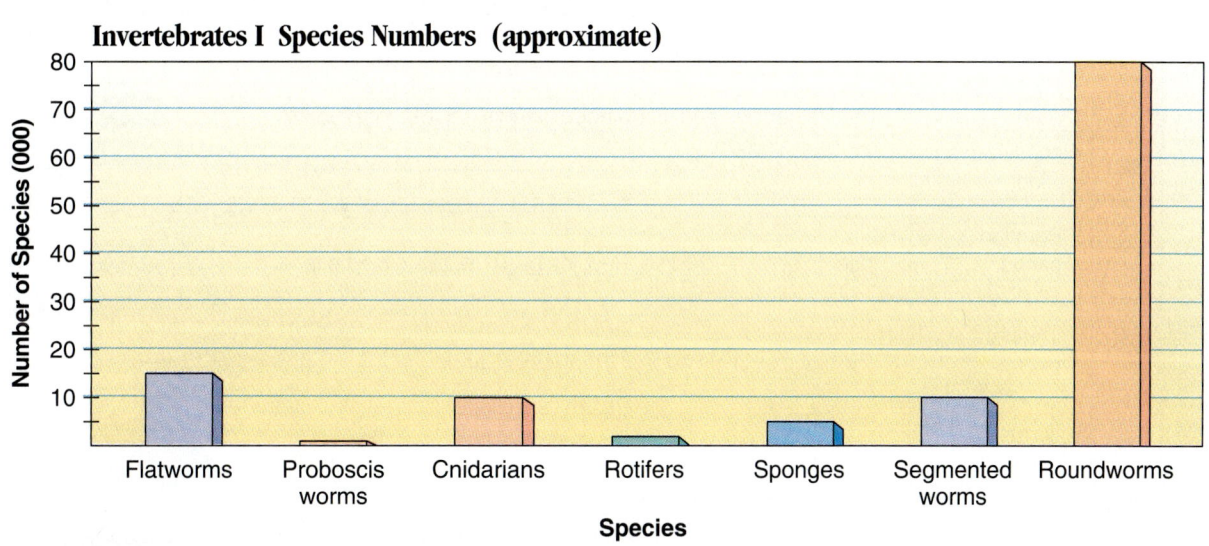

Invertebrates I Species Numbers (approximate)

Bar chart showing Number of Species (000) on the y-axis (0 to 80) for the following species:
- Flatworms: ~15
- Proboscis worms: ~1
- Cnidarians: ~10
- Rotifers: ~2
- Sponges: ~5
- Segmented worms: ~10
- Roundworms: ~80

Cancer Risk Factors

Type of Cancer	Risk Factors	Warning Signals
Breast cancer	Over age 50, first child born after age 30, personal or family history of breast cancer, never had children	Breast lumps, swelling, thickening; nipple discharge, pain, tenderness, scaliness
Uterine cancer	Early age at first intercourse, multiple sex partners, late menopause, history of infertility, failure of ovulation, and combination of diabetes, obesity, and high blood pressure	Unusual vaginal bleeding or discharge
Lung cancer	Heavy cigarette smoking, history of smoking 20 years or more, exposure to certain industrial substances, such as asbestos	Persistent cough, spitting blood, chest pain
Oral cancer	Heavy drinking and smoking, use of chewing tobacco	Sore lips, tongue, mouth, or throat that bleeds easily and does not heal; lump or thickening; difficulty chewing, swallowing
Skin cancer	Excessive exposure to the sun, fair complexion, occupational exposure to coal tar, radium, arsenic compounds, pitch and creosete	Unusual skin conditions, especially size or color changes of a mole or a darkly pigmented growth or spot

Data Bank

Evolution of Mammals (millions of years ago)

Era	MYA	Period/Epoch	Description
PALEOZOIC ERA	570	Cambrian period	Trilobites become abundant.
	505	Ordovician period	Corals, brachiopods, nautiloids, and graptolites are common.
	438	Silurian period	Fish with jaws appear. Sea scorpions present.
	408	Devonian period	Fish become abundant. The first amphibians appear.
	360	Mississippian period	The first reptiles and first winged insects appear. Amphibians become abundant.
	320	Pennsylvanian period	
	286	Permian period	Insects become diverse. Reptiles begin to take over the land.
MESOZOIC ERA	245	Triassic period	The first mammals appear. Reptiles are abundant.
	208	Jurassic period	The first birds appear. Dinosaurs dominate.
	144	Cretaceous period	Mammals and birds begin to diversify. Dinosaurs become less common and finally die out.
CENOZOIC ERA	66	Palaeocene epoch	The first primates appear. Mammals rapidly diversify, but are still unlike those alive today.
	58	Eocene epoch	Bats and early horses appear.
	37	Oligocene epoch	The first mastodons appear, and many relatives of the rhino.
	24	Miocene epoch	Apes present. More modern plant-feeding mammals become abundant.
	5	Pliocene epoch	The first hominids evolve.
	1.6	Pleistocene epoch	Ice Age mammals abundant as the ice caps advance and retreat.
	0.01	Holocene epoch	Modern mammals. Humans increase on all continents.

Large Reptile Size Comparison

1-3 m Giant tortoise

2m Leatherback sea turtle

3m Komodo dragon

1m Grass snake

2m Rattlesnake

6m Crocodile

5.5m King cobra

9m Anaconda

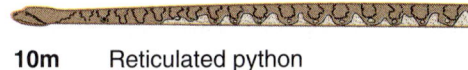

10m Reticulated python

Energy Use in the United States by Type of Fuel

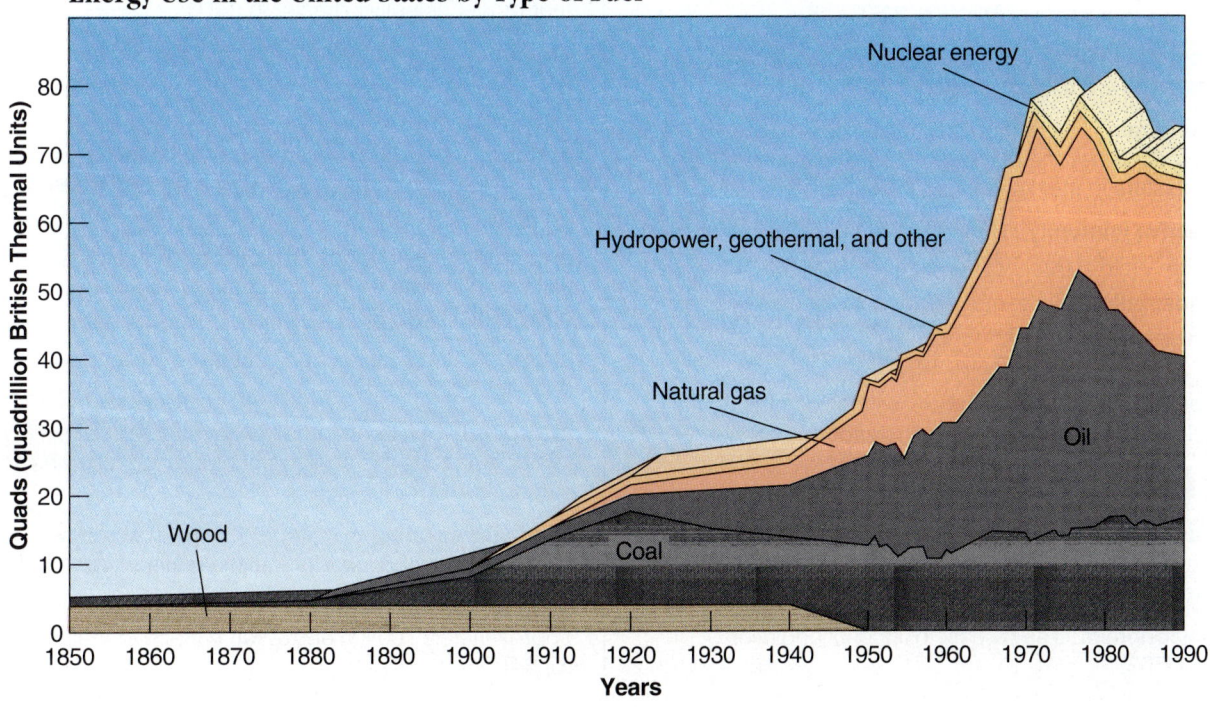

Y-axis: Quads (quadrillion British Thermal Units), 0 to 80+
X-axis: Years, 1850 to 1990

Labels: Wood, Coal, Natural gas, Oil, Hydropower, geothermal, and other, Nuclear energy

Data Bank

Glossary

A simple, phonetic spelling is given for words in this book that may be unfamiliar or hard to pronounce.

Stressed syllables are printed in capital letters. Sometimes a word has two stressed syllables. The syllable with the primary stress is printed in full capitals. The syllable with the secondary stress is printed in small capitals.

Example: *Chromosome* is pronounced KROH muh zohm.

Most of the time, the phonetic spelling can be interpreted without referring to the key. The key to the right gives the pronunciations for letters that are commonly used for more than one sound.

Pronunciation Key

a	cat	ih	pin
ah	hot	oh	grow
ai	care	oo	rule, music
ah	all	ow	now
ay	say, age	oy	voice
ee	meet	u	put
eh	let	uh	sun, about
eye	ice or by	ur	term

A

abdomen In arthropods and vertebrates, the part of the body containing many of the animal's organs. (p. 325)

abiotic factor The nonliving part of an ecosystem, such as air, soil, and water. (p. 538)

acid rain Highly acidic droplets formed when gases produced by the burning of fossil fuels combine with moisture in the clouds. (p. 613)

active immunity Body's production of its own antibodies for a particular pathogen. (p. 526)

adaptation A change in character or structure that helps an organism live and reproduce successfully in its environment. (pp. 39, 139)

addiction Physical and/or psychological dependence on a drug or other substance or behavior, usually including an overwhelming urge to continue or increase the use or the behavior. (p. 499)

aerobe (AIR ohb) A microorganism that requires oxygen to live and grow, such as an aerobic bacterium. (p. 201)

alcoholism The disease of alcohol addiction. (p. 503)

algae (AL jee) A type of protist, single-celled or many-celled, containing chlorophyll and having no true root, stem, or leaf, such as seaweed. Singular: alga. (p. 211)

alveoli (al VEE uh LY) Tiny air sacs in the lungs at the end of the bronchioles. (p. 431)

ameba (uh MEE buh) Single-celled microorganism that uses pseudopods for moving and engulfing food. (p. 214)

amino acid Simple molecule containing carbon, oxygen, hydrogen, and nitrogen; the building block of protein molecules. (pp. 64, 488)

anaerobe (AN ur OHB) A microorganism that does not require oxygen to live and grow, such as an anaerobic bacterium. (p. 201)

angiosperm (AN jee oh SPURM) A flowering plant that produces seeds enclosed in a fruit or pod. (p. 239)

animals Many-celled organisms consisting of complex cells, having membrane-bound nuclei and organelles, that can move, eat, and usually reproduce sexually. (p. 181)

antenna (an TEHN uh) One of a pair of appendages extending from the head of certain arthropods, used for balancing and sensing. (p. 328)

anther (AN thur) A saclike part of a flower's stamen that produces pollen grains. (p. 281)

antibiotic A chemical that inhibits or stops bacterial growth, such as penicillin; used to treat infectious diseases. (p. 527)

antibody A specially-shaped molecule that is part of the immune system; interacts with and neutralizes a particular pathogen. (p. 518)

arachnid (uh RAK nihd) Arthropods, such as spiders, ticks, and scorpions, that generally have two main body regions. (p. 327)

artery A blood vessel that transports blood from the heart to organs and muscles. (p. 422)

arthropod (AR throh pahd) An invertebrate with jointed appendages and segmented body, such as a spider, insect, or lobster. (p. 325)

asexual reproduction Reproduction in which a single individual copies its genetic material. (p. 145)

atmosphere An envelope of gases that extends approximately 700 km above the surface of the earth. (p. 34)

atom The building block of an element; atoms combine with other atoms to form a molecule. (p. 53)

B

bacilli (buh SIHL EYE) Rod-shaped bacteria. (p. 201)

bilateral symmetry Characteristic of certain animals, such as humans and insects, in which the body has two similar sections. (p. 300)

binary fission The process of asexual reproduction in protists and most protozoans in which the cell of the organism divides into two identical cells. (p. 202)

biomass energy Energy that is released when plants and plant products are burned. (p. 609)

biome A large community of plants and animals whose makeup is determined by soil and climate. (p. 579)

biosphere A life-supporting zone extending from the earth's crust into the atmosphere. (pp. 31, 537)

biotic factor The living part of an ecosystem. (p. 538)

bivalve (BY valv) A mollusk with two shells hinged together, such as a clam or oyster. (p. 323)

bronchi (BRAHN kee) The two main branches of the trachea that lead into the lungs. (p. 430)

bryophyte A type of nonvascular plant, such as mosses and liverworts, that grows only in moist environments. (p. 256)

budding A type of asexual reproduction in which a new individual develops from an outgrowth on the parent. (p. 224)

C

Calvin cycle The second stage of photosynthesis, in which chemical reactions produce glucose and other simple sugars. (p. 244)

cambium (KAM bee uhm) Growth tissue in a plant that makes up new xylem and phloem cells. (p. 276)

cancer A body disorder in which abnormally rapid cell reproduction causes a tissue to grow and spread. (p. 522)

capillary The smallest blood vessel in the body, where the exchange of oxygen and wastes between blood cells and the surrounding tissue occurs. (p. 422)

capsid A protein coat that makes up most of a virus and gives the virus shape. (p. 195)

carbohydrate An organic compound made of carbon, hydrogen, and oxygen, such as sugar, starch, and cellulose. (p. 487)

carcinogen A substance that causes cancer, such as some chemicals, radiation, and some viruses. (p. 523)

cardiac muscle (KAR dee ak) The striated, involuntary muscle of the heart. (p. 403)

cartilage (KART uhl ihj) Strong, flexible tissue in the skeletal system of vertebrates; also forms the skeleton in cartilaginous fishes. (pp. 349, 396)

cast A fossil formed by minerals in water that build up in a mold. (p. 163)

cell Basic unit of a living organism that can perform all the processes associated with life. (pp. 37, 75)

cell division The process in which one cell divides to produce two identical cells. (p. 103)

cell membrane The thin structure, made mostly of proteins and lipids, that encloses the cytoplasm and nucleus of plant and animal cells. (p. 79)

cell wall The stiff structure outside the cell membrane in a plant; provides support for the cell. (p. 79)

centriole A Structure in the cytoplasm that duplicates during interphase and forms the spindle. (p. 103)

cephalopod (SEF uh loh pahd) A mollusk with a distinct head and a foot divided into tentacles, such as a squid or octopus. (p. 323)

cerebrum (SUR uh bruhm) Part of the brain in vertebrates where higher brain functions occur. (p. 378)

chemical change Change in the chemical identity of a substance. (p. 59)

chemical digestion Chemical changes that take place when nutrients in food are broken down into simpler molecules that dissolve in water. (p. 415)

chemical equation Symbols and formulas that represent a chemical reaction. (p. 60)

chemical formula A combination of chemical symbols that represent a compound. (p. 55)

chemical symbol The shorthand name of an element, consisting of either a single capital letter or a capital and lowercase letter. (p. 55)

chlorophyll Green pigment in plants that captures the sun's energy for photosynthesis. (p. 83)

chloroplast An organelle in plant cells that contains chlorophyll. (p. 83)

cholesterol (kuh LEHS tuh ROHL) A fatty substance found in all animal tissues. (p. 488)

chromosome (KROH muh ZOHM) A threadlike strand in the nucleus that controls cell activity and carries genetic material to pass traits to offspring. (p. 104)

Glossary

cilia (SIHL ee uh) Tiny, hairlike structures that facilitate movement in protozoans and small worms; also move fluid in animals with greater specialization, as in the human trachea. (pp. 215, 430)

class A division of a phylum. (p. 177)

climate Average weather patterns from year to year. (p. 579)

climax community Stable, permanent ecosystem. (p. 554)

cnidarian (ny DAIR ee uhn) An invertebrate, usually marine, with a body cavity and stinging cells, such as a jellyfish or a sea anemone; also called coelenterate. (p. 305)

cocci (KAHK sy) Round or egg-shaped bacteria. (p. 201)

cochlea (KAHK lee uh) The spiral-shaped, fluid-filled structure of the inner ear. (p. 452)

coefficient The number that shows how many molecules or atoms of a substance are involved in a chemical equation. (p. 60)

coelom (SEE luhm) The fluid-filled body cavity in some animals, providing space for organ systems. (p. 313)

community Two or more groups of different organisms that live and interact in the same area. (p. 538)

compact bone The dense, smooth outer layer of bone. (p. 395)

competition Struggle among living things for the proper amount of water, food, and living space. (p. 44)

compound A substance made of more than one kind of element. Most compounds are made of molecules. (p. 54)

condensation The process in which a gas changes phase to a liquid. (p. 549)

coniferous (koh NIHF ur uhs) Referring to trees that bear their seeds in cones and do not shed their leaves at one time. (p. 587)

conifer Most common living gymnosperm, with needle-shaped leaves and seeds produced in cones. (p. 262)

conserve To protect from being lost, reduced, or damaged; to save. (p. 602)

constant A factor that is kept the same in a controlled experiment. (p. 7)

consumer An organism that eats other living things. (p. 542)

controlled experiment Experiment with two test groups: an experimental group and a control. (p. 7)

crustacean (kruhs TAY shun) An arthropod, such as a lobster or crab, with a hard outer shell and specialized appendages. (p. 328)

cubic meter The amount of space occupied by a cube: 1 m X 1 m X 1 m; abbreviated m3. (p. 14)

cytoplasm A jellylike substance that makes up all the living material in a cell except the nucleus. (p. 81)

D

data Information from which conclusions can be made. (p. 3)

deciduous (dee SIHJ oo uhs) Referring to trees that lose their leaves annually. (p. 586)

decomposer An organism, such as a fungus or bacterium, that feeds on and breaks down dead plants and animals. (pp. 203, 542)

density Measure of how much matter exists in a given volume; density = mass/volume. (p. 15)

depressant Drug that relaxes the central nervous system, causing the heart rate to slow and blood pressure to fall. (p. 500)

dermis Lower layer of skin, beneath the epidermis. (p. 408)

desalination (dee sal ih NAY shun) Process by which salt is removed from ocean water. (p. 45)

desert A biome that receives less than 25 cm of rainfall per year and has a greater rate of evaporation than rainfall. (p. 590)

diatomic molecule A molecule made of two chemically bonded atoms. (p. 55)

dicot (DY kaht) A variety of angiosperm whose seeds have two cotyledons; examples are maple trees and tomatoes. (p. 273)

diffusion (dih FYOO zhuhn) Movement of a substance from an area of high concentration to an area of low concentration. (p. 93)

digestion The process of breaking down food into simpler forms the body can use. (p. 415)

diversity The variety of life. (p. 140)

dominant gene Strong form of a gene, which is expressed even if a recessive gene is present. (p. 113)

duodenum (DOO oh DEE nuhm) First section of the small intestine. (p. 417)

E

eardrum The round, thinly stretched membrane that vibrates when struck by sound waves. (p. 452)

echinoderm (ee KY noh DURM) An invertebrate with radial symmetry that lives in the ocean, such as a sand dollar or sea star. (p. 337)

ecosystem Area in which living things interact with each other and the environment. (p. 538)

ectotherm (EHK toh THURM) An animal, such as a reptile or fish, whose body temperature varies according to the external temperature; often called cold-blooded. (p. 346) Adjective form: *ectothermic*. (p. 43)

element A form of matter that cannot be changed into simpler substances by any chemical process or by heating. Elements exist in nature as solids, liquids, or gases. (p. 53)

endocrine gland A ductless gland that releases hormones directly into the bloodstream. (p. 458)

endoplasmic reticulum (ehn duh PLAZ mihk rih TIHK yuh luhm) An organelle that transports materials throughout a cell. (p. 82)

endoskeleton (ehn doh SKEL uh tuhn) The skeleton inside the body. The backbone is the central part of an endoskeleton. (pp. 345, 393)

endospore A bacterium with a thick protective wall. (p. 202)

endotherm (EHN doh THURM) An animal, such as a dog, bird, or human, capable of maintaining a stable body temperature regardless of surroundings. (p. 345) Adjective form: *endothermic*. (p. 43)

energy pyramid A diagram showing the amount of energy available at each level of a food chain. (p. 545)

epidermis (EHP uh DUR mihs) The outer layer of protective cells in plants and animals. (pp. 276, 408)

era The largest division of the earth's history. The Precambrian, Paleozoic, Mesozoic, and Cenozoic eras are measured in millions of years. (p. 158)

esophagus (ih SAHF uh guhs) Muscular tube extending from the mouth to the stomach through which food travels. (p. 416)

eukaryote (yoo KAIR ee OHT) An organism with complex cells, having nuclei, mitochondria, and other organelles. (p. 211)

evaporation The process by which heat energy changes a liquid to a gas. (p. 549)

evolution The concept originated by Charles Darwin, explaining that organisms are the products of historical change and that new species gradually developed from previous ones. (p. 149)

exoskeleton The hard outer support structure of an arthropod. (p. 325)

extensor A muscle that straightens a joint. (p. 404)

extinct Referring to forms of life that have died out. (p. 162)

F

family A division of an order. (p. 177)

feces (FEE seez) Semisolid wastes that form in digestive organs. (p. 418)

fermentation An anaerobic process in which cells break down sugar or starch into carbon dioxide, either alcohol or lactic acid, and a small amount of ATP. (p. 100)

fertilization The process in which an egg cell and a sperm cell unite to form one cell called a zygote. (p. 469)

fetus The unborn young of an animal while still in the uterus. The human embryo is called a fetus after the eighth week of pregnancy. (p. 474)

fiber Indigestible substances in plant cell walls that help move food quickly through the intestine. (p. 489)

filter feeding The method of feeding used by certain animals, such as sponges, in which food is removed from water passing through the animal's body. (p. 302)

flagellum (fluh JEHL uhm) A whiplike structure of some cells and single-celled organisms that allows movement. (p. 200)

flagellate (FLA juh LAYT) Protozoan that uses a flagellum to move. (p. 215)

flatworm A worm with a flattened shape. (p. 311)

flexor A muscle that bends a joint. (p. 404)

food chain A sequence of organisms through which food energy passes. (p. 542)

food web A network of relationships in which the flow of energy branches out in many directions. (p. 544)

fossil Remains or traces of an organism that lived in the past. (p. 162)

fossil fuel An energy source formed from the buried remains of decayed plants and animals that lived hundreds of millions of years ago. (p. 608)

fossil record Record of life on earth provided by fossils. (p. 162)

fungus (FUHN guhs) An organism with a membrane-bound nucleus and organelles. Most fungi are many-celled; examples are mushrooms. (pp. 180, 222)

G

gamete (GAM eet) In organisms that reproduce sexually, one of the sex cells that unites with another to form a zygote. (p. 114)

gametophyte (guh MEET uh fyt) A generation of a plant that produces gametes and reproduces sexually. (p. 254)

gastropod (GAS troh PAHD) A mollusk that glides along on a foot underneath its body, such as a snail or slug. (p. 322)

gene A unit of genetic material that determines a trait. (p. 113)

genetics The study of heredity. (p. 111)

genotype (JEHN oh TYP) Gene combination that determines phenotype. (p. 113)

genus A division of a family. (p. 177)

geologic time The time scale of the history of the earth and its life. (p. 158)

geothermal energy (jee oh THUR muhl) Energy that comes from heat from the earth's hot interior. (p. 610)

gill An organ that removes dissolved oxygen from water. (p. 348)

Golgi body (GOHL jee) An organelle that manufactures and moves materials within a cell. (p. 83)

grassland A biome that is characterized by grasses, having hot summers, cold winters, rich soil, and unevenly distributed rainfall. (p. 589)

gymnosperm (JIHM noh SPURM) A plant that produces naked seeds not enclosed in fruit, such as a pine, spruce, or cedar. (p. 239)

H

habitat The area or place where an organism naturally lives in an ecosystem. (pp. 139, 305, 539)

hair follicle (FAHL ih kuhl) A structure in the dermis; each one grows a single hair. (p. 408)

hallucinogen (huh LOO sih nuh juhn) A drug or other substance that causes hallucinations, or dreamlike "trips," in which a person sees, hears, smells, or feels things that do not exist. (p. 501)

hazardous waste Waste substances that are very harmful to humans and other organisms. (p. 614)

heredity The passing of traits from parents to offspring. (p. 111)

holdfast The rootlike structure in brown and red algae that anchors the alga to a stable surface. (p. 219)

homeostasis Ability of an organism to keep conditions inside its body the same even though conditions in the environment may change. (pp. 43, 79)

hominid (HAHM uh nihd) The primate ancestor of modern humans. (p. 168)

hormone (HOR mohn) A chemical secretion made in endocrine glands that regulates a certain body function. (p. 458)

host A living thing another organism lives on or in for protection or nourishment. (p. 196)

hybrid An organism that has two different genes for a trait. (p. 113)

hydroponics (HY droh PAHN ihks) Method of growing plants without soil. (p. 242)

hyphae (HY fee) Branching, threadlike filaments that form the bodies of many-celled fungi. (p. 222)

I

immune system The complex set of cells and tissues that makes up the body's natural defenses, which attack invading pathogens and prevent them from growing inside the body. (p. 514)

impermeable Referring to a membrane through which substances can not pass. (p. 95)

incomplete dominance A condition that results when genes produce a trait somewhere in between the traits of the parents. (p. 116)

incubate To keep an egg, or other object, in an environment suitable for hatching or development. (p. 373)

infectious disease A disease that can be transmitted between people by harmful organisms. (p. 511)

ingest To take in food from the environment. (p. 299)

innate behavior Behavior that is natural for an organism and does not need to be learned. (p. 571)

inorganic compound A compound that usually does not contain carbon. (p. 64)

inorganic nutrient A substance the body needs, such as water or a mineral, that does not come from a living thing. (p. 490)

invertebrate An animal without a backbone. (p. 301)

iris The round, colored disk surrounding the pupil that controls how much light enters the eye by changing the size of the pupil. (p. 451)

J

joint A place where two or more bones meet. (p. 397)

K

kilogram Basic unit of mass; abbreviated kg. (p. 15)

kingdom One of the five major divisions into which all living things can be classified. (p. 176)

L

labor The process of human childbirth, especially the muscle contractions that push a baby out of the uterus and through the vagina. (p. 475)

Glossary

lactic acid A chemical that builds up in muscles during anaerobic reactions. (p. 406)

larva (LAHR vuh) The early stage of an organism's development which changes structurally when the organism becomes an adult; plural: larvae. (p. 303)

lens A piece of glass that can bend light in some way to make an object look smaller, larger, closer, or farther away. (p. 21)

ligament Stretchable connective tissue that joins bones and reinforces movable joints. (p. 397)

limiting factor An environmental condition that keeps a population from increasing, such as the availability of food, water, and living space, and the spread of disease. (p. 562)

lipid (LIH pihd) A type of organic compound that is insoluble in water, such as fat and oil. (p. 488)

liter A metric unit of volume; abbreviated L. (p. 14)

littoral zone (LIHT uh ruhl) The place where land and ocean meet; also called the seashore. (p. 592)

liverwort A bryophyte with either flat, leaflike structures, or a body divided into rounded sections, or lobes. (p. 257)

lysosome An organelle that breaks down food molecules, waste products, and old cells. (p. 83)

M

mandible (MAN duh buhl) A jawlike appendage in arthropods used for chewing and crushing food; the lower jaw of invertebrates. (p. 328)

marine Referring to the ocean. (p. 593)

marsupial (mar SOO pee uhl) A mammal with a pouch in which the young complete their development. (p. 381)

mass A scientific measurement for the amount of matter that an object contains. (p. 14)

matter Anything that takes up space and has mass. (p. 51)

mechanical digestion Physical changes that take place as food is broken down into smaller pieces. (p. 415)

medusa (muh DOO suh) Body form of cnidarians that is free-swimming. (p. 305)

meiosis (my OH sihs) Cell division that produces gametes or spores having one set of unpaired chromosomes. (p. 118)

melanin (MEHL uh nihn) The dark pigment in skin, hair, and parts of the eye that partially determines the color. (p. 410)

meniscus (mih NIHS kuhs) Curved surface of a liquid in a graduated cylinder. (p. 14)

menstruation (MEHN STRAY shun) The flow of blood and tissue from the uterine lining out of the body when a human egg is not fertilized; also, the period of time when this flow occurs. (p. 470)

metabolism The continual chemical and physical processes that balance the production, consumption, and storage of energy in an organism. (p. 37)

metamorphosis (MEHT uh MOR fuh sihs) The process of development in which a young organism undergoes changes in appearance to become an adult. (pp. 333, 355)

meter Basic SI unit of length; abbreviated m. (p. 13)

microscope An instrument with one or more lenses that makes very small objects appear larger so they can be seen and studied. (p. 21)

migrate To move to a different place for a certain part of every year, or simply to move from one place to another. (p. 372)

mitochondrion (myt uh KAHN dree uhn) An organelle that produces energy in the cell. (p. 82)

mitosis (my TOH sihs) The process by which a cell nucleus divides into two identical nuclei. (p. 103)

mixture A substance with two or more components that are not combined chemically. (p. 58)

mold A fossil impression left in rock by the hard parts of an organism. (p. 163)

molecule The smallest part of a compound that has all the properties of the compound; two or more atoms chemically bonded. (p. 54)

mollusk (MAHL uhsk) An invertebrate with a soft, unsegmented body, such as an oyster, snail, or octopus; many are covered with hard shells. (p. 321)

moneran A simple organism with no membrane-bound nuclei or organelles; occurs as a single cell or in colonies; examples are bacterium and blue-green bacterium. (pp. 180, 200)

monocot (MAHN oh KAHT) A variety of angiosperm whose seeds have only one cotyledon; examples are corn and tulips. (p. 273)

monotreme (MAHN oh TREEM) An egg-laying mammal, such as a platypus, with two unique characteristics: hair and mammary glands. (p. 380)

moss A bryophyte with tiny green leaflike structures arranged in a spiral on a short stalk. (p. 256)

motor neuron A specialized nerve cell that carries an impulse from the central nervous system to voluntary muscles, causing the muscles to contract. (p. 447)

Glossary

mutation A change in DNA or chromosomes; can occur in body cells and sex cells. Most mutations are not harmful. (p. 124)

N **narcotic** A drug that numbs or takes away feelings and is often addictive. (p. 501)

natural resource Material from the environment that people use to carry on their daily lives. (p. 601)

natural selection The process by which organisms that are best adapted to their environment survive and reproduce, passing their traits on to their offspring. (p. 150)

nephron (NEHF rahn) One of many tiny structures in the kidney that filter blood. (p. 436)

neuron (NOO rahn) A nerve cell; receives and sends electrical and chemical messages. (p. 445)

niche (nihch) The specific role of an organism in its habitat. (p. 539)

nicotine The mild stimulant in tobacco. (p. 500)

nitrogen fixation The bacterial conversion of atmospheric nitrogen to usable compounds. (p. 550)

nonrenewable resource A natural resource, such as oil or natural gas, that cannot be replaced by natural processes. (p. 601)

nonvascular plant (NAHN VAS kyuh lur) A plant that lacks the specialized tissues to transport water, such as a moss. (p. 238)

nuclear energy Energy released by either the splitting apart or the joining together of atoms of certain elements. (p. 610)

nuclear membrane A thin layer that separates the nucleus from the rest of the cell. (p. 80)

nucleic acid A combination of proteins and carbohydrates that controls all cell activities and carries reproduction instructions; examples are DNA and RNA. (p. 65)

nucleus The control center for most of a cell's activities. (p. 80)

nutrient A substance needed by an organism to live and grow. (p. 415)

nymph (NIHMF) A stage in certain insect metamorphosis, such as that of a grasshopper, in which the organism looks like an adult but has no wings and cannot reproduce. (p. 333)

O **oral groove** A channel in a paramecium's body that is lined with cilia; food particles are swept down the oral groove into the mouth. (p. 215)

order A division of a class. (p. 177)

organ A group of tissues that work to perform a special function. (p. 87)

organ system A group of organs working together. (p. 87)

organelle A tiny part of a cell having a special form and function; organelles carry out all life processes. (p. 81)

organic compounds A carbon compound that occurs naturally in the bodies, products, and remains of living things. (p. 64)

organic nutrient An essential compound found in living things; examples are carbohydrates, fats and oils, proteins, vitamins, and fiber. (p. 487)

organism The highest level of cell organization. All organisms carry out life processes. (p. 87)

osmosis (ahz MOH sihs) Diffusion of water or other solvent through a membrane. (p. 94)

ovary (OH vuh ree) The female sex gland, which produces egg cells and, in vertebrates, sex hormones. (p. 469)

oviduct (OH vi duhkt) The tube through which egg cells travel from ovary to uterus. (p. 469)

P **passive immunity** Temporary immunity to a disease, gained when a person is injected with antibodies produced by a donor. (p. 526)

pathogen A microorganism that causes an infectious disease. (p. 511)

pelagic zone (pih LAYJ ihk) Part of the ocean beyond the sublittoral zone, where water is deeper than 200 m. (p. 593)

penis The male external sex organ, through which semen and, in mammals, urine flows out of the body. (p. 468)

peristalsis (PEHR uh STAHL sihs) Rhythmic contraction and relaxation of smooth muscles that pushes food through the digestive tract. (p. 416)

permeable Referring to a membrane through which substances diffuse freely. (p. 95)

pharynx (FAR ihnks) The funnel that leads from the nose and mouth to the windpipe and esophogus. (p. 430)

phase Property of matter; the three phases of matter are solid, liquid, and gas. (p. 52)

phenotype (FEE noh TYP) A trait that an organism actually shows. (p. 113)

phloem (FLOH ehm) A set of linked cells inside the vascular tissue of a plant that serves as the distribution path for glucose and nutrients. (p. 275)

638 Glossary

photosynthesis (foh toh SIHN theh sihs) The process by which plants use energy from sunlight and raw materials from air and water to make glucose to obtain energy. (pp. 97, 243)

phylum (FY luhm) The largest group in the animal kingdom, or a subdivision in the plant kingdom. (p. 176)

physical change The change in a substance's physical properties, but not in its chemical identity. (p. 58)

physical property A characteristic of matter that can be observed or measured without changing the make-up of the object. (p. 51)

pioneer community A community of the first organisms to settle in an area. (p. 554)

pistil The female reproductive part of a flowering plant. (p. 281)

placenta (pluh SEHN tuh) An organ in most mammals that connects the fetus to the uterus of the mother, through which nutrients and waste products are exchanged. (p. 381)

placental mammal A mammal whose young develop in the mother's uterus, attached by a placenta. (p. 381)

plant A many-celled organism with cell walls that makes glucose by photosynthesis and does not move voluntarily. (p. 181)

polar climate A climate with a long cold season and a short warm season. (p. 580)

pollen Dustlike particles that are protective carrying cases for the sperm cells of gymnosperms. (p. 262)

pollination The transfer of pollen, containing the male gametophyte, to the pistil, containing the female gametophyte. (p. 263)

polyp (PAHL ihp) Body form of cnidarians that is usually attached to a stable surface. (p. 305)

population A group of individuals of the same species living in a particular area. (pp. 150, 539)

population density The number of individuals in an area, calculated by dividing the number of organisms in a certain area by the area's size. (p. 562)

precipitation Moisture in the atmosphere that condenses and falls to the earth as sleet, snow, hail, or rain. (p. 549)

predator An animal that obtains its food by killing and eating other organisms. (p. 563)

prey An organism that is hunted or eaten by predators. (p. 563)

primate A member of a group of mammals that includes humans. Primates appeared more than 65 million years ago. (p. 167)

producer An organism that can make its own food. (p. 542)

product A substance produced by a chemical reaction. (p. 60)

protein A large, complex molecule made of amino acids. (p. 488)

protist (PROHT ihst) A member of the kingdom Protista, most of which are single-celled organisms, with complex cells and membrane-bound nuclei and organelles, that reproduce by cell division and may make or eat food. (p. 181)

protozoan (PROH tuh ZOH uhn) A single-celled, animal-like organism belonging to the kingdom Protista. (p. 211)

pseudopodium (soo duh POH dee uhm) A temporary footlike projection of cell matter that facilitates protozoan movement and ingestion. (p. 214)

pupa (PYOO puh) The stage of metamorphosis in which an insect is often surrounded by a cocoon and many changes occur. (p. 333)

pupil The opening in the choroid layer at the front of the eye. (p. 451)

R radial symmetry A characteristic of certain animals, such as cnidarians, in which the body is organized around the center. (p. 300)

radioactive dating A method of determining an object's age based on known decay rates of radioactive elements. (p. 161)

reactant Raw material in a chemical reaction. (p. 60)

recessive gene The weak form of a gene, which is not expressed when the dominant form is also present. (p. 113)

renewable resource A natural resource, such as trees or soil, that can be replaced by natural cycles or processes. (p. 601)

replication The copying process in which new DNA is created. (p. 123)

reproduction The process by which organisms produce more organisms like themselves. (p. 38)

respiration The process by which a cell breaks down glucose into carbon dioxide and water. (p. 99)

response A reaction to a stimulus. (p. 38)

retina (REHT ih nuh) The light-sensitive layer that lines the back and sides of the inner eye. (p. 451)

rhizoid (RY zoyd) The rootlike structure of bryophytes. (p. 256)

Glossary

ribosome An organelle that produces protein for the cell. (p. 82)

roundworm A worm with a round shape and a one-way digestive tube with two openings. (p. 312)

S

sarcodine (SAR kuh deen) Type of protozoan, including the ameba, that moves using pseudopodia. (p. 214)

scrotum (SKROHT uhm) A sac that protects the testes in most male mammals. (p. 468)

sedimentary rock Rock formed from layered sediments that pile up and squeeze together, offering clues to the earth's past. (p. 160)

segmented worm A worm whose body is divided into small segments or units. (p. 313)

selectively permeable Referring to a membrane through which only some materials may pass. (p. 95)

semen (SEE muhn) The fluid mixture of sperm cells and accompanying secretions produced by male animals. (p. 468)

sensory neuron A nerve cell that carries impulses from other neurons to the spinal cord. (p. 447)

sepal (SEE puhl) A specialized structure at the base of a flower that encloses or protects the flower as it develops. (p. 281)

sex-linked trait A trait whose genes are located on the sex chromosomes. (p. 119)

sexual reproduction Reproduction through the union of a male and a female gamete, each contributing genetic material to the offspring. (p. 145)

simple sugar A substance in living things, such as glucose, that provides food for energy. (p. 65)

skeletal muscle A type of muscle attached to bones by tendons, allowing movement at the joints. (p. 403)

smog Air pollution caused by automobile exhaust, burning coal, or by other means, resulting in air that is unhealthy to breathe. (p. 612)

smooth muscle A type of muscle that makes up the walls of most internal organs. (p. 403)

solar energy Energy from the sun. (p. 609)

solution A type of mixture in which one substance is evenly mixed with another substance. (p. 59)

specialized Referring to a cell, tissue, or other unit that has certain characteristics in order to perform a particular function. (p. 86)

species The basic unit of classification, the division of a genus, made of very similar organisms that are able to mate and reproduce offspring of the same type. (pp. 142, 177)

spinal cord Nerve fibers extending from the base of the brain down the back. (p. 444)

spirilla (spy RIHL uh) Spiral-shaped bacteria. (p. 201)

spongy bone Soft, lightweight bone that contains red marrow. (p. 395)

spore A reproductive cell produced by sporozoans. (pp. 216, 254)

sporophyte (SPOH roh FYT) A plant that reproduces nonsexually by spores. (p. 254)

sporozoa (SPOH roh ZOH uh) A group of parasitic protozoan living in the bodies of another animals; produces spores, from which new sporozoans grow. (p. 216)

stamen (STAY muhn) The male reproductive part of a flower. (p. 281)

stigma The part of a flower's pistil that collects pollen. (p. 281)

stimulant A substance that stimulates the nervous system, raises blood pressure, and makes the heart beat faster. (p. 500)

stimulus A change that occurs in the environment, such as sound, light, or odor, that may affect an organism's behavior. (p.38)

stomate (STOH mayt) An opening in a plant surface through which gases are exchanged. (p. 244)

stress Emotional or mental strain or imbalance. (p. 496)

sublittoral zone (SUHB LIHT uh ruhl zohn) Part of the ocean near the littoral zone, where the water is less than 200 m deep, and the bottom slopes gradually; also called the continental shelf. (p. 592)

subscript The number written below the line in a chemical formula, that indicates the number of atoms in a molecule. (p. 55)

succession The replacement of one community by another over a period of time. (p. 553)

symbiosis (SIHM by OH sihs) The relationship between species that live together, in which at least one member benefits. (p. 567)

synapse (SIHN aps) The space between neurons. Electrical impulses must be chemically transmitted across the synapse. (p. 445)

synthetic polymer A giant molecule made of many smaller molecules that does not occur naturally, such as plastics and nylon. (p. 62)

T

taxonomy The science of classifying living things. (p. 175)

temperate climate A climate with a cold season and warm season of nearly equal lengths. (p. 580)

tendon A connective tissue that joins muscle and bone. (p. 397)

testes (TEHS teez) Male sex glands that produce male hormones and sperm. (p. 468)

thorax The part of an arthropod's body where legs are attached; in some vertebrates, the part of the body between the neck and the abdomen. (p. 325)

tissue A group of specialized cells in plants and animals that are organized to perform a certain function. (p. 86)

tolerance The body's adjustment to a drug, so that more and more must be used to obtain the same effects, possibly resulting in addiction. (p. 499)

trace element An element occurring in only a very small quantity in a certain substance. (p. 57)

trachea (TRAY kee uh) In most land vertebrates, the windpipe, leading from the larynx to the bronchi. In insects, the tubules that conduct air from the exterior. (p. 430)

transpiration A plant's loss of water through the stomates of the leaves. (p. 275)

tropical climate A warm climate with little variation in temperature during the year; most tropical climates have a lot of rainfall. (p. 580)

tropism (TROH PIHZ uhm) In plants, movement or turning in response to an external stimulus, such as light or gravity. (p. 287)

tundra The vast, treeless plains of arctic and subarctic regions. (p. 585)

U

umbilical cord A soft rope of blood vessels that connects the embryo to the placenta, through which materials are exchanged between mother and embryo. (p. 474)

ureter A tube through which urine passes from the kidney to the urinary bladder or cloaca. (p. 436)

urethra A tube through which urine passes from the bladder to the outside of the body. (p. 436)

uterus (YOOT uhr uhs) A hollow, muscular organ in female mammals where a zygote embeds and develops into a baby. (p. 469)

V

vaccine A medicine that produces immunity to a disease without causing the disease; contains weakened or dead pathogens, which induce the production of antibodies. (p. 525)

vacuole (VAK yoo ohl) An organelle that stores water, food, and wastes in a cell and helps get rid of wastes. (p. 83)

vagina (vuh JY nuh) A canal in female mammals extending from the cervix to the outside of the body; receives the penis during sexual intercourse. (p. 469)

variable The factor that is changed in a controlled experiment. (p. 7)

variation A difference within a species. (p. 144)

vascular plant (VAS kyuh lur) A plant, such as a flowering plant, having tissues with tubelike structures that transport water and other materials. (p. 238)

vegetative propagation (prahp uh GAY shun) A method of producing genetically identical plants from pieces of stems or leaves. (p. 284)

vein A blood vessel that carries blood to the heart. (p. 422)

vertebra (VUR tuh BRUH) One of the small bones that make up the backbone. (p. 393)

vertebrate (VUR tuh briht) An animal with a backbone. (p. 301)

villi (VIHL eye) Tiny, fingerlike projections that line the small intestine and absorb nutrients from food. (p. 418)

virus A piece of hereditary material with a coat of protein. (p. 195)

vitamin An organic nutrient that helps regulate the chemical reactions that convert food into energy and living tissue. (p. 489)

volume The amount of space that something occupies. (p. 14)

X

X chromosome The longer sex chromosome. Females have two X chromosomes; males have one X chromosome and one Y chromosome. (p. 119)

xylem (ZY luhm) A set of linked cells inside the vascular tissue of a plant that forms a pipeline for water and minerals. (p. 275)

Y

Y chromosome The shorter sex chromosome. Males have one X chromosome and one Y chromosome. (p. 119)

Z

zygote A cell formed by the union of two sets of genes when a male and female gamete unite; a fertilized egg. (p. 145)

Glossary

Index

Note: Boldface numerals denote definitions. Italic numerals denote illustrations.

A

Abdomen in arthropods, **325**, 332
Abiotic factors, **538**, 539
Acid rain, **613**, 618
Acne, 409–410
Acquired immune deficiency syndrome (AIDS), 197, 472, 513–515
Active immunity, **526**
Adaptation, **39**, **139**, 141
    in cnidarians, 306
    diversity of, 140
    in fish, 346–347, 351
    in flowering plants, 238, 240–241, 271–272, 278–279
    freshwater biomes, 594–595
    in insects, 331
    niches and, 540
    in nonflowering plants, 253–254, 259, 262
    reasons for, 143, 150, 167, 262, 370
    in reptiles, 357–358
    in segmented worms, 313
    specific to species, 142
    in vascular plants, 259
Addiction, **499**
Adolescence, 479
Adrenaline, 458
Adulthood, 480
Aerobe, **201**
African sleeping sickness, 215
Agricultural researcher, 606
Air pollution, 433, 612
Air pressure, breathing and, 429
Alcohol, 100, 503–504
Algae, **211**, 212, 217, 219
    in coral reef building, 308
    as food, 213, 221
    types of, 220–221
    vs. plants, 237
Algin, 213, 219
Allergen, 524
Allergy, 285, 524
Alligator, 360
Altitude and climate, 581
Alveoli, **431**
Ameba, *214*, 214
Amebic dysentery, 214, 512
Amino acids, **64**, 66, 488
Amniocentesis, 477
Amphetamines, 500
Amphibians, 347, 353–356
Anaerobe, **201**
Analyzing data, 5
Anaphase, *104*, 104
Angiosperm, **239**, 271–273
Animalcules, 75
Animal Kingdom, 181
Animal(s), **181**
    bacteria and, 203
    behavior of, 571–574
    cells, *80*, 80, 85, 103–105
    characteristics, 299
    classification of, 300–301
    diseases spread by, 513
    energy processes, 99
    testing, 378
    trainers, 572
Anorexia nervosa, 420
Ant colony, 334
Antenna, crustacean, **328**
Anther, **281**
Antibiotics, 227, **527**
Antibodies, 425–426, **518**. See also Rh factor.
Antigens, 425
Antiseptics, 206
Anus, 419
Apgar Score, 476
Appendicular skeleton, 393, 398
Aqualung, 432
Arachnids, 327
Archaeopteryx, 368
Aristotle's classification system, 175
Arteries, **422**
Arthropod, **325**, 326–329
Arthroscopic surgery, 399
Artiodactyla, 383
Aspirin, 274
Asexual reproduction, **145**
Asthma, 432
Atherosclerosis, 427
Atmosphere, **34**
Atom, models of, **53**, *53*
ATP, 99–101, 244–245
Auto-immune diseases, 521
Auxin and phototropism, 287
Axial skeleton, 393, 398

B

Bacilli, **201**
Bacteria
    classification, 201
    diseases caused by, 204–205, 512, 527
    in genetic engineering, 129
    nitrogen cycle and, 203, 550
    parasitic, 570
    reproduction, 202
Bacteriophage, 196
Balance
    in ecosystems, 541
    middle ear control of, 452
Ball and socket joints, 398
Barbiturates, 500
Bar graphs, 18–19
Barnacles, 328
Base pairs, 123
Biennial plants, 288
Bilateral symmetry, *300*, 300, 310
Bile, 417
Binary fission, **202**, *202*
Binomial nomenclature, 178
Biomass energy, **609**
Biomes, 579, *584*, 584–590, 592–595
    climates in, 579–580
Biosphere, **31**, 537
Biotechnology in agriculture, 289
Biotic factors, **538**
Birding, 185
Birds, 367–371, 373
    pair bonds in, 373
    perching, 369
    seed-eating, 369
Birth rate, population size and, 561
Bivalve, **323**
Blood, 424–427, 513
    pressure, 422, 427
    vessels, *422*, 422
Blue–green bacteria, 201
Body disorders, 521–522
Body secretions, 517
Body temperature control, 347, 358, 367, 377
Bone marrow, *395*, 395
Bones, 393–397, 400, 405
Book lungs, 327
Botanist, 257
Botulism, 205
Brain, 444, *446*, 446, 449
Breadmaking, fermentation in, 100, 224
Bronchi, **430**
Bronchioles, 431
Bronchitis, 432
Bryophytes, 238, 256–258
Budding, **224**
Bulimia, 420
Burns, treatment of, 84

C

Cacti, 240, 590
Caffeine, 500
Calcitonin, 459
Calories, 495
Calvin cycle, **244**

642   Index

Cambium, **276**
Cancer, 106, 124, 505, **522**, 523
Capillaries, **422**
Capsid, 195
Carbohydrates, 65, **487**
Carbon dioxide, 44
    formation of, 60
    global warming and, 552
    molecule, **54**
    in oxygen–carbon dioxide cycle, 551
    in photosynthesis, 97–98
    release during respiration, 431
Carbon in living things, 64
Carcinogens, **523**
Cardiac muscle, **403**
Cardiovascular disease, 427–428, 488, 505, 522
Careers in science, 10, 81, 184, 197, 220, 257, 284, 360, 403, 447, 572, 606
Carnivores, 382, 543
Carrageen, 213
Carrier molecules, 95
Carson, Rachel, *Silent Spring,* 368
Cartilage, **349**, **396**, 397
Cast, fossil **163**, *163*, 166
Cell, **37**, **75**
    culturing, 84
    parts of, 78–83
    plant vs. animal, 85
    specialized, 86
Cell cycle, 104–105
Cell division, 103, 107
Cell membrane, **79**, *79*, **95**, 95
Cell plate, 106
Cell theory, development of, 75–76, **77**
Cellulose, 79
Cell wall, **79**, *79*
Celsius scale, 16
Cenozoic era, *159*, 159
Centipedes, 329
Central nervous system (CNS), 444
Centriole, 105
Centromere, 105
Cephalopod, **323**
Cephalothorax, 327
Cereal grains, 246
Cerebellum, 446
Cerebrum, **378**, 446
Cervix, 469, 475
Cetacea, 382
Chelicerae, 327
Chemical bonds, 54
Chemical changes in matter, 59
Chemical digestion, **415**
Chemical equations, **60**, 61
Chemical formulas, **55**
Chemical reaction, 60, 63, 244
Chemical symbol, **55**
Chemical Transportation Emergency Center (CHEMTREC), international safety symbols, 11

Childbirth, 475–476
Childhood stage of development, 479
Chiroptera, 383
Chlamydia, 472
Chlorophyll, **83**, 97–98, 201, 244
Chloroplasts, **83**, 97–98
Cholera, 570
Cholesterol, 352, **488**
Chordates, 345
Choroid layer of eye, *451*, 451
Chromatid, 105
Chromosomes, **104**, 105, *117*, 117, 119
Cilia, **215**, 430
Ciliates, 215
Cinchona tree, 274
Circle graphs, 18
Circulatory system, *423*, 421–428
Class, **177**
Classification, 5
    of angiosperms, 273
    of animals, 300–301
    of arthropods, 327–329
    of bacteria, 201
    of echinoderms, 338
    of fungi, 226
    history of, 175
    of plants, 241
    systems, *176*, 176–179
    of viruses, 196
Climate zones, **579**, *580*, 580, 583
Climax community, 553, 555
Club fungi, 225
Club mosses, 239, 261
Cnidarian, 300, **305**, 306–307
Cocaine and crack, 500
Cocci, **201**
Cochlea, **452**
Coefficient, 60–61
Coelom, **313**
Cold virus, 198
Colony, 219, 334
Commensalism, 568
Common cold, 198–199, 432
Community, **538**
    pioneer, 554
    succession, 553–554
Comparative anatomy, 182
Competition, **44**, 562
Compound, **54**, 64
Computerized axial tomography (CAT), 449
Computerized tomography (CT), 449
Condensation, **549**
Conifers, **262**
Conservation of resources, 602–606, 611
Constants, **7**
Consumers in food chain, **542**, 543
Continents, *32–33*, 32–33, 161
Controlled experiment, **7**
Copepods, 328
Coral, reef-building, 307
    reefs, 308–309

Cotyledons, 273
Cousteau, Jacques–Yves, 432
Crayfish, *328*, 328
Crocodile, 360
Cro–Magnons, 169
Crustacean, **328**
Cryobiology, 96
Cubic meter, **14**
Culturing cells, 84
Cycads, 262
Cyclic behavior, 571, 574
Cystic fibrosis, 125
Cytoplasm, **81**

D

Darwin, Charles, *148*, 148–149
Data, **3**
DDT. *See* Pesticides.
Death rate, population size and, 561
Decomposers, **203**, 222, 542–543
Deforestation, 604
Dendrite, 445
Density, **15**, 51
Deoxyribonucleic acid. *See* DNA.
Depressants, **500**, 504
Dermis, **408**
Desalination process, *45*, 45
Desert, 240, **590**
Desertification, 591
Dewey Decimal Classification System, 179
Diaphragm, 378, 429
Diatoms, 218
Dicot, **273**, *273*, 278
Diffusion, **93**, 94, 431
Digestion, 299, 419
    processes of, **415**, 416–417
Digestive system, 371, 415–420, *416*
Dinoflagellates, 217
Disease. *See also* Body disorders; *particular diseases*, e.g., Infectious disease.
    natural defenses against, 516–518
    noninfectious, 511
    nutrition and, 521
    preventive medicines, 525–526
    tobacco use and, 505
Diversity, **140**, 146
DNA, 65, *123*, 123, 129, 182
Dominant gene, **113**
Drugs, 498–501, 526
Duodenum, 417
Dwarfism, 462

E

Ear, structure and hearing, *452*, 452
Eardrum, **452**
Earth, 31–34, 57
    history of life, 157
    protection of resources, 616

timeline, 158–159. *See also* Geologic time scale.
Earthworms, *313*, 313–314
Eating disorders, 420
Echinoderm, **337**, 338–339
Ecology, 537
Ecosystems, **538**, 541, 548–551, 553–556
Ectothermic, **43**
Ectotherms, **346**
Edible plants, 280, 289
Eggs, 373, 473
Electron, *53*, 53
Elements, **53**, 55–57
Embryo, 263, 283, 381
Endangered species, 605. *See also* Extinction.
Endocrine glands, **458**, 459–460
Endocrine system, *459*, 458–459, 461, 463
Endoplasmic reticulum, *82*, 82
Endoskeleton, 337, **345**, **393**
Endospore, **202**
Endothermic, **43**
Endotherms, **345**, 346
Energy, 37, 41
  alternative resources, 608–610
  conservation of, 602, 611
  food and, 542–546
  processes, 97–101
  resources, 601
Energy pyramid, **545**, *545*
Environmental Protection Agency (EPA), 433, 616
Enzymes, digestive, 415–416
EPA. *See* Environmental Protection Agency.
Epidermis, **276**, 279, **408**
Epiglottis, 430
Epiphytes, 588
Era, **158**
Erosion, 604
Esophagus, 416
Estimating skills, 4
Euglenas, *218*, 218
Eukaryote, **211**
Euphorbias, 241
Evaporation, **549**
Evolution, **149**. *See also* Natural selection.
  of amphibians, 353–354
  of arthropods, 326
  of birds, 368
  evidence for, 165
  human, 167–169
  mammals vs. birds, 376
  of nonflowering plants, 253
  of reptiles, 357
  of vertebrates, 346–347
  worms vs. mollusks, 322
Evolutionary tree, *164*, 164
Excretory system, 435–438
Exercise, 494–495, 497

Exercise specialist, 403
Exoskeleton, **325**, 326, 332, 393
Experiments, 7–8
Extensor, **404**
Extinct, **162**, 356, 379, 606
Eyes, 410, *451*, 451, 457, 520
  compound, 326, 328, 332

F Factors, genetic, 113
Family (classification), **177**
Farsightedness, 456
Fats. *See* Lipids.
Feathers, *371*, 371, 375
Feces, 418–419
Fermentation, **100**, 101, 224
Ferns, 239, 259–261
Fertilization, **469**, 473, 483
Fetal alcohol syndrome, 506
Fetal development, 474
Fetus, **474**
Fever biology, 347
Fiber in human nutrition, 489
Fibrin in blood clotting, 424
Field guide, 184
Filter feeding, 302
Fishes, 346–352
  bony, structure of, *351*, 350–351
  buoyancy in, 350–351
Flagella, **200**, 215, 217–218
Flagellates, 215
Flash distillation, 45
Flat bones, 394
Flatworm, **311**
Fleming, Alexander, 227
Flexor, **404**
Flight, 370, 372
Flightless birds, 369
Flower, structure of, *281*, 281, 290
Flowering plants, 240–241, 271–272, 281–284, 288
Flukes, 311
Food, 41
  algae in, 213, 221
  bacteria in production of, 203
  diseases carried by, 513
  energy and, 542–546
  fish and shellfish as, 324, 352
  labeling, 493
  preventing spoilage, 204–205
  staple, 492
Food chain, **542**, *543*, 547
Food groups, 491
Food plants, 246–247
Food poisoning, 204
Food web, **544**, *544*, 546–547
Foot, muscular, 321–323
Forests
  coniferous, *584*, **587**
  deciduous, *584*, **586**
  logging of, 555
  management, 604

  rain, *584*, 588-589
Fossil fuels, 165, 255, **608**
  carbon dioxide and, 551
  pollution from, 611–612, 616
Fossil record, **162**
Fossils, **162**, 163, 166, 171
Frog metamorphosis, *355*, 355, 362
Fronds, 259–260
Fruits, 282
Fuels. *See also* Fossil fuels.
  alternative, 607–610
Fungi, 180, 222–227
  diseases caused by, 512
Fungus, **222**

G Gagnan, Emile, aqualung, 432
Galápagos Islands, life forms, 149
Gametes, **114**, 145
Gametophyte, **254**, 256–257, 260, 263, 282
Gas
  atmospheric, 34
  phase of matter, 52
Gas exchange, 43–44, 354, *431*, 431
Gastric juice, 416
Gastropod, **322**
Genes, **113**
  diseases from, 125, 522
  dominant and recessive, 114, 116, 125
  inherited, 114, 121, 125
Genetic code, 122
Genetic diversity, 146
Genetic engineering, 128–130, 289
Genetics, **111**, 114. *See also* Selective breeding.
  applied, 127–131
  Mendel's experiments, 111–114, 117
Genital herpes, 472
Genome, 130
Genotype, **113**
Genus, **177**
Geologic time scale, **158**, 159–160, 171, 264
Geothermal energy, **610**
Gericke, W. F., 242
Germplasm banks, 152
Gill, 322–323, 345, **348**, 351
Ginkgo tree, 262
Gliding joints, 398
Global warming, 552
Glucagon, 459–461
Glucose
  breakdown during respiration, 99–100
  control of blood levels, 460–461
  as energy source, 487
  manufacture of, 97, 243, 279
  transport in vascular plants, 275

644 Index

Glycogen, 101
Gnetophytes, 262
Golgi bodies, *83*, 83
Gonorrhea, 472
Goose bumps, 40
Graduated cylinder, 14
Graphing, 18–20
Grasslands, **589**
Greenhouse effect, *613*, 613
Groundwater, 549, 615
Gymnosperm, **239**, 254, 262–264

H Habitat, 44, **139**, 141, 539, 605
adaptation to, 139,140, 150, 240–241
Hagfish, 349
Hair, 377, 407–408, 410
Halley, Edmund, diving bell, 432
Hallucinogens, **501**
Haversian canals, 395
Hazardous material, 11
Hazardous waste, **614**
Head in arthropods, 325, 332
Health technician, 81
Hearing, 452
Heart, 87–88, 358, 371, 403, *421*, 421
Heart rate, 422
Hepatitis virus, 198
Herbaceous stems, 278
Herbivores, 543
Heredity, principles of, **111**, 112–116
Heroin, 501
Herpes viruses, 198
Hibernation, 571
Hinge joints, 398
Histamine, 517, 524
Holdfast, 219
Homeostasis, **43**, 79
Hominid, *168*, **168**
Honeybee, 335, 573
Hooke, Robert, 75–76
Hookworms, 312
Hormones, 243, 246, 287, **458**, 459–461, 467
Hornworts, 257
Horsetails, 239, 260
Horticulturist, 284
Host, **196**
—parasite relationship, 569
viral, 196
Human Genome Project, 130
Human growth hormone (HGH), 459, 462
Human immunodeficiency virus (HIV), 197–198, 514–515
Humans
development of, 474, 478–480, 482–483, 489
elements in, 56–57
evolution, 167–169, 171
inherited disorders in, 125
as mammals, 384
place in food web, 546
population growth, 564–565
reproduction, 467–471
symbiotic behavior, 570
Huntington's disease, 125
Hybrid, **113**
Hydrogen fuel, 607
Hydroponics, 242
Hydropower, 610
Hydrozoans, 307
Hyphae, **222**, 223–225
Hypothalamus, 347, 459
Hypothesis, 5

I Immune system, **514**, 518, 523, 525
Immunity, 518, 525–526
Incomplete dominance, **116**
Incubate, **373**
Infancy, 478
Infectious disease, **511**, 512-513
Inference, 4
Inflammation, protective function, 517
Infrared sensors, 34–35
Inherited diseases, 125, 130
Innate behavior, **571**
Inorganic compounds, **64**
Insectivora, 382
Insects, 331–336
Insulin, 129, 459, 461
Interferon, 517
Interphase, *105*, 103–105
Intestine, 417–418
villi in, **418**, *418*,
Inuit Indians, high–cholesterol diets, 352
Invertebrate, **301**
Iris, of the eye, **451**
Irradiation of food, 204
Irregular bones, 394

J Jellyfish, 306–308
Jointed appendages, 325
Joints, **397**, *397*, *398*, 398

K Karyotype, 120
Kelp, 219, 340
Keratin, 407
Kettlewell, H. B., 151
Kidneys, *436*, 436–438
Kilogram, **15**
Kingdoms (classification), **176**, *180*, *181*, 180–181, 187

L Labor, stages of, **475**
Laboratory safety, 6
Lactic acid production, 100–101, 406
Lagomorpha, 383
Lake community, 554
Lamarck, Jean, 150
Lampreys, 349
Land biomes, 579
desert, *584*, **590**
forests, 586–588
grassland, *584*, 589
tundra, *584*, 585
LANDSAT I satellite, 34
Land use, 604
Larva, **303**, 333
Larynx, 430
Lasers, *24*, 24, 27
Latitude and climate, 581
Lavoisier, Anton, 56
Law, scientific, 9
Leaf, 246, structure of, *279*, 278–279,
Leakey, Louis and Mary, 168
Legumes, 550
Length measurements, 13
Lens, **21**, *22*, 23–25
concave, *22*, 22
convex, *22*, 22
for correcting vision, 455–456
development of, 76
in human eye, 451
Lever systems, *404–405*, 404-405
Library of Congress classification system, 179
Lichen, 225, 568
Life cycle
jellyfish, 306
plants, 254, 256–257, 260, 263
Ligaments, **397**
Light, image and, 451
Limited resources, 150
Limiting factors
marine biomes, 593
for population size, **562**, 563
Line graphs, 19–20
Linnaeus, Carolus, 175, 241
Lipids, 65, **488**
Liquids, 52
Liter, **14**
Littoral zone, *592*, 592
Liver, *417*, 417, 435
Liverwort, **257**
Living space, 44
Living things
characteristics of, 36–39
common ancestry, 165
elements in, 56–57
energy sources, 37, 41, 97–101
habitats, 31
needs of, 41–44
organization of, 86–87
Lizards, 359
Long bones, 394–395
LSD, 501
Lungs, 327, 351, 378, 429, *430*, 431
measuring capacity of, 434

Index 645

Lyme disease, 513, 515
Lymphatic system, *425*, 425
Lymph nodes, 425
Lymphocytes, 425, 518
Lysosomes, 83

M
Magnetic field, 161
Malaria, 216, 274, 512
Mammals
    characteristics of, 377, 380–383
    egg–laying, 380
    endangered species, 379
    evolution, 376
    humans as, 384
Mammary glands, 376
Mantle, 321
Many–celled organisms, 86, 103
Marijuana, 501
Marine biome, 340, *592*, 592, **593**
Marine life technician, 220
Marine Mammal Protection Act, 340
Marsh community, 554
Marsupials, **381**
Mass, **14**, 15–16, 51
Matter, **51**
    changes in, 58–59
    particle theory, 52
    phases of, 52, 549
    physical properties, 51
Measuring
    with scientific units, 12–17
    skills, 4
Mechanical digestion, **415**
Medicine, 274, 525–528
Medulla, 446
Medusa, 305–306
Meiosis, **118**, *118*, 145
Melanin, 410
Memory cells, 518
Mendel, Gregor, genetic experiments, 111–114, 117
Meniscus, 14
Menopause, 480
Menstrual cycle, *470*, 470–471
Menstruation, **470**, 471
Mesoderm in worms, 310
Mesozoic era, *159*, 159
Metabolism, **37**
Metals. *See* Trace elements.
Metamorphosis, **333**, *333*, 355, 362
Metaphase, *105*, 105
Meter, **13**
Metric units. *See* Système internationale d'unitès.
Microclimates, 582
Microscope, **21**, *23*, 22–24, 27, 76–77
Migrate, **372**, 571
Millipedes, 329
Minerals
    in human nutrition, 490
    transport in vascular plants, 275
Miscarriage, 476
Mitochondria, *82*, 82
Mitosis, **103**, *104*, *105*, 104–106, 474
Mixtures, 58–59
Models
    of atoms, 53
    of inheritance, 121
    in science, 10–11
Mold, fossil, **163**, 166
Molds, 46, 226
Molecule, **54**, 55, 95
Mollusk, **321**, 322–324
Molting, 326
Monerans, 180, **200**, 201
Monocot, **273**, *273*
Monotremes, **380**
Moss, 239, **256**, 261. *See also* Sphagnum moss.
Motor neurons, **447**, 448
Movement
    in mammals, 377
    of substances, 93–96
Multigene traits, 116
Muscles
    action of, 404
    in birds, 370
    in lever systems, 405
    soreness after exercise in, 101, 406
    structure, *401*, 401
    types of, 402–403
Muscular system, *402*, 401–406
Mushrooms, 223, 225
Mutagens, 124
Mutation, **124**, 146
Mutualism, 568
Mycelium, 223
Mycologist, 223

N
Nails, 407
Narcotics, **501**
National Plant Germplasm System, 152
Natural defenses, 516–518
Natural resources, **601**
    conservation of, 602–603, 616
Natural selection, 142, 148–149, **150**, 151, 153
Nautilus, 323
Neanderthals, 169
Nearsightedness, 455–456
Nematodes. *See* Roundworms.
Nephritis, 437
Nephrons, **436**, *436*
Nerves, 444–445
Nervous system, *444*, 443–449
Nestmaking, 373
Neurons, **445**, *445*, 447–448
Neutron, *53*, 53
Newts, 356

Niche, **539**, 540
Nicotine, 504
Nitrogen, 34, 55, 123
Nitrogen cycle, *550*, 550
Nitrogen fixation, 203, **550**
Nomenclature. *See* Binomial nomenclature.
Nonflowering plants, 253–255, 259–265
Nonrenewable resources, **601**
Nonvascular plants, **238**
Nose, 430, *453*, 453
Notochord, 345
Nuclear energy, **610**
Nuclear membrane, 80
Nucleic acids, **65**, 80
Nucleus, 53, **80**, *80*
Nutrients, **415**, 487–490
Nutrition, 490–491, 521
Nymph, 333

O
Observation techniques, 3
Oceans, *32–33*, 32–33
Octopus, 323
Oil glands, 408
Oils. *See* Lipids.
Old age, 480–481
Omnivores, 543, 546
Optical instruments, 25
Order (classification), **177**
Organelles, **81**, *82*, *83*, 81–83
Organic compounds, **64**
Organisms, 87, 141, 538–539. *See also* Many–celled organisms; Single–celled organisms.
    identifying, 183–186
Organ, **87**, 96
Organ systems, 87, 313, 378
Origin of Species (Darwin), 149
Osmosis, **94**
Ossification, 396–397
Otters, protection of, 340
Ovaries, **469**
Over–the–counter drugs, 498
Oviduct, **469**
Ovulatory cycle, 470–471
Ovum, 119
Oxygen, 34, 57, 245, 429, 431
Oxygen–carbon dioxide cycle, *245*, *551*, 551
Ozone depletion, 613
Ozone hole, *613*

P
Pacemaker, *88*, 88
Pain, sensory receptors, 454
Paleozoic era, *158*, 158–159
Pancreas, *417*, 417, 459
Pangea, 264
Papermaking, 264–265
Paramecium, 145, *215*, 215

Parasitism, 216, 311–312, 315, 349, 569
Particle theory of matter, 52
Passive immunity **526**
Pathogens, **511**, 513, 517–518
Peat formation, 258
Pedigree chart, 119
Pelagic zone, *593*, 593
Penicillin, 227
Penis, **468**
Peppered moth, 151
Peregrine falcon, effect of DDT on, 374
Periosteum, 395
Peripheral nervous system (PNS), 444
Perissodactyla, 383
Peristalsis, 416–417
Permafrost, 585
Permeability of cell membranes, 95
Pesticides, 334, 368, 374
Pet store owners, 360
Phagocytes, 518
Pharynx, 430
Phenotype, **113**, 115
Phloem, **275**, *275*, 279
Photochemical smog, 612
Photosynthesis, **97**, 98–99, 201, *244*, 243–245
Phototropism, 287
Phylum, **176**
Pigments, 217, 220
Piltdown Man, 170
Pistil, **281**
Pituitary gland, 459
Pivotal joints, 398
Placenta, **381**, 474–475
Placental mammals, **381**, 382
Planarian, *311*, 311
Plankton, 212
Plant(s). *See also* Flowering plants; Nonflowering plants; Vascular plants.
   alpine, 241
   cell structure, *79*, 79, 85
   characteristics of, 237
   chemistry, 243–247
   classification, 238–239, 241
   conservation of, 606
   edible, 280, 289
   effect of acid rain on, 618
   energy sources, 97–100
   evolution, 238–239
   food, 246–247
   growth of, 203, 287, 291, 581–582
   kingdom, 181
   light and growth of, 102
   stages of, 286–288
   vs. algae, 237
   water absorption, 266
Plasma, 424
Plasmid, 129
Plastics, formation of, 62
Platelets, *424*, 424

Pneumonia, 432
Polar climate, **580**
Pollen, **262**, 263, 272, 281, 285
Pollination, **263**, 282
Pollution, 524, 612–617
Polyps, cnidarian, 305–306
Pond ecosystem, *42*, 42–43
Populations, **539**, 566
   changes in, 561–565
   isolation, 151–152
   natural selection, **150**
   relationships among, 567–570
   sampling, 540
Population density, **562**
Porifera. *See* Sponges.
Positron–emission tomography (PET), 449
Pouched mammals, 381
Precambrian era, *158*, 158–159
Precipitation, **549**, 580–581
Predators, **563**,
   –prey relationships, 563
Predictions, 4
Pregnancy, 474, 476–477
Premenstrual syndrome (PMS), 471
Prescription drugs, 498
Primate, **167**, 383
Proboscidea, 382
Producers in food chain, **542**, 543
Product, **60**
Prophase, *105*, 105
Proteins, 64, 66, 95, 122, **488**
Protista kingdom. *See* Protists.
Protists, 181, **211**
Proton, *53*, 53
Protozoan, **211**, 212, 214, 512
Pseudopodia, **214**
Puberty, 467, 479
Pulmonary circulation, 423
Punnett squares, 115
Pupa, insect, **333**
Pupil of the eye, **451**, *451*

Q

Quinine, 274

R

Rabies, 513
Radial symmetry, *300*, 300, 305, 337
Radiation, 124, 126
Radioactive dating, 161
Radon, 433
Radula, 322
Rain forests, 240, 588–589
Reactant, **60**
Recessive gene, **113**
Recombinant DNA, 129
Recording and organizing data, 5
Rectum, 419
Recycling, 617

Red blood cells, *424*, 424
Redi, Francisco, 39
Red tide, 217
Reef–building coral, 307
Reflex actions, *448*, 448
Regeneration of body parts, 311, 339
Remote sensing, 34–35
Renewable resources, **601**
Replication, **123**, 196
Reproduction, 38. *See also* Life cycle; Meiosis; Mitosis.
   amphibians, 354
   asexual, 145
   bacterial, 202
   birds, 373
   cnidarians, 306
   flowering plants, 271–272, 281–284
   fungal, 223
   gymnosperms, 263
   human, 467–471
   insects, 336
   protists, 212
   sexual, **145**
   sponges, 303
   viruses, 196
   without seeds, 284
   worms, 311–312
   yeasts, 224
Reproductive system, human, *468*, *469*, 468–471, 476
Reptiles, 357–360
Respiration, **99**, 429
   in insects, 327, 332
   in plants, 245, 248
Respiratory system, 198, *430*, 429–434
Response. *See* Stimulus and response.
Retina of the eye, **451**, *451*
Rh factor, 519
Rhizoids, 224, **256**
Rhizomes in ferns, 260
Rhizopus stolonifer, 224
Ribonucleic acid. *See* RNA.
Ribosomes, *82*, 82
RNA, 65
Rock cycle, 548
Rodentia, 382
Rods and cones, 451
Root crops, 247
Roots, 276–277
Root system, *276*
Roundworm, **312**, *312*

S

Salamanders, 356
Salinity, 593
Saliva, 416, 517
Salmonella food poisoning, 204
Salt, need for, 45
Salt–marsh organisms, 547

Index   **647**

Salt water, 45, 592
Sarcodines, 214
Satellites, 34
Scanning electron microscope (SEM), 24
Scavengers, 542
Schleiden, Matthias, 76
Schwann, Theodor, 76
Science
   definition, 3
   facts, 9
   models in, 10–11
   skills, 3–5
Scientific method, 8–9
Scientific names, 178
Scientific photographers, 184
Scientific units, 12–17
Sclera, 451
Scrotum, **468**
Sea anemones, 307
Sea cucumbers, 338
Sea lilies, 338
Sea urchins, 338, 340
Seaweed, 212, 220
Sedimentary rock, relative age determination, **160**
Seeds, 254, 262, 272, 282–283
Segmented worm, **313**, *313*, 325
Selective breeding, 127, 289
Selective permeability, 95
Semen, **468**
Semicircular canals, 452
Senses
   human, 450–456
Sensory neurons, **447**
Sensory receptors, 409, 450, *454*, 454
Sepals, **281**
Sex determination, 119
Sex–linked traits, 119
Sexual intercourse, 468–469
Sexual reproduction, **145**
Sexually transmitted disease (STD), 472, 513
Sharks, 349–350
Short bones, 394
SI. *See* Système internationale d'unitès.
Sickle-cell anemia, 125, 521–522
Sight, light and image, 451
Silent Spring (Carson), 368
Simple sugars, **65**
Single–celled organisms, 86, 103, 145
Skeletal muscle, **403**
Skeletal system, 393–400
Skeleton, 393, *394*, 398. *See also* Endoskeleton; Exoskeleton.
   bird, 370
   in sponges, 303
Skin, 377, 407–408, *408*, 410, 435, 516
Sleep, 574
Slime molds, 226
Smell, sense of, 453, 455

Smog, **612**
Smooth muscle, **403**
Snakes, 359, 361
Social behavior, 573
Soil, 203, 276, 314, 603
Solar energy, **609**, house, *609*
Solar system, 31
Solids, 52
Solid wastes, 614
Solutions, 59
Sori, 260–261
Sound waves, 452
Speciation, 151
Species, **142**, 144, 151–152, **177**
Speech pathologist, 447
Sperm, 119, 468, 473
Sphagnum moss, 255, 258
Spiders, 327, 330
Spinal cord, 444, *447*, 447–448
Spinal nerves, 444
Spindle fibers, 105–106
Spirilla, **201**
Sponges, 302–305
Spontaneous generation, *39*, 39
Spore cases, 224, 261
Spores, **216**, **254**, 260
Sporophyte, **254**, 256–257, 260, 263
Sporozoa, 216
Squid, 323
Staghorn coral, 307
Stamen, **281**
Starch, 65, 280, 487
Starfish, *339*, 339
Stem, plant, 278
Stigma, **281**
Stimulants, **500**
Stimulus and response, **38**, 40
Stinging cells, 306, 308
Stomach, 416, 517
Stomate, **244**, 279
Stress, 496
Struggle for existence, 150
Subatomic particles, 53
Sublittoral zone, *592*, 592
Subscript, 55
Substance abuse, 498–499, 502. *See also* Drugs.
Succession, **553**, *554*
   in cities, 555
Sucrose molecule, 55
Sun as energy source, 41
Sunburn and DNA damage, 126
Sweat glands, 408
Sweating, 409, 435
Swim bladder, 350
Symbiosis, **567**, 570
Synapse, **445**
Synfuels, 607
Synthetic polymers, 62
Système internationale d'unitès (SI), 12–17
Systemic circulation, 423

T Tadpoles, 354–355
Tapeworms, 311, *315*
Taste, sense of, 453, 455
Taxonomic key, 183, 186
   for fungi, 226
Taxonomy, **175**
Tears, protective function, 517
Teeth in mammals, 377
Telescope, *25*, 25
Telophase, *104*, 104
Temperate, climate, **580**
Temperature, 16. *See also* Body temperature control.
   influencing factors, 580–581
   of living things, 42–43
Tendons, **397**
Termites, social behavior, 573
Testes, **468**
Testosterone, 468
Theophrastus, plant classification system, 241
Theory, scientific, 9
Thermometer, 16
Thorax in arthropods, **325**, 332
Thyroid gland, 459
Thyroxine, 459
Time measurements, 17
Tissues, **86**, 87
Toads, 355
Tobacco, 504–505
Tongue and taste buds, *453*, 453
Tortoises, 359
Trace elements, 57
Trachea, 430
Traditional medicine, 527
Traits of living things, 111–113, 119
Tranquilizers, 500
Transmission electron microscope (TEM), 24, 77
Transpiration, **275**
Transport
   cell membranes and, 95
   of substances, 93–96
   in vascular plants, 275
Trees, 183,
   making paper from, 264
Trichinella worms, 312
Trilobites, *163*, 326
Triticale, 128
Tropical climate, **580**
Tropism, **287**
Tube feet, 337, 339
Tuberculosis, 570
Tundra, **585**
Turtles, 359

U Ultrasound, 437, 477
Umbilical cord, 475
Underwater exploration, 301, 432
Urea, 435
Urethra, 437

Urinary bladder, 436
Urine, 436
Uterus, **469**

V
Vaccines, **525**, 526
Vacuoles, 83
Vagina, **469**
Van Leeuwenhoek, Anton, 75
Variables, **7**, 19
Variation
    in humans, 146
    in populations, 150–152
    within species, **144**
Vascular plants, **238**, 239
    nonflowering, 259–265
    root system, 276
    shoot system, 278
    transport of substances in, 275
Vegetative propagation, **284**
Veins, **422**
Vertebrae, **393**, 393–394
Vertebrate, **301**, 345–349
Viral replication, *196*, 196
Virchow, Rudolf, 76
Virologist, 197
Virus, **195**, 196
    diseases caused by, 197–199, 512
    vaccines for, 526
Vision, 455–456
    in birds, 370
Vitamins, 489
Volume, **14**, 14
    measurements of, 14
Von Linné, Carl. *See* Linnaeus, Carolus.

W
Water
    absorption in plants, 266
    conservation, 602
    diffusion, **93**
    diseases carried by, 513
    formation of, 61
    fresh, 594-595
    function in living things, 42
    in human nutrition, 490
    molecule, 54
    on earth, 31–33
    osmosis, **94**
    salt, 45
    transport in vascular plants, 275
Water biomes, 592–596
Water birds, 369
Water cycle, *549*, 549
Water power. *See* Hydropower.
Weight measurements, 14–15, 17
Wetland ecosystem, 538–539, 541
Wet mount slide preparation, 26
White blood cells, *424*, 424, 517
Wildlife conservation, 605
WIN 51, 199, 711

Windpipe, 430
Wind power, 610
Wing, aerodynamic properties, 372
Woody stems, 278
Worms
    characteristics of, 310
    diseases caused by, 512
    parasitic, 311–312, 315
    types of, 311–313

X
X chromosome, **119**
Xylem, **275**, *275*, 279

Y
Y chromosome, **119**
Yeasts, 100, 224, 228
Yellow River Valley (China), 596

Z
Zoo animals, 44, 380
Zygote, **145**, 469, 473-474

Acknowledgments

Photographs

Title page i-CR David M. Phillips/Visuals Unlimited; i-R M. Austerner/Animals, Animals; i-L Norm Thomas/Photo Researchers; i-CL Zig Leszczynski/Animals, Animals

Contents iii-T Roger Ressmeyer/Starlight; iii-BR Zig Leszczynski/Animals, Animals; iii-BC Zig Leszczynski/Animals, Animals; iv-R Biofotos Associates/Science Source/Photo Researchers; iv-L Photo Researchers; v-B Eric Gave/Phototake; v-T Ken Karp*; vi-C Alford W. Cooper/Photo Researchers; vi-BL Gareth Hopson*; vii-BC Brian Parker/Tom Stack & Associates; vii-BR David Scharf/Peter Arnold, Inc.; vii-T Paul Skelcher/Rainbow; vii-C Runk-Schoenberger/Grant Heilman Photography; viii-BR Breck P. Kent/Animals, Animals; viii-BL Runk-Schoenberger/Grant Heilman Photography; viii-BC Science Photo Library/Science Source/Photo Researchers; ix-T David Scharf/Peter Arnold, Inc.; ix-BR Dennis Kunkel/CNRI/Phototake; ix-BL Manfred Kage/Peter Arnold, Inc.; ix-BC Dennis Kenkel/CNRI/Phototake; x-T Fritz Polking/Peter Arnold, Inc.; x-B Grant Heilman/Grant Heilman Photography; xi-TL Calvin Larsen/Photo Researchers; xi-BL Gary Braasch; xi-BR NASA

Unit 1 xviii Gareth Hopson*; xviii-1 Will & Deni McIntyre/Photo Researchers;

Chapter 1 2 David Scharf/Peter Arnold, Inc.; 3B John Colwell/Grant Heilman Photography; 3T Craig Newbauer/Peter Arnold, Inc.; 4 Richard Shiell/Earth Scenes; 5 Larry Brownstein/The Stock Shop; 11B Livestock; 11T Ron Colby/Fran Heyl Associates; 13 Ken Karp*; 14B Yoav/Phototake; 14C Ken Karp*; 15 Ken Karp*; 16 Ken Karp*; 17 Ken Karp*; 21 Phillip A. Harrington/Fran Heyl Associates; 24B Argon Dyelaser/Denny Studios/TheStock Shop; 24C CNRI/Science Photo Library/Custom Medical Stock Photo; 24T Secchi, Lecaque, Roussel, Uclaf,CNRI/ Science Photo Library/Photo Researchers; 25 Roger Ressmeyer/Starlight

Chapter 2 30 Photo Researchers; 35 NASA; 36 Zig Leszczynski/Animals, Animals; 37B Eric V.Grave/Photo Researchers; 37T Rod Planck/Photo Researchers; 38 Gregory G. Dimijian, M.D./Photo Researchers; 39 Leonard Lee Rue III/Photo Researchers;40 Stephen J. Krasemann/DRK Photos;41 Zig Leszczynski/Animals, Animals;44 Livestock; 45 Grant Heilman/Grant Heilman Photography

Chapter 3 50 Frans Lanting/Minden Pictures;51 NASA;56 The Bettmann Archive; 57L Lawrence Migdale*, 57R The Stock Shop;58 Tom Till/DRK Photos; 59B Martha Cooper/Peter Arnold, Inc.; 59T Brian Lovett/The Stock Shop; 62 Lawrence Migdale*; 64 Fred Bavendam/Peter Arnold, Inc.; 65 Renee Lynn*; 70-71 Gary Braasch/Woodfin Camp & Associates; 70B Gerard Lacz/Peter Arnold, Inc.; 70T Gareth Hopson*

Unit 2 72 Hans Reinhard/Bruce Coleman Inc.; 72-73 Jim Foster/The Stock Market

Chapter 4 74 Manfred Kage/Peter Arnold, Inc.; 75 The Bettmann Archive; 76 The Bettmann Archive; 77 Courtesy Cornell University; 78L Manfred Kage/Peter Arnold, Inc.; 78R E.R. Degginger/Animals, Animals; 81 K. Love/Custom Medical Stock Photo; 82B Photo Researchers; 82T Science Photo Library/Photo Researchers; 83L James Dennis/CNRI/Phototake; 83R J. L. Carson/Custom Medical StockPhoto; 84 Nubar Alexanian/Woodfin Camp & Associates; 86C Don Fawcett/Visuals Unlimited; 88 Martin/Potker/Phototake

Chapter 5 92 Science Photo Library/ Photo Researchers; 93B Ken Karp*; 93C Ken Karp*; 93T Ken Karp*; 96 Peter Menzel; 97 Gary Braasch/Woodfin Camp & Associates, Inc.; 100 Barbara Pfeffer/ Peter Arnold, Inc.; 101 David Burnett/Contact Press Images/Woodfin Camp & Associates, Inc.; 104B John D. Cunningham/Visuals Unlimited; 104T Eric Grave/Visuals Unlimited; 105B John D. Cunningham/Visuals Unlimited; 105C John D. Cunningham/Visuals Unlimited; 105T Eric Grave/Phototake; 106 J. L. Carson/Custom Medical Stock Photo

Chapter 6 110 Heather Angel/Biofotos; 111 Bob Daemmrich/Stock, Boston; 116 Dan Clark/Grant Heilman Photography; 117 Biofotos Associates/Science Source/Photo Researchers; 119 CNRI/Phototake; 120 Richard Hutchings/Science Source/Photo Researchers; 120R CNRI/Science Photo Library/ Photo Researchers; 126 Bob Daemmrich/Stock, Boston; 127B Jacana/Scientific Control/Photo Researchers; 127T Frank Siteman/Stock, Boston; 128B Keith V. Wood; 128TR Grant Heilman/Grant Heilman Photography; 130 Peter Menzel; 134-135 Art Wolfe; 134TC Stan Osolinski/Oxford Scientific Films/Earth Scenes

Unit 3 136 Norman Owen Tomalin/Bruce Coleman Inc.; 136-137 Jeff Foott/Bruce Coleman Inc.

Chapter 7 138 M. P. Kahl/Photo Researchers; 139B Anup & Manoj Shah/Animals, Animals; 139T Walter E. Harvey/Photo Researchers; 140BC Stephen P. Parker/Photo Researchers; 140BL Warren Garst/Tom Stack & Associates; 140BR John R. MacGregor/Peter Arnold; 140TC Calvin Larsen/Photo Researchers; 140TL Grant Heilman/Grant Heilman Photography; 140TR G. C. Kelley/Tom Stack & Associates; 141L Tom McHugh/Photo Researchers; 141R Tom McHugh/Photo Researchers; 142 John Cancalosi/Stock, Boston; 144 Karen Tweedy-Holmes/Animals, Animals; 145R Eric Grave/Phototake; 145L Roger Wilmshurst/Bruce Coleman Inc.; 148 The Granger Collection; 150 Mark Stouffer/Animals, Animals; 150R David M. Phillips/Visuals Unlimited; 151 Nuridsany & Perennou/Photo Researchers;152B J. Cancalosi/Stock, Boston; 152T Ken Cole/Animals, Animals

Chapter 8 156 J. Koivula/Photo Researchers; 160 D. & J. Heaton/Stock, Boston; 162 Gerd Ludwig/Woodfin Camp & Associates; 163B Fred Bavendam/Peter Arnold; 167 Larry Tackett/Tom Stack & Associates; 168BR UPI/Bettmann; 168TL Cleveland Museum of Natural History

Chapter 9 174 Norbert Wu; 175B Breck P. Kent/ Animals, Animals; 175C Jacana Scientific Control/Photo Researchers; 175T Charles Palek/Animals, Animals; 177 Jim Brandenburg/Minden Pictures; 179 Ken Karp*; 180L Rod Planck/Tom Stack & Associates; 181 Michael Abbey/Photo Researchers; 183 Michael Giannechini/Photo Researchers; 184 Lynn

Funkhouser/Peter Arnold, Inc.; 185L Doug Wechsler/Animals, Animals; 185R Ken Karp*; 190-191 Stephen Frink/AllStock; 190BL Fred McConnaughey/Photo Researchers; 190BR M. Doolittle/Rainbow; 190CR Ralph Oberlander/Stock, Boston; 190T Nuridsany et Perennou/Photo Researchers; 190TR L. West/Photo Researchers; 191 Phil Dotson/Photo Researchers

Unit 4 192 Runk-Schoenberger/Grant Heilman Photography; 192-193 Doug Wechsler/Earth Scenes

Chapter 10 194 Dr. Tony Brain & David Parker/Science Photo Library/Custom Medical Stock Photo; 197 Custom Medical Stock Photo; 198BL Science Photo Library/Photo Researchers; 198BR CNRI/Science Photo Library/Photo Researchers; 198CL P. Hawtin, University of Southhampton/ Science Photo Library/Photo Researchers; 198CR Photo Researchers; 201C David M. Phillips/Visuals Unlimited; 201L David Scharf/Peter Arnold, Inc.; 201R David M. Rollins/Visuals Unlimited; 203L Pat Miller/The Stock Shop; 203R Inga Spence/LiveStock; 204 Thomas Hovland/Grant Heilman Photography

Chapter 11 210 Dr. Paul A. Zahl/Photo Researchers; 211 Pat & Tom Leeson/Photo Researchers; 212 Manfred Kage/Peter Arnold, Inc.; 212C Eric Grave/Photo Researchers; 212T Fred Bavendam/Peter Arnold, Inc.; 213L W. H. Hodge/Peter Arnold, Inc.; 213R Lawrence Migdale*; 214B Eric Gave/Phototake; 215B Michael Abbey/Photo Researchers; 215T Eric Grave/Photo Researchers; 216 Holt Studios Ltd./Animals, Animals; 217A M.I. Walker/Science Source/Photo Researchers; 217B Biofoto Associates/Science Source/Photo Researchers; 218BL E. R. Degginger/Animals, Animals; 218CL E. R. Degginger/Animals, Animals; 218TR Peter Parks/Oxford Scientific Films/Animals, Animals; 219C Manfred Kage/Peter Arnold, Inc.; 219L Dan McCoy/Rainbow; 219R Runk-Schoenberger/Grant Heilman Photography; 220B Greg Vaughn/Tom Stack & Associates; 220T Brian Parker/Tom Stack & Associates; 221 Runk-Schoenberger/Grant Heilman Photography; 222 S. Rannels/Grant Heilman Photography; 224B Dr. Paul Zahl/Photo Researchers; 224TL Ralph C. Eagle/Photo Researchers; 225B Frans Lanting/Minden Pictures; 225T Ted Streshinsky/Photo 20-20; 226 David M. Dennis/Tom Stack & Associates; 227 Dr. Jeremy Burgess/Science Photo Library/Custom Medical Stock Photo; 232-233 Renee Lynn*

Unit 5 234 Charles Krebs/The Stock Market; 234-235 Richard R. Hansen/Photo Researchers

Chapter 12 236 Rod Planck/Photo Researchers; 237B Owen Franken/Stock, Boston; 237T Norm Thomas/Photo Researchers; 238B John Gerlach/Earth Scenes; 238C Runk-Schoenberger/Grant Heilman Photography; 239BL Ed Reschke/Peter Arnold, Inc.; 239BR Coco McCoy/Rainbow; 239TC F. Gohier/Photo Researchers; 239TL Doug Wechsler/Earth Scenes; 239TR Thomas Kitchin/Tom Stack & Associates; 240BL Frans Lanting/Minden Pictures; 240BR L. L. T. Rhodes/Earth Scenes; 240C Dallas & John Heaton/The Stock Shop; 240T Greg Vaughn/Tom Stack & Associates; 241TL Patti Murray/Earth Scenes; 241TR Michael Fogden/Earth Scenes; 242 Catherine Ursillo/Photo Researchers; 246B Fred Bavendam/Peter Arnold, Inc.; 246T Stephen R. Swinburne/Stock, Boston; 247 Christiana Dittmann/Rainbow

Chapter 13 252 Matt Meadows/Peter Arnold; 253 Greg Vaughn/Tom Stack & Associates; 255B Sinclair Stammers/Science Photo Library/Photo Researchers; 255T Breck P. Kent/Earth Scenes; 256 C.W. Perkins/Earth Scenes; 257B Gary Braasch/Woodfin Camp & Associates; 257T Runk-Schoenberger/Grant Heilman Photography; 258 Addison Geary/Stock, Boston; 260 Ed Reschke/Peter Arnold, Inc.; 261 Runk-Schoenberger/Grant Heilman Photography; 262BL Patti Murray/Earth Scenes; 262C Patti Murray/Earth Scenes; 262R Zig Leszczynski/Earth Scenes; 262TL Grant Heilman/Grant Heilman Photography; 263 Runk-Schoenberger/Grant Heilman Photography; 264B Sinclair Stammers/Science Photo Library/ Photo Researchers; 264T J. & L. Weber/Peter Arnold, Inc.; 265L John Blaustein/Woodfin Camp & Associates; 265R Ted Streshinsky/Photo 20-20

Chapter 14 270 D. Cavagnaro/DRK Photos; 271B G. I. Bernard/Earth Scenes; 271T G. I. Bernard/Earth Scenes; 272 Peter Arnold, Inc.; 274 M. J. Balick/Peter Arnold, Inc.; 277C Lynwood M. Chace/Photo Researchers; 277L Jeff Foott/Tom Stack & Associates; 277R Don & Pat Valenti/Tom Stack & Associates; 278BC Manfred Kage/Peter Arnold, Inc.; 278CL Breck P. Kent/Earth Scenes; 278L W. H. Hodge/Peter Arnold, Inc.; 278T Z. Leszczynski/Earth Scenes; 280 Renee Lynn*; 282 Renee Lynn*; 282TL Jeff Lepore/Photo Researchers; 283CL Runk-Schoenberger/Grant Heilman Photography; 283CR Runk-Schoenberger/Grant Heilman Photography; 283L Runk-Schoenberger/Grant Heilman Photography; 283R Runk-Schoenberger/Grant Heilman Photography; 284B Nick Pavloff*; 284T Steve Maines/Stock, Boston; 285C David Scharf/Peter Arnold, Inc.; 285T Grant Heilman/Grant Heilman Photography; 287B Runk-Schoenberger/Grant Heilman Photography; 287T Runk-Schoenberger/Grant Heilman Photography; 288C Richard Shiell/Earth Scenes; 288T Fletcher & Baylis/Photo Researchers; 289 Peter Menzel; 290 Alford W. Cooper/Photo Researchers; 294-295 Renee Lynn*; 294BL Renee Lynn*; 294R Renee Lynn*; 294TC Renee Lynn*; 295 Jerry Jacka Photography

Unit 6: 296 M. P. Kahi/DRK Photos; 296-297 Art Wolfe

Chapter 15 298 S. Summerhays/Biofotos; 299 Anne Wertheim/Animals, Animals; 302T J. Selig/Photo 20-20; 304T Charles Hornbrook Photo Co.; 304CL Gareth Hopson*; 307BL Dave B. Fleetham/Tom Stack & Associates; 307BR Fred Bavendam/Peter Arnold, Inc.; 307C Ed Robinson/Tom Stack & Associates; 307TL Denise Tackett/Tom Stack & Associates; 307TR Runk-Schoenberger/Grant Heilman Photography; 309 Kevin Schafer/Tom Stack & Associates; 310 Runk-Schoenberger/Grant Heilman Photography; 311 Ed Reschke/Peter Arnold, Inc.; 312B Cubberly/Phototake; 314A David M. Dennis/Tom Stack & Associates; 315 J. H. Robinson/Photo Researchers

Chapter 16 320 E. R. Degginger/Animals, Animals; 321B W. Gregory Brown/Animals, Animals; 321T Neil G. McDaniel/Photo Researchers; 322B Runk-Schoenberger/Grant Heilman Photography; 323T Ed Robinson/Tom Stack & Associates; 324B Runk-Schoenberger/Grant Heilman Photography; 324T Takeshi Takahara/Photo Researchers; 326R Breck P. Kent/Animals, Animals; 326L David Scharf/Peter Arnold, Inc.; 327BC Kevin Schafer; 327BL Paul Skelcher/Rainbow; 327BR Mary Clay/Tom Stack & Associates; 328L Runk-Schoenberger/Grant Heilman Photography; 328T George H. Harrison/Grant Heilman Photography; 329B Doug Wechsler/Animals, Animals; 329T Klaus Uhlenhut/Animals, Animals; 330 Richard Kolar/Animals, Animals; 332 Billy Jones/Photo Researchers; 334B A. Griffiths Belt/Woodfin Camp & Associates; 334T Stanley Breeden/DRK Photos; 335 Mark Moffett/Minden Pictures; 337

Acknowledgments **651**

E. R. Degginger/Animals, Animals; 338BR Fred Bavendam/Peter Arnold, Inc.; 338C Denise Tackett/Tom Stack & Associates; 338TL Dave B. Fleetham/Tom Stack & Associates; 338TR Chuck Nicklin/ Nicklin & Associates; 339 Zig Leszczynski/Animals, Animals; 340 John Cancalosi/Tom Stack & Associates

Chapter 17 344 Gerard Lacz/Peter Arnold, Inc.; 345 G. I. Bernard/Oxford Scientific Films/Animals, Animals; 348 Larry Lipsky/Tom Stack & Associates; 349B Kelvin Aitken/Peter Arnold, Inc.; 349L Runk-Schoenberger/Grant Heilman Photography; 349R Andrew J. Martinez/Photo Researchers; 350 Tom McHugh/Photo Researchers; 351 Zig Leszczynski/Animals, Animals; 352 Jim Brandenburg/Minden Pictures; 353 Tom McHugh/Photo Researchers; 354B Barry L. Runk/Grant Heilman Photography; 354T David M. Dennis/Tom Stack & Associates; 355 Zig Leszczynski/Animals, Animals; 356 Zig Leszczynski/Animals, Animals; 357 Joe McDonald/Tom Stack & Associates; 359C Zig Leszczynski/Animals, Animals; 359L Don & Ester Phillips/Tom Stack & Associates; 359R Zig Leszczynski/Animals, Animals; 360B Lawrence Migdale*; 360C Brian Parker/Tom Stack & Associates; 360T Stephen J. Krasemann/Peter Arnold, Inc.; 361 David M. Dennis/Tom Stack & Associates

Chapter 18 366 Frans Lanting/Minden Pictures; 367 Mangelsen/Peter Arnold, Inc.; 368 AP/Wide World Photos, Inc.; 368BL Perry D. Slocum/Animals, Animals; 369BR Carl R. Sams II/Peter Arnold, Inc.; 369C John R. MacGregor/Peter Arnold, Inc.; 369TL Jeff Lepore/Photo Researchers; 369TR Tom McHugh/Photo Researchers; 373 Hal H. Harrison/Grant Heilman Photography; 374 Galen Rowell/Mountain Light; 377C Brian Parker/Tom Stack & Associates; 377L Gunter Ziesler/Peter Arnold, Inc.; 377R Stephen Dalton/Oxford Scientific Films/ Animals, Animals; 378 Matt Meadows/Peter Arnold, Inc.; 379B Fred Bavendam/Peter Arnold, Inc.; 379C Renee Lynn/Photo Researchers; 379T Steve Kaufman/Peter Arnold, Inc.; 380 Tom McHugh/Photo Researchers; 381 John Cancalosi/Peter Arnold, Inc.; 382BL Francois Gohier/Photo Researchers; 382BR Cris Crowley/Tom Stack & Associates; 382C J. R. McGregor/Peter Arnold, Inc.; 382TR Judd Cooney/Phototake; 383BL Mickey Gibson/Animals, Animals; 383BR Frans Lanting/Minden Pictures; 383C Fritz Polking/Peter Arnold, Inc.; 383TL M. Austerman/Animals, Animals; 383TR Gerard Lacz/Peter Arnold, Inc.; 384 Lawrence Migdale*; 388-389 Renee Lynn*

Unit 7 390 CNRI/Science Photo Library/Photo Researchers; 390-391 Michel Tcherevkoff/The Image Bank

Chapter 19 392 Pix Elation/Fran Heyl Associates; 395 Breck P. Kent/Animals, Animals; 397L Biophoto Associates/Science Source/Photo Researchers; 397R Biophoto Associates/Science Source/Photo Researchers; 399 Nicholas Secor/Medichrome Div./The Stock Shop; 402B Biophoto Associates/Photo Researchers; 402L M. I. Walker/Science Source/Photo Researchers; 402T Michael Abbey/Photo Researchers; 403 Lawrence Migdale*; 406 Focus on Sports; 407 Biophoto Associates/Science Source/Photo Researchers

Chapter 20 414 Dennis Kunkel/CNRI/Phototake; 418 Electra/Phototake; 424C Runk-Schoenberger/Grant Heilman Photography; 424L Dennis Kunkel/CNRI/Phototake; 424R Lennart Nilsson/Boehringer Ingelheim International GmbH; 427 Lennart Nilsson/The Incredible Machine; 428 Lori Adamski Peek/Tony Stone Worldwide; 432 Jose Fernandez/Woodfin Camp & Associates; 433 Day Williams/Photo Researchers; 435 Lennart Nilsson/The Incredible Machine; 437 Tom Raymond/Medichrome Div./The Stock Shop, Inc.

Chapter 21 442 Secchi-Lecaque/Roussel-Oclaf/CNRI/Science Photo Library/Photo Researchers; 443 Lawrence Migdale*; 445 Ed Reschke/Peter Arnold, Inc.; 447 Lawrence Migdale*; 448 Tim Davis*; 449 Science Photo Library/Science Source/Photo Researchers; 450 Tony Savino/The Image Works; 451 Manfred Kage/Peter Arnold, Inc.; 452 Patricia Barber, RBP/Custom Medical Stock Photo; 453 Lennart Nilsson, 'BEHOLD MAN', Little Brown and Company

Chapter 22 466 Howard Sochurek/Medical Images Inc.; 467 Dan McCoy/Rainbow; 473 David Scharf/Peter Arnold, Inc.; 474C Lennart Nilsson, 'A CHILD IS BORN', Dell Publishing Company; 474R Fran Heyl Associates; 474C Lennart Nilsson, 'A CHILD IS BORN', Dell Publishing Company; 475C Lennart Nilsson, 'A CHILD IS BORN', Dell Publishing Company; 475L Lennart Nilsson, 'A CHILD IS BORN', Dell Publishing Company; 475R Penny Gentieu/Black Star; 476 The National Foundation, March of Dimes; 477 Richard Hirneisen/Medichrome Div./The Stock Shop; 478 Anthony Bannister/Earth Scenes; 479B Peter Menzel; 479T Lowell Georgia/ Photo Researchers; 480L Susan Kuklin/Photo Researchers; 480R Susan Kuklin/Photo Researchers; 481C Nathan Benn/Woodfin Camp & Associates; 481T Frank Siteman/Stock, Boston

Chapter 23 486 D. Cavagnaro/DRK Photos; 487 Jim Brandenburg/Minden Pictures; 488 Renee Lynn*; 489 Renee Lynn*; 492C C. J. Collins/Photo Researchers; 492T D. & J. Heaton/The Stock Shop; 494 Mickey Pfleger/Photo 20-20; 496L Bob Daemmrich/Stock, Boston; 496R Joseph Lynch/Medical Images Inc.; 497 Roberto Soncin Gerometta/Photo 20-20; 498 Harry J. Przekop, Jr./Medichrome Div./The Stock Shop Inc.; 500 Gareth Hopson*; 500L Owen Franken/ Stock, Boston; 501L E. R. Degginger/Earth Scenes; 501R Deni Bown/Oxford Scientific Films/Earth Scenes; 502B Coco McCoy/Rainbow; 502T Laura Dwight/Peter Arnold, Inc.; 506 Yoav Levy/Phototake

Chapter 24 510 Lennart Nilsson, Boehringer, Ingelheim International GmbH; 511 Yoav Levy/Phototake; 512 Lennart Nilsson, Boehringer, Ingelheim International GmbH; 513BR Ken Karp*; 513C Renee Lynn*; 513L Kent Wood/Photo Researchers; 513TR Bernard Furnival/Fran Heyl Associates; 514 Alon Reininger/Woodfin Camp & Associates; 515 Dr. Durland Fish/Fran Heyl Associates; 516 Bob Daemmrich/Stock, Boston; 517 Manfred Kage/Peter Arnold, Inc.; 521B Bill Longcore/Photo Researchers; 521T Dennis Kunkel/CNRI/Phototake; 522L Dan McCoy/Rainbow; 522R Electra/CNRI/Phototake; 523B Matt Meadows/Peter Arnold, Inc.; 523T Geoff Tompkinson/Science Photo Library/Photo Researchers; 524B Dennis Kunkel/CNRI/Phototake; 524T David Scharf/Peter Arnold, Inc.; 525 Carl Purcell/Photo Researchers; 526 Barbara Alper/Stock, Boston; 527 Jim Brandenburg/Minden Pictures; 528 David M. Dennis/Tom Stack & Associates; 532 Dennis Kunkel/CNRI/Phototake; 532-533 Biophoto Associates/Photo Researchers; 533BL Dennis Kunkel/CNRI/Phototake; 533TL Dennis Kunkel/CNRI/Phototake; 533TR Institut Pasteur/CNRI/Phototake

Unit 8 534 Larry Lipsky/Tom Stack & Associates; 534-535 Tom Stack & Associates

Chapter 25 536 Stephen J.Krasemann/Peter Arnold, Inc.; 537 NASA; 540 Richard R. Hansen/Photo Researchers; 541 Arthur Gloor/Earth Scenes; 542 Grant Heilman/Grant Heilman Photography; 548 Wayland Lee*; 550 Runk-Schoenberger/Grant Heilman Photography; 552 Compoint/Sygma; 553B Gary Braasch; 553T Gary Braasch; 555 Harvey Lloyd/Peter Arnold, Inc.; 556 Mike Yamashita/Woodfin Camp & Associates

Chapter 26 560 William Townsend/Photo Researchers; 561 Michio Hoshino/Minden Pictures; 562 Francois Gohier/Photo Researchers; 563 Tom & Pat Leeson/DRK Photos; 565 Timothy Eagan/Woodfin Camp & Associates; 567 Fritz Polking/Peter Arnold, Inc.; 568BR F. Gohier/Photo Researchers; 568L Francois Gohier/Photo Researchers; 568TR Ira Kirschenbaum/Stock, Boston; 569BR Patti Murray/Animals, Animals; 569C Charlie Ott/Photo Researchers; 569L CNRI/Science Photo Library/ Photo Researchers; 569TR Robert A. Luback/Animals, Animals; 571 Frans Lanting/Minden Pictures; 572B Warren & Genny Garst/Tom Stack & Associates; 572T C. Prescott-Allen/ Animals, Animals; 573 Grant Heilman/Grant Heilman Photography

Chapter 27 578 John Gerlach/Earth Scenes; 579 Robert Caputo/Stock, Boston; 581 Stephen Frisch*; 582 John Marshall/AllStock; 585BL W. Perry Conway/Tom Stack & Associates; 585C Jack Stein Grove/Tom Stack & Associates; 585TL Len Rue, Jr./Stock, Boston; 585TR Johnny Johnson/Animals, Animals; 586B John Lemker/Earth Scenes; 586C Robert Winslow/Tom Stack & Associates; 586TL Stanley Breeden/DRK Photos; 586TR Rod Planck/Tom Stack & Associates; 587B Bob Pool/Tom Stack & Associates; 587TC Grant Heilman/Grant Heilman Photography; 587TL Stephen J. Krasemann/Peter Arnold, Inc.; 587TR Tom & Pat Leeson/Photo Researchers; 588B Alan Pitcairn/Grant Heilman Photography; 588C Luiz Claudio Marigo/Peter Arnold, Inc.; 588TL Warren Garst/Tom Stack & Associates; 588TR Kimball Derrick/Rainbow; 589B John Cancalosi/Stock, Boston; 589T Eastcott/Momatiuk/Woodfin Camp &Associates; 590BR Jeff Foott/Tom Stack & Associates; 590L Christine M. Douglas/Photo Researchers; 590TC John Cancalosi/Stock, Boston; 590TR E. R. Degginger/Animals, Animals; 591 Arthur Rothstein/The Bettmann Archive; 594B Jack Wilburn/Earth Scenes; 594TL Greg Gawlowski/Photo 20-20; 595 Leo Keeler/Animals, Animals; 596 Delta Willis/Bruce Coleman Inc.

Chapter 28 600 Woodward Payne/Photo 20-20; 601 Adam Woolfitt/Woodfin Camp & Associates; 602 Richard Kolar/Earth Scenes; 603B J. Howard/Stock, Boston; 603T Ian Lloyd/Black Star; 604B Francois Gohier/Photo Researchers; 604T Greg Vaughn/Tom Stack & Associates; 605BL Leonard Lee Rue III/Photo Researchers; 605BR Frans Lanting/Minden Pictures; 605C Cary Wolinsky/Stock, Boston; 605TL S. J. Krasemann/Peter Arnold, Inc.; 605TR Lawrence Migdale/Stock, Boston; 606 Nick Pavloff*; 607C Mazda; 607T John Marmaras/Woodfin Camp & Associates; 608 Jon Brenneis/Photo 20-20; 610L Simon Fraser/Science Photo Library/Photo Researchers; 610R Giorgio Sclarandis/Black Star; 612 Tom Stack/Tom Stack & Associates; 613B NASA; 613T Breck P. Kent/Earth Scenes; 614B Bruce M. Wellman/Stock, Boston; 614C Holt Confer/Grant Heilman Photography; 615B E. Adams/Sygma; 615C, Grant Heilman/Grant Heilman Photography; 616 Mark N. Boulton/Photo Researchers; 617 Phil Degginger/Bruce Coleman Inc.; 622-623 Thomas Bachand*

*Photographed expressly for Addison-Wesley Publishing Company, Inc.

Illustrations

Raychel Ciemma
pgs. 393, 395, 397, 401, 408

Elizabeth Morales-Denney
pgs. 158 -159, 163, 164, 165, 169, 238 -239, 243, 244, 245a

Carlyn Iverson
pgs. 4, 7, 8, 9, 22, 23, 26, 52, 53, 54, 55, 60, 446, 447, 448, 454, 456, 459, 468, 469, 470

Mapping Specialists
pgs. 32-33, 157, 580, 584

Barbara Massey
pgs. 112, 113, 118, 123, 302, 303, 322, 323, 325, 328, 331

Michael Maydak
pgs. 42-43, 134, 135, 232, 233, 345, 346, 351, 354, 355, 358, 370-371, 372, 373, 376, 388-389, 538-539, 543, 544, 545, 554

Barbara Melodia
All reflective artwork for unit and chapter openers, half-page activity bars, chapter review bars, icons for WarmUp, WorkOut, and STS feature.

Precision Graphics
pgs. 18, 19, 29, 49, 63, 69, 94, 103, 109, 114, 115a, 119, 124, 129, 133, 142, 155, 173, 189, 209, 231, 245b, 251, 269, 288, 293, 319, 343, 362, 365, 387, 413, 424, 441, 461, 465, 485, 493, 509, 531, 559, 563, 564, 566, 577, 599, 602, 603, 613, 621, 624, 625, 626, 627, 628, 629, 630, 631

Rolin Graphics
pgs. 79, 80, 82, 83, 86, 87, 214, 215, 218, 223, 224, 225, 305, 306, 308, 311, 312, 313, 504

Margo Stahl-Pronk
pgs. 273, 275, 276, 279, 281, 282, 286, 287, 327, 332, 333, 339, 394, 398, 402, 404-405, 592-593

Carla Simmons
pgs. 176, 180-181, 186, 254, 256, 259, 260, 263, 300, 573

Nadine Sokol
pgs. 195, 196, 200, 202, 416, 417, 422, 423, 436

Sarah Woodward
pgs. 38, 39, 95, 98, 104-105, 141, 143, 149, 415, 418, 421, 425, 429, 430, 431, 444, 445, 451, 452, 453, 518, 540, 549, 550, 551

Science and Literature Credits

Unit 1 From *The Gorilla Signs Love* by Barbara Brenner. Copyright © 1984 by Barbara Brenner. By permission of Lothrop, Lee & Shepard Books, a division of William Morrow & Co., Inc.

Unit 2 From *The Plant People* by Dale Carlson, Franklin Watts, 1977. Copyright © 1977 by Dale Carlson. Used with permission of Franklin Watts, Inc., NY.

Unit 3 Reprinted with the permission of Atheneum Publishers, an imprint of Macmillan Publishing Company, from *In the Night, Still Dark* by Richard Lewis. Text copyright © 1988 by Richard Lewis. "The Weasel" from *Consider the Lemming* by Jeanne and William Steig. Verse copyright © 1988 by Jeanne Steig. Reprinted by permission of Farrar, Straus & Giroux.

Unit 4 "The Great Mushroom Mistake" from *Uninvited Ghosts and Other Stories* by Penelope Lively. Copyright © 1974, 1977, 1981, 1984 by Penelope Lively. Used by permission of Dutton Children's Books, a division of Penguin Books USA Inc.

Unit 5 "Songs in the Garden of the House God," from *Journal of American Folklore,* Vol. 7, 1894, pp. 187–193, reprinted in *In the Trail of the Wind,* John Bierhorst, ed., Farrar, Straus & Giroux, 1971.

Unit 6 "The Cloud Spinner" is adapted from "The Spider Weaver," *Japanese Children's Favorite Stories,* edited by Florence Sakade. Used by permission of Charles E. Tuttle Co., Inc. of Tokyo, Japan.

Unit 7 From *Fantastic Voyage* by Isaac Asimov, © 1966 Bantam Books. Based on the screenplay by Harry Kleiner from the original story by Otto Klement and Jay Lewis Bixby.

Unit 8 "All Things Are Linked" reprinted by permission of Coward, McCann & Geoghegan from *The Crest and the Hide* by Harold Courlander, text copyright © 1982 by Harold Courlander.